Protective Relaying for Power Systems II

Protective Relaying for Power Systems II

Edited by

Stanley H. Horowitz

A Selected Reprint Volume
IEEE Power Engineering Society, *Sponsor*

The Institute of Electrical and Electronics Engineers, Inc., New York

This book may be purchased at a discount from the publisher
when ordered in bulk quantities. For more information contact:

IEEE PRESS Marketing
Attn: Special Sales
PO Box 1331
445 Hoes Lane
Piscataway, NJ 08855-1331
Fax: (908) 981-8062

Printed in the United States of America
10   9   8   7   6   5   4   3   2   1

ISBN 0-7803-0413-6

IEEE Order Number: PC0309-5

Library of Congress Cataloging-in-Publication Data

Protective relaying for power systems II / edited by Stanley
Horowitz.
    p.   cm.
    "IEEE Power Engineering Society, sponsor."
    "A Selected reprint volume."
    "IEEE order number : PC0309-5"—T. p. verso.
    Includes bibliographical references and index.
    ISBN 0-7803-0413-6
    1. Protective relays. 2. Electric power systems—Protec-
tion.
    I. Horowitz, Stanley H. II. IEEE Power Engineering
Society.
TK2861.P762   1992
621.31′7—dc20
                  92-10664
                    CIP

Contents

Part 4   Substation Equipment   **289**

Preface

THE first volume of *Protective Relaying for Power Systems* was published in 1980. It contained papers published during 1947–1979 which recognized a fundamental power system relaying problem and presented the latest solution to that problem. Applicable Power Systems Relaying Committee (PSRC) reports and summaries of standards and guides were also included which presented the dominant practice within the United States and offered additional comments from other experts in each field.

Volume I also provided practicing relay engineers with a readily available compilation of source material to assist them in discharging their responsibilities. It became evident, however, that, in reality, Volume I served a much larger purpose than simply as a reference book: It became useful as a training and teaching aid in utilities and universities as well as a document of historical value. It was, therefore, decided to reevaluate some of the papers that were not included in Volume I; papers which should be available to the serious student of power system protection.

Volume II continues the basic purpose and format of Volume I. It should not be considered an updated version of Volume I; rather it is an extension of it. We have, however, included important fundamental papers that were published before 1980, but which were not included in volume I.

As in Volume I, there is an introductory discussion preceding each part that presents a general recognition of the problems associated with the equipment covered in that section. These discussions take into account the changes that have evolved since 1980. Discussions on the reprints themselves, and, in many instances, a supplementary bibliography to direct the interested reader to further appropriate literature are also included. In addition, Volume II contains the discussions that were presented at the IEEE meetings at which each paper was presented. Significant papers that have been presented at forums other than IEEE are also included.

Two subjects that have been studied and written about extensively in recent years (and were touched upon in Volume I, but not examined in great detail) are digital devices and equipment and system monitoring. Although a separate section was devoted to digital devices in volume I, it only covered specific applications that were relatively new and just being introduced. The novelty of such equipment no longer exists. Computer-based relays are now offered by all manufacturers and are being considered by all utilities. In this volume, Part 6: Digital Relays is devoted to the fundamental concepts within the digital technology. Papers associated with specific applications of computer relays are now presented in the part related to each class of equipment. Monitoring (Part 7) has assumed a major role as part of the protection chain. It is directly related to the analysis of system disturbances and affects decisions related to restoration. Therefore it is discussed in a separate part.

Obviously, the selection of papers that encompasses the entire spectrum of system protection subjects for more than a decade cannot be performed adequately by a single editor. In determining which papers to include, or exclude, and creating or editing the discussions and comments, the support of J. W. Chadwick, Jr., J. R. Linders, and A. G. Phadke was invaluable. The thoroughness and timeliness of this volume could not have been achieved without them. In addition, the officers and members of the PSRC provided excellent suggestions and comments. In the end, however, decisions regarding both content and style must be made by one person. The inclusion and exclusion of any paper and the exact words in the part introductions and discussions are the responsibility of the editor.

Significant books and publications that cover the general theory and application of power system protection are:

J. Lewis Blackburn, *Protective Relaying—Principles and Applications*, Marcel Dekker, Inc., 1987.

GEC Measurements, *Protective Relays Application Guide*, 1983.

S. H. Horowitz, *Protective Relaying for Power Systems*, New York: The Institute of Electrical and Electronics Engineers, Inc., 1980.

S. H. Horowitz and A. G. Phadke, *Power System Relaying*, London: Research Studies Press, New York: Wiley, 1992.

C. R. Mason, *The Art and Science of Protective Relaying*, New York: Wiley, 1956.

I. T. Monseth and P. H. Robinson, *Relay Systems*, New York: McGraw-Hill, 1935.

Power System Protection—Reference Manual—Reyrolle Protection, NEI Electronics, LTD. 1982.

A. R. Van C. Warrington, *Protective Relays: Their Theory and Practice*, vol. 1, London: Chapman and Hall, and New York: Wiley, 1962.

A. R. Van C. Warrington, *Protective Relays: Their Theory and Practice*, vol. II, London: Chapman and Hall, and New York: Wiley, 1969.

Westinghouse Electric Corporation, *Applied Protective Relaying*, 1976.

I. GENERAL DISCUSSION

THE fundamental philosophy and considerations of protective relaying have not changed significantly since Volume I was published. The basic requirements of protection, i.e., reliability, speed, selectivity, and economics are still the same, although changes in hardware, particularly the use of microprocessors, have had a noticeable impact on virtually all aspects of design and application of protective relays. Part I: Fundamental Considerations, as in Volume 1, concerns itself with those areas that are basic to the overall technology of power system protection or involves relaying considerations that apply to a variety of equipments or situations. Specific applications are addressed in those parts associated with individual equipment. For instance, the method of symmetrical components is still the primary way of calculating short circuit currents. However, the proliferation of microcomputers has significantly altered the manner in which the calculation is performed. In particular, reprints involving relay settings and coordination were not included in Volume I but are considered in this volume.

II. DISCUSSION OF REPRINTS

The first two papers, ''A Survey of Relay Test Practices'' and ''Review of Recent Practices and Trends in Protective Relaying,'' sponsored by the Power Systems Relaying Committee (PSRC), update the overview of relaying practices and present a survey of test practices. These updates present the effect of electronic equipment, the increasing system voltages, fault currents, and the changing economic and environmental conditions. In particular, the paper on relay practices and trends provides valuable details on a wide variety of new problems and concepts, such as the gas insulated substation (GIS), HVDC, and fault monitoring. The paper on test practices gives the results of a survey of 125 relay engineers relative to the qualifications of test personnel, intervals, and equipment.

An area of increasing interest and concern involves the protection of distribution feeders, especially with the increasing use of small generators being connected directly to the distribution circuits. Wagner et al. in the paper ''Relay Performance in DSG Islands,'' review the performance of relays in the presence of such generators. The increased use of semiconductors in industrial, commercial, and residential installations has raised the issue of harmonics and wave distortion on the power system. This subject is discussed in ''Influence of Harmonics on Power Distribution System Protection,'' by Fuller, Fuchs, and Roesler.

Capitalizing on the advantages of computers, considerable work has been done on managing the complexities of relay settings and coordination. Two papers that trace the progress of relay coordination using an off-line computer, from the earliest attempts to the present, are presented. One is a PSRC report, ''Computer Aided Coordination of Line Protection Schemes,'' which describes the use of computers to provide coordination with inverse time current characteristics; the other is a PSRC prize paper, ''Coordination of Directional Overcurrent Relays in Transmission Systems—A Subsystem Approach,'' by Ramaswami, Damborg, and Venkata, which covers distance relays.

Perhaps one of the most exciting new concepts is the advent of adaptive relaying and control. The main thrust of this concept is to make on-line changes in relaying settings and/or characteristics and control logic in response to system changes. Three papers dealing in this area are presented. The prize paper, ''Adaptive Transmission Relaying Concepts for Improved Performance,'' by Rockefeller et al., describes the concept using the adaptive capabilities of digital relays as applied to transmission lines, on-line changes in settings, characteristics, or logic. The result is to improve reliability and better the utility of transmission facilities. A companion paper, ''Adaptive Transmission System Relaying,'' by Horowitz, Phadke, and Thorp also describes this concept as applied to transmission lines and extended transformer protection and reclosing. The third paper, ''Some Applications of Phasor Measurements to Adaptive Protection,'' by Thorp et al., discusses a new measuring technique that emerged from the digital relay algorithm studies. This technique involves the calculation of phasors to establish a variety of parameters, such as frequency and phase angle, which can be useful in power system protection and operation. This subject is also covered in some detail in Part 7: Monitoring.

III. ASSOCIATED STANDARDS

''*Recommended Practice for Electric Power Distribution for Industrial Plants*,'' Red Book, ANSI/IEEE Std 141—1986.

''*Recommended Practice for Protection and Coordination of Industrial and Commercial Power Systems*,'' Buff Book, ANSI/IEEE Std. 242—1986.

''*Sine-Wave Distortions in Power Systems and the Impact on Protective Relaying*,'' IEEE Special Publication 84 TH 0115-6 PWR, pp. 1–33.

IV. SUPPLEMENTAL BIBLIOGRAPHY

R. E. Albrecht, M. J. Nisja, W. E. Feero, G. D. Rockefeller, C. L. Wagner, ''*Digital Computer Protective Device Coordination Program—Part I General Program Description*,'' T-PAS 64, vol. 83, no. 4, pp. 402–410, April 1964.

W. K. Sonnemann, ''*Simplification of A.C. Voltage Vector Systems*,'' AIEE Trans., pp. 1320–1328, Oct. 1954.

W. K. Sonnemann, ''*A Study of Directional Element Connections for Phase Relays*,'' AIEE Trans., vol. 69, pp. 1438–1450, 1950.

D. I. Jeerings and J. R. Linders, ''Ground Resistance—Revisited,'' *IEEE Trans. on Power Delivery*, vol. 4, no. 2, April 1989, pp. 949–956.

''Fault Induced Wave Distortion of Interest to Relay Engineers,'' PSRC Committee Report, *IEEE Trans. on Power Apparatus and Systems*, vol. PAS-104, no. 12, Dec. 1985, pp. 3574–3584.

A SURVEY OF RELAY TEST PRACTICES

IEEE Power Systems Relaying Committee Report

Abstract– Results of a survey on protective relaying testing practices are presented. The responses of 125 relay engineers to questions on the qualifications, training, and work scope of test personnel, testing equipment, testing practices, test intervals, transformer fault pressure relay practices, circuit breaker trip tests, and performance checks are listed.

INTRODUCTION

The IEEE Power Systems Relaying Committee reviewed the 1956 committee report, "A Survey of Relay Test Methods"[1], for areas where changes might be of significance. A new survey was initiated to detail these changes and to cover new areas of concern. Some of the previously reported detailed test procedures were judged as sufficiently standardized to omit in the new survey. Questions regarding the qualifications and training of relay test personnel were expanded. Test practices for static relays and transformer sudden pressure relays are new areas of interest.

The results were tabulated from 125 replies. Six of these replies are from large industrial systems. Geographically, most of the United States, Canada, and Puerto Rico are included. A total of 1,157 generating stations, 5,096 transmission stations, and 17,849 distribution substations are represented.

The data given in the tables is tabulated in three categories based on relative system peak loads: Group I under 500 MW, Group II 500 to 2,000 MW, and Group III 2,000 MW and over.

SUMMARY OF SURVEY

General information regarding relay test personnel education and training is found in Table I. Some 26% of the reporting companies have college graduates with engineering degrees as test personnel. Other comments indicate that some companies are finding it more difficult to secure college graduates for test work. Most companies require a minimum of a high school education. Further training includes company courses and evening courses at the college level. On-the-job training with close supervision in the early phases is basic, as shown in Table II.

The job scope of most testers, as shown in Table I, includes related protective equipment, such as CT's and PT's, relaying communication, instrument and control telemetry. The changes brought about by static relays are apparently being successfully met. This is evidenced in Table III by the 67% who use the same people for static and electromechanical relay testing.

Table II shows technical material used for training and reference. A large majority, 78%, provide supplementary information to the manufacturers' bulletins for test personnel. The small response to the need for a text to train relay test men is perplexing. It was not until 1956 that a general text on protective relaying was available.[2] It is assumed that this relay text book and others are in wide use for both engineer and tester training. This may account for the decline to 17% in need of a relay test training text shown in this survey compared to the 34% in the 1956 survey.

Paper 71 TP 540-PWR, recommended and approved by the Power System Relaying Committee of the IEEE Power Engineering Society for presentation at the IEEE Summer Meeting and International Symposium on High Power Testing, Portland, Ore., July 18–23, 1971. Manuscript submitted December 30, 1970; made available for printing April 27, 1971.

TABLE I Relay Test Personnel

	Group I	Group II	Group III	Total Replies
Relay test personnel have the following training and/or education:				
College graduates, engineering degree	4	9	20	33
Men with limited college training in engineering	9	21	13	43
Technicians with vocational training in testing relays	20	34	13	67
Other (See Summary)	15	21	6	42
Total Replies	(48)	(85)	(52)	(185)
Relay test personnel are:				
Specialists who test protective relays only	5	10	4	19
Specialists who do other work on related protective equipment such as CT's, PT's, carrier equipment, telemetering	24	49	23	96
General test men who also work on OCB's, transformers, batteries, etc.	16	9	5	30
Employees of another company	1	--	2	3
Total Replies	(46)	(68)	(34)	(148)
Number of companies in each group	40	59	26	125

As shown in Table III, most companies use the same test crews to calibrate static relays as they do for electromechanical relays. The percentage of companies replying increased with size. This may indicate that fewer small companies have static relays.

The majority check all relays in a station at one time with one test crew. About 1/3 test relays in conjunction with the equipment they protect.

In Table IV, the opinion that relay test equipment be in readily portable units is almost universal. Commercially packaged test units usage is a relatively recent trend indicating increasing acceptance of the adequacy and convenience of these units. The only significant alternate is the use of mobile test laboratories. These are mainly recommended by Group I and II responders. A smaller number of group I & II companies indicate that the use of permanently installed test equipment in a central location with the relays brought in for testing is preferable.

Individual units of commercially available standard test devices connected as needed in the field are widely used. The use of especially designed equipment is less common and is almost confined to group II & III companies.

Table V shows most all companies test relays according to schedule. The few that report testing on installation only also test following

Reprinted from *IEEE Trans. Power App. Syst.*, pp. 1191–1196, May/June 1972.

TABLE II Training of Test Personnel

Training of test personnel uses:	Group I	II	III	Total Replies
The apprentice system	15	29	10	54
Special company courses	4	10	7	21
Close supervision in the early phases	31	41	19	91
Other	1	6	2	9
Total Replies	(51)	(86)	(38)	(175)
Relay test personnel are:				
Furnished with detailed test manuals especially written for your system	4	10	7	21
Dependent upon manufacturer's bulletins for test methods	19	14	5	38
Dependent on test procedures supplied by test equipment manufacturer	5	2	1	8
Total Replies	(56)	(73)	(36)	(165)
In regard to relay test information:				
Manufacturer's bulletins are considered adequate	17	21	7	45
A relay test guide is needed to supplement bulletins	21	39	21	81
A text is needed for training relay test men	7	7	7	21
Total Replies	(45)	(67)	(35)	(147)
Number of companies in each group	40	59	26	125

TABLE III Allocation of Test Work

Static relays are calibrated by:	Group I	II	III	Total Replies
The same people as are used for electromechanical relays	21	42	21	84
Special test crews	2	4	2	8
Total Replies	(23)	(46)	(23)	(92)
Testing protective relays in a particular station:				
All relays are tested by a single crew in one concentrated effort	27	36	11	74
Relays are tested by different crews according to voltage and complexity of relays in one concentrated effort	--	2	3	5
Relays are tested according to complexity and voltage by different crews at different times	3	19	10	32
Relays are tested in conjunction with the equipment they protect	13	21	10	44
Total Replies	(43)	(78)	(34)	(155)
Number of companies in each group	40	59	26	125

an incorrect operation. Note that none of the companies depend on a factory test without some checking on installation. The majority also test relays following an incorrect operation. Different procedures for initial and routine testing are used by a majority.

Field testing by clearing the primary and circulating current through the primary circuits is infrequently practiced.

A majority use test plugs or draw-out relays and test one relay at a time with the circuit in service except for test tripping. The trend seems to be in this direction as the clearing of the circuit and testing of relay calibrations with secondary values has less use. Load growth, heavily scheduled weekend work, and system security are factors that may account for the predominance of this practice. A large majority test the relay on the actual setting specified by the relay engineer.

The testing of input and output devices and circuits shows a large majority performing a test tripping. The establishment of C.T. continuity has a larger response than C.T. ratio tests. Response to the testing of P.T. circuits was less than that for C.T. continuity. A slightly higher number report on the testing of potential and control fuses. About 1/3 of the companies also test meters while testing relays. The smallest response was to the testing of wiring insulation.

Table VI lists the maintenance and calibration intervals of electromechanical and static relays by voltage classification. The distinction between high and low voltage circuits was made by the responders.

For time overcurrent relays on high voltage circuits, the one-year interval is predominant for both electromechanical and static relays. For low voltage circuits, the one-year interval is in the majority for both types of relays. There is however a trend towards longer intervals for overcurrent relays of both types. Note that sum of 1-1/2, 2, and 3-year intervals is almost double the one-year interval for the low voltage electromechanical overcurrent and equal to that for static overcurrent relays.

The one-year interval is predominant for both high and low voltage circuits with both electromechanical and static impedance relays. However, static relays have a greater percentage of the total responses in the 2 and 3-year intervals than do the electromechanical types. One hesitates to call this a trend as yet.

With electromechanical differential relays the one-year interval is followed by the majority for both high and low voltage circuits. The pattern for intervals greater than one year seems to be the same as for electromechanical overcurrent relays.

The one-year interval is practiced by a large majority for both high and low voltage circuit electromechanical and static carrier relays and channel equipment. Note that the 6 months or less interval is the largest for any relay function. From the questionnaire comments there is frequent testing of channel equipment. The total of 46 responses in carrier relaying is the largest for static relays on high voltage circuits.

Pilot and remote trip schemes totals also show the one-year interval is practiced by a large majority with both types of relays.

A question not included in Table VI referred to maintenance and calibration intervals: "Do you think the time intervals indicated above are correct? If not indicate what you would like to do." There were 24 responses; 13 thought the intervals should be decreased and 11 thought they should be increased.

4

TABLE IV Relay Test Equipment

	Group I	II	III	Total Replies
Relay test equipment:				
Consists of standard commercial test devices connected as needed (individual units)	26	20	11	57
Especially designed as no suitable equipment is available	1	13	10	24
Consists of both of above	13	29	14	56
Commercially packaged test unit	17	16	3	36
Total Replies	(57)	(78)	(38)	(173)
Relay test equipment should be:				
Readily portable, units can be handled by one or two men	31	57	26	114
Installed in a mobile test laboratory	10	5	1	16
Installed in a central location, relays brought in for testing	3	5	--	8
Total Replies	(44)	(67)	(27)	(138)
Number of companies in each group	40	59	26	125

TABLE V Relay Testing Practices

	Group I	II	III	Total Replies
Relays are tested:				
According to schedule	32	53	25	110
Only on installation	4	--	--	4
By manufacturer only	--	--	--	--
After incorrect operation	21	38	20	79
Using different initial and routine testing procedures	17	39	18	74
Total Replies	(74)	(130)	(63)	(267)

Relay testing is done principally by (indicate percentage of each method used):	Ave. %	Ave. %	Ave. %	Weighted Ave. %
Clearing the circuit and circulating current through the primary circuits	5 (9)	9 (14)	8 (7)	7.56 (30)
Clearing the circuit and testing calibration of relays with secondary values	31 (21)	35 (33)	56 (22)	40 (76)
Using test plugs or draw out relays to test one relay at a time with the circuit in service except for test tripping	63 (30)	68 (52)	43 (18)	62 (100)
Testing on actual setting	67 (33)	91 (55)	92 (23)	84 (111)

Secondary relay test procedures include the testing of:	I	II	III	Total
Breaker tripping	29	44	24	97
CT continuity	18	32	19	59
CT ratio	16	19	11	46
PT circuits	17	27	14	58
Fuses (potential & DC)	18	25	20	63
Meters	12	18	12	42
Insulation of wiring	9	12	14	35
None	5	3	1	9
Total Replies	(124)	(180)	(115)	(419)
Number of companies in each group	40	59	26	125

Of the 13 desiring decreased intervals, one gave no period, 5 desired 1 to 1-1/2 year periods, the other 7 chose 2 to 4-year intervals. Their present schedules are 4 with 2 to 5 years and 9 with 3 to 10-year intervals.

Of the 11 desiring increased intervals, 10 are now on a 1-year schedule. Six preferred a 2-year interval, one 1-1/2 years, one 2 to 3 years, and 3 gave no definite choice.

Table VII shows that about 62% of the companies connect transformer fault pressure relays to trip. This is indicated by the 78 companies who do installation checking of the trip circuit. It is assumed that those who do no installation checking connect to alarm only. Only 10 companies, or 8%, make laboratory installation checks of the pressure relays. There is as yet no commercially available test apparatus for gas or oil pressure relays. Hence, many companies have not built the apparatus required for complete testing. The 44 replies, or 35%, indicating field calibration on installation are probably one point checks. A majority of 52% assume the manufacturer's calibration as correct.

Periodic checks of fault pressure relays show a smaller number of companies checking trip and alarm circuits. The drop in periodic pressure testing is not as severe, some 41% as contrasted to 53% on installation testing. The not tested percentage rises to 22%.

A large majority of companies who periodically test these devices use a one-year test interval.

Circuit breaker trip test responses are shown in Table VIII. A slight majority indicate that breakers are trip tested only during relay testing and breaker maintenance. The tripping of high and low voltage circuit breakers by station operators or service personnel received more replies for high voltage, 50, than for low voltage, 41. The high voltage breakers are also more frequently exercised. Comments on special conditions include the use of weekly trip tests by supervisory control, on

suspicion of malfunction, routine tests by operators at certain stations, and at maintenance shut-down of industrial plants.

Table IX shows the practices of checking relay performance after installation. About 85% use oscillograms, various recorded charts, reported relay targets, etc. for checking performance. The 91% response to the periodic testing of relays shows the degree of dependence on this method.

Staged primary tests are reported by 17 companies; the 14% in the current survey is a significant reduction from the 31% in the 1956 survey.

Among the other performance checks reported were:
a) Coordination tests simulating fault conditions.
b) Applying secondary values of calculated fault current and voltage

5

TABLE VI Maintenance and Calibration Intervals

Relay Function	Group	High Voltage Circuits								Low Voltage Circuits							
		6 mos. or less	1 year	1-1/2 years	2 years	3 years	4 years or more	5 years or more	Total	6 mos. of less	1 year	1-1/2 years	2 years	3 years	4 years or more	5 years or more	Total
Time Overcurrent Relays: Electromechanical	I	–	21	3	5	4	–	–	33	–	14	2	7	3	–	4	30
	II	2	33	2	7	9	3	–	56	2	17	2	12	15	3	6	57
	III	1	10	2	7	3	–	–	23	1	5	6	8	4	–	–	24
Total Replies		3	64	7	19	16	3	–	112	3	36	10	27	22	3	10	111
Static	I	–	6	–	–	1	–	–	7	–	6	–	–	1	–	1	8
	II	1	4	–	3	3	–	–	11	1	4	1	2	3	1	–	12
	III	–	4	–	3	1	–	–	8	–	2	1	3	1	–	1	8
Total Replies		1	14	–	6	5	–	–	26	1	12	2	5	5	1	2	28
Impedance Relays: Electromechanical	I	1	20	3	5	3	–	–	32	–	12	1	–	2	–	2	17
	II	2	33	1	8	6	–	3	53	1	16	2	6	1	1	1	28
	III	1	16	3	4	1	–	–	25	1	9	2	4	–	–	–	16
Total Replies		4	69	7	17	10	–	3	110	2	37	5	10	3	1	3	61
Static	I	1	5	–	–	1	–	–	7	–	4	–	–	1	–	–	5
	II	2	10	–	4	6	–	1	23	1	1	–	4	1	–	–	7
	III	–	7	–	3	2	–	–	12	–	5	–	2	–	–	–	7
Total Replies		3	22	–	7	9	–	1	42	1	10	–	6	2	–	–	19
Differential Relays: Electromechanical	I	–	19	2	8	4	–	–	33	–	12	1	4	5	–	2	24
	II	1	31	3	12	6	1	2	56	1	21	1	16	4	1	2	46
	III	1	17	3	3	1	–	–	25	1	10	3	6	2	–	–	22
Total Replies		2	67	8	23	11	1	2	114	2	43	5	26	11	1	4	92
Static	I	–	4	–	–	1	–	–	5	–	4	–	–	–	–	–	4
	II	1	5	–	4	3	–	1	14	1	1	–	4	–	–	–	6
	III	–	5	–	1	2	–	–	8	–	2	–	2	–	–	–	4
Total Replies		1	14	–	5	6	–	1	27	1	7	–	6	–	–	–	14
Carrier Relays (and Channel Equipment): Electromechanical	I	5	17	1	3	2	–	–	28	–	9	–	–	2	–	–	11
	II	6	30	5	5	5	–	3	54	3	11	2	2	2	–	1	21
	III	5	12	3	2	2	–	–	24	1	4	1	2	1	–	–	9
Total Replies		16	59	9	10	9	–	3	106	4	24	3	4	5	–	1	41
Static	I	1	7	–	–	1	–	–	9	–	5	–	–	1	–	–	6
	II	7	15	–	1	4	–	–	27	1	2	–	1	1	–	–	5
	III	1	6	–	2	1	–	–	10	–	2	–	1	–	–	–	3
Total Replies		9	28	–	3	6	–	–	46	1	9	–	2	2	–	–	14
Pilot & Remote Trip Scheme: Electromechanical	I	1	17	1	3	3	–	1	26	–	9	–	1	2	–	–	12
	II	5	30	5	5	5	–	1	51	3	10	2	6	–	–	–	21
	III	4	15	2	2	1	–	–	24	2	7	1	3	1	–	–	14
Total Replies		10	62	8	10	9	–	2	101	5	26	3	10	3	–	–	47
Static	I	–	5	–	–	2	–	–	7	–	4	–	–	1	–	–	5
	II	3	9	–	2	4	–	–	18	2	–	–	2	–	–	–	4
	III	1	5	–	1	1	–	–	8	1	2	–	1	–	–	–	4
Total Replies		4	19	–	3	7	–	–	33	3	6	–	3	1	–	–	13

TABLE VII Transformer Fault Pressure Relay Testing Practices

| | Installation | | | | Periodic | | | |
	I	II	III	Total Replies	I	II	III	Total Replies
Transformer fault pressure relays are:								
Laboratory calibrated	1	4	5	10	--	1	--	1
Field calibrated	7	25	12	44	6	17	12	35
Assumed correct as received from manufacturer	28	27	10	65	19	13	6	38
Total Replies	(36)	(56)	(27)	(119)	(25)	(31)	(18)	(74)
Trip circuit checked	19	37	22	78	11	19	18	48
Alarm circuit checked	15	29	15	59	7	15	13	35
Pressure tested with gas or oil	12	39	15	66	9	30	12	51
Not tested	9	2	1	12	19	7	2	28
Total Replies	(55)	(107)	(53)	(215)	(46)	(71)	(45)	(162)

Periodic Testing Intervals

| | Calibration Checked | | | | Trip Circuit Tested | | | | Alarm Circuit Tested | | | |
	I	II	III	Total Replies	I	II	III	Total Replies	I	II	III	Total Replies
Test interval:												
Installation only	--	1	--	1	--	4	1	5	--	3	1	4
6 months or less	--	--	--	--	--	2	1	3	--	1	1	2
1 year	4	10	7	21	10	16	12	38	9	11	11	31
1-1/2 years	--	2	1	3	1	1	2	4	--	1	2	3
2 years	1	5	3	9	2	5	2	9	1	5	1	7
3 years or more	1	--	3	4	2	5	--	7	2	2	--	4
Total Replies	(6)	(18)	(14)	(38)	(15)	(33)	(18)	(66)	(12)	(23)	(16)	(51)
Number of companies in each group	40	59	26	125								

to the relay and checking overall performance with a direct writing oscillograph.

c) Verify proper phase angle relationships at time of installation.

d) Tests initiated by the field test force or required by the engineering group.

e) Special test after incorrect operation.

COMMITTEE COMMENTS

Since the 1956 survey, new designs and materials have been applied to standard types such as the time overcurrent relay. These changes in design and materials are initially evaluated by laboratory tests and compared with the previous design prior to system application. Thus, although the same relay tests for a new design may be made, the service experience for the type is not representative.

Maintenance intervals are determined by the experience with the relay type, service environment, frequency of operation, and the importance placed on the protected circuit or apparatus. The availability of circuits and equipment is often a decisive factor in scheduling relay maintenance. Coincident periods with generator and circuit breaker maintenance is common practice.

The committee members are well aware of the "best-foot-forward" response to survey questionnaires. Hence, the maintenance interval responses given may be objectives rather than reality in many cases. Several comments of this nature were received. The problem of adhering to scheduled intervals had the following causes given. Construction programs keep experienced test personnel checking new installations and construction outages often make other circuits unavailable for week-day maintenance. Overlaying these conditions are the heavy loadings resulting from load growth and the limited amount of work that can be done on week-ends. Hence, there is a stretch-out of the intervals.

TABLE VIII Circuit Breaker Trip Test

| | Group | | | Total Replies |
	I	II	III	
A trip test of all power circuit breakers is made:				
Only during routine relay testing	4	13	3	20
Only during routine breaker maintenance	12	6	3	21
During both relay testing and breaker maintenance	6	17	3	26
Total Replies	(22)	(36)	(9)	(67)
Number of companies in each group	40	59	26	125

Other trip tests by station operators or service personnel of all power circuit breakers is made at the following intervals:

| | High Voltage | | | | Low Voltage | | | |
	I	II	III	Total Replies	I	II	III	Total Replies
1 month	6	7	6	19	2	3	4	9
2 month	1	-	2	3	-	-	2	2
3 month	1	4	5	10	2	3	3	8
6 month	1	5	1	7	1	4	1	6
12 month	4	4	1	9	4	6	1	11
18 month	1	-	-	1	2	-	-	2
24-36 month	-	1	-	1	-	2	1	3
Total Replies	(14)	(21)	(15)	(50)	(11)	(18)	(12)	(41)

TABLE IX Performance Checks

In service performance of relays is checked by:	Group I	Group II	Group III	Total Replies
Analysis of relay operations during faults by means of targets, oscillograms, etc.	28	55	23	106
Periodic testing of relays	36	54	24	114
Staged primary faults	5	8	4	17
Other methods	2	2	3	7
Total Replies	(71)	(119)	(54)	(244)
Number of companies in each group	40	59	26	125

much toward wear and trouble, as to assurance of "will it or won't it operate" since relays generally operate relatively rarely.

While no formal analysis has been made, it has been a general observation that the reported high performance of relay systems is approximately the same for users who do extensive maintenance as for those who do minimum to very little maintenance.

It is noted that static relays did not show any indications toward less testing than electromechanical types. This should be an early and transient experience when a new technology needs the learning and hardware shakedown of actual field operation. This "living in the field" is an important and necessary final laboratory for any design and one that cannot be completely reproduced in the laboratory with system models.

Manuscript received August 16, 1971.

REFERENCES

[1] A Survey of Relay Test Methods (Committee Report) AIEE TRANS. (POWER APPARATUS AND SYSTEMS) vol. 75 pt. 3, pp. 254–260, 1956.
[2] C. Russel Mason, "The Art and Science of Protective Relaying," New York: John Wiley & Sons, Inc.

Members of the IEEE Working Group on the Survey of Relay Practices of the IEEE Relaying Practices Subcommittee of the Power System Relaying Committee are: V. A. Nosko, Chairman; H. L. Goodridge, H. H. Leech, O. F. Pumphrey, J. R. Turley, and D. R. Volzka.

Discussion

J. L. Blackburn (Westinghouse Electric Corporation, Newark, N.J. 07101): As the last paragraph indicates, it is very difficult to separate actual practice from established policy in a survey such as this. So it would appear that the data, particularly in Table VI, represents company policy. Maintenance is very necessary but it is very important to recognize that maintenance always runs the risks of putting as much or more trouble into the equipment as is taken out. Today with rapidly changing personnel and technology it is vital that maintenance be carefully scrutinized, and I feel extended wherever possible, particularly on established designs.

There seems to be a general tendency to maintain basic types of relays on a fixed interval schedule and perhaps this is a convenient policy. It is hard to believe that many of the modern relays designed and built with more attention to minimizing wear and deterioration, etc. should be maintained in the same period as the older designs of the same family where the materials and art were not so fully developed. In other words shouldn't maintenance be more related to actual experience on a specific design?

Perhaps a partial answer to a blanket maintenance period is this discusser's feeling that protective relay maintenance is directed not so

V. A. Nosko: Mr. Blackburn's thoughtful discussion is necessarily from the viewpoint of a manufacturer of basic sensing components in a protective system. Similarly, my remarks are based on experience with an operating company and should not be interpreted as committee opinion.

Mr. Blackburn's observation of the risk that preventive maintenance may cause trouble is valid. However, the possibility of human error exists in any endeavor and is dependent on the competence of the personnel involved.

Periodic maintenance is partially based on actual experience with a specific design. For example, the intervals given in Table VI for both high and low voltage overcurrent relay shows very few intervals extending beyond three years. Could this be due to the common experience of silver contact corrosion within this period?

The fact that some utilities trip test circuit breakers apart from protective relay tests indicates that operational assurance is a factor in the choice of maintenance periods.

The survey proper does not substantiate the general observation that performance is approximately the same for users who do extensive maintenance as for those who do a minimum to very little maintenance. My observation, is that lack of maintenance can cause embarrassing outages.

The shakedown period of static relays is now underway in the industry and this experience has disturbed a number of relay engineers. Hopefully, improved relay components and suppression of control wiring transients will result in a highly reliable, low maintenance relay system.

All operating utilities desire to reduce maintenance and testing costs to a minimum. Unfortunately, the minimum is a maintenance only upon failure program; this is precluded by the customer demand for high service reliability. Service reliability in an operating system can only be assured by an adequate maintenance program. The search for an economic optimum cost for service reliability is forwarded by a comparison of industry-wide test practices with that of the individual utility and its experience.

Manuscript received September 16, 1971.

REVIEW OF RECENT PRACTICES AND TRENDS IN PROTECTIVE RELAYING

IEEE COMMITTEE REPORT

Abstract - This paper is a review and supplement to the original paper "Recent Practices and Trends in Protective Relaying" published in 1960 and its supplement published in 1964. IEEE Trans., PAS VOL 78, 1959, Pg. 1759-79 and PAS VOL 83, 1964 Pg. 1064-69.

Most of the relaying concepts discussed in the original paper and its supplement are valid. Advances in the electronic equipment available, larger fault currents, higher voltages, more restrictive design margins in heavy power equipment, greater emphasis on reliability, greater power transfers, and changing environmental and economic conditions have forced consideration of factors not previously considered significant.

GENERAL SUBJECTS

There have been some new problems and concepts arise in the last few years in addition to the evolutionary type of changes in protective relaying.

Subsynchronous Resonance and Torsional Oscillations Due to Series Compensated Lines [1,2,3]

The application of series capacitors to EHV transmission lines is attractive from an economic and stability standpoint, but for certain conditions they can produce subsynchronous resonance in the electrical system which can excite the turbine generator mechanical systems and result in severe vibrations and shaft failures. Relays, filters and other methods have been developed to protect the turbine and generator shafts.

Electrical and Mechanical Relay Environment [4]

All of the new solid state relays must be capable of passing the Surge Withstand Capability (SWC) tests so that they may be installed in locations subject to high transient voltages. Relays also may be required to be seismically qualified to prevent misoperation during an earthquake. Standardized tests are being developed to determine a relay tolerance for other electrical transients and radio waves.

Open Power Circuit Breaker Flashover

Certain designs of EHV power circuit breakers have shown the need for protection against flashover occurring across the open contacts, especially when being used to synchronize a generator. Some form of breaker failure relaying is normally installed to protect the generator under these conditions.

81 WM 121-3   A paper recommended and approved by the IEEE Power System Relaying Committee of the IEEE Power Engineering Society for presentation at the IEEE PES Winter Meeting, Atlanta, Georgia, February 1-6, 1981. Manuscript submitted February 8, 1980; made available for printing December 12, 1980.

Gas Insulated Substations [5]

These compact substations have special problems, some of which involve relay protection. These problems are due to the gas pressure and purity, the reliability of the power supply, when the switches and breakers can be opened, when they must be opened, when they can't be opened, how to handle breaker failure, how to ground, how to inspect and how to maintain.

Relaying by Travelling Waves [6,7,8]

A new concept has been field tested whereby a fault generated wave will reveal the location of the fault by the transit direction through a current transformer, plus a comparison by communication with the other line terminal within 1/4 cycle of time.

Relay Input Levels [9,10]

Development work continues in the field of reduced energy levels for relay inputs by the use of transducers having outputs in the milliampere and millivolt range, and is a natural result of the use of solid state relays. Experiments are being made with fiber optics as a medium for the transmission of information to a relay.

AC GENERATOR PROTECTION

The application of protective relays for ac generators varies, depending on the size of the machine, the type of prime mover, the station design, past practices, etc. The size of steam turbine-generators has steadily increased to the point where a single unit may represent a substantial portion of the system's capacity and usually requires more sophisticated protection and control. Small peaking units such as gas turbine-generators are usually unattended, remotely-controlled units with complete automation rather than alarm and operator control. (also see section on computer relaying)

Stator Winding Protection [11,12]

It is common practice to ground the stator neutral through a high resistance using a distribution transformer and secondary resistor with an over-voltage relay connected across the transformer secondary to detect and trip the unit for stator ground faults. In a few instances, a distribution transformer and secondary reactor (ground fault neutralizer) have been used. Recently, ground relays have become available which will provide protection for 100 percent of the stator winding when high-resistance generator grounding is used. Backup stator ground protection is being applied using a short-time overvoltage relay connected across the broken-delta secondary of a set of wye primary-delta secondary auxiliary voltage transformers fed from the unit. This relay provides approximately the same sensitivity as the neutral ground overvoltage relay used with the distribution transformer-secondary resistor grounding method.

Although manufacturers recommend that the generator be brought up to near rated speed prior to applying field, there still exists the possibility of the generator operating at other than rated frequency during the start-up or shutdown period. In some

Reprinted from *IEEE Trans. Power App. Syst.*, vol. PAS-100, no. 8, pp. 4054-4064, Aug. 1981.

9

cases, relay sensitivity at other than rated frequency is inadequate. For this reason, an instantaneous overvoltage ground relay has been applied across the secondary resistor in the high-resistance grounding method. The relay has a relatively constant pickup with respect to frequency down to approximately 20 Hz. Sensitive phase fault protection has been provided with instantaneous overcurrent relays. These relays are extremely sensitive and blocked from operating when the generator has been synchronized to the system.

Solid-state relays have been introduced for differential protection and provide an operating time of less than one cycle. Recently, the concern for transient performance of current transformers has created interest in high-impedance relays for use as differential protection.

An ion chamber detector has been developed for early detection and warning of local overheating in the stator windings. These devices are called "core monitors" or "condition monitors" and are applied on generators where at least part of the cooling is with hydrogen. They will detect the thermal decomposition of small amounts of organic materials (epoxy paint, core-lamination, and insulating materials) used in the generator.

Field Winding Protection

The brushless excitation system has been introduced which requires that the ac exciter armature, the rectifying assembly, and the generator field winding be located on the generator shaft. Since continuous monitoring of the field circuit is not practical, a ground detection system has been developed which checks the field circuit automatically at a given time interval or upon manual initiation.

The increased synchronous impedance of the large machines has resulted in a larger circle characteristic for the loss-of-field relay setting. This has made the relay susceptible to stable power system swings where the apparent system impedance may enter the loss-of-field relay operating characteristic. A two-zone, loss-of-field relay scheme is being used whereby a restricted circle first-zone provides fast tripping for severe loss-of-field conditions. A delayed, larger circle, second-zone is used to provide enough time-delay to override transient power swings and trip the machine for less severe loss-of-field conditions.

A sudden loss of load can result in high terminal voltage which may induce a high level of magnetic flux in the generator core as well as any connected transformers. Excessive flux can also be caused by field energization with the regulator in service during start-up and shutdown of the generating unit. Volts/hertz limiting may be included in the voltage regulator; however, volts/hertz relays are also used to prevent core heating as a result of excessive flux.

Generator Backup Protection
Balanced Fault Protection

Three-phase faults external to the generator zone of protection are expected to be cleared by the system protection (bus, transmission line, transformer). In the event the fault is not cleared, backup protection is applied to detect these faults and initiate tripping of the unit. This protection

generally applied at the generator current transformers and is designed to "look" into the system.

A common form of backup protection consists of a three-phase, single-zone distance relay together with a dc timing relay for coordinating purposes. Instantaneous overcurrent and voltage-balance relays have been used to supervise the distance relay. In some cases, voltage- restrained or voltage-controlled overcurrent relays are used for backup protection.

Unbalanced Fault Protection[13,14]

Unbalanced faults or open-phase conditions may cause negative- sequence currents (I_2) to flow which produce heating in the nonmagnetic rotor wedges and retaining rings. The permissible integrated product I_2^2t for turbine-generators has decreased to values of 10 for 800 MVA and smaller machines and decreased linearly to a value of 5 for 1,600 MVA machines. Solid-state, negative-sequence current relays are available with increased sensitivity and I_2^2t matching capability as low as 2.

Abnormal Operating Conditions
Underfrequency[15]

There is a definite trend toward protection of turbine-generators against underfrequency operation. Both the turbine and generator are limited to the degree of tolerable underfrequency operation; however, the turbine is more restrictive since mechanical resonances could cause turbine blade damage in a short period of time for small departures in speed. Underfrequency protection is being applied using either electromechanical or solid-state relays. In some instances, several levels of underfrequency protection may be employed in inversely corresponding time delays.

Reverse Power Protection

Turbine pressure differential switches, turbine exhaust hood temperature switches, hydraulic oil pressure switches, or control valve auxiliary contacts indicate either motoring or loss of or reduced steam flow which may result in reverse power flow. Backup protection is an accepted practice using a power directional relay. This protection, when applied to cross-compound machines, requires special study since one of the two units may motor during startup as a normal condition.

Overload Protection

Resistance temperature detectors (RTDs) or thermocouples are used to monitor the stator winding and core iron to alarm when safe operating temperatures have been exceeded. On directly-cooled generations, loss-of-coolant protection is applied to alarm in the event of loss of coolant and then initiate a load runback scheme followed by a time-delayed trip, if necessary.

Standstill Protection

Concern exists over the possibility of accidentally energizing the turbine generator while at standstill or on turning gear and also the possibility of generator breaker main contact flashover. There are several devices in the normal generator protection that will detect this condition. Overcurrent relays distance relays, and loss-of-field relays may detect this condition. Some utilities are requiring this protection to be part of the switchyard relaying to eliminate the possibility of this protection being taken out of service when the generator is taken off line.

GENERATOR AUXILIARIES[16]

The steadily increasing size of modern generators has been accompanied by a more extensive and complex system. In the unit connected concept, this greatly increases the exposure of the generator leads and reduces unit reliability. The increase in auxiliary load, also increases the transformer sizes. These larger transformers result in greater short circuit levels which stress the momentary and interrupting ratings of the switchgear and thus also reduces system reliability. In addition, environmental concerns require that scrubbers and precipitators have the same reliability as the unit, placing greater emphasis on the configuration and protection of this auxiliary load.

The basic relaying is overcurrent type protection. These devices are applied for fault protection and do not necessarily provide overload protection.

In the auxiliary system, overcurrent relays on the secondary of unit auxiliary transformers provide bus protection and backup relaying for the feeder breakers. Overcurrent relays on the primary side of the auxiliary transformer provide backup relaying for the auxiliary bus. The philosophy of some utilities is to install bus differential relaying for their auxiliary buses. Care must be taken in the selection of all current transformers in the auxiliary system to assure proper accuracy, application ratio, thermal and mechanical withstand capability, etc.

The practice of supplying an automatic alternate source for the auxiliary buses varies with the kind of power plant and the philosophy of the particular utility. In general, the alternate sources are not to maintain unit operation but to provide a source of power for orderly and safe shutdown of the unit. If an automatic transfer is provided, it may be a fast or delayed transfer and may require a supervising relay for either synchronism check or voltage sensing depending on the type of transfer employed.

Nuclear units introduce additional factors to be considered in designing the auxiliary buses. The requirement to provide shutdown power despite a multiplicity of eventualities invariably leads to the use of on-site diesel generators and a complex configuration of auxiliary buses engineered to isolate the reactor safety systems. The need to seismically qualify all relays has apparently been met with available designs.

AC MOTOR PROTECTION[17,18,19]

The present-day power plants utilize higher voltages and larger motors than the plants of the early 1960's. Induction motors power the vital fuel, air, gas, water and coolant systems in Power Plants. Therefore, there is greater emphasis on protection of motors. Major changes since the sixties involve primarily solid state versions of older electromechanical relays, packaged solid state protective devices, and a trend to somewhat more sophisticated relays.

Motors must be protected against all abnormal operating conditions. Depending upon operating practices, the abnormal conditions may include overload, locked rotor, faults, overtemperature, unbalanced and abnormal voltages. Over the years, the type of protection applied for these conditions has become fairly standardized with some variations in specific applications. This complement of relays for motors below 1500 HP normally consists of time delay and instantaneous phase overcurrent, instantaneous ground overcurrent, thermal replica and time delay undervoltage devices. Instantaneous ground overcurrent relays are commonly supplied from a single window type CT enveloping all three phase conditions supplying the motor. Packaged solid state protective devices are commonly used on motors below 600 volts. On motors above 1500 HP, additional protection, such as resistance temperature detection, unbalanced voltage, unbalanced current and differential relaying is frequently specified.

The problem of protecting for locked rotor has been compounded with decreased design margins between starting and thermal damage times. Recently, impedance relays and directional overcurrent relays have been introduced as a means for locked rotor protection. The impedance relay senses the difference in motor impedance at locked rotor and at normal running conditions. The directional overcurrent relay senses the current and phase angle during starting and running conditions to detect a locked rotor condition.

Embedded resistance temperature detectors are the more accurate devices for measuring prolonged overtemperature conditions. However, the protection provided by these detectors, and by thermal replica and overcurrent relays is not effective for multiple or hot motor starts. A solid-state relay has been developed using analog thermal tracking techniques which purportedly accommodates all start, stall and cooling conditions.

The use of ground sensors fed from torodial current transformers encompassing all three phases provides better ground fault protection of motors than the residual ground relay configuration. The zero-sequence sensing system eliminates the misoperations of residual ground relays associated with the motor starting function.

For synchronous motors, protection against loss of field and loss of synchronism is also provided. For very large synchronous machines, an impedance type loss-of-field relay is often used to detect VAR flow into the machine.

TRANSFORMERS, SHUNT REACTORS AND SHUNT CAPACITORS

Transformers[20-24]

The past 15 years have seen no dramatic changes in the overall philosophy of transformer protection. The protection is still largely dictated by size, vulnerability, accessibility, importance to system integrity, operating practices and economics. Where these factors and good engineering judgment indicate slower clearing times are satisfactory, protection may be by fuses or by overcurrent relays. Where faster times or greater sensitivity are considered necessary, sudden pressure and high-speed differential relays are employed.

Transformers have increased in electrical size while decreasing in physical size per MVA of capacity. These design changes, along with increasing system fault current, have caused concern for the effectiveness of conventional protective practices for low level internal faults caused by winding movement or loss of life on through faults. Transformer iron reduction provides increased flux densities and may affect differential relay response during magnetizing inrush. These concerns have produced considerable discussion, but no resolution. The basic operating conditions and phenomena affecting transformer relay performance have not changed and the solution of problems arising from connection, thru faults, turn-to-turn faults, inrush, overexcitation, frequency and grounding still continue to challenge the relay engineer.

Probably the most reliable and commonly used type of transformer protection, the harmonic restrained percentage differential relay. This relay has evolved until today there are units available with multiple restraint windings. Higher system fault currents and varied differential zones which may include multiple breakers on the high or low side of transformers demand more than the two or three restraint windings incorporated in the original design if adequate and reliable protection is to be provided. Use of the solid state versions of this type of relay is increasing.

The application of gas accumulation or pressure relays to detect faults of small magnitudes in transformers is an established practice. Sudden pressure relays may be subject to false operation due to winding movement on thru faults or restart of oil pumps on restoration of station service. Whether these relays are used to trip or to alarm is largely based on individual user's historical field experience. The newer designs have eliminated most of the possibilities for malfunctioning and with proper testing and maintenance, these relays offer economical transformer protection.

In the late sixties, concern over cases of damage due to overexcitation of generator step-up transformers prompted investigations and the development of relays and schemes to prevent such damage. Overexcitation or excessive core flux, with attendant core heating, in generator step-up transformers can occur during start-up or shut-down of units with sudden load rejection. Both electromechanical and solid state volts/hertz relays are now available for this type of protection and are now specified routinely.

Research and development organizations are searching for a direct winding temperature detector, a surge discharge detector and an improved gas and oil detector (also see section on computer relaying).

Shunt Reactors[25,26]

The increased use of EHV transmission systems has necessitated the increased use of shunt reactors for voltage and reactive power flow control. Extremely long EHV lines of bundled conductors connected to isolated generation sources pose operating problems whose only solutions lie in application of reactors directly to the lines or to the tertiaries of associated transformers. The importance of these shunt reactors, both from an economical and operational standpoint, dictates that they be adequately protected from damaging faults.

The two most commonly used types of reactor are the gapped iron core and the air core. Units are available in three phase or single phase configuration, oil or air insulated.

When connected to EHV lines, reactors are usually considered a part of the line with no automatic disconnecting means except that associated with the line itself. In these cases, transmission line relaying affords some overall protection for the reactors, but dedicated reactor relaying which provides additional sensitivity is preferred. This dedicated relaying normally consists of generator type percentage differential or high impedance relaying, inverse overcurrent relaying, and sudden pressure relaying for oil filled units. These relays trip the local line circuit breaker(s), block reclosing, and transfer trip the necessary remote power circuit breaker(s) to remove the faulted reactor from the system.

Reactors connected to transformer tertiary windings are provided with their own relaying scheme including phase overcurrent relays and neutral ground protection.

A mho distance relay specifically designed for shunt reactor protection has been applied by some utilities. Negative sequence, neutral voltage, and current balance schemes are being used to some extent. A unique method of reactor fault detection using sensor coils on the magnetic circuit of the reactor is presently being tested in service.

Shunt Capacitors[27]

The use of shunt capacitors in power station switchyards and substations for reactive power control has increased dramatically in the last 15 years. Ratios of installed shunt capacitor capacity to installed generation capacity run as high as 0.7 with a somewhat lower figure of 0.5 as an industry average.

The cost of protecting these capacitors is relatively small but the technical aspects have stirred considerable debate and controversy in two areas: (a) comparison of the advantages and disadvantages of the various configurations, arrangements, and connections possible, and (b) the search for dependable and accurate overvoltage protection of remaining capacitor units after unit failures in a series group.

All individual capacitor units are usually protected by fuses supplied by the manufacturer. Fault protection of the bus section between the source circuit breaker or power fuses and the capacitor bank is accomplished with conventional time delay and instantaneous phase and ground overcurrent relaying or power fuses. The solution of the difficult problem of detecting sustained overvoltage on multi-group banks after loss of individual units is attempted with current or voltage relays on grounded, ungrounded, or double wye banks.

BUS PROTECTION[28-31]

Since the last review, EHV systems have developed beyond the embryonic stage and more power transfer capability is required in HV systems. New stations in these systems usually start with a ring bus arrangement which is converted to a breaker-and-a-half arrangement when more than four or five circuits are involved. Breaker-and-a-half type bus arrangements have simplified bus protection in that circuit connections are not capable of being switched between different buses with the associated complexities of switching the tripping and possibly the CT circuits.

Faster fault clearing required to minimize the probability of system instability and to minimize the damaging effects of higher short circuit currents has resulted in the development of bus protection relays with operating times of 8 milliseconds. In some cases, two sets of high speed bus protection relays with separate ct circuits are used on each bus for increased reliability. This includes separate auxiliary tripping relays, associated dc control circuits and possibly separate batteries. (also see section on computer relaying)

On lower voltage network type systems more loads are being tapped to these lines due to the need to defer expenditures and the difficulties in obtaining right-of-way. To reduce the probability of sustained outages to loads tapped to these lines during bus faults at the terminals with breakers, bus protection is being provided at more of these latter terminals.

Some form of differential type bus protection is being considered for auxiliary station power buses in generating plants to reduce the time duration of the abnormal conditions caused by bus faults.

TRANSMISSION AND SUBTRANSMISSION LINE RELAYS[32]

Over the years, there has been considerable change in the character of transmission systems and in the relays to protect them. These changes have been evolutionary in nature, and the hardware and techniques required for protection were developed as the need arose.

Concepts

Transmission and some lower voltage systems, especially the more complex networks, are protected extensively by use of separate primary and backup (secondary) relay schemes with the degree of separation and independence varying with the importance of the station.

The form of the primary and backup relay schemes at the transmission level is principally a form of pilot relay as primary and distance and/or directional overcurrent as backup. The various transfer trip schemes (such as underreaching, overreaching etc.) are considered pilot relaying. Exceptionally short lines, or those that are critical to stability, will often use a dual primary (sometimes referred to as primary and secondary) relaying concept e.g., phase comparison and directional comparison with the pilot channels being over different media such as microwave and power line carrier. Some use is being reported of a dual pilot scheme with timers as backup or even completely separate time delay backup schemes.

Channel and other trends listed in the 1964 report are unchanged.

Phase Relaying

Distance relays, of some form, are still the predominant relays used for transmission and subtransmission system phase fault protection because of their insensitivity to variations in fault current and their virtual immunity from operating on normal load current. Changes since 1964 have been primarily in improved products and relay characteristics.

Ground Relaying

Directional overcurrent relays are the most often used relays for ground protection, with zero sequence current and/or voltage being the most common polarizing sources. This is true for both pilot and direct trip relaying. Where mutual induction problems exist sometimes compounded by autotransformers, there is an increased usage of negative sequence directional sensing and/or a form of a phase comparison pilot relaying. Ground distance relays have also found increasing use, particularly on EHV transmission systems, as an alternate solution to these problems.

Solid State Relays

The use of solid state relays has also increased in recent years, particularly on EHV systems where the increased sensitivity and speed is important. Solid state relays are still somewhat plagued with transient problems; however, protection against this by isolation, shielding or better components is continually improving performance. The hope for reduced maintenance may not have always been achieved, but the solid state relay still appears to be the relay of the future. Most solid state relays are purchased as a packaged scheme, but the use of individual solid state relays as a replacement for the older electro mechanical relays is increasing.

Pilot Relaying

Pilot relaying most commonly used is directional comparison, with the use of phase comparison or some form of transfer trip (under- reaching permissive, direct, etc.) increasing. Short lines, particularly at the subtransmission level, use ac or dc pilot relaying over leased or private wires. The increasing use of series compensation and individual pole operation of power circuit breakers has resulted in the development of single phase fault discrimination and tripping schemes such as the single phase comparison relay system. While single phase pilot relaying uses additional channel spectrum, it solves a number of problems.

HIGH VOLTAGE DIRECT CURRENT

Several HVDC (high voltage direct current) transmission lines have been put in operation in the last 10 years beginning with the 1440 MW Pacific HVDC Intertie commissioned in 1970. There are now four commercially operating systems in the US with a combined capacity of 3040 MW.

A typical HVDC converter station consists of AC harmonic filters, ac shunt capacitor banks, converter transformers, converter valves, dc smoothing reactor (5) and dc harmonic filters. Most of the protective relaying needs of the ac portion converter stations are solved by means of conventional ac relaying schemes or slight adaptations thereof; however, the protection of HVDC systems differs in many respects from ac system protection. For instance:

a) dc systems are normally equipped with very fast-acting control systems. Therefore, in many cases short circuits do not result in high current and conventional overcurrent relays are often not suitable components of the relay protection system.

b) In a bipolar dc system, the two poles can operate independently of each other even though long-term monopolar operation may be restricted, since it results in ground currents. However, special considerations are needed for faults that inject direct current into the station ground mat, since the direct current can flow into the ac network through the converter transformer neutrals and affect ac system protection as a result of CT saturation. Therefore, these types of faults need to be cleared reasonably fast. The clearing time should not exceed about 1 second.

c) Conventional current transformers are, of course, unsuitable for direct current measurements. Current measuring transductors (a form of magnetic amplifier) are used where dc current measurements are needed. However, even in the ac circuits of a converter, the risk for CT saturation is substantial and air-gapped CTs should be used.

d) High harmonic content is present in many parts of a converter. This has to be considered where ac type protection is applied.

e) Converters are normally connected in series rather than in parallel for best overall economy. Faults many times can be cleared by means of grid control actions that temporarily allow the dc current to flow through (bypass) the converter and block the ac current flow. (This is normally referred to as a blocked condition). Restart (unblocking or deblocking) is normally a simple and fast process, and is also accomplished by means of the grid control system. In case of a permanent fault, a switch is closed that allows the direct current to bypass the converter. Sometimes, opening of the ac circuit breakers follows the bypass operation, but it is not always required.

A special problem area is the coordination between ac and dc system protection. A fault in the ac system often affects the operation of the dc system, but only rarely does a fault in the dc system effect ac system protections. Hence, it is important to allow the ac system protection time to clear any fault before a dc system protection is allowed to take the affected portion of the dc system out of service. Similar coordination problems exist between rectifiers and inverters in the dc system itself.

BACKUP PROTECTION[33]

The subject of backup protection, including actual industry practices, was thoroughly dealt with in a 1970 Power System Relay Committee report. Since that time, general concepts have changed very little although much has been done to make devices and circuitry more dependable and secure.

For EHV systems, it is common practice to provide for relay-failure backup with redundant high-speed relay protection for the various elements of the system, and some utilities have extended this practice to their HV systems where fast clearing is a stability requirement. These designs usually include separate current transformers, separate voltage sources (either entirely separate CCVT's or separate windings of common CCVT's), separate batteries, and separate power

circuit breaker trip coils. Modern generator protection is likewise divided into two separate trip paths for functional redundancy.

In the area of breaker-failure backup protection, dual timing schemes are sometimes used where extremely short critical clearing times dictate narrow margins between normal clearing and backup tripping initiation. The use of two times allows fast tripping only when necessary and slower tripping for less critical conditions. This minimizes exposure to possible misoperations due to the narrower time margin. Operation of the fast timer is controlled in various ways depending on the requirements of the particular situation. Some applications require fast tripping for high magnitude three phase faults only, others require fast tripping for any multiphase fault, while in still others, close-in faults of any type must be cleared fast. These various requirements are accomodated through the proper contact arrangement and setting of fault detecting relays. Implementation can be simplified when the dual timing scheme is applied in conjunction with individual pole power circuit breakers by the use of power circuit breaker auxiliary switches to provide indication of power circuit breaker status.

DISTRIBUTION PROTECTION[34,35]

Many distribution circuits are now operated at 25 kV or 35 kV for more economic distribution to heavier loads. Few changes in distribution relay systems are noted as a result of these higher voltages. Higher values of fault current have increased the hazards of both pole and pad mounted distribution transformer explosions. Many utilities are now using current limiting fuses especially in pad mounted transformers and underground feeders to minimize this hazard. Backup current limiting fuses are being used in areas where the fault current exceeds the interrupting capability of the fuse cutout. Pressure relief devices are also being incorporated in some distribution transformers. A few utilities are using heat sensors in vaults or pad mounted enclosures to detect arcs or fires. These sensors trip breakers, reclosers, or similar devices.

Underground residential distribution (URD) systems are being installed in many areas. These underground feeders are frequently a part of an overhead system where reclosing is used. If the underground system is extensive, it often merits the use of a sectionalizer on the source terminal of the underground section to prevent the feeder breaker at the substation from re-energizing an underground fault. If the underground system is small, fuses can be used for the same purpose provided instantaneous relays on the feeder breaker are disabled in a manner to operate the fuse. Underground feeder laterals can be isolated by the use of fuses in pad mounted enclosures. There is a tendency to use current limiting fuses in these or similar applications.

There are several computer programs available to both calculate the radial feeder fault currents and the coordination time between protective devices such as breakers, fuses and reclosures. Some companies have selected standard feeder relay schemes and have been able to use the same relay schemes and settings on a majority of their feeders. This practice not only reduces the amount of coordination work but in general makes field operations less complex.

Multiple instantaneous tripping of a feeder circuit breaker to prevent unnecessary lateral fuse operation during heavy lightning storms is employed on occasion.

Due to environmental constraints more emphasis and study is being directed toward the detection of low value ground fault current, greater dependence on advanced SCADA systems for reclosing and certain limited advantages of the fast reset characteristics of static relays.

Some utilities are using distance relays to permit greater loads on distribution lines.

AUTOMATIC RECLOSING [36-38]

It has long been recognized that a great many overhead line faults are temporary. If the fault can be cleared before serious damage occurs, the circuit may be automatically reenergized and service restored. This is particularly important where supervisory control is not available; however, the advent of widespread supervisory control has not had a well-defined impact on automatic reclosing practices.

Distribution

Automatic reclosing is used almost universally to improve service continuity in overhead distribution circuits. The first reclosure is usually instantaneous but with dead-time allowed for the majority of faults to clear. As many as three subsequent delayed reclosures may be used. However, certain reclosing practices have recently been suspected of contributing to the failure of source power transformers. One utility has reported a marked reduction in substation transformer failures following the elimination of the initial instantaneous reclosure. Not only did this modification result in fewer reclosing attempts, it resulted in a larger percentage of successful first reclosures, subjecting transformers to fewer through faults.

Transmission

During the last 15 years, the following reclosing concepts have received increasing acceptances:

Selective Reclosing - Various combinations of high-speed and time delayed reclosing can be selected with a switch depending on circuit breaker configuration and system conditions.

Single-Pole Reclosing - When single-pole tripping is used (tripping only the pole associated with a single-phase-to-ground fault) single-pole reclosing is applicable. Arc deionization time will be longer because of (a) capacitive coupling between the energized sound phases and the disconnected faulted phase and (b) inductive coupling between load current flowing in the energized sound phases and the disconnected faulted phase.

Out-of-Step Reclose Blocking - Out-of-Step relays may be applied to detect loss of synchronism between two systems and to block reclosing of lines which tripped for that condition.

Multi-Phase Fault Reclose Blocking - Reclosing into a multi-phase fault may severely aggravate the system swings caused by the initial fault. Such reclosing may also subject turbine-generator shafts to undesirable stresses as discussed

below. Multi-phase line fault detectors may be applied to block reclosing under these conditions.

Recently concern has been expressed over the torsional stresses on machine shafts during reclosing cycles. Shaft damage is a possibility when reclosing into close-in multi-phase faults. Some utilities avoid dead-line reclosing (both high-speed and time-delayed) on lines emanating from generating stations. A common practice is to reclose the line from the remote terminal first and then by synch-check from the generating station end.

Although no trends are obvious, high-speed reclosing is generally being reexamined by utilities for its impact on system stability and equipment damage.

FAULT AND DISTURBANCE MONITORING

Fault recording oscillographs using moving mirror galvanometers connected directly to instrument transformer secondaries have been in use for over forty years. The early instruments contained six or seven recording elements and the recording medium was photographic paper requiring dark room development. Starting times were in the order of 1/2 to 1 cycle resulting in the probable loss of the first current and voltage peak.

Faster clearing time for EHV lines created a need for an instrument which would capture all of the first fault cycle, and passive delay lines came into use in the early 60s but on a limited scale primarily because of their high cost. In 1965 the first commercially feasible prefault recorder came into existence. The technique involved a continuously rotating FM memory drum to provide about ten cycles of prefault. Most of these instruments were capable of recording a maximum of 32 variables all of which were delayed by the ten cycle memory. Direct print-out recording chart paper became available at about the same time simplifying and greatly reducing the time required for the chart development process. This encouraged many users to install additional instruments.

The most recent fault recorders use solid state digital or analog techniques for storing the fault data. In the case of the digital approach signals from instrument transformers are attenuated and digitized, delayed several cycles in shift registers, converted back to analog signals by D to A converters, and then recorded on a direct writing oscillograph. This approach lends itself to the transmission of the data in digital format to a central readout location, and a trend is developing to utilize this technique so as to obtain data on essentially a real time basis.

Another innovation was the introduction of continuous FM recording on magnetic tape. This technique provides several hours of system's recording prior to and after the occurrence of a disturbance or fault. Because it is a continuously running mechanical instrument, maintenance requirements may abe considered burdensome but the magnetic tape approach is the only method of providing these lengthy periods of recording.

Large scale sequential events recorders have been installed in power plant control rooms for many years, and recently these instruments have been finding many applications in transmission substations for monitoring the sequence of system protection functions. There are substation installations capable of monitoring over 1000 binary points with time resolution of one millisecond between successive operations. Although sequential events recorders do not normally provide analog or even magnitude data, sequential on-off information with high speed resolution has been of considerable value in analyzing the sequence of relay and power circuit breaker operations for incorrect trip outs as well as during system faults.

Several utilities have recently initiated programs to transmit sequential events data from several remote sites to a master location. An accurate common time base is used. The transmission of this on-off data to a central location together with analog data transmitted from remote fault recorders provides a powerful tool for rapid analysis of system faults and disturbances particularly with the addition of a central computer.

LOAD SHEDDING

The phenomena of system frequency changes as a result of load and generation imbalance and relays specifically designed to respond to a change in frequency are not new concepts. However, power system design parameters, rather spectacular outages and even acceptance of the term "load shedding" are recent changes affecting relaying practice. The study of system frequency decay rates, the coordination of trip frequencies and load dropping schedules on a large area basis are becoming general practice.

Early frequency relay designs were slow speed electromechanical tuned circuit type, sensitive to temperature, voltage and rate of frequency changes in operating characteristics. Higher speed relays of electromechanical tuned circuit type reduced the effects of temperature, voltage, and rate of frequency change; use of solid-state components within the relays reduced the objectionable effects still further. Recent relay designs using solid-state components generally incorporate a stable clock frequency source and a detector circuit to measure time between zero voltage crossings and include timing trip and other auxiliary functions. Reliability and aging characteristics of relay components are extremely important.

A recent relay type used extensively by one operating company is the rate of frequency change relay. Reports from that company indicate very satisfactory results have been obtained with its use. This type relay may be impossible or at least extremely difficult to coordinate with the fixed time relays unless all members of a grid use it.

The use of frequency relays for load restoration as frequency returns to normal is also practiced by some companies.

In order that unwanted trips be prevented, many schemes have evolved. These include a fixed time delay trip output, dual measurements before trip, mixed solid state and high speed systems, two out of three systems and various other combinations.

COMPUTER RELAYING[39-47]

Computers have been applied to, or considered for, a variety of off-line and on-line tasks related to electric system analysis and operation. Off-line tasks include such things as load flows, fault studies, stability studies, relay setting and coordination studies. On-line tasks include such things as generation scheduling and dispatching, supervisory control and data acquisition (SCADA), sequence of events monitoring, sectionalizing and load distribution.

The use of digital computers for protection of power system equipment is of relatively recent origin. The first serious proposals appeared in the late 1960's. A great deal of research is continuing in this field but there are no devices or systems commercially available at the present time.

A computer for any purpose has several inherent advantages that make it attractive and provide incentive for further work. It is always active. For relaying, this is particularly desirable since it permits constant monitoring and self-checking. It also has the ability to consolidate logical functions of many devices in one processor unit, thus, possibly avoiding duplication in situations where many separate pieces of equipment use identical inputs or perform similar functions. Finally, in an integrated station concept, there is potential for significant economics.

Computer relaying, however, is not inherently free of many of the problems that beset electromechanical or solid state relays. Input signal errors caused by transients, d-c offsets, CT saturation must still be recognized and considered. The station environment, for which much has been done regarding conventional devices, must be re-evaluated in terms of computer technology.

Most of the investigators in the field envision, in the ultimate, an integrated station using digital techniques involving protection, control and monitoring. One report concerned with the use of a central mini-computer lists station functions amenable to this technique. Reference 42 includes a feasibility study of using a computer for the protection of all classes of station equipment. The bibliography lists a sampling that cannot even approximate the number of papers that have been published in this area. It does present the scope and variety of the activities. An Electrical Power Research Institute Research Project with Westinghouse Electric Corporation is specifically directed towards a unified station concept using distributed, dedicated microprocessors. The actual activities and research to date has concentrated on individual functions, the most popular being line protection. Some theoretical work is being done on bus and transformer and generator protection but, this effort is much more limited.

It is revealing to note that the bulk of the effort has been under the auspices of universities and, as such, tends to concentrate on algorithms and software or on models that can be tested on multi-purpose mini-computers and on special purpose circuits and hardware that are laboratory-oriented. There have only been three actual station test installations and all three were primarily concerned with line protection with an impedance algorithm.

For the present, computer relaying cannot be considered as a viable protection system.

TESTING AND MAINTENANCE

Relay testing can be divided into four categories: Acceptance tests, calibration tests, installation tests, and routine tests.

Acceptance tests are generally made on new or unfamiliar relays to check that the relays meet the needs and specifications of the user. This testing covers the entire range of relay performance and may concentrate on areas of more particular interest to the user.

Calibration tests set the relay for the specific application for which it will be used. Relay performance is thoroughly checked at these settings.

Installation tests check out the entire protective relaying system before it is put in service and relied upon. This includes checking all external wiring to the relay - current transformers, potential transformers, and control circuit wiring. This invariably means tripping the breaker(s) by operating the relay and checking for correct current and/or voltage to the relay with the high side energized. Staged fault tests, where an actual fault is applied to the high voltage system, with backup clearing provided separate from the system being tested, are sometimes performed.

Routine tests are made to check that the relays remain in calibration and/or are functionally operable (key carrier, give an alarm, etc.). Relays are routinely tested at various intervals from six months to four years. The more complex relay systems are usually given functional tests at more frequent intervals. Currently there exists a trend to extend the routine testing interval. In general, the relay scheme is deactivated for testing while the power circuit remains in service protected by backup relays.

A practice to utilize data processing equipment for better record keeping and maintenance analysis has been incorporated by many companies. A greater emphasis on the use of solid state relays has resulted in the utilization of different test methods and test equipment suitable for electronic type devices.

REFERENCES

GENERAL SUBJECTS

[1] R.G. Farmer, A.L. Schwalb, E. Katz, "Navajo Project Report on Subsynchronous Resonance Analysis and Solutions", IEEE Trans., Jul/Aug 1977, Vol PAS 96 No 4 Pg 1226.

[2] M.C. Hall, R.L. Daniels, D.G. Ramey, "A New Technique for Subsynchronous Resonance Analysis and An Application to the Kaiparowitz System", IEEE Trans., Jul/Aug 1977, Vol PAS 96 No 4 Pg 1251.

[3] Committee Report, "A Bibliography for the Study of Subsynchronous Resonance Between Rotating Machines and Power Systems", IEEE Trans., Jan/Feb 1976 Vol PAS 95 No 1 Pg 216.

[4] W.C. Kotheimer, L.L. Mankoff, "Electromagnetic Interference and Solid State Protective Relays", IEEE Trans., July/Aug 1977 Vol PAS 96 No 4 Pg 1311.

[5] Committee Report, "Bibliography of Gas Insulated Substations", IEEE Trans., Jul/Aug 1977 Vol Pas 96 No 4 Pg 1280.

[6] H.W. Dommel, J.M. Michels, "High Speed Relaying Using Travelling Wave Transient Analysis", IEEE Trans., A 78-214-9.

[7] M.T. Yee, J. Esztergalyos, "Ultra High Speed Relay for EHV/UHV Transmission Lines-Installation, Staged Fault Tests and Operational Experience, IEEE Trans., F 78 222-2.

[8] M. Chamia and S. Liberman, "Ultra High Speed Relay for EHV/UHV Transmission Lines - Development, Design and Application," IEEE Trans., Vol PAS 97, No 6, Nov/Dec 1978, Pg 2104-2116.

[9] G. Manchur, R. E. Beaulien, R. D. Brown, "Line Protection Using Low Level Principles", IEEE Trans., A 78 218-0.

[10] S.C. Sun, J.C. Gambale, "An EHV Current Transducer", IEEE Trans., C 73 330-8.

AC GENERATOR PROTECTION

[11] Eric T. B. Gross and Edward M. Gulachenski, "Experience of the New England System with Generator Protection by Resonant Neutral Grounding," IEEE Trans., July/August, 1973, vol. PAS 92, pp. 1186-1194.

[12] C. C. Carson, S. C. Barton, L. P. Grobel, "Immediate Detection of Overheating in Gas-Cooled Electrical Machines," IEEE Winter Power Meeting, 1971.

[13] Standard for Cylindrical Rotor Synchronous Generators, ANSI Standard C50.13-1977.

[14] D.J. Graham, P.G. Brown, R.L. Winchester, "Generator Protection with a New Static Negative Sequence Relay," IEEE Trans., July/Aug 1975, vol. PAS 94, No. 2, Pg. 1202-1213.

[15] J. Berdy, P.G. Brown, L. E. Goff, "Protection of Steam Turbine Generators During Abnormal Frequency Conditions," General Electric Publication "Relaying The News" RN No. 75, July 26, 1974.

GENERATOR AUXILIARIES

[16] J.L. Koepfinger and K.J.S. Khunkhun, "Protection of Auxiliary Power Systems in a Nuclear Power Plant," IEEE Trdns., Vol PAS 98, No 1, pp. 290-299, Jan/Feb, 1979.

AC MOTOR PROTECTION

[17] W.A. Elmore, "Some New Thoughts on Large Motor Protection." Protective Relaying Conference, Georgia Tech, 1976.

[18] D.R. Boothman and E.C. Elgar, "Thermal Tracking - A Rational Approach to Motor Protection." IEEE Winter Meeting, New York, N.Y. January 27-February 1, 1974.

[19] "Guide for AC Motor Protection," ANSI C37.96-1975 (IEEE Std. 588-1976).

TRANSFORMERS, SHUNT REACTORS AND SHUNT CAPACITORS

Transformers

[20] K. Winick and W. McNutt, "Transformer Magnetizing Inrush Currents and Harmonic Restrained Differential Relays," General Electric Publication, Relaying The News, RN No. 86 November, 1975.

[21] C.L. Wagner, "Overexcitation Problems and Protection of Generators and Transformers," Westinghouse Electric Corp. publication Silent Sentinels, RPL 67-7, October, 1967.

[22] C.H. Einvall and J.R. Linders, "A Three Phase Differential Relay for Transformer Protection," Vol PAS 94, Part 6, Nov/Dec 1976, pp. 1971-1980.

[23] E. Boyaris and W.S. Guyot, "Experience with Fault Pressure Relaying and Combustible Gas Detection in Power Transformers," 1971 American Power Conference Proceedings, Vol 33, pp. 1116-1125.

[24] "Guide for Protective Relay Applications to Power Transformers," ANSI C37.91-1972, IEEE Std. 273-1967.

Shunt Reactors

[25] M. Christoffel, "The Design and Testing of EHV Shunt Reactors," IEEE Trans., Vol PAS 86, June 1967, pp. 684-692.

[26] J.P. Vora and A.E. Emanuel, "Sensor Coil for Internal Fault Protection of Shunt Reactors," IEEE Trans. Vol PAS 93, Nov/Dec 1974, pp. 1917-1926.

Shunt Capacitors

[27] IEEE Guide for Protection of Shunt Capacitor Banks, C37.99 (P569/D8) - 1979.

BUS PROTECTION

[28] T. Forford and J.R. Linders, "A Half Cycle Bus Differential Relay and Its Applications," IEEE Trans., July/Aug 1974, VOL PAS 93, pp. 1110-1120.

[29] Sandor Mentler, "A Half Cycle Static Bus Protection Relay Using Instantaneous Voltage Measurement," IEEE Trans., May/June 1975, Vol PAS 94, pp. 939-944.

[30] IEEE Guide for Protective Relay Applications to Power System Busses, ANSI 37.97-1978.

[31] J. Kauferle, D. Pohu, "Concepts of Overvoltage and Overcurrent Protection of HVDC Converters," CIGRE 1978, 14-08.

TRANSMISSION AND SUBTRANSMISSION LINE RELAYS

[32] IEEE PSRC Special Publication - Protection Aspects of Multi-Terminal Lines, 18 pages.

BACKUP PROTECTION

[33] IEEE Committee Report, "Local Backup Relaying Protection" IEEE Trans. Vol PAS 89, pp. 1061-1068, July/Aug 1970.

DISTRIBUTION PROTECTION

[34] R.G. Draw, C.W. Fromen, J.D. Borst and A.M. Lockie, "The Application of Current-Limiting Fuses to Padmounted XURD Systems," IEEE Trans. Vol PAS 91, No 3, May/June, 1972, pp. 946-951.

[35] E.A. Goodman, "A New Approach to Distribution Transformer Protection," Proceedings of the American Power Conference, Vol 34, 1972, pp. 984-992.

AUTOMATIC RECLOSING

[36] L. Johnston, et al, "An Analysis of VEPCO's 34.5 kV Distribution Feeder Faults as Related to Through Fault Failures of Substation Transformers," IEEE Trans., PAS Vol 97, No 5, Sept/Oct, 1978, pp. 1876-1883.

[37] R. Haun, "13 Years' Experience with Single-Phase Reclosing at 345 kV," IEEE Trans. PAS, Vol 97, No 2, March/April, 1978, pp. 520-528.

[38] A. Abolins, et al, "Effect of Clearing Short Circuits and Automatic Reclosing on Torsional Stress and Life Expenditures on Turbine-Generator Shafts," IEEE Trans., Vol. PAS 95, No 1, January/February, 1976, pp. 14-22.

COMPUTER RELAYING

[39] PSRC Report - Central Computer Control & Protection Functions. IEEE F76 317-8.

[40] G.D. Rockefeller, "Fault Protection with a Digital Computer," IEEE Trans., Vol. PAS *8, No. 4, April, 1969, pp. 438-461.

[41] J. Carr and R.V. Jackson, "Frequency Domain Analysis Applies to Digital Transmission Line Protection," IEEE Trans., July/Aug, 1975, Vol. PAS 94, pp. 1157-1166.

[42] J.G. Gilbert, E.A. Udren and J. Sackin, "Evaluation of Algorithms for Computer Relaying," IEEE A77 520.

[43] G.S. Hope, P.K. Dash and O.P Makik, "Digital Differential Protection of a Generating Unit: Scheme and Real-Time Test Results," IEEE Trans., Mar/Apr, 1977 Vol PAS 96, pp. 502-512.

[44] G.D. Rockefeller and E.A. Udren, "High-Speed Distance Relaying Using a Digital Computer, Part II - Test Results," IEEE Trans., May/June, 1972 Vol PAS 91, pp. 1244-1258.

[45] A.G. Phadke, T Hlibka and M. Ibrahim, "A Digital Computer System for EHV Substations: Analysis and Field Tests," IEEE Trans., Jan/Feb 1976, Vol PAS 95, pp. 291-301.

[46] W.D. Breingan, M.M. Chen and T.F. Gallen, "Laboratory Investigation of a Digital System for Protection of Transmission Lines," IEEE F77 052-4.

[47] W.D. Breingan, M.M. Chen and T.F. Gallen, "Field Experience with a Digital System for Transmission Line Protection," IEEE F79 252-8.

Discussion

John T. Tengdin (General Electric Company, King of Prussia, PA): The committee has done an excellent job in summarizing recent practices and trends in protective relaying. This paper will be a valuable reference for years to come. In the area of generator protection, an additional paper could be cited. With the new larger generators, an out-of-step condition may cause the impedance locus to swing through the generator or step up or transformer impedance. In such cases, out-or-step protection for generators must then be considered (Reference A).

REFERENCE

[A] J. Berdy, "Out of Step Protection for Generators", Protective Realy Conference, Georgia Tech. and Texas A&M, 1976. General Electric Publication *Relaying The News* RN No. 92, July 1976.

Manuscript received May 11, 1981.

J. M. Crockett (Westinghouse Canada Inc., Hamilton, Ontario, Canada): This paper is a commendable effort, and it would seem fitting that those responsible should be named.

In the area of stator ground protection for high resistance grounded generators, more specific mention might be made of methods in current use. These include third harmonic neutral undervoltage detection; comparison of neutral and terminal third harmonic voltage magnitudes; and monitoring of current resulting from injection of a low frequency signal into the stator circuit.

Shunt reactor shorted turn detection by means of zero sequence potential polarized directional relays might also be included.

The use of current compensation to improve the polarizing voltage available to zero or negative sequence directional relays is also worthy of mention.

Distance relay dynamic characteristics due to "memory action" have been increasingly exploited, and improved methods have been developed for analyzing distance relay response during system disturbances.

Positive sequence restrained zero or negative sequence overcurrent relays have been used to prevent relay misoperation due to asymmetrical by pass of capacitors in series compensated lines.

Manuscript received February 23, 1981.

L. E. Landoll: My apologies to the members of the Working Group in that their names were omitted during the publishing process. They are: L. E. Landoll, D. C. Adamson, W. H. Butt, R. E. Carlson, M. J. Fein, K. R. Gruesen, S. H. Horowitz, F. B. Hunt, T. L. Kaschalk, D. K. Kaushal, M. D. Limerick, T. J. Murray, T. Niessink, S. Nilsson, H. M. Paul, M. Rosen, E. T. Sage, Jr., H. S. Smith, and F. Von Roeschlaub. With regard to Mr. Tengdin and Mr. Crockett's comments on the paper, I can only make the following general statements. With the volume of items reviewed by the Working Group, it was not possible to include all possible references. We would direct any interested parties to the IEEE Relay Bibliography of published papers. It was also not possible to address details of certain types of protection except in general terms. Also, the scope of the project was to address trends; that is, items that were becoming used in some significant quantity, and practices; those things which have become generally applicable to power systems. We thank the discussers for their comments. I am sure they will be useful to those who read the paper.

Manuscript received April 3, 1981.

RELAY PERFORMANCE IN DSG ISLANDS

C. L. Wagner, Fellow
Electric Research
& Management, Inc.
Export, PA

W. E. Feero, Fellow
Electric Research
& Management, Inc.
State College, PA

W. B. Gish
Senior Member
Electric Research
& Management, Inc.
Thomaston, ME

R. H. Jones
Senior Member
Rochester Gas &
Electric Corp.
Rochester, NY

ABSTRACT

When Distribution System Generators (DSGs) involving induction and synchronous generators are separated from the utility source, they can become self-excited and drive the isolated section into ferroresonance provided: the isolated load is less than three times the rating of the DSG, there is adequate capacitance and there is at least one transformer remaining connected to the island. High overvoltages and distorted wave shapes resulting from this condition can damage utility equipment and other connected loads. Relays must be responsive to this condition and remove the DSG itself from the island.

During field tests to confirm the existence of these ferroresonant conditions, a number of voltage and frequency relays were installed and their performance monitored while exposed to these highly distorted waves. While the general performance of relay groups was satisfactory, certain problem areas became apparent. These are discussed in the paper.

INTRODUCTION

The phenomenon of self-excitation of induction generators has been known for many years [1-3]. It occurs when an isolated generator is connected to a system having capacitance equal to or greater than its magnetizing reactance requirements. Depending on the value of the capacitance and the kilowatt loading on the machine, terminal voltages as high as 1.5 to 2.0 per unit can be produced.

While recognized as a potential problem, there have until recently been very few cases where system conditions have been such to produce these self-excitation conditions. With the increase of DSGs in recent years however, the problem has become real. The possibility of DSG islands being formed when feeder faults trip utility substation breakers raises the question as to whether the capacitance and load left in the island can cause the self-excitation condition.

To compound the problem for these types of systems, studies have shown that a special case of ferroresonance, which can cause overvoltages of over 3.0 per unit [4-6], can also occur. These overvoltages are produced by the discharging and charging of the system capacitance through the highly non-linear

magnetizing reactance of the system transformers as they pass into and out of a saturated condition. The result is high overvoltage and distorted waveforms which not only contain the ferroresonance but also all the natural resonant frequencies of the distribution circuits excited by the ferroresonant pulses.

While the exact description of the self-excitation and ferroresonance phenomena is contained in the references [4-6], the following conditions must exist for it to occur:

1. The DSG must be separated from the utility source (the islanding condition).
2. The kilowatt load in the island must be less than 3 times the rating of the DSG.
3. The system capacitance must be greater than 25 and less than 500 percent of the rating of the DSG generation.
4. There must be at least one transformer connected to the island.

If all these conditions exist and ferroresonance occurs, the techniques for mitigation become paramount. The studies have shown that all types of generators (induction and synchronous, single-phase and three-phase) are susceptible. All types of transformer connections (wye-delta, delta-wye, wye-wye, delta-delta) are also susceptible. Changes in the generators or transformers will not alleviate the problem.

Surge arresters will clip the peaks of the overvoltages, but will not suppress the ferroresonance. Unfortunately the arresters may be damaged thermally in the process. The most practical solution is to trip the DSG from the system and remove the driving source. But this tripping requires protective relays which may not have characteristics completely defined for this application. Many questions arise. How do the relays react to this phenomena? Do the relays perform satisfactorily for the distorted waveforms encountered? Field tests were run in an attempt to answer these questions.

FIELD TEST DESCRIPTION

The analytical studies of the ferroresonance phenomena were conducted using the hybrid computer and power system simulator at Purdue University and a digital transient program developed by Electric Research & Management, Inc. These studies occurred in three steps; an initial general study and two succeeding studies to investigate additional system parameters. To validate the findings of the analytical studies, a series of field tests were conducted on a distribution feeder test line constructed by Rochester Gas & Electric Corporation on the campus of the Rochester

88 WM 124-0   A paper recommended and approved by the IEEE Power System Relaying Committee of the IEEE Power Engineering Society for presentation at the IEEE/PES 1988 Winter Meeting, New York, New York, January 31 - February 5, 1988. Manuscript submitted August 27, 1987; made available for printing November 4, 1987.

Reprinted from *IEEE Trans. Power Delivery*, vol. 4, no. 1, pp. 122-131, Jan. 1989.

Institute of Technology. These, too, occurred in three steps, each following its associated analytical study.

The test line, shown schematically on Figure 1, had an overall length of 2800 feet. Construction was based on 12470 - 7200 volt standards but the circuit was operated at 4160 volts (to prevent damage due to the 3.0 per unit overvoltages expected). The tests involved the use of diesel-driven generators of different types: two different three-phase induction units, a three-phase synchronous unit, and a single-phase induction unit. Loads were simulated by various sized resistors fed by both three-phase and single-phase transformers on the feeder or directly connected to the 208 volt DSG bus. For several tests an induction motor load was also connected to the 208 volt bus.

Capacitors of various sizes were connected in grounded wye to the 4160 volt circuit and/or in delta on the 208 volt bus. The DSG step-up transformer was connected either grounded wye - grounded wye or grounded wye-delta with the wye on the 208 volt side. The feeder recloser was opened to initiate the islanding condition.

The main purpose of the field tests was to verify the results of the computer studies; the tests also provided the opportunity to monitor the operation of the relays during these islanding conditions. During the first two series of tests, the measuring and recording equipment was selected to provide rapid on-the-spot display and measurement and only a few relays taken from the utility stock were tested. The resulting relay data was interesting and indicated that some of the relays could be confused by extremely distorted waveforms. The data was not sufficiently complete to draw objective conclusions.

For the final test series, performed in August of 1985, the oscillographs were chosen to allow more complete monitoring and digital analysis of relay performance. The test sample was expanded to 73 relay functions representing relays from nine manufacturers. The relays primarily monitored the four functions felt necessary for DSG interconnection protection: overvoltage, undervoltage, overfrequency, and underfrequency.

There were other relays available, such as overcurrent, negative sequence, and reverse power. Since faults on the feeder while connected to the utility were not performed, data from these relays were statistically insignificant.

Some of the relays were taken from utility stocks and some were supplied by the manufacturers. To minimize comparison problems, all relays were to have had time delays set to a minimum. However the minimum time requirements were not clearly communicated and some of the relays used could not respond in the desired time period.

Data for each test in this series was recorded on a Rochester Instrument Systems TR-1620 digital recorder. The three phase-to-ground voltages were recorded and stored in digital form to be played back when required or desired. The status of the various relay contacts was also recorded together with their times of operation. Since there were 86 contacts for the 73 relay functions and the recorder could accept only 48 contact inputs, double runs for each set of system conditions were required. Half the relays were monitored for each of the double runs.

EVALUATION METHOD

All the relays were set approximately the same as they would be on an actual system: the overvoltage relays to 1.1 per unit, the undervoltage relays to 0.9 per unit, the overfrequency relays to 60.5 Hz, and the underfrequency relays to 59.5 per unit. To determine when an instantaneous relay should have operated, three software based detectors were developed and reviewed by the manufacturers that determined when the above limits were exceeded.

The software detectors were based on test data taken from the second field test. In an attempt to understand the relaying problem, a plot of the locus of the phase voltage amplitude vs frequency was prepared from several field tests. Although the plot does not provide as clear a picture of the actual harmonic activity of the voltage, it does show the complex nature of the signals that relay detectors must utilize. Figure 2 shows paths that the voltage took for several tests. On path 1 the voltage-frequency signal exceeds trip levels for all four functions and even returns momentarily to operation inside the voltage-frequency window. On path 2, the voltage rises on phase A but falls on phase B (path 3). Path 4 demonstrates that, except for very brief periods, only the overfrequency boundary is exceeded.

Since some of the relays respond to peak voltages while others respond to rms quantities, both peak and rms voltage software detectors were required. The third type of detector was a frequency detector.

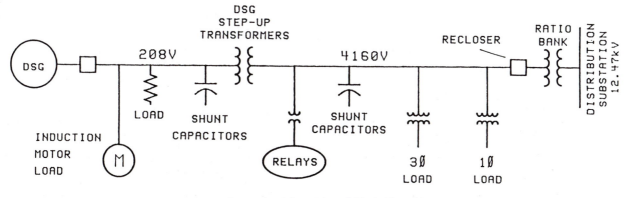

Figure 1. Schematic of Test Circuit.

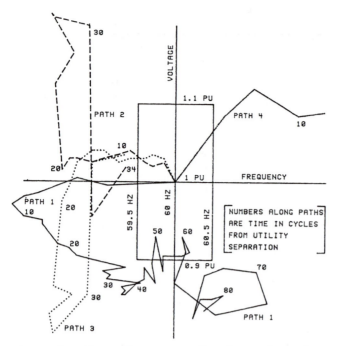

Figure 2. Voltage-frequency paths for various phase voltages measured during field testing. Path 1: 56 kw induction DSG, 64 kw load, 33 kvar capacitance, wye-wye step-up transformer, phase A. Path 2: 56 kw induction DSG, 48 kw load, 33 kvar capacitance, wye-wye step-up transformer, phase A. Path 3: same as path 2 but phase B. Path 4: 56 kw induction DSG, 0 kw load, 33 kvar capacitance, wye-delta step-up transformer, phase B.

Each of these software detectors was adjusted to pick up at the same levels at which the relays were also set. By comparing the software detector times with the actual relay times a judgment could be made about the adequacy of the relay.

The first step in the evaluation, therefore, was to analyze the oscillograms and detector times to determine which relays should have operated for each case. For the case of Figure 3 for example, all three phases show high magnitude, highly distorted, underfrequency traces. Phase "b" voltage also collapsed for several cycles at the beginning of the transient. This would indicate that the overvoltage, undervoltage, and underfrequency relays should operate. The overvoltage detectors in all three phases operated in approximately 1/2 cycle, the undervoltage detector on phase "b" operated in approximately 1 1/2 cycles, and the underfrequency detector in approximately 2 cycles. All the relays then were checked as to their operating times (if any) for this test condition and a judgment made as to whether the operation was satisfactory or not.

But how fast should the relays operate? The following system conditions were considered:
1. If high overvoltages (1.5 per unit or higher) were experienced, damage could occur to other loads on the feeder. Limiting the time to 10 cycles relay time plus up to 5 cycles of circuit breaker time should protect surge arresters and other equipment.
2. A moderate sustained voltage would not damage equipment but could be a safety

hazard to maintenance workers.
3. A decaying voltage, while not a damage or safety problem, could damage the DSG should the utility not block reclosing of its substation breakers or reclosers.
4. While the voltage may be steady or decaying initially, as it decays the load characteristics may change with frequency and subsequently go into ferroresonance.

For these reasons and from the data similar to Figure 2, it was felt that all relays should operate in 10 cycles or less. This time was used in the following as the criterion for a successful operation.

It is recognized that a problem might exist if this criterion were used for all over and undervoltage relays. A fault on an adjacent substation feeder might cause a low voltage on the faulted phase and a high voltage on the unfaulted phase. This could cause a DSG trip for many faults on adjacent feeders. Unnecessary trips may be acceptable in some cases and unacceptable in others. For the latter it would be necessary to delay the voltage relays beyond the maximum fault clearing time of the adjacent feeders. It might also be possible to use two overvoltage relays: an instantaneous (or less than 10 cycles) relay set to 1.3 to 1.5 per unit (which is above the unfaulted phase voltage), and a second set to 1.1 per unit and timed longer than the adjacent feeder fault clearing time. This would satisfy all the above criteria for overvoltage conditions. The undervoltage trip may be lengthened as long as it is still less than the utilities fastest reclose time. If the utility is certain to block reclosing, for example by means of a live line test, the undervoltage trip time might be further increased.

In the analysis, however, the 10 cycle time criterion was used because it would require no change in present utility operating practice which then would not require years of operating experience before possible undesired effects on non-generating customers might be discovered. Note that this criterion was selected after the tests and not before; so some of the relays (such as the induction disc relays) could not possibly meet this requirement.

TEST RESULTS
The performance of each relay and each relay group used during the 26 tests was analyzed separately. These results and comments on specific questionable performance were sent to each manufacturer. However, this data will not be published because these tests by themselves are not an adequate basis for relay selection, for the following reasons:
1. The 10 cycle performance criterion was not clearly communicated prior to the field tests and thus precluded more optimal relay schemes from being supplied.
2. Several manufacturers have indicated that since the tests in August 1985, relays better suited to DSG applications have been developed.
3. There is no statistical basis for performance. Only one relay of each style or model was tested. This relay may not be representative of the complete line.

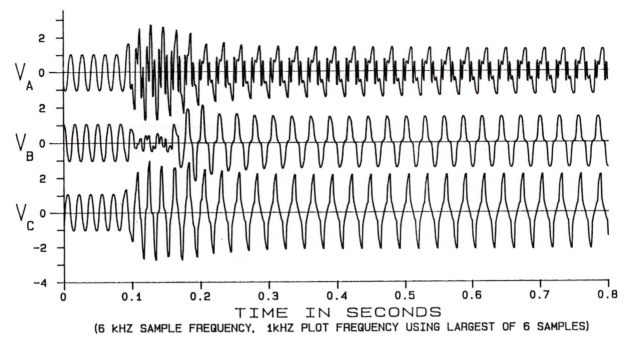

TIME IN SECONDS
(6 kHZ SAMPLE FREQUENCY. 1kHZ PLOT FREQUENCY USING LARGEST OF 6 SAMPLES)

Figure 2. Case #A7. 50 kw synchronous DSG, 9 kw load, 100 kvar capacitance, and wye-delta step-up transformer. Maximum voltages: A = 2.74 p.u., B = 2.34 p.u., C = 2.92 p.u.

4. Complete acceptance or specification verification tests were not made on all relays prior to the testing. Relays were set and given routine in-service tests by experienced relay technicians; however, time did not permit meticulous acceptance type testing for each relay.
5. The relays for this test were chosen based on availability, requests of DSG owners, and test participants and manufacturers suggestions. Such a sampling approach invalidates any manufacturer-to-manufacturer comparisons that might be attempted.

Because of the above, and because injustices might be done by using the data out of context, the data on specific manufacturer's relays is only available from that manufacturer. Questions on specific relays should therefore be directed to the manufacturer.

PROBLEM AREAS

While the data gathered was not adequate for the analysis of the performance of specific relays, it did serve the purpose originally intended. Through the analysis of a large number of different relays and a large number of test conditions, problem areas both with specific types of hardware and specific application concepts were discovered. The data also pointed out certain general problem areas that need further evaluation.

Induction Disc Relays

Several manufacturers supplied induction disc relays and as expected, their operating times were longer than the 10 cycle operating criterion. If one is to insist that these criteria be met, then induction disc relays should not be used for this application.

One possible application for induction disc relays would be as a companion to a high speed overvoltage relay. The high speed overvoltage relay would be set to 1.3 to 1.5 per unit and be used to prevent damage to other feeder loads. A slow speed relay set to 1.1 per unit would be used for all other self-excitation and ferroresonance conditions. This latter relay could be an induction disc relay.

Excessive Time Delay in Overvoltage and/or Undervoltage Relays

In addition to the induction disc relays, many of the other relays tested had time delays that prevented them from meeting the 10 cycle operating criterion. Some of these relays had adjustable delays which could have been set to a lower value. Others had built-in delays that could be reduced by a modification to the design or by selecting a different model with a shorter delay that is available from the manufacturer.

One of the problems encountered was simply that tripping occurred in times longer than 10 cycles for simple over or undervoltage conditions. Sometimes the beat frequency, having both overvoltage and undervoltage peaks, caused delayed tripping due to insufficient time of the particular over or undervoltage elements to time out.

A specific problem exists with combined over/undervoltage relays where the long relay times, combined with the pulsating nature of the wave, prevents either element from operating. An example of this is shown on Figure 4 where the voltage increases and then decreases with a period of approximately 600 milliseconds. If the overvoltage element has sufficient delay it will reset when the voltage decreases below pickup. The undervoltage element will similarly reset when the voltage again increases. Tripping may never occur.

23

Excessive Time Delay in Overfrequency and/or Underfrequency Relays

A similar situation exists with the frequency relays, especially with the highly distorted wave shapes encountered during ferroresonance. For example, in Figure 5, which was a true overfrequency condition, there were cases of either extremely long tripping times or failure to trip altogether. This was caused by the distorted wave exhibiting one or more cycles of long zero crossing times each beat cycle which allowed the overfrequency relay element to reset. Shorter relay operating times may have cured this situation.

Lack of Individual Overvoltage/Undervoltage Phase Relays

There were a number of cases where there was an undervoltage condition on one phase but no undervoltage relay installed on that same phase. Figure 6 shows such a condition. Fortunately, there was usually an overvoltage condition on another phase with an overvoltage relay connected to that phase. Therefore, correct tripping occurred. It is conceivable, however, that if only one overvoltage and one undervoltage relay were used, a situation could occur that no tripping would result. Individual relays on each phase are therefore recommended.

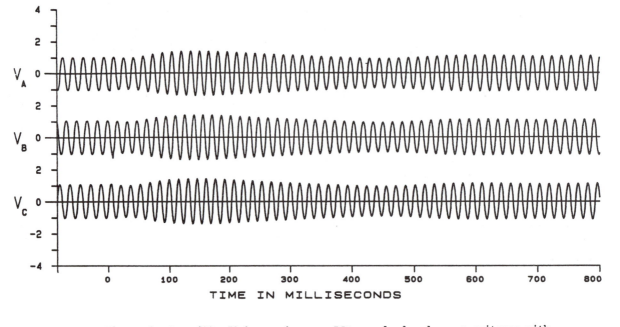

Figure 4. Case #92. 50 kw synchronous DSG, no load and no capacitance with wye-wye step-up transformer. Maximum voltages: A = 1.39 p.u., B = 1.4 p.u., C = 1.4 p.u.

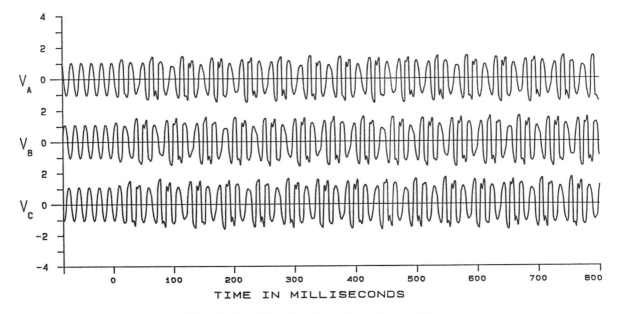

Figure 5. Case #20. 64 kw induction DSG, 12 kw load, 100 kvar capacitance, and wye-wye step-up transformer. Maximum voltages: A = 1.52, B = 1.63 p.u., C = 1.69 p.u.

Averaging of Three Single Phase Voltages

Problems can occur if the three single phase voltages are averaged to operate an overvoltage or undervoltage element. Figure 7 for example, shows Phase "a" and "c" with a definite overvoltage condition but Phase "b" with a definite undervoltage. The average of the three however, varies just slightly above and below the pickup settings of the relay. In this particular case, the over/undervoltage relay did not operate. Perhaps if the delay time of the relay had been reduced, it might have operated, but it indicates a potential problem area.

Effect of Low Voltage on Frequency Relays

Some frequency relays are designed to trip when the applied voltage is reduced below a certain level. Others are designed to prevent tripping under this low voltage condition. In the former, a false indication of a low frequency condition would be given which might be confusing to the operator. It might also trip the DSG falsely for a fault on an adjacent feeder if the time of the tripping were not delayed. The pros and cons of this were discussed in a preceding section of this report.

For those relays that block tripping, the DSG may fail to trip unless a separate undervoltage relay was applied to the same phase as the frequency relay.

While neither of these situations is intolerable, they should be recognized by the user and suitable measures taken into account for them.

Need for Individual Frequency Relays per Phase

In most cases, an over or underfrequency condition can be recognized by relays connected to only one phase of the three-phase circuit. The exception is the low voltage tripping or blocking condition just mentioned. To avoid the problems mentioned above, some users may want to install frequency relays on more than one phase.

SUMMARY

It has been clearly shown from the field test experiences that the most difficult problem encountered in using protective relays for DSG applications is the communication of relay requirements between the user, the utility and the manufacturer. The tests have provided a vehicle which can aid in understanding these requirements.

One should not get the impression that the relay problem areas represent insurmountable difficulties or are major problems. In practically all cases, at least one relay from each manufacturer's package, did or would have operated in 10 cycles or less, if the proper relay or proper setting had been used. By examining the specific problems encountered during the tests however, the relay application engineer and the manufacturers can provide means for correcting or minimizing these difficulties.

From this extensive series of tests, it can be concluded that unless careful site specific analysis has been performed, prudent relaying practice requires that a full complement of relay functions be provided at any interconnection. A full complement of functions includes:

1. At least one overvoltage element per phase to detect a condition that might damage other loads and equipment on the feeder. If only one relay is used it should be set at 1.1 per unit. If two relays are used per phase, the high speed unit should be set to 1.3 to 1.5 per unit and the slow speed unit at 1.1 per unit.

2. At least one undervoltage element per phase to detect a sustained voltage that could cause overheating of motor loads and be hazardous to utility personnel. It would also be used to prevent the DSG circuit breaker or contactor from closing until the feeder is energized from the utility source. These relays should be set at 0.9 per unit.

3. At least one overfrequency element to

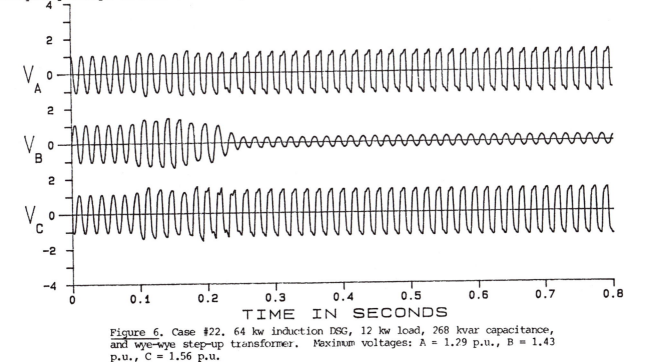

Figure 6. Case #22. 64 kw induction DSG, 12 kw load, 268 kvar capacitance, and wye-wye step-up transformer. Maximum voltages: A = 1.29 p.u., B = 1.43 p.u., C = 1.56 p.u.

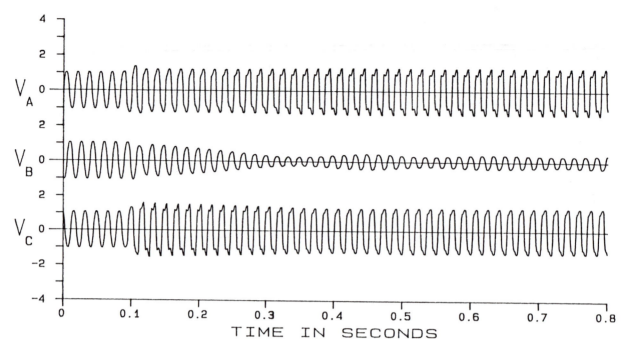

Figure 7. Case #51. 64 kw induction DSG, 11 kw induction motor load, 200 kvar capacitance, and wye-wye step-up transformer. Maximum voltages: A = 1.38 p.u., B = 1.08 p.u., C = 1.57 p.u.

detect those cases where the self-excitation voltage is between 0.9 and 1.1 per unit, and the feeder load is much less than the DSG rating. The relay, set at 60.5 Hz, would operate on the resulting increase in frequency.

4. At least one underfrequency element to detect those cases where the self-excitation voltage is between 0.9 and 1.1 per unit, and the feeder load causes a decrease in frequency. This relay would normally be set at 59.5 Hz.

ACKNOWLEDGMENT
The study and field tests of small induction and synchronous generators for DSG applications was funded by Pennsylvania Power and Light Company, Philadelphia Electric Company, and the Empire State Electric Energy Research Corporation (Central Hudson Gas & Electric Corporation, Consolidated Edison Co. of N.Y. Inc., New York Power Authority, New York State Electric & Gas Corporation, Niagara Mohawk Power Corporation, Orange and Rockland Utilities, Inc. and Rochester Gas & Electric Corporation).

The relays used and/or supplied for the tests were from the following manufacturers: Adtech, Inc., ASEA Electric, Basler Electric Co., BBC Brown Boveri, Inc., Beckwith Electric Co., General Electric Co., Square D, Westinghouse Electric Corp., and Wilmar Electronics.

Special mention should be given to D. L. Basset of Pennsylvania Power & Light Co. and T. F. Gallen of Philadelphia Electric Co. for their assistance during the tests and data analysis, and to S. Greuel of Marathon Electric, M. A. Reynolds of Continental Electric Co., E. Schrom of the New York Public Service Commission, and the Rochester Institute of Technology.

REFERENCES
1. E.D. Bassett and F.M. Potter, "Capacitive Excitation for Induction Generators", AIEE Transactions Vol. 54, May 1935, pp. 540-545.
2. C.F. Wagner, "Self-Excitation of Induction Motors", AIEE Transactions Vol. 58, February 1938, pp. 47-51.
3. J.E. Barkle and R.W. Ferguson, "Induction Generator Theory and Application", AIEE Transactions Vol. 73, February 1954, pp. 12-19.
4. W.E. Feero and W.B. Gish, "Overvoltages Caused by DSG Operation: Synchronous and Induction Generators", IEEE Transactions on Power Delivery Vol. 1, January 1986, pp. 258-264.
5. W.B. Gish, "Small Induction Generator and Synchronous Generator Constants for DSG Isolation Studies", IEEE Transactions on Power Delivery Vol.1, April 1986, pp. 231-239.
6. W.B. Gish, W.E. Feero and S. Greuel, "Ferroresonance and Loading Relationships for DSG Installations, IEEE Transactions on Power Delivery Vol. 2, July 1987, pp. 953-959.

Discussion

J. B. Bunch and **L. J. Powell** (General Electric Company, Malvern, PA and Schenectady, NY): The test results described by the authors provide interesting and helpful data for relay application engineers where small induction and synchronous generators are interconnected to distribution systems. There are several areas which should be clarified when considering these results and the relaying described in the paper.

The term DSG normally has been used in the industry to mean dispersed storage and generation, rather than distribution system generator. In this sense, DSG encompasses a wide variety of types and sizes of generation. Application may be on the utility distribution, subtransmission, or transmission system and/or industrial system. Sizes may range from the generator ratings used in the tests up to 100 MW or larger. Considering the possible range of DSG applications, the over and under voltage and over and under frequency relaying functions will often be a minimum complement of relaying rather than a maximum complement. System applications will often require additional relaying functions, depending on the size and type of DSG.

Several criteria are listed in the paper as requirements for self-excitation and ferroresonance conditions to occur in an islanding situation. It would be helpful if the authors would describe further the basis for the criteria that kilowatt load must be less than three times and system capacitance must be less than five times the rating of the DSG generation. Comments on extrapolation of the results and criteria to other system configurations and larger DSG sizes would be helpful.

Fault conditions were not considered in the tests. Would the authors anticipate any effects on the results if faults had been imposed to initiate the disturbance, rather than initiation from opening the recloser?

The paper provides a very helpful addition to the DSG literature in providing test results and describing conditions which may be encountered in protective relaying applications in this area.

Manuscript received February 12, 1988.

D. C. Dawson (South California Edison Co., Rosemead, CA): The authors are to be commended for making the results of these unique studies and tests available to the industry. Their comments on the application problems of particular relaying principles are incisive and valuable.

At Southern California Edison, we protect DSG installations with three voltage relays, but these are grouped differently than the one-relay-per-phase recommended by the authors. Our installations utilize one zero-sequence overvoltage relay and two over/undervoltage relays connected to phase-to-phase voltages. We use this approach because it separates the overvoltages caused by single line-to-ground faults from those which could be caused by DSG isolation. This technique allows optimum settings for each relay. We would therefore be interested in whether the authors have investigated the zero-sequence and phase-to-phase components (or Clarke components) of their results. Such an analysis might also lead to additional insights into the ferroresonance mechanism which produces the overvoltages.

Additionally, we would like to second the authors' caution about undervoltage blocking functions on underfrequency relays. We have already experienced one incident in which clearing under low voltage, low frequency isolation conditions were much slower than desired because the frequency relays were set to block at a relatively high undervoltage level. Such high settings are appropriate for underfrequency load-shedding relays where it is desired not to have the frequency relays operate if the area is deenergized. They are not desirable in DSG applications where the goal is to disconnect the DSG as rapidly as possible after islanding.

Our experience to date supports the authors' conclusion that the relaying problems of DSG's are not insurmountable. With many hundreds of such installations in service, we have yet to have any evidence of utility or customer equipment damage from DSG-caused overvoltages. We attribute this to adequate relaying and careful attention to neutral grounding and var supply for the DSG's.

Manuscript received February 12, 1988.

Walter A. Elmore (Westinghouse Electric Corporation, Coral Springs, FL): The authors, Pennsylvania Power & Light, Philadelphia Electric Co., and ESEERCO are to be congratulated for undertaking this sort of field investigation. Laboratory and Model Power system tests have produced marvelous advances in the art over the years, but only field tests can provide the concrete assurances of proper relay response.

Fig. 2 of the paper is particularly interesting. Part 2 is described as a case

involving more generation than load, yet underfrequency results for at least 34 cycles. This occurs even though the voltage is only moderately elevated for the first 20 cycles. How is this explained? Also on this path, an abrupt voltage elevation occurs at 20 cycles. Could the authors identify the cause of this?

Was there any unique character of the 33 kvar of capacitance except that it overcompensated all of the induction machines tested?

Manuscript received February 18, 1988.

C.L. WAGNER, W.E. FEERO, W.B. GISH, AND R.H. JONES: The authors thank Messrs. Bunch, Powell, Elmore and Dawson for their comments. The comments are a welcome addition to the paper.

To answer Messrs. Bunch and Powell, the criterion for possible ferroresonant conditions based on the size of the load and capacitance was developed in Reference 1 of the paper. This criterion is based on the induction and synchronous machine equations and on the fact that speed and voltage are not constant but may vary over significant ranges. The analysis would apply to any size machine. In practice, the larger machines would not normally be isolated with excessive capacitance as is possible with small machines on distribution systems and the larger transformers may have higher saturation voltages with less saturation slope. The larger machines would be more extensively protected by relaying systems since the investment would warrant such protection. However the analysis applies to machines in general, and ferroresonance may occur on large machines or in multi-machine systems just as it does in distribution systems if the capacitance, load and transformer saturation falls within the criterion in Reference 1.

Although no feeder faults followed by feeder isolation were attempted during field testing, faults were placed on the feeder during isolated operation. The relays were monitored during the faults and the relays detected the low voltage of one phase and the ferroresonant overvoltages on the other phases. Computer simulations of feeder faults have verified these results and have also shown that isolation after the feeder fault occurs does not significantly change the waveforms.

One of the most difficult concepts that the authors faced during the ferroresonance studies was the understanding of energy balance during a ferroresonant condition. It is an axiom of power system studies that an isolated generator increases speed when the load is less than the generation before isolation. Indeed, for normal system conditions, the axiom is true. However, when voltage and speed can significantly vary, when voltage buildup is a function of machine speed and frequency applied to the capacitors, when the currents in the machine may be very large, and when the torques in the machine involve extensive harmonics, the energy balance is extremely difficult to determine by intuitive processes. Analysis involving the induction machine is also complicated by the fact that the frequency is related to the inertia of the machine through slip rather than directly to speed, and the slip is a function of voltage and power factor as well as speed. As pointed out by Mr. Elmore however, the field test that produced Path 2 of Figure 2 of the paper showed that frequency reduced even though the load was less than the generation before isolation. Computer simulations, both analog and digital, showed the same effects for similar conditions. The authors do not have a ready intuitive explanation for the complex interaction.

The abrupt voltage elevation which occurs on Path 2 of Figure 2 of the paper however, is due to the rapid escalation of harmonic activity as the machine slip, machine inductance, transformer saturation level and capacitance all move through a point of high resonant activity. As the slip changes due to prime mover response, the harmonics reduce and lower the voltage. Path 2 should be utilized as one specific case of many cases tested and treated as typical only as an

27

indication that the frequency and voltage may make abrupt changes during ferroresonant conditions. There were cases where the resonances remained in constant activity and abrupt changes were not present.

The 33 kvar of capacitance was not significant in itself. The paths of Figure 2 were chosen from an entire series of field tests. It happened that the desired characteristics in the paths were found in four early tests and all had the same capacitance. Many field tests were run with as little as no kvar and as much as 266 kvar of capacitance. The development of ferroresonance occurred as predicted by the ferroresonant criterion.

Mr. Dawson mentions that he uses two phase-to-phase connected over/undervoltage relays and one zero-sequence overvoltage relay rather than the three phase-to-ground connected voltage relays mentioned in the paper. Actually several phase-to-phase relays and phase-to-phase detectors were also used during the tests. Since the great majority of the relays were connected phase-to-ground however, only they were mentioned in the paper. The results of the few phase-to-phase units were almost identical to the phase-to-ground units. At least one relay in the phase-to-phase package operated correctly, and if set fast enough, would have operated in 10 cycles or less.

It had been planned to also use a zero-sequence relay for the test, however it failed to arrive in time for the test. In a previous test however, the open delta voltage was monitored and its signal gave a strong indication of ferroresonance overvoltages with an even more distorted waveform. Therefore if the zero-sequence relay used by Mr. Dawson is not bothered by this distorted waveform, his technique should be very effective.

The authors wish to again thank the reviewers as such discussions greatly increase the usefulness of the paper.

Manuscript received April 19, 1988.

INFLUENCE OF HARMONICS ON POWER DISTRIBUTION SYSTEM PROTECTION

J.F. Fuller,
Fellow, IEEE
Dept. of Electrical &
Computer Engineering
University of Colorado
Boulder, CO 80309

E.F. Fuchs,
Senior Member, IEEE
Dept. of Electrical &
Computer Engineering
University of Colorado
Boulder, CO 80309

D.J. Roesler
Senior Member, IEEE
U.S. Dept. of Energy
Washington D.C. 20585

Abstract

The effects of nonsinusoidal voltages and currents on the performance of static underfrequency and overcurrent relays were experimentally studied. The tests were conducted such that the frequency, amplitude and phase shift of individual harmonics could be adjusted in a controlled manner by employing a waveform generator with a phase-locking circuit. The relation between the harmonic currents and voltages was modelled through the power system impedances within residential distribution systems. It was found that for harmonic voltage and current amplitudes, as they occur in distribution systems, underfrequency relays and the time delay operation of overcurrent relays show a marked deterioration in their performance. The instantaneous operating characteristics of overcurrent relays, however, are hardly affected by the presence of harmonic currents. The work described has been supported by the U.S. Department of Energy.

INTRODUCTION

The fast growing use of solid-state switching devices has caused an increasing content of harmonics in both the voltage and current on electrical utility systems. The switching devices are applied to a wide range of equipments including variable frequency drives for induction motors, single-and three-phase battery chargers, dc motor controllers fed from ac lines and high frequency lighting power supplies. Figure 1 shows the actual line current to a three-phase, half-wave battery charger in a power plant installation.

Figure 1   Nonsinusoidal line current of a battery charger.

During this same period static relays have been introduced which in many cases have supplanted the induction cup or disk relays. If the relay has appreciably different characteristics in the presence of harmonics, the effect on a feeder or system can be devastating if relay operation is either too soon or too late. Installations that were initially trouble free may cause problems as the total harmonic distortion grows.

Most textbooks and handbooks concerning relays as they occur in power system protection deal with the design and performance of relays under sinusoidal conditions [1,2]. The IEEE Transactions on Power Apparatus and Systems have published every three or four years since 1941 the most pertinent relay literature published in English. The latest paper of the IEEE Power System Relaying Committee is given in Ref. 3. To date only a few publi-

87 WM 108-4   A paper recommended and approved by the IEEE Power System Relaying Committee of the IEEE Power Engineering Society for presentation at the IEEE/PES 1987 Winter Meeting, New Orleans, Louisiana, February 1 - 6, 1987. Manuscript submitted August 27, 1986; made available for printing November 12, 1986.

cations are concerned with the effect of harmonics on electromechanical overcurrent relays [4,5,6]. The most comprehensive overview available on the influence of harmonics with respect to electromechanical and static relays is given in References 7 and 8.

The intent of this paper is to discuss the influence of voltage harmonics on the performance of a static underfrequency relay [9], the impact of current harmonics on the performance of two different types of solid-state overcurrent relays [10,11] and the effect of current harmonics on the performance of an electromechanical overcurrent relay [12].

At the outset of any investigation of the influence of harmonics on relays one has to define the magnitudes and frequencies of voltage and current harmonics which may have a detrimental influence on the performance of such relays. It is well known that solid-state circuits and other nonlinear devices of the power system generate harmonic currents [13] which result in a harmonic voltage drop across the power system impedance and produce amplitude modulations of the power system voltage. In addition, certain components of the power system, e.g. cycloconverters and induction motors, may cause either subharmonics, harmonics or fractional[*] harmonics. Most single and three-phase induction motors generate fractional harmonics within the rotating flux wave and the terminal currents due to the choice of the stator and the rotor slots and because of air gap eccentricity [18]. Figure 2 shows the voltages induced within two full-pitch search coils placed in quadrature within the stator of a 2 hp single-phase induction motor. Note that some of the harmonic orders are not an integer multiple of 60 Hz and therefore produce a nonstationary image on the oscilloscope screen.

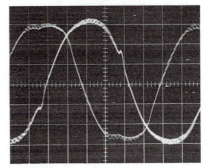

Figure 2   The generation of fractional harmonics within an induction motor.

For a given injected current harmonic amplitude the generated voltage modulations are dependent upon the power system impedance measured at the location where the injection occurs. To estimate the magnitudes of occurring harmonics within an urban residential distribution system an equivalent power system impedance has been calculated for a single-phase (Figure 3a) as well as for a three-phase (Figure 3b) network at 60 Hz [17].

From Figure 3a one obtains for the harmonic voltage drop across the line-to-line impedance

$$V_{(l-l)\nu}^{single, urban} = \frac{\sqrt{(0.102)^2 + \nu^2(0.0252)^2}}{2V_{l-N}} \cdot I_\nu \cdot 100\%, \quad (1)$$

[*]All other sinusoidal voltage and current components not being harmonic or subharmonic are called fractional harmonics.

Reprinted from *IEEE Trans. Power Delivery*, vol. 3, no. 2, pp. 549–557, April 1988.

where ν is the harmonic order. For the harmonic voltage drop across the line-to-neutral (centertap) impedance

$$V_{(l-n)\nu}^{single, urban} = \frac{\sqrt{(0.0984)^2 + \nu^2(0.0188)^2}}{V_{l-N}} \cdot I_\nu \cdot 100\%. \quad (2)$$

From Figure 3b the following harmonic voltage drops result

$$V_{(l-n)\nu}^{three, urban} = \frac{\sqrt{(0.752)^2 + \nu^2(0.0157)^2}}{V_{l-N}} \cdot I_\nu \cdot 100\% \quad (3)$$

and

$$V_{(l-l)\nu}^{three, urban} = \frac{2\sqrt{(0.051)^2 + \nu^2(0.0126)^2}}{\sqrt{3}\, V_{l-N}} \cdot I_\nu \cdot 100\%. \quad (4)$$

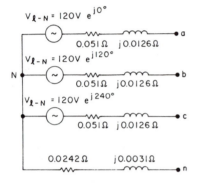

Figure 3a   Equivalent circuit of a single-phase system.

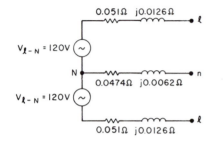

Figure 3b   Equivalent circuit of a three-phase system.

For rural distribution feeders the system impedances tend to be larger than those of urban systems, therefore for a given current harmonic amplitude I_ν the voltage harmonics will be correspondingly larger.

Example: For an assumed third harmonic ($\nu = 3$) current amplitude, $I_3 = 20\%$ one obtains from Eq. 2 for the third harmonic line-to-neutral (centertap) voltage $V_{(l-n)3}^{single,\ urban} = 1.89\%$. If one supposes that the corresponding impedance of a rural single-phase system is three times larger than that of the urban system (Eq. 2) one would get $V_{(l-n)3}^{single,\ rural} = 5.67\%$. With the recommended maximum voltage harmonic spectra of Refs. 13 and 14 one notes that harmonic currents of 20% are possible within a distribution system and therefore harmonic current amplitudes of 20% and 40% of the fundamental current will be investigated. These percentages are not out of line for actual values found in the field. In a similar manner one can arrive at percentages for voltage harmonics which are 5%, 10% and 15% of the fundamental voltage. Note that the lower percentages are within the permissible range for low harmonic orders only as far as capacitors are concerned [15,16].

The experiments were performed in a controlled manner such that only one harmonic with a prescribed amplitude, phase shift and frequency was superimposed on the fundamental. A locked-phase frequency generator that was designed and built at the University of Colorado permitted a continuous phase shifting of any harmonic with respect to the fundamental (Figure 4). Two distinct phase shifts were tested: the first one where the peak-to-

peak value of the quantity (either voltage or current) was a maximum and the other one where the peak-to-peak value was a minimum.

The papers available on electromechanical overcurrent relays indicate that the effect of harmonics on their operation is of minor importance if an instantaneous operating characteristic is considered [4,5,6]. For this reason the main thrust of this paper will be directed to static relays.

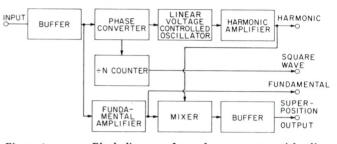

Figure 4   Block diagram of waveform generator with adjustable phase of either harmonic voltage or current.

PRESENT DESIGN CHARACTERISTICS OF RELAYS

The introduction of static relay designs required a decision in the logic circuit as to the characteristics of the quantity that would be measured. Relay testing standards for ac relays apply only to the use of single frequency sinusoids and assume no harmonic content. This, then, allows a peak measurement to be made and an appropriate multiplier used to relate to the rms value. Some devices measure very closely to a true rms value even when the wave contains harmonic orders as high as the 11th or 13th. Other circuits assume that the wave is always at 60 Hz fundamental and measure peak values only. The calibration value is stated as an rms current. Instruction books [10,11] usually contain a statement that when performing field tests on the relay the user must make certain that the test wave shape of voltage or current contains no harmonics. Others [9] discuss the "purity" of a sine wave, meaning no harmonics. Most all imply, if not state, that harmonics will alter the operation of the relay but do not necessarily state how the operation will be affected.

Some relay handbooks [1] state that the "solid-state instantaneous overcurrent unit" has a higher pickup value at other than rated frequency to minimize the effect of harmonics. However, one should not generalize unless the logic circuit design is actually known since tests indicate the pickup value can be either higher or lower than rated.

The behavior of the time delay circuit of the relay may also be subject to wide differences when harmonics are present and again it is difficult to generalize.

DISCUSSION OF EXPERIMENTS
Types of Relays Tested

This paper deals with four different applications. Relay "A" is a static underfrequency relay of the type used for load shedding at substations [9]. This relay is responsive both to the rate of change of frequency and total frequency change. In a pool operation load shedding capability is considered crucial if the system is to be maintained under loss of lines or loss of generation problems.

Relay "B" is a static time overcurrent relay normally used on distribution lines. The instantaneous overcurrent element is a hinged armature type [12].

Relay "C" is a three-phase solid-state time and instantaneous overcurrent relay that is normally applied to low voltage breakers used for power house auxiliaries or industrial plants. The relay is provided as an integral part of a three-phase breaker in ratings from 50 to 4000 amperes. The relay is self contained and can have short time, long time, instantaneous and ground protection depending on the options selected [10].

Relay "D" is a three-phase solid-state instantaneous overcurrent relay generally similar to that of relay "C" [11].

Testing Procedure

In order to investigate the effect on either the pick-up time or the pick-up value of current or voltage, a harmonic generator was inserted in series with a 60 Hz fundamental that initially contained no more than 3% harmonics. The harmonic generator was capable of having the phase shifted. This allowed the percent harmonics to be kept constant while the observed wave shape changed from a maximum peak-to-peak to a minimum peak-to-peak. Figure 5a shows a 30% second harmonic adjusted for maximum peak-to-peak while Figure 5b is for minimum peak-to-peak. Figures 6a and b are for 30% third harmonic. Relay "C" had a test point available that permitted finding the long-time delay pick-up current at 1 per unit of rating.

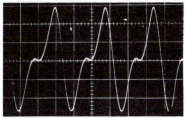

Figure 5a   Thirty percent second harmonic of the fundamental adjusted for a maximum peak-to-peak value.

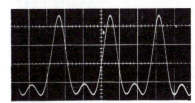

Figure 5b   Thirty percent second harmonic of the fundamental adjusted for a minimum peak-to-peak value.

A Clarke-Hess model 255 Digital V-A-W meter which is a true rms ammeter was used to measure the current. In order to check the values, a Hewlett-Packard spectrum analyzer model 3580A measured the magnitude of the fundamental and the har-

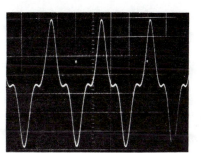

Figure 6a   Thirty percent third harmonic of the fundamental adjusted for a maximum peak-to-peak value.

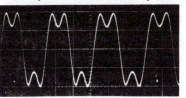

Figure 6b   Thirty percent third harmonic of the fundamental adjusted for a minimum peak-to-peak value.

monic. These values were than cross-checked against the ammeter reading. Short times were measured on a cycle counter while a stop watch was used for extended times.

The harmonic generator capable of phase shifting has been mentioned in the introduction. (See Figure 4)

TEST RESULTS FOR UNDER FREQUENCY RELAY "A"

The test circuit for the underfrequency relay "A" is illustrated in Figure 7. This circuit was designed to give varying rates of change of frequency in order to match the manufacturer's calibration curves of the relay operating characteristics. The function generator was capable of shifting frequency in response to a dc bias voltage. An RC circuit was used as shown to provide an almost linear rate of change of frequency.

Tests were performed at rates of frequency change of 1, 2, 3 and 4 Hz per second. Odd harmonic voltage contents of 0, 5, 10

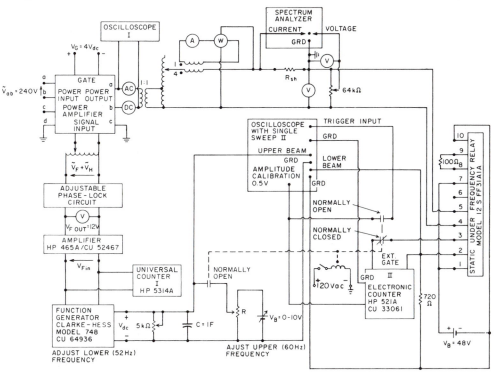

Figure 7   Test circuit for underfrequency relay "A".

31

and 15 percent of the fundamental from the 3rd through 7th or 9th were superposed on the fundamental voltage. The time delay setting, T_D, was unchanged at $T_D = 29$ cycles or 483.34 ms but two different measuring element set points were adjusted, 59 Hz (58.99) and 58 Hz. The same measuring series was performed but with a time delay setting of $T_D = 6$ cycles or 100 ms. From the measurements it can be noted that the time delay adds to the operating time T_T in an arithmetic manner. This means the largest percentage errors in the total operating time $T_S = T_T + T_D$ will occur at a time delay of $T_D = 0$. Figure 8 typically illustrates the course of the dc input voltage of the function generator of Figure 7. The oscillogram corresponds to a set point of 59 Hz (58.99) where the frequency starts the drop from 60 Hz to a minimum value of 52 Hz at a frequency rate of change of 1 Hz/sec. Shown are the values of the input voltage of the function generator at times $t = 0$ and $t = \infty$, $V_{dc}(t=0)$ and $V_{dc}(t=\infty)$, respectively. Also indicated is the relay operating time T_{S1} for a sinusoidal voltage ($\nu = 1$) and a time delay of $T_D = 29$ cycles.

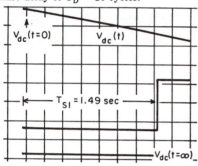

Figure 8   Input voltage of function generator as a function of time, $V_{dc}(t)$. Relay operating time T_{S1}, for a time delay of $T_D = 29$ cycles at a frequency rate of change of 1 Hz/sec. Sweep time = 0.2 sec/cm.

To show that the delay time T_D adds in an arithmetic manner to T_T, a second series of measurements was performed for a delay time of $T_D = 6$ cycles. Figure 9 corresponds to a set point of 59 Hz (58.99) where the frequency starts to drop from 60 Hz to 52 Hz at a frequency rate of change of 1Hz/sec. Marked are the values of the input voltage of function generator at times $t = 0$ and $t = \infty$, $V_{dc}(t) = 0$ and $V_{dc}(t = \infty)$, respectively. Also indicated is T_{S1} for a time delay of $T_D = 6$ cycles.

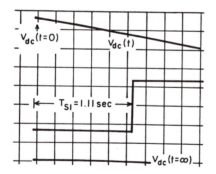

Figure 9   Input voltage of function generator as a function of time, $V_{dc}(t)$. Relay time T_{S1} for a time delay of $T_D = 6$ cycles at a frequency rate of change of 1 Hz/sec. Sweep time = 0.2 sec/cm.

Both measurement series as applied to time delays of $T_D = 29$ cycles and 6 cycles lead to the same operating time (T_T) increases due to voltage harmonics of 5, 10 and 15% of the fundamental if the time delay is subtracted from the measured T_S.

Example 1: At a set point of 59 Hz (58.99) the frequency drops from 60 Hz to 52 Hz. One obtains for a sinusoidal voltage ($\nu = 1$) exciting the underfrequency relay a time $T_{S1} = 89.5$ cycles for a time delay of $T_D = 29$ cycles. Therefore, the operating time becomes in this case $T_{T1} = T_{S1} - T_D = 60.5$ cycles.

If a 9th voltage harmonic of 15% is superimposed with the rated (100%) fundamental voltage then one obtains an increased total operating time of $T_{S9} = 130.6$ cycles for a delay time of T_D

= 29 cycles. The operating time becomes therefore, $T_{T9} = T_{S9} - T_D = 101.6$ cycles and the percentage increase of operating time

$$\Delta T = \frac{T_{T9} - T_{T1}}{T_{T1}} = \frac{101.6 - 60.5}{60.5} \cdot 100\% = 67.79\%.$$

This is identical with T_S for a zero time delay ($T_D = 0$).

The increases in T_S due to voltage harmonics of 5, 10 and 15% of the fundamental are shown in Figure 10a for a set point of 59 Hz (58.99) at a frequency rate of changes of 1 and 2 Hz/sec, at a delay of $T_D = 0$ sec. Similar total operating time increases are presented in Figure 10b for a set point of 59 Hz (58.99) and for frequency rate of changes of 3 and 4 Hz/sec, at a delay time of $T_D = 0$ sec. Corresponding time increases are shown in Figures 11a, b for a set point of $F_p = 58$ Hz.

From these T_S increases at $T_D = 0$ the T_S increase for any non zero time delay ($T_D \neq 0$) may be calculated as follows

$$\Delta T|_{T_D=0} = \frac{T_{T\nu} - T_{T1}}{T_{T1}} \tag{5}$$

and

$$\Delta T|_{T_D \neq 0} = \frac{T_{T\nu} + T_D - (T_{T1} + T_D)}{(T_{T1} + T_D)} \tag{6}$$

Introducing Eq (5) into Eq (6) gives for the percentage T_S increase for a non zero time delay ($T_D \neq 0$):

$$\Delta T|_{T_D \neq 0} = \frac{T_{T1} \cdot \Delta T|_{T_D=0}}{(T_{T1} + T_D)} \tag{7}$$

Example 2: For a time delay of $T_D = 6$ cycles and using the values of Example 1 one gets

$$\Delta T|_{T_D = 6 cycles} = \frac{60.5 \cdot 67.79\%}{60.5 + 6} = 61.67\%$$

or for a time delay of $T_D = 29$ cycles

$$\Delta T|_{T_D = 29 cycles} = \frac{60.5 \cdot 67.79\%}{60.5 + 29} = 45.82\%$$

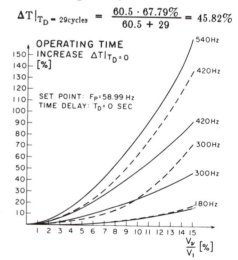

Figure 10a   Operating time increase $\Delta T|_{T_D=0}$ due to voltage harmonics for set point $F_p = 58.99$ Hz.
—— Frequency rate of change of $\Delta F = 1$Hz/sec and an operating time of $T_{T1} = 60.5$ cycles or 1.008 sec.
---- Frequency rate of change of $\Delta F = 2$ Hz/sec and an operating time of $T_{T1} = 25$ cycles or 0.417 sec.

In all cases the voltage harmonics increased operating time and at 15% of voltage harmonics, time more than doubled. When measuring the increase in operating time the phase shifts of the voltage harmonics were adjusted such that the largest possible increase in operating time resulted. This means Figures 10 and 11 represent worst cases and it might well be that for a given har-

32

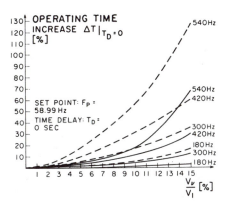

Figure 10b Operating time increase $\Delta T|_{T_D = 0}$ due to voltage harmonics for set point $F_p = 58.99$ Hz.
— Frequency rate of change of $\Delta F = 3$ Hz/sec and an operating time of $T_{T1} = 17.5$ cycles or 0.292 sec.
---- Frequency rate of change of $\Delta F = 4$ Hz/sec and an operating time of $T_{T1} = 13$ cycles or 0.217 sec.

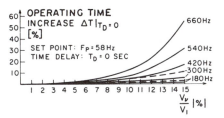

Figure 11a Operating time increase $\Delta T|_{T_D = 0}$ due to voltage harmonics for set point $F_p = 58$ Hz.
— Frequency rate of change of $\Delta F = 1$ Hz/sec and an operating time of $T_{T1} = 125.25$ cycles or 2.088 sec.
---- Frequency rate of change of $\Delta F = 2$ Hz/sec and an operating time of $T_{T1} = 56.5$ cycles or 0.943 sec.

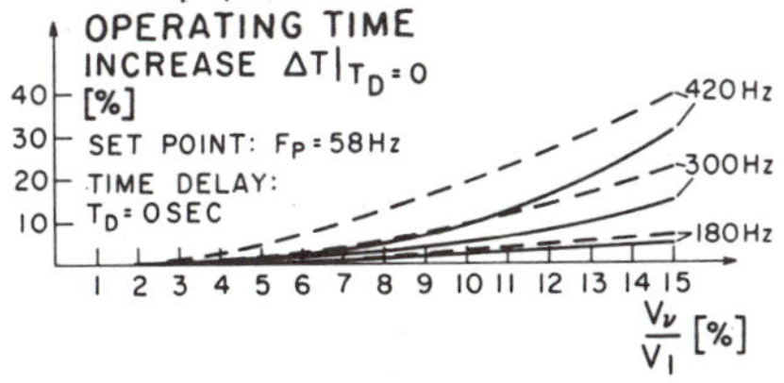

Figure 11b Operating time increase $\Delta T|_{T_D = 0}$ due to voltage harmonics for set point $F_p = 58$ Hz.
— Frequency rate of change of $\Delta F = 3$ Hz/sec and an operating time of $T_{T1} = 37$ cycles or 0.617 sec.
---- Frequency rate of change of $\Delta F = 4$ Hz/sec and an operating time of $T_{T1} = 26$ cycles or 0.433 sec.

monic amplitude of say 15% the increase in operating time is well below those of Figures 10 and 11. Finally, tests indicated that the underfrequency relay under consideration is very sensitive to fractional voltage harmonics and it is recommended that such harmonics should be limited to less than 0.5%.

TEST RESULTS FOR OVERCURRENT RELAY "C"

Two solid-state overcurrent relays of the same model number "C" were connected in series and three tests were performed on both relays to determine their

1) instantaneous pick-up rms current

2) the time delay for a given pick-up rms current,

3) and the long-time delay pick-up current at 1 per unit of rating.

The test circuit as proposed in Reference 10 was set up whereby the rms value of the current was measured with the Clarke-Hess true rms ammeter mentioned under "Testing Procedure". Since both relay devices generated very similar test data, only the test results of one relay device will be presented. For all three measuring series two distinct harmonic phase shifts were investigated. The first one resulted into a current waveshape where the peak-to-peak value is a maximum (see Figures 5a and 6a) and the other where the peak-to-peak value is a minimum (see Figures 5b and 6b).

Determination of the Instantaneous Pick-Up rms Current

At the outset of the tests the long-time delay pick-up was set at the maximum in order to exclude pick-up in this mode. Then a pick-up multiplier of 1.2 was chosen which resulted with ampere tap settings of 4, 5, 6, 8, 10 and 12 in nominal pick-up rms current values of 4.8, 6, 7.2, 9.6, 12 and 14.4A, respectively. The percentage deviations from the nominal (sinusoidal) rms current values operating the overcurrent relay at lower (-%) or higher (+%) (nonsinusoidal) rms current values are listed for various harmonic amplitudes and phase shifts in Table I. From Table I it can be seen that the current harmonics cause the overcurrent device in the worst case to pick up at $(100 \pm 10)\%$ of the nominal rms value based on a sinusoidal waveshape.

Table I
Influence of Harmonic Amplitudes and Phaseshifts on the Instantaneous Pick-Up rms Current of Relay "C"

ν	$\dfrac{I_\nu}{I_1}$ [%]	tap setting						peak-to-peak value
		4	5	6	8	10	12	
1	100	4.20A	5.20A	6.22A	8.24A	9.92A	11.2A	
2	40	-0.95*	-1.15	-2.25	-7.28	-6.85	-5.71	min
		1.43	0.38	-1.29	-6.31	-5.24	-2.86	max
	20	-0.95	-0.77	-1.61	-3.88	-4.44	-6.07	min
		-0.48	0.00	-0.64	-3.64	-3.22	-0.71	max
3	40	-1.43	-0.77	-0.64	-6.55	-6.05	-2.50	min
		16.67	15.00	14.47	7.77	6.05	5.71	max
	20	-0.95	-1.15	-0.97	-5.09	-4.03	-3.57	min
		7.14	6.92	6.11	5.58	6.45	7.14	max
5	40	8.57	12.69	4.50	-1.94	-2.42	-0.36	min
		0.95	0.77	2.26	-1.70	-4.84	-4.64	max
	20	5.71	6.92	3.54	-0.49	-0.81	0.36	min
		0.00	0.38	0.65	-6.55	-6.45	-3.57	max
7	40	3.37	3.08	0.00	-5.34	0.81	5.71	min
		4.29	5.00	5.14	-4.37	-7.26	-5.17	max
	20	0.00	0.77	0.97	-1.94	-6.85	-4.64	min
		3.81	1.92	-1.28	-5.83	-4.84	-2.86	max
9	40	6.19	5.00	4.18	-0.24	-4.03	-5.36	min
		3.33	4.23	1.61	-7.77	-6.85	-2.86	max
	20	3.81	3.85	0.96	-6.31	-6.05	-4.64	min
		0.48	-0.38	0.00	-0.49	3.23	3.93	max

* All measured values are in percent.

Determination of the Time Delay for a given Pick-Up rms Current

For the time delay test the ampere tap setting of the instantaneous current pick-up was set at the maximum value (12A) to prevent any instantaneous pick-up. The rms current value of 1.5 times pick-up of 1.2A or 1.8A ± 1% was adjusted and the delay times were measured for minimum, intermediate and maximum long-time delay settings. The percentage deviations from a time delay based on the nominal (sinusoidal) rms current value are listed in Table II. The negative (-%) or positive (+%) values represent operating times which are either shorter or longer than the nominal operating time of the overcurrent relay, respectively.

As seen in Table II the overcurrent relay under investigation

Table II
Influence of Harmonic Amplitudes and Phaseshifts
on the Time Delay for a Given rms
Current Value (1.8 A ± 1%) of Relay "C"

ν	$\dfrac{I_\nu}{I_1}$ [%]	minimum peak-to-peak value			maximum peak-to-peak value		
		T_{max}	T_{int}	T_{min}	T_{max}	T_{int}	T_{min}
1	100	493 sec	142 sec	55 sec	493 sec	142 sec	55 sec
2	40	-39.75*	-37.32	-38.18	-27.99	-24.65	-23.64
	20	-26.57	-24.65	-21.82	-13.18	-13.38	-12.73
3	40	8.52	7.04	5.45	-45.64	-42.96	-41.82
	20	26.37	19.72	18.18	-33.87	-30.98	-30.91
5	40	-36.11	-32.39	-32.73	-40.97	-35.21	-34.55
	20	-20.48	-19.01	-18.18	-27.99	-25.35	-25.45
7	40	-39.55	-36.62	-34.55	-36.92	-34.51	-34.55
	20	-23.53	-21.13	-21.82	-27.18	-24.65	-23.64
9	40	-40.36	-36.62	-36.36	-38.13	-33.80	-34.55
	20	-27.58	-25.35	-27.27	-23.53	-21.13	-23.64

* All measured values are in percent.

Table III
Influence of Harmonic Amplitudes and Phaseshifts
on the Long-Time Delay Pick-Up rms Current
at 1 Per Unit of Rating of Relay "C"

ν	$\dfrac{I_\nu}{I_1}$ [%]	minimum peak-to-peak value	maximum peak-to-peak value
1	100	1.73 A	1.73 A
2	40	-16.76%	-8.67%
	20	-8.67%	-4.62%
3	40	4.91%	-22.54%
	20	6.94%	-15.32%
5	40	-15.61%	-19.08%
	20	-6.36%	-12.14%
7	40	-18.49%	-17.34%
	20	-9.25%	-10.69%
9	40	-19.08%	-17.34%
	20	-11.27%	-9.54%

has changes at the time delay settings up to 45% shorter and up to 26% longer than the nominal time delay with no harmonics. Figure 12 is a graphical representation of Table II.

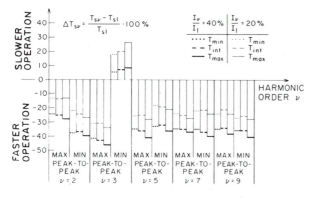

Figure 12   Influence of Harmonic Amplitudes and Phase-Shifts on the Long-Time Delay for a Current of 1.8A ± 1%.

Determination of Long-Time Delay Pick-Up Current at 1 Per Unit of Rating

According to the relay time-current characteristics of Reference 10 the operating times, if the relay operates in the long-time delay mode, are very long. The instruction book [10] offers a convenient testing method to find the exact start of minimum pick-up or the 1 per unit value.

Using this method, the 1 per unit pick-up current value was found to be 1.73A for the fundamental. Thereafter, current harmonics with 40 and 20% and two distinct phase shifts were superimposed on the fundamental of 60 Hz. Table III summarizes all measurements and lists the percentage deviations from the per unit current (1.73 A) which will eventually cause the relay to operate due to the presence of harmonics in the sensed current. Positive or negative percentages correspond to rms current values which are either lower or higher than the nominal per unit current of 1.73A, respectively.

Table III shows that the long-time delay pick-up current value is reduced for most harmonics by 20% and increased for certain harmonics by 7%. Figure 13 illustrates the data of Table III.

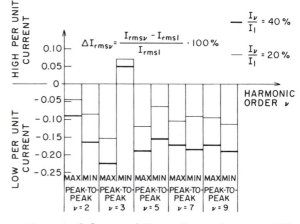

Figure 13   Influence of Harmonic Amplitudes and Phase-Shifts on the Long-Time Delay Pick-Up rms Current at 1 per unit of rating.

TEST RESULTS FOR OVERCURRENT RELAY "D"

Solid-state overcurrent relay "D" only could be operated in instantaneous pick-up characteristic. As for relay "C" the lowest available ampere tap setting of 300A was chosen and the instantaneous pick-up was set at 4. Table IV lists the percentage deviations from the nominal (sinusoidal) rms current value operating the overcurrent relay at lower (-%) or higher (+%) (nonsinusoidal) rms current values at various harmonic amplitudes and phase shifts. It can be seen that the current harmonics cause the overcurrent device to predominantly pick up at (100 + 20) % of the nominal rms value based on a sinusoidal waveshape.

TEST RESULTS FOR OVERCURRENT RELAY "B"

Although electromechanical overcurrent relays were studied in References 5, 6, 7 and 8 under the influence of harmonics it was deemed necessary to investigate at least one electromechanical relay with the testing method at hand in order to be able to compare the test results of this paper with previously published data. The relay used for testing is a hinged armature overcurrent relay with an instantaneous pick-up characteristic as well as an inverse time-delay characteristic. At the outset of the tests it was noted that the resetting of the target changed the required pick-up torque and thus has an effect on the pick-up values. Therefore, it was decided to reset the target after each operation of the relay. For all tests the relay was operated in the 4 per unit range.

Table IV
Influence of Harmonic Amplitudes and Phaseshifts on the Instantaneous Pick-Up rms Current of Relay "D"

ν	$\dfrac{I_\nu}{I_1}$ [%]	tap setting 4	peak-to-peak value	ν	$\dfrac{I_\nu}{I_1}$ [%]	tap setting 4	peak-to-peak value
1	100	4.04A		1	100	4.04A	
2	60	2.22% 10.12%	min max	9	60	8.17% 7.92%	min max
	40	0.00% 3.70%	min max		40	4.95% 3.71%	min max
	20	-0.25% 0.99%	min max		20	3.22% 0.00%	min max
3	60	-0.5% 22.28%	min max	11	60	8.17% 8.42%	min max
	40	-3.22% 18.81%	min max		40	4.46% 4.46%	min max
	20	-3.22% 7.92%	min max		20	0.50% 2.48%	min max
5	60	9.90% 5.45%	min max	13	60	7.92% 7.67%	min max
	40	9.16% 0.74%	min max		40	4.21% 4.46%	min max
	20	4.70% -1.24%	min max		20	1.98% 0.99%	min max
7	60	8.17% 7.92%	min max	15	60	7.43% 7.43%	min max
	40	2.48% 5.94%	min max		40	4.46% 4.46%	min max
	20	-0.50% 4.21%	min max		20	1.24% 1.73%	min max

Determination of the Instantaneous Pick-Up rms Current

For this test the adjustable pole piece of the instantaneous overcurrent unit was set at minimum. As before, for the solid-state relays, the percentage deviations from the nominal (sinusoidal) rms current value operating the overcurrent relay at lower (-%) or higher (+%) (nonsinusoidal) rms current values are given in Table V for various harmonic amplitudes and phase shifts. Note that the current harmonics cause the overcurrent relay to pick up in the worst case at (100 + 10)% of the nominal rms value based on a sinusoidal waveshape.

Table V
Influence of Harmonic Amplitudes and Phaseshifts on the Instantaneous Pick-Up rms Current of Relay "B"

ν	$\dfrac{I_\nu}{I_1}$ [%]	range 4 p.u.	peak-to-peak value	ν	$\dfrac{I_\nu}{I_1}$ [%]	range 4 p.u.	peak-to-peak value
1	100	2.30A		1	100	2.30A	
2	40	0.00% -3.91%	min max	7	40	6.96% 7.39%	min max
	20	2.17% -1.74%	min max		20	3.04% 2.60%	min max
3	40	10.87% 4.35%	min max	9	40	8.26% 7.83%	min max
	20	2.61% 3.04%	min max		20	2.61% 2.61%	min max
5	40	8.26% 6.52%	min max				
	20	3.91% 4.40%	min max				

Determination of the Time Delay for Given Pick-Up rms Current Values

The time-delay mode was tested with a time dial setting of 5, with a current tap block setting of 3, with the time dial Vernier at 0, and with the time target tap block at 0.2. Two test series were performed: the first one with 2.0A·3=6A and the second one with 6.67A·3=20A. For both test series the percentage deviations from a time delay based on the two nominal (sinusoidal) rms current values are listed in Table VI. The negative (-%) or positive (+%) values represent operating times which are either shorter or longer than the appropriate nominal operating time of the overcurrent relay, respectively. The current harmonics cause the hinged armature overcurrent relay to pick-up in the worst case at (100 + 43)% of the nominal rms value based on a sinusoidal waveshape.

Table VI
Influence of Harmonic Amplitudes and Phaseshifts on the Time Delay for Given rms Current Values of Relay "B"

ν	$\dfrac{I_\nu}{I_1}$ [%]	Delay-Time Setting nominal value: 2.0A·3=6A	nominal value: 6.667A·3=20A	peak-to-peak value
1	100	162.4 sec	75.8 sec	
2	40	8.56% 8.81%	1.58% 12.53%	min max
	20	2.65% 2.40%	0.26% 1.98%	min max
3	40	-4.56% 30.11%	-6.33% 43.40%	min max
	20	-4.74% 13.12%	-5.41% 13.85%	min max
5	40	11.64% 0.80%	15.17% -4.62%	min max
	20	7.14% -2.52%	11.21% -5.01%	min max
7	40	3.26% 8.37%	1.19% 4.62%	min max
	20	0.99% 5.73%	-3.30% 8.18%	min max
9	40	5.91% 5.73%	2.90% 3.30%	min max
	20	4.50% 0.56%	1.58% -2.00%	min max

CONCLUSIONS

For the relays investigated one can draw the following conclusions:

1. Solid-state, static and electromechanical overcurrent relays behave in a similar manner as far as the instantaneous and the time-delay pick-up are concerned. The (nonsinusoidal) rms value of the pick-up current depends upon the waveshape, that is the phase shifts of the harmonics with respect to the fundamental current. For a harmonics content of 40% the overcurrent devices operating under the instantaneous characteristic pick up in the worst case at about (100 ± 15)% of the nominal pick-up value. Under the time-delay characteristic both types of relays pick up in the worst case at about (100 ± 45)% of the nominal time delay.

2. Solid-state overcurrent relays have worst case long-time delay pick-up current values of about 80% of the nominal value which occur at a harmonic content of 40%. Such a large harmonic content can be caused by battery chargers, variable-speed motor controllers and high frequency lighting devices.

3. The static underfrequency relay has under the influence of voltage harmonics increased operating times and at 10% of voltage harmonics, in the worst case time almost doubled. Tests indicated that the underfrequency relay is very sensitive to fractional voltage harmonics and therefore such harmonics should be limited to less than 0.5%.

4. Relay standards may need to address the problems of harmonic content and their impact on relay characteristics.

5. It is impossible to generalize the behavior of any relay response to harmonics without testing or reviewing and understanding the actual device design. This applies to static, solid-state, induction cup or disk or plunger or clapper type relays.

6. Any one relay may have two different responses of higher and lower pick-up values to different harmonics or even to the same harmonic with the phase shifted.

7. Installations that initially cause no trouble may experience nuisance operation of relays in the future as the percentage of harmonics on the system grows.

ACKNOWLEDGEMENT

The work described was supported by the Department of Energy, Washington, D.C., under contract #DE-AC02-80RA50150. The authors are especially grateful to Mr. Tom Yohn of Public Service Company of Colorado, Denver for making relays available for testing.

REFERENCES

[1] Applied Protective Relaying, Westinghouse Electric Corp., Sections 22-1, 22-2.

[2] C.R. Mason, The Art and Science of Protective Relaying, John Wiley and Sons, 1956.

[3] IEEE Committee Report, "Bibliography of Relay Literature, 1982-1983", IEEE Transactions on Power Apparatus and Systems, vol. PAS-104, no. 5, May 1985, pp. 1189-1197.

[4] D.E. Saramaga, R.W. Menzies and G.W. Swift, "Effects of Harmonic Currents in Induction Disc Relay Operating Torque", IEEE PES Summer Meeting Vancouver, Canada, Paper #C73 351-4, July 1973.

[5] W.F. Horton and S. Goldberg, "The Effect of Harmonics on the Operating Points of Electromechanical Relays", IEEE Transactions on Power Apparatus and Systems, vol. PAS -104, no. 5, May 1985 pp. 1178-1188.

[6] A.E, Emanuel, Discussion to "The Effect of Harmonics on the Operating Points of Electromechanical Relays", by W.F. Horton and S. Goldberg, Ibid.

[7] F.A. Jost, D.F. Menzies and M.S. Sachdev, "Effect of System Harmonics on Power System Relays" Paper presented at the Canadian Association Spring Meeting 1974, March 1974.

[8] IEEE Power System Relaying Committee, "The Impact of Sine-Wave Distortions on Protective Relays", IEEE Transactions on Industry Applications, Vol. IA-20, No. 2, March/April 1984.

[9] General Electric Instruction Book GEK 49923 for Underfrequency Relay Type SFF 31 and 32.

[10] ITE - Imperial Corp. Instruction Book IB-9.1.7-21 Issue C for ITE Solid-State Trip Device Type SS 3 G4.

[11] Brown Boveri Corp. Instruction Book IB-9.1.7-22D for ITE Solid-State Trip Device Type SS7.

[12] General Electric Instruction Book GEK 36887 for Static Overcurrent Relay Type 12 SFC.

[13] "The Limitation of Disturbances in Electricity Supply Networks Caused by Domestic and Similar Appliances Equipped with Electronic Devices", European Standard EN 50 006, European Committee for Electrotechnical Standardization, 5, Boulevard de l'Empereur - Bte n° 14-1000 Bruxelles, Belgium.

[14] E.F. Fuchs, D.J. Roesler and K.P. Kovacs, "Aging of Electrical Applicances Due to Harmonics of the Power System's Voltage", IEEE Transactions on Power Delivery, vol. PWRD-1, no. 3, July 1986 pp. 301-307.

[15] IEC Publication 70/1967.

[16] ANSI/IEEE Std. 18-1980, Shunt Power Capacitors.

[17] J.-P. P. Montu, "Impact of Dispersed Photovoltaic Systems on the Design of Residential Distribution Networks", Doctoral Dissertation, University of Colorado, June 1986.

[18] J.R. Cameron, W.T. Thomson and A.B. Dow, "Vibration and Current Monitoring for Detecting Air Gap Eccentricity in Large Induction Motors", IEE Proceedings, vol. 133, Pt.B, no. 3, May 1986.

Discussion

John R. Linders (Nordon R&D, Inc. Sarasota, FL 33581):
This paper provides valuable information on a subject of considerable current interest. An additional reference which enlarges on [8] is, "Fault Induced Wave Distortion of Interest to Relay Engineers," a Report prepared by the Power System Relaying Committee, IEEE Transactions on Power Apparatus and Systems, vol PAS-104, no. 12, pp. 3574–3584, Dec. 1985. This and [8] are part of a more complete report, "Sine-Wave Distortions in Power Systems and the Impact on Protective Relaying," an IEEE Special Publication no. 84 THO 115-6 PWR. Regarding the choice of maximum peak-to-peak and minimum peak-to-peak criteria, as written, the choice is arbitrary and as such, lessens the value of the conclusions. However, these choices do give the maximum difference between rms and ac average values for single-frequency harmonic distortion and as such do provide valuable benchmarks. A next step is to further explore multifrequency distortions with square waves as from thyristors and chopped waves as from saturated CT's.

One should make a distinction between ambient distortion and fault-induced distortion. Ambient distortion is most important in the application of overload relays and remote backup relays. Fault-induced distortion (including CT saturation) is of more concern in primary fault clearing. Frequency relay applications need to consider both conditions and additionally other distortions which are possible during a severe system disturbance. There would appear to be value in classifying the various types of distortion according to the likelihood of its occurring under each of these different types of conditions. From such a classification would come understandings as to how best to cope with widely varying types of distortion and widely varying types of relay responses.

Regarding harmonic terminology, induction motor slot harmonics and similar effects, while related to the system frequency, are not integers of that frequency. Their moving appearance on an oscilloscope suggests that they be referred to as "floating harmonics" or some similar term. The terminology "fractional harmonics" used in the paper, could then more properly be used to describe full integral multiples of a subharmonic, such as 3/2's or 7/4th's, etc.

The caption on Fig. 6(a) should probably be 40-percent third harmonic and not the noted 30-percent. A 30-percent third harmonic, phased for maximum peak voltage will not cause an inflection at the zero crossing as shown. Also, with other harmonic phase relations the shape and position of the zero crossing can be significantly affected. Thus the maximum/minimum peak criteria used in the study are not as applicable to static frequency relays and other zero-crossing detectors as it is to overcurrent relays. Another frequency measuring problem occurs during an unbalanced fault. Any such fault will cause a 30° phase shift in at least one of the system voltages and if the relay is connected to that voltage it will be subjected to a major change in the apparent frequency. (Hence the need for delay in most high-speed relays which use zero crossings.) This problem can be substantially reduced by energizing the relay with positive sequence voltage. This voltage does not shift in phase angle during an unsymmetrical fault and does not go to zero except for a nearby three-phase fault. It also will have inherently less harmonic distortion in it than the single-phase voltage and thus the harmonic problem is minimized as well.

Manuscript received February 17, 1987.

J. E. Stephens (Illinois Power Co., Decatur, IL):
The description of the Relay C instantaneous unit tests indicate that the 60-Hz nominal pickup rms current values should have been 4.8, 6, 7.2, 9.6, 12, and 14.4 A on the respective taps 4 through 12. However, Table I seems to indicate that the actual 60-Hz sinusoidal pickup values were 4.2, 5.2, 6.22, 8.24, 9.92, and 11.2 A, respectively. These are from 12.5 to 22.2 percent below the nominal relay pickup settings. Could the relay calibration be corrected to bring the 60-Hz pickup values up to agree with the relay tap settings?

The values given in Table I are identified as the percentage deviations from the *nominal* pickup values of 4.8 to 14.4 A, respectively. However, it appears that they are deviations from the *actual* 60-Hz sinusoidal pickup values given in the first line of the table.

Similarly to the instantaneous unit, the long-time delay pickup was found to be 1.73 A, an odd value. We wonder what the actual tap setting was and if the relay calibration could be corrected to match the tap value. The third harmonic, minimum peak-to-peak pickup deviation values recorded in Table III are inconsistent with the remaining data and appear questionable. Did the authors repeat those tests or look for an explanation for the apparent inconsistency?

The long-time timing tests of Relay C were made at 1.8 A or 1.5 times the 60-Hz sinusoidal pickup value. At this test current of only 1.5 times the nominal pickup value, the timing is greatly affected by the *actual* pickup current of the relay. The test is more a measure of pickup value than it is of timing. Depending on the shape of the time–current characteristic curve of the relay, timing tests should be made at 5 to 10 times the tap setting to minimize the effect of actual pickup variations.

Table II shows the same inconsistency as Table III for third harmonic current with minimum peak-to-peak phasing, but the longer time is consistent with the higher pickup. The significantly faster timing of all other tests of Table II agrees with and is caused by the lower pickup as shown in Table III. The apparent unique response of the relay to the third harmonic is reason for concern. Does the relay respond to peak values rather than rms values?

The discussion of the Relay B tests states that the relay was operated with a tap block setting of 3 and with test currents of 2.0 A·3 = 6 A and 6.67 A·3 = 20 A. Does this mean that the relay sinusoidal pickup setting was 3 A and test currents of 6 A and 20 A (2 and 6.33 times pickup, respectively) were used?

Was the pickup of Relay B time-delay mode affected in the same way as the instantaneous relay given in Table V? What was the actual sinusoidal pickup current for the time tests?

Manuscript received February 17, 1987.

W. Mack Grady (Department of Electrical and Computer Engineering, University of Texas at Austin, Austin, TX 78712): As Chairman of the Working Group on Power System Harmonics, I would like to thank the authors for an important piece of work which gives some additional insight on the effects of harmonics on protective relays. This is a timely topic which will become more important in the future as the number of harmonics producing loads increases.

Activities are presently underway to establish more stringent guidelines for harmonic levels (IEEE Standard 519—pending update). However, while these new guidelines will help greatly, it is generally expected that the overall level of harmonics in power systems and the number of problem cases will continue to increase for some time.

Concerning the technical content of the paper, it is shown in Figs. 12 and 13 that the presence of out-of-phase third-harmonic currents results in slower operation of type "C" overcurrent relays. This is not surprising since the third-harmonic minimum coincides exactly with the first-harmonic peak at every half-cycle, thereby reducing the peak current magnitude significantly. Other harmonics do not have as significant an effect.

The surprising result shown is that the highest third-harmonic level (40 percent) has a less pronounced effect than does the lower level (20 percent). Could the authors please speculate on this phenomenon?

Again, I would like to congratulate the authors for an important contribution.

Manuscript received February 27, 1987.

E. F. Fuchs and **J. F. Fuller**: The authors thank all discussers for their interesting comments and constructive questions.

The queries of Mr. Linders are well justified. The authors are glad that

Mr. Linders explained the interrelation between the various reports of the Power System Relaying Committee. A classification of harmonics into those caused by ambient distortion and harmonics due to fault-induced distortion should be established by the Working Group on Power System Harmonics. Also, the authors strongly urge the above mentioned Working Group or a Subcommittee of the Power Engineering Society to find adequate words to describe these fractional harmonics so that the meaning will be clear to all. The authors have subjected Fig. 6(a) to a Fourier analysis and the fundamental and the harmonics of Table C1 have been obtained.

Table C1
Fourier Coefficients of Fig. 6(a)

ν	1	3	5	7	9	11
V_ν[pu]	21.64	9.12	0.81	0.53	0.56	0.25

From this table it is apparent that Mr. Linders is correct. Obviously the incorrect photograph for the harmonic percentage of 30 percent has been used. Note that all harmonics were checked with the Clarke–Hess meter as indicated in the paper.

Prof. Grady rightfully points out the peculiarity that for the third harmonic, if the peak-to-peak value of the wave is a minimum, Relay C exhibits a slower operation than for all other cases of Tables II and III or Figs. 12 and 13. In addition, a 40-percent harmonic content has less effect than a 20-percent harmonic content. When testing the two in series connected Relays C, these peculiar results were noted and the respective measurements were repeated several times. Here, it is important to mention that the two in series connected Relays C generated almost identical results. In order to explain this phenomenon, one has to analyze the time-delay circuit of the respective overcurrent relay. The authors chose not to investigate specific relay design characteristics, but to rely on instruction book information as to operating characteristics.

Mr. Stephens is right in pointing out the difference between the nominal and the actual 60-Hz sinusoidal pickup rms current values as they apply to Relay C. The percentage values of all tables and figures of this paper are referred to the actual (measured) 60-Hz sinusoidal pickup rms current values. No attempts have been made to calibrate any of the tested relays because no calibration circuit was available and it was assumed that the relays provided by Public Service Company of Colorado were indeed sufficiently calibrated. The long-time delay pickup of 1.73 A has been determined as proposed on p. 14 of [10]. The solid-state relay has a unique part of the circuit that allows one to find the exact pickup point as follows: "The LONG-TIME pick-up may be accurately determined by connecting a VOM from terminal 10 to 9 (10 plus). Pickup is that current at which the VOM just deflects. If allowed to persist long enough, the breaker will eventually trip after the VOM is removed." Again, the authors did not attempt to check the calibration of the long-time delay of Relay C. The authors agree with the discusser that the third-harmonic, minimum peak-to-peak pickup deviation values recorded in Tables II and III appear to be inconsistent with the remaining data of these two tables. The authors noted this inconsistency at the time of testing and repeated these measurements several times: there is no question that these values are not questionable. In order to be able to explain such an inconsistency, the complete time-delay circuit design would have to be reviewed. The authors do not know whether the relay responds to peak values rather than to rms values. The authors agree with the discusser, the long-time delay tests of Relay C should have also included higher tap settings. Indeed, for Relay B, the sinusoidal pickup setting was 3 A and test currents of 6 and 20 A were used (2 and 6.667 times pickup, respectively). Finally, one should note that the instantaneous and time-delay characteristics of Relay B are not interdependent since the instantaneous characteristic is based on a plunger-type relay and the time-delay characteristic is based on a solid-state circuit.

Manuscript received March 31, 1987.

COMPUTER AIDED COORDINATION OF LINE PROTECTION SCHEMES
IEEE COMMITTEE REPORT

Members of the Computer Aided Coordination of Line Protection Schemes Working Group of the IEEE Power System Relaying Committee are: J. M. Postforoosh - Chairman, J. Akamine, H. N. Banerjee, J. R. Cornelison, M. K. Enns, S. E. Grier, R. W. Haas, I. O. Hasenwinkle, R. W. Johnson, C. J. Mozina, R. Ramaswami, H. S. Smith, R. P. Taylor, J. R. Turley, J. A. Zipp. Former members: C. H. Castro, D. H. Colwell, E. J. Mayer, G. R. Nail.

Abstract - IEEE PSRC Working Group on Computer Aided Coordination of Line Protection Schemes was established in 1985 to investigate the application of computers to the task of setting and verifying the setting of line protective devices. This report provides the findings of the Working Group regarding functional requirements of the data base, software, and hardware for this system as well as the methods of presenting the results to the protection engineer.

1. INTRODUCTION

Digital computers have revolutionized the process of designing, analyzing and controlling electric power systems. The development of efficient computer algorithms has facilitated large scale load flow, short circuit, and transient stability studies and a variety of real-time control and management functions. However, considerably less effort has been focused on developing computer software for solving protection engineering problems.

This may be attributed to the general concerns that protection software designed for one user may not meet the needs of other companies or may be in conflict with some of their practices. Protection engineers are increasingly concentrating their efforts on improving productivity through computer applications and generally agree that a computer aided coordination (CAC) system is badly needed.

To define a computer aided system that would meet the needs of protection engineers, this Working Group was formed with experienced members from seventeen companies nationwide. The Working Group has identified the most important considerations in the design of a CAC system to be:

1. The protective devices and electrical system data that the protection engineer has available and can provide as input to the software.

2. The functional specification of the software that meets the needs of the protection engineer.

3. The results and outputs that the protection engineer expects from the software.

4. The typical computer hardware configuration suitable for this system.

The intent of the Working Group has been to specify a system that would be acceptable to protection engineers with a wide range of applications for use throughout the Industry.

90 SM 386-3 PWRD   A paper recommended and approved by the IEEE Power System Relaying Committee of the IEEE Power Engineering Society for presentation at the IEEE /PES 1990 Summer Meeting, Minneapolis, Minnesota, July 15-19, 1990. Manuscript submitted April 2, 1990; made available for printing April 24, 1990.

1.1 Summary of Past Work

The bibliography included at the end of this report lists most of the published work done in the area of computer aids in relay coordination studies. Most of the initial approaches [2,4,5,17,40] resulted in "batch" off-line computer studies for the relay coordination problem. Interactive methods were then gradually employed [6,7,9,12,14-16,18,20,22-27,41-46] to place the protection engineer in the control loop to guide the solution process. Graph-theoretical methods were applied for the system-wide coordination studies [10,11,14,20,25,29]. Performance measures to evaluate alternate solutions were suggested [26,27]. Conversion of existing relay setting files into a computerized data base was presented in [21] and the design and implementation of a relational data base for the protection system using a commercially available data base management system was reported in [22]. Recently, optimization techniques have been applied [30] for obtaining "optimal" relay settings. Subsystem coordination [25] and adaptive coordination [28,47] were proposed as natural extensions to the system-wide coordination concept.

Specific areas of the protection problem, such as coordination of relays in radial distribution lines and industrial power systems, have been considered by a number of researchers [31-39]. Mathematical modeling of overcurrent relays for computer representation has also been reported [1,3,8,48].

2. SYSTEM DATA BASE

The data base required for a CAC system is described in the following three sections:

2.1 Short Circuit Data Base
2.2 Protective Relay Data Base
2.3 Protective Equipment Data Base

These sections represent the type of data the utilities generally have available and can provide as input for the various tasks of a CAC software.

2.1 Short Circuit Data

This section provides information about the data normally required for a reliable analysis of short circuits or faults. The following types of data will be required to form the system model:

(1) Source Data
(2) Bus Data
(3) Branch Data
(4) Transformer Data
(5) Shunt Data
(6) Mutual Impedance Data

Reprinted from *IEEE Trans. Power Delivery*, vol. 6, no. 2, pp. 575-583, April 1991.

2.1.1 Source Data

The source data section will contain impedances of generators and motors and equivalent impedances to represent interconnected systems. For short circuit purposes, generators and motors are represented by connecting their subtransient reactances between the reference bus and their connected bus. Equivalent systems can be represented by impedances in each of the three sequence networks.

2.1.2 Bus Data

The bus data section will contain such information as bus number, bus name, bus type, and bus voltage.

2.1.3 Branch Data

The branch data section will contain circuit identification, such as from bus number, to bus number and circuit number, and branch impedance values. The impedances are normally in per unit. This section should also allow for change codes to indicate additions, removals, or conversions. A means of indicating an open circuit should be provided.

2.1.4 Transformer Data

In addition to the branch data, the transformer data section will contain transformer connections and grounding. Two-winding transformers can be represented by their positive- and zero-sequence impedances, along with a code to indicate the connection (delta-wye, delta-delta, or wye-wye). If a wye winding is grounded, a means of indicating this fact must be available.

Three-winding transformers and autotransformers with tertiaries, including phantom tertiaries, may require an additional midpoint bus. Impedances are entered between the midpoint bus and each terminal bus of the transformer. Connections of each winding and grounding must also be indicated.

The capability of including transformer tap positions may be required by some users. There may also be a need to be able to represent the phase shifts resulting from phase shifters, delta-wye, or other transformers.

2.1.5 Shunt Data

The shunt data section should allow incorporation of shunt impedances into the model. These impedances may represent reactors, capacitor banks, or load and would normally be shown as located at a bus. Pre-fault bus voltages should be consistent with the currents taken by the shunt devices.

2.1.6 Mutual Impedance Data

The mutual impedance data section will contain zero-sequence mutual coupling impedance between pairs of circuits. This impedance data can be either a positive or negative number and must include provisions for that information. For accurate calculation of faults between line terminals, the actual distribution of mutual coupling must be modeled.

2.2 Protective Relay Data Base

The protective relay section of the system data base contains data that describes four levels of relay data:

(1) Relay Characteristic Data
(2) Relay Identification Data
(3) Relay Settings Data
(4) Relay Input Sources Data

By assembling the information in these four sections, the data manager will be able to compile setting sheet output, or direct relay data into the coordination program.

2.2.1 Relay Characteristics

The relay characteristics section should contain data which describes each relay by type and model number. This section should be broken down by general relay classes such as overcurrent, impedance, breaker failure, etc. Other sections of the program will refer to this section to determine the input data parameters of each relay.

Following is some of the information which should be included in this section:

(1) Relay type and manufacturer's model number
(2) Relay parameters
  (a) Tap, time dial, operating characteristic, time-current characteristic, etc.
  (b) Timer set ranges
  (c) Restraint setting ranges and tap tables for impedance relays
  (d) Maximum torque/reach angles
  (e) Parameters for test data calculations
(3) Polarizing/directional requirements and quantities
(4) Operating speeds
(5) Relay burden data
(6) Thermal limits, short-time and continuous

2.2.2 Relay Identification

The second portion of the relay data base contains the relay identification information. This section should be set up with a hierarchy to reduce duplication of data when describing individual relays. The hierarchy could start at the substation level and proceed down through voltage, bus protection, transformer protection, line protection, terminal designation, and individual relay ID. The following should be included in the relay identification data:

(1) Location data - substation, kV, protective zone, associated breaker, relay ID number
(2) Primary or secondary system
(3) Relay device function number - i.e. defined by American Standard C37.2
(4) Function code - zone 1, step distance, phase overcurrent, ground overcurrent, etc.

2.2.3 Relay Settings

The relay setting portion of the data base contains the settings placed on each relay. There should be some means of tying this setting data back to the relay identification data (such as the relay ID number). This data will be divided into data structures similar to the relay characteristics structure.

There will be a number of different data categories for each existing relay, with flexibility to add more if needed. The following are some typical classifications:

(1) Time overcurrent, directional, and instantaneous overcurrent units
(2) Phase distance, single and three-phase
(3) Ground distance
(4) Line protection packages
(5) Pilot wire
(6) Breaker failure
(7) Reclosing
(8) Differential

Each of these categories would be then broken down to the individual setting data required, such as relay ID number, range, tap, lever setting, etc. There should also be room to add comments for each relay. This will add flexibility so that the user can make minor adjustments to data without making program changes.

2.2.4 Relay Input Sources

The inputs to a relay along with its setting determine its operation. Relay inputs are generally currents and voltages which are obtained from current transformers (ct's) and voltage transformers (vt's). The ct and vt locations and the relays they supply must correspond with the buses and line segments in the short circuit data base. This ensures that the modeled relays receive the correct inputs. The source input data should include the following:

(1) Available range of taps, connected ratio, wye, delta, and other connection details
(2) Accuracy class
(3) Internal resistance of ct and connecting wire
(4) Volt-ampere capability of voltage transformers

This data will be accessed by the software to provide the engineer with information on the performance and saturation as well as burdens that exceed ratings.

To determine the direction of a fault, a directional relay requires a reference or a polarization quantity against which line quantities can be compared. Relays can be polarized by voltage, current, voltage and current, or a sequence component. These polarization quantities represent additional input for which data is required. The same input information (above Items 1 through 4) for operating ct and vt sources should be provided for polarization sources.

2.3 Protective Equipment Data Base

Protective equipment refers to fuses, reclosers, sectionalizers, etc. which are involved in coordination with line relays. The data base should include both the available parameters and the actual settings for these devices. A general outline of the type of information that may be required for both is shown below.

2.3.1 Protective Equipment Characteristics

The following general information would be common to all sections in the protective equipment data base:

(1) Manufacturer
(2) Type -- The device type should include the model and style numbers used to describe the device.
(3) Interrupting Rating -- The interrupting ratings of the protective equipment, where applicable, should be included.
(4) Continuous Rating -- The continuous current and/or voltage rating of the device.

2.3.1.1 Fuses

To identify fuses, the data base should contain the following additional information:

(5) Ampere Rating and Speed Characteristics -- The fuse ampere rating should include letters (E, K, T, etc.), designating the "speed" of the fuse's time-current characteristic curve (slow, standard, fast, etc.).
(6) Time-Current Characteristic (TCC) Curves (minimum melting and total clearing)
(7) Preloading and Ambient Temperature Adjustment Characteristics -- Changes to the minimum melting curve due to preloading or elevated ambient temperatures. Current limiting fuses require a representative model of the peak let-thru current and the I^2t characteristics.

2.3.1.2 Reclosers

Recloser modeling can best be accomplished by the following additional parameters:

(5) TCC Model and Trip Rating -- The TCC model and trip rating should include the available settings (ct ratios and taps, or series trip coil ratings) for both phase and ground trip elements.
(6) Number of Trips -- Reclosers may have many trips before lockout and different curves for phase and ground trip elements.
(7) Reclosing Times -- The number of reclosures and time between reclosures.

2.3.1.3 Sectionalizers

In addition to the general information, the data on sectionalizers should contain the following:

(5) Trip Rating -- The trip rating should include the available settings (taps, coils, timing, voltage, restraints, etc.) for phase and ground.
(6) Number of Trips -- The number of upstream trips before sectionalizer opening.

2.3.1.4 Non-Fault Devices

Non-fault devices include frequency, voltage, power, and synchronizing relays, and network protectors. In addition to the general information, the data on non-fault devices should contain the following:

(5) Trip Rating -- The trip rating should include the frequency, power, or voltage ratings of the device.

2.3.1.5 Generic Devices

There may be other devices not addressed in this report that are employed on a power system. The data base should be designed so that these devices can be incorporated with relative ease. Since the generic device may emulate any of the relays, fuses, or reclosers previously described, or combinations thereof, any or all of the data described in Sections 2.2, 2.3.1, and 2.3.2 should be available to describe the generic device.

2.3.2 Protective Equipment Application

Data required to describe the settings of protective devices is as follows:

(1) Location Data -- This data should include the substation and/or line name, number, division, and district.
(2) Device ID Number -- This is the number the company uses to identify the particular device. In certain instances, this may be the switch number.
(3) Switch Number -- The breaker, recloser, or switch (fuse or sectionalizer) number used to identify the device.
(4) Device Function Number -- The device number is used to identify the function of the device. This number may or may not be according to ANSI standards.
(5) Setting -- The data base should include sufficient information on the settings of the devices for user information, as well as coordination checking by the software. This information should consist of the ct ratio and tap or coil (link) size for both phase and ground elements.
(6) Device History -- Some information describing the past locations of the device may be desirable.
(7) Maintenance History -- Information describing maintenance dates and past repairs.

3. SOFTWARE

The functional specification of the software that meets the needs of the protection engineer is described in the following sections:

(1) Data Base Management System
(2) Short Circuit Software
(3) Coordination Software

Enhancements may be desirable if they do not hinder the performance of the required features.

3.1 Data Base Management Software

Protection engineering requires work with large sets of dissimilar data. It includes information on:

(1) Network topology
(2) Structural and configuration data of transmission lines and towers
(3) Parameters of power system equipment
(4) Location, type, model, and setting parameters of protective relays
(5) Equipment failure records
(6) Maintenance history and other reliability data

The main purpose of a data base management system (DBMS) is to eliminate, or more practically to control, data redundancy. Elimination or control of redundancy, in turn, greatly eases the burden of data maintenance. Some of the considerations in the selection of a DBMS are:

(1) The type and quantity of data that must be stored and retrieved
(2) The need for flexibility in adding and rearranging data and relationships
(3) Speed requirements in accessing or storing data
(4) The kinds of relationships among the data
(5) The need for ad hoc queries and reports
(6) The degree to which there are multiple independent users that require concurrent access
(7) The availability of programming language interfaces for the more commonly used languages

The physical and logical organization of data base management systems will be reviewed before going on to their specific application to protection engineering.

3.1.1 Physical Data Organization

The physical data organization is of only secondary interest to a user of a DBMS, although of primary interest to a DBMS developer. To a user it is sufficient to know the impact of the physical data organization on the DBMS performance. Performance measures are often based on retrieval and storage speeds, storage requirements, and speed and flexibility in data base reorganization.

3.1.2 Logical Data Organization

Data base management systems typically use one of three logical data organizations or data models: hierarchical, network, or relational. The inverted file or inverted data model is sometimes considered a logical as well as a physical data organization. The hierarchical and network models have many similarities. It is fairly straightforward to convert a network model to a hierarchical model, at some cost in redundancy. Relational models can be considered as a higher logical development of file management systems. The inverted data model possesses outstanding retrieval characteristics for static data but, as a data model, has limitations that make it unsuitable for the protection engineering application. Almost all current DBMS research and development is directed toward the relational model, sometimes incorporating features of the network model. For this reason, as well as for its

inherent advantages, it is probably best to consider only relational DBMS for computer-aided protection engineering. Almost all of the popular DBMS for personal computers use the relational model, although often only a simplified form with a single table is used. The relational model is fast displacing the other models in large systems.

The relational DBMS model is based on simple "tables" or "flat files." Each column of a table contains data items of a given type; a row contains all of the data for some entity. The columns are often called "fields" and the rows "records." This arrangement is easy to understand and contributes to the ease of using relational DBMS. Different tables or relations are automatically linked by any fields they have in common. The simple table arrangement of the relational model makes it easy to expand an existing data base. New relations can be added at any time, and they are automatically linked to the existing data base by their common fields. However, the construction of the data base (the design of the "schema") and the application of the relational algebra or calculus is quite complex and requires the services of a data-base expert.

3.1.3 Specific Requirements for Protection Engineering

Specific requirements for a DBMS applied to protection engineering will now be discussed. Some of these requirements are general, but others are specific to protection engineering, or at least to engineering applications.

Programming language interface -- Most DBMS have their own programming language, often of a "nonprocedural" type, that makes it easy to perform most required data manipulation. For engineering use, however, the DBMS should have a programming language interface between the data base and a number of commonly used programming languages such as FORTRAN, Pascal, and C. This interface is usually implemented either through subroutine calls or by embedding high-level data base calls in the application programs. The latter approach is gaining in popularity and eases application program development.

Interactive data edit or forms capability -- The DBMS should include a "forms" facility through which data entry and edit are achieved in an interactive, user-friendly fashion. The forms package should offer on-line help, pop-up menus, and screen painting to help a user design, build, and modify forms used to query and maintain data base tables.

Report generator -- A report generator should be available through which the user can generate formatted reports that present and summarize selected data from the data base. It may also be useful if the data resulting from queries can be automatically converted and presented graphically in pie charts, bar graphs, and line graphs with appropriate legends and titles.

Engineering data types -- To accommodate data associated with protection engineering, the DBMS should provide for engineering data types such as vectors, matrixes, complex numbers, and double-precision numbers, in addition to the standard data types used in a commercial environment. Another requirement is the capability to include mathematical expressions for defining and reporting certain data fields.

Interactive query facility -- The DBMS should have a user-friendly interactive data base query language through which data items can be selectively displayed and manipulated. Nested queries should be allowed to combine data across a number of tables. A "complete" set of relational algebra commands should be available to ensure that different "views" of the data can be obtained directly using the query language, with no additional application language programming required.

File transfer -- The DBMS should allow loading of data from an external file and for optionally routing output to a file. For example, a utility will probably already have network data in a formatted file, and will want to load the data into the data base without manual re-entry.

Data checking -- The DBMS should include data verification and validation features to protect data base integrity. This has the added benefit of eliminating the need for extensive data checking in application programs.

Security -- The DBMS should provide for access control features to ensure data security. Access controls should be provided at data base, relation, view, and field levels.

Maintenance tools -- Routine maintenance tools such as backup and garbage collection should be provided as part of the DBMS. A means should be provided for recovering the data base after an accidental system crash.

Portability -- It is desirable that the DBMS be capable of running on a number of systems. At least partial portability is provided by using Standard Query Language (SQL), since it is being proposed as a "relational standard."

Performance -- The DBMS should have good performance in terms of storage and retrieval time, storage requirements, and speed and flexibility in data base reorganization.

Networking capability -- The trend towards using networks of workstations and personal computers for power system engineering makes it desirable that the DBMS support networks of computers. The protection engineering function does not have the highly demanding "consistency" and "concurrency" requirements of a transactions-oriented DBMS used in some commercial applications such as an airline reservation system. However, reasonable support for a number of simultaneous users will be required.

3.2 Short Circuit Software

Short circuit studies are used as a basis to determine relay settings and to check relay performance for a variety of system conditions. Using information available in the short circuit data base, the short circuit software calculates the fault currents at as many points as required to simulate the worst fault conditions. Results are then used by the coordination software to coordinate, set or evaluate the performance of the relays. The specific requirements for the fault calculations may vary according to user preference. In general, the short circuit software should meet the minimum requirements as described below.

3.2.1 Fault Types

Three-phase and single-phase-to-ground faults are required for coordination of most protective relays. Three phase faults are used for phase relay settings and to determine the response of phase distance relays. Single-phase-to-ground faults provide the necessary information to be used in setting ground relays. If relay performance for these types of faults is correct, it will usually be satisfactory for two-phase-to-ground faults.

Ideally, the software should determine the short circuit conditions required for relay settings and coordination. However in any system there will sometimes be short circuit conditions that are not covered by program routines but are necessary to consider in calculating relay settings. Therefore, the program should give the user capability to select additional fault conditions to determine the performance of relays for all possible conditions.

3.2.2 System Conditions

An important consideration in conducting short circuit studies is the configuration of the system. The amount of system generation and the dispatch of the units will affect the magnitudes and distribution of the fault currents. When determining relay settings both maximum and minimum generation cases should be considered. The user should also have the option to specify fault impedances for reduced minimum conditions. Settings for instantaneous trip units are determined at maximum system generation. Relay sensitivities and settings for supervising fault detectors are determined under minimum conditions. It is sometimes necessary to verify relay coordination for both conditions.

3.2.3 Normal System Faults

Normal system faults are those taken with all lines in service in a "normal" operating configuration. Faults are taken at each bus in the area of interest. These faults are necessary to check the coordination with bus protection and to coordinate line relaying for close-in line faults. For the latter case, the short circuit software should sum the appropriate current contributions to determine the current flowing in the line.

3.2.4 Contingency Conditions

Relay coordination is usually determined assuming at least one contingency in addition to the system fault. A contingency can include any of the following:

(1) A transmission line out of service
(2) A transformer out of service
(3) A generator out of service (other than for maintenance and refueling)

Faults under any of these contingencies are defined as "line-out" faults.

3.2.5 Line-Out Faults

For this type of fault, currents are calculated with one or more lines removed from service. Usually, faults are taken only at the bus where the line ordinarily terminates.

Line-out faults are important in that they frequently result in the highest current contribution from other lines. As a result, line-out faults can represent the worst case for relay coordination and for setting instantaneous overcurrent relays. The line-out fault is also used to simulate a "test" where a line breaker is closed into a close-in fault.

3.2.6 Line-End Faults

Line-end faults are taken at the end of a transmission line with the remote breaker open. These faults are used to determine coordination under sequential clearing or coordination with backup relays in case of a breaker failure. Line-end faults can also represent the critical case for coordinating ground overcurrent relays for parallel lines.

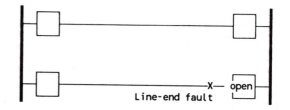
Line-end fault

42

3.2.7 Multiterminal Line Faults

Additional faults should be taken on multiterminal lines to check relay performance with one of the line breakers open. On a three-terminal line, for example, faults at the number two terminal should include contingencies with each of the number three and number one terminals open at a time.

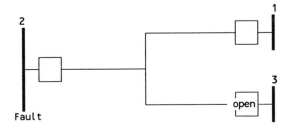

3.2.8 Instantaneous Relay Reach Faults

Close coordination between primary and backup overcurrent relays can be obtained if inverse time elements are coordinated for faults at the end of reach of the primary relay's instantaneous element. The fault currents for this condition have been traditionally obtained by assuming that the distribution factors (i.e. ratio of line currents to total fault current) are the same as for a bus or line-end fault condition. The backup relay current is then determined by multiplying its current distribution factor by the primary relay instantaneous setting. This simplifying assumption will produce satisfactory results in most cases, but a more exact solution is desired by protection engineers.

3.3 Coordination Software

The physical characteristics of each power system determine how it will behave under short circuit conditions. Many variables such as load concentration, available generation, geographical layout, grounding, and design techniques make each utility power system unique. To satisfy the protection requirements of each unique system, the utilities have developed their own protective relaying guidelines. Although the guidelines may vary, the basic techniques used adhere to the same industry practices. Therefore, the coordination software should be based on the industry practices with capability to incorporate the user guidelines.

The coordination software sections that follow describe the software requirements for coordinating and setting of line protective relays. Specific or preferred algorithmic approaches are not identified. However, the bibliography includes references of work done in this area.

3.3.1 Coordinate Relays and Determine Desired Settings

The definition of coordination is explained in the introduction of the IEEE Committee Report "Computer Representation of Overcurrent Relay Characteristics" [48], at least in terms of overcurrent devices. That definition is generalized and repeated below.

Protective devices are selected and calibrated to isolate only the section of the system affected by a fault, so that the remaining portion of the system can continue to operate normally. The basic approach is that abnormal fault currents and voltages are clearly distinguishable from normal load currents and voltages, and the protective devices are designed to respond to the abnormal currents and voltages and open the affected circuit rapidly.

Since faults can occur at any point in the system, many protective devices are required. These devices are located such that a single device, or combination of devices, operate to isolate every fault. Good practice dictates that the area isolated must be as small as possible, with only the devices nearest the fault operating. In addition, the possibility of failure of a protective device must be considered. In this situation the next upstream device, or device combination, must operate to provide backup (remote) protection. When two devices operate properly in this primary/secondary mode for any system fault, they are said to be coordinated. Proper coordination is achieved by time discrimination between successive devices.

In a protective system, each device is assigned a primary function to clear faults in a specific zone and a secondary function to clear faults in adjacent or downstream zones to the extent the range of the device permits. The operating time for the primary device for a fault in the protected zone must be sufficiently short to allow the secondary device to reset to normal without causing a trip. If the primary device fails to operate the secondary device is expected to clear the fault.

The specific zones that a protective device is assigned to protect correspond to power system apparatus. Generators, buses, transformers, and motors are very compact compared to the area covered by a power system. For these zones, communication between protective devices on zone boundaries (differential relaying) is not a problem. When there is a communication link between protective devices on zone boundaries, coordination is relatively simple. Transmission lines differ from the other elements of a power system because the distances between line terminals may be hundreds of miles. An external communication link between line terminals (pilot relaying) may not be practical, or it may be desirable to have a backup system which does not depend on communication. In this case, the protective devices on the boundaries must be applied and set to coordinate with the remainder of the system (allow sufficient time for adjacent zone protective devices to operate for faults external to the zone). The protective devices must make decisions to trip or not trip based entirely on voltage and current information available at the device location.

Software which is designed to aid the power system protection engineer must be able to determine settings that will satisfy the coordination requirements defined above. The degree to which the software can satisfy these requirements is constrained by the characteristics of the power system under consideration, as well as the protective device models, with the assumption that these have been defined. With this in mind, it becomes evident that the portion of the software that determines settings enters the realm of the "art of protective relaying." In the determination of protective device settings, there is more than one answer to the same problem, with the desirable answer being dependent on the prioritization of speed and sensitivity versus selectivity. This brings to light some distinct coordination software requirements. The methodology used by the software in the determination of settings must be apparent to the user as well as controllable by the user. In addition, this methodology must incorporate an evaluation of all significant primary/backup relationships for all relevant fault conditions.

The software must also allow application of its methodology in either a global or local fashion controlled by the user. To implement the requirements outlined above, an implicit requirement of the software arises. This is that the software must have accurate protective device models which emulate physical hardware to the extent of providing an indication of operation (including operating time) or nonoperation, given the necessary electrical quantities. The bibliography contains papers that provide specific algorithmic approaches, along with examples.

3.3.2 Check Coordination of Specified Settings

In addition to executing the software to achieve coordinated settings, it must be possible to have the software simply check the coordination of the specified settings. In this case it must generate information on where coordination is not achieved. This information must include the identification of the devices which do not coordinate, the fault condition or conditions for which they do not coordinate and the device operating times for these faults. This same information must be generated for those devices which do coordinate in the strictest sense but for which a desired time margin, generally defined as the "coordinating time interval" (CTI), is not achieved.

3.3.3 Check Coordination of Alternative Settings

The software must operate on two sets of device settings. The first, or initial, settings are specified by the user. The second, or alternative, settings may be determined in two ways. First, they may be produced by the execution of the coordination logic. The software must determine automatically what settings must change to achieve coordination. Those changes will be made to the specified settings to create the alternative settings. Second, the user may create the alternative settings by making changes to the specified settings. In either case the software must be able to do the same check on coordination as described in Section 3.3.2 and generate the same information as it applies to the alternative settings.

3.3.4 Provide Comparison of Specified and Alternative Settings

Since two sets of settings are maintained by the software, it is natural that the software be required to make comparisons between the two. These comparisons are as simple as a listing of device setting changes and as complex as comparisons of system performance for each set of settings. The comparisons of system performance should include, but not be limited to, a measure of speed of fault clearing, a measure of protection sensitivity, and a measure of system device coordination.

3.3.5 Criteria

The methodology to determine coordinated settings must be controllable by the user. The user must be able to incorporate specific company practices and policies into the coordination process. The user-specifiable coordination criteria must include but not be limited to the following items:

(1) Device setting constraints
(2) Line coverage requirements for each protective zone
(3) Maximum operating time of primary protection for which coordination is required
(4) Coordinating Time Interval for each device

The user must have an option as to whether or not sequential tripping operations are included in the coordination process.

3.3.6 Application Considerations

In conjunction with the coordination process the software should be able to evaluate proper device application. These application checks should include but not be limited to:

(1) Device directional requirements
(2) Current transformer saturation
(3) Device energy requirements
(4) Polarization problems

3.3.7 Batch or Interactive Mode

Software should allow operation in batch as well as an interactive mode. In the batch mode the coordination process should run through to completion to create a set of coordinated settings. Optionally, it should run in an interactive mode with user inputs allowed during the process. For example, the software may automatically compute device settings to achieve coordination. However, the user must be able to specify particular devices whose settings cannot change, particular sets of devices which must or must not be coordinated, and fault conditions that must or must not be considered for coordination. Additionally the criteria used in coordination must have default values to be used automatically unless the user elects to specify them.

3.3.8 Subsystem Capability

The software must allow application of its methodology in either a global or a local fashion. This means the user can access a subset of the full system data to study a localized coordination problem.

4. PRESENTATION OF RESULTS

The program executive should be able to store, print and communicate the results of the different processes to meet the needs of the relay engineer. A configuration file should exist in which the user specifies what output data should be saved and what medium it should reside on, such as the screen, disk, file, tape or paper. This will allow the relay engineer to only save data needed for tracking setting calculations and to discard working data that will not be used later.

The two basic types of coordination processes would require output files to be displayed in different ways. In the interactive mode, the relay engineer will only be coordinating a few relays. The engineer may then want the results of the short circuit calculations and the relay operating times to be displayed on the screen or on an attached printer. In the batch mode, when coordination of an area is being performed, the relay engineer may only be interested in the miscoordinations. The total results would then be stored in a file for later printing, while the problems are displayed on the screen and/or printed immediately.

4.1 Short Circuit Output

The short circuit module should provide the engineer with the results of the faults as described in Section 3.2. The output data should include the normal system as well as the contingencies considered, i.e., line-out faults, line-end faults, etc. This data could be stored in a separate file for later use or could be included in the data base.

4.1.1 Tabulation of Faults

Tabulation of the output should include information on the sequence components of the fault currents. Current flows should be included for current contributions within the user-specified depth around the faulted bus. The short circuit data may be sorted and prepared for input into the coordination module.

4.1.2 Graphics

At the user's option, the results of a fault may be displayed on a one-line graphic diagram of the system on the interactive display terminal. The diagram should include options for displaying the symmetrical component or phase component phasors. Additionally, the option should be available to the user to print the diagram along with the symmetrical or phase components of the fault.

4.1.3 Interactive Color Graphics and Interface

To perform fault studies of a certain contingency, the user should have the option to enter the contingency graphically on the one-line diagram using the cursor. Color-coding of changes should show the user the changes just made, and update of the symmetrical and phase components should follow to indicate the results of the fault with that contingency.

4.1.4 Data Base

The short circuit output results may be stored in the data base via the data base management software for later queries.

4.2 Protection Output

The output of the protection coordination should include the fault information and operating time of the relays corresponding to the faults, as well as the coordination time intervals between relays.

4.2.1 Tabulation of Results

Tabulations of relay coordination should include relay numbers, fault description, fault currents, times of operations, and the coordination intervals. When two sets of relay settings are being evaluated, the output should provide a comparison table for the settings.

4.2.2 Coordination Graphics

A color graphic terminal display of the system one-line diagram, showing faults with times of operation of each relay corresponding to the fault current and the coordination intervals, should be available from the user's command line entry. The user should be able to select the related fault and contingencies and graphically receive the coordination result updates in response. For distance relays, a diagram should show the overlapping of the distance zones with the primary and backup relays displayed in colors corresponding to each relay.

4.2.3 Display Copy

The user should have the option to print results of each display.

4.3 Polarization Results

Polarizing quantities should be available both via the data base and with the relay coordination output tabulation as an option.

4.4 Coordination Curves

The user may need to compare the time-current characteristics of different relays, reclosers, fuses, or other devices on a log-log scale at various pickup positions. This option will allow the user to interactively coordinate relays on the display for the radially operated lines with the computer coordinated relays.

5. COMPUTER SYSTEM REQUIREMENTS

In order to support the computer-aided coordination functions as outlined in this report, a variety of system configurations could be deemed suitable. The computer system requirements for any software are not only dependent on the size of the system to be studied and the number of anticipated simultaneous users, but are subjective, e.g.,

what is the maximum response time tolerated? What level of file control is implemented? What are the budgeting limitations?, etc. Therefore, the guidelines presented here are for a "representative" application with a "reasonable" man/machine interface.

The following are minimum specifications for computer system requirements under the above assumptions:

(1) CPU Configuration
  (a) 32-bit word size
  (b) 4 megabytes physical memory
(2) Peripheral Configuration
  (a) Medium resolution video graphic CRT (at least 640x480 display matrix)
  (b) Digitizing tablet (.005" resolution) or mouse
  (c) Hard disk (70 megabytes or larger)
  (d) Printer with graphics capability or plotter
(3) Operating System Configuration
  (a) On-line file handling capabilities
  (b) Virtual memory
  (c) Interactive operation

No attempt has been made to classify computers, i.e., mainframe, mini, micro, or PC, due to the gray area that exists here. The actual supporting computer system would likely consist of the generic classifications.

CONCLUSION

The definition of a computer aided system for coordination of line protection schemes that would meet the needs of a large segment of the relay protection engineering community has been completed. Those companies who have an interest in developing software of this type should obtain the complete special publication "Computer Aided Coordination of Line Protection Schemes" from the publications office of the IEEE.

REFERENCES

1. G. E. Radke, "A Method for Calculating Time-Overcurrent Relay Settings by Digital Computer," IEEE Transactions on PAS, Special Supplement, Vol. 82, 1963, pp. 189-205.
2. R. E. Albrecht et al., "Digital Computer Protective Device Coordination Program I - General Program Description," IEEE Transactions on PAS, Vol. 83, No. 4, April 1964, pp. 402-410.
3. H. Y. Tsien, "An Automated Digital Computer Program for Setting Transmission Line Directional Overcurrent Relays," IEEE Transactions on PAS, Vol. 83, October 1964, pp. 1048-1053.
4. S. S. Begian et al., "A Computer Approach for Setting Overcurrent Relays in A Network," IEEE PICA Conference Record, Vol. 31C69, May 1967, pp. 447-457.
5. W. M. Thorn et al., "A Computer Program for Setting Transmission Line Relays," Proceedings of American Power Conference, Vol. 35, Chicago, May 8-10, 1973, pp. 1025-1034.
6. R. B. Gastineau et al., "Using the Computer to Set Transmission Line Phase Distance and Ground Backup Relays," IEEE Transactions on PAS, vol. 96, No. 2, March/April 1977, pp. 478-484.
7. R. B. Gastineau and R. H. Harris, "Setting Phase Distance Relays by Computer," Transmission and Distribution, Nov 1979, pp. 51-54.
8. M. S. Sachdev et al., "Mathematical Models Representing Time- Current Characteristics of Overcurrent Relays for Computer Applications," IEEE PAS Paper A78 131-5, January 1978. Abstract PAS-97 No. 4, July-Aug. 1978, p. 1008.
9. K. Suzuki et al., "Interactive Computation System of Distance Relay Setting for A Large Scale EHV Power System," IEEE Transactions on PAS, Vol. PAS-99, No.1, Jan/Feb 1980, pp. 165-173.
10. A. H. Knable, Electrical Power Systems Engineering Problems and Solutions, McGraw Hill, New York, 1967.
11. M. H. Dwarakanath and L. Nowitz, "An Application of Linear Graph Theory for Coordination of Directional

Overcurrent Relays," Electric Power Problems - The Mathematical Challenge, Proceedings of the Society for Industrial and Applied Mathematics (SIAM) Meeting, Seattle, WA, March 18-20, 1980, pp. 104-114.

12. M. J. Damborg et al., "Applications of Computer Aids to Transmission Protection Engineering," Proceedings of the Ninth Annual Western Protective Relay Conference, Spokane WA, October 26-28, 1982.

13. J. P. Whiting and D. Lidgate, "Computer Prediction of IDMT Relay Settings and Performance for Interconnected Power Systems," IEE Proceedings, Vol. 130, Pt. C, No. 3, May 1983, pp. 139-147.

14. M. J. Damborg and S. S. Venkata, "Specification of Computer- Aided Design of Transmission Protection Systems," Final Report EL-3337, RP 1764-6, EPRI, January 1984.

15. M. J. Damborg et al., "Computer Aided Transmission Protection System Design, Part I: Algorithms," IEEE Transactions on PAS, Vol. PAS-103, January 1984, pp. 51-59.

16. R. Ramaswami et al., "Computer Aided Transmission Protection System Design, Part II: Implementation and Results," IEEE Transactions on PAS, Vol. PAS-103, January 1984, pp. 60-65.

17. J. Huisma et al., "Computer Determination of the Zone Settings Distance Relays on the Basis of Short-Circuit Calculations," Paper 34-12 presented at the International Conference on Large High Voltage Electrical Systems, Paris, Aug 29 - Sep 6, 1984.

18. D. E. Schultz and S. S. Waters, "Computer-Aided Protective Device Coordination, A Case Study," IEEE Transactions on PAS, Vol. PAS-103, No. 11, November 1984, pp. 3296-3301.

19. Y. Oura et al., "Interactive Computation System for Relay Setting and Simulation of Relay Operation in A Large Scale EHV Power Systems," IEEE Transactions on PAS, Vol. PAS-104, No. 7, July 85, pp. 1767-1773.

20. R. Ramaswami et al., "Enhanced Algorithms for Transmission Protective Relay Coordination," IEEE Transactions on Power Delivery, Vol. PWRD-1, No. 1, Jan.1986, pp. 280-287.

21. R. Zimmering and R. Allen, "Computerization of A Large Relay Setting File," IEEE Transactions on Power Delivery, Vol. PWRD-1, No. 1, Jan 1986, pp. 135-142.

22. M. J. Damborg et al., "Application of Relational Data Base to Computer-Aided-Engineering of Transmission Protection Systems," IEEE Transactions on Power Systems, Vol. PWRS-1, No. 2, May 1986, pp. 187-193.

23. J. Postforoosh, "Computer Program Coordinates Relay Operations," Transmission and Distribution, June 1987, pp. 50-52.

24. User Manual for Computer Aided Protection System (CAPS), EPRI RP 2444-2, University of Washington, Seattle, July 1985.

25. R. Ramaswami, "Transmission Protective Relay Coordination: A Computer-Aided-Engineering Approach for Subsystems and Full Systems," Ph.D Dissertation, University of Washington, Seattle, January 1986.

26. S. S. Venkata et al., "C.A.E. Software for Transmission Protection Systems: Puget Power Experience," IEEE Transactions on Power Delivery, Vol. PWRD-2, No. 3, July 1987, pp. 691-698.

27. J. A. Juves et al., "The concept of Figure of Merit Applied to Protection System Coordination," IEEE Transactions on Power Delivery, Vol. PWRD-1, No. 4, October 1986, pp. 31-40.

28. A. K. Jampala, "Adaptive Transmission Protection: Concepts and Computational Issues," Ph.D Dissertation, University of Washington, Seattle, November 1986.

29. V. V. B. Rao and K. S. Rao, "Computer-Aided Coordination of Overcurrent Relays: Determination of Break Points," IEEE Transactions on Power Delivery, Vol. 3, No. 2, April 1988, pp. 545-548.

30. A. J. Urdaneta and L. G. Perez Jimezez, "Optimal Coordination of Directional Overcurrent Relays in Interconnected Power Systems," IEEE Transactions on Power Delivery, Vol. 3, No. 3, July 1988, pp. 903-911.

31. K. S. Bajaj, "Industrial Power System Protective device Coordination by Computers," Proceedings of IEEE IAS Annual Meeting, Los Angeles, CA, October 2-6, 1977, pp. 528-532.

32. J. D. Langhans and A. E. Ronat, "Protective Device Coordination via Computer Graphics," Proceedings of IEEE IAS Annual Meeting, Cleveland Ohio, October 1979, pp. 1209-1217.

33. H. A. Smolleck, "A Simple Method for Obtaining Feasible Computational Models for the Time-Current Characteristics of Industrial Power System Protective Devices," Electric Power Systems Research, 2(1979), pp. 129-134.

34. J. K. Wagner and F. C. Trutt, "Interactive Relay Coordination Using Microcomputer Graphics," Electric Power Systems Research, 4(1981), pp. 129-134.

35. D. H. Malone and D. C. Schroeder, "Computerized Method of Coordinating System Protective Devices for Transient Fault Conditions," Proceedings of the 24th Midwest Symposium on Circuits and Systems, June 29-30, 1981, pp. 611-615.

36. U. Sachs et al., "Computer-Aided Coordination of Overcurrent- Time Protection," Siemens Power Engineering and Automation, VII(1985), No.2, pp.72-76.

37. C. R. St. Pierre and T. E. Wolny, "Standardization of Benchmarks for Protective Device Time-Current Curves," IEEE Transactions on IAS, Vol. IA-22, NO.4, July/August 1986, pp. 623-633.

38. K. A. Brown and J. M. Parker, "A Personal Computer Approach to Overcurrent Protective Device Coordination," IEEE Transactions on Power Delivery, Vol. 3, No. 2, April 1988, pp. 509-513.

39. L. L. Lai and J. G. Hadwick, "Computer-Aided Design for Protection Coordination in Industrial Power Systems with the Inclusion of Transient Phenomena," Proceedings of IEEE Power Industry Computer Applications Conference PICA, May 18-22, 1987, Montreal, Canada, pp. 453-459.

40. W. E. Feero, "Setting Ground Relays with a Digital Computer," IEEE 1967 PICA Conference Proceedings, May 1967, pp. 443-446.

41. R. H. Cauthen and W. P. McCannon, "Computer-Aided Protection Engineering - Do You Need It?," Protective Relaying Conference, Georgia Institute of Technology, April 30, - May 2, 1986.

42. J. R. Adomaitis, R. C. Wingerter, H. W. Anderl, and R. W. Johnson, "Advantages of Using Computer - Aided Protection Techniques over Conventional Methods," Pennsylvania Electric Association, September 23-24, 1987.

43. E. J. Mayer et al., "Computer-Aided Transmission Protective Relay Coordination with Interactive Graphics," Western Protective Relay Conference, Spokane, WA, October 1985.

44. E. Mayer and R. Baysinger, "Improving System Reliability through the Application of Computer Aided Techniques to the Protection Problem," Texas A&M Relay Conference April 1987.

45. R. Cauthen and W. McCannon, "The CAPE System: Computer-Aided Protection Engineering," IEEE Computer Applications in Power, Vol. 1, No. 2, April 1988, pp. 30-34.

46. R. W. Johnson, "Modeling Needs in Protective Device Coordination Software," Instrument Society Of America (ISA) Modeling and Simulation Conference, paper No. 87-2236, 1987.

47. A. K. Jampala et al., "Adaptive Transmission Protection: Concepts and Computational Issues," IEEE Transactions on Power Delivery, Vol. 4, No. 1, January 1989, pp. 177-185.

48. "Computer Representation of Overcurrent Relay Characteristics," an IEEE Committee Report, IEEE Transactions on Power Delivery, Vol. 4, No. 3, 1989, pp. 1659-1667.

49. F. L. Alvarado, M. K. Enns, and W. F. Tinney, "Sparsity Enhancement in Mutually Coupled Networks," IEEE Transactions on Power Apparatus and Systems, Vol. PAS-103, No. 6, pp. 1502-08, June 1984.

50. F. L. Alvarado, S. K. Mong, and M. K. Enns, "A Fault Program with Macros, Monitors, and Direct Compensation in Mutual Groups," IEEE Transactions on Power Apparatus and Systems, Vol. PAS-104, No. 5, pp. 1109-1120, May 1985.

COORDINATION OF DIRECTIONAL OVERCURRENT RELAYS IN

TRANSMISSION SYSTEMS - A SUBSYSTEM APPROACH

R. Ramaswami
Member
Electrocon International, Inc.
611 Church Street
Ann Arbor, MI 48104

M.J. Damborg   S.S. Venkata
Member   Senior Member
Department of Electrical Eng.
University of Washington, FT-10
Seattle, WA 98195

Abstract- A new concept called subsystem coordination for efficiently computing proper settings of directional overcurrent relays on transmission systems in response to changes in system topology and load levels is proposed. An automatic window identification and coordination algorithm has been developed, implemented and tested on a realistic transmission system. The test results have been validated through a full system coordination study and the significant reduction in computer time required by the subsystem approach has been demonstrated.

1. INTRODUCTION

The major task faced by the protection engineer is to set and maintain coordination among a number of protective relays. This task is complicated because the transmission system is very often subjected to a number of changes due to maintenance activities, network reconfigurations, switching actions after faults and load level changes. The basic methodology in selecting suitable relay settings to achieve sensitivity, selectivity, reliability and speed has been known for a long time. Traditionally, these relay setting and coordination computations have been carried out manually. This process involves substantial manpower constantly laboring with a large number of computations involving huge data sets. These time consuming relay setting and coordination calculations are best handled by a computer.

Computer aids in setting and coordinating directional overcurrent relays of a transmission system are relatively new. Recently a great deal of work has been reported in this area. Data base designs and implementations for representing protection engineering data were reported [3, 10]. Application of interactive graphics in the setting and coordination process were recommended [7, 9 and 11]. Algorithms and methodologies for a system-wide coordination were proposed [1-5, 7-9]. All of the system-wide methods, however, dealt with a "full system" approach in which each system change requires a new complete system

coordination study. These methods are not efficient for analyzing and coordinating large networks in the face of changing system conditions.

This paper proposes a new algorithm called the Subsystem Coordination Algorithm [6,14]. In the event of structural or load changes, this algorithm automatically identifies a subsystem of the original network as a "window" around the region of changes. This window is identified by studying the sensitivity of the relay setting parameters with respect to the new network conditions. Only those relays within this window will require new settings due to the system change. The settings of the relays outside the window will be unchanged. This subsystem is then coordinated for proper relay actions considering the new system configuration. This coordination process also accounts for the "boundary" conditions which are the coordination constraints imposed on the relays within the window by the relays just outside the window. The subsystem coordination algorithm has been implemented using a database management system, RIM [3, 5], for data handling. This implementation has been successfully tested on the 115-kV transmission system of Puget Sound Power and Light Co. (PSPL), a 38 bus/61 line system. This system has all the characteristics of a typical transmission network such as a large number of loops, multi-terminal lines, and parallel lines. These are the characteristics which make the coordination process a tedious and highly iterative one. The protection engineers at PSPL have had a history of difficulty in coordinating the ground directional overcurrent relays on this system manually. We have therefore selected this system for testing the proposed subsystem coordination method. The concepts and development of this subsystem approach along with the test results showing the significant advantages of this suggested methodology are presented in this paper.

2. NEED FOR A SUBSYSTEM COORDINATION APPROACH

The protection engineer, while working with an existing system, often encounters a situation where he would like to study a part of the large system for proper coordination of the protective relays. One possible reason may be that this subsystem was coordinated earlier but may not be responding properly for the present conditions due to some system changes that might have taken place after it was last coordinated. More specifically, he may also like to determine the changes in relay settings needed in response to structural changes like addition or removal of lines or of generators, or to significant

89 SM 728-7 PWRD   A paper recommended and approved by the IEEE Power System Relaying Committee of the IEEE Power Engineering Society for presentation at the IEEE/ PES 1989 Summer Meeting, Long Beach, California, July 9 - 14, 1989. Manuscript submitted January 31, 1989; made available for printing April 27, 1989.

Reprinted from *IEEE Trans. Power Delivery*, vol. 5, no. 1, pp. 64–71, Jan. 1990.

changes in system loading conditions. The
changes in system topology may be temporary
due to maintenance, or permanent because of
network reconfiguration or the result of
switching actions after a fault. For a large
transmission system, it may not be desirable
and, at times, may even be impossible due to
the system size, to carry out a complete
system coordination process for every one of
these modifications. Coordination of the full
system may be prohibitively expensive in
terms of the manual and computer effort
involved in obtaining new operating
parameters for the protective relays. A more
efficient way to solve this problem is to
adopt a subsystem coordination approach where
only that part of the network which has been
affected by the change is considered for
recoordination.

A preliminary investigation using a
small test system revealed that structural
changes generated significant changes in the
parameters of the overcurrent relays in the
neighborhood of the disturbed region but
negligible changes occured in the parameters
of distance relays. Changing load levels had
minimal effect on the overcurrent relays but
had significant effect on the third zone
settings of the phase distance relays. It can
therefore be expected that coordinating the
overcurrent and the distance relays may
require different subsystems. It is well
recognized by protection engineers that
coordination of inverse time overcurrent
relays in a looped network is much more
complicated and time consuming than that of
distance relays. It was, therefore, decided
to limit our focus on the study of
directional overcurrent relays only. This
paper, therefore, reports on a project to
study the sensitivity of overcurrent relays
to system changes and to identify and
coordinate a suitable subsystem.

3. CONCEPTS OF THE SUBSYSTEM COORDINATION APPROACH

Figure 1 will be used to define the basic
terminology which will be used to describe
the concepts and algorithms of the subsystem
coordination process. A local region of a
typical large system is shown in this figure.
Let us assume that the entire system has
earlier been coordinated. Consider the
problem of removing the line M and studying
the effect of this line removal on all the
relays in the system. The removal of this
line can be viewed as a "structural
disturbance" at the end buses of this line,
namely buses 1 and 2. This disturbance will
affect the fault current profiles on the
neighboring lines and hence affect the
coordination of the relays under the existing
settings. However, we know that the changes
in fault current profiles diminish as we move
farther from these disturbed buses. Hence, we
can reasonably expect that relays which are
far away from the disturbed zone may still
remain coordinated even with this system
change. The idea of subsystem coordination is
to identify those relays in and around the
disturbed zone which are affected by the
system change and to obtain new parameters
for these relays to ensure coordination under
the present system conditions.

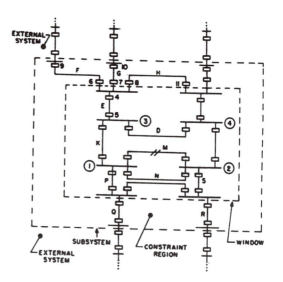

Figure 1. An Example System for Definitions of Subsystem Terminology

Definitions used in the subsystem
coordination process are given below:

1. An area called the "window" is
automatically identified around the
disturbed zone. This window contains all
the relays which are to be reset to obtain
coordination for the new system
conditions. This window is identified by
studying which relays need "significant"
changes in their parameters for proper
coordination. The criteria for identifying
significant changes will be explained
later.

2. Considering the window shown in Figure 1,
one can also observe that the relays at
the boundary of this window are
constrained in their operation by the
relays which are just outside the window.
The "constraint region" is then fixed by
the window and the "level of remote
backup" desired. Following the typical
utility practice of one level of remote
backup, we can then identify the
constraint region to consist of all relays
within one tier around the window. This
region is therefore made up of all relays
on lines immediately adjacent to the
window. It is to be noted that these
relays in the constraint region themselves
have fixed settings, that is the changes
in the system structure have not affected
their settings and they remain as before.
The relays in this constraint region
impose constraints on both the primary
and the backup operation of the relays in
the window. We can therefore classify
these constraints into the two categories,
the "primary constraints" and the "backup
constraints." The primary constraints are
those imposed by the relays in the
constraint region on the PRIMARY operation
of the relays in the window and the backup
constraints are those imposed by the
relays in the constraint region on the

BACKUP operation of the relays in the window. Again referring to Figure 1, relays 9, 10 and 11 in the constraint region backup relay 4 in the window. Hence, relay 4 is to be set for its PRIMARY operation to coordinate for the existing fixed settings of relays 9, 10 and 11 for faults on line E and bus 3. These relays 9, 10 and 11 are therefore said to impose primary constraints on relay 4. Similarly, relay 5 in the window backs up relays 6, 7 and 8 in the constraint region. Hence, relay 5 is to be set for its BACKUP operation to coordinate for the existing fixed settings of relays 6, 7 and 8 for faults on lines F, G and H respectively. Relays 6, 7 and 8 are therefore said to impose backup constraints on relay 5. Thus, in addition to coordinating all the relays in the window, the subsystem coordination procedure must also include these primary and backup constraints imposed by the relays in the constraint region.

3. The combination of the "window" and the "constraint region" is called the "subsystem." Thus the subsystem consists of all those relays which are to be considered in the subsystem coordination process though only a portion of this subsystem namely the "window", contains relays which will have changed settings.

4. Everything outside the subsystem is referred to as the "external system." The settings of relays in this external system remain unaltered.

4. SUBSYSTEM COORDINATION PROCEDURE

Identifying and coordinating a subsystem can be thought of as three basic procedures:

A. Subsystem Identification
Once the buses involved with the system changes (the "disturbed" buses) are specified, a radial search starting from these buses is carried out. Those relays whose settings need to be changed are identified. The window containing all the lines on which these relays are located is then obtained. The constraint region is then identified as explained later completing the subsystem identification.

B. Subsystem Database Setup
All the data corresponding to the topology, relays and fault study information for the subsystem are extracted from the full database and loaded into the relations pertaining to the subsystem. The subsystem coordination program accesses only these relations to carry out the coordination process.

C. Subsystem Coordination
The topological analysis process [1, 4] is then carried out on the subsystem to identify an efficient sequence of primary/backup relay pairs. A complete loop coordination is then carried out on the subsystem. This coordination process includes all the constraints imposed by the relays in the constraint region on those in the window.

A detailed flow chart indicating the steps involved in this subsystem identification and coordination process is shown in Figure 2.

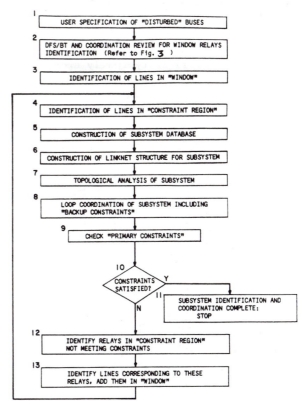

Figure 2. Subsystem Identification and Coordination Algorithm

A description of these steps of the flow chart keyed by the numbers of the individual blocks is given below:

1. The user specifies a set of buses as the "disturbed" buses. This set could contain buses where configuration changes like line removal or addition has taken place, where load changes have occured or simply those where relay coordination is in doubt.

2. The depth-first-search and back-tracking (DFS/BT) technique is used to identify the relays to be included in the window. This process is described in further detail in the flow chart shown in Figure 3. This search process is initiated from each of the disturbed buses. As a particular bus is visited in this process, coordination of relevant relays are reviewed and new settings are found if necessary. In this review process, all the primary and the backup relays for faults at the subject bus are identified first. For each of these relays, new settings are then obtained using the fault data corresponding to the modified system. Those relays whose settings have changed "significantly" from the previously existing values are included in the window. We need a precise criterion to define what we mean by a significant

change. For the overcurrent relays, the pickup and the instantaneous taps are discrete in nature while the time dial tap can be varied continuously. Therefore, a new discrete tap different from the existing value is considered to be a significant change for the pickup and instantaneous settings. A user specified precision value is used for the time dials. This value is typically about 0.1.

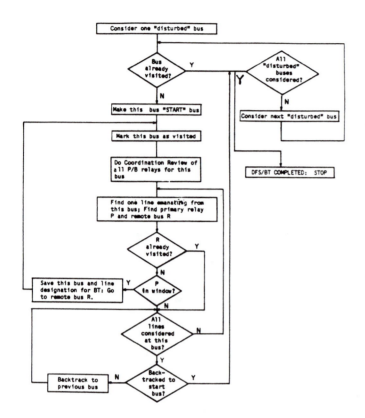

Figure 3. DFS/BT for Window Relays Identification

Once this review process is completed for all the relays for the subject bus, the depth-first-search process proceeds in the direction of those primary relays which have been included in the window. The review of relays is repeated at all the buses encountered in this search. The search in one direction terminates when none of the relays corresponding to a bus need to be changed. This bus is then a boundary bus for the window. We then back-track to the previous bus and continue with the search process. Once we back-track to the bus we started with, we consider the next disturbed bus and repeat the same procedure. Once all the disturbed buses are considered, we have identified all the relays which are to be included in the window. It is to be noted that new values of pickup and instantaneous settings for all the relays in the window have been obtained in this process and these will be the final settings. We have also obtained the time dial settings for these relays but these may not be the final values

since we have not carried out an iterative loop coordination for the subsystem. This iterative loop coordination of the subsystem is carried out using an approach very similar to that for the full system coordination algorithm described in [4]. The details are described in the following steps.

3. We have thus far identified only the relays which are to be included in the window. Since we want to carry out a complete loop coordination of the subsystem, we need to form a complete database for the subsystem. Since the topology of the window is described in terms of the lines contained in the window, we need to determine the lines on which the identified relays are located.

4. The next step is to obtain the description of the constraint region. The constraint region consists of all the lines which are one bus away from the boundary buses of the window. Since we have formed the window by identifying all the lines and the boundary buses of the window, we can now obtain the lines which should be in the constraint region. Thus we have obtained the subsystem which is the combination of the window and the constraint region.

5. We then form the database corresponding to the subsystem. This is achieved by extracting the subsystem data from the existing complete system database and forming additional relations for the subsystem. The topology of the subsystem is described in terms of the lines in the subsystem. The parameters of the relays in the subsystem are loaded into the relevant relations. The fault data needed for coordinating all the relays in the subsystem are also extracted from the entire fault set and loaded into subsystem relations. This approach of forming separate relations for the subsystem facilitates fast data access and retrieval for the coordination programs and hence permits faster execution of these programs.

6. The data base for the subsystem is now available and therefore we can proceed with a complete coordination process of the subsystem. The first step is to create the sparse data structure, LINKNET [1,2,5] for the subsystem.

7. We then carry out the topological analysis on the subsystem to obtain the efficient sequence of relay pairs.

8. We can now carry out the coordination of the overcurrent relays of the subsystem. We note that the pickup and the instantaneous settings of the relays have already been obtained in the subsystem identification process. We have also obtained preliminary values for the time dial settings. These settings are preliminary in the sense that only a "radial" coordination has been carried out on the subsystem. With these settings as initial values, we now carry out an

iterative loop coordination. We have to augment the coordination procedure performed by the overcurrent coordination program OCCORD described in [4] by including the constraints imposed by the relays in the constraint region on those in the window. We recall that there are two types of constraints, the "primary" and the "backup" constraints. Consideration of the backup constraints is relatively straightforward since these are constraints on the backup operation of the relays in the window. The time dial settings of the relays in the window can be assigned to satisfy coordination with other relays in the window as well as to satisfy these backup constraints. The loop coordination is carried out iteratively until we get the final converging solution.

9. We now need to ensure that the "primary" constraints are satisfied. We recall that these are the constraints imposed by the relays in the constraint region on the primary operation of the relays in the window. We carry out a coordination check procedure to verify that all of these primary constraints are satisfied by the relays in the window with the final settings we have obtained in step 8.

10 & 11. If all of these primary constraints are satisfied, we have solved the subsystem coordination problem. The settings of the relays in the window are the final settings correponding to the modified system.

12. If some of the primary constraints are not satisfied, we identify those relays in the constraint region which impose these primary constraints. This situation means that the settings of these relays

can not remain fixed and these relays must be included in the window.

13. We then identify the lines on which these relays are located and include these lines in the window. We therefore have an expanded window. Due to these added lines in the window, the constraint region is also now changed. We therefore go back to step 4 to identify the new constraint region and then to proceed to the subsequent steps of coordinating the subsystem.

5. IMPLEMENTATION AND TEST RESULTS

The subsystem coordination process has been implemented in a manner consistent with our previous comprehensive Computer-Aided-Engineering package, CAPS [5], for protection studies. As with that package, a database management system, RIM, is employed for efficient data retrieval and management. The resulting code has been successfully tested on the 115-kV transmission system of PSPL, a 38 bus/61 line system. Reference [6] describes this test system, a portion of which is shown in Figure 4. In this section we will consider the case of removal of the line OB-MI1 (the line #1 from O'Brien substation to Midway substation in Figure 4). Since this line is an electrically short one (low impedance), it is expected to affect the fault profiles on adjacent lines significantly and, hence, disturb the coordination of the relays in the neighborhood. The subsystem coordination procedure is applied to obtain the new setting values for the system under this modification. The results obtained from the subsystem approach will be validated by comparing them with those obtained when the full system is coordinated after this line is removed.

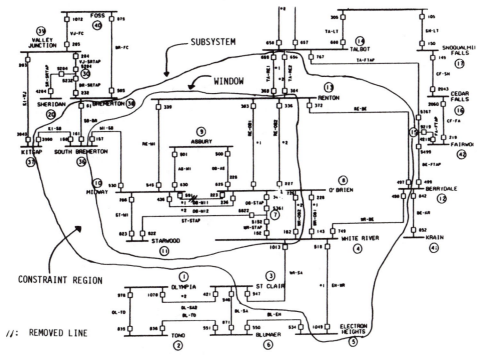

Figure 4. Subsystem Identified by the Coordination Program

51

Table 1. Comparison of Relay Settings by Full- and Subsystem Methods

RLY No.	PICKUP SETTING			INSTANTANEOUS SETTING			TIME DIAL SETTING			
	OLD	NEW		OLD	NEW		OLD	NEW		
		FSC	SSC		FSC	SSC		FSC	SSC WI	SSC WC
162	1.5	1.5	1.5	20	20	20	2.04	2.19	2.18	2.19
226	0.6	0.6	0.6	20	20	20	2.78	2.78	2.78	2.78
152	0.6	0.5	0.5	24	24	24	5.67	6.21	5.77	6.21
361	0.5	0.5	0.5	15	15	15	4.10	4.22	4.22	4.22
822	0.6	0.6	0.6	15	15	15	5.53	5.56	5.53	5.53
823	1.0	1.0	1.0	24	30	30	5.52	5.15	5.15	5.15
766	1.0	1.0	1.0	24	24	24	2.72	2.68	2.72	2.72
236	0.8	1.0	1.0	20	30	30	3.59	4.95	4.95	4.95
436	0.5	0.6	0.6	20	20	20	2.60	2.66	2.60	2.60
625	1.5	1.5	1.5	40	40	40	5.11	4.75	4.75	4.75

OLD: System before change   FSC: Full system coordination   WI: After window identification
NEW: System after change   SSC: Subsystem coordination   WC: After window coordination

The subsystem coordination algorithm has identified that a total of seventeen relays should have their settings changed for proper coordination with the new system conditions. A window consisting of 7 buses and 10 lines has been identified. The constraint region has also been identified and a subsystem of 13 buses and 19 lines has been formed. The resulting subsystem, constraint region and the window are shown in Figure 4. The loop coordination takes two passes through all the relays in the window to get the final settings. With these settings, it is determined that all the "primary constraints" are satisfied and hence no expansion of the window is necessary.

In order to verify the settings obtained through the subsystem approach, the modified system is also coordinated using the full system approach described in [4,6]. Identical precision values for checking the convergence of the relay settings are used in both the full and subsystem coordination procedures. The results of the full system coordination confirmed that the window we obtained through the subsystem approach is indeed valid. Table 1 compares the settings obtained for some of the relays in the window from both the full system approach and the subsystem approach. We observe that the settings obtained through the subsystem method are in very good agreement with those obtained through the full system method. The time dial settings obtained through the subsystem method are shown in two steps, one after the window identification process (column indicated as "WI") and the other after the loop coordination of the window (column indicated as "WC"). For our example system, we observe that for most of the relays we get the final settings at the end of the window identification process itself. Relays 162 and 152 are the only relays which changed their settings during the window coordination process to reach their final settings.

The subsystem coordination approach provides significant benefits in terms of computation time. In order to carry out the coordination process of the example system, the subsystem coordination approach needed only 337 CPU seconds in contrast with the full system method which took 1161 seconds. We thus observe a substantial reduction in the time required using the subsystem approach. We can expect even a larger reduction in computer time when we apply the subsystem coordination concept to study modifications to larger test systems. Recently, the performance of the full system coordination program has been improved by a factor of eight by limiting database access [12]. We can expect a similar reduction in the execution time of the subsystem coordination program also and this performance improvement is underway.

6. CONCLUSIONS AND SCOPE FOR FUTURE WORK

The problem of determining proper relay settings for a large system to account for changes in the system parameters has been tackled in this paper. A new subsystem coordination algorithm which will efficiently compute the relay settings for a part of the system in response to changes in system conditions in that part has been developed. This procedure automatically identifies the subsystem which is affected by the system change and carries out the coordination process to ensure correct relay operation under the new system conditions. This algorithm has been successfully implemented and tested on PSPL's transmission network. The test results have been verified to agree with those obtained when the modified system is coordinated as a single entity using the full system methods. A significant advantage of the subsystem approach is the reduction in computation time required when compared to the full system method. For the test system considered, the CPU time figures improved by about a factor of 3.5 using the subsystem method when compared to the approach of using the full system method for obtaining new settings for the modified system.

It is, however, important to recognize that the definition of the subsytem coordination problem and a possible approach to solving this problem as proposed in this paper are completely new. Hence, this problem area is still in the initial stages of research. In this first attempt, only a single test system, which the authors believe to be typical for transmission networks, has been considered to check the validity of the proposed method. More extensive testing using a variety of test systems is needed to gain confidence in the proposed methodology. In particular, we do not know if there are cases when the subsystem identification approach may fail in some sense. For example, the identified subsystem may still be too large to coordinate with existing tools.

Interactive computer aids for protection system engineering are relatively new and the work reported here is a major step toward achieving such aids. A number of possible extensions which can build on our work are foreseeable. The first enhancement is to integrate this module with the full system coordination package. This integration will involve providing a suitable user interface and managing the subsystem data and results with respect to the full system data. A second enhancement is related to the fault study capabilities which can be realized if a complete integration of the fault study into the protection package is achieved. Due to the limitations of the present interface with the fault study, we need to carry out the fault study on the entire system and store the data in the database before attempting a subsystem coordination. With the fault computation as an integral part of the protection software, we would need to analyze only faults on the buses in the subsystem.

The present subsystem coordination approach also needs to be investigated for sensitivity of the size of the subsystem to the type of relays being considered. In the work reported in this paper, we have developed a method of identifying the subsystem based on overcurrent relays. A similar approach needs to be adopted for distance relays. A more involved extension of the subsystem approach which will facilitate the full-scale analysis of a large system is to develop a piecewise coordination procedure. The idea here is to divide the entire system into a number of small subsystems and carry out the coordination of each of these subsystems separately. These subsystems are then coordinated at a higher level to meet the coordination constraints imposed by the relays in the boundaries of the subsystems. Finally, this ability to identify automatically the subsystem being affected and carry out coordination will be an important component in an adaptive protection system [12,13], where relays have to be recoordinated quickly in response to a system change.

7. ACKNOWLEDGEMENTS

This research was supported by the Electric Power Research Institute under project RP 2444-2, Mr.J.V. Mitsche, Project Manager.

8. REFERENCES

1. M.J. Damborg, et al., "Computer Aided Transmission Protection System Design, Part I: Algorithms and Part II: Implementation and Results", IEEE Transactions on PAS, Vol. PAS-103, January 1984, pp. 51- 65.

2. R. Ramaswami, et al., "Enhanced Algorithms for Transmission Protective Relay Coordination", IEEE Transactions on Power Delivery, Vol. PWRD-1, No.1, Jan. 1986, pp.280-287.

3. M.J. Damborg, et al., "Application of Relational Database to Computer-Aided-Engineering of Transmission Protection Systems", IEEE Transactions on Power Systems, Vol. PWRS-1, No.2, May 1986, pp. 187-193.

4. M.J. Damborg and S.S. Venkata, Specification of Computer- Aided Design of Transmission Protection Systems, Final Report EL-3337, RP 1764-6, EPRI, January 1984.

5. User Manual for Computer Aided Protection System (CAPS), EPRI RP 2444-2, University of Washington, Seattle, July 1985.

6. R. Ramaswami, "Transmission Protective Relay Coordination: A Computer-Aided-Engineering Approach for Subsystems and Full Systems", Ph.D Dissertation, University of Washington, Seattle, January 1986.

7. S.S. Venkata, et al., "C.A.E. Software for Transmission Protection Systems: Puget Power Experience", IEEE Transactions on Power Delivery, Vol. PWRD-2, No. 3, July 1987, pp. 691-698.

8. J.A. Juves, et al., "The concept of Figure of Merit Applied to Protection System Coordination", IEEE Transactions on Power Delivery, Vol. PWRD-1, No. 4, October 1986, pp. 31-40.

9. E.J. Mayer, et.al., "Computer Aided Transmission Protective Relay Coordination with Interactive Graphics", Western Protective Relay Conference, Spokane, WA, October.1985.

10. R. Ramaswami and P.F. McGuire, "Navigating A Protection Engineering Data Base", Accepted for publication in IEEE Computer Applications in Power, Volume 2, Number 2, April 1989.

11. R.H. Cauthen and W.P. McCannon, "The CAPE System: Computer Aided Protection Engineering," IEEE Computer Applications in Power, Volume 1, Number 2, April 1988, pp. 30-34.

12. A.K. Jampala, et.al., "Adaptive Transmission Protection: Concepts and Computational Issues", Paper 88 SM 528-2 Presented at the IEEE/PES 1988 Summer Meeting, Portland, OR, July 24-29, 1988.

13. G.D. Rockefeller, et. al., "Adaptive Transmission Relaying Concepts for Improved Performance", IEEE Transactions on Power Delivery, Vol.3, No. 4, October 1988, pp. 1446-1458.

14. S.S. Venkata, et. al., Computer-Aided Relay Protection Coordination, Final Report EL-6145, RP 2444-2, EPRI, December 1988.

ADAPTIVE TRANSMISSION RELAYING CONCEPTS FOR IMPROVED PERFORMANCE *

G.D. Rockefeller, Fellow IEEE
Electric Research & Management
St. Rose, LA

C.L. Wagner, Fellow IEEE
Electric Research & Management
Export, PA

J.R. Linders, Fellow, IEEE
Consultant
Sarasota, FL

K.L. Hicks, Sr. Member IEEE
Stone & Webster Engineering
Boston, MA

D.T. Rizy, Member IEEE
Energy Division, Oak Ridge National Laboratory
Oak Ridge, TN

Abstract - Concepts for adaptive protective relaying of transmission lines are presented. These include changes in on-line relay settings, relay characteristics or logic in response to power system or environmental changes, or as a result of operating experience. Such changes as line-out or generator-out contingencies, which affect fault current distributions and degrade the system's security level, initiate adaptive protection system (APS) responses.

Adaptive relaying is shown to be capable of improving relaying reliability and power system security plus achieving better utilization of transmission facilities. Most of the concepts require a hierarchical computer system, involving front-line, parallel processors, a substation host and remote central processors, linked by channels that transmit data or relaying changes prior to or after a disturbance.

Maximum value from adaptive relaying will result from integration with existing or planned substation control and data acquisition (SCADA) functions and interfacing with the central energy management system (EMS). Emphasis is placed on use of interim hardware approaches to allow evolutionary development, starting with just local signals and the first level of a hierarchy [1].

INTRODUCTION

Since no official definitions exist, for the purposes of this paper the following terms are defined:

Adaptive Protection An on-line activity that modifies the preferred protective response to a change in system conditions or requirements. It is usually automatic, but can include timely human intervention.

Adaptive Relay A relay that can have its settings, characteristics or logic functions changed on-line in a timely manner by means of externally generated signals or control action.

87 SM 632-3   A paper recommended and approved by the IEEE Power System Relaying Committee of the IEEE Power Engineering Society for presentation at the IEEE/PES 1987 Summer Meeting, San Francisco, California, July 12 - 17, 1987. Manuscript submitted January 30, 1987; made available for printing April 24, 1987.

Transmission system margins for contingencies are lessening, resulting from several factors such as difficulties with obtaining right of way, generation deferrals and uneven load growth, plus an increase in wheeling. With this situation transmission relaying reliability becomes increasingly important. Sympathy trips (those incorrect trips during faults) represent a major source of concern to the planner and operator, since they remove additional circuits along with the loss of the faulted line. Reclosing is one well-established ameliorating means to these sympathy trips. Single-phase tripping (opening only the faulted phase) also offers a very attractive means to reduce the impact of unneeded trips on system security, as well as offering some opportunities for implementing adaptive techniques.

Short critical switching times for maintaining stability represent an important concern to those facing these situations. One partial solution to the problem has been the use of independent-pole breaker controls to minimize the possibility of all three phases failing to open for a three phase fault. Another solution, offered in this paper, is a more sophisticated breaker failure timing scheme.

The last decade has seen widespread concern with generator shaft fatiguing due to faults and switching. This paper introduces an adaptive concept to reduce this fatiguing.

Until recently, response to operating experience or power system changes, in the form of setting changes or design improvements, has required protection equipment outages. Now, new microprocessor relays can be provided with automatic or remotely initiated setting changes. Perhaps fortuitously, fiber optic technology is maturing and being enthusiastically embraced by the utility industry, providing what promises to be reliable, wide-band communication links [2]. These developments offer opportunities for greater relaying sophistication, including use of more adaptive techniques.

The paper's adaptive concepts for transmission system relaying and their operating improvements are summarized in Table I. Most of these concepts require a hierarchical computer structure to process power system information and develop setting changes.

*Research sponsored by the Office of Energy Storage and Distribution, Electric Energy Systems Program, U.S. Department of Energy, under Contract No. DE-AC05-84OR21400 with Martin Marietta Energy Systems, Inc.

Reprinted from *IEEE Trans. Power Delivery*, vol. 3, no. 4, pp. 1446–1458, Oct. 1988.

TABLE 1 Adaptive Transmission System Protection Techniques

ADAPTIVE TECHNIQUE	BENEFITS
Adaptive system impedance model (permits calculation of fault-distribution)	Improved relaying reliability and possible avoidance of current line construction
Adaptive sequential instantaneous tripping (detection of far-end breaker openings)	Faster back-up protection and possible elimination of the need for a second pilot scheme
Adaptive multi-terminal relay coverage (accounts for changes in infeed ratios)	Improved zone 1 and zone 2 settings
Adaptive zone 1 ground distance (accounts for large apparent impedance in fault resistance	Greater sensitivity to high resistance ground faults
Adaptive response to defective relaying equipment	Minimizes need for second pilot scheme and need to take affected line out of service
Adaptive reclosing	Faster restoration following incorrect trips, reduced number of unsuccessful reclosures, reduced shaft fatiguing
Variable breaker-failure timing (detects failure to interrupt)	Improves back-up timing margins and eliminates unneeded tripping of back-up breakers
Adaptive last-resort islanding (system splitting to isolate generators with manageable load levels)	Improved probability of maintaining units in service to facilitate load restoration
Adaptive internal logic monitoring	Improved relaying reliability
Relay setting coordination checks (checks coverage and selectivity)	Coordination optimized, starting from existing power system conditions and minimizes operating constraints

The adaptive techniques presented require only slow speed responses such as those of a SCADA system, in contrast to the high-speed channels used in pilot relaying between interconnected transmission line terminals. Methods of collecting data and implementing control actions for the adaptive techniques are described in the remainder of the paper.

ADAPTIVE SYSTEM IMPEDANCE MODEL

System conditions external to the protected line influence relaying performance. An up-to-date system impedance model can provide the relays with a very useful input, including minimization of the contingencies to be considered. Later sections offer examples of the use of this model.

Fig. 1 shows the simplest form of system representation viewed by line T-U relays, with the external system represented by equivalent impedances Z_{E1}, Z_{E2} and Z_{E3}. The equivalents could be developed at the substation using a locally available system impedance data base and circuit-out information from local events and from remote computers. Alternatively, the impedance data base could be maintained and calculations could be performed centrally for each protected line and then the three equivalents transmitted to each transmission line processor. In either case the line currents

and voltages for critical fault locations would be calculated by the local relay computer, on-line, prior to a disturbance, or the impedances used directly in routines at fault inception, such as for pinpointing the fault location.

The adaptive protection system (APS) may need a more complex model than that of Fig. 1, with additional lines, breakers or other elements shown discretely, rather than lumped into an equivalent. Nevertheless, the principle is the same. The required model is adapted to system changes to provide improved relaying performance.

An approximate impedance model can be updated based solely upon conditions within the substation. Open breakers or undercurrent conditions can identify the opening of lines connecting to the station. Also, distinctive load flow patterns may identify an outage of a nearby generator. Upon detection of a configuration change, the substation host computer would initiate an update of the impedance model, using pre-loaded system equivalents and local circuit impedances. This mode would be used prior to linking to remote computers or for a failure of such a link. While the model may be relatively inaccurate, it could still provide the basis for significant benefits as described in succeeding sections.

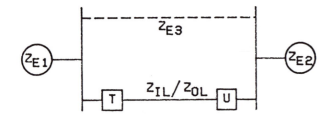

T, U CIRCUIT BREAKERS
Z_{IL}, Z_{OL} POSITIVE- AND ZERO-
SEQUENCE IMPEDANCE,
RESPECTIVELY

Z_{E1}, Z_{E2}, Z_{E3} EQUIVALENT
IMPEDANCES

Fig. 1 Equivalent impedance model for use by
transmission line T-U relaying

The accuracy of the impedance model improves
with better knowledge of the status of the
system beyond the confines of the substation.
Once the computer relaying hierarchy expands
beyond the substation, this refinement becomes
practical.

IMPROVED STAND-ALONE RELAYING

Stand-alone protection performs without the
benefit of knowledge of conditions at the
remote line terminal during the fault, in
contrast to pilot schemes which use a
communication channel to interchange data
during the fault. Stepped-distance protection
is an example of conventional stand-alone
relaying. The adaptive stand-alone schemes
described in this paper benefit from data
transmitted prior to the fault, but do not use
this link during the fault. Indeed, the
stand-alone protection would still perform,
generally with some degradation in quality,
should the pre-fault data link be inoperative,
by reverting to default settings.
Transmission relaying normally utilizes one,
and sometimes two, pilot schemes to provide
simultaneous high-speed tripping of all line
terminals, relying on high-speed, real-time
channels to distinguish between an internal
and an external fault. As with any
protection, these systems can become
unserviceable, with the communication channel
the frequent culprit. Accordingly, back-up
devices are provided which don't require a
remote communications link during the fault.

Adaptive Sequential Instantaneous Tripping

Conventional stand-alone relaying may provide
high-speed tripping for most internal faults;
however, for faults near the remote station,
the local electrical conditions immediately
after fault inception will not indicate
whether the fault is internal or external to
the protected line. For these faults
conventional stand-alone relaying is delayed
to prevent unneeded tripping for an external
fault. However, once the far terminal of the
line opens, changes in current flows may
provide the basis for immediate tripping. If
these changes are consistent, complete
sequential instantaneous response can be

achieved, reducing the clearing time from
about 0.35s to about 0.15s. In this case
installation of a second pilot scheme might be
obviated, based on the 0.15s clearing time
being tolerable when the pilot scheme is
defective. Also, stand-alone protection that
provides 0.15s maximum clearing time may
eliminate the need to remove a line from
service if its pilot protection becomes
inoperative.

Fig. 2 shows illustrative system
configurations that can utilize adaptive
sequential instantaneous tripping. In Fig.
2(a) parallel lines connect external systems
S1 and S2. Breaker M opens first for fault 1,
causing currents to change from "I" to "F"
values. Figs. 2(b) and (c) show four lines
interconnecting systems S1, S2 and S3. Strong
source S2 delivers high levels of fault
current; weak source S3, low levels of fault
current.

An instantaneous overcurrent relay at breaker
L in Fig. 2(a) may operate for a far-end fault
1 once breaker M opens, responding to the
increase in current, F vs. I, resulting from
the redistribution forced by breaker M
opening. This response may be limited with
non-adaptive relaying, if the settings are
based on worst-case conditions. The
instantaneous overcurrent unit at breaker L
must be set assuming line W-Y open, degrading
coverage with line W-Y closed; contingencies
affecting the S1 and S2 source equivalents
must also be studied. Thus, a conventional
setting will be determined by worst case
conditions, or by probable worst case
conditions, resulting in compromises and
degraded performance for all-in conditions.

With line W-Y closed in Fig. 2(a), the initial
current in line L-M will be less than with
line W-Y open for the far-end fault, so the
coverage provided by the instantaneous
overcurrent unit will be less than optimum
under all-in conditions. By adapting the
setting to changes in the power system
configuration, the unit's coverage can be
increased both initially and sequentially.
However, changes occurring during an external
fault can catch the relay with an overly
sensitive setting. For example, consider
fault 2 under all-in conditions. If a
sympathy trip occurs at breaker W or Y, the
relay at breaker L will be caught with the
wrong setting, possibly resulting in its
operation. Then, two lines will trip
incorrectly. An incorrect trip would also
occur at breaker L for fault 3 if breaker W
opens before breaker Y. These sympathy trips
could be avoided by using "transient
blocking", which would allow the more
sensitive current setting to be effective for
just the period prior to the point when line
W-Y could trip; with two cycle breakers, this
period could be the first 1.5 cycles after
fault inception. However, transient blocking
would prevent sequential tripping, so separate
logic would be needed to detect far-end
opening for an internal fault. Just sensing an
increase in current won't provide proper
security.

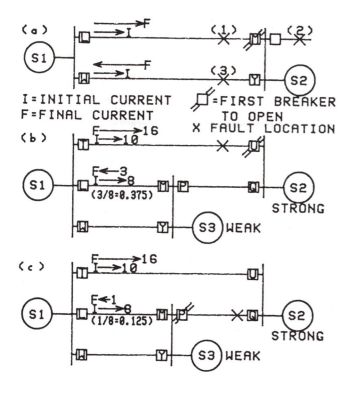

I=INITIAL CURRENT   □=FIRST BREAKER
F=FINAL CURRENT   TO OPEN
                     X FAULT LOCATION

Fig. 2 Current-flow patterns with sequential
        fault clearing

For a line-ground fault on a two-terminal
line, far-end opening will interrupt current
flow on the sound phases, except for line
charging current. A substantial drop in
sound-phase current within the expected time
window might provide a reliable indication,
although sequential clearing of external
faults might fool the logic with some power
system configurations.

In Fig. 2(a) at breaker L, a relatively simple
solution to detect breaker M tripping would be
to sense the reversal in the direction of line
W-Y current to indicate that breaker M has
opened for an internal fault. However, if
line W-Y is open, this signal becomes
unavailable.

Instead of looking at the current in an
external line, the protection scheme could
operate on the percent increase in the
protected line current as an indication of
far-end breaker opening. However, such an
increase could also occur for an external
fault, either for sequential clearing or as a
result of the fault evolving to a multiphase
fault.

Fig. 2(b) shows a more complex configuration,
with a current reversal occurring in line L-M
for a sequential clearing on line T-U. Such a
current reversal (I vs. F direction) also
occurs in Fig. 2(c) for a fault on line P-Q if
breaker P opens first, so the breaker L logic
must sense more than just a line L-M current
reversal to reliably detect the opening of
breaker U. In this example the relative
magnitude of the reversal in breaker L
pinpoints breaker U opening, since this event
yields a 0.375 ratio of final to initial
current, compared to a 0.125 ratio for breaker
P opening for a line P-Q fault. A ratio of

0.25 plus a reversal and an increase in
breaker T current would initiate tripping of
breaker T. As the power system configuration
changes, the setting or logic could be changed
adaptively to maintain secure protection. For
example, with line L-M open, relay T could use
breaker W current with an appropriate setting.

The above discussion suggests that the
preferred logic and settings may vary with
power system conditions. Different logic
could be enabled adaptively.

Adaptive Multi-Terminal Distance Relay
Coverage

Current flow from the third terminal of a line
can drastically affect distance relay reach in
another line terminal--an outfeed or infeed
effect [3]. Zone 1 protection must always
underreach any remote terminal, so it is set
assuming no infeed from another leg (i.e.
assuming a two-terminal configuration). Then,
with all terminals closed, infeed causes the
zone 1 units to underreach. Coverage can be
improved by increasing reach when infeed is
present by inputting the position of the
remote breakers into the protective relay
logic; relay settings can also adapt to
changes in the current distribution resulting
from external power system changes. Each time
the system changes, the impedance model would
be updated and infeed ratios calculated for
critical fault locations. This illustrates
the use of a slow-speed channel to modify
settings prior to a fault.

The system of Fig. 3 will be used for the Fig.
4 example of adaptive reach settings. The
line and source impedances are specified.
With all terminals closed, the zone 1 phase
distance relay at breaker T is set adaptively
for a percentage of the smaller of:

(1)
$$Z_y + Z_x + Z_x(I_s/I_t)$$

(2)
$$Z_y + Z_z + Z_z(I_r/I_t)$$

where Z_x, Z_y and Z_z are the positive
sequence line impedances to the tap and
I_r, I_s and I_t are the leg currents for a
phase fault

I_s/I_t and I_r/I_t are infeed ratios for the
present power system conditions

Then, regardless of whether the fault is at 1
or at 2, zone 1 at breaker T underreaches,
avoiding incorrect tripping.

Zone 1 at breaker T is adaptively set on the
basis of a fault at 1 in Fig. 4(a) for 90%
of the apparent impedance: 14.4 with breaker S
open and 34.4 if breakers R & S are closed.
Higher apparent impedances result for a fault
at 2; therefore, a fault at 1 determines the
breaker T zone 1 setting. While not shown in
Fig. 4(a), the settings must also adapt to the
opening of breaker R. By adapting the
settings based on the amount of infeed, zone 1
coverage is extended with R & S closed,
improving both stand-alone coverage, as well
as underreaching transfer trip pilot
protection, when used.

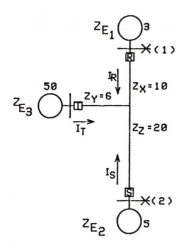

I_R, I_S, I_T RELAY CURRENTS
FAULT LOCATION

$Z_{E_1}, Z_{E_2}, Z_{E_3}$ EQUIVALENT SOURCE
IMPEDANCES

Z_X, Z_Y, Z_Z LINE IMPEDANCES
TO TAP POINT

Fig. 3 Three-terminal line example

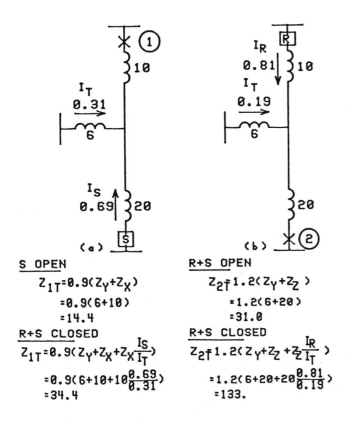

S OPEN
$$Z_{1T} = 0.9(Z_Y + Z_X)$$
$$= 0.9(6+10)$$
$$= 14.4$$

R+S CLOSED
$$Z_{1T} = 0.9(Z_Y + Z_X + Z_X \frac{I_S}{I_T})$$
$$= 0.9(6+10+10 \frac{0.69}{0.31})$$
$$= 34.4$$

R+S OPEN
$$Z_{2T} = 1.2(Z_Y + Z_Z)$$
$$= 1.2(6+20)$$
$$= 31.0$$

R+S CLOSED
$$Z_{2T} = 1.2(Z_Y + Z_Z + Z_Z \frac{I_R}{I_T})$$
$$= 1.2(6+20+20 \frac{0.81}{0.19})$$
$$= 133.$$

Fig. 4 Zone 1 (Z_{1T}) and zone 2 (Z_{2T}) adaptive settings for breaker T

Similarly, a fault at 2 controls the zone 2 setting at breaker T. Per the Fig. 4(b) calculations, zone 2 is set for 1.2 times the apparent impedances: 31 with breaker R open

and 133 with breaker R & S closed. With a conventional fixed setting of 133 ohms, two possible problems may develop. First, with breaker R or S open, zone 2 at breaker T may reach so far into the external system as to be non-selective in its delay setting with adjacent line relaying. Secondly, the source supplying breaker T may convert from a weak source to a strong one, catching zone 2 with inadequate loadability, resulting in tripping breaker T for load flow into the line.

If a fault is seen at inception as beyond either zone 1 or zone 2 their reach must be pulled back to avoid misoperation should infeed be removed before the fault is cleared by incorrect (sympathy) tripping of one of the other protected-line terminals. Also, the external fault can, through sequential clearing, modify the infeed ratio and hence the relay reaches. Multiple faults in a short space of time might also catch the relays with improper settings, so after about 1.5 cycles of detection of the first fault, the zones 1 and 2 reach should be pulled back and not re-extended until the central computer through a slow-speed channel has had the opportunity to update the settings, based on the new status of the power system.

Adaptive Zone 1 Ground Distance

A supplementary zone 1 ground-fault protection approach based on the fault location techniques of Ref. 4 offers fast tripping (e.g. 80ms) for high-resistance faults (e.g. mid-span tree fault). Because of the computational requirements of this solution, it probably won't provide the high-speed response needed for high current faults, so it would supplement an algorithm with much less fault resistance accommodation, but faster operation. The adaptive method compensates for the apparent reactance developed in the fault resistance if load current is not excluded and by out-of-phase fault current, $(1- D_A)I_F$ in Fig. 5, flowing into the fault from the far end of the line. At breaker A the total current consists of the superposition of the prefault current and the change in current caused by the fault. The voltage consists of the drop developed by the total current flow in the line to the fault plus the drop in fault resistance R_F.

In general the prefault current will not be in phase with the fault current $D_A I_F$, so use of the total current (pre-fault plus current change) will result in an apparent reactance in the fault resistance. Also, if the fault current from B, $(1-D_A)I_F$, is not in phase with the fault current from A, this will also develop an apparent reactance in the fault as viewed from A. If the current change at A is divided by the distribution factor D_A, I_F is obtained. D_A is a function of the network impedances and the fault location. The latter and R_F are unknowns, but they can be computed if the external network is completely defined, using an up-to-date system impedance model as introduced earlier. By solving for the fault location, zone 1 protection is achieved by limiting tripping to cases where the location is indicated in say the first 80% of the line.

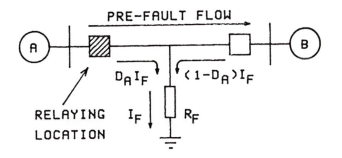

PRE-FAULT FLOW

$D_A I_F$ $(1-D_A)I_F$

I_F R_F

RELAYING
LOCATION

A, B EQUIVALENT SOURCES
D_A FAULT CURRENT DISTRIBUTION FACTOR FOR A
I_F CURRENT IN FAULT POINT
R_F FAULT RESISTANCE

Fig. 5 Total current results from super-position of prefault flow plus current caused by fault - only the latter flows in R_F

To avoid the processing of unreliable signals, the fault locator algorithm of Ref. 4--used in this solution as a zone 1 relay unit--must only be enabled for a forward fault using the appropriate voltage and current depending on the phase involved, so ancillary algorithms must be used: a fault detector, a directional check and a faulted-phase selector. In the Ref. 4 approach, conventional distance relaying with only limited fault resistance accommodation provides these functions. So a different approach is needed to detect higher resistance faults.

The fault detector checks the value of each current sample with the value one cycle earlier. Any disturbance produces a difference value which operates the fault detector, which in turn calls the directional-check routine.

The direction of the fault, if any, is determined by using the change in the fundamental components of residual current and voltage. That is, the prefault values are subtracted from the sampled readings, so the angle comparison between current and voltage uses only the change in values caused by the fault. This eliminates the effects of unbalanced load flows caused by non-transposed lines, which could be larger than the fault-generated values. Use of the fundamental components eliminates the effects of harmonics. Because of generator swings, the load current will change during the fault and begin to adulterate the current-change values; however, the directional check should be valid for 10 cycles--ample time to compute the fault location. Except for use of the change in voltage and current, the directional check uses a conventional voltage polarized approach.

The phase selector establishes that a single phase fault has occurred and determines the affected phase. The distance algorithm then processes the faulted phase quantities per Ref. 4 to determine the fault location. (If a multi-phase fault occurs, this routine is bypassed in favor of a more conventional algorithm, since resistance levels should be moderate for multi-phase faults.) For multi-phase faults a substantial change in line to ground voltage or phase-current will occur on more than one phase, providing the basis to block further processing of the distance tripping routine.

For a single-phase fault with very high resistance, voltage changes may not provide a reliable basis to select the faulted phase. Instead, a comparison of the phase angle between the change in zero-sequence current and the change in positive- plus negative-sequence current will select the faulted phase. For a single-line-to-ground fault, these two quantities will be nearly in phase on the faulted phase and almost 180 degrees out on the sound phases.

To avoid the possibility of a misoperation during a sequentially cleared external fault, output will be blocked if the distance routine locates the fault as external while processing the first 1.5 cycles of data. With the external system not properly represented once the first line terminal opens, the fault might then appear as internal.

The Ref. 4 solution is rigorous provided the external system is perfectly represented, including load impedances in the positive- and negative-sequence networks. At present, representative external impedance values are input [4], but in the future these values could be adaptive to maintain accuracy as the external system changes. Since load impedances probably produce only a secondary effect on accuracy, these could remain fixed at representative values or changed infrequently, while the series line elements would be changed with each significant contingency.

ADAPTIVE RESPONSE TO DEFECTIVE RELAYING

Adaptive responses to trouble alarms can relieve operating personnel of decision making. A channel failure represents the most likely type of alarm. The operator usually relies on the back-up protection or a second pilot scheme, but may elect to request emergency investigation, rather than wait for normal-hour action. In rare cases setting changes are implemented or the line removed from service.

Adaptation may obviate the need to install a second pilot scheme, take the line out of service or call out testers on an overtime basis. The zone 2 delay of the stand-alone protection can be reduced, either for phase faults or for both phase and ground faults, eliminating coordination with breaker-failure timers. Zone 1 reach could be extended to 120% for the first trip, with the normal underreach setting upon reclosing. (This change would have low impact on system security if single phase tripping were utilized.) Other logic judged less secure than the normal protection could be enabled; this might include means to detect far-end opening to provide a sequential instantaneous trip for a far-end fault.

If the line computer becomes inoperative, the bus differential computer could be assigned the added temporary job of providing pilot protection where the channel is functional. The bus computer would need a pilot-channel interface for each line computer to be backed up. Further, with a breaker-and-a-half station, the bus system needs to also read the center breaker current to perform the line relay function, unless this breaker is opened. The bus computer, then, has all the inputs required to perform all the line protection functions. However, the computational requirements appear excessive to provide this function for more than one line at any time.

An adaptive-response strategy could include a check for simultaneous alarms involving multiple lines in the area, perhaps alerting the operator of the need for emergency callout. Circuit outages and the power system security state could also temper the response, whether automatic or an operator aid. Adaptation minimizes the need to depart from an optimum generation schedule that might otherwise be triggered by removal of a line from service because of a relaying deficiency.

ADAPTIVE RECLOSING

Adaptive reclosing logic offers many significant opportunities, including high-speed response to a sympathy trip, minimizing unsuccessful reclosing and reducing system and equipment shock.

Sympathy Trip Response

Breaker reclosing reduces the impact of sympathy trips by promptly restoring what should not have tripped in the first place. Sympathy trips are much more serious than incorrect trips that occur with no fault on the power system, since multiple outages will result. Protection logic to detect that a sympathy trip has occurred could initiate a high-speed breaker reclose where none is normally employed, or a second high-speed reclose could be enabled where one is normal. A second reclose would handle the case of a second sympathy trip where the reclose into the faulted line is unsuccessful.

While most lines are equipped for high-speed breaker reclosing, this practice is frequently moderated by using delayed dead-line reclosing at the remote terminal and syn-check reclosing near the power plant because of generator shaft-fatigue concern. In this case high-speed reclosing could be enabled following a sympathy trip, possibly contingent on the fault location, number of phases involved, power-system security state, number and location of sympathy trips, number of generators or line outlets in service at the plant, etc.

In a directional comparison blocking scheme during an external fault the directional relays at only one end should see the fault as in the trip direction and should be blocked from tripping by the signal from the remote transmitter. In the absence of this channel blocking signal, just one end of the line usually trips for an external fault. Here, near-normal voltage will reappear following fault clearing, reliably signalling the

soundness of the line. High-speed reclosing, with or without a high-speed check for synchronism across the breaker, will restore the line with minimum hazard to the system.

Sympathy trips due to a directional sensing failure can be detected by the substation computer host (ST) if it has access to the phasor quantities for the line that tripped. ST makes an independent check on the direction of the fault, using different signal sources than those used by the line relaying. If ST senses fault-power flow from the line, it advises the reclosing logic that a sympathy trip has occurred. In this case it is likely that the remote line terminal has also tripped, since fault-power flow at its terminal would be into the line. The remote end, therefore, must await the reclose of the local end to establish that a sympathy trip occurred, unless the local logic transmits advice to the remote end.

A directional fault-power-flow check would detect a bus differential relaying sympathy trip. The program would determine the direction of flow separately on each bus connection. Use of the current change (total minus the prefault value) would avoid a wrong analysis where load current predominates on one or more connections to the bus. The substation host should use the line-relaying phasor computations for the directional check, assuming that the bus differential scheme uses different current sources than does the line relaying. Reclosing would be initiated on either a preselected breaker to test the bus, or one chosen by logic which looks at substation status, and perhaps, system status. The remaining breakers would close, supervised by a synchronism check.

Reducing Unsuccessful Reclosures

While secondary arc extinction doesn't guarantee a successful reclose, the failure to extinguish this arc guarantees lack of success. When the arc extinguishes, the voltage to ground partially recovers as a result of the coupling to adjacent phases or other lines in the vicinity. This effect allows detection of the time of extinction if it occurs [5]. The amount of recovery voltage and its wave form will be highly variable with different applications. Where three-phase tripping occurs voltage recovery can still occur because of parallel lines and from the trapped energy on the unfaulted phases of the tripped line.

Some delay in the initiation of reclosing following extinction may be needed with fast acting breakers to allow for the deionization of the air. In any event some initiation delay should be employed to allow for possible re-striking of the arc if the recovery voltage is excessive.

For three-phase tripping applications, the amount of capacitive coupling may be highly variable considering parallel line configuration and contingencies, indicating a possible need for an adaptive voltage setting.

A high resistance fault (e.g. >25 ohms) should block reclosing, based on the assumption that a mid-span fault has occurred with attendant low probability of success (and increased possibilities of human hazards). The resistance component of the apparent impedance obtained from a conventional distance algorithm might be accurate enough. Alternatively, the method of Ref. 4 could be employed. This uses a representative model of the external system in order to pinpoint the fault location, but the technique can be readily expanded to also calculate the fault resistance. Its accuracy can be enhanced by changing the source impedance model in response to actual power system changes, as discussed for the zone 1 concept.

One US utility's central computer adapts high-speed reclosing, enabling it only when lightning activity is detected, on the basis of heightened probabilities for a successful reclose.

VARIABLE BREAKER-FAILURE TIMING

Conventional dual-timer breaker-failure relays provide a shorter delay for multi-phase or 3 phase faults and a longer delay, for phase-ground faults. This provides a greater timing margin for most faults. Worst-case conditions (i.e. close-in fault) dictate the time settings. An adaptive variable breaker-failure time can provide a conservatively secure setting when events permit it and in extreme cases avoids the need to immediately trip the back-up breakers except where the critical switching time (CST) is too short to allow a check for proper clearing of the primary breaker.

Fig. 6 shows a sophisticated approach, where timing is a function of positive-sequence voltage, V_1 during the fault. This voltage embodies the effects of both fault location and the number of phases involved, indicating the degree of disruption of power flow and, therefore, the impact of the fault on stability. For a given power system pre-condition, the central computer would specify a time-voltage characteristic for each critical location on the system. Fig. 6 shows an example of a range of characteristics for different conditions for a given breaker location. The actual breaker-failure timer delay T_{bf} would depend on the severity of the fault--i.e. upon V_1.

The lower curve of Fig. 6 applies to an extreme case where V_1 values between points j and k indicate need for an immediate trip of the back-up breakers. Use of V_1 confines this drastic action to only those faults and pre-conditions requiring this response.

Fig. 7 shows an actual 138 kV power system area used to test the principle of Fig. 6. A simulation applied 3 phase faults along line 1 at or near station P to obtain the critical switching time for each location. Both all-in and line 2-out cases were run. MW flows are indicated; for line flows, the top value applies with all-in conditions and the bottom value, with line 2 out. Reclosing effects were not considered.

Fig. 8 plots the results of the simulation, critical switching time (CST) vs V_1. Two mileage points are shown to provide a feel for the effect of fault location. For the all-in

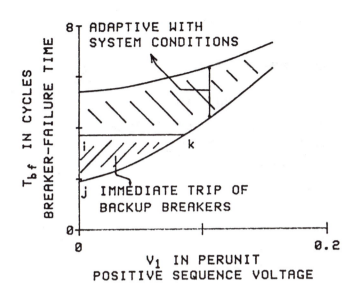

Fig. 6 Breaker-failure time T_{bf} varies with fault severity as gauged by positive-sequence voltage V_1

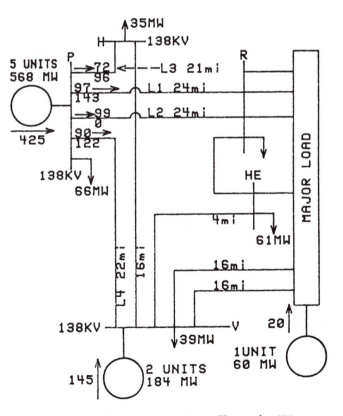

Fig. 7 Example power system flows in MW

case a 14 cycle CST applies for a 3 phase fault 5 miles from station P ($V_1=29\%$); the 19 cycle value, for a 3 phase fault 10 miles out ($V_1=41\%$). These compare to a 10 cycle CST for a close-in 3 phase fault. As the fault approaches station R, the station P voltage decreases somewhat from its maximum (not shown in Fig. 8), because of the parallel paths. However, the station P CST remains above the problem range for these remote faults.

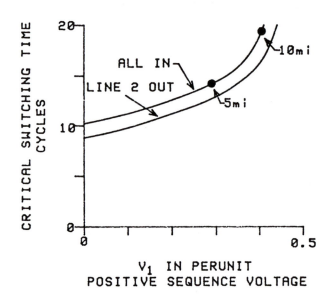

Fig. 8 Critical switching time as a function of V_1 at station P for nearby 3-phase faults on line 1

The Fig. 9 example shows an adaptive timing margin of two cycles based on a five cycle time setting and a CST of 10 cycles, which applies to a close-in 3 phase fault under all-in conditions in Fig. 8. In contrast, conventional protection would use a 4 cycle time setting, based on the 9 cycle line-out case in Fig. 8 for a close-in 3 phase fault; the resulting one cycle margin is tight. In contrast, Fig. 10 shows the application of adaptive settings for one set of pre-fault loading conditions. A fixed timer setting of 6 cycles was used at the higher values of V_1 for an indicated margin of 3 cycles. Below about 20% V_1, T_{bf} is adaptive.

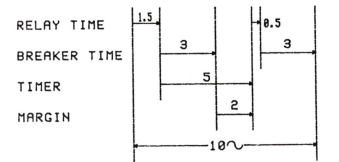

Fig. 9 Timing relations for breaker-failure protection

Initially the relay could use a fixed, worst case CST vs. V_1 characteristic. When a link to a central computer is available, an adaptable characteristic could be implemented, using predetermined curves keyed to pivotal

system conditions. Since the characteristic would be updated prior to any disturbance, slow speed communications suffices.

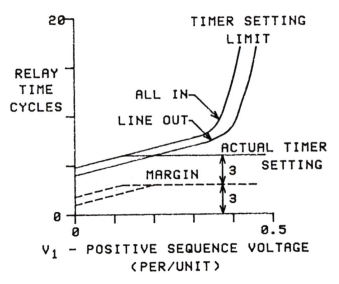

Fig.10 Breaker-failure timer settings to avoid instability

ADAPTIVE LAST-RESORT ISLANDING

Following a power system overload, if the load shedding program doesn't arrest the frequency decay, selected generators should be promptly isolated with load for the prime purpose of facilitating restoration of the system. This action must be neither premature nor too late. A prediction of insufficient shedding based on the frequency decay rates might provide faster generator isolation. A program would choose from pre-selected split points based on generating plant capability, bus loads and circuit configuration. Since the split-point selection is periodically updated and not frequently affected by the disturbance producing the overload, only slow speed communications links are needed. If the breakup producing the overload fragments the projected island, the last-resort islanding may be unsuccessful. However, if several islanding schemes are implemented in various areas of the system, probability should be high that one or more plants can be successfully isolated.

Fig. 11 shows the results of a dynamic simulation of an islanding for the system configuration of Fig. 7 for three levels of island load, with separation occurring at about 56.5 Hz and about 3 seconds after inception of a system overload. The pre-loading conditions of Fig. 7 do not apply to this simulation; in particular, the two units at station V have an output of 100 MW, vs. the 145 MW for the prior simulation. The island includes station V and its two units rated at 92 MW each. The system overload was simulated by tripping 138 kV lines 1 to 4, which were carrying an aggregate of 316 MW. Load shedding in effect was:

Frequency	%Load
59.2	15
58.8	8
58.4	7.6

With the separation of station P, the system had 180 MW of load assigned to shedding--not enough for this contingency. The islanding scheme was actuated by underfrequency detection about one second after the frequency dropped to the 57 Hz set point.

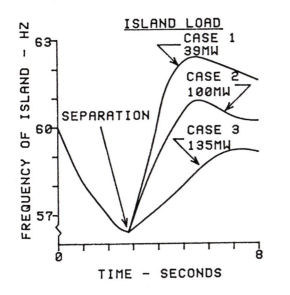

Fig.11 Frequency response of island for three cases of loading pre-disturbance generation

If the station V units in Fig. 7 are islanded with just the local load of 39 MW, compared with an output of 100MW at separation (Case 1), the machines experience a load rejection sufficient to actuate the auxiliary governors, which transfer governor control to the intercept valves. If the two units have similar characteristics, the island may survive, but there's a high probability that governor hunting will cause them to shut down. If station HE load is added to the island (Case 2), the station V units are loaded to 100MW, an approximate balance with generator output. If both stations H and HE are included (case 3), the total load on the units is 135 MW and the units level out at about 59 Hz, based on a 5% governor droop characteristic.

The likelihood of the units staying on line is greater for Cases 2 and 3, where the plant transients are less severe than for Case 1.

Depending upon station V output and loading conditions, it is possible for a central computer to select the best split points for islanding the generators.

ADAPTIVE INTERNAL LOGIC MONITORING

Internal logic event recording [6, p1665] offers important advantages over conventional macro-recording. Key software events are logged, whether or not the program initiates breaker tripping. The output of this recording provides for an open-loop adaptation based on engineering analysis. For example, the apparent impedance trajectory of a distance routine or the margin of breaker-failure timing could be recorded. In contrast to hard-wired systems, this recording can be readily accomplished in software with minimal impact on the running program. The advantages of this concept are that it:

1. helps in the identification and diagnosis of incorrect operations;

2. provides data to assess the appropriateness of security timing margins and logic;

3. provides more meaningful statistics on the number of correct operations by defining which schemes had the opportunity to misoperate during an external fault.

4. performs without significant degrading the performance of the software it monitors;

5. reduces need for routine testing; and

6. enhances analysis of any type of disturbance even though no incorrect operations occur.

This data could:

a) lead to changes in setting or application policies or use of adaptive settings, if too many schemes see some faults; and

b) provide planners and operators with better data for probability analyses (e.g. knowledge that faults in some areas are more likely to produce sympathy trips).

Ultimately relay settings or logic might adapt automatically based on analysis of these events.

RELAY SETTING COORDINATION CHECKS

Each new project, whether a circuit addition or change in basic configuration of the power system, causes ripples over a wide area as it effects the coordination of relay settings. Typically, only a limited setting review is conducted for each project with the intent to follow with a more comprehensive review on a periodic basis. However, the temptation is to postpone these periodic reviews in the face of other pressures.

A key aspect of settings is the choice of contingencies for which coordination is attempted. While all imaginable faults must be cleared, it isn't usually feasible to achieve coordination for every conceivable permutation of power system configuration. Accordingly, the engineer must choose what he considers are those contingencies with reasonable probabilities of being encountered. Even then he usually must make compromises, accepting some miscoordination or slim timing margins for some contingencies. In addition, to some extent the quality of protection for normal conditions is degraded by the need to cover contingencies. Adaptive techniques should resolve these deficiencies.

Additionally, an on-line program could periodically check relay coordination based on the actual power system conditions and selected contingencies starting with present conditions. Fault quantities would be calculated using the up-to-date system impedance model. The program would alarm if it found any inadequate or improper settings, including excessive clearing time. Settings could be either manually input or obtained via links to the line relaying microprocessors. Adaptive setting changes would also initiate coordination checks. Since a setting check over a wide area requires extensive processing time, on-demand initiation would have to be limited. The checking responsibility could be spread to involve area computers where they exist, as well as the central computer.

The program would not only indicate points of possible misoperation, but would also provide an analysis of the origins of the problem. The problem might originate with a setting at the boundary of the area under study, indicating that another area computer or the central computer should run its checking program.

These checks would not only supervise manual procedures, but would also monitor adaptive setting routines.

IMPLEMENTATION CONSIDERATIONS

The paper has attempted to show how adaptive protection principles can be implemented on an evolutionary basis. This applies not only from a hardware viewpoint, but also for the software. Confidence can be gained by starting with simpler concepts and those of a less critical nature. For example, various means of detecting a sympathy trip could be implemented with a moderate software effort and minimum risk to power system security.

Top management support will be needed not only for funding, but also because many aspects of adaptive protection systems (APS) may cross existing functional lines of responsibility (e.g. control and protection functions).

An effective way to get an APS program started on a solid economic basis is to identify those areas where the relaying can cause operating constraints today. These would include any economic dispatch restraints, stability margins, weather related hazards, overstressed equipment, maintenance constraints or manpower limitations. Any new or upgraded facility program should include an evaluation of possible benefits of including an APS.

CONCLUSIONS

Adaptive conceptual solutions have been introduced to improve the performance of transmission line protective relaying and, hence, the reliability of power system operation in the following areas:

1. Achievement of consistent high-speed sequential clearing of far-end faults without a pilot channel, possibly eliminating the need for a second pilot scheme.

2. Adaptive use of current infeed ratios on multi-terminal lines to control distance relay reaches, increasing zone 1 coverage during high-ratio conditions and avoiding excessive zone 2 reach during low-ratio periods.

3. Use of an adaptive system impedance equivalent with a fault-locator algorithm to achieve reliable zone 1 relaying for high-resistance ground faults (up to 500 ohms).

4. Responses to defective relaying alarms that obviate the need for providing a second pilot scheme or for removing the protected line from service, by automatic temporary setting changes or by enabling standby protection in the bus protection processors.

5. Means for fast detection of a sympathy trip (incorrect trip during an external fault) and selective initiation of high-speed reclosing based on power system conditions.

6. Detection of secondary-arc extinction as a permissive control of reclose initiation, to reduce unsuccessful reclosing.

7. Measurement of fault resistance, to block reclosing for mid-span faults (e.g. > 25 ohms).

8. Use of an adaptable breaker-failure timing function that varies with the positive-sequence voltage during the fault, thereby minimizing the possibility of unneeded back-up breaker tripping.

9. Adaptive last-resort system splitting, to isolate generating plants with manageable amounts of load and to facilitate load and transmission system restoration.

10 Detailed monitoring of the internal performance of relaying algorithms, permitting adaptation of settings or logic in response to operating experience.

11 Periodic area-wide relay coordination studies starting from the present power system configuration, to achieve a realistic supervision of the relay setting process.

ACKNOWLEDGEMENT

This study was conducted for the Power Systems Technology Program of the Energy Division at Oak Ridge National Laboratory (ORNL) under subcontract 86X-22012C.

REFERENCES

1 "Adaptive Power System Transmission Protection", ORNL/Sub/85-22012C/1, prepared by Electric Research and Management, Inc., for the Oak Ridge National Laboratory.

2. H.F. Parker, et al, "Testing of Composite Overhead Ground Wire Containing Optical Fibers", IEEE Transactions, Vol. PAS-104, No. 6, June 1985, pp1571-77.

3. IEEE Working Group, "Protection Aspects of Multi-Terminal Lines", IEEE Special Publication 79 TH0056-2-PWR, 1979.

4. L. Eriksson, et al, "An Accurate Fault Locator with Compensation for Apparent Reactance in the Fault Resistance Resulting from Remote-End Infeed", IEEE Transactions, Vol. PAS-104, Feb. 1985, pp424-436.

5. P.O. Geszti, et al, "Problems of Single-Pole Reclosing on Long EHV Transmission Lines", CIGRE 33-10, 1982.

6. J.S. Deliyannides and E.A. Udren, "Design Criteria for an Integrated Microprocessor-Based Substation Protection and Control System", IEEE Transactions, Vol. PAS-101, No. 6, June 1982, pp1664-73.

7. CIGRE Working Group 34.01 Report, "Final Report on Computer Based Protection and Digital Techniques in Substations", May 1984.

Discussion

George R. Nail (Public Service Company of New Mexico, Albuquerque, NM): The use of microprocessor technology to overcome the traditional boundaries in protective relaying is not a new idea. However, the use of these computers to adaptively change the characteristics of the protection packages is innovative. The traditional method, irrespective of hardware, was to design the scheme and calculate fixed settings for the relays. Changes would only be made when the configuration or system was significantly modified. To use the computer to evaluate system conditions and determine the course of action, is the ultimate. Obviously, the hardware to do many of the functions in the paper it not economically justified or is not yet developed to the point that would be necessary to accomplish these tasks. Perhaps one of the most important points that was not discussed is the need for standardization of data interchange formats and protocols. This is especially important if new equipment is going to be installed over a period of time rather than all at once. Future additions by other venders must be compatible. The health and the future of his concept is dependent on this compatibility.

Manuscript received July 31, 1987.

S. H. Horowitz (AEP Service Corporation, Columbus, OH), **A. G. Phadke** (Virginia Polytechnic Institute and State University, Blacksburg, VA), and **J. S. Thorp** (Cornell University, Ithaca, NY): Our congratulations to the authors for this creative and thought-provoking paper. This paper covers a parallel effort by the discussers on a study of adaptive protection sponsored by the Department of Energy. It is not surprising, therefore, that there are strong similarities in concepts. The discussers obviously have no fundamental disagreement with the authors' positions. There are differences, however, in emphasis and approach that are worth noting.

Since there is no industry definition on the term "adaptive protection", each investigator is free to create his own and then continue the investigation under these self-imposed constraints. The subject paper adopts a definition that specifically allows human intervention. This definition is compatible with their further specification that the adaptive techniques that are described required only slow speed responses.

In the investigations made by the discussers, and reported in references 1, 2 and 3, we adopted a definition that permitted us to explore a wider range of concepts using high speed communication. Have the authors also studied concepts using high speed communication?

Reclosing is a particularly fertile area for adaptive techniques and both the authors and the discussers have recognized this potential for significant improvements in system control. Again, the difference in emphasis is apparant as the authors have opted for an excellent adaptive scheme that can be implemented in specific situations. The discussers adaptive reclosing concept (1) is a widespread, long-range concept that addresses basic planning criteria.

This paper also discusses "sympathy trips" which are recognized as serious obstacles to optimum system performance. We invite the authors to join us in a call to the Power System Relaying Committee to begin an industry study into this phenomenon so adaptive concepts could be further enhanced.

References

[1] S. H. Horowitz, A. G. Phadke, J. S. Thorp, "Adaptive Transmission System Relaying", IEEE Summer Power Meeting 1987—Paper 87 SM 625-7.
[2] J. S. Thorp, A. G. Phadke, S. H. Horowitz, M. M. Begovic, "Some Applications of Phasor Measurements to Adaptive Protection", PICA '87, #124.
[3] A. G. Phadke, J. S. Thorp, S. H. Horowitz, "Impact of Adaptive Protection on Power System Control", 9th Power Systems Computation Conference, Lisbon, Portugal.

Manuscript received August 3, 1987.

S. S. Venkata and **M. J. Damborg** (University of Washington, Seattle, WA) and **A. K. Jampala** (ESCA Corporation, Bellevue, WA): The authors of these papers are to be commended for postulating a new and novel concept for protecting power transmission lines on an on-line and adaptive basis. The papers are well documented with thought-provoking ideas which will have far-reaching implications in the coming decades. They are bound to become classics and pivotal reference material for future researchers in this area. In our opinion, the adaptive protection (AP) concept became an attractive possibility due to the simultaneous emergency of three technologies:

(i) Availability of large, relatively inexpensive computers for on-line analysis such as the rapid computation of coordinated relay settings.
(ii) On-going development of large bandwidth communication systems, using new technologies such as fiber-optics, to link all modes of a utility network.
(iii) Computer controlled or digital relays whose performance can be influenced by remote commands through their communication lines.

We would like to make the following specific observations:

1) Both papers offer a similar definition for AP, except that Horowitz et.al did not refer to "on-line" characterization explicitly. If the adaptive protection process is to be an integral part of an EMS package, should not the definition also include "real-time"? In view of these comments it is therefore desirable suggest a more precise definition for adaptive protection.

2) It is our opinion that the adaptive protection philosophy should evolve gradually, in such a way that it would allow for utilization of the exisiting relaying architecture in the short range to the fullest advantage. This would in turn pave the way for a smooth transition to a digital relaying scheme in the next two decades. Such an approach is extremely desirable in order to convince utilities to switch over from standard electromechanical and solid-state relays to digital schemes. The utilities might find these ideas attractive not only from an improved system performance point-of-view but also from the benefits that could accrue from additional functions that the sophisticated digital relays could offer them.

3) Both papers emphasize multi-terminal lines in proposing the AP concept. We hope the readers realize that the concept is equally applicable to two-terminal as well as parallel lines.

4) Is it possible that with the adoption of the AP idea, the need for backup protection and consequently, coordination process could be eliminated if the sophisticated digital relays can be designed to protect the lines with the highest degree of dependability, security and reliability? This last point is merely posed as a philosophical question so the authors can express their views.

5) This item specifically pertains to the section titled "relay setting coordination checks" in the paper by Rockefeller et.al. We agree with one of the basic thrusts of this section which we understand to be:

> Adaptive protection should allow better performance from a protection system by allowing the setting to be made with consideration of fewer contingencies than is presently the case. That is, fewer constraints on a problem leads to a solution which is closer to the ideal.

For example, using a 38 bus, 61 line test system, we have shown that, when coordinating overcurrent relays for present system conditions rather than planned system conditions (with contingencies included), the mean operating time decreased by 18% for primary relays and 14% for backup relays [1].

The authors seem to suggest this adaptive concept in the context of *checking* relay setting on-line and providing an alarm if problems are found. We would suggest that it is possible to go further in the adaptive process. Specifically, we think it is feasible to respond to a change of system configuration (e.g. loss of a line due to a switching action) by not only identifying those relays which have inappropriate settings for the new configuration but by computing new settings. We also suggest this identification and coordination action can be done rapidly with on-line computers and the new settings can be transmitted to the digital relays quickly enough for satisfactory protection for the next contingency.

We have experimented with this question of the time required for coordination computation for the test system cited above, and we have demonstrated that the affected relays can be identified and the relay setting recalculated "promptly". The definition of "promptly" is, of course, relative but as a test of what is possible with massive computer power we have calculated all overcurrent relay settings for the 38-bus system in 6 seconds on a CRAY-XMP supercomputer [1].

We would appreciate the authors' comments on our observations on adaptive protection.

Reference

[1] A. K. Jampala, "Adaptive Transmissions Protection: Concept and Comptational Issues", Ph.D. Dissertation, University of Washington, Seattle, Washington, 1986.

Manuscript received August 7, 1987.

G. D. Rockefeller, K. L. Hicks, C. L. Wagner, J. R. Linders, and **D. T. Rizy**:
The authors are anxious to spark the evolutionary development of more sophisticated relaying, building on the widespread acceptance by utility relay engineers of "transparent" microprocessor-based commercial designs. These are stand-alone (except for the communications port in some cases), tend to look much like hard-wired designs and have setting facilities similar to their predecessors. The potential for implementing adaptive techniques within this framework is limited, since the relay is getting only conventional on-line information. This in contrast to use of a hierarchical structure. To gain acceptance an adaptive relaying plan must be evolutionary as suggested by discussers Venkata, et al. It must be evolutionary both in terms of software sophistication and in its hierarchical complexity. For example, a logical advance would be to build adaptation into the line protection using input from a substation computer with information from just the local substation (in addition to the conventional pilot signal from the remote line terminal). Experimental hardware structures are now being installed in the field which can accomodate some advanced adaptive concepts.

Aside from the acceptance hurdle, the hierarchical structure and adaptive protection must be economically driven. Because relaying represents a very low volume market, the transition from the use of transparent designs for conventional decentralized protective zones to integrated substation complexes (and later on, ties to an EMS systems, etc.) is a major quantum leap. The latter represents an even lower volume, particularly when considering the software costs. Nail's call for "standardization of data interchange formats and protocols" is a key ingredient for acceptance and economic justification.

The authors selected concepts from the larger study which seems to be shorter range to provide a base for evolution to more elaborate and more expensive implementations. The study itself included longer range considerations and the use of higher speed communications than that of a SCADA system.

Reclosing seems a logical early area for adaptation, because it is less critical than fault clearing functions and because in this area there are a number of potential improvements within relatively easy reach. For example, improved response to sympathy trips promises major benefits to network reliability.

Venkata, et al, suggest that the definition for adaptive protection include the phrase "real time". This seems unduly restrictive, since many adaptations need not stay in step with power system events.

The authors suggested use of an on-line coordination *check*, not as the ultimate, but as an evolutionary step. As Venkata, et al state, a central computer could run fault studies in response to power system changes, recoordinate and issue setting updates, all automatically. Ironically this would return us to the days of batch process computer coordination, in contrast to off-line interactive computer setting calculations. Acceptance of batch process settings was limited because the user had little control of the process, particularly in making compromises.

Manuscript received September 4, 1987.

Adaptive Transmission System Relaying*

S.H. Horowitz
Fellow, IEEE
AEP Service Corp.
Columbus, Oh.

A.G. Phadke
Fellow, IEEE
Virginia, Tech.
Blacksburg, Va.

J.S. Thorp
Senior Member, IEEE
Cornell Univ.
New York, N.Y.

Abstract - This paper describes the results of an investigation into the possibilities of using digital techniques to adapt transmission system protection and control to real-time power system changes.

The adaptive possibilities studied include transmission system protection philosophy, transmission line protection, relay settings, transformer protection and automatic reclosing.

All protective relaying decisions involve a compromise. One of the most difficult compromises to resolve is that of Reliability. Reliability has two factors; dependability and security, these factors are, in general, mutually contradictory. A relay scheme is traditionally designed with a bias towards one factor or the other. Transmission line protection embodies a wide variety of conflicting criteria. For example, multi-terminal lines involve considerations of current contributions from each terminal and relay settings must be chosen either with or without this effect. The percentage differential relays used for transformer protection are designed to accommodate errors that are possible, but not precisely known at any given moment. Automatic reclosing practices are established to cover a wide range of possible events and configurations but may not reflect the exact situation at the time a closing command is initiated.

In this paper we have examined each of these situations and developed adaptive strategies to minimize the compromises required and, thereby, optimize system performance.

INTRODUCTION

Adaptive protection is not a new concept. Time-delay overcurrent relays adapt their operating time to fault current magnitude. Directional relays adapt to the direction of fault current. Harmonic restraint relays adapt to the difference between energizing a transformer and faults within a transformer. These, however, are permanent characteristics of a relay or relay system and are included as part of the original design or installation to perform a given function.

The adaptive concept, as it is implemented in this paper, is defined as follows:

"Adaptive protection is a protection philosophy which permits and seeks to make adjustments to various protection functions in order to make them more attuned to prevailing power system conditions." [1]

The implementation of this concept is based on the availability of a microprocessor-based system of protection.

* Research sponsored by the office of Energy Storage and Distribution, Electrical Energy Systems Program U.S. Department of Energy, under Contract No. DE-AC05-84OR21400 with Martin Marietta Energy Systems, Inc.

87 SM 625-7   A paper recommended and approved by the IEEE Power System Relaying Committee of the IEEE Power Engineering Society for presentation at the IEEE/PES 1987 Summer Meeting, San Francisco, California, July 12 - 17, 1987. Manuscript submitted December 15, 1986; made available for printing April 28, 1987.

Adaptive techniques have been developed for the protection of equipment such as transmission lines, transformers and buses and for control and monitoring at the substation, regional and system levels. In this paper, we will concentrate on selected protection and control aspects of the power system. The accompanying considerations associated with the organization of computers, their hierarchy and architecture and system control considerations will be presented at other forums appropriate to those subjects. [1]

It is generally agreed that a 1969 paper by G.D. Rockefeller [2] is the correct starting point for any history of computer relaying. In his paper, Rockefeller undertook the study of protection with digital computers of all of the power equipment in a substation. Other investigators [3],[4],[5], began developing algorithms for transmission line protection. Eventually, field installations were made to prove the viability of this technology in the harsh electrical environment of the substation [6],[7],[8].

A 1984 report by CIGRE SC34,[9], includes a summary of 15 world-wide projects, primarily concerned with transmission line protection but also describing other substation functions. Today, there are investigations into computer relays by every major relay manufacturer and there are test installations in stations throughout the world.

Since there is only modest experience with digital systems for protection, control and monitoring and there is a natural tendency to perfect the primary function of the relay before advancing to sophisticated enhancements, the majority of functions are not adaptive. Most of the historic experience is to have adaptive changes made by human intervention, either operating personnel, technicians or engineers through their periodic updates of settings or characteristics.

There is a limited adaptive capability in some electromechanical and solid-state analog systems but the amount is small due to the extremely limited logic and memory capability.

Digital technology, on the other hand, is inherently programmable and has "essentially" unlimited logic and memory capability. Digital systems also have greatly enhanced capability for data transfer, analysis, communication between modules (both local and remote), data reduction and exception reporting. These expanded capabilities make adaptive functions much more feasible in a digital system. The protection, control and monitoring functions can automatically adjust their performance to match the needs of changing power system conditions and can handle changing system configurations.

With the advent of digital technology in a microprocessor-based system, the implementation of an adaptive concept is practical and straightforward. The use of adaptive features coupled with its inherent diagnostic capability can enhance protective schemes rather than degrade them as may be the case with analog systems.

Adaptive Security/Dependability

Relay engineers, almost universally, recognize that existing non-adaptive relaying schemes involve a compromise between security and dependability; dependability being a measure of the relaying equipment's ability to correctly clear a fault while security is a measure of the relaying equipments' tendency not to trip incorrectly. By utilizing redundant relaying schemes, dependability and security can be controlled.

Reprinted from *IEEE Trans. Power Delivery*, vol. 3, no. 4, pp. 1436-1445, Oct. 1988.

As an example, consider two independent line protection systems protecting the same line. If either system can cause a trip (contacts connected in parallel) then the combined system is more dependable. If both systems must operate to clear a fault (contacts connected in series) then the combined system is more secure. Most back-up and redundant relaying of transmission lines favor dependability.

The reason for the criteria for the high dependability of existing relaying schemes is that the power system itself is highly redundant. Generally, the loss of a single transmission line element will not affect the power system's ability to perform its function. There are, of course, exceptions in radial systems and at times when the system is highly stressed. In a highly interconnected transmission system, the failure to clear a fault may have severe consequences, while a false trip may have no serious consequences at all. Dependability is at a premium compared to security in such situations. As lines are tripped out, however, network redundancy is reduced and security becomes more of an issue.

A fixed (non-adaptive) relay has a given security and dependability which is a function of its type, construction and application. When such a relay is used alone, the relaying scheme has the security and the dependability of the relay itself. If additional relays are employed in a redundant connection, then either the security or the dependability is improved at the expense of the other. Once a non-adaptive relaying system has been designed and installed, its security and dependability are fixed and cannot respond to changing system conditions.

In general, dependability is obtained through additional equipment connected redundantly. The necessary degree of security will have to be obtained by testing, monitoring or self-checking [10].

This reference examines the reliability of relays and how testing, monitoring and self-checking affects both their reliability and availability. Although most of this reference is beyond the scope of this paper, several observations and conclusions are pertinent.

(1) Manual periodic testing is the most complete method of uncovering failures but the unavailability of this equipment during testing is a significant disadvantage. The length of time between tests is relatively long and a hidden failure may exist during that time.

(2) Automatic testing introduces additional equipment which, itself, adversely affects reliability. In addition, most automatic checks are functional and, while they may catch the catastrophic failures, may not reveal incipient or potential failures.

(3) Monitoring is continuous, relatively simple, with few components. In this regard, monitoring is less a concern than automatic testing, but it suffers from the same limitation of possible inability to uncover subtle, potential failures.

(4) Self-checking in a microprocessor-based relay has none of the disadvantages of 1), 2) or 3) above. It can be so designed to improve both security and dependability. This, of course, is the specific interest in our concept of adaptive relaying. Self-checking provides the ability to establish that a failure has occurred, to transmit an alarm to the appropriate personnel and to correct it before the equipment has had an opportunity to operate incorrectly.

To examine this practice from the standpoint of developing an adaptive concept, it is useful to look at two aspects of the protection system:

(1) Multiple Systems - Every protection system has some degree of redundancy. In the simplest case this will be a primary and a back-up system. Ideally, they should be completely independent of each other. Practically, there are always some shared elements such as the battery, trip coil or potential sources. In more sophisticated relaying systems, usually associated with higher voltages, there may be two primary systems and one back-up system. In such systems, there are fewer shared elements. The fundamental purpose of redundant relays and the use of limited shared elements is to emphasize dependability. Either system will initiate a trip.

2) Multiple Components within a System - To properly protect a system against all possible faults, any relay system is composed of many elements. For example, there are usually relays connected to each phase and to ground, fault detectors, directional elements, communication inputs and multiple output devices. Some of the elements are connected to provide dependability, others for security. For instance, for dependability, every fault will be seen by two or more of the phase and ground relays and any of these relays will initiate a trip. Fault detectors and directional contacts are used for security so all of these relays must operate to initiate a trip. Power line carrier communication usually operates in a mode that favors dependability over security but can be provided with a feature, known as "cross-blocking", to increase its security at the expense of dependability.

This technique is used with two independent directional comparison schemes in which the blocking signal of each system is connected to the logic of the other system. The result is that either blocking signal will prevent a trip from being initiated in either relay system.

With these two aspects in mind, it is possible to develop an adaptive protection philosophy to alter the security and the dependability of the relaying scheme as a function of the system state--i.e. whether it is in a normal, alert, emergency or restorative condition. It can be argued that, in the alert or restorative state, security has more importance than in the normal state. Since it is the entire relaying philosophy that must be altered to change the security/dependability of the protection scheme, more than the individual relay characteristics are involved in the adaptation. Figure 1 shows the logic that could be used to reconfigure two relay systems or some element within a relay system from a bias towards dependability to one favoring security. The "switches" are, of course, symbolic representations of the action required but the logic is as follows:

In the normal state, (all switches set to "N"), a trip will occur from either relay system through OR#1. Within each relay system, a trip can be initiated by any phase or ground relay through, OR#2 or OR#3, provided the appropriate supervising element, such as the fault detector, directional element and communication signal, is also present. This is the usual configuration favoring dependability.

In an abnormal state, such as Alert or Restorative, all switches can be "thrown" by a command from a central location. This would reconfigure the relay systems so that both systems must issue a trip signal through "AND 1". This favors security but retains the parallel configuration of the individual relays.

A subset of this concept could be to only change the switch positions associated with the communication signal of both relay systems. This converts the parallel trip path to a series trip path and is, in fact, the cross-blocking feature discussed above.

In an analog system, the complexity of the circuitry required to implement this concept makes it untenable. In a digital system, however, when the switches reside in software, the logic can be reconfigured by commands, making the concept viable.

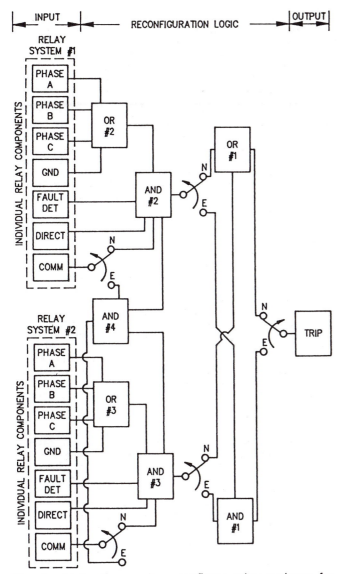

Fig. 1. Logic diagram to reconfigure relay systems for dependability or security.

ADAPTIVE RELAYING OPPORTUNITIES

The protection and control schemes that have been examined include the adaptive possibilities associated with multi-terminal transmission line protection, relay settings, transformer protection using harmonic-restrained differential relays and automatic circuit breaker reclosing control.

Multi-Terminal Protection

The settings associated with multi-terminal line protection probably result in more compromises than any other single protection situation. Multi-terminal line protection is never as straightforward as that of two terminal lines. The taps, when they are contributors of positive - or zero - sequence currents, are likely to produce erroneous estimates of fault distance, direction, or levels. The objective of an adaptive protec-

tion system would be to achieve reliable protection for a multi-terminal line that is comparable to that obtained with a two-terminal line.

The problem to be addressed stems from the current contributions from active taps, and the difference in relay response when the contribution is, or is not, present. This contribution can be either into the protected line section (in-feed) or out of the protected line section (out-feed), depending on the system parameters involved. Usually in-feed is the problem. Its effect is to increase the apparent impedance seen by a relay, i.e., for a given relay setting, infeed shortens the reach. In practice, the accommodation for this difference is conservative. Since zone 1 must never overreach its associated line section, the zone 1 relay setting is calculated without in-feed. Thus, with in-feed, the relay reach is reduced.

Conversely, since zone 2 must always cover the entire associated line section, it is set with in-feed (and appropriate time delay). The absence of in-feed will then extend the reach and increase the protected zone.

Similarly, in a directional comparison blocking pilot scheme, the blocking relay must always overreach the tripping relay to avoid an incorrect trip for an external fault. The tripping relay is, therefore, set with in-feed to obtain the maximum apparent impedance. The blocking relay is set by calculating the impedance seen by the tripping relay without in-feed and coordinating for that condition.

In general, even with these compromises, the effect on the total protection package is not great. There are, however, reductions in coverage or relay loadability, or increased coordinating times. Calculations using typical AEP multi-terminal configurations show improvement in transmission line protection coverage from 65-70% without provisions for adaptively changing the relay settings to 80-90% when these adaptive concepts are implemented.

The multi-terminal line protection problems can be summarized as follows:

(1) Distance relays used for high speed, pilot or backup protection see erroneous distances to a fault because of the in-feed from active taps.

(2) Directional relays will operate incorrectly if one of the active taps produces out-feed during an internal fault.

(3) Mutually coupled circuits compound the problem.

The compromises forced upon the relay engineer are:

(1) Acceptance of sequential fault clearing

(2) Reduced load limits on the line

(3) Acceptance of loss of remote backup protection.

(4) Longer fault clearing times.

The availability of an adaptive protective system lets us organize the settings for every situation. If we can adapt the setting to the real-time situation regarding in-feed or out-feed, we can maintain the maximum coverage and the minimum coordinating margins. Within this concept, there are several levels of sophistication possible:

(1) Transmit on-off indication from the remote breakers. If a remote terminal of a multi-terminal system is taken out of service, a pre-calculated setting can be implemented to reflect the existing contributions. Only on-off information is needed, and very fast notification of status changes is not required. If a computer system exists at each of the remote terminals, the status of the line or tap can be established locally and, since a single digit on-off

signal is all that is sent to all relays, a low bandwidth channel is acceptable. When the remote ends do not contain a substation computer, the status of all of the relevant disconnects and circuit breakers must be communicated. In addition, station 1-line diagrams of each of the remote ends must be stored in the adaptive relay if it must determine the switch and circuit breaker status information.

As each switching operation takes place, back-up zone settings (as well as instantaneous zone settings where appropriate) would be reset accordingly. This would permit the use of optimum settings relative to the relay reach, instead of accepting a compromise that is applicable for all system configurations.

(2) Transmit all analog values of current to all terminals. This concept furnishes the ultimate refinement in relaying multi-terminal lines. With this information, accurate distance calculations can be made at each terminal. This requires measurements at all terminals as input to all adaptive relays. Samples of current in each phase and the zero sequence contributions are needed. In addition, if line charging current is a significant quantity, each phase voltage must also be communicated. As a good approximation, it may be possible to use the voltages at the relay terminals and thus eliminate the need to measure and transmit voltage sample data. If a sampling rate of 720 Hz is assumed, each terminal would thus contribute either 4 or 7 16-bit words at each sampling time, depending on whether voltage measurements are transmitted or not. Furthermore, a method of synchronizing the samples at each terminal must be implemented.

(3) Adapt to system load. One of the simplest, albeit most powerful, adaptive features to incorporate in a relay, is the ability to subtract the load current from the post-disturbance current in order to assess the properties of a fault. Load current can be continuously monitored and stored. A transient detector can then trigger the procedure which removes the load current from the total current. In effect, this removes the loadability limits imposed on the line by the relay setting. No additional inputs are required.

(4) Adapt to a system Thevenin equivalent. In a power system that is monitored by a central computer that determines the actual system configuration from a system-wide data acquisition system in conjunction with State Estimation algorithms, it is possible to produce an on-line multi-port Thevenin representation for the power system as seen from each terminal of a multi-terminal line. If these equivalents are made available to the relays, all fault locations can be identified uniquely, regardless of the number of taps on the line. Note that the Thevenin representations for positive and zero sequence networks would have to be included in order to account for phase and ground faults. In this fashion, the normal relationship between the lengths of lines and the zone settings of the relay would be restored. This technique would require the update of the equivalent whenever it changes significantly from the previous value. This determination could be made at the control center and the update at the relay initiated by the center. Prefault voltages at each line terminal must also be available. This is more information than is required for 1) and 2) above, but the amount of data is modest and need not be transmitted in real-time. It only has to be made available at the relay in time for the next disturbance.

When real-time current phasor measurements from each line terminal are available as discussed above in (2), the need for the Thevenin representation disappears. The current injections at each line terminal can be used instead to calculate the fault location precisely. If line terminal voltages are also made available, the distance calculation can be made free of errors caused by the fault path resistance. This technique would permit the zone settings to be extremely narrow, thereby improving the line loadability.

The first three adaptive protection schemes depend upon modifications of the relay setting based upon system conditions as communicated to the relay location at relatively slow speeds; thus the speed and accuracy of the distance relay functions are not affected in these cases. Where load current compensation is used-in case (3) - the transient monitor and load compensation calculation are added burdens to the relay computer.

All of the adaptive features described above result in a more accurate determination of the fault location. Thus, the underreaching zones will continue to underreach under various system operating conditions. Also, an overreaching zone will continue to overreach and can be set more closely than is possible in a non-adaptive environment. The adaptive relays, therefore, do not require any changes in existing adaptive or non-adaptive relays.

Adaptive Relay Settings

Relay settings represent the ultimate synthesis of technology and art, fact and judgement. In every phase of the development of a transmission line protective scheme, a compromise is usually made between economy and performance, dependability and security, complexity and simplicity, speed and accuracy, credible vs. conceivable. It is, however, in the relay settings themselves that all of these compromises are focused. It is the settings that transform the concept to a reality and allow the relays to perform at their maximum or minimum potential.

The objective of providing adaptive relay settings is to minimize the compromises that accumulate during the engineering process and allow the relays to respond to actual conditions.

Relay settings, at the present time, are calculated from short circuit studies that include a wide variety of system configurations, generation schedules and reasonable voltage excursions. The user's setting philosophy and criteria establishes limits to assure maximum possible coverage in the fastest possible time. In the end, the settings are the result of an engineer's judgement as to the best overall protection consistent with reliable system operation.

Experience indicates that the relay engineer's judgement is, in fact, extremely effective. In practice, there are very substantial margins available and the compromises that must be made are between performance that is either excellent or performance that is very good; not between adequate or inadequate. The burden then might shift from the most desirable to one less desirable in terms of additional equipment or complexity. Digital devices, whose settings are resident in software, can automatically adjust to many changes in system conditions and, thus, eliminate the complexities that might be involved.

There are several conditions that, in extreme situations, would lend themselves to significant improvement of relay settings by using the adaptive techniques.

(1) Pre-Fault Load - The positive sequence voltages and currents at the relay location are directly affected by load. Usually, the effect is ignored. For long lines or heavy loads, however, this effect can be a problem. Overcurrent relays must be set sufficiently above load and below fault current to allow for both dependability and security. Distance relays must not encompass the load impedance. The traditional solution is to add voltage restraint, directionality or blinders, or to compromise the setting. Digital relays can recognize pre-fault load as a steady state condition and can include, in its logic statements, the ability to ignore this current. This technique was discussed more fully in the section on multi-terminal transmission line protection and has general application for most relays.

(2) Source-Impedance Ratio (SIR) The ratio of the power system source impedance to the impedance of the protected zone - A high SIR results in a low fault voltage and, more importantly, a small change in the voltage for a fault at one end of a line or the other. The changes in SIR during normal system operation are usually well within the ability of a given relay setting to operate correctly. The relay selected for a given line has the required operating characteristic. However, unusually large changes, such as those which would accompany a wide-area disturbance, can result in fault voltages that are below the normal range of the relay. The digital inputs to a digital relay are derived from the analog output of Current and Potential Transformers through analog-to-digital (A/D) converter systems. The operating range of the relay can be adjusted by rescaling the A/D converter when such a disturbance occurs.

(3) Cold-Load Pickup - This phenomenon is well known and extremely difficult to resolve. It is usually a problem on subtransmission or distribution circuits and is related to the loss of diversity of thermostatically controlled devices following an extended outage. The simultaneous reenergization of all of the connected load, aggravated by the additional current accompanying motor starting, results in currents that exceed normal relay settings. The common practice is to allow the instantaneous relays to trip once, remove them from service and allow the time-delay relays, which have enough time to override the inrush, to operate if a fault occurs. Digital relays can be provided with the intelligence to recognize a sustained outage and can adjust their settings accordingly.

(4) Line Charging and System Asymmetries - The importance of capacitance on overhead transmission lines depends upon the length of the line and the operating voltage. Normally, this effect on the relay settings can be ignored. However, long lines have experienced line charging currents of sufficient magnitude to cause false relay operations of directional-comparison ground relays. Phase comparison relays are also subject to misoperation if the line capacitance changes too much with operating voltage. Digital relays can reconfigure the logic to provide parallel zero sequence fault detectors on low capacitance lines, and series fault detectors on high capacitance lines and adjust this logic if the operating voltage changes sufficiently to effect a significant change in the capacitance.

In addition, an untransposed line presents a higher impedance between the two outer conductors than between the middle conductor and either outer one. The apparent impedance, i.e., V/I, seen by the relays is not equal between phases at the same terminal or on the same phase at the two ends of the line. Digital relays can take both the charging current and the asymmetry into account and make a setting that is customized for each relay.

All of the measurements required to adapt relay settings as discussed above are available at the relay location. If we assume that the system changes do not coincide with a fault, there are also no stringent communication requirements.

All of the adaptive measures discussed above are in the direction to make the relay settings more precise; to more accurately locate the fault. To that extent, there should be no impact on existing non-adaptive relays. If, however, any relay setting is adaptively changed in the direction of longer times or reach, coordination with all other relays must be examined.

Transformer Protection

Setting the pick-up and slope of a percentage differential relay for power transformer protection is at best a compromise. The slope and pick-up settings must be selected to account for Current Transformer

(CT) mismatch, variable tap settings and magnetizing inrush. In order to avoid misoperation under these conditions the relays must be set so that some internal faults cannot be detected. The objective of an adaptive percentage differential relay would be to adjust the pick-up and slope on the basis of information relative to actual currents that are experienced or changes in operating conditions so as to reduce the number of undetectable faults.

The details of the percentage differential protection depend on the winding type and the number of phases or windings but can be illustrated on the single-phase, two-winding transformer shown in Figure 2.

Neglecting magnetizing current,

$$N_1 I_1' = N_2 I_2' \qquad (1)$$

$$\text{if } \frac{n_1}{n_2} = \frac{N_2}{N_1} \qquad (2)$$

where,

$N_1 n_1$ - number of windings on the primary-side of the power transformer and current transformer, respectively

and $N_2 n_2$ - number of windings on the secondary-side of the power transformer and current transformer, respectively,

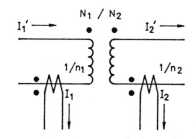

Fig. 2. Single phase, two winding transformer.

then I_1 must equal I_2 under normal operation. Even if equation (2) is valid at the nominal tap setting, a change in tap will result in $(I_1 - I_2)$ not being zero under normal load or an external fault. To account for CT mismatch and off-normal turns ratios, the percentage differential characteristic shown in Figure 3 is used.

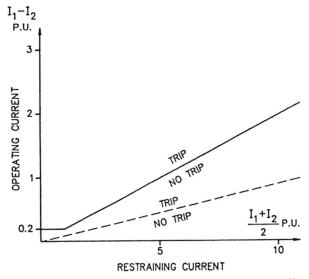
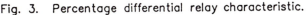

Fig. 3. Percentage differential relay characteristic.

Figure 3 is drawn for a 20% pick-up and a 20% slope and is a fixed characteristic. The dash line in Figure 3 is a locus of points for $N_1 n_1 / N_2 n_2$ = 0.9 and an external fault and indicates the effect of the conservatism built into the 20% slope/20% pick-up characteristic; i.e., an internal fault could lie above the dash line (the trip zone in the adaptive case) but below the solid line (the no-trip zone in the non-adaptive case).

The slope of the percentage differential characteristic must be large enough to account for CT mismatch even if there is no tap-changing mechanism. The problem of CT mismatch can occur because equation (2) is not exactly satisfied even at the nominal tap settings. Error current may also be due to residual magnetizing currents in the CT's or to the different magnetizing and saturation characteristics of the CTs. The relay must also operate correctly in the presence of magnetizing currents which exist during energization or overexcitation. Both of the latter conditions are partially addressed using an additional harmonic restraint element on the relay.

In summary, the selection of the pick-up and slope settings of the percentage differential relays involve a compromise between false trips and an inability to detect partial winding faults. The difficulty with existing relays is that the pick-up and slope settings are fixed and cannot respond to additional information.

A variety of adaptive techniques can be employed in power transformer protection depending on the amount of external information that is available to the relay.

(1) Adapt to tap position: By monitoring the tap setting and using the correct turns ratio in computing the trip and restraint currents a digital relay could use a smaller pick-up and slope settings than required when the tap setting is unknown. The only input, in addition to the transformer currents, is a digital indication of the tap position. As only a discrete number of tap positions are possible, a relatively simple encoding of tap positions into a digital input seems possible.

(2) Adapt to excitation level: If voltages were available to a digital relay, the settings for the pick-up and slope could be reduced when voltages were below some threshold values. The pick-up and slope settings would be increased for higher voltages when large magnetizing currents were present due to overexcitation. The minimum input requirement for adapting to excitation level is a digital indication of whether the voltages are above or below a level that will produce excessive magnetizing currents. If other digital units in the substation are using voltage measurements (e.g. for line protection) then those units could make the determination of the level and only transmit a binary level indicator to the transformer protection unit. Another possibility with a larger communication burden is to transmit the actual voltage samples from the line modules to the transformer module.

(3) Flux Restraint: If voltages were available, the harmonic restraint could be replaced by a restraint based on determining that the flux was saturated [11]. Such a relay would be more recognizable as adaptive if the pick-up and slope settings also responded to the flux calculation. The flux restraint algorithm would require the actual voltage samples be communicated to the transformer protection unit at the same sampling rate as the current samples.

(4) Remove harmonic restraint in steady state: In conjunction with (2) it would be possible to invoke harmonic restraint only during energization. That is, in steady state the only source of large magnetizing current is overexcitation. If harmonic restraint were removed then the problem of CT saturation for an internal fault would be solved. No additional inputs beyond the excitation level indicator of (2) are required. The transformer protection unit can determine from the currents themselves that the transformer is not energized. It is also possible to monitor the system so that harmonic restraint is restored when a

fault occurs or a parallel transformer is energized to accommodate harmonics due to sympathetic inrush or recovery voltage.

(5) Estimate CT ratios: In conjunction with (1) and (2) it is possible to estimate the CT ratios so that the pick-up and slope settings could be further reduced. With the tap setting accounted for as in (1) and magnetizing current eliminated as a source of apparent CT mismatch as in (2) then CT ratios necessary to obtain a zero differential current can be estimated from past data available to the relay. The process can be understood by considering a collection of points on the percentage differential characteristic of Fig. 3, corrected for tap setting and including data for which voltages were within nominal ranges. If the CT mismatch were constant then the points would lie along a straight line such as the dashed curve in Fig. 3. The tap setting from (1) and the excitation level from (2) are the only inputs required.

(6) Variable speed/security: More severe faults (further from the trip-no-trip boundary line in Figure 3) can be cleared faster than less severe faults (closer to the trip-no-trip boundary line in Figure 3). In essence the relay averages more measurements and therefore takes longer to make a decision, if the data lies close to the boundary. No additional inputs are required.

Since all of the required inputs are at the substation level, none of the adaptive transformer protection schemes involve extensive communication needs.

While the tap setting and excitation level are very low data rate signals it is important that changes in tap setting and/or excitation level be communicated to the transformer protection module in a timely manner. If the pick-up and slope settings are reduced because the tap setting is assumed to be known then the relay could trip if tap setting information were late in reaching the transformer relay. If the transformer differential characteristic is checked every T seconds then a change in tap setting or excitation level must be determined and communicated within T seconds.

A second consideration in timing is involved if Fourier methods are used to compute harmonics for harmonic restraint. If a full cycle window is used for the calculation of harmonics then a change in tap setting will take a cycle to propagate through the harmonics computed for the trip and restraint currents. If the change in tap setting is small the error in harmonics will not be too large. Nevertheless, the one cycle delay must be considered in adjusting the pick-up and slope.

If actual voltage samples are used for flux restraint then the voltages samples must be synchronized with the current samples which would imply synchronization of all sampling in the substation if both line and transformer relays are to share samples.

Adaptive Reclosing

The primary objective of automatic reclosing circuit breakers, following a fault, is to return the transmission system to its normal configuration with a minimum outage time of the affected equipment and the least expenditure of manpower.

In addition to the rapid restoration of the system to its normal configuration, the following benefits of automatic closing are generally obtained:

(1) It restores the power transfer capability of the system to its former level in a minimum of time.

(2) It adheres to predetermined reclosing restraints and times.

(3) It reduces the service interruption time to customers fed directly from the transmission system.

(4) It may relieve the operators from restoring service manually during storms and other widespread disturbances.

Automatic reclosing can be high speed (HSR), or time-delayed, single-shot or multiple-shot, supervised or unsupervised. HSR refers to automatic reclosing of a circuit breaker with no intentional delay beyond that necessary to deenergize the fault arc. All other reclosures are referred to as time-delayed.

Automatic reclosing is practiced throughout the world, based in large measure on the fact that approximately 80% of all faults are temporary. Each element of a power system, however, has its own criterion relative to the advantages and disadvantages and the type of reclosing that is applicable.

(1) Stability - The adverse effect of unsuccessful HSR is well known. Many users prefer to forego the advantages of HSR to avoid this risk. There are, however, several factors involved in this decision. Stability may be threatened by 3-phase faults and not by single-phase faults, or by certain configurations and not others, or in only certain areas and not throughout the system.

(2) Generators - In recent years considerable study and analysis have focused on the stresses to the shafts and other components of a turbine-generator set following reclosing.

(3) Transformers - Reclosing is not generally recommended where transformers are involved, unless the protective zones can clearly rule out the possibility of an internal transformer fault. Similarly, cables, reactors, gas-insulated equipment follow the same criteria.

(4) Single-phase switching - The problem involved with single-phase switching is the phenomenon known as the secondary arc. During a single-phase-to-ground fault, if only the faulted phase is deenergized, the remaining energized phases tend to reignite the arc path after the primary arc has been interrupted. HSR would then be unsuccessful. It is possible, and common in Europe, to simply delay the reclose until the secondary arc extinguishes itself due to arc instability or wind. In other instances, however, a sufficiently high current may sustain the arc or the required time delay may exceed the maximum permissive time for stability, voltage, or negative sequence considerations.

Automatic reclosing presently embodies most of the concept of adaptive control. Initiating HSR for zone 1 faults blocking HSR for out-of-step conditions or blocking reclosing for equipment failures are all commonly applied adaptive measures with existing analog relays. Use of digital devices greatly simplifies and enhances the implementation of this adaptive reclosing capability.

(1) Type and Severity of Fault - With the ability of digital relays to identify the location and type of fault, it is a minor task to include this information in the control logic. Presently, since the tripping zone is not coincident with the restrictive reclosing zone, a separate complement of relays are required for the reclose logic.

(2) Hard vs. Soft Circuit Breaker Operation - Circuit breaker operations that stress the insulation systems or mechanical structure are referred to as "hard" operations in contrast to "soft" operations where arcing, voltage stress, vibration or shock are minimized. Precisely controlling the closing and tripping impulse to a circuit breaker offers the opportunity to reap significant benefits in terms of reducing the exposure of the breaker to system transients and reducing breaker maintenance. The ideal control scheme would be to have the breaker contacts in each pole, following a fault, open when the current in that pole was zero. This would result in minimum arcing time. This can be done with the digital relay.

Similarly, if there is a permanent fault on a line, it is desirable for the first pole of a deenergized line to close at the maximum of the bus voltage.

This would minimize the d-c offset of the short-circuit current. If there is no fault, however, subsequent poles at the same terminal would be closed at the minimum voltage across the breaker.

Figure 4 shows the voltage relationship that exists on each phase for various faults after the first phase is energized. These values have been confirmed during commissioning tests of AEP 765 kV stations. Referring to this figure, the control logic is as follows:

When the first phase is energized, if no fault exists, the resulting voltage on the energized phase will be equal to the source voltage. If the line is untransposed the voltages on the other two phases will be a small percentage of the source voltage but unequal to each other.

If a line-to-ground fault exists on the energized phase, the resulting fault current will cause the breaker to trip and logic provided to lock out the closing of the other two poles.

If a line-to-ground fault exists on either of the deenergized phases. there will be no coupled voltage on the faulted phase and the first breaker pole can be tripped and further closing prevented.

If a phase-to-phase fault exists, involving the energized phase, the coupled voltage on the second phase will be equal to the source voltage.

If a phase-phase fault exists on the other two phases, the coupled voltage will be equal to each other, in phase, and a small percentage of the source voltage.

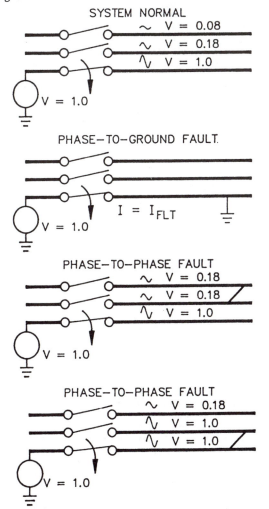

Fig. 4. Induced voltages for single phase energization (Untransposed line.)

(3) Optimizing Circuit Breaker Wear - Most high voltage transmission lines are configured in a ring bus or a breaker-and-a-half arrangement, providing two circuit breakers per element. Unless other considerations are present, it is desirable to rotate the reclosing duty to equalize contact wear, compressor operation, etc. This is normally done by maintaining appropriate records, and providing the operators with selector switches. A local or remote computer can reassign these responsibilities automatically. For instance, during a sleet or lightning storm, a line can be faulted many times in a very short period of time; each reclose being successful but all of the operations being done by the same breaker(s). Presently, even supervisory control cannot easily redirect the open and close impulses to apportion the work evenly among all of the breakers. A computer, however, would have no difficulty in following all of the operations, revising the tripping and reclosing sequence and then returning the system to its normal state when the storm has passed.

SUMMARY

The concept of adaptive protection is not new, and its introduction into the existing power system need not be revolutionary. The increasing acceptance and use of digital equipment has made these adaptive concepts viable. In this paper we have examined a representative sample of system protection situations and the possible improvements offered by their application. Specifically we have described the implementation of adaptive techniques to reliability, multi-terminal transmission line protection, relay settings, transformer protection and automatic reclosing.

Several typical situations which force the relay engineer to make compromises have been presented and adaptive strategies have been developed which minimize these compromises.

A scheme has been described which can reconfigure a total relay package so that dependability is provided during normal system conditions but security is emphasized during abnormal conditions.

The effect on relay settings by the presence or absence of infeed is eliminated through the use of adaptive relaying concepts. In addition, relay setting compromises caused by heavy load or cold-load pick-up are minimized by adaptive techniques.

A dynamic slope and pick-up characteristic for a percent differential relay is developed for protection of power transformers which allows the relay to protect more of the transformer winding.

Adaptive reclosing concepts are presented which adjust reclosing to specific fault conditions or prevent ever closing into a multi-phase fault.

ACKNOWLEDGEMENT

This investigation was conducted for the Power Systems Technology Program of the Energy Division at Oak Ridge National Laboratory (ORNL) under subcontract 86X-22005C. D.T. Rizy was the ORNL project engineer for the project.

REFERENCES

[1] A.G. Phadke, J.S. Thorp, S.H. Horowitz, "Study of Adaptive Transmission System Protection and Control," ORNL/SUB/85-2205C, Prepared by Virginia Polytechnic Institute and State University for Oak Ridge National Laboratory (to be published).
[2] G.D. Rockefeller, "Fault Protection with a Digital Computer", Trans. IEEE, Power Apparatus and Systems, Vol. PAS-88, No. 4, April 1969, pp 438-461.
[3] B.J. Mann and I.F. Morrison, Digital Calculation of Impedance for Transmission Line Protection", Trans. IEEE, Power Apparatus and Systems, Vol. PAS-90, No. 1, Jan/Feb 1971 pp 270-279.
[4] B.J. Mann and I.F. Morrison, "Relaying a Three Phase Transmission Line with a Digital Computer", Trans. IEEE, Power Apparatus and Systems, Vol. PAS-90, No. 2, Mar/Apr 1971 pp 742-750.
[5] G.B. Gilchrist, G.D. Rockefeller and E.A. Udren, "High Speed Distance Relaying Using a Digital Computer, Part 1-System Description", Trans. IEEE, Power Apparatus and Systems, Vol. PAS-91, No. 3, May/Jun 1972, pp 1235-1258.
[6] A.G. Phadke, M. Ibrahim, T. Hlibka, "Computers in EHV Substation: Programming Considerations and Operating Experience", CIGRE Study Committee No. 34 Colloquium, Philadelphia, Oct. 1975.
[7] A.G. Phadke, M. Ibrahim, T. Hlibka, "A Digital Computer System for EHV Substations: Analysis and Field Tests, Trans. IEEE, Power Apparatus and Systems, Vol. PAS-95, No. 1, Jan/Feb 1979, pp 291-301.
[8] M.M. Chen and W.D. Breingan, "Field Experience with a Digital System with Transmission Line Protection", Trans. IEEE, Power Apparatus and Systems, T-PAS, Sep/Oct, 1979, pp 1790-1805.
[9] Final Report on Computer-Based Protection and Digital Techniques in Substations, CIGRE Study Committee No. 34, Working Group 02 Report. (1986)
[10] W.G. 34.03, J. Gantner, Convener, "Use of Equipment Built-in Automatic Testing, Self-Checking and Monitoring With a View to Improving Reliability", CIGRE Study Committee No. 34 Report. (1986)
[11] A.G. Phadke, J.S. Thorp, "A New Computer-Based Flux-Restrained Current Differential Relay for Power Transformer Protection, (PICA 1983), T-Pas, Nov. 83, pp 3624-3629.

Discussion

G. D. Rockefeller (Rockefeller Associates, Inc., St. Rose, LA): On page 4, column 1, numbered paragraph (3), the authors propose to remove pre-fault current as a means to allow more sensitive reach settings for the distance protection, in this case in the context of multi-terminal applications. This concept is certainly readily implementable with stored program hardware and is used in at least one commercial relaying product for a different purpose. It deserves further study and not just for multi-terminal cases. Its use in distance protection is probably not as straightforward as might be inferred from the authors statements. The change in current (actual minus pre-fault) in combination with voltage yields an error in impedance calculation, since the voltage is the drop to the fault point produced by the *actual* current flow, not the current change. [Voltage change cannot be used, as this is the source drop.]

The error introduced by the drop caused by pre-fault current can be eliminated by compensation if we know the fault location and the current

infeed. The compensation can assume that the fault is at the end of the zone (the ''balance point''). The infeed can be based on the system conditions existing before the fault.

It seems possible that even with compensation, sufficient overreaching could occur to be of concern for zone 1, although the discusser is not aware of any analysis that would support this statement.

Sequential clearing of an external fault will change the load current distribution in the network, possibly causing a zone 1 overreach whether compensated or not. In this case the pre-fault current no longer represents the actual load flow as breakers open during the fault clearing sequence. Swinging during a protracted fault might also cause a zone 1 overreach, again because the prefault current doesn't continue to represent the actual load current. Also, blocking might be needed following fault clearing, to avoid operation on a swing.

Manuscript received July 20, 1987.

Donald R. Volzka (Wisconsin Electric Power Co., Milwaukee, WI): I would first like to congratulate the distinguished authors on the development of a well written paper that extends the horizon seen by relay and operations people concerned with protection systems. It is important that a number of views of the future be expressed to enable research and development efforts to be provided the benefit of discussion of the architecture of protection systems prior to commercial design. My comments and questions on the paper are as follows:

In order for transmission system relaying systems to be adaptive to system conditions, it appears that they would have to be provided with data on the loading, status, and notification of switching to be done from a central computer, and permission granted to change performance factors in accord with planned action resembling a form of expert systems technology. The paper is primarily concerned with needs of the protection system, but does not treat the information requirements to accomplish the objectives, nor the technology that will permit adaptive relaying systems to become working tools.

The development of relaying systems that can adapt to the needs of the power system is a worthy goal. It would appear that a corollary activity is needed, namely, to develop a method of providing feedback to system operating computers that will minimize the amount of time spent in a compromised position or of initiating analysis of alternatives by system operators that might preclude such operation altogether.

It appears that the discussion of adaptive transformer protection is based to a large extent upon technology and terminology of the past and is in need of a design philosophy that can take advantage of computer technology. Some thoughts along these lines are as follows. Why not discuss an adaptive transformer protective relay in terms of algorithms for automatic adjustment for ratio error associated with LTC operation or normal CT mismatch instead of pickup and percentage slope? This might be part of a "learn" mode whereby the relay does harmonic analysis of transients and adapts to the excitation transients and characteristics of the protected transformer instead of relying on harmonic restraint. Would it be desirable to use voltage measurement and associated excitation current measurements for better protection against overexcitation rather that as a restraint function? An adaptive transformer protection system should be able to provide much improved performance under short and long term overload conditions through feedback of actual winding temperature response to load changes. Have concepts such as these been contemplated?

The author's comments would be appreciated.

Manuscript received July 31, 1987.

S. S. Venkata and M. J. Damborg (University of Washington, Seattle, WA) and A. K. Jampala (ESCA Corporation, Bellevue, WA): The authors of these papers are to be commended for postulating a new and novel concept for protecting power transmission lines on an on-line and adaptive basis. The papers are well documented with thought-provoking ideas which will have far-reaching implications in the coming decades. They are bound to become classics and pivotal reference material for future researchers in this area. In our opinion, the adaptive protection (AP) concept became an attractive possibility due to the simultaneous emergence of three technologies.

   (i) Availability of large, relatively inexpensive computers for on-line analysis such as the rapid computation of coordinated relay settings.
   (ii) On-going development of large bandwidth communication systems, using new technologies such as fiber-optics, to link all modes of a utility network.
   (iii) Computer controlled or digital relays whose performance can be influenced by remote commands through their communication lines.

We would like to make the following specific observations.

1) Both papers offer a similar definition for AP, except that Horowitz et.al did not refer to "on-line" characterization explicitly. If the adaptive protection process is to be an integral part of an EMS package, should not the definition also include "real-time"? In view of these comments it is therefore desirable suggest a more precise definition for adaptive protection.

2) It is our opinion that the adaptive protection philosophy should evolve gradually, in such a way that it would allow for utilization of the existing relaying architecture in the short range to the fullest advantage. This would in turn pave the way for a smooth transition to a digital relaying scheme in the next two decades. Such an approach is extremely desirable in order to convince utilities to switch over from standard electromechanical and solid-state relays to digital schemes. The utilities might find these ideas attractive not only from an improved system performance point-of-view but also from the benefits that could accrue from additional functions that the sophisti-

cated digital relays could offer them.

3) Both papers emphasize multi-terminal lines in proposing the AP concept. We hope the readers realize that the concept is equally applicable to two-terminal as well as parallel lines.

4) Is it possible that with the adoption of the AP idea, the need for backup protection and consequently, coordination process could be eliminated if the sophisticated digital relays can be designed to protect the lines with the highest degree of dependability, security and reliability? This last point is merely posed as a philosophical question so the authors can express their views.

5) This item specifically pertains to the section titled "relay setting coordination checks" in the paper by Rockefeller et.al. We agree with one of the basic thrusts of this section which we understand to be.

> Adaptive protection should allow better performance from a protection system by allowing the setting to be made with consideration of fewer contingencies than is presently the case. That is, fewer constraints on a problem leads to a solution which is closer to the ideal.

For example, using a 38 bus, 61 line test system, we have shown that, when coordinating overcurrent relays for present system conditions rather than planned system conditions (with contingencies included), the mean operating time decreased by 18% for primary relays and 14% for backup relays [1].

The authors seem to suggest this adaptive concept in the context of checking relay setting on-line and providing an alarm if problems are found. We would suggest that it is possible to go further in the adaptive process. Specifically, we think it is feasible to respond to a change of system configuration (e.g. loss of a line due to a switching action) by not only identifying those relays which have inappropriate settings for the new configuration but by computing new settings. We also suggest this identification and coordination action can be done rapidly with on-line computers and the new settings can be transmitted to the digital relays quickly enough for satisfactory protection for the next contingency.

We have experimented with this question of the time required for coordination computation for the test system cited above, and we have demonstrated that the affected relays can be identified and the relay setting recalculated "promptly". The definition of "promptly" is, of course, relative but as a test of what is possible with massive computer power we have calculated all overcurrent relay settings for the 38-bus system in 6 seconds on a CRAY-XMP supercomputer [1].

We would appreciate the authors' comments on our observations on adaptive protection.

Reference

[1] A. K. Jampala, "Adaptive Transmissions Protection: Concept and Comptational Issues," Ph.D. Dissertation, University of Washington, Seattle, Washington, 1986.

Manuscript received August 7, 1987.

S. H. Horowitz, A. G. Phadke, and J. S. Thorp: Mr. Rockefeller is certainly correct in stating that load currents and voltages must be removed from post-fault quantities with care since the changes in voltage and current represent the source impedance, and not the impedance of the faulted line section. It should be noted that this effect requires different treatment for three-phase faults compared to unbalanced faults. In the case of unbalanced faults, the negative and zero sequence measurements will cancel out the source impedance, and only the faulted line impedance will remain. These concepts have been used in our Symmetrical Component Distance Relay (SCDR) design [1]. The three-phase fault calculation is done with actual voltages and currents; not with changes in these quantities.

Regarding the use of load current compensation during sequential switching operations and power swings, we propose the use of compensation up to the time when the first switching operation takes place. Allowing for circuit breaker operating times, this should be well beyond the operating period of zone-1 relays at all terminals. The transient-monitor function used in the SCDR provides the necessary delimiters within which compensation should be used. Compensation with the pre-fault load current may not be used when slower operating times - such as those related to backup operations or power swings - are called for. Even without the swing conditions, sampling clocks are likely to produce sufficient apparent phase differences so that the old values of load current are invalid. In these longer periods one would fall back on the uncompensated mode of operation.

Mr. Volzka's observation regarding the need to consider the data and communication requirements is, of course, pertinent to the ultimate goal of implementing the adaptive concept in a total system. In the DOE report, reference [1] of the paper, we examine this issue in great detail. Within the space limitations of the subject paper, however, we only alluded to these requirements. The closing paragraphs of each section of the Adaptive Relaying Opportunities in the paper discuss the type of input required, its location and the speed of transmission involved.

We agree with the discusser's appeal to improve the participation of system control computers in this adaptive concept. References [2] and [3] below are our attempts to do that and are the result of our studies related to this paper. Clearly much more must be done and we welcome Mr. Volzka's call for more active participation in this area. Hopefully, the appropriate IEEE technical committees will respond.

Similarly, the suggestion to take a quantum jump in transformer protection technology beyond replicating existing relays is exactly our intent. Again, space limitations do not permit in-depth design considerations, nor was that our studies' primary purpose. We hope, however, that we have conveyed the same thoughts às the discusser, i.e., the reduction of "error current" in the operating winding by constantly adjusting for power transformer taps and CT mismatch. The use of flux and voltage is also addressed in both this paper and in reference [11] of the paper.

We appreciate the points raised by Messrs. Venkata, Damborg and Jampala. They extend the concepts covered in our paper into areas that will have to be addressed by all serious investigators. Our specific responses to their questions are as follows.

1) The addition of "real-time" is useful and in keeping with our own interpretation. It may, however, be premature to establish more than a working definition at this time. A more rigorous definition will undoubtedly emerge as more studies illustrate additional applications.

2) We agree that the adaptive protection philosophy will evolve gradually. This, in fact, is also true or the entire digital technology and is the traditional and conservative approach. It is unlikely that there will either be the opportunity or the desire to install this equipment in any other way.

Digital equipment will be added to existing equipment to gain experience and confidence.

3) Our emphasis on multi-terminal lines is due to the fact that this configuration requires more compromise in settings than the 2-terminal line and, hence, offers greater benefits. In fact, 2-terminal lines are very well protected with existing relays. There are ample margins between trip and no-trip situations and less opportunity to demonstrate the advantages of the adaptive concept.

4) The possibility that computer relays will eliminate separate backup protective devices has· intrigued investigators from the beginning. We believe that the backup function will always be present; the exact form, however, is not clear at this time. Data sharing, diagostic and communication capabilities presents attractive possibilities for local or remote backup. The specifics will depend on the user's experience and philosophy. The elimination of backup relays will be as evolutionary as the introduction of computer relays.

References

[1] A. G. Phadke, M. Ibrahim, T. Hblika, "Fundamental Basis for Distance Relaying with Symmetrical Components", IEEE Transactions on Power Apparatus and Systems March/April, 1977 pp 635-676.

[2] J. S. Thorp, A. G. Phadke, S. H. Horowitz, M. M. Becovic, "Some Applications of Phasor Measurements to Adaptive Protection", PICA 87, Montreal, Canada.

[3] A. G. Phadke, J. S. Thorp, S. H. Horowitz, "Impact of Adaptive Protection on Power System Control" Power System Computer Conference, Lisbon, Portugal, September, 1987.

Manuscript received August 31, 1987.

SOME APPLICATIONS OF PHASOR MEASUREMENTS TO ADAPTIVE PROTECTION*

J. S. Thorp
Senior Member, IEEE
Cornell University
Ithaca, NY

A. G. Phadke
Fellow, IEEE
Virginia Tech
Blacksburg, VA

S. H. Horowitz
Fellow, IEEE
A. E. P. Serv. Corp.
Columbus, OH

M. M. Begovic
Virginia Tech
Blacksburg, VA

Abstract

The application of microprocessor based phasor measurements to new adaptive protection schemes is presented. The concepts of digital adaptive protection in which relay characteristics are modified in response to external signals and conditions in the system are examined. The importance of real-time phasor measurements in adaptive protection systems is illustrated by two examples. The first involves obtaining synchronizing information about phasor measurements to be used in a new type of current differential protection for a multiterminal line. By estimating the reference angles for the phasor measurements the effect of synchronized sampling can be obtained without the expense of synchronizing equipment.

The second use of phasor measurements in adaptive protection is the adaptive setting of out-of-step blocking and tripping. Traditional problems of setting out-of-step blocking and tripping are caused because the settings are, of necessity, compromises based on off-line studies. By reacting to real-time phasor measurements from selected buses in interconnection it is shown that improved out-of-step blocking and tripping is possible.

INTRODUCTION

Although many existing protection schemes are adaptive in a limited sense, the introduction of microprocessor based protection and improved communication systems offer the possibility of extended adaptive protection techniques. An adaptive relay can be thought of as a protection device whose characteristics or function is altered on the basis of external information. Examples can be given [1] where that information is available in the substation or where that information is obtained from adjacent substations or even from the overall system. The focus of this paper is on the use of microprocessor based phasor measurements as an information source for adaptive relaying.

---

*Research sponsored by the Office of Energy Storage and Distribution, Electrical Energy Systems Program, U. S. Department of Energy, under contract No. DE-AC05-84OR21400 with Martin Marietta Energy Systems, Inc.

This paper was sponsored by the IEEE Power Engineering Society for presentation at the IEEE Power Industry Computer Application Conference, Montreal, Canada, May 18–21, 1987. Manuscript was published in the 1987 PICA Conference Record.

The use of real-time phasor measurements in static state estimation has been presented [2], [3]. The phasor measurements are a by-product of the Symmetrical Component Distance Relay (SCDR) [4], [5] which makes accurate real-time measurement of symmetrical components of voltages and currents for a transmission line. The symmetrical components are by definition fundamental frequency quantities, and the SCDR calculates these from sampled data in the presence of harmonic distortions of the current and voltage waveforms. Thus the front end of the SCDR acts as a highly accurate symmetrical component filter.

Some experiments concerned with common-reference phase angle measurements across transmission lines have been reported in the literature [6]–[10]. These schemes have utilized radio receivers tuned to standard time broadcast systems such as the WWV transmission or to satellite transmissions. In most cases, the zero-crossing instant of a voltage wave was used as a measure of its phase angle with respect to the reference time pulse. In at least one case the intended use of the measurements was that of monitoring system response to actual disturbances [10]. In contrast to the zero crossing schemes the SCDR uses waveform information contained over a complete period (or multiples of a period).

In order to obtain phasor measurements from a number of locations on a common reference using the SCDR it is necessary to synchronize the sampling clocks at the various locations. Sampling clock synchronization to within a few microseconds seems to be possible. In addition to the satellite signals used in the zero crossing experiments there are currently available LORAN-C receivers with accuracies of one microsecond. Another alternative would be to use a dedicated synchronizing medium - such as microwave or fiber-optic links between substations. Clearly such channels cannot be installed economically for the purpose of clock synchronization, but once in place for other reasons, they become an attractive option for this purpose. Such a system has been reported as being in use in a Japanese installation [11]. If the use of fiber-optics over ground wires becomes common in transmission systems in the future, this technique of clock synchronization shows much promise. Practical experience with these and other techniques for synchronization of clocks throughout the power system are reported in [12]. The synchronization reported in [12] was primarily used to time-tag data and accuracies of 1 ms were deemed acceptable.

While 1 ms accuracies are adequate for time-tagging data, 1 ms at 60 Hz fundamental frequency corresponds to 21.6° angular error in a phasor. The various adaptive relay applications to be discussed will require more accurately synchronized phasors to be effective. As an inexpensive alternate to accurate synchronization, it is possible in some

Reprinted from *IEEE Trans. Power App. Syst.*, vol. 3, no. 2, pp. 791–798, May 1988.

77

situations to use network and measurement redundancy to improve low quality synchronization. In the next section a technique will be presented to synchronize phasors which are originally measured with millisecond timing errors. The technique is applied to a 3-phase four-terminal line. A number of possible applications of phasor measurements in adaptive protection are discussed which are related to detection of system instabilities.

PHASOR SYNCHRONIZATION THROUGH NETWORK REDUNDANCY

To obtain the necessary redundancy to estimate the timing errors a three phase (unbalanced) representation of the system will be used. Since all samples within the substation are synchronized this means that, at a minimum, six measurements (the three phase voltages and three phase currents) will have a common synchronizing error. If n buses are involved in the measurement set then the true phasor voltages and phasor currents are related by a $3n \times 3n$ complex impedance matrix with pq^{th} entry Z_{pq}. If we concentrate only on the timing errors associated with the n sampling clocks and take bus 1 as a reference then the synchronizing problem can be posed as that of estimating the $(n-1)$ angles δ_j such that

$$
\begin{vmatrix} V_1 \\ V_2 \\ V_3 \\ V_4 e^{j\delta_2} \\ V_5 e^{j\delta_2} \\ V_6 e^{j\delta_2} \\ V_7 e^{j\delta_3} \\ \cdot \\ \cdot \\ \cdot \\ V_{3n} e^{j\delta_n} \end{vmatrix} = \underline{Z} \begin{vmatrix} I_1 \\ I_2 \\ I_3 \\ I_4 e^{j\delta_2} \\ I_5 e^{j\delta_2} \\ I_6 e^{j\delta_2} \\ I_7 e^{j\delta_3} \\ \cdot \\ \cdot \\ \cdot \\ I_{3n} e^{j\delta_n} \end{vmatrix} \tag{1}
$$

where V_l and I_l are the measured phasors. Equation (1) represents $3n$ complex equations in which the only unknowns are the synchronizing errors. Equation (1) may be rewritten as

$$
\begin{vmatrix} 1 \\ 1 \\ 1 \\ e^{j\delta_2} \\ e^{j\delta_2} \\ e^{j\delta_2} \\ e^{j\delta_3} \\ \cdot \\ \cdot \\ \cdot \\ e^{j\delta_n} \end{vmatrix} = M \begin{vmatrix} 1 \\ 1 \\ 1 \\ e^{j\delta_2} \\ e^{j\delta_2} \\ e^{j\delta_2} \\ e^{j\delta_3} \\ \cdot \\ \cdot \\ \cdot \\ e^{j\delta_n} \end{vmatrix} \tag{2}
$$

where $M_{pq} = Z_{pq} I_q / V_p$. If the $3n \times n$ matrix S is defined as

$$
S = \begin{vmatrix} 1 & 0 & 0 & \ldots & 0 \\ 1 & 0 & 0 & & 0 \\ 1 & 0 & 0 & & 0 \\ 0 & 1 & 0 & & \cdot \\ 0 & 1 & 0 & & \cdot \\ 0 & 1 & 0 & & \cdot \\ & & & & 1 \\ & & & & 1 \\ & & & & 1 \end{vmatrix} \tag{3}
$$

and x defined as

$$
x = \begin{vmatrix} e^{j\delta_2} \\ e^{j\delta_3} \\ \cdot \\ \cdot \\ \cdot \\ e^{j\delta_n} \end{vmatrix} \tag{4}
$$

then (2) can be written

$$
S \begin{vmatrix} 1 \\ x \end{vmatrix} = M S \begin{vmatrix} 1 \\ x \end{vmatrix} \tag{5}
$$

or

$$
(M - I)S \begin{vmatrix} 1 \\ x \end{vmatrix} = \underline{0} \tag{6}
$$

If we partition the $3n \times n$ matrix

$$
(M - I)S = [-b \mid \dot{A}]
$$

then the estimation of the $(n-1)\delta_s$ is equivalent to the "solution" of the overdefined equations

$$
A x = b \tag{7}
$$

where A is $3n \times (n-1)$ and represents the last $(n-1)$ columns of the matrix $(M - I)S$.

If the only errors are in synchronizing then there are $(n-1)\delta s$ which solve Equation (1) and hence there is an x which solves (7). Since there are almost certainly other measurement errors involved, a least squares solution of (7) is used. The QR algorithm can be used for the solution of a set of complex overdefined equations [13]. The QR algorithm factors the matrix A into the product of an orthogonal matrix Q and an upper triangular matrix R and provides a numerically attractive solution technique for (7). The solution minimizes

$$
(x^+ A^+ - b^+)(Ax - b)
$$

where $^+$ denotes conjugate transpose. Symbolically

$$
\hat{x} = (A^+ A)^{-1} A^+ b \tag{8}
$$

It should be emphasized that (8) is not used to compute $\hat{x}$ but rather the QR algorithm. In the presense of other errors appearing linearly in (7) Equation (8) is a least squares estimate of the components of x.

To use this technique for synchronizing the phasor measurements it is only necessary to crudely synchronize the sampling so that full period ambiguities in the angles are resolved. The measured voltage and current phasors and the known impedance matrix are used to compute the matrix M, the matrix A and the vector b are formed and $\hat{x}$ is computed. The δs are taken as the angles of the complex vector $\hat{x}$. The voltage at bus i, for example, is then given by

$$V_{3l-k}\, e^{+j\delta_l} \quad k = 0, 1, 2 \tag{9}$$

The voltages given by (9) are then on a common referenced defined by bus 1.

The System Model

The model on which the algorithm was tested is a four-terminal EHV network where the synchronizing was desirable for current differential protection. The single phase diagram of the network is shown in Figure 1.

An unbalanced 3-phase representation of the system was used to provide an accurate system model. Each of the five lines was taken as a 100 mile long, untransposed, 800 kV line with a physical description shown in Figure 2.

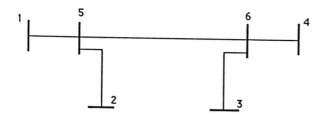

FIGURE 1. Four Terminal Line Used for Simulation. Each of the five lines is 100 miles of untransposed 800 kV line

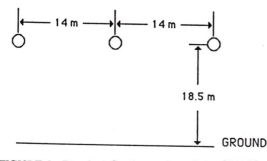

FIGURE 2. Physical Configuration of the 800kV lines in Figure 1

The data for the line was taken from [14] where

$$z = \begin{vmatrix} .1165 & .0975 & .0943 \\ .0975 & .1176 & .0975 \\ .0943 & .0975 & .1165 \end{vmatrix} + j \begin{vmatrix} .8095 & .2961 & .2213 \\ .2961 & .7994 & .2961 \\ .2213 & .2961 & .8095 \end{vmatrix}$$

$$y = \begin{vmatrix} 0. & 0. & 0. \\ 0. & 0. & 0. \\ 0. & 0. & 0. \end{vmatrix} + j \begin{vmatrix} 7.125 & -1.133 & -.0284 \\ -1.133 & 7.309 & -1.133 \\ -.0284 & -1.133 & 7.125 \end{vmatrix}$$

where z is the series impedance in ohms/mile and y is the shunt admittance in microsiemens/mile. The 12 x 12 admittance matrix for the 3-phase description of the network shown in Figure 1 was formed and inverted to form the 3-phase impedance matrix. Phasor voltage and currents (on a 3-phase basis) were computed for the network corresponding to the power flows given in Table 1. There were no injections at buses 5 and 6.

bus	Re $\{S_i\}$Mw	Im $\{S_i\}$ MVARS
1	2206	- 307
2	6	- 458
3	- 2172	- 191
4	5	- 432

Table 1: Injected Powers

The six complex phasors at bus two were then all rotated by 21.6° corresponding to a 1 ms timing error. The phasors at bus three were rotated by 32.4° (1.5 ms) and those at bus four by 43.2° (2 ms).

Error Model

The rotated complex phasors were used to create instantaneous voltage and current signals to which random measurement errors were added. The measured voltages and currents are assumed to be sampled 12 times per period and are of the form

$$v_k = \sqrt{2}\, V\, \sin(2\pi k / 12 + \theta) + \epsilon_{kv} \tag{10}$$

$$i_k = \sqrt{2}\, I\, \sin(2\pi k / 12 + \phi) + \epsilon_{kI} \tag{11}$$

where ϵ_{kv} and ϵ_{kI} are zero mean gaussian random variables which are independent from sample point to sample point and independent from each other. The variances of the voltage and current errors are taken to be

$$\sigma_{vk} = 0.0017 f_{mv} + 0.005 |v_k| \tag{12}$$

$$\sigma_{Ik} = 0.0017 f_{mI} + 0.010 |i_k| \tag{13}$$

where f_{mv} and f_{mI} are the full scale range of the measuring instrument and $|v_k|$ and $|i_k|$ are the magnitude of the instantaneous measurement. The real and imaginary parts of the computed phasors are given by

$$v_R = (\sqrt{2}/12) \sum_{k=1}^{12} v_k \sin(2\pi k / 12) \tag{14}$$

$$v_I = (\sqrt{2}/12) \sum_{k=1}^{12} v_k \cos(2\pi k / 12) \tag{15}$$

the corresponding variances in the real and imaginary parts can be computed as

$$\sigma_{v_R}^2 = (\sqrt{2}/12)^2 \sum_{k=1}^{12} \sin^2(2\pi k / 12)\sigma_{v_k}^2 \tag{16}$$

$$\sigma_{v_I}^2 = (\sqrt{2}/12)^2 \sum_{k=1}^{12} \cos^2(2\pi k / 12)\sigma_{v_k}^2 \tag{17}$$

Equations (16) and (17) produce variances which depend on the angle of the phasor in an approximate double angle form as shown in Figure 3.

Figure 3 is drawn for a one cycle calculation of a voltage with f_{mv} set equal to 120% of the nominal voltage. In addition, the estimates of the real and imaginary parts are correlated. Because of these complexities a Monte Carlo simulation was performed to obtain the performance of the synchronizing technique.

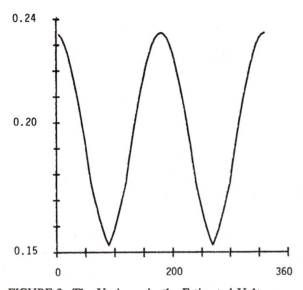

FIGURE 3. The Variance in the Estimated Voltage
Phasor as a Function of the Phasor
Angle for a One Cycle Window and
$f_{mv} = 1.2\ pu$

Simulated Results

At each trial the corrupted measurements of the instantaneous voltages and currents at all three phases at buses 1-4 of Figure 1 were formed at a sampling rate of 12 times per cycle. The phasors were formed using (14) and (15) with appropriate modifications for longer data windows. The estimated phasors were used to compute the A and b of (7) and x was computed using the complex QR algorithm. In addition to computing an experimental variance the estimated probability that the error in each of three δ_i was less than 0.1° or 0.2° was computed. The results for 100 Monte Carlo trials are shown in Table 2 for windows of 1 cycle through 5 cycles.

cycles	bus	σ in degrees	# < 0.1°	# < 0.2°
1	2	0.0917	72	98
	3	0.128	59	85
	4	0.0917	71	97
2	2	0.0675	88	100
	3	0.0922	68	100
	4	0.0669	88	99
3	2	0.0607	88	99
	3	0.0807	78	99
	4	0.0657	90	99
4	2	0.0535	91	100
	3	0.0649	97	100
	4	0.0499	95	100
5	2	0.0454	96	100
	3	0.0543	94	100
	4	0.0478	96	100

Table 2: Simulation Results

The larger errors at bus 3 are due to the error model in (12) and (13) having a component proportional to the signal size. The currents at bus 3 are considerably larger than at buses 2 and 4 so the errors are larger and the angular estimate is poorer. The error model in equations (12) and (13) also has a fixed component proportional to the full scale range of the measuring instrument. To examine the effect of this term when currents were small the algorithm was tested with one of the terminals open. In addition the algorithm was tested with comparable power flow in all the lines. In both cases the results are similar to Table 2.

DETECTION OF INSTABILITY

Detection of instability in power systems is not a relaying function, *per se*. However, many relaying functions (load shedding, for example) depend upon the premise that given the current state of a power system, it will remain stable following a transient oscillation. Conversely, the out-of-step tripping function is based upon the premise that the post-disturbance system will be unstable. If a developing transient could be judged to be either stable or unstable, then the operation of such relays could be made to adapt to this judgement. Thus instability detection becomes a precursor of the adaptive relaying process.

In a power system consisting of two synchronous machines and a connecting network over which synchronizing power can flow, the problem of instability detection can be solved in real-time. The equal-area criterion is applicable in this case, and if the machine rotor angles and speeds can be measured in real-time a prediction algorithm can be developed for the detection of instability. This algorithm is presented in the last section. When ac power is imported over relatively long distances, so that both the sending system and receiving system consist of groups of coherent generators, one of the modes of oscillations of such a power system is that of a two-machine system. Similarly, when a single machine or a single plant is connected to a strong system, the modes of oscillation of the single machine or plant are similar to those of a simple two machine system.

An example of such a two-machine system is illustrated by the history of the Florida-Georgia interface [15]-[20]. During the past ten years a series of steps have been taken to deal with instabilities caused by sudden loss of generation in peninsular Florida. A large number of system separations in the late '70's were caused by transient swings with electrical centers in the northern Florida-southern Georgia area. Advances in load shedding techniques and improvements in machine response along with construction of new transmission facilities have contributed to improving the situation. In addition out-of-step relays were installed to effect a controlled separation. In each area extensive use was made of off-line simulation programs and the outputs of Continuous Monitoring Fault Recorders. Using this information a set of tie line relays were designed that are capable of detecting power swings and determining those swings which are severe enough to force a split. It is recognized that the problem is an evolving one that will change as more load and generation is added and as even more power is imported.

Recently the Florida-Georgia interface has been investigated using the transient energy function method [21]. In that study a three-machine equivalent was derived using the E.P.R.I. "coherency-Based Dynamic Equivalencing" program [22]. In that three-machine equivalent, one machine was an actual machine (the machine to be dropped) and the other two were large equivalents. In the equivalencing the four major tie-lines between Florida and Georgia were retained.

If we examine the system after the generator has been dropped the system can be represented as in Figure 4 where the two generators are the equivalents and the only 'real" buses are those at the ends of the tie lines.

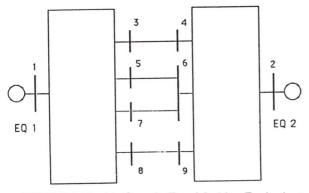

FIGURE 4. Florida-Georgia Two Machine Equivalent with retained tie-lines. The buses 3-9 are retained in the equivalent.

If synchronized real-time phasor measurements at the tie-line buses are made then the internal voltages of the equivalent machines can be determined. From the reduced Y_{bus} matrix the generator voltages and currents, V_G, I_G are related to the measured tie-line bus voltages, V_M, by

$$\begin{vmatrix} I_G \\ 0 \end{vmatrix} = \begin{vmatrix} Y_{11} & Y_{12} \\ Y_{21} & Y_{22} \end{vmatrix} \begin{vmatrix} V_G \\ V_M \end{vmatrix} \quad (18)$$

where, for example, if all 7 tie-line bus voltages are measured Y_{22} is a 7 x 7 matrix. The last partition in (18) yields

$$Y_{21}V_G + Y_{22}V_M = 0 \quad (19)$$

Equation (19) can then be solved for the generator voltages (again in the least squared sense if there are more measurements than generators).

Since line relays will produce phasors for line currents they also can be included in the measurement set. The maximum number of phasor measurements is then 15 (7 voltages and 8 currents) which implies that with a full measurement set the measurements may not need to be synchronized. If one of the tie-line buses is chosen as a reference then the problem of estimating both the six synchronizing errors and the two machine voltages given the 15 measurements would reduce to the solution of 15 overdefined complex linear equations in 8 unknowns. The procedure of the preceding section could be used on a single phase basis.

Alternately if the measurements are synchronized then all 15 measurements are unnecessary. In this case one issue is to determine the location of a more limited measurement set.

It is clear that detection of instability is a task to be performed at a central location. Consequently, all data must be transmitted from substations where measurements are made to a chosen central location. The measurements are phasors and either the phasors, or their contributions to the Thevenin phasor voltages of the equivalent generators must be sent to the central computer. In any case, these are digital data, and input needs of this procedure are confined to digital data transmission from remote locations.

In terms of output considerations, all the outputs will consist of digital control signals for relays at various substations. These would be blocking signals for some relays and tripping signals for others. Also, load shedding relays would be controlled from such a system. Any other stability-enhancement controls — such as fast valving, dynamic brakes and HVDC systems — that can be used for controlling developing instability would also receive signals from this function.

The central function of instability detection is based heavily on the ability to communicate with substations. Consider first the data rate which must be accommodated at the central computer interface. The fastest stability swings may have a frequency of approximately 2 Hz, it may therefore be necessary to have as many as 10 measurements of machine states in a quarter cycle of this oscillation to produce good instability prediction. This corresponds roughly to a snap-shot of system measurements every cycle of the power frequency (16 milliseconds). At an interval of 16 milliseconds either the measurements of voltage and current phasors, or the computed contributions to the equivalent generator internal voltages (4 quantities) must be communicated to the center. If the center is getting data from 10 key stations on the system, this corresponds to a data rate of 10 x 4 x 16 bits in 16 milliseconds. Thus, allowing for data acquisition and computation by the central computer, a data handling capability of 10 times this rate — or about 400 kb/second is indicated. If the power system oscillation frequency is lower than the assumed 2 Hz, this rate would go down proportionately. If more stations contribute data to the center, higher rate of data handling capability would be needed at the center. The communication channel capacity needed between the central computer and each of the remote transmitting stations is of the order of 40 kb/seconds. (Assuming 10 transmitting stations). This is also the data rate at the computer interfaces in the transmitting stations.

The data flow in the reverse direction for controlling the relays would require much lower band-width. If a synchronizing clock pulse is transmitted to each station, it would require the highest band width channel. In fact, in all likelihood, the synchronizing pulse would require a dedicated fiber in a multi-fiber link.

As the instability detector assumes a system configuration to remain unchanging during the prediction interval, it is imperative that all other relays be inhibited from tripping on out-of-step detection. Thus the adaptive instability detection device can not function in conjunction with devices that are non-adaptive. Of course, this restriction is relevant to critical facilities only: any line that is not significant to the flow of synchronizing power for the given disturbance may be left nonadaptive. It is thus necessary to determine key facilities as a preliminary step, and then include all of their protection and control in the adaptive system. Similar considerations apply to the load shedding relays also.

EQUAL AREA ALGORITHM

The phasor measurements of the previous section can be used to estimate the internal machine voltages for the equivalent two-machine system shown in Figure 5.

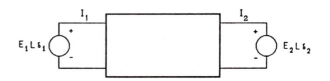

FIGURE 5. Classical Two Machine System used in the Equal Area Criterion

For simplicity in presentation the network will be assumed to be a simple series reactance. The algorithm is only somewhat more complicated for an arbitrary (known) two port connecting the two generators. With

$$M_1 \ddot{\delta}_1 = P_{m1} - P_{e1} \qquad (20)$$

$$M_2 \ddot{\delta}_2 = P_{m2} - P_{e2} \qquad (21)$$

and

$$\delta_1 - \delta_2 = \delta \qquad (22)$$

then

$$\ddot{\delta} = c - k\sin\delta = c - P_e^1 \qquad (23)$$

A loss of generation in one area can be modeled as a change of operating point in Figure 6 from the value c_0 to the value c_1.

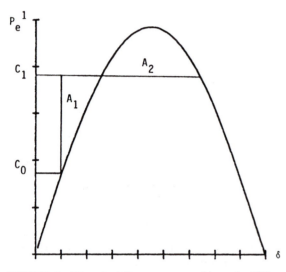

FIGURE 6. Electrical Power versus the angle difference between the two machines of Figure 5. The value c_0 represents the predisturbance operating point while c_1 represents the new equilibrium.

To predict instability it is necessary to use the estimated values of the machine voltage and current phasors to estimate the new value of c_1. Once c_1 is known the areas A_1 and A_2 can be computed and stability or instability can be determined. Since the phasor measurements can be converted to estimates of δ and P_e^1 and since these measurements can be made on a once a cycle basis, the problem can be posed as that of given estimates (or measurements) of $\delta(t)$ and $P_e^1(t)$ in the form

$$\delta_k = \delta(k\Delta) \quad k = 1, 2, .., n \qquad (24)$$

$$P_k = P_e^1(k\Delta) \qquad (25)$$

where $\Delta = 1/60$, estimate c_1 where

$$\ddot{\delta} = c_1 - P_e^1(t) \qquad (26)$$

Using a first order hold equivalent (assuming $P_e^1(t)$ is piecewise linear between sample instants) we can write

$$\delta_k = \delta(0) + \frac{\Delta^2}{2}[c_1 k^2 - \sum_{l=0}^{k-} 1\, P_l (2k - 2l - 1) \qquad (27)$$

$$- \sum_{l=1}^{k-} 1\, P_l + (k-1)P_0 - (P_k - P_0)/3]$$

or with

$$v = \begin{vmatrix} 1 \\ 4 \\ 9 \\ . \\ . \\ . \\ n^2 \end{vmatrix}, P = \begin{vmatrix} P_0 \\ P_1 \\ . \\ . \\ . \\ P_{n-1} \end{vmatrix}, \bar{P} = \begin{vmatrix} P_1 - P_0 \\ P_2 - P_0 \\ . \\ . \\ . \\ P_n - P_0 \end{vmatrix} \qquad (28)$$

$$\delta = \begin{vmatrix} \delta_1 \\ \delta_2 \\ . \\ . \\ . \\ \delta_n \end{vmatrix}, 1 = \begin{vmatrix} 1 \\ 1 \\ . \\ . \\ . \\ 1 \end{vmatrix} \text{ and} \qquad (29)$$

$$M = \begin{vmatrix} 1 & 0 & 0 & 0 & \ldots & 0 \\ 2 & 2 & 0 & 0 & \ldots & 0 \\ 3 & 4 & 2 & 0 & \ldots & 0 \\ 4 & 6 & 4 & 2 & \ldots & 0 \\ . \\ . \\ n & & & 6 & 4 & 2 \end{vmatrix} \qquad (30)$$

a sequence of n measurements can be related by

$$\delta = \delta(0)1 + \frac{\Delta^2}{2}[Vc_1 - M\,P - \frac{1}{3}\bar{P}] \qquad (31)$$

Equation (31) with the "measured" or estimated angles and powers represents n overdefined equations in the unknown $\hat{c}_1$. The estimated value of $\hat{c}_1$ is given by

$$\hat{c}_1 = \frac{2}{\Delta^2(v^T v)}[v^T \delta - v^T 1\delta(0)] \qquad (32)$$

$$+ \frac{1}{(v^T v)}[v^T M\,P + \frac{1}{3}v^T\,\bar{P}]$$

Equation (32) has some interesting properties in terms of real time implementation. Each "measured" quantity δ_k, P_k or $\delta(0)$ is multiplied by an integer and divided by another integer. The use of the "measured" electrical power in predicting instability has also been suggested in [23]. The difference between [23] and this application is that the measurements in [23] were on an actual machine while here the equivalent machine bus does not physically exist so the measurements $P_e^1(k\Delta)$ are themselves estimates.

The algorithm was tested on a simple example with a relatively fast swing frequency. The system

$$\ddot{\delta} = c - 45.48\sin\delta$$

has a swing frequency of approximately .93 Hz [24]. The predisturbance c_0 is 18.879 with a $\delta(0)$ of .4 radians. If c_1 is 37.699 then the new equilibrium value is $\delta = .8908$ radians and the swings are stable. The first swing is shown in Figure 7. The algorithm of Equation (32) was used on data with additive errors in P_e^1 corresponding to zero-mean, independent gaussian errors with a σ of 2% of the measured power and similar errors in the angles with a σ of 1°. The errors are on the conservative side since the "measured" angles and powers are themselves estimates made from phasor measurements. The experimental standard deviations (from 50 Monte Carlo trials) in the esti-

82

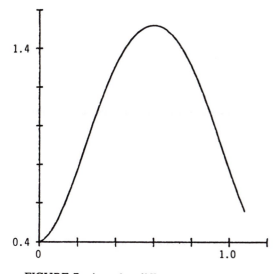

FIGURE 7. Angular difference versus time for the
system $\delta = c - 45.48 \sin\delta$ with $c_0 = 18.879$,
$\delta(0) = .4$ radians, $c_1 = 37.699$

mate of c_1 as a function of the time is shown in Figure 8.
It can be seen that in .25 sec (15 samples at one sample per
60 Hz period) the error is sufficiently small to accurately
predict that the swing will be stable. The actual σ after
15 samples is .245 or less than 1% of 37.699. It should be
observed that the swing does not reach the maximum
angle until about $t = .6$ sec.

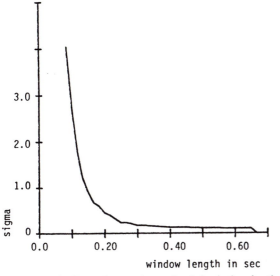

FIGURE 8. Experimental standard deviation in the
estimate of c_1 from Figure 7 versus
the length of the data window. Results are
from 50 Monte Carlo trials.

CONCLUSION

The issue of real-time phasor measurements of bus
voltages and line currents as a source of information for
adaptive relays has been proposed. The synchronization of
these phasor measurements through the use of network
redundancy is possible in a variety of situations. If the
sampling clocks are roughly synchronized (to a few milliseconds) then the synchronizing errors can be estimated.
For a four-terminal 800 kV line the synchronizing error
can be reduced to as little as 0.1 degrees by using an
unbalanced 3-phase representation of the line.

Synchronized phasor measurements can be used to
predict instability in situations where the power swings
are one coherent group of machines against another. The
instability prediction can be used to adapt load shedding
and/or out of step relays. The technique requires that
phasor measurements at real tie-line buses be used to
estimate phasor quantities at fictitious buses of equivalent
machines in a two machine system. The technique
requires communication of the phasor measurements at
roughly a 60 Hz rate to a central location and communication from the central location back to the appropriate
relays.

Investigation is underway on the extension of the
technique to multi-machine situations. The difficulty, of
course, is that the equal area criterion has no simple extension to the multi-machine case. Techniques based on the
potential energy boundary surface are being considered.

ACKNOWLEDGEMENTS

This study was conducted for the Power Systems
Technology Program of the Energy Division at Oak Ridge
National Laboratory under subcontract 86X-22005C. D. T.
Rizy was the project engineer for the project.

REFERENCES

[1] A. G. Phadke, J. S. Thorp, S. H. Horowitz, "Study of
Adaptive Transmission system Protection and Con-
trol," ORNL/SUB/85-2205C, Prepared by Virginia
Polytechnic Institute and State University for Oak
Ridge National Laboratory (to be published).

[2] J. S. Thorp, A. G. Phadke, K. J. Karimi, "Real-time
Voltage Phasor Measurements for Static State Estimation," T-PAS, Nov. 85, 3098-3107.

[3] A. G. Phadke, J. S. Thorp, K. J. Karimi, "State Estimation with Phasor Measurements," *IEEE Trans. on PS*,
vol. PWRS-1, no. 1., pp. 233-251, February 1986.

[4] A. G. Phadke, M. Ibrahim, T. Hilbka, "Fundamental
Basis for Distance Relaying with Symmetrical Components", *IEEE Trans. on PAS*, vol. PAS-96, no. 2,
pp. 635-646, March/April 1977.

[5] A. G. Phadke, T. Hilbka, M. Adamiak, M. Ibrahim, J.
S. Thorp, "A Micro-Computer Based Ultra-High-Speed
Distance Relay - Field Tests", *IEEE Trans. on PAS*,
vol. PAS-100, no. 4, pp. 2926-2936, April 1981.

[6] A. G. Phadke, J. S. Thorp, M. G. Adamiak, "A New
Measurement Technique for Tracking Voltage Phasors,
Local System Frequency and Rate of Change of Frequency", *IEEE Trans. on PAS*, vol. PAS-102, no. 5,
pp. 1026-1038, May 1983.

[7] P. Bonanomi, "Phase Angle Measurements with Synchronized Clocks - Principles and Applications",
IEEE Trans. on PAS, vol. PAS-100, no. 12, pp.
5036-5043, Dec. 1981.

[8] G. Misout, J. Beland, G. Bedard, "Dynamic Measurement of the Absolute Voltage Angle on Long
Transmission Lines", *IEEE Trans. on PAS*, vol. PAS-
100, No. 11, pp 4428-4434, Nov. 1981.

[9] G. Misout, J. Beland, G. Bedard, P. Bussieres, "Study of
Time Dissemination Methods Used on an Electric
Power System with Particular Reference to Hydro-
Quebec", *IEEE Trans. on PAS*, vol. PAS-103, no. 4,
pp 861-868, April 1984.

[10] R. O. Burnett, Jr., "Field Experience with Absolute Time synchronism between Remotely Located Fault Recorders and Sequence of Event Recorders", *IEEE Trans. on PAS*, vol. PAS-103, no. 7, pp. 1739-1742, July 1984.

[11] C. P. Dalpiaz, D. J. Hansen, "High Rate Telemetry of System Voltage Phase Angle and other Stability Related Quantities", submitted to PICA 1985.

[12] T. Takagi, Y. Yamakosi, M. Yamaura, R. Kondow, T. Matsushima, M. Masui, "Digital Differential Relaying System for Transmission Line Primary Protection Using Traveling Wave Theory - Its Theory and Field Experience", 1979 IEEE Paper No. A79096-9.

[13] J. J. Dongarra, C. B. Moler, J. R. Bunch, G. W. Steward, "LINPAC Users' Guide", SIAM, Philadelphia, 1979.

[14] "Transmission Line Reference Book/345 kV and above" Second Edition, EPRI, Palo Alto, 1982.

[15] A. N. Darlington, "Response of Underfrequency Relays on the Peninsular Florida Electric System for Loss of Generation", presented at the 1978 Georgia Tech Relay Conference, May 1978.

[16] P. B. Winston, R. O. Burnett, Jr., "Impact of Disturbances on Weak Intercompany Transmission Ties", presented at the 1980 Georgia Tech Relay Conference, May 1980.

[17] W. E. Lester, "Florida Power Corporation Out-of-Step Relay Application" presented at the 1980 Gerogia Tech Relay Conference, May 1980.

[18] T. A. Pruitt, R. O. Burnett, P. B. Winston, "Response of Plant Hatch Nuclear Power Units to Grid Disturbances", presented at the 1981 Georgia Techn Relay Conference, May 1981.

[19] J. Benson, "Development of the Present Peninsular Florida Underfrequency Load Shedding Scheme", presented at the 1983 Georgia Tech Relay Conference, May 1983.

[20] R. W. Ohnesorge, T. A. Pruitt, "Operational Aspects of the Southern-Florida Interface - Past, Present and Future", presented at the 1984 Georgia Tech Relay Conference, May 1984.

[21] A. A. Fouad et al., "Investigation of Loss of Generation Disturbances in the Florida Power and Light Company Network by the Transient Energy Function Method", IEEE Winter Power Meeting, paper 86WM067-3, February 1986.

[22] J. C. Giri, "Coherency Reduction in the EPRI Stability Program", *IEEE Trans. on PAS*, vol. PAS-102, no. 5, pp. 1285-1293, May 1983.

[23] Y. Ohura et al., "Development of a Generator Tripping System for Transient Stability Augmentation Based on the Energy Function Method", *IEEE Winter Power Meeting* paper 86WM117-6, February 1986.

[24] A. R. Bergen, *"Power System Analysis"* Prentice-Hall, Englewood Cliffs, New Jersey, 1986.

I. General Discussion

THE protection problems associated with transmission line protection has always been the keystone of power system protection. Although the possible number of faults is limited to the ten types of short circuits, e.g., phase-to-phase, phase-to-ground, double phase-to-ground, and three-phase faults, the variety of protective relays and their associated equipment is virtually unlimited. This is a reflection of the scope of the problem, i.e., the extent of exposure in miles of transmission lines, weather conditions, different system configurations, and the many compromises between dependability and security that must be accommodated. As a result, although the basic protection requirements do not differ greatly from other protection situations, transmission line protection has occupied the attention of relay engineers to a far greater extent than other classes of equipment.

Since Volume I was published in 1980, there have been changes in the relaying technology; changes that affect all classes of equipment but are particularly significant for transmission lines. The steady and dramatic developments in circuit components from electromechanical to transistors to microprocessors are reflected, not so much in new relaying concepts as in their implementation. Progress in solid-state components and digital relays have made relays more reliable and adaptable and have improved testing and self-checking ability. Transmission line protection involves interaction between many components and the new circuits allow these components to be combined in compact logical units. The ability to apply additional intelligence to the basic measuring function of a transmission line protective relay adds another dimension to this technology. In particular, adaptive relaying, i.e. the ability to alter relay performance, settings, or characteristics in response to system changes has become an active field of investigation.

II. Discussion of Reprints

Recalling our interest in providing both a ready reference for practicing relay engineers and a tutorial for newer engineers, Volume II includes several classic papers on the fundamentals of transmission line relaying. The 1963 paper, "Negative Sequence Directional Ground Relaying," by Elmore and Blackburn, describes the use of negative sequence as an alternative to zero sequence for polarizing ground relays and thus avoiding the problem of induced voltages. In the 1976 paper, "Selective-Pole Switching of Long Double-Circuit EHV Line," Kimbark introduced the concept of selective pole switching to deenergize only the faulted phase, an idea that today is becoming more attractive as transmission line

right-of-way is becoming more difficult to obtain and the loss of each circuit is more detrimental to system integrity. An updated report on this concept is presented in the section on substation control. In "Limits to Impedance Relaying," Thorp et al. examine the relationship among relaying speed, fault, clearing, and system performance and conclude that there are fundamental system parameters, particularly the transients associated with faults, that impose a limit to the speed with which relays can perform properly. This is an important conclusion in that it forces the recognition of the interaction of all elements of the protection chain instead of concentrating on any one element.

The increasing levels of transmission system voltages requires a continuous review of the associated protection practices. A PSRC report, "EHV Protection Problems," summarizes those problems that are becoming significant in light of today's environmental concerns and technological advances. This report covers the effects of line parameters, transient responses of transducers, sub-synchronous resonance and stability, and the influence of harmonics and power-line noise. In "Principles of Ground Relaying for High Voltage and Extra High Voltage Transmission Lines," Griffin reexamines the particular problems associated with ground relaying and provides an excellent summary of the ability to detect this most common transmission line fault. In "A Directional Wave Detector Relay with Enhanced Application Capabilities for EHV and UHV Lines," Giuliante et al. describe how traveling waves can be used in the protection of EHV and UHV lines. Solutions for difficult application problems, as well as economic considerations are presented.

The use of fiber optics is becoming extremely attractive in view of its capacity for a great many channels and immunity to electromagnetic interference (EMI). "A Current Differential Relay Using Fiber Optic Communications" by Sun and Ray describes the use of a unique fiber-optic channel in place of a metallic circuit in a current differential scheme. These changes increase the relay system reliability and broaden the area of fiber-optic application.

Energized distribution lines lying on the ground in built-up areas present a serious hazard to the public, and, due to the high ground resistance usually associated with such locations, are extremely difficult to detect and operate protective relays. "Feeder Protection and Monitoring System, Part I: Design, Implementation, and Testing," by Aucoin, Zeigler, and Russell examines the theory of detecting high impedance faults and the application of digital overcurrent relays, including field tests.

III. Associated Standards

"*Protection Aspects of Multi-Terminal Lines*," IEEE Special Publication 79 TH 0056-2-PWR, 1979.

"*Microwave Communication Channels for Protective Relaying Applications*," IEEE Special Publication 85 TH 0129-7 PWR, 1985.

"*Requirements for Power Line Carrier Coupling Capacitors*," ANSI/IEEE C93.1.

IV. Supplemental Bibliography

"*Downed Power Lines: Why They Can't Always be Protected*" IEEE Power Engineering Society Report 1989.

Negative-Sequence Directional Ground Relaying

W. A. ELMORE
MEMBER AIEE

J. L. BLACKBURN
MEMBER AIEE

Summary: The principles, problems, and solutions in applying negative sequence for directional ground relaying in 3-phase power systems are given in detail. Two general application areas are covered: (1) where system economics do not permit the necessary equipment for zero-sequence polarization and (2) for mutual induction on parallel lines. A practical method is included to evaluate relay operation without the knowledge of negative-sequence voltages and currents.

THE customary method of establishing a sense of direction in ground relays is through the use of zero-sequence polarizing current or voltage. These methods of polarizing have been described in references 1 and 2. Current polarizing requires that the polarizing power transformer bank contain a Y and delta winding. In the absence of a grounded neutral in the power transformer for a zero-sequence current source, zero-sequence current polarizing is unsuitable.

Zero-sequence voltage polarizing requires the use of three line-to-ground connected potential transformers or potential devices with a broken delta-connected secondary. This secondary winding generally is reserved exclusively for polarizing use.

No zero-sequence component is ever present in the phase-to-phase voltages. This is demonstrated in Appendix I. Consequently, there is no zero-sequence voltage at the secondary of a line-to-line connected potential transformer, and zero-sequence voltage polarizing is impossible with delta- or open-delta-connected potential transformers.

Paper 62-1072, recommended by the AIEE Relays Committee and approved by the AIEE Technical Operations Department for presentation at the AIEE Summer General Meeting, Denver, Colo., June 17-22, 1962. Manuscript submitted March 12, 1962; made available for printing April 13, 1962.

W. A. ELMORE is with the Westinghouse Electric Corporation, Seattle, Wash., and J. L. BLACKBURN is with the Westinghouse Electric Corporation, Newark, N. J.

Negative-sequence voltage, on the other hand, is present in the line-to-line voltages whenever an unbalanced condition exists. Thus, negative-sequence directional sensing is applicable in many cases where zero-sequence directional sensing is impossible.

Many informative papers have been written on the subject of ground relaying utilizing zero sequence, but these have only given a cursory mention of negative-sequence directional ground relaying. System economics and the wider use of parallel lines make this tool increasingly valueable and important. This paper will review the fundamentals, define the practical problems that can be encountered, and provide typical solutions. With these in hand, relay applications can be made more efficiently and with increased confidence.

Definitions

The term negative-sequence directional ground relay refers to a protective relay where the fault directional sensing unit operates from the power system negative-sequence current and voltage components. The fault detecting or discriminating units, either time overcurrent or instantaneous overcurrent, or both, operate from the power system zero-sequence current component. Thus, this type of relay contains a negative-sequence current filter and a negative-sequence voltage filter to energize the directional unit. The directional unit supervises by torque controlling the zero-sequence operated overcurrent units.

Various relays and relay systems applicable in ground fault protection are summarized in Appendix II. The general principles of operation and application of these relays and systems are covered in more detail in references 7 and 8.

Negative sequence, positive sequence, and zero sequence are those associated with the method of symmetrical components and have been covered in many textbooks and references.[3,7]

Applications

The major applications where negative-sequence directional relays are advantageous are shown in Figs. 1 through 4. In each case the system is assumed to be grounded at each end of the incoming lines, thus requiring the application of directional-type ground relays.

Case I. Open-Delta Bus Potential Transformers, No Ground Source at Station (Fig. 1)

The station does not have a ground source at the bus voltage and for economy, only two potential transformers are used. Hence, no zero-sequence polarizing source exists for the directional ground relays on the lines. However, the open-delta potential transformers provide an adequate polarizing source for application of negative-sequence directional sensing.

Case II: No Bus Potential Transformers, No Ground Source at Station (Fig. 2)

Directional ground relaying for the high-voltage lines can be obtained without high-side potential through the use of negative-sequence directional relays energized as shown from the low-side potential transformers. It should be noted that in this type of application (a) the negative-sequence voltage at the relay for high-voltage faults will be less than the high-voltage bus voltage by the amount of the I_2Z_2 drop through the transformer bank, and (b) a 30-degree phase shift exists in the negative-sequence voltage if the transformer bank is connected delta-Y or Y-delta. If the bank is connected according to ASA (American Standards Association) standards (high-side positive-sequence voltage leads low-side positive-sequence voltage by 30 degrees), the negative-sequence voltage on the low side leads the high-side negative-sequence voltage by 30 degrees.

According to principles I_{a2} always leads

Reprinted from *AIEE Power App. Syst.*, pp. 913–920, 1963.

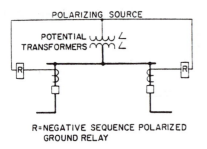

POLARIZING SOURCE

POTENTIAL
TRANSFORMERS

R = NEGATIVE SEQUENCE POLARIZED
GROUND RELAY

Fig. 1. Directional ground relaying utilizing negative sequence possible at stations without local ground source and only open-delta potential transformers

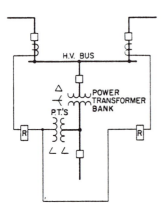

H.V. BUS

POWER
TRANSFORMER
BANK

P.T.'S

Fig. 2. Directional ground relaying utilizing negative sequence possible at stations without local ground source and only low-side potential transformers

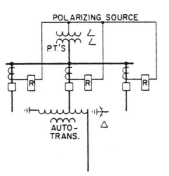

POLARIZING SOURCE

P.T.'S

AUTO-
TRANS.

Fig. 3. Directional ground relaying utilizing negative sequence possible at stations where local ground source is not suitable for zero-sequence current polarization

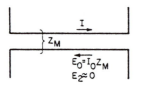

$$E_0 = I_0 Z_M$$
$$E_2 \approx 0$$

Fig. 4. Mutual induction between parallel transmission lines. Current I in top lines can induce a large E_0 in bottom line

V_{a2} by an angle of 180 degrees minus the system impedance angle for all phase-to-ground and 2-phase-to-ground faults. I_{a2} is the phase a negative-sequence current flowing instantaneously to the fault and V_{a2} is the negative-sequence voltage drop phase-a-to-neutral $(V_{a2} = 0 - I_{a2}Z_2)$. The system impedance angle is the angle of the negative-sequence impedance Z_2 from the neutral bus to the point of measurement. A typical negative-sequence directional unit by design has maximum torque when I_2 leads V_{a2} by 100 degrees. Thus, in normal applications, maximum torque occurs for a negative-sequence system angle of 80 degrees, and in the application of Fig. 2, for a system angle of 50 degrees.

CASE III. GROUNDED SOURCE NOT SUITABLE FOR CURRENT POLARIZATION (FIG. 3)

Negative-sequence ground relays can also be applied where the neutral of an autotransformer bank is not reliable as a zero-sequence current polarizing source, and the delta is not accessible. In addition, only two potential transformers are required in these applications. The criteria for polarizing ground relays from autotransformers are given in reference 1.

The zero-sequence impedance of autotransformers usually is low, which could make the negative-sequence voltage for double-line-to-ground faults low. These faults are also recognized by the phase relays so that low negative-sequence voltage does not eliminate the application of negative-sequence directional relays.

Line-to-ground faults, on the other hand, produce sizeable negative-sequence voltage when the zero-sequence impedance is low (such as in Case III when the autotransformer shown is in service), thereby overcoming the limitation of using zero-sequence potential polarizing alone in such a case. The possible switching of a bank of power transformers used for polarizing, which prohibits the use of current polarizing alone, would not limit the negative-sequence relay.

Negative-sequence relays can definitely be useful in cases where dual polar-

ized ground relays are currently applied. The predominant limitation of these relays is the adequacy of torque produced on the directional unit by the available negative-sequence quantities during line-to-ground faults. This is discussed later.

CASE IV. MUTUAL INDUCTION BETWEEN PARALLEL LINES (FIG. 4)

The mutual coupling between 3-phase transmission lines on the same tower or on adjacent towers in the same right-of-way may present relaying problems to which negative-sequence directional relays can provide an excellent solution.

With 3-phase conductors paralleled either on the same tower or adjacent towers, the negative-sequence mutual impedance is very small. For practical separations of the conductors and circuits it is less than 10% of the self-impedance of the line and usually it does not exceed 3 to 7% of the self-impedance. Consequently, the negative-sequence induction is negligible from a practical standpoint. A detailed study of this for transposed and nontransposed lines is given in reference 9.

On the other hand, the zero-sequence mutual impedance on paralleled lines can be as high as 50 to 70% of the self-impedance and must be considered both in the fault calculations and its effect on the ground relays.

Since positive- and negative-sequence induction is negligible and the zero sequence is large, the only relaying problems that might arise would involve the use of zero-sequence in ground relaying.

These can be discussed as three general types:

1. Parallel lines with common positive- and zero-sequence sources.

2. Parallel lines with common positive-sequence sources but zero-sequence sources isolated.

3. Parallel lines with positive- and zero-sequence sources isolated.

Before discussing these individually, a brief review of the techniques of handling the zero-sequence mutual impedance in the zero-sequence network is in order. A typical parallel line system is shown in

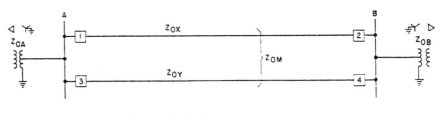

Fig. 5. Parallel line transmission system

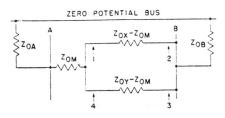

Fig. 6. Zero-sequence network for the system of Fig. 5

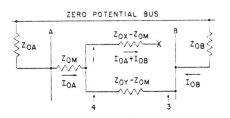

Fig. 7. Zero-sequence network for system of Fig. 5 with breaker 2 open

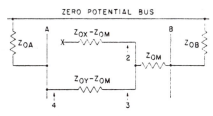

Fig. 8. Zero-sequence network for the system of Fig. 5 with breaker 1 open

Fig. 5. The lines are bussed at each station with a ground source at each end. This ground source may be at the station or at a remote station beyond A or B. Usually the self-impedances of the lines are equal, but they may be different as shown.

The equivalent zero-sequence network for Fig. 5 is shown in Fig. 6. Z_{0A} and Z_{0B} are the equivalent zero-sequence impedances behind bus A and B respectively. Either one of these sources could be ungrounded, in which case the impedance value Z_{0A} or Z_{0B} would be infinite.

The upright arrows in the network indicate the equivalent location of the line breakers for determining the current flow. All voltage would be measured at the respective bus A or B.

If $Z_{0X} = Z_{0Y}$, then the total parallel line impedance between A and B is 1/2 $(Z_{0X}+Z_{0M})$. Thus, the zero-sequence impedance between stations A and B is in the order of 150 to 170% of the value $(1/2\ Z_{0X})$ obtained by neglecting mutual.

Line-end faults can be calculated from the equivalent circuit. Assume that the redistribution of fault current is desired for a line fault outside breaker 2 at B after breaker 2 opens. The equivalent zero-sequence network would be as shown in Fig. 7.

If the line-end fault for breaker 1 at A is desired, the equivalent circuit can be redrawn by connecting the common mutual circuit to bus B instead of A. This is shown in Fig. 8.

Following the common practice today, either or both lines may be tapped or bussed. A combination arrangement is illustrated in Fig. 9. The zero-sequence network for this system is shown in Fig. 10. Many other possible combinations will be related to Fig. 9 from which the zero-sequence network can be derived by a study of Figs. 9 and 10. As long as the mutually coupled lines or sections have one common bus the mutual can be represented by the equivalent network of Fig. 6, and the sequence network can beset up on a d-c network analyzer.

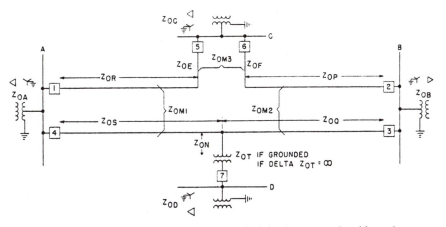

Fig. 9. Parallel transmission line system which has been tapped and bussed

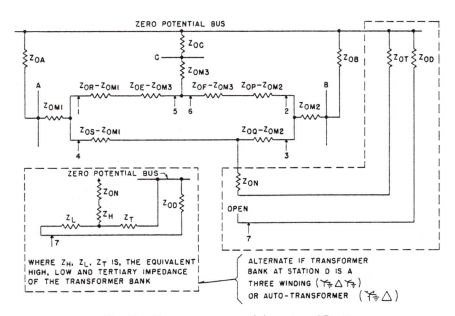

Fig. 10. Zero-sequence network for system of Fig. 9

Where the lines do not terminate in a common bus the technique of an ideal or perfect transformer can be used to represent the mutual effect. This is illustrated in Fig. 11. The ideal transformer as shown provides electrical connection isolation between the two lines, yet inserts the necessary $(I_{0X}+I_{0Y})\ Z_{0M}$ drop in both circuits. Thus, the transformer should have no loss with infinite magnetiz-ing- and zero-self-impedances. Practically, a high-quality transformer is used in the a-c network analyzer. This equivalent cannot be used on a d-c analyzer but there are techniques of introducing an equivalent voltage drop in each branch, as discussed in reference 5. This is a cut-and-try method since the voltage to be introduced is a function of the current flow which in turn depends on the volt-

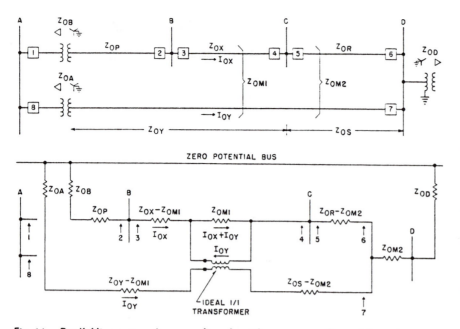

Fig. 11. Parallel line system where some lines do not have a common bus and the zero-sequence network for this system

age. Where computers are used for fault calculation the mutual effect should be written in the network equations.

Type A. Parallel Lines with Common Positive- and Zero-Sequence Sources

These are the cases most frequently encountered. Typical examples are shown in Figs. 5, 9, and 11. In these systems zero-sequence current always flows up the neutrals, no matter where the ground fault occurs or what combination of breakers are in service, as long as the zero-sequence isolation does not occur. Hence, conventional zero-sequence directional units can be applied for correct directional indication for all lines. The principles of ground relay polarization are documented in reference 1. Also, the zero-sequence current is always larger through the relay nearest the fault than for the co-ordinating relay, so that the mutual does not change the co-ordination problem. Of course, the zero-sequence mutual does change the actual magnitudes and distributions. Therefore, mutual induction does not introduce any new problems in these systems and conventional relaying systems can be applied.

There are two exceptions to this. One is a case of zero-sequence isolation which can occur by having a ground fault on line 3–4 in Fig. 9, with Z_{0T} grounded, breaker 7 tied to a power source, and breakers 3 and 4 open. Then zero-sequence current would be induced in line 1–5 for ground fault near breaker 4 and in line 6–2 for ground faults near breaker 3.

This will be discussed later in more detail.

The second exception applies to product-type relays on parallel lines. Under certain system conditions and with one breaker open, as shown in Figs. 7 or 8, the product at the remote relay can be greater than that at the near relay, thereby causing incorrect tripping. In other words, with reference to Fig. 7, the product of line current and polarizing quantity at breaker 3 can be larger than the corresponding product at breaker 1. This occurs with current polarization when $I_{0A}(I_{0A}+I_{0B})<(I_{0B})^2$ where I_{0A} is the current through the Z_{0A} branch and I_{0B} is the current through the Z_{0B} branch.

Type B. Parallel Line with Common Positive-Sequence Sources but Zero-Sequence Sources Isolated

As pointed out previously, Fig. 9 with breakers 3 and 4 open represents a case of zero-sequence isolation. Another example is the system of Fig. 11 where zero-sequence isolation occurs if breaker 7 is open. A ground fault on the line side at breaker 7 in Fig. 11 would induce zero sequence in lines 1–2, 3–4, 5–6, and conversely, ground faults on these top lines would induce zero sequence in line 7–8. The voltage would be the fault current times the mutual (Z_{0M}) impedance and the amount of current induced is equal to this voltage divided by the total Z_0 impedance. In the example of Fig. 11, with breaker 7 open and a ground fault near this breaker, fault current I_{0Y} will flow through Z_{0A} and the line to the fault.

This will induce a voltage I_{0Y} ($Z_{0M1}+Z_{0M2}$) in the top lines. The current flowing in the lines will be

$$I_{0X} = \frac{I_{0Y}(Z_{0M1}+Z_{0M2})}{Z_{0B}+Z_{0P}+Z_{0X}+Z_{0R}+Z_{0D}}$$

This current will flow up Z_{0D} neutral, from D to A and down Z_{0B} neutral and appear to be an internal fault to a directional comparison pilot relay system.

A further example is shown in Fig. 12 where zero-sequence isolation exists under all conditions.

Relaying Solutions to Type-B Problems

Several solutions are available and the best choice will vary with different systems and conditions. It is very reliable and economical to use negative-sequence current and voltage for directional sensing of the ground relays. As has been pointed out, negative-sequence induction is small. With common positive and negative sources, the normal negative-sequence current in the line would further negate or swamp out any effect of negative-sequence induction so that the very sensitive negative-sequence directional units will give correct directional indication. For example, in Fig. 9, if station D does not have a positive-sequence source, then a ground fault on line 3–4 with breakers 3 and 4 open does not have any meaning. In Fig. 11, breaker 8 supplies negative sequence through bus A for breaker 7 line-side ground fault with breaker 7 open. Similarly, in Fig. 12 negative-sequence fault current is supplied through busses A or B or both.

As indicated, negative-sequence directional zero-sequence overcurrent ground relays can be used throughout the systems described here. As an alternate, current-polarized zero-sequence directional overcurrent relays can be used at breakers 1 and 8 in Fig. 11 where the polarizing winding is energized by paralleling the current transformers in both Z_{0A} and Z_{0B} transformer bank neutrals.* Since the per-unit induced current is always less than the per-unit fault current, the sum of the two will be in the same direction as that of the fault current. The current transformer ratios must be the same on the same voltage level or properly pro-

---

* The examples illustrate a basic principle. Directional ground relays would not be required at these stations because the banks block ground current for bus faults. Such relays would be applicable with instantaneous reclosing to avoid reclosing on bank faults.

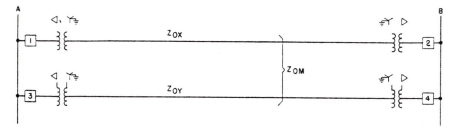

Fig. 12. Parallel line system with common positive- and negative-sequence sources but with zero-sequence source isolation

portioned if the voltage levels of the paralleled lines are not the same to maintain a positive sum. In Fig. 12 again zero-sequence current polarization can be used by paralleling the current transformers in both bank neutrals at stations A and B. If the banks are not in the same station, then this solution is not practical.

For lines where pilot relaying systems are applied, the phase comparison[7] system is applicable in all cases. Systems such as the pilot wire relay or the phase comparison carrier relay operate on current comparison only and therefore are not subject to incorrect operation on induction.

The directional comparison pilot relay systems with negative-sequence directional ground relays and the combined pilot relay system are also applicable.

TYPE C. PARALLEL LINES WITH POSITIVE- AND ZERO-SEQUENCE SOURCES ISOLATED

This type is not common with today's complex meshes and interconnections. Complete independence of the systems except at the point of mutual induction could mean that negative-sequence induction might be sufficient to operate sensitive negative-sequence directional units. Again, negative-sequence directional zero-sequence overcurrent ground relays can be used with a negative-sequence overcurrent fault detector to avoid any possible problem from negative sequence.

The phase comparison, directional comparison with negative-sequence directional ground relays and the combined pilot relay systems also can be used, as outlined previously.

An actual example of electrically isolated systems exists where two companies jointly use the same tower for one of their high-voltage lines. Usually, the companies are interconnected but it is possible to operate separately. Directional comparison carrier with negative-sequence ground relays were applied and negative-sequence fault detectors were used since

the systems could operate separately. A recent report after many years of operation indicated successful operation with no difficulties.

TYPE-D—MULTIPLE PARALLELED CIRCUITS

Where several circuits run along the same right-of-way, mutual coupling exists between each of them and must be included in zero-sequence fault calculations. This is done by determining the mutual impedance between several pairs using the techniques charted in reference 4 and including this mutual in the zero-sequence networks as outlined earlier. These equivalent circuits for three and four parallel lines are given in Figs. 6 and 8(D), pages 187 and 193 of reference 6. These complex circuits require the use of ideal transformers and should be solved by an a-c network analyzer or equivalent.

With close coupling from several parallel lines it is possible for the mutual induction to cause zero-sequence current reversal in the nonfaulted line resulting in current flowing down the delta-star grounded transformer neutrals. In these cases negative-sequence directional ground relays or the phase comparison-type relays must be used for correct relaying.

Comparison of Negative-Sequence Versus Combined Relaying Systems

Both the directional comparison pilot relay system with negative-sequence ground relays and the combined pilot relay system are available and in use for solving the mutual induction problem as indicated earlier. Phase comparison in the combined scheme solves the mutual problem in a more complex and costly manner. The combined and phase comparison systems require a wider bandwidth or more carrier spectrum than the directional comparison system.

On the other hand, the application of negative sequence solves the problems by

the substitution of a negative-sequence directional unit in place of the zero-sequence directional unit. Some evaluation is in order to make certain that sufficient negative-sequence voltage and current are available under the various fault conditions to produce adequate operating torque on the directional unit of the relay. A system grounded "too solidly" will impose a problem during double line-to-ground fault conditions and a "poorly grounded" system will impose a restriction for single line-to-ground faults.

Calculation and Evaluation of Negative-Sequence Quantities

NEGATIVE-SEQUENCE VOLTAGE AT THE FAULT

The negative-sequence voltage to be expected at the fault for various ratios of Z_0 to Z_1 is shown in Fig. 13. This is obtained from Appendix III which derives V_{A2} in terms of the Z_0/Z_1 or Z_1/Z_0 ratios for the two types of faults. This figure is based upon the assumption that $Z_2=Z_1$, the impedance angle of Z_2 equals the impedance angle of Z_0, and normal line-to-line voltage at the relay is 115 volts. In the normal range of Z_0/Z_1 ratios found on grounded utility systems (1 to 20), the line-to-ground case is the limiting one. The minimum values of negative-sequence current required for proper directional element response in one relay are plotted in Fig. 14 versus the corresponding voltages. The required V_{A2} continues to decrease as I_{A2} is increased. At $I_{A2}=45$ amperes, $V_{A2}=0.04$ volt. These values are based on the assumption that the negative-sequence current leads the negative-sequence voltage by the maximum torque angle of 100 degrees. The corresponding values for other angles are:

$$I_{2(\theta)} = \frac{I_{2(100)}}{\cos(\theta - 100)}$$

where

$I_{2(\theta)}=$ negative-sequence current required leading the negative-sequence voltage by angle θ

$I_{2(100)}=$ current required at 100 degrees

For cases where the negative-sequence impedance angle does not equal the zero-sequence impedance angle, the value of V_{A2} will be greater than the values shown in Fig. 13. Application of Fig. 13 then, will provide a conservative estimate of the negative-sequence voltage available for polarizing during close-in faults.

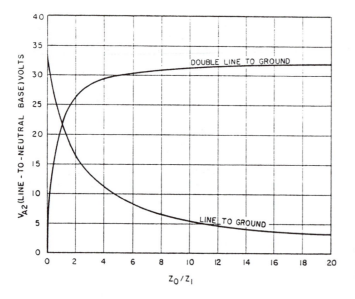

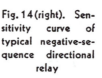

Fig. 13 (left). The negative-sequence voltage at the fault versus Z_0/Z_1 ratio for ground faults

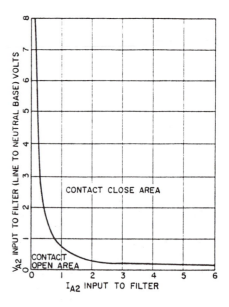

Fig. 14 (right). Sensitivity curve of typical negative-sequence directional relay

Voltage at Relay for Remote Fault

The negative-sequence voltage available at the relay is less for faults remote from the relay location. Since maximum negative-sequence voltage occurs at the ground fault location, the greater the separation between fault and relay the lower the negative-sequence voltage available for direction unit reference.

An effort is made at this point to simplify the determination of the negative-sequence voltage at the relay for various ground fault locations. In terms of the information the protection engineer customarily has at his disposal, Appendix IV shows that a simple factor can be applied to the quantity taken from Fig. 13 to obtain this voltage. The factor is the ratio of the source contribution at the relay bus to a 3-phase fault at X, to the source contribution at the relay bus to a 3-phase fault at the relay bus or I_{sx}/I_{sr}. Since Fig. 13 is established on the conservative basis that the positive-sequence and zero-sequence impedance angles are identical, the multiplication by this factor will produce a conservative estimate of the negative-sequence voltage at the relay for a line-to-ground fault at the point under consideration.

The process is simple and straightforward. (1) Obtain the Z_0/Z_1 ratio at the fault point to be investigated; (2) consult Fig. 13 for the equivalent secondary negative-sequence voltage at the fault; (3) multiply by the factor I_{sx}/I_{sr}. The resulting value should equal at least 4 volts to assure positive action of the directional unit.

Simultaneous analysis of negative-sequence current and voltage will show cases that are satisfactory where the negative-sequence voltage is less than this 4-volt figure. See Fig. 14.

Network analyzers or computers can readily supply negative-sequence quantities and these should be obtained as a matter of practice when making ground fault studies.

Comparison of Negative- and Zero-Sequence Energy

The product of sequence voltage and current is indicative of the amount of energy available to operate a relay directional unit. While this product is of little quantitative importance unless a precise phasor relationship between current and voltage is identified for a given case and unless the maximum torque angle of the particular relay element is known, it is of interest to compare the relative I_2V_2 and I_0V_0 quantities.

Zero-sequence voltage for line-to-ground faults increases as system zero sequence impedance increases. This produces a compensating effect for the decrease in zero-sequence current tending to hold zero-sequence energy within fairly narrow limits. Negative-sequence voltage, on the other hand, decreases as zero sequence impedance increases. This produces an adverse compounding effect tending to cause negative-sequence energy for operating a directional unit to decrease markedly as zero-sequence impedance increases. The ratio of zero-sequence volt-amperes to negative-sequence volt-amperes produced by a line-to-ground fault is simply Z_0/Z_1. For a double line to-ground fault, this ratio is the inverse of this, or Z_1/Z_0. The normal range of Z_0/Z_1 ratios would produce more zero sequence volt-amperes for line-to-ground faults. Since the sequence quantities are limiting only with respect to directional sensing, magnitude is of little consequence so long as adequacy is assured.

One Advantage of Negative-Sequence Directional Ground Relays

Negative sequence has a very important advantage over current polarization either when used alone or in dual combinations with voltage polarization. Negative-sequence relays can be easily checked in the field using load current to assure correct installation. This check cannot be made with zero-sequence current polarization unless primary zero-sequence current is circulated through the power system.

The check with negative-sequence relays using normal load flow is accomplished by opening and short-circuiting various current and potential secondary circuits to simulate primary fault conditions. A number of combinations can be used depending on the direction and phase angle of the load and to provide several tests for complete assurance of correct installation.

Conclusions

Negative-sequence directional sensing for ground overcurrent relays provide excellent solutions to relaying problems imposed by system economics and mutual induction. They also aid in avoiding erroneous tripping under abnormal operation conditions which upset zero-sequence polarization of the ground relays. The relaying problems, solutions, and limitations are reviewed and discussed.

It should be emphasized that the negative-sequence ground relays can be applied rather simply and a great deal of detailed analysis is not ordinarily required. A practical method is given to determine whether sufficient negative-sequence current and voltage product is available for a system where a question exists or where the fault values may not be readily available. Negative-sequence relays are easy to test and the installation can quickly be checked using load current flow.

It is hoped that this paper will assist relaying engineers in providing increased reliability and security in relaying systems —systems of minimum equipment and circuitry to give maximum protection at minimum cost.

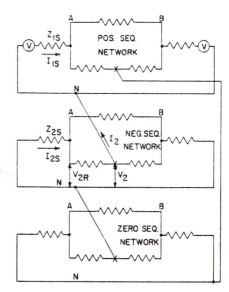

Fig. 15. Sequence networks and interconnections for a line-to-ground fault on a line between busses A and B in an equivalent system with sources at A and B. Top impedance between A and B represents the equivalent of all other ties between the two busses

Appendix I. Sequence Components in the Line-to-Line Voltages by Fundamental Definitions of Symmetrical Components[3]

$$V_a = V_1 + V_2 + V_0 \tag{1}$$

$$V_b = a^2 V_1 + a V_2 + V_0 \tag{2}$$

$$V_c = a V_1 + a^2 V_2 + V_0 \tag{3}$$

where

a = phasor operator, $1\,/120$ degrees
V_a = phase-a-to-ground voltage drop
V_b = phase-b-to-ground voltage drop
V_c = phase-c-to-ground voltage drop
V_1, V_2, V_0 = positive-, negative-, and zero-sequence components respectively of V_a
V_{ab} = phase-a-to-phase-b voltage drop

$$V_{ab} = V_a - V_b = (V_1 + V_2 + V_0) - (a^2 V_1 + a V_2 + V_0) \tag{4}$$

$$V_{ab} = V_1(1-a^2) - V_2(1-a) \tag{5}$$

V_{ab} therefore contains no zero-sequence component. By similar analysis, it can be shown that V_{bc} and V_{ca} also contain no zero-sequence components.

Appendix II. Review of Available Relays

A complete line of relays is available for transmission line relaying with or without zero-sequence induction. For reference the various types and basic characteristics are summarized here.

1. Directional time overcurrent relays for primary or back-up protection with inverse, very inverse, moderately inverse, extremely inverse or definite time curve characteristics. Nondirectional instan-

taneous trip attachments can be added for high-speed clearing of heavy faults.

2. Directional instantaneous and time overcurrent relays for primary or back-up protection. These are similar to those under item 1, with an added directionally controlled instantaneous unit.

3. Directional instantaneous overcurrent relays for primary protection. These are used separately for phase or ground protection or in the directional comparison pilot relay systems providing high-speed ground protection.

These three families have various members for:

(a). Phase fault protection.
(b). Ground fault protection, current polarized.
(c). Ground fault protection, potential polarized.
(d). Ground fault protection, dual polarized.
(e). Ground fault protection, negative-sequence directional.

These relays are completely self-contained. The sensitivity of a typical negative-sequence directional unit is 0.75 volt-ampere without the negative-sequence fault detector or 1.1 volt-amperes with the fault detector. The directional unit operating time is 1 cycle at 100 times pick-up (60-cycle basis).

4. Phase comparison pilot relay systems. These provide simultaneous, instantaneous phase and ground fault protection.

(a). Pilot wire relays for use over pilot wire channels.
(b). Phase comparison carrier relays for use over power line carrier or microwave channels.

These systems compare the phase position of the currents at the terminals to determine whether the fault is internal or external.

One pilot wire system is applicable to two or three terminal lines. It is desirable that the minimum internal fault current be approximately 1.67 maximum load current. Potential transformers are not required. There is no problem with induction in these systems.

5. Directional comparison pilot relay systems. These provide simultaneous instantaneous phase and ground fault protection. This system uses the conventional distance relays and can be used over wire, tone carrier or microwave channels. The direction of fault power flow is compared via the channel to determine the internal or external fault. The system is quite versatile and applicable to multi-terminal lines.

6. Combined pilot relay systems. This system provides a directional pilot system for phase faults and a phase comparison pilot system for ground faults and is generally used over a carrier or microwave channel. It provides the advantages of distance relays for phase fault discrimination and phase comparison for ground fault so that induction is not a problem. It is more complex and hence more expensive.

Appendix III. Deriving Negative-Sequence Voltage at the Fault Point in Terms of Total System Impedances

Case A: Line-to-Ground Faults

Consider a line-to-ground fault in a power system where the total positive-, negative-, and zero-sequence impedances to the fault are Z_1, Z_2, and Z_0 respectively. With the networks connected in series for the phase-a-to-ground fault, the total current components are

$$I_1 = I_2 = I_0 = \frac{V_{an}}{Z_1 + Z_2 + Z_0} \tag{6}$$

The negative-sequence voltage at the fault is

$$V_{A2} = 0 - I_2 Z_2 = -\frac{V_{an} Z_2}{Z_1 + Z_2 + Z_0} \tag{7}$$

Assuming $Z_2 = Z_1$, then

$$V_{A2} = -\frac{V_{an}}{1 + 1 + \dfrac{Z_0}{Z_1}} = -\frac{V_{an}}{2 + \dfrac{Z_0}{Z_1}} \tag{8}$$

Case B: Two-Line-to-Ground Fault

For a two-line-to-ground fault, the three networks are connected in parallel so that

$$I_1 = \frac{V_{an}}{Z_1 + \dfrac{Z_2 Z_0}{Z_2 + Z_0}} = \frac{V_{an}(Z_2 + Z_0)}{Z_1 Z_2 + Z_1 Z_0 + Z_2 Z_0} \tag{9}$$

$$I_2 = -I_1 \frac{Z_0}{Z_2 + Z_0} = -\frac{V_{an} Z_0}{Z_1 Z_2 + Z_1 Z_0 + Z_2 Z_0} \tag{10}$$

$$V_{A2} = -I_2 Z_2 = +\frac{V_{an} Z_0 Z_2}{Z_1 Z_2 + Z_1 Z_0 + Z_2 Z_0} \quad (11)$$

Assume $Z_2 = Z_1$, then substituting and reducing,

$$V_{a2} = +\frac{V_{an}}{2 + \dfrac{Z_1}{Z_0}} \quad (12)$$

Appendix IV. Determining the Ratio of Negative-Sequence Voltages, At Relay Versus At Fault

Compare the 3-phase and line-to-ground faults both at the same remote point X. The sequence networks and connections for the line-to-ground fault are shown in Fig. 15. For the 3-phase fault at X, only the positive-sequence network of the general system is involved. Using the customary assumption that the positive- and negative-sequence systems are identical ($Z_1 = Z_2$), the current splitting factors between the two faults are identical. Thus,

$$\frac{I_{2S}}{I_2} = \frac{I_{SX}}{I_X} \quad (13)$$

where I_{2S} and I_2 are the source A and total negative-sequence currents respectively, as shown in Fig. 15 for the line-to-ground fault at X. I_{SX} and I_X are correspondingly positive-sequence currents for the 3-phase fault at X.

$$V_{2R} = -I_{2S} Z_{2S} = -I_2 \frac{I_{SX}}{I_X} Z_{2S} \quad (14)$$

V_{2R} is the negative-sequence voltage at bus A for the line-to-ground fault.

The current components in the line-to-ground fault are defined by equation 6. Since $Z_1 = Z_2$ and $V_{AN} = 1$ per unit, this becomes

$$I_2 = \frac{1}{2Z_1 + Z_0} \quad (15)$$

$$V_{2R} = -\frac{I_{SX}}{I_X} \frac{Z_{2R}}{(2Z_1 + Z_0)} \quad (16)$$

For a 3-phase fault at X,

$$I_X = \frac{V_{AN}}{Z_1} = \frac{1}{Z_1} \quad (17)$$

Substituting this in equation 16,

$$V_{2R} = -\frac{(Z_1 Z_{2S})}{(2Z_1 + Z_0)} I_{SX} = \frac{Z_{2S} I_{SX}}{2 + \dfrac{Z_0}{Z_1}} \quad (18)$$

Equation 8 gives the fault voltage component which is

$$V_2 = -\frac{1}{2 + \dfrac{Z_0}{Z_1}} \quad (19)$$

Thus,

$$\frac{V_{2R}}{V_2} = Z_{2S} I_{SX} \quad (20)$$

Now, the current contribution from the source (Z_{1S}) for a 3-phase fault at bus A is

$$I_{SR} = \frac{1}{Z_{1S}} \quad (21)$$

Finally,

$$\frac{V_{2R}}{V_2} = \frac{I_{SX}}{I_{SR}} \quad \text{since } Z_{1S} = Z_{2S} \quad (22)$$

References

1. Ground Relay Polarization, J. L. Blackburn. AIEE Transactions, pt. III (Power Apparatus and Systems), vol. 71, 1952, pp. 1088–95.

2. Special Circuits for Ground Relay Current Polarization from Autotransformers Having Delta Tertiary, P. A. Oakes. Ibid., vol. 78, 1959, pp. 1191–96.

3. Symmetrical Components (book), C. F. Wagner, R. D. Evans. McGraw-Hill Book Company, Inc., New York, N. Y., 1933.

4. Speed Zero Sequence Impedance Calculation for Transmission Lines, J. L. Blackburn. Electrical World, New York, N. Y., vol. 153, March 1960, pp. 60–61.

5. A Device for Solving Mutual Induction Problems on a D-C Network Analyzer, Torgeir Karlsen, H. A. Wallhausen. AIEE Transactions, pt. III-A Power Apparatus and Systems vol. 78, Oct. 1959, pp. 754–59.

6. Circuit Analysis of A-C Power Systems Edith Clarke. Symmetrical and Related Components, John Wiley & Sons, New York, N. Y., vol. 1, 1943.

7. Applied Protective Relaying (book). Westinghouse Electric Corporation, Pittsburgh, Pa., Publication B 7235.

8. The Art and Science of Protective Relaying (book), C. R. Mason. John Wiley & Sons, Inc., 1956.

9. Voltage Induction in Paralleled Transmission Circuits, J. L. Blackburn. AIEE Transactions, See pp. 921-29 of this issue.

———◆———

Discussion

J. G. Hendrickson (Department of Public Utilities, City of Tacoma, Tacoma, Wash.): The authors are to be commended for preparing an excellent paper covering an area on which little correlated information has been available.

While our actual operating experience with negative-sequence relaying has not been extensive, several favorable features have become apparent. Our experience definitely verifies the authors' contention of an important advantage to the user of this type of relaying, i.e., the ease of testing and initial checkout of the installation. The fact that the installation can be completely and accurately checked by means of manipulating the load current in the secondary to present fault conditions to the relay for its recognition, means that many hours of labor can be saved. No assembling of bulky equipment is required, as is usually necessary to determine the zero-sequence polarizing current direction by direct primary methods. The possible danger from temporary wiring for direct primary testing is eliminated. In addition,

when adding another installation to an existing polarizing source, it is usually impossible to verify the polarizing direction in the new installation because the source is in operation and cannot be released for testing. One must then rely on a careful sight check of all wiring and the final test is the action of the installation during an actual fault. These uncertainties are eliminated when using the negative-sequence relay and its means of positive test.

Economy is another feature. Our applications were made because negative-sequence ground relays were more economical than purchasing additional potential transformers which would have been necessary if the conventional zero-sequence relaying had been used, since a local ground source did not exist for current polarizing.

The analyses presented for problems encountered by mutual induction between parallel lines with common positive-sequence sources, are corroborated by actual experience in several relatively short-line installations in our system. We have used pilot wire relay systems exclusively, and where the proper protection is applied to the wire, the problems introduced by mutual induction are handled by the relays as predicted. In our experience, where this type of relaying is applied within its broad limitations, it cannot be surpassed either in performance or from an economic standpoint, if a basic wire communication system is available. Consequently, a field of application which might be available to the negative-sequence relays as presented in this paper, is practically usurped by the pilot wire relay.

One measure of the merit of a relay by the utility application engineer is the susceptibility of the relay to false tripping, either under unusual operating conditions such as switching or lightning surges, or conditions accidentally imposed by actions of maintenance people, such as loss of potential or short-circuiting of current transformers. It would be interesting to know whether any such limitations exist in this type of relaying.

J. B. Gray, H. V. Beck (Puget Sound Power and Light Company, Bellevue, Wash.): The paper by Messrs. Elmore and Blackburn should be welcomed by relay engineers responsible for the design and specification of relay circuits. In this paper, and the references cited, the subject of negative-sequence polarization has been covered quite thoroughly. Also, the authors have described the various applications in which negative-sequence polarized ground relays may be utilized to advantage and have compared their characteristics with relays using other forms of polarization, which makes this paper a very valuable source of information when the engineer is faced with an unusual situation. This discussion suggests such a situation and the solution of its problems.

The problem was caused by pole top fires on an ungrounded 55-kv delta-connected transmission network. The faults were generally arcing single line-to-ground faults that were very difficult to locate

94

even by extensive switching. The ensuing outages frequently covered large areas that were disproportionate to the magnitude of the fault. It was believed that a better way of isolating faults than by manual switching could be found.

Previous studies had shown that the cost of grounding the system properly and installing conventional zero-sequence current polarized directional overcurrent ground relays was prohibitive. The system involved was composed of nine interconnected major stations and approximately 325 miles of transmission lines, the longest of which was approximately 40 miles and the shortest approximately 7 miles long. An investigation and inventory of the stations in the system involved, disclosed that there were two stations near one end of the system with transformers whose 55-kv windings were connected in ungrounded Y and the neutrals were accessible, and that all but one of the other stations were equipped with two potential transformers connected in open delta. Space limitations precluded the installations of a third potential transformer so that potential polarization could not be achieved. It appeared that only negative-sequence polarizing could be accomplished economically.

From calculations, it was determined that overcurrent relay co-ordination could be obtained using zero-sequence current

and that for negative-sequence polarizing, at two of the stations, the negative-sequence voltage would be low enough under some conditions to give marginal operation. A review of the combinations of conditions under which operation would be questionable revealed that they were far less frequent than those where operation would be ensured. On this basis, it was decided to install negative-sequence polarized ground relays at all points where zero-sequence current was not available as a polarizing source. Although wrong relay operations occasionally could be expected at the two points mentioned, correct operations on the rest of the system would greatly improve the over-all operation of the system.

Since the first installation, a small grounding bank has been installed at the far end of the system from the original wye-connected banks, not for polarizing, but for additional zero-sequence current for better relay co-ordination near the end of the system. Twenty-two negative-sequence polarized overcurrent ground relays have been installed to date, and in the past 4 years, 20 operations have been logged, with no improper operations. One of these occurred at a point where marginal operation was expected. The assumption of the calculated risk of wrong operations has been fully justified and the improvement in over-all system operation has been truly gratifying.

W. A. Elmore and J. L. Blackburn: We appreciate the interesting discussions. They contribute materially to the paper by adding practical experiences and by lending emphasis that negative sequence is another valuable tool for relay engineers.

In comment on the last paragraph of Mr. Hendrickson's discussion, the susceptibility of the relay to false tripping is extremely low and we are not aware of any operation problems in this connection. Several factors are involved to assure a high security to these relays: (1) the negative-sequence directional unit is used in the sense of a fault detector in conjunction with fault measuring or discrimination units; (2) the relays are designed to have negligible response to offset transients; (3) 3-phase loss of potential or 3-phase short-circuiting of the current transformers does not provide negative-sequence quantities; and (4) lightning and switching surges generally are not significant producers of negative sequence.

Unequal loss of potential or current and unequal pole closing can produce negative sequence and could cause directional unit operation. In these cases the fault measuring or discriminating units and these inherent, or built-in, time delays would provide security at least until the second contingency of a fault occurred. We are not aware of any field problems in this area with the present relays.

SELECTIVE-POLE SWITCHING OF LONG DOUBLE-CIRCUIT EHV LINE

Edward W. Kimbark
Bonneville Power Administration
Portland, Oregon

ABSTRACT

This paper develops the joint use of several new concepts for improving stability, reliability, and safety of a double-circuit three-phase EHV line:

1. Use of selective-pole switching; that is, only the faulted conductors are opened and reclosed.

2. Exclusive use of that arrangement of conductors that gives the least current to ground from a large vehicle capacitively coupled to the line: like phases are in diagonally opposite positions.

3. Use of bank of 3 shunt capacitors and 9 shunt reactors for neutralizing the 15 interconductor capacitances and thereby eliminating the shunt capacitive coupling that tends to maintain the secondary fault arc.

4. Sectionalizing the faulted conductor or conductors into two or more longitudinal sections by remote-controlled switches in order to reduce the longitudinal resistive-inductive coupling that also tends to maintain the secondary fault arc.

These concepts are illustrated by applying all of them to a simulated 735-kV, heavily loaded, 200-mile line having 6 bundles of 4 Chukar conductors each.

INTRODUCTION

Selective-Pole Switching

Selective-pole switching means the tripping and high-speed reclosure of the circuit-breaker poles connected to all the faulted conductors of a line, while the circuit-breaker poles connected to the unfaulted conductors remain closed. If selective-pole switching were applied to a three-phase single-circuit line, one pole would be tripped and reclosed for a single line-to-ground (SLG) fault, two poles for a line-to-line (LL) or two-line-to-ground fault, and all three poles for a three-phase fault. Such switching was proposed as long ago as 1939,[1] but it has been seldom, if ever, used, except for clearing SLG faults. Other types of fault are generally cleared by three-pole switching without automatic reclosure.

One of the reasons for current practice is shown by the stars in Fig. 1. With one open conductor and two energized conductors, the transmitted power is reduced to about 54% of its initial value (the value before occurrence of the fault, when three conductors are energized). With a transitory fault and rapid reclosure, the power system is likely to be stable. However, with two open conductors and one energized conductor, the power is reduced to only 12% of the initial value, which is probably insufficient for maintaining stability.

On double-circuit lines there is a greater opportunity for improvement of stability by use of selective-pole switching. On adequately shielded EHV lines, lightning strokes to the tower or ground wires, passing through the tower footing resistance to ground,

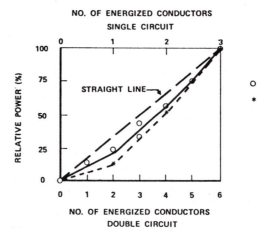

NO. OF ENERGIZED CONDUCTORS
SINGLE CIRCUIT

Fig. 1. Relative power transmitted at a fixed value of phase difference between the terminal voltages as a function of the number of energized conductors.

raise the voltage of the tower with respect to ground. The voltages of the energized conductors to ground are great enough to increase the probability that flashover, if it occurs, will involve the conductor or conductors having the highest instantaneous voltage of polarity opposite from that of the tower. Thus, on a double-circuit tower, it is likely that two conductors of like phase may flash over simultaneously, creating two SLG faults, one on each circuit. If three-pole switching were used, all six conductors would be disconnected, making instability almost certain. But, if selective-pole switching were used, only the two faulted conductors would be switched off and reclosed a fraction of a second later, constituting a very minor disturbance. For any other combination of 1, 2, or 3 faulted conductors, stability would be improved but not so much.

Residual Fault Current and Related Quantities

Residual fault current (also called secondary fault current) is that which exists after isolating the faulted conductor(s) by opening the appropriate circuit-breaker poles. It is much smaller than the short-circuit current (or primary fault current) which exists before opening the circuit breakers. It is maintained by two different kinds of coupling between the isolated, faulted conductor(s) and the energized, unfaulted conductors: (1) shunt coupling due to interconductor distributed capacitance, and (2) series (longitudinal) coupling due to mutual resistance and inductive reactance, much of which is in the common return path in the ground.

Recovery voltage is the voltage across the fault path after extinction of the secondary fault arc and before reclosure of the circuit breakers. It has two components corresponding to those described above.

Values of residual fault current and of recovery voltage cited in this paper are rms steady-state values, but this does not imply that the circuit is really in a steady state.

The ratio of the recovery voltage to the residual fault current is the Thevenin's impedance. On an uncompensated line, it is the single-phase equivalent shunt capacitive reactance between a disconnected conductor and all the energized conductors.

Thevenin's admittance is the reciprocal of Thevenin's impedance. It is proportional to the length of the line insofar as the spacing between conductors is constant and is independent of the power transmitted by the line.

Paper F 75 511-6, recommended and approved by the IEEE Transmission & Distribution Committee of the IEEE Power Engineering Society for presentation at the IEEE PES Summer Meeting, San Francisco, Calif., July 20-25, 1975. Manuscript submitted January 30, 1975; made available for printing May 1, 1975.

Reprinted from *IEEE Trans. Power App. Syst.*, vol. PAS-95, no. 1, pp. 219–230, Jan./Feb. 1976.

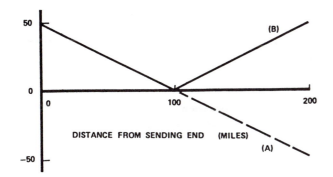

Fig. 2. *Idealized profile of voltage to ground of one open conductor of loaded, untransposed line due to longitudinal induced voltage. Plotting conventions: (A) negative ordinate denotes phase reversal with respect to positive ordinate; (B) positive ordinate denotes absolute value.*

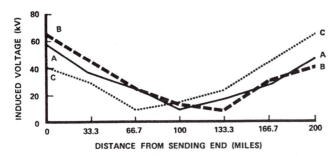

Fig. 3. *Induced voltage from one open conductor (a, b, or c) to ground, $h_1 = 0.7$. Power received on 5 conductors, 5073 MW.*

The foregoing definitions can be applied directly to single-phase (SLG or LL) faults, but they lose their precision in multiphase faults because of ambiguity of the actual arc paths and because of the likelihood of arc extinction in successive steps.

The shunt component of recovery voltage is independent of the length and loading of the line. It is proportional to the voltages of the energized conductors but these are substantially constant. Hence this component may be regarded as constant.

The series component is proportional to the currents in the energized conductors and to the mutual impedance between these conductors and the disconnected ones. Hence it is proportional to the length of the line and to its loading.

This voltage is induced longitudinally in the open conductors. The voltage from such a conductor to ground varies along the line, depending on the distribution of the shunt capacitances. If the capacitance per unit length is constant, the conductor-to-ground voltage is zero at the midpoint of the line and increases linearly to each end, having opposite phases on the two sides of the midpoint. See Fig. 2. On a transposed line, the capacitance per unit length is different in each transposition section, giving the voltage profiles illustrated in Fig. 3.

The two components of recovery voltage differ in phase by about 90°, and the actual recovery voltage at any point of the line is their vector sum.

The residual arc current is equal to the product of the Thevenin admittance and the recovery voltage. Its shunt component is proportional to length of line but is independent of line loading. The series component is theoretically proportional to the loading and to the square of the length of line. Its maximum value occurs at one end or the other, where the vector sum of the two components is also a maximum.

The residual fault arc seems to be inherently unstable and will eventually be extinguished. However, for the preservation of system

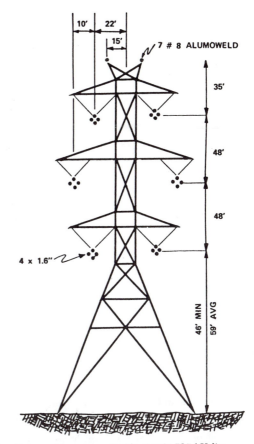

Fig. 4. *Proposed tower for double-circuit 735-kV line.*

stability, it must be extinguished rapidly so as to permit successful fast reclosure. The extinction time depends on the current and on random factors, including wind. Tests on lumped-capacitance simulations of uncompensated lines have shown that, if the residual fault current does not exceed 20 A, the extinction is so rapid that half-second reclosure is nearly always successful. If, however, the current exceeds 30 A, the extinction time is too great to permit reliable rapid reclosure.

If reliable clearing of faults by single-pole or selective-pole switching is desired and if the residual fault current exceeds 20 A, the best remedy is to add special shunt compensation which will reduce that current to a value well below 20 A at the greatest expected load.

Line Chosen as Example

For a test of the adequacy of special shunt compensation for expediting extinction of residual arcs on a double-circuit line, a long, EHV, heavily loaded, stacked double-circuit line with large bundle conductors was chosen. More specifically, the line is described as follows:

Normal voltage, 735 kV, rms, phase to phase

Frequency, 60 Hz

Length, 200 mi (322 km)

Configuration, stacked (see Fig. 4)

Bundle conductors, 4 Chukar subconductors (diameter 1.6 in. = 4.06 cm) at corners of an 18-in. (46-cm) square.

The shunt capacitances are listed in Table I. There are 21 of them, but because of symmetry they have only 12 different values.

Table I — Shunt Capacitances of Line (nF/mi)

Conductor to Ground		Interphase	
$C_{01} = C_{06} =$	10.878	$C_{12} = C_{56} =$	3.021
$C_{02} = C_{05} =$	7.623	$C_{13} = C_{46} =$	0.806
$C_{03} = C_{04} =$	9.216	$C_{23} = C_{45} =$	3.328
Avg. $C_g =$	9.239	Avg. $C_h =$	2.385

Intercircuit Unlike Phases		Intercircuit Like Phases	
$C_{15} = C_{35} =$	1.310	$C_{14} = C_{36} =$	0.551
$C_{24} = C_{35} =$	1.511	$C_{25} =$	1.918
$C_{16} =$	2.969		
$C_{34} =$	3.448		
Avg. $C_i =$	2.010	Avg. $C_j =$	1.007

SHUNT COMPENSATION OF SINGLE−CIRCUIT LINE

Purposes

Shunt compensation of long lines serves two purposes:

1. Limitation of power-frequency overvoltage, especially at an open end of the line and at buses having transformers or loads.

2. Neutralization of shunt-capacitive coupling between phases, making single-pole or selective-pole switching feasible on a longer line.

For serving the first purpose alone, one or two banks of Y-connected shunt inductors with solidly grounded neutral are commonly employed. The size of the inductors can be expressed (a) as the three-phase reactive power (megavars) consumed at rated voltage, (b) as the reactance per phase X_p, or (c) as the degree of shunt compensation,

$$h_1 = -B_{L1}/B_{C1} \quad \text{(numeric)} \qquad (1)$$

See Table II for the notation employed, including sign conventions. For serving both purposes simultaneously, the neutral point of the inductors selected for the first purpose is grounded through a fourth inductor, the neutral inductor. The bank of inductors then becomes a four-pointed star (Fig. 5). The method of determining the reactance of the neutral inductor will now be described, using a method which can be extended to double-circuit lines.

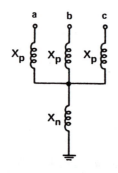

Fig. 5. Four-branched symmetrical star circuit for shunt compensation of balanced single-circuit line.

Table II — Notation

Symbol	Definition	Reference
Main Symbols		
B	Susceptance†	
C	Capacitance	
h	Degree of shunt compensation	Table IV
X	Reactance†	
Subscripts		
C	Capacitive	
L	Inductive	
1, 2, 3, 4, 5, 6, . .	Positions of line conductors*	Table I, Fig. 11
a, b, c, d, e, f . . .	Line conductors (phases)*	Figs. 9, 10
g or 0	Ground	Tables I, V
0, 1, 2, 3, 4, 5 . .	Symmetrical components* Branches of symmetrical circuits	Fig. 13
g, h, i, k	Mesh circuit	Figs. 6, 7, 14
m, n, p	Tree circuit	Fig. 12, Table IV
n, p	Star circuit	Fig. 5

* For single circuit, use first 3 characters.
† Inductive reactance and susceptance are positive; capacitive reactance and susceptance are negative.

Shunt Capacitances of Line and Their Symmetrical Components

The shunt capacitances of a transmission line are usually described by either a capacitance matrix or by a mesh circuit (Fig. 6), the two representations being related simply. For a balanced line, there are only two different values of capacitance. In the matrix the elements on the main diagonal have one value; the off-diagonal elements, another value. In the equivalent mesh circuit, the phase-to-ground capacitances have one value; the phase-to-phase capacitances, another and smaller value. We shall use the capacitive susceptances of the mesh circuit, B_{Cg} and B_{Ch}. The symmetrical components of these susceptances are found by assuming a set of voltages applied line to ground, first of zero sequence, and second of positive sequence, and finding the currents that are caused by each set of voltages. The zero-sequence voltages are equal in phase and magnitude. Therefore, zero-sequence currents exist only in the grounded branches, with none in the phase-to-phase branches. Positive-sequence voltages are equal to one another in magnitude but differ in phase by 120°. They cause currents in both the phase-to-ground branches and in the phase-to-phase branches.

The resulting expressions for symmetrical components of capacitive susceptance in terms of the capacitive susceptances of the branches of the mesh circuit are as follows:

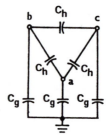

Fig. 6. Mesh circuit representing shunt capacitances of balanced single-circuit line.

$$B_{C0} = B_{Cg} \tag{2}$$

$$B_{C1} = B_{Cg} + 3B_{Ch} \tag{3}$$

The inverse set of equations is:

$$B_{Cg} = B_{C0} \tag{4}$$

$$B_{Ch} = (B_{C1} - B_{C0})/3 \tag{5}$$

The negative-sequence susceptance of this circuit is equal to the positive-sequence susceptance and need not be considered further.

Conditions for Shunt Compensation in Terms of Symmetrical Components

The conditions to be met are, for the first purpose of shunt compensation, from eq. 1,

$$B_{L1} = -h_1 B_{C1} \tag{6}$$

and, for the second purpose,

$$B_{Lh} = -B_{Ch} \tag{7}$$

where B_{Lh} is the inductive susceptance between phases in the mesh circuit (Fig. 7) equivalent to the star circuit (Fig. 5) which is used because of its simplicity. However, rather than to pass directly from mesh to star, we shall use the intermediate step of symmetrical components. By eq. 5 and the corresponding equation for the inductive mesh,

$$B_{L1} - B_{L0} = -(B_{C1} - B_{C0}) \tag{8}$$

By substituting eq. 6 into eq. 8 and solving for B_{L0},

$$B_{L0} = -[B_{C0} - (1 - h_1)B_{C1}] \tag{9}$$

and we already have (eq. 6)

$$B_{L1} = -h_1 B_{C1}$$

These are the two conditions to be met.

Inductive Reactances of Star Circuit

Now, because of the independence of the sequence circuits corresponding to a symmetrical actual circuit, these circuits are decoupled, and we have

$$X_{L0} = 1/B_{L0} \tag{10}$$

and

$$X_{L1} = 1/B_{L1} \tag{11}$$

Examination of the star circuit of Fig. 5 with pure positive-sequence currents into the terminals gives

$$X_{L1} = X_p \tag{12}$$

and, with zero-sequence currents,

$$X_{L0} = X_p + 3X_n \tag{13}$$

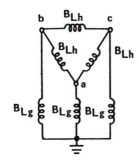

Fig. 7. Symmetrical mesh circuit equivalent to the symmetrical star of Fig. 5.

The inverse set of eqs. 12 and 13 is

$$X_p = X_{L1} \tag{14}$$

$$X_n = (X_{L0} - X_{L1})/3 \tag{15}$$

To summarize, X_n is given by eq. 15, wherein X_{L0} and X_{L1} are the reciprocals of B_{L0} and B_{L1} respectively, which are given by eqs. 9 and 6, respectively, in terms of known line capacitive susceptance B_{C0} and B_{C1} and known degree h_1 of shunt compensation. Because all the equations involved are so simple, we can derive the following expression for X_n as a function of the known quantities:

$$X_n = \frac{B_{C1} - B_{C0}}{3h_1 B_{C1}[B_{C0} - (1 - h_1)B_{C1}]} \tag{16}$$

Unbalance of Capacitances and its Remedy

Two common configurations of single-circuit lines are symmetrical with respect to the central vertical plane. Each of these has four different values of shunt capacitance, as shown in Table III. It may be observed that there is such a serious unbalance of the phase-to-phase capacitances that it is not possible accurately to compensate a long untransposed line by use of a symmetrical circuit of inductors.

Table III — Capacitances of Single-Circuit Lines

Configuration (See Fig. 8)	Shunt Capacitance per unit length (nF/mi)			
	Phase to ground		Phase to phase	
	Outers	Middle	Outer to Middle	Between Outers
Flat	14.83	13.32	3.068	0.803
Delta	13.10	12.55	3.017	2.059

The unbalance may be eliminated either by transposition or by addition of a lumped capacitance of proper value connected between the two outer conductors.

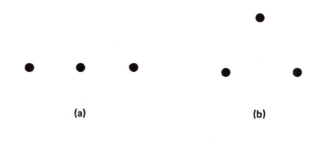

Fig. 8. Common configurations of single-circuit lines.

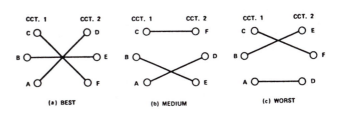

Fig. 9. Cross sections of double-circuit line showing three different arrangements of the phase conductors in which each circuit is kept on its own side of the tower. Like phases are joined by lines.

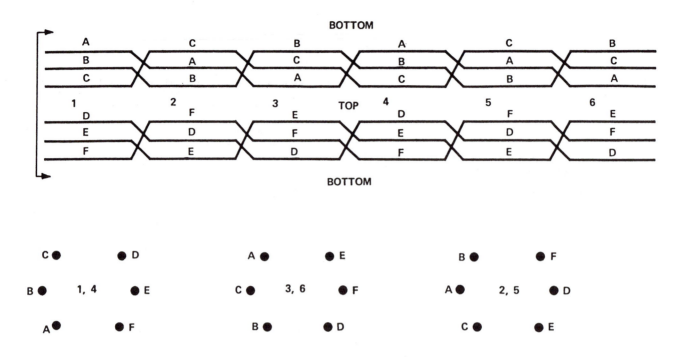

Fig. 10. *Transposition scheme of line studied. Developed plan view and the three different cross sections.*

SHUNT COMPENSATION OF DOUBLE–CIRCUIT LINE

Shunt Capacitances and Reduction of the Number of Independent Values

The double-circuit line, having six energized conductors, has, in addition to 6 phase-to-ground capacitances, $(6 \times 5)/2 = 15$ interconductor capacitances which, in the general case, might all be different. Transposition could reduce the number of different values of capacitance, and in the extreme reduction all the interconductor capacitances could have equal values.

There is, however, a restraint imposed on the transposition scheme of a double-circuit EHV line by the limitation on the capacitive current from the energized conductors to a large vehicle located on the ground beneath the line. Figure 9 shows three different arrangements of the phase conductors of a stacked double-circuit line. At 500 kV, two of these arrangements (*a* and *b*) are acceptable from this standpoint; at 700 kV, only arrangement *a* is acceptable. Then the only permissible transposition scheme is the one in which the six conductors maintain the same relative positions, as in Fig. 10. The effect of such transposition is to equalize the average value over three equal, successive line sections of all the capacitances in each of the four groups shown in Fig. 11. Numerical values for the line used as example appear in Table I.

Next let us consider some of the symmetrical polyphase circuits that might be used for the shunt inductive compensation of a double-circuit three-phase line. Three such circuits, all of tree form, are shown in Fig. 12. They correspond to three different ways of factoring the number 6 into the numbers of equal branches at each fork, proceeding from the ground to the phases: (*a*) 1×6, (*b*) $1 \times 2 \times 3$, and (*c*) $1 \times 3 \times 2$. The most suitable circuit for compensating a double-circuit line appears to be that shown in part *b*.

None of these compensating circuits, however, has more than three different values of reactance, while there are four different values of capacitances to be compensated (Fig. 11), including those to ground. If we transform the tree circuit of Fig. 12*b* to a mesh circuit, we find that the 9 intercircuit branches corresponding to the capacitive branches of Fig. 11*c* and *d*, are all equal. This suggests a remedy: connection of lumped capacitors between like phases, as in

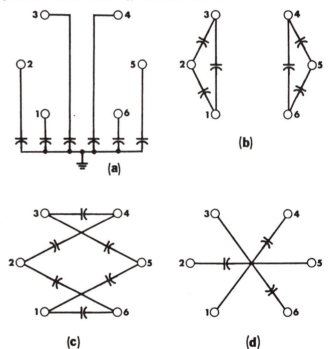

Fig. 11. *Shunt capacitances of double-circuit line:*
(a) conductor to ground,   *(b) interphase,*
(c) intercircuit, unlike phases,   *(d) intercircuit, like phases*

Figure 11*d*, which, added to the distributed shunt capacitances between like phases, raise them to equality with the capacitances between unlike phases (Fig. 11*c*). The cost of this method is expected to be moderate because the capacitance of these "trimming capacitors" is small and because the voltage across them is normally zero. In what follows, these capacitors are assumed to be connected, and they are treated henceforward as part of the line capacitance.

Equations for computing the reactances of the compensating circuit of Fig. 12*b* will now be derived. In the derivation, it is found

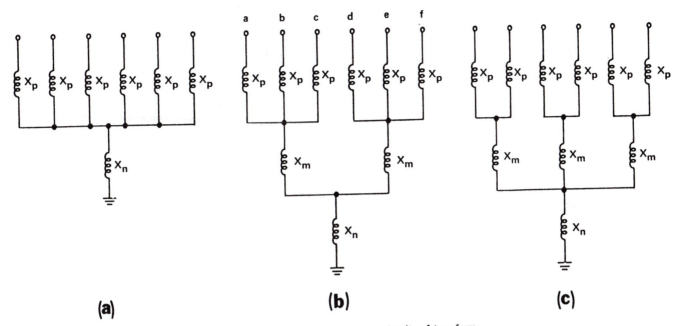

Fig. 12. Six-terminal symmetrical polyphase circuits of tree form.

simpler to introduce the intermediate step of obtaining the symmetrical components of the line capacitive susceptances and of the inductive reactances of the compensating circuit, than to pass directly from a symmetrical mesh circuit to a symmetrical tree circuit.

Symmetrical Components

This process requires that the symmetrical components for a three-phase double-circuit line be defined. They are based on the well-known symmetrical components of a single three-phase circuit, elaborated as follows. Sequences 0, 1, and 2 (zero, positive, and negative sequences) have equal quantities in both circuits. Sequences 3, 4, and 5 have quantities in the second circuit that are antiequal* to the corresponding quantities in the first circuit. The unit vectors are shown in Fig. 13. The defining equations are given here only for voltages of sequences 0, 1, and 3. The equations for current differ from those for voltage only in that V's are replaced by I's. All the V's and I's are complex (phasors).

$$V_0 = (V_a + V_b + V_c + V_d + V_e + V_f)/6 \tag{17}$$

$$V_1 = (V_a + aV_b + a^2V_c + V_d + aV_e + a^2V_f)/6 \tag{18}$$

$$V_3 = (V_a + V_b + V_c - V_d - V_e - V_f)/6 \tag{19}$$

where $a = \underline{/120°}$ and $a^2 = \underline{/240°} = \underline{/-120°}$. $\tag{20}$

A sample of the inverse equations, giving phase voltages in terms of their components, is the following:

$$V_e = V_0 + a^2V_1 + aV_2 - V_3 - a^2V_4 - aV_5 \tag{21}$$

Sequence 1 may be regarded as the normal operating condition.

Symmetrical Components of Shunt Capacitive Susceptances of the Line

These are found in the same way as for a single circuit: by assuming a set of voltages of one sequence applied from conductor to ground, taking each sequence, 0, 1, and 3, in turn. (For static symmetrical circuits, sequences 1, 2, 4, and 5 have identical admittances or impedances.) The currents in each branch of the mesh circuit from a particular terminal — say, phase a, Fig. 14 — and their sum are written. For example, for zero sequence, there are equal

voltages V_0 from every phase to ground. Hence there is no current in any branch except that from a to ground, and the current entering terminal a and every other phase terminal is:

$$I_a = I_0 = -j\omega C_g V_a = -jB_{Cg}V_0 = -jB_{C0}V_0 \tag{22}$$

Hence

$$B_{C0} = B_{Cg} \tag{23}$$

For positive sequence, the expression is similar to eq. 3 but with an additional term due to coupling between circuits:

$$B_{C1} = B_{Cg} + 3(B_{Ch} + B_{ci}) \tag{24}$$

For third sequence:

$$B_{C3} = B_{Cg} + 4B_{Ci} + 2B_{Cj} = B_{Cg} + 6B_{Ci} \tag{25}$$

The inverse equations are:

$$B_{Cg} = B_{C0} \tag{26}$$

$$B_{Ch} = (B_{C3} - B_{C0})/6 \tag{27}$$

$$B_{Ci} = (2B_{C1} - B_{C0} - B_{C3})/6 \tag{28}$$

Symmetrical Components of Inductive Reactances of Shunt Tree Circuit

The procedure for finding these is the dual of the procedure for finding the symmetrical components of shunt capacitive susceptances. A set of currents of a single sequence is impressed on each terminal instead of a voltage, the resulting branch currents and voltage drops are found by inspection, and the drops are summed from phase to ground. The results are these:

$$X_{L1} = X_p \tag{29}$$

$$X_{L3} = X_p + 3X_m \tag{30}$$

$$X_{L0} = X_p + 3X_m + 6X_n \tag{31}$$

The inverse equations are:

$$X_p = X_{L1} \tag{32}$$

$$X_m = (X_{L3} - X_{L1})/3 \tag{33}$$

$$X_n = (X_{L0} - X_{L3})/6 \tag{34}$$

*Equal in magnitude but opposite in sign (or 180° apart in phase).

101

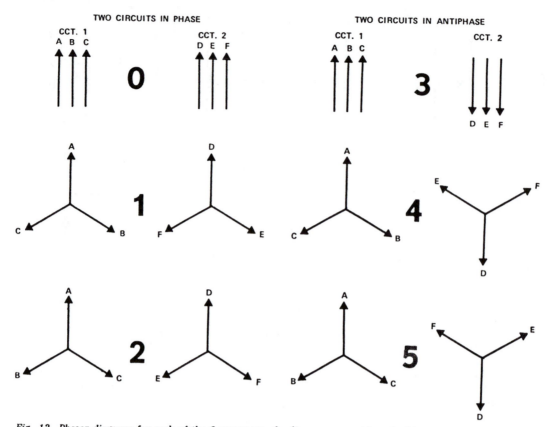

Fig. 13. Phasor diagrams for each of the 6 sequences of voltage or current in a double-circuit 3-phase line.

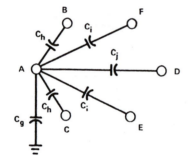

Fig. 14. Line capacitances connected to conductor a.

Required Conditions of Shunt Compensation

See footnote to Table II for convention on algebraic signs of reactances and susceptances.

For the desired degree h_1 of positive-sequence shunt compensation,

$$B_{L1} = -h_1 B_{C1} \qquad (35)$$

For obtaining complete compensation of the interphase capacitive susceptances of each circuit,

$$B_{Lh} = -B_{Ch} \qquad (36)$$

and, for the intercircuit susceptances,

$$B_{Li} = -B_{Ci} \qquad (37)$$

where B_{Lh} and B_{Li} are the inductive susceptances of branches of a mesh circuit equivalent to the desired tree circuit. Equations 36 and 37 will be put into terms of symmetrical components. By analogy with equations 27 and 28,

$$B_{Lh} = (B_{L3} - B_{L0})/6 \qquad (38)$$

and

$$B_{Li} = (2B_{L1} - B_{L0} - B_{L3})/6 \qquad (39)$$

When the right-hand members of eqs. 27, 28, 38 and 39 have been substituted into equations 36 and 37, we obtain, after clearing of fractions,

$$B_{L3} - B_{L0} = -(B_{C3} - B_{C0}) \qquad (40)$$

and, with additional substitution of eq. 35,

$$2h_1 B_{C1} + B_{L3} + B_{L0} = 2B_{C1} - B_{C3} - B_{C0} \qquad (41)$$

By taking half the sum and half the difference of the last two equations, and solving for B_{L0} and B_{L3}, we get:

$$B_{L0} = B_{C1}(1 - h_1) - B_{C0} \qquad (42)$$

$$B_{L3} = B_{C1}(1 - h_1) - B_{C3} \qquad (43)$$

while B_{L1} is given by eq. 35.

Next we take the reciprocal of each B_L in order to obtain the corresponding X_L:

$$X_{L0} = 1/B_{L0}, \quad X_{L1} = 1/B_{L1}, \quad \text{and} \quad X_{L3} = 1/B_{L3}. \qquad (44, 45, 46)$$

We now use eqs. 32 to 34 to obtain the reactances of the tree circuit.

It should be noted that all these reactances depend upon the degree h_1 of positive-sequence shunt compensation. At a certain value of this compensation ($h_1 = 0.588$ for the line used as example), $X_n = \infty$. This means that the compensating circuit has only the branches X_p and $2X_m$ and is ungrounded. The same phenomenon occurs in the compensation for a single circuit, normally at a lower value of h_1.

Numerical Values of Reactances

Computed values of the reactances of the tree-form shunt-connected compensating circuits for the line used as example are given in Table IV.

102

Table IV — Numerical Values of the Branch Reactances
of Shunt-Connected Compensating Circuits (ohms)
for 100 Miles of Line

Degree of Shunt Compensation			Tree Circuit		
h_1	h_0	h_3	X_p	X_m	X_n
.050	−1.306	.000	23,660.	∞	∞
.100	−1.184	.052	11,830.	3,975.	−4,363.
.200	− .941	.158	5,915.	661.	−1,824.
.300	− .699	.263	3,943.	264.	−1,474.
.400	− .456	.368	2,958.	141.	−1,613.
.500	− .213	.474	2,366.	88.1	−2,680.
.588	.000	.566	2,012.	62.6	∞
.600	.029	.579	1,971.	60.0	15,990.
.700	.272	.684	1,690.	43.5	1,456.
.800	.515	.789	1,479.	33.0	666.8
.900	.757	.895	1,314.	25.9	399.8
1.000	1.000	1.000	1,183.	20.8	270.9
1.100	1.243	1.105	1,075.	17.1	197.2
1.200	1.485	1.211	985.8	14.3	150.6
1.300	1.728	1.316	910.0	12.2	119.1
1.400	1.971	1.421	845.0	10.5	96.7
1.500	2.213	1.526	788.7	9.1	80.2

TESTS OF SHUNT COMPENSATION

Line Represented by Capacitors (Series 1)

In the first series of tests run on a digital computer, half of the example line (100 mi) was represented only by its shunt capacitive susceptances, and one bank of shunt inductors was connected in parallel with it. This circuit was designed for each of several degrees of positive-sequence shunt compensation. Balanced three-phase voltages of 735 kV rms line to line were applied. Many different kinds of short-circuit faults were applied to one or both line circuits.

The purpose of these tests was to discover and to correct errors made in the derivation of the compensating circuits. After correction of the errors, residual fault currents were of the order of milliamperes, and recovery voltages were of the order of volts. The fact that these quantities were not all null is the result of round-off errors.

Line Represented by Pi Sections (Series 2 and 3)

The second and third sets of tests were made with the line represented by six sextuple pi sections in cascade, each representing one transposition section 33.3 miles long. Balanced positive-sequence voltages of 735 kV$/\sqrt{3}$ (line-to-ground) were applied to each end. In the second series of tests, the receiving-end voltage was in phase with the sending-end voltage for simulating no-load conditions. In the third series, 50% series compensation was inserted at the midpoint of the line, and the delivered power was increased to 6.64 GW (1.5 SIL) by making the phase difference between sending-and receiving-end voltages equal to 34°. Shunt compensation, including capacitors between like phases, was connected at both ends. In both sets of tests, several different types of faults were applied; however, Table V shows the results only for SLG faults on circuit 1 or on both circuits. In addition, comparative cases were run with no compensation. Values recorded are the highest for any point of any phase of either circuit.

In the no-load tests (series 2) with shunt compensation, the recovery voltages were a few kilovolts, and the residual arc currents, a few amperes. These values were about 1000 times as great as in series 1. The probable reason is that the voltage is not constant along the line, being higher at the midpoint than at the ends. Nevertheless, the values are very small and would give very rapid extinction of

arcing faults. With no shunt compensation, arc extinction, even at no load, would be intolerably slow.

In the load tests (series 3), with no compensation, the secondary arc current and the recovery voltage are increased by 13 to 19%. This shows that the shunt-capacitive coupling is more important than the series-inductive coupling at a load of 1.5 SIL. With shunt compensation, the recovery voltages were reduced from 0 to 30%, depending on the degree of compensation; but the residual fault current was reduced more, by 67 to 90%. However, some of these currents were still as high as 32 to 44 A, considerably above the tentative limit of 20 A. Since the shunt coupling was neutralized, these currents were caused almost entirely by series-inductive coupling. Methods of reducing the effects of series coupling were therefore sought.

REDUCTION OF LONGITUDINAL COUPLING

Three different methods were employed for reducing longitudinal inductive coupling.

Three-pole Switching

Tests were made with SLG faults on one or both circuits. The values of residual fault current were still too high to give reliable arc extinction. In addition, three-pole switching, even with rapid reclosure, would seriously stress the system and jeopardize stability after occurrence of a double-circuit fault.

Grounding the Faulted Conductor (Series 4)

A method of quenching residual fault arcs proposed long ago[2] is to ground the faulted, disconnected conductor at one or both terminals through either a switch or a liquid spray. When this method was tried by placing a SLG fault at the midpoint of the line adjacent to a series capacitor, the residual arc current was very high, 1580 A. The high value occurs because of series resonance, as the capacitor gives 100% compensations to half the line. Still the capacitor current (3011 A) would not be great enough to fire the protective gap. With a higher degree of series compensation, resonance could occur with the fault at some other location. This method was discarded.

Sectionalizing the Faulted Conductor (Series 5)

The most promising method is to provide a switch in each of the six conductors at the midpoint of the line. Upon occurrence of a fault, the faulted conductors are first disconnected at the two terminals and are then sectionalized by the appropriate midpoint switches. After expiration of the time allowed for arc extinction, first the midpoint switches are reclosed and finally the terminal PCB's are reclosed. The relays which trip the PCB poles at the terminals send a signal to open the appropriate midpoint switches, and the reclosing relay sends a signal to reclose those switches an instant before reclosing the PCB poles. The sectionalizing switch breaks and makes only the residual fault current but must carry load current continuously.

The effect of conductor sectionalization upon residual fault current and recovery voltage of a SLG fault are shown in the lower part of Table V and are compared with the corresponding values for an unsectionalized line in Table VI.

It may be noted that the ratios of residual fault current are very nearly half the corresponding ratios of recovery voltage. This could have been expected from the fact that the shunt capacitive admittance of half the line is half that of the whole line, these being the Thevenin's admittances. Both ratios are higher than the theoretical ones (0.5 for voltage and 0.25 for current).

Voltage profiles of an open conductor, sectionalized, are shown in Fig. 15, which should be compared with Fig. 3, drawn for an unsectionalized conductor.

It is clear that more reduction of recovery voltage and of residual fault current than those obtained by division of disconnected conductors into two sections could be obtained by division into a

Table V — Residual Fault Currents and Recovery Voltages of SLG Fault
and of Simultaneous SLG Faults on Same Phase of Both Circuits
of Double-Circuit 735-kV 200-Mile Line Represented by Six Pi Sections

Test Series	Shunt Circuit	Compensation Degree h_1	Line Load (GW)	Fault on Conductors	Recovery Voltage (kV)	Residual Fault Current (A)	Thevenin's Impedance (kilohms)	Remarks
2	None	0	None	ag or bg	79.5	116.	0.685	
	Tree	0.7	None	cg	3.48	1.78	1.96	
	Tree	0.9	None	bg	5.56	1.02	5.45	
3	None	0	6.64	bg	89.7	134.	0.669	
	Tree	0.6	6.64	bg	62.9	44.5	1.41	50% series compensation at midpoint.
	Tree	0.7	6.64	bg	65.9	35.2	1.87	
	Tree	0.7	6.64	adg	60.3	32.2	1.87	
	Tree	0.9	6.64	bg	89.3	17.2	5.19	
	Tree	0.9	6.64	adg	69.2	13.2	5.25	
4	Tree	0.7	6.64	ag	44.0	1,581.	0.028	Faulted conductor grounded at both ends.
5	None	0	6.64	bg	78.7	57.4	1.37	
	Tree	0.6	6.64	bg	41.4	14.1	2.94	Faulted conductor sectionalized at midpoint of line.
	Tree	0.7	6.64	bg	41.2	10.9	3.78	
	Tree	0.9	6.64	cg	112.	10.6	10.6	

Table VI – Decrease of Recovery Voltage and Residual Fault Current
by Division of Faulted Conductor into Two Sections; SLG Fault

h_1	Recovery Voltage (kV)			Residual Fault Current (A)		
	Sectionalized	Unsectionalized	Ratio (Sec./Unsec.)	Sectionalized	Unsectionalized	Ratio (Sec./Unsec.)
0	78.7	89.7	0.879	57.4	134.0	0.428
0.6	41.4	62.9	0.657	14.1	44.5	0.317
0.7	41.2	64.9	0.625	10.9	34.7	0.310
0.9	92.0	66.6	1.38	8.8	12.8	0.69

greater number of sections. This, of course, would require more switches and would result in lower reliability of operation.

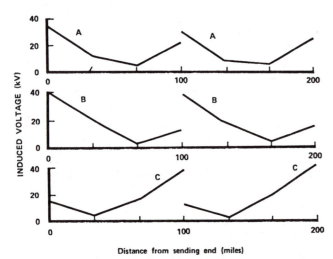

Fig. 15. Voltage profile of one open conductor, sectionalized at midpoint, to ground, $h_1 = 0.7$ Voltages across sectionalizing switch (a) 49.4 kV, (b and c) 50.3 kV.

CONCLUSIONS

It has been shown that:

1. Selective-pole switching can be used for clearing any type of transitory fault involving any 1, 2, 3, 4, or 5 of the 6 conductors of a double-circuit line.

2. The transient stability limit will be higher than that for three-pole switching, especially for faults involving both circuits.

3. For such switching to be successful on a long EHV line, the shunt capacitive susceptance must be neutralized.

4. A feasible method of neutralization consists of one or more banks having the following three types of branch:

a. Three capacitors connected between like phases;

b. Six equal inductors connected in two Y's and having a value of reactance that furnishes the desired degree of positive-sequence compensation.

c. A three-terminal circuit connected among the neutral points of the two Y's and ground. This circuit may consist of three reactors in T connection.

5. If the residual fault current and/or recovery voltage are still too high for prompt extinction of residual arcs, these quantities may be decreased by separating the faulted conductor or conductors into two or more sections soon after these conductors have been opened at the two ends of the line.

The method described above has been tested by computation on a 200-mile, 735-kV, 3-phase, double-circuit line having bundle conductors of four Chukar subconductors. It is expected to be applicable to a 300-mile line of the same or similar type. It is also applicable to 500-kV double-circuit lines.

ACKNOWLEDGEMENTS

I thank R. M. Hasibar for his discovery of errors in my derivations and for performing nearly all of the computations.

REFERENCES

1. H. Thommen, "Investigations on High-Speed Reclosing on the Occurrence of Short-Circuits in Overhead Line Systems and the Application of the Air-Blast High-Speed Breaker for the Purpose", *C.I.G.R.E.*, 1939, report 108, vol. III, 26 pp.

2. A. Amstutz, "Residual Currents and Voltages with Single-Pole Reclosing", *Brown Boveri Rev.*, vol. 35, pp. 220–226, July/August 1948.

Discussion

F. S. Prabhakara and K. R. Shah (Commonwealth Associates Inc., Jackson, Michigan): The author is to be congratulated for a very good paper on the subject of selective pole fault clearing — the problems and possible solutions.

The discussers would like to make a few observations about the paper. The first comment is about the permissible residual fault current. The author has assumed 20A as permissible limit for 700 kV. How does this limit compare with the following statement made in reference (A) which we quote, "the formula would indicate that a recovery voltage of less than 20 kV with 40 ampere would not be extinguished on BPA 500 kV lines. Single pole reclosing tests have indicated about 45 kV recovery voltage at 40 amperes as about the limit". The deadtime required for fault extinction would increase with the increase in recovery voltage and the residual current. The values in Table V also indicate that the recovery voltage and residual current increase with the prefault load carried by the line. Since the ultimate quantity of interest in the deadtime required for successful automatic reclosure, it appears that the time required for secondary arc extinction for all cases in Table V should be very useful. Could the author include this in his closure.

The nonsuitability of grounding the ends of the faulted conductor to reduce the residual fault current is very well illustrated by the results of test series #4. This study indicates that the most desirable configuration would be tree connected shunt circuit, with degree compensation $h_1 = 0.7$ and faulted conductor sectionalized at the midpoint of the line. Another question is what would be the effect of changing series capacitor location to one or both the ends. Would this change the results in Table V considerably?

Another alternative to increase the transient stability margin is to provide a switching station in the middle of the line. With this arrangement only 100 mile line section has to be disconnected for isolation of fault. In addition single pole switching could be applied to each section. Due to smaller line length being switched out, the deadtime for successful reclosure may not be very high. Would the author comment on relative technical and economical advantages and disadvantages of such an alternative?

Another important point brought out in the paper is the optimum phase configuration. The author has also discussed in the paper, the best, medium and worst double circuit phase configuration (Fig. 9) from the electrostatic ground gradient point of view. When optimum phase configuration of long untransposed multicircuit lines are to be determined, other factors such as circulating currents, high and low reactance configurations should be considered. Recently, we completed a study to determine optimum phase configuration for two 500 kV double circuit lines on the same right-of-way. Maximum electrostatic ground gradient for twenty-one phase configurations were calculated. Eight configurations having acceptable maximum gradient values were identified. For each configuration circulating current unbalance and power loss were calculated. The most suitable configuration in the order of preference is given below:

BAC–CBA–ABC–BCA	— optimum
BCA–CBA–ABC–BAC	— 2nd best
ABC–BAC–CBA–BCA	— 3rd
CBA–BAC–BCA–ABC	— 4th

Fig. A. Phase Arrangement of BCA–CBA–ABC–BAC 500 kV
2 D/C Lines in the Same R–O–W

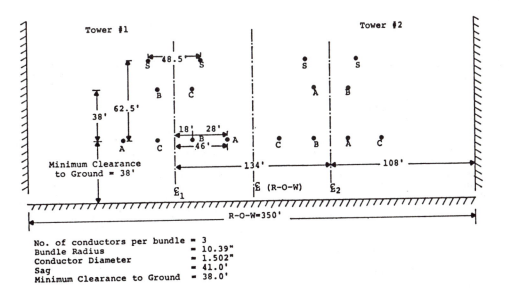

No. of conductors per bundle = 3
Bundle Radius   = 10.39"
Conductor Diameter   = 1.502"
Sag   = 41.0'
Minimum Clearance to Ground   = 38.0'

Manuscript received July 23, 1975.

ABC–BAC–BCA–CBA — 5th
BAC–CBA–BAC–CBA — 6th
BCA–ABC–BCA–ABC — 7th
CBA–BCA–ABC–BAC — 8th

However due to the structural considerations at the station, the 2nd best configuration was selected. This configuration is shown in Fig. A.

The last question is about disconnecting one of the two lines. Under such circumstances should the intercircuit capacitances shown in Fig. 11 be disconnected too. Also which type of shunt reactor connection (shown in Fig. 12) would avoid or minimize any possible resonant overvoltages (C).

REFERENCES

(A) "Electromagnetic Effects of Overhead Transmission Lines – Practical Problems, Safeguards, and Methods of Calculation", IEEE Trans. Vol. PAS-93, No. 3, May/June 1974. See disc. by D. A. Gillies and D. E. Perry, page 900.
(B) "Transmission Line Reference Book – 345 kV and Above", Electric Power Research Institute, California 1975.
(C) M. J. Pickett et al, "Near Resonant Coupling on EHV Circuits Part I – Field Investigations", Presented at IEEE Winter Meeting, 1967.

A. B. Sturton (The Shawinigan Engineering Co. Ltd., Montreal, P.Q., Canada): The discussion is intended to support this very excellent paper by actual operating experience (on unscheduled fault, except where noted) — although at lower voltages, and on uncompensated lines — on systems where we were responsible for applications and design.

At the 230 kV levels, SLG fault occurred on a 100 mile, radial line connecting 300 MW of hydro generation into a major system. The fault was cleared through one pole opening, but failed to reclose at one line terminal, leaving power transmittal over two phases. Steady state stability was attained during the approximate 2 seconds time lag required for breaker disagreement protection to operate.

The LL fault does not necessarily require that both faulted phases be opened, and indeed, single pole switching of LL fault has been applied and performed correctly on three – 230 kV circuits in Quebec, Canada. It is acknowledged, however, that there is relatively little requirement for accommodating the LL fault, and in these particular cases, it was influenced largely through simplification of relaying design.

The LLg fault has been switched two phase on single 230 kV transmission (no other interconnection) between utilities, with a preload of only 20 MW, and without instability.

Simultaneous Lg fault on the same phase of double circuit, 135 mile, 230 kV transmission has been switched correctly without instability, and with order of magnitude prefault loading of 350 MW.

The Lg fault has been switched successfully on 345 kV transmission on numerous occasions. During scheduled fault throwing tests, and with 225 miles of uncompensated line operating at 400 kV, it was demonstrated that a 55 ampere secondary arc was unstable. (Our own investigation into this problem indicates that secondary arc stability is very much contingent on the ratio of power arc/secondary arc current.)

Oscillograph records of unscheduled faults at 345 kV show clearly that secondary arc restrike does occur. Typical values are approximately 50–80 ms from breaker opening to initial secondary arc extinction, 4–16 ms to restrike, and 30–50 ms to second extinction. We have no recorded cases of more than 1 restrike. During scheduled test, however, it was demonstrated that for remote end fault on a 275 mile, 230 kV line, with a power arc/secondary arc ratio of only 14:1, time to initial arc extinction was 700 ms.

In further support of Mr. Kimbark, our own experience regarding transient SLG faults is that for:
  i) Dead times < 0.3 seconds, reclosing will probably be unsuccessful, even if secondary arc goes out in 50–100 ms, since there is a real and apparently fairly long time required for build up of the di-electric strength of the air around the arc path – unless there is a substantial wind.
  ii) Dead time > 0.3 seconds, but < 0.5 seconds, reclosing may not be successful.
  iii) Dead time > 0.5 seconds, secondary arc and di-electric recovery will not interfere with successful reclose.

We would commend the author on the content and presentation of this paper, and believe it is a major contribution to a technique which is not adequately used today.

Manuscript received July 28, 1975.

G. B. Furst (The Shawinigan Engineering Co. Ltd., Montreal, P.Q., Canada): I would like to join my colleague, Mr. A. B. Sturton, in complimenting Dr. E. W. Kimbark for a very fine paper and would like to make the following comments.

1. I assume that the residual voltages shown in the paper are the steady state values. On a line partially compensated by shunt reactors the sec. arc recovery voltage will oscillate with a beat which is determined by the degree of shunt compensation and it is the first peak of the recovery voltage which has the dominent influence on the extinction of the secondary arc. This peak can be considerably smaller than the steady state value, some studies showed values in the order of 35%. The secondary arc current also exhibits superimposed oscillations with transient peaks considerably larger than the steady state value.

It is therefore important to carry out a detailed digital transient study or TNA study to determine the correct magnitude of secondary arc current and recovery voltage.

2. The extinction of the secondary arc current has a statistical distribution which is shown by the laboratory tests on insulator strings but is, of course, even more so in real life where the success of arc extinction is affected by such factors as wind and the preceding power arc. There is little if any information available on secondary arc extinction at 500 kV and above where depending on line design the power arc can occur across the string, or from the conductor to the structure directly, giving different values for maximum permissible secondary arc current and recovery voltages.

We should try therefore, concurrently with devising methods to reduce secondary arc current and recovery voltages, to gain more knowledge of their permissible limits by lab. or system tests so that the cost benefit ratio and reliability of the various proposed secondary current limiting schemes can be properly evaluated, bearing in mind the primary purpose of the application whether it is to increase stability limits or simply minimize disturbances in the system.

3. In the same context referring to Table IV on page 8 of the paper, which shows the tree circuit parameters as a function of compensation, it would seem that the value of X_M is significant only for unusually low values of shunt compensation, thus in most cases the tree reduces to the relatively simple scheme with a neutral reactor which was previously suggested by the author for single circuit lines.

4. The parameters for the tree circuit were computed for the double circuit configuration. Have calculations been made on the quality of secondary arc current and recovery voltage compensation with one circuit temporarily out of service?

Manuscript received August 7, 1975.

Edward W. Kimbark: Messrs. Prabhakara and Shah ask how the limit of 20 A to the permissible residual fault current, assumed for a 700-kV line, compares with the 40 A cited from the earlier discussion by Messrs. Gillies and Perry of BPA (reference A) as "about the limit" for BPA 500-kV lines. Actually both of these limits are based on the same set of laboratory tests made by BPA. I would say that 20 A is a more conservative limit than 40 A and would insure that a high percentage of the transitory faults would be extinguished rapidly enough to permit 0.5-second reclosures. This does not deny that some of the faults having a higher residual current — up to 40 A or more — would be extinguished just as rapidly and that others of them would be extinguished more slowly.

The same discussers ask how the placement of the series capacitors would affect the residual fault current if the method of arc extinction is to ground the faulted conductor. Assuming that the degree of series compensation (50 percent) is unchanged but that it is placed at one end of the line, a fault at the midpoint of the line would give the same amount of resonance, whereas if the series compensation were divided into two equal parts, one part placed at each end, the resonant fault locations would be approximately one-fourth of the line length from each end.

The discussers next propose a switching station at the midpoint of the line instead of the six single-pole circuit switchers assumed to be there in the study; and they ask for the advantages and disadvantages. The disadvantage is a considerable increase in the cost, because the substation would have a bus and at least 12 single-pole circuit breakers, each more expensive than one of the circuit switchers, if a plain bus or a ring bus were chosen, and perhaps even more — three more for a bus sectionalizing breaker between the two circuits and six more if the breaker-and-a-half arrangement were adopted. In addition, the number

Manuscript received September 26, 1975.

of protective relays would be doubled. The advantage is some improvement in transient stability limit of the system because a lower net reactance between line terminals results from the fact that only one 100-mile section of the faulted conductor or conductors is opened instead of two such sections.

I would advise against the provision of a full switching station unless it were found necessary to feed a load or to make a connection with a lower-voltage transmission system.

I am interested in the problem of the best arrangement of conductors on *two* 500-kV double-circuit lines and note that in the arrangement selected no two adjacent conductors of the lower tier of eight conductors are energized at the same voltage phase; instead, conductors of like phase are three spaces apart. The arrangement chosen for our own study of one double-circuit 735-kV line (Fig. 10) also gives balanced currents and minimum reactance.

If one of the two circuits is disconnected but physically intact, the compensation scheme chosen for double-circuit operation and including the shunt capacitors between like phases is still correct. This is true because the compensating circuit is correct for *any* voltage of each conductor, including zero voltage if the conductor is grounded or a voltage determined by the capacitive network if the conductor is open at both ends. If, however, one circuit has not yet been strung, the other circuit should be operated with four-reactor shunt compensation (Fig. 5) in which the neutral reactor X_n has a value determined for single-circuit operation and different from that used for double-circuit operation. This difference is not economically important because such reactors, having low reactance and low BIL and being energized only intermittently, are much cheaper than the phase reactors. Also, one could specify a tapped reactor having two values of reactance depending on the tap used.

The final question asked by these discussers is which type of shunt reactor connection shown in Fig. 12 would avoid or minimize any possible resonant overvoltage. I regret that the paper did not make it clear that only the circuit of Fig. 12(b) was proposed for use with selective-pole switching. The circuits of parts (a) and (c) of Fig. 12 were presented merely as additional examples of a class of circuits which can be described as six-terminal symmetrical polyphase tree circuits. However, the circuit of Fig. 12(a) can be used in some cases, as suggested by Mr. Furst.

In the study we did use two other compensating circuits which were omitted from the paper to conserve space. They are modifications only of the central part of the circuit of Fig. 12(b), where reactors X_m and X_n are T-connected. The six phase reactors X_p are unchanged. The fact that any T circuit has its equivalent π circuit is familiar to most of us. This π, together with the phase reactors, constitutes the first alternate circuit. The second alternate circuit has two reactors, X_t and X_z, and one transformer, as shown in Fig. 16. If the transformer is considered to be ideal, the reactances of this circuit in terms of those (X_m

and X_n) of the equivalent T circuit and again, in terms of the symmetrical components X_0, X_1, and X_3 of all these circuits, are as follows:

$$X_t = 2X_m = (X_3 - X_1)2/3$$
$$X_z = X_n + X_m/2 = (X_0 - X_1)/6$$

We noted no difference among the performances of the three circuits nor any appearance of resonance, except with $h_1 = 0.9$.

The beats of recovery voltage between the forced frequency and the natural frequency, which Mr. Furst mentions, are very conspicuous in computed curves in which the fault is assumed to behave like an ideal switch, having zero impedance when closed and infinite impedance when open. The actual secondary arc, however, has a resistance which increases from some low value, which it has at the clearing of the primary arc, to a much higher value just before extinction of the secondary arc. This resistance damps the natural frequency rapidly, with the result that the beats are much less conspicuous. Unfortunately, we cannot simulate the phenomena of arc extinction by TNA or otherwise unless we know how to correctly represent the increase in arc resistance as a function of time.

I agree heartily with Mr. Furst on the necessity of laboratory and system tests on extinction time of residual arcs. I feel that we have reasonably satisfactory test values from tests on lines without shunt reactors and voltages up to 700 kV. However, because of the need for special shunt-reactor banks for reducing the residual arc current and recovery voltage on lines of 500 and 700 kV (except on very short lines), we need also tests on residual arc extinction time as a function of these two quantities on reactored lines. I see no reason to believe that the results of tests on unreactored lines will be valid for reactored lines.

The laboratory test circuit can be a single-phase Thevenin's equivalent, which, in its simplest form would have one capacitor, one inductive reactor, and one source of power-frequency emf, all adjustable. Since the natural frequencies of such a circuit are low, the test quantities from such a circuit of lumped elements should not differ significantly from those using a smooth line.

Point 3 of Mr. Furst's discussion is well taken. We have made computations on the simulated test line with X_m set equal to zero, and the results are shown below, in comparison with those where X_m has its theoretically correct value. Both the residual fault current and the recovery voltage are still acceptable if they were acceptable before and unacceptable if they were unacceptable before.

TABLE VII. Change of Recovery Voltage and Residual Fault Current by Setting $X_m = 0$ in Fig. 12(b); SLG Fault

Degree h_1 of Shunt Compensation	Faulted Conductor Sectionalized?	Recovery Voltage (kV)		Residual Fault Current (A)	
		$X_m = 0$	$X_m > 0$	$X_m = 0$	$X_m > 0$
0.7	No	66.5	65.9	34.6	35.2
0.9	No	95.8	89.3	16.9	17.2
0.7	Yes	40.8	33.2	10.5	8.8
0.9	Yes	122.3	92.0	10.7	8.8

Point 4, concerning shunt compensation while one circuit is temporarily out of service, has already been answered above. The compensation need not be altered.

I am grateful to Mr. Sturton for his summary of his extensive experience with single-pole switching on uncompensated transmission lines rated at 230 and 345 kV. From his previous publications and correspondence, I know that he is very enthusiastic about single-pole switching. I hope that future experience on long lines of 500 kV and above, with four-reactor shunt banks for aiding fault extinction, will show an equally good record.

He corroborates that simultaneous line-to-ground faults on the same phase of both circuits of a double-circuit line do occur.

It is interesting to know that single restrikes can occur during single-pole switching. I expect that restriking may not occur — or, at most, will occur less often — on lines with suitable shunt compensation.

His summary — based on experience — of necessary dead times is interesting and valuable. If a similar summary were to be made for lines with special shunt compensation, after enough experience has been accumulated, I hope and expect that a shorter dead time may prove satisfactory.

I thank him for his complimentary remarks.

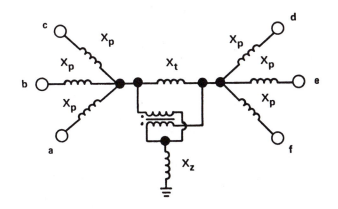

Fig. 16. Shunt compensating circuit with central part having a transformer and two reactors. Dots opposite transformer winding indicate subtractive polarity.

LIMITS TO IMPEDANCE RELAYING

J.S. Thorp
Member, IEEE
Cornell University, Ithaca, N.Y.

A.G. Phadke
Senior Member, IEEE

S.H. Horowitz
Senior Member, IEEE

J.E. Beehler
Fellow, IEEE

American Electric Power Service Corporation, New York, N.Y.

ABSTRACT

The paper reviews the role of impedance relaying in modern transmission line protection systems. The speed and reach of the first zone of an impedance relay are identified as being the most important attributes of a relay. The paper develops a mathematical model of the relaying process in the presence of transient phenomena which accompany faults. It is shown that there is an uncertainty associated with the impedance estimated by a relay and this uncertainty depends upon the speed with which these estimates are made. A quantitative relationship between the relay speed and its reach is obtained.

INTRODUCTION

Impedance relays have always been a significant element in EHV transmission line protection. They have, in fact, been the major common element in a wide variety of protection schemes designed to meet specific and often unique, situations. The speed and reach of an impedance relay are interrelated. We begin our examination of this relationship by first reviewing the present day relaying practices and the available relaying characteristics. Next, the influence of relaying speed upon the power system is discussed. An attempt is made to go beyond the simplistic statement that "the faster the relay, the better it is". The relative value of a gain in the speed of a relay is discussed in terms of overall fault clearing times.

Having discussed the influence of relay speed and reach upon power system performance, we examine in detail the relaying process to obtain the limits of the relay reach - speed relationship.

It will be shown that an attempt to make an impedance relay faster inevitably makes for a less accurate impedance measurement because of the transient phenomena accompanying a fault. The nature of these transient phenomena is studied with the aid of a small scale model of the power system. Based upon this quantitative description of the transients, the optimum speed vs. reach relationship for an impedance relay is obtained. Through the analytical expressions given, the influence of different types of transients on this optimum relationship can

be obtained. Finally, digital relaying is shown to be comparable to analog relaying in terms of its optimum performance curves.

RELAYING CHARACTERISTICS AND PRACTICES:

In 1963, R.A. Larner reporting on Western European relaying practices noted that for short circuit protection for phase faults on the 380-kV systems, impedance relays were used in an underreaching mode. [1] H.J. Sutton described the relaying of a 500-kV system in Louisiana that had a particularly difficult mutual ground current problem requiring the use of phase comparison relaying. Nevertheless, for 3-phase fault protection a mho-type distance element was required. [2] Similarly, the BC Hydro 500-kV system, [3] the Keystone 500-kV system [4] and the AEP 765-kV system [5] all incorporate impedance relays although the overall protective schemes are widely disparate. Our interest here is in the accuracy, speed and reliability of the impedance element itself, recognizing that the overall protection package may compensate for any limitations.

Relay Reach:

In applying and setting an impedance relay we must consider both the extent of the zone of protection and the relay operating time. In electromechanical relays there is a rough relationship between the two that must be considered when setting the protective zones of the relay. In solid state relays this relationship is more precise. The boundary between trip (contacts close) and no-trip of electro mechanical relays is not well-defined. For faults just beyond the desired reach, the relay operating time is infinite and for faults just within the desired reach, the time is finite. In actual practice the relay setting is accepted as the point at which the contacts just close. This is a judgement affected by the skill and patience of the relay tester.

Solid state relays on the other hand have a very well-defined boundary. The trip output is a voltage that is either present or absent and the pickup point is usually very repeatable under steady state test conditions. In practice there are sufficient inaccuracies in the impedance calculations, current transformer and potential transformer response and relay manufacturing tolerances so that all 1st zone impedance relay settings are calculated to reach 85-90% of the line. At this setting one should not expect the relays to overreach although the electromechanical relay operating time at the end of the protected zone is much longer than for close-in faults.

Operating Speed:

Electromechanical impedance relays will operate between these minimum and maximum times:

F 78 219-8. A paper recommended and approved by the IEEE Power System Relaying Committee of the IEEE Power Engineering Society for presentation at the IEEE PES Winter Meeting, New York, NY, January 29 – February 3, 1978. Manuscript submitted September 1, 1977; made available for printing November 11, 1977.

Reprinted from *IEEE Trans. Power App. Syst.*, vol. PAS-98, no. 1, pp. 246–260, Jan./Feb. 1979.

- For faults at the relay location, with operating currents greater than 50 amperes the relay will operate in 0.5 cycles on a 60 Hz basis.

- For faults between 85-90% of the relay reach setting and at an operating current of 10 amperes the relay will operate in 2-3 cycles on a 60 Hz basis.

- Below 10 amperes or between 90% and 100% of the relay reach setting, the relay operating time is indeterminate.

On the other hand the operating time of solid state relays is practically constant for all faults within its reach setting but is very dependent upon the angle of the voltage wave when the fault occurs. This time is nominally between 4 and 8 ms. for the measuring unit of the relay (depending on fault current asymmetry). The effect of filter circuits and the output relay may introduce some delay, particularly for faults between 90% and 100% of the reach setting.

Both electromechanical and solid state relays are subjected to transients coupled from the primary system due to lightning and switching, directly from the secondary wiring due to interruption of auxiliary relay coils, arcing battery grounds and control circuit switching. The relay manufacturer takes special pains to insure accuracy under these conditions. Inevitably the measures taken adversely affect relay operating times, particularly at low levels of fault current or at the relay reach setting.

Pilot Relaying and High Speed Reclosing:

Generalizing for all impedance relays, the fastest and most consistent operating time is, therefore for faults less than 80% of the reach setting which itself is less than 90% of the total line length. To protect 100% of the line, the usual practice is to employ a pilot scheme, that is, a scheme using a communication channel between the relays at the two ends of the line. Information from both ends is then used to determine whether to trip or not. The use of a power line carrier or microwave pilot channel for HV or EHV transmission lines is practically universal in the United States. It is the use of this scheme that permits virtually all U.S. utilities to justifiably claim they achieve 1 cycle (16.67 ms) relay protection on 100% of the line length.

It is interesting to review the need for such protection from both the historical and present-day point of view. In 1935 Sporn and Muller [6] proclaimed "one-cycle carrier relaying accomplished" and gave as the reason for this development "material increases in the power limits of combination generator and transmission line systems running to as much as 20% when switching time was decreased from 6 to 3 cycles (in terms of 60 cycles)". As further explanation their article describes the increasing use of single circuit transmission lines to reduce the investment per kilowatt of capacity and the advantages, in such configuration, of high speed reclosing (HSR) to maintain continuous service. HSR requires simultaneous clearing at both ends of a faulted line, hence the need for a pilot scheme.

In modern systems a single path on which to transmit power from one point to another is a rarity. Far more common are double circuit lines or parallel transmission lines at different voltage levels. And, as discussed below, stability on today's systems of long lines and strong sources can be maintained by quickly clearing a fault from the closest terminal.

This, of course, does not mean HSR and pilot relaying are not required. It means they are not required for the same reasons that existed in 1937. HSR is still a valuable tool. It restores the system to a normal configuration as rapidly as possible, minimizing the effects of errors, reducing abnormal contingencies and reestablishes the margins upon which system integrity is based. Pilot relaying is a necessary prerequisite for successful HSR. It also clears faults from remote ends in first zone times which reduces the fault damage and the effects of reduced voltage on the remainder of the system. Of course on short line, weak source systems where fault magnitude is not dependent on location, pilot relaying provides constant clearing time for the entire length of the line.

Influence of Relay Speed on Power Systems:

A justification for attempting to reduce the time to clear a fault is the beneficial effect such an action has on system stability. Although it is difficult to generalize such a complex phenomenon, there are, nevertheless, existing situations and studies of future trends which give us a good basis for determining the required operating time of impedance relays.

(a) On the PJM Conemaugh 500-kV Project stability studies established 12 cycles for single-phase-to-ground faults and $7\frac{1}{2}$ cycles for multi-phase faults. To accomplish these times and still provide adequate breaker failure protection a unique 2-step timing sequence was introduced. [7] Both primary and backup relays energize a 4-cycle timer for multi-phase faults. This timer is deenergized by high set fault detectors as soon as enough circuit breaker poles have opened, or the fault current has decreased to less than a critical value. An 8-cycle timer is energized by all primary and backup relays and does not reset unless all phases indicate zero current.

(b) Two independent studies by Concordia, Brown [8] and Lokay, Thoits [9] examined the effects of future Turbine Generators on System Stability. Lokay and Thoits established minimum fault clearing times of 3 cycles for the most severe probable future conditions. The Concordia, Brown paper established an index of stability (Ks) derived from a mathematical expression that includes critical clearing time (CCT), inertia constant, short circuit ratio, and subtransient and synchronous reactance for probable future machine design. They determined a ratio of present to future of 0.6 for the most probable future designs. A

CCT of 7 cycles would therefore reduce to 4.2 cycles.

(c) In a general discussion of fast relaying using a 2 machine, 2 line equivalent of representative systems, Andrichak and Wilkinson [11] established that:

(i) For a system with long lines and strong sources, the most serious threat with respect to stability is a 3-phase fault at the terminals of one of the lines. Thus it is imperative that the fault be cleared rapidly at the end near the fault. For faults closer to the center of the line, the system becomes capable of transferring some power across the remaining lines, thus the stability margin will increase and an increase in relaying time can be tolerated.

(ii) For a system with short lines and weak sources or a short line in parallel with a weak interconnection, a 3-phase fault is still the most onerous to the system. However, the threat to stability does not diminish to any great degree as a function of fault location.

The above findings are similar to our own studies in this area. On the AEP System we have calculated critical switching times of 4-6 cycles during periods of construction and the resultant abnormal system configuration. When the planned system improvements were completed, this time increased to 8-10 cycles. We find that we can remove the threat to stability with first zone protection and local breaker failure operating in times attainable today where present-day relays and breakers are capable of clearing a fault in 3 cycles from the primary protective package and 6-7 cycles in "local breaker failure time". The term "local breaker failure" includes any failure of the fastest protective equipment, not just the breaker. It implies a backup scheme at the same location as the primary scheme, initiated by all of the primary relays and resulting in clearing the fault by tripping additional breakers in the station. It is of course, desirable to establish as a criterion for stability a 3-phase fault that is maintained until it is cleared by the backup breakers. Using nominally-rated 1 cycle relays and 2 cycle breakers the minimum time would be 1 cycle for the relays to initiate trip and start a timer, 2 cycles for the breaker to operate, 1-2 cycles coordinating time to stop the timer if the fault is cleared and finally, 2 cycles for the backup breaker to operate, or 6-7 cycles total time.

Faster clearing times may be required in unique or severe situations such as in the 500-kV AC and 800-kV DC parallel system in the northwestern U.S. In such cases, totally new relay or equipment concepts must be employed.

Total Fault Clearing Time:

Figure (1) is derived from reference (11) and shows the minimum and maximum fault duration time for the entire range of short circuit current offsets for the present and future system X/R ratios as breaker and relay times vary. In this analysis the assumption is made that the breaker will interrupt at the first current zero after the contacts part and that the breaker will not force a current zero. Breakers whose contact theoretically part at a current zero will not be able to interrupt until the next current zero because the contact must part a sufficient time before current zero to allow arc contaminants to be dissipitated and the breaker recovery voltage withstand strength to become high enough so that this arc will not re-ignite. This, minimum time is in the order of .1 to .2 cycles and is included in the relay time.

Inspection of this figure indicates that for a given circuit breaker, the decrease in fault duration is not directly proportional to a reduction in relay time. For instance with a System X/R of 34 and a 2-cycle breaker (1.2 cycle contact opening time) reducing the relay time .5 cycle from 1.5 to 1.0 results in only a .2 cycle reduction in fault clearing but an additional .5 cycle reduction to 0.5 cycle relay time produces a 0.75 cycle reduction in fault clearing. The next additional .5 cycle reduction in relay time will again result in only a .15 cycle reduction in fault clearing. The effect of the system X/R can also be examined. As the system X/R ratio increases the relationship between relay time and fault duration becomes less flat so that a small increment in relay time for a given circuit breaker can produce large changes in fault clearing time. For system X/R ratio of 34, it can be seen also that reducing a two cycle breaker opening time by 0.2 cycles and the one cycle relay time by 0.2 cycles reduces the fault duration time 0.75 cycles.

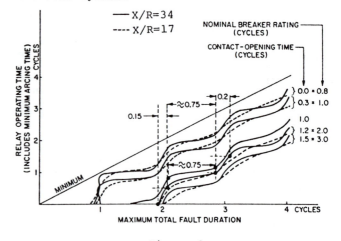

Figure 1

Maximum Total Fault Duration vs. Relay Operating Time

In the final analysis, using present day technologies, the combined action of both the relays and the circuit breaker will determine fault duration. The fault duration time in a growing system must be decreased to maintain stability margins. Referring to the 4.2 cycles established in reference [8] the only way stability can be maintained today would be to disregard breaker failure as a criterion or to resort to the independent-pole, 2 timer scheme of reference

110

[7]. In such a situation relay times would have to be those associated with close in faults and the time settings would provide virtually no margin for error. Breakers having 16 ms contact parting time are beginning to appear on the market.

The next logical subject for analysis is the ultimate speed that an impedance relay is capable of, and an examination of the consequences of achieving this high speed. These limits to impedance relaying are established in the following.

LIMITS TO IMPEDANCE RELAYING, WHAT IS POSSIBLE:

Impedance is a steady-state concept. In the relaying sense it refers to the ratio of fundamental frequency components of a selected voltage and current signal pair. During normal system operating conditions, the voltage and current signals associated with a transmission line are reasonably close to being constant fundamental frequency waveforms. However, every change in the state of the power system (especially the occurrence of faults) gives rise to transient phenomena. During the first 10 or 20 milliseconds immediately following the inception of these transients, there are significant non-fundamental frequency components in the current and voltage waveforms.

Formally, the objective of a high speed impedance relaying system is to extract the fundamental frequency components from the corrupted voltage and current signals, and thence to determine the impedance to the fault. There are benefits to be had if this function is performed quickly, while the desire for a secure overlapping first zone demands that there be a certain degree of confidence in the estimated values of the fundamental frequency components. We will try to determine the relationship between the speed and accuracy of an impedance relay in the presence of non-fundamental frequency components in current and voltage waveforms. The attribute of accuracy can be related to a permissible reach for a secure first zone, and consequently we will obtain a relationship between the speed and the first zone reach of a high speed impedance relay.

The present investigation was motivated by our studies of impedance relaying with a computer. A number of conclusions we draw below are therefore expressed in the terminology of digital signal processing. However, a key result of this paper - the relationship between the speed and first zone reach of an impedance relay - is independent of any digital considerations.

We begin with a functional description of the relaying process. Only one voltage and current signal pair is considered below, although a complete relaying system must deal with several such signal pairs. Figure (2) shows the major functional blocks relevant to this discussion. The transmission line is an integral part of a large network, and the entire ensemble is responsible for the nature of voltage and current signals at the

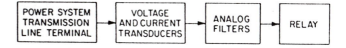

Figure 2

Relaying System Functional Blocks

transmission line terminals. These **signals** are transformed to a low power level by suitable transducers. In normal relaying practices the transducers would be current or voltage (PT or CCVT) transformers. Analog signal filters are generally low pass filters and may include such special circuits as mimic circuits for current signals (to suppress the transient dc offset). Low pass analog filters are a necessity for digital computer based relays to avoid the introduction of aliasing [14] errors in the sampled data. However, even in a purely analog environment, a low pass filtering effect is always present because of the nature of the transducers, signal cables, and surge suppression circuitry.

The performance of the system shown in Figure (2) is examined in three stages:

. Nature of non-fundamental frequency components in voltage and current signals.

. Best estimates of the fundamental frequency components as function of relay speed.

. Digital computer relaying considerations.

Nature of non-fundamental frequency components in voltage and current signals:

In the context of this discussion, we will use the term "error signal" to denote the non-fundamental frequency components of a signal. The transient phenomena accompanying the occurrence of a fault are a prime source of the error signals. The magnitude and frequency of the dominant component of the error signals is determined by the prevailing structure of the power system, and the existing load levels at the instant of fault occurrence. Given a specific configuration of a power system, the composition of the error signals can be determined through small scale modeling of the network or through the computer simulation of the network. However, from the relaying viewpoint, the power system cannot be assumed to be in a known state, and consequently the error signals cannot be treated as having a known or a determinate structure. An attempt was made to obtain a spectrum of the error signals for different operating states of a representative transmission network. Clearly the details of the structure obtained in this manner are not valid for all other situations. However, it can be reasonably expected that the order of magnitude and frequency content of the spectrum of the error signal give us a basis from which valid results can be obtained.

A laboratory model suitable for transient analysis of three-phase networks

[13] was used to obtain the error signal data. Figure (3) is a one-line diagram of the network that was represented in detail using three-phase pi-sections for transmission lines and multiwinding transformers as needed. The two 138-kV lines

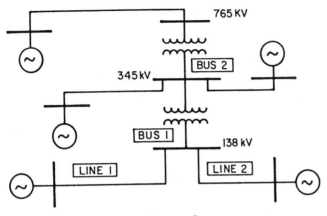

Figure 3

One-Line Diagram of the Laboratory Model

(lines 1 and 2) were subjected to various phase and ground faults at varying distances from bus 1. Loading of the lines was changed, as was the configuration of the connected network. Voltage and current signal spectra were analyzed using a high speed sampling rate (3840 Hz). Standard Fast Fourier Transform techniques [15] were used to find the frequency components of the signals from dc to 1920 Hz in steps of 15 Hz. No analog antialiasing filters were used in these experiments since the transmission line models used have a cut-off frequency of about 1800 Hz which is well below the Nyquist [14] rate for the chosen sampling rate.

A set of 320 faults were placed on the two lines with following variations:

(a) Four types of faults: 3-phase, phase to phase, phase to phase to ground, phase to ground.

(b) Ten Fault Locations: At 10% to 100% line length from bus no. 1.

(c) Eight System Configurations: Length of lines connected to bus no. 2 varied between 30 and 300 miles. No single line longer than 180 miles.

Each of the eight system configurations produced 40 cases. It was observed that for a given system configuration, the error signal composition was almost independent of fault types and fault location. Line loading was varied in some of the cases, but its effect on the error signals was minimal. Half the faults were placed at the maximum voltage and the remaining at about 87% of the maximum.

The voltage error signal for each of the 40 cases was decomposed into its components, and the squared magnitude at each frequency (which is a measure of the energy content of that component) was normalized using the average-squared fundamental frequency components of the post-fault signals. The resulting spectrum for one of the 8 system

configurations tested is shown by a dotted line in Figure (4). It can be seen that for this system most of the error signal energy of the post-fault voltage waveforms is concentrated at 255 Hz. For different system configurations the spectrum of the voltage signals is generally similar to Figure (4), with the exception that the frequency at which most of the error signal energy is concentrated is different. For the eight system configurations considered, this frequency changed from 225 Hz to 555 Hz, the lower frequency corresponding to a system configuration with the most extensive 345-kV network. The results obtained for the eight different system configurations were further combined and normalized to produce the composite spectrum shown by a solid line in Figure (4). Note that this spectrum shows a marked attenuation at higher frequencies. It is believed that this is due to higher losses present in the power system model at higher frequencies, and in actual power systems the spectrum can be expected to be relatively flat.

Similar experiments were performed with fault current waveforms. To simplify the digital signal analysis, the fault incidence angles were adjusted to produce no dc offsets in the current analyzed. The spectrum for the current signals is also shown by a heavy line in Figure (4). As expected, the magnitude of the normalized error signals in the current waveforms is smaller than in the voltage waveforms.

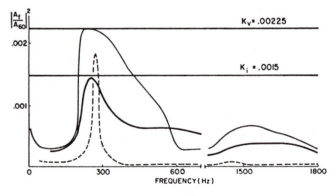

Figure 4

Spectrum of post-fault voltage error signals

For faults at or near the line terminal the normalization procedure described above (which uses the post-fault 60Hz component as a per-unit base) becomes unusable for voltage signals, since the post-fault 60 Hz component of the voltage signals are very small for such faults. Hence the pre-fault (i.e. normal system condition) voltages are used for normalization for near fault cases, and the resulting spectra for these cases are also similar to Figure (4).

We will stipulate that the error signals in current and voltage waveforms as described above are the only error signals present. Admittedly this procedure underestimates the error signals, since the contribution of other sources to the error signals is not taken into account. The transducers, especially the current transformers, are known to saturate, thereby producing error components in the current signals. The CT saturation phenomenon is difficult to assess as it depends substantially upon the design of the CT's, CT selection criteria, and prevailing shortcircuit levels on the power system. Saturation, if it occurs, would be significant for near faults which are well within the first zone of an impedance relay. Also, CT saturation is less likely to occur within 1/2 cycle to 1 cycle duration after the occurrence of a fault. CT saturation effects are therefore not considered to be significant in the present context. Coupling Capacitor voltage transformers (CCVT) are also likely to produce significant error signals for near faults, when the faulted phase voltage undergoes a sudden change of a relatively large magnitude. Once again, since these faults are well within the first zone, the CCVT error contributions to the error signals for near-faults are also not considered to be significant.

Best estimates of the fundamental frequency components as function of relay speed:

Considering the results of the experiments described in the preceding section, it is possible to model the error signals as a random process. That is, the error signal is an unknown time function which depends upon certain parameters which are not known to the relay. The fact that the transient phenomenon is a deterministic phenomenon (knowing the system parameters, the transient could be predicted accurately) does not preclude a random process description for the error signal [14]. The experimental results obtained in the previous section were spectra of squared Fourier components, which are not the same as power density spectra. Since the present discussion deals with relaying applications, we will be considering the current and voltage signals over a relatively short time interval after the occurrence of a fault. During this interval (the data window), the error signal describing the transient can be modeled as a noise signal with a power density spectrum resembling the experimental results described earlier.

It can be seen that the presence of a random component in the current and voltage signals will cause random errors in the impedance calculated from them. Given a characterization of the random noise component, it is possible to obtain bounds for the estimates of the fundamental frequency component of a signal. Consider a current or voltage signal given by

$$y(t) = Y \sin(\omega_o t + \beta) + e(t); \quad 0 \leqslant t \leqslant T$$
$$= Y_r \sin(\omega_o t) + Y_i \cos \omega_o t + e(t) - (1)$$

where the estimates of Y_r and Y_i in the presence of the error signal $e(t)$ are needed, T being the data window over which observations are made. The random process $e(t)$ may be assumed to have zero mean, and the best estimates of Y_r and Y_i must also be designed to have a mean equal to the correct values of Y_r and Y_i. Denoting the best estimates of (Y_r, Y_i) by $(\hat{Y}_r, \hat{Y}_i)$ the

following two random variables are therefore required to have zero mean:

$$p = Y_r - \hat{Y}_r$$
$$q = Y_i - \hat{Y}_i \qquad\qquad -- (2)$$

The variance of p and q which is a measure of the deviation of the estimates from their true value, must be minimum when $\hat{Y}_r$ and $\hat{Y}_i$ are optimum estimates of Y_r and Y_i. These conditions of optimality are expressed by

$$E\{p\} = E\{q\} = 0$$
$$E\{p^2\} = \sigma^2_p \quad \text{minimized} \qquad -- (3)$$
$$E\{q^2\} = \sigma^2_q \quad \text{minimized}$$

It is possible to calculate the optimal estimates for an error signal whose spectral structure matches the experimentally observed results shown in Figure (4). However, the intended use of Figure (4) is to specify a quantitative measure for the possible magnitudes of error signals, and not to provide a detailed model for the spectrum. Consequently, we assume that the actual error signal spectrum to be a constant as shown by the heavy lines in Figure (4), which encompasses the experimental data.

Denoting the constant magnitude of the error signal spectrum by K (K_v for voltage and K_i for current), the optimal estimates $\hat{Y}_r$ and $\hat{Y}_i$ which satisfy the conditions set forth in equation (3) are given by:

$$\hat{Y}_r = \frac{A}{F} \int_o^T y(t) \sin\omega_o t \, dt - \frac{B}{F} \int_o^T y(t) \cos\omega_o t \, dt$$
$$\qquad\qquad -- (4)$$
$$\hat{Y}_i = \frac{D}{F} \int_o^T y(t) \cos\omega_o t \, dt - \frac{C}{F} \int_o^T y(t) \sin\omega_o t \, dt$$

$$\sigma_p^2 = \frac{AK^2}{F}, \quad \sigma_q^2 = \frac{DK^2}{F}, \quad F = AD-BC \qquad -- (5)$$

$$A = \frac{1}{\omega_o}\left[\frac{\Phi}{2} + \frac{\sin 2\Phi}{4}\right]$$

$$B = C = \frac{1}{\omega_o} \frac{\sin^2\Phi}{2} \qquad\qquad -- (6)$$

$$D = \frac{1}{\omega_o}\left[\frac{\Phi}{2} - \frac{\sin 2\Phi}{4}\right]$$

and $\Phi = \omega_o T$

which reduces to the following expressions when the data window T is a multiple of half-period of the fundamental frequency

$$\hat{Y}_r = \frac{2}{T} \int_o^T y(t) \sin\omega_o t \, dt$$

$$\hat{Y}_i = \frac{2}{T} \int_o^T y(t) \cos\omega_o t \, dt \qquad -- (7)$$

For Φ = multiple of π

113

Equation (7) represents the familiar Fourier coefficient expressions.

Using the optimum estimates $\hat{Y}_r$ and $\hat{Y}_i$ of equations (4) and (7) respectively, the variance of the error of the estimates is given by

$$\sigma^2 = \sigma_p^2 + \sigma_q^2 = \frac{4\omega_o K^2 \phi}{\phi^2 - \sin^2 \phi} \qquad -- (8)$$

for all ϕ

$$\sigma^2 = \frac{4K^2}{T} \qquad -- (9)$$

for ϕ = multiple of π

the standard deviation of the optimal estimate of the fundamental frequency component of the voltage and current signals as given by equation (8) depends upon the data window as shown in Figure (5):

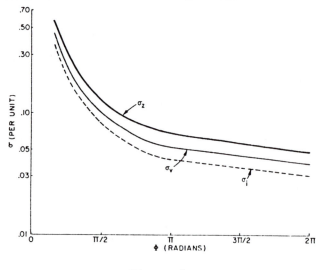

Figure 5

σ^2 as a function of data window ϕ

If an impedance is calculated as a ratio of the voltage estimate and the current estimate:

$$\hat{Z} = (\hat{R}, \hat{X}) = \frac{(\hat{V}_r, \hat{V}_i)}{(\hat{I}_r, \hat{I}_i)} \qquad -- (10)$$

the per unit variance of the error in impedance calculation for small σ_v and σ_i is given by

$$\sigma_z^2 \cong \sigma_v^2 + \sigma_i^2 \qquad -- (11)$$

The standard deviation of the impedance estimate is also shown in Figure (5).

Digital Computer Relaying Considerations:

In digital computer based relaying systems, the basic process involved is that of sampling the analog signals (voltages and currents) at a suitable sampling rate and performing the necessary computations on the sampled data. Fourier Transform type calculations are often needed for relaying applications. A cardinal principle for such a procedure is to assure that the analog signals are band-limited in frequency below the Nyquist rate so that aliasing errors are avoided. [14] The higher the sampling rate chosen, higher is the bandwidth of these anti-aliasing filters, and consequently higher is the noise or error signal presented to the relay by the anti-aliasing filters. We will assume that the selected sampling rate and the anti-aliasing filter are matched to meet the Nyquist criterion (filter cut-off frequency equal to one-half the sampling rate).

Consider the sampled data generated by sampling the signal of equation (1) at a sampling frequency N/T. The n'th sample is given by

$$y(n\frac{T}{N}) = Y_r \sin(\omega_o \frac{nT}{N}) + Y_i \cos(\frac{\omega_o nT}{N}) + e(\frac{nT}{N}) \qquad -- (12)$$

Using the straight-line characterization of the error signal spectrum shown in Figure (4) we will define the error signal samples to have an auto-correlation function [14]

$$R\left\{(m-n)\frac{T}{N}\right\} = \frac{NK}{T} \delta mn \qquad -- (13)$$

Notice that this is the sampled data analog of the continuous signal case shown in Figure (4). Also note that equation (13) quantitatively displays the effect of higher sampling rates; a larger N, which represents a higher sampling rate, leads to an auto-correlation function of larger magnitude. Defining p and q as before for the sampled data case [see equation (2)], the optimal estimates $\hat{Y}_r$ and $\hat{Y}_i$ are obtained by minimizing the variance of p and q:

$$\hat{Y}_r = \frac{A}{DA-BC} \Sigma y(\frac{nT}{N}) \sin \frac{\omega_o nT}{N} - \frac{B}{DA-BC} \Sigma y(\frac{nT}{N}) \cos \frac{\omega_o nT}{N}$$

$$\hat{Y}_i = \frac{D}{DA-BC} \Sigma y(\frac{nT}{N}) \cos \frac{\omega_o nT}{N} - \frac{C}{DA-BC} \Sigma y(\frac{nT}{N}) \sin(\frac{\omega_o nT}{N})$$

$$\qquad -- (14)$$

$$\sigma_p^2 = \frac{AK^2}{DA-BC}$$

$$\sigma_q^2 = \frac{DK^2}{DA-BC} \qquad -- (15)$$

$$\sigma^2 = \sigma_p^2 + \sigma_q^2 = \frac{(A+D)K^2}{DA-BC}$$

where

$$A = \frac{T}{N}\left[\frac{N-1}{2} + \frac{\cos^2 \omega_o T}{2} + \sin^2 \omega_o T \cdot \frac{\cos \frac{\omega_o T}{N}}{\sin \frac{\omega_o T}{N}}\right]$$

$$B = C = \frac{T}{N}\left[\frac{\sin \omega_o T \cdot \sin(\omega_o T + \omega_o T/N)}{2 \sin(\omega_o T/N)}\right] \qquad -- (16)$$

114

and

$$D = \frac{T}{N}\left[\frac{N+1}{2} - \frac{\cos^2\omega_o T}{2} - \frac{\sin^2\omega_o T}{4} \cdot \frac{\cos\omega_o T/N}{\sin\omega_o T/N}\right]$$

For a data window which is a multiple of half cycle:

$$\hat{Y}_r = \frac{2}{N}\Sigma y\left(\frac{nT}{N}\right)\sin\left(\frac{nT}{n}\right)$$

$$\qquad\qquad\qquad -- (17)$$

$$\hat{Y}_i = \frac{2}{N}\Sigma y\left(\frac{nT}{N}\right)\cos\left(\frac{nT}{N}\right)$$

$$\sigma_e^2 = \frac{4K^2}{T} \qquad\qquad -- (18)$$

These results are sampled data analogs of equations (4)-(9). Equations (17) represent the fundamental frequency component of the Discrete Fourier Transform of the sampled data.

It is interesting to compare equations (8) and (15) and study the degradation of the optimum estimate resulting from the sampling process. The standard deviations for continuous case (infinite sampling rate) and discrete case estimates (sampling rate of 12 times the fundamental frequency) are plotted in Figure (6):

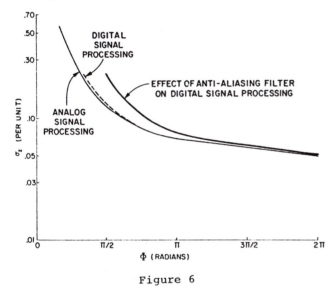

Figure 6

σ_z for discrete and continuous signal processing

It should be noted that optimum solutions possible with sampled data techniques are somewhat more error prone than the corresponding optimum solutions in the analog processing case. However, with data windows of multiple of half cycles, the optimum solutions for the two cases are equally accurate. Furthermore, at these data windows the number of samples obtained per half cycle do not affect the accuracy of the computation. The sampling rate selected does however affect one important feature of the relaying process. Recall that the anti-aliasing filters selected to meet the Nyquist condition add a delay (phase shift) in the signal processing path. To maintain the recursive properties of the algorithms

based on equations (14) a half cycle data window (or its multiple) is necessary. An anti-aliasing filter which adds little delay to the total relaying process while still maintaining the required filtering property is found to produce a delay of about one sample period.

Consequently, the sampled data optimum estimates are available one sample time later, and the overall curve for this case taking into account the delays of the anti-aliasing filter is as shown by the shifted curve in Figure (6). It should be clear that lower sampling rates with matched anti-aliasing filters are not equal in performance to higher sampling rates from an overall system consideration.

At extremely high sampling rates, another aspect of the anti-aliasing filter becomes operative. If the error signals are band-limited at their source (all physical analog signals are of course band-limited), then at some sampling rate the matched anti-aliasing filter has a bandwidth wider than that of the analog signal. In this case, the autocorrelation function for the error signal can no longer be assumed to be proportional to the sampling frequency. Instead of equation (13), the corresponding equation for such a case is

$$R\left\{(m - n) \frac{T}{N}\right\} = K \cdot \delta mn \qquad -- (19)$$

From our laboratory experiments described above, we feel that such a description of the error signals cannot be justified at sampling frequencies lower than 2000 Hz.

The exponentially decaying dc offset present in the current signals leads to certain variations on the procedures described above. A pure dc offset if allowed to remain in the signal, would lead to gross errors at a data window of half cycle (and at odd multiples of half cycles). A full cycle data window would be immune from such errors, although an exponentially decaying dc offset would produce errors even with a full cycle data window. The use of mimic circuits is well known in relay engineering literature. This circuit (which may be an inductive burden on the CT, on an operational amplifier realization of the inductive burden) eliminates the offset from the current signal; the offset removal is complete if the time constant of the burden matches that of the faulted network. With the use of such a mimic circuit, the objection to a half cycle data window is removed. The mimic circuit does in fact affect the error signal spectrum, it amplifies the higher frequencies in proportion to the fundamental frequency. This can be corrected easily by considering the mimic circuit and the anti-aliasing filter as a single entity whose cutoff frequency satisfies the Nyquist condition. This introduces a slight additional delay (of the order of one-half sampling period) in the overall relaying process.

Alternate forms of digital signal processing:

We will briefly consider two other forms of digital techniques discussed in the relaying literature. A number of papers propose the use of least-square polynomial fits to the sampled data to achieve data smoothing from which the fundamental frequency components are calculated. [16] These techniques are reasonably close to the optimum estimates shown in Figure (6) at relatively narrow data windows (less than 1/6 of the fundamental frequency period). At longer data windows, their performance deteriorates rapidly and is not acceptable. It is possible to use a moving narrow window over a longer period and use averaging techniques to improve the performance of these methods, but the overall performance is less accurate than the corresponding optimum estimate over a data window spanning the entire data used in the averaging process.

The other digital technique discussed in the literature models the fault circuit as a series R, L circuit and estimates R and L (and hence the distance to a fault) from the differential equation representation of the phenomenon. [17] An apparent advantage of this technique is that all signal components (including the exponentially decaying dc offset) which satisfy a single differential equation tend to aid in the estimation process. However, in the presence of error signals (for example, those components which do not fit the differential equation model) the numerical performance of the technique is considerably degraded when an exponentially decaying dc offset is present. In the absence of a dc offset, and when an averaging procedure is used over several narrow windows, the results produced by the differential equation method are comparable to the optimum estimates discussed above.

Optimal Speed vs Reach Characteristic for the Impedance Relay:

The variance of the impedance calculation as a function of the data window shown in Figure (6) is one of the principal results of this paper. The variance σ_z is a measure of the expected dispersion of the calculated impedance value from its true value. It is possible to estimate the size of a region within which the impedance calculation would lie with a certain degree of confidence level. As an example, if a confidence level of 99% is desired, a statement about the size of the region (expressed in terms of σ_z) could be made if the details of the distribution of the impedance calculation were known. Normal (or Gaussian) distributions are usually assumed

because they lead to a simple mathematical formulation. However, a Gaussian distribution in the present context is not to be expected. Other possibilities are a rectangular distribution (equal likelihood within a band) or a triangular distribution (likelihood of occurrence weighted towards the center of the band). These three distributions and their corresponding 99% confidence level ranges are given in Figure (7). It can be seen that a range of about 2.5 σ_z can be used to describe a 99% confidence limit for all distributions given above. The radius of this zone

$$r_z = 2.5\sigma_z \qquad \text{-- (20)}$$

as a function of the data window for the optimum analog relay and the optimum digital relay using a sampling rate of 720 Hz is shown in Figure (8):

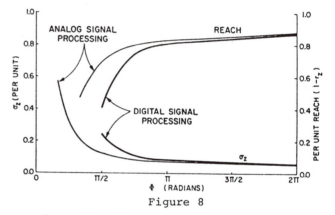

Figure 8

Optimum Data Window (speed) vs Accuracy and Reach of the Impedance Calculation

Note that the data in Figure (8) is based upon an error signal characterization as obtained experimentally from a model system. Although it is expected that this is a representative error signal spectrum, the techniques described here are applicable to any other error signal spectrum that can be determined.

For example, the optimum reach corresponding to a speed of one cycle is .83 per unit line length (see Figure (8)); this corresponds to the constants K_V and K_i of Figure (4) and $Y_z = 2.5\sigma_z$. This choice of K's and r_z is on the conservative side. It may be argued that particularly in digital estimation procedure one should use the average value of the power density spectrum rather than its peak to determine the K's. Also, the uniform distribution rather than the Gaussian or the triangular [see Figure (7)] is likely to be more realistic. These considerations would lead to $K_V = .00112$ and $K_i = .00075$ respectively, and $Y_z = 1.8\ \sigma_z$. Use of these parameters leads to the solid curve in Figure (9), where for the sake of comparison the speed-reach curve of Figure (8) is redrawn with a heavy line. It is apparent that such an interpretation of the error analysis leads to a more accurate expected impedance estimate.

Finally, we make some comments about the performance of an impedance relay using symmetrical components [18]. Calculation of

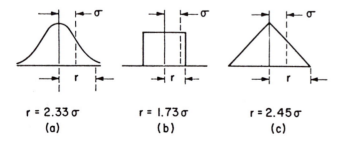

Figure 7
Possible Distribution of calculated impedance values; and zones for 99% confidence level
(a) Gaussian   (b) Rectangular   (c) Triangular

symmetrical components involves using information from all three-phase voltages to produce the symmetrical components of the voltages. In such a procedure, there is an additional smoothing effect upon the computations, with the result that the σ_z for calculations using symmetrical components of voltages and currents is about 40% smaller than σ_z for calculations using phase currents and voltages. Thus, for the error signals having the spectrum shown in Figure (4), a symmetrical component based impedance calculation shows a performance as shown in Figure (9):

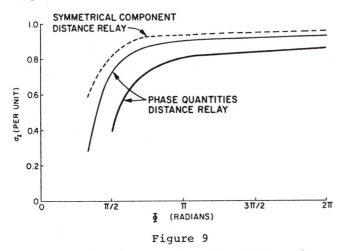

Figure 9

Optimum Speed vs Reach of Impedance Relays and Symmetrical Component Relays

We have not considered in this discussion the effect of Analog Digital conversion errors upon the performance of a digital computer relay - neither have we considered the effect of finite word length computation. Both these effects are truly random in nature, and in a well designed system their effects should be far smaller than the error levels discussed above.

CONCLUSIONS:

(1) The speed of the direct trip function of an impedance relay has a significant influence on the total fault clearing time. Although it is true that a higher relay speed always leads to a shorter total fault duration, the incremental reduction in the total fault duration due to an increase in relay speed depends upon the circuit breakers associated with the relay.

(2) For an impedance relay there is a relationship between the accuracy of its impedance estimate and the speed with which that estimate is obtained. This relationship can be determined from a knowledge of the energy content of non-fundamental frequency components of line voltage and current waveforms. Furthermore, this relationship holds irrespective of the actual hardware used by the relay - whether digital or analog.

(3) Since the reach setting of an impedance relay is determined by the error of its impedance estimate, an impedance relay with a specified reach setting cannot operate at arbitrarily high speeds. There is a well

defined limit to this speed which can be determined from a knowledge of the current and voltage waveforms.

(4) It is possible to design relays with adaptive reach-speed characteristic within the direct trip zone. That is to say an impedance relay can be made to operate faster for close in faults, and slow its speed for faults approaching the boundary of the direct trip zone. Digital computer relays can be designed with such adaptive features.

REFERENCES:

(1) R.A. Larner, "Protective Relaying Practices in Western Europe", IEEE Transactions on Power Apparatus & Systems, Feb. 1963, pp. 1020-1025.

(2) H.J.Sutton, "Application of Relaying on an EHV System ", *Ibid*, April 1967, pp. 408-414.

(3) S.R. Hayden, K.H. Engelhardt, J. Crockett, W.P. Duggan, "Relaying the BC Hydro 500 kV System", *Ibid*, May/June 1971, pp. 1190-1200.

(4) Committee Report, "Relaying the Keystone 500 kV System", *Ibid*, June 1968, pp. 1434-1439.

(5) S.H. Horowitz, H.T. Seely, "Relaying the AEP 765 kV System", *Ibid*, Sept. 1969, pp. 1382-1389.

(6) P.Sporn, C.A. Muller, "One Cycle Carrier Relaying Accomplished", Electrical World, Vol. 105, Oct. 1935, pp. 26-28.

(7) D.R. Holland, G.D. Paradis, G.E. McNair, J. Senuta, W.H. Van Zee, "Conemaugh Project: New Concepts for 500 kV System Protection", IEEE Transactions on Power Apparatus & Systems, March/April 1971, pp. 852-858.

(8) C. Concordia, P.G. Brown, "Effects of Trends in Large Steam Turbine Driven Generator Parameters on Power System Stability", *Ibid*, Sept./Oct. 1971, pp. 2211-2218.

(9) H.E. Lokay, P.O. Thoits, "Effects of Future Turbine Generator Characteristics on Transient Stability", *Ibid*, Nov./Dec. 1971, pp. 2427-2431.

(10) Committee Report, "Local Back Up Relay Protection", *Ibid*, July/Aug. 1970, pp. 1061-1068.

(11) J.G. Andrichak, S.B. Wilkinson, "Considerations of Speed, Dependability and Security in High Speed Pilot Relaying Schemes", Minnesota Power Systems Conference, Oct. 12-13, 1976.

(12) J.E. Beehler, "One Cycle Breakers-Are They Required", IEEE paper no. C 74 175-6.

(13) A.G. Phadke, M. Ibrahim, T. Hlibka, "Computer in an EHV Substation:

Programming Considerations and Operating Experience", CIGRE Study Committee No. 34: Protection, Colloquium in Philadelphia, Oct. 1975.

(14) W.B. Davenport Jr., W.L. Root, Random Signals and Noise, McGraw-Hill, New York, 1958.

(15) A.V. Oppenheim, R.W. Schafer, Digital Signal Processing, Prentice-Hall, Englewood Cliffs, N.J., 1975.

(16) P.J. Mann, I.F. Morrison, "Digital Calculation of Impedance for Transmission Line Protection", IEEE Transactions on Power Apparatus & Systems, Feb. 1971, pp. 270-279.

(17) A.D. McInnes, I.F. Morrison, "Real Time Calculation of Resistance and Reactance for Transmission Line Protection by Digital Computer", IEE Transactions, Institute of Engineers, Australia, EE7, No. 1 (1971), pp. 16-23.

(18) A.G. Phadke, M. Ibrahim, T. Hlibka, "Fundamental Basis for Distance Relaying with Symmetrical Components", IEEE Transactions on Power Apparatus & Systems, March/April 1977, pp. 635-646.

APPENDIX (List of Symbols:)

A, B, C, D, F	Constants in optimization procedure defined in equation (6).
A_f, A_{60}	Amplitude of a signal component corresponding to frequency f and 60 Hz respectively.
$E\{\cdot\}$	Expected value of the quantity within brackets.
K, K_v, K_i	Magnitude of a constant power density spectrum for a general variable $y(t)$, voltage $v(t)$, and current $i(t)$ respectively.
N	Number of samples per period of a 60 Hz wave.
T	Data window; a time duration over which signal can be observed.
Y, Y_r, Y_i	Amplitude, real part and imaginary part of a 60 Hz sinusoid.
$\hat{Y}_r, \hat{Y}_i$	Optimum estimates of Y_r and Y_i.
σ	Standard deviation of a random variable. Variable appears as a subscript (e.g. v,i,z,p,q).
ω_o	System frequency: 377 radians/second.

Discussion

W. A. Lewis (Consulting Electrical Engineer, Sunnyvale, CA): In this paper the authors consider the total clearing time for a fault that is cleared by the action of a relay system and circuit breakers and undertake to establish a limiting total clearing time that can be used in stability calculations for system reliability. In this analysis the time assigned for a circuit breaker to open is the time specified for the circuit breaker, whether two cycles, three cycles, or some other specific figure.

However, all circuit breakers considered depend upon the zero of the current wave to open. When the line in question is close to a generating station, so that the subtransient generator reactance is a major part of the reactance from the generator to the fault, it is possible for the short-circuit current contribution from the nearby generating station not to pass through zero for several cycles. This phenomenon results from the fact that the decrement of the substransient component may be much faster than the decrement of the dc component, so that, for the phase that has maximum offset, zero will not be reached for several cycles, if the short circuit persists.

This phenomenon has been investigated by Owen and Lewis [1]. The analysis shows that, if the offset is near maximum, the current in one phase will not go through zero for several cycles, but the current in the other two phases will pass through zero each cycle. Therefore, if a circuit breaker attempts to open during this time, two phases can be expected to open as the current reaches zero in each of them. As these currents are interrupted, a sudden change occurs in the remaining phase, causing that current to go to zero very quickly thereafter. The third phase will then be in a position to open. The net effect is that the final opening of the circuit breaker is delayed by perhaps a cycle or so. Thus, if reliability criteria are based on "worst case" analysis, the circuit-breaker opening time should be increased by perhaps a cycle when the fault is at a location that will cause the current not to go to zero during the first few cycles.

Even when breaker-failure protection is allowed for, the total clearing time may have to be increased for faults in locations where delayed opening can occur, because the breaker failure relay should not operate when the primary circuit breaker is going to clear the fault, with some delay for offset, and so the total time of clearing should be correspondingly increased.

REFERENCE

[1] Robert E. Owen and William A. Lewis: Asymetry Characteristics of Progressive Short Circuits on Large Synchronous Generators. T-PAS 71 Mar/Apr. 1971, pp. 587-596.

Manuscript received February 23, 1978.

C. A. Mathews (General Electric Co., Philadelphia, PA): The authors are to be commended for attacking the question of the maximum speed that can be attained with relays that measure impedance to determine the distance to a fault. The fact that they do not come up with a simple answer indicates that the problem is not a simple one and does not have one absolute answer.

This paper covers several different subjects, and much of the error analysis is developed around digital relaying techniques. My experience has been entirely with the analog type of solid-state relaying presently in use on many EHV lines, and my comments will be confined largely to this portion of the paper.

I agree in principle with two of the major conclusions of this paper, (1) that an impedance relay with a specified reach setting and used in an underreaching mode cannot operate at arbitrarily high speeds, and (2) that the presence of non-fundamental frequency components in the current and voltage signals just after the fault occurs requires some delay in operation for faults near the balance point if overreach is to be avoided.

There are a few other points with regard to solid-state relaying where I feel some additional emphasis, or some clarification, is needed.

First, there is a statement that solid-state relays have a constant operating time for faults anywhere within their reach. This is true for simpler relay designs and was true for some of the first solid-state transmission line relays built, but it is not correct to make a general statement that it is true for all solid-state relays. The designer of a first-zone solid state impedance relay has many design techniques at his disposal to achieve an optimum speed-reach characteristic, balancing speed against security, and the relays I am familiar with have the "adaptive" characteristic described in conclusion 4 where the relay operates faster for close-in faults and slows its operation for faults approaching the boundary of the direct trip zone. This approach to impedance relaying is described in detail in Reference 11, the paper by Andrichak and Wilkinson.

A second point is that high speed relaying with a pilot scheme does not require high speed for faults near the reach setting of the relay. In conventional solid-state relays accuracy of impedance measurement is associated only with faults near the reach setting. The "adaptive" characteristic mentioned above can provide high speed for the fault

118

locations where it is needed without sacrificing accuracy for faults near the relay reach. Therefore the question of the limit of speed for an accurate impedance measurement is of concern only for the first-zone, direct-trip function.

A third point involves CVT errors. It is stated in this paper that they are ignored on the assumption that they are only significant for faults that are well within the first-zone reach setting. This is probably true for the system shown in Fig. 3, but our model power systems tests have shown that where a short line is supplied by a high source impedance, these errors can be quite significant for faults near the reach setting of the relay. Our experience has shown that in some cases this may be the deciding factor in determining the limits of speed for faults near the reach setting of the relay.

I would like to ask some questions regarding these points:

1. The statement in conclusion 2 that the ''speed-reach relationship holds regardless of the hardware used'' seems to contradict the results shown in Fig. 8, where analog and digital processing give different results, and in Fig. 9, where symmetrical component processing appears to be better than phase quantity processing. Would the authors elaborate on the meaning of conclusion 2?

2. Fig. 4 shows magnitudes of the ratio of the non-fundamental frequency signals to the fundamental frequency. In practice, the error signals are a decaying transient and the ratio decreases with time. Were the measurements made at some fixed time after fault inception, and if so what was the time? Are they the ratios of instantaneous values, of average values, or of some other value?

3. A commonly used reach setting for phase relays is 90% of the line. Fig. 8, based on one assumption of error signal distribution, indicates that more than one cycle is needed. Fig. 9, based on another assumption, which is called more realistic, indicates that only a quarter cycle is needed. Is this a correct interpretation of Figs. 8 and 9?

Manuscript received February 23, 1978.

R. P. Sood and J. N. Sinha (Regional Engineering College, Kurukshetra, India):
We wish to compliment the authors for presenting a comprehensive review of the present day distance relay practice and critically analyzing the effect of relay speed on the accuracy and performance of distance protection.

As mentioned by the authors, the error signal energy is concentrated at a particular frequency which depends upon system configurations. For the eight system configurations considered, they have found this frequency to lie between a spectrum of frequencies from 225 to 555 Hz. Could the authors please comment on the possibility of any correlation between it and the sampling frequency?

The error signal has been described by the authors as a random process and an unknown time function depending on certain parameters which are not known to the relay. We feel that error function distributions can be obtained for different types of faults and then an overall distribution could be determined assigning different weightage to the respective faults depending on their frequency of occurence. Probably this may introduce more complexity in the relaying process and the authors comments on this point would be appreciated.

As shown in Fig. 5, the dispersion in current, voltage and impedance data indicate a finite minimum values for even a very large data window (i.e. 2π). Does it mean that the amount of error is unavoidable? Comparing from Fig. 5, the ζ, increases roughly from 0.05 to 0.15 if the data window is decreased from 2π to $\pi/2$. It would be appreciated if the authors could suggest an optimum value of data window and its dependence on the type of transmission system.

The authors have selected a range of about 2.5ζ, for describing a 99 percent confidence limit for impedance. This is based on the comparison of the three distributions (i.e. Gaussian, Rectangular and Triangular) and the maximum value of the range has been selected. However, no valid reason has been given for the selection of these distributions. Usually a Gaussian distribution is assumed but the authors have rejected the possibility of such a distribution in the present context without any explanation. Could the authors suggest an unique distribution?

The parameters m and δ_{mn} appearing in equations 13 and 19 have not been defined anywhere in the text.

We once again commend the authors for their contribution in advancing the art of distance protection.

Manuscript received February 24, 1978.

W. D. Breingan, M. M. Chen, and J. G. Jewell (General Electric Co., Philadelphia, PA):
We would like to congratulate the authors on their paper which applies the powerful technique of harmonic analysis of random functions to achieve a general approach to impedance relaying. Our immediate interest in the application stems from our own development of a digital-relaying system based upon estimation of the parameters of a differential equation representing the transmission line. With a sample rate of 20 per cycle and a window of 3 samples, we have correctly calculated resistance and inductance in 2.5 milliseconds. We therefore share the authors' interest in establishing the limits of impedance measurement and related methods of relaying. We have the following comments and questions.

The method chosen by the authors to estimate impedance at fundamental frequency forces them to discard much useful information that is contained in the transient. The differential equation technique for determination of the circuit parameters of R and L makes good use of the information from the exponentially decaying dc offset to obtain a rapid estimate of impedance. An assumption implicit in the authors' analysis is that the transient can be modeled as a *stationary* random process. Whatever the validity of this assumption for determining impedance when information from the transient is discarded by filtering or other means, the stationary assumption must be abandoned when the transient information enters into the impedance calculation. Have the authors analyzed such a calculation, and by what means? We have not been able to verify their statement that the differential technique is degraded in the presence of the transient, either in our laboratory investigation of a digital system (See reference cited with this discussion) or in the on-going field test of the two terminal version of the same system.

The accompanying figure shows an example from the laboratory test in which convergence occurred at about the same rate with a symmetrical transient (no offset) as when a significant offset was present. The figure shows the error in the inductance calculation for a 3 phase fault located 64 miles from the relaying point on the laboratory model power system. In the field test of the two terminal digital system, an external fault occurred near voltage zero resulting in a significant d.c. offset. The system, based on its R and L calculation, diagnosed the fault as external and started carrier blocking within 5 milliseconds.

The authors mention an additional smoothing effect when sym-

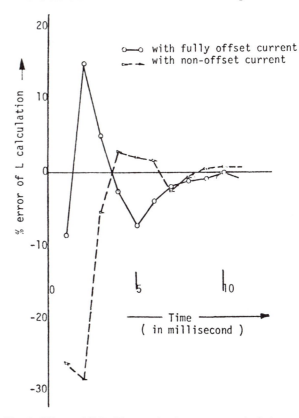

Fig. 1. Effects of DC offset on circuit parameter calculation

metrical components are used. They attribute the improved performance of a symmetrical component relay to the use of all three-phase voltages. What effect would the use of delta quantities or zero sequence compensation which also use information from several phases have on performance?

The authors present two methods of estimation, one using continuous integration, the other, a discrete summation. Is not the difference in Figure 6 between analog and digital processing this mathematical distinction? Is this same difference also the basis for the reference in conclusion 2 to analog and digital hardware?

The attempt by the authors to apply mathematical techniques to establish general criteria for protective relaying is laudable. It is hoped that continual dialogue will eventually begin to establish limits to many of the techniques used in relays.

REFERENCE

W. D. Breingan, M. M. Chen, T. F. Gallen, "The Laboratory Investigation of a Digital System for the Protection of Transmission Lines", IEEE PES Winter Meeting, February 1977 (F77 052-4).

Manuscript received February 24, 1978.

J. S. Thorp, A. G. Phadke, S. H. Horowitz, and J. E. Beehler: We welcome these discussions which help identify critical factors involved in understanding the behaviour of impedance relays and their influence on the operation of a power system.

Mr. Mathews makes several key points in his discussion. We are pleased to have his clarification that modern solid state impedance relays do indeed use an inverse-time type of characteristic within the first zone. We agree with him that the speed-reach characteristics of our figures (8) and (9) are of concern mainly for the direct trip function of the first zone. A pilot scheme, which only requires a secure directional decision (and therefore can tolerate a higher relative error in the impedance estimate) can operate much faster to clear faults at the end of the line. Mr. Mathews is quite right in pointing out that CVT errors can be serious at zone-one boundary of a short line connected to a weak system. These CVT errors will contribute to the error spectrum of voltage signals, and consequently will make the optimum estimates less accurate. We offer the following comments in response to Mr. Mathews' numbered points:

(1) Conclusion (2) of our paper states that the optimum speed-reach characteristic for an impedance relay is completely determined by the non 60-Hz components of the current and voltage waveforms. This relationship is thus independent of the hardware used to implement the relay. However, to the extent that a specific hardware realization falls short of the optimum, the characteristics of that particular hardware will be hardware dependent. In figure (8), a realization of the relay using digital techniques with a sampling frequency of 720 Hz is shown, and it is indeed poorer (slower for comparable accuracy) than the optimum labeled 'Analog Signal Processing'. Similarly, figure (9) shows that a digital relay based upon symmetrical components is more accurate for the same operating speed than a phase-quantity relay. Both of these realizations of course fall somewhat short of their respective optima.

It should be pointed out that the primary reason for the slower response of the digital relay (see figures 6 and 8) is the delay caused by the antialiasing filter. This delay is of the order of one sample interval. Thus a digital relay using a higher sampling rate would approach the speed of the optimum more closely. Incidentally, the differences in relay speed caused by different sampling rates are most noticeable in the 1/4 to 1/3 cycle speed region. Beyond a 1/2 cycle operating speed, these differences are not significant.

(2) The non 60 Hz components do decay after the occurrence of the fault. The error signal spectra of our figure (4) were obtained by analyzing the first cycle of the waveforms immediately following the fault.

(3) Mr. Mathews' interpretation in this instance is generally correct. (Although, the 90% point in figure 9 corresponds to little over 1/2 cycle speed for the phase quantity relay, and to about 1/3 cycle speed for the symmetrical component relay). The general thrust of this portion of our paper is that the distribution of the error signals—if known precisely—can lead to a more refined estimate of the impedance inac-

curacies. If, by model studies or by observation of system transients, error distributions for the waveforms could be specified, then the relay reach-speed setting can be specified commensurate with the known error distribution.

Professor Lewis points out that for faults near a generating station, the total fault duration may be longer than that obtained under the assumption that a natural current zero occurs every cycle of a fault current. This is true, and must be considered in evaluating the performance of the protection system for such faults. The conclusions of our paper are applicable to faults of this type also, when the appropriate addition to the fault duration is taken into account.

Messrs. Sood and Sinha would like to know the relationship between the observed frequencies of our model studies and the sampling rate chosen. We would like to point out that the spectrum of the non 60 Hz components was obtained from the model power system by using a sampling rate of 3840 Hz. Frequency components in multiples of 15 Hz were computed from the sampled data. However, the spectrum itself (which was obtained by interpolating between the 15 Hz interval computed values) is in no way related to the 3840 Hz sampling frequency. It should also be remembered that our relaying algorithm uses a sampling rate of 720 Hz, and this latter rate determines the performance of our algorithm in the presence of noise defined by the observed spectrum.

We should re-iterate our observation that if the precise noise structure is known, a relay will perform within accuracy limits which can be determined far more precisely. Alternatively, a special algorithm can be devised which attenuates that specific noise to a greater extent, thereby improving the accuracy of the relay. However, in a practical system where a relay must operate under changing system conditions, the relay setting must accommodate the noise spectrum that is an ensemble average over all possible system configurations.

Figure (5) of our paper illustrates that for the noise spectrum observed a one cycle window relay has an unavoidable error component in its impedance estimate. Our figures are based on phenomena observed on our model system. It would be wrong to assume that the same numbers are applicable for other systems. However, our paper provides equations for calculating relay performance criteria for other types of systems if the noise spectrum for those systems could be specified.

In our opinion it is not possible (nor is it worth the effort) to determine the exact distribution of error signals. If the noise component is a single non-decaying sinusoid, its distribution can be easily specified. Since the noise usually consists of several sinusoids each decaying at a different rate, at best a reasonably shaped distribution may be postualated. In our view, a rectangular or triangular distribution is adequate in the present context.

The δmn symbol used in our paper is the Kronecker delta; 0 when $m \neq n$ and 1 where $m = n$.

Messrs. Breingan, Chen and Jewell have indicated that their laboratory and field test results have been very accurate for a relatively narrow data window. This in our view says something about the nature and the amount of noise present in their signals. The conclusions of our paper are that the noise in the signals determines the optimum speed-reach relationship obtainable by any relay; a specific realization of the relay—whether analog or digital—can only approach this optimum, and perhaps equal it for some realizations. Certainly if the power system is such that the noise components of the signals are weaker than those shown in our figure (4), the corresponding of the signals are those shown in our figure (4), the corresponding optimum speed-reach relationships as illustrated in our figures (5) through (9) should be changed as specified by our equations (8), (9), (15) and (18).

We would like to re-emphasize that our noise description of figure (4) is an ensemble average. All types of faults on all possible system configurations on our model system were analyzed to produce figure (4). Some faults—on a single, simply structured system—may produce fault waveforms having much less energy in the noise components. For such a system, the relaying algorithm may well produce a very precise result. However, the reach setting of the relay must be so adjusted that it will operate securely under all possible system configurations in a system of representative complexity, and consequently must accommodate impedance errors compatible with expected system noise.

A fault location calculation on the other hand—since it can use a much larger data window—is capable of far more precision.

The stationary random process description of noise used in our paper specifically excluded the decaying dc offset used by the discussors in their differential equation algorithm.

The discussors state that by discarding the dc offset (through the use of a mimic circuit), we apparently discard a signal component which

120

is capable of aiding the impedance estimation process. On closer examination it is found that this is not true. Indeed, using the differential equation algorithm on a fully offset fault current is equivalent to using a digital realization of the mimic circuit.

We offer the following brief error analysis of the differential equation algorithm. Let the differential equation

$$v = Ri + Li' \qquad (21)$$

describe the fault process simulation. In the response of error signals (i.e. those non-60 Hz components which do not satisfy the differential equation; the decaying dc offset is NOT such a component, and is included in i and i'), equation (21) is perturbed as follows:

$$v + n_v = (R + \Delta R) \cdot (i + n_i) + (L + \Delta L) \cdot (i' + n_i') \qquad (22)$$

Where the n's are the noise components in each of the observables, ΔR and ΔL are the corresponding errors in the estimates of R and L. Equation (22) can be simplified to include first order terms only:

$$n_v - (Rn_i + Ln_i') = \Delta Ri + \Delta Li' \qquad (23)$$

With assumed independence between (n_v) and $(Rn_i + Ln_i')$, their covariance and power density spectra are additive. We will assume (as in our paper) an upper bound K_v for the spectrum of (n_v) and K_i for the spectrum of $(Rn_i + Ln_i')$, such that $K_v + K_i \equiv K$.

Now consider a fault incidence angle θ, such that the fault current and its derivative are given by

$$i = i_0 [\sin (\omega t + \theta) - \sin \theta \cdot e^{-\alpha t}]$$
$$\qquad (24)$$
$$i' = \omega i_0 [\cos (\omega t + \theta) + (\alpha/\omega) \sin \theta \, e^{-\alpha t}]$$

We will assume a half-cycle data window for the sake of compatibility with our results, although calculations for other window lengths can be carried out just as easily. Over the half-cycle window, the minimum variances of ΔX and ΔR ($X \equiv \omega L$), using the differential equation estimator (23) are given by

$$\sigma^2 \equiv \sigma_X^2 + \sigma_R^2 \cong K (A + D)/(AD - BC) \qquad (25)$$

Where

$$A = T/2$$
$$D = (T/2) + T \sin^2 \theta - (2/\omega) \sin 2\theta \qquad (26)$$
$$B = C = (2/\omega) \sin^2 \theta$$

Equation (25) becomes exact when α, the power system decrement factor, is 0; and is a good approximation for realistic α's found on a power system. If there is no dc offset in the current wave, $\theta = 0$, and equation (25) becomes

$$\alpha^2 = 4K/T \qquad (27)$$

which agrees with equation (9) of our paper. With a fully offset current ($\theta = \pi/2$), the corresponding result is

$$\alpha^2 = (1.45) \, 4K/T \qquad (28)$$

The multiplier (1.45) appearing in equation (28) is the amplification factor on the covariance of the error estimate produced by the differential equation estimator when compared to the optimum (Fourier Type) estimator discussed in our paper. In our simulation of the differential equation algorithm, this conclusion has been borne out.

It is interesting to note that the term $(Rn_i + Ln_i')$ appearing on the left-hand side of equation (23) is identical to the noise contribution of an analog type mimic circuit. Thus, the quality of the current signal presented to the Fourier Type algorithm by analog type mimic circuit is comparable to the quality of the current and current derivative terms used by the differential equation type algorithm.

On narrower data windows, the covariance of impedance error estimates for the differential equation algorithm is also poorer than the corresponding Fourier estimates. It should also be clear that in discrete time domain, the optimum estimates achievable by either technique are further degraded.

With regard to the discussors' figure, we have already stated that the performance of any algorithm is determined by the noise components of the fault waveforms.

We have previously commented in our response to Mr. Mathews' discussion on the role of impedance estimates in a pilot scheme.

The use of delta quantities for relaying would provide more smoothing than the use of single phase quantities. The delta quantity smoothing in turn is less than that obtainable with a symmetrical component calculation.

The difference between analog and digital signal processing in figure (6) is caused by the discretization process which does not use as much information as is used by the analog (or continuous) process. Although it is not clear from figure (6) because of its scale, the two curves—analog and digital—coincide at multiples of half cycle.

We wish to thank all the discussors for their contribution to our paper.

Manuscript received May 8, 1978.

EHV PROTECTION PROBLEMS

An IEEE Power Systems Relaying Committee Working Group Report

ABSTRACT

This paper presents a broad overview of system considerations affecting protective relaying applications on extra-high voltage (EHV) systems. Discussions include the effects of line parameters on relaying, transient response of current transformers and coupling capacitor voltage transformers, control circuit and relay surge protection, subsynchronous resonance problems, stability considerations, influence of harmonics and power line noise on carrier relaying, automatic reclosing considerations and simulation of transmission line transients for relay applications.

INTRODUCTION

This paper discusses problems and possible solutions affecting protective relaying and controls on EHV (extra high voltage) transmission systems. Although some of these problems may also occur on HV (high voltage) systems, they are more pronounced on EHV systems.

EFFECTS OF LINE PARAMETERS ON RELAYING

The application of line relaying systems at EHV levels is complicated by line shunt capacitance, untransposed lines, shunt reactor compensation, and series capacitor compensation. The effect of each of these parameters on relaying is discussed in the following sections.

Line Shunt Capacitance

The use of bundled conductors at EHV levels results in substantially larger line charging reactive power than found on HV systems. For example, charging kVA for 500 kV bundled conductor lines is approximately 2000 kVA/mile, vs. approximately 270 kVA/mile for single conductor 230 kV lines. During light load conditions this can cause system overvoltage, requiring compensation by shunt reactors, spinning reactive power, deenergizing lines, or a combination of these.

Line charging current may produce different positive, negative and zero sequence quantities at the two ends of an unfaulted line during system faults. This can result in unequal sequence network outputs in relay systems utilizing sequence quantities. Consequently, the relay systems may misoperate during external faults if not properly applied and if this factor is not taken into account in the relay setting. Surge currents and voltages with frequencies up to 1.5 MHz are produced during switching or system faults. Some of these surges may be several milli-seconds in duration and in some of the initial relay designs caused sensitively set high speed solid state relays to misoperate. Relay manufacturers have recognized these problems and have designed relay systems accordingly.

Untransposed EHV Transmission Lines

Untransposed lines have unbalanced impedances which may adversely affect the relay systems applied. Unbalanced line impedances

EHV Protection Problems Working Group membership: J. J. Bonk (Chairman), D. C. Adamson, J. R. Boyle, A. D. Cathcart, D. H. Colwell, R. W. Dempsey, C. W. Fromen, F. O. Griffin, J. M. Intrabartola, J. R. Latham, R. W. Pashley, R. C. Stein, D. Zollman. Former members: F. E. Newman (past Chairman), R. F. Arehart, J. E. Benning, C. W. Cogburn, J. B. Dingler, R. W. Hirtler, H. M. Paul, A. C. Pierce, A. B. Webb.

81 WM 120-5   A paper recommended and approved by the IEEE Power System Relaying Committee of the IEEE Power Engineering Society for presentation at the IEEE PES Winter Meeting, Atlanta, Georgia, February 1-6, 1981. Manuscript submitted November 6, 1980; made available for printing November 10, 1980.

result in negative and zero sequence current flow within the system. These sequence quantities may become large enough to cause false response in line relays. There is also a risk of damage to rotating machines. The magnitude of the negative sequence current component in some machines on the system could exceed the continuous negative sequence current carrying capability.

The problem is further aggravated on long untransposed EHV lines which are series capacitor compensated. In this case the capacitors cancel a large percentage of the inductive reactance of the line, but do not reduce the magnitude of unbalance between phases. Therefore, the effect of the unbalance becomes more pronounced on the series compensated line. If unbalanced series compensated lines are in parallel, there is a risk of having high circulating currents which may further complicate the application of relays on these lines.

As an example, one west coast utility had a 500 kV, 150 mile, 70% series compensated, untransposed transmission line paralleled by two 500 kV series, compensated transposed transmission lines. Under light load, negative sequence currents on the untransposed line exceeded 30% of the positive sequence currents. As the load increased, the percentage of negative sequence currents decreased, but the actual magnitude increased. A loading of approximately 300 MW resulted in enough negative sequence current to pick up the low-set negative sequence fault detectors on the phase comparison relays, which turned on the pilot channel. Emergency overloads could have picked up the high-set fault detectors, greatly increasing the possibility of the protection tripping incorrectly. Because of these problems, this line was transposed.

Shunt Reactor Compensation

Shunt reactors are applied to compensate and control the charging current of an EHV line to keep steady state voltage within normal range. These reactors reduce the adverse effects of the line shunt capacitance of long lines on relaying, but also create some problems.

Shunt reactors may be connected to a line, to a bus, or to transformer tertiary windings. If line-connected they are usually protected by line circuit breakers and are included in the zone of protection of the line relays. Where bus VT's are used for relay potential, line relays will not see the oscillatory discharge of the line and reactors when the line is deenergized. However, if line potential devices are used, the relay potential circuit will be subjected to an oscillatory potential and may produce an undesirable trip output at the instant of reclosing. In like manner, relays connected to reactor bushing CT's will be subjected to oscillatory currents and there may be some danger of undesirable line lockout caused by misoperation of reactor overcurrent relays due to the oscillatory discharge current when the shunt compensated line is deenergized. This oscillation may be sustained for several seconds.

Problems can also arise due to induced voltage from a parallel line if system parameters are right. In one case a shunt compensated line was deenergized and sectionalized at a series compensation station about one fourth of the distance to the next bus. This section of line with reactors connected provided the parameters to allow rated reactor current, in phase in all three phases, to be induced by a parallel line carrying load. These currents, adding in the reactor neutral connection, caused the reactor neutral relay coil to be destroyed by a sustained current over the coil rating.

Various relay schemes have been used for reactor protection, including differential, distance, overcurrent, negative sequence, sudden pressure, or some combination of these relays. Users report sudden pressure relays as one of the best means of reactor protection. Direct transferred trip is needed for protection of reactors when they are line-connected. Distance protection can be applied to single phase EHV shunt reactors to detect shorted turns which involve greater than 5% to 10% of the total turns. This type of reactor protection is possible due to the significant reduction in its 60 Hz reactance under shorted turns conditions.

Reprinted from *IEEE Trans. Power App. Syst.*, vol. PAS-100, no. 5, pp. 2399–2406, May 1981.

The use of distance relays for reactor protection has been discontinued by one utility because of misoperations during reactor switching. However, one user has reported satisfactory experience with a type of distance relay designed specifically for reactor protection. Tuned circuits cause the reach of the relay to be appreciably shortened at other than rated frequency. This reduces the risk of operating incorrectly on the natural frequency oscillation of a deenergized line.

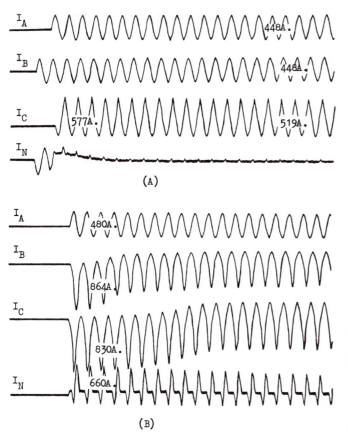

NOTE: Calibration not the same for each trace

Fig. 1. 230 kV 180 MVA shunt reactor energization (A) with circuit switcher, (B) with circuit breaker. (Courtesy of Bonneville Power Administration)

An example of the energization transient of a shunt reactor is shown in Figure 1. This oscillogram shows the charging current to a 230 kV, 3 phase, 180 MVA, gapped core reactor during energization with a circuit switcher (upper record) and a gas blast circuit breaker (lower record). The oscillograph calibration on the two records is the same. Note that the rms measurement of currents in B and C phase respectively, on the lower record are 192% and 194% of the nominal rated current. This is due to the circuit breaker contacts closing at a near zero point in the voltage wave.

Series Capacitor Compensated Lines

Short and medium length transmission lines at voltages below the EHV level usually exhibit waveforms which are smooth and regular enough to be predictable before, during, and after a fault. However, EHV made longer lines economically attractive and series and shunt compensation a technical necessity, and introduced additional relay problems due to the transients and distorted waveforms caused by this compensation. One of the causes of transients is the firing of the gap across the capacitor bank to protect it from overvoltage. In addition to the effect on relays, transients associated with series capacitor gap firing can affect power line carrier operation. The voltage generated in the capacitor discharge circuit can produce an erroneous power line carrier signal, resulting in misoperation of carrier relaying. Figure 2, which is an oscillogram of a staged fault test, illustrates this condition. Note the trace labeled "Bank #1, E_{CO}" and the effect of this

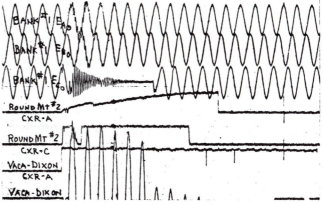

Fig. 2. Series capacitor gap firing transients (Table Mt. staged fault tests - courtesy Pacific G & E Co.)

voltage on the power line carrier signal, "Round Mtn #2, CXR-C." This effect was noted only on this one occasion by this utility, and they could not produce the same effect by secondary testing.

The presence of series capacitors on the system can produce wave form distortion when the protective gaps do not flash. Figure 3 shows subsynchronous currents in the order of 30 Hz during and after clearing a one-line to ground fault on a series compensated system where the protective gaps did not flash. Waveform distortions shown on these oscillograms can cause improper relay operation if not recognized and compensated for when the relay system is designed and applied.

EHV lines with series compensation are difficult to protect with conventional directional comparison relaying schemes. If series capacitors are located at the ends of the transmission line, and fault currents are not large enough to flash the capacitor protective gaps, faults just beyond the capacitor bank on either the protected line or on external lines appear to be in the reverse direction on a steady state basis.

Two different relay schemes have been predominantly used to protect series compensated lines. One scheme uses phase comparison and the other scheme uses directional comparison designed specifically for series compensated lines.

Phase comparison has the advantage that only currents are used and therefore it is not affected by voltage reversals that can be introduced by the series capacitor. However, special consideration must be given to schemes which use weighted sequence currents generated by

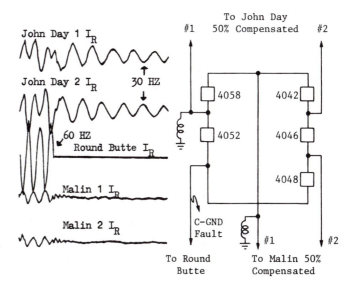

Fig. 3. Subsynchronous line currents on series compensated lines after clearing an external fault. (Courtesy of Bonneville Power Administration)

sequence networks. Phase impedance unbalance due to unsymmetrical gap flashing can result in angular differences between the sequence quantities at the two ends of the line and can cause failure to trip or unnecessary tripping if not properly allowed for. Schemes which employ a separate relay for each phase do not have this problem, but require special channel considerations.

Directional comparison schemes designed especially for series compensated lines usually rely on the transient rather than the steady-state relay characteristic to give correct directional sensing. In one particular scheme, the tripping relay operates faster than the blocking relay at the same location for an internal fault, and the blocking relay operates faster than the tripping relay for external faults. A seal-in circuit uses this information to maintain the trip or block signal, even if the steady state output is wrong.

Additionally, a scheme utilizing wave detection techniques has been developed. Limited field experience is available at present.

Solid state relays are usually applied to series compensated lines. However, electromechanical phase and ground relays may, under certain circumstances, also be successfully applied.

TRANSIENT RESPONSE OF CURRENT TRANSFORMERS[1,2]

Response of current transformers is becoming increasingly important in EHV systems because of the increase in fault current magnitudes, the increase in the system time constants, and the increased incidence of asymmetrical faults because of high closing velocity of air and gas blast breaker contacts. The stringent protection reliability requirements for EHV systems can be met only if the current transformers reproduce the primary current within required accuracy limits. Factors that can be controlled by the user to provide best performance of a given CT are (1) secondary burden and (2) CT ratio. Flux density in the CT core is directly proportional to secondary burden and inversely proportional to the square of the secondary turns. Therefore, it is important to keep the burden low and to use the full secondary winding if possible. Long cable runs associated with large EHV stations tend to make the cable resistance an appreciable part of the secondary burden. Use of relay houses in the substation yard is one method of shortening cable lengths.

Careful selection of the proper full winding rating of CT's in each station is required to attain best utilization of the voltage output capability of those CT's. For example, high ratio CT's should not be applied where relay sensitivity would require a much lower ratio unless ultimate development of the system will allow eventual use of the full winding. On the other hand, a low ratio CT should not be applied in a developing system which will ultimately require a higher ratio. Although it is desirable to use the full winding of a CT, it is usually permissible to use lower ratio taps during development of the power system because fault currents are relatively low at this time.

For angle sensitive relays, such as directional overcurrent or directional distance relays, current distortion from CT saturation may distort the steady state characteristic or, in extreme cases, cause incorrect directional sensing. On distance relays, CT saturation may cause relays to under-reach. Protection schemes utilizing zero crossings may be adversely affected by CT saturation.

The configuration of most EHV stations requires paralleling of CT's to provide line relay current. The CT's to be paralleled should be identical in full winding ratio and taps if taps must be used. Use of the full winding on one CT and a tap or another CT to provide equal ratios of the two should be avoided. The difference in transient performance characteristics of two CT's which are paralleled may produce false differential current due to unequal saturation for through fault current. Two CT's with the same relay accuracy rating may not be identical in performance. More practically, one of the two paralleled CT's will usually have higher current than the other because of a source connected between them. This will cause more error in the one with higher current if conditions are conducive to poor CT performance.

Use of auxiliary CT's should be avoided if possible, but if used should not degrade the overall performance of the relaying scheme.

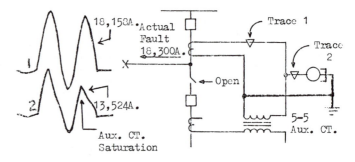

Fig. 4. *Effect of auxiliary current transformer saturation on relay current for a close in line fault. (Courtesy of Bonneville Power Administration).*

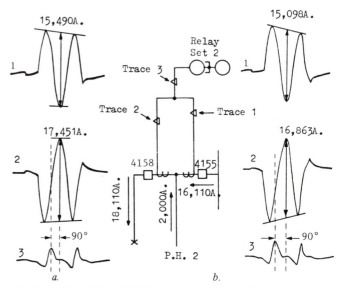

Fig. 5a and 5b. *Effect of difference in performance between current transformers. (Courtesy of Bonneville Power Administration)*

Examples of inadequate CT performance are shown in Figures 4 and 5. These figures show results from a series of staged fault tests on the Bonneville Power Administration system. Figure 4 shows the effect of inadequate auxiliary CT's used to sum currents for a line in a breaker-and-a-half scheme. The total fault current was 18,300 amperes. This was measured by a CT (not shown on the diagram) having a C2000 accuracy rating. As indicated on Trace 1, the C800, 3000/5 CT reproduced the current with good accuracy, but the relay current, Trace 2, was only about 75% of this magnitude. Saturation of the 5/5 auxiliary CT's (C200) caused the appreciable reduction in current to the relay.

Figures 5a and 5b show the effect of difference in performance between relaying CT's. Both sets are C800, 3000/5, connected 1600/5. Poor CT performance caused distortion of the secondary currents, particularly in breaker 4158. This caused the relay current to be badly distorted with the positive peak being shifted approximately 90° from the expected P.H. #2 relay current, and much higher in magnitude. This kind of CT performance can cause a low set directional ground relay to misoperate. For both these cases, a C2000 CT (not shown) was used to measure total fault current.

One important factor affecting CT operation is that of residual magnetism in the core. Refer to Figures 5a and 5b. Note that the measured magnitudes of current through breakers 4155 and 4158 are lower in Figure 5b than in Figure 5a. This is due to a difference in residual flux in the CT cores caused by successive asymmetrical faults in the same direction. Some manufacturers will provide antiremnance air gaps in the cores of their CT's, essentially eliminating this problem. Core gaps in the order of 0.0001 per unit of the core length will provide adequate relaying accuracy. Bracing of gapped core CT's must be adequate to withstand the forces caused by fault current, shipping, and temperature cycling.

An important consideration in the design of an EHV station is the number of CT cores required in order to avoid the use of auxiliary CT's. Two sets of relays on separate CT's will require 4 cores per breaker on a ring bus. On a 1-1/2 breaker scheme, 4 cores are required on the center breaker and 3 are required on the bus breakers, assuming one is used for bus differential relays. If a metering accuracy core is required, CT's with 5 cores should be procured in order to avoid use of auxiliary CT's in relay circuits.

TRANSIENT RESPONSE OF COUPLING CAPACITOR VOLTAGE TRANSFORMERS (CCVT'S)[3]

Because CCVT's are widely used as the source of a.c. voltage input to protective relays on EHV transmission lines, their transient response, and the effect of this response on the relays, is of great importance.

High speed relay operation is essential in order to obtain high speed fault clearing. Solid-state relays which can operate in less than one cycle are being used, and it is desirable that the transient response of the CCVT should neither delay the operation of these relays nor cause them to operate incorrectly.

There are many parameters that affect the transient response of CCVT's. The most significant are the amount of capacitance in the device itself, the voltage of the divider section, the type of ferroresonant suppression system used, the magnitude and power factor of the burden on the CCVT, and the point on the primary transmission voltage wave where the fault occurs. Figure 6 shows a ferroresonant condition on a CCVT. The second oscillograph trace from the top depicts a ferroresonant condition occurring upon reclosing approximately 25 cycles from the start of the fault. The condition, which lasted for approximately 9 cycles, was not one that existed on the primary system, but a secondary phenomenon in the potential device. A blowup of this condition is shown in the lower right-hand corner. The poor performance of the carrier signal, trace 4, was probably caused by the ferroresonant condition and the saturation of the neutral CT, trace 6. Ferroresonant conditions do not normally occur if proper suppression devices are installed.

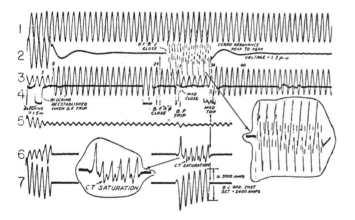

Fig. 6. CCVT ferroresonance
(Courtesy of TVA)

A study[3] indicated that the CCVT error voltage is maximum for a fault that occurs at zero on the voltage wave, and is minimum for a fault that occurs at the peak of the voltage wave, if the burden is resistive. It also showed that CCVT's with larger values of capacitance produce less magnitude of error, but the time constant of the error decay is greater and therefore the error voltage is present for a longer time. The ferroresonant circuit and the makeup of the burden add other distortions to the CCVT output voltage.

Figure 7 shows the A and B phase-to-neutral voltages on both 500-kV lines to the Sullivan Substation. An A phase-to-ground fault on the 500-kV line between the Bull Run and Sullivan terminals cleared by the opening of breakers 1, 2, and 4 in approximately 3 cycles. Upon reclosure the output voltage of the B phase CCVT on the Bull Run line (fourth element from the top labeled B-ϕ NEUT: POT. Bull Run 500-kV line) was 1.85 times higher than its correct

counterpart on the Broadford line which constituted a good reference because it was never deenergized.

There are standards which define the method of testing the transient response of a CCVT, but there are no IEEE or ANSI standards which place a limit on how much CCVT error signal is acceptable.

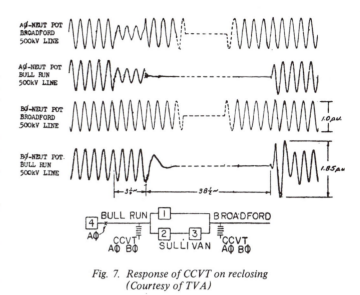

Fig. 7. Response of CCVT on reclosing
(Courtesy of TVA)

While it is difficult, if not impossible, to make any general statements about the effect of CCVT errors on relay performance, one possible relay problem that has been identified is the overreach of solid-state first zone distance relays. This has not been a serious problem in actual installations, but it can be shown by laboratory tests, including a model of the CCVT, that overreach may occur if the change from prefault to fault voltage is large.

Several solutions have been devised to overcome the CCVT transient error problem as far as overreach is concerned. In one case, the operation of phase relays was delayed and the reach of ground relays was reduced. In another case, the operating signal inside the relay was modified in such a way as to insert filtering only when needed to prevent overreach. Another solution is the use of a CCVT with more accurate transient response.

The effect of the CCVT error on any particular relay depends on how the relay is designed, and the manufacturer of the relay should be consulted for his recommendations with respect to a particular CCVT design.

SUBSYNCHRONOUS RESONANCE AND TORSIONAL OSCILLATIONS DUE TO SERIES COMPENSATED LINES[4,5,6]

The application of series capacitors to EHV transmission systems is very attractive from an economic and stability standpoint. However, for certain system configurations and conditions, application of series capacitors can result in subsynchronous resonance of the electrical system with the turbine-generator (T-G) mechanical system. This can produce damped or undamped mechanical oscillations corresponding to the natural frequencies of the components of the T-G, which may result in shaft failure. Figure 8 shows a simplified circuit for explaining the electrical subsynchronous oscillation problem.

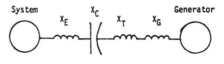

X_E - Net System Inductive Reactance
X_C - Series Capacitive Reactance
X_T - Transformer Reactance
X_G - Generator Subtransient Reactance

Fig. 8

This circuit has an electrical resonant frequency

$$f_{er} = f_o \sqrt{\frac{X_C}{X_L}}$$

where f_o is the synchronous frequency defined by rotor average speed (60 Hz in the U.S.); X_C, the series capacitive reactance of the system at f_o, and X_L, the total inductive reactance at f_o, the sum of $X_E + X_G + X_T$ in Figure 8.

Since less than 100% series compensation is used, f_{er} will always be less than the synchronous frequency, f_o. Under this condition, positive sequence component of stator current produce rotor current at subsynchronous frequency f_r, equal to $f_o - f_{er}$. Also, the interaction of the rotor magnetic field and the rotating subsynchronous mmf in the armature at frequency f_{er}, produces a subsynchronous torque of frequency equal to $f_o - f_{er}$.

These can result in two forms of self-excited oscillations: 1) an induction generator effect, where the rotor resistance to the subsynchronous current viewed from the armature terminals is negative and exceeds the sum of the generator armature and network resistance at f_{er}; 2) a torsional interaction, where a generator rotor oscillation at a torsional mode frequency close to $f_o - f_{er}$ induces armature voltage components of frequencies which are phased so as to sustain subsynchronous torques equal or in excess of the inherent mechanical damping torque of the rotating system.

In a compensated system, electrical shocks such as faults, although damped, may also result in a build-up of torsional forces that may exceed the shaft elastic limit.

By properly selecting the amount of series compensation, the subsynchronous resonance problem can be mitigated. The more serious problems occur when series capacitors and generation are electrically close (with insufficient load). The most serious case occurs when generation remote from load is connected to the system through a series compensated line. Steam units are more likely to be damaged than hydro units, as the shaft torsional frequencies of steam units are closer to the torsional stress frequency caused by $f_o - f_{er}$.

The Southern California Edison Company has developed a current operated subsynchronous relay which has been installed on the two generators at Mohave.[6] This relay will trip the units from the system in the event of a sustained subsynchronous oscillation which could cause shaft damage. The relay has two tripping elements. The low-set element is set at 5% of the generator current rating and timed for 2.0 seconds. The high set element has been set at 30% of the generator current rating and timed for 0.33 seconds. A manufacturer has also supplied a subsynchronous current relay of different principle for generator units of the Arizona Public Service Co.[7,8]

One manufacturer has supplied a protection package to the Salt River Project including (1) a three-phase blocking filter in the transformer neutral circuit, (2) a supplementary excitation damper control, and (3) a relay which directly measures any change in shaft rotational speed (from magnetic pickups on the shaft) and operates through several coordinated magnitude pickup and time delay settings to trip the generator breaker. In this application, the relay is backup protection to cover the possibility of component failure in the filter or in the damper control.

Another manufacturer has supplied a dynamic compensating device that also is controlled from magnetic pickups on the generator shaft, also for application in the Southwest.[9]

Various filter circuits and capacitor bypass and reinsertion schemes have also been developed to minimize the shaft torsional stress due to transient subsynchronous frequencies.

SURGE PROTECTION OF STATIC RELAYS[10,11]

The growing use of solid state semiconductor control equipment, including static relays, on electric power systems has accentuated the problem of transient electromagnetic influences on the control or protective system itself. These influences are commonly called "surges," EMI or control circuit transients. The influencing transients from various sources have been recorded in a number of field tests and sufficient data is available to identify their probable nature and severity. Also, methods have been evolved which both the user and the equipment manufacturer can employ to reduce the effect of these surges on static relays and other control equipments. Metal Oxide Varistors, Zener diodes, and surge capacitors have been used to suppress transients and protect control equipment.

Field measurements have shown that, in general, surges in control circuits have an oscillatory nature, the oscillation frequency being in the range 200 kHz to several MHz or higher. The voltage amplitudes measured range from 100 volts to as high as several kV. The effective source impedance may range from 10 ohms to over 1000 ohms depending on the particular mode of coupling and equivalent circuit parameters. They appear on just about all low voltage wiring in a station, but are not always generated within the same circuits. The source can be in the EHV circuits in many cases. The surge voltage can appear between two parts of the same circuit, i.e., the transverse mode, or between the circuit as a whole and a common reference, e.g., ground. This is the so-called common mode. The oscillatory bursts can occur singly or be repeated several hundred times per second. Single oscillatory bursts may last from 1 to 100 microseconds.

Surges can be generated outside or inside the control circuit. External sources can be in other low voltage circuits, which are then coupled to the control circuit via various stray mutual impedance effects. They can also be generated in the EHV power circuits and coupled similarly by stray mutual effects into the control circuit. This is a very common situation.

The use of secondary cables to CCVT's, CT's, and d.c. circuits which have good conducting shields, properly grounded, greatly reduces the secondary surge voltage levels. Shields are only effective for both the electric and magnetic components if both ends are grounded.[12] Filtering, such as surge capacitors and battery clamps, is not always adequate for circuit protection; therefore, surge protection devices may be required at the source of the transient, such as across d.c. coils.

EFFECTS OF STATIC RELAY SURGE CAPACITORS ON DC CONTROL BUSES

The application of static relays generally requires surge protection capacitors. This adds to the already present stray capacitance to ground on the plus and minus station battery leads. In large stations, the energy available from the lumped and distributed capacitance of battery negative to ground is enough to cause circuit breaker and auxiliary relay operation. An accidental ground in the control house, or an insulation failure in a cable, may cause discharge of this capacitance through a breaker trip or close coil, causing the breaker to operate. Auxiliary relays may also be operated due to d.c. grounds.

The breaker trip and close coils and some auxiliary relays are normally connected only to the negative bus. Connection to the positive bus is usually through an open contact. Therefore, some means of holding the negative bus near ground potential will eliminate the possibility of unwanted operation of these devices due to d.c. grounds. Various schemes are being used to hold the negative bus near ground potential. A resistor in the order of 1000 ohms or a light bulb will effectively ground the negative lead and still not cause short circuit problems if the positive bus becomes grounded. A means of switching the grounding device is desirable to allow detection of other grounds on the circuit.

INFLUENCE OF HEAVY LOAD CAPABILITY OF EHV LINES

Load growth in an area served by power transmission facilities generally requires continued additions to these facilities. An overlay of an existing high-voltage system by an EHV system is common practice today. EHV ties between large systems are becoming more common. Loss of these high capacity lines may lead to overloads, power swings, or instability of the underlying system. It is advisable, if not mandatory, to look at the effect on the system of the loss of these EHV facilities by running load flow and transient stability studies. The results of these studies, along with system planning criteria, must

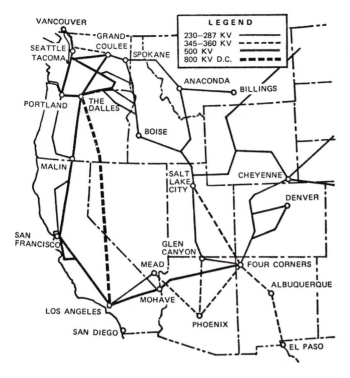

Fig. 9. Western states major transmission system

be analyzed to determine whether or not the transmission system meets the expected system reliability.

Loss of heavily loaded EHV ties between systems can cause severe disturbances that require remedial actions such as generator dropping, dynamic braking, and load shedding to meet reliability criteria. It may also require intentional islanding of systems, particularly if these systems form a parallel tie. Transient stability studies must be made in order to evaluate the system operation and plan the control facilities required.

An example illustrating the above problems is the Pacific Northwest-Pacific Southwest Intertie, consisting of two parallel 500-kV lines rated 1250 MW each and a d.c. line rated 1400 MW. The lines are loaded to capacity at certain times of the year with power flow from north to south. Figure 9 shows the basic system with parallel HV lines. Out-of-step relaying was part of the initial design and was installed at Malin, where intentional separation of the 500-kV systems occurs for unstable swings. After operating experience, which included some 500-kV intertie separations, it was found that large power transfers through the paralleled HV system to the east of this 500-kV intertie (HV system through Idaho, Utah, and Arizona) caused those systems to break up uncontrollably on separation of the 500-kV ties. These effects are reduced to a tolerable level by transmitting an islanding signal to the southwest at Four Corners, separating the eastern HV system upon loss of the 500-kV intertie. Generator dropping up to 1400 MW in the Northwest is also required. The sensing schemes also separate the 230 kV from the 500 kV system at appropriate points to prevent overloading the 230 kV system.

This has proved to be a satisfactory solution. Several separations have occurred with successful operation of the islanding and generator dropping schemes. Underfrequency load-shedding is required in the islanded areas because of the varying patterns of power interchange during the year.

Since the 500-kV intertie first went into service, system studies have necessarily been expanded to the whole western United States rather than smaller areas previously studied by each company.

Loss of both poles of the d.c. line (1400 MW) and the consequent transfer of this power to the 500 kV a.c. intertie causes a stable swing, but the impedance seen by the out-of-step relays at Malin comes marginally close to the operating point in the worst case. The severity of this swing is decreased by dropping up to 800 MW of generation in the Northwest upon loss of the d.c. transmission.

AUTOMATIC RECLOSING

Automatic high-speed reclosing of transmission lines after fault clearing is an accepted practice on most HV lines. The practice is also extended to EHV lines when the system has developed to the point that a second fault can be tolerated immediately following the original. Also, as in lower voltage systems, the system must have developed to the point that synchronism is maintained by other parallel lines as the reclosing takes place.

The benefits expected from automatic reclosing include the following:

(1) Improved system reliability - lines are immediately returned to service, thereby reducing risk of multiple line outages.

(2) Reduced consequences of incorrect relay operations.

(3) Relieves operators of unnecessary manual switching duties.

The EHV system has presented new but not unique problems in applying high-speed reclosing. None of these has proved to be unsurmountable.

High-speed reclosing is most often done without dead line checking of voltage conditions on the line; the trapped charge that will exist when the line is interrupted at other than voltage zero will decay through any direct current path to ground or through insulation leakage paths. In the latter case, the decay time will be long and there will be a remaining trapped charge when the line is reclosed. This may produce switching surge overvoltages but is usually within normal design parameters. On EHV systems and some HV systems, the normal circuit breakers are equipped with pre-insertion resistors for control of these switching surge overvoltages. Another means of controlling these voltages is the use of synchronous closing devices on circuit breakers.[13,14,15]

When shunt reactors are connected to the line, as is commonly done in EHv systems, the trapped charge will decay as an oscillation at the natural frequency of the line and shunt compensation inductance, and the line shunt capacitance. There may also be a system frequency component due to mutual coupling from parallel lines. If the shunt compensation of the line is near 100%, the natural frequency will be near system frequency and the mutual coupling may cause the line to

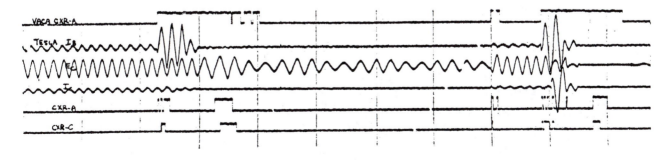

Fig. 10. Oscillating voltage decay of a compensated line. (Courtesy Pacific G & E Co.)

remain at near normal voltage, even though disconnected from the system.

As example of an oscillating voltage decay of a compensated line is shown in Figure 10, an oscillogram of a successful reclose on a 134 mile, 40% compensated 500-kV line which, subsequent to the reclose, refaulted. The natural frequency in this case was about 37 Hz, and at the time of reclose the residual voltage was about 35% of normal. The apparent increase in voltage at the beginning of the decay is due to CCVT performance at 37 Hz.

When it is necessary to limit switching surges and the decay of line potential is oscillatory, the line condition can be monitored with voltage relays or frequency relays and used to block the reclosing control until the line has decayed to a reasonable level of voltage. It is desirable for this application to use voltage relays with flat frequency characteristics and frequency relays with flat voltage characteristics.

Successful high-speed reclosing can be expected in a high percentage of cases, perhaps 70% or higher, which has been attained in HV systems. The remaining 30% must be assumed to subject the system to a second or third fault condition. Depending upon the type of fault and the location, this second fault may cause separation of systems or interruption of a large area. This consideration is particularly important to EHV systems due to the large blocks of power being transmitted and the low impedance ties to large generating plants.

Stability studies of the system are necessary to properly evaluate the effect of refaulting the system upon reclosing. In most cases reclosing into three-phase faults is not justified. Most often the three-phase fault is not caused by natural insulation failure and is usually persistent. The presence of three-phase or multiphase faults, which may present stability problems, can be detected and the reclosing blocked or delayed for a sufficient time to remove the risk of instability. Very seldom does the single-phase fault create stability problems and luckily this is by far the most probable fault condition.

On one 500-kV system in which reclosing is permitted at 30 cycles after all except three-phase faults, the reclosing has been 72% successful; (41 of the 57 faults where reclosing was attempted). On another 500-kV system, with 35 cycles reclosing on one-line-to-ground faults, reclosing has been 73% successful; (91 of the 124 faults where reclosing was attempted).

The three-pole tripping and reclosing of lines that are the only or principal ties to large generating units present special problems, due to the transient torques that may be produced in the turbine generator shaft even if the reclosing is successful. These problems may also exist with single-pole tripping and reclosing but generally are less severe. Due to load distribution, large phase angles may appear across a PCB to be reclosed even though the unit is in synchronism with the system. These problems warrant special studies to evaluate the effects. The alternative is not to use reclosing or to use more sophisticated reclosing schemes until the undesirable effects are reduced to a tolerable level by normal system expansion and integration.

In many EHV station designs, the bus arrangement will be a ring or breaker-and-a-half. It is, of course, undesirable to permit high speed reclosing of all four breakers, since this could result in four additional faults if reclosing is unsuccessful. Various switching schemes, either automatic or manual, or high-speed voltage interlocks are used to avoid this problem.

In applying external single-shot reclosing relays, some of which utilize capacitor discharge for operating energy, the presence of large amounts of surge protection capacitors in EHV static relaying systems must be considered carefully. The surge capacitors may interfere with or cause unexpected operation of this type of reclosing relay.

METHODS USED TO RETAIN STABILITY FOLLOWING MULTIPHASE FAULTS

On those systems where line loads and system configuration pose stability problems, multiphase faults close to large generating plants may trigger instability. Fault clearing time is an important factor in the severity of the resulting swing. Reference 16 gives the difference in relative phase angles of power plant generators with four cycle clearing of three-phase faults close to large generating plants.

A 1400 MW braking resistor is in service in the Pacific Northwest to damp system swings caused by faults in that area. The brake is applied for most 500-kV multiphase faults, some 230-kV multiphase faults and some 500-kV ground faults. Application is controlled by rate-of-change-of-power relays connected to generator lines at certain locations, supervised by fault detecting relays.[17]

DIGITAL COMPUTATION AND ANALOG SIMULATION OF TRANSMISSION LINE TRANSIENTS

On EHV systems, faults produce electromagnetic transients which contain oscillations not only at power frequency but at other frequencies as well. Such transients may cause misoperation of relays. It is therefore important to simulate transmission line transients.

Simulation by hand calculation is only possible for very simple cases. This led to the development of transient network analyzers (TNA's) in the late 1930's. In recent years, digital computer programs have been developed[18,19,20], which make simulations by digital computers.

Most programs use travelling wave methods for the line simulation, which are straightforward for transposed lines but can also be used for untransposed lines. Some programs can also handle cascade connections of multiphase π-circuits, similar to TNA simulations, but with greater flexibility in the representation of electrostatic and electromagnetic coupling in single and parallel circuits. It is beyond the scope of this paper to describe the modelling of other devices, such as lightning arresters, transformers, and generators. Reference 20 attempts to give a summary of modelling techniques as well as solution methods.

General-purpose transients programs have been primarily used for switching surge studies, where good agreement among various programs and TNA's has been established[21]. There is reason to believe that such programs will also become useful for relay application studies. Such studies may require certain refinements, however. For instance, the frequency dependence of the zero sequence parameters may have to be modelled to obtain correct damping of the high frequency oscillation[22]. The initial conditions in relay application studies will often be a.c. steady-state load flows. Therefore, subroutines for the automatic computation of a.c. steady-state conditions as initial conditions for the transients calculations are helpful, even though superposition techniques can sometimes be used to circumvent the a.c. steady-state computation. Also, some devices which are not normally represented in switching surge studies may have to be modelled in relay studies, such as the relays themselves and voltage and current transformers.[23]

Preliminary work in applying digital techniques to transient analysis for relay application has shown excellent correlation with results from model power systems. This work is proceeding and is being used to investigate problems associated with a large 500 kV development which includes series compensation and single-pole reclosing.

Valuable use has been made of analog model power system testing to develop transmission line relays for EHV line protection. In these facilities, all of the transients described above are simulated at the same energy level as the actual CT's and PT's would produce, and the outputs are applied to the actual relays to check their response. This has been a valuable and essential design tool both to develop the relay circuits themselves and to demonstrate correct performance on simulations of actual power system configurations. They are used by most relay manufacturers, and their contribution to the solution of EHV relay problems has been invaluable.

INFLUENCE OF HARMONICS AND POWER LINE NOISE ON CARRIER RELAYING

Steady-state currents with appreciable harmonic content have been encountered on EHV systems during light load conditions. These harmonics are produced by equipment having non-linear characteristics, such as transformers, and are noticeable during light load periods due to high voltage and overexcitation of this equipment. Distorted fields in rotating machines also produce harmonics. These harmonics are predominantly third, but higher order odd harmonics are also present. Third harmonic currents in each phase add directly to produce significant current in line residual current and in transformer neutrals. Distribution of these third harmonic currents in the system is dependent on system parameters and is affected by transformer tertiary windings, shunt capacitors and shunt capacitance of transmission lines.

Sensitively set relays, such as carrier ground, can be picked up by harmonic currents. An example of the type of ground relaying problem which can be caused by excessive harmonics in residual currents is illustrated by one company's investigation of a case of undersirable continuous carrier transmission during light load periods, which disappeared during heavier load periods. In this case, a solid-state carrier start ground relay, calibrated for 144 amperes primary, continued to stay picked up even though an ammeter connected to the residual circuit measured only 79.2 amperes. An oscillograph trace showed that the residual current contained a large degree of harmonic current which could not be measured by the ammeter. Subsequent measurements and analysis indicated that the residual current consisted of approximately 47% fundamentals, 48% third harmonic and 5% higher harmonics. The trace of this residual current is shown in Figure 11.

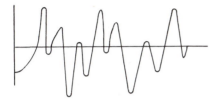

Fig. 11. Harmonic currents in residual circuit

Power line noise interference with carrier channels is caused by arcs which range from low energy level corona discharges to high energy level fault arcs.

Corona discharges are very common on EHV systems and are particularly troublesome during fog, rain, snow, and ice conditions. It is usually necessary to apply 100 watt amplifiers, in multi-function carrier systems, and multiphase coupling techniques to insure reliable carrier operation during adverse weather conditions, particularly on long (100 miles or more) EHV lines.

Arcing noise due to disconnect switching in a substation has caused false tripping of a direct trip dual channel transfer trip carrier system. Fault arcs within a substation, or an adjacent line, can also cause noise interference in carrier channels that could result in false tripping or failures to trip.

REFERENCES

1. IEEE Special Publications 76 CH 1130-4, "Transient Response of Current Transformers". Presented at the IEEE PES 1977 Winter Meeting.

2. Summary and discussion of IEEE Special Publication 76 CH 1130-4, "Transient Response of Current Transformers". Paper F 77 177-9, IEEE Transactions on Power Apparatus and Systems, Vol. PAS-96, No. 6, November/December 1977.

3. A. Sweetana, "Transient Response Characteristics of Capacitive Potential Devices". Paper 71 TP 197-PWR, IEEE Transactions on Power Apparatus and Systems, Vol. PAS-90, No. 5, September/October 1971.

4. IEEE Committee Report, "A Bibliography for the Study of Subsynchronous Resonance Between Rotating Machines and Power Systems". Paper F 75 515-7, IEEE Transactions on Power Apparatus and Systems, Vol. 95, No. 1, January/February 1976.

5. IEEE Committee Report, "First Supplement to a Bibliography for the Study of Subsynchronous Resonance Between Rotating Machines and Power Systems". Paper F79 266-8, IEEE Transactions on Power Apparatus and Systems, Vol. 98, No. 6, November/December 1979.

6. IEEE Special Publications 76 CH 1066-0, "Analysis and Control of Subsynchronous Resonance". Presented at the IEEE PES 1976 Winter Meeting and Tesla Symposium.

7. B. L. Agrawal, R. G. Farmer, "Application of Subsynchronous Oscillation Relay - Type SSO". Presented at the Joint Power Generation Conference, Phoenix, Arizona, September 28 - October 2, 1980.

8. S. C. Sun, S. Salowe, E. R. Taylor, Jr., C. R. Mummert, "A Subsynchronous Oscillation Relay - Type SSO". Presented at the Joint Power Generation Conference, Phoenix, Arizona, September 28 - October 2, 1980.

9. IEEE Committee Report, "Countermeasures to Subsynchronous Resonance Problems", Paper F79-754-3, IEEE Transactions on Power Apparatus and Systems, Vol. 99, No. 2 March/April 1980.

10. ANSI Standard C37.90a-1974, "Surge Withstand Capability".

11. W. C. Kotheimer and L. L. Mankoff, "Electromagnetic Interference and Solid State Protective Relays". Paper F 77 057-3, IEEE Transactions on Power Apparatus and Systems, Vol. PAS-96, No. 4, July/August 1977.

12. D. A. Gillies, H. C. Ramberg, "Methods for Reducing Induced Voltages in Secondary Circuits". Paper No. 31 TP 66-329, IEEE Transactions on Power Apparatus and Systems, Vol. PAS-86, No. 7, July 1967.

13. E. Maury, "Synchronous Closing of 525 kV and 765 kV Circuit Breakers, A Means of Reducing Switching Surges on Unloaded Lines". Report 143, CIGRE 1966 Session.

14. G. E. Stemler, "BPA's Field Test of 500 kV PCB's Rated to Limit Line Switching Overvoltages to 1.5 Per Unit". Paper F75 542-1, IEEE Transactions on Power Apparatus and Systems, Vol. PAS-95, No. 1, January/February 1976.

15. R. N. Yeckley, R. E. Friedrich and M. E. Thuot, "EHV Breaker Rated for Control of Closing Voltage Switching Surges to 1.5 Per Unit". Paper 71 TP 571 PWR, IEEE Transactions on Power Apparatus and Systems, Vol. PAS 91, No. 2, March/April, 1972.

16. Berglund, Mittelstadt, Shelton, Barkan, Dewey and Skreiner, "One-Cycle Fault Interruption at 500 kV: System Benefits and Breaker Design". Paper T 74 176-4, IEEE Transactions of Power Apparatus and Systems, Vol. PAS-95, No. 5, September/October 1974.

17. Shelton, Mittelstadt, Winkelman, Bellerby, "Bonneville Power Administration 1400-MW Braking Resistor". Paper T 74 433-9, IEEE Transactions on Power Apparatus and Systems, Vol. PAS-94, No. 2, March/April 1975.

18. H. B. Thoren, K. L. Carlsson, "A Digital Computer Program for Calculation of Switching and Lightning Surges on Power Systems", Paper 69 C2-PWR IEEE Transactions on Power Apparatus and Systems, Vol. PAS-89, No. 2, February 1970.

19. D. E. Hedman, J. D. Mountford, "Traveling Wave Terminal Simulation By Equivalent Circuit", Paper 71 TP 149-PWR, IEEE Transactions on Power Apparatus and Systems, Vol. PAS-90, No. 6, November/December 1971.

20. H. W. Dommel, W. S. Meyer, "Computation of Electromagnetic Transients". Proceedings of the IEEE, 1974.

21. A. Clerici, "Analog and Digital Simulation for Transient Overvoltage Determinations", ELECTRA, No. 22, pp. 111-138, May 1972.

22. W. S. Meyer, H. W. Dommel, "Numerical Modelling of Frequency Dependent Transmission Line Parameters in an Electromagnetic Transients Program", Paper T 74 080-8, IEEE Transactions on Power Apparatus and Systems, Vol. PAS 93, No. 5, September/October, 1974.

23. G. C. Kothari, K. Parthasarthy, B. S. Ashok Kumar, H. P. Khincha, "Digital Transient Analysis of Power System and Transducers for Relay Analysis". Paper No. C 73 349-8, presented at the IEEE PES Summer Meeting, Vancouver, Canada, July 15-20, 1973.

PRINCIPLES OF GROUND RELAYING FOR HIGH VOLTAGE AND EXTRA HIGH VOLTAGE TRANSMISSION LINES

C. H. Griffin
Senior Member, IEEE
Georgia Power Company
Atlanta, Georgia

Abstract - This paper is a tutorial discussion of the basic principles of ground relaying for high voltage and extra high voltage transmission lines. Three different HV configurations are considered: Long lines, lines with a weak mid-point station, and mutually-coupled lines. Application criteria for EHV circuits are also discussed, and specific setting calculations are included where appropriate.

INTRODUCTION

The purpose of this paper is to discuss the basic principles of ground relay protection for high voltage (HV) and extra high voltage (EHV) transmission lines. "High voltage" lines are generally defined as those circuits operating at voltages from 115 kV up to and including 230 kV. In North America, the definition "EHV" is usually applied to 345 kV and 500 kV circuits. Approximately 90% of all faults that occur on HV and EHV lines involve ground. At the 500 kV level, single-phase-to-ground faults predominate to the extent that on many well designed circuits, no other type of fault has ever occurred, even after years of service. In this paper, methods for ground fault protection of transmission lines of various voltages and configurations are described, including illustrative examples of actual installations on the Georgia Power Company system.

LONG HIGH VOLTAGE LINE CONNECTING STRONG SOURCES

Directional Ground Overcurrent Protection

Two basic principles that are applied to the protection of transmission lines on the Georgia Power Company system are first, to install only those relays that are required to properly protect the line, and second, to provide redundancy in the form of two completely independent relay schemes at each line terminal. The first principle implies that faults should be cleared with all necessary speed, as opposed to the highest possible speed the state of the art will allow. It also follows the maxim, "If a relay is not installed, it cannot misoperate and cause an unnecessary outage".

The simplest high voltage transmission line to protect is the long line connecting two strong sources. Figure 1 shows a 230 kV line, 76 miles long, connecting the Bonaire 500/230/115 kV Substation with the North Tifton 500/230/115 kV Substation. Note that the source behind the Bonaire bus is 21,000 amperes, while the far-end ground fault at North Tifton, with all breakers closed, is only 700 amperes. Note also that the ground fault current fed from North Tifton to Bonaire is reduced from 19,600 amperes to less than

82 SM 403-4   A paper recommended and approved by the IEEE Power System Relaying Committee of the IEEE Power Engineering Society for presentation at the IEEE PES 1982 Summer Meeting, San Francisco, California, July 18-23, 1982. Manuscript submitted January 14, 1982; made available for printing April 28, 1982.

850 amperes. Actually, on most long two-terminal HV lines, the current fed through the line for single-phase-to-ground faults at the remote ends will be about the same at both terminals over a wide range of source impedances. This is because of the high zero-sequence impedance of a long line, and it may not be true of some mutually-coupled circuits.

The Bonaire-North Tifton 230 kV line is constructed of 795 MCM ACSR conductor. Two ground relays are installed at each terminal - a dual-polarized directional overcurrent ground relay connected to the primary relaying current transformers, and a non-directional overcurrent ground relay (device 50/51) connected to the secondary relaying current transformers. C. T. ratio is 800/5 (160/1) for all relays.

All overcurrent ground relays installed to protect transmission lines on the Georgia Power Company system are very inverse. Ground relay tap is generally selected to give a relay pickup value of from one-fifth to one-tenth of the far-end fault with all breakers closed. Consider the Bonaire terminal. If we select a tap of 0.8 for the primary ground relay, the pickup value will be 0.8 x 160 = 128 amperes, so that the far end fault of 700 amperes is approximately 5.5 times the tap setting. A lower tap may be selected, with the understanding that the additional burden may result in C. T. saturation and degrade relay action for high current, close-in faults. In general, unless a high C. T. ratio can be employed (above 1000/5), it is wise to avoid using the minimum taps on most electromechanical ground relays. This is because the small additional sensitivity may be gained at the expense of poor relay performance for faults approaching 100 secondary amperes.

Time-dial settings for primary ground relays are selected to allow about 0.4 second coordination time interval (CTI) for line-end faults. For ease of coordination, all Georgia Power overcurrent ground relays are calibrated at ten times pickup. For the Bonaire terminal, a time dial calibration of 18 cycles at 1000% (8 relay amperes) will provide a CTI of 27 cycles for the 700 ampere fault at North Tifton, which is 550% of pickup.

The instantaneous overcurrent (IOC) unit on the Bonaire directional ground relay is set for 120% of the far end fault. The calculated value is (1.2 x 700)/160 = 5.3 amperes. To provide margin, and to allow for convenience of application in the field, this should be rounded up to the nearest integer, which is 6 relay amperes and 6 x 160 = 960 primary amperes. This will provide instantaneous protection for about 95% of the distance from Bonaire to North Tifton, without danger of overtripping.

Faults in the last 5% of the line near North Tifton will be cleared by a "sequential-instantaneous" operation. The sequence of events will be as follows:

(1) Ground fault occurs.
(2) North Tifton PCB tripped by ground IOC unit(s) at North Tifton. (Elapsed time = 5 cycles.)

Reprinted from *IEEE Trans. Power App. Syst.*, vol. PAS-102, no. 2, pp. 420-432, Feb. 1983.

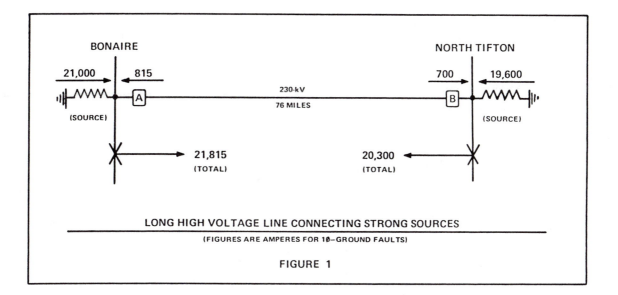

LONG HIGH VOLTAGE LINE CONNECTING STRONG SOURCES

(FIGURES ARE AMPERES FOR 1Ø-GROUND FAULTS)

FIGURE 1

(3) Current fed to line-end fault through Bonaire PCB increases from 700 amperes to 1240 amperes.

(4) Directional ground relay IOC unit at Bonaire (set for 960 amperes) detects fault and trips Bonaire PCB. (Total elapsed time = 13 cycles.)

Settings for the directional ground relay at North Tifton will be slightly higher than those at Bonaire, due to the higher value of fault current for a line-end fault. However, the difference will not be significant; and this scenario indicates that the directional ground relay time overcurrent (TOC) units on this line will seldom be called upon to operate, but will mainly serve as backup protection.

Non-Directional Ground Overcurrent Protection

The most reliable, secure, and effective ground relay for long HV transmission lines is the non-directional overcurrent relay. When properly set, the IOC unit is inherently self-directional, is very fast (considerably faster than directional overcurrent relays), and has the advantages of not being dependent on external polarizing sources. Except for two-terminal buses, long HV transmission lines should be protected by a non-directional overcurrent ground relay at both ends. For most faults, this will be the first, if not the only relay to operate, no matter what other more sophisticated or exotic types of protection are installed.

Since a non-directional relay will operate for fault current flowing in either direction, it must coordinate for faults behind the relay, as well as faults off the downstream bus beyond the protected line. For this reason, the protection engineer must examine the line-end fault current at both terminals in order to properly set the relay. On the Bonaire-North Tifton line, these values are 700 amperes at North Tifton, and 815 amperes at Bonaire, with the relay settings always determined by the higher value. Note that for proper coordination on external faults, the non-directional ground relays must be set exactly the same at both ends.

The TOC unit will serve principally as backup protection. In order not to overtrip for faults behind the relay, it should be set to allow a CTI of about 0.6 second. At Bonaire and North Tifton, this will require a tap setting of 0.8 ampere (= 128 amperes

primary), with a time-dial calibration of about 28 cycles at 1000% of pickup.

The important element of a non-directional ground overcurrent relay is the IOC unit. In order to prevent this unit from misoperating on the asymmetrical value of fault current, it must be set for at least 170% of the maximum line-end fault in either direction. With this setting, it is converted into a high speed (one cycle) directional ground relay. The calculated setting for both terminals is (1.7 x 815)/160 = 8.66 amperes (use 9 amperes). Ground faults in the middle 75% of the line will be cleared by the non-directional IOC units at Bonaire and North Tifton. Faults near either terminal will be cleared by the non-directional IOC unit at that terminal, and the directional IOC unit at the remote terminal.

In order to provide redundancy, the non-directional ground relay at each terminal should be connected to a separate set of current transformers and separate PCB trip coil(s) from the directional ground relay; and at EHV substations, the ground relays on all lines should be connected to separate control batteries.

Pilot relays are not required for proper relaying of long HV transmission lines, except in rare instances, and indeed are undesirable in most cases. Such relays are expensive to install, expensive to maintain, and tend to reduce the security of the transmission system. They should only be installed on long lines when stability or other considerations make it essential that both terminals be tripped simultaneously. This will usually be true only when the long line is connected to a very weak bus, so that the source impedance is of the same order of magnitude as the line itself.

HIGH VOLTAGE LINE WITH WEAK MID-POINT STATION

Figure 2 illustrates a somewhat more complex ground relaying problem. This figure shows the 115 kV line from Porterdale to East Social Circle, with the Hercules station load connected near Porterdale. There is no source at Hercules, so that this installation is an example of "breakers in series". That is, any fault that occurs in either line will cause fault current of the same magnitude and polarity to flow in two line breakers simultaneously. For example, relays in PCB "D" and PCB "F", facing Porterdale, will both receive the same 2100 amperes for a fault near PCB "C". This may make it difficult for directional ground relays to select without excessive "pile up" of the TOC unit curves. Directional relays on PCB "E", facing East Social Circle, will restrain, and will not have to be coordinated for this fault. It is generally wise to permit only one mid-point station between any two source buses. For example, the temptation to install breakers on either side of station "G" should be firmly resisted, unless the local transformer provides a source of zero-sequence current, in which case they will be required. Otherwise, due to curve "pile up", directional over-current relays will provide poor protection for most faults, and expensive alternatives will be necessary.

Primary Ground Relaying

The Porterdale-Hercules 115 kV line is fairly short, and can only be properly protected by some type of pilot relaying system. Several types may be applied, according to user preference. This line happens to be protected by a directional-comparison blocking scheme, but other types, such as over-reaching transferred trip, or ac pilot wire could be used. On directional-comparison systems, proper coordination for external ground faults is achieved by setting the ground carrier start units the same at both terminals, and the ground carrier trip units the same at both terminals, generally 200% of the value of the starting relays. A usual setting might be 0.5 ampere for G1 (start) and 1.0 ampere for G2 (trip).

After careful study, it was decided that pilot relaying was also required for the line from Hercules to East Social Circle, although it was correctly relayed for a number of years without it. Actually, this line is an ideal candidate for two-zone ground distance relaying. Zone 1 on PCB "E" can be set for 75% of the line, and will push delayed clearing of ground faults well away from the mid-point station. Zone 2 can be set for 175% of the line, with a 25 cycle delay. It will not have to be coordinated with anything, due to the heavy infeed at East Social Circle. Facing the other direction, the same settings may be used, except for the time delay on Zone 2. This will have to be increased to 30 or 35 cycles, since failure of the carrier pilot relays to trip for a fault near "C" requires that the ground relays on PCB "F" be coordinated with the backup relays on PCB "D". The protection engineer must keep this possibility in mind at all times.

Secondary Ground Relaying

If directional overcurrent relays are to be applied to this type of line configuration, setting calculations should always commence with the mid-point station. The settings for PCB "E", followed by those for PCB "C" may be determined first, for faults moving from left to right. Next, an entirely independent set of calculations for PCB "D", followed by PCB "F" may be taken in hand. Non-directional overcurrent relays cannot be applied on these lines, except for PCB "F", which has weak back-feed. It also has a fairly strong source, and a fairly good fault current slope toward Hercules (14,000 down to 2800 amperes). There is no way a non-directional overcurrent relay can be made to select on either breaker at Hercules and such a device will have only minimal value at Porterdale. For instance, the IOC unit at the latter location would have to be set for (1.7 x 2600) = 4420 primary amperes. Since the source behind the relay is only 3900 amperes, this setting effectively removes the IOC element from service.

The Porterdale - Hercules - East Social Circle line is constructed of a mixture of conductor sizes, principally 336 MCM ACSR. For ease of calculation, consider the current transformer ratio on all circuit breakers to be 600/5. The coordination of directional ground relays on PCB's "C" and "E" will be fairly routine, due to the great difference in line-end faults. (PCB "C" is looking at 2600 amperes, while between "E" and "F", the current slopes down to 780 amperes.) An "E" relay can be set as follows:

Tap 0.8 (= 96 primary amperes), 22 cycles at 1000%.
IOC = (1.2 x 780)/120 = 7.8 Use 8 amperes (= 960 primary amperes).

For directional ground relay on PCB "C":

Tap 2.0 (= 240 primary amperes), 30 cycles at 1000%.
IOC = (1.2 x 2600)/120 = 26 amperes (= 3120 primary amperes).

These settings are plotted on Figure 3, and indicate a CTI of 0.36 second for a fault just off the Hercules bus beyond PCB "E". This is probably adequate. However, since the source behind PCB "C" is only 3900 amperes, it is evident that fast clearing of ground faults on the Porterdale - Hercules line

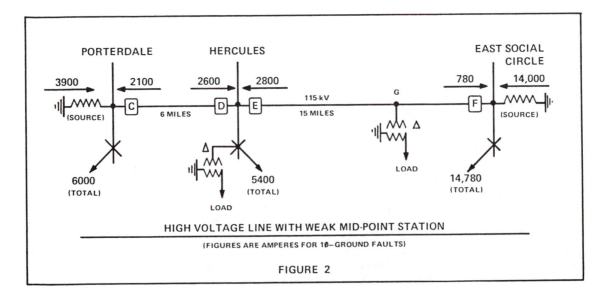

HIGH VOLTAGE LINE WITH WEAK MID-POINT STATION

(FIGURES ARE AMPERES FOR 1Ø-GROUND FAULTS)

FIGURE 2

is almost completely dependent on proper action by the carrier relays. This becomes even more apparent when we examine the settings that may be applied to relays on PCB's "D" and "F":

First a conscious decision must be made to reduce the CTI for line-end faults near "C" from a desired value of 0.4 second to 0.3 second. This is done in order to exert some control over the clearing time for the relay at East Social Circle, and will permit the following setting for the "D" relay:

Tap 1.5 (= 180 primary amperes), 22 cycles at 1000%.
IOC = (1.2 x 2100)/120 = 21 amperes (= 2520 primary amperes).

For PCB "F" at East Social Circle:

Tap 2.0 (= 240 primary amperes), 39 cycles at 1000%.
IOC = (1.2 x 2800)/120 = 28 amperes (= 3360 primary amperes).

These settings are plotted on Figure 4. Again note the fact that the IOC unit on the "D" relay has very little current to work with, so proper carrier relaying action is all important. Also note that even though the coordination time interval between the two relays is squeezed down to (0.62 - 0.32) = 0.3 second, we are still faced with a 0.6 second relay time for PCB "F" for the 2800 amperes fault near "E".

Consideration of these problems has led Georgia Power Company to install a directional-comparison carrier relaying system on the Hercules - East Social Circle line. This has two very beneficial effects. First, it reduces maximum ground relay time from 36 cycles to less than 3 cycles, and second, it permits a relaxation of the very tight directional overcurrent relay settings, since these relays now are relegated to a back-up role. Unfortunately, these improvements are somewhat offset by substantial additional installation and maintenance costs, and by a reduction in security against over-tripping.

MUTUALLY-COUPLED LINES

Primary Ground Relaying

Figure 5 is an example of an elementary mutual coupling problem. Three 115 kV lines are constructed close together on the same right-of-way between Plant Arkwright and South Macon 230/115/12 kV transmission substation. Due to the zero-sequence mutual impedance between the circuits, unbalanced current (normally 3Io) flowing in any one circuit induces a current of opposite polarity in the other two circuits. This causes the ground fault currents in each line to vary over a wide range, according to fault location and circuit breaker status.

Most protection engineers will prefer some type of pilot relaying as the primary protection for this type of configuration. Four types may be recommended:

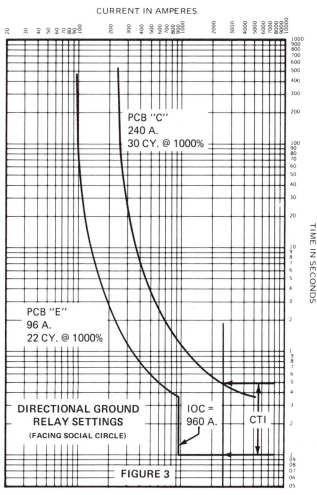

FIGURE 3

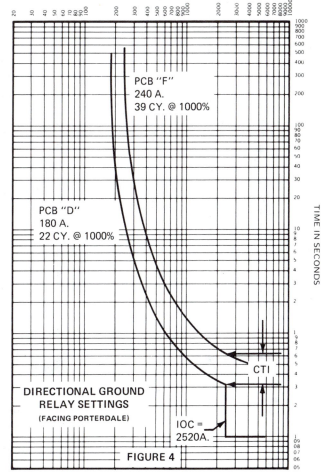

FIGURE 4

133

(1) AC Pilot Wire: Pilot-wire (PW) relays are appropriate for mutually coupled lines because they are not dependent on external polarizing sources. They should be applied only over short distances (generally less than 10 miles) and only when a well shielded, low impedance, company-owned pilot wire circuit is available. Most leased circuits are not sufficiently reliable for secure pilot-wire relaying.

(2) Phase-Comparison Relaying: This type of relaying system is made-to-order for mutually coupled circuits, because, like PW relays, it does not require external polarization for ground faults. Either power-line carrier or microwave channels may be used. The main restriction on the phase-comparison scheme is that for any line there must be an adequate margin between the maximum expected load current and the minimum phase-to-phase fault current. If there is, and the relays are properly set in accordance with the manufacturers' instructions, ground-fault sensitivity will usually prove adequate.

(3) Over-Reaching Transferred Trip or Carrier Unblocking: Ground distance relays should usually be used as the ground fault detector for these schemes, although the relay reach may be adversely affected by the line mutuals. This can be compensated for by auxiliary current transformers, if the circuits terminate in the same station, but this is not preferred. In most cases, setting the ground distance fault detectors for 175% of the protected line will permit full coverage, and will not cause coordination problems, particularly on lines connecting strong buses. Microwave, powerline carrier, and audio tones on hard-wire communications circuits may serve as a relaying channel, according to user preference or local constraints.

(4) Directional-Comparison Relaying: Directional-comparison systems may be successfully applied to mutually-coupled lines if adequate ground fault current is available with all breakers closed. As noted previously, if the ground carrier start element is set for 0.5 ampere, the trip should be

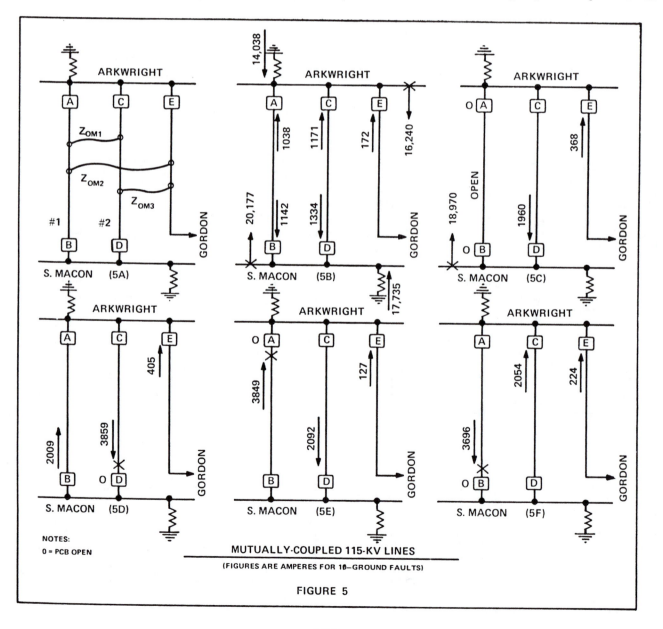

NOTES:
O = PCB OPEN

MUTUALLY-COUPLED 115-KV LINES

(FIGURES ARE AMPERES FOR 1Ø-GROUND FAULTS)

FIGURE 5

134

set for 200%, or 1.0 ampere. With these settings, the minimum secondary ground fault should be 2.0 amperes for fast relay action. Also, the user must verify that there is adequate energy available to drive the ground directional unit under all conditions.

Completely secure directional-comparison relaying for ground faults on mutually-coupled transmission lines can best be achieved with negative-sequence voltage polarizing. Polarizing the ground relays on these circuits with zero-sequence current is extremely risky, due to the possibility of polarizing current reversal, and with the negative-sequence-voltage option available, there is no need to take this risk. Georgia Power Company operates 126 negative-sequence-polarized carrier ground relay terminals, with a record of nearly 100% correct operations over many years of service; and no case of reverse throw has ever been recorded. If zero-sequence current must be used, extensive short circuit studies will be required in order to verify the magnitude and direction of the polarizing quantities under all conceivable fault locations and circuit configurations. Even then, due to Murphy's Law, complete satisfaction cannot be guaranteed.

Secondary Ground Relaying

In order to apply directional or non-directional overcurrent relays to mutually-coupled lines, short circuit studies must be conducted for the following fault conditions:

(1) Ground faults at each line terminal with all breakers closed.
(2) Line-end faults at each end of each line with local PCB open.
(3) Bus faults with each line out of service, one by one, showing fault current delivered over the remaining lines.

Consider the case of two parallel lines terminated at the same stations at both ends. In order for ground overcurrent relays to properly select for any fault location, the TOC units must be set the same at all four terminals. This means that the relay requiring the highest setting will control the setting on the other three relays. Figure 5 shows the fault current flows on the Arkwright - South Macon 115 kV lines for five possible fault conditions. The highest through-fault current is the 2092 amperes flowing in Line #2 from Arkwright to South Macon ("C" to "D", Figure 5E) for a line-end fault off PCB "A". This value is substantially higher than the amount for a fault at South Macon with all breakers closed (1334 amperes, Figure 5B), or the amount for a fault at South Macon with Line #1 out of service (1960 amperes, Figure 5C). If the C. T. ratio for breakers "A", "B", "C", and "D" is 1000/5 (200/1), the TOC units on all four directional ground relays may be set on tap 1.0 (= 200 primary amperes), 24 cycles at 1000% of pickup. The IOC unit for PCB "C" should be set for (1.2 x 2092)/200 = 12.6 amperes. Rounding up to 13 amperes gives a setting of 13 x 200 = 2600 primary amperes. This setting will permit a sequential-instantaneous operation, as shown by Figure 5D, which indicates that the line-end fault near "D" increases to 3859 amperes after "D" opens.

The maximum through fault in breakers "A", "B", and "D" is slightly less than for "C", but not

enough so that the setting required for "C" results in unacceptable settings at the other locations. This is not always true, however, particularly in cases where three or four lines of different voltages are involved. For example, if two identical 115 kV lines are closely coupled to two parallel 230 kV lines, the fault currents in the two 115 kV lines may differ considerably from each other, according to the degree of coupling with the two higher voltage circuits.

An interesting effect of zero-sequence mutual coupling is shown by the fault current in the line from Gordon to Arkwright. A fault near Arkwright on the line itself produces 172 amperes (Figure 5B). Only when the fault is moved away, toward PCB "D", does the maximum through-fault of 405 amperes appear (Figure 5D). This result occurs because of the transformer action between adjacent circuits. The 3859 amperes flowing from "C" to "D" induces a current in phase with the current entering PCB "E" so that the total current in the Gordon - Arkwright line is increased. For the Arkwright fault in Figure 5B, however, the Line #2 current is reversed, and this produces a "choking" effect on ground fault current entering PCB "E". This fact must be taken into account when setting a ground overcurrent relay at Gordon, or overtripping may occur. It could also cause overreach of a zone 1 ground distance relay, unless a conservative setting is applied.

Directional overcurrent relays should be polarized by negative-sequence voltage for the same reasons as the ground relays in directional comparison systems. This will prevent incorrect relay action during unexpected reversal of zero-sequence polarizing quantities. Large systems with many strong grounding sources are especially vulnerable to this phenomenon, particularly during line-end fault intervals caused by unequal PCB operating times.

EXTRA HIGH VOLTAGE LINES

In designing ground relay protection for EHV transmission lines, the engineer must not only strive for reliable, high-speed protection, but carefully consider the need for security against false operation. This latter requirement is much more important for EHV lines than for lower voltage circuits, due to their very large power transfer capability. (Some 500 kV lines on the Georgia Power Company system regularly carry over 1000 megawatts under normal load conditions.) An incorrect relay operation that causes the simultaneous loss of two or more EHV circuits due to a single line fault could result in a widespread system disruption. In general, for EHV lines, given the necessity of making a choice between slightly higher ground relay speed, and increased security, it is nearly always wise to opt for the more secure system. In most instances, a two-cycle ground relay will provide just as satisfactory protection as a one-cycle relay will, and the slower relay will be much less likely to trip incorrectly. Note, however, that this rule may not apply to phase faults, where very high speed may be required due to stability consideration. Fortunately, phase faults are rare on well designed EHV circuits; and since such faults almost never occur, the security problem is not acute.

The four watch-words for relaying on EHV systems are RELIABILITY, SECURITY, SPEED and REDUNDANCY. These criteria also apply to lower voltage lines, of course, but are much more important on EHV circuits due to the high power transfer levels and very large short circuits involved. These four criteria are closely related, and all depend on proper design of the instrument transformer circuits. The very best relaying systems available today will not perform properly unless supplied with accurate replicas of the primary current and voltage quantities.

EHV relays must be connected to high accuracy, high ratio current transformers, with the primary and secondary systems coupled to separate C.T.'s. On 3000 ampere 500 kV circuit breakers, the C800 multi-ratio C.T.'s will frequently be connected on the 2000/5 ampere tap for line protection. This reduces the relaying accuracy to approximately 2/3 of the value for the full winding. Improved relaying action may be obtained by ordering the C.T.'s with C800 accuracy on the ratio that will actually be used in the field.

Figure 6 is a diagram of the 500 kV transmission lines connected to Plant Bowen on the Georgia Power Company system. For proper ground relaying, the immense difference in ground fault magnitude between close-in faults and far-end faults must be carefully considered. On the Sequoyah line for example, the Bowen ground relays will receive about $40,000/400 = 100$ amperes for faults near Bowen, but only $1200/400 = 3.0$ amperes for faults at Sequoyah. This is somewhat mitigated by the fact that the level will increase to $3000/400 = 7.5$ amperes after the Sequoyah PCB opens.

Capacitor voltage transformers may not produce a correct relaying signal for several milliseconds after fault initiation. This is not a problem for electromechanical relays, but must be taken into account in the design of solid-state systems. This generally requires that the relay "go-no go" decision be delayed about one-half cycle until the secondary transients have subsided.

All relaying on EHV systems must be designed so that primary and secondary protection is provided for all circuit elements, with the primary relays connected to separate PCB trip coils and a separate control battery from the secondary protection. Otherwise, adequate redundancy cannot be obtained.

Pilot Relay Protection

Reliable pilot protection for ground faults on EHV lines is achieved by designing highly redundant schemes that are fast, easy to apply, and do not require a "juggling act" when making the relay calculations. Pilot schemes should be self-monitoring. Probably the best overall system for EHV lines is the over-reaching transferred trip scheme, using a microwave channel, or an unblocking scheme using power line carrier. The guard frequency should be continuously monitored by a high resolution digital event recorder with supervisory alarm. If solid-state phase and ground distance relays are used for fault detectors, very high speed relaying will be achieved. Using this system, Georgia Power Company clears most 500 kV line faults in two cycles, including breaker time. Distance-supervised phase comparison schemes will also provide reliable protection to EHV circuits. However, such schemes are not as secure against false trips, and the setting calculations are considerably more difficult. The reduction in security comes about due to the very precise format and timing required of the blocking square wave which is transmitted on alternate half-cycles - any hash or low area under the curve may be falsely interpreted as a trip command. Also, the setting calculations may become quite onerous on systems that are arranged so as to convert three phase faults into what appear to be phase-to-phase faults as far as the pilot relays are concerned.

Transferred-trip schemes are made secure by the design of the relay logic, which requires that the following conditions be satisfied for a trip output:

(1) Guard signal must <u>not</u> be received from remote terminal (unblock).
(2) Trip signal must be received from remote terminal.
(3) Local relay must send trip signal.

Both phase and ground distance fault detectors should be supervised by overcurrent units. The phase unit should be set above maximum load current, if possible. The ground unit should be set less than 50% of the far-end fault, but not below 0.6 ampere secondary. This rule also applies to carrier unblock schemes. Carrier unblock is installed as primary protection on the Bowen-Sequoyah line. On this circuit, sufficient margin is available so the ground overcurrent unit at Bowen is set for 400 primary amperes, which is 33% of the far-end fault at Sequoyah.

The Sequoyah line is also protected by a directional comparison carrier-pilot scheme, which operates on "hot standby". The GD unit is a ground distance relay set for 275% of the line impedance to accommodate fault resistance. The G1 unit, which starts carrier, is set for 0.5 ampere, and the G2 unit, which supervises GD, is set for 1.0 ampere (= 400 primary amperes). For proper coordination on external faults, the same settings are used at both terminals. The ratio between the far-end fault and the G2 setting is $1200/400 = 3$, which exceeds the required minimum of 2/1.

Secondary Protection

Directional overcurrent relays, non-directional overcurrent relays, and two-zone ground distance relays may all be profitably used for secondary ground fault protection on EHV lines, and all three methods are widely used on the Georgia Power Company EHV system. The following guide-lines are suggested:

Ground Distance: Both solid state and electromechanical types work well. The solid state versions are faster, the electromechanical types are considerably more secure. If the Zone 1 reach is limited to 70% of the line, accidental overreach will be rare, even with strong mutuals. The Zone 2 setting can range from 175% of the protected line for long lines, to 300% for short ones, in order to accommodate fault resistance. Caution No. 1: Ground distance relays may not respond to some very high resistance faults (green trees in line, for example). Therefore, whenever they are installed on a line breaker, sensitive ground overcurrent relays are also required. Caution No. 2: All solid state relays on EHV circuits should have a <u>contact</u> <u>output</u>. Relays with a thyristor output are subject to false operations in EHV substations, even when connected to well shielded cables and protected by a properly designed capacitor "surge fence".

Ground Overcurrent: All EHV circuits should be protected by a "stand-alone" directional ground overcurrent relay at each terminal. For utmost security, these relays should be polarized with negative sequence voltage. They will serve as reliable backup for the pilot relays and will protect for high resistance faults that may not be detected by ground distance relays. A coordination time interval of 0.3 second will prove satisfactory in most instances.

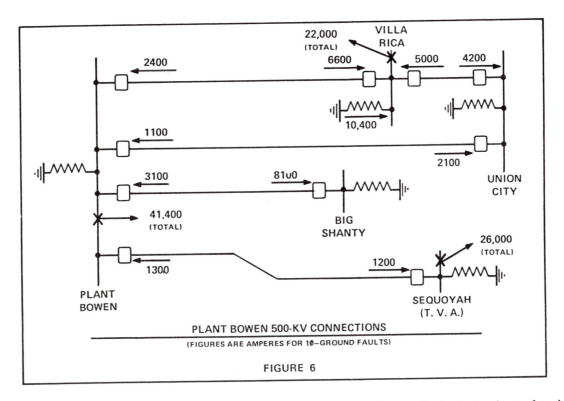

PLANT BOWEN 500-KV CONNECTIONS
(FIGURES ARE AMPERES FOR 1Ø—GROUND FAULTS)

FIGURE 6

As on lower voltage circuits, the effectiveness of EHV directional overcurrent relays is generally directly proportional to the length of the protected line. However, this effect is more pronounced at the 500 kV level due to the very low per unit line impedance for these circuits. Refer to Figure 6. On the fairly long line from Bowen to Union City, the fault current slopes from 40,000 amperes to 2100 amperes. An IOC unit setting of 2800 amperes at Bowen will provide instantaneous tripping for about 90% of the line length, and sequential instantaneous protection for the remainder. The TOC unit can be set on tap 0.6 (= 240 primary amperes), and will give sensitive backup for faults not cleared by the primary relays. An IOC will not be as effective on the Villa Rica - Union City line, since the current slope is not so steep (17,000 down to 4200 amperes for a far-end fault). However, an IOC setting at Villa Rica of 5200 amperes will still reach over 60% of the line, and will overlap the IOC unit at Union City. As before, a sensitive TOC setting is required for stand-alone service.

Non-directional ground overcurrent relays can be very useful on long EHV lines, such as the Bowen - Sequoyah and Bowen - Union City circuits. They should not be installed at locations such as the Villa Rica terminal of the Bowen - Villa Rica line, where the back-feed fault (6600 amperes in this case) is substantially higher than the line-end fault in the forward direction. Also, the usefulness of non-directional relays on EHV systems will be somewhat limited by the prevalence of three-circuit buses (two lines with one transformer, or three lines). In this configuration, the TOC units on the non-directional ground relays must be set exactly the same in order to properly select for all faults. This is usually impractical, and will normally require that this type of relay be omitted as a hazard to system security.

CONCLUSION

This paper has given a brief outline of some of the more important principles of ground fault protection for high voltage and extra high voltage transmission lines. Emphasis has been placed on four basic criteria: reliability, security, speed, and redundancy. The methods described for achieving these criteria have been developed on a very large electric system over many years of service. Consistent application of these guidelines will result in proper and timely clearing of transmission line ground faults, and will enhance the security of the bulk power network.

Discussion

C. E. Ojanen (Detroit Edison Company, Detroit, MI): The author is to be commended for preparing a report on the important principles when designing ground fault protection. Several questions and comments arise from some of the example calculations. Setting the instantaneous overcurrent (IOC) unit 120% of the far end fault does not appear to provide a sufficient safety margin against overtripping. What tolerance for errors in circuit parameters, settings, etc., is recommended to prevent tripping for faults beyond the far end? We are concerned that reclosing of breakers on other lines at North Tifton could result in false tripping at Bonaire due to appearance of end-of-line fault current to the Bonaire relay.

Setting of IOC to provide instantaneous or "sequential instantaneous" tripping appears to jeopardize security for fast clearing of low fault current. Want benefits does the author expect by not waiting for the TOC unit to operate in 25-30 cycles for these low currents? The 800/5 current transformer ratio on the 230 kV lines must also be used for the load carrying phase relays. Is the loading on the 230 kV lines restricted by the phase relay thermal limits?

On the EHV system, is the two cycle clearing time you obtain on the 500 kV system a design criteria? If so, what breaker failure time do you require? Are these operating times required on all the 500 kV system?

The Detroit Edison EHV system (345 kV) consists of mostly short lines (less than 30 miles). Protection for these lines are provided by duplicate pilot systems. The pilot systems used are directional comparison carrier, directional comparison permissive overreaching unblock on micro-wave channels and pilot wire relaying. These schemes are generally designed as you describe. Directional IOC and TOC ground relays are used as added backup for the duplicate pilot systems.

Redundancy is maintained by duplicate batteries, breaker trip coils, current and potential sources to the protective relays.

Manuscript received August 2, 1982.

137

T. Niessink (Commonwealth Associates Inc., Jackson, MI): The author is to be complimented for preparing a very detailed paper on the Georgia Power Company ground relaying practices. Papers of this nature incite the interest of practicing engineers and should be encouraged.

Protective relaying is an art, not an exact science. Consequently, protective relaying practices vary according to experience, unique system requirements, and practices of other divisions such as system planning and operation within a company. Some of the practices discussed in the paper which fall into this judgemental and experience category are:

1. Coordinating Time Interval (CTI) between relays regardless of relaying operating time and "multiple of times pickup."
2. Selection of ground relay tap setting which determines sensitivity to high resistance faults, backup protection and to what extent coordination is obtained by "multiples of times pickup" versus lever setting characteristics of an inverse-type overcurrent relay. The author's philosophy of generally selecting "a relay pickup value of from one-fifth to one-tenth of the far-end fault with all breakers closed" would appear to result in reliance on lever setting rather than the inverse characteristic of the relay for selectivity.
3. Extent to which negative sequence polarized ground relays are used.
4. Extent to which impedance ground relays are used.
5. Need for contact output on all static relays on EHV circuits.

Because of these and other variations in practice, it is recommended that the IEEE Power System Relaying Committee update their 1966 survey on ground relaying practices in the U.S.A., taking into account the judgemental factors presented in Mr. Griffin's paper.

In the section entitled "Secondary Ground Relaying," a statement is made that "the TOC units must be set the same at all four terminals" in the case of two parallel lines terminating at the same stations at both ends. This should be expanded upon. Consider as an example a case as in Figure 5A of the paper where there is a ground source at Arkwright but none or a very weak ground source at S. Macon. Probably nondirectional ground relays would be provided on breakers "A" and "C" and directional ground relays at "B" and "D." Would the TOC units be set the same at all four terminals in this case? If not, what are the conditions for setting the relays the same at all four terminals?

REFERENCE

[1] "Ground Relaying Practices and Problems, A Power System Relaying Committee Survey," IEEE Committee Report, *IEEE Transactions on Power Apparatus and Systems*, Vol. Pas-85, No. 5, May 1966.

Manuscript received August 9, 1982.

George D. Rockefeller (Rockefeller Associates, Inc., NY): There are some apparent discrepancies regarding setting criteria. The directional IOC is set for 120% of the maximum far-end current while the nondirectional IOC is set for 170% of the maximum line-end current in either direction. I assume that the 120% criterion is based on use of unit not subject to transient overreach resulting from dc offset (typically a cylinder).

Also three CTI criteria for TOC units are stated:

Line	Directional?	Time-s	Fault
HV	YES	0.4	Line End
HV	NO	0.6	Not Stated
EHV	YES	0.3	Not Stated

Would the author explain his reasons? Also, why is the line-end fault used, since operation is desired for this fault?

The author's company uses ground distance relays on EHV, a common practice. I agree with his statement that a ground overcurrent relay should complement the distance relay to clear high-resistance faults. The failure of the line relaying to detect a high-resistance fault could be disastrous if back-up overcurrent relays elsewhere can detect the fault. For example, it is common to apply a neutral TOC relay on generation step up transformers. These must not be allowed to operate for a transmission line fault barring a total failure of the line relaying. The author's recommendation of using negative-sequence polarization is a good one to minimize the possibility of incorrect polarization from mutual induction.

Ground distance relays have a number of advantages compared to overcurrent. These include: more stable reach and less tendency for the settings to become obsolete as the power system is re-configured. One disadvantage of some distance-relay designs is their susceptibility to operation on swings. It is common practice to rely on a ground overcurrent fault detector to block swing tripping. This is dangerous, because a sound-phase relay can operate during a swing and an external ground fault. For example, a swing could result from an external ground fault; an unsuccessful reclosure could catch one of the sound-phase relays in the operate condition.

Manuscript received June 25, 1982.

C. J. Pencinger (Brown Boveri Corporation, North Brunswick, NJ): The author is to be complimented for presenting a useful and practical paper on the application of ground fault protection for HV and EHV lines. For the protection of mutually-coupled HV lines, we agree that a pilot system is preferred. Ground distance relays using cross polarized mho characteristics can provide effective protection for these lines without the need for compensation (1). Our studies show that the error in distance measurement due to the mutual impedance is not more than $\pm 15\%$ in most cases with a $\pm 30\%$ error remotely possible in exceptional cases.

In the case of short lines, particularly with strong sources on both ends, ground distance relays equipped with a polygon characteristic can offer advantages of detecting ground faults with high fault resistances (2). High immunity to false operation on heavy load currents can still be maintained with this characteristic.

For solid state EHV line protection system, we agree that the output should be a contact and not a solid state device. The advantages of contact output are galvanic separation, higher insulation test voltage, greater immunity to dc control power transients. We have not found a suitable commercially available solid state device which meets the strict requirements of the application.

When pilot relay protection for EHV lines is used, it is our opinion that not only the pilot channel should be supervised but the relay system itself as well. Modern solid state EHV relay systems can incorporate self monitoring and diagnostic provisions which enhance the reliability and security of the protection scheme (3).

We disagree with the use of supervision by high set overcurrent relays of phase and ground distance measuring units. We prefer to use low set over-current fault detectors set at 0.15 x rated current to prevent false operation due to line charging current and CVT de-energizing transients. However, non directional impedance type fault detectors are applied in addition to the distance measuring units (4). The use of the impedance fault detector (called starting elements) offers the advantages of better security because three criteria must be met for tripping: starting, measuring and low set overcurrent. In addition, the starting elements provide phase indication and selection where necessary, line pick up protection, non directional back-up protection (with time delay) and preparation of carrier or microwave equipment to receive or send a trip signal.

Finally, as would like to ask the author why he does not consider the use of single pole tripping and reclosing for EHV line protection. With single pole relaying, temporary phase to ground faults are cleared without a major power system disruption (5) (6) (7). Particularly as the author points out, three-circuit busses prevail at EHV. Since, at EHV single phase faults predominate, a station with a three-circuit bus can benefit greatly when single pole relaying is applied. Consider the case where one line is out of service. The remaining line can provide continuous service to the load even in the event of a ground fault. With two lines in service and three pole relaying, a fault on one line and a relay malfunction on the ground healthy line would leave the station connected only with the third circuit (line or transformer). With single pole relaying only one phase is temporarily interrupted.

REFERENCES

1. H. Ungrad, V. Narayan, "Behavior of Distance Relays Under Earth Fault Conditions on Double-Circuit Lines", Brown Boveri Review, Vol. 56, October, 1969, PP. 494-501.
2. F. Ilar, "Detection by Line Protection of the Earth Fault Resistance", Brown Boveri Review, Vol. 65, June, 1978, PP. 371-378.
3. W. Kolbe, F. Ilar and G. Bacchini, "A New Distance Relay for HV and EHV Systems Type LZ 96", Brown Boveri Review, Vol. 67, October, 1980, PP. 599-607.

4. V. Narayan, "Distance Protection of HV and EHV Transmission Lines", Brown Boveri Review, Vol. 58, July, 1971, PP. 276-286.
5. J. Esztergalyos, "The Application of High Speed Grounding Switches on EHV/UHV Power Systems to Enhance Single Pole Reclosing - Control and Protection", Western Protective Relay Conference, Spokane, Washington, October 27-29, 1981.
6. E. Kimbark, "Selective-Pole Switching of Long Double-Circuit EHV Lines", IEEE Trans., PAS-95, No. 1, Jan./Feb., 1976.
7. M. Mueller, F. Gygax, C. Hahn, P. Baltensperger, "Protection of EHV Systems, Taking Into Account Single Phase Automatic Reclosure on Very Long Lines", Brown Boveri Review, Vol. 45, June, 1958, PP. 243-253.

Manuscript received August 2, 1982.

W. A. Elmore (Westinghouse Electric Corp., Coral Springs, FL): Mr. Griffin has provided us with another of his clear, useful and learned expositions on the subject of protective relaying. With basic tenents so well described, and with the paper sprinkled with well-chosen actual case histories, one can add little by discussion, but there are a few points I would like to make and others I would like clarification on.

I agree with the comments suggesting the avoidance of use of minimum tap on electromechanical ground relays in this application. On higher taps, the effective sensitivity of a ground relay (in primary current terms) may well be *better* than that on a lower tap. On the lower tap exciting requirements are higher for the current transformers that are not associated with the faulted phase because they are back-fed by the higher voltage appearing across the higher burden.

In the Bonaire - North Tifton example why was 27 cycles used as a CTI rather than the 0.4 second the author recommends. Our investigation of the need for CTI indicated that 0.3 seconds plus breaker time is suitable considering ct error, relay deviation from typical curves, virtually negligible relay overshoot plus margin. Even tighter CTI is realizeable with proper evaluation of all these factors.

To avoid transient overreach by the instantaneous overcurrent relays due to the dc component of fault current, a factor of 1.2 is generally used for some Georgia Power applications and 1.7 is used for others. It is assumed that this is because Georgia Power uses cylinder type directional overcurrent units and clapper type non-directional overcurrent units.

Our experience indicates that a cylinder unit will display an overreach of no more than 10% for a circuit having an angle typical of EHV lines. A clapper unit will have a 50% maximum overreach on a comparable circuit. The multipliers being used by the author provide a comfortable margin. Does the author agree that in many cases, the increased coverage by a cylinder unit instantaneous relay will justify its use in a non-directional overcurrent application?

Under "Non Directional Ground Overcurrent Protection", the author indicates a CTI of 0.6 should be used. Why is this different from the 0.4 seconds recommended earlier?

Does the author feel that pilot relaying is justified for systems smaller than the Georgia Power System where no substantial 500 kv overlay exists and the 230 kv system performs the bulk transmission function?

The author mentions that the Hercules - East Social Circle line is an ideal candidate for a two-zone ground distance scheme. Others should be cautioned that the zone 1 setting suggested is dependent on the tapped load bank having substantially higher impedance than either line section. A longer line, or other than mid-tap configuration or a very large bank may force other settings to be used.

Comparing figures 3 and 4, a much wider separation between curves is noted in 3. This is partly due to the smaller CTI in figure 4, but it is also due to the ratio of the settings chosen for C and E and for D and F. Could not a much lower tap setting have been chosen for C while still retaining the 0.36 second CTI.

It is interesting to note that Mr. Griffin correctly defines CTI differently in figures 3 and 4. In Figure 3 it is defined at the maximum current level "seen" by "C" and "E" (2600 amperes). In figure 4, it is defined at the instaneous trip pick up level (2520 amperes). Coordination must exist at both levels, of course. The 0.1 second pickup time used for the IT (Mr. Griffin's IOC) is quite conservative and assures some additional margin.

The experience described for the negative sequence polarized directional ground relays is gratifying. In addition to the absence of a "reversal" problem, negative sequence directional units afford another ad-

vantage. Negative sequence voltage is generally lower at the fault point for phase-to-ground faults than the zero sequence voltage, but the product $V_2 I_2$ tends to vary less at a given relay location than $9V_0 I_0$ as fault location varies and will in general be higher for far-end terminal faults. What is meant by the term "reverse-throw"?

"Dependability" has come to a wider acceptance than "Reliability" as used in describing the ability of a device to operate when it is required to. "Reliability includes both "Dependability" and "Security".

Under "Pilot Relay Protection" the author recommends a ground unit setting of 0.6 ampere or more. With the burden of solid state relaying systems being so very low, irrespective of settings, this criterion is probably not based on similar considerations to that expressed earlier in the paper. Is the lack of transmission line transpositions, and the zero sequence current resulting from positive sequence load flow through the unbalanced phase impedances, behind this recommendation?

There is no question that time overcurrent relays have a place in HV and EHV relaying and can be coordinated but Mr. Griffin will probably agree that it is not an inconsequential task to maintain the settings as system variations occur and as fault magnitudes change over a period of years. Distance relays, with or without pilot support, minimize this problem while at the same time decreasing tripping times to a low and reasonably fixed value.

Manuscript received August 2, 1982.

Frank B. Hunt (New England Power Service, Westborough, MA): The author has written an excellent paper on Ground Relaying of Transmission Lines to go along with his paper on "Steam Generating Stations" and I compliment him.

Some minor points and matters of opinion are:
  1. We try to avoid having more than 100 A secondary, so at Bonaire, we would try to have a higher ratio than 800/5. Of course, this may be a problem with the NORTH TIFTON fault.
  2. The author has not said what he would use for a tree fault current. We try to have a back up ground relay (long time) pickup set no more than 240 A for these faults. In fact, we try to limit ground relay pickup to 240A, for primary protection, at all voltages. This often makes our ground relay curve cross our phase relay curves, so high current ground faults give a misleading target. We seldom set ground relay pickup below 90 A because of CT errors.
  3. We have had reasonably good experience with leased telephone lines using a monitor. They are not as good as our own of course.

Manuscript received August 4, 1982.

D. R. Volzka (Wisconsin Electric Power Company, Appleton, WI): The author is to be congratulated on the development of a very fine paper that provides excellent insight into the reasons relaying is considered both an art and a science. The experience factors contained in this paper make it worthy of study as opposed to casual reading.

I would, however, raise several issues for clarification. The author has mentioned two different figures for the setting of the instantaneous overcurrent element (IOC); 120% for the Directional Overcurrent unit and 170% for the nondirectionsl overcurrent element. The treatment of assymetrical fault current varies across the industry with some using the maximum offset factor of 1.62 and others essentially ignoring its effect. Proper application lies in between and is a function of system X/R ratio, the speed of the relay, and the design of the relay element. Is there a specific reason for the different settings on your system?

The author has also stated that it is wise to avoid using the minimum taps on most electromechanical ground relays. Perhaps a better statement would be to avoid using a lower tap than is absolutely necessary and to assure that the total burden connected to the current transformers is within their capability. Verification that burden is within limits is critical to proper relay performance.

The usage of current transformers with a C800 rating on the 2000/5 tap of a 3000/5 full winding is quite interesting. Were there specific reasons or problems that prompted such CT's to be used? Are any special test procedures or equipment required?

Mr. Griffin also advocates several concepts with which I fully agree and wish to emphasize. The first is usage of continuous type carrier systems as opposed to quiescent systems. Carrier failures are detected much more reliably resulting in improved security. The second area I wish to emphasize is in utilizing sequence of events recorders to analyze relay/communications system performance. We do not utilize this technique at present and are missing extremely valuable information in analyzing relay operations and in trouble shooting. I feel that one cannot really do a complete analysis without information of this type.

Manuscript received August 4, 1982.

J. E. Stephens (Illinois Power Co., Decatur, IL): This paper will serve as a very good reference and guide. It brings out many important points that may be easily overlooked by the less experienced.

Although Mr. Griffin does not specifically state, the discussion and examples are for two terminal lines. Some of the points raised may not apply to 3 terminal lines. For example, it is correctly stated that long (2 terminal) HV lines may, and should, have a non-directional overcurrent ground relay at each end. However, this is usually not practical with multiterminal lines even though each line section is "lomg". Pilot relays are not required for long two terminal HV lines but may well be required for a 3 terminal line.

Mr. Griffin has indicated many of the practices of the Georgia Power Co. Other operators may use different practices. For example, in setting the IOC relay, there is a significant difference in rounding up to the next integer at 6 amperes than at 30 amperes. Some users may prefer to choose the desired margin and set at that value. Also, calibration of the TOC unit at 10 times pickup may be above the IOC unit setting at which the TOC unit is not needed. For instance, in the example of Bonaire terminal of Figure 1, the setting is based on a time requirement of 27 cycles at 700 primary amperes. The operating time of the TOC unit at higher currents is of no concern but calibration is specified at 1280 primary amp. (8A. sec.), or nearly double the current at which timing is of interest. Other users may prefer to calibrate the TOC unit at an integer multiple of pickup and integer amperes (to minimize curve and ammeter reading errors) just under the IOC setting or approximately at the critical coordination current on which the required time setting is based.

The 0.4 second coordination time interval (CTI) suggested by Mr. Griffin is quite conservative and may not be justified with short network lines unless it is essential to coordinate with breaker failure clearing of the remote fault. CTI consists of three nearly independent factors:

   a. Breaker interrupting time - 2, 3, or 5 cycles. (Double if reclosing time is under 1 sec.)
   b. Effective relay coasting time - from .02 to .05 sec. depending on relay type, tap setting, and magnitude of current.
   c. Safety margin - 20% or more of the sum of all known time factors depending on accuracy of the available data.

If a time delayed relay is backing up instantaneous tripping of a remote breaker, the breaker rated interrupting time is known, the instantaneous relay time is known and very small, and the local relay effective coasting time is known. Therefore, only a small safety margin, and CTI, is required to insure coordination. If coordination is required where the remote breaker relaying is time delayed rather than instantaneous, then additional uncertainty exists in the much longer remote breaker relay time, requiring a greater CTI for the same assurance of coordination.

A greater CTI in the setting of a TOC relay to coordinate with the remote station breaker failure clearing time may not be necessary since the risk of a breaker failure operation is small, coordination is still obtained for a more remote, less critical fault, the remote end breaker may be locked out by the breaker failure operation, and automatic reclosing will reenergize the line. If, however, the line must remain in service without interruption during the breaker failure operation at the remote station, and switching does not open the remote end of the line, then the TOC relay should be set accordingly.

Mr. Griffin suggests a CTI of 0.6 second for the non-directional TOC relay in order to not overtrip for faults behind the relay. If the TOC relay setting is based on coordination for faults behind the relay, as it must be, then the same CTI should be applicable as for faults beyond the remote bus.

However, it must be emphasized that coordination and settings must be based on the reasonable outage condition for which coordination is required and which results in the maximum line fault current flow. For example, removing a strong zero sequence current source from the remote bus, or opening a parallel line, will increase the line ground current flow. It is assumed that the figures given are for conditions which produce maximum line flows. If an acceptable CTI is chosen for the

most critical coordination condition, than a greater CTI will exist for all other conditions.

Another factor affecting the CTI of a TOC relay setting is the reclosing time of the remote breaker. If it is fast, not allowing the TOC relay to reset, then the CTI must allow for two operations of the remote breaker.

In the example of the Bonaire - N. Tifton line, the primary advantage of the directional relay is that the induction cup type IOC unit is set for 120% of maximum through fault currents as compared to 170% for the non-directional IOC relay. With the proper settings as discussed, the directional relay can be operated in a non-directional mode to eliminate being dependent on the external polarizing source and directional unit.

In Figures 3 & 4, it would be desirable to show the PCB "C" and "F" IOC settings since the TOC curve does not apply above those points. The actual CTI of Figure 3 appears to be somewhat more than that shown and stated in the text since the PCT "E" IOC operating time would be 3 cycles or less at 2 times pickup rather than the flat 6 cycles shown. The directional ground relay at PCB "C" may be made non-directional to eliminate the external polarizing source and directional unit since the settings are higher than those at the other end of the line at PCB "D".

Directional comparison carrier relay settings

Mr. Griffin suggest (for 2-terminal lines) identical settings for both terminals, usually 0.5A. carrier start and 1.0 A. carrier trip. If possible, the trip setting should be not more than 50% of the line end fault under minimum fault conditions. It must be at least 125% of the remote terminal start setting. However, if fault currents are adequate, it is recommended that trip settings be made well above the minimum requirements for coordination in order to minimize the probability of a false trip due to failure to properly block for an external fault.

For multiple terminal lines, the trip setting must be at least 125% of the *sum* of carrier start settings of all remote terminals where fault current flow is OUT for a single external fault. Because of this, the desire to minimize false tripping upon blocking failure, different CT ratios, and differences in ground fault current sources, there may be differences in the settings for each terminal. Of course, on any multiple terminal line, carrier must be stopped by a high speed ground directional relay. If the trip relay is not able to operate initially due to fault current distribution, the directional unit must operate to remove blocking and allow remote terminals to trip, after which the local trip relay can operate for sequential instantaneous tripping.

Manuscript received July 16, 1982.

C. H. Griffin: These discussions raise several interesing questions and point up the need for clarification in some areas. With regard to Mr. Ojanen's concern that IOC units are set for 120% of the maximum far-end fault, we feel that this is a conservative value, and we have not had any problems with over-tripping. This setting is used only on directional ground relays with cylinder units, which, as noted by Mr. Elmore, have a maximum transient overreach of 10% or less. It must also be kept in mind that this type relay is relatively slow, since the IOC unit is not activated until after the directional unit operates. I must confess, however, that on very short lines, we would probably increase the IOC unit margin to 125%, even though we have considerable confidence in our short circuit studies.

"Sequential instantaneous" clearing of ground faults is not a goal, but a result, which does not always occur, and probably would not occur on short lines; and, as Mr. Ojanen remarks, we would have no problem with a 25 cycle clearing time for low current ground faults on the Bonaire - North Tifton line. With regard to the 800/5 current transformer ratio on this line, he is correct in deducing that this ratio was selected to accommodate the phase relays and line meters. However, this line has now been upgraded to 100°C operation, and the C. T. ratio has been increased to 1000/5. I should note at this point that the Bonaire - North Tifton 230-kV line has been in service for 20 years, that over two dozen ground faults on this line have been cleared during this period, and that no over-tripping has ever occurred.

With regard to Mr. Ojanen's final questions, our design criteria for fault clearing on the 500-kV system is a *maximum* of three cycles. Our breaker-failure timers are set for 5.5 cycles at generating plants and seven cycles at 500/230-kV step-down substations. This results in breaker-failure clearing times of 9.5 to 11.5 cycles, according to location.

Mr. Niessink's point that many relay setting criteria are judgemental and based on the configuration of the overall system is well taken. I agree that the five that he lists fall into this category. However, the methods employed by Georgia Power Company in these areas are based on close study and long and favorable experience with fault clearing on our system. In 1980, for example, out of 3602 transmission line breaker operations, 3573 were correct and 29 incorrect, or 99.2% correct. Of the 29 incorrect, most were due to relay or circuit breaker failure (open trip coil, etc.), and only three to incorrect relay settings. I agree with Mr. Niessink that it would be appropriate for the IEEE Power System Relaying Committee to conduct an up-to-date survey on transmission line ground relaying practices.

The statement that "the TOC units must be set the same at all four terminals" in the case of two parallel lines terminating at the same station at both ends needs clarification. This statement applies only to relays of the same type. That is, all four *directional* ground relays should be set the same. If each terminal is also equipped with a *non-directional* ground relay, these may be set differently (less sensitively or slower) than the directional relays, but all four must be set alike. Otherwise, they will not select. Over 30 years of hard experience has proven this theory time and again.

I wish to thank Mr. Rockefeller for pointing out some poor wording with regard to IOC units. The *far* end fault is used to calculate both the directional and non-directional settings. This applies to the TOC units as well. Our standard coordination time (which includes breaker time) is 0.4 second for high voltage lines, which has been reduced to 0.3 second for EHV. This reduction is made possible by faster relays and circuit breakers at the EHV level. Since non-directional ground relays may operate for faults behind a line breaker, and since the TOC units in these relays serve mainly as emergency backup devices, we feel that it is prudent to provide ample margin (0.6 second), particularly at the 230 and 500-kV level.

I appreciate the interesting comments by Mr. Pencinger, particularly with regard to the distance measuring errors of ground distance relays on mutually-coupled lines. This is valuable information and probably not widely known. With regard to supervision by ground distance fault detectors, we do not use "high set overcurrent relays" as Mr. Pencinger remarks. The paper states that these fault detectors should be set *"less than 50% of the far-end fault, but not below 0.6 ampere secondary"*. Mr. Pencinger recommends a setting of "0.15 x rated current". Since rated secondary current is usually five amperes, this works out to $(0.15 \times 5) = 0.75$ secondary ampere. This is the same setting normally used by Georgia Power Company. My objections to an impedance fault detector are that its operation depends on the same voltage signal as the relay measuring element, and that it may not be immune to operation on loss of potential. With a properly set overcurrent relay, accidental interruption of the relay potential supply will not cause a false trip.

Single-pole tripping and reclosing is not used on the Georgia Power Company system because it is complex, expensive, and not needed. For weak systems, where the loss of one or two lines could cause grave problems, single-pole tripping has proved valuable in maintaining service. Georgia Power Company has an extremely strong system, which can accommodate the simultaneous loss of several high capacity transmission circuits without causing disruption of the network. It is certainly possible, however, that we may have to consider single-pole tripping for some of the new 500-kV lines now in the planning stages, particularly in the case of proposed ties to Florida.

I appreciate the thoughtful discussion by Mr. Elmore, and I will try to answer the questions he has raised. First, the paper says that GPC policy is to allow *about* 0.4 second coordination time. 27 cycles is 0.45 second. I believe that this would qualify as "about" 0.4 second. Also, I agree that 0.3 second plus breaker time is a suitable CTI. Since there are a large number of 5 cycle breakers in service, this works out to $(18 + 5) = 23$ cycles = 0.4 second total time.

Both Mr. Elmore and Mr. Volzka invite discussion on the setting of instantaneous overcurrent units. Most clapper-type non-directional IOC units operate in less than one cycle. During this interval, the assymetrical value of the fault current on HV and EHV systems can range up to 1.5 times the symmetrical value. The ground IOC unit must be set high enough to ride through this transient. The IOC cylinder unit on directional overcurrent ground relays cannot trip until after the directional unit operates (see above). On an electromechanical relay, this takes from 2 to 5 cycles, during which time the assymetrical offset substantially decays. Since, as Mr. Elmore points out, a cylinder unit is much less susceptible to false operation on dc offset anyway, it is usually perfectly safe to set these devices for 120% of the maximum far-end fault; and, as Mr. Elmore suggests, there definitely are cases where it

would be advantageous to have cylinder-type non-directional IOC units, particularly on lines with a shallow end-to-end fault current slope. A cylinder unit would permit considerably improved high-speed coverage on these lines.

Georgia Power Company feels that pilot relaying is required on many transmission lines below the 230-kV level, with or without a 500-kV overlay. Mutually-coupled lines, three-terminal lines, and very short lines are typical examples. The general rule is that *"all buses must appear as instantaneous buses when viewed from all adjacent buses"*. This means that there must be no zone-two overlap on phase or ground line relays. Also, system stability frequently requires that all line terminals be tripped simultaneously. This can only be accomplished with some type of pilot relaying.

I certainly agree with Mr. Elmore's reminder that the impedance of transformers tapped to lines protected by phase distance relays should be substantially higher than the impedance of the line itself. This is almost always the case. I suspect that this may also be true for ground distance relays, although I am not sure how these relays will respond to ground faults on the low side of a delta-wye transformer bank.

With regard to the tap setting on the PCB "C" ground relay, the far-end fault seen by this relay is 2600 amperes. A 240 ampere tap is about one-tenth of this value, which is usually a desirable setting. It might be possible to reduce this tap to 1.5 (= 180 amperes primary), but the time-dial setting would have to be increased to compensate.

I am glad to have Mr. Elmore's comments on negative-sequence versus zero-sequence polarizing quantities. "Reverse throw" is utility relay engineer's jargon that indicates that a directional relay trips incorrectly for a fault behind the relay and not on the protected line.

The recommendation for using 0.6 ampere as a minimum setting for pilot relay overcurrent units is, as Mr. Elmore suspects, based on the fact that most HV and EHV transmission lines are not transposed. As a result, residual currents of 0.4 ampere and higher may be present in the ground relays, particularly during heavy load periods. A setting of 0.6 ampere for the pilot ground fault detectors will usually provide security against accidental trip during testing, etc., and will prevent spurious carrier alarms due to unbalanced phase currents.

Georgia Power Company is well aware of the advantages of ground distance relays, particularly the fact that they do not have to be reset as system conditions change. However, they are expensive, and they are complex; and, as stated in the paper, ground overcurrent relays are still required "to protect for high resistance faults that may not be detected by ground distance relays".

We agree with Mr. Hunt's rule of trying to limit maximum relay current to 100 amperes. As already noted, we have now increased the C. T. ratio on the Bonaire - North Tifton line to 1000/5, which reduces the maximum relay current to 105 amperes. We also agree with Mr. Hunt's practice of limiting ground relay TOC unit settings to a maximum of 240 amperes. We have always followed this rule on lower voltage circuits. We have not been as careful about this at the 230-kV and 500-kV level, probably banking on the fact that the same fault that will barely be detected by a 240 ampere setting at 25-kV should produce nearly 10 times as much current at 230-kV. I see no problem with the fact that a ground relay curve may cross the phase relay curves. This is a common experience on all circuits where sensitive ground settings are required, but, as Mr. Hunt notes, it will result in phase relay targets for some high current ground faults. We have long since learned to ignore statements from the field like "I know it was not a ground fault because the only flag showing was on the phase-three TOC unit".

In reply to Mr. Volzka, I believe that the practice of using 3000/5 current transformers with C800 rating on the 2000/5 ampere tap was pioneered by Bonneville Power. Such C. T.'s are only required where ratios less than the full winding are to be used. They should be ordered from the manufacturer when the associated circuit breaker is ordered, and statements such as "we can't supply C. T.'s with that rating" can be expected. I am not aware that any special equipment is required to test these C. T.'s in the field.

I agree with Mr. Stephens that some of the setting criteria outlined in the paper would have to be modified in the case of three-terminal lines. I also agree that pilot relaying may be required for long three-terminal lines, particularly where one leg is considerably shorter or weaker than the other two.

There are certain advantages to varying relay calibration points, according to fault current levels, as Mr. Stephens points out. Georgia Power Company feels that these advantages are outweighed by the advantage of using a standard calibration point for all ground relays (in our case, ten times tap setting), since this reduces the possibility of error in comparing one setting with another. Which will give the longer tripping time at "x" amperes - 30 cycles at 1000% or 42 cycles at 600%? One must plot the curves to find out.

The Georgia Power Company 0.4 second coordination time rears its head again in Mr. Stephens' discussion. We are not trying to sell this number to anybody; we like it; it works. It probably originally derived from the fact that Georgia Power Company uses high speed reclosing on most transmission line breakers. Mr. Stephens states that this type of reclosing will have the effect of doubling circuit breaker clearing time. This seems reasonable for breakers that are new or just overhauled. On our system, where a single transmission line breaker may operate six or eight or more times a week during lightning season, the effects of high-speed reclosing into a permanent fault may be much more severe. For this reason, we find that allowing 15 cycles clearing time for a 5 cycle breaker is prudent, particularly on those occasions where a breaker recloses on a close-in bolted fault. With the number of tree cutters in Georgia, this is a common occurrence. I should also note that since most ground faults on medium and long lines are cleared by sequential instantaneous operations, shortening the CTI will not speed up the fault clearing time, but will reduce the security of the bulk power network. Finally, all our short lines are protected by pilot relays. On these lines, the directional and non-directional overcurrent ground relays serve as backup protection only, so that again, reducing the CTI from 0.4 to 0.3 second would not reduce the clearing time for most faults.

I agree with Mr. Stephens that there are some locations, such as Bonaire, where it is possible to convert a directional relay into a non-directional relay, and thus eliminate dependence on an external polarizing source. We used to do this. We stopped doing it about 15 years ago, because Company operating procedures require that non-directional ground relays be turned off during disconnect switch operation. Since a directional ground relay will not operate on the residual of unbalanced phase currents, it may be left in service at all times.

Mr. Stephens brings up an important point in regard to the setting of carrier ground relays, where he reminds us that an adequate margin is required for proper setting of the G2 (trip) unit. We try to follow his recommendation that G2 be no more than 50% of the minimum far-end fault. However, we have some very long mutually-coupled 230-kV lines, where this is not possible. These lines terminate in large grounding transformers. The shunting effect of these sources reduces the remote current for terminal faults, so that the carrier ground trip element at the remote end may not operate until after the local breaker opens. This is an example of a sequential operation where the near terminal is tripped by the non-directional IOC unit, and the far terminal by the carrier ground relay. On circuits of this type, both carrier ground relays will usually operate only for faults in the middle 80% of the line. We think Mr. Stephens is well advised when he states that "if fault currents are adequate, it is recommended that trip settings be made well above the minimum requirements for coordination". As he points out, this will help reduce the possibility of a false trip, if one terminal fails to block.

Manuscript received September 20, 1982.

A DIRECTIONAL WAVE DETECTOR RELAY WITH ENHANCED APPLICATION CAPABILITIES FOR EHV AND UHV LINES

A. T. Giuliante, Member, IEEE
R. R. Slatem

G. Stranne, Member, IEEE
Carl Öhlén

ASEA Inc.
Relay & Control Division
One Odell Plaza
Yonkers, New York 10701

Abstract - A directional wave detector relay capable of fulfilling the high-speed or ultrahigh-speed (UHS) protection requirements for EHV and UHV transmission lines is described. The results of combining directional wave detection with a sequential measurement distance relay and a sensitive ground overcurrent unit are discussed. Multiple functions operating in series and parallel enhance the security and dependability by mutual support whereby the inherent strengths of each measurement technique are used to improve the characteristics of others. Solutions for difficult application problems are presented. In addition, economic considerations are noted for increased power transfer and reduced circuit breaker costs.

INTRODUCTION

Relay engineers constantly strive to develop line protection having greater speed, selectivity, and sensitivity to meet the increasing demands of modern EHV and UHV transmission systems. A step in this direction was taken when the directional wave detector (DWD) relay [1, 2] was developed. This relay satisfied the ultrahigh-speed requirements for one cycle fault clearance and has been in successful service for over 4 years. Being a pure pilot scheme, it provided no remote back-up and its maximum sensitivity could be limited for certain system conditions.

A new relay is described which combines the features of the high-speed directional wave detector relay with a sequential measurement distance relay [3] for remote back-up, plus neutral current monitoring for maximum high-resistance ground-fault sensitivity without loss of security. With these features, the relay is able to provide fast fault clearance for the protected line, remote distance back-up, a relatively fast zone 1 independent of the communication channel, excellent high-resistance fault sensitivity with clearance in 20-40ms even with in-zone surge arrestors. It is also able to cope with weak infeeds and in-zone reactor switching. The relay measures on a per phase basis which provides an enhanced adaption to system parameters compared to a pure negative-, positive-, or zero-sequence directional comparison scheme. A phase selector is provided for single-phase tripping applications or where breaker failure times or the reclosing mode must differ for different fault types. In addition, distance type switch-into-fault protection independent of breaker-closing signals is provided. Inherent in the directional wave detection principle is the ability to protect series compensated lines

and to be relatively independent of power swings. This capability has been used to control the distance measurements thus retaining these desirable characteristics.

Problems such as transient saturation of current transformers, transient phenomena of CCVT's, and the detection of broken loaded phase conductors are also solved by the relay. Because the directional wave detector controls the zone 1 and 2 distance measurement, high load transfer capability is provided and there is no need for high-speed blocking for a VT fuse failure.

DESCRIPTION

Directional wave detectors are combined with three distance functions and a neutral current unit to provide complete phase and ground protection for transmission lines. The relay utilizes a single communication channel in each direction. Also included are back-up functions which operate independent of the communication channel. System signals are obtained from conventional current and voltage transformers.

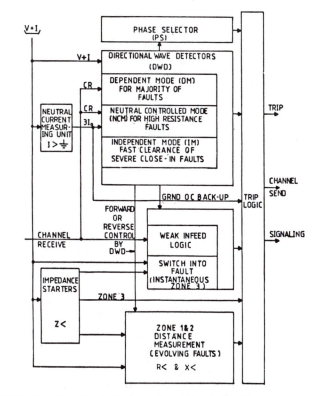

Fig. 1 Overview block diagram of combined directional wave detector (DWD) and distance relay.

Directional wave detectors (DWD) operate if sudden changes in voltage and current exceed their settings. The relay utilizes three different directional wave detecting modes, (Figure 1). The dependent mode (DM) works in

83 WM 097-3   A paper recommended and approved by the IEEE Power System Relaying Committee of the IEEE Power Engineering Society for presentation at the IEEE/PES 1983 Winter Meeting, New York, New York, January 30-February 4, 1983. Manuscript submitted September 8, 1982; made available for printing November 11, 1982.

Reprinted from *IEEE Trans. Power App. Syst.*, vol. PAS-102, no. 9, pp. 2881-2892, Sept. 1983.

conjunction with the communication channel to provide complete pilot protection for the majority of faults. The directional decision, forward or reverse, is made independently at each line terminal and then compared over the channel to provide a trip signal for internal faults or blocking for external faults. The neutral current controlled mode (NCM) is the most sensitive and uses criteria of neutral current and a selectable time delay of either 20 or 40ms, combined with the directional wave detection principle. This mode provides high-resistance ground-fault detection capability without degrading security. The additional delays are acceptable for the low level faults intended to be covered by this mode. The independent mode (IM) is comparable to a high-set directional overcurrent relay and operates independent of the channel to provide fast detection (1-2 ms) of severe close-in faults.

Novel use of filters results in the independence of the directional wave detector to system loading, sensitivity to low level faults, speed, and insensitivity to power swings. Instead of filtering the components of fault currents and voltages after a fault has occurred as in conventional relays, the DWD suppresses the steady state voltages and currents and reacts only to sudden changes. These sudden changes pass through the steady-state suppression filters without delay. Filtering speed is not critical provided that it is fast enough so the relay does not react to changes resulting from power swings or load variations.

Faults occurring on the protected line are detected by the directional wave detector. Once the direction to the fault is established, all subsequent information from the CT's and VT's is disregarded by the DWD until it resets. For the unlikely event of a subsequent internal fault occurring during this reset time, the zone 1 and 2 distance measuring functions will be enabled by the wave detector start, either forward or reverse.

By controlling the zone 1 and 2 distance measurement with the wave detectors, the inherent and highly desirable DWD features of insensitivity to power swing and VT fuse failure are retained for these measurements. The zone 1 and 2 distance measurements are of the sequential measurement type using 3 impedance and 1 neutral current starters and 1 dual measuring element. Depending on the fault type as determined by the starters, optimum measuring quantities are selected by a static signal selector and fed into the measuring element for determination of direction and distance to the fault. Highly selective directional sensing is ensured by unfaulted phase polarization for unbalanced faults and memory for three-phase faults. The memory function utilizes a phase locked loop oscillator which locks-in on the phase of the voltage signal immediately prior to the faults and is unaffected by the fault voltages.

The zone 1 reach in the resistive and reactive axes can be set independently. After expiration of the set zone 2 time delay, the reach is extended by a set constant, see Figure 2.

The impedance starters operate independent of the wave detector section and provide three functions:
1. Instantaneous switch-into-fault protection when the line is being energized. This always gives a three-phase trip output.
2. Time delayed non-directional zone 3 back-up distance protection.
3. Starting (determination of fault-type) for the zone 1 and 2 distance measurements.

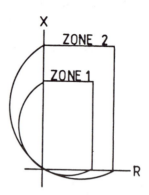

Fig. 2 Zone 1 and 2 distance characteristics

The switch-into-fault protection is necessary since the directional wave detector with the normal line side VT's may not detect an adequate voltage change because of the presence of maintenance grounds or a line fault. This protection operates if, after the line voltage has been low for a set time, the impedance starters operate. It remains enabled as long as any line voltage is below a set value and is disabled immediately when all line voltages exceed this value. This eliminates the need for breaker-closing signals.

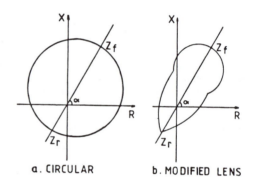

Fig. 3 Zone 3 distance and starting characteristics

The impedance starters have two possible characteristics, circular or modified lens, Figure 3. These starters can be set to have the desired forward and reverse reaches. Generally, a modified lens characteristic with minimum reverse reach would be selected for a long, heavily loaded line. High-resistance fault detection capability is not required by these starters since this function is covered by the neutral current controlled mode of the wave detector. This is a classic example of the mutual support roles provided in the relay whereby the inherent strengths of one relaying principle are used to improve those of the complementary principle.

In some applications one line terminal may at times have a weak source. Since the operation of each end in a permissive scheme is dependent on receiving a communication signal from the opposite end, the weak-infeed end must be arranged to echo the signal received from the stronger end. The logic available for achieving this is shown in Figure 4. If a communication signal is received from the remote end and there is no local general trip or start of the wave detector in either direction, forward or reverse, the local end retransmits the channel received signal, thus enabling the remote end to trip. If, in addition, one or more phase voltages are low at the weak-infeed end, the relay will initiate a local trip. The necessary signal level coordination between blocking at the weak-infeed end and tripping at the strong-infeed end to prevent misoperation for marginal external fault conditions is built into the relay.

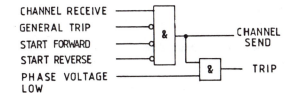

Fig. 4 DWD weak-infeed logic

The current monitor for the neutral current controlled mode can be arranged to provide definite-long-time delayed ground overcurrent back-up.

For applications such as single-phase tripping, phase selectors are included with the directional wave detector. The simplified phase selection logic is shown in Figure 5. The maximum of the peak current changes of the three phases is extracted and halved and, in an individual summing amplifier per phase, the difference between this maximum and the individual phase peak change is determined. If this difference is positive and either the DWD dependent mode (DM) or neutral current controlled mode (NCM) has started in the forward direction, the corresponding AND gate is armed and sets its flip-flop to give an output indicating the phase involved.

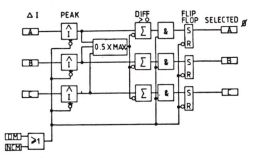

Fig. 5 DWD phase selection logic

Further logic ensures a three-phase trip for internal faults involving more than one phase. Therefore, a phase will only give an output to the indication and trip logic if there has been a neutral or dependent mode start and the peak current change signal of that phase is greater than 50% of the maximum peak current change of the three phases. With these phase selectors the use of current changes, rather than current signals exceeding an absolute level, enhances the reliability of the phase selection appreciably.

The relay conceptual design is based on ensuring optimum reliability, both dependability and security, by series and parallel combinations of functional criteria. This is illustrated by Figure 6 which indicates in simplified form the inter-relation between the various measurement functions together with typical operating times. The use of multiple measurement techniques such as continuous steady state measurement, sudden changes, and sequential measurement also contribute significantly to reliability.

APPLICATION

The relay is applied on EHV and UHV lines. It must be emphasized that its application is not limited to systems requiring ultrahigh-speed fault clearance although it is well suited for such applications when combined with a communication channel having an appropriate speed. In general, overall operating time will be determined by the channel since the directional wave detector will make its directional decision independently in 1-2ms. Consequently, by suitable choice of the channel, overall operating time can be controlled by the user to suit system needs.

Besides speed, other reasons for applying this relay are:
- o Different measurement principle required for redundancy.
- o High-resistance faults.
- o High load transfer.
- o Power swing insensitivity.

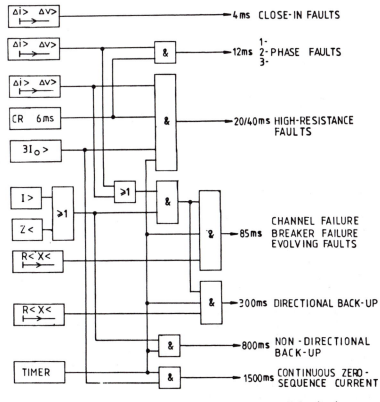

Fig. 6 Enhanced reliability using series and parallel criteria

o Series compensated line.
o CT or CCVT transient response problems .
o Zero sequence mutual coupling.
o Single-phase tripping and reclosing.
o Phase selection for breaker failure, reclosing, or fault locators.
o Broken loaded-conductor detection.

High-Speed Advantages

Greater emphasis is being placed on ultrahigh-speed fault clearance because of the increasing need to transfer larger blocks of power over existing or planned lines. This results from economic and environmental considerations. The principal advantage of UHS relaying is increased stability for a given power transfer, or conversely a greater power transfer for a given stability limit (Figure 7), but other benefits are provided because of reduced fault damage to insulators, conductors and hardware; reduced safety hazards to substation personnel and line crews from step and touch potentials; reduced stresses on major substation equipment, generators and turbines; and the possibility of faster reclosure due to less arc ionization.

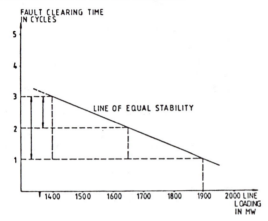

Fig. 7   Increased transmission capability possible with faster fault clearance

An important contribution that can be made by using an ultrahigh-speed relay is in reduced circuit breaker costs. By reducing relaying time less expensive and possibly more reliable breakers can be used for a given fault clearance time and this could also reduce breaker maintenance costs. When faster fault clearance is required it is generally less costly, particularly on an existing system, to provide ultrahigh-speed relaying than to retain conventional relaying and use faster breakers.

Sensitivity to High-Resistance Faults

In many areas high-resistance ground-faults can occur due to broken conductors falling onto rocky ground, trees under the line, bush fires, or where there are no ground wires and high tower footing resistances represent a high resistance fault for all single-phase to tower faults. The described relay provides sensitive (as low as 20% of rated current) pilot protection for these faults and an operating time of the order of 20-40ms. Such an operating time ensures that there is no loss of security during in-zone surge arrestor discharge of power-follow current after a lightning surge. A definite-long-time delayed function (alarming or tripping) for these high-resistance faults has been added for redundancy.

Effects of Shunt Reactor Switching

Independently switched shunt reactors should preferably be excluded from the protected zone by locating current transformers in the reactor circuits. Where switched shunt reactors cannot be excluded from the zone, provision has been made to block the sensitive neutral controlled mode output and desensitize or block the dependent mode output of the directional wave detector, as desired, for a preselected time. This action is initiated from the reactor switching controls. The independent mode output is never blocked nor desensitized since its settings are chosen such that it cannot respond to changes resulting from reactor switching in-rush current. This option is therefore always available for rapid clearance of severe close-in faults. Directional wave starting is not blocked so that the zone 1 and zone 2 functions are enabled and provide the line with direct local tripping and stepped distance protection during the reactor switching period.

Sensitivity with Heavy Loading

Applying relays on modern transmission systems is often made more difficult because heavy load flow situations must be considered. Conventional relays often require less sensitive settings as a result of heavy loading. The DWD relay overcomes this difficulty by responding only to sudden current and voltage changes. The distance measuring functions in the relay have characteristics which can be set insensitive to load flow since high-resistance faults are covered by the neutral current controlled mode. Hence the relay retains its high sensitivity, even for heavy load flow conditions.

Power Swing Insensitivity

Power swings do not cause the DWD to respond because the associated changes occur sufficiently slow for suppression by the filters. Since the zone 1 and 2 distance functions are controlled by the DWD they also are unaffected by power swings and, unlike conventional distance relays, do not require power swing blocking. The zone 3 impedance starters are not controlled by the directional wave detectors and therefore their tripping output must be blocked by power swing detection relays if the swing can enter their characteristic and remain there for 800 ms.

Series Compensated Lines

Applying the relay on series compensated lines poses no problem since the DWD responds only to the changes in the voltage and current and, unlike conventional distance relays, is not affected by bus voltage reversals. Figure 8 shows how the change in voltage, ΔV_A retains its sign even though the bus voltage has reversed due to the capacitors. Thus, the directional wave detectors provide high-speed protection for internal faults.

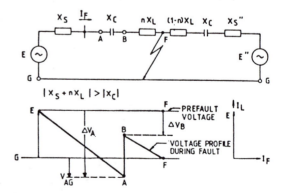

Fig. 8   Effect of series capacitors. Bus voltage reversal but no reversal in voltage change

146

For evolving faults the zone 1 and 2 functions will provide backup protection. For unbalanced faults, the zone 1 and 2 directional measurements are unaffected because of unfaulted phase polarization. This, in effect, means that they obtain their directional polarizing quantity from the voltage behind the source impedance which does not suffer reversal. For balanced three-phase faults, the zone 1 backup function is delayed to allow capacitor gap firing to take place so it makes its directional decision after the bus voltage reversal has been removed. Zones 2 and 3 provide a non-directional backup as do other types of distance relays on series compensated lines for such faults.

Instrument Transformer Response

Near modern large power plants the system primary time constants tend to increase. Also, fault levels are rising steadily and this combination can lead to rapid transient saturation of current transformers. Remanent magnetism left by previous faults can aggravate the problem. Since some distance relays cannot measure correctly with the distorted secondary currents, appreciable delays in operation could occur with consequent adverse effect on system stability. The directional wave relay makes its directional decision in 1-2ms, before transient saturation can occur, and is unaffected by the distorted waves. The zone 1 distance measurement, which backs the DWD, requires that the current transformers reproduce the primary waves faithfully for only 50ms.

At EHV and UHV levels distance relays often require extra high capacitances for the coupling capacitor voltage transformers (CCVT's) to ensure correct distance and directional measurement for faults which result in low voltage at the relaying point. For such faults, the low frequency transient of CCVT's can have several times greater amplitude than the signal to be measured, resulting in a signal/noise ratio less than unity. The directional wave relay measures change in voltage and for these conditions the change is many times greater than the low frequency transient amplitude. This implies that the signal/noise ratio is always much greater than unity so correct measurement is ensured, even in the presence of CCVT transients (Figure 9).

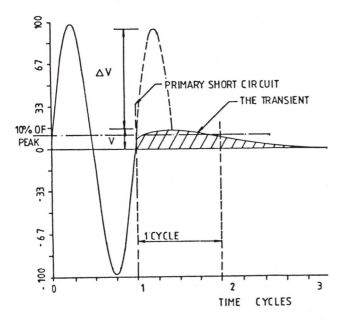

SECONDARY VOLTAGE (%)

Fig. 9   Influence from CCVT transient errors minimized by measuring voltage change

Effects of Mutual Coupling

Zero sequence mutual coupling between parallel lines normally gives difficulty of reach with ground distance relays. The directional wave detector, which provides the primary protection function, is unaffected by this coupling. Only the distance measurement, which is secondary in function, requires settings which take mutual coupling into account.

Phase Selection Capability

The special phase selectors discriminate between single-phase-ground and other faults to make the relay particularly suitable for single-phase tripping and reclosing applications or for situations where different breaker failure times are desired for single-phase and multi-phase faults. It can also be used where different reclosing policies are applied for single-phase and multi-phase faults or where phase selection is required for fault locators.

Broken Loaded-Conductor Detection

One of the most difficult faults to detect on a line is a series fault or broken phase conductor even if this also results in an actual shunt fault since the latter can usually be detected only from one end. Because it measures the changes in voltage and current, the DWD can often detect the series interruption of one phase of a loaded line at both ends, whether or not it results in a shunt fault. The change in current is equal to the lost load current and, provided the changes in voltage at each end of the line (which depend on load magnitude and source impedances) exceed the setting, the relay will detect the condition and could actually clear the line before the broken phase results in a shunt fault.

Other Effects

High-speed VT fuse failure detection is required by distance relays to prevent misoperation from the resulting voltage loss. The DWD, requiring simultaneous sudden changes in both voltage and current for its operation, does not misoperate from this cause. The zone 1 and 2 distance functions are controlled by the directional wave detectors (in either direction) thus they are also unaffected. Slow VT fuse failure detection is required to block the zone 3 impedance relays and to give an alarm so that the fuses can be replaced before any system fault occurs. Miniature circuit breakers can be used instead of VT fuses to eliminate the need for fuse failure detection. Blocking of the zone 3 impedance relays is accomplished by auxiliary contacts in the miniature circuit breaker.

In a conventional permissive overreaching pilot scheme, overreaching distance relays covering an unfaulted line require blocking during sudden power reversals to prevent them from misoperating. Such blocking is built into the DWD since the directional wave detector will have blocked at one end. Subsequent internal faults occurring on the previously unfaulted line during the blocking of the directional wave detector are covered by the zone 1 and zone 2 functions.

The directional wave detector relay responds correctly to transients so line transients due to switching or faults are no problem. For internal faults they tend to increase the desired signal change and for external faults they can only cause the relay to block temporarily if they exceed the relay setting.

Setting and Testing

It is of fundamental importance to appreciate that the setting and testing of the relay, including the directional wave detector modes, is based entirely on 60Hz quantities in spite of the fact that the relay responds to sudden changes.

It is not necessary to have knowledge of the system transient behavior to apply, set, or test the relay. Allowance for these phenomena is built into the design. For setting the directional wave detectors the only departure from normal system studies is that the rms values of the changes instead of the actual rms current and voltage values must be determined. Determination of the settings and testing of the distance functions follow standard practice.

Adjustment of the relay to the required settings is simple since it merely involves the setting of precision calibrated potentiometers or thumb-wheel switches located on the front panels. No special tools are required for making these adjustments.

Built-in functional test facilities are provided for simulating the various fault types and checking the logic behavior of both the directional wave and distance functions. These tests do not check calibrations. Calibration is performed during secondary injection testing which is facilitated by a built-in test switch. The relay type designation is RALZA and it is shown in Figure 10.

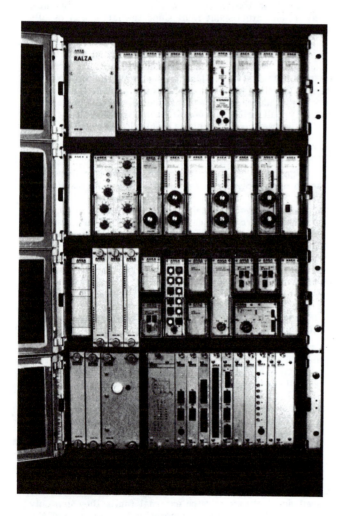

Fig. 10   Type RALZA Relay

CONCLUSIONS

The described relay employs proven multiple measurement techniques and functions to ensure the largest possible fault coverage at speeds commensurate with the power system requirements. Optimum sensitivity and reliability, immunity from system and transducer error

phenomena, phase selectivity, and modest demands on the communication channel are features achieved by the unique combination of measuring techniques. Multiple functions are integrated in the relay to provide superior performance by mutual support whereby the inherent strengths of each measurement technique are used to improve the characteristics of others.

Wide setting ranges and flexibility of characteristics enable the relay to be used to protect any EHV or UHV line which may be short or long, series compensated, heavily loaded, or equipped with in-zone switched shunt reactors.

REFERENCES

[1]   M. Chamia, S. Liberman, "Ultra High Speed Relay For EHV/UHV Transmission Lines - Development, Design and Application". IEEE Transactions on Power Apparatus and Systems, vol. PAS-97, pp. 2104-2116, No. 6 Nov/Dec 1978

[2]   M. T. Yee, J. Esztergalyos, "Ultra High Speed Relay for EHV/UHV Transmission Lines - Installation, Staged Fault Test and Operational Experience". IEEE Transactions on Power Apparatus and Systems, vol. PAS-97, pp. 1814-1825, No. 5 Sept/Oct 1978

[3]   F. Andersson, J. R. Linders, R. Slatem, G. Stranne "An Adaptable Static Distance Relay for Transmission Lines and Cables" Western Protective Relay Conference Oct. 27-29, 1981

Discussion

H. Stanley Smith, (Southern Company Services, Birmingham, AL): The authors have presented a very interesting relay scheme which has distinct advantages expecially from a system stability standpoint. By utilizing the ultra high speed relaying to reduce the fault clearing time at a generating plant by as little as 1/2 cycle could increase the power transfer capability of the tranmission system enough to be the difference of maintaining the plant stability under certain critical stability situations.

Would the authors expand on the filters used for "steady-state suppression" described on page 2?

Also on page 2 the authors state that..."the zone 1 reach in the resistive and reactive axes can be set independently. After expiration of the set zone 2 time delay, the reach is extended by a set constant..." Would the authors clarify the last statement?

The authors state on page 4 "An important contribution that can be made by using an ultra high speed relay is in reduced circuit breaker costs." I would agree with the statement if the authors are meaning, for example, the cost savings of not having to use IPO breakers because of the faster clearing that could be achieved with the ultra high speed relaying which would tend to increase the stability limit.

Manuscript received April 15, 1983.

W. A. Moncrief (Georgia Power Company Atlanta, GA): The misoperation of a relaying system on an EHV line can have a variety of effects on the Electric Utility, ranging from a poor statistical record to significant financial loss. Efforts to improve the security of relaying systems without compromising reliability are welcomed by all.

The new relay described has, in addition to the traveling wave pilot system, dependent and independent relaying functions. Would the authors comment on the common failure modes, if any, that would affect all of the relaying functions?

The choice of communication channels should be addressed. As the speed of the channel determines the overall speed of the pilot system, the reliability of the channel also determines the reliability of the pilot system. Would the authors recommend anything other than an unblocking scheme for this relay?

In ring-bus arrangements, the line and relaying potentials are typically removed from service while the circuit breakers and the stub or bus section remain energized. Do the zero sequence and switch-into-fault circuits provide continuous protection for the bus section?

The paper states that the relay settings may be determined knowledge of the system or line transient behavior, yet the relay instructions caution that dependent mode settings must be higher than voltage and current changes due to internal switching operations and surge arrester discharges. Would the authors address this inconsistency?

The paper recommends substracting the currents of independently switched line reactors from the relays. Is this practical for a relay that operates so quickly on changes in voltage and current? How closely do the CT characteristics have to match for this approach to work?

Manuscript received March 2, 1983.

Evan T. Sage, Jr., (Baltimore Gas and Electric Company, Baltimore, MD): The additional functions incorporated in the new relay result in a very comprehensive protection system. One of the more valuable features is the possibility of ultra high speed (UHS) detection of faults caused by closing into maintenance grounds, which represent a significant portion of those faults for which UHS clearing is desirable.

There are several points which the paper was unclear on and which the authors will hopefully clarify.

1. The text indicates a detection time of 1-2 ms for severe, close-in faults, whereas Figure 6 indicates 4 ms. Which is correct? Also, what are the limits of applicability of this feature with respect to short lines?

2. If line-side CCVT's are employed and a breaker is closed into a bolted, three-phase fault at the station, detection will depend on the operation of the impedance starters. What is the approximate detection time in this case?

3. Figure 6 indicates an 85 ms (5 cycle) detection time for zone 1 faults. is that what is meant by "a relatively fast zone 1"? Also, it is stated that "The zone 1 distance measurement...requires that the current transformers reproduce the primary waves faithfully for only 50 ms." Why should this figure not be *85 ms*?

4a. The authors state that "Because the directional wave detector controls the zone 1 and 2 distance measurement, high load transfer capability is provided..." This would seem to imply that the directional wave detectors (DWD) will prevent a false trip under conditions in which the distance measuring units would otherwise have operated due to heavy load flow. This being the case, how would the relay behave if an external fault, or even a remote switching operation (e.g. a reactor) appearing to the DWD as an internal fault, were then to occur? Presumably the DWD would enable tripping of the distance measuring units until the DWD reset. (What causes reset?) Are not the distance unit settings therefore subject to the same load limitations suffered by more conventional relays?

4b. The authors state that "The distance measuring functions in the relay have characteristics which can be set insensitive to load flow since high-resistance faults are covered by the neutral current controlled mode." Why would this be significant given the statement quoted in (4a)? In any event, even setting a zone 2 *phase* relay on a long, heavily-loaded line can present a problem due to arc-resistance and its apparent impedance to the far-terminal relays.

Manuscript received January 5, 1983.

W. A. Elmore, (Westinghouse Electric Corp., Coral Springs, FL): The authors have described an interesting extension of their existing relaying system, and it would be helpful to the reader if certain apparent contradictions were clarified.

1. Improvement in dependability over the previous pilot-only DWD system is described and yet the DWD directional unit seems to supervise everything-the pilot trip, the zone 1, and the zone 2 trips. Further from the descriptions of the zone 1 and zone 2 elements it appears that only one distance measuring unit is used for all fault combinations and for both zones. Would not dependability be enhanced more by keeping these functions completely independent of one another?

2. Increased power transfer *and* reduced circuit breaker costs are suggested by the authors where a faster relaying system is used. It is unlikely that both could be realized in any realistic case. Thorp, Phadke, Horowitz and Beehler in their paper "Limits to Impedance Relaying" [1] showed that relay operating time vs. maximum total fault duration is distinctly non linear. Considering the requirements for breaker current-zero interruption and differing system dc time constants, faster relaying speed, in and of itself, does not assure faster interruption and therefore may enither increase permissible power transfer nor reduce breaker costs.

3. From observation of figure 6, it is not clear what the speed of this system is being compared to when the term "faster" is used. Are these minimum, maximum, typical? Are they times from fault inception to trip coil energization?

REFERENCE

[1] J. S. Thorp, A. G. Phadke, S. H. Horowitz, J. E. Beehler, "Limits to Impedance Relaying" IEEE Transactions on Power Apparatus and Systems Jan./Feb. 1979 pp 246-260.

Manuscript received February 10, 1983.

J. M. Crockett (Westinghouse Canada Inc. Hamilton, Ontario, Canada): Since the presentation of the authors' references 1 and 2, considerable interest has been shown in these relaying concepts and further elaboration on some of the material in this present paper would be welcome.

Analyses by Vitins[1] and Bignell[2] illustrate significant reach variation with source impedance and fault initiation angle. Figure 2 of the paper would be more informative if it also illustrated the range of close in fault coverage and the range of DWD reach for some small range of source/line impedance ratio, say 0.5 to 2.5. Bignell[2] suggested the use of a replica impedance approach to minimize these reach variations, and to eliminate any need to store the first directional indication resulting from a disturbance. The authors' Figure 6 indicates 300 msec directional back up. Does this indicate a time duration of DWD seal in on forward or reverse faults?

Figures 1(a), 1(b), 1(c) are copies of 500 KV line automatic oscillograph records. I understand 735 KV records have illustrated similar phenomena. Figure 1(a) illustrates a non fault event, presumed to be a lightening disturbance, 22 $\frac{1}{2}$ cycles before an actual SLG fault. Figure 1(b) illustrates 2 such events following line opening on a fault and prior to successful reclose.

In the cases of Figures 1(a) and 1(b) there were actual SLG faults, cleared in 24 cycles. I assume the automatic oscillograph starts were due to actual fault current. However in the case of Figure 1(c) the non fault event disturbance was severe enough to start the oscillograph.

On October 14, 1982 in Quebec, Canada there were two faults on a 735 KV system within a few cycles.

It would appear that, since the DWD disregards subsequent information from CT's and VT's, and that since the phase locked loop oscillator locks in on the phase of the voltage signal immediately prior to the faults, there may be cause for some concern about reliability. Do the authors have any concern for possible dependability or security problems due to such non fault events close in time to actual faults or due to near simultaneous faults on different circuits?

An open phase of a loaded line should be detected by zero or negative sequence directional comparison before the broken conductor hits the ground. However, this does not apply if only one conductor of a phase bundle breaks. Bundled conductors will be the norm on EHV and UHV systems. Figure 2 shows that an open conductor is a special case of impedance inserted in one phase. Another, and seemingly more possible, case would be the by-pass of capacitors in one phase due to excess line current during a power swing. A sudden change would result and an undersirable line trip might result. Are the authors concerned about this possibility? Zero and negative sequence directional comparison schemes must include provisions to prevent sound line tripping due to asymmetrical capacitor bypass.

A power swing may very well be preceded by a fault which will be detected by the DWD elements on sound lines adjacent or parallel to the faulted line. Would the authors explain why the sudden change associated with the fault does not defeat the described power swing insenstivity? Similiarly, I would expect that the sudden change due to the fault would block the directional scheme to avoid any undersired tripping due to sound line capacitor by pass resulting from the fault or subsequent swing.

Would the authors explain the deliberate 20 or 40 msec delay and also clarify the 20% of rated current as regards sensitivity to high resistance ground faults? Is this 20% figure the current in each terminal of the faulted line or; is it total current in the fault with appropriate allowance for unfavourable distribution factors? The sensitivity limit is due to the DWD, and there are probably applications where zero or negative sequence directionals would significantly improve sensitivity. Channel independent protection for high resistance ground faults not detected by distance elements might be faster and more secure if provided by an inverse time zero sequence current relay rather than definite time. Would the authors give their opinion of these comments?

A high resistance line to ground fault could be located in Figure 3 so as to reduce receiving end faulted phase current to near zero. I believe the phase selector principle described in the paper remains effective. Do the authors agree? If the phase selector is not included, will the sequential relaying properly identify the fault type in a similar situation?

Zero sequence current compensated ground distance relays utilize the ratio of line zero and positive sequence impedance. This ratio increases when the line is series compensated and this should be considered.

Ground faults predominate on EHV lines, and the speed and selectively possible with relatively simple instantaneous channel independent overcurrent relays should not be overlooked.

100 to 300 MVA of shunt reactors may be used on a 500 KV 3000 MVA line. Can one relate the 20% sensitivity figure and maximum expected reactor inrush to determine whether switched reactors would be a problem?

Would the authors elaborate on the requirement for 50 msec faithful CT response? Is the concern Zone 1 underreach, overreach, or loss of directional integrity? This does not seem to be an onerous requirement when current is limited by the impedance of 100 or more Km of line.

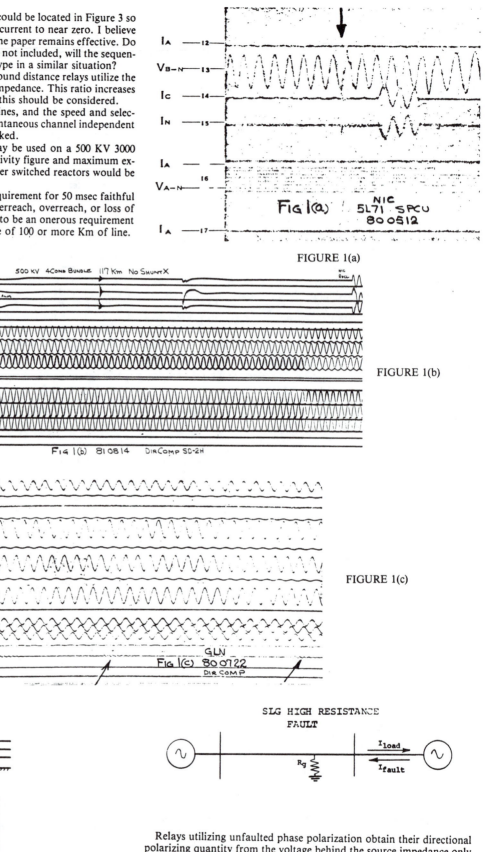

FIGURE 1(a)

FIGURE 1(b)

FIGURE 1(c)

FIGURE 2

Relays utilizing unfaulted phase polarization obtain their directional polarizing quantity from the voltage behind the source impedance only if there is no pre fault load current. Load may be a significant factor in a fault, swing, unsuccessful reclose sequence of events. Based on steady state analysis of a Zone 1 distance relay applied to a series compensated line, dependability not security is enhanced by use of a proper polarizing voltage. How are three phase faults identified and what is the concern? Delaying Zone 1 operation until capacitor gaps flash suggests

Zone 1 settings based on the uncompensated line impedance. Does the user have to ensure flasing of protected line gaps in order to prevent overreach on external faults? This would seem to defeat the intent of the series capacitors, and in fact there is a trend to use ZnO limiters in place of gaps to ensure that sound line capacitors remain in service during disturbances.

Have the authors ever considered the potential advantages and problems of building a distance relay incorporating a true replica line impedance including replica capacitor protection?

On a transient basis, the integral and time derivative of current initially have the same polarity and hence the voltage across capacitive and inductive reactances are initially the same polarity. This suggests that faults on series compensated lines may briefly appear to be further away rather than closer as indicated by steady state analysis. Is this significant to the independent or dependent mode DWD settings?

As a final comment, I would suggest that relaying of series compensated systems is an overreaching channel dependent proposition. A Zone 1 with a secure and useful reach through capacitor banks to coordinate with adjacent Zone 2 relays is desirable but appears difficult to impossible if Zone 1 settings must be based on compensated line impedances and if adjacent line Zone 2 seetings must be based on uncompensated line impedances. Logic to enhance the selectivity of channel independent protection in the event of a channel failure would be a distinct asset. Would the authors see any advantage in adjusting the Zone 2 time downwards or enabling Zone 2 time trip if there is an indication of channel failure?

REFERENCES

[1] M. Vitins, "High Speed Transmission Line Relaying" 1980 Western Relay Conference, Spokane Washington
[2] A. M. Bignell, "Further Thoughts on Ultra High Speed Relaying of Transmission Lines" 1982 Western Relay Conference, Spokane, Washington

Manuscript received February 10, 1983.

C. J. Pencinger (Brown Boveri Corporation North Brunswick, NJ):
I would assume that the application of this relay would be for an EHV line 230 kV and above. However, the combination of a UHV relay with a switched distance relay is a sharp contrast. The switched relay is normally applied at lower voltage being an economic compromise because of its slower speed (2 cycles) and low cost. Why doesn't the author consider using a non-switched relay for backup and thereby improve operating times for zone 1 and high resistance faults to one cycle or less? Also, the dependability of the backup would be considerably improved with a non-switched distance relay having separate measuring systems for each phase.

Manuscript received February 23, 1983.

M. Haenggli (Brown Boveri & Company, Ltd. Baden, Switzerland):
The authors have presented a promising protection systems which takes advantage of multiple, partially redundant functions to provide superior performance. The selected measuring principles allow severe faults to be detected in a few milliseconds while only complex fault situations, which in general, cause less harm to the power system, require more time to be detected reliably.

The discusser believes that problems can arise in the coordination of the different measuring systems. It is not clear from the paper how the relay will react in the following situation of an evolving fault:

Assume that a fault occurs on a parallel line. The directional wave detectors (DWD) on the faulted line then detect the fault correctly. On a healthy line, the fault is detected as being outside the protected zone. Now the fault evolves and causes a single phase fault on the hitherto healthy line, thus producing an internal fault on this line. Could the authors explain by which criteria the phases to be tripped are selected?

The authors mention that the relay system is suitable for automatic reclosing. If upon a reclosure a fault continues to be present, do the undirectional impedance starters lead to a trip as in the case of manual closing when the line is being energized?

A very interesting feature of the directional wave detector is its ability to detect a broken loaded conductor before a shunt fault results. The presented relay seems not distinguish between a broken phase conductor and any ohter kind of fault, thus an automatic reclosing will occur. In this case, the broken line will be tripped definitely only if the impedance starters or the neutral current measuring unit detect the resulting shunt fault (or an asymmetry), despite the directional wave detector initially has detected the broken conductor.

The relay presented by the authors utilizes favorable the information contained in current and voltage signals by using different measuring systems that support mutually.

It seems, that for line protection, the integration of multiple measuring principles could offer significant advantages for the future.

For this kind of a relay system, the discusser suggests to use a microprocessor instead of several different analogue meausuring units. Until now, the realization of microprocessor controlled relays was too expensive because fast and highly resolving A/D converters are needed to digitize the incoming analog data. In the presented relay, the same data is used to perform multiple measurements and a relatively expensive data aquisition is justified. It is our experience that the high speed needed for the directional wave detector can be achieved by using parallel microprocessors (for example one per phase). The costs of such a relay need not to exceed the costs of a conventional static relay because all further measuring algorithms can be implemented without any additional hardware.

Manuscript received February 23, 1983.

A. T. Giuliante, G. Stranne, C. Ohlen, and **R. R. Slatem:**
The authors appreciate the discussors' thoughtful comments and their questions relating to various aspects of this new relay. the described relay in a complete, integrated relay terminal providing UHS and conventional distance protection plus redundant backup protection for all types of faults and system configurations. The apparent complexity as reflected in several of the discussors' questions should thus be viewed in terms of the application problems associated with a complete protection terminal. In this respect, the discussions illustrate the very broad range of competence required of today's relay application engineer.

Since the initial testing in 1975, ASEA has accumulated more in-service experience of relays based on travelling wave measurements than any other manufacturer and today appears to be the only manufacturer with commercial offerings of travelling wave relays. These designs are based on extensive power system and relay computer-modelling that proved the validity of the concept. Staged fault testing by the Bonneville Power Administration and others further verify these designs. Additional installations on the BPA and other systems have confirmed the advantages of the DWD principle used in pilot scheme arrangements. With regard to back-up protection, another operating principle is required which has been considered in this new relay system.

Several of the discussors commented on various aspects of dependability and possible disabling failure modes of this new relay system. Loss of auxiliary power or failure of the output electromechanical relay are the only common total failure modes visualized. Failure of other individual components will not result in complete failure because of the mutally supporting measuring techniques.

This may not be fully evident from Figs. 1 and 6 because of the simplified illustration used for the sake of brevity and clarity of main purpose. A design objective has been to provide more than one type of measuring system in the relay so as to implement in one relay package, the dependability philosphy of providing dissimilar, redundant protection for EHV and other strategically important circuits. The single measuring element in the sequential measurement distance relay section does not represent a lack of dependability as is occasionally suggested. In fact, the overall reliability of this technique is perhaps better than some other commonly used solutions. In a parallel measurement system utilizing three directional distance relays for phase-phase faults and three additional directional distance relays for L-G faults, only one relay responds for most single fault conditions. An exception is the L-L-G fault where more than one relay may see the fault. However, for this fault, overreaching may occur for certain system fault locations, while on the other hand, a fault at another location may not be seen by any ground relay. Thus, a multiplicity of individual relays does not necessarily enhance either the dependability nor the security aspects of relay reliability. When selective single pole tripping is required, additional demands are placed on the relay system and relay reliability becomes still more important to the continued operation of the power system. The described relay includes facilities for selective single pole tripping. It also includes protection against maloperation due to current reversal upon clearing an external fault, saturating CT's, and a solution to the weak infeed problem. With these features included in the manufactured product, system hazards during commissioning are reduced because the relay system can be thoroughly checked out before actual installation in the field. the performance of the communication channel also affects relay reliability. The conventional blocking mode can be used with this relay. However, security is enhanced by using the permissive or unblocking modes which are therfore preferred. Any loss in dependability in these modes, compared for example with the blocking mode (there is always a trade-off), is compensated for by redundant features in this new relay design.

In respect to questions on operating times, those quoted are typical and adjustable over wide ranges. The mentioned 85 ms for a "1st zone" time is suggested for series compensate lines to prevent overreaching before the gaps flash over. This can be reduced to 35 ms for uncompensated lines. The 4 ms shown in Fig. 6 for the DWD includes the 3 ms output tripping relay whereas the text mentions a 1-2 ms "decision time" for the DWD section that does not include the output relaY. The indicated 12 ms for the dependent mode will, of course, be a function of the communication channel speed. The shown delay for high resistance faults is not critical and is needed only when surge arrester power follow current within the protected zone·or series capacitor gap flashing are considerations.

Several questions related to settings. These are determined based on conventional 60 Hz (50 Hz) calculations. There is no travelling wave theory required in the application of the relay. The characteristics of any component within the protected line section, however, must be considered. This would include energizing inrush to a shunt reactor or a Tee'd transformer, the power follow current of a surge arrester, or the gap flashing across a series capacitor. When determining the ΔV's and ΔI's for any of these internal non-fault events, or for any fault condition, it is the *phasor* change in V and I which should be determined since the relay responds to this change and not to just the change in magnitude. It is this phasor characteristic which makes the relay sensitive to high resistance ground fault since these faults may result in negligible change in voltage magnitude. The settings in the impedance section of the relay do need consideration of maximum load flow as noted by some of the discussors. When this is a problem resulting from a power swing due to an external fault condition, an optionally available power swing blocking function may be required. However, the conventional conflict between impedance relay loadability and fault sensitivity is minimized as the result of two features in the relay. First, as shown in Figs. 2 and 3, the characteristic has more loadability for a given allowance for fault resistance than provided by some other commonly used characteristics. Also, this reach in the resistive direction is fixed, it is not a function of source impedance, thus no allowances need be made for a variation in reach during a system disturbance. Secondly, the impedance section need not be used for low current ground faults. The DWD section provides the primary protection and the equally sensitive back-up is adequately provided in the "$3I_0$" function as shown in Fig. 6.

The swithc-into fault (S-I-F) protection feature used with line side VT's, is armed upon loss of three phase voltage and remains armed until after voltage restoration. Any fault during this period beyond the CT location will be recognized by the starters and give a trip output within 10 ms. When bus side VT's are used, the DWD section will automatically provide the S-I-F protection.

"Evolving faults", "sequential faults" and "cross-country faults" are of concern to protection engineers. These are not new protection problems, they are just discussed more frequently today. Electromechanical relays appear to have given good performance over the years during these infrequent events for three reasons: 1) These events do not occur very often, 2) The use of automatic reclosing has minimized the consequences of a malfunction due to these more complex fault situations, and 3) Many of these faults are not as complicated to detect properly as appears at first glance. To the extent that these faults can be specified and yield to calculations, the new relay can be set to respond correctly. However, priorities for other conditions must also be considered. For example, security against overtripping due to current reversal upon breaker opening on an ex·ernal fault may take precedence over fast clearing of the second fault of a sequential fault sequence.

Mr. Smith's comments regarding improved system stability are pertinent both from the ultra-high speed capability standpoint and the selective single pole trip capability of the relay. The cost advantages discussed could be equal to the cost for a 1 cycle breaker speed difference, e.g., 3 cycle breakers with the ultra high speed relay compared to 2 cycle breakers with conventional relays. More pronounced is the most advantage when 1 cycle breakers are otherwise required from a system stability standpoint, i.e., with a fast relay, the same degree of stability is ob-

tained with a slower, lower cost and possibly a more relible breaker. We offer an additional reference, "Feasibility and Economics of Ultra High Speed Fault Clearing" (1) by K. L. Hicks and W. H. Butt, which evaluates in more detail these economic considerations. The steady state suppression filters are based on the theory that any change in an electrical network requires additional higher and lower frequencies than the fundamental for a brief period of time to effect the change.

These nonfundamental frequency components completely describe the change taking place in the fundamental component. A slow change, as in a power swing or a small change such as a load variation, results in negligible "sidebands". Thus, by suppressing the steady state component, the relay can be made responsive only to changes (eg. faults) and not to loading regardless of its level. In regards to the R and X reaches for the zones of the distance relay in the sequential measurement system, these are established for one of the zones and the other zone reach is determined by changing the percentage of system voltage applied to the measuring element. Thus the ratio of the X to R reach is the same for both zoneș. The time at which the setting is changed from zone 1 to zone 2 is set in a timer controlled by the starting elements.

Mr. Moncrief's questions on switch-into-fault protection coverage, common failure modes, settings and communication preferences were treated above in response to those questions asked by more than one discussor. In response to Mr. Moncrief's other question on CT matching on reactors, there is no critical requirement since it is generally only necessary to reduce the influence of the reactor currents, not eliminate them entirely. One would generally pick a reactor CT with the same turns ratio and voltage classification as the main line CT's. Routine calculations as made for any CT application should suffice to confirm the adequacy.

Mr. Sage's questions regarding operating times were answered above. Improved nonmenclature is admittedly needed to clarify some of the operating features of this new relay. One generally thinks of a first zone as having no intentional delay. In this case the 85 ms "first zone" of the distance relay is actually a backup function and as such the operating time has been deliberately extended 50 ms to accommodate series capacitor gap flashing. This time delay can thus be eliminated for nonseries-compensated lines. The 50 ms CT accuracy requirement is not directly related to this "first zone" operating time. Other features in the relay come into play after the 50 ms which reduce the adverse effects of CT saturation.

The use of the independent DWD mode (4 ms in Fig. 6) for close-in faults is generally set with a 100% non-trip margin for symmetrical faults in the next line section. Its effectiveness for seeing close-in faults will then be a function of the source-to-line impedance ratio (SIR) and can be calculated conventionally. Concerning heavy line loading, Mr. Sage is partially correct in regard to setting limitations. The DWD section is not influenced by loading, since it responds only to the changes in voltage and currents brought on by the fault. The distance measurement is affected, but the shape and stability of the quadrilateral characteristic minimizes the loading effects which he notes as common to all distance relays.

We beg to differ with some of Mr. Elmore's comments. The purpose of the new relay is not to enhance the dependability of the authors' DWD relay, the RALDA. As noted above, this relay has a remarkable in-service record since its initial field testing in 1975. The purpose of the new relay, the RALZA is to integrate the DWD concept with distance relaying concepts to provide a complete factory wired and tested multizone UHS relay terminal. The resulting mutually supportive characteristics provide a relay system with dependability and security comparable to any hand tailored scheme for any application. Specifically, the DWD unit, as shown in Fig. 6, consists of three sets of level detectors, Not shown is that these are on a per phase basis and that each responds to phase *or* ground faults involving the respective phases. Thus, while Mr. Elmore's comment is correct. the redundancies inherent in the system makes his point moot. In respect to the single measuring unit in a sequential measurement scheme, as noted above, this approach is *not* inherently less reliable than a multiplicity of measuring units. Attention to engineering detail and manufacturing quality are more important to assuring a reliable system. We agree with Mr. Elmore's second point that faster relaying does not assure faster interruption, per se. But we disagree with the conclusion he reaches. In the time frame of interest, Thorp, *et al* in their referenced paper clearly point out the benefit of faster relaying. In discussing their Fig. 1, they state, with 2 cycle breakers," —reducing relay time 0.5 cycles from 1.5 to 1.0 cycles results in only a 0.2 cycle reudction in fault clearing—". However, the last part of this same sentence states, "but an additonal *0.5 cycle reduction (from 1.0) to 0.5 cycle relay time produces a 0.75 cy-*

cle reduction in fault clearing.'' They further state, ''as the system X/R ratio increases—a small increment in relay time for a given circuit breaker can produce large changes in fault clearing time''.

Several of Mr. Crockett's comments have been responded to above. Points not covered include suggested use of replica impedance in the DWD section of the relay. A replica impedance approach, which has been carefully studied by the authors' company, invovles complexities such as series capacitors and gap-flashing characteristics which may be difficult to implement, but are to be considered in the replica principle to enable its use. The replica approach for travelling wave relaying to provide a distance measurement does not appear to provide a practical solution at this time.

In the three actual oscillograph examples Fig. 1 a, b, and c given by Mr. Crockett, the RALZA relay would have no problems in correctly discriminating the internal and external faults. The reason being that there are time, amplitude and sequence (TAS) criteria applied to the three levels of DWD measurements plus the impedance back-up elements. It should be pointed out that the neutral current element supervises the most sensitive DWD mode and does not prevent operation of the other DWD modes, the settings of which are higher. The DWD requires simultaneous changes in phase-current and phase-voltage for operation. These are not present in Fig. 1c. In Fig. 1a, the B-N flash-over 2.5 cycle sprior to the C-N fault would probably start the most sensitive DWD elements but tripping would not be affected since neutral current does not last 20/40 ms. The C-N fault that follows is heavier and would be cleared by the dependent DWD mode. In Fig. 1b the changes in the voltage traces after breaker opening do not coincide with any simultaneous current-changes, which would make the DWD not respond to this non-fault event.

The phase-locked loop memory oscillator will not lock in on the spurious events shown in the oscillograms and will not cause reliability problems as suggested by Mr. Crockett, since the memory is only used for three phase faults for the impedance portion of the relay, not for the DWD's. Evolving fault elements are included in RALZA to handle sequential faults. Thus, Mr. Crockett's concern as to the reliability of the relay response to situations illustrated by his oscillograms is due largely to the brevity of detail in our paper.

Series capacitor gap flashing from faults, post-fault power swings, or simple overloads are all accommodated within the relay logic. The necessary relay setting determinations to avoid undesired tripping for these unsymmetrical line conditions are based on straight forward 60 Hz calculations. The transient build-up of current in a faulted line with series capacitors does not complicate the application or settings. The noted 20% senstivity is for each relay terminal. This can be supplemented by the weak-infeed logic mentioned in the paper. Thus the sensitivity is limited only by the need for a 20% rated current change at only one terminal. Most transmission relays use fixed time delays rather than inverse time and this new relay is based on that philosophy because of coordination advantages. We question if a break in one conductor of a bundle (without a fault) could activate any relay. The change in impedance of that phase would be negligble in view of the frequent paralleling of conductors by the mechanical bundle spacers. We also question the comment that the relay sensitivity is limited by the DWD section and that zero or negative sequence relays are more sensitive. The DWD responds only to changes in phase current and voltage which includes all symmetrical components. It is difficult to visualize how use of the total current which includes synchronizing currents, load currents, chaging currents, and reactor exciting currents could result in a more fault sensitive and secure relay. The problem posed with Fig. 3 where-in at one terminal the sum of the high resistance fault current and the load current is nearly zero, fails to recognize that the DWD responds to the *phasor change* in current (and voltage) and the magnitude of the resulting total current is of no concern. We do agree with the stated concepts of simplicity. However, protection with only simple overcurrent relays can seldom provide complete protection and one must then make the decision of how to provide for the remaining protection needs. the described relay has the needed features for practically all EHV situations and the flexibility for any specific situation. Mr. Crockett and Mr. Hanggli suggest other approaches to reliable high speed line protection. Obviously all manufacturers are constantly examining new methods for improving their product line. Mr. Crockett's suggestion for reducing the relay time settings to compensate for loss of channel is practical and easy to implement when desired. The distance relay section of the new RALZA is based on the RAZOA distance relay in which this adaptability to external conditions includes both time and reach.

We agree with Mr. Pencinger's comment that switched relays are not generally used on EHV because of their two cycle repsonse, but disagree on his dependability comments as we have noted above. We distinguish between electromechanical ''switched'' relays and ''sequential measurement'' relays in the degree of sophistication incorporated in the logic made possible by a 100% static approach. The described relay uses a sequential measurement technique in the distance relay section for backup functions only. As such, the slightly longer operating times are not objectional and are, in fact, needed in most cases as noted in the paper. Simultaneous measurement relays are also offered by the authors' company with operating times as short as ½ cycle. These can be used to augment the DWD for the backup function if one so desires. However, time delay for high resistance faults may still be necessary if a tripout due to power follow current in a surge arrester within the protected line section is to be avoided.

Mr. Hanggli poses an interesting sequential fault problem which was discussed above. But we wish to further clarify that the selective single pole logic results in a single pole or a three pole tripping. There is never a two pole tripping. Thus any fault or fault sequence involving more than one phase as seen by the DWD will be tripped three phase. Clarifying other comments, the switch-into-fault feature is effective regardless of what control action causes the breaker to close. the broken conductor reclosing could be solved by supervising the reclosing initiation with a zero sequence overcurrent relay; no severe ground fault current during a single phase abnormality, no reclose. Mr. Hanggli's other comments are appreciated and confirm that the evolution of protective relays is proceeding in all countries.

REFERENCE

[1]Hicks K. L. and Butt W. H. ''Feasibility and Economics of Ultra-High-Speed Fault Clearing'' presented at the IEEE PES Winter Power Meeting, New York, N.Y., February 3-8, 1980, F80 239-4.

Manuscript received April 4, 1983.

A CURRENT DIFFERENTIAL RELAY SYSTEM USING
FIBER OPTICS COMMUNICATIONS

Shan C. Sun, Senior Member, IEEE

Roger E. Ray, Member, IEEE

Westinghouse Electric Corp.

4300 Coral Ridge Drive

Coral Springs, FL. 33065

ABSTRACT

Relay systems have made use of metallic circuits for communications entering the power station environment. These circuits have at times affected the system reliability due to extraneous voltage influences. The system described makes use of a unique fiber optic channel in place of the metallic circuit. The system takes advantage of the glass fiber's virtual immunity to the effect of extraneous voltages. The relay system also uses new techniques for its sequence network and its current comparison method. These improvements increase the relay system reliability and broaden the area of application.

INTRODUCTION

Many relay systems in the past have made use of a pilot wire over which information is transmitted to determine if the relay system should trip for a fault condition. [1] These systems use a continuous metallic circuit between the substations. This makes the channel susceptible to extraneous voltages, such as longitudinally induced voltage and station ground mat rise.

These outside influences have at times impaired the reliability of pilot wire systems. The system described in this paper, known as the LCB, makes use of a fiber optics channel between the line terminals to solve these problems. The system not only makes use of a new channel, but new techniques are applied to the sequence network and current comparison circuits.

The new relay system should be more reliable because the optics channel is not affected by outside voltage influences caused by the fault. The sequence network used is simpler and more accurate than any previously known designs using either passive or active components. Therefore, the system should be more secure for faults external to its zone of protection, and more dependable for internal faults. Another factor which will help increase the overall performance is that the relay system design considered the total problem of relaying as well as communications.

BACKGROUND AND RELAY REQUIREMENTS

The years of experience with pilot wire relaying have made it evident that if the problems of the pilot wire could be solved then the overall system performance would be improved. Fully 90% of the causes of false operations or failures to trip are in

82 SM 397-8   A paper recommended and approved by the IEEE Power System Relaying Committee of the IEEE Power Engineering Society for presentation at the IEEE PES 1982 Summer Meeting, San Francisco, California, July 18-23, 1982. Manuscript submitted February 2, 1982; made available for printing April 26, 1982.

one way or another related to the pilot wire. The fiber optics solution then presents itself as the solution to be considered for a new type of line protection system using a differential current technique. In the development of the new system the use of fiber optics for the channel was to be the prime consideration; however, it should not be the only aspect to consider. Some of the other items to consider in the design of a new system are:

a) The relay system should have very low burden since burden affects current transformer performance. It is, of course, very important that the current transformers not saturate during external faults when using a current differential system.

b) The sequence network used in the relay must have an accurate response and more importantly a repeatable response from unit to unit. Also the transfer angle, that is, the angle between the input current and the output voltage should be independent of input current magnitude. Also it is known that a sequence network designed for the power system frequency will have an error in its output due to other frequencies. This error cannot be removed, but it is very necessary that the sequence network have a consistent error for relays at different ends of the same line.

c) The range of linear comparison should be increased over the present pilot wire systems. Although this is not a main cosideration it would be desirable.

d) Due to the type of channel (other than fiber optics) which may be used with this relay system the channel delay may be significant, and a phase delay circuit must be considered. The type of delay circuit used must have equal time delay for a wide range of frequencies other than the power system frequency. This is to say the delay circuit must delay a transient but also accurately reproduce it at its output. Also an arrangement must be provided by which two incoming channels can be equalized in time for three terminal line applications.

Items a to d above were some of the points considered in the design of a new relay system. Items a and b point out that the input network design should be solid state since all sequence networks in the past that used iron have suffered from these problems. Also it appeared that a new technique should be considered so that positive, negative, and zero sequence sensitivity be controlled independently. The following sections of this paper will dicuss the unique portions of the relay system.

RELAY SYSTEM DESCRIPTION

A functional block diagram of the relay system is shown in Figure 1. A low burden, wide range (100 p.u. symmetrical, 1 p.u. = 1A or 5A) current transformation package supplies the local three phase line current information in voltage form to a precision composite sequence network. The sequence network output characterizes the three phase current in a single

Reprinted from *IEEE Trans. Power App. Syst.*, vol. PAS-102, no. 2, pp. 410–419, Feb. 1983.

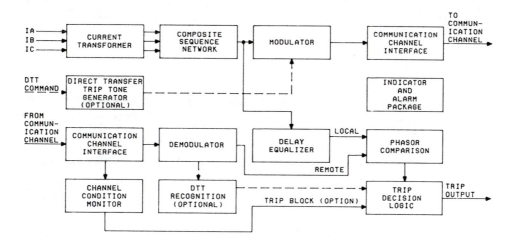

FIGURE 1 FUNCTION BLOCK DIAGRAM OF LCB RELAY

phase form. The modulation circuit converts this information into an FM like signal which is the input to the optical transmitter. At the receiving end, the optical signal is first converted back to an electrical signal before the demodulation process. The delay equalizer circuit further conditions the local sequence network output to reflect the same transfer characteristic of the modem process of the remote quantity. The two sequence quantities are therefore very closely matched in terms of the frequency and time responses prior to the comparison process. The comparison circuit performs a true phasor evaluation of the local and remote quantities and outputs the results to the trip decision circuit.

A direct transfer trip (DTT) feature is also incorporated in the design as an option. The DTT is a priority function and when activated, the local quantity transmission is disabled. In its place, a coded tone signal signifying a DTT command is transmitted.

The overall relay system has been designed to accommodate either two or three terminal lines. For three terminal applications, a second modem channel can be added.

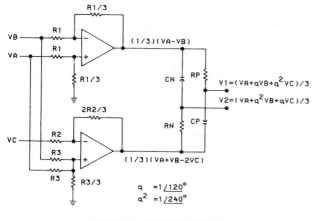

$$V1=(VA+qVB+q^2VC)/3$$
$$V2=(VA+q^2VB+qVC)/3$$

$$(1/3)(VA+VB-2VC)$$

$$q = 1\underline{/120°}$$
$$q^2 = 1\underline{/240°}$$

FIGURE 2 SEQUENCE NETWORK

SEQUENCE NETWORK

The sequence network (Figure 2) implemented in the LCB relay uses a simple circuit theory developed recently [2]. The sequence network completely eliminates the phase and magnitude dependency and sequence purity problems found in the traditional design. Furthermore, this design supplied both

the positive and negative sequence detections independently from one basic circuit. Figure 3 shows a diagram of the overall composite sequence network.

The sequence network will provide accurate linear sequence responses for current magnitudes up to 25 p.u. The performance is fully repeatable and consistent from unit to unit.

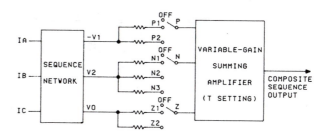

FIGURE 3 COMPOSITE SEQUENCE NETWORK

DELAY EQUALIZATION

A reliable performance of a differential comparison relay depends heavily on the ability of the design to bring the remote and local signals together with a response agreement, especially during transient conditions. In the LCB relay the local signal prior to comparison is conditioned by a series of circuits nearly identical to the ones needed to process the remote signal. This design practice ensures the integrity of the differential comparison, because the local signal now has the same transient response as the remote signal. However, the real time delay occurring in the communication channel must also be provided for in the local signal by an adjustable, distortion-free delay equalization circuit. Designs that provide ideal time delays are very involved and not practical for this application. A much simpler approach (Figure 4) consisting of sectionalized all-pass delay networks has been developed to provide the time delay compensation. The design supplying an adjustable delay time up to 8 ms is similar to a

155

lumped-parameter delay line circuit. It exhibits a linear phase (constant time delay) characteristic over a wide frequency range. A similar design is also used for equalizing the remote signals of a three terminal line application.

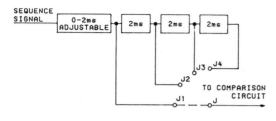

FIGURE 4 DELAY EQUALIZER

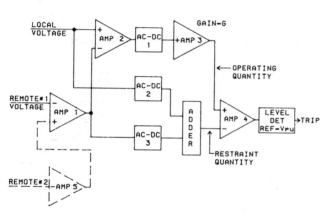

FIGURE 5 COMPARISON CIRCUIT

COMPARISON TECHNIQUE

A block diagram of the comparison circuit is shown in Figure 5. For the comparison process, two quantities are generated from the local and remote voltages. The first is called the operating quantity and is derived by the phasor addition of the local and remote voltages. This addition is performed by differential amplifier 2 in Figure 5. The inputs to the amplifier are the local voltage into the non-inverting input, and the remote into the inverting input after passing through inverting amplifier number 1. Thus, the output of amplifier number 2 is the phasor sum of the local and remote voltages. The operating quantity is given by equation 1.

$$V_{op} = |\bar{V}_L + \bar{V}_R| \qquad (1)$$

where V_{op} = the operating quantity
  V_L = the local voltage
  V_R = the remote voltage

The magnitude of the operating voltage is obtained by full wave rectifying the ac voltage and filtering to obtain a dc voltage proportional to the incoming ac. This is accomplished by the AC-DC circuits number 1. The restraint quantity is obtained by adding the local and remote voltages on a magnitude basis, thus phase angle does not enter into the result. In Figure 5, AC-DC

circuits numbers 2 and 3 are used to obtain the magnitude of both the local and remote voltages. The sum is then obtained in the adder amplifier. The restraint quantity is given in equation (2).

$$V_{res} = |\bar{V}_L| + |\bar{V}_R| \qquad (2)$$

where V_{res} = the restraint voltage

The quantities V_{op} and V_{res} must now be compared in some way in order for the relay to determine if the fault is internal or external to the protected line. Equation (3) shows how this comparison is accomplished.

$$V_{op} \times G - V_{res} \geq V_{pu} \qquad (3)$$

where G = a constant: G = 1.4
  V_{pu} = a preset pickup threshold

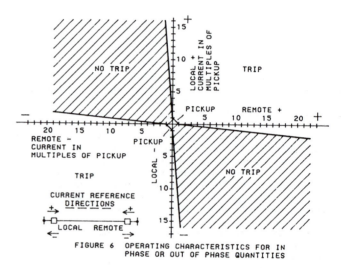

FIGURE 6 OPERATING CHARACTERISTICS FOR IN PHASE OR OUT OF PHASE QUANTITIES

Amplifier numbers 3 and 4 and the level detector shown in Figure 5 perform the function of equation (3). The level detector has its reference set at a value equal to V_{pu}. Figure 6 represents the characteristic of the comparison circuit if the fault currents at the two ends of the line are either in phase (internal fault) or out of phase (external fault). Since many times the currents at the two ends are not exactly in or out of phase, Figure 7 shows the relay characteristic when the local (V_L) is at three times pickup, and the remote varies in phase and magnitude. A family of curves similar to Figure 7 could be drawn for different magnitudes of local voltage. It can be seen from Figure 7 that the relay system will operate for small amounts of out-feed at the remote terminal. This is the differential aspect of the device, and in weak terminal applications this out-feed trip capability is valuable. Also Figure 7 shows that the relay has about an 82 degree characteristic, that is, the two quantities can be up to 82 degrees out of phase and the relay will trip for any set of magnitudes. This angle characteristic can be varied by changing the value of G in equation (3).

Three terminal lines can be protected by the addition of the dotted blocks shown in Figure 5. Again the operate quantity is the phasor sum of the local plus both remote voltages, and the restraint quantity is the magnitude sum of the local quantity and the vectorially combined remote quantities.

MODULATION TECHNIQUE

Optical fibers are know to have superb characteristics suitable for signal transmissions of any nature. However, precision

156

amplitude modulation designs using electro-optical components and fiber optical links are still difficult to achieve. This is due to the lack of consistent transfer characteristics in electro-optical components presently available as well as the losses in the fiber. In this respect, fiber optics offer no prevailing advantage over conventional communication channels. For applications such as in the LCB relay, where the sequence quantities have to be accurately transmitted and recovered, selections of a suitable modem scheme are limited to either a digital or an FM scheme. In either case, the resultant signal transmission in the optical fiber is an on-off light. A digital approach would have been ideal if the communication channels for the LCB were exclusively fiber optical. To enhance the application flexibility, communication links such as leased data channels had to be considered. These other channels, unfortunately, do not permit the data rates needed for sending signals with fidelity by digital schemes. This limitation leaves FM as the only choice to meet the need. Common FM designs, however, are also inadequate to provide the needed speed and accuracy. This is again due to the bandwidth limitation in the standard leased data circuit (3002, unconditioned or equivalent channel). To circumvent the difficulty, a pulse period modulation (PPM) scheme was developed. The process is done entirely in the time domain. All design parameters are precisely under control producing a performance that meets all the objectives.

PPM, where the carrier time period is varied linearly to the modulating signal amplitude, is similar to FM when the deviation ratios are small. A prominent merit of the PPM scheme is the comparatively miniscule filter requirements in the demodulator. A sample-and-hold process implemented in the demodulator virtually removes the carrier frequency content without the need of a filter. This feature greatly enhances the relay speed and also simplifies the circuit that is needed to compensate the local sequence signal.

Diagrams illustrating the PPM operation are shown in Figure 8 and 9. The modulator formation is based on the charging time of capacitor Cl under a constant current I_1. The voltage on Cl is compared to the modulating voltage V_1. When the voltage on Cl reaches V_1, flip-flop FF1 is toggled and Cl is reset, commencing another charging cycle. This encoding operation is performed at 27.2 kHz center frequency which is subsequently counted down by 16 to 1.7 kHz for transmission. Performing modulation at a higher frequency permits a more practical design. This choice also produces a digitally selectable center frequency feature which may be beneficial for applications in the future.

The PPM modulation can be characterized by the equation,

$$T(t) = T_0 + K_1 \times V_1(t) \qquad (4)$$

where
$T(t)$ = period of modulated carrier
T_0 = period of unmodulated carrier
$K_1 = Cl/I_1$, a design constant
$V_1(t)$ = the modulating voltage
$T(t)$ and $V_1(t)$ are time functions.

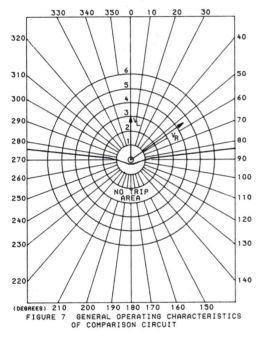

FIGURE 7   GENERAL OPERATING CHARACTERISTICS OF COMPARISON CIRCUIT

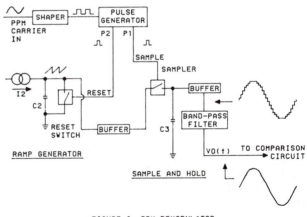

FIGURE 9 PPM DEMODULATOR

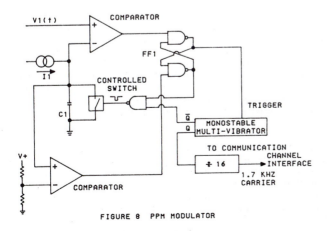

FIGURE 8 PPM MODULATOR

In the demodulator, the incoming carrier is first shaped to a square wave. Two pulse signals, P1 and P2, generated from transitions of the square wave are used respectively for the sample-and-hold operation and capacitor C2 reset function. Constant current I_2 charges C2 to produce a linear ramp voltage. Prior to reset (P2 function) of the ramp, C2 voltage is sampled (P1 function) and stored in capacitor C3. C3 holds the sampled ramp voltage until the next P2 pulse appears. Assume that the modulating signal in the carrier is a 60 Hz sinusoidal wave then the recovered signal at the holding circuit output will be a stepped 60 Hz sinusoidal wave carrying very little residual carrier frequency components. A simple filter will remove the steps accurately recreating the original signal. The demodulator operation is given by the equation,

$$V_0(t) = K_2 \times T(t) \qquad (5)$$

where $K_2 = I_2/C2$, a design constant.

157

To demonstrate the overall PPM process, one can substitute $T(t)$ by the expression given by equation (4). Then

$$V_o(t) = K_2 \times T_o + K_1 \times K_2 \times V_1(t) \qquad (6)$$

$K_2 \times T_o$ is a constant voltage term and can be removed by a simple highpass filter leaving $V_o(t) = K \times V_1(t)$. Where $K = K_1 \times K_2 = C1 \times I_1 / C2 \times I_2$, a constant term representing the gain of the PPM process.

It is noted that the demodulation is performed on each half of the carrier signal. This yields an effective sampling rate of 3.4 kHz, or approximately 55 samples per cycle of a 60 Hz frequency. This is an adequate rate to ensure a good reproduction of the original modulating signal.

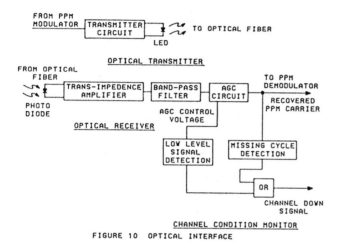

FIGURE 10 OPTICAL INTERFACE

OPTICAL COMMUNICATIONS INTERFACE

The optical channel interface comprises an optical transmitter, an optical receiver, and a channel monitor as shown in Figure 10. The transmitter circuit obtains the driving signal from the PPM modulator circuit. This signal controls a transistor switch to turn a light emitting diode on and off at the PPM carrier rate.

The key element in the receiver circuit is a photo PIN diode. This diode implemented in a trans-impedance amplifier circuit operates in the photoconductive mode. It has a conversion gain of about 0.5 ampere per watt of optical power. The trans-impedance amplifier consists of two cascaded non-linear amplification stages providing a maximum gain of 4.4×10^7 volt per ampere. The output from this circuit is conditioned by a bandpass amplifier followed by an automatic gain control (AGC) circuit to supply a relatively constant amplitude carrier signal to the PPM demodulator circuit.

The channel condition monitoring section comprises two detection circuits. The low carrier signal detection uses the AGC control voltage to detect carrier signals that are below a preset level. A low carrier signal signifies a malfunction in the channel. Since the AGC voltage is a relatively slow responding signal, this detection is implemented primarily to provide an early warning indication that problems are developing in the channel. Adequate margin has been given in the design to accommodate the time delay effect in the AGC voltage. The missing cycle detector, on the other hand, is a fast responding detection circuit. By sensing the carrier signal directly, any fast interruption in the channel lasting for more than one carrier cycle will activate this circuit instantly. The two detection outputs are combined to produce a channel malfunction signal sent to the relay circuit. Upon receiving this signal the relay circuit immediately discards the remote signal and trip blocking is imposed. The relay under this condition can still be used as an overcurrent function using only

the local sequence quantity if desired. Further discussions concerning this feature are given in the Application Consideration section.

DIRECT TRANSFER TRIP

The direct transfer trip (DTT) function is supplied as an option. The design consists of a transmitter and a receiver. When activated, a tone signal of 400 Hz synchronized with the carrier is produced to substitute the sequence signal as the new modulating signal. While the DTT function is being sent the sequence signal is blocked from the modulator. The modulation deviation during DTT is greater than the maximum deviation allowed for the normal sequence signal operation. This implementation requires that the DTT detection in the receiver must statisfy both the magnitude and frequency of preset values to be recognized as a correct signal. Furthermore, the correct signal must last a minimum of 12 ms to allow a trip.

The DTT trip and the differential comparison trip are mutually exclusive functions. Separate trip indicators are provided for the two functions.

APPLICATION CONSIDERATIONS

GENERAL

The relay system described can be applied to either two or three terminal lines. It is of no concern as to whether or not a terminal is weak since the relay system does not require a contribution from all terminals for an internal fault. The only requirement to trip all line breakers is that one terminal of the line be a strong source which produces current above the pickup of the relay.

It is always preferable that the transmission line current transformers have the same ratio at both ends of the line. However, if this is not possible the relay system may be set to overcome as much as a three to one difference in current transformer ratio. Careful consideration must be given to the current transformer with the lower ratio, since it may saturate before the current transformer with the higher ratio. As in all applications of a current differential system, the current transformers should not saturate for a maximum external fault. The very low burden of the relay aids in solving this problem.

The relay philosophy is independent of the length of the transmission line, but the length of the fiber optics channel must be considered.

CHANNEL CONSIDERATIONS

The most common fiber being used today for communication systems is a graded index fiber with a core diameter of 50 micrometers. This type of fiber attempts to limit pulse dispersion and, as a result, has a very large bandwidth for high density communication systems. If fibers are being installed for high density communications along with fibers in the same cable for use with the LCB, then the 50 micrometer graded index fiber is the best choice. If on the other hand, a fiber is being installed only for the purposes of the LCB then the 50 micrometer fiber is not the best choice. In the latter case, a fiber with a larger diameter would be preferred.

The LCB does not need a large bandwidth, in fact, its bandwidth is limited to about 2 kHz, and all optical fibers can transmit this small bandwidth. The most important items of concern are the amount of light that gets into the fiber and the total loss of the fiber. Since the LED used has a light emitting spot of about 200 micrometers in diameter any fiber with a smaller diameter will cause a significant loss of light when coupling to the fiber. When the LED is coupled to a 50 micrometer fiber a light loss of 14 db is encountered primarily due to the diameter difference of the LED spot and the cable diameter. If a larger cable is used then the loss is less and more light is available for transmission.

TABLE 1
SYSTEM LOSS CALCULATION

ITEMIZED LOSSES		LOSSES IN dB
FIBER LOSS	4 dB/Km (50um) 8.5 Km	34
SPLICE LOSS	0.5 dB/SPLICE✦ 8 SPLICES	4
CONNECTOR LOSS	1 dB/CONNECTOR 2 CONNECTORS	2
TOTAL SYSTEM LOSS		40

✦ FUSION SPLICE

The 50 micrometer fiber is the reference in this paper when considering channel loss capability, and the maximum allowable channel attenuation for this fiber is 40 dB. The 40 dB is based on the given light source and detector selected. The coupling losses have alredy been taken into account and the 40 dB is just losses in the fiber optic cable itself, splice losses, and connector losses. The channel attenuation limit listed above allows for a 3 dB system degradation and a minimum of a 20 dB signal to noise ratio at the receiver. The following is an example of the loss calculation. Assume that the fiber being used has an average loss of 4 dB per km and the optical channel length were 8.5 km with 8 fusion type splices. A su;mmary of the losses is shown in Table 1. The example described represents the maximum channel loss for this system. If the cable had been a 125 micrometer cable with the same loss per km then the system would handle an added length of two km. This is because of the extra light which would be coupled to the larger fiber. For any size fiber then the maximum allowable channel loss is given by the formula:

$$L = 40 + 20 \times LOG(D/50) \tag{7}$$

Where L= maximum channel loss in dB and D= fiber diameter in micrometers. The above is true for fibers up to 200 micrometers. Above this size very little gain is realized due to increasing fiber diameter.

SETTING CRITERIA

The relay is set based on the same criteria as used in previous pilot wire systems with some minor changes. The power system information required is as follows:

I_{3p}=The minimum three phase fault current from the strongest terminal.

I_g=The minimum phase to ground fault current as fed from the strongest terminal.

I_L=The maximum expected load current through the protected line.

The above quantities are always the secondary current magnitudes. The settings to be made on the relay are as follows:

Positive sequence sensitivity-P jumper to P1,P2, or OFF
Negative sequence sensitivity-N jumper to N1,N2,N3, or OFF
Zero sequence sensitivity-Z jumper to Z1, Z2, or OFF
Current sensitivity adjustment T
Phase Delay-J jumper to J1,J2,J3, or J4
Channel block configuration-L jumper to B or U

The sequence network voltage output referenced to secondary current quantities is shown in equation (8).

$$V_F = (14.14/T)(-C_1 \times I_{a1} + C_2 \times I_{a2} + C_0 \times I_{aO}) \tag{8}$$

where
V_F=voltage output of network
T=current setting of the relay
C_1=positive sequence network constant
C_2=negative sequence network constant
C_0=zero sequence network constant
$I_{a1}, I_{a2},$ and I_{a0}= A phase positive, negative, and zero sequence current components respectively (phasor quantities)

TABLE 2
SEQUENCE NETWORK CONSTANTS

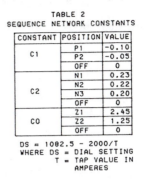

CONSTANT	POSITION	VALUE
C1	P1	-0.10
	P2	-0.05
	OFF	0
C2	N1	0.23
	N2	0.22
	N3	0.20
	OFF	0
C0	Z1	2.45
	Z2	1.25
	OFF	0

DS = 1082.5 - 2000/T
WHERE DS = DIAL SETTING
T = TAP VALUE IN AMPERES

Table 2 shows the actual constant values of C_1, C_2, and C_0 for various jumper settings. The T setting may vary between 2 to 20 amperes. The network output voltage needed to operate the system is 1.414 volts by design. Therefore, at system pickup equation (B) becomes:

$$0.1 = (C_1 \times I_{a1} + C_2 \times I_{a2} + C_0 \times I_{a0})/T \tag{9}$$

Equation (9) may be used by those desiring to check detailed pickup for various fault conditions. The right side of equation (9) must be equal to or greater than the left side in order for the system to operate.

For the vast majority of applications the above detailed calculations need not be made, and the following general criteria will be adequate. Table 3 shows the preferred combinations of jumper settings, and these should be used unless special circumstances dictate otherwise.

The criteria are as follows:

$$I_{3p} \geq 0.1 \times T/C_1 \tag{10}$$
$$I_L \leq 1.25 \times (0.1 \times T/C_1) \text{ See note below} \tag{11}$$

$$I_g \geq 0.3 \times T/(C_1 + C_2 + C_0) \tag{12}$$

NOTE: If the system is strapped to block on loss of channel then criterion (11) may be ignored.

TABLE 3
PREFERRED JUMPER COMBINATIONS

COMB #	JUMPER POSITION
1	P1,N1,Z1
2	P1,N1,Z2
3	P2,N2,Z1
4	P2,N2,Z2
5	OFF,N3,Z1
6	OFF,N3,Z2
7	P1,OFF,Z1
8	P1,OFF,Z2
9	P2,OFF,Z1
10	P2,OFF,Z2

The first step in picking the relay settings is to first pick the smallest value of C_1 and calculate the value of T in order to satisfy criterion (10). Then check to see if criterion (11) is met. If criterion (11) is not satisfied then C_1 and T must be varied in order to satisfy both criteria (10) and (11) or jumper L must be set to B position. When a value for current setting T and C_1 has been selected then the setting for C_2 is obtained from Table 3. The next step is to select jumper C_0,

159

and this is done by using criterion (12). Criterion (12) assumes that $I_{a1}=I_{a2}=I_{a0}$. This then completes the current sensitivity setting of the relay.

The next selection to be made is to select the L jumper position. It has two positions. Position B means the relay system will block during a loss of channel condition, and position U means that each terminal operates as an independent overcurrent relay during loss of channel. Position U is the preferred position unless criterion (11) above cannot be satisfied. The last item to consider is the phase delay adjustment. The jumper J should be set to J1 and the time delay potentiometer set to a minimum. This sets the phase delay to almost zero time since the optical channel delay is very small for the lengths of fiber being considered.

If the current transformers at the two ends of the line are not the same ratio then use the higher of the two ratios to make all the calculations described above. Raise the T current setting at the terminal with the lower ratio current transformer by a factor equal to the higher ratio divided by the lower ratio.

MODEL POWER SYSTEM TESTS

The relay system was tested on a model power system in order to verify the design and philosophy prior to placing the system in service. Due to the limited space the details will not be presented in this paper but a summary of the tests is presented.

There were four power system configurations used for the tests, and they are listed below:

a) The first configuration was a single transmission line with balanced source impedances at the two ends of the line. The difficult aspect of this configuration was that the line charging current was high with respect to the minimum three phase fault current. The ratio of three phase fault current to line charging current was 6.25. For these tests the load current was zero amperes.

b) The second configuration was identical to the first one except source impedance at one end was raised to a high value to make it a weak source. The source/line impedance ratio was 62.5 for the weak terminal.

c) The third configuration again was the same as in a), but the source impedance at one end was lower and was shifted in angle by 30 degrees. This caused 3.1 amperes of load current flow on the line prior to the fault.

d) Configuration four included a parallel line. This set up was to test the effects of external faults which cause power reversals in the protected line.

For the first three configurations internal faults were placed on the line at each end and the center of the line. All types of faults were applied at each location. Each type of fault was applied at various fault incidence angles. In some cases, the angle was varied only 15 degrees; and, in other cases, the angle was varied 45 degrees based on previous tests as a guide to which angles produced the slowest trip times. Table 4 gives a summary of the trip time results for the internal faults, and the table also relates the fault current magnitude to trip time.

There·were all types of external faults applied to the system at each bus. Each fault was applied 10 times with a record taken of only one test for each type of fault. In configurations 1 through 3 a total of 220 external faults were applied and the relay system did not operate. In fact the oscillographs show that the relay system did not have any trend towards the tripping state. A few special situations were tested for the external faults. The first was configuration four which tested the relay for its ability to block during a power reversal in the protected line when clearing a fault on a parallel line.

Again the relay system showed no tendency to operate during this transient condition. The last type of external fault that was tried was an arctype fault which made the through fault current rich in harmonic content. Three phase and phase to phase arcing faults were run with no load current flowing on the line. Zero load current is a worst case situation since load current puts the relay system in a stron restraint state prior to the fault, whereas with just charging current on the line the relay is looking at a low level of internal three phase fault current. The relay system again had no tendency to trip.

The results of the tests were that the relay system performed very well, and the design was continued to the manufacturing stage. The next step is to place the relay system in sevice on an actual power system. This is scheduled to be accomplished by the time this paper is presented.

TABLE 4
LCB SYSTEM TRIP TIMES

CONFIGURATION	TYPE OF FAULT	TRIP TIME RANGE IN ms	MINIMUM MULTIPLES OF PICKUP
SINGLE LINE NO LOAD $I_{3P}/I_{CH}=6.25$	THREE PHASE	17-26	3.75
	PHASE TO PHASE	16-20	3.75
	GROUND FAULTS	10-14	28.7
SINGLE LINE LOAD CURRENT	THREE PHASE	21-29	4.12
	PHASE TO PHASE	21-24	4.12
	GROUND FAULTS	16-22	34.3

CONCLUSIONS

The LCB relay system has been designed and tested on a model power system, and it has performed to its initial design requirements. The use of a new solid state sequence network, a new comparison technique, and a fiber optic channel has enabled the relay system to overcome many of the problems associated with present day pilot wire relay systems. There are problems which have not been solved. The main problem, of course, is the limited distance associated with the fiber optic channel. This distance may be increased by using more sophisticated light sources and detectors at the expense of cost, circuit simplicity, and component life. Since the fiber optic field is new and ever changing, the future may allow longer distances without the use of repeaters or other complexities. On the other hand, if repeaters were used and they were of the regenerative type the distance would be practially unlimited.

The next step is to place the relay system in an actual field installation to obtain field experience. This will be done at a Duquesne Light Company site during the first quarter of 1982. Both the fiber optic channel and a version for audio tones will be placed in service. The response of the relay system to these field trials will determine the final relay design, if in fact changes are required.

REFERENCES

[1] "Applied Protective Relaying", Chapter 14, Westinghouse Electric Corporation, 1976

[2] S. C. Sun, "A Simplified Sequence Filter Circuit", presented at Western Protective Relay Conference, October 16-18, 1979.

Discussion

George D. Rockefeller (Rockefeller Associates, Inc., NY): A significant feature of this development is the transmission of current *magnitude* intelligence. Its electromechanical cousin, the pilot wire differential relay, is a percentage differential relay at lower levels of fault current; at the higher currents it becomes a phase-comparison system. Energy is transmitted over the pilot wire from a strong-source station to a weak-feed one, allowing both stations to clear at high speed. The pilot-wire relay can also clear an internal fault on a three-terminal line with moderate outfeed at one terminal; a pure phase comparison system will be blocked by an outfeed. These advantages of the pilot-wire system are retained and extended in the new solid-state system.

A substantial market may exist for replacement of pilot-wire systems using either user-owned pairs or leased circuits. A relay-case package might substantially reduce the cost of retrofitting.

Creeping pilot-wire resistance increases occur on leased circuits as defective sections of larger gauge cable are replaced with smaller wire. The day of reckoning can be postponed by usng a tapped 4/1 insulating transformer which can extend the range from 2000 to 2500 ohms. However, suitable metallic circuits will not always be available.

The authors infer that the design is suitable for 3002-conditioned voice-grade circuits. For retrofits of pilot-wire systems, particularly over leased circuits, such an offering would be attractive, obviating the need to install a user-owned fiber optic system.

Manuscript received June 25, 1982.

J. Miller (Long Island Lighting Company, Hicksville, NY): For many years the 60 hz. ac pilot wire relay system and its carrier cousin, the phase comparison relay system using a combination positive, negative and zero sequence input filter have been the workhorse pilot relay systems for the protection of the urban/suburban electric utility high voltage transmission systems. Unfortunately, modern communication technologies, including the shrinking of common carrier cable pair wire sizes, have forced the ac pilot wire relay system into a state of virtual retirement unless the electric utility is willing to build and maintain its own pilot wire communication links. A need exists for a low burden, solid-state protection system as a replacement for the former ac current differential pilot wire relay system.

The authors have presented a well-written and concise paper describing a rather sophisticated and clever approach using solid-state technology to develop a relay system (LCB) that could rejuvenate the use of current differential pilot protection in urban/suburban areas.

The methods of developing the operating and restraining voltages and the comparison techniques are quite novel as are the PPM modulation/demodulation techniques.

Question: Am I correct in assuming that the unmodulated 1700 hertz carrier signal generated in the Fig. 8 PPM modulator is a square wave?

One of the major advantages of fiber optics is that the cables employing optical transmission neither pick up nor emit electromagnetic radiation and relaying provides total electrical isolation. The use of fiber optics with pilot wire is ideal. The pilot wire relays have proven to be reliable, but the weak link is the wire circuit. The results of the field installation of the LCB will help determine its reliability. We are also investigating the use of fiber optics from the substation to a point outside the influence of the ground potential rise and then converting back to metallic wire. This eliminates the need for protection for equipment and personnel from GPR. The telephone company has been putting increasing pressure on us to provide this protection. Other companies in our area have tried fibers for GPR protection and have had good success. *A utility user comment concerning the fiber optic link:* While the concept of an end-to-end optical communication link would greatly help to solve most of the electrical noise and ground potential rise problems to which the older pilot wire relay systems are prone, it is doubtful that the common carriers will readily install fiber optic subscriber cables into the electric utility substation. On the other hand, if the electric utility chooses to install its own end-to-end fiber optic communication channel it places itself in the business of installation, maintenance and repair of the communications link with its accompanying problem set. An ideal approach would be a fiber optic link from the substation to some point beyond the substation ground rise zone of influence and then an optic-electric interface into the conventional telephone network.

The authors imply that the LCB system is adaptable to an audio tone version which could overcome the distance limitations of both the older

pilot wire system and the new fiber optic system described herein. Could the authors enlighten us further on this aspect and also describe briefly the testing being done on the Duquesne Light Company system?

Manuscript received August 2, 1982.

Raymond VanDusen (Public Service Company of Oklahoma): I would like to commend the author on a well written paper and the information presented.

We have experienced many of the problems associated with pilot wire that was pointed out in this paper. Many mis-operations have been caused by extraneous voltages.

These extraneous voltages sometimes are caused by faults on feeders that are adjacent to the feeders protected by the pilot wire. The LCB Pilot Wire Relay and fiber optic should eliminate this problem. One advantage of using your own fiber optics is that you have control over the circuit. We have had several false trips due to the phone company disturbing the circuit wire.

TOne of the major advantages of fiber optics is that the cables employing optical transmission neither pick up nor emit electromagnetic radiation and relaying provides total electrical isolation. The use of fiber optics with pilot wire is ideal. The pilot wire relays have proven to be reliable, but the weak link is the wire circuit. The results of the field installation of the LCB will help determine its reliability. We are also investigating the use of fiber optics from the substation to a point outside the influence of the ground potential rise and then converting back to metallic wire. This eliminates the need for protection for equipment and personnel from GPR. The telephone company has been putting increasing pressure on us to provide this protection. Other companies in our area have tried fibers for GPR protection and have had good success.

The LCB relay is similar to HCB pilot wire relay. The LCB make use of advances in solid state electronics and fiber optics technology. The pilot wire for the HCB relay wire requires an end-to-end metallic circuit over which 60 Hz ac voltage and dc current must be carried.

The design procedure for a system that uses optical fibers should take into account the signal to be transmitted, the length of cable, and the SNR. After these parameters are known it is a matter of selecting the best source - fiber - detector combination. As in any system some consideration should be given to economics of the system.

The LCB pilot wire system uses a composite filter to convert the three phase current at each terminal of a two or three terminal line into a signal that is compared at the remote terminal to determine whether the fault is inside the protected line. The comparison technique produces two quantities, an operating and restraint voltage. These voltages are compared and if the fault is internal to the protected line then the line is tripped.

Optical fiber modulation can be either analog or digital. Analog modulation has large signal to noise ratio which limits it to low bandwidth and short distance. The pulse position modulation has improved noise immunity and bandwidth. The light source can be an LED or laser diode. LED's are normally used for short distance but are more economical than lasers. Two types of detection are available which match the existing light source. These are the pin diode and APD detector. The APD detector has an inherent internal gain. There are a variety of fiber cables available. New fibers are being discovered that will carry light over a much longer distance before amplification is needed.

The LCB relay system takes a new approach to an old problem and should provide improved reliability in pilot wire relaying.

Manuscript received August 9, 1982.

J. S. Benton, D. W. Hand, and **J. W. Pope** (Georgia Power Company, Atlanta, GA): The authors' paper describes a current differential relaying scheme using fiber optics which can readily accommodate a weak source terminal. The current differential technique has application advantages of both security and reliability over phase comparison techniques because of the current differential restraint characteristic. Phase comparison relays operate like differential relays without restraint.

Current differential relaying requires that both amplitude and phase information be conveyed accurately and reliably over the communications channel. Generally leased circuits are unreliable for relaying applications because of poor performance during fault conditions. What

161

would be the effect of phase jitter and phase delay variations occurring over leased circuits? It seems that phase errors tend to cause failure to operate, since the operate voltage is derived from the vector sum of local and remote signals. Do the application problems associated with excessive pilot wire shunt capacitance affect the LCB scheme in the same manner as the conventional ac pilot wire scheme?

The problems associated with large tapped transformers on the protected line were not discussed. Shouldn't the relay pickup be above the three phase fault value for a low side transformer fault? Is distance supervision recommended in those cases where it is impractical to meet this setting constraint? The setting criteria outlined in this paper appears simple and straightforward.

Are the tripping times shown in Table 4 for a fiber optic channel? What was the delay for this channel? The tripping time seems to be relatively long compared to power line carrier tripping times.

Would the authors comment on the relative speed of this system if digital modulation techniques using fiber optics channels were used exclusively. It seems that digital modulation electronic circuitry would be more stable and easier to trouble-shoot than the pulse position modulation sample and hold electronic circuitry. The authors allude to this fact in the paper when they state that the digital approach would have been ideal if the communications channel were exclusively optical.

Manuscript received August 12, 1982.

A. T. Giuliante (ASEA Inc., Yonkers, NY): The several sections to this paper bring out the many areas for consideration in the design of a pilot wire relay system. It shows the detail engineering which must be given to each area and to the interfaces between these areas. This discussion is limited to the relay requirements and to some of the application considerations. Three of the four states design requirements need clarification if they are to have the implied general applicability. For example:

a. The stated need for non-saturating ct's during external faults is true only for some pilot wire systems. The discusser's company offers a wire-pilot differential system that does not have this limitation. Other current differential systems in the discusser's product line are similarly not so limited.

b. The statement that all sequence separating networks have frequency errors is misleading. Those errors due to variations in system frequency are a matter of filter design. The sequence network used by the discusser is substantially free of these errors, having a separation error of not more than 1.5% within a band of +/− 5Hz of the rated frquency.

c. The statement that a large linear dynamic current range is a desirable requirement is not true for all pilot systems. The discusser's wire-pilot system is designed to specifically limit the linear range to about three times the pick-up value.

The stated conclusions of this part of the paper are much broader than the facts justify with respect to the need for a solid state solution to separation networks. Networks are available with all passive components which can satisfactorily perform all of the needs of current processing in a pilot differential system. Further the reference to poor separation network performance due to use of iron, and the comparison of Figure 2 to "traditional designs" should be limited to clearly exclude the discusser's relay. The sequence segregating aspects of the circuit of Figure 2 appear to be the pair of typical R-C voltage separation networks, back-to-back so as to provide both sequence components. As such, it will have the traditional severe sensitivity to variations in system frequency. The solid state elements in Figure 2 appear to provide only the needed signal scaling factors for the input voltages which could be provided readily by passive components.

The section on ct application needs clarifying. The lower ratio ct may not be the limiting component in regards to saturation. Saturation depends on ct accuracy class, burden, and on fault current distributions as well as on the simple turns ratios. And as we noted above, it is not a universal pilot relay requirement that ct's not saturate for external faults as erroneously stated in the paper.

Manuscript received August 13, 1982.

S. C. Sun and **R. E. Ray**: The authors would like to thank all of the discussers for their comments and recommendations. Since there are a great deal of commonalities in the areas addressed by the discussers, these items will be considered first, and then some of the more specific subjects will be discussed.

An audio-tone interface for the new relay has indeed been developed to supplement the applications where the use of fiber optical cables is not practical or conventional communication channels are already in place. This interface, as briefly mentioned in the paper, allows the LCB scheme to be implemented on a 3002 unconditioned circuit over microwave, leased lines and single-side band power line carriers. Literature providing detailed description of the audio-tone interface will be made available in the near future.

As for the optical-to-audio-tone and audio-tone-to-optical repeaters, the subject has been under study by us for some time. These repeaters could further broaden the application of the LCB scheme in the areas addressed by the discussers as well as a number of attendees of the Summer Meeting. An actual product of this nature will not come out until some minimal field experience of the new relay has been acquired.

The PPM technique is considered as the most suitable modem method to implement a practical design when taking all factors into consideration. If the design were to consider fiber optics exclusively, the PPM would still be chosen for its simplicity. The practical design of the PPM modem provides us a performance which is more than adequate for the LCB application. For a digital system to surpass this performance, the design would be quite involved and there is really no justification for such a complication.

Mr. Miller, in his discussion, is absolutely correct in assuming that the 1.7 KHz signal is a square wave.

Several specific questions were asked by Messrs. Benton, Hand, and Pope. First, slow long term phase variations on a channel will have no effect on the demodulated 60 Hz wave. However, phase jitter will manifest itself as a higher frequency ripple added to the 60 Hz demodulated signal. If the phase jitter is of a magnitude such as to cause sidebands outside the channel bandpass, the frequency monitoring circuit can detect the phase jitter and block the system for this condition. In cases where the phase jitter does cause this blocking action then the system security can be affected. However, the 60 Hz demodulated signal is passed through a low-pass filter which will greatly attenuate frequencies above 120 Hz thus reducing the effect of phase jitter. If the phase jitter becomes large enough to seriously affect security, the system will be clamped by the frequency monitoring circuit. It should be pointed out that the above discussion applies only to the audio-tone version of the relay.

The effect of channel shunt capacitance on the relay system is not the same as in the case of standard pilot wire relays. In the case of the LCB, the effect of shunt capacitance will increase channel attenuation and reduce the effective bandwidth. These factors are taken into account by specifying a minimum of a 3002 unconditioned line.

The paper does not discuss the use of the LCB relay to tapped transformer loads. However, the application and setting criteria would be the same as discussed in other literature published by the authors' company under the subject of HCB relay. Certainly, as suggested by the discussers, the LCB pick-up would be set above the maximum load out of the tap and low side transformer faults.

It is the authors' opinion that an average trip of one to one and a half cycles is a very respectable trip time for a pilot relay system operating at 3 to 4 times pick-up. However, it should be pointed out that the major objective in the relay design was not speed but security and dependability.

Mr. Giuliante's discussion is welcome because it provides an opportunity to discuss some of the points in the paper in more detail.

Regarding ct (current transformer) saturation, the references in the paper were directed toward extreme saturation. A completely saturated ct delivers only short duration spikes to the relaying burden, and therefore, provides no usable information to any relay requiring current as a decision making quantity. The point was that the burden of the relay is very low, minimizing any saturation tendency of the cts due to the relays themselves. External faults will produce similar distortion of the cts at both ends of the line because the primary currents are identical except for distributed capacitance effects. If the ct response is not identical due to, for example, residual flux, there may be an interval of unequal relay currents at the two line terminals. In this case, security constraints contained in the design will allow the correct performance of the system as demonstrated by model power system testing. The method of forming the restraint quantity (sum-of-magnitudes), the bandpass filtering to permit essentially only 60 Hz quantities to be compared, and the energy integrating coincidence principle (not just time coincidence) all contribute to this security.

The discussion on the sequence network suggests that more detail of the design should have been presented. However, the subject was thoroughly treated in a previous paper as referenced. The frequency-

162

dependent error in a sequence network is a well-known fact as long as the design uses a phase-shift network. Unless Mr. Giuliante's design employes a constant-phase shift network, a term that exists on paper but not easily realized in a practical design, the frequency-dependent error would have to be in his design as well. Furthermore, the errors referred in the paper are the errors caused by the harmonic frequencies of 60 Hz, not just the errors due to the variation of the power system frequency. Sequence networks using components such as iron-core transformers and/or inductors, which are potentially non-linear when subjected to a dc offset or high fault currents, will suffer additional errors in repeatability and, of course, absolute accuracy. The design shown in Figure 2 avoids this problem since all of the components used are absolutely linear. Repeatability is therefore not a problem. The merit of the proposed design, however, is more than just a good repeatability, many other operational amplifier based designs using linear components would also have this feature. The prominence of this design is its simple form and the function it performs. In addition to the scaling function, the outputs of the two amplifiers supply a pair of phasors which were sought by many sequence network designers in the past. This unique phasor pair permits a connection of an RC network between them to achieve both the positive and negative sequence component detections. By making the resistors adjustable, one can use wide tolerance capacitors to achieve the needed phase shift. There is no interaction or dependency between the magnitude and phase response in this phase-shifter design whereas any prior design, active or passive, would suffer this problem. Because of this unique feature, the calibration of this network is very simple. By supplying a balanced negative sequence to its input, one only needs to adjust the variable resistor in the positive sequence branch to produce a zero voltage at the positive sequence output. A similar procedure applies to the negative sequence calibration. This and other features are hidden in a seemingly very simple circuit. Readings of Reference 2 given in the paper and also U.S. Patent No. 4146913 are recommended.

The comments may not have been totally clear concerning unequal ct ratios. An attempt was made to point out that there are problems associated with using different ratio cts at both ends of the line. Of course, there are many factors involved where ct saturation is concerned, and the paper may have over-simplified the problem. However, with all other factors being equal the ct with the lower ratio will saturate first. It is a universal fact that if a ct at one end of a line saturates and the other end does not any current differential relay system will be subject to a possible mis-operation during external faults. It is also a universal fact that it is good relay engineering practice to attempt to apply cts which do not saturate for maximum system fault conditions when applying any relay system.

Manuscript received September 28, 1982.

FEEDER PROTECTION AND MONITORING SYSTEM, PART I:
DESIGN, IMPLEMENTATION AND TESTING

Mike Aucoin
Member

John Zeigler
Member

B. Don Russell
Senior Member

Electric Power Institute
Texas A&M University

Abstract – This paper describes the development of a microcomputer-based Feeder Protection and Monitoring System at Texas A&M University. The Feeder Protection and Monitoring System includes an overcurrent relay to provide overcurrent protection for a distribution feeder and it includes an arcing fault detector which identifies some low current faults which are not cleared by overcurrent protection. The system also provides a monitoring capability which supports data storage and remote interaction with a user. The paper describes the design of the system, how the design was implemented, and the testing of three prototypes which were built for research purposes. These units properly demonstrated the overcurrent protection, arcing fault detection and monitoring functions of the system, as well as many of the features of power system automation.

INTRODUCTION

Much research has been done in recent years on the application of microcomputer technology to power system control, monitoring and protection. The Electric Power Institute at Texas A&M University has actively participated in this research, focusing on two areas of distribution protection: the detection of low current faults, and the microcomputer implementation of the overcurrent relay. We combined these two protection functions and additional monitoring capabilities into one package called the Feeder Protection and Monitoring System (FPMS). This paper describes the design, implementation and testing of the Feeder Protection and Monitoring System.

The microcomputer-based overcurrent relay (OVC) performs all the functions that an electromechanical relay does, but includes enhancements offered by microcomputer technology. Some of these enhancements include the implementation of any shape relay curve, easily changeable settings, remote interaction, and data storage.

Some distribution faults are not overcurrent faults, and the arcing fault detector (AFD) improves the detection of such low current faults which are typically caused by downed distribution primary conductors. The AFD identifies changes in high frequency feeder current to detect a fallen conductor.

84 SM 621-9 A paper recommended and approved by the IEEE Power System Relaying Committee of the IEEE Power Engineering Society for presentation at the IEEE/PES 1984 Summer Meeting, Seattle, Washington, July 15 - 20, 1984. Manuscript submitted February 15, 1984; made available for printing May 7, 1984.

The AFD is an example of a new type of protection offered by microcomputer technology. The AFD also allows easily changeable settings, remote interaction and data storage.

The data storage and remote interaction capabilities in both the OVC and the AFD provide the monitoring portion of the package. These capabilities enable a user to remotely have access to information similar to that provided by the feeder ammeter and a strip chart (fault) recorder.

We performed previous work on the overcurrent relay under a contract with DOW Chemical, U. S. A. [1]. A prototype overcurrent relay has been installed at their Freeport, Texas plant since 1981. The previous work on the arcing fault detector was performed under a contract with the Electric Power Research Institute and resulted in a prototype demonstration at Texas Electric Service Company [2,3]. A second AFD prototype has been installed at Public Service Company of New Mexico (PNM) since 1981.

The natural evolution in our research was to combine the overcurrent relay and arcing fault detector with their monitoring functions into one package. Houston Lighting and Power (HL&P) and PNM provided Texas A&M with grants to develop and demonstrate prototype Feeder Protection and Monitoring Systems. The goal of the project was to build prototype devices which would demonstrate:

o Improved detection of low current faults, or those faults with current below the pickup level of overcurrent devices

o Improvements in overcurrent protection brought about by microcomputer technology

o Additional monitoring functions such as data storage and remote interaction

Concisely, the motivation for the project was to improve the protection and monitoring of distribution feeders. One prototype FPMS was installed at HL&P in July, 1983, and two were installed at PNM in August, 1983. The performance of the units has been monitored over several months.

DESIGN GOALS

Our overall goal for this project was to integrate the overcurrent relay and the arcing fault detector with their monitoring functions into one unit called the Feeder Protection and Monitoring System.

The integration of functions refers to using one device to perform several protection and monitoring functions which would otherwise be performed by several discrete mechanical devices. The advantages of integration are that protection and monitoring can be performed more effectively and ultimately more

Reprinted from *IEEE Trans. Power App. Syst.*, vol. PAS-104, no. 4, pp. 873–880, April 1985.

economically. For example, the OVC could replace the two or three phase relays, the ground relay, and the ammeter on a feeder. Additionally, it can perform breaker failure detection, store data from faults, and communicate information to a central office (functions which are typically not available on a distribution feeder). The integration of all these functions with the addition of arcing fault detection makes an attractive package.

Combining the AFD and the OVC into one package made sense from the standpoint of relaying and monitoring. The FPMS could offer protection for the feeder for high current faults and some low current faults. Integrating these functions enables one to correlate data from the two functions, making it easier to identify operations on the feeder. We felt that the demonstration of the AFD and the OVC along with the demonstration of the automation capabilities of the system would make a substantial step in the development of power system automation.

Although we had demonstrated the OVC and AFD at previous installations, both functions were still in the research stage. We sought to design a unit which would enable us to gather data and adjust settings or algorithms easily and remotely. This approach would help us to gain experience with the algorithms and hardware, and regularly make evaluations of system performance. The inclusion of these interactive functions also fulfilled the project goal of demonstrating some of the benefits of power system automation.

Overcurrent Relay

The overcurrent relay provides integrated overcurrent protection and monitoring for all three phases and the neutral of a distribution feeder. The instantaneous and time overcurrent trip functions provide primary feeder protection equivalent to that offered by their electromechanical counterparts. Breaker failure detection has been added to these conventional protection functions to provide enhanced capabilities. This is accomplished by allowing the relay to continue monitoring the line currents after a trip signal has been issued. If the line currents do not fall below an open circuit threshold within a specified time period, a secondary trip signal is sent. This signal could be connected to an annunciator panel or to backup protection on the source side of the failed breaker.

The flexible time-current coordination curves stored within the relay incorporate the standard tap, time dial, and curve type specifications of conventional devices. These curves may be defined as any single valued function of current, and are not restricted to the inverse characteristics of electromechanical devices. The relay uses one curve for the protection of all three phase circuits and a second curve for the protection of the ground circuit. Additional curves may be downloaded to the relay to update its coordination characteristics as feeder conditions vary.

The relay provides a data recorder function which stores several types of information about each disturbance seen by the overcurrent functions. Each storage file includes the date and time of the disturbance, the maximum phase and ground currents, the settings in effect at the time, and the classification of the disturbance. If no trip is output, the disturbance is classified as an event, and the duration of the event is added to the file. If either an instantaneous or time overcurrent trip is output, the disturbance is classified as a fault. For

these files, the trip type is stored along with the identification of the first phase requesting a trip, the fault currents at the time of the trip, the time to trip, and the breaker operation time. Oscillographic data storage is provided for all disturbances regardless of their classification. The storage process operates continuously so that several cycles of data prior to the disturbance may be retained. Two fixed length recording windows are provided to capture the beginning and end of the disturbance. For long duration disturbances, the intermediate data is typically repetitive and is discarded to conserve memory.

Arcing Fault Detector

We developed the arcing fault detector to improve the detection of low current distribution primary faults. Such faults typically occur when a primary conductor breaks and contacts earth. If the impedance at the point of the fault is high enough, fault current may remain too low for an overcurrent device to clear the fault. Arcing commonly occurs with fallen conductor faults, and the arcing produces high frequency transients in the feeder current which the AFD uses as an indicator of a fault. Although the AFD cannot detect all low current faults, it does improve the chance of detecting many of them. Limitations of the detection concept are still being investigated.

Previous research showed that arcing on the feeder primary could be detected by identifying an increase in the 2-10 kHz components of feeder neutral current. Arcing also occurs with normal switching operations but is limited in time. The arcing which occurs with a downed conductor fault often lasts for as long as the conductor is grounded, which may be several minutes or even hours. In a general sense, the arcing fault detection technique is the identification of an increase in the 2-10 kHz feeder current which lasts for a sufficiently long time. The basic design goal for the AFD is to implement this technique.

Another important function of the AFD is storing high frequency data from the feeder current. We still need more experience with the AFD to determine the algorithm and settings to use for optimum results, and this effort is substantially helped by the monitoring function. Whenever the detection threshold is exceeded, the AFD stores the average 2-10 kHz signal magnitude for each cycle of 60 Hz and continues to do so for a fixed length of time. At the end of this time it determines whether or not a fault exists.

Coverage Regions

Figure 1 provides a sketch of the approximate coverage regions of the FPMS for a typical feeder. This graph plots the possible range of fault currents which could be observed for different faults on this feeder, and the graph shows which devices may be expected to detect a fault exhibiting a given fault current. The coverage regions are marked for the AFD, the OVC, the existing electromechanical substation relay (marked "SUB. RELAY") and the existing downstream protection on the feeder (fuses and line reclosers). The OVC will cover approximately the same range of fault currents as the existing substation relay. The AFD covers a range of fault currents extending below that of the substation relay and the downstream protection. The designation of the region of very low fault currents with a question mark means that no present commercial or research device can detect faults with only a few amperes of current, and it is very unlikely that any substation-based device ever will detect such low fault currents.

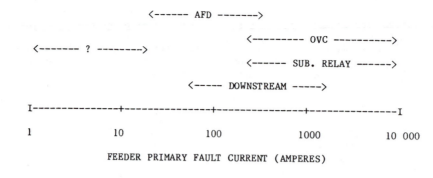

```
                        <------ AFD ------->

            <------- ? -------->        <---------- OVC ---------->

                                        <------ SUB. RELAY ------>

                        <----- DOWNSTREAM ----->

        I--------------+--------------+--------------+--------------I

        1           10           100          1000        10 000
```

FEEDER PRIMARY FAULT CURRENT (AMPERES)

Figure 1. Coverage Regions of FPMS.

Communications Function

The other major function of the FPMS is its ability to communicate with a local or remote user. We deemed this design feature important for two reasons. First, this capability demonstrates some of the advantages of power system automation. Second, we needed to communicate easily with the units to determine their performance and to retrieve data from them regularly for research purposes. Because some units were to be several hundred miles away, we chose to use telephone access to communicate with each FPMS. We desired to perform six types of operations over the communications link.

o Check system status and time.

o Retrieve stored data.

o List present 60 Hz load currents or high frequency currents.

o Change and verify settings.

o Download programs or curves.

o Use a base level monitor for diagnostics.

Our experience has been that utility personnel are particularly excited about the remote communications capabilities of the FPMS. For example, they find it worthwhile to determine the present load on a feeder while they are at a computer terminal in a downtown office.

IMPLEMENTATION

We chose for research purposes to concentrate on demonstrating the functions which could be performed by the system. We did take some steps to help the unit survive in the substation environment, but we did not have sufficient resources to design for high reliability. High reliability and survivability can be achieved in a straightforward but more costly manner.

One implementation problem involved integration. We originally intended to highly integrate the OVC and AFD functions, especially their signal conditioning and conversion operations. However, due to limited resources, we chose to separate the functions according to their data bases. The data and operations associated with the OVC are entirely separate from those of the AFD. Likewise, the transducers, signal conditioning and conversion functions of each are separate. The OVC and AFD do share some functions, namely power and communications, which do not materially affect the unit's complexity. We also connected status signals between the AFD and OVC to allow them to share some information.

While we chose to keep the OVC and AFD functionally independent, we also allowed expansion of the system. We allowed for the inclusion of a third microcomputer within the system called the supervisor (SPV) which could handle communications, data storage, and trip/reclose logic. The SPV was not included in our present implementation.

We will now discuss some of the details of the hardware and software implementation. A functional diagram of the FPMS is shown in Figure 2 for reference.

Hardware System Implementation

We chose to use commercially available boards to implement the generic microcomputer system in both the AFD and OVC sections. The word "generic" here describes the microprocessor, memory, serial and parallel interfaces, and signal conversion functions. We had some unique design problems which required that we custom design boards for signal conditioning and for operating the front panel display. For all the boards, we used commercial grade components which have an operational temperature range of 0-70 degrees C.

Microcomputer System

We chose the Adaptive Science Modulas One microcomputer system to implement the FPMS. This system consists of 4-1/2 inch by 6-1/2 inch boards which are inserted into a card cage with a backplane. The Modulas One system is based on the Motorola 6809 microprocessor. The AFD and the OVC each had a separate microcomputer system which included several boards which are now described.

The single board computer has 8K of EPROM and 4K of static RAM. For data storage we included a memory board with 32K of dynamic RAM. We also included a serial/parallel interface board. The serial interface portion of the board is RS-232C compatible, and it can connect with a local terminal or modem. The parallel interface portion of the board is TTL compatible and is used to convey signals between the AFD and OVC subsystems. We used the Modulas One A/D board for signal conversion. This board uses a Datel ADC-HZ12BGC converter which has 12 bit resolution and an 8 microsecond conversion time.

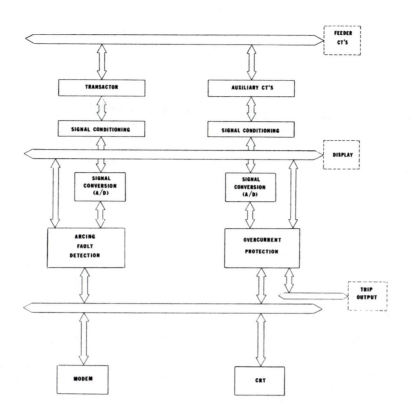

Figure 2. Functional Diagram of FPMS.

OVC Signal Conditioning

The input to the OVC is from the secondary of the existing feeder current transformers used for metering or relaying. These CT's typically have a 5A rms output for rated load on the feeder. Under fault conditions, these currents may reach 100A momentarily. We used an auxiliary current transformer with a load resistor to further step down these currents and to convert them to voltages for input to the FPMS signal conditioning board.

We wished to measure normal feeder load currents to an accuracy of 10 A. Since the feeder currents could reach 10 000 A during a fault, the signal conversion system must have a large dynamic range. We chose to implement an auto-gain circuit to provide the necessary dynamic range for the OVC.

A block diagram of the signal conditioning board containing the auto-gain circuit is given in Figure 3. The fixed gain amplifier stage buffers the analog input voltage from the auxiliary CT and provides proper voltage levels to the remaining circuitry. The full-wave peak detector and the set of comparators classify the voltage magnitude into one of five voltage bands. The digital results of this classification are fed through a multiplexer to the gain control inputs of the variable gain amplifier stage. This stage provides gains of 1, 2, 4, 8, and 16 V/V. In the automatic mode, the gain of the final amplifier stage is determined by the magnitude of the input voltage at each sample period. The gain setting is read by the microprocessor so that it can scale the digitized data accordingly. The board also supports a manual mode, where the gain of the final stage is set

by the microprocessor independently of the input voltage magnitude.

The signal conditioning board also supports three filter stages which can be configured for the desired transfer function. The first two stages are designed for switched capacitor filters which can provide lowpass filter characteristics. These two stages were not required for the OVC signal conditioning, however, because the passive lowpass filter in the third stage provided adequate anti-aliasing filtering.

AFD Signal Conditioning

The input to the AFD is the feeder neutral current, which also comes from the secondaries of the existing metering or relaying CT's. We want only the 2-10 kHz component of this signal. The steady-state 2-10 kHz feeder current, or base noise level, is on the order of 0.01-0.1A, and during an arcing fault, it may be 1.0A on the feeder primary. The 60 Hz component of neutral current often reaches 100-200 A. The signal of interest is therefore 60-80 dB below the level of 60 Hz current. The AFD signal conditioning must sufficiently reject 60 Hz while also amplifying the 2-10 kHz component to a level which is easily measurable.

General Electric supplied a transactor with a suitable transfer function to accomplish the desired conditioning. In the 2-10 kHz passband, this transducer gives a 100 V/A output. At 60 Hz, the output is 7.5 V/A. We constructed a passive highpass filter on the transactor output to effectively eliminate this remaining 60 Hz signal.

167

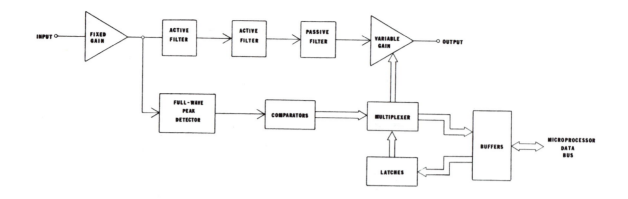

Figure 3. Functional Diagram of Signal Conditioning.

The output of this passive filter is' then fed into the AFD signal conditioning board, which is very similar to the OVC signal conditioning board shown in Figure 3. Because the AFD signal does not have a wide dynamic range, the auto-gain circuit was not used and both amplifiers on the board were set to unity gain. The board was designed to allow the use of switched capacitor lowpass and highpass filters to further accentuate the 2-10 kHz band. These switched capacitor filters were not needed because the passive filters sufficiently conditioned the signal.

Communications

Communications between a terminal user and the FPMS is over an RS-232C compatible link locally, and over a telephone modem remotely. Both the AFD and the OVC share the same communications link which requires a user to enable communications with each one as desired. For the default case, neither processor will echo input, but after one receives the proper enable string, it will echo input and perform commands until disabled. We used a Racal-Vadic VA-355P 300 baud modem for communications via telephone. Because a dedicated telephone line could not be provided at any of our installation locations, we programmed each unit to wait ten rings before answering its telephone, allowing utility employees to answer the phone first.

OVC Software

The OVC software is based on a cross-correlation algorithm for estimating the magnitude of the primary feeder currents. Each of the four currents is sampled at 240 Hz and correlated with reference sine and cosine values. The resulting sine and cosine component magnitudes are then used in a Walsh magnitude estimation procedure [4]. This final estimate has been modified to give the rms magnitude of the 60 Hz component of the currents.

A table lookup algorithm continuously determines the time to trip as long as an estimated magnitude exceeds a fault pickup threshold. The table provides 256 points partitioned into four unequally-sized sections for each coordination curve. This partitioning provides greater resolution at low fault currents where the curve is generally steep. The table lookup procedure permits custom coordination curves to be used, so that the relay is not restricted to the limited inverse curve characteristics of electromechanical relays.

The data recorder portion of the OVC software

samples each of the four currents at 480 Hz to provide oscillographic waveform data. Whenever the estimated magnitudes exceed an event pickup threshold, this function stores 14 cycles of data for each channel, beginning two cycles before the disturbance was detected. For disturbances longer than this initial storage window, a second storage window is recorded when the disturbance is cleared. This window provides eight cycles of data extending three cycles beyond the end of the disturbance.

AFD Software

The AFD software primarily consists of event detection and event identification algorithms. The AFD samples the 2-10 kHz signal at a rate of 3840 Hz. It takes the average of 64 such samples, which corresponds in time to one cycle of 60 Hz. Event detection consists of monitoring this average 2-10 kHz signal and detecting when this signal exceeds a threshold (usually a 50% increase in the signal) for three out of five consecutive cycles. Once an event has been detected, the AFD then performs event identification to classify the occurrence as a normal operation or as an arcing fault. If 48 of the 260 cycles after an event indicate a high frequency signal above the threshold, the AFD determines that a fault has occurred. We chose sensitive settings for the AFD during our research so that it could detect as many faults as possible.

The threshold for event detection adapts with the base level high frequency signal on the feeder. Our previous research has shown that normal "noise" levels on the feeder may vary quite a bit. The algorithm tracks slow changes in the noise level, but it detects a sharp increase in the signal as an event.

We programmed the AFD to store data from every event. It also captures the data for the ten cycles prior to the event. The data stored is the average 2-10 kHz feeder current for each cycle of 60 Hz.

Power, Packaging and Display

Power to each FPMS is supplied by 120 VAC station service. We accepted the fact that station service is occasionally lost. DC power would have been more continuously available but much more costly. Furthermore, our experience shows that commonly available low power DC supplies are not suitably reliable. We installed a surge filter on the AC line into each unit. The unit consumes a maximum of approximately 100 W.

168

We constructed each prototype in a standard 19" cabinet. A continuously-operating fan at the rear of the unit exhausts air which is drawn in through vent holes underneath the card cage.

The front panel of the FPMS provides a visual display of the status for both the OVC and AFD functions. Refer to Figure 4 which is a photograph of the front panel. LED targets are provided for the status of each function. A 32 character fluorescent alphanumeric display also provides numerical data. For example, Figure 4 shows a display of the present feeder load in amperes. Each prototype also displays the date and time, the 2-10 kHz signal level, the cumulative events, faults and trips, and the unit identification.

FIELD TESTING

After the units were installed, we monitored their performance for a period of several months, during which time several overcurrent faults occurred normally. Figure 5 shows the oscillographic plot of data stored by the OVC during one of these faults. This event was a phase C to ground fault which was cleared by a line recloser in approximately four cycles. Notice that the vertical scale is in kiloamperes. This plot shows that the OVC picked up and gathered data properly during the event.

Downed conductor fault tests were also staged at each location to test the operation of the AFD in each system. Figure 6 shows the plot of the data stored by the AFD during one of the staged fault tests. Each vertical line gives the average magnitude of the 2-10 kHz signal over one cycle of 60 Hz. The vertical scale is in amperes of 2-10 kHz feeder current. The fault begins at time zero and arcs intermittently thereafter. The AFD correctly identified this activity as a downed conductor fault while the overcurrent devices on the feeder did not detect this fault.

A companion paper, "Feeder Protection and Monitoring System, Part II: Staged Fault Test Demonstration," provides an in-depth discussion of the performance of the FPMS units during the staged fault

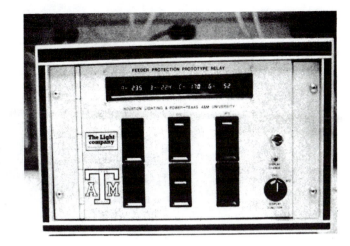

Figure 4. FPMS Display of Feeder Load Currents.

tests.

Because our goal was to demonstrate the functions of the system for research purposes, the prototype units may not achieve the reliability needed for commercial use, and some component failures may occur during the testing period.

CONCLUSIONS

We have described the design of the prototype Feeder Protection and Monitoring System. The FPMS demonstrates some of the advantages obtained by applying microcomputer technology to power system protection and monitoring. The FPMS detects both overcurrent and some low current faults. We designed the unit with these functions and its monitoring capability to demonstrate how these features might be applied to the operation of a distribution feeder. We have also discussed the implementation of the system and its performance during testing and field use. The

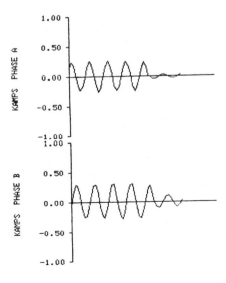

UNIT: HLP001   DATE: 08-24-83   TIME: 17:31:50.326   FILE TYPE: EVENT   TAPE ID: FILE #: 0001

Figure 5. Data Plot of Overcurrent Fault Seen by OVC.

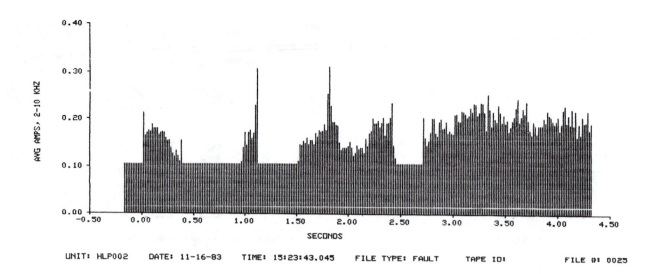

Figure 6. Data Plot of Arcing Low Current Fault Seen by AFD.

experience gained with these prototypes should help utilities learn how to best apply the new technology and functions on their systems.

ACKNOWLEDGEMENTS

We are grateful for the support of Houston Lighting and Power and Public Service Company of New Mexico on this project. We also thank the General Electric Company for providing some of the transducers for the prototypes, and we thank Mr. Gary Gerloff and Mr. Les Perry for help with printed circuit board layout.

Several persons and groups at Texas A&M were instrumental in supporting this effort. In particular, we thank the Electric Power Institute, the Electrical Engineering Department and the Texas Engineering Experiment Station.

Finally, we are grateful to the former members of the Power System Automation Laboratory at Texas A&M for their continued help.

REFERENCES

[1] B. M. Aucoin and B. D. Russell, "Distribution High Impedance Fault Detection Utilizing High Frequency Current Components," IEEE Transactions on Power Apparatus and Systems, Vol. PAS-101, June, 1982, pp. 1596-1604.
[2] "Detection of Arcing Faults on Distribution Feeders," EPRI Report EL-2757, Prepared by Texas A&M University, December, 1982.
[3] J. W. Horton, "The Use of Walsh Functions for High Speed Digital Relaying," IEEE Summer Power Meeting, San Francisco, July 20-25, 1975, Paper No. A75 582-7.

Discussion

Walter A. Elmore, (Westinghouse Elec. Corp., Coral Springs, FL): The authors are to be congratulated on their efforts to develop a more responsive ground detecting concept for distribution circuits and should be heartened by their progress so far.

We agree with the question mark of Fig. 1 in the 1 to 20 amp. range. It is unlikely that ground faults of these values will ever be recognizable by a practical distribution substation device. Presently available ratio-ground relays are able to recognize fault magnitudes well below the ap-proximately 200 amp. level indicated in Fig. 1, with or without arcing and inherently detect many open conductor conditions. With a 400:5 ct, 40 amp. sensitivity is achievable at light load.

Though the microprocessor affords remarkable flexibility in the choice of curve shapes, one still must fall back on the same system requirements which dictated the curves in present use such as cold load pickup, thermal characteristics of circuits and devices, and of course the need to coordinate with other existing devices on both sides of the feeder protective device. While sensitivity with security are vital goals for any new device, the need for a different curve shape is not apparent.

Manuscript received August 6, 1984.

M. Aucoin, J. Ziegler, and **B.D. Russell:** The authors are grateful to Mr. Elmore for his interest and worthwhile comments on this paper.

It may be helpful to point out again that there are two distinctly different subsystems within each FPMS unity. Each subsystem is responsible for the detection of a different class of fault. The OVC subsystem detects high current faults, the AFD detects low current (high impedance) arcing faults and is more sensitive than overcurrent protection. The OVC duplicates and enhances existing overcurrent protection. The AFD in this demonstration did not use time-current curves in the detection algorithm. The discussion of time-current curves is applicable only to the OVC subsystem.

There are numerous reasons for taking advantage of the flexibility in curve shapes afforded by microcomputer technology. The relay engineer is given the capability to allow the system coordination requirements to dictate the necessary relay curve for a device instead of having to accept the given curves of a particular device which dictate the effectiveness of coordination. Additionally, a utility would no longer need to stock or order a separate electromechanical relay to provide each needed curve characteristic. There is need for only one generic piece of microcomputer hardware; different curves are generated by different memory chips. Likewise, to change a relay curve shape, one only replaces a memory chip instead of replacing three or four electromechanical relays.

The ratio-ground relay (RGR) referred to by Mr. Elmore is helpful for improving the detection of high impedance faults. The AFD and the RGR each address different portions of the high impedance fault problem. As Mr. Elmore implies, the sensitivity of the RGR is dependent on load current. The relay must be set so that the operation of the largest fuse on the feeder is not interpreted as a broken conductor condition. This coordination practice implies that the coverage of the RGR is limited to that portion of the feeder between the substation and the first fuse on each phase. The most effective high impedance fault detection system will likely require a combination of detection techniques, one of which has been presented in this paper.

Manuscript received September 4, 1984.

Part 3
Rotating Machinery

I. GENERAL DISCUSSION

THE protection of generators, e.g., stator phase and ground faults, rotor faults, motoring, and loss-of-excitation are still primary concerns. The reprints included in Volume I presented an overview of the existing generator and motor protection practices before 1980. Since Volume I was published, however, there have been significant changes within the power industry that have forced a slightly different focus on the industry's activities relative to rotating machinery protection. The advent of independent power producers without the large engineering staff or experience of the more traditional utility, consulting, and manufacturing organizations has resulted in a need to publish a variety of protection guides. In addition, several incidents have occurred which have forced recognition of other areas not previously considered as probable sources of damage to machines.

Torsional vibration studies resulted from two incidents of subsynchronous resonance in systems involving series capacitors and were extended to the general effect of switching and reclosing on lines close to generators. Several cases of inadvertent energization of generators, i.e. paralleling a unit to the system while it is still on turning gear, resulted in catastrophic damage to the generator-turbine shafts and spurred investigations into protective relays and circuits to prevent a recurrence. Also, protection of the entire winding of a generator was not considered a major cause for concern because the area closest to neutral is not highly stressed electrically. Nevertheless, several failures have occurred in those portions of the winding and this subject has been studied quite extensively. The advent of computer-based relays had been particularly useful in motor protection and has resulted in the development of digital relays that are providing protection, control, and monitoring that is superior to that provided by electromechanical or solid-state relays.

II. DISCUSSION OF REPRINTS

One of the primary responsibilities of the PSRC has been to develop standards and guides related to generator protection to assist the many agencies involved in generator design and operation as discussed above. The PSRC has published three standards on generator protection: ANSI/IEEE C37.102—1987 examines the overall protection problems; ANSI/IEEE C37.101—1985 covers ground protection; and ANSI/IEEE C37.106—1987 presents a guide for protection during abnormal frequency operation. In addition to publishing the complete documents, the practice of summarizing these publications at appropriate IEEE meetings has been well received, thus publicizing the documents and receiving the benefit of widespread review and discussion. Summaries of the standards mentioned above are found in the first three papers presented in this part.

To supplement these PSRC documents, three additional papers are also included. In "A Comparison of 100% Stator Ground Fault Protection Schemes for Generator Stator Windings," Pope reviews the problem of protecting the entire generator stator winding and compares several alternative schemes. Schlake, Buckley, and McPherson present the specifics of one particular scheme in "Performance of Third Harmonic Ground Fault Protection Schemes for Generator Stator Windings," and Griffin and Pope describe ground fault protective relaying schemes in "Generator Ground Fault Protection Using Overcurrent, Overvoltage, and Undervoltage Relays." Two other PSRC reports are also included. In "Inadvertent Energizing Protection of Synchronous Generators," the possible causes and results of this catastrophic event are discussed, and in "Out-of-Step Relaying for Generators," the need to implement this feature is described as well as the impedance characteristics that would detect this condition.

Following the torsional vibration incidents mentioned above, Bowler et al., have written a pioneering paper, "Evaluation of the Effect of Power Circuit Breaker Reclosing Practices on Turbine-Generator Shafts." This paper applies statistical methods to evaluate the effects of reclosing on turbine-generator shafts. Together with several discussors, this paper compares the different reclosing practices and the prevailing philosophies regarding restrictions on reclosing, particularly high-speed reclosing.

The practice of using resonant grounding for unit-connected generators is not very popular in the United States, although it is common in Europe. The paper by Khunkhun, Koepfinger, and Haddad, "Resonant Grounding (Ground Fault Neutralizer) of a Unit Connected Generator," describes this grounding method and its advantages over the more common resistor grounding scheme. A method of evaluating its performance is presented.

Computer relaying is uniquely suitable for motor protection. This idea is presented in "Complete Motor Protection by Microprocessor Relay," by Elmore and Kramer. A complete analysis of motor starting, identification of the source data, and application to a personal computer (PC) to determine the most important protection criteria is covered by Zocholl in "Motor Analysis and Thermal Protection."

III. ASSOCIATED STANDARDS

"*Guide for AC Motor Protection,*" ANSI/IEEE C37.96—1988.

"*Guide for Generator Ground Protection,*" ANSI/IEEE C37.101—1985.

"*Guide for AC Generator Protection,*" ANSI/IEEE C37.102—1987.

"*Guide for Abnormal-Frequency Protection for Power Generating Plants,*" ANSI/IEEE C37.106—1987.

SUMMARY OF THE
"GUIDE FOR AC GENERATOR PROTECTION" ANSI/IEEE C37.102-1987

The AC Generator Protection Guide Working Group of the IEEE Power System Relaying Committee:
L. E. Landoll, Chairman, R. F. Arehart, D. M. Clark, L. H. Coe, R. W. Haas, J. Latham,
H. O. Ohmstedt, A. C. Pierce, D. E. Sanford, C. L. Wagner

ABSTRACT

This paper serves as an introduction to the Guide for AC Generator Protection (ANSI/IEEE C37.102-1987). It briefly describes the material contained in the guide and presents some examples. Those interested in the material presented in the summary paper should obtain the complete document.

INTRODUCTION

This new guide presents a review of the generally accepted forms of protection for synchronous generator and their excitation system. It describes the proper application of relays and other devices and serves as a guide for the selection of equipment to obtain adequate protection. The guide is primarily concerned with protection against faults and abnormal operating conditions for large hydraulic, steam, and combustion-turbine generators.

This guide is not a standard and does not purport to detail the protective requirements of all generators in every situation. Standby and emergency use generators are specifically excluded. The suggestions made pertain to typical generator installations. However, sufficient background information relating to protection requirements, applications, and setting philosophy is given to enable the reader to evaluate the need, selection and application of protection for most situations.

DESCRIPTION OF GENERATORS, EXCITATION SYSTEMS AND GENERATING STATION ARRANGEMENTS

The intent of this section is to present information on typical generator designs and connections, grounding practices, excitation systems, and generating station arrangements as it affects the protection and selection of protective relays. For example, single and multiple parallel winding configurations connected in wye or delta are described, since they affect the type of phase fault protection discussed in the following section.

Since the method of grounding of the generator neutral determines the protection schemes applicable for stator ground faults, the guide discusses the most commonly used methods for the various arrangements used to connect the generator to the system. For unit-connected generators both high resistance distribution transformer grounding and ground fault neutralizer grounding are discussed. For two or more generators bussed at generator voltage and then connected to the system through a step-up transformer, or connected directly to a three-wire distribution system, low resistance grounding is discussed. Low reactance grounding is discussed for those generators connected to four-wire distribution systems. Various grounding-transformer schemes are described for those systems using delta-connected generators and other special connections where the other grounding methods are not applicable. The detailed specifications for these grounding schemes are covered in other ANSI and IEEE standards and guides. Only those characteristics of each scheme, such as the ground fault current and voltage levels, that affect generator protection systems are discussed in this guide.

The type of excitation system affects the type of rotor field protection that can be used, so the four basic types of excitation systems are briefly described. These include the DC generator commutator exciter, the alternator rectifier exciter with a stationary rectifier or rotating rectifier, and the static excitation system.

Finally, this section of the guide describes and shows the system diagrams for a number of the most common generator connections and station arrangements: the unit generator-transformer configuration, the unit generator-transformer with generator breaker configuration, the cross compound generator, multi-generators sharing a unit transformer, and generators connected directly to a distribution system.

PROTECTION REQUIREMENTS

This section briefly describes the damaging effects of faults and abnormal operating conditions and the type of devices and their settings commonly used to detect these conditions. A clear understanding of the effects of abnormalities on generators will assist the reader in evaluating the need for and the means of obtaining adequate generator protection in specific situations. Areas covered are:

Generator Stator Thermal Protection

Thermal protection for the generator stator core and windings may be provided for the following contingencies:

1. Generator overload.
2. Failure of cooling systems.
3. Localized hot spots caused by core lamination insulation failures or by localized or rapidly developing winding failures.

88 SM 526-6   A paper recommended and approved by the IEEE Power System Relaying Committee of the IEEE Power Engineering Society for presentation at the IEEE/PES 1988 Summer Meeting, Portland, Oregon, July 24 - 29, 1988. Manuscript submitted March 21, 1988; made available for printing April 15, 1988.

Reprinted from *IEEE Trans. Power Delivery*, vol. 4, no. 2, pp. 957-964, April 1989.

Field Thermal Protection

Thermal protection for the generator field may be divided into two categories:

1. Protection for the main field winding circuit.
2. Protection for the main rotor body, wedges, retaining ring, and amortisseur winding.

This section is primarily concerned with the first category, protection for the main field winding.

Generator Stator Fault Protection

Generator faults are always considered to be serious since they can cause severe and costly damage to insulation, windings, and the core; they can also produce severe mechanical torsional shock to shafts and couplings. Moreover, fault currents in a generator do not cease to flow when the generator is tripped from the system and the field disconnected. Because of the trapped flux within the machine, fault current can continue to flow for many seconds, thereby increasing the amount of fault damage.

As a consequence, for faults in or near the generator which produce high magnitudes of short-circuit currents, some form of high-speed protection is normally used to trip and shut down the machine as quickly as possible in order to miminize damage. Differential relays and turn fault relays are used to detect phase and turn-to-turn faults, while a number of relay schemes are available for ground faults, depending upon the method used to ground the generator(s). Where external impedances are used to limit ground fault currents to a few amperes, slower forms of protection may be justified.

Generator Rotor Field Protection

This section is primarily concerned with the detection of ground faults in the field circuit. Other protection for the field circuit is covered in the section on loss of field and the section on excitation system protection.

Generator Abnormal Operating Conditions

This section describes those hazards to which a generator may be subjected that may not necessarily involve a fault in the generator. It discusses the typical means for detecting these abnormal operating conditions and the tripping practices.

Protection is normally provided for the following abnormal operating conditions:

1. Loss of field
2. Unbalanced currents
3. Loss of synchronism
4. Overexcitation
5. Anti-motoring
6. Overvoltage
7. Abnormal frequencies

System Backup Protection

The protective relaying described in the preceding sections provides protection for all types of faults in the generator zone and for generator abnormal operating conditions. In addition to this protection, common practice is to provide protective relaying that will detect and operate for system faults external to the generator zone that are not cleared due to some failure of system protective equipment. This protection, generally referred to as system backup, is designed to detect uncleared phase and ground faults on the system.

Generator Breaker Failure Protection

When the protective relays detect an internal fault or an abnormal operating condition they will attempt to trip the generator breaker and at the same time initiate the breaker-failure timer. If the fault or abnormal condition is not cleared in a specified time, the timer will trip the necessary breakers to remove the generator from the system. This section describes why a breaker "a" switch must be used to indicate that the breaker has failed to open, and describes a scheme to detect an open generator breaker flashover.

Excitation System Protection

The excitation system has many similarities to the generator it supplies, and hence requires much of the same type of protection. Although the consequences from equipment damage are less serious, adequate protection of the excitation system is important for reasons of continuity of service.

Excitation system protection may be provided for the following:

1. Exciter phase unbalance: Symptomatic of a serious problem, such as a phase-to-phase or turn-to-turn fault, this condition may lead to winding damage. Detection may be accomplished by comparing three-phase voltages with their average or by differential relaying.

2. Exciter ground fault: This condition may be detected by generator field ground protection.

3. Overcurrent: Detection is provided by generator field protection.

4. Loss of rectifier cooling: To prevent rectifier thermal damage, some form of loss-of-flow detector should be present, backed up by an overtemperature alarm.

5. Alternator armature winding overtemperature: This condition may indicate stator winding damage due to failure of the stator cooling system. Detection is provided by imbedded thermocouples or resistance temperature detectors.

6. Alternator air cooler loss-of-water flow: Thermal damage is prevented by a flow detector alarm.

7. Bearing vibration: Severe mechanical damage may be prevented by using vibration detectors/recorders.

Loss of Excitation Example

If a generator is operating at full load when it loses excitation, it may attain a speed of 2% to 5% above normal. The level of kVARS drawn from the system can be equal to or greater than the generator kVA rating. If a generator is initially operating at reduced loading (for instance 30% loading), the machine speed may only be 0.1% to 0.2% above normal and it will receive a reduced level of vars from the system.

In general, the most severe condition for both the generator and the system is when a generator loses excitation while operating at full load. For this condition, the stator currents can be in excess of 2.0 per unit and, since the generator has lost synchronism, there can be high levels of current induced in the rotor. These high current levels can cause dangerous overheating of the stator windings and the rotor within a very short time. In addition, since the loss-of-field condition corresponds to operation at very low excitation, overheating of the end portions of the stator core may result. No general statements can be made with regard to the permissible time a generator can operate without field; however, at speeds other than synchronous, it is very short.

With regard to effects on the system, the var drain from the system can depress system voltages and thereby affect the performance of generators in the same station, or elsewhere on the system. In addition, the increased reactive flow across the system can cause voltage reduction and/or tripping of transmission lines and thereby adversely affect system stability. When a lightly loaded machine loses field, the effects will be less damaging to the machine but the var drain may still be detrimental to the system.

The most widely applied method for detecting a generator loss of field is the use of distance relays to sense the variation of impedance as viewed from the generator terminals. It has been shown that when a generator loses excitation while operating at various levels of loading, the variation of impedance as viewed at the machine terminals will have the characteristics shown on the R-X diagram in Figure (1). In this diagram, curve (a) shows the variation of impedance with the machine operating initially at or near full load. The initial load point is at C and the impedance locus follows the path C-D. The impedance locus will terminate at D to the right of the (-x) ordinate and will approach impedance values somewhat higher than the average of the direct and quadrature substransient impedances of the generator. Curve (b) illustrates the case in which a machine is initially operating at 30% load and underexcited. In this case, the impedance locus follows the path E-F-G and will oscillate in the region between points F and G. For a loss of field at no load, the impedance as viewed from the machine terminals will vary between the direct and quadrature axis synchronous reactances (X_d, X_q). In general, for any machine loading, the impedance viewed from the machine terminals will terminate on or vary about the dashed curve (D-L).

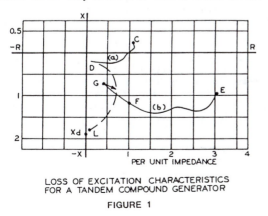

LOSS OF EXCITATION CHARACTERISTICS
FOR A TANDEM COMPOUND GENERATOR

FIGURE 1

There are two types of distance relaying schemes used for detecting the impedances seen during a loss of field. One approach is shown in Figure (2) where one or two offset mho units are used to protect a machine.

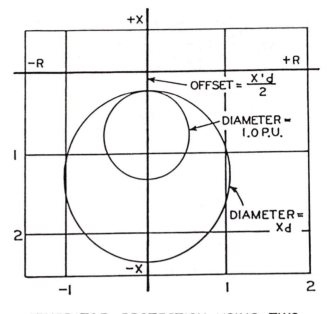

GENERATOR PROTECTION USING TWO
LOSS OF EXCITATION RELAYS

FIGURE 2

These relays are applied to the generator terminals and set to face into the machine. On small or less important units, only a single relay may be used, with the diameter of its circular characteristics set equal to the synchronous reactance of the machine (X_d) and with an offset equal to one half transient reactance (X'_d). A time delay of 0.5 to 0.6 second may be used with this unit in order to prevent possible incorrect operations on stable swings.

Depending upon machine and system parameters, two relays are sometimes used. The relay with 1.0 per unit impedance diameter will detect a loss of field from full load down to about 30% load. This relay is generally permitted to trip with 0.1 second time delay and provides fast protection for the more severe conditions in terms of possible machine damage and adverse effects on the system. The second relay is set with a diameter equal to X_d, and with a time delay of 0.5 to 0.6 second. Both units are set with an offset equal to one half transient reactance.

The second distance relaying approach is illustrated in Figure (3). This scheme uses a combination of an impedance unit, a directional unit and an undervoltage unit applied at the generator terminals and set to look into the machine. The impedance (Z2) and directional units are set to coordinate with the generator minimum excitation limiter and its steady state stability limit. During abnormally low excitation conditions, such as might occur following a failure of the minimum excitation limiter, these units operate and sound an alarm, allowing a station operator to correct the condition. Should a low voltage condition also exist, indicating loss-of-field, the undervoltage unit operates and initiates tripping after a time delay of 0.25 to 1.0 second.

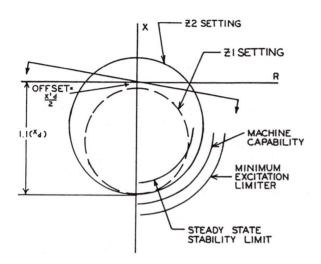

LOSS OF EXCITATION RELAYING SCHEME

FIGURE 3

Two relays may also be used in this scheme, with the second (shown as Z1 on Figure (3)) set with an offset equal to $X'_d/2$ and with the long reach intercept equal to 1.1 times X_d. In this case, the relay with the Z1 setting should trip without any external time delay, with the other relay delayed by approximately 0.75 second to prevent operation on stable swings.

In both of the above schemes, where two relay units are used, one may be considered primary protection and the second as backup. However, there is a widespread practice of using no backup for the loss of field relay. Dependence is placed upon an operator to trip the machine before it is damaged if the primary protection and that included in the excitation system both fail.

For machines paralleled at their terminals, such as cross-compound units, the terminal voltage will be maintained by the output of the good machine. In this case the undervoltage unit of the loss of excitation relay should be deactivated.

On small generators, loss of field may be detected by sensing the magnitude of field current, or by a power relay connected to sense var flow into the generator, or by sensing power factor in excess of some angle, such as 30 degrees underexcited. These devices tend to be less secure than the distance relay approach and therefore are often used to alarm only.

Loss-of-field protection is generally connected to trip the main generator breaker(s) and the field breaker and to transfer the unit auxiliaries. The field breaker is tripped to minimize damage to the rotor field in case the loss of field is due to a rotor field short circuit or a slipring flashover. With this approach, if the loss of field is due to a

some condition which could easily be remedied, a tandem compound generator could be quickly resynchronized to the system.

This approach may not be applicable with once-through boilers, cross-compound units, or those units that cannot transfer sufficient auxiliary loads to maintain the boiler and fuel systems. In these cases, the turbine stop valves should also be tripped. Cross-compound units with directly interconnected stator circuits can be resynchronized with the system only if the units are in synchronism with each other. If the units are out of synchronism, normal starting procedures must be used to return the units to the line. However, recent developments in the industry have established that it may be possible to resynchronize particular cross-compound generators after an accidental trip without returning the two generators to turning gear speed. This procedure should be established only after consultation with the manufacturer.

OTHER PROTECTIVE CONSIDERATIONS

This section of the guide discusses additional types of generator protection and describes other factors that should be considered in the generator zone.

Current Transformers

The overall performance of current transformers (ct) is an important factor in the performance of protective relays used in the generator zone. The guide stresses two areas of particular concern regarding ct performance. These are the adverse effects of residual flux and stray external flux fields (proximity effects).

Voltage Transformers

The guide enumerates and presents solutions for two problems affecting the output of voltage transformers (vt) used in the generator zone.

1. A blown fuse can cause misoperation of relays and voltage regulators by changing the magnitude and phase angle of the secondary voltage signal.

2. Ferroresonance can occur whenever grounded vts are connected to an ungrounded system. If permitted to continue, the resulting high exciting currents will cause vt failure in a relatively short period of time.

Protection During Start-up or Shut Down

During start-up or shut-down, a generator may be operated at reduced frequency. At frequencies below 60 Hz, the sensitivity of some generator relays may be adversely affected. Figure 4 shows the effects of frequency on the sensitivity (pickup) of relays associated with generator protection. To ensure machine protection during start-up and shut-down, supplementary protection using relays not greatly affected by frequency may be provided. This sensitive supplementary protection is usually deactivated when the unit is connected to the system.

Protection for Accidentally Energizing a Generator on Turning Gear

If a generator is accidently energized while on turning gear, it will accelerate as an induction

176

motor. During the period the machine is accelerating, high currents will be induced in the rotor and the time-to-damage may be in the order of a few seconds. High speed protection for this contingency is desirable. The guide lists those relays in the generator zone that may detect this condition and then describes the limitations of each type in providing protection for inadvertent energizing.

Other concerns affecting the generator and its protection that are addressed in this section of the guide are:

1. Subsynchronous Resonance
2. Transmission Line Reclosing
3. Synchronizing Equipment
4. Incipient Fault Detection

This section presents detailed one-line diagrams and their associated control logic diagrams which are classified according to the method by which the generator is connected to the system. These diagrams show the combination of relays (and their control function) generally applied for generator and excitation system protection in accordance with good engineering practices. These diagrams also consider the protective devices on other equipment in or adjacent to the generating station which are connected to trip or shut down the generator.

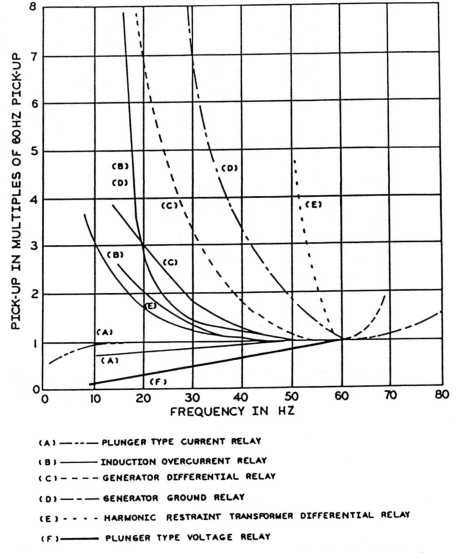

(A) —————— PLUNGER TYPE CURRENT RELAY

(B) ——————— INDUCTION OVERCURRENT RELAY

(C) — — — — GENERATOR DIFFERENTIAL RELAY

(D) ——— — GENERATOR GROUND RELAY

(E) - - - - HARMONIC RESTRAINT TRANSFORMER DIFFERENTIAL RELAY

(F) ——————— PLUNGER TYPE VOLTAGE RELAY

NOTE: CURVES WITH LIKE LETTERS INDICATES COMPARABLE RELAYS OF DIFFERENT MANUFACTURERS.

RELAY PICK-UP VS FREQUENCY

FIGURE 4

The following is typical of the one-line diagrams shown.

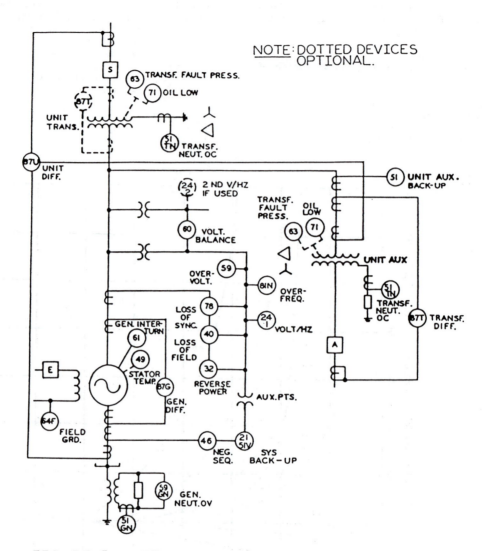

NOTE: DOTTED DEVICES OPTIONAL.

FIGURE 5 UNIT GENERATOR-TRANSFORMER CONFIGURATION.

This particular diagram illustrates the unit generator-transformer configuration with single generator breaker. Other configurations illustrated are:

° Unit generator-transformer configuration with dual generator breakers

° Cross-compound generators

° Several generators sharing a unit transformer

° Generators connected directly to a distribution system.

Accompanying each of the one-line diagrams is a d-c tripping logic diagram similar to the following.

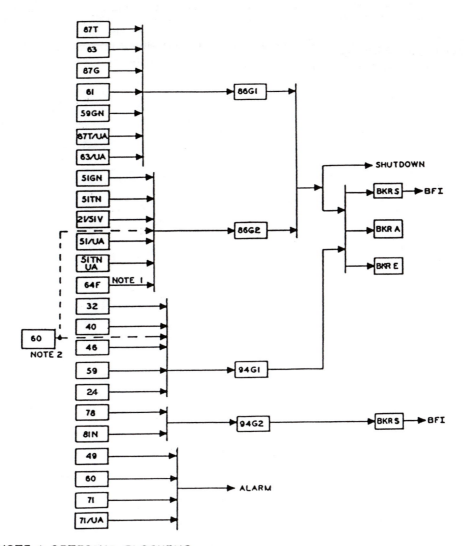

NOTE 1-OPTIONAL BLOCKING.
NOTE 2-DEVICES SHOW TRIPPING APPROPRIATE AUXILIARY
        IF TRIPPING IS THE SELECTED OPTION.
NOTE 3-UA INDICATES UNIT AUX

**GENERATOR-TRANSFORMER
CONFIGURATION DC
TRIPPING LOGIC.**

FIGURE 6

This particular diagram accompanies the one-line diagram for the unit generator-transformer configuration with single generator breaker shown above. These d-c tripping logic diagrams illustrate suggested methods for grouping the outputs of the protective relays shown on the one-line diagram to achieve such functions as shutdown of the generator, tripping associated circuit breakers, initiation of breaker-failure schemes, actuation of fire protection schemes and alarming.

CONCLUSION

The "Guide for AC Generator Protection" should be of considerable value to the industry. This paper has attempted to cover the highlights of this new guide. Discussion of generator protection methods is encouraged to provide the Power System Relaying Committee with feedback that can be used to further improve the guide in the future.

EA688/024

179

J. E. Stephens, (Illinois Power Co.) Decatur, Illinois): Figure 6 of the Summary paper is a d-c tripping logic diagram, one of several included in the guide. Suggested methods of grouping the outputs of the various protective relays are illustrated. Two lockout relays, 86G1 and 86G2, are shown with 87T (87U), 87G, and 63 operating the same lockout relay. Since 87T is somewhat redundant to 87G and 63, it may be preferred that they operate different 86G relays for greater reliability. The same would apply to the 87T/UA and 63/UA relays.

Since breaker failure relay initiation is a parallel function to tripping the breaker, I believe it would be more appropriate to show the BFI as a parallel function from the trip bus line rather than a series function from each breaker box.

Manuscript received June 29, 1988.

R. J. Fernandez: The Working Group would like to thank Mr. Stephens for his interest and comments pertaining to the paper.

Mr. Stephens makes two very good points that will enhance the reliability of the protection and control schemes presented in the guide. These points will be included when the guide is scheduled for reaffirmation or revision.

Manuscript received September 9, 1988.

GENERATOR GROUND PROTECTION GUIDE

A working-group* report for the Power System Relaying Committee
A. C. Pierce, Senior Member IEEE
Factory Mutual Engineering, Norwood, MA

ABSTRACT

The "Generator Ground Protection Guide" ANSI/IEEE Standard C37.101 has been prepared recently by the Power System Relaying Committee to aid in the application of relays and relaying schemes for the protection of synchronous generators for single-phase-to-ground faults in the stator winding. The various protective schemes along with recommended and alternate methods of protection for the most commonly used generator connections and grounding practices are covered. The purpose of this paper is to review some of the highlights of the Guide, to familiarize the industry with the scope and usefulness of this new document, and to invite considered discussion for the purpose of improving this guide in the future.

INTRODUCTION

Historically the ground fault is the most common type of fault to which a generator is subjected. The "Generator Ground Protection Guide" ANSI/IEEE C37.101 provides detailed application information on the protection schemes most commonly used in North America. Protective schemes used elsewhere in the world and which are just starting to be used in North America are described in an appendix. The application of a specific scheme is dependent upon the grounding method, the generator connection and the owner's protection philosophy. Sections are provided in the Guide which describe the various generator connection and grounding methods. A summary of the recommended protective schemes is given in a matrix table. The main body of the Guide provides application information on each protective relay scheme. A second appendix in the Guide provides examples of how to calculate ground overcurrent and overvoltage relay settings for the various protective schemes and how to coordinate them with voltage transformer secondary fuses. The third appendix is a bibliography of available literature on the ground fault problem from which source material was drawn.

SUMMARY OF PROTECTIVE SCHEMES

The summary of recommended protective schemes is given in Table 1, which is a matrix of generator connections, generator grounding methods, and the scheme numbers which identify the protective schemes. Across the top of the table, above each column, are one line diagrams covering the significant variations of generator-transformer-bus-circuit breaker arrangements that are encountered. Vertically along the left side, heading the rows, are one line diagrams of approved grounding methods, covered in the IEEE Application Guide for Grounding of Synchronous Generator Systems, IEEE 143. The neutral point (N) of the generator dia-

84 WM 108-7   A paper recommended and approved by the IEEE Power System Relaying Committee of the IEEE Power Engineering Society for presentation at the IEEE/PES 1984 Winter Meeting, Dallas, Texas, January 29 - February 3, 1984. Manuscript submitted October 25, 1983; made available for printing November 11, 1983.

grams connects to the Point N of the grounding method diagram. Where a generator neutral point does not exist or is not brought out a distribution transformer bank can be connected to the associated generator main leads to provide grounding. The individual boxes in Table 1 list by scheme number (1, 2, 3, etc.) the different applicable ground fault protective schemes that apply for a given generator connection and a given grounding method. Schemes within the brackets are the most widely used and recommended. The suffix S on the scheme number (5S, 8S) indicates that the protective scheme is suitable for use only when the machine is running and disconnected from the system but with field excitation applied. The protective scheme numbers in the boxes refer to protective schemes that are completely illustrated in the Guide section entitled, "Single-Phase-To-Ground Fault Protective Schemes." The X-Y points in the grounding methods diagrams connects to voltage relays illustrated on the protective scheme diagrams. Likewise, the R-S and W-Z points connect to current relays illustrated in the protective scheme diagrams.

GENERATOR CONNECTIONS

The six different classes of generator connections illustrated in Table 1 are intended to be representative of connections in common use today. Although two different connections are shown in Column A, the same protective schemes may be applied to both. The criteria for these connections is that a single-phase-to-ground fault in the generator will not produce significant zero sequence currents or voltage in the system, nor will a similar fault in the system produce any significant zero sequence quantities in the generator circuit. When two units are paralleled on one transformer delta winding only one neutral is normally grounded. Where machines are connected to separate low voltage transformer windings, each unit is grounded separately and has its own protective scheme.

In Column B an autotransformer is shown with either a wound or a phantom tertiary. In either case the autotransformer provides a zero sequence connection between the generator and the system. With the generator neutral ungrounded, the tertiary is a source of ground fault current even when the main circuit breaker is open for a ground fault in the stator with excitation applied.

Column C is similar to A except that the generator is connected in delta. Grounding is obtained by connecting wye-broken delta distribution transformers to the generator main leads (Grounding Method VI).

Columns D & E, respectively, indicate delta and wye wound generators connected directly to the system bus without any interposing step-up transformer. These are normally small generators connected to a grounded (in contrast to ungrounded) power system.

*Members of the Working Group are A. C. Pierce, Chairman, J. Berdy, H. DiSante, E. J. Emmerling, C. H. Griffin, I. R. Harley, G. P. Stranne, C. L. Wagner, R. C. Zaklukiewicz.

Reprinted from *IEEE Trans. Power App. Syst.*, vol. PAS-103, no. 7, pp. 1743-1748, July 1984.

TABLE 1 Generator Connections, Generator Grounding. Methods and Protective Scheme Numbers.

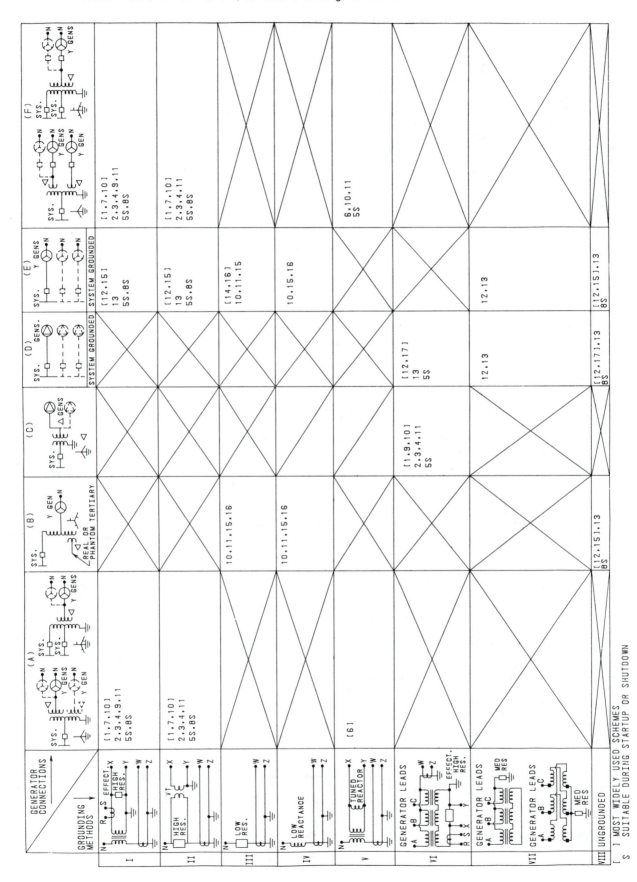

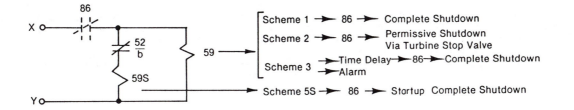

Ground Overvoltage Schemes 1, 2, 3 & 5S
Figure 1

Column F generator connections are the same as A except that individual generator low side circuit breakers are utilized. Each generator will normally have its own grounding and protection. Different time delay settings are required on ground relays to sequentially trip generators to a common delta transformer winding in order to determine the ground fault location.

GROUNDING METHODS

The Guide lists eight grounding methods which are subsequently described. The Guide describes protection for six of the seven grounding categories covered in the IEEE Application Guide for Grounding of Synchronous Generator Systems, IEEE 143. Protection for solidly grounded generators is not considered.

Grounding Method I – Distribution Transformer Grounded.

This is a high resistance grounding method which utilizes a distribution transformer connected to the generator neutral with a primary voltage rating equal to or greater than the line-to-neutral voltage of the generator. Transient overvoltages are controlled by selecting the secondary resistor so that its KW loss during a single-phase-to-ground fault at the terminals of the generator, is equal to or greater than the zero sequence capacitive KVA to ground. The maximum single-phase-to-ground fault current is normally limited to 3-25 amperes.

Grounding Method II – Neutral Resistor Grounded.

This is a high resistance grounding method which is functionally equivalent to Grounding Method I. The same criteria is used in sizing the directly connected neutral resistor to limit the ground fault current as in I.

Grounding Method III – Neutral Resistor Grounded.

This is a low resistance grounding method which limits the ground fault current to values high enough to operate the standard generator differential relay.

Grounding Method IV – Neutral Reactor Grounded.

This is a low inductive reactance grounding method which permits many times higher fault current than the previous methods described.

Grounding Method V – Ground Fault Neutralizer Grounded.

This is an inductively tuned grounding method which uses a distribution transformer selected as per Method I, with a secondary reactor. The reflected value of the secondary reactor is equal to one-third of the zero sequence capacitive reactance to ground. The ground fault current is limited to values that will not sustain an arc. This method may not be easily applied to units where low side breakers are applied (Column F).

Grounding Method VI – Grounding Transformer Grounded.

This is a high resistance grounding method used for delta connected generators which utilizes three distribution transformers connected grounded wye-broken delta to the generator main leads. The distribution transformers and the secondary resistor across the broken delta are selected per Method I.

Grounding Method VII – Grounding Transformer Grounded.

This is a medium resistance grounding method which permits sufficient current for selective ground relaying. The primary winding of a wye-delta or zigzag transformer is connected to the generator leads with a resistor connected from the transformer neutral to ground.

Grounding Method VIII – Ungrounded.

No grounding is employed on the leads or neutral of the generator.

PROTECTIVE SCHEMES

The protective schemes listed in Table 1 are fully described and illustrated in the guide. A brief description of the 17 protective schemes most commonly used in North America follows. Protective schemes Nos. 1 through 4 and 7 are used for detecting single-phase-to-ground faults on high resistance grounded generators that are connected to the system through delta-wye connected transformers. For these connections a ground fault at the generator terminals will produce a full neutral shift with a resultant voltage appearing across terminals X-Y illustrated in Table 1. This voltage will be roughly proportional to the distance from the neutral as a percentage of the total winding. Fault detection in these applications is achieved by connecting an overvoltage relay, device 59, across terminals X-Y as indicated in Figure 1. In protective schemes 1 through 4 the fault is sensed by the voltage across the neutral grounding device, whereas in Scheme 7 the voltage is sensed across the broken delta secondary winding of the voltage transformers. The voltage pickup setting of device 59 must be high enough so that it will not operate on fundamental frequency voltages produced by normal system unbalances, or third harmonic voltages generated by the machine under full-load conditions. Relays are available with third harmonic filters and can be safely set to within 10 percent from the neutral end of the stator winding. A short time delay may be required to prevent false generator trips for faults on the transmission system. Time delay is also required to coordinate with voltage transformer fuses for phase-to-ground faults in the voltage transformers or their secondary leads.

183

The overvoltage relay employed in protective schemes 1 through 4 and 7 will not provide sensitive protection at frequencies significantly below rated frequency. If field excitation is applied while the machine is being brought up to speed or being shut down, a startup scheme similar to 5S or 8S should be used in addition to scheme 1 through 4 and 7.

During a ground fault device 59 used in protective scheme 1 operates and energizes a lockout relay, device 86. The lockout relay initiates a complete shutdown which includes tripping the main and field breakers and closing the turbine stop valves or gates.

Protective schemes 2, 3, 4 and 7 are variations of protective scheme No. 1 and utilize the same overvoltage relay, device 59. Scheme 2 is a permissive shutdown scheme where tripping of the main and field circuit breakers is supervised by position switches on the turbine stop valves as indicated in Figure 1. Scheme 2 prevents full load rejection with its accompanying overspeed condition. A contact on the lockout device 86 is employed to interrupt the overvoltage relay operating coil circuit since higher than rated coil voltage may be applied for a prolonged period of time.

Scheme 3 also indicated in Figure 1 provides an immediate alarm and a time delay complete shutdown. A further variation of this scheme would be for the lockout relay device 86 to trip the turbine stop valve for a permissive shutdown.

Protective scheme 7 utilizes wye-broken delta voltage transformers connected to the generator leads as shown in Figure 3. The resultant voltage appearing at the relay terminals will be equal to 3 times the phase-to-neutral voltage divided by the voltage transformer ratio for a ground fault on the generator terminals. Scheme 7 is normally used where two or more machines, each with its own low side circuit breaker, are connected to the same transformer delta winding. Scheme 1 is used for individual generator protection while scheme 7 is used for protection of the delta transformer winding and associated bus.

Protective schemes 5S and 8S are startup ground overvoltage schemes for stator ground fault detection during the time that the machine is disconnected from the system and running with field excitation applied. Scheme 5S is shown in Figure 1 and is applied in conjunction with schemes 1 through 4 previously described for high resistance grounded machines. It is also applied to high resistance and ground fault neutralizer grounded machines where low side breakers are used. Scheme 8S is applied in conjunction with scheme 7 as indicated in Figure 3 and ungrounded machines. Device 59S, used in schemes 5S and 8S, is an instantaneous overvoltage relay and has a relatively constant volts-per-hertz response down to its dc pickup. The same level of protection can therefore be provided while the generator is brought up to speed or shut down. The operating coil of device 59S in series with the associated circuit breaker auxiliary switch (52/b) is connected across terminals X-Y to provide protection when the circuit breaker is open.

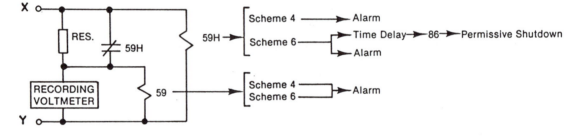

Ground Overvoltage Scheme 4 & Ground Fault Neutralizer Scheme 6
Figure 2

Scheme 4 indicated in Figure 2 provides only an alarm. If the overvoltage relay device 59 cannot withstand the applied voltage to its operating coil for a prolonged period of time an additional less sensitive, higher rated overvoltage relay, device 59H is also employed. Device 59H is set below the continuous rating of device 59. The circuitry is arranged so that the 59H during a ground fault will operate to insert a resistor in series with the 59 coil and therefore reduce the voltage on device 59 to a safe value.

Protective scheme 6 is applied to machines that are grounded by means of the ground fault neutralizer method. Scheme 6 is a variation of scheme 4, and employs the same 59 and 59H devices previously described and shown in Figure 2. Ground fault neutralizers limit the ground fault current in the stator winding and connected equipment to magnitudes so low that an arc cannot be maintained. This method permits orderly shutdowns since fault damage is severely restricted. A recording voltmeter is used to monitor the small zero sequence voltage always present across the neu-

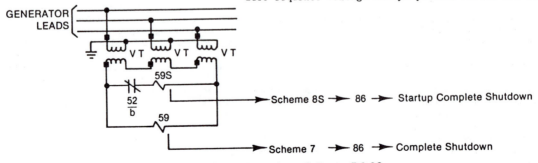

Wye-Broken Delta VT Ground Overvoltage Schemes 7 & 8S
Figure 3

tralizing reactor. Reduction in this voltage indicates a fault-to-ground near the generator neutral terminal. An increase in voltage indicates insulation deterioration.

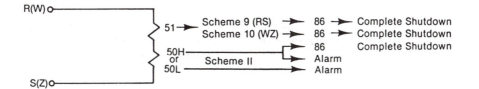

Ground Overcurrent Schemes 9, 10 & 11
Figure 4

Protective schemes 9, 10 and 11, shown in Figure 4, are overcurrent relaying schemes used for detection of single-phase-to-ground faults. All three schemes can be used for ground fault detection on generators that are high resistance grounded and that are isolated from the system by the delta winding of the generator step-up transformer. Schemes 10 and 11 can also be used for low impedance grounded systems.

Scheme 11 is an instantaneous over-current relay that may be used in conjunction with either scheme 9 or 10. When used in conjunction with scheme 9, this device, 50H, will provide for high speed detection of all ground faults in the transformer delta winding, main leads, and 50% to 70% of the generator stator winding. Device 50H can be used for alarm or trip and will help determine fault location. When used in conjunction with scheme 10, this device, 50L, generally alarms since it is set very sensitively to detect faults near the generator neutral which may not be sensed by device 51 of scheme 10.

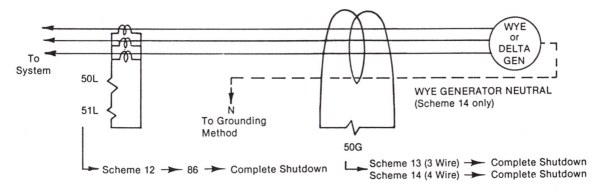

Ground Overcurrent Schemes 12, 13 & 14
Figure 5

Scheme 9 uses an overcurrent relay connected to terminals R-S of a secondary connected CT. The CT ratio is selected so that the relay current is equal to the generator neutral current. Scheme 9 is essentially a variation of scheme 1 and the application discussion for scheme 1 also applies to scheme 9. Since the scheme 9 overcurrent relay is sensitive to harmonics it must be set less sensitively than the scheme 1 overvoltage relay which is insensitive to harmonics.

Scheme 10 is a variation of scheme 9 except that the overcurrent relay is connected to terminals W-Z of a primary current transformer in series with the generator grounding impedance. For high resistance grounded generators isolated from the system by the step-up transformer the same consideration as described under schemes 1 and 9 are applicable. A 5/5 CT ratio is used to match the relay current to the generator neutral current. For low impedance grounded generators a relatively high CT ratio (typically 400/5) must be used and the generator ground relay must be coordinated with system ground relays. On low impedance grounded machines, serious generator damage may occur from the resultant high levels of ground fault current.

Protective relay schemes 12, 13 and 14 shown in Figure 5 are ground overcurrent relay schemes for generators that are connected to a grounded system either directly or through an autotransformer. The overcurrent relay detects ground fault current flowing through the generator leads from the system to the generator. For ungrounded and high resistance grounded generators the instantaneous ground overcurrent relays used in schemes 12 and 13 must be set above the generator zone zero sequence capacitance current contribution for external system ground faults.

Scheme 12 uses both time and instantaneous overcurrent relays residually connected to current transformer in each phase of the generator leads. The instantaneous overcurrent relay must be set above any false residual current caused by unequal current transformer characteristics.

Scheme 13 is a variation of scheme 12 but makes use of a window type current transformer which surrounds the phase leads to the generator. The window CT measures the zero sequence current in the generator leads during a ground fault. The window type CT and its associated relay must be tested as a system to determine time current characteristics and sensitivity.

Scheme 14 is similar in principle to scheme 13 and makes use of a window type CT and an instantaneous overcurrent relay. Phase leads and the generator neutral lead passes through the CT and the neutral "N" is then connected to the particular method of generator neutral grounding. This arrangement results in a self-balancing differential ground relay scheme with the CT output to the relay for an internal fault being the sum of ground fault current from the system and the generator. For an external fault no current will flow in the relay.

coil in the neutral of the differential relay circuit and its polarizing coil energized from a CT in the generator neutral. Scheme 16 provides additional sensitivity for internal faults and security against external faults. Scheme 17 is the application of the differential relay to delta connected generators for ground fault protection.

CONCLUSION

A guide for ground fault protection of generators is now in print and should be of considerable value to

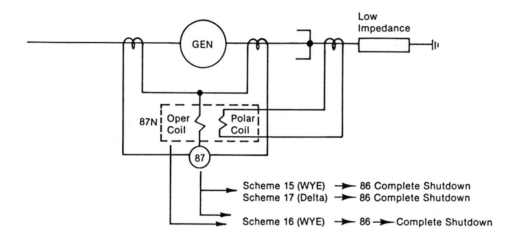

Generator Differential Schemes 15, 16, & 17
Figure 6

Protective relay schemes 15, 16 and 17 indicated in Figure 6 are conventional generator percentage differential relay schemes that can be used for ground fault protection if the generator is connected to a solidly grounded system either directly or through an autotransformer. Scheme 16 also uses a current polarized product type overcurrent relay with its operating

the industry. This paper has attempted to cover the highlights of this new guide. Discussion of generator ground fault protection methods is encouraged to provide the Power System Relaying Committee with feedback that can be used to further improve this guide in the future.

SUMMARY OF THE
"GUIDE FOR ABNORMAL FREQUENCY PROTECTION FOR
POWER GENERATING PLANTS" ANSI/IEEE C37.106-198X

Prepared by:

Generating Plant Abnormal Frequency Working Group × (See Page 5)
Rotating Machinery Protection Subcommittee
IEEE Power System Relaying Committee

Abstract - The paper serves as an introduction to the "Guide for Abnormal Frequency Protection for Power Generating Plants" soon to be published. The summary paper briefly discusses the hazards of operating generation equipment at abnormal frequencies and describes acceptable protective schemes. If the reader is interested in the material presented the summary paper should lead him to seek out the complete document as a protection guide.

INTRODUCTION

Following is a summary of the "Guide For Abnormal Frequency Protection for Power Generating Plants," prepared by a working group under the sponsorship of the Power System Relaying Committee. The Guide covers relay applications for the protection of generating plant equipment from damage caused by operation at abnormal frequency including over-excitation. Emphasis is placed on the protection of the major generating station components at steam generating stations, nuclear stations and on combustion turbine installations. Consideration is also given to the effect of abnormal frequency operation on those associated station auxiliaries whose response can affect plant output.

The Guide presents background information regarding the hazards to generation equipment from operation at abnormal frequency. It documents typical equipment capabilities and describes acceptable protective schemes. Recommended methods for coordinating the underfrequency protective scheme with system load shedding schemes are also included. Sufficient information is provided to apply suitable coordinated protection for given specific situations.

STEAM GENERATING PLANT-ABNORMAL FREQUENCY OPERATION

General Background

There are two major considerations associated with operating a Steam Generating Station at abnormal frequency. They are:

1. Protection of equipment from damage which could result from operation at abnormal frequency.

2. Prevention of a cascading effect that leads to a complete plant shutdown as long as limiting conditions are not reached during abnormal frequency operation.

86 SM 369-3   A paper recommended and approved by the IEEE Power System Relaying Committee of the IEEE Power Engineering Society for presentation at the IEEE/PES 1986 Summer Meeting, Mexico City, Mexico, July 20 - 25, 1986. Manuscript submitted February 24, 1986; made available for printing April 29, 1986.

Printed in the U.S.A.

The major components within a steam plant which are affected by abnormal frequency are the generator and unit step-up transformer, the turbine and the station auxiliaries.

Generator Over/Underfrequency Capability

While no standards have been established for abnormal frequency operation of synchronous generators, it is recognized that reduced frequency results in reduced ventilation; therefore, operation at reduced frequency should be at reduced output.

Operating precautions should be taken to stay within the short time thermal rating of the generator rotor and stator. The underfrequency limitations on the generator, however, are usually less restrictive than the limitations on the turbine.

Operation within the allowable overfrequency limits of the turbine will not produce generator overheating as long as operation is within rated kVA and 105% of rated voltage.

Turbine Over/Underfrequency Capability

A turbine blade is designed to have its natural frequencies sufficiently displaced from rated speed and multiples of rated speed (i.e., the rated fundamental frequency and its harmonics) to avoid a mechanical resonant condition which could result in excessive mechanical stresses in the blade. Excess stress can occur if the system damping is insufficient to overcome the excitation stimulus produced by turbine steam flow. For a resonant condition, the vibratory stress can be 300 times greater than the stress during non-resonant operation conditions. Operation at frequencies other than rated or near rated speed is time restricted to the limits shown for the various frequency bands published by each turbine manufacturer for various blade designs.

The abnormal frequency limits are generally based on worst-case conditions because (1) the natural frequencies of blades within a stage differ due to manufacturing tolerances, (2) the fatigue strength may decline with normal operation for reasons such as pitting corrosion and erosion of the blade edges, and (3) the limit must also recognize the effect of additional loss of blade life incurred during abnormal operating conditions not associated with underspeed or overspeed operation.

It is recommended that the manufacturer be consulted in order to obtain the applicable frequency limits of the turbine to be studied. These limits are usually presented in the form of allowable time vs. frequency curves.

The abnormal frequency capability curves are applicable whenever the unit is connected to the system. These curves also apply when the turbine generator unit is not connected to the system if it is operated at abnormal frequency while supplying its auxiliary load. During periods when the unit is being brought up to speed, being tested at no-load for operation of the overspeed trip device, or is being shut down, blade life is not significantly affected if the manufacturer's procedures are followed.

Reprinted from *IEEE Trans. Power Delivery*, vol. 3, no. 1, pp. 153-158, Jan. 1988.

Power Plant Auxiliaries - Underfrequency Considerations

The ability of the steam supply system to continue operating during an extended period of underfrequency operation is a function of the margin in capacity of the auxiliary motor drives and shaft-driven loads.

The most limiting auxiliary equipments are generally the boiler feed pumps, circulating water pumps, and condensate pumps, since each percent of speed reduction causes a large percent loss of capacity. The critical frequency at which the performance of the pumps will affect the plant output will vary from plant-to-plant. However, tests and experience have shown that the plant capability will begin to decrease at 57 Hz, and that frequencies below 55 Hz are critical for continued plant operation due to the reduction in the output of the pumps. The effects of operating at below rated voltage on the performance of station auxiliary equipment are not covered in this Guide.

UNDERFREQUENCY PROTECTION METHODS FOR STEAM TURBINES

This section describes possible protection methods to prevent turbine operation outside the prescribed limits. The discussion is limited to underfrequency protection since overfrequency relay protection is generally not applied because governor runback controls or operator action are counted upon to correct the turbine speed.

Since it has become common practice to drop load by automatic underfrequency relays to maintain a load-to-generation balance during system overload, system load shedding is considered the primary turbine underfrequency protection. Appropriate load shedding can cause the system frequency to return to normal before the turbine abnormal limit is exceeded. When designing underfrequency load shedding systems, it is necessary to make assumptions about the degree to which a normally interconnected system may break up into "islands" during a major disturbance. Conditions may be such that during a system breakup, the system may break into islands different from those assumed in the load shedding design study and sufficient load shedding may not be attained to permit frequency recovery. Under these conditions, underfrequency protection of turbine generators should be considered a means to reduce the risk of steam turbine damage in the islanded area. Also, if it is possible that the load shedding system could fail, turbine generator frequency protection could provide backup protection against such a failure. However, generator underfrequency tripping should be considered as the last line of defense since it is likely to cause an area blackout.

A turbine underfrequency protection scheme must be adaptable in order to protect turbines with different operating frequency limits and flexible enough to be adjusted in case underfrequency operating limits are revised by designers due to new technological discoveries and improved material applications.

The following design criteria are suggested as guides in the development of an underfrequency protection scheme:

1. Establish trip points and time delays based on the manufacturer's turbine abnormal frequency limits.

2. Coordinate the turbine generator underfrequency tripping relays with the system automatic load shedding program.

3. Failure of a single underfrequency relay should not cause an unnecessary trip of the machine.

4. Failure of a single underfrequency relay to operate during an underfrequency condition should not jeopardize the overall protective scheme.

5. Static relays should be considered as their accuracy, speed of operation, and reset capability are superior to the electro-mechanical relays.

6. The turbine underfrequency protection system should be in-service whenever the unit is synchronized to the system, or while separated from the system but supplying auxiliary load.

7. Provide separate alarms to alert the operator for each of the following:

• A situation of less than the nominal system frequency band on the electrical system.

• An underfrequency level detector output indicating a possible impending trip of the unit.

• An individual relay failure.

To avoid unnecessary generator trips during a disturbance from which the system could recover and to minimize stresses on the turbine, a protection scheme providing 5 or 6 frequency bands is desirable. The ideal protection system would accumulate time spent in each underfrequency band and preserve it in a nonvolatile memory. Two protective schemes are presented which provide different levels of protection and utilize the turbine abnormal frequency capability to different degrees.

SCHEME 1:. Multi-setpoint scheme with frequency band logic and accumulating counters (See Figure 1)

This scheme is designed to closely follow the turbine manufacturer's limit curves for underfrequency operation. This scheme can be applied where it is desired to fully protect the turbine and allow as much underfrequency operation as possible before tripping. Scheme 1 takes into account the following factors in its application:

1. It utilizes six underfrequency setpoints in addition to supervision steps and takes into account the cumulative time spent in each underfrequency band (multi-stage underfrequency relays can be used to reduce the number of relays required).

2. It accumulates the time spent in each underfrequency band independently and stores it in a nonvolatile memory.

3. A time delay of 10 cycles should occur before accumulation begins to allow the underfrequency blade "resonance" to be established to avoid unnecessary accumulation of time.

4. When the time for the particular band is used up, an output is given which can be used for alarms or tripping.

5. All frequency steps are supervised so that two static relay steps in series are required for an output.

6. All outputs are event recorder monitored.

7. Where possible, units are tripped to station service load if the turbine and boiler limits allow.

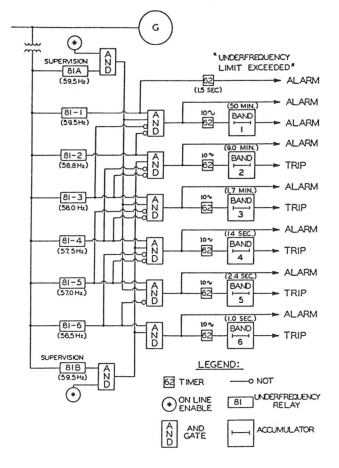

FIGURE 1
Protection Scheme 1 Block Diagram

Scheme 2: 4 setpoint scheme - two alarms, slow trip and fast trip (See Figure 2)

Scheme 2 is a simple protection system using two frequency steps. It may be considered for applications where it is deemed acceptable to have less than full time vs. frequency relay protection for the turbine. Extremely low probability of underfrequency occurrence and high reliance on the system load shedding program are factors which would be evaluated in considering use of this scheme.

Protection Scheme 2 takes into account the following factors in its application:

1. For the severe frequency decays (below 58.5 Hz) automatic relay action is taken. For the less severe frequency decays (59.5 - 58.5 Hz) the system operator will be relied on to take corrective action.

2. The protection engineer must consider an acceptable percentage of available abnormal frequency operating time for the turbine.

3. Scheme 2 does not accumulate time for multiple underfrequency events. Therefore, this scheme does not protect the multiple underfrequency events whose sum is greater than the turbine abnormal frequency capability.

4. If the frequency relay timer is set for 50% of the abnormal frequency capability, one event using 50% of the capability will cause a relay operation.

5. Two static relays must operate for a trip output.

The Guide presents an applicational example showing how each of the two protective schemes may be applied to protect a turbine.

TYPICAL UNDERFREQUENCY PROTECTION SCHEME
FOR TWO STEPS

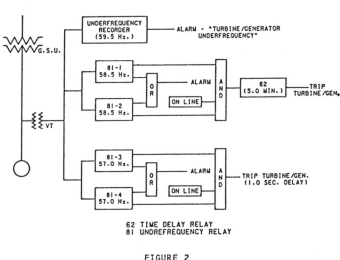

62 TIME DELAY RELAY
81 UNDERFREQUENCY RELAY

FIGURE 2
Protection Scheme 2 Block Diagram

COORDINATION WITH SYSTEM LOAD SHEDDING SCHEME

Introduction

The turbine underfrequency protection scheme should coordinate with other underfrequency protection schemes used on the connected system. It should not operate until the automatic load shedding scheme has operated to maintain the system frequency at an operable level.

Requirements

In order to coordinate a steam-turbine underfrequency protection scheme with a system automatic load shedding scheme, the following information is required:

1. A frequency response characteristic of the system for the conditions to be considered, including the effects of the automatic load shedding scheme used.

2. A time-frequency characteristic of the proposed turbine underfrequency scheme.

System Frequency Response

For a large interconnected system, modeling of loads, turbine and boiler control, area load-frequency controls, and spinning reserves are difficult to include in a simple representation of the system. An imbalance of load and generation will cause the system frequency and voltage to vary. The system load itself may vary with changes in frequency and voltage which affects the load to generation imbalance. The generation inertia constants may vary throughout the mix of steam turbine, hydro, and combustion turbine generators in the island under consideration. If the system has islanded, each island will probably have a different mixture of the above types of generation.

189

The islanded system frequency response is determined primarily by the system inertia constant, and the magnitude of the internal overload. The smaller the inertia constant the faster the frequency decline for a given overload. Newer generating units may have inertia constants of 2 to 3 since the trend in turbine generator design is toward larger outputs with smaller rotor masses. Older generators with massive rotors have inertia constants as large as 10.

Large generating units with the smaller inertia constants will tend to dictate the composite system inertia constants.

A utility must identify the condition or conditions of overload or loss of generation from which it can reasonably expect to recover. A load shedding scheme can then be applied to optimize the system frequency response.

When a scheme for turbine underfrequency protection is determined and relay setting tentatively determined, a time versus frequency curve can be drawn for the scheme similar to Figure 3.

Coordination

The system-frequency response should be compared with the turbine protection characteristic to determine if coordination between the two exists. For example, in Figure 3, the relay characteristic is a series of frequency bands. In order to trip, the system frequency must remain inside a particular band (that is, below the pick-up point) in excess of the time setting for that particular band. The time in a band may be cumulative depending on the scheme used and the previous history may have stored sufficient time such that this incident could cause the relay to trip. Precaution should be taken to assure that the turbine protection frequency relay has sufficient time delay to prevent tripping on a recoverable swing.

If there is insufficient margin between a recoverable swing and tripping a turbine for a projected system-frequency response, there are several options available:

1. Modifying the load shedding scheme by increasing the number of steps or the amount of load to be shed. This should change the frequency response sufficiently to allow the desired margin to be obtained.

2. Modify the turbine protection scheme to take additional risk of loss of blade life for the turbine.

3. Accept that, for worst-case conditions, coordination is not possible for the degree of turbine protection desired.

Options 1 or 2 could be employed to add margin to the scheme. The user must decide what is the most important for his application and make a decision accordingly. In general, Option 2 would be the least desirable compromise to make. The best choice would be to go as far as possible or practical with Option 1 before having to resort to Options 2 or 3.

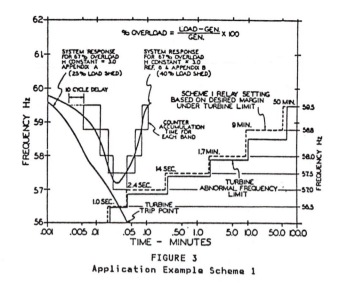

FIGURE 3
Application Example Scheme 1

GENERATOR - TRANSFORMER OVEREXCITAITON (VOLTS/HERTZ) CONSIDERATIONS

Magnetic flux in generator and transformer cores is directly proportional to voltage and inversely proportional to frequency. When the volts/hertz ratio exceeds the equipment design limit, thermal damage to the core and adjacent structures can result due to high eddy currents. High volts/hertz conditions for either the generator or the Generator Step Up (GSU) transformer can occur during unit shutdown with the generator voltage regulator left in service, following a load rejection with the voltage regulator out of service, and during startup with improper manual voltage adjustment.

The Guide discusses limiting (volts/hertz) conditions both for generators and transformers. Included is a sample calculation showing the limitations in effect when the generator and GSU low-voltage ratings are different, as is often the case.

Keeping the volts/hertz level at a safe value can be achieved with limiting devices associated with the generator voltage regulator; however, if these devices fail to operate properly, it will be necessary to trip the generator and its excitation. This can be achieved automatically with volts/hertz protective relaying, which the Guide discusses in detail. Relay setting examples are illustrated graphically for the different types of volts/hertz relays now available.

NUCLEAR GENERATING PLANTS

In addition to considering the effects of underfrequency on turbine blades for nuclear generating plants, consideration must also be given to the effect frequency changes have on the nuclear steam system. The output of electrical pumps in the system will vary with frequency and will cause various coolant flows in the nuclear steam system to change. In some cases, reduced coolant flow may be detrimental to equipment.

Some Boiling Water Reactor (BWR) units employ non-seismically qualified motor-generator sets to supply power to the reactor protection system. Redundant underfrequency protection is provided to both the primary and alternate power sources to ensure that these systems are capable of performing their intended safety function during seismic events. Consideration must be given to the tolerance of the underfrequency relay, the slip of the motor-generator set and the characteristics of the power system load shedding scheme when determining settings.

190

In Pressurized Water Reactor (PWR) units, reactor design requires the coolant flow rate to be proportional to the rate of heat production in the reactor. Abnormal system frequency will affect the reactor coolant pump flow since the flow rate is proportional to the pump speed which varies with the power system frequency.

Underfrequency relays are applied to trip the reactor at a frequency above a specified Departure from Nucleate Boiling Ratio (DNBR) level to prevent fuel rod cladding damage. In applying underfrequency protection at a PWR plant consideration must be given to: the designed DNBR, the size of the coolant system with respect to the reactor core, the rating of the core with respect to loading, the power system maximum frequency decay rate, coordination with power system load shedding and, power system voltage conditions during abnormal frequency conditions.

COMBUSTION-TURBINE UNDERFREQUENCY OPERATION

In most respects, abnormal frequency limitations of combustion turbine generators are similar to those for steam-turbine generators. However, protective requirements may differ due to application and some design differences.

In general, for combustion turbines, the turbine blades limit continuous operation to a range between 57 Hz and 60 Hz. Most combustion turbine controls unload the units as speed decreases thus offering protection to both the blades and the generators during off-frequency operation. Additional underfrequency protection may be considered depending on whether the units are used for peaking purposes and restoration following a black-out or supply a major portion of the system load.

Combined cycle generating installations, which are a combination of a combustion-turbine unit and a steam turbine unit, will require separate underfrequency protection for each unit.

CONCLUSION

The Guide discusses the fundamental reasons for potential turbine blade failure and suggests alternate remedial measures which can be taken to prevent damage. The working group feels that the Guide will be a valuable aid to the protection engineer in deciding how to approach the generating plant abnormal frequency problem.

* Generating Plant Abnormal Frequency Working Group

Members Are:

Chairman - C. J. Pencinger

J. Berdy	G. R. Nail
J. J. Bonk	G. C. Parr
D. C. Dawson	J. W. Pope
R. J. Fernandez	E. T. Sage, Jr.
S. E. Grier	H. S. Smith
R. W. Haas	R. C. Stein
T. L. Kaschalk	R. D. Stump
L. E. Landoll	F. Von Roeschlaub
J. R. Latham	

Past Members Are:

D. C. Adamson	W. M. Strang
W. A. Elmore	J. E. Waldron
C. H. Griffin	F. Wolf
J. A. Imhof	R. C. Zaklukiewicz
C. M. Shuster	

Discussion

G. D. Rockefeller (Rockefeller Associates, Inc., St. Rose, LA): This discussion focuses on the blade-fatiguing aspect of the report: whether or not a user applies automatic underfrequency protection and the trip delays he uses may be a function of the number of system underfrequency conditions he expects during the lifetime of the unit. Some areas experience several per year, while once or twice a decade may be experienced elsewhere. The objective may be to avoid fatiguing or it may be to limit the amount of fatiguing. If the generator underfrequency protection introduces a considerable risk of a system shutdown, some blade fatiguing seems a reasonable price to pay in exchange for a longer delay setting (or no automatic tripping). One needs to ask the turbine supplier the right question to meet the specifics of the installation and philosophy of the user.

Scheme 2 is very elaborate and doesn't seem to reflect existing practice. Is the Working Group aware of applications similar in scope to that of scheme 2?

Manuscript received August 5, 1986.

John R. Linders (Consulting Engineer, Sarasota, FL): The purpose of the Guide summarized in the present paper is to enhance the protection of turbogenerators and thereby the ultimate reliability of the power system. The purpose of this discussion is to further this objective by alerting the reader to several areas of possible misunderstanding or misinterpretation. To illustrate: In section 2.5.5.2, (3) of the Guide it is implied that if the relay settings which protect the turbine blades do not coordinate with the load-shedding relays, it is acceptable to modify (i.e., forego) the turbine protection. The reverse is of course to be preferred, i.e., modify the load-shedding scheme to coordinate with the turbine capabilities. Section 4.3.1 concerning CTGs concludes with "...consideration should be given to a facility for manual defeat of the protection under this condition." "This condition" refers to quick restoration of the system frequency. The reader is left confused by two ideas implied in this sentence: a) it is within the philosophy of this Guide to ignore blade protection; b) quick frequency restoration increases the risk of blade damage. The methodology of Figs. 8 and 10 (Fig. 3 in summary paper) illustrating the protection schemes described in section 2.5.5 is not sound and can result in failure to attain the desired result. In the system frequency plot, time is the independent variable and frequency the dependent variable (H constants and the extent of the overload are other parameters). These are all shown properly. For the relay performance, the dependent variable is the time of operation. However, this is plotted to the same time coordinate as the independent variable for system frequency. It is fallacious to plot these two functions on the same coordinate and then assume coordination depending on the relative position of the two plots. This is particularly true when the reset ratio of an instantaneous relay is large and also when inverse time relay characteristics are involved.

Once the proposed protection has gone through a disturbance, the initial margin between the load-shedding relays and the blade protection may no longer be valid. To maintain these margins, the accumulated time in the blade protection timers (Fig. 1 in the paper, Fig. 7 in the Guide) needs to be read out and used to recalibrate the load-shedding relays either manually or automatically. The critical resonant frequency curve, Fig. 4, in the Guide would be meaningful if typical values had been placed on the coordinates. The proposed turbine underfrequency protection results in ultimately tripping (manually or automatically) for any frequency below 59.5 Hz. Yet Figs. 3 and 4 in the Guide show only discrete critical blade resonant frequencies for a given machine. One thus wonders why the Guide proposes to trip all machines at all accumulated frequencies below 59.5 Hz regardless of criticality of a given machine.

In sections 2.7 and 2.8 on overexcitation protection, the excitation is assumed to come only from the local generator. In practice, overexcitation resulting from an underloaded island can be the result of external excitation (line capacitance and PF capacitors). The corrective needs are the same, but the resulting voltage profile may warrant additional analysis for a satisfactory solution.

Manuscript received August 11, 1986.

191

R. F. Arehart: *Concerning Discussion Submitted by Mr. G. D. Rockefeller*

In applying generator underfrequency protection, a concerted effort must be made to obtain coordination with the underfrequency load-shedding scheme and thus avoid a system shutdown during an underfrequency condition. If coordination is not obtainable and the generator underfrequency protection thus introduces considerable risk of a system shutdown, then, as Mr. Rockefeller states, the only reasonable option is to save the system by delaying the tripping of the generator underfrequency protection and accept the risk of some blade fatigue.

Mr. Rockefeller states that the generator underfrequency protection labeled as scheme 2 is very elaborate and doesn't seem to reflect existing practice. Since scheme 2 has only two frequency setpoints and scheme 1 has six frequency setpoints in addition to cumulative timers, it is believed that Mr. Rockefeller is referring to scheme 1. Scheme 1 is elaborate and is being used by some U.S.A. utilities.

Concerning Discussion Submitted by John R. Linders

Remarks on 2.5.5.2(3)—This third setting criteria states that the turbine underfrequency settings may be modified in some cases to provide coordination with the system load-shedding scheme. Mr. Linders is concerned that the turbine blade protection will be sacrificed if this protection is modified rather than the settings of the load-shedding scheme.

In 2.6.5 (Coordination) — The Guide states that the most desirable option is to modify the load-shedding scheme in order to obtain a margin between a recoverable swing and a turbine tripping. If coordination cannot be obtained through this option, only then must consideration be given to saving the system at the expense of risking possible turbine blade damage.

Remarks on 4.3.1 (Combustion Turbine Protection Philosophy & Relay Settings) — This section states that with combustion turbine generators (CTGs), consideration should be given to a facility for manual defeat of the protection "under conditions of a quick restoration of the system."

When the preponderance of weight falls on quick restoration of the system versus possible blade damage on a small machine such as a CTG, it would seem that the option involving possible turbine blade damage is a considered risk that is worthy of taking.

Remarks of Figs. 8 and 10 (Applications of Turbine Underfrequency Protection Relay Schemes) — Mr. Linders is concerned that by plotting the system response and turbine underfrequency protection using the same coordinates, the desired goal of obtaining coordination will not be obtained if the relative position of the two plots on these figures is the prime consideration used in determining that coordination.

Mr. Linders is correct, the relative position of the two plots in Figs. 8 and 10 should not be used in determining the existence of coordination. The figures were included to illustrate that with adequate load shedding and the existence of coordination with the turbine protection, the system can experience a recoverable swing. Mr. Linders, the relay performance plot should have time as the ordinate (dependent variable) and frequency as the abscissa (independent variable). Figs. 8 and 10 were plotted as shown to correlate with the turbine operating limitations of Figs. 5 and 6. It is with these coordinates that manufacturers present the user with their turbine operating limits.

It is true, as Mr. Linders states, that after a disturbance the coordination margin between load shedding and blade protection no longer exists since the blade protection timers have accumulated time. It would be desirable to know and track the accumulated time in the blade protection scheme, but it seems highly unlikely that the setting of load-shedding relays would be changed either automatically or manually, based on the accumulated timer readings alone. The sheer numbers of load-shedding relays would preclude multisetting changes.

Mr. Linders states that the proposed turbine underfrequency protection results in tripping *all* machines at *all* accumulated frequencies below 59.5 Hz regardless of a given machine's critical frequency. I believe Mr. Linders is referring to Fig. 6, which is a composite representation developed from the capability curves of five of the world's turbine manufacturers. The composite was developed in order to evaluate the performance of different relay schemes. In Section 2.3 (Turbine Over/Underfrequency Capability) it is strongly pointed out with underlining that the Fig. 6 curve is not to be used as a standard for any specific steam turbine. The Guide recommends that the turbine manufacturer be consulted in order to obtain the applicable curve for a specific machine.

Manuscript received September 11, 1986.

A Comparison of 100% Stator Ground Fault Protection Schemes
for Generator Stator Windings

J. W. Pope
Senior Member, IEEE
Georgia Power Company
Atlanta, Georgia

ABSTRACT

This paper describes three different schemes for detecting single-phase-to-ground faults in 100% of the generator stator winding. The operating theory, a setting example, and field measurements are provided for each of the three schemes. A chart comparing the significant application characteristics of the three schemes is presented.

INTRODUCTION

Single-phase-to-ground faults in the generator stator are by far the most common faults to which generators are subjected. The practice of the author's Company is to ground all system generators through a distribution transformer with a resistance loaded secondary. This type of grounding is, in effect, very high resistance grounding resulting in maximum ground fault currents in the order of 5 to 10 amperes. With high resistance grounding, the possibility of fault damage is greatly reduced, and high speed relaying, field breakers, or neutral breakers are not required to de-energize the machine before serious damage occurs. Even so, it is recommended that the generator be immediately tripped off with the occurrence of a ground fault, rather than delay this action until the unit can be shut down more conveniently. As long as the generator is operating with one phase grounded, the possibility exists for a phase-to-phase fault occurring if a second phase goes to ground. This could result in very high fault current and major equipment damage. Serious damage may also result if a fault occurs very near the generator neutral, and is then followed by a second ground higher up in the same phase. Standard overcurrent or overvoltage ground relaying will not detect this second fault, since the stator neutral is shorted. The generator differential relays will not detect this fault either if both grounds are inside the differential current transformers.

Standard overcurrent generator ground fault relaying consists of a current transformer connected in series with the distribution grounding transformer secondary resistor. The current transformer supplies current to one or more overcurrent relays. When properly set, these relays will provide sensitive protection for 90 to 95 percent of the generator stator winding and will not operate incorrectly for external faults. Standard overvoltage protection consists of a relay tuned to 60 Hz and connected across the secondary grounding resistor.

Even though the overcurrent or overvoltage generator ground fault protective scheme is straightforward and dependable, it suffers from two disadvantages. First, it will not detect ground faults near the generator neutral, and second, it is not self-monitoring. For example, an open or short circuit anywhere in the relay, primary or secondary of the current transformer, or an open or shorted grounding resistor, may not be detected before a fault occurs.

In addition to the overcurrent or overvoltage generator ground fault protection schemes, it would be prudent to protect all large generators with an additional ground fault protection system that is completely independent of the overcurrent scheme, gives reliable protection to 100% of the generator, and continuously monitors the generator grounding system.

This paper describes the ground fault protection installed on three Georgia Power Company generators. The operating theory and a setting example is provided for each type protection scheme. A chart comparing the significant application characteristics of the three protective schemes is presented.

Type 1 injects a voltage at a subharmonic frequency and trips on an increase in current. The current is caused by the shorting out of generator capacitance resulting from a single-phase-to-ground fault. Type 2 employs two overlapping voltage relays—an overvoltage relay that protects the high voltage end of the machine, and an undervoltage relay tuned to respond to the third harmonic protecting the neutral. Type 3 is a harmonic voltage comparator relay that operates on a change in the ratio of terminal to neutral end third harmonic voltage. This scheme must be used in conjunction with the standard overcurrent or overvoltage scheme to obtain 100% coverage.

TYPE 1—SUBHARMONIC INJECTION SCHEME

Operating Theory

The neutral voltage injection scheme detects ground faults by injecting a 15 Hz voltage between the generator neutral and ground, and measuring the resultant 15 Hz current. When a ground fault occurs, the 15 Hz current increases and causes the relay to operate. The 15 Hz injection signal is synchronized to the 60 Hz generator terminal voltage. Since the 15 Hz current measurement is done by integrating during a complete half wave (15 Hz), all other system signals existing at the harmonics of 15 Hz (i.e., 30 Hz, 60 Hz and higher) will be integrated to zero and will not influence the measurement. The use of a subharmonic injection frequency will improve sensitivity due to the higher impedance path of the generator and connected equipment capacitance at this low frequency. Refer to **Figure 1** for a simplified schematic of the Type 1 scheme.

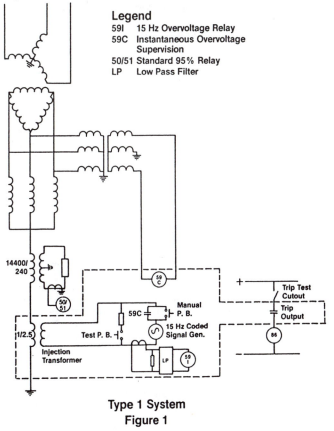

Legend
59I   15 Hz Overvoltage Relay
59C   Instantaneous Overvoltage Supervision
50/51 Standard 95% Relay
LP   Low Pass Filter

Type 1 System
Figure 1

83 SM 416-5   A paper recommended and approved by the IEEE Power System Relayed Committee of the IEEE Power Engineering Society for presentation at the IEEE/PES 1983 Summer Meeting, Los Angeles, California, July 17-22, 1983. Manuscript submitted February 4, 1983; made available for printing May 10, 1983.

Reprinted from *IEEE Trans. Power App. Syst.*, vol. PAS-103, no. 4, pp. 832–840, April 1984.

Security against misoperation is achieved by coding the 15 Hz injection voltage. The signal consists of alternating intervals when the transmission of the 15 Hz signal is either "on" or "off." Each test period of 470ms includes an "on" interval of about 200ms (3 cycles) followed by a 270ms "off" interval when no transmission takes place. The "on"-"off" sequence takes place continuously with measurements made during each test period. **Figure 2** is a timing chart showing the measurement technique used during each test period. Measurements are made during three successive "on" interval half-cycles of the 15 Hz signal. The measurement technique makes independent integrations of two negative and one positive half-cycles (T_1-T_3 on Figure 2). These integrals for each "on" interval are compared against a reference set by the pickup potentiometer. The results of these three comparisons are stored in memory. During the "off" interval, similar integrations are made of six successive 15 Hz half-cycles (B_1-B_6, Figure 2). Only 15 Hz noise and leakage current in the neutral circuit contribute to these integrals. For increased security, sensitivity of the pickup circuit is automatically increased to 40-60% of nominal pickup during the "off" interval. If any of the six integrals in each "off" interval falls above the threshold setting in the relay, the relay trip signal is blocked and an alarm sounds. Two conditions must be satisfied to produce a trip output. First, **all** three half-cycle integrations during the "on" interval must be above the pickup setting. Second, **none** of the "off" interval integrations during the same test period can be above the blocking threshold setting. The memory is reset at the end of each test period. When these two conditions are satisfied during one test period, a trip output is given.

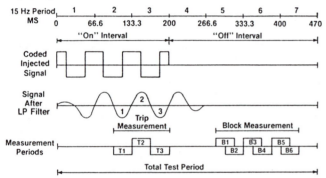

Type 1 System—Timing Chart
Figure 2

Operation time varies between one and two test periods (470-940ms) depending on when during the test period the ground fault occurs. When the generator is on turning gear and not excited, the injection voltage must be turned off for personal safety reasons. A voltage supervision relay (59C in **Figure 1**) is used to deactivate the injection scheme when the generator voltage is below 40% of rating. The injection scheme may be put in service with the generator on turning gear, by using the manual pushbutton shown in **Figure 1**. This feature allows a check for grounds after maintenance and before energizing the generator.

The Type 1 scheme can readily be tested while in service by using the test pushbutton shown in **Figure 1**. A stator ground fault is simulated by pressing the test pushbutton, which places a short circuit across the 15 Hz signal generator output. The trip test cutout switch must be opened during this test to prevent tripping the generator. The logic and integrators are functionally tested using this feature. The Type 1 scheme will operate with the same sensitivity for phase-to-ground faults at the generator neutral or terminals. A significant flaw in this scheme is its inability to detect and trip or alarm for open circuits in the grounding transformer primary or secondary. An open grounding transformer will result in a decrease in the 15 Hz current level. The Type 1 system is looking for an increase in current level caused by shorted capacitance. An undercurrent relay with time delay may be applied with the Type 1 scheme to detect a loss of the injected 15 Hz signal and alarm indicating a grounding system problem or relay failure.

Setting Example
The following setting example is based on an existing Type 1 installation on Georgia Power Company's Plant Hatch Unit No. 1. This is an 810 MW, 24-kV nuclear unit connected to the 230-kV network in South Georgia.

The application of a Type 1 scheme requires that four specific values be calculated to permit specification of components and settings:

1. **Minimum theoretical setting** corresponds to the 15 Hz current that will flow under normal (no fault) conditions.

2. **Maximum theoretical setting** corresponds to the 15 Hz current that will flow for a bolted phase-to-ground fault at the generator neutral.

3. **The maximum 60 Hz ground fault current** determines the coupling transformer and padding capacitors rating.

4. **The pickup setting during the "off" interval** has 40%, 50% or 60% taps available.

The data required for these application calculations is:

Generator winding capacitance to ground per phase.	C_G = 0.43000	μF
GSU transformer low voltage winding phase-to-ground capacitance.	C_T = 0.00975	μF
Sum of phase-to-ground capacitance of generator leads, potential transformer primary and other incidental equipment.	C_V = 0.00245	μF
Phase-to-ground capacitance of unit auxiliary transformer(s) high voltage winding.	C_A = 0.00710	μF
Total single phase capacitance to ground of the protected zone.	C_{TOTAL} = 0.44930	μF
Leakage impedance of the generator grounding transformer referred to the generator side.	$Z_{GT} = R_{GT} + X_{GT}$ ohms = 15.6 + j35.0 (60 Hz) = 15.6 + j8.75 (15 Hz)	
Leakage impedance of the injection transformer.	$Z_W = R_W + jX_W$ ohms = 36 + j125 (60 Hz) = 36 + j31.2 (15 Hz)	
Turns ratio of injection transformer (V_1/V_2)	$\dfrac{V_1}{V_2} = \dfrac{1 \text{ (GEN.)}}{2.5 \text{ (RELAY)}}$	
Grounding resistor ohms.	R_x = 0.4316 ohms	
Generator phase-to-phase voltage.	V_{L-L} = 24-kV	

1. Minimum Theoretical Setting. Calculate the normal 15 Hz leakage current through the circuit of **Figure 3**.

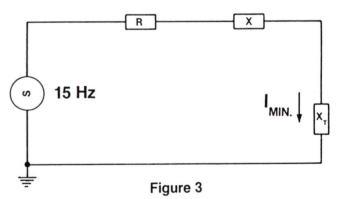

Figure 3

$$R = R_S \text{(primary)} + R_{GT} + R_W = 1605 \text{ ohms} \quad (1)$$

$$X = X_{GT} + X_W = +j39.95 \text{ ohms (15 Hz)} \quad (2)$$

X_T = Three phase capacitive reactance to ground of protected system.

$$X_T = \frac{1}{3\omega C_{TOTAL}} = -j7872 \text{ ohms (15 Hz)} \quad (3)$$

$$I_{MIN.} = \frac{V}{R + j(X + X_T)} = 7.0 \text{ mA} \quad (4)$$

140 volts is the peak 15 Hz relay output voltage specified by the relay manufacturer.

The current flowing in the injection transformer secondary (relay input) is:

$$I_{MIN. RELAY} = \frac{V_1}{V_2} I_{MIN.} = 2.8 \text{ mA} \quad (5)$$

To avoid false tripping, the relay setting must be above 2.8 mA.

2. Maximum Theoretical Setting. Calculate the 15 Hz current that will flow during a bolted phase-to-ground fault at the generator neutral, which is, the current flowing in **Figure 3** with a short around X_T.

$$I_{MAX.} = \frac{V}{R + jX} = 34.9 \text{ mA} \quad (6)$$

The current flowing in the injection transformer secondary (relay input) is:

$$I_{MAX. RELAY} = \frac{V_1}{V_2} I_{MAX.} = 13.95 \, mA \qquad (7)$$

To be sure that the relay will trip for ground faults, the relay setting must be below this value.

The relay should be set for approximately 75% of $I_{MAX. RELAY}$.[2] The setting must be higher than $I_{MIN. RELAY}$.

$$Setting = (75\%)I_{MAX. RELAY} = (75\%) \, 13.95 \, mA = 10.46 \, mA \qquad (8)$$

10.46 mA is significantly higher than $I_{MIN. RELAY} = 2.8 \, mA$. The selected settings will accommodate a fault resistance of:

$$10.46 \, mA = \frac{140 \, (\frac{1}{2.5}) \, (1000) \, volts}{1605 + R + j39.95 \, ohms}; \; R = 3748 \, ohms \qquad (9)$$

3. Calculate the Maximum 60 Hz Ground Fault Current.

$$I_{MAX. 60} = \frac{V_{L-N}}{R_S + R_{GT} + R_W + j(X_{GT} + X_W)} = 8.60 \, amperes \qquad (10)$$

$I_{MAX. 60}$ on the secondary side of the injection transformer (relay input) is:

$$I_{MAX. 60 RELAY} = \frac{V_1}{V_2} I_{MAX. 60} = 3.44 \, amperes \qquad (11)$$

The maximum allowed value is 3.5 amperes due to the relay design. The coupling transformer and padding capacitor should be specified for thermal capability of no less than 8.60 amperes.

4. Calculate the pickup setting during the "off" interval.

Percent dead interval pickup

$$\leq 80 \, (\, 1 - \frac{I_{MIN. RELAY}}{RELAY \, SETTING}) \; \leq 58.6\% \qquad (12)$$

Available taps are 40%, 50%, and 60%.
A setting of 40% or 50% is acceptable.
Select 40%.

Field Measurements

The Type 1 relay system was installed on Hatch Unit No. 1 with the relay settings as calculated in the previous example.

Field measurements were made to verify relay performance and the accuracy of data used in the setting calculations. **Table 1** shows measurements taken for different fault locations. Bolted single phase-to-ground faults were placed at the neutral and terminal ends of the generator with the generator de-energized.

Relay Current (15 Hz—Peak)	Calculated	Measured
No Fault, $I_{MIN. RELAY}$	2.8 mA	2.2 mA
Neutral Fault, $I_{MAX. RELAY}$	13.95 mA	17.1 mA
Terminal Fault, $I_{MAX. RELAY}$	13.95 mA	17.1 mA
Relay Setting	10.46 mA	12.8 mA

Table 1

The relay setting was modified to reflect the actual measured value.

$$Setting = 75\% \times 17.1 \, mA = 12.8 \, mA \qquad (13)$$

The variations between calculated and measured values can be attributed to a number of factors.

1. The 140 volt peak output specified by the relay manufacturer was measured to be 150 volts peak.
2. Since the no-fault calculated value was larger than the no-fault measured value, the capacitance values supplied by the plant designers and equipment manufacturers must be larger than the actual protected zone capacitance.
3. Since the neutral fault calculated value was lower than the neutral fault measured value, the generator grounding resistor value, injection transformer impedance and grounding transformer impedance may be lower than specified. Some variation could be attributed to the "cold" temperature of this equipment during test.

The manufacturer does not specify that field measurements are required to verify the Type 1 installation. The Hatch Unit No. 1 measurements indicate the calculated and measured values may differ significantly. Field measurements as described are recommended.

TYPE 2—60 HZ OVERVOLTAGE AND 180 HZ UNDERVOLTAGE

Operating Theory

The overlapping overvoltage/undervoltage scheme provides 100% protection for generator stator ground faults by using two measuring functions that cover different portions of the machine winding. The voltage relays are connected as shown in **Figure 4**.[3]

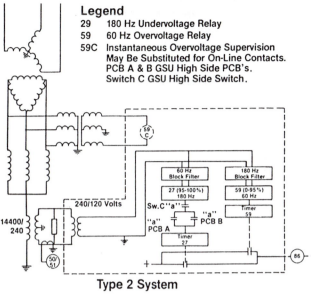

Legend
29   180 Hz Undervoltage Relay
59   60 Hz Overvoltage Relay
59C   Instantaneous Overvoltage Supervision
     May Be Substituted for On-Line Contacts.
     PCB A & B GSU High Side PCB's.
     Switch C GSU High Side Switch.

**Type 2 System
Figure 4**

The overvoltage protection path consists of a 240/120 volt isolating transformer across the neutral resistor, a 180 Hz blocking filter, an overvoltage level detector (59) and a timer.

The 240/120 volt isolating transformer limits the voltage to the continuous rating of the relay, provides isolation of the protection schemes and allows the grounding of one input at the relay location. An isolating transformer with a D-C resistance of less than 2.5 ohms should be used in order to prevent excessive voltage drop. A 500 VA transformer with a resistance of 0.89 ohm was selected.

The 180 Hz blocking filter and overvoltage device 59 are calibrated as a unit. The pickup voltage at 180 Hz is about 18 times the 60 Hz value.

The connection of the overlapping undervoltage relay is also shown in **Figure 4**.[3] This relay is tuned to operate on the 180 Hz component of the voltage appearing at the generator neutral. The undervoltage protection path consists of a 240/120 volt isolating transformer across the neutral resistor, a 60 Hz blocking filter, an undervoltage level detector, on-line logic, and a timer. The 60 Hz blocking filter and the undervoltage device labeled 27 are calibrated as a unit. The frequency response of this unit is such that the pickup voltage at 60 Hz is about 100 times the 180 Hz value. This results in a 1.0% leakage of the fundamental voltage. This filter has recently been redesigned to provide a 1.0% leakage instead of 3.3% that was provided in the past. A significant flaw in this scheme is its inability to detect and trip or alarm for open circuits in the secondary grounding resistors.

Since device 27 responds to an absence of third harmonic voltage, a supervision scheme must be provided to block false trips when the generator is out of service. For example, overcurrent supervision may be necessary with generator designs that do not produce significant third harmonic voltage until loaded. In this case, an alternate protective scheme should probably be considered, since a third harmonic undervoltage relay would be out of service under light load conditions. **Figure 4**[3] shows an alternate overvoltage supervision scheme which could also be considered in place of the on-line supervision switch.

Setting Example

The following setting example is based on an existing Type 2 installation on Georgia Power Company's Plant Bowen Unit No. 2. This is a 789 MW, 25-kV coal fired unit.

The overvoltage relay (59) setting is controlled by the maximum voltage applied to the relay for a phase-to-ground fault on the machine terminals, and the neutral voltage resulting from a phase-to-ground fault on the high side of the generator step-up transformer due to capacitive coupling between the transformer windings.

In the case of Bowen Unit No. 2, the maximum voltage applied to the relay is:

$$V_R = \frac{V_{L-N}}{N \times 240/120} = \frac{25000/1.73}{60 \times 2} = 120V \qquad (14)$$

In order for the overvoltage relay to detect ground faults in the first 95% of the stator winding, measured in from the generator high voltage terminals, it must be set for $(0.05 \times 120) = 6.0$ volts.

The effect of a high side phase-to-ground fault on the generator neutral voltage may be calculated using the circuit of **Figure 5**.

V_{HO} = high side zero sequence voltage for a 500-kV phase-to-ground fault = 147-kV

C_{H-L} = generator step-up transformer high-to-low winding capacitance = 0.012 μF

C_O = generator system zero sequence capacitance per phase = 0.254 μF

R_N = equivalent resistance in the generator neutral = $0.78 \times 60^2 = 2809$ ohms

Z_O = parallel combination of $3R_N$ and X_{CO}

X_{CO} = zero sequence capacitive reactance per phase

The voltage appearing across the primary of the generator grounding transformer is:

$$V_R = \frac{V_{HO}Z_O}{Z_O - jX_{H-L}} = 4280 \text{ volts} \tag{15}$$

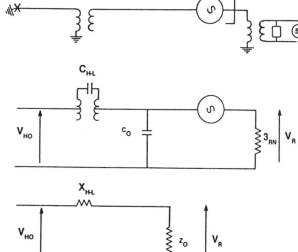

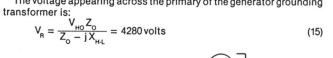

Figure 5

This voltage is $\frac{4280}{25000/1.73} = 30\%$ neutral shift and results in 36 volts supplied to device 59 during the time that a phase-to-ground fault exists on the high voltage bus. The pickup of device 59 may be set larger than this value or device 59 time delay setting must be coordinated with the clearing time of 500-kV backup relaying. In this case, timer 59 will be set for 75 cycles to eliminate possible overtripping for ground faults on the transmission system. If a setting larger than 36 volts were used (now set for 6.0 volts), device 59 sensitivity would be reduced drastically.

Determination of device 27's pickup level requires calculations of third harmonic voltage distribution, field checking of this value, and an analysis of the variation of third harmonic voltage at the neutral for ground faults at various points in the machine. The minimum third harmonic voltage generated by Bowen Unit No. 2 is given by the manufacturer as 200 volts at no load. The third harmonic voltage distribution may be analyzed by grouping the terminal end capacitance and the neutral end resistance and capacitance and performing a voltage drop calculation to determine the portion of the 200 volts third harmonic that will appear across the grounding transformer primary. Performing this calculation for Bowen Unit No. 2 yields:

V_N = voltage at neutral = 103 volts

The ratio of the neutral voltage (V_N) to the generated voltage is (103/200), so that 52% of the third harmonic voltage is present at the neutral.

The third harmonic voltage at the relay is:

$$V_R = 103 \times \frac{240}{14400} \times \frac{120}{240} = 0.86 \text{ volt at no load.} \tag{16}$$

The dropout of the third harmonic voltage relay should provide a 2/1 margin for the minimum no-fault voltage appearing at the neutral.[4] In this

case, a relay setting of $(\frac{1}{2})(0.86) = 0.43$ volt may be considered. Thus, when the third harmonic neutral voltage decreases to 0.43 volt, the 27 relay will drop out and start the timer. The unit will be tripped after a delay of about three seconds.

Voltage variations during faults are significant in determining an appropriate undervoltage setting, and then calculating the overlap of the two voltage relays. Phase-to-ground faults at the machine terminals will cause the fundamental frequency voltage at the neutral to rise to the phase-to-neutral voltage. The third harmonic voltage will also increase above the unfaulted level, since the terminal capacitance is shorted. A phase-to-ground fault at the neutral will short circuit the grounding transformer, causing the fundamental and third harmonic voltages to equal zero. Previous calculations show that the fundamental frequency overvoltage relay (59) protects 95% of the generator winding, measured from the high voltage terminals. Since the third harmonic voltage increases for faults at the generator terminals, and equals zero for neutral faults, there is some point in the winding where a phase-to-ground fault will have no effect on the third harmonic voltage at the generator neutral. For Bowen Unit No. 2, this point is 52% up from the neutral end of the machine.

It is now necessary to confirm the adequacy of the overlap of the 0-95% relay (59) and the 95-100% relay (27). At the 5% pickup point of device 59, the third harmonic voltage at the neutral will be:

$$V_N = 103 \times \frac{5}{52} = 9.9 \text{ volts} \tag{17}$$

The voltage applied to the relay will be:

$$V_{NR} = 9.9 \text{ volts} \times \frac{240}{14400} \times \frac{120}{240} = 0.083 \text{ volt} \tag{18}$$

The operational margin of the third harmonic relay is $0.43/0.083 = 5.18$ to 1. This margin will be reduced by the 60 Hz voltage which will appear at the third harmonic relay filter output. This voltage is 6 volts $\times 1.0\% = 0.06$ volt. This 60 Hz voltage reduces the margin to $0.43/\sqrt{(0.06)^2 + (0.083)^2} = 4.20$ to 1 at **no load**.

From the above discussion it is evident that the 60 Hz blocking filter output affects relay operate margins. Considering the filter leakage, the maximum reach of the relay may be described by the following equations:

The no-load third harmonic voltage at the relay (V_{R3}) as a function of (S), the distance in percentage of winding measured up from the neutral, will be:

$$V_{R3} = 103 \times \frac{S}{52} \times \frac{1}{60} \times \frac{1}{2} = 0.0165S \tag{19}$$

The fundamental frequency voltage at the relay (V_{R1}) due to 60 Hz blocking filter leakage of 1.0% is:

$$V_{R1} = \frac{S}{100} \times \frac{25000}{1.73} \times \frac{1}{60} \times \frac{1}{2} \times 1.0\% = 0.0120S \tag{20}$$

at any load.

The total operate voltage applied to the relay is $\sqrt{V_{R3}^2 + V_{R1}^2} = 0.0204S$. At the relay setting of 0.43 volt, the reach up from the neutral at **no load** is:

$0.43 = 0.0204S$, and

$S = 21\%$

$$V_{R3} = 103 \times \frac{7.8}{1.8} \times \frac{S}{52} \times \frac{1}{60} \times \frac{1}{2} = 0.0715S \tag{21}$$

at full load.

(7.8 and 1.8 are the full load and no load voltages measured across the resistor—see field measurements.)

The total full load operate voltage applied to the relay is $\sqrt{V_{R3}^2 + V_{R1}^2} = 0.0725S$.

At the relay setting of 0.43 volt, the reach up from the neutral is:

$0.43 = 0.0725S$, and

$S = 5.9\%$

The 27 relay operate ratio for a ground fault 5.0% up from the neutral is 1.2 to 1, and this means that the overlap of the 27 and 59 relays is marginal for a 27 relay setting of 0.43 volt. For this reason, the preferred 2/1 third harmonic requirement must be modified. Reducing this margin to 1.5/1 establishes an undervoltage setting of $(0.86/1.5) = 0.57$ volt. At this setting the reach up from the neutral at **full load** is:

$0.57 = 0.0725S$, and

$X = 7.9\%$

This setting increases the coverage at no load to $0.57/0.0204 = 28\%$. Another option is to reduce the 6.0 volt setting of the overvoltage relay

59 to 5.0 volts. This setting would provide 59 device coverage for all except the last 5.0/120 = 4.17% of the stator winding.

Field Measurements

The application of harmonic sensitive voltage relays to the generator neutral requires an analysis of the frequency content of the neutral voltage.

Figure 6[3] shows the frequency spectrum of the voltage appearing across the resistor on the secondary side of a generator grounding transformer. This plot is for Bowen Unit No. 2 loaded to 600 MW and 90 MVAR (out). The plot was made by passing a 3 Hz frequency window across the frequency spectrum and plotting the voltage passed by this window. Note that the peak voltage was the third harmonic, which reached 6.3 volts (RMS). 6.3 volts across the resistor corresponds to:

$$\frac{60 \times 6.3}{25000/1.73} \times 100\% = 2.6\% \qquad (22)$$

of the generator phase-to-neutral voltage rating. The 60 Hz component of the neutral voltage was 0.07 volts (RMS) across the resistor.

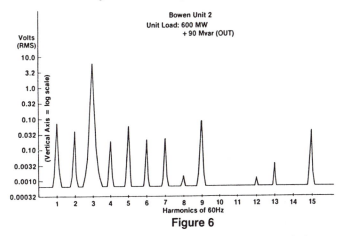

Figure 6

The third harmonic content of the neutral voltage may vary between 1% and 10% of the generator terminal voltage, depending on the design philosophy. One can expect at least 50% more third harmonic at full load than at no load.[4] A two-month survey of third harmonic voltage variations across the neutral resistor of this unit showed that the third harmonic voltage varied from 5.5 volts to 7.8 volts under normal load conditions. The minimum third harmonic voltage of 1.8 volts was observed prior to a routine unit trip and was observed again when the unit was resynchronized to the line. For this machine, the no-load third harmonic voltage is:

$$\frac{1.8}{7.8} = 23\%$$ of the maximum value monitored. The calculated third harmonic voltage across the resistor is 103 × 240/14400 = 1.72 volts, which compares favorably with the measured value of 1.8 volts at generator minimum load.

Measurements were also made on this unit to determine the variation of third harmonic voltage at the neutral as a function of power factor angle. The unit reactive component was varied from 200 MVAR "out" to 80 MVAR "in" while holding the real power constant at 630 MW. No variations in the third harmonic voltage were noted.

TYPE 3—180 HZ VOLTAGE COMPARATOR

Operating Theory

The Type 3 generator stator ground detector is a third harmonic voltage comparator relay. The relay compares the magnitude of the third harmonic voltage at the generator neutral to the third harmonic voltage at the generator terminals. Stator ground faults near the neutral or terminal end of the generator will upset the normal third harmonic distribution resulting in a relay operation. The Type 3 relay supplements the normal 90% overvoltage or overcurrent relay to provide total stator winding ground fault protection coverage.

Figure 7 shows a simplified schematic for a Type 3 installation. The relay measuring circuit consists of a d'Arsonval type dc contact making milliammeter (0.75-0-0.75 mA dc), two bridge rectifiers, two 180 Hz pass filters and one isolating/matching transformer. The isolating/matching transformer provides a step-up ratio from primary (input) to secondary (comparator voltage) of 1:1 to 1:4.1 depending on the taps selected for S, C and F. The step-up ratio is determined by the equation S(C + F).

The basic operating principle of the Type 3 scheme is a differential third harmonic approach. The matching transformer is used to balance the

third harmonic voltage from the generator neutral connection with the third harmonic voltage from the geneator terminal connection. The difference voltage will cause a current to flow in the 59D relay of **Figure 7**. This approach assumes that the ratio between the terminal and neutral end voltages will remain constant for varying unit load levels. If the ratio of terminal to neutral voltage changes, the difference voltage will appear as an operate voltage to the relay. Various laboratory tests were performed on the Type 3 relay to determine its operating characteristics.

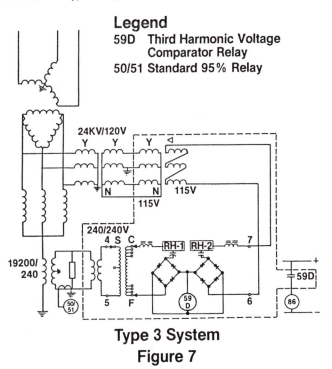

Legend
59D Third Harmonic Voltage
Comparator Relay
50/51 Standard 95% Relay

Type 3 System
Figure 7

The contact making milliammeter sensitivity was measured by reading the deflection (labeled in milliamperes where 0.75 mA = contact closure) resulting from a voltage on one input only. The matching transformer was tapped for a ratio of 1:1.

With RH-1 and RH-2 calibrated per the manufacturer's recommendations, a trip output was obtained with 2.2 volts (180 Hz) applied to one input with no voltage applied to the other.

The operating time for contact closure varies from 528 ms at 2 times pickup to 190 ms at 10 times pickup.

Setting Example and Field Measurements

Setting calculations for the Type 3 scheme are simple and straightforward. The settings are directly dependent on field measurements and do not require involved calculations unless one desires to calculate the exact portion of the stator winding protected by this scheme.

The only setting to be calculated is the isolating/matching transformer ratio. This ratio may be determined as follows:
1. With the generator at or near full-load, measure the third harmonic voltage at the appropriate P.T. secondaries.
2. Calculate the appropriate tap setting for the ratio matching transformer to provide a balance.
3. After setting the isolating/matching transformer, it may be necessary to adjust RH-2 to balance the resistance of RH-1 plus the impedance of the isolating transformer.
4. Connect the relay and observe that the moving contact is at the center position and that it remains within the balance tolerance of ± 0.075 mA deflection over the entire range of **operation of the generator.** (This requirement can only be realized if the ratio of terminal to neutral end voltage does not change over the operating range of the generator. The following example shows that this requirement cannot be attained in some cases.)

The following example is based on a Type 3 installation on Georgia Power Company's Plant Yates Unit No. 6, 350 MW, 22-kV unit.
1. Columns 2 and 3 of **Table 2** provide measured values for this unit at various loadings. The full-load 180 Hz voltages are 8.0 volts from both relay inputs.
2. The tap setting for the ratio matching transformer should be:

$$\frac{\text{TERMINAL VOLTAGE}}{\text{NEUTRAL VOLTAGE}} = \frac{8.0V}{8.0V} = 1.0 \qquad (23)$$

Column 4 of **Table 2** shows that the terminal voltage to neutral voltage ratio varies from 1.48 at no-load to 1.0 at full-load. The Type 3 instruction book indicates that this ratio should be constant over the operate range of the unit. Reference 5 provides additional field measurements for various size generators.

3. With full-load third harmonic voltages applied to the relay, RH-2 was adjusted to 500 ohms and RH-1 was adjusted to achieve a balance.

4. With the unit load reduced to zero, a **trip output was obtained.** The matching transformer ratio was changed to 1.15 to allow the relay trip contact to float open with the generator at no-load.

Unit Load MW, MVARS	180 Hz (RMS) Voltage		180 Hz Voltage Ratio	60 Hz (RMS) Voltage	
	Gen. Neut. (Studs 4-5)	Gen. Term. (Studs 6-7)	Term. V./Neut. V.	Gen. Neut. (Studs 4-5)	Gen. Term. (Studs 6-7)
0,0	2.5	2.7	1.48	1.8	4.7
7,0	2.5	3.7	1.48	1.9	4.9
35, −5	2.7	3.8	1.41	1.9	4.9
105, −5	4.2	5.0	1.19	2.0	5.0
175, −25	5.5	6.2	1.13	2.0	4.9
340, −25	8.0	8.0	1.00	4.7	4.7

Table 2

It is significant to note that the **range** of mA deflection and the **centering** of the meter movement about the null point are dependent on the matching transformer tap.

The matching transformer ratio that will result in centering the meter movement about zero may be calculated as follows:

N = Desired matching transformer ratio

A = Minimum load generator neutral voltage (180 Hz) = 2.5 volts (Table 2)

B = Minimum load generator terminal voltage (180 Hz) = 3.7 volts (Table 2)

C = Maximum load generator neutral voltage (180 Hz) = 8.0 volts (Table 2)

D = Maximum load generator terminal voltage (180 Hz) = 8.0 volts (Table 2)

$$(AN - B) = -(CN - D) \qquad (24)$$

$$(2.5N - 3.7) = -(8.0N - 8.0)$$

$$N = 1.12$$

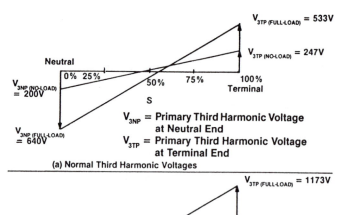

(a) Normal Third Harmonic Voltages

V_{3NP} = Primary Third Harmonic Voltage at Neutral End

V_{3TP} = Primary Third Harmonic Voltage at Terminal End

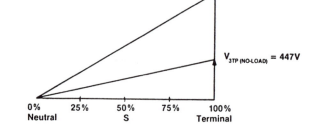

(b) Third Harmonic Voltages for Fault at Neutral

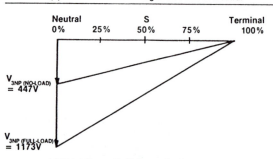

(c) Third Harmonic Voltages for Fault at Generator Terminal

Figure 8

	Type 1—Subharmonic Injection	60 Hz Overvoltage Type 2-180 Hz Undervoltage	Type 3—180 Hz Comparator
1. **Percentage Coverage** Portion of generator bus, connected transformer windings, stator winding, neutral connections protected.	Covers 100% of stator winding and connected equipment. No additional protection required for 100% coverage.	The combination of overvoltage and third harmonic undervoltage covers 100% of the stator winding and connected equipment. Overlap calculations are required.	Covers the terminal and neutral ends of the stator winding and connected equipment. Conventional stator protection is required to cover center portion of stator winding.
2. **Self-monitoring** (a) Detects short circuits in stator grounding system. (b) Detects open circuits in stator grounding system.	(a) Operates for shorts across the primary or secondary of the grounding transformer. (b) Will not detect or alarm for open grounding transformer primary or secondary.	(a) Operates for shorts across the primary or secondary of the grounding transformer. (b) Operates for open grounding transformer primary or secondary. Will not detect or alarm for open grounding resistor.	(a) Operates for shorts across the primary or secondary of the grounding transformer. (b) Operates for open grounding transformer primary or secondary.
3. **Off line supervision required.**	Yes	Yes	No
4. **Field measurements required for setting.**	Yes	Yes	Yes
5. **Setting calculation/field commissioning difficulty**	Moderately difficult/Easy	Moderately difficult/Easy	Easy/Easy
6. **Ground detection on turning gear or at standstill.**	Yes	No	No
7. **Sensitivity affected by machine loading.**	No	Yes	Yes
8. **Connection to neutral of generator P.T.'s.**	No	No	Yes
9. **Requires blocking for generator P.T. blown fuse.**	No	No	Yes
10. **Tripping time**	940 mS max.	59 Unit = 1250 mS, 27 Unit = Adjustable	528 mS @ 2 times P.U.
11. **Test features**	Yes	No	No
12. **Power supply required**	Built-in	No	No
13. **Approximate price**	3x	X	X
14. **Applicable to multiple units on same bus.**	Yes	Maybe	No

Table 3
Comparison of 100% Stator Ground Fault Schemes

The final value of 1.12 for the matching transformer ratio gives a range of meter deflection from + 0.57 mA at no-load to − 0.50 mA at full-load. This setting provides margin against misoperation at no-load and full-load by centering the range of meter movement about the null point. If the ratio of terminal to neutral voltage varies so much from no-load to full-load that a secure setting cannot be obtained, the contact that trips for faults near the terminal end may be disabled allowing a secure setting for neutral end protection.

The variation in the ratio of terminal to neutral third harmonic voltage was monitored as a function of power factor. No variation was detected over the normal operating range of this unit.

The portion of the stator winding covered with this relay setting can be calculated by determining the magnitude of the terminal and neutral voltage as a function of fault location. **Figure 8** demonstrates how the terminal and neutral voltages vary with fault location. Calculations show that a matching transformer ratio of 1.12 will provide the following coverage for single phase-to-ground faults:

The no-load protection range is 0-50% and 67%-100% of the stator winding measured up from the neutral. The no load dead band is from 50%-67% of the stator winding.

The full load protection range is 0-46% and 53%-100% of the stator winding measured up from the neutral. The no load dead band is from 46%-53% of the stator winding.

The standard overcurrent (or overvoltage) 60 Hz 0-95% relay will overlap the no-load or full load dead band resulting in 100% stator protection.

SUMMARY COMPARISON CHART

Table 3 provides a summary of the more important application and operating characteristics of the three types of 100% stator ground fault protection discussed.

REFERENCES

(1) P. G. Brown, **Generator Neutral Grounding**, General Electric Co. Application Enegineering Information GET 1941A, General Electric Co., Schenectady, NY, p. 5.

(2) "Protection Against Ground Faults Covering 100% of the Stator Windings Type GIX103," **Brown Boveri Corporation Relay and Protection Schemes Publication CH-ES 31-40A.**

(3) C. H. Griffin and J. W. Pope, "Generator Ground Fault Protection Using Overcurrent, Overvoltage and Undervoltage Relays," **IEEE Transactions on Power Apparatus and Systems,** Vol. PAS-101, No. 12, December 1982, pp. 4490-4497.

(4) M. Stein and J. R. Linders, "Ground Fault Protection of the Complete Generator Winding," **Proceedings of the Pennsylvania Electric Association Relay Committee,** October 6-7, 1977.

(5) J. Pilleteri and J. R. Clemson, "Generator Stator Ground Protection," **Proceedings of the May 7-8, 1981 Protective Relaying Conference,** Georgia Institute of Technology, Atlanta, GA.

Discussion

Gunnar Stranne, (ASEA Inc., Yonkers, NY): It is valuable to the industry to get users' input as to how various relaying concepts are viewed in terms of applicability and complexity. The use of the third harmonic zero sequence voltage component for 100% stator ground fault relaying offers simplicity in that no additional HV apparatus is required. The injection principle offers a solution to applications where no third harmonic voltage is availble for relaying, but depends on the availability of a reliable sub-harmonic source. The influence of the injected signal on the generator system may also require studies. Service experience is also a factor that would be of interest. Some comments in this regard would be appreciated, i.e., how prevalent are faults in the neutral region of the stator?

The described type 2 system is a non-standard version. The standard relay rcommended by the manufacturer uses a terminal voltage relay to supervise the third harmonic undervoltage relay. This offers the advantage of providing protection also during start-up when the supervised voltage level is reached prior to synchronization. The fundamental overvoltage relay provides increased sensitivity to ground faults during start-up when the frequency is between 20 and 60 Hz. Since generator voltage would be required to permit fault current to flow, this appears to be a more desirable approach than supervision via auxiliary contacts.

One advantage of the type 2 relay is that the third harmonic of the machine terminal does not need to be supervised. This means that additional PT's or auxiliary PT's are not required to provide the third harmonic terminal quantity. This often results in a less expensive installa-

tion, particularly for small hydro-installations that do not have a three-phase potential transformer. Since only the neutral third harmonic voltage is used, there is no need to evaluate the level and ratio of the neutral to terminal third harmonic, which results in a somewhat less complex setting and calculation procedure than for type 3 relays with regard to the coverage and overlap with the 95% relay obtained thru the use of the third harmonic voltage. Point 5 in Table 3 should therefore reflect this difference, particularly since there is no difference in the need to provide the 95% overvoltage function with these two relay principles.

The ratio of terminal to neutral end third harmonic voltage is not linear with loading, and the setting of the type 3 relay therefore has to be determined at the time of installation. If this ratio difference varies to such an extent that the pick-up value for the relay is exceeded at some load, the relay will misoperate. The effects of generator over-excitation should also be considered with regard to variations in the third harmonic ratio and absolute values, to ensure that no false operation occurs, under any generator service condition. Can the author comment on these effects?

The third harmonic sensitvity of the type 2 scheme is better than stated for the type 3 relay (0.15−0.45 V setting range alternatively 0.4−1.2 V). The type 3 relay requires a fixed 22. V difference to operate. This sensitivity difference is not reflected in the paper with regard to the application limitations. Could this be expanded upon?

If only the third harmonic detection relay is required to supplement existing relays for full stator ground faults coverage, this function is available as a separate unit providing selective neutral end coverage. This variant of scheme 2 is more comparable to the scheme 3 described.

With regard to the "redesigned" filter, the 100 to 1 damping of 60 Hz is obtained by an optional external transformer-resistor combination that replaces the 240/120 V auxiliary PT indicated in the Georgia Power scheme. The basic relay is therefore not changed, but the applicability is expanded.

Also, ferrorsesonance PT conditions should be considered when broken delta voltage is used. The paper does not address this problem with regard to the type 3 scheme.

The open grounding resistor problem is a deficiency in both type 2 and 3 standard schemes when distribution grounding-secondary resistor is used. The type 2 relay can be provided with a resistor current monitor that will detect and open resistor, when this is a concern. This deficiency does not exist with either scheme 2 or 3 relays in installations with primary resistor and PT operated relays.

Manuscript received August 11, 1983.

R. E. Nienaber (The Cincinnati Gas & Electric Company, Cincinnati, OH): The author is to be complimented for a clear and interesting presentation of the three types of 100% stator ground protection schemes. The Cincinnati Gas & Electric Company has installed two of the type 3 relays which are discussed in the paper. Two more will be installed in the near future. These are the latest model relays with an additional temperature-compensating adjustable copper resistance "RT" in series with stud 7, L2, and the RH-2 resistor which are shown in Figure 7 of the paper. The "RT" resistor is not shown in early instruction books or in the paper. The relay setting procedure which was developed may be of some interest since it uses a simulation procedure to obtain the setting.

Setting Example:

1. Generator Voltage Measurement-Equipment Required:
   a. An accurate frequency selective voltmeter for 180 Hz, accurate to two decimal places, and a voltmeter that is not frequency selective.
   b. An oscilloscope for observing waveforms.
   c. A type 3 relay.

An accurate set of 180 Hz voltage readings from the open delta auxiliary P.T. at the generator terminals, and from the auxiliary P.T. at the neutral grounding resistor should be obtained by connecting the frequency selective voltmeter to studs 6-7 and 4-5 in Figure 7 of the paper. Readings should be made for generator loadings varying from zero megawatts to full load. The terminal voltage/neutral voltage ratio will likely not be constant over the range of generator loadings. Also the voltage ratio might be slightly affected by whether the auxiliaries are carried by the unit auxiliary transformer or by a reserve auxiliary transformer, and by the tap in use on the main step-up transformer.

The relay should be installed in the circuit with an initial rough setting; for example, balanced at full load, so that the proper burden will be reflected in the voltage readings. The meeter deflection should be recorded for the various loadings. The initial setting could be done by selecting "S", "C", and "F" taps along with the associated matching RT taps. Initially the RH-1 and RH-2 resistors can be set fully counterclockwise. Adjust the "S", "C", and "F" taps for approximate balance (o ma.). Do not be concerned if the meter moves to full scale (+0.75 ma.) for some load points. The recorded data will be used in a test set-up in the shop and the relay will be readjusted there to give balanced deflection with adequate margin (at least ±0.20 ma. to trip). The shop simulation will be based on a 180 Hz voltage source without any harmonics present. Therefore knowledge of the actual harmonics present at the machine is important. The voltage waveform should be observed on an oscilloscope. The frequency selective voltmeter can be used to measure other harmonics present, mostly 60 Hz. Voltage read on a voltmeter which is not frequency selective could be used for comparison. The filters in the relay reject the majority of frequencies which are other than 180 Hz. However, if the 60 Hz voltage present is much larger than the 180 Hz quantity, the relay will be affected.

The relay should be installed in the circuit with an initial rough setting; for example, balanced at full load, so that the proper burden will be reflected in the voltage readings. The meeter deflection should be recorded for the various loadings. The initial setting could be done by selecting "S", "C", and "F" taps along with the associated matching RT taps. Initially the RH-1 and RH-2 resistors can be set fully counterclockwise. Adjust the "S", "C", and "F" taps for approximate balance (o ma.). Do not be concerned if the meter moves to full scale (+0.75 ma.) for some load points. The recorded data will be used in a test set-up in the shop and the relay will be readjusted there to give balanced deflection with adequate margin (at least ±0.20 ma. to trip). The shop simulation will be based on a 180 Hz voltage source without any harmonics present. Therefore knowledge of the actual harmonics present at the machine is important. The voltage waveform should be observed on an oscilloscope. The frequency selective voltmeter can be used to measure other harmonics present, mostly 60 Hz. Voltage read on a voltmeter which is not frequency selective could be used for comparison. The filters in the relay reject the majority of frequencies which are other than 180 Hz. However, if the 60 Hz voltage present is much larger than the 180 Hz quantity, the relay will be affected.

2. Bench Simulation in the Shop-Equipment Required:
   a. Two sine wave generators capable of about 10 to 15 volts at 180 Hz without clipping when connected to the type 3 relay, or one sine wave generator plus an adjustable 10 turn resistor.
   b. An accurate frequency selective voltmeter and an oscilloscope.

Connect one 180 Hz source to stud 6-7 and the other source to stud 4-5 on the relay. A variable resistor may be used if only one 180 Hz source is available. A stable sine wave with line trigger should be present on the oscilloscope.

To verify the bench set-up, the relay deflection should be able to be reproduced within about +0.10 ma. when the same voltages are applied to studs 6-7 and 4-5 as were recorded at the generator, with the same setting on the relay.

Initially set both RH-1 and RH-2 to full counterclockwise which produces zero ohms in the bridge circuit and maximum sensitivity. Apply the "full load" voltages (maximum megawatts) to studs 6-7 and to 4-5. Note the meter reading. Apply the "no load" (zero megawatts) voltages and note the meter reading. Repeat while adjusting S, C, and F (along with RT) to give approximate equal deflection from zero for each case. Turning RH-1 and RH-2 clockwise inserts resistance into the bridge circuit which reduces sensitivity. If the meter deflects too much (more than ±.55 ma.) add RH-1 and RH-2 resistance. Since one of the lugs on RH-1 and RH-2 is unused, an accurate reading of the setting is possible by measuring from the center (wiper) to the unused lug. The ohms in the circuit is the difference between the maximum value in the full counterclockwise position and the set value.

Several "no load" to "full load" simulated "runs" can be made for various sensitivity settings. A neutral fault can be simulated by reducing stud 4-5 voltage. If stud 6-7 voltage is not increased, a conservative estimate of the sensitivity is obtained. During an actual fault the amount of decrease in stud 4-5 voltage will tend to be added to stud 6-7 voltage. Some larger values of RH-1 and RH-2 may be found which do not produce sufficient sensitivity for the simulated fault case.

Presently, an "alarm" output will be obtained if the relay operates. An added benefit of this relay is that it monitors the integrity of the generator's neutral grounding system. At this time, the relay's response to extreme values of reactive generation or to close-in external faults is not known. When the generator is taken off line, the relay's meter

returns to zero. No separate "off line" logic is required. Since relay operation can occur if a fuse blows in the potential circuit, the relay's operating contacts should be supervised by the "60" relay in the generator's potential circuits.

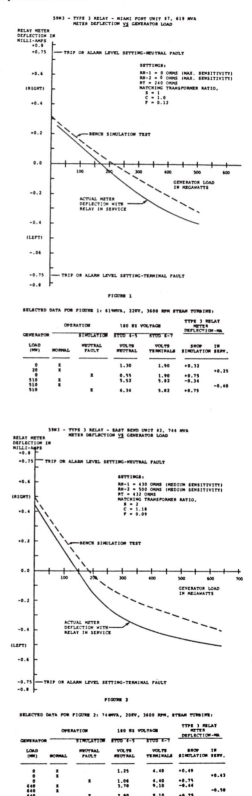

SELECTED DATA FOR FIGURE 1: 619MVA, 22KV, 3600 RPM STEAM TURBINE:

GENERATOR LOAD (MW)	OPERATION		180 HZ VOLTAGE		TYPE 3 RELAY METER DEFLECTION-ma
	NORMAL	SIMULATION NEUTRAL FAULT	VOLTS NEUTRAL STUD 4-5	VOLTS TERMINALS STUD 6-7	SHOP / IN SIMULATION SERV.
0	X		1.30	1.90	+0.32
20	X				+0.25
0		X	0.55	1.90	+0.75
510	X		5.52	5.82	-0.34
510	X				-0.40
510		X	4.34	5.82	+0.75

SELECTED DATA FOR FIGURE 2: 744MVA, 20KV, 3600 RPM, STEAM TURBINE:

GENERATOR LOAD (MW)	OPERATION		180 HZ VOLTAGE		TYPE 3 RELAY METER DEFLECTION-ma
	NORMAL	SIMULATION NEUTRAL FAULT	VOLTS NEUTRAL STUD 4-5	VOLTS TERMINALS STUD 6-7	SHOP / IN SIMULATION SERV.
0	X		1.25	4.40	+0.49
0	X				+0.43
0		X	1.06	4.40	+0.75
640	X		3.70	9.10	-0.44
640	X				-0.50
640		X	2.80	9.10	+0.75

A plot of the meter deflection for the simulated test and the actual deflection when installed on the generator is shown for each generator in Figure 1 and Figure 2. In both cases, the bench simulation procedure produced an acceptable relay setting.

Manuscript received December 2, 1983.

200

John W. Pope: The author appreciates the thoughtful and pertinent comments of the discussors, which have added considerably to the value of this paper. With regard to the points raised by Mr. Stranne, we have no statistics on the number and location of ground faults experienced by other systems. Acutally, "stator" ground faults is probably a misleading term. Many faults detected by stator ground fault relays are not in the generator itself, but in the isophase bus network. To my knowledge, Georgia Power Company has experienced only one stator ground fault which occurred near a generator neutral and was not cleared by the "90%" relay. On June 2, 1980, a fault occurred near the neutral of Unit No. 2 at Wallace Dam. This is a 52 MW, 14.4-kV pumped-storage generator/motor. The unit was under-going its final full-load rejection test when a turn-to-turn fault occurred approximately 12.8% up from the machine neutral. The neutral ground relay was set to cover 88.5% of the stator winding. The machine ran for about five minutes before being tripped by the generator differential relays. Inspection showed evidence of a sustained ground fault where a coil comes out of the stator iron. Arcing at this point eventually damaged an adjacent coil, and this caused sufficient current to flow to operate the differential. The arcing (high resistance) nature of the original fault, and the fact that the operating quantity supplied to the relay was very near its pickup setting, resulted in no ground relay operation. A 100% stator ground fault protective scheme would have cleared the initial fault promptly before the adjacent coil became involved.

Figure 4 for the Type 2 system shows an optional 59C instantaneous overvoltages supervision relay that may be used in place of the on-line supervision contacts. Initially we were unsure as to how the third harmonic voltage at the generator neutral will vary, paticularly at no load. For this reason, on-line contacts were used instead of overvoltage supervision. Field measurements taken since the original installation indicate that overvoltage supervision would be secure.

Item 5 in Table 3 attempts to compare the relative difficulty in calculating settings for the three types of 100% stator protection. Each system requires that field measurements be taken to insure the security of the calculated setting. It was determined that overlap calculations are not routinely reqired for the Type 3 scheme, but are required for the Type 2 scheme. For this reason, the Type 3 setting calculations were considered slightly less difficult than those required for the Type 2 scheme.

In response to Mr. Stranne's concern over variations in the third harmonic voltage ratio (generator terminals to neutral) and absolute value with generator over-excitation, I can offer no experience other than measurements discussed in reference (3) which shows no variation in neutral third harmonic as the one unit's excitation is varied over its operating range.

The statement that "the third harmonic sensitivity of the Type 2 scheme is better than stated for the Type 3 relay (0.15.-0.45V setting range alternatively 0.4-1.2V)" is not correct. The operating principle of these two schemes is so different that a simple comparison of relay ranges is misleading. The Type 2 scheme is more sensitive at no load than at full load. The higher the setting on the Type 2 scheme, the more sensitive and less secure is the setting. The Type 3 scheme is more sensitive at full load than at no load. A comparison of Type 2 and Type 3 scheme sensitivity can be obtained by application of the Type 2 scheme to Plant Yates Unit No. 6 as described in the Type 3 example. Following the Type 2 setting guidelines and using the no load and full load measurements in Table 2 results in the following stator coverage where the percentage of stator covered is measured up from the neutral.

	NO LOAD	FULL LOAD
Type 2	0-19%	0-6.2%
Type 3	0.-50% 67-100%	0.-46% 53-100%

These calculations indicate that in this example the Type 3 scheme is more sensitive and provides greater stator coverage than the Type 2 scheme.

Mr. Nienaber has provided a detailed setting calculation method for the Type 3 scheme based on field measurements followed by test lab simulations. This method appears well thought out and should result in the same relay response as the method described in the paper. Mr. Nienaber correctly makes the point that RH-1 and RH-2 control the relay sensitivity. Using his method, RH-1 and RH-2 must be equal.

Manuscript received October 18, 1983.

PERFORMANCE OF THIRD HARMONIC GROUND FAULT PROTECTION SCHEMES FOR GENERATOR STATOR WINDINGS

R. L. Schlake
Westinghouse Electric Corp.
East Pittsburgh, Pa.
Member

G. W. Buckley
Detroit Edison
Detroit, Michigan
Member

G. McPherson
University of Missouri-Rolla
Rolla, Missouri
Senior Member

ABSTRACT

The paper shows how the normally generated third-harmonic voltage can be used to protect the lower ten to twenty percent of generator stator windings against ground faults. A method of determining the applicability of the scheme to a given machine is described which takes into account the design of the generator and its externally connected apparatus. The effects of finite resistence ground faults are also investigated and other advantages of the scheme, such as protection of the neutral, are also highlighted.

INTRODUCTION

The reliability requirements placed on modern power systems demand increasingly complete protection of the generator. Because the major causes of stator winding ground faults are electromechanical stresses and local insulation deterioration, the locations of ground faults on the stator cannot be predicted a priori. Thus the entire stator winding (neutral end included) must be considered when designing schemes for protection against ground faults.

The conventional unit type generator has the neutral stabilized by a resistance loaded distribution type transformer, see Figure 1. The load resistor is sized to limit the in-phase component of line-to-ground fault current to a value not less than the total charging currents of the capacitances-to-ground of the generator winding and the associated isolated phase bus, transformers and leads. This value of neutral effective resistance has evolved over the years and results in moderate ground fault currents, generally under fifteen amperes for a solid phase-to-ground fault at the generator terminals. Currents of this magnitude generally do not cause serious damage to the core steel. However, while it is unnecessary to use high speed detection for ground faults, good industry practice dictates prompt tripping of the generator with the minimum delay practical.

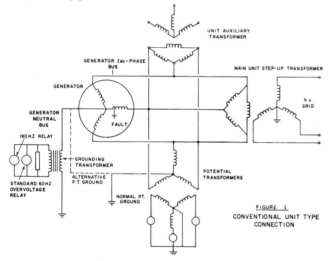

FIGURE 1
CONVENTIONAL UNIT TYPE
CONNECTION

For a single ground fault near the neutral end of the winding, there will be proportionately less voltage available to drive current to ground, resulting in a lower fault current and a lower neutral bus displacement voltage. At the lower limit, a fault on the neutral bus will result in no fault current or displacement voltage at all.

Small 60 Hz ground currents may flow under normal conditions, possibly due to generator winding imbalances or to faults on the secondaries of the generator potential transformers. Under these conditions the generator

81 WM 028-0 A paper recommended and approved by the IEEE Power System Relaying Committee of the IEEE Power Engineering Society for presentation at the IEEE PES Winter Meeting, Atlanta, Georgia, February 1-6, 1981. Manuscript submitted June 13, 1980; made available for printing October 29, 1980.

should not be removed from service. To allow for these small ground currents, trip settings for over-voltage ground relays are generally set to detect neutral displacement voltages in excess of 5-10% of the phase-to-neutral voltage. The potential hazard of an undetected ground fault in the lower portion of the winding, where the neutral bus displacement voltage is below the ground relay trip setting, therefore arises. Moreover, a fault at the neutral end of a generator phase winding will effectively short circuit the neutral grounding transformer. Under these conditions the conventional ground fault protection scheme will be totally inoperable, with potentially catastrophic results.

If a ground fault occurs and remains undetected because of its location or otherwise, the probability of a second fault occurring is much greater. This second fault may result from insulation deterioration caused by transient over-voltages due to erratic, low current, unstable arcing at the first fault point. This second fault may yield currents of devastating magnitudes.

The present paper describes an investigation into the feasibility of utilizing the third harmonic (180 Hz) emf, induced in the stator phase windings of all large generators, to give ground fault protection in the lower 10-15% of the winding. In conjunction with the conventional 60 Hz scheme, which relies on a resistance loaded distribution transformer to limit ground fault currents, this can be designed to give ground fault protection over 100% of the stator phase winding. In addition to the above, the paper also highlights a number of other operational advantages of the scheme and investigates its range of validity. A further major advantage of the scheme is its simplicity and low cost. The relay can also be easily retrofitted on existing installations.

DISCUSSION

Typical Unit Connected System

A three-phase circuit model of a typical unit connected generating station is shown in Figure 1. Normally the potential transformer (P.T.) neutral is directly grounded. However, some power companies connect the neutral point of the P.T. primary winding to the generator neutral, although this practice is deprecated by an IEEE Committee Report[1].

When providing protection for the generator stator winding, it is important to consider the winding neutral for several reasons. During high potential testing of the stator winding it is necessary to attach safety grounds to the generator neutral. Other conditions occasionally require disconnecting the generator neutral and leaving the neutral open. There are many recorded cases of these grounds not being removed when the generator is returned to service or alternatively, the generator neutral to grounding transformer connection is not made. This is a potentially dangerous situation, both for the machine, and also for plant personnel. In addition, the presence of the cable which connects the generator neutral bus and the potential transformer primary neutral increases the exposure of this neutral to ground faults. If a ground occurs in this cable, the neutral grounding transformer is short circuited.

The third harmonic relay will detect all of these conditions.

The Problem

The relay scheme consists of two relays in parallel with the load resistor, as shown in Figure 1. A regular over-voltage relay, operating at the fundamental generator frequency of 60 Hz, protects the upper 90% to 95% of the stator winding as explained earlier. The lower 10% (neutral end) of the winding is protected by a relay which detects a reduction in the third harmonic, or 180 Hz, component of the neutral bus voltage following a fault.

No 180 Hz component of line-to-line voltage appears in either delta or wye connected machines. However, most generators are connected in wye to eliminate 180 Hz circulating currents which can cause undesirable heating, but a phase to ground 180 Hz voltage does exist. This voltage causes 180 Hz current to flow, through the various leakage capacitances to ground, returning to the generator neutral bus through the neutral grounding transformer. This current will cause a 180 Hz voltage drop across the load resistor. If a ground fault occurs in the lower end of the generator winding or on the neutral bus (see Figure 1), the grounding transformer will be effectively short circuited and this 180 Hz current will bypass the ground transformer. This will cause a reduction in the 180 Hz load resistor voltage. Thus a ground fault in the lower 10% of the generator winding may be detected by a marked reduction in the 180 Hz neutral bus displacement voltage.

In developing a method for determining the applicability of this relay scheme for a typical unit-connected generator, the standard 60 Hz scheme of Figure 1 is assumed appropriate for the upper 90 to 95% of the winding. The applicability of the package thus depends on the proper operation of the 180 Hz relay. There are two problems associated with the operation of the 180 Hz relay.

180 Hz Neutral Bus Voltage: The applicability of the protection scheme depends on the magnitude of the 180 Hz neutral bus voltage under all ex-

Reprinted from *IEEE Trans. Power App. Syst.*, vol. PAS-100, no. 7, pp. 3195–3202, July 1981.

pected operating and fault conditions. The unfaulted 180 Hz neutral bus voltage must be within the relay's operating range. In addition, for a fault in the lower 5 to 10% of the winding, the resulting reduction in the 180 Hz neutral bus voltage must be sufficient to cause the 180 Hz relay to trip and indicate the presence of the fault.

A method had to be developed to calculate the 180 Hz neutral bus voltage under all expected load and fault conditions. These estimates could then be used to determine the resulting 180 Hz neutral bus displacement voltage range for a given machine.

Effect of the P.T. Connections: The second objective of this project was to determine the effect of the P.T. connection described previously on the applicability of the relay package. The performance of the 180 Hz relay depends on the generated 180 Hz current returning to the generator neutral through the neutral bus grounding transformer. This P.T. connection provides a second current path in parallel with the neutral bus grounding transformer. This current path might bleed enough of the 180 Hz current around the grounding transformer to limit the range of applicability of the relay package. Thus a method for determining the 180 Hz currents flowing through the P.T. primary windings was also desired.

The 180 Hz neutral bus displacement voltage is directly proportional to the 180 Hz current flowing through the grounding transformer primary winding. Thus both of the problems associated with the operation of the 180 Hz relay can be solved by determining the distribution of the 180 Hz currents in the third harmonic, generating station impedance network.

Method of Solution

As part of the present work, a simple, but extremely versatile computer program named GTHAP was written to fully model the machine on a coil by coil basis, together with its externally connected apparatus such as the isolated phase bus, main generator step-up transformer, etc. This program is used to estimate the 180 Hz circulating ground current for any given machine based on either a calculated or measured third harmonic voltage value. Thus the applicability of the third harmonic scheme, and its range of protection for a given machine, can be determined.

Since the third harmonic voltage acts as a zero sequence component, no third harmonic currents flow in the low voltage delta windings of the unit and auxiliary transformers. This means no 180 Hz voltage appears on the wye connected, high voltage side of the unit transformer. Thus the main power system appears as an open circuit in the third harmonic current network.

Using the assumed method of connecting the potential transformer (P.T.) primary neutral to the generator neutral, see Figure 1, there are two parallel paths external to the generator through which 180 Hz current is driven. One of these paths is through the various machine capacitances to ground, returning to the generator through the neutral grounding transformer. The second path is trhough the Y-Y connected P.T. primary windings and back to the generator neutral. However, initial calculations and subsequent simulations with GTHAP have shown that the 180 Hz series impedance of these P.T. windings is very large. Thus the effect of this second current path on the relay performance is almost undetectable and it can safely be neglected in evaluation of the relay scheme

Large generators often contain several parallel paths in each phase winding, so the precise path for the third harmonic current depends on the type of generator connection. It also depends on whether the machine winding is faulted or unfaulted. The precise equivalent circuits for each case are shown in Figures 2 through 5 and described later in this section.

General Outline: Each of the circuit models yields a set of mesh current equations and GTHAP was developed to solve any of these sets of equations for any fault point and fault resistance.

In carrying out a solution, the 180 Hz emfs and the various 180 Hz system impedance values are calculated first. The 180 Hz system impedances are then used to fill the appropriate impedance matrix. A complex matrix inversion subprogram is then called to determine the resulting system admittance matrix. The system voltage vector and admittance matrix are then multiplied to calculate the desired currents.

180 Hz Circuit Models: The unfaulted system is of primary importance in determining the normal 180 Hz current distribution. Under faulted conditions, two types of circuit models are relevant. A faulted, single path per phase model must be used when there is only one current path per phase. When there are two or more parallel current paths per phase, another circuit model is needed. For these reasons three cases must be simulated.

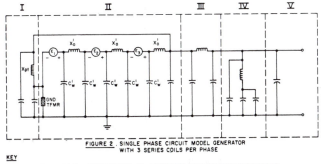

FIGURE 2. SINGLE PHASE CIRCUIT MODEL GENERATOR WITH 3 SERIES COILS PER PHASE

KEY
I = GENERATOR POTENTIAL TRANSFORMER II = GENERATOR AND GROUNDING TRANSFORMER
III = ISOLATED PHASE BUS IV = UNIT AUXILIARY TRANSFORMER
V = UNIT STEP-UP TRANSFORMER

The general circuit model for one phase of the generating station system is shown in Figure 2. In the circuit models developed the following assumptions are made:

1. All distributed windings, bus bars and cables are represented by Pi-line equivalents.

2. The generator neutral bus grounding transformer has been approximated by the load resistor reflected to the primary of the transformer.

3. The driving voltages, E_1, E_2, E_3, in Figure 2, are the emf's induced in the coils of the generator stator winding for a 3 coil machine. Additional coils would lead to further subdivision in the same manner.

4. No 180 Hz current flows in the line and the delta primary winding of the generator step-up transformer, and the h.v. grid appears as an open circuit to the third harmonic. This renders the equivalent impedance of the unit and auxiliary transformers equal to the phase-to-ground capacitance of the delta winding only.

5. For simplicity, an equivalent 180 Hz impedance, Z_{eq}, will be substituted for the circuit elements external to the machine terminals. These components are the generator isolated phase bus, the station auxiliary bus and the unit and station auxiliary transformers.

6. Since a Pi-line equivalent has been made for the P.T. windings, one-half of the capacitance to ground of each phase of the P.T. appears in parallel with the grounding transformer. An equivalent impedance, Z_{gnd}, is then made for these parallel circuit elements.

7. Interturn capacitances between adjoining stator slots are negligible due to the electrical shielding properties of the core steel. The winding capacitance-to-ground in a given slot outweighs the interturn capacitance between the conductors in that slot. Assuming a typical conductor height-to-width ratio of 3:1, the total ground capacitance will be at least 26 times the interturn capacitance between two windings in the same slot. Thus the interturn capacitive reactance will be 26 times larger than the impedance-to-ground and can safely be neglected.

A further reason for neglecting the interturn capacitance is that very little voltage is available to drive leakage current through it. Since the three third harmonic phase voltages are in phase, the highest possible voltage between turns of different phases is equal to the third harmonic phase voltage. This is not a large voltage. Also, a practical winding will be short pitched, resulting in only one or two slots where the potential difference exceeds 80% of the third harmonic phase voltage. Thus, there is generally less voltage available to drive interturn leakage currents than there is to drive leakage currents to ground.

Mutual inductances under zero sequence excitation are non-existent. By definition, zero sequence impedance is measured by applying the same voltage and current to all phases of a machine. Thus, each of the phase windings will develop an MMF in opposition to the others. No flux will link any two windings and only leakage flux will flow. Thus, only leakage inductances are involved in the zero sequence impedance of the unfaulted generator windings. These (zero sequence) impedances should be used in modelling the generator stator winding, although their overall effect is small.

Under fault conditions, one phase is partially shorted to ground. This fault will therefore cause a smaller opposing MMF in that phase. Such an imbalance in the phase MMF's will cause a mutual flux to flow and the zero sequence impedance will no longer involve only the leakage impedance. This imbalance will depend on the magnitude of the current in the faulted section of the winding. However, the fault current path includes the grounding transformer which appears as a very high impedance. This will keep the fault current small and the MMF imbalance will be negligible. Therefore, three times the normal 60 Hz zero sequence impedance should be used for the 180 Hz winding impedance under fault conditions.

Cases to be Considered

Case 1: Unfaulted Systems: The circuit model for the unfaulted machine is shown in Figure 3. When the machine is unfaulted, all three phases are identical. As explained earlier, the three single phase third harmonic induced emf's are equal and in phase.

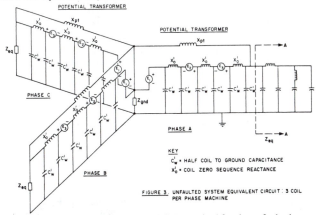

FIGURE 3. UNFAULTED SYSTEM EQUIVALENT CIRCUIT : 3 COIL PER PHASE MACHINE

KEY
C_w^i = HALF COIL TO GROUND CAPACITANCE
X_o^i = COIL ZERO SEQUENCE REACTANCE

To minimize the computer core and time required for the unfaulted case, a single phase equivalent circuit can, and has been developed for the three phase network. To adjust for the effects of coupling to the other phases, the grounding impedance in the single phase equivalent is three times the actual grounding impedance, Z_{gnd}.

Case 2: Faulted system with a Single Current Path per Phase: The circuit model for the faulted case with one current path per phase is shown in Figure 4. This system is not symmetrical, so all three phases must be included in the vector-matrix voltage equation.

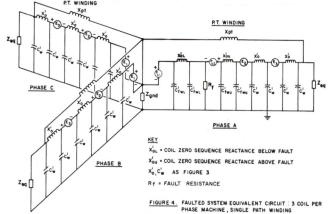

FIGURE 4. FAULTED SYSTEM EQUIVALENT CIRCUIT : 3 COIL PER PHASE MACHINE , SINGLE PATH WINDING

KEY

X_{oL} = COIL ZERO SEQUENCE REACTANCE BELOW FAULT

X_{ou} = COIL ZERO SEQUENCE REACTANCE ABOVE FAULT

X_o, C'_w AS FIGURE 3

R_f = FAULT RESISTANCE

Only one fault is considered at a time. This fault is placed in Phase A as shown in Figure 4 and the fault resistance denoted R_f. This phase is then modelled by a voltage source and a Pi-line coil equivalent below the fault. The remainder of the faulted coil above the fault and all succeeding coils are modelled individually. The winding capacitance and series impedance are distributed according to the proportion of the winding faulted. The 180 Hz voltage induced in each section of the winding is discussed later. See Case 3 below for an explanation of the coil voltages e_{a1} and e_{a2}.

Case 3: Faulted System with Two Parallel Current Paths per Phase: The circuit model for the faulted case with two parallel current paths per phase is shown in Figure 5.

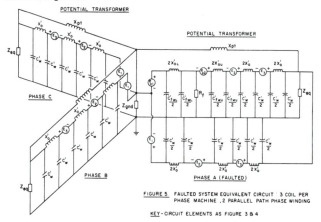

FIGURE 5. FAULTED SYSTEM EQUIVALENT CIRCUIT : 3 COIL PER PHASE MACHINE , 2 PARALLEL PATH PHASE WINDING

KEY - CIRCUIT ELEMENTS AS FIGURE 3 & 4

Again, only one fault is considered at a time. This fault is placed in Phase A as shown. Phases B and C are modelled as in Case 2. The voltage sources in the unfaulted current path are the normal 180 Hz induced voltages, E_1, E_2, E_3, etc. The sum of the voltages e_{a1} and e_{a2} in the faulted path is equal to the coil voltage, the individual voltages being obtained from the ratio of unfaulted to faulted coil length as in Case 2.

If the generator winding contains two parallel current paths per phase, the faulted and unfaulted paths in Phase A must be modelled separately. The capacitance per path will be one-half of the total capacitance per phase. The series impedance per path will be twice the total series impedance per phase. Again, simple circuit theory is used to solve for the current flowing through the grounding transformer.

A faulted machine with more than two parallel paths per phase may be modelled as a two path machine by approximating the unfaulted paths by one equivalent winding. The faulted path must then be modelled separately.

Calculation of 180 Hz Faulted Phase emf's

In order to calculate the individual third harmonic coil emf's for a particular machine, the number of slots, poles and parallel winding paths must be known. Either a measured or calculated 3rd-harmonic rms line to neutral voltage must also be known. A conventional coil voltage phasor diagram for the particular machine can then be constructed as shown in Figure 6. The relationship between the nth harmonic phase-to-neutral voltage, ^nE_r, (measured or calculated) and the nth harmonic coil voltage, ^nE_c, is given by

$$q^nE_c = {^nE_r}/{^nK_d} \qquad (1)$$

where ^nK_d, is the nth harmonic distribution factor. For a three phase winding:

$$^nK_d = \sin(n\pi/6) / \sin(n\pi/6q) \qquad (2)$$

Where q = number of slots/pole/phase for the fundamental.

To illustrate the manner in which the partial coil voltages, e_{a1} and e_{a2}, shown earlier in Figures 3-5, are obtained, Figure 6 shows a case for a fault part of the way into the first coil. It should be noted that each coil emf, including those in the faulted coil, is used with its correct in-phase and quadrature components to determine the correct ground path and leakage current. Similar constructions are used for other fault points and winding configurations.

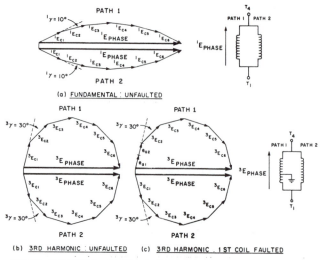

(a) FUNDAMENTAL : UNFAULTED

(b) 3RD HARMONIC : UNFAULTED   (c) 3RD HARMONIC : 1ST COIL FAULTED

FIGURE 6 FUNDAMENTAL AND 3RD-HARMONIC COIL VOLTAGE PHASOR DIAGRAMS 2 POLE, 2 PARALLEL PATH , 36 SLOT MACHINE

Effect of Load and Power Factor on Generated Third Harmonic Voltage

The current paths for the 180 Hz (third harmonic) stator currents, discussed earlier, present very high impedances. Therefore, the 180 Hz stator currents are very small in relation to the 60 Hz load currents. These currents produce a very small 180 Hz armature reaction stator MMF, resulting in a negligible effect on the 180 Hz line-to-neutral terminal voltage.

A major source of third harmonic line-to-neutral voltage is the effect of the rotor ampere-turn distribution. The fact that the field winding coils are in distinct slots around the rotor surface and also that the pole face may be asymmetrical, gives rise to a stepped rotor MMF wave. This stepped MMF space wave may be Fourier analyzed to yield many odd harmonic components. The effect of this source of third harmonic air-gap flux depends on the rotor winding configuration, the field current and the core steel saturation level, which in turn determines the reluctance of the magnetic circuit in the machine. The portions of the third harmonic MMF caused by the rotor turn and airgap permeance distributions are space stationary with respect to the rotor direct axis (the center of the pole face).

A further source of third harmonic voltage is the third harmonic component of the air-gap flux wave due to saturation of the rotor and stator teeth. Little saturation occurs in the stator yoke (the area of iron behind the stator teeth). This is because the stator yoke is sized mainly from mechanical considerations. Hence, this component of the third harmonic flux magnitude in the machine depends mainly on the saturation level in the rotor and stator teeth and the rotor core.

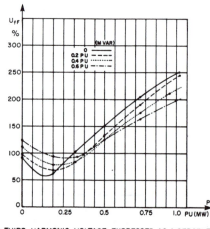

U_{fF} = THIRD HARMONIC VOLTAGE EXPRESSED AS A PERCENTAGE OF NO LOAD VALUE.

PU = PER UNIT POWER OUTPUT.

<u>FIGURE 7</u> VARIATION OF THIRD HARMONIC VOLTAGE WITH REAL AND REACTIVE POWER. (REF. 2)

With a fixed terminal voltage, the flux density in the air-gap will remain nearly fixed. This results in a nearly constant level of saturation in the rotor core and stator teeth. However, as the load angle increases, the angle between the rotor d-axis and the peak of the air-gap flux wave increases. Thus more of the air-gap flux passes through the rotor teeth instead of the pole face. This renders the rotor teeth saturation component of the 180 Hz flux wave dependent on the power factor of the generator load.

This rotor tooth saturation component is not space stationary with respect to the pole face. Rather, for a symmetrical pole face machine, it can be expected to be space stationary with respect to the position of peak air-gap flux, an : is therefore load dependent.

The combination of these effects renders the third harmonic terminal voltage dependent on the excitation level, the load and the power factor at the generator terminals. The curves of Figure 7, taken from Reference 2, illustrates the behavior of the third harmonic terminal voltage for a typical machine under varying load conditions. The exact relationships governing the interactions of the sources of third harmonic terminal-to-neutral voltage, discussed above, are extremely complicated, and beyond the scope of this present paper.

EXPERIMENTAL VERIFICATION OF THE MODEL

In order to verify the reliability of the computer program developed for this project, three experimental investigations were made.

Model Compatibility

The first of these was an attempt to determine the compatibility of all three models, i.e., whether or not they predict the same grounding transformer current when no fault exists. First, the program was run for the unfaulted case using impedance data from a typical machine on the Detroit Edison system. The unfaulted system was then approximated by running the faulted system of Cases 2 and 3 with 0.01% of the winding faulted through a nine megohm resistance. The results of this investigation are shown in Table 1. These three models gave the same results with negligible error.

TABLE 1 - COMPARISON OF PREDICTED GROUNDING TRANSFORMER CURRENTS UNDER UNFAULTED CONDITIONS

Case	180 Hz Line-to-Neutral Volts	Fault Position	Fault Resistance	Transformer Load Resistance	Grounding Transformer Primary Current	% Error
1	730	Unfaulted		0.498Ω	0.138A	—
2	730	0.01%	9MΩ	0.498Ω	0.138A	0.0%
3	730	0.01%	9MΩ	0.498Ω	0.137A	0.7%

The single phase unfaulted model is effectively equal to the three phase faulted system shown in Figure 5, with 0.01% of the winding faulted through a nine megohm resistance. Since the three phase faulted circuit with such a fault agrees with the single phase, Case 1 circuit, the validity of the single phase model is proven.

Reliability of the Unfaulted System Model

The second investigation was an attempt to verify the reliability of the unfaulted system model of Figure 3.

Measurements of the neutral grounding transformer and potential transformer 180 Hz secondary voltages were made at various output power levels on the chosen test unit.

The 180 Hz neutral grounding transformer secondary voltage was measured under generator start-up conditions. The voltage was recorded for 75 hours during which time the machine was off line five times. The maximum power output attained was 400 MW, which was 0.5 p.u. on the generator MW base.

On another occasion, the 180 Hz generator terminal line-to-neutral voltage and the generator output power were recorded, beginning in the evening and running until the morning. This experiment was run overnight to record the 180 Hz terminal voltage behavior at reduced load.

Knowing the 180 Hz voltage on the secondary of the neutral grounding transformer and the load resistance, the transformer's 180 Hz secondary current was calculated at various generator output levels. The unfaulted system simulation was then computed using the 180 Hz generator terminal-to-neutral voltage measured at each power level.

The tested generator's load resistor is rated at 0.588 ohms when a maximum current of 307 amperes is flowing through it. This current corresponds to a solid phase-to-ground fault at the machine terminals. The resistor was measured at 0.498 ohms, prior to installation, with no current flowing through it. To determine the effect of variations in this resistance on the error encountered in calculating the grounding transformer current, two computer simulations were run. The load resistors were set equal to the limiting cases of 0.498 ohms and 0.588 ohms.

The comparisons of the estimated and measured 180 Hz neutral grounding transformer primary currents are shown in Tables 2 and 3. Table 2 assumes the grounding transformer load resistance to be 0.498 ohms, while Table 3 assumes it to be 0.588 ohms.

As indicated in Tables 2 and 3 the error encountered in predicting the 180 Hz grounding transformer current was approximately 20 to 30%. There are several possible sources of the error but most relate to either experimental or data errors. Fortunately errors of this magnitude can be tolerated in making a determination of the applicability of this protection scheme to a particular generator. It is believed that the errors arise from faulty data as to winding-to-ground capacitances, and to the somewhat uncontrolled conditions under which third harmonic voltages were measured.

In Tables 2 and 3 the effect of the grounding transformer load resistor on the 180 Hz grounding transformer current magnitude is evident. However, the percent error is not greatly affected since the load resistance appears in the calculation of both the measured and estimated currents.

Predicted Reduction of the 180 Hz Grounding Transformer Currents Under Various Fault Conditions

The third area of investigation is an example of the method used to determine the applicability of the relay package to a specific generator. This method entails determining the expected reduction of the 180 Hz neutral grounding transformer current under various fault conditions.

The 180 Hz terminal line-to-neutral voltage is assumed to be 730V in this experiment (730V was an actual measured voltage, see Table 3). In an actual application of this program, the smallest expected voltage would be used since

TABLE 2 - ERROR IN PREDICTING THE 180 Hz GROUNDING TRANSFORMER CURRENT
Assuming the Load Resistor = 0.498 Ohms

Load		Terminal-to-Neutral 180 Hz r.m.s. Voltage	Load Resistor	Grounding Transformer			Computer Estimated 180 Hz Current	Error	
MW	MVAr			Secondary 180 Hz Volts	180 Hz Current	Primary 180 Hz Current		mag.	%
607	59	730V	0.498Ω	4.6V	9.24A	0.111A	0.138A	0.027A	20
404	30	605V	0.498Ω	3.3V	6.63A	0.080A	0.114A	0.034A	30
224	6	437V	0.498Ω	2.4V	4.82A	0.058A	0.083A	0.025A	30
234	-2	440V	0.498Ω	2.5V	5.02A	0.060A	0.083A	0.023A	28
380	25	590V	0.498Ω	3.2V	6.43A	0.077A	0.111A	0.034A	30

TABLE 3 - ERROR IN PREDICTING THE 180 Hz GROUNDING TRANSFORMER CURRENT
Assuming the Load Resistor = 0.588 Ohms

Load		Terminal-to-Neutral 180 Hz r.m.s. Voltage	Load Resistor	Grounding Transformer			Computer Estimated 180 Hz Current	Error	
MW	MVAr			Secondary 180 Hz Volts	180 Hz Current	Primary 180 Hz Current		mag.	%
607	59	730V	0.588Ω	4.6V	7.82A	.094A	.119A	0.025A	21
404	30	605V	0.588Ω	3.3V	5.61A	.068A	.099A	0.031A	31
224	6	437V	0.588Ω	2.4V	4.08A	.049A	.071A	0.022A	31
234	-2	440V	0.588Ω	2.5V	4.25A	.051A	.072A	0.021A	29
380	25	590V	0.588Ω	3.2V	5.44A	.066A	.096A	0.030A	31

this would produce the smallest magnitude of unfaulted grounding transformer current and the smallest reduction in this current following a fault. This experiment considers both single path per phase and a two parallel path per phase winding machines.

First, the unfaulted case was run to determine the expected 180 Hz grounding transformer current with a given 180 Hz terminal line-to-neutral voltage level. Simulations were then run with the faulted system models of Case 2 and Case 3. Six different fault resistances were placed at each of six fault locations in the lower 10% of the winding. The cases were all run with the grounding transformer load resistor equal to the resistor's value measured prior to installation. The resulting 180 Hz grounding transformer current magnitudes and percent reductions for the Case 3 model are shown in Figure 8. The results for a Case 2 model agreed so closely with those for the Case 3 model that Figure 8 can be used for both cases. Figure 8 shows that the scheme is most sensitive for fault resistances in the range 0 to 1000Ω.

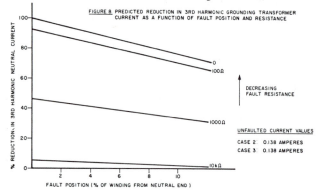

FIGURE 8. PREDICTED REDUCTION IN 3RD HARMONIC GROUNDING TRANSFORMER CURRENT AS A FUNCTION OF FAULT POSITION AND RESISTANCE

Percentage of Winding Protected

The only questions which now remain to be answered are, firstly, what percentage of the stator winding can be effectively protected by the third harmonic scheme?, and secondly, how can we determine this?

Under normal circumstance with no faults present, a certain amount of third harmonic voltage will be measurable at the neutral bus. If the neutral bus suffers a fault, then this voltage will fall to zero because the neutral grounding transformer is short-circuited. At the other extreme, if a zero resistance short circuit is applied to the machine terminals, then the level of third-harmonic current flowing through the grounding transformer will increase because the current can flow through this fault as opposed to flowing through the leakage capacitances of the circuit. This increased current will raise the neutral bus voltage above the unfaulted level. Between these two extremes, there exists a position on the winding at which, if a zero resistance ground fault were to occur there, there would be no reduction in the third harmonic bus voltage, thus the third harmonic relay would not detect the fault. It is obviously important that this position is above the lower point protected by the conventional 60 Hz relay.

The above questions can be answered very simply by simulating a fault at progressively different points along the stator winding. Zero resistance faults in the range 10 to 80% in steps of 10% were simulated and the results are shown in Table 4. As we move out from the neutral bus, zero resistance faults will result in a reduction in the third harmonic neutral bus voltage, until some position is reached where an increase is noticed. A finite resistance short circuit will alter this position somewhat and it is possible to construct a family of curves from an extension of Table 4 which would depict the variation of third harmonic neutral voltage as a function of fault location with various resistance values as a parameter.

On a practical note however, because we are attempting to protect the lower portion of the stator winding, the likelihood of a high or medium resistance short caused by insulation cracking, surface tracking or discharges is small because the available voltage is small. It is much more likely that such a fault would be as the result of mechanical damage such as fretting caused by relative motion of the stator winding and the stator core and therefore be of very low resistance. According to Table 4, theoretically, a zero resistance ground fault at the 45% position would be undetectable on this particular machine. Practically though, the relay would typically be set to detect a 50% reduction in neutral bus displacement voltage[3], then Table 4 shows that the 180 Hz scheme will protect at least 20% of the winding giving good overlap with the conventional 60 Hz scheme.

TABLE 4 - DETERMINATION OF RANGE OF PROTECTION

Driving Voltage	Fault Resistance Ω	Fault Position %	I_3 Amps	% Reduction in Current
730	0	10	0.033	76
730	0	20	0.065	53
730	0	30	0.095	31
730	0	40	0.123	11
730	0	50	0.150	-9
730	0	60	0.169	-22
730	0	70	0.187	-36
730	0	80	0.199	-44
Unfaulted base case. V_3 = 730, I_3 = 0.138 amps				

CONCLUSIONS

The following conclusions can be reached as a result of this investigation.

1. Calculations indicate that the additional loop provided by connection of the P.T. neutral to the generator neutral does not bleed sufficient current away from the ground path for the effectiveness of the scheme to be impaired.

2. Good coverage, typically 20% of the winding, can be obtained with the third harmonic relay package.

3. Preliminary studies given in Figure 8 show that the most significant reductions in third harmonic grounding transformer current are produced for fault resistances in the range 0 to 1000 ohms. This applies for all fault positions between 0.01 and 10.0% of the stator winding.

4. The combined relay package can provide 100% stator ground protection if sufficient third harmonic voltage is generated. However, it should be pointed out that, even though 100% of the generator winding is protected, no backup protection is provided.

5. As shown in Figure 7, because the third harmonic voltage passes through a minimum value at low loads, as the generator is brought up to full load, the 180 Hz relay may indicate an invalid alarm if the ratio of minimum to maximum third harmonic voltage is less than the relay setting. For this reason, if this ratio is not known, the 180 Hz should only be used to alarm and the normal 60 Hz relay used to trip the machine in the usual way.

ACKNOWLEDGEMENTS

The authors are indebted to their respective organizations for the provision of support and facilities in the course of the present work. In particular, the authors would like to thank Mr. Tom Roberts and other members of the Relay Division, Electrical Systems Dept. of Detroit Edison for their help and encouragement. The authors would also like to thank Mr. Lon W. Montgomery of the Generator Stator Systems Dept. at Westinghouse Electric Corporation for his consistent help, and support during the course of the present work. The authors are grateful to the authors of Reference 2 for permission to include Figure 7, taken from their work, in this paper.

REFERENCES

1. "Potential Transformer Application on Unit-Connected Generator", IEEE Power System Relaying Committee Report, IEEE Trans., Vol. PAS-91, 1972, pp 24-28.
2. Ilar, M. et al, "Total Generator Ground Fault Protection", *Protective Relaying Conference*, Georgia Institute of Technology, May 3-4, 1979.
3. Stien, M. and Linders, J.R., "Ground Fault Protection of Complete Generator Windings", *Fourth Annual Western Protective Relaying Conference*, October 18-20, 1977.

Discussion

K. H. Engelhardt (British Columbia Hydro and Power Authority, Vancouver, Canada): The authors have done an excellent and thorough job in analyzing, on the basis of generator design data, the third-harmonic voltage and current distribution for normal conditions and for stator ground faults at various winding locations, and applying a 100% ground fault detection system on these principles.

The utilization of third-harmonic voltages for the detection of ground faults at or near the neutral end of the stator (being a "blind" zone to conventional 60 Hz voltage relays) poses some practical problems and constraints to relay application engineers. These are briefly described in the following.

1. Determination of Third Harmonic Voltages

Details of generator winding design, and the kind of computer program described in the paper are generally not available to utility relay engineers. On the other hand, instrumentation for *measuring* the third harmonic voltages under various unit output conditions is relatively easy to come by. Having such *measured* voltages obtained at the neutral end and the bus end of a generator, it is then possible to determine the *approximate* 180-Hz voltage reduction as a function of fault distance from the neutral. This need not be accurate, as long as it can be established that the 180-Hz detection zone extends beyond the 60-Hz blind spot near the neutral.

2. Application of 180-Hz Undervoltage Relay

The typical relay setting for detection of 50% reduction in neutral-end 180-Hz voltage as mentioned in the paper, may be very difficult to accomplish in practice with the wide operational variations of this voltage sometimes found. Figure 1 is a striking example.

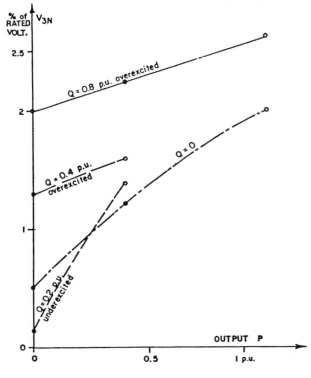

Fig. 1 Example of Measured Neutral 180-Hz Voltages

Here the 180-Hz voltage is so low under certain generator output conditions that the secure application (i.e. without possible *false* operations) of such a scheme is barely practical, or if so, then only by supervising it with judiciously set minimum current or MW/MVAR detectors. Such supervision would of course limit the unit's operating range within which neutral end ground faults are detectable.

3. 180-Hz Voltage Comparator Method

An alternative detection method is offered by the fact that for stator ground faults, a 180-Hz voltage *reduction* at one end is accompanied by an *increase* at the opposite end, except for the 180-Hz "blind" area (given at the 45% position in the example of Table 4 of the paper). The discussor has applied a third-harmonic voltage *comparison* relay based on this principle, a simplified schematic of which is given in Figure 2, on a number of generators.

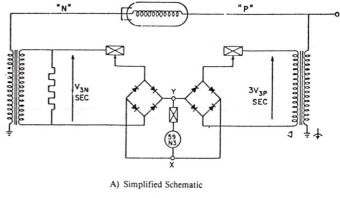

A) Simplified Schematic

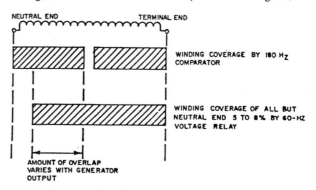

B) Third-Harmonic Secondary Voltage Distribution

Fig. 2 Principle of Third-Harmonic Voltage Magnitude Comparator

As the third harmonic voltages at the neutral end and at the terminal end will shift their relative magnitudes for faults in either of the end zones (in respective opposite directions of course), such relays will see ground faults in both these zones, as shown in Figure 3.

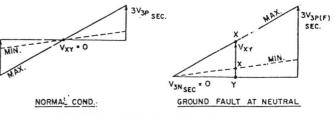

Fig. 3 Stator Winding Coverage Provided by 180 Hz Comparator and 60 Hz Overvoltage Relay

The blind spot of the 180-Hz relay has a different location than the blind spot of the 60 Hz relay. Thus, a considerable portion of the winding will be covered by both relays, thereby providing some degree of mutual backup. The calculations described in the paper, and/or third-harmonic voltage field measurements at both ends of the generator stator, can be used as basis for the application of the 180-Hz voltage comparator.

Manuscript received February 20, 1981.

R. J. Marttila (Ontario Hydro, Toronto, Canada): The authors have presented an interesting paper on the performance of a third harmonic ground fault relay for generating units.

The relay principle discussed by the authors is based on detecting a 50 per cent reduction in the third harmonic voltage at the neutral of the generator. Could the authors elaborate on the criteria used in distinguishing between a 50 per cent reduction in the third harmonic voltage caused by a ground fault and that caused by normal variations in

a) real power output
b) reactive power output?

207

In the case of a), Figure 7 of the paper indicates that reduction of more than 50 per cent in the third harmonic source voltage can occur as the unit loading is reduced from the maximum to the minimum. Also, in the case of b), our measurements indicate that a reduction exceeding 50 per cent can occur in the neutral third harmonic voltage in response to a normal change in the reactive power output.

To have an accurate measure of the third harmonic voltage reduction, the 60-Hz component needs to be filtered out. What rejection ratio is required of the filter to limit the 60-Hz component to an acceptable level under fault conditions, to allow the necessary measurements to be made on the third harmonic component? What is the minimum unfaulted value of third harmonic voltage required to make the necessary measurements?

Alternate approaches to that used by the authors have been discussed in the references cited below. Both of these approaches require as input the neutral voltage and the residual terminal voltage and depend on the ratio of the third harmonic components of these voltages to be relatively constant over the full operating range of an unfaulted machine. For a ground on the stator at or near the neutral, the relationship between the two voltages changes sufficiently from the quiescent to allow detection of the ground condition. Have the authors considered and what opinion may they have of this type of an approach? We have several relays of the type described in reference [1] in service to give an alarm with output from the relay. Some relays have been in service for more than three years. So far, no alarms have been recorded. We would like to point out, however, that according to our measurements, the radio of the two voltages used by the relay is not as constant as would be expected under normal operating conditions of some machines and could, therefore, result in nuisance alarms from this relay. We are currently investigating alternate implementation techniques of the "two-voltage" approach in an effort to accommodate the possible variations in the ratio of the voltages, as well as provide sensitivity for detection of grounds with the relatively small no-load third harmonic voltages.

REFERENCES

[1] K. H. Engelhardt. "A Composite Ground Fault Detection Scheme for High Resistance Grounded Generator Stators". Presented at the meeting of Canadian Electrical Association. September 1973. Quebec, Canada.

[2] L. Pázmándi. "Stator Earth - Leakage for Large Generators". *IEEE Transactions on Power Apparatus and Systems*. Vol. PAS-94. No. 4. July/August 1975. pp. 1436-1439.

Manuscript received February 20, 1981.

M. S. Baldwin (Westinghouse Electric Corporation, East Pittsburgh, PA) and **W. A. Elmore** (Westinghouse Electric Corporation, Coral Springs, FL):
1. This paper affords considerable insight into the modeling and analysis of third harmonic effects in generators. Much of the difficulty described by the authors in setting a relay that will be both secure and sensitive can be overcome by utilizing a relay that compares the third harmonic voltage, neutral-to-ground, with that at the machine terminals (as derived from a wye-ground broken-delta voltage transformer connection). These voltages "track" one another (rise and fall together) reasonably well as loading and/or excitation is changed. A fault at one end (either end) of the winding will cause the third harmonic voltage to drop at the faulted end and rise at the other. The relay using this concept complements the 60 hertz relay conventionally used for generator ground detection and overlaps its coverage for a large part of the machine winding. A much larger part of the winding can be covered by a relay using this concept than by one responding to third harmonic alone.
2. Does Detroit Edison use the "alternate" vt connection with the vt primary neutral connected to the generator neutral rather than to ground? If so, why?
3. Page 1 - Middle of right hand column "During high potential-it is necessary to attach safety grounds to the generator neutral." Actually, to hi-pot the generator winding it is necessary to *isolate* the neutral from ground. Following the hi-pot testing the entire winding should be grounded to drain off any charge. However, it is agreed that the absence of third harmonic should detect the conditions stated in the paper.

Manuscript received February 20, 1981.

Modolf Stien (ASEA AB, Vasteras, Sweden) and **John R. Linders** Sarasota, FL):
This is an interesting paper which in a clear and straight forward way presents the basic facts about the Third Harmonic Voltage Scheme for complete stator ground protection. As noted by the references, third harmonic voltages, measured on the neutral of the machine, have been providing this protection for some years. This paper extends the understanding of third harmonics, including, as shown in Figure 6, their use in the proteciton of parallel winding machines.

The correlation between the computer program and the field data would be more meaningful if all of the component capacitances were given. For example, it is not evident from Table 4 that the refinement of using the third harmonic phase displacements of Figure 6 and PI sections of Figures 2-5 add to the accuracy of the calculations. Also what percent of rated L-N voltage is the 730 V of third harmonic?

The third harmonic load profile of Figure 7 should be used with caution since it applies to a specific machine. The paper notes that several sources of third harmonics are developed in the generator. With minor changes in machine parameters, these sources may come into phase reinforcement of opposition at significantly different (relay and reactive) loading conditions. Thus Figure 7 should not be taken as typical of a given machine without confirming tests or calculations. From the application point of view it would have been of interest if the authors had included some information on how the 3rd Harmonic Voltage of the generator can be estimated from basic design data of the machine.

In Tables 2 and 3, on what is the estimated error based? Does the computer program calculate the third harmonic voltage (Column 3) or is this the starting point of the program? In the reference to Figure 8, has any work been done to represent the high resistance fault as a nonlinear resistance which is most likely in a real situation?

The philosophy reflected in the final conclusion that the third harmonic relay should only alarm because it may falsely indicate a stator ground fault is not favored by most operating utilities. If a given relay is not secure it would probably not be used except under very unusual cases. An essential reason for only alarming on third harmonic is that the condition is not *immediately* critical (as contrasted to a differential relay operation, for example). A procedural shutdown is thus quite acceptable and surely less hazardous to the system. But the "alarm only" concept should not be used to transfer the responsibility from the protection engineer to a plant operator. False signals from third harmonic relays cen be avoided by use of suitable arming relays or otherwise adapting the system automatically to the prevailing non-fault conditions.

R. L. Schlake, G. W. Buckley, and **G. McPherson:**
The authors are grateful for the kind comments made and the interest shown by each of the discussors.

So far as making measurements of the third-harmonic voltage at the neutral bus are concerned, Mr. Englehardt is correct in stating that such measurements are relatively easy to make and, provided that the more difficult values of third harmonic terminal voltage can be obtained, some estimate of the 180Hz voltage reduction as a function of fault distance from the neutral can be obtained. It is of course vital to successful operation of this relay scheme that a significant reduction in the 180Hz voltage occurs for fault positions past the blind point of the conventional 60Hz relay. As shown in the paper, the fault resistance can have a significant effect on the amount of winding protected. Thus caution needs to be exercised when using these simpler methods of evaluation.

Mr. Englehardt makes an important point when he speaks of the variation of the third harmonic voltage with real and reactive power output from the generator. A user unfamiliar with this variation could potentially design a scheme which would generate an unnecessary sequence of nuisance trips or alarms. There is a definite need for further measurements of the variation of this voltage on typical units, because relatively few published records of such measurements exist. There is also a need for a simplified method of calculation to help steer potential users of this scheme around possible pitfalls. We are pleased to see the additional measured results supplied by Mr. Englehardt, but still advice they be interpreted with caution because of the small number of observation points in his results.

When considering the practicality of using a relay setting of 50% reduction in neutral end voltage, one has to remember the length of time that a given unit is to operate at reduced load. In reality, most modern steam turbine generators operate at or close to full load for the major part of their operating life and only pass through these lower loads during unit loading or unloading. We therefore do not consider this point a major problem area, provided steps are taken to defeat false alarms at low loads.

The relay scheme considered by the authors is only one of several possible alternatives. Our primary purpose in studying the particular scheme we did was that it was to be applied at two new generating stations on the Detroit Edison system. The 180Hz Comparison Method is a admirable way of achieving the same result and our calculations and computer programs can also be used to assist in the application of this alternate method.

In response to Mr. Marttila's questions, we have considered three possible solutions to the problem of distinguishing between genuine reductions in third harmonic neutral voltage caused by winding faults and those resulting from changes in load. Firstly, the 50% reduction was selected based on a combination of suggestions made in the paper by M. Stien and J. R. Linders (see Reference #3 of the paper) and the need to obtain a good range of protection. This 50% figure is not unchangeable and can be altered to reflect the particular users needs and conditions.

Ideally, the voltage at which the relay is set to trip should be not more than 80% of the minimum value of voltage through which the machine passes during normal operation. This would completely avoid spurious trips or alarms due to load changes. However, this has a penalty. Based upon a limited number of published test results, plus our own measurements, the 180Hz neutral bus voltage does not fall lower than about 60% of the open-circuit value, and this figure is readily available from the generator manufacturer. If we were to select a relay setting of 80% of this value, say 50% of open-circuit value, using a typical ratio of full load to open-circuit 180Hz voltages of 2:1, this means the relay is set to detect a 75% reduction in 180Hz voltage when at full load. Reference to our Table 4 and Figure 8 shows that, for the machine studied in our paper, we could only protect 7-10% of the lower part of the winding for fault resistances up to 100Y. This range of protection barely covers the blind spot of the conventional 60Hz relay and is not felt to be satisfactory. The measurements necessary to determine the minimum voltage must obviously come from field tests during commissioning of the unit.

An alternative method would to use an arming relay to arm the device only for loads greater than, say 40 to 50% full load and to set the relay at 50% of full load and to set the relay at 50% of full load or near full load value.

To aid our understanding of the use of this relay, we are very seriously considering adding a requirement to future generator specifications that the manufacturer supply curves depicting the variation of 180Hz terminal voltage with changes in real and reactive power. Using the kind of modelling described, the range of protection could be definitively calculated.

We are surprised to hear that machines on which Mr. Marttila has made measurements have changes in neutral bus 180Hz voltage of more than 50% for practical changes in reactive power only (i.e. at constant real power). We strongly encourage Mr. Martilla to publish these measured results at the first convenient opportunity. They may aid the industry significantly.

With regard to filter requirements, the unit used is a six (6) pole active filter with a roll off of 36 dB with a corner frequency set between 120 and 150Hz octave. Thus the filter attenuates the 60Hz feed through by about 40dB at the point of connection of the 180Hz relay.

In answer to the questions by Mr. Elmore and Mr. Baldwin, Detroit Edison did use the ''alternate'' vt primary neutral connection until about five years ago when design of our latest power plant started. Our reasons for using the alternate connection were primarily because it eliminated the need to coordinate the setting of the 60Hz ground relay and the fusing of the vt secondary winding. Our latest practice though is to directly ground the vt primary neutral. We were always aware of the potential dangers of this alternate connection but did not actually change our practice until publication of the IEEE Committee Report listed as Reference #1 in the paper.

In order to emphasize the dangers of this alternate connection, it may be of interest for you to know that a Detroit Edison machine did in fact suffer a fault where the cable connecting the vt primary neutral to the generator neutral failed. The fault originated with failure of a single leaf of a multi-leaved flexible connector on the center phase of the isolated phase bus. The failed leaf touched to ground, initially causing a single line to ground fault at a relatively low current. The fault arc traversed the length of the isolated phase bus (forced air cooled type) and blew out the de-ion grids at the transformer end of the bus, creating a three phase short-circuit. Fortunately, the hv breakers had already tripped the unit from the system but sufficient flux still remained in the machine for the fault to cause considerable damage. During the single line to ground phase of the fault, the generator neutral is raised to line to ground potential, as is the cable connecting the two neutrals. This cable failed and flashed over to ground, effectively short-circuiting the grounding transformer. Fortunately, the cable withstood line potential long enough for the normal protection to have time to operate. Had this not been the case, damage which was already severe, could conceivably have been worse. We learned three major things from this fault; firstly, the quality of the cable installation (if it is to be used) is very important; secondly, all multi-leaved flexible connectors in isolated phase buses are to be taped to prevent possible grounds on failure; and thirdly, all de-ion grids are to be bolted in place.

The authors are, of course, very much aware of the procedure for high potential testing and we appreciate the time taken by Mr. Baldwin and Mr. Elmore to improve on our poor wording.

In answer to the questions by Mr. Stein and Mr. Linders, we believe it is important to correctly include the phase of the third harmonic voltage if accurate results are to be obtained. We see no benefit in further simplifying what is already a fairly simple precedure and it did not occur to us because to leave this out because unnecessarily large errors would otherwise have resulted. The error in the coil voltages that would occur if we assumed no phase shift in the coil voltages, can be obtained from the inverse of the distribution factor, and is about 55% in the 6 slot/pole/phase case studied. We must here apologize for the omission of the term q in equation (2) which should read $nK_d = \sin(n\pi/6)/q\sin(n\pi/6q)$. 730 volts is equal to 2.8% of the rated line voltage of 26kV.

Mr. Stien and Mr. Linders are quite correct in stating that Figure #7 of our paper should be used with caution. It is explicitly stated in the paper that Figure #7 contains only typical results and is certainly not applicable to all machines. So far, we have not perfected a simple method which can be relied upon to give acceptable accuracy for a wide range of machine designs. However, initial work in this area has been done and we hope to complete this in the close future.

The errors quoted in Tables 2 and 3 are derived from the differences in the values in columns 7 and 8 of these same tables, i.e. it is the percentage difference in the measured and computed results. At the present time, the computer program uses measured values of third harmonic as input. When the previously mentioned mathematical technique is perfected, calculated values could also be used.

We do not feel it is necessary to represent any nonlinearity in the fault resistance because we are considering faults close to the neutral which, because of the distribution of potential across the winding, are likely to be low resistance faults caused by mechanical damage as opposed to carbonizing of the insulation by slot or internal discharges. Hence, even if a severe nonlinearity existed, say a resistance change of 5:1, this value is still likely to be small compared to roughly 4000Ω in the grounding transformer circuit plus the winding capacitive reactance.

We are indebted to Mr. Stein and Mr. Linders for clearly explaining a benefit of choosing an alarm only condition for the delay. We were aware of this and it was in fact incorporated as one of the reasons for recommending an alarm only condition to our Company. This recommendation has been neither accepted nor rejected at this time.

The authors would once again like to thank the discussors for their interest and stimulating questions.

Manuscript received April 2, 1981.

GENERATOR GROUND FAULT PROTECTION USING
OVERCURRENT, OVERVOLTAGE, AND UNDERVOLTAGE RELAYS

C. H. Griffin
Senior Member, IEEE
Georgia Power Co.
Atlanta, Georgia

J. W. Pope
Senior Member, IEEE
Georgia Power Co.
Atlanta, Georgia

Abstract - This paper describes the protective relaying schemes employed by Georgia Power Company to protect synchronous generators from single-phase-to-ground faults. Three types of relays are connected in the secondary of a distribution grounding transformer. These include a conventional electromechanical overcurrent relay with time overcurrent unit and instantaneous overcurrent unit, a solid-state overvoltage relay (with timing module) tuned to reject frequencies near 180 hertz, and a solid-state undervoltage relay (with timer) tuned to reject frequencies near 60 hertz present at the generator neutral. The proper method of selecting the generator grounding components and protective relays is described, and detailed setting instructions for all relays are also included.

INTRODUCTION

Experience shows that by far the most prevalent fault to which generators are subjected is a short-circuit to ground. For over 30 years, Georgia Power Company has grounded all system generators through a distribution transformer with a resistance-loaded secondary. A current transformer is then connected in series with the secondary resistor to supply current to one or more overcurrent relays. When properly set, these relays will provide sensitive protection for 90 to 95 percent of the generator stator winding, and will not operate incorrectly for external faults. This system is now installed on 126 generating units, ranging in size from 15 to 900 MW. In the past 25 years, nearly 20 ground faults have been correctly cleared with minimal equipment damage, and no incorrect operations have occurred.

An overcurrent (or overvoltage) generator ground fault protective scheme is straight-forward, secure, and reliable. However, it suffers from two disadvantages. First, it will not detect ground faults near the generator neutral, and second, it is not self-monitoring. That is, an open circuit anywhere in the relay, primary or secondary of the current transformer, or an open grounding resistor, may not be detected before a fault occurs.

In 1977, Georgia Power Company concluded that it would be prudent to protect all large generators with an additional ground fault protection system that was completely independent of the existing overcurrent scheme, would give reliable protection to 100% of the generator, and would continuously monitor the generator grounding system. Two types of systems have been installed. One type injects a current at a subharmonic frequency, and trips on an increase of current caused by the reduction in generator capacitance that results from a single-phase-to-ground fault. The other type employs two overlapping voltage relays - an overvoltage relay that protects the high voltage end of the machine, and an undervoltage relay, tuned to respond to the third harmonic, that protects the neutral. This paper describes the ground fault protection installed on Georgia Power Company generators, particularly Unit No. 2 at Plant Bowen. This generator is protected by the overvoltage/undervoltage system, as well as the conventional overcurrent scheme. It should be noted now that both "100%" schemes have performed extremely well, and the "overvoltage/undervoltage" scheme has already properly cleared a ground fault on Unit No. 1 at Plant Bowen. It should also be noted that in 1980, a Georgia Power Company generator on which a "100%" scheme had not yet been installed, was badly damaged by a ground fault that occurred very near the neutral and was not detected by the "90%" relay.

FUNDAMENTALS OF DISTRIBUTION TRANSFORMER GROUNDING

Any generator connected to the delta winding of a generator step-up transformer, as shown in Figure 1, may be grounded through a distribution transformer and protected for ground faults by overcurrent relays connected to a current transformer in the distribution transformer secondary.

This type of grounding is, in effect, very high resistance grounding. Maximum ground fault currents are in the order of 5 to 10 amperes. Under this system, the possibility of fault damage is greatly reduced, and high-speed relaying, field breakers, or neutral breakers are not required for the purpose of getting the machine de-energized before serious damage occurs. Nevertheless, it is recommended that the generator be immediately tripped off with the occurrence of a ground fault, rather than to delay this action until it can be shut down more conveniently. This is because as long as the machine is operated with one phase grounded, the possibility exists of a phase-to-phase fault occurring if a second phase goes to ground. This could result in a very high fault current and major damage to equipment. Serious damage may also result if a fault occurs very near the generator neutral, and is then followed by a second ground higher up in the same phase. Standard overcurrent or overvoltage ground relaying will not detect this second fault, since the stator neutral is shorted. The generator differential relays will not detect this fault either, if both grounds are inside the differential current transformers.

The distribution transformer grounding system is normally used with unit-connected machines as shown in Figure 1. It may also be used when two or more units are connected to a common transformer winding as shown in Figure 2. In this case, however, it may be necessary to trip off both generators for a fault in one machine. This is a small disadvantage, however, since a fault in the main step-up transformer, or low-side bus connections, will re-

82 WM 171-7   A paper recommended and approved by the IEEE Power System Relaying Committee of the IEEE Power Engineering Society for presentation at the IEEE PES 1982 Winter Meeting, New York, New York, January 31-February 5, 1982. Manuscript submitted October 2, 1981; made available for printing November 16, 1981.

Reprinted from *IEEE Trans. Power App. Syst.*, vol. PAS-101, no. 12, pp. 4490–4501, Dec. 1982.

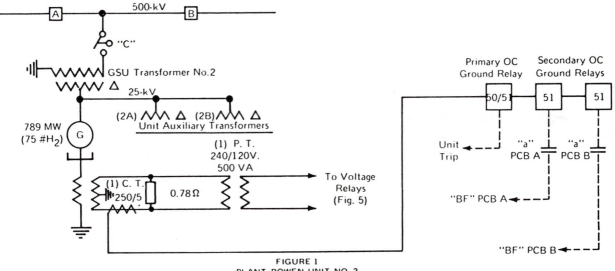

FIGURE 1
PLANT BOWEN UNIT NO. 2
GROUND OVERCURRENT PROTECTION

quire both units to be tripped in any event. It is important to note that with this generator grounding method, appreciable ground-fault current is not required for positive and fast relay protection. This is because the presence of any current or voltage between the machine neutral and ground may be used to relay the generator off the system.

RATINGS OF SECONDARY RESISTOR AND GROUNDING TRANSFORMER

Where the effective generator neutral resistance is extremely high, undersirably high transient overvoltages may result from circuit breaker restrikes, or arcing grounds. The magnitude of these transient overvoltages will be largely determined by the neutral resistance and the capacitance to ground of the generator and of the connected equipment at generator voltage.

Transient overvoltages are kept low by selecting the secondary resistor so that its kW loss during a line-to-ground fault at the generator terminals is equal to the capacitive kVA to ground. Stating it another way, the effective resistance in the machine neutral should be equal to $(1/3)$ (X_c) of the entire machine-voltage network. The value of this resistance is not critical; however, it should be noted that the conservative direction for lower transient voltages is with the greater I^2R loss - that is, higher generator ground fault current. Note also, however, that reducing the ohmic value of the secondary resistor in order to reduce transient overvoltages for external faults may tend to increase any damage that may occur due to ground faults in the protected zone.

In calculating the total capacitance to ground, the important items are the capacitance of the machine stator winding, the unit step-up transformer, the unit auxiliary transformers, the generator bus duct, and the surge protective capacitors (if used). The capacitance to ground of generator lightning arresters is relatively small and need not be considered.[2]

For the 789 MW generator shown in Figure 1, the capacitance values are as follows:

Stator winding -	0.224 uF (C_G)
Isophase bus -	0.002 uF (C_B)
GSU transformer -	0.020 uF (C_H)
UAT's (2) -	0.008 uF ($2C_A$)

Total (per phase) -	0.254 uF
Total (3ø) -	0.762 uF

The distribution transformer primary voltage rating is normally selected as a standard rating equal to or greater than generator line-to-neutral voltage. For a generator rated 25-kV, a 14.4-kV distribution transformer may be used. The secondary resistor should be connected across the 240 volt secondary winding, with the center tap on this winding solidly grounded to the station ground mat.

The following equations may be used to calculate the resistor and transformer ratings based on making the resistor kW loss equal to the charging kVA to ground. Let:

KV_{TR} = distribution transformer primary kV rating

N = distribution transformer turns ratio = primary volts/secondary volts

C = microfarads capacitance to ground of all three phases

R_s = secondary resistance

$$R_s = \frac{10^6}{N^2 \times 2\Pi fC} \text{ ohms} = \frac{153}{(KV_{TR})^2 \times C} \text{ ohms} \quad (1)$$

for 240 volt secondary and 60 Hz.[3]

If $C = 0.762$ uF, then $R_s = \dfrac{153}{(14.4)^2 (0.762)}$

$$= 0.97 \text{ ohm (at 415°C)} \quad (2)$$

To provide margin, and allow for heating effects, a standard cast grid resistor with an ohmic temperature coefficient of 7% per 100°C, and rated 0.78 ohm at 40°C, is selected.

After the secondary resistor ohmic rating has been determined, the maximum generator ground fault may be calcualted as follows: Let

I_p = maximum generator ground fault amperes

E_p = generator phase-to-neutral volts

R_s = secondary resistor rating in ohms

V_r = transformer voltage ratio = primary volts/ secondary volts

Using these relationships, it can be shown that

$$I_p = E_p/R_s(V_r)^2 \qquad (3)$$

The generator of Figure 1 is grounded through a distribution transformer rated 14,400/240 volts. Since the generator line-to-line voltage is 25,000 volts, and the secondary resistor is rated 0.78 ohm at 40°C, then the maximum generator neutral current during a fault will be

$$I_p = \frac{25,000/1.73}{0.78(60)^2} = 5.15 \text{ amperes} \qquad (4)$$

I_s, the maximum current in the secondary resistor, may now be calculated:

$$I_s = (I_p)(V_r) \qquad (5)$$
$$= (5.15)(60) = 309 \text{ amperes}$$

Either a one-minute or ten-minute resistor rating may be used. However, with the resistor sizes normally applied, there is little or no cost difference between the two, and the more conservative ten-minute rating should usually be selected.

The kVA rating of the grounding transformer for continuous duty may be derived by multiplying the generator voltage by the maximum fault current. Thus:

$$(25/1.73)(5.15) = 74.4 \text{ kVA}$$

A standard 75 kVA unit should be applied.

SELECTION OF CURRENT TRANSFORMER

In order to supply current to the ground overcurrent relays, a current transformer of high relaying accuracy is installed with the C.T. primary winding in series with the secondary resistor. A 5kV or 15kV C.T. with C100 accuracy will provide a conservatively rated current source.

The C.T. ratio is selected so that the maximum current in the generator ground relays is approximately the same as the maximum generator ground fault amperes. If the maximum secondary current is 309 amperes, a C.T. ratio of 250/5 (50/1) may be selected. This will produce a maximum current in the ground relays of 309/50 = 6.2 amperes.

APPLICATION OF THE PRIMARY OVERCURRENT GROUND RELAY ON UNIT-CONNECTED GENERATORS

Each unit-connected generator should be protected by one Primary Overcurrent Ground Relay, plus one Secondary Overcurrent Ground Relay for each breaker used to connect the generator to the transmission system. Thus, the generator of Figure 1 requires that three overcurrent ground relays be installed. The coils of all three relays are connected in series, so that each relay receives the same current. Relays operating with an inverse or very inverse time-overcurrent curve are generally applied.

With the unit on the line, there will be a continuous flow of current in the overcurrent relays, caused by the stray capacitances of the system being protected. This current will consist mostly of harmonics of the fundamental frequency, principally the third. It will vary directly with the load on the machine, so that the maximum current flow will occur with the unit fully loaded. If the secondary resistor is properly selected, this value will seldom exceed 0.5 ampere, where the maximum fault current (in the generator and in the relays) approaches 10 amperes. Actual field measurements on 29 hydro and 59 thermal units on the Georgia Power Company system, ranging in size from 15 to 900 megawatts, showed relay current from 0.1 to 0.6 ampere, with a mean value of 0.3 ampere.[4]

It is important that the current in the operating coils of the overcurrent relays be measured with the unit running at full load. This value should not exceed 75% of the relay tap setting. The maximum steady-state current in the relays protecting Plant Bowen Unit No. 2 is approximately 0.3 ampere. If the Primary Overcurrent Ground Relay is set on tap 0.6, this will provide protection for all but the last (0.6/6.2)(100%) or 9.7% of the stator winding.

Since a voltage may exist at the generator neutral when a fault occurs on the high-tension side of the GSU transformer, some time delay must be provided for the TOC unit. Otherwise, the machine may be incorrectly tripped for a transmission system fault. A time-dial setting of 3.5 to 4.0 will usually prove adequate if very inverse relays are used.

The Primary Overcurrent Ground Relay should generally be equipped with an instantaneous overcurrent unit. This device will provide for ultra-high-speed clearing of ground faults in the transformer delta windings and the bus-work connected to the generator terminals. It will also give high-speed protection for all faults in the first 50 to 70 percent of the generator, measuring in from the high voltage end of the machine. Thus, the IOC unit is extremely valuable in limiting machine damage, particularly in the case of nearly simultaneous ground faults on two different phases. However, if it is desired to coordinate the generator neutral ground relays with the generator potential transformer fuses, the IOC unit will have to be connected to alarm only. This will still prove of considerable value, since the action or inaction of this unit will aid in the frequently difficult task of determining fault location.

To prevent incorrect operation of the Primary Overcurrent Ground Relay IOC unit for faults on the high voltage side of the GSU transformer, a setting of not less than three times the setting of the TOC unit, and not less than 2.5 relay amperes should be selected. For Plant Bowen Unit No. 2, this will permit the IOC unit to protect the entire machine-voltage network, plus 100% - (2.5/6.2)(100%) = 60% of the stator winding.

APPLICATION OF THE SECONDARY OVERCURRENT GROUND RELAYS ON UNIT-CONNECTED GENERATORS

In order to protect against failure of the generator power circuit breaker(s) to trip, a Secondary Overcurrent Ground Relay should be installed for each generator PCB that is installed. As shown in Figure 1, Plant Bowen Unit No. 2 has two generator PCB's so two Secondary Overcurrent Ground Relays are required, and it will be noted that these relays receive the same current as the Primary Overcurrent Ground Relay. Each secondary relay is armed by an "a" auxiliary switch of the PCB for which it is providing breaker-failure protection. This is important, since on a unit-connected generator, ground fault current will continue to flow for as long as 10 to 15 seconds after a primary relay operation. This is due to the relatively slow decay of the generator voltage after the field has been de-energized.

In a modern breaker-failure scheme, each generator circuit breaker will be protected against failure to trip by an overcurrent fault detector connected to a timing relay. This system is armed by the GSU transformer differential relays, and other relays, such as the generator differential relays, that may be called on to trip the generator high-side bank breaker(s). Since the distribution transformer grounding method limits a ground fault on the generator side of the GSU transformer to a few amperes, such a fault will not be detected by the fault detectors which measure current flowing in the PCB's on the high-tension side. The Secondary Overcurrent Ground Relays, therefore, serve as replacement for these fault detectors when a ground fault occurs at the generator voltage level. Each Secondary Overcurrent Ground Relay should be connected to the timing relay that provides breaker-failure timing for the PCB for which the secondary relay is providing ground-fault breaker-failure protection. This will ensure that the proper PCB's are tripped in the stuck-breaker mode.

The Secondary Overcurrent Ground Relays should contain a TOC unit only. They should be set on the same tap as the Primary Overcurrent Ground Relay, with a time-dial calibration of 6.5 to 7. This will provide proper coordination with the primary relay, particularly considering the fact that the breaker-failure timing relay setting will be added to ground relay time.

APPLICATION TO TWO UNITS CONNECTED TO A COMMON DELTA WINDING

Figure 2 shows an example of two machines connected to the delta winding of GSU transformer through individual generator breakers. These units may also be protected against ground faults by overcurrent relays, with the understanding that a fault in one machine may require that both units be removed from the system. This may not be a serious disadvantage, as long as a scheme is devised that will determine in what part of the generator-voltage network the fault has occurred.

With generators connected as shown in Figure 2, the selection of the secondary resistor will require some compromise. Normally, the value selected should be the same for both units, and should be calculated by using the capacitance of the system with one generator breaker open. With both breakers closed, the parallel resistance to ground will now be less than would usually be selected for a single unit-connected machine. However, as already noted, this value is not critical, and proper relaying may be accomplished over a wide range of neutral resistance. It will also be recalled that the conservative direction for lower transient overvoltages is with higher fault current.

When a fault occurs, it is usually desirable to know its location as quickly as possible. If this is the case, the IOC unit in the Primary Overcurrent Ground Relays will have to be omitted, or connected to alarm only. Both primary relays should be set on the same tap, with the time-dial for the first unit set on division 3, and the second unit set on 4.5. If the ground is removed by separation of the first unit, there will be no further relay action, and the fault location is established. If the ground is removed only after both units are tripped, the location is established as being on the generator side of the second unit PCB. If the ground is on the transformer side of the generator breakers, a fault will still exist. To detect this fault, a zero-sequence overvoltage relay may be installed. It should be connected across the broken-delta secondary of a set of wye-delta potential transformers installed between the GSU transformer and the generator breakers. A ground in this area will impress a zero-sequence voltage, derived from the broken delta, on the ground overvoltage relay. To avoid improper coordination, this relay should be armed only when both generator PCB's are open, and should be connected to trip the same as the GSU transformer differential relays.

To prevent ringing, the potential transformers, although connected line to ground, should carry full line-to-line rating. The overvoltage relay should be rated to withstand a maximum zero-sequence voltage of 199 volts. This will permit a relay with a range of 16 to 64 volts to be applied, and this relay should be set on the minimum tap and maximum time dial.

A Secondary Overcurrent Ground Relay should be installed on each unit to protect for a stuck generator PCB. Each secondary relay should be set on the same tap as the primary relays, with a time-dial setting of 6.5 to provide coordination. Both secondary relays should be connected to trip the GSU transformer differential lockout relay.

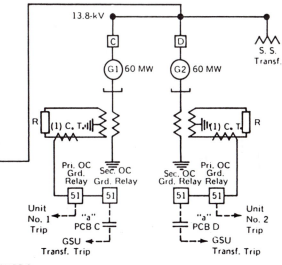

FIGURE 2
TWO UNITS CONNECTED TO ONE TRANSFORMER
GROUND OVERCURRENT PROTECTION

213

APPLICATION CONSIDERATIONS FOR
GENERATOR NEUTRAL VOLTAGE RELAYING

The application of voltage relaying to a generator neutral requires an analysis of the frequency content of the neutral voltage. Conventional generator designs strive to keep harmonics that appear in the phase-to-phase voltage to a low value, since this distortion adversely affects the electrical efficiency of the generator. On a balanced three-phase system, the fifth and seventh order harmonics appear in the phase-to-phase voltage only. The third, ninth, and fifteenth harmonics generated in a three-phase machine show up only in the neutral voltage as a zero sequence component, while the fifth and seventh behave as negative and positive sequence components respectively.[5]

Figure 3 shows the frequency spectrum of the voltage appearing across the resistor on the secondary side of a generator grounding transformer. This plot is for Bowen Unit No. 2 loaded to 600 MW and 90 MVAR (out). The plot was made by passing a 3 hertz frequency window across the frequency spectrum and plotting the voltage passed by this window. Note that the peak voltage was the third harmonic, which reached 6.3 volts (RMS). 6.3 volts across the resistor corresponds to $(\frac{60 \times 6.3}{25000/1.73} \times 100\%) = 2.6\%$ of the generator phase-to-neutral voltage rating. The 60 hertz component of the neutral voltage was 0.07 volts (RMS) across the resistor.

The third harmonic content of the neutral voltage may vary between 1% and 10% of the generator terminal voltage, depending on the design philosophy. One can expect at least 50%[6] more third harmonic at full load than at no load.[6] A two-month survey of third harmonic voltage variations across the neutral resistor of this unit showed that the third harmonic voltage varied from 5.5 volts to 7.8 volts under normal load conditions. The minimum third harmonic voltage of 1.8 volts was observed prior to a routine unit trip and was observed again when the unit was resynchronized to the line. For this machine, then, the no-load third harmonic voltage is $\frac{1.8}{7.8} = 23\%$ of the maximum value monitored.

Figure 4 shows the variation of third harmonic voltage at the neutral of this unit as a function of power factor angle. The unit reactive component was varied from 200 MVAR "out" to 80 MVAR "in" while holding the real power constant at 630 MW. No variations in the third harmonic voltage were noted.

APPLICATION OF OVERVOLTAGE GROUND RELAYS
ON UNIT-CONNECTED GENERATORS

The voltage relaying system employed by Georgia Power Company for ground fault protection of unit-connected generators provides for two relays with overlapping zones. These relays are connected as shown in Figure 5. The overvoltage protection path consists of a 240/120 volt isolating transformer across the neutral resistor, a 180 hertz blocking filter, an overvoltage level detector (59), and a timer.

The 240/120 volt isolating transformer limits the voltage to the continuous rating of the relay, provides isolation of the protection schemes, and allows the grounding of one input at the relay location. An isolating transformer with a D-C resistance of less than 2.5 ohms should be used in order to prevent excessive voltage drop. A 500 VA transformer with a resistance of 0.89 ohm was selected. Although the relay burdens would not appear to require a transformer this large, these burdens are at very low voltages and would be substantially higher at the rated voltage of the transformer.

The 180 hertz blocking filter and overvoltage device 59 are calibrated as a unit. The pickup voltage at 180 hertz is about 18 times the 60 hertz value. The setting of the 59 unit is controlled by the maximum voltage applied to the relay for a phase-to-ground fault on the machine terminals, and the neutral voltage resulting from a phase-to-ground fault on the high side of the generator step-up transformer due to capacitive coupling between the transformer windings. In the case of Bowen Unit No. 2, the maximum voltage applied to the relay is:

$$V_R = \frac{V_{L-N}}{N \times \frac{240}{120}} = \frac{25,000/1.73}{60 \times 2} = 120 \text{ V} \qquad (6)$$

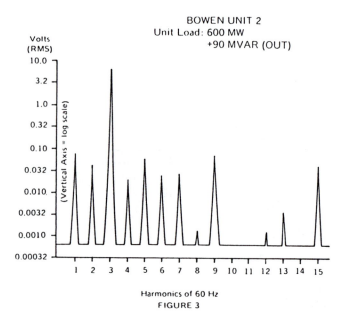

BOWEN UNIT 2
Unit Load: 600 MW
+90 MVAR (OUT)

Harmonics of 60 Hz
FIGURE 3

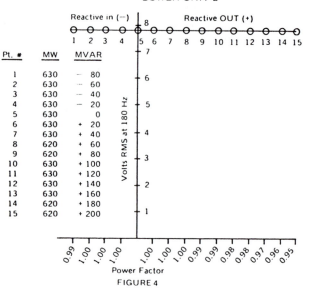

BOWEN UNIT 2

Pt. #	MW	MVAR
1	630	− 80
2	630	− 60
3	630	− 40
4	630	− 20
5	630	0
6	630	+ 20
7	630	+ 40
8	620	+ 60
9	620	+ 80
10	630	+ 100
11	630	+ 120
12	630	+ 140
13	630	+ 160
14	620	+ 180
15	620	+ 200

Power Factor
FIGURE 4

In order for the overvoltage relay to detect ground faults in the first 95% of the stator winding, measured in from the generator high voltage terminals, it must be set for $(0.05 \times 120) = 6.0$ volts. Calculations indicate that with this setting, device 59 will probably not detect faults on the high voltage side of the GSU transformer. However, absolute security may be obtained by setting timer 59 on 75 cycles. This will provide adequate coordination, and eliminate any possibility of generator overtripping for ground faults on the transmission system.

APPLICATION OF THIRD HARMONIC UNDERVOLTAGE GROUND RELAYS ON UNIT-CONNECTED GENERATORS

The connection of the overlapping undervoltage relay is also shown in Figure 5. This relay is tuned to operate on the 180 hertz component of the voltage appearing at the generator neutral. The undervoltage protection path consists of a 240/120 volt isolating transformer across the neutral resistor, a 60 hertz blocking filter, an undervoltage level detector, on-line logic, and a timer. The 60 hertz blocking filter and the undervoltage device labeled 27 are calibrated as a unit. The frequency response of this unit is shown in Figure 6. Note that the pickup voltage at 60 hertz is about 30 times the 180 hertz value, and this results in a 3.3% leakage of the fundamental voltage.

Since device 27 responds to an absence of third harmonic voltage, a supervision scheme must be provided to block false trips when the generator is out of service. Georgia Power Company employs an on-line supervision switch, but other methods may be used. For example, overcurrent supervision may be necessary with generator designs that do not produce significant third harmonic voltage until loaded. In this case, an alternate protective scheme should probably be considered, since a third harmonic undervoltage relay would be out of service under light load conditions.

Determination of the pickup level of device 27 requires the calcuation of third harmonic voltage distribution, field checking of this value, and an analysis of the variation of third harmonic voltage at the neutral for ground faults at various points in the machine. The minimum third harmonic voltage generated by Bowen Unit No. 2 is given by the manufacturer as 200 volts at no load. The capacitance values for this unit are given on page 2, and the third harmonic voltage distribution may be analyzed using Figure 7. The capacitances per phase may be grouped at the neutral and terminal ends:

$$\text{Neutral} - 0.5C_G = 0.112 \text{ uF} \qquad (7)$$

$$\text{Terminal} - (0.5C_G + C_B + C_U + 2C_A) = 0.142 \text{ uF} \quad (8)$$

The neutral end impedance at 180 hertz is the parallel combination of $3(0.5C_G)$ and R_N, where $R_N = 0.78(60)^2 = 2808$ ohms.

$$1.5\, C_G = 0.336 \text{ uF}$$

$$X_{CG} = \frac{1}{2\Pi\, fC} = 2632 \text{ ohms at } 180 \text{ hertz.} \qquad (9)$$

Let

Z_N = parallel combination of X_{CG} and R_N.

$$Z_N = \frac{(2808)\,(-j2632)}{2808 - j2632} = 1920 \;\underline{/-46.85°}$$

$$= 1313 - j1401 \text{ ohms.} \qquad (10)$$

The capacitance to ground at the generator terminals is 0.142 uF.

The capacitive reactance per phase is:

$$X = \frac{1}{2\Pi\, fC} = 6227 \text{ ohms at } 180 \text{ hertz.} \qquad (11)$$

The three phase capacitive reactance at the generator terminals is $-j6227/3$ ohms $= -j2076$ ohms, and the third harmonic voltage drops at no load will be:

V_T = voltage at terminals

$$= 200 \left(\frac{-j2076}{1313 - j1401 - j2076}\right) \qquad (12)$$

$$= \frac{2076 \;\underline{/-90°}}{3716 \;\underline{/-69.3°}}(200) = 112 \;\underline{/-20.7°} \text{ volts.}$$

V_N = voltage at neutral

$$= 200 \left(\frac{1313 - j1401}{1313 - j1401 - j2076}\right) \qquad (13)$$

$$= \frac{1920 \;\underline{/-46.85°}}{3716 \;\underline{/-69.3°}}(200) = 103 \;\underline{/22.5°} \text{ volts.}$$

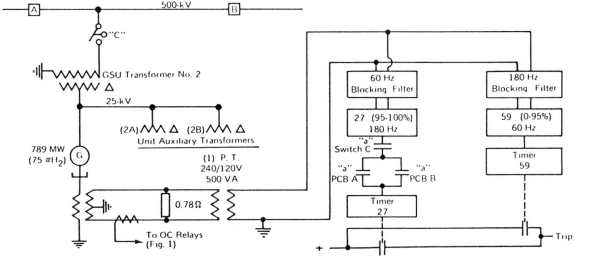

FIGURE 5
PLANT BOWEN UNIT NO. 2
GROUND VOLTAGE RELAY PROTECTION

The third harmonic voltage at the relay is:

$$V_R = 103 \times \frac{240}{14400} \times \frac{120}{240} = 0.86 \text{ volt at no load.} \tag{14}$$

The calculated third harmonic voltage across the resistor is $103 \times \frac{240}{14400} = 1.72$ volts, which compares favorably with a measured value of 1.8 volts at generator minimum load.

The ratio of the neutral voltage (V_N) to the generated voltage is (103/200), so that 52% of the third harmonic voltage is present at the neutral.

Voltage variations during faults are significant in determining an appropriate undervoltage setting, and then calculating the overlap of the two voltage relays. Phase-to-ground faults at the machine terminals will cause the fundamental frequency voltage at the neutral to rise to the phase-to-neutral voltage. The third harmonic voltage will also increase above the unfaulted level, since the terminal capacitance is shorted. A phase-to-ground fault at the neutral will short circuit the grounding transformer, causing the fundamental and third harmonic voltages to equal zero. Previous calculations show that the fundamental frequency overvoltage relay (59) protects 95% of the generator winding, measured from the high voltage terminals. Since the third harmonic voltage increases for faults at the generator terminals, and equals zero for neutral faults, there is some point in the winding where a phase-to-ground fault will have no effect on the third harmonic voltage at the generator neutral. For Bowen Unit No. 2, this point is 52% up from the neutral end of the machine.

The dropout of the third harmonic voltage relay should provide a 2/1 margin for the minimum no-fault voltage appearing at the neutral. In this case, a relay setting of $(1/2)(0.86) = 0.43$ volt may be considered. Thus, when the third harmonic neutral voltage decreases to 0.43 volt, the 27 relay will drop out and start the timer. The unit will be tripped after a delay of about three seconds.

It is now necessary to confirm the adequacy of the overlap of the 0-95% relay (59) and the 95-100% relay (27). At the 5% pickup point of device 59, the third harmonic voltage at the neutral will be:

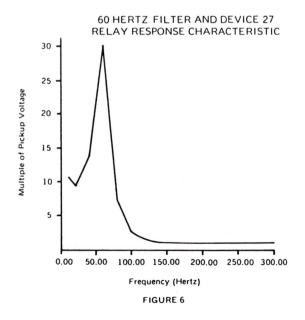

60 HERTZ FILTER AND DEVICE 27
RELAY RESPONSE CHARACTERISTIC

FIGURE 6

$$V_N = 103 \times \frac{5}{52} = 9.9 \text{ volts} \tag{15}$$

The voltage applied to the relay will be:

$$9.9 \text{ volts} \times \frac{240}{14400} \times \frac{120}{240} = 0.083 \text{ volt} \tag{16}$$

The operate margin of the third harmonic relay is 0.43/0.083 = 5.18 to 1. This margin will be reduced by the 60 hertz voltage which will appear at the third harmonic relay filter output. This voltage is 6 volts x 3.3% = 0.20 volt. This 60 hertz voltage reduces the margin to $0.43/\sqrt{(.2)^2 + (.083)^2} = 1.99$ to 1 at no load.

From the above discussion it is evident that the 60 hertz blocking filter output establishes relay operate margins. Considering the filter leakage, the maximum reach of the relay may be described by the following equations:

The no-load third harmonic voltage at the relay (V_{R3}) as a function of (S), the distance in percentage of winding measured up from the neutral, will be:

$$V_{R3} = 103 \times \frac{S}{52} \times \frac{1}{60} \times \frac{1}{2} = 0.0165S \tag{17}$$

The fundamental frequency voltage at the relay (V_{R1}) due to 60 hertz blocking filter leakage of 3.3% is:

$$V_{R1} = \frac{S}{100} \times \frac{25000}{1.73} \times \frac{1}{60} \times \frac{1}{2} \times 3.3\% = 0.0397S$$

at any load $\tag{18}$

The total operate voltage applied to the relay is $\sqrt{V_{R3}^2 + V_{R1}^2} = 0.0430S$.

At the relay setting of 0.43 volt, the reach up from the neutral at no load is:

$$0.43 = 0.0430S, \text{ and}$$

$$S = 10\%$$

$$\text{At full load, } V_{R3} = 103 \times \frac{7.8}{1.8} \times \frac{S}{52} \times \frac{1}{60} \times \frac{1}{2}$$

$$= 0.0715S \tag{19}$$

The total full load operate voltage applied to the relay is $\sqrt{V_{R3}^2 + V_{R1}^2} = 0.0818S$.

At the relay setting of 0.43 volt, the reach up from the neutral is:

$$0.43 = 0.0818S, \text{ and}$$

$$S = 5.3\%$$

The 27 relay operate ratio for a ground fault 5.0% up from the neutral is 1.1 to 1, and this means that the overlap of the 27 and 59 relays is marginal for a 27 relay setting of 0.43 volt. For this reason, the preferred 2/1 third harmonic requirement must be modified. Reducing this margin to 150% establishes an undervoltage setting of (0.86/1.5) = 0.57 volt. At this setting the reach up from the neutral at full load is:

$$0.57 = 0.0818S, \text{ and}$$

$$S = 7.0\%$$

This setting increases the coverage at no load to 0.57/0.043 = 13.3%.

CONCLUSION

Unit-connected synchronous generators should be protected for ground faults by two completely separate relaying schemes. The trip output from these schemes should be connected to separate lock-out relays, and on large units, separate control batteries should be employed. It is also important that one of the schemes be self-monitoring, so that a failure of the neutral grounding equipment, or of a protective relay, will be detected before a fault occurs. Higher reliability is achieved by using two entirely different types of relaying systems, preferably one using current detection and one responding to voltage. For all types, careful calculations are required in order to determine the precise degree of protection being provided.

C_G = Total winding capacitance to ground per phase.

C_B = Total lead capacitance to ground per phase.

C_U = Total GSU transformer capacitance to ground per phase.

C_A = Total station service transformer capacitance to ground per phase (each of 2 transformers).

R_N = Effective neutral resistance.

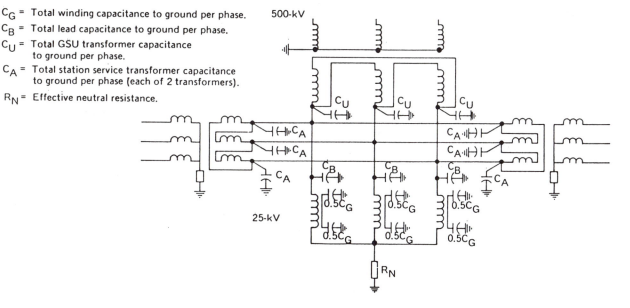

FIGURE 7
PLANT BOWEN UNIT NO. 2
DISTRIBUTION OF LUMPED CAPACITANCES

REFERENCES

(1) P. G. Brown, Generator Neutral Grounding, General Electric Co. Application Engineering Information GET 1941A, General Electric Co., Schenectady, NY, p. 5.

(2) Ibid, p. 12.

(3) Ibid, p. 13.

(4) Generator Ground Protection Guide, IEEE Power System Relaying Committee Project P490, Institute of Electrical and Electronic Engineers, New York, NY, Appendix B. (To be published as ANSI C37.101 by American National Standards Institute).

(5) Electrical Transmisssion and Distribution Reference Book, Fourth Edition, Westinghouse Electric Corporation, Pittsburgh, PA, p. 758.

(6) M. Stein and J. R. Linders, "Ground Fault Protection of the Complete Generator Winding", Proceedings of the Pennsylvania Electric Association Relay Committee, October 6-7, 1977.

Discussions

E. J. Emmerling (Camargo Associates Ltd., Cincinnati, OH): Generator ground protection has been discussed in numerous papers for over 30 years, but these authors have produced a practical guide of special interest. The worked examples and the account of extensive experience with 126 units will enable the average relay engineer to apply generator protection readily.

I would like to ask about your practices on the grounding transformers:

1. A standard overhead 75kVA transformer would have about 31 gallons of oil. Located near the generator or in the plant, this may be rated a hazard by you or the insurance company. Would you recommend:
    a. askarel filled - not PCB
    b. cast apoxy dry type, may require 112.5kVA rating
    c. dry type with space heater; 95kV BIL
    d. gas filled
2. You show the grounding transformer with the secondary mid-tap grounded. This ground has not been observed in other papers, except the new IEEE Guide [4] has a secondary terminal ground. It appears that a primary-secondary breakdown, before or during a stator fault could cause failure of the ground protection with the mid-tap ground. Since the generator is to be protected regardless of any other consideration, except personal hazard, the ungrounded mid-tap is expected to be more reliable.

The number of stator ground faults seems high, even for this large system. How does this compare with other systems? If your experience has led to the development of improved installation design and maintenance of generators, we would be interested in your comments.

Referring to the 1980 trouble with the "90%" relay that did not clear a near-neutral fault, how was the fault detected? Was there burning due to the 5A current and arcing effects which resulted in a phase to phase fault?

The authors are to be commended for making their excellent relay practices available to the industry.

Manuscript received January 11, 1981.

George D. Rockefeller (Consolidated Edison Company of New York, Inc., New York, NY): The authors and their company have made a significant contribution with this excellent, practical paper. The discusser's prime interest is to request amplification of various aspects.
Self-Monitoring vs: 100% Coverage
Should both of these objectives be equally weighted? A significant number of applications don't detect ground faults near the neutral end, with the rationale that the low insulation stress in this region makes a ground most unlikely. However, the authors report such a case where the generator was "badly damaged". Would the authors provide some details? Would a ground fault alarm and a prompt unit shutdown have avoided severe damage? Are they aware of other utilities that have experienced neutral-end faults?
How important is "self-monitoring" to protect against a defect in the grounding circuitry or associated relays? Are the authors aware of a failure to trip resulting from a defect? Would a periodic manual check suffice (e.g. a third-harmonic voltmeter reading)? Admittedly, weak insulation in the neutral transformer could go undetected until stressed by a generator ground fault. The neutral-transformer fault could significantly increase the current and attendant generator-iron damage.

Subharmonic Current Injection
The authros state that "both '100%' schemes have performed extremely well", referring to the "overlapping voltage relay" and subharmonic current schemes. Why did they confine the paper to the overlapping voltage relay method? To what extent do they consider the subharmonic technique "self-monitoring"?
Reliability of Undervoltage Protection
The authors' calculations highlight the critical importance of the Figure 6 filter rejection characteristic and its stability (and the desirability of improving it). Use of solid-state circuitry which rejects the *fundamental* rather than a filter which rejects *60 Hz* voltage would seem to be in order.
The relay output is enabled by "an on-line supervision switch". What is meant by this? Does this require human intervention? Breaker "a" switch contacts also supervise the relay trip; this requires lead runs to the switchyard. Why not use a generator phase current detector instead of the supervision switch and "a" contacts?
Could a relay with a dropout voltage that varies with generator load

current be beneficial? The authors state that harmonics "vary directly with the load on the machine . . . ." What did they intend by the use of "directly"? Did they merely mean that the loaded level is higher than the unloaded level?
Because of a coincidence, the origin of the "52" in Eq (15) may be misconstrued. Earlier the authors state that the no-effect point on third harmonic voltage for a winding fault is "52% up from the neutral . . ." However, the "52" comes from the ratio "103/200" specified earlier in the text.
Sensitivity to High Side Faults
Ground voltage shifts occur on the low side of the delta-wye GSU transformer for high-side ground faults because of the interwinding capacitance coupling. The authors state their setting practice but do not support their statements with any calculations. The value of the paper would be enhanced with an example or other support.
Breaker Failure Timing Initiation
Will the authors clarify the reason for the "secondary oc" relays in Figure 1. Why not initiate breaker-failure timing by the unit trip lockout and the breaker "a" contact in series? Isn't this needed anyway for non-electrical trips? If the breakers and associated bus are in service with the generator isolated by a disconnect switch, a "unit trip" will unnecessarily open the breakers. Alternatively, the trips from the plant can be interlocked with a disconnect switch auxiliary-switch contact to avoid nuisance trips; however, this seems to be very poor practice because a phone auxiliary switch position could wipe out all plant trips with the generator on line. These considerations are not obviated by the application of the "51" relays in Figure 1.

Manuscript received January 11, 1981.

J. L. Koepfinger (Duquesne Light Co., Pittsburgh, PA): The authors have presented useful information with regards to the application of generator ground protection. It is noted that the calculations method used avoided the application of symmetrical components, but instead used logical reasoning to arrive at the values of currents and grounding resistance. The examples using this method should be useful to those not versed in symmetrical components.

Neglect capacitance at neutral and lump with other capacitance

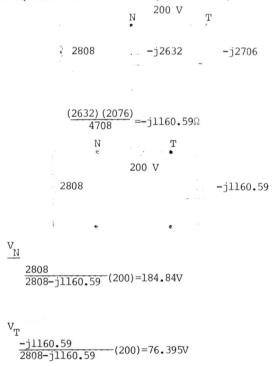

$$\frac{(2632)(2076)}{4708} = -j1160.59\,\Omega$$

$$V_N$$

$$\frac{2808}{2808 - j1160.59}(200) = 184.84V$$

$$V_T$$

$$\frac{-j1160.59}{2808 - j1160.59}(200) = 76.395V$$

Third harmonic across the resistor is (184.84) 1/60 = 3.08 volts

In the calculation of the third harmonic voltages appearing at the generator terminals, it was observed that the authors chose to assume that ½ the stator capacitance appears at the neutral of the stator. It is felt that this is an important assumption. It is this value, which in parallel with the generator grounding resistance, can have a significant effect upon the voltage applied to the third harmonic relays. If it were assumed that all of the stator capacitance were lumped at the generator terminals, the third harmonic voltage across the neutral resistor would be 3.08 volts based upon the impedances and third harmonic source voltages given in the text. Perhaps the authors would comment on the bases of their assumption.

This relay scheme presented is unique in that it provides protection for a failure of the grounding transformer as well as generator ground fault protection.

Manuscript received January 29, 1981.

Joe Pilleteri (Westinghouse Electric Corp., Coral Springs, FL): The paper offers a profound insight into ground-fault protection for synchronous generators. The methods and calculations discussed provide a thorough understanding of basic concepts used in Generator ground-fault protection. However, there are some points which may merit additional discussion and clarification.

(1) The authors (s) stated the third harmonic voltage at the neutral of a particular generator, Bowen Unit #2, was monitored while varying the reactive output from 200 MVAR "out" to 80 MVAR "in". It was also noted that the real power output was held constant throughout the investigation. The results showed no change in third harmonic voltage. Do the authors feel that as a general rule the reactive output has no effect on the residual third harmonic voltage?

(2) The author (s) should expand on the comments concerning the difficulty encountered with a third harmonic undervoltage relay for some operating conditions. The problem involves setting the relay with a minimum amount of third harmonic voltage generated under light load conditions. Figure 1 can be used to analyze the problem. Note that the relay is monitoring the third harmonic voltage at the generator neutral. It appears that the setting requirements for both light load and maximum generator output conditions cannot be met in some cases. With a ground-fault near the neutral point, tripping would occur for the light load case, but with maximum loading, there may not be enough reduction in third harmonic content to trigger relay operation.

(3) Under "Selection of Current Transformer", there is a question on the selection of the 5 KV or 15 KV rated C.T. with C100 accuracy. From an insulation standpoint, a 600 volt rated C.T. would be adequate for this application.

THIRD HARMONIC SETTING AT NEUTRAL END

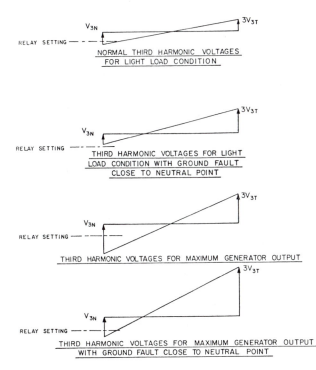

FIG. 1

Manuscript received February 3, 1981.

John R. Boyle (Tennessee Valley Authority, Chattanooga, TN): The authors are to be congratulated for an excellent critique of ground fault protection for synchronous generators. TVA shares many of the concerns expressed by the authors about the disadvantages of a generator overvoltage ground fault protective scheme.

Recently TVA experienced a serious (explosive) generator neutral transformer failure when a phase-to-ground fault occurred on the bus of a large generator. The overvoltage relay connected to the secondary of the generator neutral transformer failed to operate because one of the secondary fuses in the transformer secondary was found to be open.

I noticed that Georgia Power Company utilizes this protection, but there was no mention of the use of secondary fuses. If fuses are used, where are they located (the load side of the resistor, the load side of the undervoltage relay, etc.)?

Does Georgia Power Company utilize indicating voltmeters to measure third harmonic voltages and if so, so operating procedures require operator intervention when a voltage decrease is observed? At what level of voltage reductions is an operator instructed to take action (80 percent, 60 percent, etc.)?

Manuscript received February 16, 1981.

Walter L. Hinman (Gibbs & Hill, Inc., New York, NY): With reference to Figure 1, it is suggested that the Secondary Overcurrent Ground Relays be allowed to retrip breakers PCB A and PCB B directly (without 52a switch supervision) while at the same time starting the breaker failure timers of PCB A and PCB B via 52a switch supervison. This will provide ground relay backup and at least a measure of breaker trip circuit backup (particularly where there are dual trip coils) which is independent of the 52a switch and the possible problems associated with erroneous auxiliary switch indication.

Manuscript received February 26, 1981.

John R. Linders (Consultant, Sarasota, FL) and **Gunnar P. Stranne** (ASEA Inc., Yonkers, NY): This is a good paper for 3 reasons:
(1) It extends protection philosophy into new and desirable areas.
(2) It provides specific details for implementing sound protection philosophy.
(3) It shows the practicability of using a third harmonic undervoltage relay for protecting the low end of the generator winding and the entire neutral system.

At first glance, it is a bit awesome to consider setting a relay at a fraction of a volt and connecting it to trip a 25kV, 789 MW generator. However, if one observes on an oscilloscope this third harmonic signal as used in the described relay scheme, one finds a reliable solid signal, free of noise and distortion, and quite suitable for use in a "100%" ground fault protection scheme for unit type generators.

The self-monitoring feature of the described protection is certainly valuable, and will, undoubtedly become an important concept also in the protection of other system components. In this case, it is interesting that the monitoring provided by one set of (voltage) relays also enhances the reliability of the other set of (current) relays.

This monitoring of the integrity of the neutral system external to the generator also implements a new concept in relay protection. To date, the art of relaying has been largely limited to identifying faults and initiating segregating switching action to protect the system and the components from further damage.

It would appear that relaying is now moving towards identifying incipient fault conditions, or abnormal conditions which have the potential of serious consequences. The authors are to be commended for not only moving in this direction, but for also biting the bullet and tripping the unit rather than just alarming upon reduction in third harmonic voltage.

With regard to arming the 100% relay, schemes are available to accommodate practically any third harmonic profile with respect to machine excitation or watt or var loading. The scheme used in Figure 5 of the paper provides no protection from the 100% relay until the machine is connected to the system. Can the authors tell us why this method was chosen over a machine voltage monitoring relay which would permit arming the system before actual connection of the machine to the system?

Two machines are shown on a common bus in Figure 2, but no details are given as to the use of the 100% relay in this configuration. We would like to add that the third harmonic relay can provide ground fault protection of these machines: Generally, the calculations are more complicated when more than one machine is involved and the arming sensors may also be more involved depending on the operating requirements for load and reactive sharing. Do the authors have any experience in this area?

In this context we would like to mention the successful application of two sets of the described relays to protect the Consumers Power Company, Campbell dual winding generator. The rating of this machine is 1025 MVA at 18kV, PF 0.85, 3600 RPM.

Each of the two windings per phase is brought out to a separate neutral point. One winding is grounded through a high resistance using the typical distribution transformer connection. The other is connected to ground through a potential transformer. One of the described relays is connected to measure each neutral end voltage. The relays have been protecting the machine successfully since October 1980 with no incorrect operations.

When one can effectively arm the hundred percent relay so as to preclude nuisance relay operation under all operating conditions, tripping on the first fault regardless of location as done by the authors is to be preferred to alarming only. The discussors have no documented statistics, but considering the nature of a low end winding ground fault, it would seem reasonable to expect a higher probability of a second fault developing shortly thereafter or simultaneously. Such a second fault can result in considerable machine damage due to stored energy in the machine field sustaining the fault after all breakers are open even (although at a lesser extent) when forced field demagnetizing is used. Thus it appears desirable to trip upon the occurence of the first fault regardless of location.

Should a second fault develop on the same phase, conventional protection will be slow in responding, if it responds at all. However instantaneous tripping of such a fault is possible by a combination of this new neutral voltage relay system and a low set instantaneous negative sequence current relay. This latter function is available in some generator negative sequence relays as an alarm output. This would further enhance the protection of large machines against catastrophic failures.

With regards to breaker failure protection, we note that this is provided for neutral overcurrent relay tripping, but not for the neutral voltage tripping. Breaker failure for the latter could be added readily by monitoring the generator load current, with the breaker failure relay current detectors. A setting of 10% of rated current or some other value below any expected loading should be satisfactory. These relays are not normally picked up. Operation is enabled after initiation by any protective relay trip operation. The relay connections would be conventional, ie., any trip output from any protective relay including the 100% ground fault voltage relay would initiate the BF sequence. Reset of the BF current detectors then stops BF timing after successful tripping of the breaker.

Would the authors comment on these two suggestions for improving the generator protective scheme.

Manuscript received March 1, 1981.

Clayton H. Griffin and **John W. Pope:** The authors appreciate the thoughtful and pertinent comments of the discussors, which have added considerably to the value of this paper. With regard to the points raised by Mr. Emmerling, we agree that distribution transformers used for generator neutral grounding should not contain a flammable fluid, and Georgia Power Company has never used oil as an insulting medium for this type of application. PCB material was originally applied, but either dry-type or silicone-filled transformers are now employed. Also, it is important that the grounding transformer be located such that a violent explosion will not damage the generator or critical switchgear, and not present a hazard to plant personnel.

The secondary of the distribution transformer must have a ground for personnel safety reasons, and the mid-tap is the ideal position. It is a highly visible location, and will not be inadvertently removed during routine maintenance of the secondary resistor.

We have no statistics on the number of stator ground faults experienced by other systems. Actually, "stator" ground faults is probably a misleading term. Many faults detected by stator ground relays are not in the generator itself, but in the iso-phase bus network. Georgia Power Company has developed a routine test to verify the integrity of the generator grounding and protective system. This test is performed on a regular basis, since the results of an undetected fault anywhere in the generator low voltage network can be catastrophic. (See discussion by Mr. John Boyle above.)

Mr. Emmerling and Mr. Rockefeller have requested information on the details of a stator ground fault that occurred near a generator neutral and was not cleared by the "90%" relay. On June 2, 1980, a fault occurred near the neutral of Georgia Power Company's Unit No.

2 at Wallace Dam. This is a 52 MW, 14.4 kV pumped-storage generator/motor. The unit was undergoing its final full-load rejection test when a turn-to-turn fault occurred approximately 12.8% up from the machine neutral. The neutral ground relay was set to cover 88.5% of the stator winding.

The machine ran for about five minutes before being tripped by the generator differential relays. Inspection showed evidence of a sustained ground fault where a coil comes out of the stator iron. Arcing at this point eventually damaged an adjacent coil, and this caused sufficient current to flow to operate the differential. The arching (high resistance) nature of the original fault, and the fact that the operating quantity supplied to the relay was very near its pickup setting, resulted in no ground relay operation. A 100% stator ground fault protective scheme would have cleared the initial fault promptly before the adjacent coil became involved.

Mr. Rockefeller's comments concerning self-monitoring versus 100% coverage may be addressed by stating that both are probably equally important. The Wallace Dam failure is an example of need for a 100% scheme, and two very serious generator failures in neighboring utilities, one of which has been noted by Mr. Boyle, indicate the imperative need for continuous monitoring of the entire neutral protective system. In both instances, extensive damage resulted from the fact that the conventional "95%" neutral overvoltage protection was (unknowingly) out of service. Mr. Rockefeller is absolutely correct in stating that the characteristics of the 60 hertz blocking filter need improving; and we understand that an improved filter will soon be available from the relay manufacturer.

The "on-line supervision switch" described in this paper is an auxiliary switch arrangement verifying that the proper high-side disconnect switches and PCB's are closed, thus placing the unit on the line. No human intervention is required. (See Figure 5.) This method of arming the undervoltage relay was selected due to our lack of experience with third harmonic voltage variations at the generator neutral during off-line excitation and speed conditions. Recent field measurements indicate that generator voltage supervision would probably be secure and provide a superior arming system. Voltage supervision would not only reduce lead runs to the switchyard, but would also provide ground fault protection prior to synchronizing.

The origin for (1) the "52" in Equation 15, and (2) the statement that a fault "52% up from the neutral" will have no effect on third harmonic voltage distribution is the same. Equation 13 indicates that of the 200 volts third harmonic generated by this unit at no load, 103 volts appears across the grounding transformer. If we assume that the third harmonic is generated as a linear function of the stator winding, a fault $103/200 \times 100\% = 52\%$ up from the neutral will not cause a change in the voltage at the neutral. Using this linear relationship, Equation 15 indicates that a fault 5% up from the neutral will produce 9.9 volts third harmonic.

In response to Mr. Rockefeller's request, the following method may be used to calculate the voltage shift that occurs on the low side of a delta-wye GSU transformer due to a high-side ground fault. First, the circuit of Figure 7 may be reduced to the circuit of Figure 8 and then Figure 9. Next calculate:

V_{HO} = high side zero sequence voltage for a 500 kV phase-to-ground fault = 147 kV

C_{H-L} = generator step-up transformer high-to-low winding capacitance = 0.012 μF

C_O = generator system zero sequence capacitance per phase = 0.25 μF

R_N = equivalent resistance in the generator neutral = 0.78 $\times$ 60^2 = 2809 ohms

Then, the voltage appearing across the primary of the generator grounding transformer is:

$$V_R = \frac{V_{HO}\, Z_O}{Z_O - j\, X_{H-L}} \quad (20)$$

Z_O and X_{H-L} are calculated as follows:

X_{CO} = zero sequence capacitive reactance per phase

$$= \frac{1}{(2\pi 60)(0.254\,\mu F)} = 10443 \text{ ohms} \quad (21)$$

Z_O = parallel combination of $3R_N$ and X_{CO}

$$= \frac{(3R_N)(-j\, X_{CO})}{3R_N - j\, X_{CO}} = \frac{3(2808)(-j10443)}{3(2808) - j10443} = 5103 - j\,4117 \text{ ohms} \quad (22)$$

$$X_{H-L} = \frac{1}{(2\pi 60)(0.012\,\mu F)} = 221049 \text{ ohms} \quad (23)$$

$$V_R = \frac{147000(5103 - j4117)}{5103 - j4117 - j221049} = 4280 \text{ volts} \quad (24)$$

This voltage causes 4280/25000/1.73 = 30% neutral shift and results in 4280 × 240/14400 × 120/240 = 36 volts supplied to device 59 during the time that a phase-to-ground fault exists on the high voltage bus. Since the 59 relay is set for 6.0 volts, its associated timer is set for 75 cycles in order to eliminate possible overtripping for ground faults on the transmission system.

GENERATOR NEUTRAL VOLTAGE DISPLACEMENT FOR A PHASE-TO-GROUND FAULT

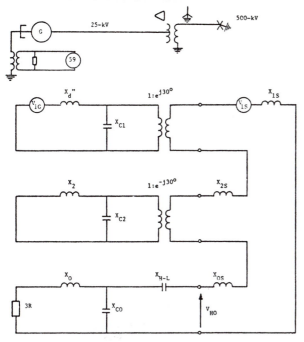

FIGURE 8

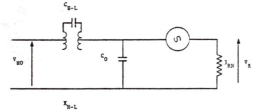

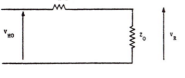

$$V_R = \frac{V_{HO} \, Z_0}{Z_0 - j \, X_{H-L}}$$

V_R = 4280 volts (36 volts to device 59)

FIGURE 9

The secondary overcurrent ground relays provide important backup protection for failure of the primary overcurrent relay and the unit lookout relay, as well as the power circuit breakers. It would not be appropriate to use the unit lockout relay to initiate breaker failure, since this device has an unacceptably high failure rate itself. The secondary overcurrent relays stand in place of the high-side PCB breaker-failure fault detectors for low-side ground faults, since such faults cannot be detected by overcurrent elements connected to current transformers in the high voltage circuit breakers. In this regard, we agree with Mr. Hinman's suggestion that the secondary overcurrent relays be allowed to retrip PBC A and PCB B directly (without 52a switch supervision), while at the same time starting the breaker-failure timers via 52a switch supervision.

In response to Mr. Koepfinger, the assumption that half the stator capacitance appears at the neutral and half at the terminals is based on Reference 6. Calculations using this assumption indicate that 1.72 volts should appear acorss the neutral resistor at no load, which compares favorably with the 1.8 volts that was actually measured.

The effect of generator reactive output on the third harmonic voltage at the generator neutral, as discussed by Mr. Pilleteri, is still under investigation by Georgia Power Company engineers. Recent field measurements indicate that generator terminal third harmonic voltage may vary significantly with excitation, while the third harmonic at the neutral may not. Additional testing is planned to address this question. With regard to Mr. Pilliteri's comment about the neutral current transformer, high voltage CT's are installed in order to provide additional insulation and mechanical strength. This may be important in case of a high-to-low winding failure in the GSU transformer, or the neutral grounding transformer, and the cost difference is trivial.

We would like to thank Mr. Boyle for his comments, as they provide important evidence for the need of continuous supervision of the generator ground protection scheme. Georgia Power Company does not fuse the grounding resistor or the voltage relays, as we prefer to accept the risk of relay damage in order to gain reliability.

Third harmonic voltage measurements are made during cut-in to verify relay setting calculations. We do not meter third harmonic voltage and, therefore, do not have procedures requiring operator action. In any case, if the third harmonic voltage falls into the danger zone, this is indication that a fault exists, and the unit will be tripped.

The discussion by Messrs. Linders and Stranne points out a number of interesting undervoltage relaying application considerations. Neutral third harmonic undervoltage relaying of two or more units on the same bus would certainly require careful calculations and field measurements under various operating conditions. Alabama Power Company has some experience with neutral undervoltage relaying of multiple hydro units connected to the same step-up transformer. They have experienced unexplained relay operations and have concluded that this is not a good application.

The two final suggestions by Messrs. Linders and Stranne may prove valuable for some installations. In our case, however, the "second fault on same phase" scenario does not appear to be a problem. If the first fault occurs in the high voltage end of a machine, the unit is tripped with high speed by the primary overcurrent IOC unit. Neutral-end faults are promptly cleared by the 100% under-voltage relay. If a second fault does occur, it would probably be after the machine was tripped, during the voltage decay period. Finally, we agree that breaker-failure protection should be provided for neutral-end faults, and the scheme proposed by these discussors appears to be suitable for this application.

Manuscript received May 26, 1981.

INADVERTENT ENERGIZING PROTECTION OF

SYNCHRONOUS GENERATORS

This report was prepared by the Inadvertent Generator Energizing Protection Working Group of the IEEE Power System Relaying Committee: C. J. Mozina, Chairman; R. F. Arehart, J. Berdy, J. J. Bonk, S. P. Conrad, A. N. Darlington, W. A. Elmore, H. G. Farley, D. C. Mikell, G. R. Nail, H. O. Ohmstedt, A. C. Pierce, E. T. Sage, D. E. Sanford, L. J. Schulze, W. M. Strang, G. Stranne, F. Tajaddodi and T. E. Wiedman

Abstract - Inadvertent energizing of synchronous generators has been a particular problem within the industry in recent years. A significant number of large machines have been damaged or, in some cases, completely destroyed when they were accidentally energized while off-line. The frequency of these occurrences has warranted the investigation of the problem by the "Inadvertent Generator Energizing Working Group" under the sponsorship of the IEEE Power System Relaying Committee. This report describes the problem of inadvertent generator energization, the hazard to the generator and turbine, and the major dedicated protection schemes employed within the industry to detect this condition.

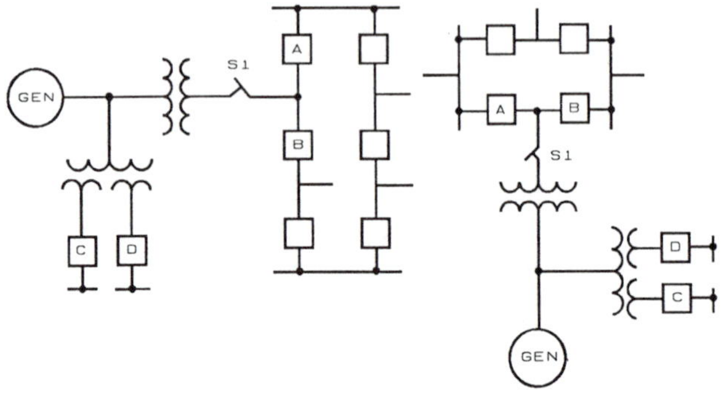

FIGURE 1 A) TYPICAL BREAKER-AND-A-HALF STATION   B) TYPICAL RING BUS STATION

INTRODUCTION

Inadvertent or accidental energization of large turbine-generators has occurred frequently enough within the industry in recent years to warrant concern. When a generator is energized while off-line and on turning gear, at a standstill or coasting to a stop, it behaves as an induction motor and can be damaged within a few seconds. Turbine damage can also occur. A significant number of large machines have been severely damaged and, in some cases, completely destroyed. The cost to the utilities for such an occurrence is not only the cost of repair or replacement of the damaged machine, but the substantial cost of purchasing replacement power during the period when the unit is out of service. Operating errors, breaker head flashovers, control circuit malfunctions or a combination of these causes have resulted in generators becoming accidentally energized while off-line. The frequency of these occurrences has prompted major U.S. turbine-generator manufacturers to recommend that the problem of accidental energization be addressed through dedicated protective relay schemes.

Operating errors have increased within the industry as high voltage generating stations have become more complex with the use of breaker-and-a-half and ring bus configurations. Figure 1 shows typical one-line diagrams for two such stations.

These station designs provide sufficient flexibility to allow a single high voltage generator breaker (A or B) to be taken out of service without also requiring the unit to be removed from service. Breaker disconnect switches (not shown) are available to isolate the breaker for repair. When the unit is off-line, however, generator breakers (A and B) are generally returned to service as bus breakers to complete a row in a breaker-and-a-half station or to complete a ring bus. This results in the generator being isolated from the system through only an open high voltage disconnect switch (S_1). Additional isolation from the power system can be provided by removing generator straps or other sectionalizing devices in the generator isophase bus. Generally, these isophase bus devices are opened to provide safety clearances or isolation for extended unit outages. There are many instances in which the high voltage disconnect switch (S_1) provides the only isolation between the machine and the system. Even with extensive interlocks between the generator breakers (A and B) and the disconnect switch (S_1) to prevent accidental switch closure, an increasing number of cases have been recorded of units being accidentally energized through this disconnect switch while off-line. Compounding this problem is the possibility that some or all unit protection, for one reason or another, may be disabled during this period.

Another path for inadvertent energizing of a generator is through the unit auxiliary system by accidental closure of unit auxiliary transformer breakers (C or D). Because of the higher impedance in this path, the currents and resulting damage are much lower than those experienced by the generator when it is energized from the power system.

The flashover of contacts in an open high voltage generator breaker (generally one or two poles) is yet another method by which generators have been inadvertently energized. The risk of a flashover is greatest just prior to synchronizing or just after the

88 SM 527-4   A paper recommended and approved by the IEEE Power System Relaying Committee of the IEEE Power Engineering Society for presentation at the IEEE/PES 1988 Summer Meeting, Portland, Oregon, July 24 - 29, 1988. Manuscript submitted April 11, 1988; made available for printing April 15, 1988.

Reprinted from *IEEE Trans. Power Delivery*, vol. 4, no. 2, pp. 965–977, April 1989.

unit is removed from service. During this period, the voltage across the open generator breaker can be twice normal as the unit slips in phase angle with respect to the system. A loss of air pressure in some types of high voltage breakers during this period can result in the flashover of breaker pole(s), energizing the generator and causing a significant flow of damaging unbalance current in the generator windings. This unique breaker failure condition must be quickly detected and isolated to prevent major generator damage.

Other large machines which are connected to the system through low voltage generator breakers have also been inadvertently energized. The use of these low voltage generator breakers is a recent practice for large generators which allows more operating flexibility than the traditional unit-connected configuration. Generally, the high voltage switchyards are breaker-and-a-half or ring buses identical to those previously described and differ only in the addition of the low voltage breaker (E). Figure 2 below shows a typical one-line for this design.

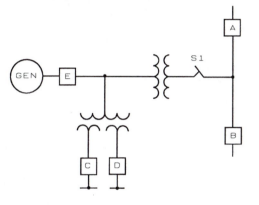

FIGURE 2   STATION WITH LOW VOLTAGE
GENERATOR BREAKER

When the generator is off-line, Breaker E is opened to provide isolation from the system. This allows the unit auxiliary transformer to remain energized, carrying load, when the generator is out of service and also provides start-up power when the generator is brought on-line. There have been cases reported of both accidental closures of Breaker E and pole flashovers due to the loss of dielectric strength which have inadvertently energized generators.

GENERATOR RESPONSE TO INADVERTENT ENERGIZING

Generator Response to Three-Phase Energizing

When a generator is accidentally energized with three-phase system voltage while at standstill or on turning gear, it behaves as an induction motor. During three-phase energization at standstill, a rotating flux at synchronous frequency is induced in the generator rotor. The resulting rotor current is forced into subtransient paths in the rotor body and damper windings (if they exist) similar to those rotor current paths for negative sequence stator currents during generator single phasing. The machine impedance during this high slip interval is equivalent to its negative sequence impedance ($R_{2G} + jX_{2G}$). The resistive component of the impedance is usually neglected. The negative sequence machine reactance is approximately equal to the average of the machine's direct and quadrature subtransient reactances ($X_d'' + X_q''$)/2. The machine terminal voltage and current during this period will be a function of the generator, unit step-up transformer, and system impedances. When a generator is inadvertently energized, the generator stator current induces high magnitudes of current in the

machine rotor, causing rapid thermal heating. This rotor current is initially at 60 Hz, but decreases in frequency as the rotor speed increases due to induction motor action.

If the generator is connected to a strong system, the machine's initial stator currents will be in the range of 3 to 4 times rating and the terminal voltage will be in the range of 50-70% of rated, for typical values of generator and step-up transformer impedances. If the generator is connected to a weak system, machine stator current may only be one to two times rating and the terminal voltage only 20-40% of rated. When the generator is inadvertently energized from its auxiliary transformer, stator current will be in the range of 0.1 to 0.2 times rating because of the high impedance in this path. The equivalent circuit shown in Appendix I can be used to determine approximately the initial machine currents and voltages when a generator is energized from the power system.

Generator Response Due to Single-Phase Energizing

Single-phase energizing of a generator from the high voltage system while at a standstill subjects the generator to a significant unbalanced current. The unbalanced current causes negative sequence current flow and thermal rotor heating similar to that caused by three-phase energizing. There will be no significant accelerating torque if the voltage applied to the generator is single-phase and the unit is essentially at standstill. Both positive and negative sequence currents will flow in the stator and each will induce approximately 60 Hz currents in the rotor. This produces magnetic fields in opposite directions essentially producing no net accelerating torque. If single-phase voltage is applied when the unit is not at standstill but, for instance, at half rated speed, the accelerating torque due to positive sequence current will be greater than the retarding torque due to negative sequence current and the unit will accelerate.

Breaker head flashover is the most frequent cause of single-phase inadvertent energizing. This situation is most likely to occur just prior to synchronizing or just after the unit is removed from service when the machine and system voltage are 180 degrees out of phase. The magnitude of stator current can be calculated using the symmetrical component equivalent circuit shown in Appendix II for a generator connected to the power system through a delta-wye grounded step-up transformer.

DAMAGE DUE TO INADVERTENT ENERGIZING

Generator Damage

The initial effect of inadvertent energizing of a generator from standstill or on turning gear is rapid heating in iron paths near the rotor surface due to stator induced current. These paths primarily consist of the wedges, rotor iron, and retaining rings. The depth of current penetration is a fraction of an inch, considerably less than the depth of the rotor windings. The contacts between these components are points where a localized, rapid temperature rise occurs, due mainly to arcing. Wedges, for example, have little "clamping" load at standstill, resulting in arcing between them and the rotor iron. The arc heating begins to melt the metals, and may cause wedges to be weakened to the point of immediate or eventual failure, depending upon the tripping time to clear the inadvertent energization incident. Damage to rotor windings, if it occurs, would result from mechanical damage due to loss of wedge support, rather than heating. Because of the low depth of current penetration, the rotor windings would not likely experience an excessive temperature rise and, therefore, would not be thermally damaged.

Generalized heating of the rotor surface to an excessive temperature takes longer than the localized areas described, but if tripping is delayed the rotor will be damaged beyond repair. The current magnitudes in the stator during this incident are generally within its thermal capability; however, if continued rotor heating occurs, wedges or other portions of the rotor may break off and damage the stator. This may result in total loss of the entire generator.

The time after which rotor damage will generally occur can be approximated by using the equation for the short time negative sequence capability of the generator - $I_2^2 t = K$. When the machine is at or near standstill and is inadvertently energized from either a single or three-phase source, the value of I_2 used in this formula should be the per unit magnitude of generator phase current flowing in the machine windings. If the generator is energized from a single-phase source at or near synchronous speed, the negative sequence component of current should be used. The equivalent circuits in Appendix I and II can be used to determine the value of current for these situations.

In the case of a cross-compound unit, sufficient field is applied at a very low speed to keep the generators in synchronism as they come up to speed. Inadvertent application of three-phase voltage will attempt to start both generators as induction motors. The thermal hazard to the rotor is the same as when no field is applied and is aggravated by the presence of current in the rotor field winding.

Turbine Damage

During an inadvertent energizing incident, the generator acts as a motor to drive the turbine, which is the reverse of normal operation. Generator, turbine, and exciter shaft inertia are the mechanical loads the generator must overcome to accelerate. The turbine represents a relatively high inertia load, two to three times that of the generator and exciter. As it comes up to speed, the unit passes through its natural (torsional) frequencies. If the generator is rotating when single-phase energizing occurs, it may take considerable time to pass through these frequencies, compared to normal start-up, because the accelerating torque may be relatively low. Therefore, torsional torque build-up can be very high when the speed corresponds to a natural frequency and can cause shaft and coupling damage to the unit. The torque pulsations are more severe if the unit has field applied, but is not yet in synchronism with the system. In the course of being subjected to these stresses, vibration, blade distortion and rubbing may cause turbine damage if the energizing source is not removed soon enough. A steam turbine, when operating at high speed, requires steam flow for cooling purposes to protect the turbine rotor and to maintain adequate clearances. If steam flow does not exist at the time of inadvertent energizing of the turbine-generator unit, it may contribute to the damage mechanisms described. Bearing failure due to insufficient lubrication can occur at the time of an inadvertent generator energization if the turbine-generator unit is at standstill and there is no oil flow to the bearings. This is not the case when the unit is on turning gear because oil pumps are running and sufficient oil flow is provided to lubricate the bearings adequately as the unit accelerates.

Hydro Machine Damage

Hydro-generators are salient-pole machines and are usually provided with damper windings on each pole. These damper windings may or may not be connected together. Inadvertent energizing may create sufficient torque in the rotor to produce some rotation. More important, the thermal capacity of the damper winding, especially at the point of connection to the pole steel, would not be adequate for the resulting currents. The heating of the connecting points, combined with the lack of proper ventilation, will create damage quickly. Since hydro-generator design is unique, each unit needs to be evaluated as to the detrimental effects of inadvertent energization. With no water passing through the turbine, vibration and friction may cause damage. Bearing failure can also occur due to low bearing oil pressure.

RESPONSE OF CONVENTIONAL GENERATOR PROTECTION TO INADVERTENT ENERGIZING

There are several relays used as part of the standard complement of conventional generator protection that may detect inadvertent generator energizing. When assessing whether these relays will provide adequate protection, great care should be taken in determining their status when the unit they protect is off-line. It is the operating practice of some utilities to remove generator voltage transformer fuses or links as a safety practice when a generator is removed from service, thus disabling most voltage dependent generator protection. Some utilities have elected to remove DC control power to generator protection when the unit is off-line to minimize the possibility of false tripping of generator high voltage breakers which have been returned to service as bus breakers in ring bus and breaker-and-a-half stations. Others elect to use auxiliary contacts of disconnect switches and/or circuit breakers to disable generator protection when the unit is off-line. The inadvertent energizing event may occur in such a manner as not to enable these relays. There have been a number of cases in which all generator protection was inoperative when the machine was accidentally energized. It appears prudent to reconsider the practice of disabling all generator protection when the unit is off-line in light of industry experience with inadvertent energizing.

Three-Phase Energizing

There are several relays used as part of the typical complement of generator protection that may detect, or can be set to detect, three-phase inadvertent energizing. They are:

- Loss of field protection
- Reverse power relay
- System backup relays

Loss of Field Protection: The most widely applied method for detecting loss of field is the use of offset mho impedance relay(s) to sense the variation of impedance as viewed from the generator terminals. There are two types of relaying schemes widely used for detecting the impedance seen during loss of field conditions.

In the first scheme, one or two offset mho units are applied to the generator terminals and polarized to "look into" the machine. Both mho units are normally offset in the trip direction from the origin by one-half of the machine transient reactance ($X_d'/2$). Since the machine appears as an impedance whose value is equal to its negative sequence impedance when inadvertently energized, this impedance must be inside the loss of field relay characteristic for the relay to operate. With normal settings, the machine negative sequence impedance is generally just inside the top of the loss of field relay characteristic. Depending upon the specific values of impedances and the tolerance of the loss of field relay, operation may be marginal. Therefore, each application should be carefully checked. It is not recommended that the offset be reduced to obtain more margin, since the loss of field relay may operate incorrectly on stable power swings.

The second scheme uses a combination of a mho unit, a directional unit and an undervoltage unit applied at the generator terminals. The mho unit is offset in the nontrip direction to incorporate the origin. Its response is supervised by both a directional and an undervoltage relay. During inadvertent three-phase machine energizing, this type of loss of field relay should operate and trip through a time delay (generally one second) since it is normally set so that the machine negative sequence impedance would be well within the relay's operating characteristics. A two-zone arrangement may also be used with this scheme with the added zone offset from the origin in the trip direction by one-half of the machine transient reactance ($X'_d/2$) as in the previously described scheme. This more restricted zone trips with a shorter time delay, but is less sensitive to the detection of inadvertent energizing.

Both loss of field relay schemes are voltage dependent. If the potential source is disconnected when the unit is off-line, the loss of field relay will not operate. It should also be noted that the loss of field relay is often removed from service by a disconnect switch and/or breaker 52/a contacts when the machine is off-line. Therefore, depending upon how the inadvertent energizing occurs, the loss of field protection may be disabled.

Reverse Power Relay: The power into a machine that is inadvertently energized from a three-phase source can be calculated using the equivalent circuit presented in Appendix I of this paper. The resulting power level will generally be within the pickup range of the reverse power relay. Tripping by reverse power relays is substantially delayed (usually 30 seconds or longer) which is much too long to prevent generator damage.

In some types of reverse power relays, this time delay is introduced through an AC voltage operated timer whose pickup level requires that 50% of rated terminal voltage be present. If the generator terminal voltage is below this level, the relay will not operate. If the potential supply is disconnected, the reverse power relay is also inoperative.

System Backup Relays: Impedance and voltage restrained or controlled overcurrent relays, used to provide backup for generation protection, can be adjusted to provide detection of three-phase inadvertent energizing. Their operation, however, should be checked by comparing their setting with expected machine terminal conditions for inadvertent energizing. These backup relays have time delay associated with tripping which may be too long to prevent the generator from being damaged. Attempts to reduce this time delay may result in false tripping for stable power swings or loss of coordination under fault conditions. Also, operation of the particular type of relay used should be reviewed for the condition when polarizing or restraining potential has been disconnected.

When a generator is accidentally energized through its station auxiliary transformer, auxiliary system overcurrent protection will generally detect the condition. Because of the low magnitude of generator current involved, time delayed tripping usually does not result in generator damage.

Single-Phase Energizing

Detection of the application of unbalanced or single-phase voltage to an off-line unit resulting from a breaker pole flashover or the closing of one pole of an isolating disconnect switch poses additional problems. For the flashover of a generator breaker pole, retripping of the breaker will not de-energize the machine. The initiation of breaker failure relaying is required to trip additional local and possibly remote breakers to de-energize the generator. The tripping of remote breakers, usually by transfer trip, can still leave one phase of the high voltage line connected to the generator/unit transformer combination through the failed pole of the breaker. If the line is long enough, the line capacitance may be enough to maintain the voltage at nearly rated values until the machine slows down, aggravating the damage to the machine and failed breaker.

There are relays used as part of the typical complement of generator, unit transformer and generator circuit breaker protection which can detect single-phase inadvertent energizing. They are:

. Generator negative sequence relaying

. Ground backup relaying

. Generator breaker failure protection in conjunction with other relays

Negative Sequence Relaying: It is common practice to provide protection of the generator from external unbalance conditions that might damage the machine. This protection consists of a time-current relay which responds to negative sequence current. Two types of relays are used for this protection: an electromechanical time overcurrent relay and a static relay with a time overcurrent characteristic which matches the I_2^2t capability curve of the generator. The electromechanical relay was designed primarily to provide machine protection for uncleared, unbalanced system faults. The negative sequence current pickup of this relay is generally 0.6 per unit of rated full load current. The static relays are much more sensitive and are capable of detecting and tripping for negative-sequence currents down to the continuous capability of the generator. The static negative sequence relay will, therefore, detect single-phase inadvertent energizing for most cases. The response of the electromechanical relay should be checked to ensure that its setting is sufficiently sensitive, especially in applications in which the unit is connected to a weak system. The tripping of these relays may be supervised by a high voltage switch or breaker 52/a contacts which could render them inoperative for a particular inadvertent energizing case. For the single pole flashover of a low voltage isophase bus generator breaker, no significant amount of negative sequence current will flow because it must flow through a high impedance path consisting of phase to ground capacitances and the generator neutral grounding impedance.

225

The generator stator ground protection may operate for this condition, depending on its setting.

Ground Backup Relaying: Generator transformer ground fault backup for system faults can be provided by a time overcurrent relay. The relay is supplied from a current transformer in the neutral of the generator step-up (GSU) transformer. When the generator is accidentally energized by a single-phase source, substantial GSU transformer neutral current will flow. The GSU neutral backup relay will operate in response to this condition if its setting is sufficiently sensitive. This backup relay is time delayed and may not prevent the generator from being damaged. The value of current, which this relay receives for a single-phase inadvertent energizing condition, can be calculated using the equivalent circuit presented in Appendix II of this paper.

Generator Breaker Failure Protection: A functional diagram of a typical generator breaker failure scheme is shown in Figure 3.

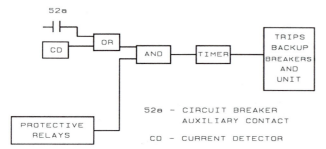
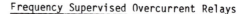

52a — CIRCUIT BREAKER
        AUXILIARY CONTACT

CD — CURRENT DETECTOR

FIGURE 3   GENERATOR BREAKER FAILURE LOGIC

When the generator protective relays detect an internal fault or an abnormal condition, they will attempt to trip the generator breakers and at the same time initiate the breaker failure timer(s). If the breaker(s) do not clear the fault or abnormal condition in a specified time, the timer will trip the necessary breakers to remove the generator from the system. The current detector (CD) or the breaker auxiliary contact (52/a) are used to detect that the breaker has successfully opened. The breaker 52/a contact must be used in this case since there are faults and/or abnormal generator conditions which will not produce sufficient current to operate the current detector (CD). If one or two poles of a breaker flash over to energize a generator, two conditions must be satisfied to initiate breaker failure:

1. The flashover must be detected by a generator protective relay that would initiate the breaker failure relay.

2. The breaker failure current detector (CD) must be set with sufficient sensitivity to detect the flashover condition.

DEDICATED PROTECTION SCHEMES TO DETECT INADVERTENT ENERGIZING

Although there are relays used as part of the normal complement of generator protection that can detect inadvertent generator energizing, these relays may:

. Be marginal in their ability to detect the condition.

. Be disabled when the machine is inadvertently energized.

. Operate too slowly to prevent damage to the generator and turbine.

For these reasons, dedicated protection schemes have been developed and installed specifically to detect inadvertent energizing. Unlike conventional protection schemes, which provide protection when equipment is in-service, these schemes provide protection when equipment is out of service. Thus, great care should be taken when implementing this protection so that DC tripping power and relay input quantities to the scheme are not removed when the unit being protected is off-line.

This section of the report describes a number of dedicated inadvertent energizing protection schemes for units without low voltage generator breakers. The judicious selection of input sources allows most of these schemes to be applied to generators with low voltage generator breakers. Whatever scheme is used to provide protection for accidentally energizing a generator, the protection should be connected to trip the generator high voltage and field breakers, trip the unit auxiliary breakers, initiate generator high voltage breaker failure backup, and be implemented so that it is not disabled when the machine is out of service.

Frequency Supervised Overcurrent Relays

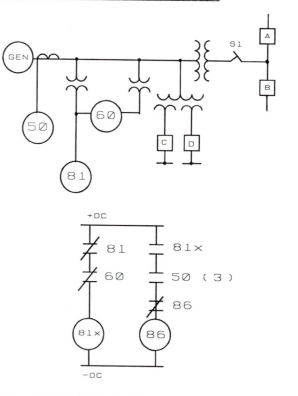

50 — THREE INSTANTANEOUS OVERCURRENT RELAYS

60 — VOLTAGE BALANCE RELAY

81 — FREQUENCY RELAY

81x — AUXILIARY RELAY (TIME DELAY DROP OUT)

86 — LOCKOUT RELAY

FIGURE 4   FREQUENCY SUPERVISED OVERCURRENT
            LOGIC

Figure 4 depicts a frequency-supervised overcurrent scheme specifically designed to detect accidental energization. The scheme utilizes a frequency relay to supervise the trip output of sensitively set instantaneous overcurrent relays. The overcurrent relays are automatically armed by the frequency relay as the unit is taken off-line and remain armed while the unit is shut down. To ensure reliable high speed tripping, the overcurrent relays should be set at 50% or less of the minimum current seen during accidental energizing. The frequency relay (81) used to identify when the generator is off-line should have a setpoint well below any emergency operating frequency. Its output contacts should also remain closed as the applied voltage goes to zero. The voltage balance relay (60) prevents incorrect operations should the frequency relay lose potential under normal operating conditions.

When the generator is taken off-line, the machine frequency will drop below the frequency relay setpoint. The frequency relay will energize auxiliary relay 81X through a normally closed contact of the voltage balance relay. An 81X auxiliary relay contact then closes to enable the trip output circuit of the overcurrent relays. The protective scheme is thereby armed and remains armed as long as the unit is shut down. Even if the a-c potential supply is disconnected while the generator is down for maintenance, the frequency relay contact should remain closed, thus allowing high speed overcurrent tripping. Should the generator be accidentally energized, the frequency relay will open its contact, but the time delayed dropout of the 81X auxiliary relay will permit instantaneous overcurrent tripping.

As the generator is accelerated in order to be brought on-line, the machine frequency exceeds the frequency relay setpoint. The frequency relay operates and de-energizes the 81X auxiliary relay. The 81X auxiliary relay, after its time delayed dropout, disarms the trip output circuit of the overcurrent relays. The frequency supervised overcurrent scheme will not provide protection for a high voltage generator breaker pole flashover just prior to synchronizing when the machine is at or near rated speed with field applied. Additional protection, as described in the next section of this report, should be considered for this situation.

Voltage Supervised Overcurrent Relays

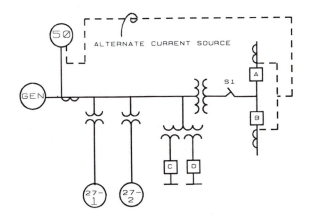

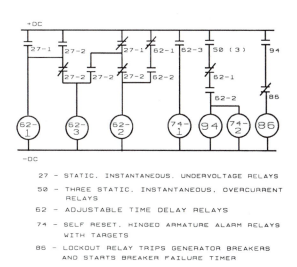

27 – STATIC, INSTANTANEOUS, UNDERVOLTAGE RELAYS

50 – THREE STATIC, INSTANTANEOUS, OVERCURRENT RELAYS

62 – ADJUSTABLE TIME DELAY RELAYS

74 – SELF RESET, HINGED ARMATURE ALARM RELAYS WITH TARGETS

86 – LOCKOUT RELAY TRIPS GENERATOR BREAKERS AND STARTS BREAKER FAILURE TIMER

94 – HIGH-SPEED, HINGED ARMATURE TRIP RELAY

FIGURE 5   VOLTAGE-SUPERVISED OVERCURRENT LOGIC

Figure 5 depicts a voltage-supervised overcurrent scheme which is designed to detect accidental energization. This scheme utilizes voltage relays (27-1&2) to supervise the output of static, high-speed instantaneous phase overcurrent relays (50) to provide inadvertent energizing protection. The overcurrent relays are automatically armed when the unit is taken off-line and remain armed while the unit is shut down. They are automatically removed from service when the unit is brought on-line. The overcurrent units are set to respond to current of 50% or less of minimum current seen during an accidental energization. Undervoltage relays (27-1&2) enable and disable the current detectors (50) through time delay relays (62-1&2). Two 27 relays are supplied from separate voltage transformers to prevent misoperations which can result from loss of one potential supply. A time delay (62-3) and a voltage detector alarm relay (74-1) are used to alarm for this situation. The 27-1&2 voltage relays are generally set at approximately 85% of rated voltage. The 62-1 timer disables tripping by (50) overcurrent relays after the unit voltage returns to normal prior to synchronizing. The 62-2 timer enables overcurrent tripping when voltage falls below 85% of normal when the machine is removed from service. The 62-2 timer is set with sufficient delay (generally two seconds) to prevent enabling the overcurrent relays for power system or unit auxiliary system faults which could drop machine terminal voltage below the 85% level. The scheme will reset when the generator field is applied to develop rated voltage prior to synchronizing. Thus, high voltage generator breaker pole flashovers just prior to synchronizing will not be detected. Supplemental protection, as described in the next section of this report, should be considered for this situation.

In order to improve the integrity of this scheme, some utilities have chosen to install it in the high voltage switchyard using CT's and DC power located in the switchyard. Others have chosen to locate it in the plant and install it in such a manner that it will not be disconnected when the unit is off-line. By locating the CT inputs at the machine terminals, the relay can be set to detect inadvertent energizing through the unit auxiliary transformer. Reference 4 provides a detailed description of the voltage supervised overcurrent scheme.

227

Directional Overcurrent Relays

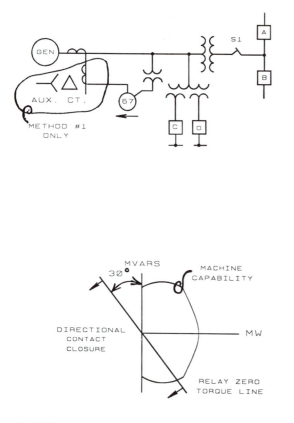

FIGURE 6   DIRECTIONAL OVERCURRENT LOGIC

Impedance Relays

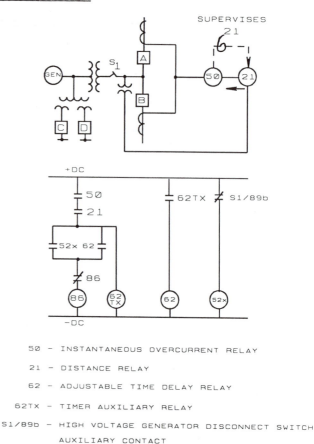

50 – INSTANTANEOUS OVERCURRENT RELAY

21 – DISTANCE RELAY

62 – ADJUSTABLE TIME DELAY RELAY

62TX – TIMER AUXILIARY RELAY

S1/89b – HIGH VOLTAGE GENERATOR DISCONNECT SWITCH
         AUXILIARY CONTACT

52x – AUXILIARY RELAY – TIME DELAY DROPOUT

86 – LOCKOUT RELAY TRIPS GENERATOR BREAKERS
     AND STARTS BREAKER FAILURE TIMER

FIGURE 7   IMPEDANCE RELAY LOGIC

The scheme depicted in Figure 6 employs three directional inverse time-overcurrent relays. Voltage and current sensing are obtained from the generator terminals. Two different methods are used. Method 1 uses a relay having maximum sensitivity when the current applied to the relay leads the voltage by 30°. To assure that the underexcited loading capability of the machine is not impaired appreciably, the 60° connection (I_A-I_B and V_{AC}) is used. Delta connected CT's or auxiliary CT's are required, or line-to-ground connected VT's may be applied. The setting used may involve a compromise between desired sensitivity and a setting at which the relay will not be endangered thermally by the maximum continuous load current. Method 2 uses a relay having maximum sensitivity when the current applied to the relay leads the voltage by 60°. A 90° connection to the relay (I_A and V_{BC}) allows adequate underexcited operation to be achieved. Some relays of this type have a fixed sensitivity of 0.5 ampere and a continuous rating of 5 amperes. They are generally set to operate in 0.25 second at 2.0 times rated generator current. The directional overcurrent relays (67) should trip the generator breaker(s) and start breaker failure timing. This scheme is dependent upon potential being present for proper operation. Thus, if company operating procedure requires removal of generator voltage transformer fuses or links as a safety procedure when the unit is removed from service, this scheme should not be applied.

There have been a number of schemes developed which use impedance relays located in the high voltage switchyard which are polarized to "look into" the machine as shown in Figure 7. The impedance relay is set to detect the sum of the reactance of the unit step-up transformer and the machine negative sequence reactance ($X_{IT} + X_2g$) with appropriate margin. In some cases, the impedance relay is supervised by an instantaneous overcurrent relay to prevent false operation on loss of potential. Some utilities connect the impedance relay to trip the high voltage generator breakers and initiate unit shutdown without delay whether the unit is on or off-line. The impedance relay will generally operate for unstable power swings and requires a thorough stability analysis to ensure that the scheme will not trip on stable swings. Other utilities choose to enable the scheme for high speed tripping only when the unit is off-line and add time delay to provide security when the unit is on-line. Figure 7 is an illustration of such a scheme. This approach provides a measure of protection even if generator disconnect auxiliary contacts fail to enable high speed tripping. This scheme will trip the unit if the field is applied when accidental energizing occurs provided that the unit is substantially out of phase with the system at the time of energizing. Additional protection is required for single-phase energization, since an impedance relay has limited capability to detect this condition.

Auxiliary Contact Enabled Overcurrent Relays

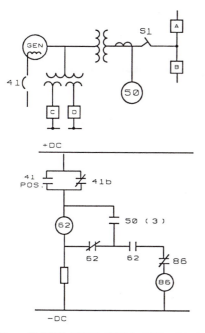

41 POS - CLOSED WHEN FIELD BREAKER IS RACKED OUT

41b - FIELD BREAKER AUXILIARY CONTACT

50 - THREE INSTANTANEOUS OVERCURRENT RELAYS

62 - AUXILIARY RELAY WITH TIME DELAY (CYCLES)
PICKUP & DROPOUT

86 - LOCKOUT RELAY TRIPS GENERATOR BREAKERS
AND STARTS BREAKER FAILURE

FIGURE 8   AUXILIARY CONTACT ENABLED OVERCURRENT
RELAY LOGIC

The scheme shown in Figure 8 uses the generator field breaker auxiliary contact to enable and disable an overcurrent relay to detect inadvertent energizing when the unit is off-line. This scheme consists of three nondirectional instantaneous current fault detectors which are armed to trip whenever the generator field breaker is open or racked out. Either of these conditions will energize a time-delay pickup and dropout auxiliary relay (62) enabling the scheme. The overcurrent relays are set at 50% or less of the minimum current seen during an accidental energization. To avoid false operation when the unit is in service, as well as misleading target information, the scheme is designed such that it is not armed unless the overcurrent relays are first reset. If the unit is on line, and the overcurrent relays are picked up due to load, the 62 relay coil is shorted out to prevent its operation, which enhances the security of the scheme. The scheme is designed such that no unit-tripping function associated with a fault or mechanical trip will activate the scheme.

As with some other schemes described, the scheme shown in Figure 8 will reset when the field is applied to the unit prior to synchronizing. The scheme will not provide protection at low turbine RPM with field on. Although field is generally not applied below synchronous speed on modern tandem units, cross-compound units require synchronization between units at very low RPM. To assure protection during the pre-synchronization period, it is necessary to use the 41b contact from the main exciter field breaker and not the startup exciter field breaker. In addition, excitation must be transferred from startup to main

exciter prior to synchronization to preclude a false trip on synchronizing.

DEDICATED PROTECTION SCHEMES TO DETECT GENERATOR BREAKER HEAD FLASHOVER

For the flashover of a generator high voltage breaker pole, retripping of the breaker will not de-energize the machine. The initiation of breaker failure relaying is required to trip additional local and possibly remote breakers to de-energize the generator. Some of the schemes discussed in the preceding section of this paper can be set to detect breaker head flashovers and provide protection in conjunction with generator breaker failure protection. Other schemes are inoperative when the generator is near rated speed and voltage prior to synchronizing and must be supplemented by additional protection.

Unbalanced currents associated with breaker head flashover will generally cause the generator negative sequence relay to operate. Breaker failure will be initiated if the breaker failure current detectors are set with sufficient sensitivity to detect the situation. Specifically designed schemes to speed up the detection and isolation of this unique form of breaker failure are described below.

Modified Breaker Failure Scheme

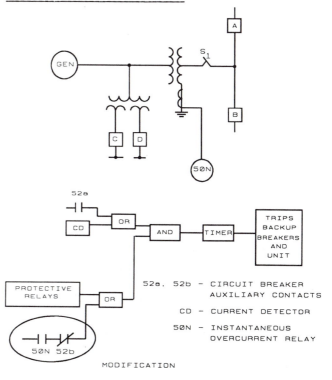

52a, 52b - CIRCUIT BREAKER
AUXILIARY CONTACTS

CD - CURRENT DETECTOR

50N - INSTANTANEOUS
OVERCURRENT RELAY

FIGURE 9   MODIFIED BREAKER FAILURE LOGIC

One approach used to speed up the detection of a breaker flashover is to modify the breaker failure scheme as shown in Figure 9. An instantaneous overcurrent relay (50N) is connected in the neutral of the generator step-up transformer and is set to respond to an EHV breaker pole flashover. The relay output is supervised by the generator breaker "b" contact providing an additional start to the breaker failure scheme. When the generator breaker is open and one or two poles of the breaker flash over, the resulting transformer neutral current is detected by the 50N relay without the delay that would be associated with negative sequence or some of the previously described inadvertent energizing schemes. The current detectors (CD) associated with the generator breaker failure scheme must be set with sufficient sensitivity to detect this flashover condition.

Breaker Pole Disagreement

It is general practice that high voltage breakers are designed with independent pole operating mechanisms. For unsymmetrical pole closures, these breakers are protected by an interconnection of auxiliary contacts. If any pole is closed at the same time that another is open, a path is provided to initiate tripping of the breaker. Since breaker auxiliary contact indications do not provide positive indication of pole position, these schemes can be augmented by a relay which monitors three-phase current flowing through the breaker and senses whether any phase is below a certain low threshold level (indicating an open breaker pole) at the same time that any other phase is above a substantially higher threshold level (indicating a closed or flashed over pole). For breaker-and-a-half and ring bus applications, zero sequence voltage across the breaker is used to supervise the relay tripping. This prevents false operation due to unbalance currents caused by dissimilarities in bus phase impedances. Thus, this current monitored pole disagreement relay provides a method of detecting generator breaker head flashovers, but tripping is generally delayed .5 seconds. Reference 5 provides a detailed description of this relay.

CONCLUSIONS

Inadvertent energizing of large synchronous generators has significantly increased within the industry in recent years as generating stations have become more complex. The widespread use of breaker-and-a-half and ring bus schemes has added significant operating flexibility to high voltage generating stations. These configurations have also increased complexity and the risk of the generator being inadvertently energized while off-line. Operating errors, breaker head flashovers, control circuit malfunctions or a combination of these causes have resulted in generators becoming accidentally energized. Because the damage to the machine can occur within a few seconds, it must be detected and isolated by automatic relay action. Although there are relays used as part of the normal complement of generator protection that may detect inadvertent generator energizing, their ability to do so is generally marginal. These relays may be disabled at the time when the machine is inadvertently energized, or operate too slowly to prevent damage to the generator and/or turbine. For these reasons, major U.S. turbine-generator manufacturers have recommended, and many utilities are installing, dedicated inadvertent energizing protection schemes. The major schemes in service in the U.S. have been described in this paper. These schemes vary because the operating practices and protection philosophies of utilities using them are different. Protection engineers must assess

the risks and determine the impact on protection of their own company's operating practices prior to deciding which scheme best suits their particular needs. It is hoped that this paper will assist in that task.

REFERENCES

1. E. R. Detjen, "Some Additional Thoughts on Generator Protection", presented at The Pennsylvania Electric Association Relay Committee, May 29, 1981.

2. IEEE "Guide for A.C. Generator Protection", ANSI/IEEE C37.102-1988.

3. J. G. Manzek and J. T. Ullo, "Implementation of an Open Breaker Flashover and Inadvertent Energization Protection Scheme on Generator Circuit Breakers," presented to the Pennsylvania Electric Association Relay Committee, September 14, 1983.

4. M. Meisinger, G. Rockefeller, L. Schulze, "RAGUA: Protection Against Accidental Energization of Synchronous Machines," presented to The Pennsylvania Electric Association Relay Committee, September 14, 1983.

5. W. A. Elmore, C. L. Wagner, "Pole Disagreement Relaying," presented to 10th Annual Western Relay Conference, Spokane, Washington, October 24-27, 1983.

APPENDIX I

Calculation of initial currents and voltages when a generator is energized from a three-phase source.

Approximate Equivalent Circuit

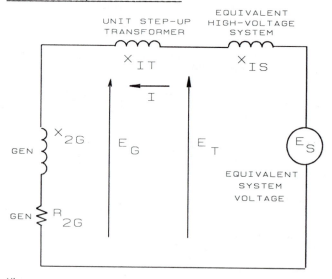

Where:

X_{1S} = System Positive Sequence Reactance

X_{1T} = Transformer Positive Sequence Reactance

X_{2G} = Generator Negative Sequence Reactance

R_{2G} = Generator Negative Sequence Resistance

E_S = System Voltage

E_T = Transformer High Side Voltage

E_G = Generator Terminal Voltage

I = Current

$P_{3\emptyset G}$ = Generator 3 Phase Power

$I = \dfrac{E_S}{X_{1S}+X_{1T}+X_{2G}}$

$E_G = (I)(X_{2G})$

$E_T = (I)(X_{2G}+X_{1T})$

$P_{3\emptyset G} = 3I^2 R_{2G}$

APPENDIX II

Calculation of initial currents and voltages when a generator is energized from a single phase source such as a breaker head flashover just prior to synchronizing.

Open Breaker Flashover

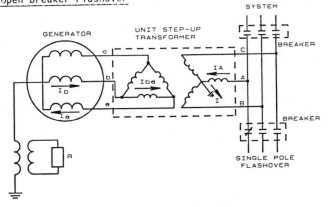

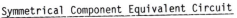

Symmetrical Component Equivalent Circuit

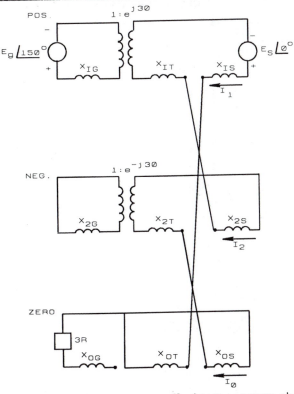

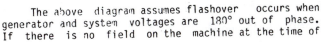

The above diagram assumes flashover occurs when generator and system voltages are 180° out of phase. If there is no field on the machine at the time of inadvertent energizing, the Eg voltage source is zero in the positive sequence equivalent circuit.

Where:

X_{1G}, X_{2G}, X_{0G} = Generator Positive, Negative and Zero Sequence Reactances

X_{1T}, X_{2T}, X_{0T} = Unit Step-Up Transformer Positive, Negative and Zero Sequence Reactances

X_{1S}, X_{2S}, X_{0S} = System Equivalent Positive, Negative and Zero Sequence Reactances

E_g = Generator Voltage

E_S = System Voltage

I_1, I_2, I_0 = Positive, Negative and Zero Sequence Currents

Simplified Equivalent Circuit

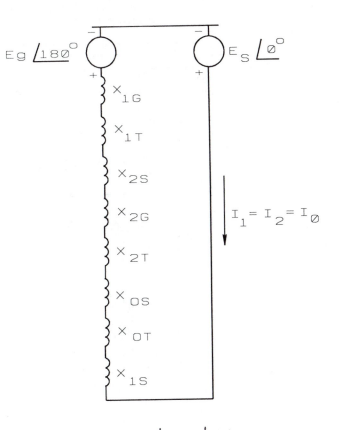

$$I_1=I_2=I_0 = \frac{E_S\;\angle 0 - E_g\;\angle 180°}{X_{1G}+X_{1T}+X_{2S}+X_{2G}+X_{2T}+X_{0S}+X_{0T}+X_{1S}}$$

The current flow from the high voltage system assuming a single pole (A∅) open breaker flashover:

$I_A = I_1+I_2+I_0 = 3I_0$

$I_B = a^2 I_1+aI_2+I_0 = 0$

$I_C = aI_1+a^2 I_2+I_0 = 0$

Where $a = 1\;\angle 120°$ and $a^2 = 1\;\angle 240°$

The current at the generator terminals assuming a single pole (AØ) open breaker flashover:

$$I_a = I_1 \angle{-30°} + I_2 \angle{30°} + I_0 = \sqrt{3}\, I_1 = I_A/\sqrt{3}$$

$$I_b = a^2 I_1 \angle{-30°} + a I_2 \angle{30°} + I_0 = -\sqrt{3}\, I_1 = -I_A/\sqrt{3}$$

$$I_c = a I_1 \angle{-30°} + a^2 I_2 \angle{30°} + I_0 = 0$$

Where $I_0 = 0$, $a = 1 \angle{120°}$, $a^2 = 1 \angle{240°}$

The 30° phase shift is due to the wye-delta generator step-up transformer.

Discussion

A. H. Ayoub and **U. C. Roumillat**, (Georgia Power Company, Atlanta, GA): The paper presented a good review of existing generator backup protection, and several schemes for providing dedicated protection against accidental energization of a generator. The two main disadvantages of using backup relays were reported to be the slow tripping inherent in backup relaying, and the likelihood of these relays being removed from service when the unit is off line. The dedicated schemes presented in the paper were all improvements over relying on backup relays, but had various short comings.

The dedicated scheme shown in figure 4 has the following disadvantages:

1. If the generator P.T.'s are jacked in with one phase not making contact, the scheme is automatically removed from service by the voltage balance (No. 60) relay.
2. The scheme will not protect the generator if the generator is being placed in service while at normal speed and voltage, but is out of synchronism with the system when the generator breaker is closed. This is due to the underfrequency relay setting being exceeded and blocking the operation of the scheme. The primary protection for this condition is a good synch check relay to monitor manual and automatic synchronizing. An adequate accidental energization protection scheme should offer backup protection against closing out of synchronism.
3. The scheme will not protect against breaker pole flashover just prior to synchronizing since the frequency of the generator will be very close to system frequency, and the frequency relay will block the operation of the scheme.

The remaining schemes are shown in figures 5, 6, 7, and 8. The voltage-supervised scheme shown in figure 5 does not provide protection against breaker pole flashover or out-of-synch condition just prior to synchronization, since the generator voltage will be close to normal and the voltage relays will block the operation of the scheme. The directional overcurrent logic scheme shown in figure 6 would not be fast enough to adequately protect the generator, and it would not detect a breaker pole flashover which is behind the 67 relay. The impedance relay logic scheme may be slow due to the 62 time delay. If zero sequence compensation is used with the 21 relay, the impedance relay will not detect a breaker pole flashover. The field breaker auxiliary contact controlled scheme shown in figure 8 will only be in service when the field breaker is open. This would not adequately protect the generator since the generator breaker could be accidentally closed regardless of the field breaker position.

Figure 1 of this discussion shows an improved scheme that is used by the Georgia Power Company. The timing relay (62) is energized by the generator off-line logic. After 62 times out, an auxiliary timing relay (62X) is energized to remove the 62 relay supply voltage. This removes the heat build-up in the timing relay during long generator maintenance outages. The heat build-up may not harm the relay, but the continuous energization is an

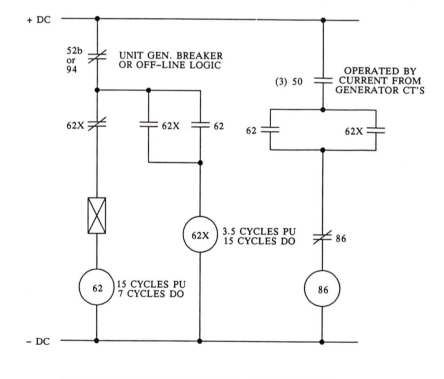

52b – STRAIGHT BUS BREAKER AUXILIARY CONTACT

94 – RING BUS OR BREAKER-AND-A-HALF OFF-LINE LOGIC:
UNIT DISCONNECT OPEN OR BOTH GENERATOR BREAKERS OPEN

62 – TIME DELAY RELAY - ADJUSTABLE PICKUP (PU)

62X – AUXILIARY TIME DELAY RELAY - ADJUSTABLE DROPOUT (DO)

50 – INSTANTANEOUS PHASE OVERCURRENT RELAY

86 – LOCKOUT RELAY TRIPS UNIT AND STARTS GENERATOR BREAKER
FAILURE TIMER

Figure 1. Off-line supervised overcurrent relay logic.

unnecessary drain on the battery. The main reason for adding the 62X relay is to provide redundancy and a longer period of protection (longer dropout time) following synchronization of the generator. If the 62X relay coil were to burn open, the 62 relay will become energized and take over. The instantaneous overcurrent relay (50) can be set slightly above the generator rated current. This offers sensitive protection against accidental generator breaker closing at stand-still, or out-of-synchronism. It will also provide protection during generator warm up at subsynchronous speed. Since the 62 or 62X relay is already energized, the 50 relay can energize the generator lockout relay in a fraction of a cycle. If a generator breaker pole flashes over when the unit is at stand-still or just prior to synchronization, the 50 relay will start breaker failure timing on the failed breaker, through the residual current breaker failure fault detector. The calculated magnitude of the open breaker flashover current shown in Appendix II of the paper is more than enough to pick up device 50.

The scheme described above has been proven in the field. A 950 MW generator was accidentally energized while rotating at a subsynchronous speed. This inadvertent energization scheme tripped and cleared the fault in approximately two cycles. No equipment damage resulted due to fast clearing. A more complicated scheme would not be as reliable.

We would like to thank the authors for presenting good reasons for having inadvertent energization protection. All of the protection ideas presented in the paper give the reader a better understanding of the various problems encountered in providing reliable inadvertent energization protection.

Manuscript received August 15, 1988.

J. E. Stephens, (Illinois Power Co., Decatur, Illinois): As discussed in the paper, one of the most common causes of inadvertent energizing of a stationary generator is the inappropriate closing of the high voltage generator disconnect switch. A simple control scheme may be applied which will trip the generator breakers if the disconnect switch starts to close while either breaker is still closed. The system is armed when the disconnect switch is opened and bus energized. The system is disabled when the generator bus is deenergized prior to closing the disconnect switch. The activating disconnect switch auxiliary contact should be one connected directly to the switch operating pipe, not a contact of the operating mechanism which could be disconnected from the switch. The auxiliary contact should be adjusted to open early in the disconnect switch closing travel which will permit tripping the breakers (if necessary) before the switch actually gets closed. A contact of the tripping relay inserted in the green light circuits permits monitoring the operation of the tripping relay. Additional contacts may be included in the breaker failure relaying input.

Manuscript received June 29, 1988.

Charles J. Mozina, (Centerior Service Company, Independence, Ohio 44131): C. J. Mozina Working Group Chairman: The authors would like to thank Messrs. Ayoub, Roumillat and Stephens for their interest in our paper and their comments. We would first like to discuss Ayoub's and Roumillat's comments concerning some of the dedicated protection schemes presented in the paper. It was the intent of the Working Group to present the schemes for detecting inadvertent energizing which have the most widespread and successful use within the U.S. and not to develop new and untried methods which have little or no in-service experience. We believe the schemes presented in the paper are the principal ones in use today in the U.S. The paper addresses the two major causes of inadvertent generator energizing: 1) accidental energizing while the machine is off-line with field removed, 2) breaker pole flashovers. Our research into the problem indicated that these are by far the most prevalent causes of inadvertent energizing. Out-of-phase synchronizing mentioned by Ayoub and Roumillat is a minor contribution to the total number of inadvertent energizing cases and there are effective protection practices in widespread use to prevent such synchronizing. The Working Group felt it was not necessary to address this problem since it was adequately covered in industry standards (ANSI/IEEE C37.1.2-1988).

The paper presents dedicated protection schemes associated with both off line energizing and breaker head flashover in two separate sections. As pointed out by Ayoub and Roumillat some of the schemes specifically designed to detect off line energizing (Figures 5, 6, 7 & 8) have limitations in the ability to detect breaker pole flashover. These limitations are clearly pointed out in the paper, and the reader is advised to read the next section of the paper which described a number of schemes designed to detect breaker pole flashovers. It is the practice of many utilities to use dedicated schemes for each from of inadvertent energizing.

We concur with Ayoub's and Roumillat's comments concerning the frequency supervised overcurrent scheme (Figure 4) being blocked by the voltage balance relay (60) if sound voltage is not present. This is a security feature associated with this scheme to prevent false operations when the generator is in service. It should also be pointed out that the voltage balance relay also generally alarms to notify the operator of this condition. When the unit is off line and inadvertently energizing does occur, this scheme will properly function even if P.T. jacks in one phase are not making contact. We disagree with the comment concerning the assertion that the directional overcurrent scheme shown in Figure 6 is not fast enough to adequately protect the generator or detect a breaker pole flashover. The section of the paper entitled, "Damage Due to Inadvertent Energizing" and Appendixes I & II provide a means of calculating the time to damage which can be used to check the directional overcurrent response time. One directional relay will operate for a breaker pole flashover condition. Comment made concerning the limitation of the impedance relay scheme to detect pole flashover are fully discussed in the text of the paper.

We would like to thank both discussors for sharing with us the scheme used by their companies. The scheme presented by Ayoub and Roumillat

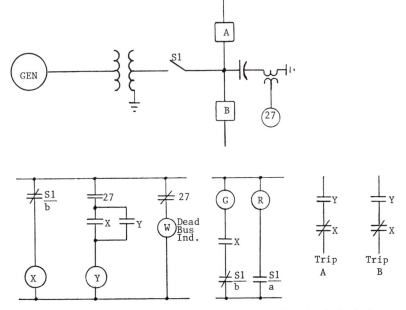

Logic to trip generator breakers if a disconnect switch is closed with the bus energized.

uses breaker auxiliary contact logic to determine whether the unit is on or off line. In our investigation of the problem of inadvertent energizing, we had found a number of cases where breaker auxiliary contacts have not provided a true indication of the position of generator breakers and high voltage disconnect switches (Figure 1, A, B & S_1).

Since most modern EHV generator breakers have independent pole mechanisms, there are a total of six auxiliary contacts for the two generator breakers. Thus, the failure of any one contact in addition to the auxiliary contacts associated with the high voltage disconnect switch could potentially disable the scheme or enable it improperly. This is one of the reasons many in the industry favor the use of frequency or voltage relays to determine whether the generator is on or off line.

The scheme present by Mr. Stephens uses a voltage relay in conjunction with an auxiliary contact of the generator high voltage mechanized disconnect switch to detect improper closure of the mechanized disconnect to inadvertently energize the generator. The scheme will not detect improper energizing if Switch S_1 is closed and breaker A or B close to inadvertently energize the generator.

As stated in the conclusion section of the paper, schemes used for detection of inadvertent energizing vary because the operating practice and protection philosophies of utilities are different. Protection engineers must assess the risks and determine the impact on protection of their own Company's operating practices prior to deciding which scheme best suits their particular need. The goal of this paper is to bring to the attention of the industry the need to use some form of dedicated protection for inadvertent energizing.

Manuscript received October 17, 1988.

OUT OF STEP RELAYING FOR GENERATORS
WORKING GROUP REPORT

J. A. Imhof   Commonwealth Edison Co. (Chairman)

J. Berdy   W. A. Elmore   L. E. Goff
W. C. New   G. C. Parr   A. H. Summers   C. L. Wagner

ROTATING MACHINERY PROTECTION SUBCOMMITTEE - IEEE POWER SYSTEM RELAYING COMMITTEE

ABSTRACT

This paper discusses the need for the methods of accomplishing out-of-step protection of generators. The report describes the loss-of-synchronism impedance characteristics of large generators in a modern EHV system. It is demonstrated that conventional out-of-step relaying schemes as applied to transmission lines can also be used to detect the out-of-step condition at the generator.

I. INTRODUCTION

In the aftermath of the 1965 Northeast Power Failure and other subsequent power system disturbances, attention was focused more acutely than ever on such system problems as low voltage and frequency, and generators operating out-of-synchronism. The need for purposeful automatic sectionalizing of systems into self-sustaining islands was recognized as being desirable for severe system swings for which recovery is not possible. Although out-of –step protection systems were available before 1965, more sophisticated schemes have been developed for application to transmission systems. These schemes provide line relays with sensing elements and control logic to detect the out-of-synchronism condition, the objective being the opening of certain transmission lines or the inhibiting of tripping of these ties to effect the maximum preservation of the system.

In 1970 the Rotating Machinery Subcommittee of the IEEE Power System Relay Committee became aware that some relay engineers were becoming more concerned with sensing the out-of-step condition electrically at the generator terminals or the high voltage terminals of the associated step-up transformer. It was recognized that sensing elements were not generally applied to cover the gap that may exist from the station transmission buses electrically back through the unit transformer and into the generator. This gap usually occurs because relays in the generator zone such as differential and most time delayed back-up relays cannot operate for an out-of step condition. This, along with the possibility of seriously damaging a generator under an out-of-step condition, prompted the Subcommittee to form a Working Group to study this problem. The assignment of the Working Group

was defined as "To investigate the need for and the methods of accomplishing out-of-step protection of generators." The results of this study including a survey letter of inquiry are presented in this report.

II. LOSS OF SYNCHRONISM CHARACTERISTICS

General

The conventional relaying approach to visualize and detect a loss of synchronism condition is to analyze apparent impedance variation as viewed at the terminals of a line or generator. This variation is a function of the system voltages and the angular separation between the systems. The apparent impedance locus also depends on the type of governor and voltage control, and the type of disturbance which initiated the swing. Depending on the rate of slip this variation in impedance can usually be detected by distance-type relaying, and the systems or generator separated before the completion of one slip cycle.

Simplified graphical procedures have been developed (2) and used to determine the variation in apparent impedance during a loss of synchronism condition. These procedures derive an impedance locus which can be plotted for study purposes with the system characteristic on an R-X diagram. Typical impedance loci obtained with this procedure are illustrated in Figure 1. The three impedance loci shown are plotted as a function of the ratio of the system voltages EA/EB which is assumed to remain constant during the swing. Moreover, in this simplified approach, the following assumptions are made: initial transients (dc or 60 Hz components) and effects of generator saliency are neglected; transient changes in impedance due to a fault or clearing of a fault (or due to any other disturbance) have subsided; effects of shunt loads and shunt capacitance are neglected; effects of regulators and governors are neglected; and the voltages E_A and E_B behind the equivalent impedances are balanced sinusoidal voltages of fundamental frequency.

When the voltage ratio $E_A/E_B = 1$, the impedance locus is a straight line PQ which is the perpendicular bisector of the total system impedance between A and B. On this diagram, the angle formed by the intersection of lines AP and BP on line PQ is the angle of separation (δ) between the systems. If system B(E_B) is taken as reference, and if it is assumed E_A advances in phase angle ahead of E_B, the impedance locus moves from point P toward Q and the angle (δ) increases. When the locus intersects the total impedance line AB the systems are 180 degrees out of phase. This point is both the electrical and impedance center of the system. As the locus moves to the left of the system impedance line, the angular separation increases beyond 180 degrees and

F 77 013-6. A paper recommended and approved by the IEEE Power System Relaying Committee of the IEEE Power Engineering Society for presentation at the IEEE PES Winter Meeting, New York, N.Y., January 30-February 4, 1977. Manuscript submitted October 22, 1976; made available for printing October 25, 1976.

Reprinted from *IEEE Trans. Power App. Syst.*, vol. PAS-96, no. 5, pp. 1556–1564, Sept./Oct. 1977.

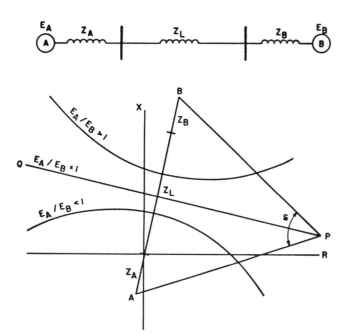

Fig. 1 Typical Out-of-Step Impedance Loci Using Simplified Graphical Procedures

Generator Loss of Synchronism Characteristics

Ten to twenty years ago, system and generator impedance characteristics were such that the electrical center during a loss of synchronism condition generally occurred out in the transmission system. The impedance loci generally intersected transmission lines and could readily be sensed by line relaying or out-of-step relaying schemes and the system could be separated without the need for tripping generators.

With the advent of EHV systems, large conductor-cooled generators, with fast response voltage regulators and with the expansion of transmission systems, system and generator impedance characteristics have changed appreciably. Generator and step-up transformer impedances have increased in magnitude while systems impedances have decreased. As a result, on many systems today, the system impedance center and the electrical center during swings can and does occur in the generator or in the generator step-up transformer.

In general, the loss of synchronism impedance loci, as viewed at the generator terminals, follows the swing characteristic where the ratio of generator to system voltage (E_A/E_B) is less than one ($E_A/E_B<1$). This is due to the fact that for most machine loadings, the equivalent internal machine voltage will be less than 1.0 per unit and less than the equivalent system voltage. This will generally be true for leading, unity and even for slightly lagging power factor loadings. Most generators are operated in this power factor range today.

Figures 2 to 4 illustrate the type of loss of synchronism impedance loci that may be encountered for both tandem and cross-compound generators. The impedance loci shown are given as a function of system impedance and were determined in a digital computer study using a comprehensive dynamic model of a turbine generator. Representations of the excitation system and governor response were included, but it was assumed the voltage regulator was out of service. With the omission of voltage regulator response, the internal machine voltages remain low during the disturbance, and therefore, the electrical center of the swings are more likely to fall within the generator zone.

In all cases, it was assumed that the disturbance and instability were caused by the prolonged clearing of a nearby three phase fault on the high voltage side of the generator step-up transformer. All impedances given are in per unit on generator MVA base and all initial loadings and system voltages are shown for each case.

Figure 2 demonstrates the loss of synchronism impedance loci for a tandem generator for system impedances of .05, .2 and .4 per unit, respectively. It should be noted the circle formed by the impedance locus increases in diameter, and the electrical center shifts from within the machine out into the step-up transformer as the system impedance increases. This increase in circle diameter and shift is due to the fact that as system impedance is increased, a higher internal machine voltage and increased reactive power output are required to compensate for the increased losses and to maintain 1.0 per unit voltage on the machine terminals and system bus. All of these loss of synchronism characteristics usually can be detected by available out-of-step relaying schemes, as discussed in Section IV.

eventually the systems will reach a point where they will be in phase again. If the systems are not separated, system A can continue to move ahead of B again and the whole cycle may repeat itself. When the locus reaches the point where the swing started, say at point P, one slip cycle has been completed.

If system A slows down with respect to B (decreases in phase angle), the impedance locus will move in the opposite direction from Q to the right to P.

When the voltage ratio E_A/E_B is more than or less than one (1), the impedance loci will be circles with their centers on extensions of the total impedance line AB. For the case of $E_A/E_B>1$, the electrical center will be above the impedance center of the system and for $E_A/E_B<1$, the electrical center will be below the impedance center.

It should be noted that the electrical centers of the system are not fixed points. The location of these centers will vary as the system impedances behind the line terminals vary and the equivalent internal generator voltages vary. Therefore, when determining the impedance locus for a system it will be necessary to consider changes in system conditions (that is, variations in Z_A and Z_B of Figure 1) and determine a locus for each condition.

The rate of slip between the two systems is a function of the accelerating torque and inertias of the systems. An estimate of the slip can be obtained from transient stability studies where angular changes of the system voltages may be plotted as a function of time. From these plots, an average rate of slip either in degrees/sec. or cycles/sec. can be determined. The rate of slip between the systems will not be constant during any given slip cycle. However, the general practice in simplified studies is to assume a constant rate of slip for the portion of the first slip cycle which is of interest; namely, the starting point of the swing locus and the point of 180 degrees separation between systems.

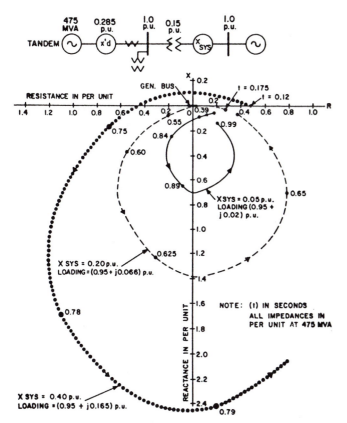

Fig. 2 Loss of Synchronism Characteristic – Tandem Unit

Figures 3 through 4 show the loss of synchronism impedance loci for a typical cross-compound generator as a function of system impedance. Figure 3 gives the impedance loci as viewed at the terminals of high pressure (HP) and low pressure (LP) units for a system impedance of .05 per unit. In this case the impedance loci are similar to those for a tandem generator. The high pressure unit with its lower inertia has completed one slip cycle while the high inertia low pressure unit has only completed a small portion of the slip cycle.

Figure 4 shows the pattern for the impedance loci when the system impedance falls in the range of .2 to .4 per unit. In this case both loci, as viewed from the machine terminals, are above the R axis. Because of the irregularity of these loci, it would be difficult to detect a loss of synchronism at the terminals of either unit with conventional relays. However, Figure 4 also shows the composite impedance locus for this case, as viewed at the low voltage terminals of the step-up transformer. The impedance locus at this point follows the same pattern as for tandem generators and could be used to detect a loss of synchronism.

As noted previously, the above loss of synchronism characteristics do not include the effect of voltage regulators. If a slow response voltage regulator is in service, the impedance locus circle will be larger in diameter but will still fall below the R axis. The increase in circle diameter is due to the increase in machine internal voltage produced by the voltage regulator. While the effect of fast response voltage regulators has not been studied, it is possible that the rapid increase in internal voltage produced may shift the electrical center out of the generator zone into the system.

The advent of the modern generator and the EHV transmission system has also resulted in significant changes in the slip characteristics of the generator during swing conditions. For the tandem generator, the average rate of slip during the first half slip cycle will usually be in the range of 250 to 400 degrees/second while for the cross-compound units, the average initial rate of slip will be 400 to 800 degrees/second. For both types of generators, the average rate of slip during the remainder of the slip cycle will fall in the range of 1200 to 1600 degrees/second. These rates of slip, it should be noted, are approximate values. The actual rate of slip will be a function of the machine inertias, machine load, type of fault and the length of time it takes to clear the fault. In general, the longer the fault clearing time, the higher the initial rate of slip.

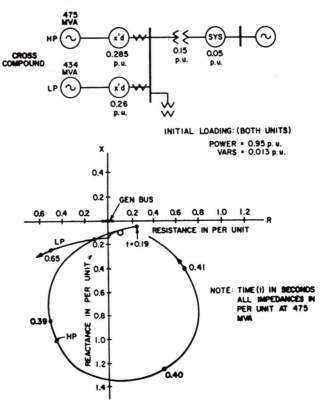

Fig. 3 Loss of Synchronism Characteristic – Cross Compound Unit

III. EFFECT ON GENERATORS OPERATING IN OUT-OF-STEP CONDITION

With the increased probability of incurring an out-of-step operating condition with the modern machine, the problem of possible damage to such a unit needs to be examined.

During an out-of-step condition as the swing angle between the generated voltage of a machine changes with respect to that of other units in the system, the current in any such unit varies in magnitude. The current surges that result are cyclical in nature; the frequency being a function of the relative rate of slip of the poles in the machine. The resulting high peak currents and off-frequency operation can cause winding stresses, and pulsating torques which can excite mechanical resonances that can be potentially damaging to the generator and to the turbine generator shafts. Therefore, it is recommended that for an out-of-step condition the generator be tripped with no intentional delay within the first slip cycle.

237

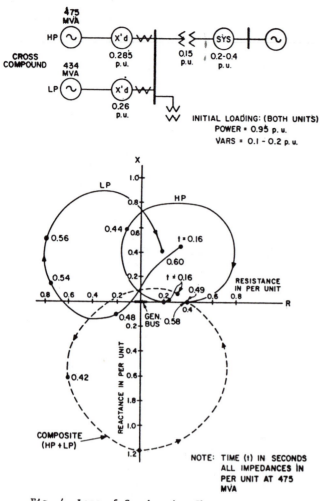

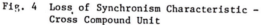

Fig. 4 Loss of Synchronism Characteristic –
Cross Compound Unit

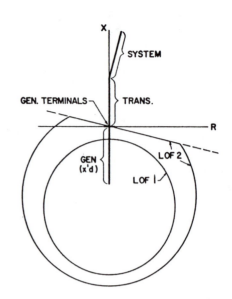

Fig. 5 Typical Loss-of-Field Relay Characteristic

IV. RELAYS FOR OUT-OF-STEP TRIPPING OF GENERATORS

General

The basic schemes that are available to the relay engineer to detect locally the generator out-of-step condition are essentially the same as those utilized to detect this condition at transmission line terminals. The methods, as discussed here, are applied to obtain a tripping function.

Loss-Of-Field Relaying

Although loss-of-field relaying is applied primarily to protect a generator for a loss-of-field condition, the conventional distance relays used to detect such a loss may provide a measure of generator out-of-step protection. The typical setting characteristics of two different relays commonly used for loss-of-excitation relay protection are shown in Figure 5. It is apparent that tripping of the generator can only occur for out of step conditions that appear electrically in the generator, i.e., swing characteristics that pass electrically thru the generator step-up transformer would not be sensed by these relays. It should be noted that because of the time delay in the protection scheme, tripping will occur only for swings that dwell within the characteristics for a sufficient period of time.

Mho Element Scheme

The mho element scheme is the simplest method of detecting an out-of-step condition that appears electrically in a generator or the step-up transformer. The concept as illustrated in the R-X diagram of Figure 6 requires the use of an impedance element sensing current and voltage at the high-side terminals of the unit transformer, and reaching electrically into the local generator. If for a loss of synchronism, the swing impedance characteristic should enter the circle, immediate tripping will occur. Without any supervision this scheme may well result in tripping for recoverable swings unless the setting sensitivity is restricted to detecting swings in excess of 120°.

The 120° criterion is the typical maximum angular separation of machines in a power system which is likely to occur without loss of synchronism. The scheme has the added disadvantage that tripping can occur when the angle approaches 180°. This subjects the circuit breaker to a maximum recovery voltage during interruption.

It should be noted that the mho circle element is sometimes applied to sense current and voltage at the terminals of the generator and be directionally oriented to look through the transformer into the system. With some reverse offset, electrically into the generator, the relay can be set to provide a characteristic similar to that shown in Figure 6. However, to prevent misoperation for faults or swings appearing beyond the high side terminals of the transformer either the reach must be set short of these terminals or tripping must be time delayed. This may nullify the use of the scheme.

An example of the mho circle scheme applied at the high side terminals of a generator step-up transformer in a typical system is shown in Figure 7. The setting objective is to provide relay operation for any out-of-step characteristic that passes electrically through the generator or step-up transformer and which cannot be sensed by a loss-of-excitation relay. The angle of swing (δ), in the sample in Figure 7 is

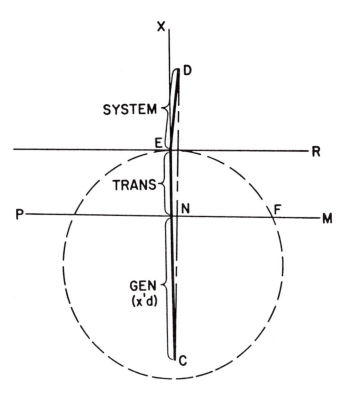

Fig. 6 Mho Circle Scheme

approximately 112° at the point where the swing impedance characteristic comes into the mho circle characteristic. At this angle of swing, recovery may be possible for some system disturbances. This dictates that the setting sensitivity must be carefully studied to insure that operation of the mho element will not occur for stable swings. As the mho circle is made smaller in size, however, it is apparent that tripping will occur at a less favorable angle to effect interruption of the swing current. The rate of angular change also becomes more critical for out-of-step sensing.

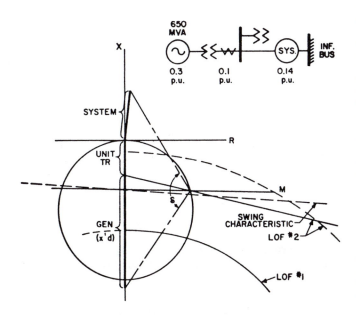

Fig. 7 Application of Mho Circle Scheme

Also note in Figure 7 that LOF #1 Relay would not operate and that LOF #2 Relay would trip, only if the swing impedance characteristic remains within the relay characteristic for the time delay inherent in the scheme.

To minimize the probability of false tripping of a generator for a loss-of-potential condition, some users may prefer to install an overcurrent fault detector in series trip logic with the mho relay. This application necessitates careful system swing studies to insure that fast operation of the over-current element will occur for nonrecoverable swings.

Single Blinder Scheme

The sensing elements of this scheme consist of two impedance elements, usually referred to as blinders, and a supervisory relay. The characteristics are shown in Figure 8 with the dashed circle representing the supervisory function. To restrict operation to swings appearing in the generator or transformer and to prevent operation for recoverable swings that pass through both blinder elements, the relay must be supervised by a mho element. This supervision also precludes tripping for the oscillatory reactive flow that predominates after synchronizing of a generator before load pickup. The blinder units have opposite polarity such that an impedance value falling between the A and B characteristics as at N will cause both units to pick up. The angle of the blinder units can be adjusted so that the characteristics can be made parallel to the equivalent system impedance represented by line DC.

Control logic is provided to aid in detecting the out-of-step condition. The differentiation between an internal fault and an out-of-step condition is made in the scheme by sensing the origin of the swing and whether there is a change from the +R to the −R region on the R-X plot. This variation corresponds to the actual real power reversal that occurs during an out-of-step condition but usually not for the fault period, i.e., from a few cycles before to a few cycles after the fault.

For the example in Figure 8, when a loss of synchronism occurs, the swing impedance having progressed to H will first cause the mho element to pick up and will also cause the A blinder unit to pick up if it is not already picked up due to load. As the swing progresses, it will cross the B unit characteristic at F and the B unit will pickup. Finally the swing impedance will cross the A unit characteristic at G, for example, and the A unit will drop out. If this sequence is followed, the breaker trip circuits will be completed when the impedance is at point G, or following reset of the supervisory unit, depending on the particular scheme being used. The reach setting of the blinder unit controls the impedance NF and NG and thus for a given system condition controls the angle, such as DFC. Since this is a measure of the angle between the source voltages, tripping can be controlled to effect circuit breaker opening when the angle is less than a particular value, such as 90°. This limits the voltage appearing across the opening breaker poles to a more favorable value for arc interruption.

It should be noted in Figure 8 that the scheme can initiate tripping for swings passing through lines terminating at the high voltage terminals of the transformer. A line distance relay will detect a swing as soon as it enters its characteristic while the blinder scheme will not provide tripping until the swing has left the supervising mho circle. This will allow line relays to clear on instability where these relays are not blocked by out-of-step detection schemes.

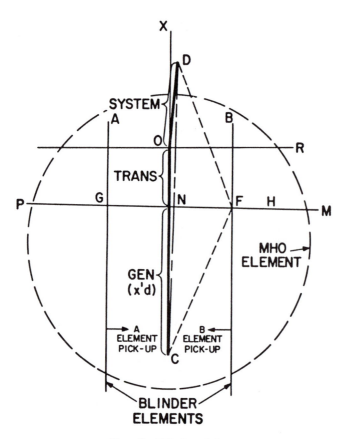

Fig. 8 Blinder Scheme

The advantages of the blinder relay scheme, as compared to the mho element scheme, for generator protection can be seen by studying the setting examples in Figures 7 and 9. As the diameter of the mho circle indicated in Figure 7 is increased to provide greater sensitivity for out-of-step swings in the generator, it is possible for tripping to occur for the recoverable swing indicated in Figure 9. With the addition of the blinder elements, however, tripping would not occur. For the out-of-step condition shown in this figure, it is clear that the scheme will provide not only tripping of the generator, but will effect interruption at a favorable swing angle. The general application of this scheme to generators, however, requires a careful study of the mho element setting to prevent tripping for swings that cross both blinder elements and fool the inherent logic of the scheme.

Double Lens or Double Blinder Schemes

The double lens and double blinder systems perform in a manner similar to the systems previously described. The supervisory mho element is included in the double blinder system to obtain the same security features covered in the discussion on the single blinder scheme. Referring to Figures 10 and 11, the outer element operates when the swing impedance enters its characteristics, as at F. Note that in the double blinder scheme the mho element will pick-up before the outer blinder element. If now the swing impedance remains between the outer and inner element characteristics for longer than a pre-set time, it is recognized as an out-of-step condition in the logic circuitry. When the swing impedance now enters the inner element characteristics, a portion of the logic circuitry is sealed in after a short time delay. Then as the swing impedance leaves the inner element characteristic, its traverse time must exceed a pre-set time before it

reaches the outer characteristic. Tripping does not occur until the swing impedance passes out of the outer characteristic, or for the double blinder scheme until the reset of the supervisory mho element depending on the particular logic being used. This is to provide for the case where sequential clearing of a fault inadvertently sets up the first two steps of logic in the scheme. In the case of a fault, the inner and outer elements reset practically simultaneously and no incorrect tripping results.

Again, the swing angle DFC is controlled by the settings to limit the voltage across the opening poles of the breaker. Once the swing has been detected and the impedance has entered the inner characteristic, the swing can now leave the inner and outer characteristics in either direction and tripping will take place. Therefore, the setting of the inner element must be such that it will respond only to swings from which the system cannot recover. This restriction does not apply to the single blinder scheme because the logic requires that the apparent ohms enter the inner area from one direction and exit toward the opposite. The single blinder scheme may for this reason be a better choice for the protection of a generator than either the mho scheme, the double blinder scheme, or the concentric circle scheme.

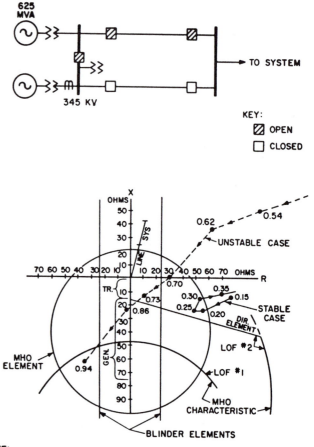

NOTE:

1. ALL IMPEDANCES IN OHMS AS VIEWED FROM 345 KV BUS.
2. ALL TIME VALUES INDICATED ON SWING CHARACTERISTICS ARE IN SECONDS.

Fig. 9 Example of Blinder Scheme for a Stable and an Unstable Case

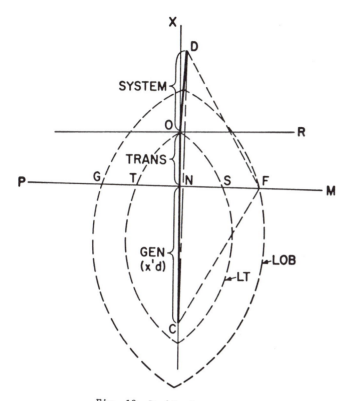

Fig. 10 Double Lens Scheme

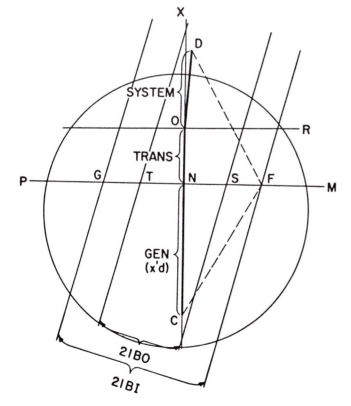

Fig. 11 Double Blinder Scheme

Concentric Circle Scheme

The concentric circle scheme uses two distance units and operates essentially the same as the double lens scheme.

When this method is employed, particular care must be exercised to insure that the inner sensing element responds only to non-recoverable swings. This requirement may well necessitate the use of the double lens or the double blinder scheme when both a fault tripping function as well as out-of-step protection are desired, i.e., those schemes are characteristically equipped to provide fault sensitivity more easily with minimum response to swing conditions.

Triple Lens Scheme

This out-of-step tripping scheme uses a relay consisting of three lens units: an outer, a middle, and an inner lens, which have a common maximum reach at the greatest sensitivity angle. This scheme is functionally more secure than the other schemes since it uses four steps in the out-of-step tripping sequence in order to discriminate between the out-of-step condition and recoverable power swings and faults. To obtain such reliability, the logic circuitry is inherently more sophisticated than with the other systems.

V. GENERATOR OUT-OF-STEP PROTECTION PRACTICES

In 1970 a survey was conducted by the Working Group to determine what out-of-step protection was being applied by relay engineers to detect such a condition at the unit location. To aid in this study, a letter of inquiry was sent to all members of the Power System Relay Committee. In 1973 a follow-up letter was sent to all who responded to the survey to aid in clarifying current protection practices.

The replies to this survey are summarized in Table I. The types of protection in use or authorized can be classified as a form of mho or blinder (or in combination) scheme. Some place dependence on the loss-of-field relaying with an ohmic sensing element for out-of-step detection.

A number of companies are employing the simple mho element scheme to provide both high-speed out-of-step protection and a measure of back-up protection for transformer faults. As generally applied, a mho element sensing current and voltage on the high side of the unit transformer is directionally oriented to electrically look into and through the unit transformer. The reach of the relay is set to obtain overlap with the impedance or mho element with which these companies obtain loss-of-excitation protection. The reach is limited to preclude pick-up during recoverable system swings. This achieves the objective of electrically bridging the gap between the phase relays of the lines terminating on the high side bus with the unit transformer and the ohmic characteristic of the loss-of-excitation relay.

Several forms of the blinder scheme are in use by a number of companies. Most of these feature mho supervision of the blinder units and are applied to function only as out-of-step trip protection.

A majority of the companies that replied to this inquiry are using conventional loss-of-field relaying to protect for both loss-of-excitation and loss-of-synchronism conditions. Several users indicated their loss-of-field relays are set so as not to operate for recoverable swings. A number of the reporting companies were among those experiencing loss-of-field relay operations for out-of-step conditions that developed during the course of the Northeast Power Failure of 1965.

It is also of interest to note that a number of companies switch a generator electrically from the system, but do not shut it down completely for an out-of-step condition.

Several companies indicated a preference for manual control to be exercised by operators for a loss of synchronism condition. One company stated that out-of-step protection is not used because loss of excitation protection and automatic speed and voltage control are applied to generators in their system to minimize the probability of incurring the out-of-step condition.

TABLE I

Summary of Replies to 1970 and 1973 Letters of Inquiry

A.   Number of companies responding
to letter of inquiry. 25

B.   Number of companies not employing
out-of-step protection for generators
(other than loss-of-field). 9

C.   Number of companies that apply various
impedance relay schemes for the protection of generators:

Type of Scheme

Mho Element		Blinder		Loss-of-Field	
E	P	E	P	E	P
8	3	8	4	13	-

Note: E-existing; P-planned

Remarks:

1. Number of companies employing out-of-step protection for generators (other than loss-of-field)

    Blinder relay only 1
    Mho relay only 2
    Blinder relay with
    overcurrent supervision. 2
    Blinder relay with mho
    element supervision. 3

2. Four of the reporting companies use mho ele schemes primarily as fault protection -- out-of-step protection for the generator is incidental. Three of the four companies use no intentional time delay in the scheme.

VI. CONCLUSIONS

Although the 1970 opinion survey conducted by the Working Group indicates light interest in local out-of-step protection for generators, some relay engineers apparently regard this zone of coverage as vital to system protection.

The pertinent points to be considered in the application of out-of-step relaying to a generator can be summarized from this study as follows:

1. There are some out-of-step characteristics that will pass electrically thru a generator or its associated step-up transformer. This tends to occur when a generator pulls out of synchronism in a relatively tight system. Such a characteristic will also exist due to a low excitation level on a generator.

2. The out-of-step characterisitics can be most simply sensed by relay schemes with a mho type of distance element oriented to look electrically into the generator and its associated step-up transformer.

    It is recommended that no intentional time delay be used in the schemes other than that required to logically sense the out-of-step conditon. Otherwise, the swing may go thru the distance characteristic before tripping can occur. The mho element can be connected to sense current and voltage quantities on either the high voltage side or the machine side of the transformer. The setting should not reach into the high voltage transmission system to avoid tripping for system faults and swing conditions that should be detected by other relay systems.

3. The simple mho circle scheme provides both out-of-step protection and fast back-up relaying for multi-phase lead faults on the generator and transformer. It has the disadvantage of circuit interruption at an unfavorable angle of swing, and is subject to tripping for recoverable swings. The more sophisticated schemes such as the blinder and lens types minimize the probability of tripping for recoverable swings and permit controlled switching of the generator at a preferred angle of swing. The conventional loss-of-field relay provides limited out-of-step protection because the associated time delay may preclude tripping for swings passing thru its characteristics.

4. It should be emphasized that the data presented in Section II - LOSS OF SYNCHRONISM CHARACTERISTICS are the results of generalized studies that do not consider the effects of all combinations of generator designs, voltage regulator characteristics, system parameters or the interaction effect of other generators. These effects can only be completely determined by the study of a generator connected to a specific system. It is strongly recommended that the user determine the actual loss-of-synchronism impedance loci for each generator considering the overall effects of the system before selecting a protection scheme. These loss-of-synchronism characteristics can readily be obtained with a high degree of accuracy using available computer programs.

REFERENCES

1. "The Art and Science of Protective Relaying," (book) by C. R. Mason, John Wiley and Sons, New York, N. Y.

2. "Impedances Seen By Relays During Power Swings With and Without Faults," by Edith Clarke, AIEE Transactions, Vol. 64, 1945, pp. 372-384.

3. "Circuit Analysis of A-C Power Systems, Volume II," (book) by Edith Clarke, John Wiley and Sons, New York, N. Y.

4. "Electrical Transmission and Distribution Reference Book," Chapter 13. Westinghouse Electric Corp., E. Pgh., Pa.

5. "A Power Swing Relay for Predicting Generation Instability," Brown & McClymont, IEEE Transactions, Power Apparatus and Systems 1965, pp. 219-294.

6. "A Modern View of Out-of-Step Relaying," by G. D. Rockefeller and W. A. Elmore, IEEE Conference Paper 31 CP 66-34.

7. "Effects of Future Turbine Generator Characteristics on Transient Stability," by H. E. Lokay and P. O. Thoits, IEEE Transactions, Power Apparatus and Systems 1971, pp. 2427-2431.

8. "The Fundamentals of Out-of-Step Relaying," by W. A. Elmore, 1974 Georgia Institute of Technology Protective Relaying Conference.

9. "Application of Out-of-Step Blocking and Tripping Relays," by J. Berdy, 1967 Texas A & M Conference of Protective Relay Engineers.

10. "Loss of Synchronism Protection for Modern Synchronous Generators," by J. Berdy, IEEE Transactions, Power Apparatus and Systems 1975, Number 5, pp. 1457-1463.

Discussion

F. H. Birch (Central Electricity Generating Board, Leeds, England); After four years study of this subject a CIGRE Working Group, of which I have been Convenor, has completed a report which will be published in ELECTRA early in 1977. The report reaches the following main conclusions.

If out-of-step operation occurs with the electrical centre passing through the transmission system, out-of-step blocking and out-of-step tripping relays can be used to prevent undesired tripping of lines and sectionalise the system into self-sustained islands. However, as system designs differ considerably, studies are necessary in each case in order to determine optimum relaying strategy. Generators should not be tripped, except where the local load (including the power station auxiliaries) is insufficient to allow the generators to operate satisfactorily after the transmission connections have been broken. In this case the load would be secured by maintaining the transmission connections and tripping the generators.

If, on the contrary, the electrical centre is liable to pass through one or more large generators or their step-up transformers, there is a growing opinion in Europe in favour of the addition of out-of-step protection, capable of disconnecting the affected units without delay. Such protection would be justified for the reasons stated by the authors and also because there is the risk of losing the auxiliaries before automatic re-synchronising takes place. Furthermore, fast removal from the system of a large generator running out of step avoids the risk of loss of auxiliaries to other generators which remain in step and minimises disturbance to other system loads. Out-of-step protection of large pumped-storage units is definitely recommended, because these are particularly prone to fall out of step when operating underexcited in the pumping mode.

A fast three pole overcurrent relay is a simple and effective means for protecting a generator against damage during very severe out-of-step conditions, such as may occur after slow clearance of a nearby three-phase fault. A relay set to operate when the current in all three phases exceeds, for example, 85% of the a.c. subtransient value for a three phase short-circuit at the stator terminals, would be stable with H.V. faults beyond the step-up transformer and under the more likely out-of-step conditions associated with low system infeeds through long transmission lines.

Manuscript received February 9, 1977.

Joseph A. Imhof, Chairman: The authors wish to thank Mr. Birch for his summary of the conclusions reached in the report of his CIGRE Working Group on out-of-step relaying for generators.

Although the authors did not address such problems as the need for sectionalizing a system into self-sustaining islands or the tripping of generators for a case of partial load rejection, they are in agreement with the CIGRE Working Group that in-depth studies are necessary to determine the optimum relaying strategy.

Our Working Group concurs with Mr. Birch's comments concerning fast tripping of generators to minimize the risk of losing auxiliary systems for a loss-of-synchronism condition. This philosophy of relatively fast tripping to maintain the auxiliary power systems can also be extended for out-of-step characteristics that appear electrically in the transmission lines connected to the station bus. This approach obviously necessitates more sophisticated sensing schemes that inhibit line tripping via the line relays and permit generator tripping via the generator out-of-step relays.

The use of a fast three-pole overcurrent relay to detect an out-of-step condition is certainly a simple and effective means of protecting a modern generator. However, to effectively discriminate between a three phase fault close to the station high voltage bus and a severe out-of-step condition it appears that this application would be limited to systems that are electrically very potent. The statement by Mr. Birch that such a relay would be stable for high voltage faults in a relatively long-line system environment tends to invalidate the application, i.e., in obtaining security for system faults, relay sensitivity for a loss-of-synchronism condition is unduly compromised. Perhaps the relay scheme as visioned by Mr. Birch is somewhat more sophisticated than his remarks seem to indicate.

Manuscript received April 11, 1977.

REFERENCES

[1] "The Art and Science of Protective Relaying," (book) by C. R. Mason, John Wiley and Sons, New York, N. Y.
[2] "Impedances Seen By Relays During Power Swings With and Without Faults," by Edith Clarke, AIEE Transactions, Vol. 64, 1945, pp. 372-384.
[3] "Circuit Analysis of A-C Power Systems, Volume II," (book) by Edith Clarke, John Wiley and Sons, New York, N. Y.
[4] "Electrical Transmission and Distribution Reference Book," Chapter 13. Westinghouse Electric Corp., E. Pgh., Pa.
[5] "A Power Swing Relay for Predicting Generation Instability," Brown & McClymont, IEEE Transactions, Power Apparatus and Systems 1965, pp. 219-294.
[6] "A Modern View of Out-of-Step Relaying," by G. D. Rockefeller and W. A. Elmore, IEEE Conference Paper 31 CP 66-34.
[7] "Effects of Future Turbine Generator Characteristics on Transient Stability," by H. E. Lokay and P. O. Thoits, IEEE Transactions, Power Apparatus and Systems 1971, pp. 2427-2431.
[8] "The Fundamentals of Out-of-Step Relaying," by W. A. Elmore, 1974 Georgia Institute of Technology Protective Relaying Conference.
[9] "Application of Out-of-Step Blocking and Tripping Relays," by J. Berdy, 1967 Texas A & M Conference of Protective Relay Engineers.
[10] "Loss of Synchronism Protection for Modern Synchronous Generators," by J. Berdy, IEEE Transactions, Power Apparatus and Systems 1975, Number 5, pp. 1457-1463.

EVALUATION OF THE EFFECT OF POWER CIRCUIT BREAKER RECLOSING PRACTICES ON TURBINE-GENERATOR SHAFTS

C.E.J. Bowler
General Electric Company
Schenectady, N.Y.

P.G. Brown
General Electric Company
Schenectady, N.Y.

D.N. Walker
General Electric Company
Schenectady, N.Y.

ABSTRACT

Several circuit breaker reclosing practices following transmission system electrical faults are evaluated, and recommendations presented, that minimize the fatigue duty on turbine-generator shafts, while still maintaining the commonly considered advantages of high speed reclosing (H.S.R.). Probability distributions were created for appropriate parameters, so that the statistical nature of high speed reclosing could be treated by Monte Carlo simulation. These include distributions for the timing of opening and reclosing of circuit breakers, the type and location of the fault, the timing of the fault relative to the position on the voltage wave, and the scatter in shaft torsional fatigue properties. For each reclosing practice analyzed, and for each of several turbine-generator designs, the probability distribution of shaft cumulative fatigue expenditure was estimated, based on a forty year fault exposure period, and on an assumed electric system model. These data give a meaningful basis for comparing different types of reclosing practices that are presently in use. This paper advances the state-of-the-art by being the first to apply statistical methods to the analysis of this subject.

INTRODUCTION

Background

High speed reclosing of transmission lines has been the subject of many papers over the past thirty years. These papers generally deal with breaker relaying techniques, with emphasis on reduction of reclosing times to the extent permitted by current breaker technology, and by de-ionization time requirements. Machine duties associated with reclosing occupied little concern, and it was generally assumed that, if excessive, such duties would be readily apparent by inspection of the stator end windings. However, during the past several years there has been an increasing recognition [1,2,3] that shaft fatigue damage, which is not an observable quantity prior to crack initiation, may be a greater limitation than stator winding duty. The two shaft failures at the Mohave Generating Station [4], while caused by unstable torsional oscillations (subsynchronous resonance), rather than fault induced transient vibration, prompted new studies of shaft torsional response and fatigue, and the damping of torsional oscillations. The studies clearly indicated the need to consider the expenditure of fatigue life on shafts, as well as on other components, when deciding on appropriate switching limits, even for systems without any series capacitor compensation. In the area of steady state planned line switching, screening levels and study procedures have been proposed [2,5]. The main purpose of this paper is to provide similar guidance with regard to circuit breaker reclosing practices.

F 80 199-0   A paper recommended and approved by the IEEE Rotating Machinery Committee of the IEEE Power Engineering Society for presentation at the IEEE PES Winter Meeting, New York, NY, February 3-8, 1980. Manuscript submitted September 6, 1979; made available for printing November 6, 1979.

The potential for shaft fatigue damage is greatly amplified with high speed reclosing because of the inherently light damping of turbine-generator torsional oscillations. Shaft oscillations created by a fault disturbance and circuit breaker clearing remain relatively undiminished in the time frame of the usual high speed circuit breaker reclosing. Thus, the added disturbance accompanying reclosing, depending on its timing, may directly reinforce the oscillations of the original disturbance. The resulting effect on shaft fatigue is very non-linear, as the addition of two similar events (such as unsuccessful reclosing) results in much more than just a simple doubling of the fatigue damage. In the past, viewing only stator winding duties as significant, one might have assessed the damage potential of an unsuccessful reclosure as simply that of two faults instead of one. It is now recognized that shaft fatigue is generally more critical, and from that standpoint, the effect of the second fault may result in one or more orders of magnitude increase in the shaft fatigue life expenditure.

While it is clear from analysis that some of the worst case reclosing scenarios (such as unsuccessful reclosure onto multi-phase faults with switching at the most unfavorable instants) can cause a significant loss of fatigue life of turbine-generator shafts, the severity of actual reclosing events varies over a wide range. A true appraisal of high speed reclosing should therefore consider the probable lifetime duty which the unit will experience, including the variables of fault frequency, fault type, fault location, ranges of clearing and reclosing times, the timing of the fault relative to the position on the voltage wave, and scatter in shaft torsional fatigue properties.

Reclosing Practices

Since experience indicates that over 80% [6] of transmission line faults are of a temporary nature, it has become common practice to automatically reclose the circuit breakers of a line tripped out due to a fault. If the fault persists, the circuit breakers are tripped a second time, usually with no further attempts to reclose automatically, but in a few cases with an automatic delayed reclosure attempt.

There are a number of forms of reclosing practice, ranging from the fastest possible automatic reclosing of both ends of the line to manual operation of breakers with synchro-check relay supervision. Between these extremes are various other practices, resulting in different degrees of turbine-generator shaft fatigue duty. The four practices listed below were evaluated in this paper.

- "Unrestricted high speed reclosing" refers to the practice of reclosing transmission line circuit breakers at both ends of the line as rapidly as possible following a fault tripout, regardless of the fault type. Successful reclosing cannot occur until the fault arc has extinguished and the dielectric strength has recovered. This may require more than a second for lines having a large amount of shunt capacitance. For short transmission lines, reclosing times may be as short as 15 electrical cycles.

- "Delayed reclosing" refers to automatic reclosing of the transmission line breakers where additional time delay is used beyond the minimum required for arc extinguishing and dielectric recovery. A ten second time delay was used with this type of reclosing in the study. The main reasons for adding time delay include the following:

  1. To improve the percentage of successful reclosings.

Reprinted from *IEEE Trans. Power App. Syst.*, vol. PAS-99, no. 5, pp. 1764–1779, Sept./Oct. 1980.

2. To reduce the probability of system transient instability following an unsuccessful reclosure back into a major fault [2]. Time delays of one-half to several seconds are typically used.

3. To avoid the risk of major shaft fatigue damage associated with unsuccessful reclosing into multiphase faults. Delay times of at least 10 seconds have been recommended for this purpose [1,2].

- "Sequential reclosing" refers to the practice of reclosing the line first from the end which is remote from the plant, after which, on the basis of check relays (usually voltage and phase angle) to assure that the fault no longer exists, the plant end breaker is automatically closed. One to several seconds time delay is inherent in this type of reclosing. This type of reclosing is therefore particularly applicable where a short delay is desired anyway for reducing the risk of transient instability [2]. The effectiveness of sequential reclosing in avoiding the compounding of fatigue damage when reclosing into permanent faults is of course a function of the electrical network. Sequential reclosing would be of little value in tight networks where the "remote" line terminal is electrically close to the plant through other connections, or if another generating plant was located close to the "remote" end of the line.

- "Selective reclosing" as used in this paper is the practice of distinguishing the type of fault and permitting high speed reclosing only for single line to ground and line to line faults. Selective reclosing might also be applied on some other basis, such as distance from the plant, to screen out the more severe close-in faults.

There are many possible variations and combinations of these techniques, for example, selective high speed reclosing for single line to ground faults with delayed automatic reclosing (of 10 seconds or more) for all other faults. In general, techniques such as these which might be applied to reduce the risk of shaft fatigue damage tend to be helpful from the system aspect also. This is because unsuccessful reclosure of severe multi-phase faults not only results in the greatest risk of major fatigue damage but also the greatest risk of transient instability, particularly during extreme and possibly unforeseen emergency conditions. Such severe faults also statistically appear to be much more likely to be permanent.

Recognition of the potential for excessive shaft fatigue duty from high speed reclosing has led many electric utilities to change to a more conservative reclosing practice, pending completion of industry studies. An IEEE working group, composed of representatives from several electric utilities, turbine-generator manufacturers, universities, and consultants, is presently working toward development of guidelines on reclosing practices. The data contained in this paper have already been reviewed by this working group, and are expected to give guidance for developing recommendations.

Torsional Response

System electrical faults cause a sudden change in the average electrical air-gap torque of nearby turbine-generators, stimulating the torsional modes of vibration. When the fault is cleared by tripping the line, a second step-change in torque occurs. In addition, 60 Hz electrical torques result from dc offset currents of the fault, and unbalanced faults produce 120 Hz torques due to negative sequence current components. These additional electrical torque components also influence the magnitude of the lightly damped multimodal torsional oscillations in the turbine-generator shafts [5].

Electrical faults in conjunction with high speed reclosing subject the turbine-generator to a series of impact torques within a short period of time; first the initial fault, then fault clearing, then reclosure, and finally (if unsuccessful) when the

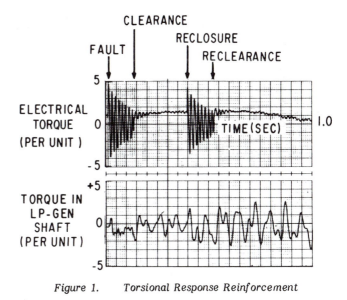

Figure 1. Torsional Response Reinforcement

fault is again cleared. This is shown in Figure 1 for the case of a three phase fault with unsuccessful reclosure. This figure shows the torsional response in the shaft that connects the low pressure turbine to the generator of a 660 MVA two-pole turbine-generator. The timings of the circuit breaker opening and reclosing in this example were selected to maximize the peak-to-peak shaft response, to illustrate the potential for torsional response reinforcement. It is seen that the peak shaft response (expressed in per unit on the rated generator MVA base) almost doubles following fault removal, and almost doubles again following the combined result of unsuccessful reclosure and final fault removal. Conversely, under the most favorable circumstances of the timing of circuit breaker opening and reclosing, the original oscillations arising from the fault may be significantly reduced. With present state-of-the-art, preselection of breaker operation times is not considered a practical method of controlling shaft torques due to variability in circuit breaker clearing and reclosing times, as well as transient stability demands for the fastest possible fault clearing.

MODELLING AND SIMULATION

Evaluations of circuit breaker reclosing practices are presented based on results obtained on seven different turbine-generator designs. Five of this population are 3600 RPM units, the remainder being 1800 RPM units. The turbine-generator configurations that were analyzed are summarized in Table I, in terms of their rating, number of turbine rotors (excludes generator and exciter rotors), number of shafts, and first torsional natural frequency. This population of machines was selected to span the spectrum of torsional duty obtained in the planned switching evaluation described in [5], which analyzed 84 significantly different turbine-generator designs.

Preliminary electrical fault torque screening studies generally indicated that whereas the damping of turbine-generator torsional oscillations is maximum at full load (tending to reduce the number of fatigue cycles), so also is the magnitude of the electrical stimulus (tending to maximize

TABLE I. Turbine-Generator Configurations

Unit	Operating Speed (RPM)	Rating (MVA)	Number of Turbine Rotors	Number of Shafts	Number of Subsynchronous Modes	First Torsional Frequency (Hz)
1	3600	890	4	5	5	15.7
2	3600	660	5	6	6	13.0
3	3600	690	3	4	4	17.5
4	3600	910	4	5	5	14.8
5	3600	815	4	5	5	16.5
6	1800	1010	3	4	6	10.0
7	1800	1280	4	5	8	7.1

245

the peak shaft response). These two effects were found to generally offset one another, and in this study all simulations were conducted at rated power output of the turbine-generators. Table II defines the modal damping values, for the first five torsional modes, that were assumed for each turbine-generator configuration at rated output. These values were obtained by extrapolations on modal damping data obtained in many station torsional vibration tests conducted over the past six years.

TABLE II. Full Load Modal Damping Values

| Unit | Damping (LOG-DEC) | | | | |
| | Mode # | | | | |
	1	2	3	4	5
1	0.0119	0.0107	0.0163	0.0025	*
2	0.0097	0.0103	0.0067	0.0051	0.0017
3	0.0081	0.0110	0.0075	0.0049	*
4	0.0117	0.0102	0.0083	0.0025	*
5	0.0216	0.0198	0.0124	0.0033	0.0130
6	0.0104	*	0.0060	0.0169	0.0010
7	0.0125	*	0.0076	0.0070	0.0109

* Negligible responses in these modes.

Vibration and Fatigue Models

In general, the transient torsional response of the turbine-generator shafts, following an electrical disturbance of short duration, consists principally of the subsynchronous torsional modes of vibration of the turbine-generator. This occurs because, in the higher modes of vibration, the generator rotor (where the stimulus is applied) generally tends to be an ineffective location to stimulate mechanical response [5]. As a consequence of the shaft response being principally in the low order modes, relatively simple torsional vibration mathematical models of the turbine-generator are adequate in fault torque evaluation studies, as described in [5].

Figure 2 is a typical point inertia torsional vibration mathematical model and corresponds to one of the 1800 RPM turbine-generators that was evaluated in this paper. In the vibration analysis, a modal transformation was utilized to convert the point inertia model to the mode spring mass model which is defined in [7]. This transformation converts from the variables of torsional motion at each inertia to the amount of response in each torsional vibration mode. This mode spring mass model reduces the computational effort by uncoupling and reducing the number of equations to be solved (by discarding inactive modal coordinates), and enables modal damping values to be used directly.

In order to estimate the amount of shaft fatigue life consumed following an electrical fault, cumulative fatigue models are required. Electrical faults generally produce complex multi-modal shaft responses. The cumulative fatigue procedure utilized must therefore be able to analyze this complex response in terms of data obtained in small specimen fatigue tests conducted in the laboratory. The procedures utilized in developing the results shown in this paper account for stress concentration, plasticity, size, mean stress, complex torque history, periodic overstrain, and several other parameters [5].

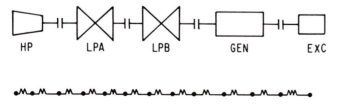

Figure 2.   Lumped Inertia Model

HP   LPA   LPB   GEN   EXC

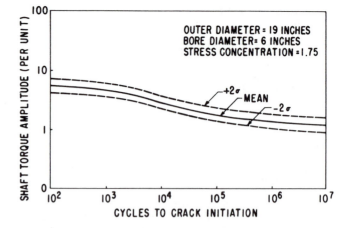

OUTER DIAMETER = 19 INCHES
BORE DIAMETER = 6 INCHES
STRESS CONCENTRATION = 1.75

+2σ
MEAN
-2σ

Figure 3.   Typical Torque-Cycles to Failure Fatigue Curve

Figure 3 shows a typical torque-cycles to failure fatigue curve that was utilized in the studies which corresponds to a 19" diameter shaft at the location of the limiting stress concentration. The definition of fatigue failure in this study corresponds to crack initiation. Here, crack initiation corresponds to 100% shaft fatigue life consumption. Linear damage rules are used so that, for example, if 1000 fatigue cycles caused crack initiation, then it is assumed that 1 cycle at the same amplitude would consume one thousandth of the fatigue life. The middle curve in this figure represents the assumed average fatigue properties for a shaft of this size based on tests on fatigue specimens, and the other curves represent $+2$ standard deviations from the mean. It is seen that the scatter in the fatigue properties, which is an inherent material attribute [5], is very significant and for this reason, the effect of scatter was included in the probabilistic evaluation.

Electrical System Models

To make fault torque computer simulation practical for generalized study purposes, it is necessary to reduce a representative electric transmission system to a simple equivalent mathematical model. The model analyzed is shown in Figure 4, and represents a two line system to an infinite bus through a reactance. The turbine-generator is connected to the station bus through a conventional delta-wye transformer. The two transmission lines shown are of unequal reactance corresponding to a two-to-one difference in length. For this system equivalent, in which both transmission lines serve the same receiving system, three phase faults at the remote end of the line are nearly as severe as those occurring at the turbine-generator end.

The transmission lines and receiving system were modelled in direct phase quantities to allow precise current zero interruption of faulted lines, and the correct zero sequence response to unbalanced faults. The step-up transformer was modelled with Clark components to achieve correct representation of sequence impedance reflection as seen by the generator or by the system, as well as achieving the thirty degree phase shift of the delta-wye transformer. The system reactance with all lines in service as viewed from the generator is 0.2 per unit on the rated MVA machine base, which corresponds to a relatively strong system representative of many United States plants.

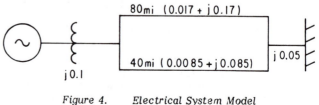

80mi (0.017 + j 0.17)

40mi (0.0085 + j 0.085)

j 0.1

j 0.05

Figure 4.   Electrical System Model

In the simulations, only one line was faulted at a time, with the other representing the synchronous connection to the system during the fault. Fault locations were at one of six evenly spaced points on each line (including the end points). In all fault cases three pole breaker operation was used.

Hybrid Computer Simulation

Many computer simulations are required to account for the variables that have probability distributions associated with them. The economic choice of simulation equipment for this study was a hybrid digital/analog computer. This computer system contains a central processing unit with 32K memory together with tape handling and disk storage capability, and interfaces with a differential electronic analog computer capable of representing up to seven transmission lines and two turbine-generators. The digital computer serves to control the run time variables and initial conditions in the analog computer associated with fault type, fault location, fault point on the voltage wave, circuit breaker clearing and reclosing times, and fault status (e.g., whether reclosure was successful or not). The digital part of the system also performs the following specific functions:

1. Initiates the analog computer runs.

2. Ensures removal of faults at normal current zeroes.

3. Captures the positive and negative peak values of the transient torque waveform in each turbine-generator shaft.

4. Calculates the loss of shaft fatigue life.

5. Stores shaft response and fatigue data for subsequent analysis.

6. Automatically initiates the next run for another set of conditions.

For a particular type of fault and location, simulations were conducted over the range of clearing and reclosing times of +3 standard deviations from the assumed nominal values, and shaft response and fatigue data were stored on magnetic tape. For each turbine-generator design, approximately 1500 computer simulations were required to effectively analyze the probabilistic loss of life at one fault location. Because of the large number of simulations required, it was necessary to structure the hybrid computer so that it could run in an unattended mode. The basic data stored on magnetic tape were later processed to estimate the probability of shaft loss of life exceeding defined values over an assumed forty year exposure period, for several circuit breaker reclosing practices.

Probability Models and Simulation

The calculation of shaft fatigue life expenditure for an assumed exposure period of a turbine-generator involves consideration of many parameters whose magnitudes or states are probabilistic. The method that is utilized in this paper to account for the probabilistic nature of circuit breaker reclosing is Monte Carlo simulation. It allows random selection of parameter values weighted by their underlying probability distributions. Following the computation of many randomly selected forty year exposure periods of the machine, the associated shaft fatigue life expenditures (% loss of life) can be plotted as a histogram. The histogram may then be cast into the form of a cumulative probability distribution, which expresses the probability that the loss of fatigue life in the shafts exceeds specified values over the assumed exposure period of the turbine-generator. Sufficient forty year periods were simulated to ensure that this distribution did not change significantly with an increase in the number of periods analyzed. Cumulative probability distributions were created for several different high speed reclosing practices.

TABLE III. Fault Statistics

	Fault Type		
	3ϕ	L–L–G	L–G and L–L
Fault type - %	3	3	94
Percent of faults that are temporary	60	60	87
Percent success of initial automatic reclosures	55	55	80
Percent success of subsequent delayed reclosures	5	5	7

Line Fault Statistics - Table III shows the assumed percentage of initial faults that are three-phase (3ϕ), double line to ground (L–L–G), and single line to ground (L–G) or line to line (L–L). L–L faults, which represent a relatively small percentage of the total (5%), and which produce a similar range of fatigue duty as L–G faults (shown in Figure 7), were included in the L–G fault category. This table also shows the assumed percentage of successful reclosures (temporary faults), both for the initial automatic reclosure, and for a subsequent delayed automatic or manual reclosure. These assumptions are based on information in [8], and represents average industry statistics. This reference is comprehensive and has the data in the form required for this study.

In addition to the fault type statistics, data on frequency of fault occurrence is required. Data available [9] indicate levels ranging from 1 to 12 faults/100 mi/yr depending on line design, tower footing resistance, and line location with respect to isokeraunic level (thunderstorm days per year). Results presented in this paper are based on an assumed mean rate of faults of 4/100 mi/yr (total for all types). A turbine-generator on-line exposure period of forty years was also assumed. It is a simple procedure to scale the study results given later for different assumptions of mean rate of fault occurrence and turbine-generator exposure period. This is discussed in more detail in the discussion of results section of the paper.

The three fault categories are assumed to occur independently at mean rates equal to the total mean rate multiplied by the proportions given in Table III. It was assumed that the fault occurrence rate (λ) is constant. The number of faults that occur in a specific time (T) will then follow a Poisson probability distribution with mean λT, and the time between fault events will follow an exponential distribution with mean $1/\lambda$. Table IV shows that, based on these assumptions, whereas it would be expected that, on average, approximately 5 three-phase faults would occur on a 100 mile line over a forty year period (0.03 x 4 x 40), there is almost a 1.5% chance of ten of these faults happening. This further reinforces the need for a statistical treatment of this subject.

Once the type of a particular fault occurrence has been identified by random sampling it is next necessary to define its location in the equivalent system. This was done by assuming a uniform distribution of faults along each transmission line, at one of six evenly spaced locations (including the end points).

TABLE IV. Three-Phase Fault Statistics

(100 miles of line, forty year period; mean rate of faults = 4/100 mi/yr, 3% three phase faults)

Probability of N faults occurring - %									
N= 1	2	3	4	5	6	7	8	9	10
3.9	9.5	15.2	18.2	17.5	14.0	9.6	5.8	3.1	1.5

Using the above distributions it was possible by random sampling to define for each forty year simulation period the number of faults, fault types, locations, and reclosing outcomes. For the purpose of this study, if a fault was identified in the simulation as being permanent, then following the unsuccessful automatic reclosure it was assumed that one unsuccessful delayed reclosure and one unsuccessful manual reclosure occurred. In the case of a temporary fault, which, based on the statistics in Table III, is unsuccessfully reclosed at the first attempt, it was assumed that a manual reclosure subsequently takes place, which is successful.

Circuit Breaker Clearing and Reclosing Times - The torsional response of a turbine-generator depends strongly on the times to clear the fault and to reclose the circuit breakers. For particular relaying and circuit breaker equipment these times are quite variable, requiring that they be treated in a statistical manner. No simple rule exists to determine the fault duration that maximizes or minimizes the torsional response, due to the multi-modal nature of the torsional response and the complex nature of the electrical torque.

The fault clearing time is the sum of the relaying and the breaker operating times. Both of these times are variable. The relaying time depends on many factors, which include: 1) relay type and design, 2) fault location and 3) fault current levels. The breaker operating time is affected by component wear and ambient conditions. The above factors influence the time to initial separation of the breaker contacts following a signal to actuate. Current interruption will occur after contact separation at the instant of a current zero condition. The arcing time may therefore span from practically zero to almost the period of one electrical cycle. The appearance of a current zero following contact separation is random, as it depends on the point on the voltage wave when the fault occurs. The statistics of fault clearing time are therefore complex and are generally unknown. It was assumed in this study that the deviation from the nominal clearing time will be ± half the period of one electrical cycle, and that these extremes represent ±3 standard deviations of a normal distribution about the nominal clearing time.

There is a lack of published information on statistics of the variability of reclosing time. However, since circuit continuity through a breaker immediately follows contact closing, it would appear that any variability in time to reclose would only depend on timer tolerance and tolerances associated with the moving components of the breaker. Breaker reclosing time is defined here as the time interval between fault clearing and breaker reclosing. Lower variability would therefore be expected for the reclosing time in comparison to the fault clearing time. It was assumed that this variability could be modelled by a normal distribution with three standard deviations corresponding to one quarter of the period of an electrical cycle.

Shaft Fatigue - With the assumptions that have been stated to this point, all important electric system related variabilities have been defined. Fatigue tests, as described earlier, exhibit significant amounts of scatter, making a statistical treatment of it desirable in this study. Shaft fatigue loss of life results can vary by more than an order of magnitude, depending on whether a fatigue curve is used based on average data or the lower bound of the scatter. The fatigue model that was utilized assumes a log-normal distribution of shaft stress about the average value for the given cycles to failure, as shown in Figure 5. This assumption is based on reference [10], which analyzed the scatter in alternating bending and steady state torsion test data. It was further assumed that the scatter in torsional fatigue test data spans ±2 standard deviations from the mean. With these assumptions, the discrete probability of a given cycles to failure curve is defined. Figure 5 shows, for example, that based on the above assumptions there is a 38% probability of the fatigue properties of a randomly selected shaft ranging between μ + 0.5σ, where μ and σ are the mean and standard deviation of the data, and only a 6% chance of fatigue properties being between ± 0.5σ of the lower bound of the test data.

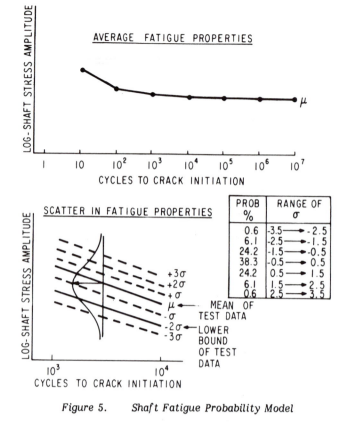

Figure 5.   Shaft Fatigue Probability Model

DISCUSSION OF RESULTS

Ranges of Loss of Fatigue Life for Various Types of Close-in Single Incident Faults

Before discussing the probability distribution of shaft fatigue expenditure, it is useful to first consider the ranges of this expenditure for a single fault incident. Figure 6 shows ranges of estimated shaft fatigue life expenditure for the population of seven machines described in Table I, assuming −2σ shaft fatigue properties. These ranges may be considered to be based on a significantly larger population, because the seven turbine-generators were selected to straddle the complete spectrum of torsional duty obtained in a planned switching study [5], involving many machine configurations. Figure 6 demonstrates that unsuccessful high speed reclosing on three phase or L-L-G faults under worst case conditions results in excessively high shaft fatigue duty. The worst case values of 30% and 100% should be interpreted as levels that are very damaging to turbine-generator shafts, since they also correspond to levels of shaft torque that result in gross yielding (permanent deformation) of shaft cross-sections. For these high overload torques, state-of-the-art fatigue models become invalid, due to violation of the linear-elastic assumption in the shaft nominal section, and also to shaft heating as a result of strain cycling. Simulation results which predict very high levels of fatigue life expenditure must therefore be subjected to additional engineering interpretation, and could easily result in the need for rebalancing, realignment, or a variety of other maintenance operations to the shaft system.

Figure 6 shows that sequential high speed reclosing reduces the maximum fatigue duty for major faults by a factor of about 6, in comparison to unrestricted high speed reclosing. The shaft fatigue damage under worst case conditions for a L-G fault is also shown to be reduced by at least an order of magnitude in comparison to other fault categories. It is seen, however, that for worst case reclosing conditions for the system configuration modelled, there is the potential for extracting as much as 2.6% of the shaft life for one L-G fault. This figure also shows that for each fault type and reclosing category the fatigue duty spans several orders of magnitude. This is due partly to differences in machine

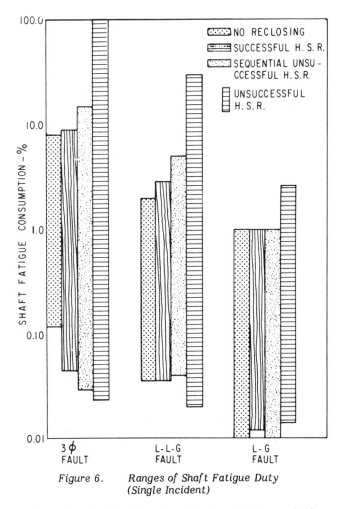

Figure 6. *Ranges of Shaft Fatigue Duty (Single Incident)*

configuration, but is also strongly dependent on whether torque cancellation or reinforcement occurs for the ranges of clearing and reclosing times that were analyzed. These ranges were chosen such that the mid-span clearing and reclosing times corresponded to peak values of fatigue duty, for each machine analyzed. Each time was varied independently in increments of one sixth of an electrical cycle.

Effect of Reclosing Time

Figure 7 illustrates the high variability in shaft fatigue life consumption for machine 3 as a function of reclosing

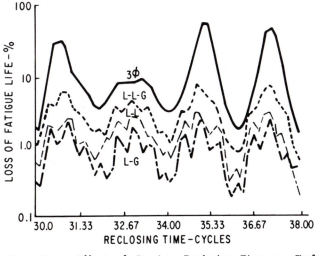

Figure 7. *Effect of Breaker Reclosing Time on Shaft Fatigue Duty*

time, for unsuccessful reclosing on all fault types. Reclosing time is defined here as the time interval in electrical cycles between fault clearing and breaker reclosure. In these simulations, the fault was cleared at the most unfavorable instant (in this example, 3 cycles), and the reclosure time varied over a wide range of 30 to 38 cycles to fully characterize possible variations in fatigue duty. The breakers were then recleared after three cycles. It is seen from Figure 7 that, for an assumed variability in reclosing time of one electrical cycle, the fatigue duty for each fault category may vary by over one order of magnitude. It is also seen that L–G and L–L faults produce similar ranges of fatigue life expenditure.

Cumulative Probability Distribution of Loss of Fatigue Life for a Close-in 3ϕ Fault

Figure 8 is a cumulative probability distribution of shaft fatigue consumption that was constructed for the case of unsuccessful high speed reclosing on a close-in three phase fault on machine 2. Seven dashed curves are shown that correspond to results using seven different fatigue curves, spanning ±3 standard deviations from average shaft fatigue data as shown in Figure 5. To construct each curve, 245 simulations were conducted, corresponding to various combinations of fault clearing time, reclosing time, and reclearing time. The time ranges selected for these variables, which were analyzed in increments of one sixth of an electrical cycle, are shown on Figure 8. Superimposed on these curves, and shown as a solid curve, is the cumulative probability distribution that has been weighted for the effects of scatter in fatigue data. This was constructed from the other curves by multiplying the probability values from each curve at a fixed fatigue life expenditure by the corresponding probabilities of having these fatigue properties (using the values in Figure 5), and then summing the results. This produces one point on the weighted probability curve. This process was repeated at several other fixed levels of fatigue expenditure to define a curve. An example of the use of this weighted distribution is that it indicates there is approximately a 5% chance of the shaft fatigue life expenditure exceeding 75% for one close-in three-phase fault that is unsuccessfully reclosed. However, the probability of experiencing a three phase fault in the first place is low, with the even lower probability that it is experienced close to the unit and unsuccessfully reclosed. The inclusion of all these statistical effects at one time will be discussed later.

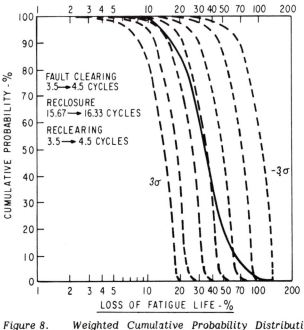

Figure 8. *Weighted Cumulative Probability Distribution (Unsuccessful H.S.R. on Close-in 3ϕ Fault on Machine 2, single incident)*

Cumulative Probability Distribution of Loss of Fatigue Life for Other Close-in Types

The cumulative probability distribution that has just been discussed, associated with fault clearing and reclosing times and shaft fatigue strengths, is shown in Figure 9 for a number of other close-in single incident cases; namely, L–G, L–L–G, and three-phase faults for both successful and unsuccessful reclosing. These curves illustrate the large differences in fatigue life expenditure for the various fault types and between successful and unsuccessful reclosures for machine 2. The L–G fault duties for this machine are shown to be nearly two orders of magnitude less than those of three-phase faults, and more than an order of magnitude less than those of L–L–G faults. Similarly, unsuccessful reclosure produces damage more than an order of magnitude greater than successful reclosure.

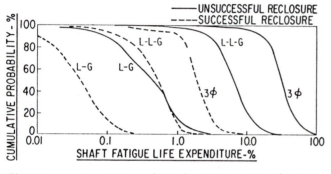

Figure 9.   Comparison of Shaft Fatigue Duty for Successful and Unsuccessful H.S.R.

For unrestricted high speed reclosing, some contributions from all of the events shown would be expected. The effect of implementing a delayed reclosure practice would be, in effect, to delete all of the solid curves from consideration. Similarly, a selective reclosing practice, blocking reclosing for the major faults, eliminates all of the multi-phase fault curves shown on Figure 9 from consideration, replacing them with delayed reclosures.

Effect of Fault Location

The curves shown in Figure 9 are worst case events from the standpoint of fault location, becoming less severe as the fault locations become more remote. This is illustrated in Figure 10 for machine 1, where the fault location is expressed

in per unit of the total length of each line. It is evident that for a uniform distribution of faults on the lines, the majority of faults will occur with much lower loss of life than the worst case near the plant. The L–G fault duties drop off faster than the three-phase as the fault moves away from the station, because of relatively higher zero-sequence line impedance for single line to ground faults. Another effect noticeable in this figure is the increasing fatigue duty for far-end faults. This effect is characteristic of this type of system, in which a three-phase far bus fault is nearly as effective as a near-end fault in causing a step change in machine average torque output. Other system types not having in effect a remote busing of all lines out of the station would experience less fatigue duty for remote faults.

Shaft Fatigue Over Forty Year Exposure Period

While the foregoing curves portray great differences in duties for different events, their significance depends on their frequency of occurrence over the assumed on-line exposure period. The assumptions as to fault type distribution and proportion of initial automatic and subsequent delayed reclosures that are successful are given in Table III and are critically important in assessing the potential for shaft fatigue life expenditure. Assuming a uniform distribution of fault occurrences along the lines at the rate of 4 faults/100 mi/yr, Monte Carlo simulations of loss of shaft fatigue life over a forty year exposure period were made for the turbine-generator units of Table I. The results for the unit in the study population which generally had the highest fatigue duty are given in Figure 11 for unrestricted high speed reclosing and several other reclosing practices.

The various reclosing practices are seen to result in widely different machine duties. If it is assumed, for example, that 30% of the shaft fatigue life is reserved for other possible events such as bus faults, out-of-phase synchronizing accidents, and the more remote line faults, then the probabilities, from Figure 11, of there being 70% or greater shaft fatigue expenditure after forty years of on-line exposures, are as follows:

Unrestricted H.S.R.	21%
Selective H.S.R.	0.9%
Sequential H.S.R.	0.1%
Delayed reclosing (10 sec.)	0.03%
No reclosing	<0.01%

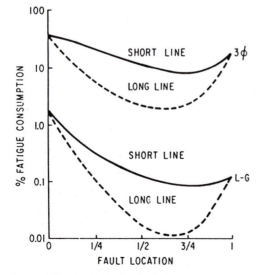

Figure 10.   Effect of Fault Location on Study System on Shaft Fatigue Duty (single incident).

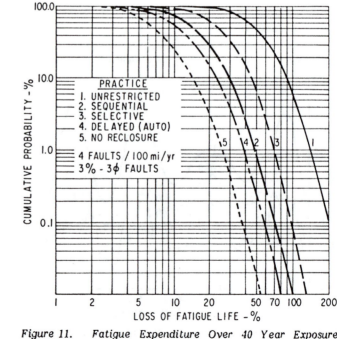

Figure 11.   Fatigue Expenditure Over 40 Year Exposure Period (3% 3ϕ Faults)

250

In general, unrestricted high speed reclosing results in a relatively high probability of expending more than the available shaft fatigue life and is therefore not recommended. Other factors discussed earlier such as possible gross yielding and permanent deformation of the shaft from worst case torque levels also reinforce this recommendation.

Effect of Different Fault Frequency and Exposure Period Assumptions

The curves of Figure 11, if properly rescaled, are equally applicable to a wide range of assumptions as to fault frequency and exposure period since these linearly affect the number of events and the shaft fatigue expenditure. For a doubling of the assumed faults/100 mi/yr to 8, for example, one needs only to double the horizontal scale values. Similarly, a reduction in assumed exposure period to thirty years and an assumed increase in fault frequency to 5/100 mi/yr are approximately offsetting effects, so that Figure 11 would be directly applicable.

The effect of fault frequency can therefore be seen to be quite important. For example, a doubling of the levels assumed to 8/100 mi/yr increases the probability of exceeding an allotted 70% fatigue life to the following, as compared to 4/100 mi/yr:

	8/100 mi/yr	4/100 mi/yr
Unrestricted H.S.R.	70%	21%
Selective H.S.R.	20%	0.9%
Sequential H.S.R.	4%	0.1%
Delayed reclosing	1.5%	0.03%
No reclosing	0.1%	< 0.01%

The foregoing points out the sensitivity of high speed reclosing, particularly to a high mean fault rate. Delayed reclosing is the obvious preferred practice from a shaft fatigue standpoint for a high rate of fault occurrence. Selective reclosing is also seen to be much less potentially damaging than unrestricted high speed reclosing, and might be considered acceptable for other assumed conditions, such as thirty years of on-line service. In some cases, very high fault rates are a result of reductions in grounding, counterpoise, and line shielding considered acceptable by virtue of the employment of high speed reclosing. The foregoing indicates the direct relation between transmission line design and the potential for shaft fatigue duty, and indicates a need for careful consideration of the trade-offs involved.

Effect of Proportion of 3φ Faults

Another parameter investigated was the assumption for the proportion of 3φ faults. The data for Figure 11 assumes 3% of the faults are three-phase and of these, 60% are temporary. The effect of increasing the number of three-phase faults to 4% is shown in Figure 12. It raises the probability of exceeding 70% shaft life to 30% for unrestricted high speed reclosing (for a fault rate of 4/100 mi/yr). The probability of exceeding 70% shaft fatigue life expenditure over a forty year exposure period is seen to be 1.4% for selective H.S.R., which is considered acceptable. This and other information is summarized in the following table:

	3% 3φ faults	4% 3φ faults
Unrestricted H.S.R.	21%	30%
Selective H.S.R.	0.9%	1.4%
Sequential H.S.R.	0.1%	0.2%
Delayed Reclosing	0.03%	0.09%
No Reclosing	< 0.01%	< 0.01%

Sequential reclosing produces the lowest fatigue duty for the H.S.R. practices and the electrical system that were evaluated. However, the sequential type is least amenable to conclusions on a general basis since its applicability is highly dependent on the specific network arrangement and particularly to locations of other generating stations which might be affected.

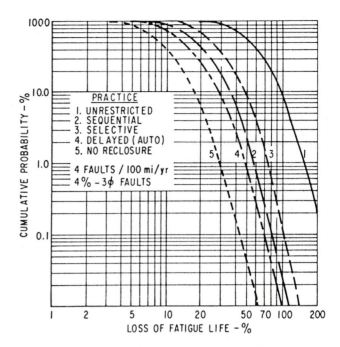

Figure 12. Fatigue Expenditure Over 40 Year Exposure Period (4% 3φ Faults)

FURTHER WORK

The potential for producing fatigue damage to turbine buckets by selective H.S.R., as a result of stimulating them in supersynchronous bucket-rotor vibration modes, has not yet been fully evaluated. However, preliminary results indicate that the likelihood of causing significant turbine bucket fatigue damage as a result of selective H.S.R. is very slight.

Further study on turbine bucket fatigue as a result of electrical disturbances is underway, as well as extending this shaft fatigue study to encompass a wider population of machines and several other transmission system configurations.

CONCLUSIONS

1. Turbine-generator shaft fatigue duty resulting from high speed reclosing is highly sensitive to many parameters, among these being the precise fault clearing and reclosing times, fault types and locations. The random variability of these requires a probabilistic approach for estimating the shaft fatigue expenditure over the assumed exposure period of the turbine-generator.

2. The calculated probabilities of expending more than 70% of the shaft fatigue life after a forty year on-line exposure period, for the turbine-generator that generally exhibited the highest shaft fatigue duty in the population studied, are 21% for unrestricted H.S.R., 0.9% for selective H.S.R., 0.1% for sequential H.S.R., and 0.03% for delayed reclosing. These values are based on 4 faults/100 mi/yr, the fault statistics in Table III, and on the assumed electrical system model.

3. High speed reclosing for nearby three-phase or L–L–G faults can result in loss of a high proportion of the shaft fatigue life in a single incident, as shown in Figure 6. While the probability of three-phase and L–L–G faults occurring under worst case conditions is very low, the consequences not only for high shaft fatigue life expenditure but also for gross permanent shaft deformation should be weighed against the expected benefits of reclosing for major multi-phase faults. These benefits appear marginal considering the high proportion of such faults which are permanent.

4. High speed reclosing for L-G or L-L faults (selective H.S.R.) does not pose an excessive risk of turbine-generator shaft damage for the assumed electrical system configuration and fault statistics, and is generally considered an acceptable practice.

5. Of the reclosing practices studied, delayed reclosing results in the least penalty in terms of added shaft fatigue duty. As such, it would be more generally applicable over a broad range of system configurations and fault statistics.

RECOMMENDATIONS

1. The following are three recommended circuit breaker reclosing practices which significantly reduce the risk of excessive turbine-generator shaft fatigue duty, and appear to be generally applicable and relatively simple to implement:

   (a) Delayed ($\geq$ 10 seconds) automatic reclosure for all fault tripouts.

   (b) Sequential high speed reclosing if the remote end of the line is not electrically near turbine-generator units.

   (c) Selective high speed reclosing for L-G or L-L faults only, supplemented by delayed ($\geq$ 10 seconds) automatic reclosure for other multi-phase fault types.

2. Delayed reclosing suggested in 1(a) should be considered particularly for systems having unusually high rates of fault occurrence.

3. The selective reclosing practice suggested in 1(c), while generally insuring against excessive turbine-generator shaft fatigue duty, also has desirable characteristics from the system security standpoint in two respects: first, in providing the fastest possible circuit restoration for single phase to ground faults which comprise the majority of temporary faults; and second, in eliminating the potential for transient instability from unsuccessful reclosing into major multi-phase faults, particularly during extreme system emergency conditions.

4. If, in considering the application of selective H.S.R., it is judged in a particular application that the assumptions made in this study are unrepresentative and unconservative, then an individual study is recommended.

5. Unrestricted H.S.R. is not generally recommended because of the potential for high fatigue life expenditure in a single incident. If, however, this practice is being considered in a particular application or is in use, it is recommended that it be studied on an individual basis.

6. To minimize shaft torsional fatigue duty, testing of a line after an unsuccessful automatic reclosure should be from the remote terminal. A similar philosophy should also be considered for the delayed reclosure practice of 1(a).

ACKNOWLEDGEMENTS

The authors are indebted to many individuals who assisted in developing the results for this paper. We particularly wish to acknowledge the contributions of the following: Paul Albrecht, Shirley Merkert, and Robert Stickle.

REFERENCES

1. J.B. Tice and P.G. Brown, "Circuit Breaker Reclosing Stability and System Considerations"; Pennsylvania Electric Association Relay Committee Meeting, Stroudsburg, Pennsylvania, October, 1973.

2. P.G. Brown and R. Quay, "Transmission Line Reclosure - Turbine-Generator Duties and Stability Considerations"; Texas A&M Relay Conference, April, 1976.

3. J.S. Joyce, T. Kulig, D. Lambrecht, "Torsional Fatigue of Turbine-Generator Shafts Caused by Different Electrical System Faults and Switching Operations"; 1978 IEEE Winter Power Meeting.

4. J.W. Ballance, et. al., "Subsynchronous Resonance in Series Capacitor Compensated Transmission Lines"; Paper No. T73 167-4, PAS Vol. 92, Sept./Oct. 1973.

5. D.N. Walker, S.L. Adams, R.J. Placek, "Torsional Vibration and Fatigue of Turbine-Generator Shafts", Paper No. F79 756-8, presented by title at IEEE-PES Summer Meeting, 1979.

6. AIEE Committee report, Report of Joint AIEE-EEI Subject Committee on Line Outages; AIEE Transactions Vol. 71, Pt.III, January 1952, page 43.

7. IEEE Subsynchronous Resonance Working Group, "Proposed Terms and Definitions for Subsynchronous Oscillations"; Paper #F79 254-4, presented at IEEE-PES Winter Meeting, 1978.

8. "Extra High Voltage Line Outages"; IEEE/EEI Committee Report 1966, Transactions Power Apparatus and Systems, Vol. 86, #5, May, 1967, page 547.

9. J.E. Beehler, "Seven Year Summary of EHV Power System Outages"; CIGRE Committee Group 13 report, Sept. 1977.

10. D. Kececioglu, L.B. Chester, T.M. Dodge, "Combined bending - torsion fatigue reliability of AISI 4340 steel shafting with K_t = 2.34;" Paper #74-WA/DE-12, presented at the Winter Annual Meeting of ASME, Nov., 1974.

Discussion

R. J. Harrington (The George Washington University, Washington, D.C.): The paper describes a method of determining the probable remaining shaft life of turbine-generator sets subsequent to their exposure to a series of nearby fault transients and gives certain recommendations for high speed auto-reclosure practice.

The authors' claim that it (the paper) advances the state-of-the-art, of presumably justifying some form of high speed reclosure, may well be reasonable. However, a method of obtaining really useful information regarding the metallurgical state of the shaft(s) of a particularly important machine or group of machines has yet to be described. By the time crack initiation has been observed, the shaft-life may reasonably supposed to have been unacceptably reduced.

All utilities will surely agree that the unscheduled loss of more than one out of a handful of critical sets within their region will cause them considerable, if not impossible, operating difficulties. It is not necessarily true, as is asserted in the paper, that shaft fatigue damage is only observable after crack initiation.

Reduction in shaft life is probably apparent in the change in the internal crystalline structure and crystal boundaries from their initial states in the original unstressed shaft. While microscopic values of crystal binding energies can be defined, they are not easily measurable. Macroscopic effects which reveal themselves in such phenomena as, typically, magnetostriction can be observed and could be used to determine the changes in the crystal structure which take place during the life of the shaft. Due to the hardness of the shafts used, magnetrostriction is not especially pronounced in turbine generator sets, so the use of this effect would require measurement techniques of an advanced nature.

Referring to Table II of the paper, the values of the log-decrement damping factors were obtained from station torsional vibrational tests conducted over the past six years. Could the authors comment on both of the following:

   (i) if the usual type of chart recorder was used to obtain these results, the inaccuracy and probably relatively poor frequency response of the recording equipment, and

   (ii) The approximate nature of the extrapolation process mentioned in the paper

would lead to doubt concerning the values quoted particularly for modes #3, 4, and 5?

What effect, if any, would the operation of a fast-acting governor system or the removal of the operating point from full to, say, half-load have on the values for the modal damping? How does the level of the excitation, i.e. over/under-excited operation contribute to the damping values at full load?

Manuscript received February 4, 1980.

J. S. Edmonds (Electric Power Research Institute, Palo Alto, CA): Debate has ensued over turbine-generator shaft system integrity vs. the various switching and reclosing techniques practiced on electrical systems today. This paper has served to bridge a very significant gap between conclusions drawn from the rash of formal calculations that have predicted catastrophic shaft failures, and the excellent, relevant operating history of these large turbine-generators.

Why did the authors choose to make recommendations on specific circuit breaker practices? If the review techniques for utility-specific situations had been recommended, would this then, have been an effective way for eliciting individual system-specific studies?

How much is actually known about turbine-generator/system interactions on an incident-by-incident basis?

At Electric Power Research Institute (EPRI) a program has been established to gain actual turbine-generator response and electric system incident data from a large number of machines and at a wide variety of locations.

Manuscript received February 20, 1980.

A. F. Armor and **J. B. Parkes** (EPRI, Palo Alto, CA): This paper introduces probabilistic methods into the analysis of shaft fatigue life expenditure following a power circuit breaker reclosure. The use of Monte Carlo analysis in predicting probable shaft fatigue is novel and overcomes to a large extent the limitation of significant scatter in the many parameters involved in the analysis. A good qualitative assessment of the dangers inherent in a selection of reclosing procedures has been derived by the authors, from which certain conclusions and recommendations are drawn.

It is both the strength and the weakness of statistical methods of analysis, that parameters are assumed to vary with a given probabilistic distribution. On the one hand such widely varying data as the measured fatigue properties of shaft materials can be factored into the analysis effectively, on the other hand the conclusions reached are best estimates rather than fully definitive and lead the reader to review the early axioms made in the analysis. Such initial assumptions as the use of 1.0 p.u. load over the machine lifetime, a 40 year life for the unit, an assumed mean rate of faults of 4/100 mi/yr. all have an effect on the final conclusions and the authors have recognized this and drawn our attention to it. The mean rate of faults appears to be particularly important and moreover is highly variable with geographical location and probably from year to year also. In light of this, the prediction of future fatigue life expenditure for any particular turbine-generator shaft-system would seem tenuous and the authors' comments on this would be appreciated.

Whatever the absolute conclusions on fatigue life expenditure, the problem certainly warrants the attention of both manufacturer in design and utility in operation. The recommendations relating to modified reclosing practices may only be one aspect of the solution to this problem. Questions relating to advanced monitoring techniques are also appropriate. Monitoring techniques for turbine-generator shaft systems fall into two general categories which may be designated indirect and direct methods. Indirect methods do not attempt to monitor crack formation per se. Instead system disturbances are translated into mechanical response using spring-mass models of the shaft system and in turn this leads to predictions on fatigue life expenditure as described, for instance, by this paper. On-line torsional monitors for example, provide an estimate of the probable shaft damage for each incident without ever being able to assess actual damage (if any). This is still a very valuable function and EPRI projects RP1531 (Determination of Torsional Fatigue Life of Large Turbine-Generator Shafts) and RP1746 (Turbine-Generator Shaft Torsional Field Monitoring Program) recognize the value of indirect monitoring to the Utilities. Direct methods, on the other hand, observe and record actual shaft deterioration and crack formation. One example involves the monitoring of acoustic emissions (sounds of low level but with frequency components into the MHZ range) from turbine-generator shafts. EPRI project RP734 in the Fossil Plant Performance and Reliability Program has for more than 1 year, been recording ultrasonic data from 41 piezoelectric transducers, installed throughout the Brayton Point plant of New England Power Service Company. Nine of these sensors were located on the bearing caps of the turbine-generator-exciter shaft systems. While this data is still bring reduced, there is encouragement that acoustic emission listening devices will eventually provide an early warning system for crack formation in a rotating shaft system. Considerable characterization work is required and is being supported by EPRI projects RP734 and RP1266-14 in conjunction with on-going tests at TVA designed to detect and locate active crack flaws in the rotors of TVA steam turbines.

In drawing our attention to the probabilistic nature of fatigue life expenditure prediction the authors have emphasized the importance of on-line monitoring systems. This in no way detracts from this paper which has arrived at significant conclusions related to high speed reclosing practices, quite valid for the parameters stated in the paper. The authors' comments relative to indirect and direct methods of monitoring shaft fatigue would be appreciated.

In summary, this is a timely and well written paper and a valuable contribution to the literature on machine/system interactions.

Manuscript received February 21, 1980.

C. A. Mathews and **S. B. Wilkinson** (General Electric Co., Philadelphia, PA): The authors are to be congratulated for their statistical approach to the subject of shaft fatigue damage, and for the very thorough study they have conducted. The results indicate that there is a very definite risk involved in the use of unrestricted high speed reclosing. The results also indicate the desirability of having available a relay scheme for reclosing that can discriminate between those conditions where high speed reclosing may be used and those where it should not be used.

In the final section of the paper the authors have recommended three alternative reclosing practices – delayed, sequential or selective. The method of presenting these recommendations gives the impression that these are the only practices which should be used. It is possible to devise other reclosing schemes which provide adequate security against shaft damage without being quite as restrictive in blocking high speed reclosing. This discussion will review the recommended practices, and will present an alternative proposal.

A. Delayed Reclosing

The first proposal in the paper, delayed reclosing for all types of faults, provides maximum protection for the machine, but sacrifices all of the advantages of high speed reclosing. This practice is more restrictive than necessary, but it is easy to implement. It is logical to favor other approaches which will permit high speed reclosing under some conditions.

B. Sequential Reclosing

The procedure described in the paper permits high speed reclosing at terminals remote from the generator, with the breaker at the generator end being reclosed only after check relays (usually voltage and phase angle) indicate that the fault has been cleared. The authors state that "one to several seconds delay is inherent in this type of reclosing." This is apparently based on the fact that it has been common practice to use synchronism-check relays that are slow to operate. However, relays are now available which will make a phase angle check in four to seven cycles, provided that the voltage is near normal. When this faster relay is used, the duration of a line outage due to a transient fault is not much longer than it would be if both ends were reclosed high speed.

C. Selective Reclosing

The selective reclosing scheme described in this paper is one that bases the selection on fault type. It permits high speed reclosing for two fault types (L-G and L-L) and inserts a delay of at least 10 seconds for other fault types.

The authors have pointed out in their conclusions that shaft fatigue is sensitive to many different parameters, and fault type is only one of those parameters. Fault location and system configuration are parameters that may have an equal effect on determining the severity of the fault. Because it is the combined effect of all these factors that determines the fault severity, and consequently the amount of damage to the machine, a relay scheme that responds to both fault type and fault severity would appear to be superior to one that responds only to fault type.

Another condition to consider is that of an evolving fault, where the fault at the time of reclosing involves more phases, or conductors, than it did when the fault was cleared. The use of a reclosing scheme based on fault severity offers the potential of blocking reclosing on some close-in single phase faults which may evolve during the reclosing period to LLG or three phase faults.

Proposed Alternative Reclosing Scheme

A reclosing scheme is proposed which responds both to fault type and fault severity. It has the following features:
(1) It blocks high speed reclosing for all three phase faults, and permits either delayed automatic reclosing or manual reclosing.
(2) It determines the severity of a fault at each terminal, for other fault types, as indicated by voltage and current measurements at that terminal.
(3) It permits high-speed reclosing at the terminal(s) if the fault is less severe than that established by the relay setting.
(4) It permits sequential reclosing at the terminal where the fault was determined to be more severe than the relay setting, assuming that the remote end has reclosed successfully to establish voltage on the line.

This proposed reclosing scheme can be implemented using protective relaying technology that is presently available. It combines some of the features of each of the schemes listed in the paper and permits high speed reclosing under more conditions while still providing an adequate level of machine protection. Do the authors agree that an expanded scheme such as this could be used in place of their recommended schemes?

We have one additional question that is not related to reclosing. Could the authors comment further on the effect of fault duration on shaft fatigue? To be more specific, the results of some other investigations have indicated that one cycle fault clearing is considerably more severe than three cycle fault clearing for one particular machine. Did the test program conducted by the authors contain any tests which would verify this observation?

Manuscript received February 22, 1980.

C. W. Taylor (Bonneville Power Administration, Portland, OR): The authors have made a significant contribution to evaluating the effects of reclosing on turbine-generator shafts. Hopefully, the benefits of using probabilisitic methods in evaluating shaft fatigue and transient stability will become widely appreciated in the future.

I have several questions regarding the effect of generator loading on shaft fatigue. The authors state that both the damping and the magnitude of electrical stimulus are maximum at full load — these factors tending to offset one another. Another factor not mentioned is the effect of mean shaft stress on fatigue. Would not this tend to produce less fatigue at lower generator loading? In this regard, would fast valving have a beneficial effect?

The recommendations for selective and sequential reclosing are very reasonable. For stability reasons, BPA has used both methods extensively for about the last 10 years. For selective reclosing, a simple inexpensive three-element plunger type overcurrent relay is used. Elements are connected in each phase and pick-up of more than one element due to multi-phase faults will block reclosing. Sequential reclosing is used near generating plants with the remote end reclosed first (for single-phase faults only). Generally, only a "hot line" check is necessary at the near end.

One reason for high-speed reclosing is the possibility of misoperation of protective relaying on an unfaulted line adjacent to a faulted line. Opening of two lines integrating generation for a single temporary fault could result in instability. In contrast with several other utilities, BPA experience in this regard has been excellent. Over a 5½-year period, we have had only 5 relay misoperations of this type out of a total of 1401 230-kV and 500-kV system faults. This success is probably primarily due to use of microwave radio transfer trip protective relaying of transmission lines.

Since BPA and several other utilities are considering more extensive use of single-pole switching, we look forward to the future work mentioned on supersynchronous vibration modes.

We would also be interested in possible future probabilistic evaluation by the authors of weaker systems that are unstable for some faults without reclosing or other stability enhancement measures. This would be representative of many situations in the Western North American interconnection where discrete stability controls, such as fast valving, unit dropping, dynamic braking, or single-pole switching are sometimes necessary.

Manuscript received March 3, 1980.

R. D. Dunlop and **M. C. Jackson** (American Electric Power Corp., New York, NY): This paper represents a significant contribution to the growing body of literature on this subject of vital concern to the industry. By taking into account: the statistical nature of the disturbance (fault type, location, persistance and incidence rate over a selected period of years); the clearing operations (timing of opening and reclosing of circuit breakers); and the scatter in the torsional fatigue properties of the shaft material, the authors provide a more meaningful basis for carrying out sensitivity studies as an aid in comparing generating unit designs, examining power system reclosing practices and in exploring the effects of configuration on torsional duties. With regard to this latter area of system configuration, it is most unfortunate that the authors have chosen a single, somewhat unrealistic configuration on which to base the conclusions and recommendations in this paper.

The authors have indicated that further work is in progress to explore the effects of system configuration. In light of the particular configuration chosen for this study, these discussors would urge the reader of this paper to exercise genuine caution in interpreting the section on "Recommendations". We would further urge the authors to insert this qualification, with regard to configuration, in the recommendations section. In fact, we feel that the value of the paper would have been greater if the authors had stated their conclusions — noting other work in progress and stating problems to be resolved — and left their recommendations for consideration by an appropriate industry committee where the views of other manufacturers and the broader concerns of the users, for overall system security and reliability, could be taken into account.

For the particular configuration chosen in this study, we agree that the parameters chosen for the transformer and line impedances and the short circuit strength of the receiving-end system are reasonably representative. The real problem with this configuration, consisting of one generating unit and only two transmission exits, is that in actual operation high speed reclosing of any type probably would not be permitted. In the absence of a high speed synchronizing device, it would be prudent to forego HSR on either line so as to preclude an out-of-phase reclosing incident in the event of a fault on one line and the remaining circuit is tripped due to relay and/or communication problems. Beyond this, even in the case of a power plant with more than two exists, it has been a longstanding practice on the AEP system to inhibit HSR whenever the number of circuits is reduced to two by prior switching events. Exceptions to this are permitted where single-phase clearing is employed to enhance stability for temporary single line-to-ground faults on a plant with limited exits or where high speed synchronizing relays, currently under review, may be utilized in the future.

The results of this paper are still very informative as a basis for exploring, in the context of a "worst-case" type configuration, the relative influence of various assumptions concerning machine design practices, disturbance characteristics and system switching practices. In our earlier work in this area, we purposely chose a similar severe configuration – based on an actual situation – because we wished to investigate the methodologies of fatigue assessment and explore the influence of switching practices in the context of a severe configuration.[A,B,C] However, such results – being based on a configuration in which HSR would not generally be permitted in actual practice – should not be used as a basis for recommendations in this area.

In discussing reclosing practices, the authors cited several advantages to "delayed reclosing". The reader is referred to our earlier work in which the advantages of "unrestricted HSR" are discussed at some length.[B] These advantages include the rapid restoration of overall system integrity, minimizing the loss of generating units due to instability, and the rapid restoration of an adjacent circuit which may be undesirably tripped at one end due to a variety of possible relay and communication mis-operations. (Incidentally, the authors point regarding enhancement in transient stability performance for delayed reclosing must be viewed with regard to the probability of successful versus unsuccessful reclosure. Since – as the authors note – over 80% of transmission line faults are temporary in nature, it would seem that unrestricted HSR would lead to an even greater overall transient stability enhancement.)

We would appreciate the author's comments on a number of matters raised in this paper. In the fatigue model used, did the $\pm 2\sigma$ deviations only relate to the scatter in the test data? In taking other effects—such as stress concentration, size scaling of test data, surface effects, etc.–into account, was the assumed mean stress lowered unilaterally or were these other effects also treated in a statistical manner? How do the authors treat couplings in their analysis? Wouldn't it be the case that in many designs the limiting piece of equipment is the coupling bolts, and would they not actually help to relieve some torques from being transmitted to the shaft system?

Extrapolation of the results to a 40-year life should be questioned in more than one request. First of all, it is unlikely that the same severe configuration or switching practices would apply over such a long period. Frequently, the system changes over time and generally in the direction to reduce switching duties. Also, the size of a new unit in relation to system load may make it of such critical importance in the early period of its life that overall system reliability considerations may imply a greater need for high speed reclosing. Secondly, it seems as though the fatigue expenditure curve, given in figures 11 and 12 of the paper, assume that the turbine-generator is operated throughout its 40-year life without being inspected. What would happen to these curves if a systematic inspection process would be carried out periodically to identify and remove any surface cracks before they are allowed to propagate?

Some manufacturers believe that 1800-rpm machines are, or can be designed to be, more capable of withstanding system switching transients than 3600-rpm machines. Have the authors found this to be true in these particular studies or in their broader experience in general?

Again, we congratulate the authors on this excellent paper and the greater application of statistical methods for which it should be primarily noted. Since this paper will be a matter of permanent record in the industry, we believe the future readers of this paper should be aware of many other references to work in this area. In particular, a bibliography in this area, which was presented during 1979 and will eventually be published in the transactions, mentions many other references in addition to some of those cited by the authors.[D] Also, a paper which has just been prepared for the 1980 CIGRE Conference by a team of international authors, includes an abundance of references. The purpose of this paper is: to summarize the state of the art in assessing the fatigue impact of electrical system transients on steam turbine-generators; to illustrate how various machine and system factors can influence torsional oscillations; to highlight the most significant industry issues and concerns; and to indicate the highest priorities for future work in the industry with regard to research and development, engineering studies, field testing and long term monitoring needs.[E]

REFERENCES

[A] M. C. Jackson, et al., "Turbine-Generator Shaft Torques and Fatigue: Part I – Simulation Methods and Fatigue Analysis." *IEEE Trans.* PAS, Vol. PAS-98, No. 6, Nov./Dec. 1979, pp. 2299-2307. (Discussion pp. 2314-2328).

[B] R. D. Dunlop, et. al., "Turbine-Generator Shaft Torques and Fatigue: Part II–Impact of System Disturbances and High Speed Reclosure." *IEEE Trans.* PAS, Vol. PAS-98, No. 6, Nov./Dec. 1979, pp. 2308-2314. (Discussion pp. 2314-2328).

[C] M. C. Jackson, S. D. Uwans, "Turbine-Generator Shaft Torques and Fatigue: Part III–Refinements to Fatigue Model and Test Results." 1979 IEEE Summer Power Meeting paper F79 607-3.

[D] P. A. Rusche, "Turbine-Generator Shaft Stresses Due to Network Disturbances–A Bibliography with Abstracts" IEEE Paper F79 752-7.

[E] R. D. Dunlop, S. H. Horowitz, J. S. Joyce, D. Lambrecht, "Torsional Oscillations and Fatigue of Steam Turbine-Generator Shafts Caused by System Disturbances and Switching Events", 1980 CIGRE Paper No. 11-06. (To be presented at the 1980 CIGRE Session; Paris, France; August 27-September 4, 1980).

Manuscript received March 4, 1980.

T. J. Hammons (Glasgow University, Glasgow, G12 8QQ, U.K.): The authors are to be commended for their effort in developing statistical methods in evaluating the effect of circuit breaker reclosing practices on fatigue life expenditure of turbine-generator shafts.

It should be noted that calculation of rotor fatigue life expenditure demands the determination of time responses for mechanical torque over an extended period of time (e.g. over several seconds). Comparing the results of calculations using constant and variable damping indicates it is essential to employ variable damping in estimating rotor fatigue life expenditure on the shaft. This is because the number of vibrations affecting rotor fatigue life expenditure depends inversely on the mechanical damping of the vibrating system. Damping increases significantly when shaft oscillations become large.

Grid network and machine electrical damping, including voltage regulator and governor damping is at least an order of magnitude greater in typical networks than steam damping. Steam damping is due to the hyperbolic nature of the turbine torque/speed characteristic. Material damping, including damping due to micro slippage at couplings and blade fixings and movement of lacing wires which stiffen LP blades is less than steam damping except when the vibrations become excessively large. Material damping is due to the energy loss per twist cycle and is greatly dependent on stress of the outer fibres of the turbine shaft.

Turbine generators should be designed to withstand clearance of 3-phase faults close to the HV transformers in addition to being designed to withstand mal-synchronization at 120° displacement angle without significant expenditure of rotor fatigue life (i.e. 5%) (A). Transmission systems should be operated in such a way that stress resulting from the most-severe switching event would not exceed the stress for which the machine was designed. Coupling bolts are often designed to be weaker than the adjacent shaft and their degree of life fraction expenditure will be greater than that of the shaft. Presence or absence of coupling bolt distortion by bending or stepping can serve as a measure of whether significant torques have been experienced during the operating life of the machine.

Could the authors compare rotor fatigue life expenditure obtained by a step-by-step solution of the complete differential equations and using variable damping with that obtained using the modal approach with (i) constant, and (ii) variable damping?

The step-by-step solution yields maximum and minimum torque per stress cycle which can be used using rain flow techniques and S/N curves to estimate rotor fatigue life expenditure provided the damping is large (i.e. there is significant grid network and electrical machine damping). It is precluded where the overall damping is low (viz. following generator disconnection) on account of integration error and the longer computation in real time this would incur. Under these circumstances a modal solution using variable damping is more appropriate.

Do the full-load damping coefficients reproduced in Table II apply only for the amplitude of torque vibration for which they were measured? Could the authors' indicate for what amplitude of torque vibration the respective coefficients were measured? Could they also indicate how the overall damping coefficient for each mode of vibration would vary as the amplitude or torque vibration is raised?

A comparison of range of shaft fatigue damage using alternative methods of simulating damping would be timely and appropriate for the more severe events. Cumulative probability distributions allowing for selection and simulation of damping weighted by their probability distribution and which express probability that loss of fatigue life exceeds specified values might be significantly different in this case.

REFERENCE

[A] Daltry, J. H., Hammons, T. J. and Muscroft, J., "Stressing and fatigue of turbine shaft couplings and blade roots due to electrical system disturbances", presented at the IMechE Steam turbines for the 1980's conference, London, England, October 9-12, 1979, paper C229/79, pp. 225-232.

Manuscript received March 4, 1980.

John S. Joyce (Utility Power Corporation, West Allis, WI) and **Dietrich Lambrecht** (Kraftwerk Union AG, Muelheim/Ruhr, Germany): Congratulations are due to the authors for their clear summary of the results of their extensive investigations into the highly complex subject of the torsional fatigue of steam turbine-generator shafts due to automatic reclosing practices. The paper makes a significant contribution to the subject by studying the relative shaft torsional impact associated with high-speed versus delayed reclosing techniques. In evaluating different methods of reclosing the authors omitted consideration of selected-pole circuit-breaker operation which is used on a standard basis in European networks. Single-pole high-speed reclosing is gaining increased acceptance for EHV systems in this country. One of its advantages is that it helps to mitigate the torsional impact on nearby turbine-generators.

The value of the paper lies mainly in its careful differentiation between the faults being combated by different triple-pole tripping and reclosing practices. We are gratified by the fact that on adopting this approach the authors have arrived at the same basic conclusions as we published two years previously in the paper which the authors kindly list as their Reference 3. In agreeing that high-speed reclosing for line-to-ground and line-to-line faults, for which they introduce the new term "selective" HSR, does not generally pose an excessive risk of turbine-generator shaft damage, whereas high-speed reclosing for nearby three-phase faults does, the authors appear to concur with our earlier published position that high-speed reclosing should normally only be blocked in the case of close-in three-phase faults [A]. Since such three-phase short circuits occur very infrequently, well over 90% of the discussion on the risk of serious turbine-generator damage as a result of high-speed reclosing is settled when manufacturers recognize that high-speed reclosing in conjunction with all the much more common system faults does not contribute to unjustifiably high cumulative shaft fatigue increments.

Despite the agreement in principle between the authors' and our own conclusions and recommendations, the results of the two independent investigations into the torsional implications of high-speed reclosing seem to differ appreciably in several respects.

While the ranges of shaft fatigue duty due to unsuccessful high-speed reclosing shown in Fig. 6 exhibit substantially the same relationship for three-phase, line-to-line and line-to-ground faults as published by us [A], the maximum potential for fatigue consumption indicated by the top of the vertical bars in the case of line-to-line and line-to-ground faults is more than three times and an order of magnitude greater respectively than determined by us. Would the authors please confirm that this is due to some of the seven machines studied by them possessing an overaverage sensitivity to 60 Hz torsional excitation?

In regard to the shaft torsional impact of unsuccessful reclosing for line-to-line faults the authors indicate a considerable difference in Fig. 7 depending on whether such short circuits are grounded or not. The magnitude of this difference in potential severity depends on various system parameters, e.g. short-circuit capacity, line reactance and zero-sequence reactance. It would be of interest to know what value of zero-sequence line impedance was assumed by the authors in their study.

Mechanical damping is strongly a function of the magnitude of the occurring shaft torsional amplitudes. Measurements show that damping decreases by orders of magnitude as the torsional oscillations become less severe after a disturbance [B]. It would appear that the authors applied constant modal damping values, namely the logarithmic decrements listed in Table II, to their calculations of shaft fatigue. Were these values selected on the basis of large torsional amplitudes typical of the initial cycles following a disturbance or of smaller amplitudes that occur as the excited oscillations decay to the endurance limit?

The employment of statistical methods to quantify the torsional hazards of automatic reclosing practices offers the prospect of obtaining concrete figures on which engineers can base their system operating decisions. The adoption of statistical scatter bands is a valuable approach to defining the torsional fatigue properties of turbine-generator shafts and the switching times of circuit breakers provided the base data are valid. Statistical assumptions of the frequency of various types of faults are obviously only as good as they accurately predict the actual fault occurrences. The critically important inherent torsional characteristics of turbine-generators, however, are not subject to random distribution, but vary greatly and uniquely from one machine design and configuration to another, depending largely on the rotor fundamental torsional natural frequency or the lower modes of vibration. The first torsional natural frequencies of the seven machines studied by the authors extend over the very wide range from 7.1 to 17.5 Hz. We have shown how very strongly the favorable and unfavorable fault clearing and reclosing times differ for machines with different torsional characteristics [3,A,B]. The authors presented the relationship between reclosing time and shaft fatigue in Fig. 7 for just one particular 3600 rpm machine. It must be remembered that slight deviations in fault clearing and reclosing times can result in several orders of magnitude higher shaft fatigue in the case of three-phase faults if the breaker timing happens to coincide with an unfavorable combination of fault clearing and reclosing times for the specific affected machine. For this reason it is essential to know whether the authors assumed in their analysis the same fault clearing and reclosing times subject to deviation bands of ±0.5 and ±0.25 cycles respectively for all seven torsionally very different machines.

Due to widely varying discrete torsional characteristics of turbine-generators, the generalized probability figures developed by the authors cannot be applied to any particular machine, not even to one of the seven studied units. It is necessary to analyze each individual machine with its own torsional characteristics and resultant special sensitivity to torsional excitation on the basis of realistic assumptions in regard to other parameters to allow a meaningful evaluation of the shaft fatigue that can be expected with different system operating practices. In other words, the statistical method of analysis advocated by the authors can only yield useful results if it is applied to a specific turbine-generator with known torsional characteristics.

Finally, we believe that the reasons that the authors concur with our own conclusions and recommendations is that they rightly do not base them on their own generalized statistical results of their study but rather, like us, consider the real despite unlikely possibility of severe turbine-generator shaft damage being incurred under unfavorable conditions by high-speed reclosing in the event of a close-in three-phase fault.

REFERENCES

[A] J. S. Joyce, T. Kulig and D. Lambrecht, "The Impact of High-Speed Reclosure of Single and Multi-Phase System Faults on Turbine-Generator Shaft Torsional Fatigue", *IEEE Transactions*, Vol. PAS-99, No. 1, pp. 111-119, Jan./Feb. 1980.
[B] J. S. Joyce and D. Lambrecht, "Status of Evaluating the Fatigue of Large Steam Turbine-Generators Caused by Electrical Disturbances", *IEEE Transactions*, Vol. PAS-99, No. 1, pp. 279-291, Jan./Feb. 1980.

Manuscript received March 4, 1980.

P. A. Rusche (Consumers Power Company, Jackson, MI): The statistical analysis of the probabilistic amplification in torsional oscillations due to discrete electrical system disturbances and related material fatigue over the life of a typical turbine generator is a commendable contribution to the state-of-the-art. Applied to the time invariant network of Figure 4, the method ranks quantitatively three-pole automatic reclosing options relative to shaft fatigue life expenditure. As qualitatively indicated in other publications, unrestricted high speed reclosing (HSR) poses a substantially higher risk than other options.

Care must be used in appraising the absolute risks for summaries like Figure 11. This figure was obtained for a generalized network and a general population of fault statistics. Individual experiences may differ significantly from these norms. Furthermore, simplifying assumptions had to be made about the fault distribution for purposes of this analysis. Specifically, constant fault rates, exponential distribution of time between faults and equal distribution of faults along affected lines was assumed. As illustrated by the following these assumptions may be conservative.

In 1974 two major disturbances occurred in the network of the East Central Area of the United States within less than one week. On March 28 and 29 a major ice storm caused 31 permanent faults on the 345 kV network of lower Michigan [a]. On April 3 of the same year a series of 147 tornadoes, which killed over 300 people, struck in ten states, including Michigan. In a matter of hours this series of tornadoes had caused structural damage at over 50 locations in principal power supply facilities.

Both events caused hundreds of automatic reclosing operations. They occurred in rapid succession. Many were successful; others involved the spectrum of permanent faults. Transmission performance during these and other disturbances are embedded in the statistics used by the authors. The analysis of such disturbances suggests the majority of faults occur in compressed time spans. Fault succession often occurs in a matter of seconds. Hence, affected generators may suffer multiple torsional excitations without returning to their steady state operating point.

The following table of fault statistics for the 345 kV system in lower Michigan illustrates other potential causes for multiple torsional excitation. At first glance the fault statistics appear less evere than those assumed by the authors. Fault frequency was only 2.5 per hundred miles per year and there were no recorded three-phase faults. However, 6% of the faults simultaneously involved adjacent circuits on common structures. More significantly, 38% of all tripouts were the result of relay error. The most common relay error is overtripping. Switching operations to clear the simultaneous faults and restore overtripped circuits induce perturbations on generator shafts in addition to those considered by the authors. The effect would be an increase in the absolute fatigue life expenditure computed by the authors. However, the ranking between automatic reclosing options would remain unchanged. Due to the exponential relationship between amplitude of torsional oscillations and fatigue expenditure, relative difference between unrestricted and selective HSR would probably increase.

TABLE I

Tripout Record for 345 kV Transmission Lines on the Consumers Power Company System Jan 1, 1970 (875 km) to Dec 31, 1978 (3,100 km)

Fault Type	No	%	Tripouts Caused By	No	%
Line-to-Ground	322	87	Lightning	122	23
Line-Line-Ground	17	4	Ice/Wind	78	15
Line-Line	33	9	Station Apparatus	23	4
Three Phase	0	0	Line Material	8	2
			Operator Error	68	12
Subtotal	372*	100	Interference	19	3
Unknown	2	–	Relay Error	202	38
No Fault Current	159	–	Unknown	17	3
			Instability	2	

*13(6%) Involved Coincident Faults on Adjacent Circuits

Average Line Tripouts

From All Causes	2.4/100 km-Year (3.9/100 Mile-Year)
From Faults	1.6/100 km-Year (2.5/100 Mile-Year)

About 87% of the line tripouts from lightning were successfully automatically reclosed.

REFERENCE

[a] "Turbine-Generator Shaft-Related System Planning Criteria, Operating Experiences and Selected Study Results" by P. A. Rusche, et al, (f 79 751-9) presented at the 1979 IEEE PES Summer Power Meeting.

Manuscript received March 7, 1980.

D. N. Walker, P. G. Brown, and **C. E. J. Bowler**: The authors wish to thank the many discussers for their thoughtful comments, suggestions and questions. In order to minimize the length of the closure, the questions will be addressed on a topical basis, and our comments on the opinions and suggestions made by the discussers will be limited to those for which there is significant disagreement.

Recommendations

Several discussers commented on the particular reclosing practices studied and the recommendations given in the paper. We would like to emphasize that there are several reclosing practices besides those studied in the paper which would be expected to result in a much lower risk of high turbine-generator shaft fatigue duty in comparison to unrestricted HSR. One such practice referred to in Messrs. Mathews and Wilkinson's discussion is a combination of selective and sequential reclosing. Recent studies which evaluated this reclosing practice (which suggests utilizing a measurement of the fault severity rather than simply fault type as used in the paper) indicates it should also be an effective selective reclosing procedure. With regard to sequential reclosing, we thank Messrs. Methews and Wilkinson for pointing out that with high speed check relays this may be accomplished within the usual definition of HSR.

With regard to the question by Messrs. Edmonds, Dunlop and Jackson on why we chose to make specific reclosing recommendations, we feel it is important that we identify, as a result of our work, several available reclosing procedures which generally seem to satisfy electrical network/turbine-generator reliability requirements. However, unrestricted HSR may risk major turbine-generator shaft fatigue duty and, as we point out in the paper, should not, in our opinion, be applied without the conduct of studies to assure acceptable machine duty. Each utility must, of course, determine the procedure best suited to its requirements and balance all the risks involved. A key purpose of our paper was to describe a probabilistic method for helping achieve this goal. In response to Mr. Edmonds' question, we would encourage individual studies by utilities, based on the probabilistic method, especially when unrestricted HSR is being applied or being considered for a particular system. For this type of study, a utility could use distributions for parameters which best fit their own experience.

System Stability

The relation of HSR to "stability," which Messrs. Dunlop and Jackson mention, is subject to confusion since this term means different things to different people. HSR may be either helpful or harmful to stability, depending on the system scenario being considered. Automatic line reclosing is a stability aid in reducing the risk of multiple outages of circuits critical to the stability of the plant, and from this viewpoint the faster the line restoration the better. On the other hand, the system must also be able to maintain transient stability for fault occurrences, and here high speed reclosing tends to be detrimental. For those occasions when the high speed reclosing is unsuccessful (i.e., closure is back into a permanent fault), it should be recognized that this second fault may cause instability by aggravating the initial electrical swings which have not had a chance to die out.

Torsional Response

We do not share the position of Messrs. Joyce and Lambrecht that only close-in three-phase faults have the potential for significant shaft fatigue damage from unsuccessful HSR. These discussers have called attention to differences in shaft loss-of-life estimates for a single incident between their paper and our own, particularly with regard to faults involving ground. In particular, Messrs. Joyce and Lambrecht report the same shaft duties for both line-line faults not involving ground and double-line-ground faults. We believe, and our studies indicate, that double-line-ground faults are inherently more severe than line-line faults, and would question results which did not reflect this. If these discussers used an unusually high zero-sequence line impedance in their studies, or omitted the grounding of the step-up transformer, this would be expected to result in similar fatigue duties for the two fault types as they have reported (reference 3 in paper). This would also result in substantially lower levels of line-ground fatigue duty. We believe that this is the primary reason for our differences in shaft fatigue results and it is not due to our study population of units possessing an over-average sensitivity to 60 Hz torsional excitation, as these discussers suggest. Other possible reasons for differences in fatigue response levels are the values of torsional damping assumed and the fatigue model.

In response to Messrs. Joyce and Lambrecht's question on the values of zero-sequence reactance used in our simulation model, we used an X_0/X_1 ratio of 3 for the transmission lines which are representative of EHV lines. We also assumed grounded-wye step-up transformers with an X_0/X_1 ratio of 1, which we also feel is representative. X_0 and X_1 represent the zero sequence and positive sequence impedances, respectively.

We agree with Messrs. Joyce and Lambrecht that the torsional response of a turbine-generator is highly dependent on the precise instant of power circuit breaker operation. In response to their question on fault clearing and reclosing times, we would like to reassure them that worst case timings were determined for each of the seven machine configurations. These worst times were then used to represent the center of the distributions for fault clearing and reclosing, and were different for each machine as they depend on their specific torsional vibration characteristics. We agree that for the statistical method of analysis to be tractable it must be applied to a specific unit in a particular Monte Carlo simulation. As we describe in the paper, Figs. 11 and 12 (the probability distributions for varying reclosing practices) correspond to the unit in our study population which generally exhibited the highest fatigue duty. For the record, and reiterating the information supplied in the paper, the probabilistic method was used and is advocated as a powerful method for evaluating and defining acceptable reclosing practices. We therefore disagree with Messrs. Joyce and Lambrecht's final remark that our conclusions and recommendations are not based on the study given in the paper.

Messrs. Mathews and Wilkinson ask what is the ffect of one-cycle clearing versus longer clearing times on shaft fatigue duty, as some investigators have found that one-cycle clearing was worse than, say, three-cycle clearing. The shaft response is highly dependent on the precise torsional vibration characteristics of the particular machine as well as the breaker operating time. Therefore, in our opinion, no general conclusion is possible.

With regard to Messrs. Dunlop and Jackson's question on whether 1800 rpm machines are torsionally more rugged than 3600 rpm units, we have not evaluated a sufficient number of 1800 rpm units to make a general statement. The two 1800 rpm units in our study population did, however, exhibit below-average levels of fatigue duty.

Study System Configuration

Messrs. Dunlop and Jackson raised some interesting points regarding our use of only two transmission lines in the model study system. There are obviously many variations in plant and transmission system configuration which, along with other variables such as fault types and their frequency and location, affect the machine duties. To keep a generalized study such as this to a reasonable size, it was necessary to use a single, relatively simple configuration which is representative of many sites. However, our simplified network can be viewed more broadly on the basis of a system with a total of 120 miles of line exposure, where turbine-generator duties are compared for various reclosing options. Alternatively, one could consider the system model as consisting of three 80-mile circuits, substituting two 80-mile circuits for the 40-mile circuit of the model. In this latter case, we think these discussers would agree that HSR would be practiced, considering the relatively high probability of relay error taking out a second line following a fault on an adjacent line. This three-line network was looked at on the same Monte Carlo basis, considering the added line exposure, and was found not to affect the general conclusions and recommendations of the paper.

We do not agree with Messrs. Dunlop and Jackson's assumption that the study system used is a worst case system. A weaker receiving system, for example, results in more severe turbine-generator duties. We explored this by considering a receiving system equivalent impedance of 40% rather than the 5% shown in Fig. 4. These results, while more severe, also did not affect the general conclusions and recommendations of the paper. In addition, relatively little difference in turbine-generator duties was found where a second unit was added to the single unit plant of Fig. 4.

We believe there are other potentially more severe system configurations, such as that comprising a number of short lines from the plant resulting in a high proportion of "close-in" faults. As stated in our recommendations, we believe that one must use judgment in extrapolating our results to other system configurations and would recommend individual studies in some cases.

Statistical Method of Analysis

The need for a statistical approach, which Messrs. Joyce and Lambrecht question is, of course, dependent on whether there is significant fatigue duty for several different fault categories over a defined turbine-generator exposure period. As previously discussed, we do not agree

with Messrs. Joyce and Lambrecht that unsuccessful HSR on three-phase faults is the only way of accumulating significant shaft fatigue life expenditure (L-L-G faults are also potentially damaging in a single worst case incident—Fig. 6). The results given in Fig. 11, for example, indicate that even though HSR on all three-phase and L-L-G faults is eliminated (curve 3), there still remains a 10% chance of the remaining, less severe faults consuming at least 50% of the shaft fatigue life over an assumed 40-year turbine-generator exposure period.

With reference to Messrs. Armor and Parkers' comments, the probabilistic method gives considerably more information than deterministic approaches and, in our opinion, is an ideal vehicle for contrasting the probability of defined levels of fatigue damage being experienced for various reclosing strategies. As described in the paper, we agreed that the most difficult step in the application of the method is the selection of parameter distributions, particularly for a generalized study. In our work, we used as much as possible published average industry statistics. This limitation, of course, becomes less severe when the method is applied to a particular machine on a particular network and, in our opinion, will become a powerful tool for assisting individual utilities to judge the reclosing practice which best meets their overall power generation and transmission reliability requirements.

We certainly agree with Messrs. Armor and Parkes that the assumption for the mean rate of faults and the turbine-generator exposure period will significantly affect the results obtained. It is important to recognize, however, that different assumptions for these paramters are easily accounted for in the data presented in Figs. 11 and 12. For example, doubling the mean rates of faults from 4 to 8 per 100 miles per year and reducing the assumed turbine-generator exposure period from 40 to 30 years would only require multiplying the horizontal scale values by a factor of 1.5 (2 x 3/4).

We would especially like to thank Mr. Rusche for the data he provided and for his comments on the dangers of relying too heavily on simple average fault statistics. This points out the need to apply margins for the rapidly occuring disturbances he describes which are beyond the norms, as well as for bus faults, out-of-phase synchronizing, and the more remote line faults. In the paper, we proposed a margin of 30% to allow for these contingencies. For some systems for which the distributions that we chose appear particularly unrepresentative, we recommend individual studies, as stated in recommendation 4.

With reference to a question from Messrs. Dunlop and Jackson, we agree that a 40-year turbine-generator on-line exposure period is probably a rather conservative assumption for many plants. However, it should be recognized that a few of our units have been in operation for more than 50 years. These discussers ask about the effect on the curves shown in Figs. 11 and 12 of an inspection process conducted periodically to identify and remove surface cracks. This would involve one machine disassembly, as cracks will initiate at points of stress concentration which are often not visible. In addition, the time between initiation of a detectable crack, and propagation by torsion or other mechanisms such as bending to a depth which would affect turbine-generator operation, may well be less than a normal inspection interval. If initiated cracks are discovered and are small enough to be safely removed, then it is possible that a major proportion of the shaft fatigue life will be restored. This use of a torsional vibration monitor (1) would give guidance to the utility as to the need for disassembly and inspection of a unit.

Torsional Damping

Mr. Harrington asks several questions about the values of torsional damping that were used in our study and how these data were recorded. Vibration data have been captured on brush recorder charts in many power station tests to determine approximate damping levels as the test proceeds. Test data have generally also been recorded on magnetic tape and analyzed with Fast Fourier Transform equipment in our data reduction facility, to yield the most reliable damping information. The results from this equipment and from analysis of information on the brush charts have generally been in good agreement. Hence the problems of poor frequency response of brush recording equipment, referred to by Mr. Harrington, are not evident in torsional damping and frequency determinations.

In response to Mr. Harrington's comments, our test experience has shown that the torsional damping of turbine-generators is highly dependent on the machine configuration, the power output, the transmission system configuration, the mode that is stimulated, and the vibration amplitude. The torsional damping levels are very low and different values have been observed on nominally identical units at a power station under the same operating conditions. These facts indicate that damping values of an untested turbine-generator cannot be predicted reliably. For our study purpose, the damping values were estimated for each unit based on test data obtained on approximately a dozen units of varying rated power output and configuration.

In response to another question by Mr. Harrington on the effect of a fast-acting governor system that reduces load following detection of a fault, this would result in some reduction of the torsional damping values as well as applying additional stimulus to the turbine-generator. This reduction in damping would not, however, be expected to occur until several seconds had elapsed after governor action because of the volume of entrapped steam between the intercept valves and the turbines. In response to the question on generator electrical operating conditions, the generator electrical damping is believed to be a minor component, particularly at full load. Consequently, we do not believe that over/under-excited operation of the generator significantly affects the torsional damping values.

We disagree with T. J. Hammons that machine electrical damping is at least an order of magnitude greater than steam damping. Our calculations indicate that the electrical damping component is a small fraction of levels that have been measured in power station tests at full load. At light loads, the electrical and mechanical components have been separated by first obtaining data with the machine unsynchronized and running at rated speed, and second following synchronization at very low load. At approximately no load, the electrical and mechanical damping components have been determined by these means to be of similar magnitude. We believe that the dramatic increase in damping with load [2] observed in many station tests is directly attributed to increased steam forces on the turbine buckets. It is our opinion that, at full load and levels of torsional oscillation below the shaft endurance limit, the dominant damping mechanism is from steam forces.

With regard to the T. J. Hammons and Joyce/Lambrecht question on the levels of torsional vibration that are reflected in the damping values given in Table II, these data were based on test data at vibration levels slightly less than the endurance level of the limiting shaft in each mode. In our study simulations, it was assumed that the modal damping values were independent of vibration amplitude. In this regard, our analyses may be conservative, because the damping values for disturbances which produce damaging levels of shaft vibration may be significantly higher than the Table II values [1]. Unfortunately, no damping data are available at damaging vibration levels. One significant benefit of implementing torsional vibration monitoring systems is that these data will be obtained relatively quickly for a variety of severities of fault and on several configurations of turbine-generator.

Fatigue

In response to Mr. Taylor's question on the effect of mean shaft steady state stress on fatigue, we have the following comments. The torsional fatigue testing program that General Electric has conducted to date has shown that there is only a small influence of torsion mean stress on torsional fatigue strength. In the shaft that spans the generator and the shaft-driven exciter (if one exists), the stress state at the limiting stress concentration causes a uniaxial stress state. However, even in this case the effect of the steady state mean stress is not significant because of its very low magnitude in this span.

Messrs. Dunlop and Jackson ask, "In the fatigue model used, did the $\pm 2\sigma$ deviations only relate to the scatter in the test data?" Our fatigue model is derived from a large data base of fatigue tests conducted over many years. It effectively includes in the statistical bands of fatigue strength the effects of size, stress concentration, surface finish, environment, and many other parameters. As our data base becomes more extensive as a result of our ongoing work, the scatter band will tend to broaden and hence our fatigue distribution may change in the future.

Monitoring

With reference to Messrs. Armor and Parkes' question on commenting on direct and indirect shaft fatigue monitoring methods, this is really outside the scope of our paper and we refer them to reference [1] which discusses this subject in some depth. We note with interest EPRI's support of development of acoustic emission listening devices intended to provide early warning of crack formation in a rotating shaft system. We agree that a direct monitoring method may be a desirable approach if it can be made to work. Development of acoustic emission devices is still in its infancy and, in our opinion, has a long way to go before these devices have the reliability and sensitivity needed for long-term monitoring of large steam turbine-generators.

REFERENCES

[1] E. E. Gibbs, D. N. Walker, "Torsional Vibration Monitoring," Pacific Coast Electrical Association Engineering & Operating Conference, San Francisco, California, March 1980.

[2] D. N. Walker, S. L. Adams, R. J. Placek, "Torsional Vibration & Fatigue of Turbine-Generator Shafts," Paper No. F79 756-8, presented by Title at IEEE-PES Summer Meeting, 1979.

Manuscript received April 22, 1980.

RESONANT GROUNDING (GROUND FAULT NEUTRALIZER) OF A UNIT CONNECTED GENERATOR

K. J. S. Khunkhun
Stone & Webster Engineering Corporation
Boston, Massachusetts

J. L. Koepfinger
Duquesne Light Company
Pittsburgh, Pennsylvania

M. V. Haddad
Duquesne Light Company
Pittsburgh, Pennsylvania

ABSTRACT

This paper outlines the advantages of resonant grounding over resistor grounding of a large unit connected generator in a modern power plant. It provides complete application procedures for this type of grounding and illustrates the means of evaluating its performance compared with resistor grounding.

INTRODUCTION

The purposes of generator neutral grounding through an impedance are to limit the damage at the point of fault, limit transient overvoltages and provide sensitive means of detection for insulation deterioration.

In the past, resistor grounding has been a popular way of grounding; however, resonant grounding fulfills the grounding objectives superbly for a large unit connected generator. In a nuclear power plant, it even provides a possibility of delayed trip for a safe orderly reactor shutdown without exposing the unit equipment to unnecessary damage during a ground fault. The reluctance to use resonant grounding is partly due to lack of both information regarding the application procedures and clear understanding of its advantages.

This paper provides a brief discussion of resonant grounding and application guidelines for its implementation. Calculations involved in comparing the damage at the point of fault and detection sensitivity are outlined in the last part of the paper, followed by an actual field experience in Appendix 1. The data for this paper are obtained from a resistor grounded unit; a similar second unit will be resonantly grounded based on techniques discussed in this paper.

DISCUSSION

The advantages of resonant grounding and the effects of high voltage system disturbances are highlighted in this section. Graph 1 shows these advantages over the entire range of resistive faults. This graph is extremely helpful in evaluating the performance of resonant grounding and provides an insight into the possible ground fault protection schemes as shown in Appendix 1, as well.

(1) The damage at the point of fault is greatly reduced when the unit is resonantly grounded. For example in a resistive fault of 5,000 Ω, energy into the fault and consequently the damage at the point of fault is 19 times more with resistor grounding compared with resonant grounding.

(2) Resonant grounding provides an increased sensitivity of detection for localized deterioration of insulation at any location in the generator system well before the imminence of solid ground. As shown in Graph 1, with resonant grounding, a developing fault of over 400,000 Ω resistive at the generator terminal will be detected by the ground fault relay set to see ninety-five percent of generator winding. A similarly set relay for resistor grounding will not see anything even as low as 40,000 Ω resistive fault. The damage for this sensitivity would be 14 times more with resistor grounding compared with resonant grounding. For a resistive fault of 5,000 Ω, sensitivity improvement of 330 percent is obtained.

(3) Surges on the high voltage system will not appear between generator neutral and ground because of the wye-delta transformation of main transformer.

(4) Transient overvoltages on the unfaulted phases caused by arcing ground are limited by resonant grounding, because arcing or current surges are reduced.

(5) The appearance of zero sequence voltage on the transmission system for a line to ground fault impresses an overvoltage on the generator system. This voltage is a function of the capacitive coupling reactance in the main transformer and zero sequence impedance of the generator system. Impressed over-voltage on the generator system will be higher in magnitude with resonant grounding than resistor grounding and must be controlled within acceptable limits by the proper selection of coil constant of the reactor. Relay coordination should be provided so that the relays monitoring generator ground may not misoperate for a fault at the transmission system.

FUNDAMENTALS OF RESONANT GROUNDING

Equipment Arrangement: The equipment arrangement shown in Figure 1 is typical of a modern power plant. A small reactor across the secondary of a distribution transformer offers a most convenient and economical way of implementing resonant grounding. The cost of resonant grounding is comparable to a similarly designed resistor grounding.

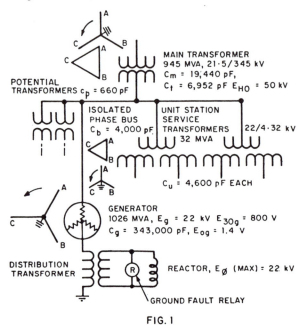

FIG. 1

Required Data: The major part of the data required for resonant grounding are system capacitances. All the capacitive values of main transformer, unit station service transformers, isolated phase bus generator and any other capacitance in the system, must be available before any attempt could be made to design the resonant grounding equipment. Calculated values of these capacitances which are readily available have been successfully used and are sufficient for this purpose. Small deviations from these values are taken care of by the provision of taps at the distribution transformer and the reactor. Typical values of the data required for resonant grounding are shown in Fig. 1 and are defined below:

F 76 330-1.   A paper recommended and approved by the IEEE Power System Relaying Committee and the IEEE Surge Protective Devices Committee of the IEEE Power Engineering Society for presentation at the IEEE PES Summer Meeting, Portland, OR, July 18-23, 1976. Manuscript submitted January 19, 1976; made available for printing April 26, 1976.

Reprinted from *IEEE Trans. Power App. Syst.*, vol. PAS-96, no. 2, pp. 550–559, March/April 1977.

Generator:

E_g = Phase to phase voltage

C_g = Single phase to ground capacitance

E_{og} = Fundamental frequency zero sequence normal voltage

E_{30g} = Third harmonic zero sequence normal voltage

Main Transformer:

N_m = Number of transformers

C_m = Single phase to ground capacitance (gen. side)

C_t = High to low side winding capacitance (per phase)

E_{HO} = Maximum zero sequence voltage due to a fault at the high side

Unit Station Service Transformer:

N_u = Number of transformers

C_u = Single phase to ground capacitance (gen. side)

Isolated Phase Bus:

C_b = Total single phase to ground capacitance

Surge Capacitors: None

Potential Transformers:

C_p = Single phase to ground capacitance (gen. side)

Maximum Acceptable Voltage:

E_ϕ (max) = Maximum phase to ground voltage the system equipment can withstand without damage

Resonant Grounding: In a unit connected generator with neutral grounded through high impedance, two sources of currents are available to a ground fault. One source is the system capacitance, which is a system function and cannot be controlled, the other is the generator neutral. The neutral current can be controlled in both magnitude and phase. In the case of resistor grounding, system capacitive current "I_c" is 90° out of phase with neutral current "I"$_{Req}$ and these two currents add up to flow through a ground fault. The magnitude of this fault current "I_{rf}" is definitely greater than the capacitive current for resistor grounding. On the other hand, with resonant grounding, system capacitive current can be completely neutralized by the provision of equal but 180° out of phase neutral current "I_L" leaving theoretically no current to flow through a ground fault. Fig. 3 shows an ideal vector relationship for both types of grounding during a solid fault.

A solid single phase to ground fault can be represented by a symmetrical component network as shown in Fig. 2(b), where r_{co}, a very small equivalent resistance is introduced in series with the zero sequence capacitance "C_o", compensating for dielectric or any other losses which might be present in the parallel circuit and R_L is a small equivalent resistance in the inductive circuit. The positive sequence elements X_1, X_{c1}, negative sequence elements X_2, X_{c2}, and zero sequence element X_o of the generator system may be ignored because $X_{c1} \gg X_1$, $X_{c2} \gg X_2$ and $X_{co} \gg X_1$, X_2 & X_o. The other elements of the network are defined below:

C_o = Total zero sequence capacitance = $(C_g + N_m C_m + N_u C_u + C_b + C_p)$ = 376,300 pF

X_{co} = Zero sequence capacitive reactance = $\dfrac{10^{12}}{2\pi(60)C_o}$ = 7050Ω

r_{co} = Resistance due to dielectric losses etc.

X_L = Equivalent inductive reactance in the generator neutral

R_L = Equivalent resistance in the inductive circuit

I_{co} = Capacitive zero sequence current contribution to a solid ground fault = $\dfrac{E_g}{\sqrt{3}\ X_{co}}$ = 1.802A

I_{LO} = Inductive zero sequence current contribution to a solid ground fault = $\dfrac{E_g}{\sqrt{3}\ \ 3X_L}$ = 1.802A

I_c = System capacitance current contribution to a solid ground fault = $3 I_{co}$ = 5.4A

I_L = Neutral current contribution to a solid ground fault = $3I_{LO}$ = 5.4A

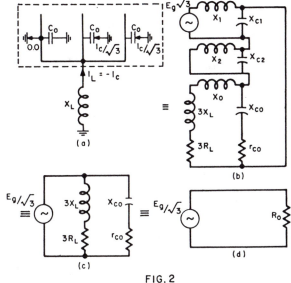

FIG. 2

To neutralize the system capacitive current, an equivalent inductive reactance is provided in the generator neutral. During a ground fault, system capacitance and neutral inductance will form a parallel tuned circuit leaving only a pure high resistance in the generator neutral as shown in Fig. 2(d). This resistance is calculated as follows:

R_o = Zero sequence equivalent impedance in the generator neutral during a ground fault.

$$= (j3X_L + 3R_L)(-jX_{co} + r_{co})\ /\ (j3X_L + 3R_L - jX_{co} + r_{co})\ \Omega$$

For resonant grounding $I_{co} = I_{LO}$

or $\dfrac{E_g}{\sqrt{3}} \times \dfrac{1}{X_{co}} = \dfrac{E_g}{\sqrt{3}} \times \dfrac{1}{3X_L}$ or $X_L = \dfrac{X_{co}}{3}$ = 2350Ω

Therefore $R_o = \dfrac{9X_L X_L}{3R_L + r_{co}} = 3\ \dfrac{X_L}{R_L + r_c}\ X_L$ for $r_{co} = 3r_c$ and R_L & r_c

very small compared with X_L

I_{nf} (max) = Maximum ground fault current with resonant grounding = $3\ \dfrac{E_g}{\sqrt{3}} \times \dfrac{1}{R_o}$ A

Resistor Grounding: A good deal of literature[1] is already available for resistor grounding. This paper assumes (as is typically done) an equivalent resistance "R_{eq}" in the generator neutral to be equal to the system capacitive reactance, ($X_c = \dfrac{X_{co}}{3}$ = 2350Ω) which gives a maximum fault current $(I_{rf}$ (max) $= I_{Req} + jI_c = 7.6 \underline{/45^\circ}$ A) to a solid ground fault as shown in Fig. 3.

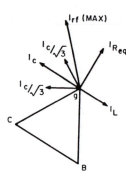

FIG. 3

Impressed Overvoltage: During a ground fault on the transmission system, high side zero sequence voltage E_{HO} is impressed on the generator system [2] due to the capacitive coupling of high to low side winding of the main transformer and the generator zero sequence impedance as shown in Fig. 4(a). In addition, neutral is also displaced by the fundamental frequency zero sequence normal voltage acting on the series resonant circuit as explained under no fault condition. As can be seen from Fig. 4(c), maximum phase to ground voltage "E_ϕ (max)" will occur when E_R (neutral displacement due to high side fault); E_P (neutral displacement due to normal fundamental zero sequence voltage) and E_L (generator phase to neutral voltage) are all in phase shown by dotted line.

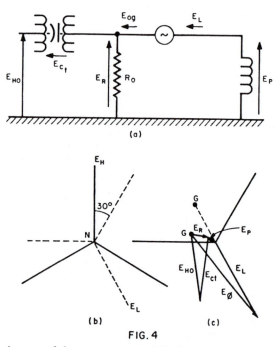

(a)

(b)   (c)

FIG. 4

The major part of the generator neutral displacement during a fault at the high side of the main transformer would be due to the capacitive coupling of high to low side windings. Assuming this voltage E_R in phase with E_L and E_P very small, it can be written:

E_ϕ (max) $= E_R + E_L = 22$ kV

X_{ct} = Capacitive coupling reactance $= \dfrac{10^{12}}{2\pi 60 C_t} = 381,557\Omega$

Also from Fig 4(a) $E_R = \dfrac{E_{HO}}{R_0 - jX_{ct}} R_0 = E_\phi$ (max) $- E_L = 9.3$ kV

or $R_0 = E_R X_{ct} / \sqrt{E^2_{HO} - E^2_R} = 72,230\Omega$

Also $R_0 = 3 \dfrac{X_L}{R_L + r_c} X_L = 72,230\Omega$

From this relation for R_0 in which r_c is constant but unknown, the value of R_L must be determined if the reactor design is to be specified to control the neutral displacement. So an assumption has to be made for the value of r_c, which if assumed so small, to be ignored in the above relation, might lead to a reactor design of so low coil constant that the expected advantages of resonant grounding would be greatly reduced. On the other hand, if the value of r_c is assumed so high that R_L could be ignored in the above relation, this may lead to a reactor design of highest available coil constant, which may cause undesirable neutral displacement, so as to require a further modification of adding a small resistor in series with the reactor. Either way, there are certain shortcomings in specifying the coil constant, which lead to a compromising decision of assuming the values of R_L and r_c equal. The experience outlined in Appendix I further justifies this assumption and provides a relation for the equivalent estimated coil constant $K = \dfrac{X_L}{R_L} = \dfrac{2R_0}{3X_L} = 20.5$

The actual effects of resonant grounding of course should be evaluated by using the actual value of R_0, which if unacceptably high, could be reduced by slight detuning of the ground fault neutralizer. Also the true value of E_R should be verified by connecting all high side bushings together and applying 120 Vac. between the phases and ground and measuring the neutral displacement as shown in Appendix 1.

EQUIPMENT SELECTION AND INSTALLATION

Selection of Distribution Transformer: The distribution transformer in the generator neutral offers not only the advantage of sizing a small low voltage reactor across its secondary, but also helps in providing effectively a variable reactor in the generator neutral through its taps for tuning purposes. Five taps, two at 2.5 percent above the normal voltage and two at 2.5 percent below the normal voltage are sufficient. The following calculations outline the distribution transformer requirements and the associated data:

E_{tp} = Primary voltage $> \dfrac{E_g}{\sqrt{3}} = 15,000$ V

E_{ts} = Secondary voltage = 120/240 V

TR = Thermal rating = $I_L \times E_{tp} \times 10^{-3} = 100$ kVA

Z = Impedance = 5 %

L = Copper losses = 1,350 W

I_{ts} = Secondary full load current = $\dfrac{TR \times 10^3}{E_{ts}} = 416.7$A

R_{ts} = Equivalent secondary resistance = $\dfrac{L}{(I_{ts})^2} = 0.0078\Omega$

Z_{ts} = Equivalent secondary impedance
$= \dfrac{Z \times 10^{-2} (E_{ts})^2}{TR \times 10^3} = 0.0288\Omega$

X_{ts} = Equivalent secondary reactance = $\sqrt{Z^2_{ts} - R^2_{ts}} = 0.0277\Omega$

Taps = $E_{tp} \times \dfrac{100 + (3-n)p}{100}$ V, where n = 1,2,3,4 & 5 and p = 2.5 for

the required taps (15750, 15375, 15000, 14625, 14250)

A dry type distribution transformer which will not saturate for

$1.5 \times \dfrac{E_g}{\sqrt{3}} = 19$kV would be safe for this application.

Selection of Neutralizing Reactor: The inductive reactance in the generator neutral for resonant grounding is provided by an indoor type, air cooled reactor with three taps, having a continuous current

262

I_n (max) and 600 V insulation rating across the secondary of the distribution transformer. The major problem of designing this reactor, of course, is its coil constant, which once the reactor is built, cannot be changed. This problem can be handled by calculating the value of R_o which will not impress dangerous voltages during a fault at the high side of main transformer and hence the value of equivalent coil constant. This is illustrated in the earlier part of the paper under impressed overvoltage. The reactor requirements are outlined below:

X_{LS} = Equivalent inductive reactance across the secondary of distribution transformer = $X_L \times (E_{ts})^2 / (E_{tp})^2 = 0.6016\,\Omega$

X_n = Actual reactance of the reactor = $X_{LS} - X_{ts} = 0.5739\,\Omega$

R_n = Actual resistance of the reactor to be determined.

K = Equivalent estimated coil constant = $\dfrac{X_L}{R_L} = \dfrac{X_n + X_{ts}}{R_n + R_{ts}} = 20.5$

From this relation, the required value of $R_n = 0.0216\,\Omega$ can be determined, because all the other values are known or can be calculated. Hence, the actual coil constant K_n is equal to $X_n / R_n = 26.6$

I_n (max) = Maximum current through the reactor

= $I_L \times E_{tp} / E_{ts} = 337.5A$

Taps = $X_n \times \dfrac{100 + (n-2)b}{100}$ where n = 1,2 & 3 and b = 15 for three taps (0.4878, 0.5739, 0.6600)

In Appendix 1, five reactor taps at 7½ percent are shown to provide the same overall range as obtained above with 15 percent, but taps 2 and 4 provide repitition of reactances in the generator neutral with no extra advantage for tuning.

The above reactor taps combined with distribution transformer taps, will provide a range of inductive reactance available for tuning in the generator neutral to cover deviation from system capacitance "C_o" of about 75 to 125 percent.

Installation and Tuning Procedures: Once all the unit equipment is in place and connected, neutralizer can be tuned using a low voltage supply.[3] A final check should, however, be made, as the machine is brought up to and above rated speed with field applied. A peaking of the neutralizer voltage at 60 Hz indicates that the ground fault neutralizer is properly tuned. The temporary connections for tuning are shown in Fig. 5 below:

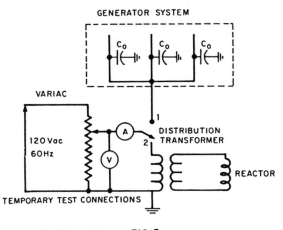

FIG. 5

The following steps should be carried out and the tabulation shown in Appendix 1 would be helpful:

a) Calculate and tabulate the available reactances in the generator neutral with different tap combinations.

b) With all the unit connections made and ground fault neutralizer removed by opening the connection at the distribution transformer, apply a low voltage at connection 1, measure the system capacitive current accurately with digital ammeter. One hundredth of generator phase to neutral voltage (127V) is most convenient.

c) With ground fault neutralizer still removed from the system, apply exactly the same voltage as in step (b) at connection 2. Measure the available reactive currents for all the tap combinations and tabulate with step (a).

d) Match the capacitive current measured in step (b) with the nearest reactive current measured in step (c) and set the taps accordingly.

e) Now with neutralizer connected to the system, measure the current applying the same voltage. Try other taps to verify, that the selected taps give the minimum current signifying proper tuning.

PERFORMANCE OF RESONANT GROUNDING

Damage at the point of fault: It is logical to assume that the damage at the point of fault would be equivalent to the energy dumped into a fault. Calculations for the percentage reduction in damage achieved by resonant grounding over resistor grounding are illustrated in this section. Even though these calculations are done for faults at the generator terminal, similar reductions will be accomplished if the faults are to develop somewhere in the generator winding. An equivalent zero sequence circuit is shown in Fig. 6, for a resistive fault of $r_f = 5000\,\Omega$.

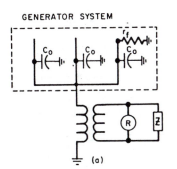

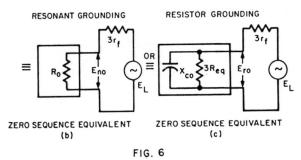

FIG. 6

E_L = Generator line to neutral voltage = $\dfrac{E_g}{\sqrt{3}} = 12{,}702\,V$

I_{nfo} = Zero sequence current with resonant grounding

= $\dfrac{E_L}{R_o + 3r_f} = 0.1456A$

I_{nf} = Fault current with resonant grounding = $3\,I_{nfo} = 0.4368A$

W_n = Energy into the fault with resonant grounding

= $I_{nf}^2 \times r_f = 954W$

I_{rfo} = Zero sequence current with resistor grounding

= $E_L / (3r_f - j3R_{eq}X_{co} / (3R_{eq} - jX_{co})) = 0.64A$

I_{rf} = Fault current with resistor grounding = $3\,I_{rfo} = 1.92A$

263

W_{rf} = Energy into the fault with resistor grounding

$\quad = I_{rf}^2 \times r_f = 18{,}432\text{W}$

W = Percentage reduction in damage with resonant grounding over resistor grounding $= \dfrac{W_{rf}}{W_{nf}} \times 100 = 1932\%$

Graph 1 shows the values of W_{nf} and W_{rf} for different values of r_f

Sensitivity of detection: Again referring to Fig. 6, the improvement in sensitivity of fault detection obtained with resonant grounding can be evaluated for similar faults as stated under damage at the point of fault. Here, the neutral voltage displacement is calculated, which actually shows up across the secondary of distribution transformer for ground fault detection. The sensitivity of fault detection for both types of grounding is calculated below:

E_{no} = Neutral voltage displacement with resonant grounding

$\quad = I_{nfo} \times R_o = 10{,}517\text{V}$

E_n = Percentage neutral voltage displacement with resonant grounding $= \dfrac{E_{no}}{E_L} \times 100 = 82.8\%$

E_{ro} = Neutral voltage displacement with resistor grounding

$\quad = I_{rfo} \times -j3\,R_{eq}X_{co} \,/\, (3\,R_{eq} - jX_{co}) = 3190\text{V}$

E_r = Percentage neutral voltage displacement with resistor grounding $= \dfrac{E_{ro}}{E_L} \times 100 = 25\%$

E = Percentage improvement in detection sensitivity with resonant grounding over resistor grounding $= \dfrac{E_n}{E_r} \times 100 = 330\%$

The values of E_n and E_r are plotted in Graph 1 for different values of r_f.

I_{op} = Fundamental frequency zero sequence normal current,

$\quad = E_{og} \,/\, (-jX_{co} + r_{co} + j3X_L + 3R_L) = E_{og} \,/\, (3r_c + 3R_L)\ \text{A}$

I_p = Fundamental frequency normal current through generator neutral $= 3\,I_{op}\ \text{A}$

V_n = Normal voltage of fundamental frequency across the reactor

$\quad = I_p \times \dfrac{E_{tp}}{E_{ts}} \times X_n\ \text{V}$

V_p = Normal neutral voltage displacement of fundamental frequency $= I_{op} \times (j3X_L + 3R_L) \approx \dfrac{K}{2} E_{og} = 14.4\text{V}$

$\quad$ for r_c and R_L very small and equal

Similar evaluation can be done for third harmonic zero sequence voltage as shown below:

X_{c3o} = Third harmonic zero sequence capacitive reactance = $\dfrac{X_{co}}{3}\ \Omega$

X_{L3o} = Third harmonic zero sequence reactance in the generator neutral $= 9X_L\ \Omega$

I_{3op} = Third harmonic normal neutral zero sequence current

$\quad = E_{3og} \,/\, (j9X_L + 3R_L - jX_{co}\,/\,3 + 3r_c)$

$\quad = E_{30g} \,/\, (j8X_L + 6X_L\,/\,K)\ \text{A}$

I_{3p} = Third harmonic normal neutral current

$\quad = 3\,I_{30op}\ \text{A}$

V_{3n} = Third harmonic normal voltage across the reactor

$\quad = I_{3p} \times 3X_n \times E_{tp}\,/\,E_{ts} = 13.7\text{V}$

APPENDIX I

Experience with Ground Fault Neutralizer — 600MVA Fossil Power Plant

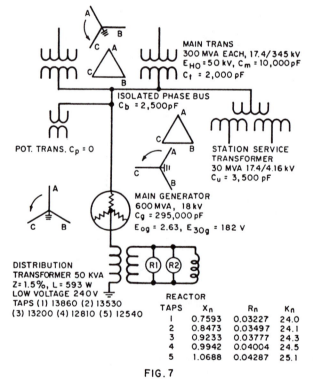

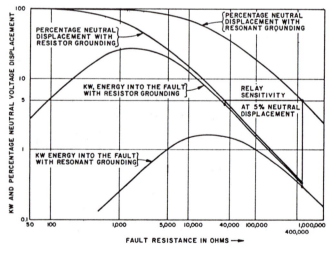

GRAPH 1

No Fault or Normal Conditions: — During normal conditions, the system capacitive reactance and generator neutral impedance are in series. Generally, fundamental frequency zero sequence voltage is small but amplified due to resonance. Third harmonic zero sequence voltage is of considerable magnitude and other harmonic voltages are extremely low and may be ignored.

For the fundamental frequency, the system capacitance and neutral inductance form a series resonant circuit and thus amplify its voltage magnitude as follows:

REACTOR TAPS	X_n	R_n	K_n
1	0.7593	0.03227	24.0
2	0.8473	0.03497	24.1
3	0.9233	0.03777	24.3
4	0.9942	0.04004	24.5
5	1.0688	0.04287	25.1

FIG. 7

Voltage applied for tuning tests = 104 V 60 Hz.
System capacitive current (step b) = 41.0 mA.

Tuning Tables:

Transformer No.	Tap	Reactor Tap	Calculated "X"$_L$ Reactance ohm (a)	Inductive current available mA (c)	Neutralized current mA (e)	
1	5	1	2101	55.8	15.0	
2	4	1	2192	52.9	12.2	
3	3	1	2328	50.3	9.8	
4	5	2	2341	50.2	9.6	
5	4	2	2442	47.6	7.4	
6	2	1	2446	47.8	7.6	
7	5	3	2549	45.9	6.0	
8	1	1	2567	45.5	5.5	
9	3	2	2594	45.2	5.3	
10	4	3	2660	43.5	4.5	
11	2	2	2725	43.0	4.3	
12	5	4	2742	42.8	4.2	
13	3	3	2824	41.3	3.8	
14	1	2	2860	41.0	(3.7	Resonance
15	4	4	2862	40.6	3.7)	
16	5	5	2946	39.9	3.8	
17	2	3	2967	39.3	4.1	
18	3	4	3039	38.6	4.4	
19	4	5	3074	37.8	5.1	
20	1	3	3114	37.5	5.3	
21	2	4	3192	36.7	5.7	
22	3	5	3264	35.9	6.4	
23	1	4	3350	35.0	7.2	
24	2	5	3429	34.2	7.9	
25	1	5	3599	32.6	9.5	

Equivalent estimated coil constant = K = 17.5

Estimated value of $R_o = \frac{3}{2} \times 2860 \times 17.5 = 75{,}000\Omega$

Actual zero sequence impedance in the generator neutral during a fault condition $= R_o \text{ (actual)} = 3 \times \frac{104}{0.0037} = 84{,}400\Omega$

Estimated value of $E_R = \frac{E_{HO} \times R_o}{R_o - jX_{ct}} \times \frac{100}{E_{HO}} = 5.6\%$ of E_{HO}

Measured value of $E_R = 5\%$ of E_{HO}

Estimated value of $E_p = \frac{K}{2} \times E_{og} = 23V$

$E_\phi \text{ (max)} = \frac{50{,}000 \times 5.6 + 23 + 10{,}400}{100} = 13{,}229$ V

I_{nf} (max) = 0.37 A actual and 0.42 A estimated.

Graph 2 below shows the performance of resonant grounding on this unit for the entire range of resistive faults.

Estimated value of I_{rf} (max) = (5.0 + j3.7)A

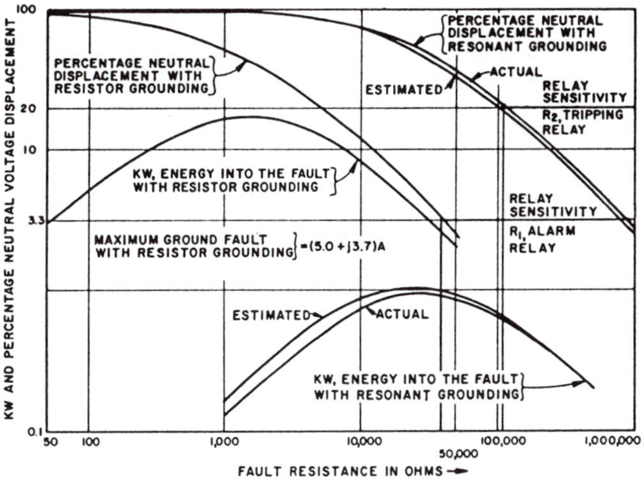

GRAPH 2

REFERENCES

(1) Westinghouse Transmission and Distribution System. Grounding Report 1, High-Resistance Grounding for unit Connected Generators Report No. 59-58 February 27, 1959 Revised April 22, 1974.

(2) Ground-Fault Neutralizer, Grounding of Unit connected Generators H.R. Tomlinson, AIEE Transactions, Vol. 72 Pt. III 1953 pp. 953-960.

(3) Experience of the New England System with Generator Protection by Resonant Neutral Grounding by E.T.B. Gross and E.M. Gulachenski IEEE Transactions, Vol. PAS92 No. 4, August 1973 pp. 1186-1194.

Discussion

H. C. Couch and N. E. Nilsson (Ohio Edison Company, Akron, Ohio):
The discussors wish to compliment the authors on an excellent, well prepared paper on resonant grounding techniques for large generators. The paper is well conceived and demonstrates the capability of resonant grounding to limit the amount of ground fault current.

In the process of limiting the amount of ground fault current, the ground fault relay selectivity is being sacrificed. Consider a single line-to-ground fault on the low side of the potential transformers used for unit relaying, metering, and voltage regulation. Assume that the potential transformers are wye-grounded and have an impedance of 2500 ohms (Z = 2000 + j1500 ohms). Figure 1 illustrates this case. For the resonant grounding scheme described in this paper, the current in the neutral of the generator for a fault on the low side of the potential transformers would be approximately 92% of the neutral current in the generator for a bolted line-to-ground fault at the generator terminals.

Manuscript received August 4, 1976.

If, on the other hand, the unit described in this paper had been equipped with a distribution transformer-resistor scheme that limited the generator neutral current to twelve amperes for a bolted line-to-ground fault at the generator terminals, then the previously described fault on the low side of the potential transformers would produce about 3.7 amperes in the neutral of the generator. This is illustrated in Figure 2. Note that this scheme can easily be modified to limit the fault current to a greater or lesser degree by simply selecting a different size equivalent resistor. In any event, the distribution transformer-resistor scheme affords adequate ranges of generator neutral current so that a critical tripping condition can be differentiated from an alarm condition, whereas the resonant grounding method does not. The discussors would be interested in the analysis of the selectivity problem that the authors used with the resonant grounding method.

The damage to a faulted machine is related to the magnitude of the fault current and the duration of the fault current. If the relaying associated with the two schemes tripped the machines at the same point in time, then the resistor grounding scheme will indeed input 19 times more energy to the fault than the resonant grounding scheme used for

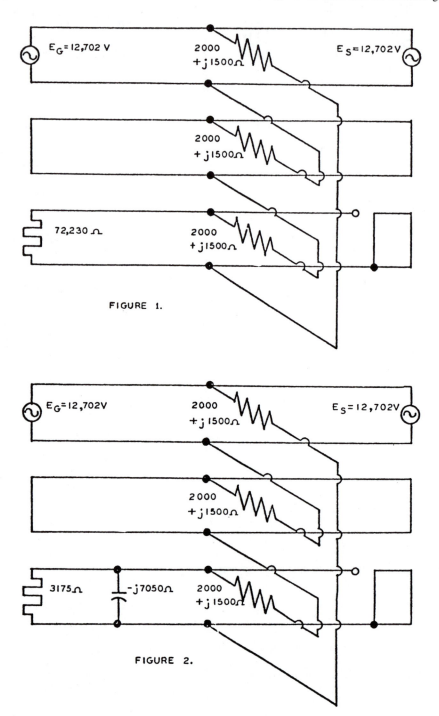

FIGURE 1.

FIGURE 2.

comparison in the paper for the 5000 ohm fault. The discussors would like to ask the authors if they expect any significant difference in the "down time" and/or repair cost for a faulted machine which employed resonant grounding versus one which used resistor grounding. There is evidence that there is no significant difference in the amount of damage caused in the cases described above. Hans Teichman, in his paper "Improved Maintenance Approach for Large Generator Armature Windings Subject to Insulation Migration (Transactions on PAS, July/August 1973, Number 4, Volume PAS-92, pp. 1234-1238)", indicates that no iron damage will occur for a single line-to-ground winding fault below ten amperes.

S. G. Whisenant (Duke Power Company, Charlotte, NC): The authors are to be commended for a well-organized paper describing the application of resonant neutral grounding to a unit connected generator. Theory of operation and the advantages of resonant grounding are briefly discussed and emphasis is placed on the design procedures for ground fault neutralizer implementation.

Several ground fault detection schemes are applicable to resonant grounded generators. Some of these schemes are a low voltage pick-up alarm and a higher voltage pick-up trip relay, a high voltage pick-up delayed trip relay, or an overvoltage relay with no trip. The authors mentioned in the introduction that resonant grounding provides a possibility of delayed trip for an orderly reactor shutdown. It would be informative for the authors to discuss their particular philosophy of operation during sustained ground faults.

Fault detection near the generator neutral is another advantage of resonant grounding which deserves more emphasis. The increased sensitivity provides more complete protection of the generator windings. However, slight system unbalances during normal conditions create an amplified zero sequence voltage because of series resonance in the generator neutral. Relays would have to be set as to prevent false alarms or false trips. There is a device on the market (manufactured by S & C Electric Co.) which compensates for system unbalances. The device contains an unbalance compensation module which compensates for error voltages by means of a summing amplifier. The amplifier combines the following three signals:

1. The neutral-to-ground voltage as sensed by a potential device.
2. The inverse component of neutral voltage caused by system voltage unbalances. This component is sensed at the bus through three wye-connected potential devices with a broken delta-connected secondary.
3. A compensating voltage which is adjusted in magnitude and phase to offset capacitor and inductor manufacturing tolerances.

An active filter removes harmonics before the signals are fed into a summing amplifier. The signals are added vectorially and the amplifier output is proportional to the neutral voltage displacement with extraneous components removed. An adjustable level detector is preset to operate a control circuit when the neutral displacement reaches a magnitude which requires relay operation. False trips caused by transient voltages can be avoided by a timing circuit which requires the signal to persist for a selectable period before a switching operation is initiated. An optional alarm module is available which gives early warning of system problems. This device can provide nearly 100% generator winding protection.

Manuscript received August 4, 1976.

Eric T. B. Gross (Rensselaer Polytechnic Institute, Troy, New York): A discussion of a paper dealing with resonant grounding would be incomplete without some historical facts added. Petersen invented resonant grounding some 60 years ago, recognizing from the beginning that fine tuning is not at all significant; he stated in his early patent application that the "grounding coil should be so adjusted that it is near to resonance;" he indicated also that iron core-air gap coils should be used. The first application[1] to generator protection was made by Tomlinson some 25 years ago and it has proved as successful as all other such installations.[2] New interest in resonant grounding of unit connected generators developed some 6 years ago because of the need for greater sensitivity emphasized and required by the ever increasing rating of the units.[3]

This paper is a valuable contribution to ground fault protection. It supplies pertinent information on the values of the various relevant capacitances to ground of all the equipment contributing to the current in a single phase to ground fault. Figures 1 and 7, for a 1000 MVA and a 600 MVA unit respectively, show the values of the pertinent capacitances to ground. The capacity of the generator stator winding is about ten times larger than the total capacity to ground of all other equip-

Manuscript received August 11, 1976.

ment. If generator breakers, provided to separate the generator from its bus, are open the change in degree of tuning will become about 10% over compensation; this is fully permissible since arcs to ground remain self-extinguishing for even ± 30% deviation from resonance.[4,5] It must be realized that the generator will rarely, if ever, operate for any length of time with such breaker open, so that it would not be worthwhile to change the coil tap setting for this unusual and temporary mode of operation. For the same reasons, the effect of resonant grounding is not impaired if the unit does not operate at its rated speed and frequency; such operation will also be unusual.

Small unbalances of the capacities to ground of the three phases produce a small system zero sequence voltage, which is amplified by resonant tuning.[6, 7] Any change of this unbalance will also change the magnitude and phase of the zero sequence voltage. Monitoring the zero sequence voltage results, therefore, in very sensitive indications of incipient ground faults,[2,8] and total stator winding, including its neutral will be protected. A recently developed relay[9] for the indication of unbalances in three-phase shunt capacitors is equally applicable here for complete protection.

With a number of apparatus connected to the generator bus of these large units, it is entirely possible that larger unbalances are established when one of these apparatus becomes accidentally disconnected on only one or two phases. As a result, undesirably larger zero sequence voltages may be developed with an air core coil of constant value.[8] The best remedy is self-detuning of the coil by saturation, one of the long and well recognized advantages of iron core coils.[10] A further advantage is the reduction of losses and the elimination of requirements for increased losses in the design of air core reactors. Saturation is equally useful in controlling the effects of ground faults on the high side of the unit.

REFERENCES

[1] See ref. (2) of the paper.
[2] See ref. (3) of the paper.
[3] Generator Ground Fault Protection by Resonant Grounding, Eric T. B. Gross, Proceedings of the Relay Conference, ECNE, Williamstown, Mass., Oct. 1970, and PEA, Reading, PA., Oct. 1971.
[4] The Why and How of Resonant Grounding, Eric T. B. Gross, Electric Light and Power, Vol. 25, 1957, July and August issues.
[5] Application of Resonant Grounding in Power Systems in the U.S., Eric T. B. Gross, E. W. Atherton, AIEE Transactions, Vol. 70, 1951, pp. 389-396.
[6] Practical Experience with Resonant Grounding in a Large 34.5 kV System, H. H. Brown, Eric T. B. Gross, AIEE Transactions, Vol. 69, 1950, pp. 1401-1408, Appendix.
[7] Neutral Grounding in High Voltage Transmission (book). R. Willhein, M. Waters, Elsevier Publishing Co., New York 1956, pp. 351-380.
[8] Sensitive Generator Ground Fault Protection, Eric T. B. Gross, Proceedings of the American Power Conference, Vol. 36, 1974, pp. 1031-1035.
[9] Improved Protection System Increases Capacitor-Bank Utilization, J. R. Cooper, J. A. Zulaski, Proceedings of the Relay Conference PEA of Oct. 27, 1972. Also: Technical Paper 740-T37, S&C Electric Company, Chicago, Ill., 60626.
[10] See discussion in ref. (2) of the paper.

P. G. Brown and **I. B. Johnson** (General Electric Company, Schenectady, N.Y.): The authors have significantly enlarged upon the published information on resonant grounding of generator neutrals by discussing some fundamentals thereon, equipment selection and installation. As the authors have stated, some important factors for consideration in the application of a particular method of grounding include the following: (1) potential overvoltage hazard; (2) fault damage; and (3) protection sensitivity. We would like to comment on these items.

In the 60 Hz resonant grounding scheme for unit-connected generators there is a question concerning overvoltages accompanying generator faults when the source frequency deviates from normal 60 Hz. Generated voltage frequencies above or below may arise from the following:

1. Generator startup.
2. Generator shutdown.
3. Sudden overload and underspeed.
4. Load rejection and resulting overspeed.

To explore the question, a generalized system was studied on the TNA as shown in Fig. 1. Three conditions of grounding were evaluated, namely: (1) isolated, (2) high resistance (distribution transformer and resistor) and (3) reactance (resonant).

Manuscript received August 12, 1976.

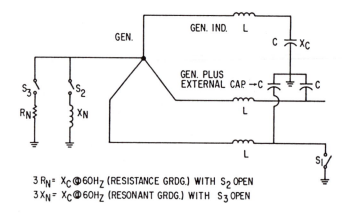

$3R_N = X_C$ @60Hz (RESISTANCE GRDG.) WITH S_2 OPEN
$3X_N = X_C$ @60Hz (RESONANT GRDG.) WITH S_3 OPEN

Fig. 1. TNA circuit studied.

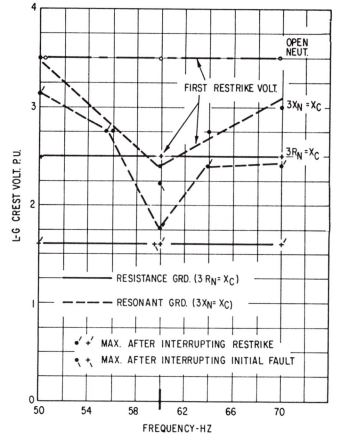

Fig. 2. Overvoltage on generator terminal of an unfaulted phase versus frequency of generator voltage for various grounding methods.

Base voltage: Prefault L-N crest voltage.
Reactances at normal frequency of 60 Hz.

The switch S_1 in Fig. 1 was adjusted to apply a fault in the vicinity of a voltage crest. It was either left closed or opened at the first fault current zero. Following interruption, S_1 was closed to simulate a restrike, maximizing for the highest voltage on a unfaulted phase. For the cases involving resistance grounding, the highest recovery voltage across S_1 prior to a restrike involved a single high frequency component. In resonant grounding, there is a second slow frequency component involving a beat frequency between the 60 Hz natural frequency of the tuned zero sequence circuit and the frequency of the generated voltage when it is other than normal 60 Hz. The switch S_1 was adjusted in some cases to obtain data with two restrikes.

The results on the maximum p.u. voltage on an unfaulted phase versus the generated frequency for one restrike and for fault interruptions at the fault are shown in Fig. 2. It is to be noted, assuming all other factors equal, that resonant grounding potentially may yield higher overvoltages than high resistance grounding during fault occurrences particularly at other than nominal frequency.

With regard to fault damage, field data obtained on a ground of asphalt-mica insulated machines which had experienced winding failures are shown in the following table for various fault current magnitudes as limited by various grounding methods:[1]

Ground Current Magnitude [A]	No. of Failures	No Repair		Partial Grinding		Partial Restacking	
		No.	%	No.	%	No.	%
0 to 10	6	6	100	0	0	0	0
100 to 1000	4	2	50.0	1	25.0	1	25.0
Above 1000	6	1	16.7	1	16.7	4	66.7
Total Failures	16	9	56.3	2	12.5	5	31.3

In none of these cases where the fault currents were in the 10 amp. range (characteristic of distribution transformer with resistor grounding) was core damage experienced.

In light of this type of service record, a question is raised on the discussion by the authors on the relative fault damage associated with 0.4368 amperes with resonant grounding versus 1.92 amperes with resistor grounding. Published service data quoted indicates there would be negligible core damage in either case. Would the authors have documented service data to the contrary?

While extended operation with low values of fault current might cause core damage there are more significant reasons for tripping a machine promptly on the inception of a ground fault rather than delaying tripping. This is the risk of a second fault which would result in fault currents of many thousands of amperes. This risk is present for any high impedance grounding; and the greater neutral displacement of resonant grounding portrayed by Graph 1 in the authors' paper and by the previously discussed TNA data suggests that these risks may be greater with resonant grounding.

The decision to continue operation after a fault condition has been detected must consider significant hazards such as the following:

1. Many failure modes involve more than one stator bar and a second point of failure can cause expensive damage from a double line to ground fault.

2. The stator winding ground is most apt to be the first signal of some other trouble within the generator. If this happens to be core melting or a breakage of a rotor part, tripping the generator immediately may minimize damage or avoid a catastrophe.

3. If a second ground occurs within the same phase, it will not be detected by the phase current differential relays. If two such grounds occur at the opposite ends of a phase, a line to neutral fault would exist with fault currents exceeding those from either a line to line or a three

phase fault. This could persist until it became a line-to-line fault that could be detected by the phase current differential relays. In either case, the fault currents would damage the stator core so severely that some restacking would most likely be required. This would require the removal of the complete winding and an outage of up to six months.

In the interest of safe, reliable generator operation, minimum generator damage when faults occur, and minimum repair down-time and costs, it would be most prudent to automatically trip the generator and excitation at the first ground occurrence. Since forced outage rates due to stator faults are responsible for only a very small portion of the over-all forced outage rate of a boiler – turbine – generator unit, the risk of continued operation would not be justified.[2]

Resonant grounding of generator neutrals is another form of high impedance grounding in the category of the distribution transformer with secondary resistor or the potential transformer. While it appears to be a satisfactory method of grounding, it is not apparent that it provides significant practical advantages to justify its special design and tuning requirements. From the standpoint of winding protection, all three methods are equivalent, seeing roughly within 3-5% of the neutral. Historically this appears to have been satisfactory.

Also, new techniques for the early detection and warning of local overheating in large steam turbine generators known as "core monitors" are now being applied.[3] This is an ion chamber detector sensing the presence of very small particles, of the order of condensation nuclei, resulting from the thermal decomposition of organic materials such as coatings and insulations. While not in any way replacing the need for good ground detection relaying it does provide a valuable supplemental means of protection sensitivity.

In summary, we have the following additional questions:

1. What is meant by "field experience" in Appendix I and how did this experience specifically verify the curves in Graph 2?

2. What is meant by the term fault damage where the fault impedance is in thousands of ohms? That is, how much of the calculated energy is in the failed insulation and how much of it can be considered as directly contributing to core damage which might affect repair time?

3. Since generators are operated occasionally at off-nominal frequencies as in load rejection and during startup and shutdown with excitation applied, would the authors comment on the safety and sensitivity of resonant grounding for these conditions?

4. With reference to plants utilizing a low voltage breaker or load break switch (as now seen in some large nuclear plants) what provisions in the resonant grounding system are required to insure adequate tuning prior to synchronizing?

REFERENCES

[1] Hans T. Teichmann, "Improved Maintenance Approach for Large Generator Armature Windings Subject to Insulation Migration." *IEEE Transactions on Power Apparatus and Systems,* July/Aug. 1973, pp. 1234-1238.

[2] M. L. Crenshaw and J. A. Massingill, "Discussion of authors' Ref. 3." *IEEE Transactions on Power Apparatus and Systems,* July/Aug. 1973. p. 1192.

[3] C. C. Carr.., S. C. Barton and F. S. Echeverria, Immediate Warning of Local Overheating in Electric Machines by the Detection of Pyrolysis Products," *IEEE Transactions on Power Apparatus and Systems,* March/April 1973, pp. 533-542.

K. J. S. Khunkhun, J. L. Koepfinger, and **M. V. Haddad:** Dr. E. T. B. Gross has advocated resonant grounding for the past many years and has written a significant number of papers on this subject. We are grateful to Dr. Gross for his discussion, which actually provides answers to many questions raised by other discussors. The use of an iron-core air-gapped coil as proposed by Dr. Gross has considerable merit. He had indicated several advantages of using an iron-core air-gapped reactor over an air-core reactor. Its application would require slightly more study than is needed to apply an air-core reactor, since one would want to assure that saturation does not occur beyond acceptable limits.

H. C. Couch and N. E. Nilsson have raised the question of ground fault relay selectivity during a single phase to ground fault at the secondaries of potential transformers (PT's). The easiest way to overcome this problem is by leaving the neutral of the secondary PT's ungrounded and by establishing a ground at B-phase.[1] The advantage of grounding one phase is that it prevents operation of the generator ground relay for a secondary phase to ground faults. In this method, the phase-to-neutral potential to burdens such as three element watt-meters, can be obtained by delta-wye connected auxiliary PT's. Answering the second question these discussors raised, we certainly believe that the reduction in fault current by resonant grounding not only will reduce the damaging effects of a ground fault but also minimize the generator ground faults by reducing the stress on insulation.

Steven G. Whisenant's discussion about the possibility of improving different sensing schemes and providing nearly 100 percent generator winding protection should be encouraging to many utilities which might like to improvise similar schemes and take full advantage of resonant grounding. We appreciate his suggestions.

P. G. Brown and I. B. Johnson have provided a very interesting discussion. We reiterate here a comment made by Dr. Gross, that there is a distinction between what a study shows can happen and what experience shows does happen. New England Electric experience, for the past 25 years, confirms the effectiveness of resonant grounding.[2] Ground fault neutralizers, as indicated in Dr. Gross's discussion, are effective even if the deviation from resonance is as much as ±30%; the fault is self-extinguishing. Since the current flowing during the fault is primarily resistive, the recovery voltage at the time of current extinction should be in phase with this current. Thus, the probability of a restrike would be low. It is not stated in the discussion whether the TNA study recognized the Q of the reactor. This varies from 15 to 50 and could have an appreciable effect upon the results, since the effective impedance of the neutral ground is R = Q (wL). Future TNA studies should investigate the effects of the Q upon the transient overvoltages during faults. If the

Manuscript received October 5, 1976.

practice of applying generator field at less than rated speed for generator preheating is not used, there is little exposure to operating a generator with field applied at other than rated frequency. In the case of a declining frequency condition such as occurs during major system separations, most modern units cannot operate in a stable manner at less than 57 Hz. According to the discussors' Figure 2, there is a little significant difference in the overvoltage by either resistance or tuned reactance grounding.

We ran an actual frequency sensitivity study on our Brunot Island Unit No. 4, which is resonant grounded to determine the effect of frequency on the resonant tuning. This was done after achieving the optimum tuning available with the taps on the reactor and transformer. During operation of the generator, a 60 Hz and 180 Hz voltage appear across the tuning reactor. The test gave the following results:

Frequency in Hertz	60 Hertz Voltage	% off 60 Hertz	180 Hertz Voltage
58	3.8	−5	2
58.4	3.9	−3	1.9
58.8	4.0	0	1.85
59.2	4.1	+3	1.85
59.6	4.1	+3	1.8
60.0	4.0	0	1.81
60.4	4.0	0	1.80
60.8	3.9	−3	1.80
61.2	3.8	−5	1.80
61.6	3.8	−5	1.80
62.0	3.7	−8	1.79

It was observed during these tests, from a scope, that the system was in maximum tuning at 59.4 Hz. Thus, it might be said that the tuning used was 1% off optimum. The most significant observation to make is that as frequency varies from 58 to 62 Hz, the tuning apparently changed only a maximum of 8%. This shows that there is relatively little effect of a change in frequency upon the tuning of a ground fault neutralizer since 8% is well within the acceptable limits of application of this device.

Authors have not made any firm recommendation for relaying schemes, but we believe that it is always good practice to remove a fault as soon as it is detected, as is indicated by the discussors. It is recognized that some operators of generation equipment have had good experience in not following this practice. When applying ground fault neutralizer, both experience and design limitation must be considered. The Fossil Unit mentioned in Appendix 1 has three relays across the reactor; two relays are sensitive only to fundamental frequency, one alarms for a resistive fault of about 1,000,000 ohms, and the other trips for a resistive fault of about 100,000 ohms. There is also a third relay which responses to all harmonic voltages. This is an undervoltage relay and alarms for a floating neutral, in case the neutral equipment is accidently disconnected. Mr. Whisenant has suggested another good protection scheme in his discussion. The answers to the last four questions are as follows:

(1) The actual performance curves shown in Appendix 1 are for the tested value of R_0 (zero sequence impedance in the generator neutral), which was obtained by applying 104 V 60 Hz and measuring the current at resonance, which was 0.0037 amp. The estimated curves were obtained as outlined under Performance of Resonant Grounding in the paper.

(2) For comparison purposes, all of the calculated energy could be assumed to be causing damage at the point of fault, anywhere in the unit equipment.

(3) A small expected deviation during load rejection does not effect the sensitivity appreciably.

(4) Ninety percent of the system capacitance is due to the generator; the remaining ten percent would not offset the tuning so as to cause any concern. We refer again to the discussion offered by Dr. Gross; he has some other good suggestions.

REFERENCES

[1] Potential Transformer Application on Unit-Connected Generators, IEEE Committee Report 70 TP 18 PWR. Presented at IEEE Winter Power Meeting, New York, 1970.

[2] See Reference 3 of the paper.

COMPLETE MOTOR PROTECTION
BY MICROPROCESSOR RELAY
by W.A. Elmore and C.A. Kramer

Introduction

Microprocessor technology has afforded us the opportunity to re-examine the basic requirements of motor protection. Traditional fault detection needs had to be fulfilled, and insulation temperature restrictions dictated their usual constraints. However, new innovations in setting procedures, data storage and access, alarm practices, and the nature and scope of protection practices were open to careful review during the development of the MPR, Motor Protection Relay.

With the logical power of the microprocessor, even the simple process of determining what to do when tripping is called for required new analysis. Do you close a contact? Open a contact? Do both? Do you always seal the trip, never seal the trip, sometimes seal the trip? Flexibility of application dictated our final choice.

Traditionally only a single RTD (resistance temperature detector) has been connected for relaying purposes. The power of the microprocessor allows essentially continuous examination of several RTD's at various points in the motor.

Another key difference from traditional practice made possible by the hardware employed, was more careful tailoring of heating and cooling response and allowing tripping characteristics to be adaptive to previous loading conditions.

A protection engineer is generally given limited information to work with in establishing protective characteristics desired for a particular motor. He knows permissible continuous allowable temperature rise, full load current, locked rotor current, permissible locked rotor time and possibly accelerating time. He may even be blessed with a complete thermal damage time curve for the motor derived from a combination of science and the motor designers best estimate. From this fairly restricted viewpoint, the user must develop settings for complete protection of the motor. The relay described here allows a wide choice of settings and functions.

Locked Rotor Protection

With the rotor blocked, losses in a motor are divided between the stator and rotor in accordance with the relative resistances of the stator and the 60 hertz resistance of the rotor. Extreme heating takes place, particularly in the rotor, and this locked rotor condition can be tolerated for only a very limited time. This time varies with the level of voltage applied to the motor terminals. It is an I^2t limit. The operating time of the protective device, then, must have a characteristic that follows an I^2t function and have the ability to be set for all reasonable values of permissible locked rotor time over the range of expected locked rotor current values.

Reprinted with permission from *Georgia Tech Protective Relaying Conference*, Atlanta, GA, pp. 1–13, April 29–May 1, 1987.

Thermal Protection Without RTD's

The heating associated with motor loading can be approximated by a consideration of $I_H^2 = I_1^2 + K I_2^2$ where I_1 is stator positive sequence current, K is a weighting factor to describe both the increased rotor voltage generated by the counter-rotating flux associated with negative sequence current and the increase of rotor resistance due to skin effect in the rotor bars at 120 hertz minus slip frequency. I_2 is stator negative sequence current in per unit of rated full load current.

Using I_H as a criterion for tripping allows the motor to be protected throughout the full range of stator current with and without unbalance. The time-current characteristic is a constant $I_H^2 t$ relationship as shown in Figure 1. The curve position is set for the locked rotor condition, slightly below the permissible (full-voltage) locked time. Also, it is cut off at the current pickup level chosen by the user, typically 115% to 125% of full load.

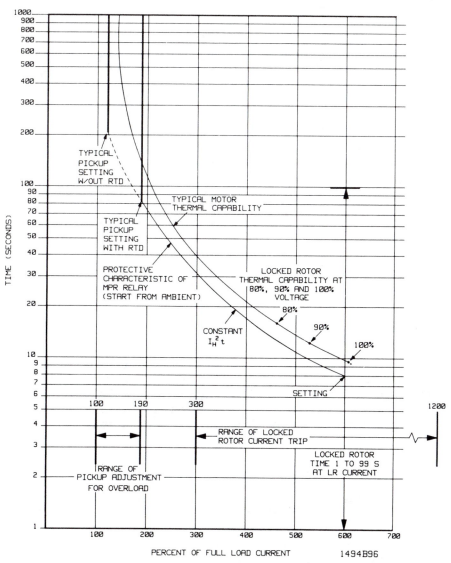

Fig. 1 Typical MPR Characteristics.

Positive and negative sequence currents are defined as:

$$3I_{A1} = I_A + aI_B + a^2I_C$$

$$3I_{A2} = I_A + a^2I_B + aI_C$$

where I_A, I_B and I_C are phase currents and "a" is operator $1/\underline{120}$.

The normal analog process for extracting $3I_{A1}$ from the three phase currents is to rotate I_B by 120°, I_C by 240° and add both to I_A. Using digital techniques we can take a sample of I_A, add that to a sample of I_B taken 120° later, then add the sum to I_C taken 240° later, giving the instantaneous value of $3I_{A1}$ that existed at the time of the I_A sample. Figure 2 describes this process. A similar process may be used to extract $3I_{A2}$ from the individual samples of I_A, I_B and I_C that are taken at 5.55ms intervals.

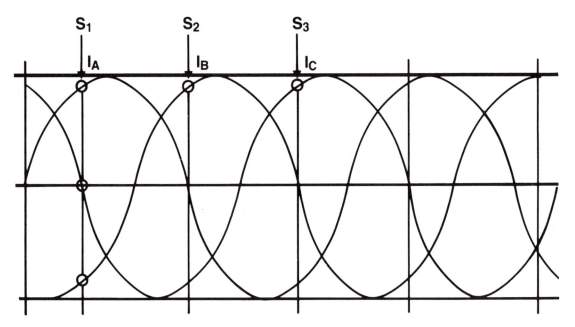

Fig. 2 Sum of Samples Taken 120° Apart Equivalent to $I_A + aI_B + a^2I_C$ Taken at S_1

The weighting factor used with I_2 allows the exaggerated heating effect associated with I_2 to be taken into account. A conservative value for K is $230/(I_{LR})^2$. This large value allows tripping to take place when the motor is in jeopardy even though all three phase currents are less than full load level.

Thermal Protection With RTD's

Where the machine is equipped with RTD's (resistance temperature detectors), highly refined protection can be applied. The RTD is an excellent indicator of the average temperature of the insulation of the stator winding because of its proximity to it. It is influenced by the recent loading history, the effects of ambient temperature and ventilation variations in a manner similar to that of the motor insulation itself.

Six winding RTD's (along with 4 bearing RTD's and 1 load casing RTD) can be used with the MPR relay. Copper (10 ohm), platinum (100 ohm) or nickel (either 100 or 120 ohm) RTD's can be accommodated, with the only restriction being that they must be the same type in any particular group. Any trip temperature can be set from 10° to 190° to cover the full range of continuous temperature levels allowed for insulation classes likely to be encountered.

The relay is intended to be used with 3 wire RTD's. Figure 3 shows how this configuration is used to assure a high degree of accuracy in measurement. A precision current is circulated by the microprocessor from point A through the RTD, returning to point C. The resulting voltage drop is measured. Then the same current is circulated from B through the two leads and back to point C. Voltage is again measured. By subtracting the latter voltage from the former, simple V/I calculations allow the RTD resistance to be calculated by the microprocessor and the corresponding temperature to be stored or displayed. The error associated with the resistance of the leads is removed and no concern need be given about the temperature variations of the leads. Simple translation from resistance to temperature is made possible by a knowledge of the physical properties of the particular material used for the RTD.

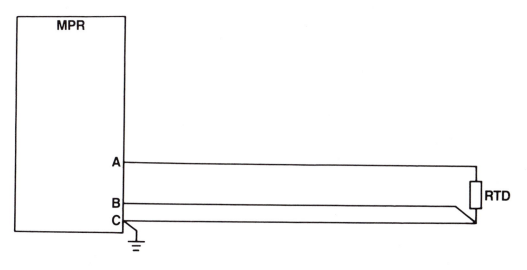

Fig. 3 Example of Connection to 3-Wire RTD's.

An alarm level is set by the user 2° to 20^oC below the selected trip temperature to alert an operator to the need for unloading the motor to avoid tripping. If the temperature indicated by the RTD exceeds the trip setting, the motor is removed from the line immediately. Tripping is, of course, delayed for this case by the time required for the temperature of the motor and the RTD to rise to a dangerous level.

For the locked rotor case, the RTD temperature is inadequate alone. For such high currents, there is a recognizable delay between the rise in insulation temperature and the rise in RTD temperature. Therefore, "I_H-sensing" is used to complement the RTD temperature measurements. A much higher setting for current pickup of the I_H element may be used in this application. This allows the RTD to provide moderate-overload protection and the I_H characteristic to provide locked rotor and unbalance protection.

273

The heating (and cooling) algorithm for the MPR relay was chosen so that the rate of rotor heating is related to I_H and to stator temperature, as indicated by the RTD resistance. I_H^2 feeds an accumulator up to the level identified by the locked rotor time setting and the locked rotor current setting, where tripping occurs. The RTD temperature establishes the starting level in the accumulator, thus acknowledging previous loading history. The analog equivalent of this is described in Figure 4.

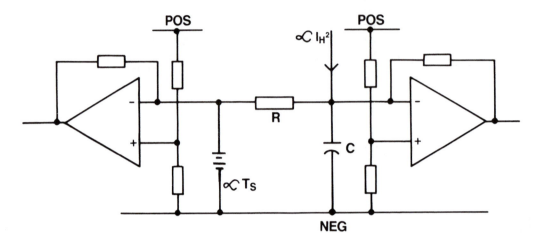

Fig. 4 Analog Equivalent of Thermal Algorithm.

For prolonged steady state operation, there is a fixed differential between the RTD temperature and the equivalent rotor temperature determined by the algorithm. Tripping will result when excessive temperature is recognized for either the stator or the rotor.

A shorted or opened RTD is recognized by the fact that its resistance is determined to be outside of the "reasonable" band. Detection of this causes the display to flash off and on. The RTD that is in trouble is identified in the "Function" display and EEEE in "Value" display indicates the measured value to be out of bounds. "Resetting" removes the EEEE error message, the faulty RTD is no longer sampled and other RTD's continue to be sampled. Whenever power is removed and restored, the faulty RTD is again sampled, and the error message is displayed to acknowledge the RTD deficiency.

Phase Reversal Protection

Reverse rotation of a motor can produce disastrous effects on the driven load. Immediate recognition of the occurrence and corrective action is mandatory in such cases. One cause of reverse rotation is reverse phase sequence of the supply, often produced by the exchange of two phase leads. This is readily detectable by an examination of the sequence of "positive-going" current in the three phases. Phase sequence ABC is normal. Detection of phase sequence ACB (detected twice for security reasons) is sufficient cause for immediate tripping with "Phase Reversal" indication.

Jam Protection

"Jam" is referred to as any phenomenon which causes binding action of the motor, of the bearings, or of the driven load resulting in excessive torque requirements and overcurrent in the motor. This is detectable by the relay but must be differentiated from other phenomenon that produce overcurrent.

High current is not recognized as a possible "Jam" condition unless the motor has been up-to-speed. The motor is determined to have been up-to-speed only if current in excess of 20% of full load has persisted for twice the "Locked-Rotor Time" setting. During starting, high current exists for only the time required to accelerate the load. This time invariably will be less than twice allowable locked rotor time. Starting current then, is prevented from causing an incorrect sensing of "Jam".

Another high current case that must be screened out occurs when the motor "contributes" to a nearby fault. This will last for only a few cycles for an induction motor. The "Jam Time Delay" setting must be chosen to override this.

Thus, if the motor is up-to-speed and experiences current in excess of the "Jam" setting (0.7 to 12 times full load) for a time in excess of the Jam delay (1 to 10 seconds), tripping occurs and the "Locked Rotor/Jam" LED becomes illuminated. No other targets are lighted for this condition.

Load Loss

Load loss refers to the sudden reduction of shaft load caused by such conditions as the shearing of a drive pin, shaft breakage or loss of prime on a pump. Any such condition requires immediate recognition and taking the motor off-line to eliminate or minimize damage to the driven load.

Detection, with security, requires recognition of the difference between no load _following_ the application of load (load loss) and no load _preceding_ the application of load. This function is not usable in applications where the motor can have its load removed without the motor being taken from the line. This function can be easily selected or deleted in service by simple push-button action and observation of the display.

A minimum operating load level is selected by setting a current value for this function. If the motor current exceeds 20% of full load current, a reasonable approximation of exciting current, and it persists for more than twice locked rotor time, the motor is presumed to be loaded. If current now reduces to a value between the minimum load setting and 20% of full load current and this persists for the pre-set load loss delay, this is identified as load loss and tripping takes place. If the motor were to have been simply removed from the line, the stator current would have immediately dropped to zero, a level being easily distinguishable from exciting current level.

Fault Protection

Faults are somewhat more prevalent in motors than other power system devices, probably because of the violent manner and frequency with which they are started. Ground faults are more likely than phase faults, but both types are expected and must be detected and cleared. Instantaneous phase and ground settings can be chosen, independent of one another, and also phase and ground fixed time delays can be set independently.

Ground current is usually sensed in motor protection using a "doughnut-type" ct, surrounding all three of the phase conductors, but residual current may alternately be used. The "doughnut-type" ct scheme may also be used in a phase differential arrangement where a 6-lead motor is involved as shown in Figure 5. This allows comparison of current "in" with current "out" of each phase winding. If they differ, of course, there is a phase or ground fault present in the winding and sensitive detection (0.1 to 1.0 amps) is possible.

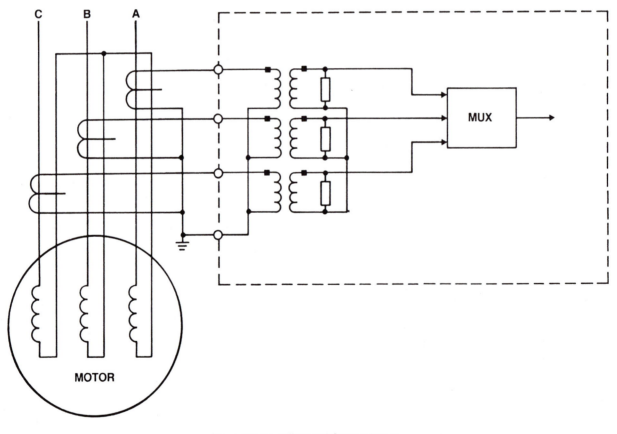

Fig. 5 Phase Differential Current Inputs.

All of the overcurrent units have a "speed-up" circuit to provide faster operation in the event the current level is well above instantaneous pickup. This allows a much higher value of current to be sensed sooner by the microprocessor for currents above this level.

Value and Setting Display

It is useful to have access to information regarding the status of various quantities in the motor. By selecting a function number, 1 through 19, by "Raise" and "Lower" buttons, various quantities may be identified and displayed. These are % load current, phase current, RTD temperature, percent of trip value for ground current and phase differential.

The remaining function numbers that may be selected, 20 through 56, allow trip levels, alarm levels and operating options to be chosen and entered into non-volatile memory. Each function may be disabled as the application dictates.

Hardware Architecture

The performance and versatility of a microprocessor allows for the design of a relay which can provide several protection functions for motors. Using techniques of converting real-time analog signals to digital data for calculation by a microprocessor, yields a complete protection system which is uniquely tailored to motor operating conditions.

Figure 6 shows a block diagram of the microprocessor motor protection relay. Inputs to the relay are switched by multiplexing circuitry (MUX), fed to the analog to digital (A/D) converter, then output directly to the microprocessor. The microprocessor controls this multiplexing, as well as the setting module, the display/indication module and outputs to the trip and alarm relays.

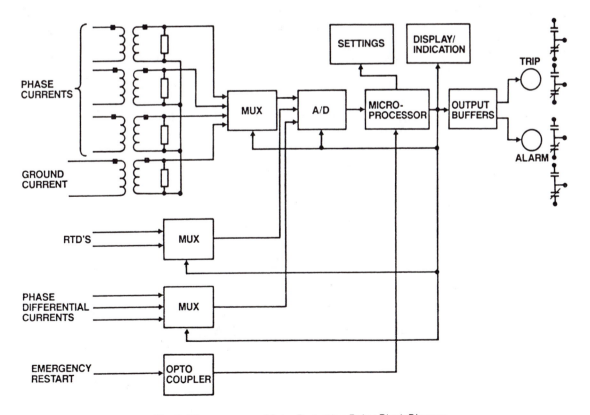

Fig. 6 Microprocessor Motor Protection Relay Block Diagram.

Phase currents, ground currents and phase differential currents are supplied to individual current transformers. RTD input is a voltage level, proportional to the resistance of the RTD, and, hence, temperature. Each of the signals is multiplexed and individually analyzed by the microprocessor.

The heart of the relay is an integrated circuit "chip-set", shown in Figure 7. It consists of a Westinghouse custom linear IC, an A/D converter and an INTEL 8-bit microprocessor. The custom linear IC performs several functions: 1. Multiplexes the analog phase current signals, 2. provides an autoranging gain to control the current input signals, 3. provides a reset function to the microprocessor during power-up conditions and during the microprocessor's self-diagnosis routine, 4. regulates the power supply voltage.

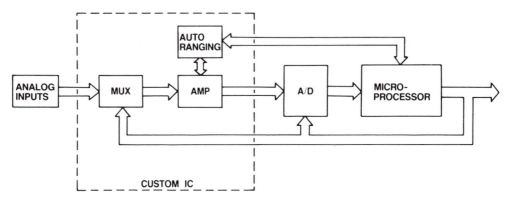

Fig. 7 Integrated Circuit Chip Set.

The interface between the microprocessor and the settings, display/indication and outputs is shown in Figure 8. The setting function of the microprocessor relay provides the capability to tailor the relay to the motor characteristics. Through the use of pushbutton switches, parameters can be programmed for each function of the relay, including primary full load current, ct ratio, frequency, types of RTD's, as well as trip and alarm levels and time delays. Each of these values are stored in a non-volatile memory until changed through a setting procedure, and are retained on "power-down" of the relay.

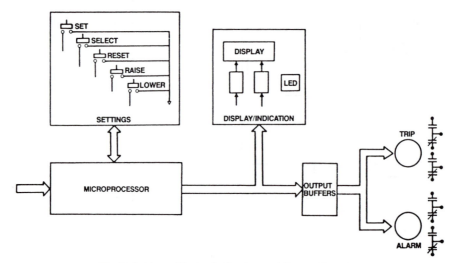

Fig. 8 Settings, Display/Indication and Output Functions.

The display/indication module includes a vacuum fluorescent display interfaced with the microprocessor through two microcontroller display drivers. Through the use of pushbutton switches, shown in the photograph of the relay in Figure 9, each motor operating value, shown in Table 1, can be displayed by incrementing or decrementing to each function.

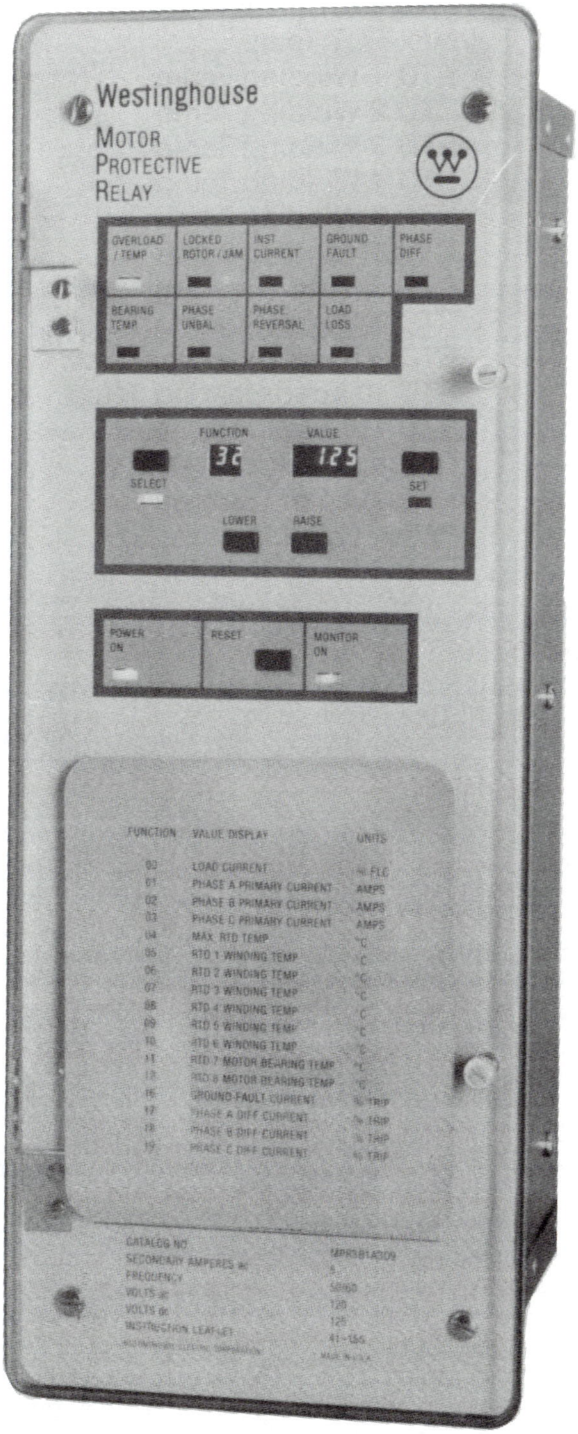

Fig. 9 MPR Microprocessor Relay.

TABLE 1
MOTOR OPERATING VALUES

FUNCTION	VALUE DISPLAY	UNITS
00	Load Current	% FLC
01	Phase A Current	Amps
02	Phase B Current	Amps
03	Phase C Current	Amps
04	Max RDT Temp	°C
05	RTD 1 Winding Temp	°C
06	RTD 2 Winding Temp	°C
07	RTD 3 Winding Temp	°C
08	RTD 4 Winding Temp	°C
09	RTD 5 Winding Temp	°C
10	RTD 6 Winding Temp	°C
11	RTD 7 Motor Bearing Temp	°C
12	RTD 8 Motor Bearing Temp	°C
13	RTD 9 Load Bearing Temp	°C
14	RTD 10 Load Bearing Temp	°C
15	RTD 11 Load Case Temp	°C
16	Ground Fault Current	% Trip
17	Phase A Diff Current	% Trip
18	Phase B Diff Current	% Trip
19	Phase C Diff Current	% Trip

When a trip has occurred, the value of current or temperature that caused the trip will remain in the display until reset. LED's provide indication for trips and alarms. Trip LED information is also stored in a non-volatile memory and retained until reset.

Trip and alarm signals from the microprocessor are fed through buffers to the output relays. The normally energized alarm relay will drop out for an alarm condition and the trip relay will pick up when trip action is called for. The trip output is an electrical reset relay with two form-C contacts. Three trip modes are selectable. Where conventional dc tripping is desired, the "normally open" contact closes for a trip condition to energize the breaker trip coil, and the reset coil is energized automatically by the microprocessor when the trip condition is removed, resetting the trip relay (Mode 3).

Where ac control is used, the normally closed contact is caused to open for a trip condition, causing the contactor to dropout. For 3-wire ac control, no provision is included for automatic restart. The trip relay stays operated until the relay is reset (Mode 1). If automatic restarting is desired when the motor "cools down" and 2-wire ac control is used, Mode 2 is selected.

A signal input to initiate a motor emergency restart will reset the I^2t tally of motor heating in memory to a half-level point, allowing a restart of the motor with the premise that it is a "hot" start. The trip relay, LED and value display will also be reset.

Through integration of a microprocessor, its support hardware devices and software programming techniques, a programmable multifunction relay has been developed to provide the complete motor protection relay shown in Figure 9.

Unique Characteristics

Unique characteristics made possible by the use of the microprocessor are:

1. Display many readings, not just a limited few, all on the same display element.

2. Protective functions in smaller space.

3. Retain current magnitude at time of trip or temperature at time of trip.

4. Retain target information following loss of power to the relay.

5. Allow all types of RTD's to be accommodated with the same relay.

6. Examine continually the temperature of all RTD's, not just the one determined to be hottest.

7. Minimize error in RTD reading by sampling technique and subtraction.

8. Detect faulty RTD, by out-of-band determination.

9. Allow emergency restart with possible sacrifice of insulation life through pre-set value in accumulator.

10. Disallow invalid settings.

11. Provide several self-checking functions.

Conclusions

This paper has described the MPR motor protection relay which incorporates the logic for detecting all of the well recognized malfunctions that can occur to a motor as well as providing easy access for setting, indication and display. It easily provides several functions that are difficult to achieve by conventional methods such as "Jam" and "Load Loss" and, further, it provides several functions uniquely suited to the microprocessor.

References

1. ANSI/NEMA Standards Publication/No. MG-1-1978 (R1981) Paragraphs MG-1-14.34 and MG-1-21.81.

2. C.F. Wagner, R.D. Evans "Symmetrical Components" Robert E Krieger Publishing Co. Malabar, Florida.

MOTOR ANALYSIS AND THERMAL PROTECTION

S.E. ZOCHOLL, IEEE FELLOW

WESTINGHOUSE ABB POWER T & D COMPANY
35 NORTH SNOWDRIFT ROAD
ALLENTOWN, PA 18106

Abstract - A motor starting study can be completed in a matter of minutes using a PC program written for the purpose. The study, run using readily available nameplate, load, and thermal limit data, determines the most significant motor protection criteria--rotor temperature rise plotted in perunit of thermal limit. The parameters for the electrical, mechanical, and thermal models are determined from the source data and used by the program to produce plots of current, torque, speed, and rotor temperature verus time. The paper identifies the necessary source data and explains the method of analysis using the example of a 6000 hp high inertia draft fan motor.

Key Words: Motor starting analysis, high inertia starting, rotor protection, thermal protection.

INTRODUCTION

In order to gain a detailed understanding of a specific application, a protection engineer should conduct a complete motor starting study. Whereas there is a tendency is to think only in terms of starting or locked rotor current duration, the primary purpose of the study is to determine the ultimate protection criteria--the temperature rise in the motor.

The general perception is that source data is difficult to obtain when in fact the study can be run using readily available nameplate rating, torque and thermal limit data. It is practical for the protection engineer to run the study since it can be completed in a matter of minutes using a PC application program (1) written to calculate and plot the voltage, current, speed, torque, and temperature as a function of time.

The benefit is that the study accesses the data usually supplied by others including the output plots which provide the same information as an oscillogram from a full scale field test. The duration of the starting current is needed to coordinate overcurrent relays for locked rotor protection. In addition, the slip variations of current magnitude and phase angle are vital in the application of impedance relays used for speed supervision of the overcurrent locked rotor protection in high inertia motor applications.

The study will be most useful in the application of advanced microprocessor based protection employing rotor and stator thermal models. In this type of protection the relay monitors continuous estimates of temperature due to the starting condition. The relay takes as its characteristic the dynamic thermal limit of the motor itself and the rise of temperature in perunit of the thermal limit determined by the study is also the dynamic response of the relay.

This paper reviews the method of analysis and

90 WM 247-7 PWRD   A paper recommended and approved by the IEEE Power System Relaying Committee of the IEEE Power Engineering Society for presentation at the IEEE/PES 1990 Winter Meeting, Atlanta, Georgia, February 4 - 8, 1990. Manuscript submitted July 5, 1989; made available for printing January 5, 1990.

identifies the required source data. The analysis is demonstrated using the example of a 6000 hp high inertia induced draft application.

Analysis Using Analytic Models

Induction motor starting can be analyzed using electrical, mechanical, and thermal models which interact as diagrammed in Figure 1. The electrical model consists of the steady state current equations for the motor equivalent circuit shown in appendix A. In the electrical model variables voltage, V, and slip, S, determine the rotor current.

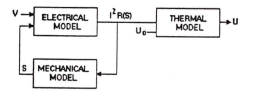

Figure 1 Block diagram of interactive models

The summation of all torques acting on the motor shaft comprise the mechanical model. The driving torque developed by the motor is resisted by the load torque and the accelerating torque of the moment of inertia of the rotating elements which are all slip dependent. In the mechanical model, slip is determined by the torque developed by a component of the total power in the rotor.

The thermal model is the equation for temperature rise due to current in a conductor determined by the thermal capacity, the thermal resistance, and the slip dependent rotor $I^2 R$ watts. Here, as well as in the electrical model, it is important to model the slip dependence of the rotor resistance as explained in reference (2). This is because the I^2R of the rotor decreases with decreasing slip and is the property which accounts for the difference in heating between starting current and locked rotor current so important in the high inertia case. As the ultimate protection criteria, the rotor thermal model is used to estimate the rotor temperature, U, resulting from the starting condition with initial temperature U_0.

A recursive solution using a finite time increment is used because the rotor impedance changes continually with slip. During each increment the input and initial conditions are held constant to calculate the output and parameters for the next time increment. At the end of the increment the output quantities are plotted and the initial conditions are updated. This process is repeated until the run is completed. This process is also called a simulation because the evolving plots emulate the actual starting process. The PC program (1) can be written to determine the model parameters from the following source data:

Reprinted from *IEEE Trans. Power Delivery*, vol. 5, no. 3, pp. 1275-1280, July 1990.

Source Data Of the Electrical Model

The motor application data can be considered complete when it can be used define the electrical, mechanical, and thermal models. For example, assuming the typical values of 83 degrees and 13 degrees for the phase angles of the locked rotor current and the full load current respectively, the impedances of the electrical model are defined by the following four items of nameplate data:

FLA - rated full load current   735 amps
LRA - locked rotor current   5000 amps
FLW - rated full load speed   713 rpm
SynW - synchronous speed   720 rpm
Q_{LR} - perunit locked rotor torque   1.2 pu

The specific values shown are for a 6000 hp, 4160 volt, 10 pole induced draft fan motor.

The correlation between the source data and the parameters of the electrical model can be established from the motor curves of the perunit current, torque, and power factor versus slip shown in Figure 2. The current curve can be thought of as the result of calculating the rotor current and phase angle using the electrical model with constant voltage and slip values ranging from 1 down to rated slip and the torque curve as having been derived from current, slip and rotor resistance in the relation:

$$Q_M = I^2 R/S \qquad (1)$$

where Q_M is the motor torque, I is rotor current, R is the rotor resistance, and S is the slip.

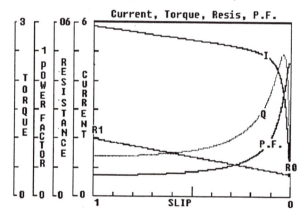

Figure 2 Motor Current, torque, and power factor versus slip.

This process can then be reversed and the total perunit impedance at locked rotor and at rated slip are the reciprocals of the plotted current values. The R and X values are then determined from the equivalent circuit using the corresponding values of slip and the phase angle values given above.

The slip dependent rotor resistance can then be separated using equation (1) in the form:

$$R = (Q_M/I^2)S \qquad (2)$$

As shown in Figure 2, the rotor resistance calculated from equation (2) is virtually a linear function of slip decreasing from a high locked rotor value to as little as one third of that value at rated slip. Consequently, it can be expressed in terms of the calculated values of the locked rotor resistance, R1, and the rotor resistance at rated slip, R0, calculated from equation (2) as follows:

At locked rotor: I_L = LRA/FLA = 6.8   S = 1

$$R1 = (Q_{LR}/I_L 2) = 1.2/(6.8)^2 = .026 \quad (3)$$

Because both the perunit current and the perunit torque equal one at rated slip:

$$R0 = S0 = (SynW - FLW)/SynW \qquad (4)$$
$$R0 = (720 - 713)/720 = .010$$

The rotor resistance, R4, at any slip is then given by the equation:

$$R4 = (R1-R0)S + R0 \qquad (5)$$

Source Data of the Thermal Model

Appendix B shows how the thermal limit curve of an induction motor is derived from the equation for the temperature rise due to current in a conductor and is the basis for the thermal model. Figure 3 shows the thermal model in the form of an electrical analog where the thermal capacitance and thermal resistance are direct analogies. The heat source is represented by a constant current numerically equal to the watts in the rotor and the temperature rise above ambient is represented by the capacitor charge voltage.

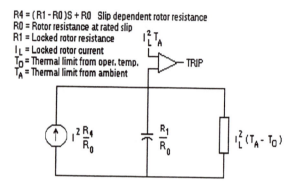

Figure 3 Rotor thermal model

It can now be seen that the slip dependent rotor resistance and the rotor current define the heat source in the rotor thermal model to be $I^2 R4/R0$. The heat source is expressed in perunit of watts at rated slip by dividing by R0. With this convention, the thermal capacity is R1/R0. As explained in Appendix B, the thermal resistance of the rotor is:

$$I_L^2(T_A - T_0) \qquad (6)$$

Figure 3 shows that parameters of the thermal model are defined by the following thermal limit times:

T_{ar} - rotor thermal limit starting from ambient 46 sec.
T_{or} - rotor thermal limit from operating temp. 40 sec.

together with the following calculated parameters:

I_L - perunit locked rotor current   6.8 pu
R1 - Locked rotor resistance   .026 pu
R0 - rotor resistance at rated speed(rated slip).010pu

Source Data of the Mechanical Model

The parameters of mechanical model is defined by the following items of source data:

RHP-rated horsepower of the motor   6000 hp
WR^2-moment of inertia (all rotating elements)695000 lbs -ft2
L-initial load torque (perunit of full load)   .4 pu
F-perunit load torque at rated slip   .6 pu

The full load and synchronous speed, FLW, and SynW, already listed and used for the electrical model will also be used here.

The equation of the total shaft torque including the motor resisted by the load torque and that of the moment of inertia comprise the mechanical model. A time discrete form of the equation useful in the

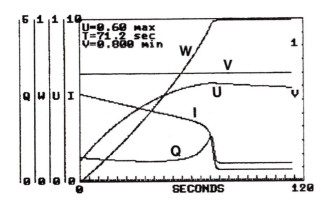

Figure 7   80 percent voltage starting condition

Locked Rotor Protection in the High Inertia Case

Overcurrent relays are routinely applied for locked rotor protection where the time duration of the starting current is short compared to the thermal limit time. However, in the high inertia motor application the starting time approaches or exceeds the thermal limit time as shown in Figures 6 and 7. Consequently, an overcurrent relay providing locked rotor protection trips on starting current and must be switched off as soon as acceleration is detected.

An impedance relay can be applied for this purpose since the apparent impedance at the machine terminals increases typically from a 0.15 perunit locked rotor impedance at 83 degrees to a rated speed impedance of 1 perunit at 13 degrees. However this application is difficult because the distance relay must detect a small variation in impedance early in the acceleration cycle where the impedance changes very slowly through the balance point of the relay. Farley and Hajos (3) give and excellent account of the tedious application procedure noting that the pickup to dropout ratios of the individual relays and slight parameter variations between machines must be considered. Since the application parameters cannot be accurately predetermined, calculated settings must be verified by test trials. Also, there is no protection during a failure to accelerate as pointed out by Eliason (4). The problem is that the magnitude of the starting current deviates little from that of the locked rotor current but has a lesser heating effect as shown in Figure 7. This occurs because the effective resistance of the rotor changes with slip from a high value at locked rotor decreasing to as little as a third of that value at rated slip as does the I^2R heating effect. Unfortunately, an overcurrent relay responds only to current and trips on starting current because it cannot detect the difference.

Microprocessor Based Rotor Thermal Protection

As a remedy the real time computational power of the microprocessor has been used to implement the thermal model as the rotor protection algorithm. The algorithm is the time discrete form of the differential equation for the temperature rise due to a current in a conductor using the slip dependent heat source.

It is not possible using overcurrent relays to account for varying slip, which is the key to the slip dependent heat source. However, in the microprocessor system a digital filter converts three phase analog voltage and current signals to digital quantities representing the rectangular form of each individual phasor. The real time computing power of the microprocessor is then used to calculate the positive sequence quantities from which the slip and the corresponding rotor resistance can then be de-

termined.

It can be shown that the positive sequence resistance of the motor is:

$$R= (|V_{a1}|/|I_{a1}|)\cos\theta = (V_{a1x}I_{a1x}+V_{a1y}I_{a1y})/|I_{a1}|^2 \quad (8)$$

Where V_{a1} and I_{a1} denote positive sequence voltage and current respectively and subscripts x and y denote the real and imaginary components of these quantities. Reference (1) shows that slip, S, and the rotor resistance, R4, are determined using:

$$S = R0/[A(R-R3) - (R1-R0)] \quad (9)$$

$$R4 = (R1-R0) S + R0 \quad (10)$$

where: R is the observed total resistance from Eq.(8)
S is the calculated slip
R4 is the calculated pos. seq. rotor resistance
R3 is the stator resistance
R0 is the rotor resistance at rated slip
R1 is the rotor resistance at locked rotor
A is the factor relating rotor to stator current.

In this algorithm R is the quantity determined by measurement. Of the remaining parameters R3 is calculated at startup while R1 and R0 are predetermined constants entered in the setting procedure.

In this manner the microprocessor based protection use motor current and terminal voltage to determine the slip dependent rotor heat source. The algorithm responds to the slip dependent I^2R rather than just to current, producing and estimate of rotor temperature. The protection is monitored to protect the motor not only for the locked rotor condition but for the entire starting cycle taking account of the heating of prior operation. The relay in effect takes on the dynamic thermal limit curve of the motor itself as its operating characteristic.

Applying Rotor Thermal Protection

The motor starting analysis is particularly useful in applying the advanced protection since the settings for the relay correspond directly to the parameters of the thermal model as shown in Figure 3 and are determined directly by the analysis. The following table shows the source data and settings calculated by the PC program(2) for the induced 6000 hp induced draft fan motor and settings calculated by the PC program(2):

Virginia Power Co. Chesterfield #6

Application: Induced Draft Fan Motor

Volts : 4160	Appl. pu V:	0.8
FLA : 735 amps	source X:	0
LRA : 5000 amps	start:	71.2sec
LRQ : 1.2 X rated torque		
HP : 6000		
Speed: 713 rpm	Synch speed: 720 rpm	
WK^2 :695000 lb-ft^2	ct rating : 1200/5	

Rotor Settings:

I_L Locked rotor current -------- setting 05   6.8
T_{or} Rotor thermal limit from
         operating Temp.   ---- setting 06   40.0
T_{ar} Rotor thermal limit starting
         from ambient-------------- setting 07   46.0
R0 Rotor resistance at rated slip-setting 08   0.010
R1 Rotor resistance at locked
         rotor-------------------- setting 09   0.026
U_{tr} Rotor Trip Threshold-------- setting 10   0.66
U_{bt} Rotor Block Start Threshold-- setting 11   0.13
FLA Full Load Current in ct amps- setting 14   3.1

The trip threshold U_{tr} setting is in perunit of the rotor thermal limit $I_L{}^2T_{AR}$. The setting

recursive solution is:

$$W = (Q_M - Q_L)(DT/M) + W_0 \qquad (7)$$

$$S = (1 - W)$$

Where W is the speed and is determined in the interval starting from the initial value W_0, DT is a discrete time interval, and Q_M and Q_L are the motor and load torque respectively. Since all the model parameters are perunit values, the moment of inertia WR^2 is converted to the inertia constant, M, using:

$$M = (WR^2/g)(2\pi/60)(SynW/Q_R) \qquad (8)$$

where g is the acceleration due to gravity and where Q_R is the rated torque of the motor calculated as follows:

$$Q_R = 5252(RHP/FLW) \qquad (9)$$

The motor torque, Q_M, already stated in equation (1) is expressed in terms of the rotor current I_4 and the slip dependent rotor resistance R_4 as follows:

$$Q_M = I_4^2 R_4/S \qquad (11)$$

The load torque is defined by the moment of inertia and initial and final torque values L and F listed above. The load of a fan or pump is slip dependent and in general increases from a low initial value to a larger final value. A convenient representation for the load torque is:

$$Q_L = L(1-W)^{E1} + FW^{E2}$$

where: W is the speed
L is the initial torque at zero speed
F is the final torques at rated speed
E1, E2 are exponents used to shape the curve

Parameters L and F set the end points and the exponents shape the load torque versus speed curve. The equation allows the user to match the load curve for a given application. Useful curves are produced with E2 equal to 2 or 3 while E1 can range from 0 to 100. The load torque for the induced draft fan (vanes closed) is emulated in Figure 4 with the following parameters in equation (11):

L=.4   F=.6
E1=5   E2=2

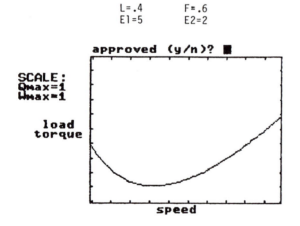

Figure 4   Induced draft fan torque with vains closed

Motor Starting Simulation

Plots of voltage, current, torque, speed, and rotor temperature versus time in seconds from the recursive solution of the models are shown in Figure 5 through 7. The plots were produced by a PC program (2) designed to determine the model parameters from the source data and perform the solution.

In each figure the voltage, current, and torque, are plotted in perunit of rated values with speed plotted in perunit of synchronous speed. Full scale

for the rotor temperature plot equals the thermal limit $I_L^2 T_{ar}$. Each plot lists the time and magnitude of the maximum temperature and also the minimum voltage for the case.

Locked Rotor Case

Figure 5 shows the 100 percent locked rotor case with the rotor initially at operating temperature. The severe heating effect of locked rotor current is caused by the high locked rotor resistance. The temperature in the case increases in a ramp from the normal operating temperature to the thermal limit in the warm thermal limit time of 40 seconds. The current is terminated at this point to show the rate of rotor cooling.

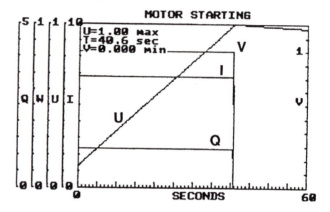

Figure 5   Rated voltage locked rotor case

Normal and Reduced Voltage Start

In contrast, Figure 6 shows the normal 100 percent voltage normal start case with a run up time of 39.1 seconds indicated by the time to reach maximum rotor temperature and the peak torque.

This case shows the dramatic reduction in the heating effect of starting current as the rotor resistance, R4, reduces with slip as indicated by equation (5). Whereas locked rotor current over heats the motor in 40 seconds, 39.1 seconds of normal staring current produced only 54 percent of the thermal limit.

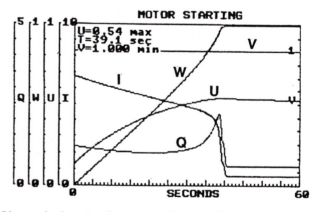

Figure 6   Rated voltage starting condition

Figure 7, shows the extended starting time of 71 seconds with 80 percent voltage the temperature reaching 60 percent of the thermal limit.

U_{bt} prevents start until the rotor cools to a predetermined percent of thermal limit to prevent overheating on frequent starts. The remaining setting is the relay tap setting, FLA, the full load current in ct secondary amps.

A 1200/5 ct is chosen so that the FLA of 735 amps is approximately 2/3 of ct rating. FLA in secondary amps., setting 14, is then 3.1 amps. Settings 05,06, and 07 are obtained directly from the motor data. Settings 08 and 09 were calculated above using equations (3) and (4).

The trip threshold, U_{tr}, setting 10, is set 10 percent above the temperature reached for the reduced voltage start case Figure 7.

Conclusions

1. A personel computer analysis of motor starting conditions can be carried out using readily available nameplate and thermal limit data. The analysis provides the protection engineer with complete understanding of the application and the relation of motor data to protective relay settings.

2. The study accesses vital application data including output plots providing the same information as an oscillogram of a full scale field test. The study also provides the ultimate criteria for protection, the temperature rise in the rotor.

3. Little analysis is needed in the routine application of locked rotor protection where the starting time is short compared to the thermal limit. However, the analysis is critical in high inertia applications where the starting time approaches or exceeds the rotor thermal limit and protection requiring speed supervision.

4. The analysis is particularly useful in the direct determination of settings for advanced microprocessor relays providing continuous rotor thermal protection.

Appendix A - Induction Motor Electrical Models

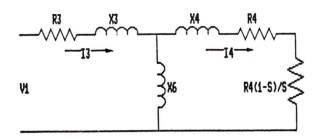

Figure A1. Positive Sequence Electrical Model

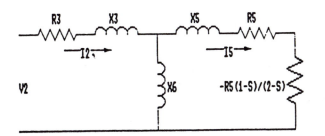

Figure A2. Negative Sequence Electrical Model.

V_1	Positive Sequence Voltage
I_1	Positive Sequence Stator Current
I_4	Positive Sequence Rotor Current
V_2	Negative Sequence Voltage
I_2	Negative Sequence Stator Current
I_5	Negative Sequence Rotor Current
$R3+jX3$	Stator Impedance
$R4+jX4$	Positive Sequence Rotor Impedance (slip dependent)
$R5+jX5$	Negative Sequence Rotor Impedance (slip-dependent)
$X6$	Mutual Reactance
S	Slip
$2-S$	Negative Sequence Slip

Although not used in the above of motor starting, negative sequence heating of the rotor is given by:

$$I_5^2 \ (R5/R0)$$

where: $$R5 = (R1-R0)(2-S)+R0$$

and used as an additional input to the rotor thermal model to account for heating due to unbalanced current.

Appendix B - The Equation of the Thermal Limit Curve

The following derivation shows how an induction motor thermal limit is constructed and how it represents a specific limit temperature.

The differential equation for the temperature rise due to current in a conductor not considering heat loss is:

$$I^2r = C_T(du/dt) \qquad (1b)$$

Where: I^2r are the input watts(loss in the conductor)
C_T is the thermal capacity (watt-sec/deg.C)
du/dt is the rate of change of temperature (deg.C/sec.)

The equation can be integrated from time zero to time t to find the temperature rise. The result for constant input watts is:

$$u = (1/C_T)I^2rt \qquad (2b)$$

Where: u is the temperature in deg. C
t is the time in seconds
I is current in amps.
r is the electrical resistance of the conductor.

The temperature can be expressed in perunit and plotted versus current as a time-current characteristic. To do this let I be expressed as a multiple M of a rated current I_{11} ;

$$I = MI_{11} \qquad (3b)$$

and substitute for I in equation (2b)

$$u - 1/C_T \ (MI_{11})^2rt \qquad (4b)$$

Divide equation (4b) through by I_{11}^2r/C_T :

$$uC_T/I_{11}^2r = M^2t \qquad (5b)$$

Which can be written as simply:

$$U = M^2t \qquad (6b)$$

Where, since M is in perunit, temperature is in seconds

This shows that an I^2t curve represents a thermal limit. The curve represents a specific limit temperature .and its plot on Log-Log paper is a straight line of negative slope 2.

The locked rotor limit starting from ambient is expressed in terms of the locked rotor current, M_L, and the locked rotor limit, T_a as follows:

$$U_L = M_L^2 T_a \qquad (7b)$$

The locked rotor limit starting from operating temperature is:

$$U_L = M_L^2 T_0 + U_0 \qquad (8b)$$

Where T_0 is the thermal limit time starting from operating temperature and U_0 represents the operating temperature.

From equation (8):

$$M_L^2 T_a = M_L^2 T_0 + U_0$$

Therefore the operating temperature in terms of the locked current and the thermal limit times is:

$$U_0 = M_L^2 (T_a - T_0) \qquad (9b)$$

References:
1. Zocholl, S.E., "Motor Simulation Software", A PC Compatible Induction Motor Starting Application Program, ABB Power Transmission Inc., Protective Relay Division, Allentown, PA.

2. Zocholl, S.E., Schweitzer III, E.O., Aliaga-Zegarra A.S., "Thermal Protection of Induction Motors Enhanced by Interactive Electrical and Thermal Models", IEEE Transactions, Vol. PAS-103, No. 7, July 1984, pp 1749-1756.

3. Farley, H.G., Hajos, L.G., "Large Induction Motors-Field Test on Locked Rotor Protection", 87 WM 107-IEEE/PES Winter Power Meeting, New Orleans, LA., February 1-6, 1987.

4. Eliasen, A.N., "The Protection of High Inertia Motors During Abnormal Starting Conditions", Vol. PAS-99, No. 4, July/August, 1980.

Part 4
Substation Equipment

I. General Discussion

VOLUME I contained separate parts on transformers and buses, reactors, and capacitors. Volume II combines those parts as well as new categories on substation control and substation design. By combining these categories, the majority of equipment that is protected by differential schemes is found in a one central location.

The differential relay has evolved from a simple one-element relay connected to the sum of the current transformers (Cts) around the protected device, to sophisticated instruments which not only evaluate the state of the protected device, but also compensate for the inadequacy of the Cts to perform its intended function. A new technology has emerged, the concept of a "fault signature," which in contrast to an error signal due to imperfect Cts or voltage transformers (Vts), provides new insight into this important problem. The number of papers in each of the categories attests to the growing importance of the entire substation as an element in power system protection.

IEEE standards, guides, special publications, and committee reports, in general, are not included in this volume. However, many of these publications are publicized in the summary-type *Transactions* papers, which have been included. These papers also broaden the value of these guides and other publications through the discussions written by knowledgeable IEEE members.

Substations are evolving into an area of system protection, which includes coordinated control of protective functions. This is resulting in the substation becoming a relaying communication center. Although the former practice of independent coordination continues, the trend is toward adaptive control of protective relay functions to better recognize the changing priorities of the continually changing power system. This evolution is occurring along two parallel paths; philosophy and hardware development. Modern technology has opened the door to new protection and control concepts while these concepts have encouraged the development of hardware and firmware to satisfy the new philosophy. This evolution is pervading the entire protective industry and the substation is the focal point for much of it.

American and European practices still differ on the use of auxiliary Cts in bus differential schemes and on switching Cts to obtain agreement with main circuit bus rearrangements. Generally, Ct switching entails separate Cts for those functions requiring this flexibility. Current transformer space constraints may then result in the need for smaller Cts with an attendant lower accuracy.

The nature of the protected device generally determines the details of the differential relay. Percentage restraints were used initially in both bus differential and transformer differential protection to minimize the effects of Ct inaccuracies. The bus relay then evolved into the voltage type in use today and a variation using a lower, though still substantial, impedance in the measuring element. Both relays, when used within their application limits provide high-speed differential protection. Both restrain on external faults even with Ct saturation. This has resulted in a secure $\frac{1}{2}$ cycle bus differential relay with no significant limitation on the number of feeders on the protected bus section. At present there are no new analog-type bus protective relays to include in Volume II and efforts at digitizing this function have not resulted in any additional breakthroughs to report.

Substation design is being influenced by the increasing application of new technologies in areas such as solid-state relays, microprocessor-based relays, and computers. Techniques have been developed by the protection engineer to provide secure operation of all these low energy devices in the harsh electromagnetic (EM) fields encountered in most substations. Four of the papers in this part deal with some aspect of this EM environment. The first three papers (which were not included in Volume I) present fundamental considerations which are of historical value and provide excellent material for training purposes. The fourth paper deals with protection aspects of new compact substation designs using gas insulated busways.

Protection requirements of large capacitor banks continues to dictate how the individual capacitor cans are arranged. Improvements have been made in minimizing the effects of system phase unbalance on the can unbalance detection needs. Capacitor protection against excessive harmonics is not yet a general problem and no major effort in this direction appears in the literature at this time.

The preferred protection for shunt reactors depends on the reactor location on the system and whether it is oil immersed or a dry type. Shunt reactor differential protection is still limited in its ability to detect turn-to-turn faults. The use of dual windings and split-phase differential relays that recognize turn-to-turn faults, as used on some generators, has had limited application on oil immersed reactors.

Transformer protection still uses predominantly harmonic restraint differential relays. However, the problem of losing fault detection sensitivity in the presence of harmonic inrush currents continues to be an area of investigation. Three of the papers in this part address this problem.

II. Discussion of Reprints

The first paper, "Summary Update of Practices on Breaker Failure Relaying," by the PSRC provides a detailed update on a 1970 paper on local backup. The paper describes the need for different schemes depending on system stability requirement, system configuration, breaker type, and bus configuration.

"Automatic Reclosing of Transmission Lines," a PSRC report, was prepared after an extensive survey was conducted on experiences and practices with high-speed reclosing in transmission lines. The report includes data from 99 respondent utilities as well as analysis of these data as they relate to the various system needs. "Single Phase Tripping and Auto Reclosing of Transmission Lines," also a PSRC report that identifies those system conditions that profit from this type of protection, describes the unique system performance during a single phase sequence, and provides a discussion of the various protective schemes applicable to this type of operation.

The first paper on substation design, "Transients Induced in Control Cables Located in EHV Station," by Sutton, is a 1970 paper that was one of the first studies to define the magnitude of transients in relaying and control circuits in EHV stations. The effectiveness of various techniques to minimize the transients are included. The second paper, "Proposed IEEE Surge Withstand Capability Test for Solid-State Relays," by Chadwick, also published in 1970, describes the rationale for the initial surge withstand capability (SWC) test standard. A circuit for generating the 2.5 kV, megahertz system design test wave is also presented. The third paper, "Electromagnetic Interference and Solid State Protective Relays," by Kotheimer and Mankoff, published in 1977, shows the need to broaden the scope of the original SWC test to accommodate the transients generated by components in low voltage control circuits. The original SWC test was concerned primarily with the surges resulting from events occurring on the high voltage system. The paper also discusses the destructive ability of radiation from portable transmitters on sensitive electronic relay components.

The fourth paper, "Design Criteria for an Integrated Microprocessor-Based Substation Protection and Control System," by Deliyannides and Udren, shows the need to revise specific past installation practices. The desired security of these newer, lower energy processors is shown to require built-in status monitors and information display. The flexibility available with these new tools requires close analysis of the system architecture to fully develop its potential. The paper describes a coordinated plan for providing all the relay protection and data acquisition for a substation using microprocessors. The economic impact is also addressed. "Pros and Cons of Integrating Protection and Control in Transmission Substations," by Nillsson et al., summarizes several field installations using a variety of configurations. In addition to describing the architecture and equipment involved, there is a discussion on the advantages and disadvantages, particularly of the communication requirements.

The protection problems of gas insulated substations (GIS) is discussed in the PSRC report, "Protection of SF_6 Gas Insulated Substations—Industry Survey Results." This paper includes an analysis of an industry survey of protection practices for GIS including recommendations. The paper shows the need to include monitoring information for detection of incipient fault conditions. Reclosing practices are also shown to be seriously impacted by this type of station design.

The only paper included in this volume on shunt capacitor protection is the PSRC report, "Shunt Capacitor Bank Protection Methods," which summarizes the capacitor protection standard, C37.99—1980. This summary includes charts that show the relay requirements for the protective relays with various can arrangements. Methods are described for negating the reduced sensitivity formerly required due to can capacitance unbalance and system voltage unbalance. Both voltage and current protection schemes are discussed.

The paper, "Shunt Reactor Protection Practices," a PSRC report, is the forerunner of a summary paper on the shunt reactor protection standard C37.109—1985. It includes substantially all the material in the summary paper and goes into more descriptive detail. The nature of reactor failures is described. Different protection schemes, which are dependent on the type of reactor and its location in the system are shown. Dry-type reactors are the most difficult to fully protect for turn-to-turn faults. A differential scheme using reactor neutral voltage and grounding transformer voltage is described for this application.

"A Three Phase Differential Relay," by Einvall and Linders, describes a relay with a single sensing element that provides improvement in fault sensitivity and minimizes misoperation due to inrush or poorly performing Cts. This relay also uses fifth harmonic restraint to prevent operation on excessive exciting currents due to overvoltage conditions. The second paper on this topic, "A Microprocessor-Based Three-Phase Transformer Differential Relay," by Thorp and Phadke, builds on the three-phase detection concept by digitizing the three-phase signals which then permits controlling the typical race between the analog restraint and trip functions with software. A third paper, "Advances in the Design of Differential Protection for Power Transformers," by Giuliante and Clough, returns to the single-phase approach, which permits attacking the problem of reduced sensitivity during inrush by utilizing a previously unused facet of inrush currents. For a small part of each cycle, transformer inrush current is substantially zero, but fault current flows for the entire cycle when there is no saturation. This difference in the current profile provides a signature needed to promptly recognize a transformer fault in the presence of inrush currents.

A fourth paper, "Development and Application of Fault-Withstand Standards for Power Transformers," by Griffin, provides much needed information on the thermal and other fault-withstand characteristics of transformers. This paper is unique in that it summarizes two standards that were developed by different Power Engineering Society (PES) committees. The first is C57.109—1984, "Transformer Through

Fault Current Duration Guide,'' which was developed by the Transformer Committee. The discussion of this standard points out information in C57.12 which is updated by this new standard. The second is C37.91—1984, ''Guide for Protective Relay Applications to Power Transformers,'' which was developed by the PSRC. It presents detailed information for coordinating relay and fuse characteristics with transformer capabilities. The four discussions to this paper are also significant in pointing out additional background information as well as noting areas in the paper which may be misinterpreted.

III. ASSOCIATED STANDARDS

''*Standard Surge Withstand Capability (SWC) Tests for Protective Relays and Relay Systems*,'' ANSI/IEEE C37.90.1.

''*Guide for Protective Relay Applications to Power Transformers*,'' ANSI/IEEE C37.91—1984.

''*Guide for Protective Relay Applications to Power System Buses*,'' ANSI/IEEE C37.97—1979.

''*Guide for Protection of Shunt Capacitor Banks*,'' ANSI/IEEE C.37.99—1980.

''*Guide for Protection of Shunt Reactors*,'' ANSI/IEEE C37.109—1984.

''*Guide for the Design and Installation of Cable Systems in Substations*,'' ANSI/IEEE 525—1987.

SUMMARY UPDATE OF PRACTICES ON BREAKER FAILURE PROTECTION

An IEEE Power System Relaying Committee Report

ABSTRACT

In the July/August 1970 issue of IEEE Transactions on Power Apparatus and Systems, an IEEE committee report was published entitled "Local Backup Relaying Protection." That report compiled the results of a survey on the current preferences on duplicate relays, dc and ac sources, control power, and breaker failure protection. Since that time, the industry has gained experience with breaker failure schemes and practices are much better defined. These practices have been compiled and are submitted in this report.

INTRODUCTION

Circuit breakers occasionally do fail to interrupt or trip for various reasons and the need for breaker failure relaying depends upon the consequences of such failures. Remote terminal relays and breakers may be able to provide backup for a failed breaker, but on many power systems this is no longer possible because of the apparent impedance caused by strong infeed from parallel sources. If backup by remote terminal relays is possible, the clearing times may be unacceptable because of system stability or excessive damage considerations. Backup by remote terminal relays may also be unacceptable because a large segment of the system may be removed from service. (Note: Breaker failure relaying should not be considered as a substitute for good system design and equipment maintenance.)

Breaker failure protection provides for the tripping of backup breakers if a fault is detected by protective relays and the associated breaker or breakers do not open after trip initiation. For example, if a fault between Stations B and C (Figure 1) is not cleared by breaker 2 within a predetermined length of time, it will be necessary to trip breakers 1, 3, and 4 locally in order to clear the fault.

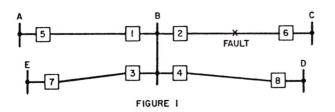

FIGURE I

Members of the "Breaker Failure Practices Working Group" of the IEEE Power System Relaying Committee are: J. R. Boyle, Chairman; R. F. Arehart, J. Berdy, B. Bozoki, G. A. Colgrove, C. W. Fromen, A. T. Giuliante, R. W. Haas, J. A. Imhof, B. L. Laird, J. R. Latham, J. Miller, T. J. Murray, G. R. Nail, G. C. Parr, R. W. Pashley, H. M. Paul, A. C. Pierce, R. C. Stein, J. E. Stephens, W. H. Van Zee, C. L. Wagner, R. C. Zaklukiewicz, and D. Zollman.

81 SM 363-1   A paper recommended and approved by the IEEE Power System Relaying Committee of the IEEE Power Engineering Society for presentation at the IEEE PES Summer Meeting, Portland, Oregon, July 26-31, 1981. Manuscript submitted April 9, 1981; made available for printing April 21, 1981.

Similar considerations must be given to multi-breaker arrangements such as ring buses or breaker-and-a-half configurations. Figure 2 illustrates the opera-

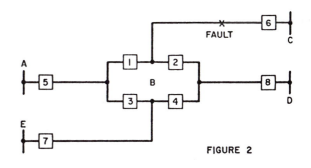

FIGURE 2

tion of a local breaker failure scheme applied to a ring bus station, showing that the equipment which must be tripped is located contiguous to the failed breaker.

A fault on any line requires tripping of two breakers at Station B. If any breaker fails to clear the fault, breaker failure protection would initiate tripping an additional local breaker and a remote breaker. For example, if a fault occurs on the line between Station B and Station C (in Figure 2), breakers 6, 1, and 2 would be tripped by the line protective relays. Assume that breaker 2 will not trip properly. The failure of breaker 2 necessitates the tripping of breaker 4 at Station B and breaker 8 at Station D. Similarly, if a fault occurs on the line between Station B and Station D, breakers 8, 4, and 2 would be tripped by the line protective relays. Again, the failure of breaker 2 would necessitate the tripping of breaker 1 at Station B and breaker 6 at Station C. Hence, irrespective of which of these two lines is faulted, the end result must be the tripping of breakers 1, 4, 6, and 8. For this reason, the breaker failure scheme may be designed to trip symmetrically all four breakers, no harm being incurred by initiating a trip command to those breakers which are already open. In all cases, all breakers tripped by the breaker failure scheme should be locked out.

A remote breaker must also be tripped either by its own relays or by transfer trip initiated by the local breaker failure protection. Remote backup clearing of the breakers may be preferred over direct transfer trip using a communications channel if relays at remote terminals provide adequate coverage and clearing times. However, direct transfer trip may be used if it is provided for other functions such as transformer or reactor protection. Apparent impedance is not a consideration in remote backup relaying since opening all breakers around a failed breaker removes the infeed effect.

Figure 3 is a logic diagram depicting a basic breaker failure protection scheme.

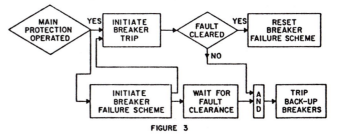

FIGURE 3

Reprinted from *IEEE Trans. Power App. Syst.*, vol. PAS-101, no. 3, pp. 555-563, March 1982.

APPLICATION AND SETTING PHILOSOPHY

General

1. Circuit breaker auxiliary switches should not be used to indicate whether or not a circuit breaker is carrying current unless there is no other way to do the job. This includes the primary and secondary line relaying schemes as well as the associated breaker failure backup protection. Such auxiliary switches may be used in conjunction with current detectors in an "OR" function. The fact that an auxiliary switch has operated is not sufficient proof that a circuit breaker has interrupted a fault. The auxiliary switch may be opened because: (a) its operating linkage is broken, (b) it is out of adjustment, (c) the breaker mechanism has operated but the main breaker contacts have failed to interrupt the current. A current detector gives a more positive indication that fault current has stopped flowing through a circuit breaker. However, it may not be able to be set to detect low magnitudes of current.

When protective relays are being tested, the breaker failure scheme should be properly blocked or isolated to prevent misoperation. Periodic relay testing is often coordinated with other equipment maintenance outages. When an auxiliary switch is used in an "OR" function, it is desirable to provide a means to disable that part of the "failure to trip scheme." Some users have found it desirable to install a maintenance switch in a breaker to open or short "52a" or "52b" contacts as required whenever the breaker is out of service for maintenance. However, unless an interlock is provided to prevent the breaker from being placed in service while the maintenance switch is in the test position, an operator could inadvertently leave the maintenance switch in the wrong position.

2. Since breaker failure relays must be initiated by a fault sensing relay, it may be desirable to provide independent breaker backup for ground faults by use of a relay in the neutral of the main transformer of a unit-connected generator. This installation would be a part of the main breaker failure scheme. It trips the high-side breakers and usually the unit lockout relay for system and transformer secondary protection. It would not necessarily deenergize the entire station if breaker failure relaying functioned properly to trip only those breakers required to clear a fault.

3. A direct transfer trip signal received at a remote terminal should trip the breaker (or breakers) and initiate breaker failure tripping because remote protective schemes may not be able to detect the fault. Since a breaker may fail to open by transfer trip, as well as by its normal protective relays, all relays should initiate the breaker failure tripping.

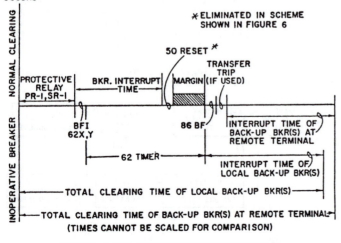

FIGURE 4 BKR. FAILURE TIME COORDINATION

Breaker Failure Timing

The breaker failure protection should be fast enough to maintain stability but not so fast as to compromise tripping security. This is particularly important on bulk transmission lines where stability is critical. A typical timing chart for the breaker failure schemes shown in Figures 8 and 9 is shown in Figure 4.

The shaded margin time provides security and should accommodate the following variables:

1. **Excessive breaker interrupting time.** At low magnitudes of fault currents, breaker interrupting times may be longer than those experienced at higher magnitudes of fault currents. (This is especially true of large oil circuit breakers whose large interrupting chambers do not permit fast clearing at low magnitudes of fault current.) Also, one must consider the increased interrupting time during a close-open duty cycle.

2. **Timer overtravel.** Normally, static timers have less than one millisecond overtravel time. However, if static or electromechanical timers are used that are known to have excessive overtravel, the excess time must be included in the overall timing sequence.

3. **Instrumentation.** Time setting errors, calibration errors, potentiometer resolution, etc.

4. **Safety factors.** In order to increase security and avoid a false breaker failure operation, an adequate margin is essential. The degree of margin desired is a direct function of the confidence level of the total protective scheme. Since interphase faults have far greater effect on stability, some users have found that the confidence level of a protective scheme can be improved by developing a breaker failure system that distinguishes between interphase faults and single phase-to-ground faults. A two-step timer arrangement which will provide fast clearing for interphase faults and slower clearing for ground faults is shown in Figure 5. Note that F2 supervises F3 via "AND" so 62-2 picks up only on multiphase faults. F3, which usually has a faster reset time than F2, is used to drop out 62-2 to increase the margin time which was reduced by the faster setting of 62-2 timeout.

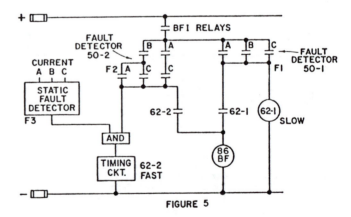

FIGURE 5

If faster time is desired only for three-phase faults, the three-phase fault detectors for the fast timer can be connected in series. The faster timer is cut off by the clearing of one or more poles which indicates that the fault will resolve from a three-phase fault to a fault involving one or two phases. This philosophy is predicated on complete pole isolation so that all poles can trip independently of one another.

As illustrated in Figures 4, 10, 11, 13, and 14, breaker failure relay timing must include the BF fault detector reset time. Another scheme illustrated in Figure 6 eliminates the device 50 reset time by allowing the device 50 to operate only after the timer has timed out. The BF timer 62 is operated directly by the BFI (breaker failure initiation) relays. The fault detector

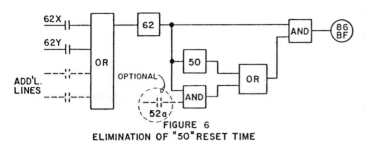

FIGURE 6
ELIMINATION OF "50" RESET TIME

in the dc control circuit is controlled by the timer so that only its operating time of 8 ms or less needs to be considered.

Fault Detectors

1. Fault detectors that have high dropout and whose dropout time is minimally affected by CT saturation and dc offset in the secondary circuit should be used. Examples of this type are induction-cup relays and static relays with suitable filtering. Hinged armature and plunger-type relays that can have a significant dropout delay should be considered carefully before use. If such devices are used, their dropout time could be ascertained under the worst conditions and these times should be considered when setting the breaker failure timer. Determining the worst condition may be difficult. When a current transformer saturates, the secondary current may not have its current zero at the same time as the primary current. Therefore, when the primary current is interrupted by a circuit breaker at current zero, the secondary current may be at some positive or negative value. This current would then have to decay through the connected relays, increasing their dropout time. It is recommended that the user contact the manufacturer when contemplating the use of plunger or hinged armature relays as fault detectors.

2. Fault detector relays should always be located in the main CT secondary circuit. Transformers, generators, and some transmission lines may be served from two breakers. Frequently CT's associated with each of these breakers are common to one set of relays. However, breaker failure protection is initiated through separate fault detector relays associated with each breaker. Thus, it is possible that a sensitively set fault detector relay associated with a breaker that has successfully operated can remain picked up from the driving voltage associated with a CT on the failed breaker. This may cause an undesired operation of the breaker failure scheme.

For example, in Figure 7 if breaker 1 fails to open and breaker 2 opens for a fault on the line, the current from the main CT-1 through the line relays can cause current to flow in the CT-2 circuit and hence may cause fault detector 50-2 to pick up. For this condition, breaker failure initiation for breaker 2 may take place even though breaker 2 is open.

Of greater concern is the possibility of CT saturation when auxiliary CT's are used. Extremely high secondary currents in the auxiliary CT associated with 50-1 may occur after breaker 2 opens. Saturation of this auxiliary CT may prevent 50-1 from operating correctly to initiate breaker failure. Therefore it is important that the CT ratios, the excitation characteristics, and the fault detector ratings be adequate for the maximum load and fault currents through each breaker. Both CT's and associated auxiliaries should have the same ratings and have adequate capacity to handle the circuit burden.

3. Permissible setting limits for fault current detectors and overall margin times must be considered when setting breaker failure relays. Factors that should be considered when setting fault detector relays are:

    (a) load current
    (b) fault current
    (c) charging current
    (d) breaker opening resistor current (which can last 1-1/2 cycles after main contact interruption and be of significant magnitude)
    (e) reset time

The main consideration for setting the current level of the fault detectors is the setting sensitivity of the protective relays and the reset time of fault detectors. The fault detector should be as sensitive as the protective relays to assure that for any fault for which the protective relays will function, the breaker failure fault detector will also operate. The thermal capability of fault detectors must be taken into consideration. Some users feel that repeated operation of the fault detectors on load current should be avoided because it may cause the contacts to stick or weld closed.

4. On transmission lines served from two breakers, even though the fault detectors are set as sensitively as the protective relays, it is possible that the breaker failure timer will not start until one of the breakers trips because, while both breakers are closed, fault current may divide unequally causing each fault detector relay to see only a portion of the total fault current.

BREAKER FAILURE CIRCUITS

Straight Bus Configuration

For reliable protection, there should be a breaker failure scheme and at least two independent protective relay schemes that cover all lines or equipment being protected. Breaker failure protection is usually initiated by both primary and secondary relays. Where one fault detector is used in conjunction with a breaker failure scheme it should be placed in the CT circuit with the least chance of CT saturation (minimum burden). For example, in Figure 8 if PR-1 were electromechanical relays and SR-1 were static relays, the fault detector 50 would be placed in the SR-1 circuit because of its lower burden. If both sets of line relays are inoperative, backup relays at a remote terminal must detect and clear the fault.

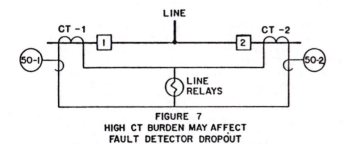

FIGURE 7
HIGH CT BURDEN MAY AFFECT
FAULT DETECTOR DROPOUT

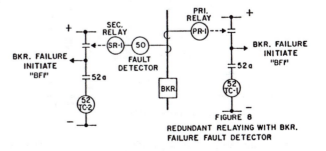

FIGURE 8
REDUNDANT RELAYING WITH BKR.
FAILURE FAULT DETECTOR

A typical single breaker trip scheme utilizing a primary set of protective relays (PR-1) and a secondary set of protective relays (SR-1) is shown in Figure 9.

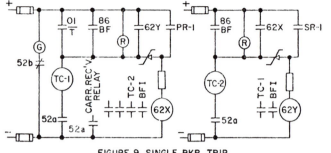

FIGURE 9 SINGLE BKR. TRIP

Each set of relays should be connected to separate trip coils if they are available. PR-1 is shown to trip through trip coil (TC-1) and SR-1 is shown to trip through trip coil (TC-2). The relays also energize auxiliary relays 62X and 62Y, respectively. The 62X auxiliary relay not only serves to actuate the breaker failure protection but also attempts to trip the breaker via TC-2 (in case TC-1 trip coil is open). It can also be used to bypass any contact circuit which is dependent on a circuit breaker "52a" switch. The 62Y relay also trips the breaker via TC-1 and actuates the breaker failure protection.

Figures 10a and 10b illustrate two schemes, one (10a) utilizing a common bus timer and the other (10b) utilizing one timer per breaker. In Figure 10a, contacts of breaker failure initiation (BFI) relays 62X and 62Y actuate the BF timer 62 only if the BF fault detector relay 50-1 is picked up (50-1 is usually comprised of two phase and one residual unit to obtain greater sensitivity to ground faults). If the breaker fails to clear the fault, the BF timer 62 then would actuate the 86BF auxiliary relay which would trip the adjacent breakers and block their closing. Other contacts of the 86BF auxiliary relay may be used to initiate transfer trip and to stop carrier for blocking pilot type relaying so that the remote terminal can trip if the remote terminal relays detect the fault. The latter two functions require independent initiation for each breaker. If the breaker clears the fault, the BF fault detector 50-1 resets and drops out the BF timer 62.

In Figure 10b, the sequence of operations would be identical to that of 10a. However, when the BF fault detector 50-1 resets, there is no possibility that the BF timer 62 can remain picked up through another BF

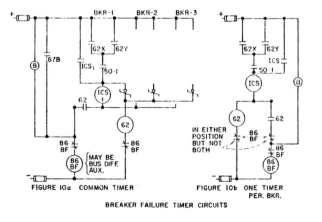

FIGURE 10a COMMON TIMER
FIGURE 10b ONE TIMER PER. BKR.

BREAKER FAILURE TIMER CIRCUITS

fault detector 50-2. The separate timer is required or preferred for the following applications:

1. Ring bus breakers.
2. All breakers of a breaker-and-a-half arrangement.
3. Breakers of double-circuit lines connected to a common bus where a line fault may spread from one circuit to a second circuit before the first circuit is cleared.
4. Bus breaker of a multiple terminal line where remote time delay relaying is not considered adequate backup for a bus fault and stuck breaker. The bus lockout relay cannot initiate remote breaker tripping to back up a stuck breaker. An independent circuit is required to selectively disable the blocking of the line pilot relay system or initiate transfer tripping.
5. Breakers arranged for single pole operation. Separate 62, 50, and 62X devices should be used with each pole.

A common timer, operating the bus lockout relay, may be used for two or more breakers connected to a single bus if conditions 3, 4, and 5 above do not apply. The following undesirable features should be considered:

1. Sequential faulting of two lines could cause sequential operation of two BFI relays and undesired continuous timer operation and bus tripping.
2. If breaker interruption times are not the same, the common timer should be set for the slowest breaker.

All breakers connected to a common bus, whether using a common timer or separate timers, may use the same circuits and lockout relay as the bus differential relays since the bus tripping functions are identical and the two input functions are not redundant. This avoids unnecessary duplicate wiring, contacts, and test switches. Some utilities have found it desirable in a breaker-and-a-half scheme to provide redundancy for device 86B (bus differential lockout relay) to prevent a single auxiliary relay failure from defeating the scheme.

Separate BF timers are preferred for bus breakers of a breaker-and-a-half arrangement to selectively trip the local center breaker and remote breakers for a bus fault. This may be done by a second contact of the BF timer or by use of a blocking diode and auxiliary relay. However, this may also be accomplished with a common bus BF timer by use of a blocking diode and auxiliary relay with each bus breaker BF input to selectively trip the local center breaker and remote breakers. Transferred tripping of a remote breaker for a bus fault and breaker failure may not be required if it is decided that remote breaker time clearing by zone 2 or ground relays is adequate.

The 86BF contact in the 62 coil circuit of Figure 10a prevents seal-in of BF timer 62 through the dc monitor lamp "B" after 86BF operates. A diode in the 62 contact circuit may be used for this purpose (see Figure 11).

A breaker may fail to open even though the trip coil is energized. To avoid trip coil damage on some breakers, a normally closed contact of 86BF may be used to open the trip coil circuit. This can be done only where a separate 86BF relay is used for each breaker. The control switch trip contact 01/T must permit manual tripping of the breaker independent of 86BF. Refer to Figure 11.

The operational testing of either the primary or secondary set of relays (with the breaker and the other set of relays remaining in service) should include disabling the BFI circuit to the timer, along with breaker tripping, to avoid operation of associated breaker failure schemes if a fault detector is picked up by load current. If this precaution has been missed, 62X and 62Y contacts, as shown in Figure 9, will trip the primary breaker to avoid backup breaker tripping. Another approach is to place a high-speed auxiliary tripping relay across the breaker failure timer with contacts to trip the primary breaker. However, this is not normally recommended because it requires additional hardware, is not applicable to a common bus timer, and when it is interrupted, inductive energy stored in the relay coil may dissipate into the timer logic and cause it to operate incorrectly.

Where target seal-in units of both BF fault detector 50-1 and BF timer 62 are used in series, their tap setting or rating should be selected to coordinate with the requirements to operate the auxiliary trip relay 86BF. The resistance of these target seal-in coils should be selected to minimize the voltage drop in the lockout relay coil circuit.

Low Air Pressure

Some air-blast breakers automatically close upon low air pressure. Some utilities elect to initiate breaker failure protection as soon as low air pressure is detected (see Figure 11). However, other utilities

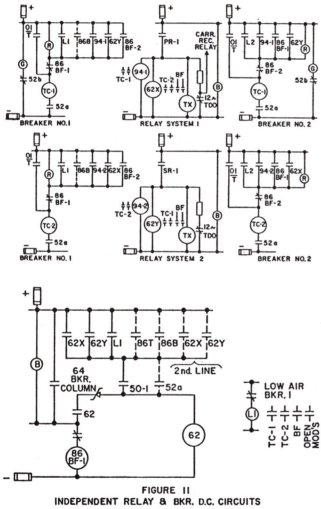

FIGURE 11
INDEPENDENT RELAY & BKR. D.C. CIRCUITS
FOR TWO BKR. TERMINAL

prefer to initiate the tripping of a breaker losing air at a higher level of air pressure (than low air pressure close) and initiate opening of isolating disconnect switches associated with the defective breaker at the same time. If the breaker opens properly, the BF fault detector relay (50) will drop out and deactivate breaker failure tripping. If, however, the BF fault detector relays stay picked up for a current in excess of their setting, breaker failure tripping will take place after

device 62 times out. This arrangement is especially suitable for supervisory controlled substations where a breaker losing air is automatically tripped and sectionalized, leaving the bus and station energized.

Ring Bus and Breaker-and-a-Half Configuration

Several schemes have been used for tripping two or more breakers of a line terminal from a single set of relays. These involve tripping diodes in either single or separate dc circuits and numerous auxiliary relay schemes. When a line terminal has two or more breakers, paralleled "52a" switches of each breaker are usually required in each relay system to drop out target and seal-in circuits. This may involve considerable wiring and exposure of the relay circuit. Wiring can be minimized by using a single "a" switch per breaker trip coil for the relay systems of both lines associated with the breaker. The "52a" switches may be connected to the respective breaker trip coil negative supply and diodes should be used to maintain independent circuit function. However, if one of the two breakers is out of service and closed, which may be common for air-blast breakers, the relay circuit for the remaining in-service breaker is sealed in after a trip and reclosing will not be successful.

One dc schematic for a two breaker arrangement is shown in Figure 11. The 12-cycle time-delay-to-open relay (TX)[1] replaces the usual paralleled "52a" switches and maintains the relay system sealed in long enough to complete breaker failure tripping. This arrangement eliminates all the problems associated with paralleled "52a" switches and the need for seal-in to maintain breaker failure timing by carrier receiver relay, nonmaintained transfer trip contacts, or an impedance relay for a zero voltage fault. All relay system wiring is then contained on the switchboard and the relay system functions are independent of the breakers. However, it is necessary to provide a seal-in for any contact not

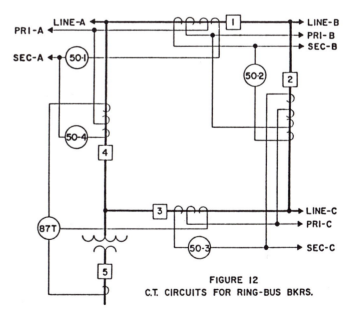

FIGURE 12
C.T. CIRCUITS FOR RING-BUS BKRS.

so equipped, such as the indicating (mechanical) instantaneous trip contact of an overcurrent relay, which will open when the fault is cleared normally and would otherwise attempt to break the resistor current.

1. Reference page 7, item 6, last paragraph.

It should be re-emphasized that with a multi-bus, multi-breaker scheme such as a breaker-and-a-half scheme it is desirable to initiate breaker failure relaying by bus relays for the breaker connected to the bus. Failure of a breaker to clear a bus fault would result in the loss of the entire station if breaker failure relaying is not used.

A typical ring bus is shown in Figure 12. This arrangement requires paralleling two sets of CT's for each set of line relays. The current detector should always be connected directly in the main CT secondary circuit and not through an auxiliary CT (refer to Figure 7). The CT ratio should be based on the maximum load flows through the bus and the breaker. The usual practice is to select the highest ratio available to avoid CT saturation with maximun fault current flowing through it. Where this ratio is too high for the line or transformer rating, auxiliary CT's or a lower ratio tap of the multiratio primary CT's not feeding the current detector, may be used to feed the terminal relays and meters.

Scheme for Generator Protection

Since transformer and generator faults may not provide sufficient current to operate a fault detector (50), a transformer differential relay, sudden pressure relay, etc., can be used to initiate breaker failure relaying, but it will require supervision by a breaker "52a" switch to control the breaker failure (BF) timer (62). This optional arrangement is shown in logic diagram Figure 13 and also in a schematic diagram in Figure 11. (Note that the "52a" switch does not monitor or control the overcurrent fault detector.) Where the breaker may be disabled and closed when it is out of service, the "52a" switch circuit should also be disabled in order to prevent undesired breaker failure tripping.

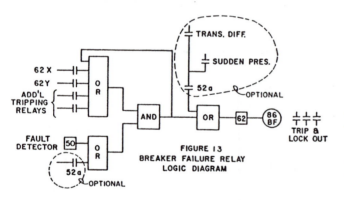

FIGURE 13
BREAKER FAILURE RELAY
LOGIC DIAGRAM

The heavy lined circuit as shown in Figure 14 shows a basic breaker failure relaying scheme with auxiliary contact logic added to provide a means of detecting and tripping for a generator breaker main contact flashthrough. This protection is intended for generator breakers only and utilizes a basic breaker failure scheme current detector and timer to achieve security. A separate fault detector (50A-1) is utilized to provide the desired sensitivity. A simpler form of the standard fault detector (50-1) in parallel with an optional "52a" switch (fully shown in Figure 11) could be retained to provide backup protection through conventional protective relays.

Generator and transformer breaker failure relaying could be initiated by lockout relays. However if the lockout relay fails, it could result in the breaker not tripping. An auxiliary relay (not shown) with both coil and contacts wired in parallel with the lockout relay coil and breaker failure initiate contacts, provides redundancy for the failure of either.

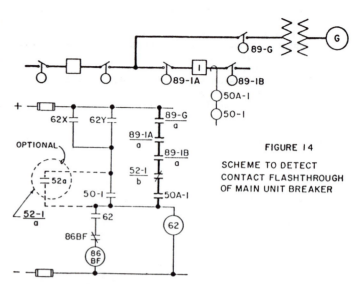

FIGURE 14

SCHEME TO DETECT
CONTACT FLASHTHROUGH
OF MAIN UNIT BREAKER

Scheme for Pole Disagreement

Pole disagreement tripping should be provided on circuit breakers which are capable of individual pole operation. This protection is provided so that after a close operation, any pole failing to close will, through appropriate circuitry, trip the poles that successfully closed.

There are no standard pole disagreement circuits since each circuit breaker design has its own specific requirements. Each scheme, however, consists of some form of series-parallel combination of "a" and "b" circuit breaker auxiliary contacts and time delay relay to provide for normal synchronization of the circuit breaker poles. The exact arrangement of this circuitry and its connection into the tripping circuits varies between the different circuit breaker designs. It is important that the protection engineers review these circuits with the manufacturer to ensure that both the system and the circuit breaker operating requirements are met.

It should be noted that these internal disagreement circuits only trip the affected circuit breaker with no tripping of backup circuit breakers provided. If the disagreement problem is caused by a jammed mechanism or operating linkage or internal flashover, the unbalanced currents will continue to flow until some other device trips the backup circuit breakers. Phase disagreement relays that operate on the current or lack of current in the individual phases are available that will perform this function.

General Design Philosophy

1. Separation must be maintained between the protective relay and the circuit breaker trip coil dc circuits so that a short circuit or blown fuse in the circuit breaker trip circuit will <u>not</u> prevent the protective relays from energizing the breaker failure backup scheme.

When two or more breakers are involved, the dc separation may be best obtained by using independent dc circuits for the protective relays and the breaker trip circuits with the relay system made independent of the breaker operation and confined to the switchboard (Figure 11). When only one breaker is involved, the relay system and breaker trip coil could be on a common dc circuit (Figure 9). With two trip coils and two relay systems, each system may provide the backup for the other. The two relay and dc systems should be made as independent as possible including separate cables and dc supply buses.

If a breaker with only one trip coil uses a common dc circuit for the trip coil and one set of protective relays, the trip coil circuit should be fused (15 or 20 amperes) to coordinate with the dc circuit supply protective fuse or breaker. This will ensure operation of the breaker failure auxiliary relay even though the trip coil may be defective. The trip coil fuse may be furnished with the breaker or located at the switchboard so as to protect the cable. (Red and green trip circuit monitoring lamps will monitor the continuity of the fuse and the power supply.)

2. The breaker failure scheme should be designed with the minimum number of trip test devices (to be determined by the user) to minimize the possibility of errors being made during normal testing.

3. Design a scheme that will not trip out a bus if some point in the dc circuit is inadvertently grounded. Sensitive auxiliary trip relays can operate as a result of capacitor discharge currents generated by long secondary dc cable leads when some point in the dc circuit is grounded. Surge capacitors on the output of SCR's may contribute to inadvertent tripouts if their capacitive discharge currents flow through the trip coils when dc buses are grounded. Battery ground lamps may also contribute to the problem by providing a path for undesired currents to flow when inadvertent grounds are applied. The installation of diodes and bleeder resistors can help eliminate this problem.

4. The physical location and mounting of breaker failure relays is an important consideration in the overall design of a breaker failure scheme. This is especially true when testing a set of protective relays and the proper care is not taken to isolate the associated breaker failure relay schemes. Electromechanical and solid state breaker failure relays may have a sensitive electromechanical output relay which could close its contacts if the panel on which it is mounted vibrates.

5. Some users have found it desirable to energize the power supply of a static breaker failure relay with fault detectors or 62X and 62Y contacts to prevent inadvertent breaker failure operations caused by dc transients.

6. Protective relays employ an internal seal-in to maintain the trip circuit once it is initiated until the breaker opens. If the breaker fails to interrupt, the protective relays maintain the trip circuit and, since the BFI relay remains energized, will complete breaker failure tripping. There are three exceptions where protective relays may not remain closed:

a. A zero voltage three-phase fault with the resulting loss of memory action of the distance relays.

b. Tripping initiated by a nonmaintained transfer trip circuit.

c. Relaying which is dependent upon a "52a" switch (carrier receiver relay).

In most cases the application of a high set overcurrent relay to trip directly will maintain the BFI for the first condition.

Where a local breaker failure scheme operation is dependent upon the reception of a transfer trip signal, direct or permissive, some means must be provided to keep the trip circuit energized and therefore the BFI relay picked up when the remote breaker opens successfully. Permissive overreaching schemes can have this feature built in through the use of time delay auxiliary relays and keying of the transmitter by "52b" contacts. A lockout relay can be used on direct transfer trip schemes at either end of the line where reclosing is not desirable, such as for transformer protection. For a permissive underreaching scheme, reclosing is desirable so neither a lockout relay nor "52b" keying of the transferred trip transmitter for a three-terminal line is applicable.

In such cases where the transfer trip signal is lost following successful opening of a remote breaker,

the time delayed auxiliary relay (TX Figure 11) could be used to seal-in the trip circuit long enough to operate the breaker failure timer.

7. An acceptable option would permit the BFI relay to trip the protected breaker, thereby avoiding the tripping of more than the one breaker if the relay were accidentally operated during testing with the current detector picked up. If the protective relay tripping were direct, then the BFI relay tripping should be by the alternate trip coil for redundancy.

8. Circuit breaker "52a" auxiliary switches should not be used in series with any of the protective relay auxiliary circuits. These could reset if a breaker were to open and a breaker interruptor were to fail. Care must also be exercised that the target and seal-in unit is set on a tap whereby it will not remain sealed in by the current drawn by an auxiliary tripping and/or BFI relay. A dc seal-in function is not required to keep the BFI relay picked up because it should remain energized as long as ac fault current flows in the fault detecting relays. The BFI relay coil should not be interrupted by anything except the protective relays. Therefore, it is important to use a BFI relay which draws low current so that the protective relay contacts can interrupt the circuit without damage.

Where only a single breaker is involved and a "52a" switch is used in the carrier receiver relay circuit, the carrier receiver relay contacts can be bypassed by contacts of the BFI auxiliary relay in order to maintain breaker failure timing if the "52a" switch should open. If the BFI relay is fast, a resistor is required in the BFI contact bypass of the phase carrier receiver relay contact to permit the carrier relay target to operate. (Without the resistor, the BFI contact shunts the target coil before it can operate.) If other relays also operate into this same trip bus and BFI relay, resistors are required in series with both BFI relay contacts in order to prevent false carrier relay targets for a non-carrier relay trip.

When two or more breakers are involved, the protective relay system should be entirely independent of either breaker operation. A time delay relay (set greater than BF timer) can be used in place of the "52a" switches to drop out targets, seal-in elements, and other auxiliaries after maintaining breaker failure initiation long enough to operate the breaker failure timer (refer to Figure 11).

The resistor must provide enough current to operate the seal-in units of all relays which may operate simultaneously. If any relay contacts are not equipped with a seal-in unit, such as the indicating instantaneous trip unit, a separate seal-in must be provided to prevent the protective relay contact from attempting to interrupt the resistor current when the fault is cleared normally before the time delay relay deenergizes the circuit.

9. Breaker failure protection cannot operate without proper initiation. It is considered good practice to provide redundant trip outputs and breaker failure inputs where some other form of redundancy does not exist. Redundant outputs can be inexpensively provided by paralleling the bus, transformer, or generator lockout relay coil with the coil of an auxiliary tripping relay.

10. The relay circuit should be examined for sneak circuits which can cause misoperation of the BFI relay when the dc is turned off or when one pole is opened (in isolating a dc system ground). A reverse polarity current may flow through the auxiliary relay coil due to the discharge of capacitance or backfeed through relay power supplies or lamps. A diode across the relay coil (with series resistor) will block operation on reversed polarity. To increase security, some users have found it desirable to determine that none of the dc auxiliary relays in a breaker failure scheme operates below half of maximum battery voltage.

BACKUP FOR BREAKER FAILURE SCHEME

If local breaker failure tripping fails, additional local backup protection can sometimes be provided at reasonable cost by a station time delay ground relay (shown at 51G in Figure 15). This assumes that independent pole tripping will cause phase faults to evolve into ground faults and all breakers (1, 2, 3, 4, and 7) can be opened without jeopardizing system security. If independent pole tripping is not provided, then additional backup phase protection may be employed.

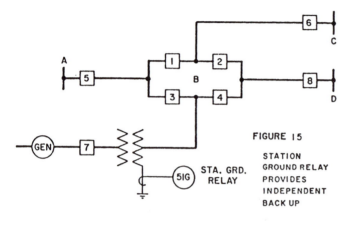

FIGURE 15

STATION
GROUND RELAY
PROVIDES
INDEPENDENT
BACK UP

STA. GRD.
RELAY

Where two or more bus sections are connected together through tie breakers, a local breaker failure backup protection for ground faults can be employed. Refer to Figure 16. For example, if breaker 2 fails to clear for a line fault, total isolation can be obtained by tripping breakers 1, 3, and 4. If a breaker failure scheme associated with breaker 2 fails to function properly, a relatively inexpensive independent backup residual ground relay can be used to detect the flow of current from bus 1 to bus 2. In Figure 16, this is

depicted as current flow through breaker 4. Directional characteristics will ensure that bus 2 is isolated. Similar schemes would be used on other bus sections. When breaker 5 is used as a substitute breaker when breaker 2 is out of service, disconnects B and C must be opened and disconnect D closed. Under this arrangement, a transfer trip to breaker 9 may be desirable to protect for transformer faults if breaker 5 malfunctions. (Care should be taken to ensure that the directional ground relays are less sensitive than the ground relays on each feeder breaker but more sensitive than the station unit neutral ground overcurrent relay.)

CONCLUSION

This report summarizes and updates practices on breaker failure protection. Experience has indicated that certain schemes perform better than others. The application of breaker failure relaying and the choice of the best scheme must reflect the application and design philosophy peculiar to each utility.

Reference

1. A. Paullow, E. T. Gray, S. H. Horowitz, A. G. McConnell, W. H. Van Zee, and P. Zarakas, "Local Backup Relaying Protection," IEEE Working Group on the Survey of Local Backup Practices. S. H. Horowitz, C. L. Wagner, and K. Winick; IEEE Working Group on the Symposium on Local Backup Practices. *IEEE Transactions on Power Apparatus and Systems*, paper No. 69TP602-PWR, vol. PAS-89 No. 6, pp. 1061-1068, July/August 1970.

Discussions

William K. Newman (Georgia Power Company, Atlanta, GA): The authors have performed a very valuable service by summarizing and updating the preferred practices in failed breaker protection. Of the many recommendations made in this paper, I believe that among the most important were those that emphasized the desirability of making the protective relaying schemes operate independently of the breaker auxiliary contacts whenever possible. This is especially important with EHV live-tank breakers.

The authors mention one advantage of the scheme shown in figure 6 as being the elimination of the device 50 reset time. I believe this to be a distinct advantage. Another advantage of this scheme is that it eliminates the problem mentioned in item 4 under the heading "Fault Detectors". Since the 62 timer does not depend on the 50 device to allow it to complete its timing cycle, by the time the 50 device is enabled by the 62 timer, the other breaker serving the line will have tripped and all of the fault current will be in the failed breaker.

I believe that the scheme shown in figure 6 would be more reliable if connected as:

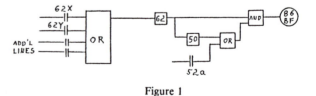

Figure 1

If this connection is made, this scheme could be used for generator breaker failure protection by connecting the generator protective relays to one of the lines marked "add'l lines". This would accomplish the function shown in your figure 13.

One feature that offers increased security is the design feature mentioned in item 7 under the heading "General Design Philosophy". I believe that this feature would reduce the instances of tripping the breaker failure lockout relay for accidental operation during testing. The "retrip" feature should be connected to energize both the primary and secondary trip circuits of the protected breaker.

Manuscript received May 14, 1981.

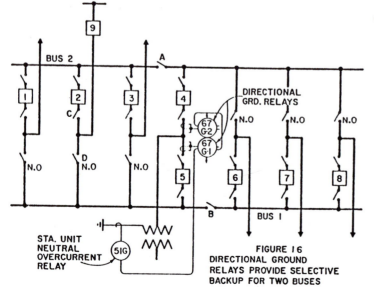

STA. UNIT
NEUTRAL
OVERCURRENT
RELAY

FIGURE 16
DIRECTIONAL GROUND
RELAYS PROVIDE SELECTIVE
BACKUP FOR TWO BUSES

R. W. Whitford (San Diego Gas & Electric Co., San Diego, CA): This is an interesting paper and provides insight into various methods to obtain breaker failure protection. It points out problem areas that one should watch for when designing a scheme and makes a good reference paper. The following is offered for consideration.

1. Figure 9 - (Single breaker trip) shows that the PR-1 relay operates to trip the TC-1 coil and also energize the 62X coil. The ICS for the PR-1 relay will pickup but should not be sealed in by the 62X coil current if the 52A contact opens properly. This should be checked. SDG&E would normally use a 50FD current relay auxiliary contact in series with the 62X coil to make sure that the PR-1 and 62X circuit is not sealed in should the breaker interrupt correctly. The 50FD which is a current relay, prevents the chance of a seal-in.

2. Figure 10B - This shows a possible seal-in path when the 50-1 ICS operates due to current from the 86BF and staying sealed up if the 62 relay draws enough current.

3. Figure 11 - This shows that with coils 94-1, 62X and TX in parallel that probably a seal-in of the PR-1 and SR-1 will occur since once picked up only about 40 percent of the pickup current is required to keep it that way. This means that the breaker could never be closed and also the 62X and 62Y relays would cause the 62 to time out.

The seal-ins described above may not be a problem if the coils are chosen so that their currents are below the PR-1 and SR-1 ICS coil maintaining current.

Manuscript received September 23, 1981.

John R. Boyle: I would like to thank the authors for their appropriate comments and suggestions. Mr. Newman is correct in his observation that one of the distinct advantages of the breaker failure scheme shown in figure 6 is the elimination of the reset time of device 50. It would have been appropriate for the paper to have mentioned this in item 4 under the heading, "Fault Detectors," because the elimination of the 62 timer's dependency on device 50 to complete its timing cycle would certainly resolve the problem of a breaker failure timer not starting until one of two breakers serving a transmission line opens.

We agree that Mr. Newman's figure 1 is probably more reliable than figure 6 shown in the paper because of the elimination of an additional "and" logic circuit. However, the elimination of this additional "and" circuit may make the scheme less secure. We concur that Mr. Newman's connection could be used for generator breaker failure protection by connecting generator protective relays to one of the lines marked "additional lines." However, it would appear that either scheme would accomplish the desired results. Of prime consideration is that circuit breaker auxiliary switches should not be used to initiate breaker failure through device 50. Both schemes accomplish this objective.

Mr. Newman's suggestion that the "retrip" feature mentioned in item 7 under "General Design Philosophy," should be connected to energize both the primary and secondary trip circuits of a protected breaker is a good one and should be a part of all design philosophies.

Mr. Whitford is correct in stating that the "ICS for the PR-1 relay will pick up but should not be sealed in by the 62X coil current if the 52A contact opens properly." The seal-in should pick up when trip coil current flows and should be set to drop out when only 62X current flows. The resistor in series with the 62X and 62Y coils should be sized to ensure that the seal-in circuits associated with PR-1 and SR-1 drop out.

Mr. Whitford's comments in item 2 concerning a possible seal-in path in figure 10B is appropriate and should have been covered in the text of the paper. The note in figure 10B that the contact of 86BF could be located in either the 62 coil circuit or the 62 contact circuit is incorrect in light of Mr. Whitford's comments. It appears now that it should be located only in the 62 coil circuit. It does not matter that the ICS picks up when the 86BF coil operates, but it is important that the ICS drops out with current flow through the 62 coil; otherwise the 86BF lockout relay could not be reset.

Mr. Whitfore's reference to figure 11 is also appropriate if the seal-in circuit associated with PR-1 and SR-1 picks up when coils 94-1, 62X, and TX are energized. However, it was the intent that the resistor associated with TX 12-cycle dropout contact be sized to permit picking up the seal-in coil but (would) drop out when the TX contact opens. This should have been mentioned in the text of the paper.

Manuscript received September 23, 1981.

AUTOMATIC RECLOSING OF TRANSMISSION LINES

AN IEEE POWER SYSTEMS RELAYING COMMITTEE REPORT

ABSTRACT

The Automatic Reclosing of Transmission Lines Working Group of the IEEE Power System Relaying Committee reviewed the application, experiences, benefits and problems associated with automatic reclosing of transmission lines. These facets of automatic reclosing are described and discussed in this paper. The Working Group conducted a three-part survey questionaire on the practices and experiences with high speed automatic reclosing of transmission lines. The results of this survey from the 99 U.S. electric utilities and a bibliography of papers for further information on various aspects of reclosing are included.

INTRODUCTION

Automatic reclosing of transmission line circuit breakers is a long accepted practice of electric utilities in this country. The intent of this practice has been to return the transmission system to its normal configuration with minimum outage of the transmission line and with the least expenditure of manpower. The increased operating costs of attended stations, and the improved reliability of automatically controlled breakers, accelerated the interest and use of automatic reclosing for first contingency situations. Individual utility policy and system requirements dictate the complexity and variety of automatic reclosing schemes in service today.

Experience in operating overhead transmission systems has indicated that approximately eighty percent of line faults are temporary in nature. Thus, with automatic reclosing on the transmission system, one or more of the following benefits are generally achieved:

1. It restores the power transfer capability of the system to its former level in a minimal period of time.
2. It adheres to predetermined reclosing restrictions and times.

Members of the Automatic Reclosing of Transmission Lines, Working Group of IEEE Power System Relaying Committee are: T. L. Kaschalk, Chairman; R. F. Arehart, T. Beckwith, J. L. Blackburn, J. J. Bonk, J. R. Boyle, M. K. Brammer, A. D. Cathcart, S. N. Chin, D. H. Colwell (previous Chairman), I. R. Harley, M. D. Hill, J. A. Imhof, R. H. Jones, D. K. Kaushal, R. J. Kuhr, L. E. Landoll, T. J. Murray, K. K. Mustaphi, G. R. Nail, R. W. Pashley, J. M. Postforoosh, M. S. Sachdev, D. E. Sanford, H. S. Smith, W. M. Strang, R. D. Stump, A. H. Summers, W. H. Van Zee, E. J. Weiss, R. C. Zaklukiewicz and D. Zollman.

3. It reduces the service interruption time to customers directly fed from the transmission system.
4. It may relieve the system operators of restoring service manually during storms or other system disturbances.

HISTORICAL BACKGROUND

Automatic reclosing was first applied in a limited fashion in the early 1900's on radial feeder circuits protected by instantaneous overcurrent relays and fuses. These schemes reclosed the circuit two or three times prior to lockout. The experience indicated from 73 to 88% success on the first reclose actions and covered both radial and looped circuits, predominantly at distribution voltages, but also including 154 kV.

By the 1930's, inverse-time relays were introduced with instantaneous trip elements that aided coordination with fuse schemes. Automatic reclosing schemes at this time reclosed the circuit once after a preset time delay for arc path deionization and relay reset, and provided lockout if a second trip occurred within 30 seconds of the first tripout. These early reclosing applications were directed at circuit restoration for service continuity only.

By the late 1930's, the industry began to develop transmission power circuit breakers with high speed mechanisms. The faster operating speeds of these new circuit breakers reduced clearing time, permitted high speed reclosing, and improved system stability. Probability studies of insulator flashover were initiated to determine minimum reclosing times that still permitted enough time for arc deionization. Early applications of automatic reclosing on multiterminal lines reclosed the circuit high speed at one terminal and if the reclosure was successful, the remaining terminals reclosed with time delay to complete the through circuit. Multishot reclosing was and still is predominantly used on radial lines for purposes of service continuity.

Concern has been created in recent years by the failures of transformers associated with reclosing feeders in distribution stations, and the possibility of exceeding stress limits in turbine generator shafts. It is interesting to note that, as early as 1944, in a paper on single pole switching, the problem of mechanical shock to generator shafts during fault clearing and reclosing was discussed. The authors concluded that the calculation of stresses "may dictate single pole switching, regardless of transient power limits."

The power circuit breaker has undergone many improvements in design, mechanism speed, and operating reliability from the 1940's to date. These improvements, as well as protective relay development and scheme sophistication, have led to the high speed reclosing applications employed today.

83 SM 411-6   A paper recommended and approved by the IEEE Power System Relayed Committee of the IEEE Power Engineering Society for presentation at the IEEE/PES 1983 Summer Meeting, Los Angeles, California, July 17-22, 1983. Manuscript submitted Aptil 4, 1983; made available for printing May 23, 1983.

Reprinted from *IEEE Trans. Power App. Syst.*, vol. PAS-103, no. 2, pp. 234-245, Feb. 1984.

AUTOMATIC RECLOSING METHODS

There are various automatic reclosing systems in use today and these may be classified as follows: high speed, time delayed, single shot or multiple shot, and single pole or three pole. These reclosing methods will be discussed in the following sections.

High Speed

High speed reclosing is the automatic closing of a circuit breaker with no intentional time delay, other than de-energized time necessary to deionize the fault arc. All other reclosures are designated time delayed.

High speed reclosing may provide the following advantages:
1. Improved system integrity and security.
2. Preservation of machine stability.
3. Minimized customer outage time.
4. Higher probability of some recovery from multiple contingency outages.

In the application of high speed reclosing, there are, however, constraints which must be considered along with the effects on system configuration, operating practices and equipment. A number of the prominent constraints are:

1. High speed relaying must be provided to initiate tripping at all transmission line terminals simultaneously for all internal fault conditions, to insure adequate fault arc deionization time.
2. System stability must be maintained following a high-speed reclosure into a permanent line fault.
3. Faults must not be sustained by induced voltages from adjacent parallel circuits.
4. Possible damage to rotating electrical equipment connected to the system must be evaluated.
5. System configuration must be capable of holding the phase angle across the open line within acceptable limits.

For a high speed reclosure to be successful, the fault arc path must deionize sufficiently to re-establish the insulating properties before the transmission line is re-energized. The factors that affect the success of high speed reclosing are:

1. Transmission line design and construction.
2. Fault clearing time.
3. Fault magnitude.
4. Location of the transmission line in relation to natural or man-made contamination sources.
5. Weather conditions.
6. Multiple component lightning strokes.
7. Capacitive and inductive coupling from parallel circuits.
8. Point on the source voltage wave at the instant the transmission line is re-energized.
9. Tapped loads, shunt reactors, series capacitors, etc.

Time Delayed

Time delayed reclosing is the automatic reclosing of a circuit breaker after a time delay which is intentionally longer than the user's designated deionization time for a specified operating voltage level. This intentional time delay usually varies from one to sixty seconds.

When applying time delayed reclosing, the following factors should be considered:
1. The preferred minimum time to re-establish the transmission circuit, recognizing the number of line terminals, system requirements, stability considerations, system damping factors and the types of customer load.
2. Minimizing the probability of re-establishing the fault arc by increasing the intentional time delay.
3. Time delayed reclosing may be desirable if high speed protective relaying is not feasible for all terminals of a transmission line.

With time delayed reclosing, only one terminal of a transmission line is usually reclosed at a time, in contrast to the restoring of the through line with most high speed schemes. Voltage monitor relays are often installed to determine the presence of system voltage on either side of the circuit breaker. The user can then preselect the system voltage conditions which must exist before a circuit breaker is permitted to reclose. Some users permit a time delayed reclosure only after verifying that the angular difference of the system voltages on each side of the open circuit breaker is within prescribed limits and in synchronism for a specified period of time. Synchronism-check relays are used for this purpose. Stability and load-flow studies should be conducted to determine the need for synchronism-check of automatic time delayed reclosing, following an abnormally long fault-clearing time, or for an unusual multiple contingency operating condition. In some systems, the number of other transmission paths in parallel with the open line, makes it unnecessary to check for phase angle difference and synchronism prior to reclose initiation.

Single Shot or Multiple Shot

Single shot reclosing is the automatic reclosing of a circuit breaker one time, within a prescribed reclosing period. This could be either high speed or time delayed. Multiple shot reclosing is the automatic reclosing of a circuit breaker more than one time, in a predetermined reclosing sequence or duty cycle.

A typical reclosing sequence at a given terminal might consist of a high speed reclosure, unsupervised by either dead line or synchronism-check relaying, followed by a time delay reclosure, supervised by dead line voltage relaying. A successful reclosure would then permit the other terminal(s) to close after a time delay, supervised by synchronism-check relaying.

In general, modern HV and EHV circuit breakers are capable of operating in any practical multiple shot reclosing duty cycle. Certain constraints, with respect to circuit breaker components, such as closing or opening resistors, the available air, gas, or fluid pressures, or a derating factor based on the prospective fault current and possible total number of fault interruptions in a given time interval, may be a limiting influence on repetitive operations or duty cycles. Reference should be made to pertinent sections of American National Standards C37.04-1979 and C37.010-1979 for circuit breakers in making an application.

Multiple shot reclosing is not generally used where system stability may be jeopardized by the shock of reclosing into a permanent fault, or when other restrictions may exist. Where it can be used, the second reclosing attempt is delayed long enough to permit post-fault transients to diminish. Likewise, multiple or single shot reclosing is not used where mechanical damage to turbine generators or large motors may result.

Over the years, transmission line reclosing schemes have been predominantly single shot; however, multiple shot reclosing schemes are receiving greater acceptance. A significant factor in this trend is that most substations are no longer manned. Generally, a transmission line may be re-energized within a few minutes following an unsuccessful automatic reclosure, if the substation is manned or supervisory control exists. However, at those locations where an employee must be dispatched, a delayed outage may unnecessarily result if a single-shot reclosure is unsuccessful. Utilities have experienced varied rates of successful second and third reclosures, sometimes exceeding 25 and 10 percent, respectively. Circuits experiencing galloping sleet-laden conductors or tree contact resulting in repeated faults, have been restored successfully because of the use of multiple reclosing. However, reclosing relay reset times and breaker constraints must be considered to avoid problems due to excessive breaker operations in a given period because of intermittent faults. If a utility does not wish to replace single-shot reclosing schemes with multiple-shot reclosing schemes, it may achieve comparable success by adding additional time delay to the single shot reclosure.

Single Pole or Three Pole

Single pole operation differs from three pole in that at each terminal of the line, only the faulted phase is tripped and reclosed high speed. While the single phase is open at each terminal, power is transmitted over the remaining two phases and through ground between the neutral points of the synchronous systems, or over other transmission circuits. This flow of synchronizing power reduces the rate of rotor angle drift between the synchronous machines and tends to maintain the stability of the system. However, during the time that the single phase is open, ground currents are being circulated in the system, and may adversely affect system ground relaying.

With single pole fault clearing, a voltage will be induced in the isolated phase due to capacitive coupling, and to a lesser degree, by the electromagnetic coupling between the two energized phases and the de-energized phase. The magnitude of the induced voltage is a direct function of the phase-to-phase and phase-to-ground capacitances which are directly related to the physical configuration and length of the transmission line. The induced voltage may sustain the arc for an extended period of time following the operation of the single pole breakers. This current is usually referred to as the secondary arc current.

The minimum dead-time required before allowing high speed single pole reclosing is determined by analytical studies. This time, without line compensation, has been found to be appreciably longer than that required for three phase tripping and reclosing. In those instances where a sufficiently high voltage may be induced to sustain the arc, either permanently or for a period greater than the maximum time permissible to maintain system stability, the capacitances must be compensated if a reclosure is to be successful. One method of neutralizing the induced voltage is by the application of shunt reactors (including a neutral reactor) on the transmission line. Another proposed method is the closing and re-opening of a single phase, high speed ground switch on the open phase conductor.

There are design considerations with single pole tripping and reclosing not associated with three pole tripping and reclosing. Special protective relays and control circuitry must be provided to distinguish between phase-to-ground and multi-phase fault conditions. In addition, intelligence must be provided to trip and reclose only the faulted phase. Generally, three pole tripping without reclosing is initiated for:

1. A multi-phase fault.
2. An unsuccessful reclosure into a single-phase-to-ground fault.
3. A second phase being faulted while a single phase-to-ground fault is being cleared.
4. A second phase being faulted during the open pole time.
5. The failure of the automatic reclosing scheme to operate in a prescribed period of time with only two phases intact.

APPLICATION CONSIDERATIONS

A number of specific reclosing practices are reviewed in this section of the paper.

Lines With Transformers

It is normal practice not to re-energize the faulted transformer until the unit has been inspected and repaired. Thus, when a transformer is connected directly to a transmission line, without a local fault interrupting device, the line terminal circuit breakers should not be allowed to reclose automatically if the fault is in the transformer. Therefore, automatic reclosing of the breakers for transformer faults will be locked out by the operation of transformer protective relays. These relays will initiate a lockout relay for the local breaker, and send a direct transfer trip signal to trip and block automatic reclosing of the remote breaker. However, automatic reclosing of the terminal breakers for a line fault should be permitted.

Some transformer installations use a motor operated air switch on the high side, for isolation of the transformer under normal switching or fault conditions. Generally, the switch is used in conjunction with a direct transfer trip scheme and/or an automatic ground switch. Opening of the air switch may be initiated directly by operation of the transformer protective relays, or it may be supervised by a voltage relay so that it opens only after the line is de-energized. In either case, the reclosing time of the remote line breakers must be coordinated with the switch opening time to prevent re-energizing when the switch is partially open.

Shunt and/or Series Compensation

Reclosing considerations for lines provided with shunt reactors are similar to those required for lines that include transformers. For reactor faults, a direct transfer trip scheme is normally required to initiate opening of the remote line breaker, since dedicated breakers are seldom provided for line connected reactors. In many cases, the line connected reactors are intended to be connected to the line whenever it is energized. Where this is the case, a reactor fault requires that the line not be reclosed until the reactor problem is resolved. However, if the system design permits operation of the line without the reactors, automatic reclosing can be used after automatic reactor isolation.

Shunt reactor compensation, together with a transmission line's natural distributed capacitance, forms a resonant circuit that produces an oscillatory decay of any trapped charge on the line after opening of the line. The frequency of this oscillation may approach system frequency and may require several seconds to dissipate. As a consequence, it may be necessary to delay high speed reclosing to control switching surges when reenergizing the line. Also, line connected voltage and frequency relays may misoperate on the oscillatory decay, and the performance of the capacitor voltage transformers may require consideration.

For lines with series capacitors, the line is usually reclosed automatically without regard to the operation of the series capacitor protective relay. Automatic controls of series capacitor banks may initiate a bypass of the bank during a line fault, followed by automatic reinsertion of the capacitor bank after successful high speed reclosing. High speed reclosing can be applied, since the system design is usually adequate to withstand the switching surge voltage that can occur with series compensation.

Multi-Terminal Lines

High speed reclosing can be applied to all terminals of a multi-terminal line, if the line protection scheme provides high speed clearing at all terminals, and the reclosing of all terminals into a permanent fault are not detrimental to system stability. High speed reclosing requires a minimum dead time that may be more difficult to obtain on multi-terminal lines than on two terminal lines. The reclosing time of all terminals must take this into account, and usually, high speed reclosing will only be permitted if the pilot relaying channel is in service, and it has been possible to set the relays at each terminal to "see" all line faults at their inception. This gives reasonable assurance that clearing will be simultaneous on all terminals. It may be more desirable to reclose one terminal high speed and provide live line checking at the remaining terminals, to eliminate multiple reclosing into a permanent fault.

Multi-Breaker Terminations

On EHV systems, it is common practice to terminate each end of the line to one or more breakers, such as in ring bus, breaker-and-a-half, or double breaker arrangements. These configurations require tripping more than one breaker to clear the fault. Automatic reclosing of all breakers simultaneously may expose them to additional fault duty. It is, therefore, desirable to test the transmission line by reclosing only one breaker and, if successful, reclose the other breaker(s). Successful reclosing can be verified either by restoration of line voltage or by synchronism-check. The initial reclosure may be high speed.

Transmission Cables

Automatic reclosing is rarely used on transmission cable. However, if a transmission line consists of part overhead line and part cable, automatic reclosing may be desirable if the cable forms only a small part of the transmission line. In some instances, separate relaying is provided for the cable section, in which case, sensing is available to block reclosing for cable faults.

Gas Insulated Bus

Normally, automatic reclosing would not be used on a gas insulated bus to minimize contamination and burn through of the containment wall due to the fault. However, if a portion of the bus is included in the transmission line protection, such as the portion of bus between the breakers on a breaker-and-a-half scheme or ring bus configuration, the reclosing practice for the transmission line would normally be followed. The gas-insulated bus section may have separate protection, in which case, sensing would be available to block reclosing for faults within the bus section.

Generators

In recent years, considerable research and analysis has been focused on the stresses in the shafts and components of turbine-generators due to switching operations. There is little documentation of actual damage to, or failure of, turbine-generators resulting from reclosing or switching. The effects of the stresses induced in the turbine-generator are cumulative and may be caused by normal switching operations or system faults; therefore, automatic reclosing may be a contributing factor to machine failure, but not necessarily the sole contributor.

The operation of closing or opening a breaker in the power system can result in the creation of power transients and current oscillations, which can stress or damage generating units located electrically close to that breaker. The power and current transients will affect various components of the turbine-generator. These components include such items as the unit shaft and all rotating components in the exciter, generator and turbine, and generator stator elements consisting of core members and winding members. The concern is the average initial power change, Delta P, which occurs when the breaker is closed, and its effect in producing torsional stresses, primarily in the rotational members of the turbine-generator. For this condition, proposed permissible limits for Delta P or Delta I at the generator terminals are .5 per unit based on rated load and power factor.

These limits should be applied only for steady-state conditions. Reclosing times in excess of 10 seconds appear to be long enough to allow oscillations from the initial disturbance to die out. Therefore, a subsequent switching operation can be assumed to occur under a steady-state condition.

The other concern is the effect of high speed reclosing following the clearing of faults on lines connected to a generating station (or electrically close to a generating station). Initial studies indicate that reclosing into a 3ϕ fault is a worst case condition and would most likely result in some loss of life to the rotational members of the machine. If a fault is electrically close to the machine, all the useful life of a rotational member can be consumed and a failure can occur. High speed reclosing into all other multi-phase faults produces stresses lower than 3∅ faults but of magnitudes still sufficient to cause some loss of life. High speed reclosing into single-line-ground faults produces stresses which are generally less than multi-phase faults. While the likelihood of damage is low for this type of fault, it is still recommended that its effect be studied.

It is probable that the effects of all types of reclosing will have to be studied to determine whether excessive stresses will occur in the turbine-generator components. These studies should also include the effects of high speed line reclosing at remote stations which are electrically close to the generating plant, since excessive stresses could occur in machines located one or more busses away from the location of the reclosing breaker.

The study of turbine-generator shaft damage, due to torsional oscillation, is complex and involves the following areas:

1. Modeling of the shaft system (rotational members)
2. Coverting the torque-time response to material fatigue values
3. Determining an acceptable screening criteria

Each of these items is a major subject which will require much additional study, yet they are not mutually exclusive. The most needed item at present is the establishment of a screening criteria which can be used in conjunction with conventional system planning studies to identify situations of concern. Some of the more recent papers indicate that shaft fatigue (as a result of high speed reclosing) cannot be directly correlated to simple measures, such as Delta P or Delta I, as indicated above.

As a result of all the uncertainties, practices amongst the various utilities differ widely. Some companies have questioned the traditional justifications for high speed reclosing, particularly at generating stations, and have changed their reclosing practices to some form of the following:

1. Synchronism check
2. Adequate time delay (10 seconds or more) to allow decay of oscillations
3. Single pole operation or other type of relaying designed to avoid reclosing on multiphase faults.
4. No automatic reclosing at all.

If automatic reclosing is not used, an underlying transmission system may be damaged or tripped by a relay due to overloads. The choice then becomes one of adjusting generation in order to reduce the phase angle across the breaker, accepting load reduction, or retaining automatic reclosing with the risk of machine damage. The selection of the reclosing relay and type of reclosing scheme should receive careful attention, in order to minimize excessive torsional stresses in the turbine-generator. The determination of more precise setting criteria is necessary before a standardized reclosing relay scheme can be applied, which will recognize potentially hazardous reclosing events and abort them.

Motors

The stresses in shafts and components noted with turbine generators, may also be found in synchronous and induction motor drive system. The effect of electrical transient disturbances on motors connected to the power system warrants concern, especially as motors and their associated drive systems become ever larger. It is not unusual in today's industrial complexes, to find motors of up to 30,000 horsepower driving fans, compressors, or pumps. In other instances, process equipment may be driven by dual motors of 10,000 h.p. via gear reducers and ring gears. It must be recognized that significant mechanical stresses can be induced in such drives as a result of transient torque produced by a switching operation.

As with generators, the effects of reclosing varies greatly depending upon the configuration of the system.

The impedance of interconnecting transformers, acting as a buffering impedance, mitigates the effects of most switching disturbances. Also where the configuration maintains a synchronizing tie to the power system, motors will remain in synchronism with little difficulty. When high speed reclosing is used, the relaying is generally high speed, providing 4 - 6 cycle clearing times. The high speed clearing can be very significant in minimizing the disturbance to large motor installations.

The system configuration that connects the large motor installation to a system may be a tapped arrangement such that the motors become isolated from the utility during the dead time of reclosing. In this case, reclosing even at high speeds such as 20 - 30 cycles, the voltage and phase angle conditions may be unfavorable. Reclosing of source breakers can and is done in these situations, but usually only with supervision by high speed relays sensing voltage, frequency or phase angle. Depending upon the specific primary connection, it may be best to ensure separation of the motor load by tripping or allow closing of the interconnecting breakers only within certain limits of voltage and angle.

It is difficult to generalize on the limits of voltage and angle for re-energizing separated motor loads by either manual or automatic reclosing means, however, delaying re-energizing until residual voltages are below 25% or the angle is within 60° is in general use. It should be noted that high speed reclosing times used in transmission systems are usually too slow to affect high speed transfer of motor loads without supervision and are too fast to permit decay of residual voltages to tolerable levels.

It is important that the system engineer be aware of the potential for damage from transient torques that may result from any of the above factors. To this extent, the above aspects should be studied in detail and their potential effect on motors connected to the system should be evaluated.

Large synchronous and induction motors, as indicated above, are potential problems in reclosing applications. However, it should be pointed out that in many industrial processes, principally where large number of smaller induction motors are in use, high speed reclosing may offer improved service to the customer. Textile mills and oil line pumping are two such situations. It is well that the customer be advised of the possible benefits and that he should consider the effects on his particular process.

It is essential that both user and supplier of electric power be aware of what takes place under switching conditions. Only when the effect of such transients on both parties is understood, can the potential risk be evaluated and acceptable solutions sought.

Transformers

Because transformer failures have been increasing dramatically, particularly on transformers manufactured in the 1960's, there is considerable activity in the area of short circuit testing of power transformers and in revising the IEEE short circuit testing standard. All power system faults impose two stresses on the power transformer with one being thermal and the other mechanical. However, the effects of these two stresses are not mutually exclusive. The strength of the transformer, depending on the materials and the clamping forces, is affected by the temperature experienced by the transformer, and in turn, is directly related to the current densities in the windings.

Recently, there has been research on the impact loading of transformers. The forces from asymmetrical currents cause much higher stress in the transformer than do symmetrical currents. Consequently, each fault does impose some deformation on the various transformer components which may or may not be permanent. If a permanent deformation occurs, the clamping forces are then lower, and subsequent faults may cause greater deformation which could lead to component fatigue, and ultimately, failure.

The majority of the available literature, including results of transformer short circuit capability tests, primarily reference the small power transformer used for feeding distribution systems.

The transmission system transformers also have similar problems to those of the smaller transformers. The specifications for, and the analysis of, the effects of reclosing on transmission substation transformers is more complicated because these larger transformers are generally multiple winding transformers, in which the system parameters affect the distribution of current in the various windings.

Other than the fatigue problem, high speed reclosing affects the thermal capability of the transformer for through faults. In general, the thermal capability is not a serious problem; however, it should be reviewed where high speed reclosing is used. The other factor that affects transformer duties is the loading practices of the utility. Overloading of transformers can seriously affect the ability of the transformer to withstand through faults and could conceivably impose some limitations on reclosing practices.

FUTURE OF AUTOMATIC RECLOSING

Previous sections have referred to single and multiple shot reclosing relays which perform the automatic reclosing of transmission line circuit breakers under pre-planned timing settings and local conditions. The number of reclosing attempts to lockout, the specific time setting of each attempt, the coordination with automatic reclosing of circuit breakers at other stations, and the locally sensed restrictions and conditions for reclosing, have largely been a matter of individual company policy, arrived at over many years of experience.

With recent advances in digital and solid state technology, it is apparent that the "one reclosing relay per breaker" application can be re-examined. For example, one dedicated reclosing microprocessor or logic unit can be programmed to intelligently perform the automatic reclosing control for all breakers at a substation. It can be programmed differently for each breaker and, as an on-line device, it can logically check the energized or de-energized status of all buses and lines, positions of all breaker control and selector switches, circuit breaker positions, closing control voltages, etc. If communication with adjacent stations is available, the knowledge of conditions at other stations could also be factored into the local automatic reclosing decisions.

With a high-speed data acquisition and communication system, it may be feasible to centralize the automatic reclosing operation of transmission line breakers on an area-wide basis, where many other system conditions could be taken into account. However, from a pragmatic view, it appears that automatic reclosing control, per se, should be maintained locally at the substation level. Any additional information from remote stations, or from an area operations office, can be additional inputs to the local decision as to when, if, and how to reclose after an automatic trip.

EXPERIENCE SURVEY

The working group conducted a three-part survey questionnaire on the practices and experiences with high speed automatic reclosing (HSAR) of transmission lines. This was undertaken because of the limited data available in the literature. The summaries which follow are the results from this three-part survey questionnaire by 99 U.S. electric utilities.

Summary of Part I

Part I covered automatic reclosing practices for all transmission lines on the system, for line voltage ratings between 60 kV and 800 kV. It included a form (AR-1) on reclosing practices and a series of questions on application considerations. The replies to the questions are summarized first, below, and a summary of Form AR-1 follows. (The numbers correspond to the number of the question in the survey.)

1.2 For faults on the line, under what conditions is HSAR intentionally not initiated or blocked

1.2.1	None	9
1.2.2	Breaker trips not initiated by pilot schemes	52
1.2.3	Cable faults	21
1.2.4	Specific types of faults	
	Three-phase	6
	Multi-phase	7
	Underfrequency trips	2
	Time delay trips	15
1.2.5	Other conditions	
	Zone 1 trips	4
	Breaker Failure Relay (BFR)	8
	Out of Step (OOS)	4
	Ground switch for transformer protection	3
	Transfer trip received (Transformer Protection)	8
	OOS and BFR	25

1.3 Does your application of HSAR include lines with tapped loads?

    1.3.1 Yes   63
    1.3.2 No   19

Under 1.3.1, 2 responded with HSAR is used on one end and sync-check on the other end. Two responded with additional time delayed reclosing is used after automatic sectionalizing takes place.

1.4 Reclosing practices on lines serving industrial customers.

    1.4.1 HSAR is used for industrials with the following types of loads.

        1.4.1.1 Both generation and large motors   11
        1.4.1.2 No generation, but with large motors   33
        1.4.1.3 No generator or large motors   21

    1.4.2 HSAR is not used on lines serving industrial customers   27

    1.4.3 Delayed reclosing is initiated after permissive signal or message is received from customer.
               Initiated automatically 11
               Initiated manually   2

    1.4.4 Other:
          Close only on request from customer   4

1.5 Where HSAR is used at one end only and manual or time delayed reclosing is applied through a synchroverifier at the other end, what angle is used?

    1.5.1 For lines terminating at generating plants:
          10° to 30° depending on system   3
          20° to 45° depending on system   1
          20°   23
          25°   1
          30°   11
          45°   2
          60° with long looped lines   4
    1.5.2 All others:
          10° to 30° depending on system   2
          20°   15
          25° to 30°   1
          25° to 55°   2
          30°   14
          40°   11
          45°   3
          60°   7
          60° @ 500kV, 40° @ 230kV, 20° lower voltages   1

1.6 Is there a planned or anticipated change by your company in application of HSAR?

    1.6.1 Yes, refer to Part II   11
    1.6.2 Yes, from – to
          HSAR with pilot trip to HSAR without pilot trip at 115kV   2
          No HSAR @ power plants to HSAR @ power plants   1
          Delete reclosing at power plants   2
          HSAR to delayed reclose on tapped lines   2

          HSAR to 5 seconds or less delayed   3
          HSAR to 5 seconds or more delayed   2
          HSAR to live line sync check   1
          No HSAR to HSAR   1
          Add sync check at power plants   1
          Three pole trip and HSAR to single pole trip and HSAR on some 500kV lines   1
          Develop uniform practice with neighbors   1
          No anticipated changes   55

1.7 If HSAR is not used presently (and is not planned in the foreseeable future), check reasons below:

    1.7.1 Dissatisfaction with HSAR used in past   3
        Other:
          Concern for stress problems in turbine-gen shafts   2
          Concern for system stability   3
          Subtransmission circuits do not have high speed tripping for all faults   4
          Advantages not sufficiently clear   2

          Low lightning frequency   1

1.8 Additional comments on philosophy, need, practices, performance, etc.

SCADA systems will change HSAR needs   2
Need to define sync-check angle requirements   2
Need to define generator and transformer capabilities in regards to HSAR   2
HSAR needs more study where customers are adding generation sources   3
HSAR adds to breaker fault duty   1
Delayed reclosing is more successful than HSAR   1
On three-terminal lines, 2 terminals HSAR and third terminal by delay or sync check   1

Summary of Form AR-1

Form AR-1 of the survey questionnaire covers reclosing practices of U.S. electric utilities, as presently applied to power systems at each nominal voltage level from 69 kV through 765 kV.

The results tend to be dominated by the utilities having the greatest number of lines at a particular voltage. This makes the reclosing practices of those utilities appear as the general practice, whereas a significant number of companies with only a few lines at a particular voltage may have different practices. For this reason, the fourth column of Table 1 indicates the utility having the maximum number of lines at that voltage, of those responding to the survey. From 69 kV through 500 kV, the minimum number of lines at each voltage having reclosing on a particular utility system was one, indicating the wide range involved.

Table 1 shows that many utilities use both single-shot and multi-shot reclosing at the same voltage level, as indicated by the last column. This is more common at the lower voltages.

Summary of Part II

Part II of the HSAR survey covers U.S. electric utility practices for transmission lines terminating at generating plants. Because of recent concerns about possible damaging shaft torques due to switching of these lines, utilities were asked to report on present and past practices, and on anticipated future practices, if changes were expected due to these concerns. Table 2 is a summary of this part of the survey.

Examination of the data presented in Table 2 indicates there is no predominant practice among U.S. utilities for lines terminating at generating plants for any voltage level, nor is there an indication of a significant change in practices in the future. The summary indicates a relatively even split between the three categories: HSAR not applied at either end, HSAR applied at the remote end only, and HSAR applied at both ends.

It is possible that many utilities may not have reviewed their reclosing practices in this area and/or reached a decision on whether or not to change. One utility representative requested to be on record that his company does not plan to change switching practices at generating stations as suggested by T-G manufacturers.

Summary of Part III

Part III rated the performance of high speed automatic relaying, covering experience, results and the utilities' assessment of reclosing on the power system. Based on the returns received, the information requested is unavailable or, at least, not in a readily retrievable form. In a few cases, this data was very complete. Generally, however, the type of fault, its cause and reclosing history was not available.

A total of 80 voltage level responses from 39 companies was tabulated. The remaining companies did not provide enough information. In general, the 39 companies appear to be satisfied with their reclosing results and component performance. The success percentages range from 62% to 85%, with 345 kV being the most successful.

Table 3 summarizes the data reported in Part III of the survey questionnaire.

TABLE 1

SUMMARY OF FORM AR-1

Voltage Class (kV) (Nominal)	No. of Utilities	Total No. of Lines - All Utilities	Max. No. Of Lines - One Util.	No. Util. With Both S.S. & M.S. Reclosing
69	47	2999	588	20
115	34	2883	343	13
138	34	1950	323	5
161	18	602	318	4
230	47	1269	136	4
345	36	534	78	4
500	21	212	61	1
765	2	18	15	0

Voltage Class (kV) (Nominal)	Number of Utilities and Number of Lines Having:									
	Single Shot Reclosing						Multi-Shot Reclosing			
	H.S. Both Ends		H.S. One End		T.D. Both Ends		One End		Both Ends	
	Util.	Lines	Util.	Lines	Util.	Lines	Util.	Lines	Util.	Lines
69	19	329	18	122	23	422	18	207	20	1443
115	21	395	13	66	16	772	16	319	19	976
138	20	520	10	134	12	401	5	118	12	643
161	8	336	8	49	7	99	3	30	7	74
230	20	372	11	87	14	142	9	61	16	403
345	18	146	15	67	15	135	5	24	9	144
500	8	103	5	37	6	27	1	3	5	22
765	0	0	0	0	1	3	0	0	1	15

TABLE 2

HIGH SPEED AUTOMATIC RECLOSING (HSAR) PRACTICES RELATIVE TO LINES TERMINATING AT GENERATING STATIONS

Reclosing Practice	Voltage	Number of Companies Indicating This As:		
		Present Practice	Past Practice	Future Practice
2.1 HSAR NOT APPLIED AT EITHER END OF LINE	69	17	15	12
Note 1. For these cases, essentially all utilities close the re-	115	14	13	8
mote end first, using line/bus voltage supervision control schemes	138	13	9	10
and/or varying degrees of time delay. Either manual/supervisory	161	5	1	5
or automatic delayed reclosing is used for the remote end, fol-	230	16	15	11
lowed by manual or automatic reclosing of the generator end of the	345	14	9	11
line via synchro-check supervision. A variety of control schemes	500	6	6	5
are used by utilities to accommplish the above, with varying degrees of sophistication.	765	1	1	1
2.2 HSAR APPLIED ONLY AT REMOTE END OF LINE	69	13	5	1
Note 2. For these cases, there is nearly an even split between	115	17	3	5
utilities whose practice is to close the generator end by operator	138	15	10	11
action, with synchro-check supervision in most cases, and those	161	5	1	3
whose practice is to close the generator end automatically after	230	21	8	13
a specified time delay, following a "live line" indication, with	345	21	10	19
synchro-check supervision. A few utilities employ both the above	500	7	2	6
methods, depending apparently on the particular plant and voltage involved.	765	0	0	0
2.3 HSAR APPLIED AT BOTH ENDS OF LINE	69	13	13	10
	115	14	17	6
	138	19	18	13
	161	4	5	3
	230	17	20	10
	345	14	22	7
	500	6	5	5
	765	1	1	1
2.4 OTHER: No responses				

TABLE 3

SUMMARY OF PART III

Voltage (kV)	Number of Companies	Recloser Success (%)	Number of Lines	Number of Miles	Average Line Length (miles)
44 & 69	12	79.41	334	4833.84	14.47
115, 138, & 161	32	62.27	1795	40152.17	22.37
230	10	71.20	406	12193.95	30.03
345	17	85.35	328	16559.6	49.69
500 & 525	8	75.56	171	11485.55	67.17
765	1	67.34	15	1359.	90.60

TABLE 3 (Cont'd)

Voltage (kV)	Transmission System Performance			Performance of Components		
	Excellent	Adequate	Poor	Excellent	Adequate	Poor
44 & 69	5	7	0	9	3	0
115, 138, & 161	15	16	1	18	12	1
230	4	4	1	4	3	0
345	10	5	1	10	6	0
500 & 525	3	5	0	2	6	0
765	1	0	0	1	0	0

BIBLIOGRAPHY

I. Historical and General

1) Anderson, A. E.; Automatic Reclosing of Oil Circuit Breakers; AIEE Transactions, Vol. 53, 1934, pages 48-53, 466 & 614.
2) Pierce, R. E.; Powers, R. E.; Stewart, E. C. and Heberlein, G. E.; Carrier Relaying and Rapid Reclosing at 110kV; AIEE Transactions, Vol. 55, 1936, pages 1120-29.
3) Sporn, Philip & Prince, D.C.; Ultrahigh-Speed Reclosing of High-Voltage Transmission Lines; AIEE Transactions, Vol. 56, 1937; pages 81-90, 100 & 1033.
4) Sporn, Philip & Muller, C. A.; Experience with Ultrahigh-Speed Reclosing of High-Voltage Transmission Lines; AIEE Transactions, Vol. 58, 1939, pages 625-31.
5) Sporn, Philip & Muller, C. A.; Five-Year's Experience with Ultrahigh-Speed Reclosing of High-Voltage Transmission Lines; AIEE Transactions, Vol. 60, 1941, pages 241-6 & 690.
6) Crary, S. B.; Kennedy, L. F.; Woodrow. C. A.; Analysis of the Application of High-Speed Reclosing Breakers to Transmission Systems; General Electric Review, Vol. 45, March 1942, pages 173-185.
7) Jancke, G.; Experience in Operation with Automatic Reclosing of Breakers After Line Faults; CIGRE, Paris, 1948, Report No. 306.
8) Parks, C. E. & Brownlee, W. R.; Consideration of Requirements and Limitations of Relaying and High-Speed Reclosing on Long and Heavily-Loaded Transmission Lines; AIEE Transactions, Vol. 69, 1950, pages 103-107.
9) Gillies, D. A.; Operating Experience with 230 kV Automatic Reclosing on the Bonneville Power Administration System; AIEE Transactions, Vol. 74, pages 1692-96.

II. Arc Extinction Technology

10) Maury, E.; Arc Extinction in the Ultra-Rapid Single-Phase Reclosure of 220 kV Lines; Revue Generale de l'Electricite, Paris, France, Vol. 53, 1944, pages 79-80.
11) Harrington, E. J. & Starr, E. C.; Deionization Time of High-Voltage Fault-Arc Paths; AIEE Transactions, Vol. 68, 1949, pages 997-1004.
12) Boisseau, A. C.; Wyman, B. W. & Skeats, W. F.; Insulator Flashover Deionization Times as a Factor in Applying High-Speed Reclosing Circuit Breakers; AIEE Transactions, Vol. 68 Part III, 1949, pages 1058-67.
13) AIEE Tranmission & Distribution Committee Working Group & Ellis, H. M.; Arc Dionization Times on High-Speed Three Pole Reclosing; IEEE Transactions, PAS Vol. 82, 1963, pages 236-53.

III. Single Pole Trip-Reclose Technology

14) Goldsborough, S. L.; Hill, A. W.; Relays and Breakers for High-Speed Single Pole Tripping & Reclosing; AIEE Transactions, Vol. 61, 1942, pages 77-81, & 429.
15) Trainor, J. J.; Hobson, J. E.; Muller Jr., N. H.; High-Speed Single-Pole Reclosing; AIEE Transactions, Vol. 61, 1942, pages 81-87 & 422.
16) Trainor, J. J.; Parks, C. E.; Experience With Single-Pole Relaying & Reclosing on a Large 132kV System, AIEE Transactions, Vol. 66, 1947, pages 405-411.
17) Dobson, W. P.; Mason V. V.; Relaying Problems & Arc Extinction Times, CIGRE, Paris, France 1952, paper 316.
18) Kimbark, E. W.; Suppression of Ground Fault Arcs on Single-Pole Switched EHV Lines by Shunt Reactors, IEEE Transactions, Vol. PAS 83, 1964, pages 285-290.
19) Edwards, Leo; Chadwick, J. W. Jr.; Riesch, H. A.; Smith, L. E.; Single-Pole Switching on TVA's Paradise-Davidson 500kV Line - Design Concepts and Staged Fault Test Results, IEEE Transactions, Vol. PAS 90, 1971, pages 2436-50.
20) Kimbark, E. W.; Selective Pole Switching of Long Double-Circuit EHV Lines: IEEE Transactions, Vol. PAS 95, 1976, pages 219-230.
21) Shperling, B. R. & Fakheri, A.; Compensation Scheme for Single-Pole Switching on Untransposed Transmission Lines, IEEE Transactions, Vol. PAS 97, 1978, pages 1421-29.

22) Shperling, B. R. & Fakheri, A.; Single-Phase Switching Parameters for Untransposed EHV Transmission Lines, IEEE Transactions, Vol. PAS 98, 1979, pages 643- 654.

23) Peterson, W. G.; Konel, H.; King, H.; Comegys, G.; Dorland, P.; Single Pole Reclose Test with High Speed Ground Switch Application, Bonneville Power Administration Test Report, October, 1979.

24) Hasibar, R. M.; Legate, A. C.; Brunke, J.; and Peterson W. G.; The Application of High Speed Grounding Switches for Single-Pole Reclosing on 500 kV Power Systems, IEEE Transactions, Vol PAS 100, Pages 1512-15.

25) Kappenmam, J. G.; Sweezy, G. A.; Koschik, V.; and Mustaphi, K. K.; Staged Fault Tests with Single Phase Reclosing on the Winnipeg -Twin Cities 500 kV Interconnection, Paper 81 SM 366-4 Presented at the 1981 Summer PES Meeting, Portland, Oregon.

IV. Reclosing With Rotating Equipment-Generators

26) Batchelor, J. W.; Whitehead, P. L.; Williams, J. S.; Transient Shaft Torques in Turbine Generators Produced by Transmission Line Reclosing; AIEE Transactions, Vol. 67, 1948, pages 159-164.

27) Cogswell, S. S.; Mauser, S. F.; Weeks, L. H.; Booth, W. H. & Markley, I. E.; Generator Shaft Torques Resulting from Operation of EHV Breakers; IEEE Transactions, Vol. PAS 93, 1974, pages 1020 Abstract: Paper C74-087 with discussion.

28) Abolins, A.; Lambrecht, D.; Joyce, J. & Rosenberg, L.; Effect of Clearing Short Circuits and Automatic Reclosing on Torsional Stress and Life Expenditure of Turbine Generator Shafts; IEEE Transactions, Vol. PAS 95, 1976, pages 14-25.

29) ASME/EEE Joint Working Group of Power Plant/Electric System Interaction, State of the Art Symposium – Turbine Generator Shaft Torosionals, IEEE - PES Special Publication No. 79TH0059-6PWR.

30) Joyce, J. S.; Kulig, T.; Lambrecht, D.; Torosional Fatigue of Turbine-Generator Shafts Caused by Different Electrical System Faults and Switching Operations; IEEE Transactions, Vol. PAS 97, 1978, pages 1965-1977.

31) Jackson, M. C.; Umans, S. D.; Dunlop, R. D.; Horowitz, S. H. & Parikh, A. C.; Turbine-Generator Shaft Torques and Fatigue Part I, Simulation Methods & Fatigue Analysis, IEEE Transactions, Vol. PAS 98, 1979, pages 2299-2307, 2314-2328.

32) Dunlap, R. D., Horowitz, S. H., Parikh, A. C.; Jackson, M. C. & Uman, S. D.; Turbine-Generator Shaft Torques and Fatigue, Part II, Impact on System Disturbances and High-Speed Reclosures, IEEE Transactions, Vol. PAS 98, 1979, pages 2308-2328.

33) Rusche, Peter A. E.; Shaft Stress Due to Switching Three-Phase Faults in Close Proximity to Generators Connected to a Strong System: Network Alternatives for Reducing Stress, IEEE Transactions, Vol. PAS 98, 1979, pages 408-415.

34) Undrill, J. M.; Shaft Impact Torques Due to Normal and Fault-Initiated Switching Sequences: IEEE Transactions, Vol. PAS 98, 1979, pages 618-628.

35) Joyce, J. S.; Kulig, T. & Lambrecht, D.; The Impact of High-Speed Reclosure of Single and Multi-Phase System Faults on Turbine-Generator Shaft Torosional Fatigue, IEEE Transactions, Vol. PAS 99, 1980, pages 279-291.

V. Reclosing With Rotating Equipment – Motors

36) McGinnis, F. J. & Schultz, W. R.; Transient Performance of Induction Motors; AIEE Transactions, Vol. 63, 1944, pages 641-646 & 1458.

37) Morton, W. R. & Shepard, R. V.; High- Speed Reclosing of Circuit Breakers on Systems Utilizing Synchonous Motors; AIEE Conference Paper CP59-881, 1959.

38) Walsh, G. W.; Effects of Reclosing on Industrial Plants; American Power Conference Proceedings, 1961, pages 768- 778.

39) DeMello, F. P. & Walsh, G. W.; Reclosing Transients in Induction Motors with Terminal Capacitors; IEEE Transactions, Vol. PAS 80, 1961, pages 1206-13.

40) Chidambara, M. R. & Ganapathy, S.; Transient Torques in Three-Phase Induction Motors During Switching Operations, IEEE Transactions, Vol. PAS 81, 1962, pages 47-55.

41) Hauck, T. A.; Motor Reclosing and Bus Transfer, IEEE Petroleum and Chemical Industry Conference Paper PC-69-8, 1969.

42) Weston, G. L.; Gareis, G. E. & Krause, P. C.; Comparison of Machine Behavior During Out-of-Phase Reclosing and a Three--Phase Terminal Fault, IEEE Conference Paper C75-016-1, 1975: Special Publication 75CH0990-2 PWR with Discussion in 75CH0991-0 PWR.

43) McFadden, R. H.; Re-energizing Large Motors After Brief Interruptions, IEEE-IAS Pulp and Paper Industry Conference, 1976, page 84.

44) Bishop, J. A. & Mayer, C. B.; A Case for High Fidelity Analysis of Non-Linear Electromechanical Torosional Dynamics - "Are You Really Sure it Won't Destroy Itself?", ASME Design Engineering Conference & IEEE/IAS Annual Meeting Conference 1976.

Discussion

D. R. Volzka (Wisconsin Electric Power Co., Milwaukee WI: The Working Group is to be congratulated on the development of a fine paper which provides a great deal of the history and philosophy behind reclosing practices and a very good summary of experiences.

In the section entitled, "Future of Automatic Reclosing," the Working Group has stated that it may be feasible to centralize the automatic reclosing operation of transmission line breakers using a high speed data acquisition and communications system. Wisconsin Electric has had such a system in service for approximately three years in one of our six operating divisions. This system provides direct, closed loop automatic reclosing of the 138 kV system in the district utilizing the central dispatch computers and existing remote terminal units. We are in the process of expanding usage of this technique to three additional divisions this year in light of favorable operating experience.

There were four basic reasons behind our development of this technique: 1) We were not satisfied with the reclosing relays available, 2) We were not using high-speed automatic reclosing, 3) Our remote control and data acquisition system was close to or represented state of the art, and 4) Our experience with communications has been excellent and is closely monitored.

Our goals in developing the system were to provide automatic reclosing during storms and other such periods to alleviate operator-required action. This feature is taken out of service from the CRT during normal periods when automatic reclosing may be undesirable. Logic checks within the program add several other refinements. The typical time from tripping to having the line back in service has been eight (8) seconds.

Manuscript received August 8, 1983.

George N. Lester (Chas. T. Main, Inc., Boston, MA 02199): The Working Group has prepared a very interesting report of the results of their survey into reclosing practices, and their efforts are appreciated.

One comment of possible interest can be added to the section on historical background. It is mentioned that in the 1930's the industry began to develop circuit breakers with higher speed mechanisms, permitting high speed reclosing. Actually, from a physical, or hardware, standpoint, it is believed to have been the development of pneumatic operating mechanisms that allowed really fast reclosing to be put into practice. Until that time. solenoid mechanisms for closing were commonly used, necessitating high control currents, a requirement that was alleviated with stored pneumatic energy, while also making possible faster operation.

The Working Group's comment would be appreciated on one question. Did the results of the survey identify rates of success, etc. (as shown in Table 3) specifically with single pole opening, and reclosing?

Manuscript received August 19, 1983.

IEEE Working Group on Automatic Reclosing of Transmission Lines: The working group (WG) thanks Messrs. Lester and Volzka for their comments and discussions.

Mr. Lester's comment on the section of historical background adds more insight into the development of the practice of high speed automatic reclosing. The published survey results dealt only with three phase reclosing practices. The original survey requested identification of lines with single pole reclosing, and only 3 companies responded in this area—two companies had one line each and the third company had 10 lines. Based on this response, the W. G. decided that it did not have sufficient data to compile meaningful results on single pole reclosing.

Mr. Volzka's discussion on automatic reclosing practices at the Wisconsin Electric Power Company (WE) are very informative and interesting. The WE practice of disabling the reclosing scheme "during normal periods when automatic reclosing may be undesirable" is of interest. The experience gained by WE, as well as by other utilities which are installing "futuristic" reclosing schemes, will be very beneficial to the utility industry. The W. G. suggests that these efforts be documented in industry publications with emphasis on why such methods were implemented and any special considerations that must be evaluated in such applications.

Manuscript received September 26, 1983.

SINGLE PHASE TRIPPING AND AUTO RECLOSING OF TRANSMISSION LINES
IEEE COMMITTEE REPORT

Members of the IEEE Power System Relaying Committee Working Group:
J.Esztergalyos-Chairman, J.Andrichak, D.H.Colwell, D.C.Dawson,
J.A.Jodice,T.J.Murray, K.K.Mustaphi, G.R.Nail, A.Politis,
J.W.Pope, G.D.Rockefeller, G.P.Stranne, D.Tziouvaras, E.O.Schweitzer

Key words: Single Phase Trip, single phase relay, auto-reclose, secondary arc current, recovery voltage, four reactor scheme.

ABSTRACT

This paper has been prepared to aid in the effective and uniform application of single-phase tripping and auto-reclosing of transmission lines. The benefits of application, relaying techniques, performance, and statistics will be discussed. The paper covers areas such as system stability, single-phase tripping and auto-reclosing, methods of secondary arc extinction, and related system requirements.

Descriptions of the various devices, definitions of terms, and references to other technical publications have been included to make the paper useful not only to relay engineers, but also to other technical people who are responsible for the installation and operation of such systems.

INTRODUCTION

The benefits of single-phase tripping and auto-reclosing are:

A.   Improvements in transient state stability

B.   Improvements in system reliability and availability, especially where remote generating stations are connected to load centers with one or two transmission lines

C.   Reduction of switching overvoltages

D.   Reduction of shaft torsional oscillation of large thermal units

Single-phase relaying application takes advantage of the fact that most faults on HV transmission lines are phase-to-ground faults. Some representative statistics are shown on Table I.[9]

TABLE I.
Relative Number of Different Types of Faults
on HV Transmission Lines

Fault Types	Percent
Single Phase-to-Ground Faults	70
Phase-to-Phase Faults	15
Double Phase-to-Ground Faults	10
Three Phase Faults	5
Total	100

91 SM 360-9 PWRD   A paper recommended and approved by the IEEE Power System Relaying Committee of the IEEE Power Engineering Society for presentation at the IEEE/PES 1991 Summer Meeting, San Diego, California, July 28 - August 1, 1991. Manuscript submitted April 2, 1991; made available for printing May 2, 1991.

On EHV/UHV lines the conductor spacing is increased, therefore, the percentage of multi-phase faults decreases. Some representative statistics for the relative number of different type of faults on a 525 kV Transmission Lines are shown on Table II.[9]

TABLE II.

Relative Number of Different Types of Faults
on 500 kV Transmission Lines

Fault Types	Percent
Single Phase-to-Ground Faults	93
Phase-to-Phase Faults	4
Double Phase-to-Ground Faults	2
Three Phase Faults	1
Total	100

DEFINITIONS

For the purposes of this paper, the following definitions apply.

Single-phase tripping: The opening of the faulted phase during a single-phase to ground fault.

Single-phase auto-reclosing: The reclosing of the faulted phase following a single-phase trip.

Primary arc current: The current in the phase to ground arcing fault prior to single-phase tripping.

Secondary arc current (Is): The current which flows in the arc after single-phase tripping is completed. (Is) is the sum of two currents derived from the electrostatic (Isc) and electromagnetic (Ism) coupling from the two energized phases and adjacent lines after the primary arc current is cleared via the line circuit breakers. The secondary arc if persistent, may prevent successful reclosing.

Recovery Voltage (Vr): The voltage which appears across the secondary arc path as soon as the arc is extinguished.

Cross Country Fault: The occurrence of simultaneous faults on the same or different phases on double circuit lines.

Neutral Reactor: A reactor used in combination with the three line-connected shunt reactors to create a high impedance against the flow of the secondary arc current.

SYSTEM REQUIREMENTS

A.   Transient Stability Criteria

Conventional studies of transient stability are based on the subject system being able to withstand a three-phase bus fault at critical locations.

Reprinted from *1991 IEEE Power Engineering Society Summer Meeting*, San Diego, CA, pp. 1-11, July/Aug., 1991.

Although three-phase bus faults are a convenient way to test the transient stability of a system, experiences and statistical data indicate that double or triple contingency faults are more common than three-phase faults.

One double contingency event more common than a three-phase fault is a fault on one line and a simultaneous trip of another line due to relay misoperation.

For example, Figure 1 shows a system with three parallel lines where a single phase-to-ground fault is applied on Line 1 and cleared in 3 cycles. Assume a simultaneous false trip of Line 2.

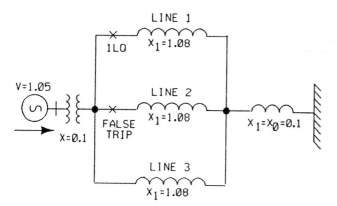

Figure 1. Single Machine-Infinite Bus Three Line Model System.

Figure 2. shows the results of the transient stability studies with three-phase and single-phase trip. The machines go unstable for a three-phase trip of line 1 (curve C), but remain stable for a single-phase trip (curves A and B). Reclose time is not critical. Note, that the units settle faster for the 90 cycle reclose than the 30 cycle reclose.

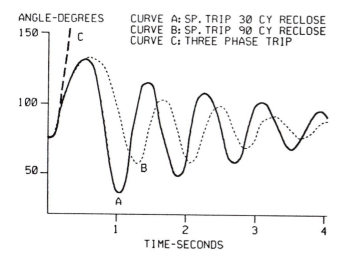

Figure 2. Rotor Angle Curves-Three Cycle Fault on Line 1 with False Three-Phase Trip on Line 2

Ideally, transient stability studies should include the statistical data of past system performance during multiple contingency events as well as three-phase bus faults. When realistic probabilities, statistical data, and cost of disturbance data are used, single-phase tripping will be cost effective in many cases.

B.   Design Requirements

Some of the obvious design requirements are:

1) A relay system which is capable of selecting and tripping only the faulted phase for single-phase-to-ground faults.

2) The circuit breakers must have independent pole operating capability to execute a single-phase trip properly.

3) The transmission line tower footing resistances must be within the range of the relay single-phase selector logic.

Other design requirements of the transmission line are less obvious namely, the effect of the electrostatic and electromagnetic coupling between the still energized phase conductors and parallel line(s). This phenomenon:

4) Tends to sustain a secondary arc current (I_s) in the primary arc path and lengthens the time of de-ionization which in turn can prevent a successful auto-reclose.

5) Causes recovery voltage (V_r) to appear across the secondary arc path as soon as the arc is broken. The magnitude of V_r and/or its rate of rise can initiate a re-strike, which can prevent a successful auto-reclose.

C.   Calculation of the Secondary Arc Current

The secondary arc current (I_s) on a single, symmetrical, fully transposed transmission line is basically the phasor sum of two currents maintained by electrostatic (I_{sc}) and electromagnetic (I_{sm}) coupling from the two energized phases.

$$I_s = I_{sc} + I_{sm} \qquad [1]$$

1)   Calculation of the Secondary Arc Current via Electrostatic Coupling

The calculation of the secondary arc current via electrostatic coupling on a single, symmetrical, fully transposed transmission line was developed by Kimbark, Peterson, Dravid, et. al. [1,2,3]. For untransposed line analysis refer to [11, 12, 13]. Figure 3a. represents a single, symmetrical, fully transposed transmission line with phase A in an open condition with a capacitance C_Δ between each pair of phases and a capacitance C_G from each phase-to-ground. The phase A-Gnd fault is represented by SW_F.

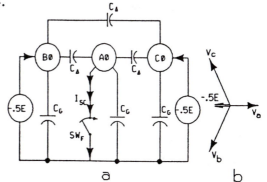

Figure 3. Electrostatic Coupling Diagram of a Single, Symmetrical, Fully Transposed Transmission Line.

The phasor of the effective voltage is shown on Figure 3b. The Thevenin equivalent circuit derived from Figure 3. is shown on Figure 4a. It is achieved by folding phase C to phase B.

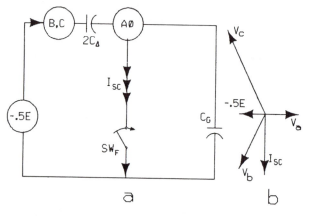

Figure 4. Electrostatic Coupling Thevenin Equivalent Diagram of a Single, Symmetrical, Fully Transposed Transmission Line.

The magnitude of the secondary arc current due to electrostatic coupling is in direct proportion to the line voltage and the line length. From inspection of Figure 4a with SW_F closed:

$$I_{sc} = -.5E \times \frac{1}{1/-j\omega 2C_\Delta} = E \, j\omega C_\Delta \qquad [2]$$

A typical secondary arc value for 500-kV lines is 20 A per 100 miles. The phase relationship between the effective phase voltages and I_{sc} is shown on Figure 4b.

2) Calculation of the Secondary Arc Current via Electromagnetic Coupling

When the transmission line is equipped with shunt reactors, there is a component of secondary arc current induced by the electromagnetic coupling from the unfaulted phases.

The simplified diagram depicting secondary arc current (I_{sm}) due to electromagnetic coupling to the open-phase A is shown on Figure 5.

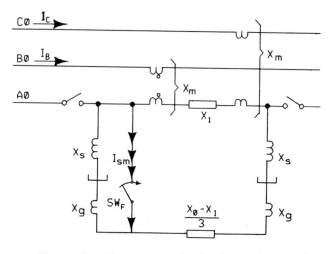

Figure 5. Electromagnetic Coupling Diagram of a Single, Symmetrical, Fully Transposed Transmission Line.

Accurate calculation of I_{sm} requires Electro Magnetic Transient Program (EMTP) studies due to the fact that mutual induction is the sum of many dynamic variables involving the line currents flowing in the sound phases, adjacent line loading, the method of secondary arc extinction, etc.

A simplified calculation of I_{sm} for a single, symmetrical, fully transposed line using a four-reactor scheme is shown in Appendix-B.

3) Calculation of the Recovery Voltage

The magnitude of the recovery voltage (V_r) is directly proportional to the line voltage and the relative values of C_Δ and C_G. Consequently, V_r does not vary with line length.

From inspection of Figure 4a. the recovery voltage on phase A with SW_F open:

$$V_r = -.5E \times \frac{1/-j\omega C_G}{(1/-j\omega 2C_\Delta)+(1/-j\omega C_G)} \qquad [3]$$

$$V_r = -E \times \frac{C_\Delta}{2C_\Delta+C_G} \qquad [4]$$

Typical V_r values are 10-25% of the line voltage without shunt reactors.

The oscillograph of a typical recovery voltage and the secondary arc current on a line without shunt compensation is shown on Figure 6.

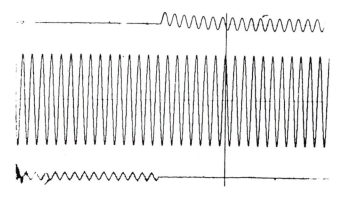

Figure 6. Recovery Voltage (top trace) and secondary arc current (bottom trace) from the 500 kV Malin-Round Mountain staged fault test.

If solidly grounded reactors are applied to the line for shunt compensation, they reduce the effective capacitance of C_G without affecting C_Δ, thus leading to higher values of recovery voltage. For shunt compensated lines where C_G represents less than one-half of the total line charging MVAR, the shunt reactors may overcompensate C_G. In these cases, the effective phase-to-ground impedance is inductive and the recovery voltage may exceed normal line to neutral voltage, limited by saturation of the shunt reactors.

The oscillograph of a typical recovery voltage and secondary arc current on a line with shunt compensation is shown on Figure 7. The voltage wave low frequency envelope is attributed to the beat frequency of the 60 Hz source to the off tuned circuit created by the open phase capacitance and the shunt compensating reactors.

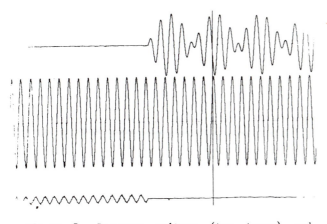

Figure 7. Recovery voltage (top trace) and secondary arc current (bottom trace) from the 500 kv Malin-Round Mountain staged fault test.

D. Methods of secondary arc extinction

The magnitude of the secondary arc current and the recovery voltage are the most important factors which determines whether or not the secondary arc will be self-extinguishing.

Briefly, the high current, high energy primary arc heats and ionizes the arc path until the faulted phase is tripped. Afterwards, the heated, ionized arc path can support the smaller secondary arc current induced by the electrostatic and electromagnetic coupling. All methods of secondary arc extinction are directed toward reducing the magnitude of the secondary arc current, which in turn will reduce the extinction time of the arc. Table III. indicates the following probable performance based on line voltage for lines without supplemental arc extinction measures.

TABLE III.

Line lengths for single-phase auto-reclosure without supplemental arc extinction devices. e.g. shunt reactors

Line Length (Mi)

Line to Line Voltage (kV)	Successful Range	Doubtful Range
765	0- 50	50 - 80
500	0- 60	60 - 100
345	0-140	140 - 260
230	0-300	300 - 500

If the line is longer than that given in the above table and single-phase tripping and auto-reclosing is to be applied, then additional measures must be taken to reduce the secondary arc as outlined below:

1) Transmission Lines with Four-Reactor Bank

Most long EHV lines require three, single-phase reactors to provide shunt reactor compensation for voltage control. A four-reactor bank is created by adding a fourth reactor in the neutral of the three single-phase reactors. The fourth reactor is used to reduce the secondary arc currents. The reactance value of the neutral reactor needed to neutralize the secondary arc current can be calculated by using the formulas given below.

Figure 8. represents a symmetrical transmission line with four reactors. The phase A to ground fault is represented by SW_F.

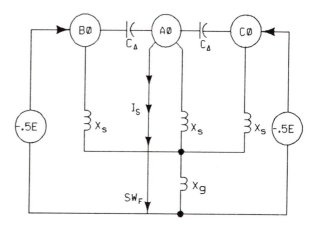

Figure 8. Transmission Line with Permanently Connected Four-Reactor Bank

The Thevenin equivalent circuit derived from Figure 8. is shown on Figure 9a. It is achieved by folding phase C to phase B.

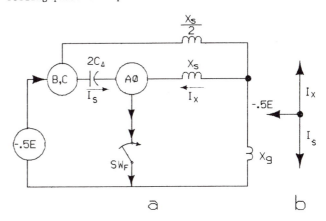

Figure 9. Permanently Connected Four-Reactor Bank Thevenin Equivalent One Line Diagram

From inspection of Figure 9a. with SW_F closed, if the magnitude of the current through the shunt reactor I_X is equal, and opposite to the magnitude of the secondary arc I_S, then the effective current through SW_F is zero.

The phasor diagram for the current I_S and I_X is shown on Figure 9b.

The voltage Eg necessary at the fourth reactor to provide I_X equal to I_S is calculated.

$$Eg = -I_X \cdot X_S \qquad -I_X = I_S \qquad [5]$$

Xs:the reactance value of the phase shunt reactor.

The current through the reactors (Xs/2) of the unfaulted phases is

$$I_T = \frac{.5E - Eg}{X_S/2} = \frac{E - 2Eg}{X_S} \qquad [6]$$

The current through the neutral reactor is

317

$$I_g = I_T - I_x \qquad [7]$$

The reactance value of the neutral reactor is

$$X_g = \frac{E_g}{I_g} \qquad [8]$$

2) Modified Selective Switched Four-Reactor Scheme.

This scheme is recommended on untransposed lines where the four-reactor banks are not effective to extinguish the secondary arc.

On long untransposed 765kV EHV lines built in a horizontal configuration, the ratio of outer-to-middle phase capacitance (C_{A-B} or C_{B-C}) to outer-to-outer phase capacitance (C_{A-C}) varies from 3.5 to 3.9. For these lines, one four-reactor bank and one modified four-reactor bank may be specified [10, 11].

The four-reactor bank at one end of the line is designed to compensate the inter-phase capacitances by a value equal to C_{A-C}. The modified four-reactor bank at the other end of the line is switched to compensate the unbalanced capacitance ($C_{A-B}-C_{A-C}$ or $C_{B-C}-C_{A-C}$). The actual reactor value selected allows for the electromagnetic component of the secondary arc current [10, 11].

Figure 10a. shows a modified four-reactor bank with a neutral switching design.

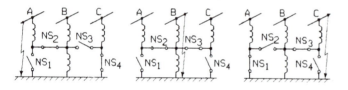

a. Neutral Switch Scheme

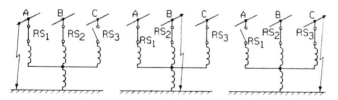

b. Reactor Switch Sheme

Figure 10. Modified Four-Reactor Bank with Neutral Switches (NS) and High Side Reactor Switches (RS) and Their Positions for Different Phase-to-Ground Faults.

All switches are normally closed but a specific pair of neutral switches operate when a particular faulted phase is identified open.

Figure 10b. shows a modified four-reactor bank with different high side reactor switch positions. This design can provide the necessary compensation for faults on various phases. The modified four-reactor bank for untransposed 765-kV lines was studied for a wide range of line lengths and reactor arrangements. The results show that the secondary arc current can be reduced to 25 A rms for 765kV lines up to 350km having practical levels of shunt reactive compensation [12].

3) High Speed Grounding Switch Scheme

This scheme involves the application of a high speed grounding switch in each phase at each end of the line. The ground switch is closed onto the faulted phase after the circuit breaker pole on that phase opens and vice versa for reclosing. In principle, the ground switch removes all voltage from the open phase on the line and, therefore, removes the driving voltage behind the secondary arc. [8, 9]

4) Hybrid Single-Phase Scheme

The scheme trips the faulted phase first. The two unfaulted phases are tripped with time delay 50-60 cycles later. The trip cycle is followed by a fast three-phase auto-reclose (within 10-15 cycles). The hybrid scheme has a dual purpose:

Keeping the two unfaulted phases closed for 50-60 cycles after the fault is cleared, which significantly reduces the system power swing.

The three-phase trip and fast auto-reclose eliminates the secondary fault current without additional hardware requirement.

E. Summary of secondary arc extinction methods

1) Permanently connected four-reactor bank:

Advantages	Disadvantages
Proven design	High cost, neutral end of phase reactors must have higher insulation, possible operational problems due to lack of flexibility in placement and lack of switching ability. May not work on closely coupled parallel lines.

2) Selectively switched four-reactor bank:

Advantages	Disadvantages
More flexibility than method 1).	Higher cost than method 1) plus complexity of high speed switches and controls.

3) High speed ground switch scheme:

Advantages	Disadvantages
Lower cost, does not require neutral reactors.	More complex protection and control scheme.

4) Hybrid single phase trip scheme:

Advantages	Disadvantages
No additional hardware is required to eliminate the secondary arc.	Cannot be used where single phase is required to maintain synchronism between the system and a remote generating plant. The torsional duty due to the multiple switching events must be evaluated when the scheme is used close-in to a thermal generating plant.[6]

318

Other Schemes

Two other schemes have been proposed in the references but have never been seriously considered. They are the series capacitor scheme [3] and the line sectionalizer scheme.

5) Series capacitor scheme:

Advantages	Disadvantages
Flexibility of reactor placement.	Complexity of design, unproven in practice.

6) Line sectionalizing scheme:

Advantages	Disadvantages
May negate need for shunt reactor.	Sectionalizer must be located in middle of line.

SINGLE-PHASE RELAYING

A. General Requirements

The installation of a single-phase relay scheme can be justified only if there is assurance that a large percentage of the phase-to-ground faults will in fact clear with single-phase trip that is followed by a fast auto-reclose.

1) Three-Phase Trip Logic

In general, the single-phase relay scheme should have the following logic that initiate three-phase trip and blocks auto-reclose:

Out-of-Step blocking logic; prevents the scheme from tripping during swing conditions.

Close-into-fault detection logic; detects a fault when the circuit breaker is being closed.

Multi-phase fault detection logic; detects multi-phase faults.

Evolving fault detection logic; detects fault on one phase, evolving into another phase during the dead time of the auto-reclose.

Directional ground overcurrent relay; provides a time delayed backup for high resistance faults.

Time delayed backup relaying.

2) Single-Phase Trip Logic

In general, the single-phase trip scheme should have the demonstrated capability to detect phase-to-ground faults on the entire length of the transmission line under the following conditions:

Heavy line loading. Heavy loading can be one of the justification to install single-phase relaying.

Tower footing resistance. Successful single-phase operation requires that a large percentage of the towers have the tower footing resistance within the range of the phase selector logic.

Series compensation.

Parallel line(s). Mutual coupling of adjacent line(s) increases the apparent resistance.

Three terminal line.

3) Verification by Testing

Model line tests by digital or analog simulations provides a check of the performance. It is necessary that extensive factory tests are run on the complete relay system and its components to judge performance over a wide range of test conditions, such as maximum and minimum source impedance, shunt reactor compensation included or excluded, and series compensation included or bypassed. Non-linear elements such as MOV's shunting series capacitors provide recent complexity additions. Staged fault testing provides a check of the complete installation and is an excellent way of obtaining proof of the performance.

B. Single-Phase Relaying Techniques

Any of the commonly applied relaying schemes can be selected in accordance with the user's philosophy with the understanding that each scheme may have drawbacks as well as advantages.

1) Distance Relays

The most common type of line protection, the distance relay, is based on the measurement of the fundamental frequency positive sequence impedance of the line. The impedance measurement has many advantages over other concepts and a few disadvantages. The distance relay measuring elements perform four different tasks:

Fault zone detection
Directional discrimination
Load discrimination
Phase selection

Trip decisions are generally carried out through logic circuits combining information from several measuring elements.

One important benefit is that the fault detection, the directional phase selection, and the trip decisions are made at the local line terminal. The relay "reach" is relatively well defined for all fault types, and system source impedance conditions do not affect the relay settings except when series capacitor compensation exists.

The ground distance relay resistive reach is restricted by the maximum line loading. The resistive reach is also influenced by the method of polarization and the system source to line impedance ratio. Modern ground distance relays have circular or other composite characteristics which enable improved resistive reach for phase-to-ground faults. These characteristics permit better discrimination between maximum load and fault impedance. System requirements, such as "Out of Step" swing conditions may reduce permissible resistive reach even further. It is therefore, necessary to add a negative or zero sequence directional current relay that will detect high resistance faults. With conventional directional overcurrent relays it is not possible to obtain phase selectivity. High resistance faults must be tripped three-phase.

Phase distance functions can operate for some phase-to-ground faults. These conditions occur when the system zero sequence impedance is small relative to the system positive sequence impedance (both viewed at the point of fault). Provisions must be made to block the phase distance functions from a false trip for a close-in phase-to-ground fault. One method utilizes a restraining signal that is proportional to the zero sequence current to supervise the phase functions. This method may not be satisfactory for remote faults if the zero sequence current is very small. This type of supervision may also block the phase functions from initiating a three-phase trip during close-in, phase-to-phase-to-ground faults.

2) Directional Comparison Relays

Directional comparison relaying is usually a hybrid scheme. The most common types use mho type directional distance relays to detect three-phase and/or phase-to-phase faults. A combination of positive, negative and zero sequence measuring techniques are used to detect phase-to-phase or phase-to-ground faults. Negative or zero sequence measurement is used provide a better sensitivity for phase-to-ground faults with high fault resistance. High resistance ground faults can be detected through separate directional comparison relay logic based on zero sequence current. The communications channel can often be shared with the pilot distance relay. It is also common to supplement this scheme with a sensitive directional ground overcurrent relay with long time delay (1-2 sec.) for back up tripping upon loss of communication.

Various methods are available to eliminate or minimize the effect of the operation of the ground functions associated with the unfaulted phases during a phase-to-ground fault. One method uses the negative and zero sequence current components to form a phase selector to minimize the area of coverage of the function associated with the unfaulted phases. Another solution is to allow the ground overreaching functions to take priority over the phase overreaching functions to prevent them from initiating a three-phase trip.

The phase selection logic that is based on the measurement of symmetrical components suffer from being dependent on source and line impedances, mutual coupling and their sensitivities.

3) Wave Comparison Relays

In addition to being fast (1/4 cycle), one of the advantages of traveling wave comparison relays is that they act on the changes in phase voltage and current, eliminating the pre-fault quantities. The simplicity of the change analysis, as compared to a more complex analysis of, for example, symmetrical components can also result in improved sensitivity by virtue of the fact that all symmetrical components are available in the phase quantities. The ability to operate on the change and not the absolute quantity of voltage and current enhances the reliability by an improvement in the signal-to-noise ratio as obtained from the instrument transformers.

This eliminates problems with CVT transients and CT saturation. Since the traveling wave relay uses the fault generated transient, the relay cannot provide a measurement after the transient subsides. The system, therefore, requires back up, usually from a more conventional distance relay.

Lightning arrester discharge, as well as switched reactors require desensitizing or delay of operation. The system is also affected by long-line/strong source limitations. Inherent phase selection is a strong feature, as well as being unaffected by fuse failure conditions, or sudden loss of A-C supply.

4) Phase Comparison and Current Differential Relays

Phase comparison or current differential relays require only line current as an input signal. One advantage of this type of protection is that it is inherently immune to tripping on a power swing or inadvertent loss of potential. The disadvantages are the critical dependency on properly functioning communications and the lack of any remote backup capability for external faults.

There are three principal types of systems:

Mixed Excitation Phase Comparison Relays: Mixed excitation phase comparison relays use a phase sequence component filter to combine three-phase currents into a single-phase signal which is compared with a similarly combined signal from the remote end terminal(s). Mixed excitation systems are not well suited for single-phase relaying since the sequence filter arrangement does not inherently identify the faulted phase. To provide single-phase relaying additional phase selection logic must be added.

If the phase selectors are current only type, the operation of the scheme depends upon having reliable sources of negative and zero sequence current at both end terminals. Mixed excitation phase comparison has possible problems when applied on series compensated lines. On such lines the sequence filters may develop a zero output for certain combinations of capacitor gap bypassing, leading to "blind spots" in the protection.

Segregated Phase Comparison Relays: Segregated phase comparison relays use three systems, one for each phase. Segregated phase comparison systems are inherently phase selective and can operate correctly for faults on another phase during the dead time of the auto-reclose. They are also immune to the problems of evolving faults on parallel lines. Weak in-feed and some out-feed conditions for internal faults can be detected with segregated phase comparison, using the offset keying technique. A fourth sub system may be added in the neutral circuit to provide three-phase tripping for high impedance ground faults.

Current Differential Relays. Current differential relays are either phase segregated or use filters and/or summation current transformers to derive an operating and restraint quantity that is the vector function of the three-phase current. Phase segregated type current differential relays are inherently capable for single-phase trip. The current values are transmitted to the remote end via metallic or optical channels, carrier or microwave radio. In one method, the operating quantity is the vector addition of the single-phase current phase and amplitude, while the restraint quantity is the arithmetic sum of the currents phase and amplitude from each end end terminal.

Current differential systems are available that use time tagged sampling of data, which therefore, make the system usable for long, multi-terminal lines.

C. Operation During the Open Pole Period Following a Single-Phase Trip

Due to the capacitive and inductive coupling, the voltage associated with an open phase can vary significantly, and may even be reversed. Therefore, when one phase has been opened, care must be taken to assure that no misoperations occur during and immediately following the (.5 to 1.5 sec.) open-phase period. This is often done via sensitively set overcurrent functions which are set to drop out during the open phase. Overcurrent supervision is also used to avoid any problems that might occur because of load flow.

Positive sequence voltage that is referenced to the associated phase will not reverse when a phase is open. It is used sometimes to provide a stable polarizing signal. The "sound" phase voltages are also often used to derive polarizing signals.

D. Double Circuit Lines

In regions where large blocks of power are being transferred over double circuit EHV transmission lines, the occurrence of a "cross country" fault, could initiate serious system stability problems, if the fault is followed by a three-phase trip of both lines. For example, consider the double circuit EHV transmission lines shown in Figure 11.

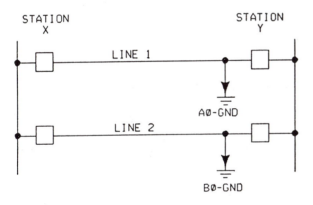

Figure 11. Double Circuit EHV Transmission Lines With "Cross Country" Type of Fault on Different Phases.

At station X, the cross country fault will appear as an A-B-Gnd fault in both lines with an attendant operation of the A-B phase functions and possibly the A and B ground functions. Thus, at station X, this condition will be recognized as a multi-phase fault.

At station Y, the fault will appear as an A-Gnd fault in line 1 and as a B-Gnd fault in line 2. Thus, at station Y, correct identification of the fault type will be made.

Segregated current phase comparison or the segregated current differential schemes are inherently phase selective. Other single-phase tripping schemes on each of the lines will initiate three-phase tripping at station X and single-phase tripping at station Y if precautions are not taken.

The problem is not insurmountable, and techniques are available to provide correct tripping of the faulted phases only. One method utilizes ground distance functions that are designed not to operate for double-line-to-ground faults, multiple communications channels, and appropriate scheme logic to recognize the fault type, and so initiate correct tripping.

E. Back Up Relaying and Redundancy Considerations

The commonly used backup involves using a second primary relay system and dual trip coils. In addition, two local breaker failure systems may be used. Single-phase reclose failure should also be considered (i.e., three-phase trip should result if the opened breaker pole in the faulted phase will not reclose). This can be taken care of by a breaker pole disagreement logic.

F. Breaker Failure

Breaker Failure protection (BF) with single-phase trip is more complex than three-phase trip.

The majority of the BF schemes use the dropout of the fault current detector logic to sense that the breaker poles have opened. In a three-phase trip scheme, it is only necessary to detect that all of the fault current detectors have dropped out.

In a single-phase trip scheme, only the breaker pole associated with the faulted phase will trip while the two remaining phases stay closed. The load current in the unfaulted phases may be above the pickup of the BF current logic. Therefore, in single-phase trip schemes, a segregated phase type BF scheme should be used.

The effect on the performance of the BF needs to be considered when a single-phase-to-ground fault evolves into other phase(s). If the BF logic uses a single timer, the overall time delay will have to be based on clearing of the evolving fault. This overall time may be excessive relative to the overall time required to clear a multi-phase fault. A possible solution for this case is to use two timers, the first timer set for single-phase-to-ground faults only, with a long enough time delay to override an evolving fault, the second timer can be set with a shorter time delay to meet system stability requirements for multi-phase faults. An alternative is to use separate timers for each phase.

OPERATIONAL REQUIREMENTS

A. Evaluation of Current Unbalance During Open Phase Conditions

When one phase of a transmission line is opened, load current still flows through the other two phases. The unbalance generate negative and zero sequence currents and voltages. Their effect should be carefully analyzed. The negative and zero sequence impedances of the power system help to reduce the amount of load current flowing through the healthy phases. The negative sequence current is usually larger than the zero sequence current flowing through the opened phase transmission line because of higher zero sequence line impedance. However, weaker negative and stronger zero sequence sources may produce higher zero sequence current.

Large generators close to the transmission line may have high negative sequence current flowing. Similarly, strong zero sequence sources will carry large amounts of zero sequence current (e.g., delta tertiary of an auto-transformer). Also, in adjacent transmission lines, negative and zero sequence directional relays, having enough polarizing signal, may try to operate the directional overcurrent relay or the pole disagreement logic unless they are properly coordinated. A sample calculation is shown in Appendix-A.

B.   Reclosing

The reclosing relay required for single-phase trip is normally different than that used for three-phase trip. It is usually preferable to have different dead time settings for single-phase-to-ground faults versus multi-phase faults. It is also desirable to be able to select whether reclosing should occur after multi-phase fault or three-phase tripping operations. On some conventional reclose schemes a mode selector switch is provided.

More advanced reclose schemes use adaptive reclosing where the line relay logic is interlocked with the recloser to delay, block or override reclose. A single-phase reclose attempt should always lead to a three-phase trip if the fault persists. A logic circuit or a "prepare three-phase trip" signal is required from the reclosing relay.

Sequential (or evolving faults) during the first phase open period should normally lead to three-phase tripping. Ring bus or breaker and half applications require special consideration regarding reclosing. It may be simpler to always trip one breaker three-phase and the other single-phase.

Synchro-check reclosing is often combined with the single-phase and three-phase reclosing relay for three-phase trip operations.

APPENDIX-A

Approximate Calculation of Current Unbalance

Calculation of the current unbalance is difficult because an open phase action involves many dynamic variables which change with respect to time from the fault inception to the phase opening. However, with some judicious assumptions the engineer can estimate the maximum negative and zero sequence current unbalance.

During single phase tripping, the reactance values of the system change with respect to time depending whether:

1) The moment of the fault is taken as the criterion.
2) The moment just after the fault is the criterion.
3) Some time after the fault has taken place.

In case 1) the reactance is the subtransient reactance X''_d

In case 2) the reactance is the transient reactance X'_d

In case 3) the reactance is the synchronous reactance X_d

o   by definition $X''_d \leq X'_d \leq X_d$ [9]

Therefore, for the "worst case" type of calculation the X''_d values should be selected on the assumption that if there is no problem with negative and zero sequence current magnitudes with the X''_d reactance then the X'_d and X_d values will always provide less negative and zero sequence current flow.

The positive, negative and zero sequence subtransient reactance values for the sources and the line are readily available from system impedance data.

For example: A simplified two machine diagram is shown on Figure 12.

The base quantities 1 per unit (pu) on a 100 MVA base are:

$$E_S = E_R = 525kV = 1 \text{ (pu)}   [10]$$

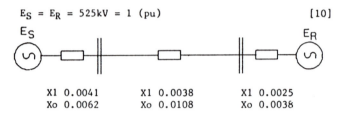

| X1 0.0041 | X1 0.0038 | X1 0.0025 |
| Xo 0.0062 | Xo 0.0108 | Xo 0.0038 |

$$Z_1(pu) = 0.0041 + 0.0038 + 0.0025 = 0.0104 \text{ (pu)}$$

Figure 12.   Two Machine Diagram

For the "worst case" condition assume that the positive sequence MVA is equal to the maximum steady state power flow P_S.

Assume:

$$P_S = 1000 \text{ MVA}   [11]$$

1000 MVA is 10 per unit on a 100MVA Base.

by definition $$P_S = \frac{E^2(pu) \times \sin \delta}{X_1(pu)}   [12]$$

Therefore $$10 = \frac{E^2 \times \sin \delta}{0.0104}   [13]$$

for E = 1 (pu) $\sin \delta = 0.104$ [14]

The sequence network interconnection for phase A open is shown on Figure 13.

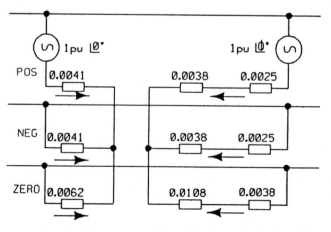

Figure 13.   Two Machine Diagram with One Phase Open

322

Assume that the power angle δ does not change between the two sources when one phase is opened. The system symmetrical component configuration is positive sequence plus a parallel configuration of negative and zero sequence network, which with one phase open will reduce the power transfer to:

$$P_s = \frac{E^2 \times \sin \delta}{X(pu)} \qquad [15]$$

$$\text{where } X(pu) = .0104 + \frac{1}{\frac{1}{.0104} + \frac{1}{.0208}} \qquad [16]$$

$$X(pu) = 0.0104 + 0.0069 = 0.0173 \text{ (pu)} \qquad [17]$$

From inspection of Figure 13. the per unit magnitude of the positive sequence current during open phase is the same as the MVA pu.

$$I_1 = \frac{E^2 \times \sin \delta}{X(pu)} \qquad [18]$$

From which

$$I_1 = \frac{1 \times 0.104}{0.0173} = 6.0 \text{ (pu)} \qquad [19]$$

% distribution of negative sequence current

$$-I_2 = \frac{0.0069}{0.0104} \times 100 = 66\% \qquad [20]$$

or

$$I_2 = -.66 \times 6.0 = -4.0 \text{ (pu)} \qquad [21]$$

% distribution of zero sequence current

$$-I_0 = \frac{0.0069}{0.0208} \times 100 = 34\% \qquad [22]$$

or

$$I_0 = -.34 \times 6.0 = -2.0 \text{ (pu)} \qquad [23]$$

$$100\text{MVA at } 525\text{KV} = 110 \text{ A} \qquad [24]$$

Under the "worst case" criteria, during one phase open:

$$\text{Maximum positive sequence amps: } I_1 = 660 \text{ A} \qquad [25]$$

$$\text{Maximum negative sequence amps: } I_2 = -440 \text{ A} \qquad [26]$$

$$\text{Maximum zero sequence amps } : I_0 = -220 \text{ A} \qquad [27]$$

It is advisable to study the effect of 440 A negative sequence and 660 A ($3I_0$) ground current flow on the power equipment and the relaying during open phase condition.

APPENDIX-B

Approximate Calculation of Secondary Arc Current via Electromagnetic Coupling

Reference the example presented in Appendix-A. The MVA power transfer during open phase are:

$$I_1 = 6.0\text{(pu)} \quad I_2 = -4.0\text{(pu)} \quad I_0 = -2.0\text{(pu)} \qquad [28]$$

from which

$$I_B = a^2 I_1 + a I_2 + I_0 = -3.0 - j8.66\text{(pu)} \qquad [29]$$

$$I_C = a I_1 + a^2 I_2 + I_0 = -3.0 + j8.66\text{(pu)} \qquad [30]$$

The relationship between I_B and I_C and the driving current I_{dr} is shown on Figure 14.

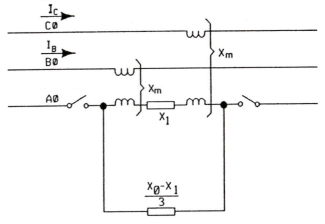

Figure 14. Driving Current One-Line Diagram.

From which the driving current:

$$I_{dr} = I_B + I_C = 2 \times (-3.0) = -6.0 \text{ (pu)} \qquad [31]$$

Note that the driving current I_{dr} is the same as $3I_0$.

$$I_{dr} = 3I_0 \qquad [32]$$

The driving voltage V_{dr} induced in phase A via electromagnetic coupling X_m and driving current $3I_0$.

$$V_{dr} = 3I_0 \times X_m \qquad [33]$$

The Carson equations define X_m for symmetrical three-phase circuits.

$$X_m = 1/3(X_0 - X_1) = 1/3(.0108 - .0038) = 0.0023 \qquad [34]$$

from which

$$V_{dr} = -6.0 \times 0.0023 = -0.0140 \text{ (pu)} \qquad [35]$$

The magnitude of the secondary arc current depends on the driving voltage V_{dr} and the location of the fault.

1) If, for example, the open phase A is shorted at each end we have the maximum I_{sm}. (As shown on Figure 14.).

$$I_{sm} = \frac{V_{dr}}{1/3(X_1 + X_2 + X_0)_{1ine}} \qquad [36]$$

$$I_{sm} = \frac{.0140}{1/3(.0076 + .0108)} = 2.28 \text{ (pu)} \qquad [37]$$

$$I_{sm} = 251 \text{ A} \qquad [38]$$

2) If the fault is in the midpoint of the line, the circulating I_{sm} currents cancel as shown on Figure 15.

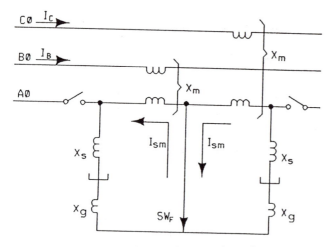

Figure 15. Midpoint Fault One-Line Diagram.

3) If the fault is at one end of the open phase with a 4 reactor scheme at the other end.

$$I_{sm} = \frac{V_{dr}}{1/3(X_1+X_2+X_0)+X_s+X_g} \qquad [39]$$

BIBLIOGRAPHY

1. E. W. Kimbark, "Suppression of Ground-Fault Arcs on Single-Pole Switched EHV Lines by Shunt Reactors," IEEE Transactions on Power Apparatus and Systems, vol PAS-83, pp. 285-290, March/April 1964,

2. E. W. Kimbark, "Selective-Pole Switching of Long Double-Circuit EHV Lines," IEEE Transactions on Power Apparatus and Systems, vol PAS-95, pp. 1, January/February, 1976.

3. H. A. Peterson, N. V. Dravid, "A Method for Reducing Dead Time for Single-Phase Reclosing in EHV Transmission," IEEE Transactions on Power Apparatus and Systems, vol. PAS-88, pp. 286-292 March/April 1969.

4. N. Knudsen, "Single Phase Switching on Transmission Lines Using Reactors for Extinction of the Secondary Arc," CIGRE Report 31-06, 1962.

5. J. G. Kappenmann, "Planning, Design and Application of a 500 kV Single Phase Reclosing Scheme," 1980 Minnesota Power Systems Conference, University of Minnesota, St. Paul, Minnesota, October 12-13, 1980.

6. A. J. Gonzales, G. C. Kung, C. Raczkowski, C. W. Taylor, "Effects of Single-and-Three Pole Switching and High-Speed Reclosing on Turbine-Generator Shafts and Blades," IEEE Transactions on Power Apparatus and Systems, vol. PAS-103, pp. 3218-3228, November/December 1984.

7. A. Amstutz, "Residual Currents and Voltages with Single Pole Reclosing," Brown Boveri Review vol. 35, pp. 220-226 July/August 1948,.

8. R. M. Hasibar, A. C. Legate, J. H. Brunke, and W. G. Peterson, "The Application of High-Speed Grounding Switches for Single-Pole Reclosing on 500-kV Power Systems," IEEE Transactions on Power Apparatus and Systems," vol. PAS-100, pp. 1512-1515, March/April 1981.

9. J. Esztergalyos, "The Application of High Speed Grounding Switches on EHV-UHV Power Systems to Enhance Single Pole Reclosing-Control and Protection," Western Protective Relay Conference, Washington State University, Spokane, Washington, October. 27-29, 1981.

10. V. Narayan, C. Pencinger, "A New Distance Protection System for Single Pole Relaying," Minnesota Power Systems Conference, University of Minnesota, St. Paul, Minnesota, October 12-13, 1982.

11. B. R. Shperling, A. J. Fakheri, B. J. Ware, "Compensation Scheme for Single-Pole Switching on Untransposed Lines," IEEE Transactions on Power Apparatus and Systems, vol. PAS-97, pp. 1421-1429 July/August 1978.

12. B. R. Shperling, A. J. and Fakheri, "Single Phase Switching Parameters for Untransposed EHV Transmission Lines" IEEE Transactions on Power Apparatus and Systems, vol. PAS-98, pp. 643-654 March/April 1979.

13. A. J. Fakheri, J. Grazan, B. R. Shperilng, B. J. Ware, "The Use of Reactor Switches in Single Phase Switching," CIGRE Report 13-06, 1980.

14. L. Edwards, J.W. Chadwick Jr, H.A Riesch, L.E. Smith, "Single Pole Switching on TVA's Paradise-Davison 500kV Line. Design Concepts and Staged Fault Test Results." IEEE Transactions on Power Apparatus and Systems, vol. PAS-90, pp. 2436-2444 November/December 1971.

TRANSIENTS INDUCED IN CONTROL CABLES LOCATED IN EHV SUBSTATION

Howard J. Sutton
Gulf States Utilities Company
Fellow IEEE

ABSTRACT

This paper was prepared from data collected in tests made to evaluate the effectiveness of shielding control cables that are laid in concrete trenches located in an EHV environment. Particular attention is given to the effectiveness of shielding against high frequency transients generated by the opening and closing of 500 kV disconnect switches. Transients or spurious signals caused by induction from 60 Hz fault currents, or by the energization or de-energization of DC contactor coils, or relay coils has fairly well been documented. Because of this, the main attention of this paper will be given to the subject of high frequency transients generated by multiple restrikes in an arc or in corona. This paper will discuss the various tests made, and the results of the tests are tabulated.

The conclusions arrived at pertain particularly to control cables laid in concrete trenches. Upon the basis of tests and late references on this subject it is the author's opinion that shielded control cables with the shield grounded on both ends should be used in EHV environment. To protect the shield and obtain further protection against transients, a heavy current carrying conductor properly grounded should be run along the same path as the shielded control cable.

INTRODUCTION

In the last several years, considerable amount of semi-conductor or solid state devices have been used in connection with protective relaying and controls. Shortly after the initial installation of this equipment, it became known that there were failures of this equipment which apparently could only be explained by spurious signals or transients in the control circuits and the wiring connected to the semi-conductor devices. This trouble has not been confined to semi-conductors or solid state devices, but it has also been noted in connection with the conventional electro-mechanical devices. This latter trouble can be attributed to the use of EHV and the presence of much higher short circuits that had been previously experienced in substations. Because of these reported troubles, a very careful investigation of the phenomena of transients has been carried on by a subcommittee of the IEEE Power System Relay Committee to investigate this phenomena.

As a result of the work of this subcommittee and others, there have been several papers prepared, and the author would like to refer to references 1, 2, 3, and 4. These references cover transients caused by 60 Hz induction, and from the energization and de-energization of DC coils. The subject of the difference in ground potential has also been considered and has been presented in several papers, one of which is reference number 5.

A third source of transients or spurious signals is that generated by high voltage arcs. This source results in high frequency transients and while this has been covered in some references, as for example 6 and 7, there still needs to be more test work and studies made of this phenomena and its influence on control circuits. On the basis of published matter, references 1 to 7, and because of trouble that had been experienced by our company with semi-conductor failures our company decided to run tests in an effort to determine the effectiveness of shielded cables as a means of protection against these high frequency transients generated by high voltage arcs. As a result, tests were performed just before the circuits were placed in service at our 500 kV substation to obtain data which would serve as a basis for future decisions as to how control cables could be protected in EHV substations.

When a high voltage arc occurs as a result of opening a disconnect, a breaker, or because of fault conditions breaking down insulation, voltages are induced in adjacent or nearby control circuits by:

(1) electrostatic coupling or induction
(2) electromagnetic coupling or induction
(3) through both electrostatic and electromagnetic coupling between the primary and secondary winding of: potential transformers, capacitor potential devices, or current transformers.

In addition to these three methods of coupling transients to the control circuits, there is a fourth means by which high voltage transients can be introduced in control circuits. This latter source is caused by a difference between ground potential at the location adjacent to the arc and the ground potential in the control house where the protective relaying is located. When considering protection against transients caused by high voltage arcs, all of these means of coupling should be considered.

THE EFFECTIVENESS OF SHIELDED CABLES

In an effort to determine the nature of the transients generated by high voltage arcs, several motor operated double break disconnects were used in the energization of and the de-energization of short 500 kV bus sections. Reference should be made to figure 1 which gives the layout of the control cable trenches with refe-

Reprinted from *IEEE Trans. Power App. Syst.*, vol. PAS-89, no. 6, pp. 1069–1081, July/Aug. 1970.

rence to the EHV busses and disconnect switches. Shown on this same sketch is the parallel path of the control circuits, or test specimen, with reference to the bus being energized or de-energized.

Since the main interest of this investigation was in reference to the voltage above ground on a conductor at the control house, all the data from tests made for this report show the measured voltage to ground of a specimen conductor with the remote end floating or disconnected from any equipment or ground. This is true with few exceptions, and in those cases the exception is noted, see Table I. In comparing the results of the tests included in this report it should be kept in mind that there is this difference with tests as reported in reference Nos. 6 and 7 in which case the remote and local ends of the conductors were both grounded.

Nearly all of our control cables for EHV substations are laid in trenches made of concrete blocks. A cross sectional view of the trench and the grounding arrangement of the trench is shown in figure 2. The general layout of the control cables in these trenches and their position in the substation is shown by figure 1. This drawing is laid out to scale, and the one line diagram is centered with the center conductor of the bus section. The bus conductors are spaced 35 ft. apart. From this you can obtain the average distance between the conductor and the trench. The bus height is at two different elevations, the lowest bus being 30 feet above ground and the higher bus being 55 feet above ground. Most of the tests were made in the EHV substation located just south of Baton Rouge, Louisiana, and called Willow Glen Substation.

Since our interest was directed towards the nature of these transients, an oscilloscope having a fast rise and high gain calibrated pre-amplifier was used for the measurements. This oscilloscope had a rise time of 23 nano-seconds and a voltage range of 50 millivolts to 20 volts per centimeter. There was an attachment on the oscilloscope which enabled the use of a Poloroid camera to record the pictures of the various shots. Care was exercised to minimize extraneous signals that might enter the scope through the power supply. For this purpose a choke coil similar to the one described in reference 8 was used.

RESULTS OF TEST

Table I has been prepared to indicate the results of the various tests performed as the basis of this paper. In Case 1 there was approximately 377 feet of parallel coupling between the control cables and the bus being energized and de-energized. You will note that with the shield grounded both in the control house and at the remote end, a peak to peak voltage of 9 volts was obtained on closing the disconnect switch and 4 volts on opening the switch. The fre-

quency was approximately 300 kHz, refer to Case 1, Shot 1. When the ground on the shield at the remote end was removed, the voltage increased slightly. However, when the ground was removed from both ends, the peak to peak voltage measured 102 volts on closing the switch, refer to Case 1, Shot 5, and 23.5 volts on opening.

This shows a possible reduction of better than 11 to 1 in the measured voltage by grounding the shield at both ends of the control cable for the closing condition.

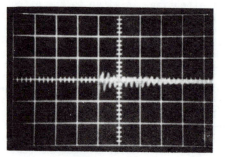

CASE 1   SHOT 1
Vert 10 volts/cm   Hor 10 μ sec/cm

CASE 1   SHOT 5
Vert 20 volts/cm   Hor 10 μ sec/cm

Case 2 and Case 3 were tests of different specimens, and as you will learn from Table I, the coupling distance was much lower than for Case 1. In Case 3, you will also notice that there was approximately 12 to 1 reduction in the pickup of voltage upon closing the disconnect switch when the shield was grounded at both ends of the cable. The reduction upon opening was not quite as great.

Since the pickup voltages measured for the specimen in the trenches was much less than recorded in Reference 6 and 7, it is believed that this much lower magnitude of transient is due to the shielding effect of the earth surrounding the trenches. Some data is included in Reference 7 on buried cables, and the range of measured voltages given are comparable to those reported in this paper. It should be pointed out that while the ground from the shield of the test specimen was removed, the grounds from the shields of all the other cables in the trench were not removed. It is believed that the shields being grounded on the other cables together with the fact that a 4/0 copper conductor was run in the trench and grounded at numerous points to the ground grid was effective in reducing the magnitude of transients in the specimens. Also from figure 2, you will note how the trench is tied into the station ground.

Since all of the measured voltages were much lower than indicated by other reports, a twin conductor No. 14, type TW 600 volt wire was run above ground next to the trenches as shown in figure 1, for Case 5. From the results of this test you will see that there was voltage in excess of 1200 volts recorded in the specimen conductor with the parallel conductor not grounded. Refer to Table I. When the parallel conductor was grounded, the voltage was reduced approximately 50% as also noted for Case 9.

In an effort to determine the effectiveness of shielded cables versus unshielded cable in a trench, a neighboring utility was visited and tests shown as Cases No. 7, 8, and 9 were performed at one of their EHV substations, see figure 3 and Table I. Their control cables have no shields, but they are run in trenches similar to our trenches except there is no heavy 4/0 grounding conductor run at the top of the trench. The configuration of the bus at this substation is quite similar to the configuration of the buses used for the first three cases shown on Table I. A bank of capacitor potential devices was connected to the bus section being energized and de-energized for the tests. This resulted in a lower frequency than that recorded at Willow Glen, see reference 9. Also the disconnect switches were single break instead of double break.

In Case 8, tests 1 and 2, you will notice a recorded peak voltage of 138 volts on closing and 168 volts on opening of the disconnect switch. The coupling was greater than shown in Case 1, and the induced voltages were higher. When comparing Case 8 test 1, with Case 1, test 1, there is nearly a 15 to 1 reduction in the peak to peak voltage. This example gives a comparison between an unshielded cable and one with the shield grounded on both ends as in Case 1, test 1. See Case 8, Shot 2 and Shot 3 shown below.

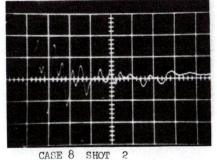

CASE 8 SHOT 2
Vert 50 volts/cm Hor 10 u sec/cm

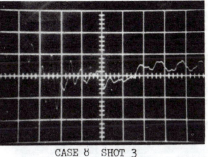

CASE 8 SHOT 3
Vert 20 volts/cm Hor 10 u sec/cm

In Case 9, with non shielded control cable laid in trenches, there was a spare conductor available in the cable being used as a specimen, and this spare conductor was grounded at both the remote end and the control house end. Case 9, tests 1 and 2, see Table I, shows the results of pickup in the specimen conductor when the spare conductor in the same cable is grounded.

In test 2, the measured voltage was low, and it is believed this was caused by the fact that only the very tail end of the arc was caught. It should be more nearly the same as Case 9, test 1. This test, Case 9, test 1, shows that by grounding a spare conductor of the specimen cable on both ends the transients could be reduced by 50%.

As a further check of the possible induced voltages present in control cables, a length of RG 8 coaxial cable was run as shown for Cases 10 to 16. The results are shown in Table I. This cable was also laid in approximately the same location as the twin conductor No. 14 TW 600 volt wire, see figure 1, and notice coupling shown for Case 10. From Case 10 with the shield ungrounded at both the control house and the remote end, you will note that a voltage in excess of 1290 volts was measured on closing the switch, and approximately 1200 volts for opening the switch, see Case 10, Shot 1. The recorded voltages shows both 300 kHz and a 900 kHz signal present during this operation.

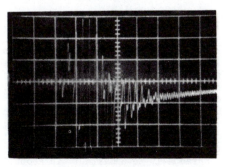

CASE 10 SHOT 1
Vert 200 volts/cm Hor 10 u sec/cm

In Case 12, tests shot 1 and 2, the switch was opened and closed with the shield grounded on both ends of the coaxial cable. In this case the voltage was reduced to a peak to peak value of 5.09 volts for closing and 1.64 volts for opening. Upon the basis of 5.1 volts the reduction in pickup was 253 to one. This is a regular communication cable, and this reduction in pickup is not an unreasonable amount. Case 12 Shot 1 is shown below.

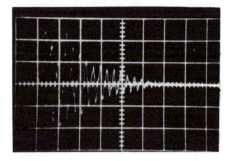

CASE 12   SHOT 1
Vert 1 volts/cm Hor 10 μ sec/cm

As a check on the amount of 60 Hz pickup that can be obtained, tests Case 7, 11, and 17 were made. Under the conditions of these tests the 500 kV bus was energized, and voltage checks made of the 60 Hz pickup on the specimen conductor. The values of pickup can be obtained from Table I. A comparison between Shots 1 and 2 of Case 11 will show the advantage of shielding for this type of pickup, which was electrostatic, see Case 11 Shots 1 and 2 below.

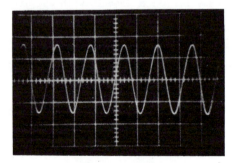

CASE 11   SHOT 1
Vert 10 volts/cm Hor 10 millisec/cm

An examination of the results shown in Table I will reveal variations in the recorded values in some of the tests such as Case I. In Mr. L. E. Smith's paper, reference No. 7, he mentioned that it could be presumed that there is one burst of induced voltage for each half cycle of the system frequency. Quoting again from Mr. Smith, "from observation, the envelope of the peaks of the bursts on switch openings increases gradually to a maximum in about 0.25 seconds; and then rises to about one half peak value where it remains until arc extinction."

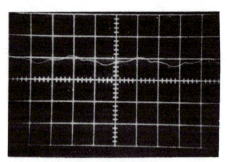

CASE 11 SHOT 2
Vert 0.5 volts/cm Hor 5 millisec/cm

The oscilloscope used in recording the data for this report with few exceptions, was triggered by an impulse of a transient of one burst and thus showed only one of many bursts of induced voltages. Consequently, the values recorded can not be considered the highest values of induced voltages present under the condition tested. Reference 10 shows that a surge counter can record values ranging from 2 to 10 times that recorded by the oscillograph. However, it is believed that the results recorded are representative of results that can be obtained by the different conditions of shielding or not shielding.

The oscillograph does provide a means of determining the frequency and nature of the wave form of the transients. This information is needed to determine preventive means of negating the effects of the transients.

In reviewing the various recordings of oscilloscope shots, you will notice that the burst starts from a high magnitude and decays to a low value in 40 to 60 microseconds. The frequency of the induced voltage was in the range of 200 to 400 kHz during the first part of the burst, and then in several cases you will note after decaying to a low peak to peak magnitude, the frequency increases to the range of 1 Megahertz. Case 10, Shot 2 shows multi-burst of voltages, thus indicating a series of burst during the opening or closing of switches, and this is as indicated in Mr. Smith's paper.

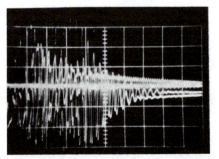

CASE 10   SHOT 2
Vert 200 volts/cm Hor 10 μ sec/cm

The tests under discussion up to now have been induced by electrostatic and electromagnetic coupling. Turning back to the introduction let us refer to Item 3, coupling between windings of current transformers. For this consideration, refer to Case 6 shown in Table I. For this series of shots the breaker CT's together with the bus between the disconnect and breaker were energized and de-energized by the opening or closing of a disconnect switch. For Case 6, Shots 29 and 31, you will notice that we recorded both a 35 kHz and a 300 to 400 kHz transients across the incoming CT lead and the ground return lead. The burden of the protective relays was also across this circuit. Reproductions of the traces Case 6, Shots 29 and 31 are shown below. The reduction of the signal strength is not as great between the shielded and unshielded specimen as for some of the other cases. However, there is some reduction by the shielding of the cable at both the control house and the remote end. The presence of two distinct low and high frequencies in this test can be explained by the discussion given in reference 11.

CASE 6   SHOT 29
Vert 1 volt/cm   Hor 10 μ sec/cm

CASE 6   SHOT 31
Vert 5 volts/cm   Hor 10 μ sec/cm

No attempt was made to measure the high frequency transients present in the secondary voltage circuits of the 500 kV capacitor potential devices. This subject is covered in reference 12. In this reference attention is called to the high potential difference between the earth in the substation and that of earth at the control house. The conclusion of this paper is that screened or shielded secondary cables limit the effects of the over voltage on the secondary circuits, the over-voltage being caused by the arcing disconnect switch. In a later paper, reference 10, where steel wire armored cable was used for the capacitor potential secondary leads, tests showed a decided decrease in the measured transients voltage on the secondary leads when the armor was earthed or grounded on both ends.

CONCLUSION

The means of inducing or coupling transients into a control circuit should be considered when deciding on protective schemes. In general, electrostatic coupling was the main cause of the transients as measured in the tests recorded in Table I, as the magnitude of the arcing current was low. The test showed grounding of the shield as being very effective, and not too much difference was noted when the shield was grounded on one end as compared with both ends. As pointed out in reference 13 as well as other references, grounding the shield is effective in reducing transients induced electrostatically.

Protection against transients induced electro-magnetically is more difficult to provide. Transients caused by electromagnetic coupling can be lessened by:

1. Use of radial circuits and the avoidance of loops in control circuits. In other words, leads going to common controls or protective circuits should take the same path out to equipment in substation and not be looped with leads to other equipment in substation, see reference 6 and 13.

2. Use of a heavy ground cable following the same path as control cable. Circulating current in this cable helps neutralize transients that might be induced in control cables, see references 4 and 13.

3. Magnetic armoring of cable can help, but in case of heavy currents, magnetic material can be saturated thus reducing its effectiveness, see reference 13.

4. Use cable with copper, lead or aluminum shield, and then ground the shield at both local and remote end. The current which will flow through the shield during transient conditions will have a neutralizing effect on transient voltage induced in the control leads, reference 13. If this method is used, a heavy current path should be provided as listed in item 2 above.

5. Avoid the causes that creates the transients. As for example, use preinsertion resistors in breakers. Apply breakers having a minimum possibility of restrike. Use care in the selection of switching devices to be used for shunt capacitor banks. Also

exert care in the connections of two or more shunt capacitor banks in a substation. Use preinsertion resistors with disconnect switches.

Protection against transients, introduced through the winding of capacitor potential devices is very difficult. It appears that the best protection is to endeavor to confine the transient to the capacitor potential secondary cable circuit. This can be done by making use of the methods of protection outlined under electromagnetic induction. Heavy parallel current paths will also tend to neutralize the transients present in the secondary cable of the capacitor potential device.

Protection against transient is not a simple problem in modern substation where heavy fault currents or high voltages are present. It is believed that protection against transient voltages induced in control circuits is justified. It is further believed that control cables with shielding and with the **shielding** grounded at both the remote and **local end** should be a part of a program of protection against induced transient voltages. With shielded cable having the shield grounded on both ends, it is essential to provide a heavy current carrying conductor or conductors parallel to the shielded cable and with this heavy conductor tied into the ground grid at numerous places including the terminating ends.

The justification is based upon the extremely high reliability factor and security factor that the user should place on relaying for EHV transmission lines. Using an old colloquial statement, "the chain is no stronger than the weakest link," it is believed that control cables should receive the same attention as the highly sophisticated relaying which is usually applied on EHV systems.

ACKNOWLEDGEMENTS

The author wishes to express his appreciation to the Baton Rouge Division of Gulf States Utilities and particularly to Mr. Wm. P. Tucker and his assistants' help in operating the Tektronix oscilloscope and the taking of the Poloroid shots. The author would also like to thank Messrs. Ray Toler and Jim Erskine of Arkansas Power Company for making it possible to conduct some tests on their EHV system.

REFERENCES

1. W. K. Sonnemann, Transient Voltages in Relay Control Circuits, AIEE Transactions, vol. 80, Pt. III. February 1962, pp. 1155-1162.

2. H. T. Seeley, Protection of Control Circuit Rectifiers Against Surges from DC Coil Interruptions, AIEE Transactions, vol. 80, Pt. III, December 1961, pp. 871-879.

3. Guide for Safety in Alternating - Current Substation Grounding, AIEEE No. 80, March, 1961.

4. E. A. Baumgartner, Transient Protection of Pilot Wire Cables Used for High Speed Tone and AC Pilot Wire Relaying, Twentieth Annual Conference for Protective Relay Engineers, Texas A&M University, April 24 - 26, 1967.

5. A Guide for the Protection of Wire Line Communication Facilities Serving Electric Power Stations, IEEE Wire Line Subcommittee of the IEEE Power System Communication Committee, IEEE Transactions, on Power Apparatus and Systems, vol. PAS - 85, No. 10, October 1966, pp. 1055-1083.

6. D. A. Gillies, H. C. Rambert, Methods of Reducing Induced Voltages in Secondary Circuits, IEEE Transaction on Power Apparatus & Systems, vol. PAS-86, No. 7, July, 1967, pp. 907-916.

7. Voltages Induced in Control Cable From Arcing 500 kV Switches, IEEE Preprint 31, pp. 66 - 467, presented at New Orleans Summer Power meeting, July 10 - 15, 1966.

8. W. K. Sonnemann, R. J. Felton, Transient Voltage Measurement Techniques, IEEE Transactions PAS - 87, No. 4, April 1968, pp. 1173-1179

9. W. C. Kotheimer, Control Circuit Transients in Electric Power Systems, Twenty First Annual Conference for Protective Engineers, Texas A&M University, April 22 - 24, 1968.

10. F. H. Birch, G. H. Burrows, H. J. Turner, Experience With Transistorised Protection in Britain, Part II: Investigation Into Transient Overvoltages on Secondary Wiring at EHV Switching Stations, CIGRE paper 31-04, Session 1968.

11. E. W. Boehne, EHV Surge Suppression on Interrupting Light Currents With Air Switches - I Capacitive Currents, IEEE Transactions on Power Apparatus & Systems, vol. PAS - 84, No. 10, 906-923, October 1965.

12. J. A. Callow, K. W. Mackey, Impulsive Overvoltages on Secondary Circuits of 330 kV Capacitor Voltage Transformers, CIGRE Paper 136, Session 1962.

13. Dietch, G. Dienne, R. Wery, Progress Report of Study Committee No. 4, (Protection and Relaying), Appendix II Induced Interference in Wiring Feeding Protective Relays, CIGRE paper 31-01, Session 1968.

TABLE 1

TRANSIENT VOLTAGES INDUCED IN CONTROL CABLES

Case No.	Shot No.	Disconnect Switch Operated — Opened	Disconnect Switch Operated — Closed	Approx. Ft. of 500 KV Bus Picked or Dropped	Conductor Tested (Note 1 & 2)	Length of Conductor with Close Coupling (See Fig. 1)	Shield Grded. at Control House	Grounded Grded. at Remote End	Measured Peak-Peak Voltage	Approx. Freq. of Induced Voltage	Transient Caused By
1	1		13019	523 FT.	Spare in Cable 515-34	377 FT.	Yes	Yes	9 Volts	300 kHz	A-C Arc
	2	13019		"	"	"	Yes	Yes	4 Volts	"	"
	3		13019	"	"	"	Yes	No	11 Volts	"	"
	4	13019		"	"	"	No	No	6.9 Volts	"	"
	5		13019	"	"	"	No	No	102.0 Volts	"	"
	6	13019		"	"	"	No	No	23.5 Volts	"	"
2	7		13011	523 FT.	Spare in Cable 510-37	30 FT.	Yes	Yes	Too Low	300 kHz	A-C Arc
	8	13011		"	"	"	No	Yes	8.0 Volts	"	"
	9		13011	"	"	"	No	No	5.5 Volts	300 kHz	"
	10	13011		"	"	"	No	No	12.4 Volts	300 kHz	
3	11		13011	523 FT.	Conductor SYP-5, Cable 510-17	125 FT.	Yes	Yes	1.05 Volts	300 kHz	A-C Arc
	12	13011		"	"	"	Yes	Yes	1.05 Volts	"	"
	13		13011	"	"	"	Yes	No	1.05 Volts	"	"
	14	13011		"	"	"	No	No	1.73 Volts	"	"
	15		13011	"	"	"	No	No	12.00 Volts	"	"
	16	13011		"	"	"	No	No	9.00 Volts	"	"
4	17		13011	See Note 3	Conductor SYP-5, Cable 510-17	140 FT.	No	No	2.4 Volts	Non-Uniform	Note 3
	18	13011		"	"	"	No	No	2.76 Volts	"	"
	19		13011	"	"	"	Yes	Yes	1.73 Volts	"	"
	20	13011		"	"	"	Yes	Yes	1.73 Volts	"	"
5	21		13011	523 FT.	Twin Conductor, TW, B1-Lead (See Note 4)	480 FT	No	No	1200 Volts	Non-Uniform	A-C Arc
	22	13011		"	"	"	No	Yes Br.	1200 Volts	"	"
	23		13011	"	"	"	Yes Br.	Yes Br.	620 Volts	300 kHz	"
	24	13011		"	"	"	Yes Br.	No	673 Volts	300&900 kHz	"
	25		13011	"	"	"	Yes Br.	No	1175 Volts	"	"
	26	13011		"	"	"	No	Yes Br.	827 Volts	Non-Uniform	"
	27		13011	"	"	"	No	Yes Br.	1200 Volts	"	"
	28	13011		"	"	"	No	Yes Br.	1200 Volts	"	"
6	29		13009	40 FT.	CT Lead 9LC-1 to 9LCO Grd.	50 FT.	Yes	Yes	5.52 Volts	35 & 400 kHz	A-C Arc
	30	13009		"	"	"	Yes	Yes	3.10 Volts	35 & 300 kHz	"
	31		13009	"	"	"	No	No	13.75 Volts	35 & 400 kHz	"
	32	13009		"	"	"	No	No	17.20 Volts	1 Meg. Hz	"
7	1	No Operation		All Busses Hot	Spare in Cable 125-19	700 FT.	No	No	3.92 Volts	60 Hz	60 Hz
8	1		5905	550 FT.	Spare in Cable 125-19	435 FT.	No	No	136 Volts	230 kHz	A-C Arc
	2	5905		"	"	"	No	No	168 Volts	230 kHz	"
	3	5905		"	"	"	No	No	103.5 Volts	230 kHz	"
	4	5905		"	"	"	No	No	98.5 Volts	230 kHz	"
	5	5905		"	"	"	No	No	92.6 Volts	230 kHz	"
9	1		5905	550 FT.	Spare in Cable 125-19	435 FT.	See Note "	See Note 9	46.8 Volts	230 kHz	A-C Arc
	2	5905		"	"	"	See Note "	See Note 9	* 6.8 Volts	230 kHz	"

FOR CASE 1 - 5 SEE NOTE 5

CASE 6 SEE NOTE 6

CASE 7 - 9 SEE NOTE 7

CASE 10 - 17 SEE NOTE 8

* CAUGHT ONLY TAIL OF ARC.

TABLE 1 CONTINUED

TRANSIENT VOLTAGES INDUCED IN CONTROL CABLES

CASE NO.	SHOT NO.	DISCONNECT SWITCH OPERATED OPENED	DISCONNECT SWITCH OPERATED CLOSED	APPROX-FT. OF 500 KV BUS PICKED OR DROPPED	CONDUCTOR TESTED NOTE 1 & 2	LENGTH OF CONDUCTOR WITH CLOSE COUPLING SEE FIG. 1	SHIELD GRDED. AT CONTROL HOUSE	GROUNDED GRDED. AT REMOTE END	MEASURED PEAK-PEAK VOLTAGE	APPROX. FREQ. OF INDUCED VOLTAGE	TRANSIENT CAUSED BY
10	1		13011	523 FT.	RG-8 COAXIAL CABLE	500 FT.	No	No	1290 VOLTS	300&900 KHZ	A-C ARC
	2		"	"	"	"	No	No	1200 VOLTS		"
11	1	No Operation		Busses Hot	RG-8 COAXIAL CABLE	500 FT.	No	No	32 VOLTS	60 HZ	60 HZ
	2	"		"	"	"	Yes	Yes	0.227 VOLTS	60 HZ	"
12	1		13011	523 FT.	RG-8 COAXIAL CABLE	500 FT.	Yes	Yes	5.09 VOLTS	300 KHZ	A-C ARC
	2		"	"	"	"	Yes	Yes	1.64 VOLTS		"
13	1		13011	523 FT.	RG-8 COAXIAL CABLE	500 FT.	No	Yes	6.4 VOLTS	200&900 KHZ	A-C ARC
	2		13011	"	"	"	No	Yes	6.0 VOLTS	30&81 MEGHZ	"
14	1		13011	523 FT.	RG-8 COAXIAL CABLE	500 FT.	Yes	No	4.9 VOLTS	300 KHZ	A-C ARC
	2		"	"	"	"	Yes	No	5.09 VOLTS	300&800 KHZ	"
15	1		13009	40 FT.	RG-8 COAXIAL CABLE	UNKNOWN	Yes	No	2.74 VOLTS	400&800 KHZ	A-C ARC
	2		13009	"	"		Yes	No	4.00 VOLTS	400&800 KHZ	"
16	1		13011	523 FT.	RG-8 COAXIAL CABLE	500 FT.	Yes	No	8.00 VOLTS	300&900 KHZ	A-C ARC
	2		"	"	"	"	Yes	No	3.46 VOLTS	600 KHZ	"
17	1	No Operation		Busses Hot	SPARE IN CABLE 515-34	377 FT.	Yes	Yes	1.45 VOLTS	60 HZ	60 HZ

NOTE No. 1: Case 1 to 4 inclusive, Case 6 & 17 - Multi conductor control cable of following specification was used: Individual conductor to have 20 mils of clear polyethene insulation and 10 mils of colored PVC jacket. Cable to be made up with suitable filler and bound with .001 inch Mylar tape. A copper metal braid shield to be in accordance with IPCEA Publication S-61-402, Part 4, Paragraph 4.1.1.2 and is to be applied over the assembled conductors. Cable is to have an overall outer sheath of black PVC, 4/64 inches thick. Conductor No. 9, 19/22 strand.

NOTE No. 2: Case 7, 8 and 9 - Test specimen similar to cable listed under Note 1 except no shield was provided for this cable.

NOTE No. 3: Case No. 4 - Measured effect of D-C pickup upon test conductor when closing & opening Sw. 13011.

NOTE No. 4: Case No. 5 - Test conductor run next to trench but on surface of ground. Black (BL) conductor used for ground to test shielding effect.

NOTE No. 5: Case 1 - 5, and 7 - 9 - Measured voltage between test conductor and control house ground. Remote end of test, conductor floating or for Case 10 - 16 terminated in 10 megohm resistor connected to remote ground.

NOTE No. 6: Case No. 6 - With CT connection normal measured induced voltage between 9LC-1 and 9LCO, - CT leads from ACB 13010. Note 9LCO grounded to control house ground.

NOTE No. 7: Case 7 - 8 and 9 - Were run at another Utility's EHV Substation, see Fig. 3. All other cases run at GSU Company's Willow Glen EHV Substation. Conductors Cases 7 - 9 in trench without shield conductor.

NOTE No. 8: Case 10 - 16 - Similar to Case 5 except RG-8 cable with shield used and run along trench on surface of ground.

NOTE No. 9: Case No. 9 - Used another spare conductor in same cable 125-19 with test conductor and grounded the second spare both on remote end and in control house.

NOTE No. 10: All dimensions given in English System. To convert to metric system: Multiply values in inches by 2.5400 to convert to centimeters; multiply values in feet by 0.3048 to convert to meters.

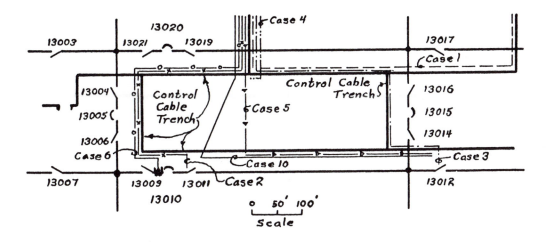

FIGURE 1: POSITION OF CONTROL CABLE TRENCHES WITH REFERENCE TO EHV BUSSES

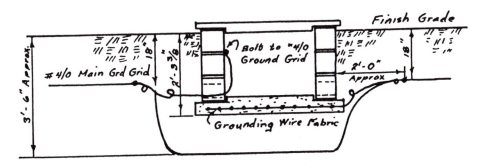

FIGURE 2: CROSS SECTIONAL VIEW OF CABLE TRENCH IN EHV SUBSTATION

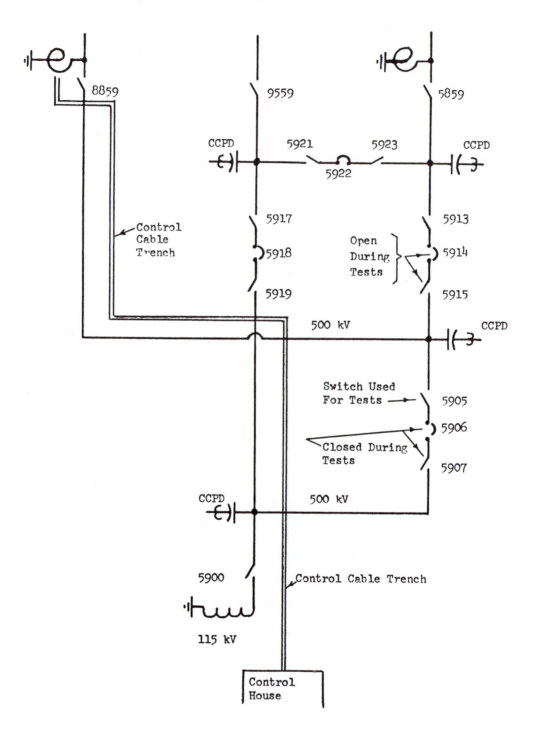

FIGURE 3: ONE LINE DIAGRAM SHOWING APPROXIMATE POSITION OF BUSSES WITH
REFERENCE TO TRENCH AND CABLE INVOLVED IN CASE 7, 8, AND 9

Discussion

W. C. Kotheimer (General Electric Company, Philadelphia, Pa. 19142): The author presents very interesting new data on the nature of the control circuit transients caused by arcing disconnect switch operation. The results in many ways are similar to those obtained by others in staged field tests. On this basis, it seems definitely established that these induced transients, in general, have lightly damped oscillatory waveforms with a dominant frequency ranging from a few hundred kilohertz to several megahertz. The level of the induced voltage on the control cable, as these tests show, is strongly influenced by the location of the cable, use of a shield, grounding methods, etc. I fully agree with the author's conclusions pertaining to cable installation, i.e., the use of shielded control cables with the shields grounded at both ends and closely paralleling these cables with heavy ground conductors also firmly tied to ground at both ends.

The author's tests show a dominant frequency content of the transients somewhat lower in value than others have obtained from similar tests. In considering the conditions of his tests, this seems reasonable because the bus length involved in the oscillation in case 1 (including the length of bus to the transformer bank) approaches a quarter-wavelength at 300 kHz. Also, in the tests, in which an EHV circuit breaker was included in the bus length, the breaker was always closed. This raises the question of how an open breaker would affect the transient frequency. The voltage distributing capacitors across its open contacts would then be in series with the bus and would increase transient frequency. This would apply particularly to cases 7, 8, and 9 where breaker 5906 was closed. Since it is very common practice to open a circuit breaker before operating the disconnects on either side, it would seem very desirable to evaluate this condition, also. For the bus structure shown in Fig. 3 (cases 7, 8, 9), if we assume that its dimensions are similar to case 1, the bus ground loop between the two coupling capacitor potential devices (CCPDs) would have about 500 μH inductance. Assuming each CCPD to have about 0.00167 μF (Hi C type), the loop resonance with the breaker closed, calculates to about 250 kHz which is close to the measured value 240 kHz. It appears that the loop oscillates at this frequency and then impresses this potential on the extended 500 kV bus structure paralleling the cable trench. However, if we assume that the breaker is open and has a capacitance of 200 pF across its contacts, calculation shows the loop resonance would increase to about 560 kHz.

Some time ago, this discusser participated in similar tests on a 500 kV station where the breaker was open and there were free-standing CTs with capacitive bushings immediately adjacent to each breaker so that a loop, much shorter than that shown in Fig. 3, was formed. The dominant measured transient frequency in these tests was 1.6 MHz, mainly due to the smaller loop and the breaker being open. The results of the various field tests to date, and also theory, seem to agree that the bus configuration and dimensions, shunt capacitance of CCPDs and CT bushings, and series capacitance across EHV breakers, if open, all taken in combination, determine the dominant frequency of the transient produced during disconnect switch operation. For a given station, with all its possible switching arrangements, this frequency could vary over at least a two to one ratio.

With the exception of case 6, the author's measurements were made on control cables which were not connected to any EHV apparatus such as CCPDs or free-standing pedestal-mounted CTs. The remote end was floating. This type of test evaluates only the capacitive coupling effect between the bus transient and the control cable. His data amply illustrate this point. In general, they show that unshielded cables located in the trench and thus partially shielded by being nearly in the ground plane, had coupled potentials in the range of 100 volts maximum. Grounding the shields on these cables reduced the coupled potentials further, to about the 10-volt range. On the other hand, an unshielded cable lying on the surface had a coupled potential in the range of 1000 volts.

Manuscript received June 12, 1969.

When control cables are connected to EHV apparatus having high capacitance to the EHV bus, particularly, pedestal-mounted CCPDs and CTs with capacitive bushings, other modes of coupling are introduced: magnetic coupling due to transient flux linkages adjacent to the apparatus ground conductor, conductive coupling due to common ground impedances, and additional capacitive coupling internal to the apparatus. These modes of coupling can also introduce transient potentials and currents into the control cable. The methods listed by the author in the conclusion, especially point 2, are quite effective in reducing this type of interference.

S. H. Horowitz (American Electric Power Service Corporation, New York, N. Y. 10004): Mr. Sutton has presented a comprehensive review of an extremely timely, and complex, subject. The specific test details are revealing, and the list of references should prove very useful. It is interesting to note that Mr. Sutton's conclusions are not based specifically on the test results but rather upon three of his references and upon the belief that special attention should be given to the cables associated with EHV relays.

I agree that special attention is required. I do not agree, however, that this attention must necessarily result in the use of shielded control cable. Referring to Table I, the highest peak-to-peak voltage for buried cable without shield grounds was 103.5 volts (case 8). On the other hand, over 600 volts appeared on cable with the shield grounded at both ends when the cable was run on the surface of the ground (case 5).

These results parallel tests we have run in which we had indications of better shielding on unshielded cable in the ground than on shielded cable on the surface. In no case did we experience induced voltages which were high enough to be of concern.

Certainly, no one can argue against the position that the control cable system must not degrade reliability. And it is equally certain that high transient voltages can exist and can be induced in cable. On the basis of our own studies, however, corroborated now by the tests reported in this paper, we believe that the portion of the cable that goes underground does not have to have any special shielding or construction. The portion from the ground up to a given device such as current transformer or potential device does not receive the benefit of ground capacitance, ground wires, or other shielded or grounded wires and in all probability should be properly shielded. Did Mr. Sutton's tests include cases related to this aspect of the problem?

If the preceding analysis is correct, the problem that is then presented is to decide if it is technically or economically more advantageous to run all shielded cable, if any part of the cable must be shielded, and if all stations should be cabled similarly. Note that 1 of Table I gives the construction of the test cable. Can the author give us the details of the unshielded cable and relative cost between the shielded and unshielded cable?

Manuscript received June 24, 1969.

Fred Chambers (Tennessee Valley Authority, Chattanooga, Tenn.): The measurement of transients in control cables is at best a tricky business and often frustrating to the engineer who seeks this information. Consequently, published papers on this subject are relatively scarce, and the author deserves credit for pursuing his investigations to a successful conclusion.

It is gratifying to note that in general the conclusions reached by the author agree very well with those of Tennessee Valley Authority engineers from several field tests performed on the TVA 500 kV system. However, a few differences are also noted. For example, the author concludes that electrostatic coupling accounted for the main portion of the induced voltages, as the magnitude of

Manuscript received June 24, 1969.

arcing current was low. We agree that the 60 Hz component of current is small, but we believe that the high-frequency component of current which results from resonant oscillations in the circuit being energized or interrupted is quite substantial, possibly reaching 1000 amperes or more on initial strike. Since structures and earth connections may have relatively high impedance at these frequencies, significant high-frequency overvoltages may result.

The author also observes that "not too much difference was noted when the shield was grounded on one end as compared with both ends." Our tests show this to be true only if the measurement is made at the end where the shield is grounded. Voltages from conductor to ground at the remote end may still be very high, approaching those which would occur on an unshielded cable. The supporting theory for this condition is that a ground at both ends of a shield having a substantially high conductivity provides in effect a single turn neutralizing transformer in which the induced voltage between conductor and shield at any location within the cable is substantially cancelled. This does not ensure that high voltages will not occur between shield and earth at intermediate points along the cable, and in order to prevent this possibility, TVA's practice is to ground the sheath at convenient intermediate points, such as in manholes. The cable shield provides considerably better coupling than a separate ground conductor, provided that it is of relatively high conductivity, is well bonded at each end, and is continuous from the terminal enclosure at the equipment in the yard to the switch house ground bus or other similar reference ground.

The author mentions the possibility of using preinserted resistors with disconnecting switches for suppressing induced voltages in control cables. TVA has some experience with such resistors and has performed several tests to determine their effectiveness. Previously we have found that such resistors are effective to some degree when they perform properly. However, recent tests have resulted in frequent flashovers due to inadequate impulse insulation of a particular design. Also some designs permit arcing from blade to jaw before the first step of resistance is cut in, and we believe that the effect of this arcing needs to be evaluated.

One source of induced voltages which has been discussed rather widely is the coupling capacitor potential device, or capacitor voltage transformer. If the distributed capacitance of the winding of such a device is fairly uniform, most of the transient voltages from this source are common mode. These voltages can be effectively suppressed by a combination of shielding and a π-section filter, consisting of toroidal ferrite cores through which all conductors pass, with capacitors to reference ground on each side of the toroid.

We wholeheartedly concur with the author that control cables should receive fully as much attention as that given to all other components of the protective relay schemes.

It cannot be overemphasized that measurements of this type are difficult, and that meaningful results are extremely elusive. To obtain results which are in error is usually worse than not making the tests. All available checks and aids should be used to the utmost degree. It is TVA's practice to check the response of as much of the measuring circuit as possible. This is sometimes done by using a square wave generator to apply a signal to the input of the measuring circuit and observing voltage at the measuring location in order to evaluate the frequency limitations of the measuring circuit. It is standard practice to perform tests to check the integrity of the measuring circuit in order to ensure that it does not have excessive pickup and error due to fields in the vicinity of the measuring cables or instruments. This is accomplished by making a test with the input leads of sensitive measuring instruments disconnected from the measuring point and connected to the reference bus. Transient voltages resulting from circuit pickup or differences in ground potential between the reference bus and the power supply source for instruments will show up in this test. A ferrite toroid with the ac supply cord wound about it for a few turns provides very good isolation for instrument power supplies.

Some rather interesting tests relating to this subject have been performed by TVA quite recently, and it is hoped that the results can be made available at a later date. However, a preliminary review of the data does not indicate a need to modify conclusions stated in [7], which are substantially the same conclusions presented by the author. Refinements in measuring methods and techniques have given firm support to these conclusions and have removed some reservations which existed concerning previous measurements.

J. L. Koepfinger (Duquesne Light Company, Pittsburgh, Pa. 15219): It is evident that the author planned these very valuable tests with meticulous detail. The author indicated that the purpose of his tests was to evaluate the coupling of induced voltages into control cables resulting from the production of switching surges in EHV circuits near the control cables. It is noted that in most cases of this test program, the specimen conductor was not terminated at either the remote location or the control house. In actual practice, the conductors in a control cable are usually terminated in equipment such as transducers and coils in the switchyard and batteries and relays in the control house.

The point that we wish to make is that relays can be subjected to both common-mode and transverse surges. These can appear across the ac inputs to the relay as well as across the dc input. In case 6, the author indicated that measurements were across relay current input between phase conductors and the ground return circuit of the control cable. Is it correct to assume that the switching surge voltage induced in the CT lead is equivalent to that produced in case 2 where the specimen conductor was not terminated? The results appear similar. If this assumption is true, it would tend to indicate that cable termination has little effect on the induced surges. Perhaps the terminating impedance of the relays appears as an open circuit to the traveling wave in the control cable.

It is also indicated that static components in relays are stressed by the surges which appear in the control cable. These tests indicate the magnitudes and frequency of these surges. We would like to know whether the author has any information which would permit us to determine the energy content of the surges in the control cable that are impressed on a relay. Was any attempt made to measure the current of the surges?

If definitive tests for static relays are to be developed, it is essential that controlled tests such as those described be made by others. Still lacking is concrete information on the amount of energy a static relay should be expected to dissipate without failure or incorrect operation. It is evident by the result of this paper and others that shielding is desirable, but is it essential in all cases? Is there a compromise between partial shielding such as ground of a conductor in the control cable and surge suppression built into the relay?

The author also indicates that tests were made on different types of shielded cable. Would he care to indicate where differences existed in the shield? If so, would he wish to rate the effectiveness of these different shields in reducing the surges in the cable?

Manuscript received July 3, 1969.

Howard J. Sutton: I appreciated the interest that is shown in this subject and would like to thank each of the discussers for their comments and questions relative to this subject.

Mr. Kotheimer's brought out the point of how the frequency of the damped oscillatory transient varies with the circuitry of the section being switched. He also raises the question of how an open breaker would affect the transient frequency. The only case pertaining to switching a bus section involving an open breaker was case 6 of the paper. In this case we picked up a bus section involving the breaker current transformers with the breaker open. This we believe would be similar to normal procedure for opening or closing the bus section by means of disconnects. Referring to Fig. 1 for case 6, we opened and closed disconnect 13009 with breaker 13010 open. One reason that we did not run more cases of energizing the bus to an open breaker is that we selected the cases where we would have the maximum coupling between the control cables and the bus being switched. As Mr. Kotheimer pointed out, case 6 was the only measurement made on control cables which were connected to EHV apparatus such as current transformers or capacitor potential devices. It is regretable that our tests did not include the recording of transients induced in the secondary of capacitor potential devices.

In reference to Mr. Horowitz's discussion, I can readily agree with Mr. Horowitz's observation that my conclusions are based as much on the references as they are on the test values included in the paper.

Manuscript received August 20, 1969.

In Mr. Horowitz's discussion he refers to Table I and points out that the highest peak-to-peak voltage for buried cable without the shield grounded was 103.5 volts (case 8). Actually, his reference should have been to shot 2 of case 8 in which the highest voltage was recorded as 168 volts rather than 103. There is a discrepancy in his reference to case 5 in which he points out that over 600 volts appeared on the cable with the shield grounded. Actually this is not a conductor with a shield but merely a specimen conductor with a parallel conductor. By grounding the parallel conductor on both ends, we reduced the pickup voltage in the specimen conductor down to approximately 50 percent of what it would be if the parallel conductor were floating. These results agree with results indicated in the various references of the paper. If Mr. Horowitz will compare case 12, shots 1 and 2, and case 10, shots 1 and 2, he will notice that the pickup voltage is reduced from a value in excess of 1200 volts down to a value of 5 volts which is a considerable reduction in favor of shielding. The specimen cable for cases 10 and 12 is an RG8 type of shielded cable.

We do agree with Mr. Horowitz that the values we recorded of the unshielded cables in trenches below the ground level were not severe enough to present a serious problem. He did indicate concern of the possible magnitude of induced transients in the portion of the control cable which was in runs above the ground level and where the exposure could be considerably greater. We believe the results as given in cases 5, 10–16 are indicative of pickup voltages that can be induced in control cable run above ground. The results of case 6 involve both the coupling between primary and secondary windings of current transformer as well as voltage induced from the coupling between high-voltage bus and control cable. The same would have been true of voltages induced in the secondary leads of capacitor potential devices.

It is questionable as to how effective it would be to shield only the portion of control cable that runs above ground and then leave the balance of cable run below the earth surface unshielded. This is because for electromagnetic coupling the current flowing through the shield has a neutralizing effect on the specimen conductor, and the neutralizing effect should be applied for the full run for maximum benefit.

Mr. Horowitz also brought up the problem of deciding whether it is technically or economically advantageous to run all shielded cable if it is technically feasible to shield part of the cables. It is roughly estimated that the shielded cables of the type used for the station considered ran approximately 26.5 percent more than unshielded cables. These are rather rough figures, but on the basis of 26.5 percent this means that we spent about $6300 more for shielded cable than we would have for unshielded cable. Comparing this combined figure for the cost of the switchboards and relaying and control cables, the increase in cost is a little over 3 percent more than if we had used unshielded cables.

I particularly appreciated Mr. Chambers' comments since I am aware that TVA has made considerable tests in reference to transients that might be present in control cables. I also agree that the measurements of these transients in control cables is a very tricky project and is sometimes very frustrating to the engineers making the tests. It is very difficult to determine whether a transient is predominantly created electrostatically or electromagnetically. Our basis for believing that most of our tests were predominantly electrostatic was the small difference noted in the results

when the cable was grounded at one end versus both ends. We agreed with Mr. Chambers that if the transients are induced by electromagnetic coupling, then the grounding of the shield on both ends will provide a neutralizing effect because of the induced transient current flowing in the shield. However, if the transient is induced by electrostatic coupling, then the induced voltage in the specimen conductor is affected by the capacitance between the specimen conductor and the shield, and between the specimen conductor and the EHV conductor from which the disturbance is emitted. This effect would not be materially changed if the shield is grounded on one end or both ends.

Mr. Chambers' reference to the preinsertion resistors in the disconnect switch test is welcomed particularly since our reference to this was drawn from the experience of others. We are in complete agreement with Mr. Chambers that extreme care must be made in making these tests to avoid erroneous results. Mr. Chambers mentioned the use of a ferrite toroid-type choke coil for the ac supply and it can be noted from our papers that we used a similar-type choke coil in our ac power supply to the oscilloscope.

These comments are directed to the discussion by Mr. Koepfinger. Mr. Koepfinger is correct that with the exception of cases 10–16 we did not terminate the cable in any impedance. In the case of 10–16 we did terminate the cable in a 10-MΩ resistance. Our reason for terminating the cable in a very high or no impedance is that our main concern was that we might pick up a transient voltage on a control wire connected to a diode or a transistor and the resulting voltage across this semiconductor to ground could damage the device. It was our understanding that the back resistance of germanium diodes and silicon diodes is in the range of 100-kΩ to 1 MΩ. It is on the basis of this high impedance that we thought it best to run our tests terminating the cable in very high impedance or open circuit. As mentioned by Mr. Koepfinger, in case 6 we did take the measurements across the relay current circuit, and the voltages are the voltages that appear across this particular circuit. Incidentally, these were coupled into the CT circuit by the coupling between primary winding and the secondary winding.

I have endeavored to obtain a power level that would cause the failure of diodes and transistors for current flow in the back direction; however this power level is so minute that the manufacturers are hesitant to furnish this information. On this basis we have assumed that if you exceed the reverse voltage rating of a diode or transistor, you can expect failure no matter how little the power level is. It is believed that the comments relative to Mr. Horowitz's discussion will answer Mr. Koepfinger's comments relative to the partial shielding of control cables.

The final question raised by Mr. Koepfinger was relative to the comparison of shielding effects of different types of shields. I would like to refer him to [14].

In summary, it is evident from the discussion and from the interest in the subject that more work needs to be done to establish good practices as to when shielding should be used and when it is not required. It is hopeful that this paper will be helpful as a basis for further work.

References

[14] B. E. Klipec, "Reducing electrical noise in instrument circuits," presented at the IEEE 13th Ann. Petroleum Industry Conf., Chicago, Ill., September 12–14, 1966.

PROPOSED IEEE SURGE WITHSTAND CAPABILITY TEST
FOR SOLID-STATE RELAYS

JOHN W. CHADWICK, JR.
Electrical Engineer
Electrical Engineering and Design Branch
Tennessee Valley Authority
Chattanooga, Tennessee

INTRODUCTION

Control circuit surges have been with us for a long time, but did not become a real problem until solid-state devices began to appear as elements in control circuits. Along with tripping diodes and blocking rectifiers came flashovers in control circuits, malfunctions of relays, and failures of circuit components, such as coils, diodes, and insulation. For a time, little was done about the unexplained phenomena, except to hide them under the mysterious and undefined label of "surges." The first real progress came in 1959, when the AIEE Relay Committee formed a Subcommittee on Surge Phenomena in Relay Control Circuits. Although its main function was to exchange information and to document phenomena that occurred in relay control circuits, trends and patterns began to emerge from the data. A paper[1] was published by the Subcommittee in 1966, enumerating 72 incidents of transient overvoltages in control circuits.

Two additional factors were beginning to have their impact in this area: the very rapid expansion of EHV and the increased acceptance of solid-state protective relaying schemes. In fact, interest in the subject progressed so rapidly that in 1966 the IEEE Power System Relaying Committee formed a Working Group on Static Relay Surge Protection. This working group is developing a "representative" test wave and a test procedure for applying it to solid-state protective relay equipment. Work is well under way on a report of this working group, to be presented at a future IEEE meeting. After all the discussions are considered, the report may then be made a part of the relay standard ANSI C37.1. All of this will take time for, as you all know, development of standards is not a speedy process, and indeed it should not be.

The purpose of this paper is to give the background behind the selection of the various parameters of this test wave, and how it is to be applied to solid-state protective relay schemes.

Because this is a "representative" wave, the equipment under test is subjected *only* to a specified level of surges, and *not* to the maximum that could be present in substations. The user, therefore, has a definite responsibility to limit the incoming surges impressed on the equipment below this test level. As you have seen from the previous papers,[2,3,4] this can be done if reasonable care is exercized. In simple terms, the manufacturer builds a "fence" around his equipment, and says that the equipment is good up to this level. It is then up to the user to be sure that he does not permit surges higher than the "fence" to get to the equipment. With reasonable care and attention to such things as grounding practices, arrangement and location of cable runs, and shielding, the integrity of the "fence" can be maintained.

SURGE GENERATOR WAVESHAPE AND CHARACTERISTICS

The *Surge Withstand Capability* (SWC) test wave is an oscillatory wave of 1.5-MHz nominal frequency (1.0- to 1.5-MHz range), having a 2500-V (-0, $+20$ percent) crest value for the first half cycle, with the envelope decaying to 50 percent of crest value in not less than 10 microseconds. The source impedance of the surge generator is 150 ohms. The test wave is applied to the test specimen at a repetitive rate of not less than 50 tests per second, for a period of not less than 2.0 seconds (Figs. 1 and 2). Note that:

1. The characteristics of the test wave, namely the crest value of the voltage and the envelope decrement, are open circuit values, i.e., without the test specimen connected.

2. The repetition rate and period of application were chosen to include 50-Hz as well as 60-Hz systems.

Reprinted with permission from *Proceedings of the American Power Conference*, vol. 32, pp. 1070–1075, 1970.

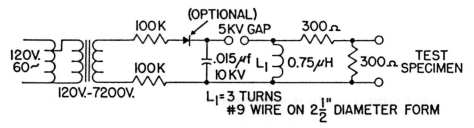

Fig. 1—Typical SWC test circuit.

Frequency

Many frequencies were discussed and investigated. Some members of the working group were strong proponents of the unidirectional wave (d-c decrement), some single polarity, some both polarities. Other members wanted multifrequency testing. Some wanted tests at lower frequencies, but a number of reliable field observations reported an oscillatory wave in the neighborhood of 1.5 MHz. Finally, most agreed than an oscillatory wave of 1.5 MHz (1.0 to 1.5 MHz range) was representative.

Magnitude

The 2500-V (−0, +20 percent) crest value was selected because the manufacturer can reasonably build protection for this value into the equipment. At the same time, it is a value to which the user can quite adequately limit the surges by the use of shielding, good grounding practices, and some simple construction and installation precautions.

Higher as well as lower values were discussed. However, 2500 volts crest certainly is not excessive, when one realizes that it is equal to only 1770 volts rms for the first crest. The wave then decays very rapidly, producing no appreciable volt-time stress on the insulation.

Decrement

The decrement selected was a reasonable value that maintained the test wave for durations well beyond those observed in the field.

Source Impedance

Field measurements indicated that the surges have a source impedance range of from 5 to 250 ohms. A compromise value of 150 ohms was specified.

Application Time

The test wave application time of two seconds was selected to represent the arcing time of a disconnecting switch, a good source of 1.5-MHz surges. Therefore, if the equipment under test can pass a two-second exposure to the test wave, it is reasonable to expect that it will perform correctly in the presence of an arcing disconnecting switch.

Repetition Rate

The repetition rate of not less than 50 times per second was set to include 50-Hz equipment. This rate also permits the insertion of the optional diode. (See Fig. 1.) The diode, in turn, limits the gap operations to one per cycle, minimizes restrikes, and improves the gap performance.

APPLICATION OF TEST WAVE

The application of the test wave should duplicate, as nearly as possible, conditions the equipment will have to face when it is placed in service. This means that the test wave should be superimposed on the normal inputs to the equipment, i.e., currents, voltages, power supplies, etc. Coupling the test wave to the input terminals can be done fairly simply by use of a 1000-V, 0.1-μF capacitor. The isolation or decoupling of the test wave from the normal source can be accomplished by inserting, in the source lead, a suitable choke: for example, J. W. Miller Company, Model 7827, rated 10 A, 2500 V, 370 μH (Fig. 3).

With the equipment to be tested set up and functioning normally with all its inputs present and isolated, the test wave is applied via suitable coupling between each terminal and the "surge reference" of the equipment. In the case of similar inputs, such as current inputs 1, 2, 3, N; voltage inputs 1, 2, 3, N; and power

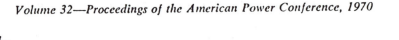

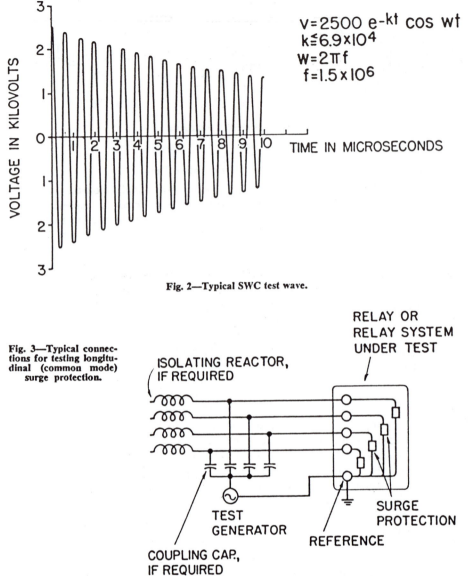

$$V = 2500\ e^{-kt}\cos wt$$
$$k \leq 6.9 \times 10^4$$
$$w = 2\pi f$$
$$f = 1.5 \times 10^6$$

Fig. 2—Typical SWC test wave.

Fig. 3—Typical connections for testing longitudinal (common mode) surge protection.

supply inputs positive, negative, etc.; these are tested individually and then the group is tested as a single input. This tests the equipment for longitudinal (common mode) surges (Fig. 3). Note that:

1. *Longitudinal (common mode) voltage* is the voltage measured at a given point between the wires of a signal pair on the one hand, and on the other, local earth (the voltage between the two wires themselves being often relatively small).

2. *Transverse (differential mode) voltage* is the voltage measured at a given point between the two wires of a signal pair.

Most surges observed in substations are longitudinal (common mode) voltages. This is understandable when you consider that EHV equipment contains fairly respectable sizes of capacitance. Capacitance bushings on current transformers and power transformers have about 800 pF; coupling capacitors have about 4000 pF, etc. These sizes of capacitors provide very good paths to "ground" for the large, high-frequency currents resulting from primary surges generated by arcing disconnecting switches, energizing lines, etc. Some pieces of equipment are mounted on pedestals or bases, which offer a relatively high impedance to high-frequency currents. The result is thousands of volts at high frequencies appearing between the case of the equipment and

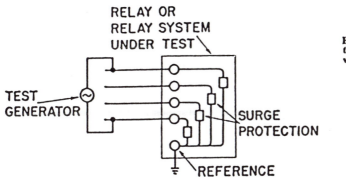

RELAY OR
RELAY SYSTEM
UNDER TEST

TEST
GENERATOR

SURGE
PROTECTION

REFERENCE

Fig. 4—Typical connections for testing transverse (differential mode) surge protection.

the station ground mat. This appears in the switch house as a longitudinal (common node) voltage between the cable conductors and the local switch house ground. Since most surges observed have been longitudinal (common mode) voltages, it follows that the main emphasis should be placed on testing the equipment for this type surge.

Additional tests probably will be specified to test the transverse (differential mode) protection of the equipment. These tests can be made by coupling the test generator to two inputs of a group, such as current inputs 1 and 2. However, this might not be as severe as the longitudinal (common mode) test because two surge units would be in series across the test generator (Fig. 4). Discussions are continuing on this test.

GENERAL COMMENTS

This SWC test is a design test for protective systems and is not intended as a routine test to be performed on each protective system. While individual unit tests may be made by the manufacturer, they do not take the place of overall system tests.

A test is successful when no erroneous output is present. An erroneous output is an output that presents false information, i.e., target lights, trip pulses, carrier bursts, carrier interruptions, etc. Several other committees both in this country and abroad are working on similar tests. IEC TC41 Working Group 4 (Static Relays) is specifying a destructive test as well as malfunction test. The BEAMA Publication 219[5] is the test that the IEC is considering for the destructive test. It was the opinion of the IEEE working group that erroneous outputs would be its criteria and that destructive tests

would not provide additional useful information.

The working group realizes that some of these tests may be difficult to interpret and apply as system tests particularly when a communications link is part of the system. The link probably is located at a remote point, and may be supplied by a manufacturer other than the relay manufacturer. It must be realized that the surges do not differentiate between the relay components and the communications link components; therefore, both must be tested. This means that the connections between the remotely located communication link and the protective relay units must receive special attention. Some form of isolation, such as buffers or conditioners, must be present at the interface or connection between these units. At this junction, assuming that the buffer or conditioner is located at the remote communications link, the input must be subjected to the SWC test. At the protective relay unit output, these terminals must also be subjected to the SWC test. Difficulties in interpretation will surely arise, but patience and experience should eventually conquer this obstacle.

CONCLUSIONS

A surge test is needed for solid-state protective equipment so that a specific level of protection can be provided and certified by the manufacturer. With a known level of surge protection built into the equipment, it then becomes the responsibility of the user to take the necessary precautions to ensure that this level is not exceeded.

The test in its present form is a design test to be applied to protective systems. When these systems include communication links, particular attention should be given the interface between the protective relays and the communication link.

The report of this working group will first appear as an IEEE paper. After incorporating the comments and discussions generated by this paper, the report may then become a test guide for a surge test to be added to the Relay Standard ANSI C37.1.

REFERENCES

1. IEEE Power System Relaying Committee. Surge Phenomena Subcommittee, "Voltage Surges in Relay Control Circuits, Interim Report," *IEEE Conference Paper 31 PP 66-314*, presented at the 1966 Summer Power Meeting, New Orleans, July 11-15, 1966.

2. Seeley, H. T. and Emmerling, H. T., "The Why and How of Surge Control in Station Relaying," Paper presented at the American Power Conference, Chicago, April 21-23, 1970. (See Authors' Index, this Volume.)

3. Dietrich, R. E., Ramberg, H. C., and Barber, J. C., "BPA Experience with EMI Measurement and Shielding in EHV Substations," Paper presented at the American Power Conference, Chicago, April 21-23, 1970. (See Authors' Index, this Volume.)

4. Sutton, H. J., "Transient Pickup in 500-kV Control Circuits," Paper presented at the American Power Conference, Chicago, April 21-23, 1970. (See Authors' Index, this Volume.)

5. British Electrical and Allied Manufacturers' Association, "Recommended Transient Voltage Tests Applicable to Transistorized Relays," *BEAMA Pub. No. 219*, (1966).

ELECTROMAGNETIC INTERFERENCE AND SOLID STATE PROTECTIVE RELAYS

W. C. Kotheimer   L. L. Mankoff

General Electric Company
Philadelphia, Pennsylvania

ABSTRACT

This paper presents new information on the nature of the electromagnetic environment in an electric power station which can influence the performance of modern solid state protective relays and other sensitive electronic control and data processing equipments. Present industry standard tests do not simulate the full range of this environment and may need to be revised. The frequency spectrum of the electromagnetic interference is shown to be significantly different from the spectrum of the voltage and current signals used for relaying. Therefore, filtering techniques can be used as a primary remedy for E.M.I. problems. The nature of transients generated in low voltage control circuits is shown in high speed oscillograms. Data are presented which show that these transients can damage avalanche type semiconductors which are usually considered to be self protecting. Finally, portable radio transmitters are shown to be capable of causing misoperation of sensitive solid state equipments if operated nearby. Remedies for this type of interference are discussed.

INTRODUCTION

Early in the 1960's, activity in the IEEE was begun which led to the development of ANSI C37.90a, the Surge Withstand Capability Test Standard for protective relays.[1] This effort was motivated by the concern of protective relay engineers about the effect of the transient electromagnetic environment in power stations on new solid state relay equipments. Since that time over a decade of field experience has revealed much new information about this environment and the methods necessary to obtain secure and reliable operation of solid state relays. In retrospect, it now appears that although the SWC test of C37.90a adequately represented the then known types of interference, it does not properly simulate all types of interference now known to exist in power stations.

Field experience has since shown that transients generated within the low voltage dc control circuits have waveform characteristics quite different from the SWC test, being typically of very short duration and having very high instantaneous power levels. They have been referred to in recent discussions as Fast Transients

Another source of interference encountered more recently is not caused by any component of the power system itself. It is due to the operation of small portable or mobile VHF and UHF radio transmitters for communications purposes. Field experience

has shown that the high frequency electromagnetic fields produced near these transmitters can directly influence certain solid state circuits and possibly cause a false or erroneous operation of an electronic system such as a solid state protective relay. Some solid state components used in these circuits have been found to be sensitive to this interference over a wide frequency range extending above 400 megahertz.

The EMI Frequency Spectrum and Relay Input Signals

Transient phenomena in a power station which give rise to interference are mostly due to intentional switching or accidental voltage breakdowns which occur on either the high voltage bus structure or the low voltage control and auxiliary power circuits. In general, their waveforms are characterized by high frequency content. A comparison of EMI frequency spectra with that of the input signals for protective relays is shown in Figure 1. Also shown are the approximate frequency bands where portable radio equipment for industrial use are operated. Figure 1 indicates that the total EMI spectrum, transients and radio, extends from about 100 kilohertz to over 500 megahertz. In comparison the frequency bandwidth needed to accommodate the transient and steady state waveforms of protective relay input signals extends from a few hertz to less than 10 kilohertz. There is a significant difference in frequency content and this is important because it provides an excellent basis for discrimination between unwanted EMI and desired input signals by using filtering techniques. Low pass filters are used for this purpose and are called surge or EMI filters. A diagram of a typical surge filter is shown in Figure 2, and Figure 3 is its frequency response. This surge filter rejects frequencies above 100 kilohertz by a factor of at least 1000 times, 60 DB, and passes frequencies below 10 kilohertz with very little attenuation.

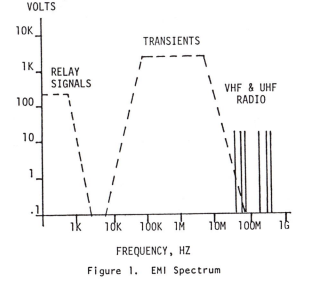

Figure 1. EMI Spectrum

F 77 057-3.   A paper recommended and approved by the IEEE Power System Relaying Committee of the IEEE Power Engineering Society for presentation at the IEEE PES Winter Meeting, New York, N.Y., January 30-February 4, 1977. Manuscript submitted September 1, 1976; made available for printing November 2, 1976.

Reprinted from *IEEE Trans. Power App. Syst.*, vol. PAS-96, no. 4, pp. 1311–1317, July/Aug. 1977.

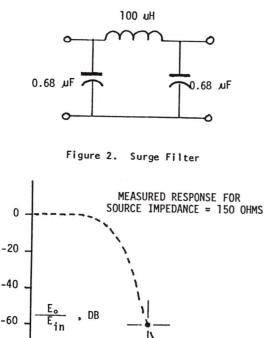

Figure 2. Surge Filter

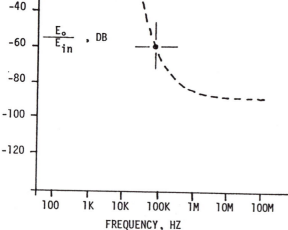

MEASURED RESPONSE FOR
SOURCE IMPEDANCE = 150 OHMS

$\frac{E_o}{E_{in}}$, DB

FREQUENCY, HZ

Figure 3. Surge Filter Response

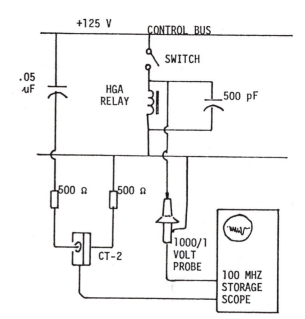

Figure 4. Laboratory setup to measure the
fast transient.

The design of a surge filter for protective relay current input circuits is complicated by the requirement to handle ct secondary currents with peaks as high as a few hundred amperes. This affects the design of the inductive elements in the filter so as to avoid saturation effects where possible.

EMI From Switching Inductive Devices In Control Circuits

It is well documented in the literature that very high voltages are generated when the current in an inductive device such as an auxiliary relay coil is interrupted suddenly.[2,3,4] A transient voltage oscillation appears across the coil. Its nature is determined mainly by the coil inductance L, the stray capacitance of the coil and wiring C, the level of current interrupted I, and the breakdown voltage/time characteristics of the switching contacts. In the limit, the maximum crest value of this voltage approaches

$$V_{max} = I \sqrt{\frac{L}{C}} \, , \qquad (1)$$

but is usually less than half this value due to losses in circuit resistance and the arc resistance of the

contacts. Even so, crest values exceeding 2.5 KV are common and values approaching 10 KV have been measured. This same voltage also appears across the stray capacitance of the coil and wiring. The oscillation frequency is not high, only about 1000 hertz. By itself the voltage oscillation appearing only across the coil is not a significant EMI problem. However, in combination with the restriking action of the switch contacts, it causes quite severe transients to be generated and conducted to the control bus. This happens during the time the contacts are sparking. The current flowing to the coil from the control bus is repeatedly extinguished and re-established by the contact interruption and subsequent restriking when the oscillation exceeds the breakdown voltage for the instantaneous contact spacing. When this occurs, the coil and wiring stray capacitance, now charged to a high potential, is suddenly and quickly discharged into the control bus as a very intense but short duration surge which then propagates along the bus.

As the contacts move farther apart, the breakdown voltage increases until the final and highest voltage discharge takes place followed by a voltage oscillation across the coil which then decays to zero. In some cases, where there is much damping, a subsequent oscillation will not occur and the voltage across the coil decays exponentially to zero.

Figure 4 shows a diagram of a laboratory experiment for studying fast transients. An HGA125 auxiliary relay has about 500 picofarads of capacitance, typical of switchboard wiring, connected in parallel with its coil to simulate the stray wiring capacitance. The coil current, 0.062 amperes, was interrupted by a simple toggle switch. In Figure 4, two methods for coupling the high voltage transient signals to the oscilloscope are shown. The voltage across the relay coil was sensed by a standard 1000/1 high impedance divider probe. This probe did not prove suitable for sensing the fast transients across the control bus. Errors resulted from transient currents in the ground return. A high frequency current transformer and an RC network were used to sense the fast transients. The isolation provided by the current transformer eliminated the offending ground currents.

344

Figure 5 shows the voltage across the coil which illustrates, at a slow sweep speed, the mechanism of voltage build-up and breakdown generating the fast transient. Figure 6 shows, at the same slow speed, the train of fast transient spikes appearing across the control bus. Figure 7 shows, at a very fast sweep speed, the detail waveform of a single fast transient. It is important to remember that there are very many of these spikes generated and impressed on the control bus, one for each restrike of the contacts as shown in Figure 6. The spike shown in Figure 7, 2600 volts in 40 nanoseconds, is not necessarily the highest value. It was the first spike to exceed the scope trigger level.

The conducted EMI traveling along the control bus can be damaging to semiconductor devices connected to the bus. This is due to the extremely high instantaneous power that can be delivered. A

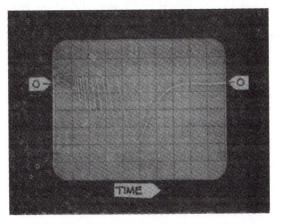

Figure 5. Voltage across the coil of an HGA-125 relay. Vertical calibration is 1000 volts per large division. Horizontal is 200 microseconds per large division.

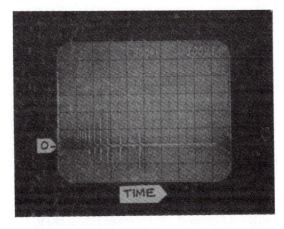

Figure 6. Train of fast transients impressed on the control bus during switching of an HGA-125 relay coil. Vertical calibration is 500 volts per large division. Horizontal is 100 microseconds per large division.

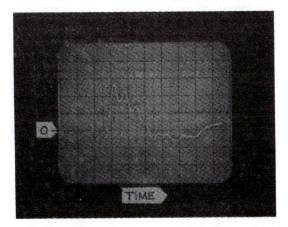

Figure 7. Detail of a fast transient. Vertical calibration is 500 volts per small division. Horizontal is 50 nanoseconds per small division.

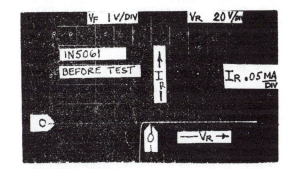

Figure 8. Avalanche diode reverse current characteristics before subjected to fast transient test.

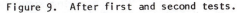
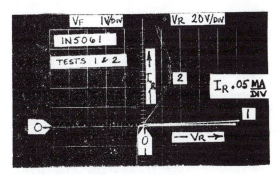

Figure 9. After first and second tests.

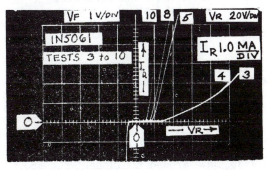

Figure 10. After tests 3 through 10.

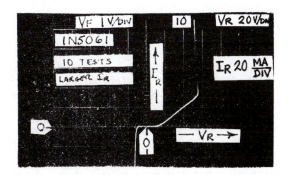

Figure 11. After test 10, larger current calibration.

5 KV fast transient on a bus with a surge impedance of 200 ohms can deliver about 20 kilowatts instantaneous power into a device such as a zener diode or an avalanche diode that supports a 1000 volt drop in the avalanche mode. This is far in excess of the short duration reverse surge power rating of most avalanche protected diodes and rectifiers. Laboratory tests have shown that avalanche devices are gradually degraded by repeated exposure to the fast transient. The damage is characterized by an increase in the reverse leakage current of the diode after each exposure. Then when the rated voltage is applied to the diode the increased reverse power dissipation causes a temperature rise which further increases the reverse leakage current. This can lead to a thermal runaway condition and catastrophic failure. The photographs displayed in Figures 8, 9, 10, and 11, show the progressive failure of a type 1N5061 avalanche protected diode when subjected to the fast transient test. The short time reverse power rating of this diode is given as 1000 watts by the manufacturer.

One author [5] has written that this type of electromagnetic interference, the fast transient phenomena, also produces a radiated field with frequency components up to 1000 megahertz. This could affect sensitive circuits nearby.

Protection Against the Fast Transient

Ideally, it is best to protect nearby equipments against the fast transient by preventing generation of the offending noise. In many cases, this can be accomplished by connecting a metal oxide varistor or other suppression means across the offending coil. This prevents generation of sufficient voltage to cause the contacts to restrike and initiate a fast transient. The varistor must be able to dissipate the energy stored in the coil safely. Other suppression means include paralleling the coil with a resistor with a reversed diode in series. In this case, the diode must be protected from surges or fast transients from other sources. When applying suppression at the source, the designer must consider the possible increase in drop out time of the suppressed relay. Even though a particular fast transient source is suppressed, there are others nearby and it may not be possible to suppress them all.

In situations where it is not practical to prevent generation of a fast transient, the affected devices - semiconductor diodes, rectifiers, etc., can be protected against damage by connecting small ceramic disc capacitors of about .01 to .05 microfarad across the component. It is important to keep the connecting leads as short as possible, preferably less than one inch. The protecting capacitor allows the fast transient current to flow around the diode. The short-time duration of the transient produces only a few volts across the diode as a result of the integrated charge on the capacitor.

VHF & UHF Radio Transmitters as Sources of EMI

Incidents have been reported recently in which protective relays in an electric power installation have been activated by operation of nearby portable rapid transmitters.[6] Others have reported incidents with solid-state control equipments associated with materials handling and continuous production processes. There have been reports of some automotive solid-state fuel control systems caused to malfunction by radio transmissions nearby.[7] The sensitivity of electronics heart pacers to microwave fields is well recognized.[8] Increasing use of radio communication and of solid-state devices for protection and control, suggests that the electric power industry, utilities, and manufacturers, need to study the phenomena to provide the best balance of hardware designs and operating procedures to minimize interference problems.

The results of preliminary tests with "walkie-talkie" units by the authors indicated that sensitive circuits can be affected over a wide band of frequencies ranging from 25 megahertz to over 500 megahertz. Any particular circuit can also exhibit a degree of selectivity over this range and be sensitive at only a number of specific frequencies.

Table I shows the approximate frequency bands allocated for public utility use.

TABLE I

Approximate Frequencies Below 550 Mhz. for Electric Utility Use in the United States (1975)

Frequency Band Mhz.	Notes
27	Citizen Band Class D
37	
47 - 48	
158 - 173	
216 - 220	
220 - 225	Citizen Band Class E
450 - 470	Citizen Band Class A
470 - 512	Land Mobile, Urban Areas

Preliminary Study - Sensitivity Factors

A preliminary study was initiated in an effort to obtain basic information for use as design guides for protective relays and for operating procedures for portable transmitters. The study included both analytical and experimental approaches. It became apparent that the potential for misoperation of a protective or control device when subjected to radiated electromagnetic interference is a function of:

1. Field intensity and frequency of radiation

2. Sensitivity of the affected circuitry to radiation.

3. Coupling efficiency resulting from device construction, wire lead configuration, etc.

An analytical approach to developing design guides and operating procedures proved to be very complex and difficult to apply except for overly simple physical arrangements. Experimental approaches, therefore, were used to evaluate sensitivities and to develop techniques to minimize the radio frequency sensitivity of protective relay designs. These experiments evaluated the following approaches to reduce sensitivity:

1. Shielding and filtering

2. Circuit design and component selections

3. Printed circuit card and relay wiring configuration.

A combination of these approaches has been adopted for the design of new protective relays to achieve practical, low levels of sensitivity to radiated electromagnetic interference.

Shielding and Filtering

A totally closed metallic shield will prevent even very intense high frequency fields from penetrating to its interior. However, such an enclosure is not generally a practical housing for electronic circuitry for protective relays since they require circuit connections to the external world. In addition, the desirability of drawout construction and relay test facilities as a part of the package, make total shielding complex. External connections provide a path to conduct induced high frequency interference into the interior. Therefore, filters are required on all these connections to protect the circuitry inside. For study, this combined approach, total shielding with filtering, was evaluated experimentally using a very sensitive high gain broad band amplifier/level detector as the sensitive circuit.

The amplifier was first tested on the bench outside the shield, unprotected. It was found to be sensitive to a small "walkie-talkie" transmitter located at the other end of the laboratory as indicated in Table II.

The shielded enclosure was made of 50 mil thickness copper sheet. All seams were soldered and the cover was screwed in place and had close fitting 1/2 inch overlapping edges. The input/output connections, for power supply and measurement, were provided by coaxial feed-through type EMI filters soldered in 1/4 inch diameter holes in the rear wall. With the circuit located inside this shield it was not possible to affect its operation even with the transmitter only a few inches from the enclosure.

The enclosure used for these tests was not a practical design for protective relays since it had no openings in the cover to permit viewing of targets, contacts, or tap settings, etc. Tests were made subsequently using standard relay cases mounted in steel panels to simulate actual installations. The glass window in the front cover was recognized as an entrance for interference so that experiments were made with closed metal covers, covers with transparent conducting layers over the glass, and metal covers with various size openings. None of these approaches was found to be satisfactory and efforts were directed towards other ways to reduce sensitivity.

Circuit Design and Component Selection

It was found by experiment that certain components, especially linear devices with high gain and broad frequency response, can be sensitive to high frequency fields.

The sensitivity can be reduced by designing UHF by-pass filters into the circuit and locating these filters very close to the sensitive device. In general, low signal levels contribute to sensitivity so that circuits should be designed to operate at as high a level as practical for the particular application. Limited tests indicate less tendency for misoperation with digital circuits as compared with the high gain linear circuits studied.

Printed Circuit Card and Relay Wiring Configurations

All circuit conductors exposed to high frequency radiated fields behave as simple receiving antennae for sensitive circuits. Our experiments showed that physical arrangement of the printed circuit conductors is important, especially the size and location of the "ground" conductor near sensitive devices. Long conductor runs connected to sensitive inputs are to be avoided.

Wiring external to the circuit boards is more exposed and harder to control since it may be routed in three dimensions. Here shielding of the conductors in conductor bundles helps. Routing, even for shielded wires, should be to avoid exposure to fields.

CONCLUSION

Much has been learned about the electromagnetic environment in a power system and its effect on solid state electronic control and protection equipment in the past decade. The complete spectrum of this interference is now becoming known. Experiments have yielded new information on extremely fast transient phenomena in low voltage control circuits which explain certain types of semiconductor failures and indicate the effective corrective measures.

The growing use of portable VHF and UHF radio equipment by power system personnel has added a new dimension to the electromagnetic environment which electronic designers and system protection engineers must now consider.

TABLE II

Tests of Shielded Enclosure

Test Conditions	Transmitter Distance Which Causes Malfunction (450 MHZ 5 watt Walkie-talkie)
1. Amplifier only	0 - 30 feet
2. Amplifier inside shielded enclosure with filtered connections	0 - No Malfunction
3. Amplifier inside shielded enclosure but no filtered connections	0 - 10 feet
4. Amplifier with filtered connections - outside shielded enclosure	0 - 6 feet

REFERENCES

[1] American National Standard C37.90a, 1974, Guide for Surge Withstand Capability (SWC) Tests.

[2] Taylor, J.R. and Randall, C.E., Voltage Surges Causes by Contactor Coils. Proceedings of the I.E.E. (London), 1942.

[3] Wilcken, J.A., Transient Phenomena at the Breaking of an Inductive Circuit. Philosophical Magazine, S7, Vol. 13, No.87, May 1932, pp. 1001-1006.

[4] Seeley, H.T., Protection of Control Circuit Rectifiers Against Surges from DC Coil Interruption. AIEE Trans. Vol.80, Pt.111, 1961, pp.871-879.

[5] Mowatt, A.Q., RFI Generation is a Factor When Selecting A-C Switching Relays. Electronics, July 8, 1976, pp. 101-105.

[6] General Electric Company Service Advice Letter No. 150.1, concerning the STD Transformer Differential Relay, November, 1975.

[7] Middleton, AFISI - Automatic Fuel Injection System Interference. Technical Correspondence, QST, June, 1972, p.38.

[8] "Say Transmitter May Halt Pacemaker", (News item), Electronic News, Monday, June 3, 1974, p.27.

Discussion

R. E. Dietrich (Bonneville Power Administration): This paper is a welcome addition to the list of references on EMI. It undoubtably will provide the necessary groundwork for future revisions of ANSI C 37.90a-1974.

It is of particular interest because it identifies problems and presents some test data on the effect of radio transmission on solid state devices. Table II indicated that perhaps satisfactory performance could be obtained without shielded enclosures but with adequate filtering in all input and output connections.

This tends to confirm experience at BPA where equipment misoperation due to a "walkie talkie" was corrected with improved equipment grounding and input filters. Shielding was not necessary.

Would the authors, other equipment manufacturers, or users care to expand on the need for shielded enclosures versus careful equipment design and filtering?

The authors have described the nature of the fast transient and methods of measurement of the wave form. There is, however, a need for a standardized wave form and a test circuit or testing device to apply the test. Have the authors considered a test circuit or a testing device? Further would they care to recommend a fast transient wave shape (crest voltage, repetition rate, etc)?

Manuscript received January 10, 1977.

H. H. Sheppard (Relay Associates, Inc., Philadelphia, PA): The paper presents a most comprehensive review and analysis of many types of electromagnetic interference. The authors report "it is well documented in literature that very high voltages are generated when the current in an inductive device such as an auxiliary relay coil is interrupted suddenly. A transient voltage oscillation appears across the coil" ... and ... "crest values exceeding 2.5 KV are common and values approaching 10 KV have been measured." "By itself the voltage oscillation appearing only across the coil is not a significant EMI problem. However, in combination with the restriking action of the switch contacts, it causes quite severe transients to be generated and conducted to the control bus." The paper then reviews the various problems these severe transients can create. One could assume from the paper that these problems are inherent in all electromagnetic auxiliary relays used in power system relaying.

Manuscript received January 31, 1977.

I believe it should be noted that there are means of greatly reducing such transients by using techniques as reported in a paper entitled "THE HIGH-SPEED AUXILIARY STORY" presented by Mr. C. R. Pope, Senior Engineer, Philadelphia Electric Company, at the Pennsylvania Electric Association, Relay Committee, at State College, Pennsylvania on October 2, 1975.

Mr. Pope's paper describes a high-speed electromagnetic dc auxiliary relay having a high resistance coil of many turns being placed in series with the operating coil as the armature approaches its picked-up position. This results in a very low steady-state energizing current, which combined with the low inductance of the coils, results in a small quantity of energy ($\frac{1}{2} LI^2$ joules) being stored in the relay. Consequently there is less energy to be dissipated when the coil current is interrupted.

In addition, heavy copper plates are molded around the ends of the coils but insulated from them, as shown in the photograph of a cross-section of the coil assembly. When the coil current is interrupted, these plates act as heavy short-circuited turns on a transformer and reduce the transient peak voltage generated by the collapsing field to a maximum value of only about 600 volts. As a result, the relay is essentially clean as a source of EMI and when the relay is operated in close electrical proximity to a "stiff" power source, no coil suppression may be required.

Mr. Pope's paper also reports that the minimum capacitive energy required for a momentary closure of this relay's contacts is 0.38024 joules (45 μF at 130 V dc) whereas a comparative value for a sensitive low inertia relay is 0.0025 joules (0.3 μF at 130 V dc). The ratio of 45 μF to 0.3 μF indicates a 150 to 1 improvement in the ability of the relay to ride through the severe transients so well described in Mssrs. Kotheimer and Mankoff's paper.

J. L. Koepfinger (Duquesne Light Company, Pittsburgh, PA): The authors have addressed two separate, but not new types of interference which can affect solid state relays and controls. It was well known to the developers of the SWC test that this test did not represent all types of interferences to which a relay could be subjected. Studies done as early as 1932 indicated that the interruption of relay coil currents could impose severe transients on the d.c. control buses.

In developing the SEC test consideration was given to using a test similar to the fast transient test presented in this paper. The SWC test as it is now presented in C 37-90a was a result of numerous compromises. At the time of its development much of the test work indicated that transients created in the relay circuits by switching the power facilities were of more importance than those transients generated within the control circuits. It was recognized that perhaps the SWC test did not fully encompass the field conditions, but was a simplified screening type test.

The Duquesne Light Company has found it desirable to use for many years both the SWC test and a chattering auxiliary relay test. It has been observed that LED isolators can be susceptible to false signal output when exposed to the chattering auxiliary relay test, yet they will successfully withstand a SWC test.

Similarly it has been well known that r.f. signals could cause misoperation of electronic equipment.

It may be desirable to expand the Surge Withstand Test to incorporate more encompassing tests. Sweden's relay standards include a radiation test for solid state relays as well as a high voltage transient surge test.

Manuscript received February 18, 1977.

W. C. Kotheimer and **L. L. Mankoff**: The authors want to thank Mr. Sheppard for calling attention to the fact that not all electromagnetic auxiliary relays are necessarily sources of severe transients. We agree that if the relay incorporates in its design features such as extra short circuited windings and copper end plate or slugs, the energy stored in the magnetic field can be dissipated in resistance losses when the circuit is opened thus considerably reducing the coil terminal voltage. Almost always, however, such techniques result in longer drop out times which the user will need to consider.

In his final comments Mr. Sheppard seems to imply that extremely fast transients impressed on the control bus by contact restriking can cause false operation of electromagnetic relays. We do not believe this is likely to happen. Because of their extremely short duration, 100 nanoseconds or less, these transients are effectively bypassed by the distributed capacitance of the coil winding and thus produce practically no transient force on the relay armature. In addition, the inertia of the

Manuscript received February 18, 1977.

armature precludes its responding to such short duration and small forces. It is conceivable that fast transients so generated could cause insulation breakdowns in the first few end turns of other relay coils connected to the control bus. We have not been able to detect such damage in our laboratory tests, however.

The authors appreciate Mr. Koepfinger's constructive comments. We agree that the existence of transients impressed on d-c buses by relay coils has been known for many years. It is only lately, however, that their extremely fast nature and the hazard this represents for solid state devices has become recognized.

Mr. Koepfinger's report of false operation of LED type opto-isolators due to transients generated by relay coils is very timely because these devices are being offered specifically for applications requiring signal transfer between electrically isolated circuits especially where transient problems exist. It appears the fast transients high dV/dt may be causing displacement currents to cross the optical barrier and affect the phototransistor directly. Perhaps a dV/dt rating is needed for these devices.

We agree with Mr. Dietrich that a standardized waveform and test generator circuit is needed to apply the fast transient test. Recently, we have made some progress in this area and hope to be able to suggest a suitable but simple circuit soon. For the waveform we would suggest the following:

Max Voltage	4000 volts ± 20 percent
Rise Time	20 nanoseconds or less, from 10 percent to 90 percent of maximum voltage level
Fall Time	100 nanoseconds or less to half value
Source Impedance	100 ohms or less, resistive
Repetition Rate	50 to 100 per second
Burst duration	2 seconds

On the question of shielding versus other measures to control radio frequency effects we note that both shielding and filtering were needed to give the best results in our tests. Shielding alone did not suffice. The degree to which measures other than shielding can achieve good results depends on the radio frequency field strength and the sensitivity of the device in question. Here again the need for a reasonable standard test is indicated.

DESIGN CRITERIA FOR AN INTEGRATED MICROPROCESSOR-BASED
SUBSTATION PROTECTION AND CONTROL SYSTEM

J. S. Deliyannides, Member, IEEE   E. A. Udren, Member, IEEE

Westinghouse Electric Corporation

Pittsburgh, PA

Abstract - Design requirements and approaches are described for the EPRI-sponsored development of an integrated, microprocessor-based system for relaying and control of transmission substations. The paper summarizes the architecture, hardware and software design, functional capabilities, and economics. Many important user benefits inherent in this dramatically different approach are listed.

INTRODUCTION

Protection and control systems for electric power substations are similar to control systems used in most other industries. Traditionally, in both cases, the bulk of process information from the plant travels from transducers through many wires to a central control area. Panels there are laden with instruments, manual controls, and automatic control packages which have been individually selected, installed and wired. In recent years, however, manufacturers have combined evolving electronic technologies to create equipment which has gradually altered this control approach. Information is converted to digital form and multiplexed onto a single trunk or highway leading to the control area, where computers record data, control process operations, and provide data records for operators. The impetus for this changeover is the potential for more sophisticated capabilities at lower installed cost.

Critical differences in protection and control tasks for electric power substations have mitigated the enthusiasm of utility engineers for direct adoption of such equipment. The speed required for the most important substation functions can't be attained with these new process control systems, and the consequence of unreliability in critical systems such as protective relaying are too catastrophic to encourage indiscriminate use of general-purpose devices.

With the goal of using the new technology while addressing specific utility concerns and problems, EPRI has sponsored a program for the development of an integrated protection and control system for transmission-level substations [1], [2], [3], [4]. Several divisions within Westinghouse Electric Corporation have combined personnel and capabilities to produce a design incorporating the newest in microprocessors, large-scale integrated circuit support, digital communications, multiplexed hierarchical architecture, and application software for relaying and other important control tasks.

This paper portrays the requirements, goals, and design approach for the substation protection and control system development. Emphasis is placed on user requirements and criteria for the usefulness of such a system, and how the design team has met the challenge of fulfilling these requirements in a specific design described below.

REQUIREMENTS FOR SUBSTATION PROTECTION AND CONTROL

Certain distinctive features of transmission substations demand primary attention in the design of a system of control equipment for use there. The starting point, then, is to list those features and to consider their implications.

Need for Sophisticated Relaying and Control

Perhaps the foremost demands made by utility engineers on the protection and control scheme are those of security and dependability for a wide range of power-system effects. Rapid discrimination of events by subtle differences in measured signals is required for many of the functions. Accordingly, sophisticated relaying and control devices are presently installed at the substation to maintain overall reliability of the power system. Communications facilities to other substations and to the utility control center are needed in conjunction with these devices.

Need for Centralized Control Functions

To achieve the discrimination which is essential for effective protection and control, the variety of sophisticated sensing functions must respond in a coordinated way. This implies that all information from the switchyard must be collected at a central location, the control building.

Large Distances

The sheer physical size of a typical air-insulated EHV substation is perhaps its most striking feature. The consequence of concern here is that the utility must run many wires hundreds or even thousands of feet between switchyard apparatus and the control building for signal transmission and for power equipment control.

Withstanding the Electromagnetic Environment

The power apparatus produces an environment of severe transient electromagnetic interference (EMI) which threatens any low level signal electronic system one might consider installing in the switchyard. Even if all such electronics are installed in the control building, the transient fields induce large voltage surges in the long wires which are conducted back to the control building and to the terminals of the electronic devices.

Need for Event Recording

The operation of control or protection elements has a direct effect on the security of the transmission network. Therefore, users justifiably insist on adding equipment to monitor the behavior and operating margins of these critical devices. The recording gear must be automatic to capture response to important transient events such as faults, power swings, and switching.

81 TD 621-2   A paper recommended and approved by the IEEE Substations Committee of the IEEE Power Engineering Society for presentation at the IEEE PES 1981 Transmission and Distribution Conference and Exposition, Minneapolis, Minnesota, September 20-25, 1981. Manuscript submitted April 1, 1981; made available for printing June 30, 1981.

Reprinted from *IEEE Trans. Power App. Syst.*, vol. PAS-101, no. 6, pp. 1664-1673, June 1982.

Time relationships of some events must be preserved with a resolution of milliseconds. Transient information thus recorded is used for post mortem analysis which leads to further adaptations of operating practice and control strategies.

Need for Unattended Operation

The remoteness of some substations, the cost of having operators on site, and the need for the speed of automatic control actions have led to the design and construction of substations which are not routinely attended. This implies, of course, that all ordinary control activities must be performed either automatically, or at the command of a remote system operator acting through the Supervisory Control and Data Acquisition (SCADA) system.

The difficulty and expense of sending a service person to the substation site imposes additional requirements for minimizing maintenance of relays and controls. It would be desirable to have some sort of self-diagnosis or external monitoring so that system maintenance personnel are alerted when something does go amiss. Finally, the equipment should be able to carry on at some level, even though crippled, until help arrives.

Need for System-Wide Control

Many modern power systems operate far closer to stability limits than was the case years ago. Consequently, comprehensive centralized monitoring and control of the system is an essential practice of modern utilities. Data from around the system is collected for detailed analysis and modeling; the results lead to reconfiguration or readjustment commands which improve the security of the system.

At the substation the system-wide control function must be supported by interface equipment which supplies digested data to the control center, and which receives and implements control commands from the master station.

Need for Periodic Modification

Rarely is a major transmission substation designed and built from the outset in its ultimate configuration. More typically, it begins as a node for only a few lines, and evolves in steps over a period of years into a major junction with many lines and transformers linking different levels of the grid.

Each step in the expansion requires additional control equipment. However, existing buses may be rearranged so that a major reconfiguration of the existing controls and relays is needed. These devices should be flexible enough to facilitate such changes with a minimum of effort and expense.

SHORTCOMINGS OF EXISTING DESIGNS

The previous sections considered the features of the substation environment which have led to the particular protection and control schemes users now install there. These represent a compromise between the demands of the application and the capabilities of the available electromechanical and electronic technology. To clarify understanding of how newer technological advances might help, this section considers the deficiencies and problems of the approaches presently used. The following lists the salient shortcomings:

Intensive Wiring

Separate terminals are provided for each input to each protection and control device or system. Each input and output is individually connected during installation or modification. Since the individual functional packages are not generally able to share inputs or outputs, field signals and control circuits must be connected to each device using them. For example, a single current may be connected to an ammeter, an oscillograph, one or more transducers for the SCADA system, and a number of relays or relaying systems. The number of connection points plus the large distances in the switchyard reflect a substantial installation cost.

What is needed to help this situation is an integration of functions so that the equipment requires a minimum of interconnections. Also, the myriad signal wires should be replaced by some multiplexed means of conveyance so that only a few long signal carriers need be pulled. And finally, it would be especially beneficial if these signal carriers could resist the surges which presently contaminate wired control circuits during switching operations.

Expense of Event and Waveform Recording

A previous section established the reasons why users want detailed records of equipment response to transient events. Monitoring conventional protection and control systems requires the purchase and installation of additional equipment which is expensive, high in maintenance, and is itself highly prone to failure. Furthermore, it can only be connected to a limited number of the possible points, making equipment monitoring a matter of guessing where problems will occur. And finally, oscillographs and event recorders rarely reveal why something didn't work as it should.

Users would fare better with relays and controls with built-in, comprehensive monitoring facilities. They would want a system which logs all events of possible interest, with related analog values and accurate time tags appended. Logs of lesser events which don't produce a control response would show whether security margins are adequate.

Failures and Periodic Maintenance

Conventional relays and controls may fail without giving any indication. Accordingly, most utilities have undertaken periodic maintenance programs in which the equipment is tested and calibrated at intervals of six months to two years. Such programs are costly, and may still fail to catch many problems until long after the failure. In addition, some users have found that such periodic human tampering may cause problems and failures of its own.

To overcome these failure-related difficulties, the equipment must have some means of monitoring its own ability to function, automatically and frequently. This self-checking scheme must not disrupt normal sensing long enough to delay critical responses; the design must minimize additional, failure-prone hardware for switching to and from the testing mode. It must be complete enough to eliminate the need for all but the most infrequent manual checking; calibration as well as gross functioning must be assured. When some function does fail, a remote alarm must be raised at once so that a repair person can be dispatched. And finally, the equipment must continue working in a reduced-capability or backup mode whenever possible in case a power-system problem arises before repairs are made.

Information Display

The only readily-available local displays of transient information on control equipment operation in present-day substations are target or indicator lights and oscillograms showing connected points. All other data are lost, or are stored on the magnetic tape of a transient event recorder which requires special playback equipment. The information is delivered as several uncorrelated records, making the task of reconstructing incorrect or suspicious operations difficult.

The task of monitoring equipment performance and understanding response to events would be more successful and less time-consuming if the equipment itself could generate complete operation records, and then filter and orgainize them for quick presentation to an operator or engineer, either at the substation or at a remote location. All important events, decisions, and responses should be placed in a single, chronological, English language display with accurate resolution of timing. Data not directly related to the operation of interest should be kept in the background.

Refinement of Characteristics

Most electromechanical or solid-state analog relays and controls have operating characteristics which are, at best, an approximation of what the designer really would like these devices to do. Improved equipment would treat inputs more analytically, apply more refined characteristics and decision criteria, seek out new relationships in the data, and apply more sophisticated logic to handle exceptional situations in specialized ways.

NEW TECHNOLOGICAL TOOLS

Having developed a list of useful attributes which are not fufilled by existing protection and control equipment, it is profitable to investigate the latest in component hardware technology to see if these new tools can offer the desired capabilities. The following sections summarize recent technological advances which could contribute toward an optimal design of the substation protection and control system.

Fiber Optics Communication

The technology now exists for producing low loss optical waveguides in the form of fine glass or plastic fibers. These fibers are encased in durable cladding so that they can be buried or pulled where conventional wiring would be run. The waveguides are capable of carrying wide band signals over long distances with minimal attenuation. Optical fibers are virtually immune to electromagnetic interference surges and transients, or ground potential differences between the terminations.

Microprocessors

Advances in large scale integration (LSI) technology have now made it possible for integrated circuit manufacturers to make complex logic circuits with the equivalent of more than 50,000 active devices on a single chip, with prospects for improving density even more in the future. Thus complete 16-bit processors are now available on a single chip and 32-bit processors are in final stages of design. As the scale of integration increases the cost per unit function decreases and the reliability per unit function also improves. The availability of such microprocessors as chips or assembled on complete processor boards makes it economically feasible to configure powerful multi-processor distributed systems. Widely available 16-bit processors provide sufficient accuracy for fixed point arithmetic operations needed in the protection algorithms. New microprocessors are becoming available which incorporate floating point arithmetic capabilities within the processor chip.

LSI Support

Along with microprocessors has come a host of LSI devices which can be used in compact digital products. Such devices include communication interface processors,

input-output interfaces, and a large variety of memory devices such as PROM (Programmable Read-Only Memory) for storage of programs, RAM (Random Access Memory) for fast handling of changing data and bubble memories for long-term bulk read-and-write storage.

Display Devices

Although these are not very recent developments, they represent a technology that is different from that used in conventional substation protection and control. The display devices include color CRT's with keyboard control as a means of operator interface. Also low cost serial interface printers could be included in this category.

The new technologies described above provide economical solutions to many of the shortcomings described in the previous section. For example, fiber optics facilitate high speed multiplexing of signals in the presence of a hostile EMI environment. Microprocessors offer the computational power needed for high-speed execution of sophisticated protection algorithms for a large number of protection terminals within the substation. LSI support processors provide economical communications interfaces and large memory capacity without moving parts. Finally, the color CRT display technology provides a means of displaying digested and organized information about many devices and events at a single operator station.

SYSTEM ARCHITECTURE

The application requirements and the problems associated with present practice, as described in the previous sections, affect the system architecture in several ways. Specifically, the criteria which affect architectural structure are volume of data flow, and system availability which depends on such factors as equipment reliability, maintainability and redundancy. These are examined in the following sections.

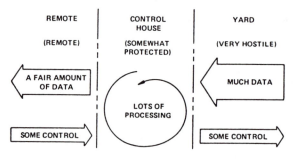

Figure 1. Data Flow in a Substation

Data Flow

Figure 1 represents one way of viewing a substation on the basis of origin and flow of data and control signals. A large amount of data originates in the switchyard where current transformers and capacitor potential devices sense instantaneous values of power system voltages and currents. These data must be digitized and transmitted from the hostile environment of the switchyard to the somewhat protected environment of the control house where processing takes place to determine the presence of fault conditions and take appropriate control action. The amount of data transferred into the control house is large whereas the number of control signals sent back to the switchyard is relatively low. There is another path of communication of data to and from the control house. It leads to one or more remotely located masters from which power system operators monitor and control the operation of substation equipment or where protection engineers monitor fault conditions. The volume of data flow to such locations is much lower than that from the switchyard to the control house. The number of control signals received is also low.

Such a distribution of data flow suggests a hierarchical arrangement as shown in Figure 2. At the top of the hierarchy is a central station computer whose functions include supervision, coordination, analysis and reporting on the basis of short term and archival data. Considerable storage is required at this level. Fortunately the volume of data flow at this level is low so that the overall computational power required is practical even for a large station. The next level down in the hierarchy consists of distributed clusters of microprocessors where the protection functions are carried out by programs operating on immediate data. The lowest level in the hierarchy consists of a number of data acquisition units, distributed throughout the switchyard, and performing the functions of synchronized sampling of power system signals and status indicators, and actuating control circuits of the power apparatus. Between each level of the hierarchy are interfaces whose bandwidth must correspond to the data volume requirements described above.

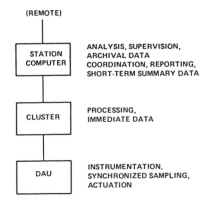

Figure 2. Hierarchy of Functions in a Substation

System Availability

System availability and security are the most important system requirements to the protection engineer. Availability is affected by hardware reliability, by the ability of the user to detect and repair malfunctioning components and by the use of redundant equipment. Security is also affected by redundancy both of hardware and of logic programmed into the computer algorithms.

The authors have compared reliability of conventional relays with that of digital systems. Although these studies have been handicapped by limited availability of well documented field data, the results indicate that the new equipment must be designed and built with extreme care to equal the hardware reliability of existing relays. Even though considerable improvement in reliability is anticipated as the scale of circuit integration increases, the authors believe that the real improvement in system availability will come about from other factors such as automatic error detection through self-checking, better equipment maintainability and subsystem redundancy.

The ability of digital relays to continuously monitor their own operation and detect any degradation in performance or outright malfunction is widely recognized. Unlike conventional relays, where a malfunction may go unnoticed until some disturbance causes either a false trip or a failure to trip, digital relays can give an immediate indication of their malfunction. During the interval between failure detection and repair trip, outputs are blocked, and tripping can only be activated when a redundant backup system is in service.

Repair time in digital relays is minimized by means of self diagnostic procedures. Maintainability is greatly affected by configuration: the higher the degree of independance among various subsystems, especially between primary and backup subsystems, the greater the maintainability. Conversely, as subsystems are interconnected to increase cost effectiveness through sharing of components, maintainability is decreased.

Redundancy

The need for redundancy and its effect on system availability have already been discussed. Different levels of redundancy are possible. Completely redundant backup is the approach most like that presently used. The backup system may be either an identical system or one of different design, (e.g. digital primary and conventional relay as backup). Extremely high availability is obtained with this approach since any relaying zone is unprotected only when both primary and backup have failed at the same time and a fault occurs before either is repaired. A somewhat less redundant but more cost effective scheme is failover backup. In this scheme data needed by each protection subsystem are also brought into a single backup subsystem which is normally in idle state. When a hardware fault in one of the protection subsystems is detected, the functions of the faulty subsystem are taken over by the backup subsystem. This has the disadvantage that the failure of a second protection subsystem will not be backed up until the first subsystem is repaired. Another aspect of redundancy is the use of multiple transducers. In present protection schemes each of the redundant protection systems is connected to its own signal sources. For example, as many as four sets of current transformers at the same location supply redundant signals to primary and backup relays for multiple zones, as shown in Figure 3a. Figure 3b shows how transducer redundancy can be minimized if protection subsystems can share input data in a secure way. Note that there are still two sources of data for each protection zone.

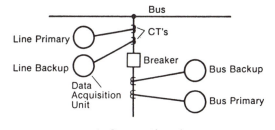

a.) Conventional

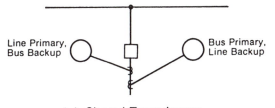

b.) Shared Transducers

Figure 3. Transducer Redundancy

Configuration Ground Rules

Consideration of hardware cost only leads to an architecture with minimum hardware. However, other factors affect cost indirectly. For example, modularity, ease of reconfiguration and software features can have a far greater impact on overall system cost than hardware cost. It is therefore necessary to perform trade-off analysis which minimizes the overall cost of the system.

Such an analysis was made for a variety of configurations taking into account the criteria discussed above. Some general configuration ground rules and a preferred configuration resulted from this study. The general ground rules are:

- Three levels exist in the hierarchy.
- Fast communication channels needed at lower level, slower ones will do at the top.
- Need for infrequent, but high priority communication between protection subsystems.
- Protection functions should be able to operate stand-alone.
- Function processors configured in clusters.
- Data acquisition units located in the switchyard.
- Flexibility should be provided in location and number of data acquisition units used.
- Flexibility should be provided in options for methods of backup.

The configuration found by the project investigators to offer the best mix of these desirable traits is shown in Figure 4 and is described in the next section.

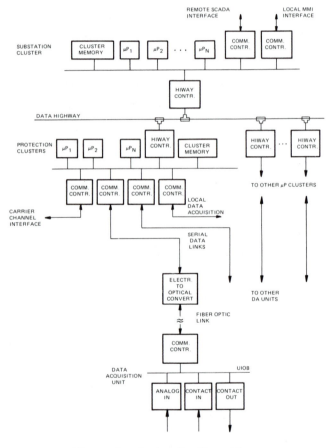

Figure 4. System Configuration

HARDWARE DESCRIPTION

The configuration shown in Figure 4 consists of a hierarchy with three levels. At the lowest level is one of many data acquisition units, located in the switchyard, which samples instantaneous values of currents and voltages and status of power apparatus contacts at the rate of 16 times per power system cycle. It also has provision for digital outputs for control of breakers and other devices. A 1 MHz serial link connects the data acquisition units to the next level in the hierarchy, the protection microprocessor clusters. Protection logic and computational algorithms are executed at this level using simultaneous current and

voltage samples received from the data acquisition units. When a fault is located in the zone of a particular relay program, a trip signal is sent to the breaker through the serial link and the data acquisition unit. The third level of the hierarchy, shown at the top, is the substation cluster of microprocessors. At this level reside the system data base, the man-machine interface system and interfaces to remote SCADA masters. Each protection cluster is connected to the substation cluster and to other protection clusters through the serial data highway. The protection clusters act as data concentrators, which convert the large numbers of instantaneous samples into stable RMS values for use by the substation processor cluster.

In the following paragraphs, the equipment at each of the three levels will be described in more detail and then an explanation of the overall operation of the system will be given.

Data Acquisition Units (DAU)

The data acquisition unit controller receives and transmits data over the serial data link and controls the flow of data on the parallel Universal I/O Bus (UIOB). When a synchronization message is received from the protection cluster the DAU controller initiates the process of sampling and holding analog inputs, gathering status inputs, converting analog inputs into digital form, and sending all the information in sequential message form up to the corresponding protection cluster via the serial data link. Three I/O cards are shown attached to the UIOB. The high speed analog input card has sample and hold circuitry, anti-aliasing filters, an A/D converter with wide dynamic range (14-bit) and a relatively fast conversion time (60 µs per point). The contact input card performs isolation, level shifting and digital contact debounce logic functions. The contact output card provides output relay driving with a high degree of security which incorporates a select, checkback of latched information and an execute signal.

Serial Data Links

The serial data link interface provides the communications between the data acquisition units and the protection clusters. The transmission medium is either coaxial cable (when the DAU's are located in the control house) or fiber optic cable (when the DAUs are located in the switchyard and noise immunity is of concern). Modems at either end of the data link modulate the signal using Frequency Shift Keying (FSK) with 3 MHz carrier. Data bandwidth is 1 MHz. The message protocol used is a subset of IBM Synchronous Data Link Control (SDLC) protocol. When the fiber optic medium is used, electrical to optical and optical to electrical converters are provided.

Protection Microprocessor Clusters

The protection cluster is a multiprocessor configuration incorporating INTEL 8086 16-bit microprocessor boards and external shared memories all connected together through the INTEL Multibus™. The clusters also include communication controllers for communication with the serial data links, the serial data highway, carrier channel equipment and with local data acquisition units for handling signals which are available within the control house. The communication controllers are attached to the Multibus and have capability for direct memory access to the cluster shared memory.

INTEL was selected as the microprocessor equipment vendor after a thorough study evaluating available 16-bit microprocessor products. The evaluation criteria included present performance of equipment, availability of a wide range of support products both at the chip and the board level, software support with special emphasis on the Pascal high level language, vendor acceptability, and long range commitment to supply state-of-the-art microprocessor systems.

Serial Data Highway

The serial data highway interface is the communications medium among the protection clusters and the substation cluster. In some back-up relaying situa-tions, such as failure of circuit breaker to clear a fault, the highway is used to communicate control signals from one protection cluster to another. The highway medium is coaxial cable. Electrically the equipment is the same as that used for the serial data links. The protocol, however, is a subset of the High Level Data Link Control (HDLC) protocol since it is necessary to specify both source and destination in a multi-drop situation. The data highway is normally used in the master-slave mode with the station processor acting as the master and periodically polling each protection cluster.

Substation Cluster

The substation cluster is configured from the same types of computing equipment as the protection clusters. Its function is to maintain the central system data base and provide interfaces to the outside world - locally to station operators through the local man-machine interface (MMI) subsystem and remotely to system operators and protection engineers through SCADA communication interfaces. Controllers for such interfaces are attached to the INTEL Multibus. The local MMI consists of a printer and a color graphic CRT.

System Operation

The protection algorithms used for high-speed relaying require simultaneous samples of currents and voltages from all transducers connected to a zone of protection. Therefore, synchronization of the sampling instants in different DAU's is a critical requirement of this system. Synchronization is achieved by tracking line frequency and dividing each cycle into 16 intervals. Elimination of timing skew between clusters is achieved by periodically broadcasting a synchronization signal over the data highway. At the beginning of each synchronized sampling interval each protection cluster sends a message to its data acquisition units through the corresponding data links. The DAU's immediately sample and hold analog values, and sequentially transmit back to the protection clusters the converted analog and digital signals. The communication controllers at the clusters deposit these data in predefined shared-memory buffers. When all the data have arrived, the protection programs commence processing the new information. The outputs produced by the protection programs are control signals which are sent to the DAU's to initiate breaker operations, plus event messages oscillographic data collected during fault conditions, and general update of the data base, all of which are sent to the substation processor for storage or display. The substation processor receives and stores all this information and in turn sends it, either automatically or on request, to remote operators. It also displays information locally and allows the local operator to initiate manual operations.

FUNCTIONAL CAPABILITIES

The hierarchical hardware configuration described above acquires data and efficiently moves it to locations where it can be processed into relaying and control decisions. The intelligence which actually defines the responses of the system to substation and power-system phenomena is embodied in the applications

software, or computer programs, which reside in the processing clusters and in the station computer.

Given below is a list of the functions to be included in the most complete version of the substation protection and control system. The most important of these functions will be programmed in the initial demonstration system now under construction.

The list of functions was compiled in discussions among the project investigators, EPRI, the host utility, and the five other utility advisors acting on EPRI's behalf. For each function, a complete list of specific capabilities, operating characteristics and logic, setting ranges, operator outputs, and other user-oriented specifications has been compiled. This Requirements Document [5], provides critical guidance in writing of the applications software.

The functions are grouped into four categories below.

Protection Functions

Under this category fall all the high-speed protective relaying programs. They are designed to work with signals from conventional ct's and cvt's, but can accomodate other transducers such as the EPRI/Westinghouse EHV current transducer [6].

Line Fault Protection. Circuit breakers at each terminal trip at high speed for a fault anywhere along the protected transmission line. Both pilot and non-pilot protection programs are included. Time-delayed remote-backup protection trips the line in case of a relay or breaker failure at a remote terminal. The demonstration will use distance-measuring algorithms plus pilot logic which is compatible with present-day directional-comparison blocking schemes.

A very substantial base of software algorithm technology [7], developed within Westinghouse and elsewhere, is available for effective, fast, and secure transmission line protection.

Transformer Fault Protection. This program quickly trips breakers to isolate the transformer bank for any fault within the zone. The primary protection scheme employs percentage-differential measurement. Discrimination is made between internal faults and through faults or inrush currents. Tripping logic also includes effects of overexcitation, time overcurrent, and sudden pressure; plus overtemperature and gas analysis if appropriate transducers are provided.

Bus Fault Protection. Phase percentage differential protection trips for faults, with ground differential overcurrent as backup. The programs are designed to use signals from conventional iron-core ct's, which need not have wired interconnections or matched ratios. Automatic reclosing is provided as an option.

Breaker Failure and Fault Protection. When the trip circuit of a breaker is energized by relays, and the current flow through the breaker does not collapse within the expected time, tripping of all adjacent backup breakers is initiated to isolate the fault and the malfunctioning breaker. Current flow in a breaker ground ct, if available, also triggers backup tripping.

Shunt Reactor Protection. Overcurrent, differential, and impedance measurement programs detect faults in line-connected or tertiary reactors. Special turn-to-turn fault protection is also included.

Transfer Tripping. Requests for remote tripping are routed to transfer-trip transmitters of existing types. Where dual channels are used, receiver outputs are compared for inconsistencies. Transfer tripping is backed up by closing a high-speed grounding switch if provided.

Automatic Control Functions

Local Control of Voltage and VAR Flow. Programs control voltage or VAR flow as measured at the station or at a remote point by operating the transformer bank LTC or by switching capacitor banks, reactor banks or synchronous condensers.

Tie Tripping. A trip signal or alarm is produced when a transmission tie to a neighboring utility shows sustained heavy power outflow and declining frequency.

Load Shedding. Preselected blocks of load are dropped at successively lower frequency levels. These loads can be automatically restored when system frequency returns to normal.

Automatic Reclosing. After high-speed clearing of line faults, reclosing of the breaker is initiated, and repeated if necessary for a specified number of attempts. Additional supervising checks such as live-bus/dead-line and syncrhonism are selectively incorporated. Reclosing capability for bus faults is also provided.

Synchronism Checking and Synchronizer Closing. The magnitude, phase angle and frequency differences between voltages on opposite sides of any open circuit breaker are measured to determine when it is safe to close the breaker. Optionally, an automatic closing program considers slip frequency, phase position and breaker contact closing time so that contacts make at the instant phase difference passes through zero; control of tap changers, exciters, governors, or phase angle regulators is provided.

Automatic Switching Sequences. In some configurations, tripping breakers to clear line or bus faults may result in needless isolation of unfaulted equipment. In such cases, post-fault automatic control sequences may be programmed which will bring sound equipment back into service. The sequences may be initiated either automatically or on operator request. Examples of such sequences are bus transfers or sectionalizing, isolation of a faulted transformer, or of a stuck breaker.

Monitoring Functions

Pilot and Transfer-Trip Channel Monitoring. Checkback tests of the pilot protection and transfer-trip channels are initiated either manually or automatically. Unique test sequences are provided for each channel type; they are designed not to interfere with communications during faults.

Load Monitoring and Out-of-Step Protection. Excessive line current is monitored and alarmed. Also, the locus of apparent line impedance is monitored and recorded. Finally, when line voltage-angle computations show that machines of the power system are slipping out of synchronism, a preprogrammed strategy of splitting the system into balanced islands is implemented.

Monitoring and Control of Breakers and Switches. The position of each breaker and switch is monitored and its operation is controlled and coordinated. Monitoring checks include agreement between auxiliary contacts and current flow or voltage data, timing of lagging poles, and detection of sustained pole disagreement. Control operations include single-pole tripping, arbitration of simultaneous or conflicting requests (automatic vs. manual), closing supervision through voltage or synchronism checks, and lockout/tag setting and removal logic.

Inferential Measurement Check (Substation State Estimation). Each ac signal reading is checked for consistency by comparing it with other related readings. Currents are checked using differentail relationships on each phase. Voltages are checked by confirming that the potential at two connected points on the same phase are the same. Currents and voltages on transformer windings should be related according to transformer ratio and tap position. Measurements found to be erroneous are pinpointed and/or corrected.

Transformer Overload Monitoring. Transformer temperature, pressure, gas analysis and current are monitored. The program accumulates hourly peak MW and MVAR values and overload history data. Additional calculations include estimate of remaining loading capability of transformer within normal or emergency limits and estimate of transformer life expectancy.

Self Checking. The integrity of each hardware element in the system is monitored by executing checking programs at the lowest priority during the processor idle time. If redundant equipment exists, an orderly transfer is executed following a primary-system failure. Checking programs fall into two categories: those which check the system hardware through equipment diagnostics, and those which check the validity of the incoming data through application-related known characteristics, such as magnitude and balance of a three-phase voltage set.

System Interfaces, Data Acquisition and Displaying

Under this category are grouped all the functions which pertain to the interfaces between the protection and control system and the external world -- SCADA masters, and local operators.

Local Man-Machine Subsystem. This subsystem enables the local operator to display and enter data through a CRT with keyboard or pushbutton console, to perform manual operations, and to initiate and monitor diagnostic and maintenance test functions.

Remote SCADA Interface. Through these programs, the system communicates with one or more remotely-located masters. It responds to requests from the master for data retrieval, control selection and activation, and return of special data. Communication line and interface equipment integrity and error statistics are monitored; detection of a problem leads to failover action.

Alarming. All events in the system (including limit violations, equipment operations, alarm resets, operator actions, and key software decisions) are recorded. From that record, different categories of events and alarms can be extracted and either displayed locally or transmitted to a remote operator.

Data Logging. All recorded operating information is formatted and presented on a CRT or printer. Information for reports to the masters is prepared for transmission through SCADA or dial-up interfaces.

Revenue Metering. High-accuracy samples are used for computation of real and reactive power. The system records kWh, kVARh, kVAh and energy demand. Pulses from conventional kWh meters are accumulated as well.

Recording and Indication of Sequence of Events (SOE). All relaying decisions and events, and all operations of station equipment, are accurately timestamped and stored. Upon request from an operator or on a prespecified automatic condition (occurrence of fault, filling of a buffer, elapsed time) the sequence-of-events files are displayed locally and/or transferred to a SCADA master.

Oscillography. Selected raw power system ac signal samples are collected and stored during faults. The raw data are delivered to the substation processor where they are recorded, displayed or logged and/or are passed on to an external interface. Waveforms may be displayed either in raw form or as smoothed curves.

Line Fault Location Estimation. The raw voltage and current signal data stored during line faults are utilized to calculate apparent physical distance from relaying location to the fault.

SOFTWARE DESIGN

The importance of software in the implementation of any digital system cannot be overemphasized. Whereas hardware costs have been decreasing consistently, software costs, i.e. the cost of specification, design, implementation, testing and documentation of the processor programs necessary for the operation of the system, have been skyrocketing. It is, therefore, necessary to select and use those tools and procedures which will help reduce overall software costs.

It has been widely recognized that the use of high level languages (HLL) is one way to reduce software costs. The use of HLL not only increases coding efficiency but also produces programs that are easy to debug, document and maintain. One high level language that has received wide attention in industry in connection with microprocessors is Pascal. Pascal has a clear syntax contained within a rigorous block structure which supports all the needed control statements and which permits programmer-defined data structures. As all high level languages, however, Pascal has its limitations and must be used judiciously and only in those parts of the application where its limitations will not be detremental to system performance. Some of these limitations are run-time inefficiency and inadequate resolution when doing 16-bit fixed point arithmetic.

Another possible contributor to software productivity is the availabity and use of the proper software services, both on-line and off-line. On-line system services are somewhat different for processors of the protection cluster and of the substation cluster. At the protection cluster there are services for handling the clock and synchronization, data highway and data link handling routines, procedure sequencing and time delay routines and system initialization. The substation cluster contains most of these routines plus some routines needed to handle the special devices (e.g. CRT, printer) which are connected to this cluster. In addition to the on-line system services, off-line program development and maintenance facilities are needed. These include file editing facilities, language processors (Pascal and Assembler), link loaders and debug packages. In the implementation of this project two development facilities are used. One is a large scale computer accessed through remote terminals. This facility is used for source editing and for compiling and execution of stand-alone programs. The second facility is an INTEL microprocessor development system with in-circuit emulation, which enables the execution of programs in the target machine environment.

A final area where improvements in software efficiency are possible is in the methodology of design, documentation and testing. This project is using many of the practices and techniques recommended by structured programming advocates [8]. First, a "top down" approach is followed during the design process. Each function is defined in broad functional terms; the definition identifies the classes of inputs to and outputs from that function. Next, each function is decomposed into one or more levels of subordinate modules. Each such module is described, its inputs and outputs are identified and the interaction among modules is shown through a sequence block diagram. The flow of logic of each module is charted on structured charts of the form proposed by Nassi and Shneiderman [9]. At each step of the process the designer produces a document. The overall program design is completed prior to writing computer programs. Coding and individual module testing follows the design. Testing uses the "bottom up" approach with individual programs integrated into functions, functions integrated into clusters and clusters integrated with one another and with the substation processor to form a system. In order to test the dynamic response of the protection algorithms the investigators are planning to test the digital relaying system with a power system simulator.

The project team expects that through the use of the techniques and procedures described above, the system can be implemented and tested within the scheduled time. At the end of this effort, the collected documentation will ease the task of maintaining and updating the design in the future.

ECONOMIC ISSUES

Perhaps the primary concern of future buyers of such a system will be the comparison of lifetime costs for the integrated versus conventional approaches. The investigators and advisors have studied and evaluated a variety of cost factors and have concluded that the new system will be economical if used in sufficient volume. Listed below are the issues which comprise such a comparison.

Direct Hardware Costs

For both conventional solid-state and new microprocessor-based hardware, individual components are low in cost, while any assembled systems are far more expensive. The trends, however, are moving in favor of the microprocessor.

Software Development and Maintenance

This is the dominant expense for the manufacturer, and is amortized over the number of identical or similar systems which can be sold.

User Engineering

Utilities now spend considerable resources on the custom engineering of a control and protection system using available component packages. This expense should be largely displaced by the manufacturer's software effort. The user's task is reduced to selection of option parameters and functional features.

Installation and Commissioning Expense

There are two key components here: field wiring costs and system integration. The integrated system should yield substantial reduction in the wiring cost, since most of the long wiring runs, plus the interconnection of the different functions in the control house, are eliminated. However, each field control, status, and analog point must still be wired to terminals in the DAU. At present, it appears that for average distances in excess of 100 feet from the control building, the multiplexed fiber-optic approach is more economical than point-to-point wiring. The size of this economic-crossover circle is expected to shrink during the time the equipment is being developed.

The commissioning effort for the new system should be very much streamlined since the new system will be assembled and tested as a complete system on the factory floor, prior to shipping. Testing in the field need only be sufficient to locate damage in shipping or installation.

Savings should also result from the replacement of erecting, punching, and outfitting control panels with the installation of a few completed modules of equipment.

Lifetime Maintenance Costs

The self-checking capabilities of the new equipment virtually eliminate the need for periodic maintenance checks and calibration; this promises a significant saving. There is also some saving of troubleshooting time if the equipment successfully diagnoses its failures to the faulty module.

The investigators have spent considerable effort in predicting the hardware reliability of the new system. Although it appears to be competitive with exisitng equipment, only field experience will provide final assurance of reliability.

Note that the hardware reliability and maintenance cost issue is distinct from the functional reliability issue; the latter is handled by redundant backup equipment and backup modes of operation in failed equipment, combined with self-checking abilities of the system.

Performance Evaluation

The new functions and improvements made possible through this system, could be of significant value to the user. For example, organized, complete, remotely-sent operation records, can save a great deal of engineers' and technicians' time and travel. Improved, refined, or faster relaying and control can extend the life of power apparatus or even reduce the number or complexity of large, expensive power devices. Savings in these areas can easily exceed the entire cost of the control and protection system, even if other savings are ignored. However, the designers have not assured that users will necessarily pay more for these new or improved functions.

CONCLUSION

This paper has presented the rationale of the design specifications for an integrated protection and control system for transmission substations. First, the authors reviewed the nature and characteristics of transmission substations, the shortcomings of existing practices, and the pertinent new technology available. With this background a set of design criteria and functional specifications have been developed. Next, the design details of the proposed system have been described in terms of architectural considerations, the hardware used, the software methodology followed, and key economic considerations. The authors believe that the system proposed and described in this paper represents a significant step in the evolution and advancement of relaying and control practice. Through the new technology available today it is possible to build systems which display many economic and technical advantages over equivalent conventional systems. To use this technology, however, requires a fresh view of the application. One crucial question remains: Can the demonstration of such an integrated digital protection and control system be convincing enough so that the system can find acceptance by the relaying community?

ACKNOWLEDGEMENTS

The authors wish to acknowledge with appreciation the support of the Electric Power Research Institute (EPRI) in this investigation. Special thanks are due to the EPRI Project Manager, Mr. Stig L. Nilsson. The host utility, Public Service Electric and Gas Company of New Jersey, and five other advisory utilities (Pennsylvania Power and Light, Public Service Company of New Mexico, Tennessee Valley Authority, Texas Electric Service Company and Utah Power and Light Company) have provided invaluable assistance in establishing system requirements. Finally, the authors wish to acknowledge the contribution of each member of the Westinghouse project team.

REFERENCES

[1] S. L. Nilsson, "EPRI Research and Development of New Substation Control and Protection Equipment," IEE (U.K.) Conference Publication 185, Developments in Power System Protection, London, June 1980, pp 88-92.

[2] E. A. Udren and M. Sackin, "Relaying Features of an Integrated Microprocessor-Based Substation Control and Protection System," Ibid, pp 97-101.

[3] J. S. Deliyannides, M. Kezunovic, and T. H. Schwalenstocker, "An Integrated Microprocessor-Based Substation Protection and Control System, " IEE (U.K.) Conference Publication 187, Developments in Power System Monitoring and Control, London, June 1980, pp 50-55.

[4] Digital Techniques for Control and Protection of Transmission Class Substations, Electric Power Research Institute (EPRI), Palo Alto, CA, Report WS-79-184, August 1980.

[5] Substation Control and Protection Project - System Requirements Specification, EPRI, Palto Alto, CA, Report EL-1813, May 1981.

[6] L. E. Berkebile, S. L. Nilsson, and S. C. Sun, "Digital EHV Current Transducer," Presented at the 1980 IEEE-PES Summer Power Meeting, Minneapolis, MN, Paper No. 80 SM 647-8.

[7] Computer Relaying, IEEE-PES Tutorial Course 79 EHO148-7-PWR.

[8] D. McCracken, "Revolution in Programming: An Overview," Datamation, December 1973, pp 50-51.

[9] I. Nassi and B. Shneiderman, "Flowchart Techniques for Structured Programming," ACM SIGPLAN Notices, August 1973, pp 12-26.

Discussion

Dorel Damsker (Gibbs & Hill, Inc., NY): The authors' closure requires a more detailed discussion about the main issue: "Hierarchical" as opposed to "Democratic" configurations. In my opinion, most industrial control and data communications systems need a functional hierarchical organization.

In the light of advanced computer networking technology, this hierarchical organization can and should be logically structured and not physically implemented. Referring to 1S0-0S1 Reference Model, we can logically superimpose the application processes on the application layer No. 7, as functionally required. The data-communication function is better serviced by a multi station, peer-to-peer multi access body, layers 2 to 1. (See Reference [1, 2]).

The three level hierarchy, described in the paper under discussion (figure No. 2), should be considered as a logical structure. The architecure of the system can be distributed implemented by only one redundant local area control network, with the necessary degree of reliability, and complete assurance of satisfactory functionality. The DAUs should work as concentrators, using scanning and/or interrupts methods, with 1 ms cycle (for scanning). This short cycle time should be confined only at the input of DAUs. The DAUs will report by exception and rate of change to the protection microcluster onto a priority physical communications medium and to the substation clusters onto a redundant communications medium. Some bridges will provide cross tying for reliability purposes. The reporting frames on these media will obey a statistical Poisson law, with an average rate of 1 second (according to the authors' estimates). A medium access control such as CSMA/CD can be successfully used because of the small ratio of protocol data units as compared to the large bandwidth of 10 Mbps (as considered by Ref. 1). A collision like error detection, caused by EMI, is avoided by the use of fiber optic media with star couplers. The back-off procedure should be favorably employed in the case of non priority frames. (See Reference [3]).

Both LAN IEEE 802 and Proway International Standards eliminate polling media access control, together with all kinds of centralized administration of protocol data units, and consider only CSMA/CD and token passing peer access methods for open interconnected systems.

As regards to redundancy issue, I am pleased to see that the authors removed the reliability weakness pointed out by my comments, and introduced a back-up station microcluster in their closure for the first time.

The time has come for the power control and data communications systems to comply with the advanced computer networking technology, as well as national and international standards.

REFERENCES

[1] IEEE Local Area Network, Draft B Standard, October 19, 1981.
[2] D. Damsker, "New Hierarchical Controls and Data Communications," Mini-Micro-Data-Com. International Conference, Geneva, Switzerland, June 17-19, 1980, Proceedings pp. 253-270.
[3] D. Damsker, "Distributed, Redundant, Local Computer Control Network" IFIP, Technical Committee No. 6 International Working Conference on Local Computer Network, Florence, Italy, April 19-21, 1982.

Manuscript received November 25, 1981.

J. S. Deliyannides and **E. A. Udren:** This second discussion adds little new information. It is a reiteration of the discusser's strong commitment to a single multi-drop communication medium properly backed up for reliability. The discusser acknowledges the fact that the nature of the application is hierarchical, but arbitrarily maintains that this hierarchical organization should be logically structured and not physically implemented. This conclusion is not supported and must be challenged. The approach that the authors have taken is to start with the application requirements and to derive an implementation architecture from them rather than presuming a given network configuration and attempting a forced fit to the application requirements. Digital protection and control combine speed, reliability, security, and maintainability requirements in a most demanding way. The authors believe that the three level physically implemented hierarchy yields a more effective system than the single distributed multi-drop data highway.

The discusser displays a misunderstanding of the application requirements when he suggests that the DAUs should only transmit data to the protection clusters by exception. The function of the protection clusters is the high speed detection of fault conditions through the execution of complex algorithms with one millisecond samples of power system currents and voltages from several DAUs. This necessitates the periodic transmission of these samples from each DAU to the protection clusters every millisecond at all times. The data transfer at an average rate of one second pertains to the interface between protection cluster and substation cluster - a different level in the hierarchy.

The authors are not averse to using accepted and established standards when they are compatible with the application. Since neither of the proposed standards mentioned by the discusser have that status, they feel no obligation to force a procrustean fit just for the sake of complying with a proposed standard.

Finally, the authors must reluctantly discredit the discusser's assertion that the backup station cluser was introduced at his suggestion, as an afterthought. The fact is that the design for such a cluster has existed for some time prior to publication of the paper. This and a host of other concepts which are not included in the paper have been studied extensively during the execution of this project over the past three years. The discusser can gain insight into some of the early design decisions by consulting Reference [4] of the paper.

Manuscript received January 15, 1982.

359

PROS AND CONS OF INTEGRATING PROTECTION AND CONTROL IN TRANSMISSION SUBSTATIONS

Contributing Authors: S. L. Nilsson, Senior Member
Electric Power Research Institute

D. F. Koenig, Member
American Electric Power Service Corp

E. A. Udren, Member
Westinghouse Electric Corp

B. J. Allguren, Member
ASEA

Edited by K. P. Lau, Senior Member
Duke Power Company

This report has been prepared by Automatic and Supervisory Systems Subcommittee, Substations Committee and is co-sponsored by Relaying Practices Subcommittee, Power Systems Relaying Committee

Abstract - It is generally accepted that the technology for integrating protection and control functions in transmission substations is here today, and many demonstration projects can testify to that fact.

However, this new technology has been implemented with caution. Many concerns have been raised regarding, reliability, security, economics, long-term availability, etc.

In this report, an assessment of the advantages and disadvantages of integrated systems are presented. Three projects with different approaches are described with particular emphasis on system criteria, architecture, justification and field experience. The Westinghouse Electric Corp. project under a contract with the Electric Power Research Institute (EPRI) and the American Electric Power Service Corp. (AEP) project are two integrating systems to be discussed. In addition a hybrid project by ASEA is presented as a separated but cooperating system.

INTRODUCTION

Parallel developments of digital control and protection systems for transmission substations are being pursued by the power industry. The development of the digital control system started when the minicomputer seemed to become economically feasible for these types of applications 10 to 15 years ago. Although some minicomputer based protection systems were built, they were mainly for technical feasibility demonstrations only. However, the microprocessor has changed the cost equation for digital protection, and several major developments of digital protection have been started and some have reached the prototype demonstration stage. Also, some of the protection systems are being built to incorporate the control system functions of the substations.

84 SM 638-3   A paper recommended and approved by the IEEE Substations Committee of the IEEE Power Engineering Society for presentation at the IEEE/PES 1984 Summer Meeting, Seattle, Washington, July 15 - 20, 1984. Manuscript submitted March 30, 1984; made available for printing May 29, 1984.

This report discusses the merits and concerns of integration of the control and protection functions in a single system.

The following section presents an overview of design and architectural issues which arise when integration of control and protection is evaluated. Following this, three specific development efforts now going on are described as examples. In each project description, selected concerns affecting this new technology are addressed.

FACTORS AFFECTING ARCHITECTURE AND DESIGN

Presently, control and protection of a typical substation is performed by means of a set of stand-alone subsystems as shown in Figure 1. Substations today may be under remote control from a central location, but only limited information about the state of the station is available to the remote operator because the cost of providing full information has been prohibitive. Detailed information about disturbances or equipment problems can therefore only be obtained locally at the station.

The stand-alone subsystems have the characteristic of inherent redundancy in the data acquisition and control circuits. Each subsystem has its input and output circuits, where possible, separated from the other subsystems.

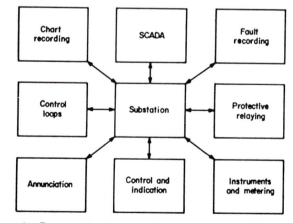

Figure 1. The conventional control and monitoring of a substation is now handled by a multitude of separate sub-systems.

Reprinted from *IEEE Transactions on Power Apparatus and Systems*, vol. PAS-104, no. 5, pp. 1207–1224, May 1985.

Some of the reasons for the adoption of digital techniques in protective relaying are:

a. Reliability improvements from use of self-checking techniques.
b. Increased relaying intelligence to minimize the incorrect operation of power circuit breakers.
c. Perceived potential operational benefits when managing critically loaded transmission systems.
d. Power apparatus protection requirements increasing as a result of the push towards incipient fault detectors:
e. Availability of cost effective base technologies (microprocessors and fiber optic communication links).

The use of self-checking techniques should reduce the number of false trips and improve the probability for correct operation of relays because relays failures will be detected as they occur. This leads to a significant improvement in availability for the protective relays.

Digital protective relaying devices can also be controlled and monitored from a remote location with suitable modems added. This can increase the operating capability because relay settings can be centrally modified if needed to handle critical load conditions. This is not something that will find ready acceptance, but it is available and can be very effective.

A major facet of the development work is the selection of a cost effective system architecture for the protective relaying system which can incorporate many protective relaying functions. The criteria for selection of the system architecture should include:

• The ability to perform all critical protection functions with any single failure in the system.
• Critical relaying functions must be operable even if the substation computer fails.
• The system must avoid a fully redundant data acquisition and control output subsystem in order to satisfy the single failure criterion.
• LSI (large scale integrated) circuit technologies should be used to the greatest extent possible.

These criteria assume that there will be a station computer for coordination of the different subsystems of the digital system. The system should be modular, such that it can be expanded from a small system to a larger substation later. Furthermore, it should be possible to use system modules as stand-alone, cost effective line protective relaying devices.

As the criteria are applied to a system for protection of a substation as illustrated in Figure 2, one notices that there is a significant overlap of protective zones in a substation. But such overlap only provides insignificant backup protection capabilities. That is, the bus protection does not protect the transformer in case the transformer protection fails. The line protection may provide backup protection only if the 4th zone (reverse looking function) distance protection function is used to trip the bus for prolonged fault indications.

To satisfy the single failure criterion for any backup protection system, redundancy is required where different designs may be used for the primary and the backup protection. But duplication of the protection is only one way to solve this problem.

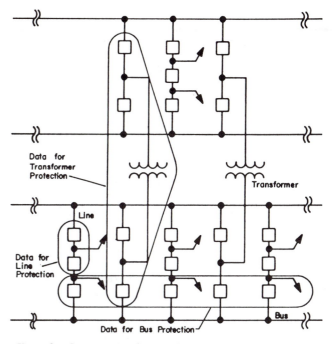

Data for Transformer Protection

Transformer

Line

Data for Line Protection

Data for Bus Protection

Bus

Figure 2. Data collection for protection functions typical substation.

Presently, different primary input sources are used for many of the protective functions. (See Figure 3a). The bus protection often relies on special, dedicated current transformers for its operation. Therefore, in order to share input/output devices to minimize the input/output subsystem cost, the digital system should be able to accommodate the functional needs of the different protection functions preferably through one and the same data acquisition module. This is not a stringent requirement because, as is shown in Figure 3b, one can make allowances for different data acquisition modules in the system. It does however, reduce the total system cost if input/output channels can be shared as shown in Figure 3c.

In the case of EPRI/Westinghouse project, a multitude of different system architectures were studied in an attempt to identify the one with acceptable performance for the lowest cost. The analyses showed that the system with the lowest cost for the input/output subsystems would be the system with the lowest cost.

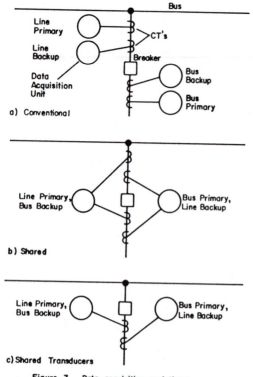

Figure 3. Data acquisition variations

A hierarchical system as shown in Figure 4, eventually emerged as the best system given the constraints above. It is assumed that the data acquisition is oriented along the breaker bay concept, because most substation cable runs are bay oriented and substation expansions are bay oriented as well. If redundant protection and input/output units are provided, the system meets all of the stated criteria (Table I). Each critical protection function will be able to function without the station computer and the highway, because all of its data and control lines for breaker trip operations are directly connected to the protection processor. Only for breaker failure protection is the highway needed, but this requirement is met without having the station computer in operation.

| Quantity | | Usage | |
	Protection	Control	Monitoring
Currents	X	X	X
Voltages	X	X	X
Breaker trip	X	X	X
Breaker close	(reclose)	X	(event)
Breaker alarm	1)	(interlocks)	X
Disc. switch control	2)	X	
Control mode	(lockout)	X	(event)

Table I: Commonality of input and output quantities for a typical breaker bay.

Note 1: The breaker is normally self-protecting, but the protection system could use information about breaker failures to initiate different protective actions or to isolate the failed breaker.

Note 2: Automatic control on disconnect switch could be beneficial in restoration after a breaker failure.

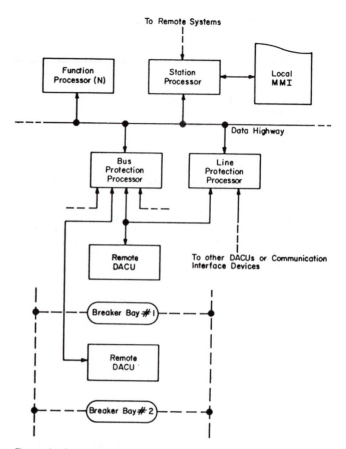

Figure 4. One Possible Architecture of an Integrated Protection and Control System. (Non-redundant System)

Integrating the control functions into a protection system

There are strong institutional reasons for separation of the control system from the protection system. These barriers to integration have to be addressed individually.

But for technical and economic justifications, the prospects for data sharing between protection devices and control systems can be a significant cost saving in the evaluation of integrated versus separate systems. This is because communication links can be effectively used to channel data and control commands between different functional modules at a lower cost than one can achieve with hardwired circuits if the communication flow is non-trivial. For the EPRI project it has been assumed that data acquisition units may be placed in the yard or close to the equipment being monitored and controlled. This is not a requirement but a system option. It is consistent with the trend to develop improved equipment monitoring devices for incipient fault detection and also permits replacement of conventional cables with fiber optic links where this is feasible. This option has litte value if SF6-insulated substations were to be the future norm, but would be cost effective if air insulated substations continue to be built.

362

A functional view of the data and control command flow is illustrated in Figure 5. Usually the protection system is the most demanding function. There are exceptions to this. The revenue metering function requires higher accuracy than the protection functions do and the oscillographic function typically has a higher bandwidth requirement than for the protection. However, in most cases the input and output channels of the protection function will meet the requirements of the control functions. Furthermore, all of the control system needs can be met with a modest expansion of the data acquisition and control submodules of the protection system. This expansion will not significantly add to the failure probability of the protection system, because the added hardware can be isolated relatively well from the hardware that the protection relies on for its operation.

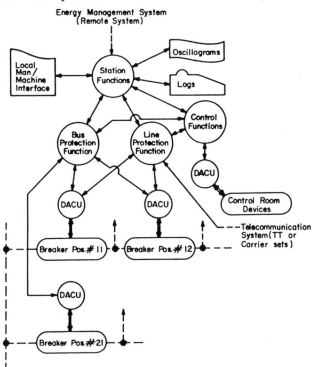

Figure 5. Signal Flow Diagram for an Integrated Protection and Control System.

Now, even insignificant compromises of the protection system integrity might not be acceptable. The alternative is full separation between the control and protection systems. An example of such a concept is illustrated in Figure 6 which also assumes full separation between the input/output subsystems of different protective relaying functions. It is apparent that the data acquisition and control subsystem can grow to unacceptable levels when full redundancy is introduced to meet the single failure criterion. Such integration can be accomplished without significantly compromising the integrity of either system.

EPRI/WESTINGHOUSE PROJECT

Westinghouse Electric Corp. with the support of EPRI has developed and is demonstrating an integrated substation protection and control system (SPCS).

SPCS employs clusters of multiple microprocessors, multiplexed digital communications, and optical-fiber data transmission media to replace the conventional complement of discrete relays, instruments, and controls with the associated interconnecting wiring.

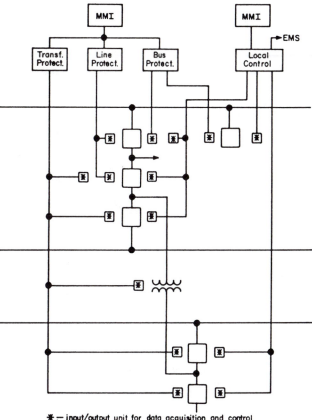

✳ — input/output unit for data acquisition and control
MMI — Man/Machine Interface
EMS — Energy Management System

Figure 6. Nonredundant control and protection system for substation with control system separate from protection system and separate input/output units for each function.

Design Criteria

Transmission Substations

As a starting point, the designers looked at the unique requirements of transmission-level substations which impact the architecture, design requirements, and economics for an integrated protection and control system.

- Need for sophisticated relaying and control functions

- Need for coordinated substation control from centralized or remote locations

- Need for unattended operation, with convenient manned operation on occassion

- Need for detailed operating data and history - event recording and oscillography

- Large distances over which information and control signals are conveyed

- Hostile electrical environment - conducted surges, RFI, EMI

- Lack of standardization in design or configuration of power apparatus, or in control and protection equipment

- Likelihood of growth or reconfiguration of the station over time

Functional Requirements

One of the first and most critical tasks for the project designers and utility participants was to compile a list of functions which the system must support in its ultimate form, with specific behavior and performance requirements for each function.

The resulting System Requirements Specification [7] describes 26 major functions identified by the participants, and has served as the starting point for the design of application programs included in the demonstration system. The same document also lists general system design requirements which have impacted architecture, hardware selection and design, testing approaches, and documentation.

Design Approach

The first phase of the development focused on evaluation of alternative design approaches in light of criteria or requirements listed above and elsewhere in this paper [6] provides much insight regarding the depth of this study. In the following we present only an overview of the end results.

Architecture

Figure 7 shows the three-level hierarchy used to meet the design requirements. The key operational features of this scheme are:

- High speed sampling of power system voltages, currents, and contact status to provide sixteen samples during each cycle. These samples are used by the protection algorithms.

- Synchronized data sampling at all data acquisition units throughout the system.

- Upward flow of power system status information and downward flow of manual, supervisory, and automatic control commands.

- Faster rate of information flow at protection level, slower at station processor level.

- Infrequent but high priority communication between protection subsystems.

- Stand alone operation of critical protection functions.

The following sections describe each element in Figure 7, starting at the bottom.

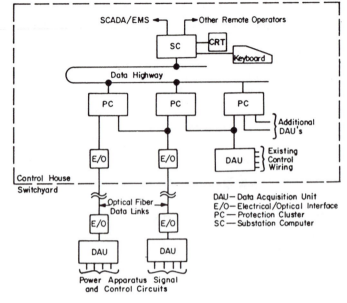

Figure 7. SPCS Architecture

1. **Data Acquisition Unit (DAU)** The DAU is the interface between the power apparatus and the integrated control system. Multiple DAU cabinets are installed in the switchyard, near the apparatus to minimize expensive wiring runs. The DAU takes instantaneous samples of ac voltage and current signals from ct's and potential devices, along with status of contact inputs such as those indicating breaker state. The samples are digitized and transmitted to the next level of the hierarchy, the protection cluster (PC), via an optical-fiber link to the control house. The link also carries power apparatus control requests from the control house PC's to the DAU's.

Additional DAU's may be installed in the control house, where they connect to conventional wiring associated with a preexisting, conventionally-controlled section of the station.

Coordinating pulses are communicated to each DAU so that all input points in the station are sampled at the same instant. The DAU's capture, digitize, and transmit sixteen sample-sets per power cycle.

2. **Protection Cluster (PC)** The protection cluster comprises a group of microprocessor boards on a common bus with shared memory, plus interfaces to DAU and substation computer (SC) data links. One such cluster is typically associated with each line, bus, or transformer protection zone. The PC receives masses of raw data samples from one

364

or more DAU's. High-speed relaying and control programs constantly evaluate these sequences of instantaneous samples, looking for symptoms of power system faults and extracting accurate phasor measurements of ac signals for transmission to the SC level.

Relaying and control programs in a PC can issue control outputs directly to DAU's, and can also check and pass along control requests sent down from the SC.

Multiple DAU's may supply a single PC with information, but it is also important to note that one DAU may connect to multiple PC's requiring signals from a particular switchyard or control house location. Also, multiple PC's can send control requests to the common DAU. This cross-linking is accomplished in the control house where link signals are in electrical form.

3. Data Highway A multidrop, serial, electrical data communications highway interconnects all of the protection clusters and the substation computer. For normal operating conditions, the substation computer is the master of the highway, sending control commands to the PC's and requesting upward transmission of a variety of data types which it collects. However, an individual PC can temporarily seize control of the highway in order to send an emergency control message to one of its peers, as it may need to do in a backup-relaying situation.

4. Substation Computer (SC) The substation computer (SC) provides a central data base and control point for the entire station. Data compiled includes periodically updated phasor values of ac signals, status of contact inputs to DAU's, internal status of programs and hardware modules of the SPCS, sequence-of-events (SOE) messages generated within programs, and lists of fault data samples used to support oscillography and other fault-evaluation measurements. The SC also performs selected low-speed control functions. Finally, the SC interfaces to local operators through color CRT, printer, and keyboard; and to a remote control center by emulating a remote terminal unit (RTU) for the utility SCADA or energy management system. The SC can send data to additional remote locations as required by individual utilities.

5. Stand-Alone Configuration The designers have configured a single transmission-line PC and associated DAU as a stand-alone protection terminal. All digital equipment is installed in a single cabinet with conventional wiring terminations. A special man-machine interface includes a keypad panel with alphanumeric data display, plus a small thermal printer, all interfaced through a dedicated microprocessor module to the protection cluster data bus.

Hardware

The designers imposed the following requirements during selection of the hardware used in the SPCS:

- Use industry-standard, proven hardware modules and designs from well-established suppliers wherever possible. This minimizes development time and cost, maximizes reliability, and provides a base for long term support of the system.

- Insure that design of this standard equipment is suitable or adaptable for use in hostile environments like that of the substation.

- Select the latest hardware available consistent with other needs, to achieve cost and performance benefits. Evaluate the ability of the manufacturer to supply even newer, yet compatible, designs as hardware technology evolves.

For the protection clusters and station computer, the designers selected the Intel Multibus (TM) and 8086-based microprocessor modules. Because of wide acceptance of both products, many manufacturers offer a broad range of compatible processor, memory, interface/communications, and test modules. New higher-performance microprocessor modules have been introduced, during the SPCS development work, with hardware and software compatible so that they have been readily integrated by the project team.

The DAU uses a proprietary Westinghouse I/O bus which has been widely applied and proven in industrial and utility environments. For the unique high-speed, high-resolution, data input and output needs of the SPCS, several new interface cards have been designed and built as well.

Software and Application Programs

Development of application and data-communications programs have consumed the biggest share of project resources. Because of the cost of software, certain factors become pivotal to the economic success of the new approach:

1. Programs must remain useable for a long time. This goal is likely to be met with a high-level language (HLL) which is not closely tied to any particular processor hardware. Accordingly, many of the SC application programs are written in Pascal. At the PC level, Pascal or other HLL's could not meet the real-time requirement for processing the masses of raw data. In this case, the programs are written in assembly language. This underscores the need to select processors from a manufacturer who will continue to support existing programs with new, higher-performance hardware as it is developed.

2. Application programs should be created from design specifications created by designers and users working together. This minimizes the chance that a program, once created, will prove to be lacking critical features. Such a specification is used in the present development [7], with additional periodic reviews by utility advisors.

3. The software should combine flexibility of features with automatic program-package generation capability, to minimize the cost of creating a program package for each new job after the initial design work has been done. Each substation installation imposes a unique combination of function and performance needs. Writing new programs to accommodate each job is too expensive.

The designers have already provided highly modular programs. They are now beginning the creation of a software system which will automatically configure these modules into program packages for each cluster and for the SC, based on user inputs and needs. Such a system-generation package is a major effort in itself, to be implemented in stages as equipment is produced for field installation.

Specific Concerns

The appeal of integrated protection and control systems is based on meaningful features and performance benefits, but designers and proponents of this new approach must not overlook or neglect legitimate concerns which make potential users cautious. Some of these have been addressed in design work already done, while others will be answered only with field experience. Surely, there are unanticipated idiosyncracies lurking in the new technology as well.

The following addresses some of the key concerns:

Security

There are some features of this integration which have the potential for increasing the chance of misoperation. For example, a number of control outputs, possibly affecting multiple relaying zones, may be connected to a single electronic interfacing package like the DAU of the SPCS described above. Furthermore, a flexible architecture results in interconnection of each such interfacing package to a number of processors performing the actual protection and control functions. What is the risk that a single hardware failure will cause false tripping of several breakers, or other widespread incorrect operation? Does this factor add to the already-existing possibility of false operation because of confusing or distorted signals from the power system?

Fortunately, the digital implementation also provides the tools not only for alleviating these new security concerns, but for actually reducing the risk of misoperation as compared to the conventional approach. Some of the measures used by the designers of the EPRI/Westinghouse system are:

- Communications and control protocols - A conventional, hard-wired breaker trip circuit may be inadvertently energized by shorting of wires or terminations. The digital communications links used in the integrated system, on the other hand, send breaker trip commands as complex streams of serial data with error-checking codes and checkback repetition. It is virtually impossible to fool the DAU into performing the trip by any disruption of link communication - the system is secure in the face of opens, shorts, tampering, fires, surges, and noise.

  Similar handshaking or protocol measures are used throughout the system, even among processors in one DAU or processing cluster, so that if malfunctions arise, the resultant behavior does not include undesired control outputs.

- Self-checking - The ability of processor-based systems to constantly test themselves, and also to evaluate the accuracy or reasonableness of incoming data, helps to alleviate many performance concerns, including concerns over security. Self-test and data-checking programs, combined with application programs which can be inhibited when a test-failure occurs, reduce the likelihood that either a system hardware failure or a failure or disruption of an input source can cause a false operation. The particular SPCS design described above will even suppress relay operation for many cases of input wiring shorts, opens, or misconnections of transducers - a capability few conventional relays possess.

- Software design - Many computer-relaying investigators have demonstrated the power and sophistication of relaying and control methods embodied as computer programs. This offers the potential for reduced chance of misoperation in the face of distorted power sytsem inputs.

- Minimization of hardware interfaces - In a conventional substation, each relay and control device has its own control connections to the power apparatus. Malfunction in any one device or its wiring causes false tripping. In the integrated system, a single, substantial SCR or contact provides a control point for all primary functions - two at most are used for primary and redundant backup control. The risk of false energization is lower.

Electrical Environment

Design of equipment, interfaces, and enclosures for the integrated system must incorporate protection against all the presently identified or standardized electrical environment influences. Most notably, these include protection against multiple categories of conducted transients [8], and radio-frequency electromagnetic fields from communications gear [9].

However, placing microprocessor-based electronics in the switchyard very near the power apparatus has raised new questions about the intensity and spectral content of radiated electromagnetic transients there. The problem is not only one of characterizing the environment, but also of devising practical and meaningful qualifying tests for the equipment.

EPRI and Texas A & M University confronted this problem to lay the groundwork for the design and testing of the SPCS. Transient EMI associated with the most offensive switching operations was actually recorded in the 345 kV and 500 kV switchyards of five utilities. Similar recordings were made using gaps and simulated buswork at the Westinghouse High-Voltage Testing Laboratory. Texas A & M researchers assembled a sophisticated mobile EM transient measurement and recording system which has a bandwidth of 100 megahertz. The results of these investigations have been reported [10] and provide a valuable base of design data for the entire industry.

With regard to qualification testing of the SPCS, system elements are being subjected to EMI fields approaching those actually found in switchyards, using the high-voltage laboratory and the same mobile instrumentation system.

The designers of the SPCS believe that the effort of accurately characterizing the environment, and of testing the ability of the equipment to withstand that environment, reinforces rather than reduces confidence in the surge-withstand capability of the new equipment.

Long-Term Availability

For all bought-outside components and modules, the key to long-term support is to use industry-standard, multiply-sourced parts, for which manufacturers provide compatible higher-performance equivalents as the technology advances.

Even though integrated control and protection systems are just emerging as commercial offerings, designers are looking at the possibility of standardizing the interfaces among major system elements. This provides the utility user with the ability to incorporate new technology in an existing system design with long-term support, and reduced inventory of spare system elements.

Reliability

Hardware Reliability - As explained above, the design of an integrated system can be arranged so that the consequence of a hardware failure is virtually always the loss of either primary or backup protection of some zone, along with other, less critical functions. With redundant backup coverage of relaying, concern focuses on how often service personnel will need to visit a normally unattended station, and

on what repairs need to be performed. A fair comparison is frustrated by the lack of complete or consistent failure data for conventional relays and substation control devices.

The project team collected and analyzed data on the reliability of microprocessors and associated components at the outset of the project, and projected that hardware reliability for the new hardware generation to be employed should be as good as that of the conventional equipment replaced. Industry-wide field experience with newer generation microprocessor-based systems accumulated since then shows that equipment to be better than originally projected. Only field experience with many utility installations can answer all hardware reliability questions, but with careful design the prospects are bright.

Relaying Reliability - Reliability in protection means something quite different: What is the likelihood that a relaying function will trip when a fault occurs in its zone? To the extent that a conventional relay may fail to respond because of a design weakness, the software approach associated with the integrated system provides more sophisticated design tools and can be expected to do a better job.

An even bigger benefit may result, however, in the case of failures to trip because of hardware malfunction. Conventional relays are idle virtually all of the time; a disabling failure may become apparent only when the relay fails to trip for a fault. Accordingly, most utilities test relays periodically, but even this will not catch all problems before they lead to trip failures. Digital processorbased, integrated protection, with multiple levels of self-testing, will alarm the user (at remote locations if needed) within seconds of a failure, so repairs can usually be made before a power-system fault occurs. The need for periodic testing by technicians is eliminated or greatly reduced, and the relaying reliability will improve even if the hardware reliability of the equipment is only the same as that of present relays.

Cost

The cost of an integrated system is a primary concern for most users. Will it cost more than a conventional system? If it does, will additional features and lifetime-cost savings justify the price differential? Unfortunately, the savings or monetary value associated with the additional features of the integrated system are hard to quantize. A one-to-one cost comparison is not conclusive.

A detailed, preliminary cost comparison was performed in the early phases of the project. The cost study focused on a small transmission-to-distribution substation where the integrated, multiplexed approach offers the least apparent cost benefit. The results indicated that the integrated and conventional system costs were about the same. However, an important factor favoring the integrated system is that its equipment to labor cost ratio is higher. Since equipment costs are decreasing and labor costs are increasing, the trend definitely favors the digital approach.

All of this assumes that the large cost of software development is amortized over a significant number of sytems sold. The cost benefit is reduced if major program writing is required for each new application. Accordingly, the availability of a semiautomatic software package generation system has a major impact on cost. This program-generation tool can be used either by the manufacturer or by the utility at his own facility. Thus, when the substation expands or changes, the software is modified to suit with a minimum of costly engineering effort.

AMERICAN ELECTRIC POWER'S INTEGRATED MODULAR PROTECTION AND CONTROL SYSTEM (IMPACS)

AEP began to investigate digital protection in the early 1970's to better understand the capabilities and limitations of digital relaying. A digital scheme for line protection, known as the Symmetrical Component Distance Relay (SCDR), was developed [11, 12, 13]. A microprocessor-based SCDR has been operational in the Matt Funk substation since the summer of 1980. Encouraged by the performance of the SCDR, AEP has evolved to the present Integrated Modular Protection and Control System (IMPACS).

IMPACS Configuration

The system configuration for IMPACS is illustrated in Figure 8. IMPACS is primarily a substation-based system that can take advantage of an existing Supervisory Control and Data Acquisition (SCADA) Master Station. The IMPACS Substation Host can replace the SCADA Remote Terminal Unit that would normally be in that substation. The Protection Modules of IMPACS are individual microprocessors with associated data acquisition and control interface hardware. The SCADA module interfaces data and control points external to the digital protection modules with the Substation Host.

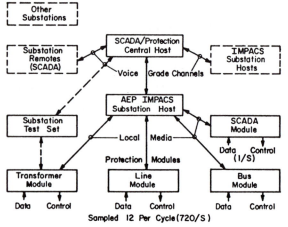

Figure 8. IMPACS System Configuration

To place benefits and cost in a known context, our focus at the protection level will be on only the line protection module where AEP has had the most experience.

IMPACS Central Host

The Central Host is basically a SCADA Master Station that can, within its communication protocol, receive power system disturbance data tables from one or more IMPACS equipped substations and send set point type information to selected IMPACS equipped substations. The SCADA Master Station must also accommodate one or two protection related computer programs

- DISTURBANCE REPORT PROGRAM, and
- REMOTE SETTINGS PROGRAM (Optional).

All other substation information, such as alarms, breaker health and IMPACS diagnostic messages, are handled by the SCADA system as SCADA-like information.

The IMPACS disturbance report program, now functional in a minicomputer at the AEP central office, satisfies the following objectives.

- Provides information about a disturbance in a clear, concise and useful report for operations and relay personnel

- Identifies

  - Line/Equipment Effected
  - Type of Fault (Ø - Ø, Ø - G, etc.)
  - Approximate location (miles)
  - Severity (Fault Current Magnitude)
  - System Response (Reclose/Lockout)
  - Sequence of events (1.4 MS Resolution)

A remote protection settings program that operates in the SCADA Master Station can be supported by the IMPACS configuration. Software now in the Substation Host accepts down-loaded information; therefore, protection criteria could be managed from a central location. Changes in settings associated with maintenance activities and seasonal influences could be accommodated.

The SCADA Master Station, could be in communication with conventional SCADA Remote Terminal Units (RTU's) as well as with other IMPACS equipped substations. The system level considerations (radial vs. party line communication channels, etc.) typical of SCADA applications would need to address the occasional need to transfer disturbance data from IMPACS equipped substations to the SCADA Master. A disturbance data table for a line event includes ten cycles of voltage and currents (960 values) and a number (usually less than ten) of sequence-of-event messages (4 bytes each). Logic in the Substation host Computer could identify which Protection Module's disturbance data table is "closest" to the event and only transmit that table to the SCADA Master Station.

For a line event, disturbance data tables from the IMPACS Protection Modules at each end of the line would be available for transmission to the SCADA Master Station and subsequent analysis by the Disturbance Report Program. The communication burden imposed by IMPACS equipped substations on the SCADA channels can be estimated based upon the frequency of disturbances and the volume of data desired for analysis. Clearly, if tables for non-faulted lines are also desired, then an additional communication burden must be accommodated. Assuming a communication channel operating at 1200 bits per second and a message protocol that has an overhead equal to the information content of the messages, a disturbance data table would be transmitted in approximately 30 seconds. The Disturbance Report Program could begin in its analysis approximately one minute after the disturbance. As a low priority program, it would be interrupted during its execution; however, a report would be printed or displayed within a few minutes. The total elapsed time between the field disturbance and the availability of a report at a central location is in the 5 to 10 minute range.

IMPACS Substation Host

The Substation Host module performs the following major functions

- Communication Processor
- Manage SCADA Type Data
- Report by Exception Processing
- Man Machine Interface
- IMPACS Diagnostics

which are each briefly described in order.

As a Communication Processor, the Substation Host buffers messages and commands between the SCADA Master Station and other IMPACS Modules in a substation. The SCADA communication protocol is accomodated on a single 1200 bits per second channel with the SCADA Master Station while local media, each operating a 9600 bits per second, are managed within the substation.

The SCADA Module and Protection Modules are read once each second and the data is stored by the SCADA Data Acquisition Function of the Substation host for subsequent retrieval by the SCADA Master Station.

The Report by Exception Processing function emulates a SCADA Remote Terminal Unit in identifying those data that have changed since last reported to the SCADA Master Station. The changed data are flagged for use by the Communications Processor in response to a data acquisition command from the SCADA Master Station. The Report by Exception Processor is also responsible for identification of which Protection Module(s) have disturbance data tables that should be sent to the SCADA Master Station for analysis. The movement of disturbance data tables is managed by the Communication Processor.

A Man-Machine Interface capability with IMPACS substation equipment is provided by the Substation Host. A permanent hard copy logger is used for a continuous sequence-of-events report of (at least) protective relay actions and IMPACS diagnostic messages. A portable hard copy computer terminal can also be used in support of installation activities and to read or modify protective relay settings in one more IMPACS Protection Modules. If the substation is manned, a computer terminal could be part of the permanent complement of equipment.

The IMPACS System continuously performs diagnostic functions. The Substation Host and Protection Module diagnostics are designed to complement each other. The Substation Host, in addition to acquiring SCADA data from all modules every second, performs sample count and data comparison checks. A sample count that was not advancing in a Protection Module would mean that a malfunction had occurred in the data sampling circuitry. When two or more IMPACS Line Protection Modules are used in the same substation, certain currents and voltages are likely to be read by two adjacent Line Protection Modules. The Substation Host, by comparing these redundant measurements, can determine when the difference is greater than normal. An alternative to the data comparison diagnostic is to read a known analog value. Each Protection Module can periodically use a known analog input as a calibration check of the data acquisition function. This last diagnostic would be useful in detecting drift in an analog to digital conversion circuit. In the presence of analog input drift, the IMPACS Protection Modeule would still perform its protection function; however, SCADA type data would have some inaccuracy.

IMPACS Line Protection Module

The Line Protection Module, with its own dedicated data acquisition subsystem, is subordinate to a local Substation Host. Alternately, a stand-alone Line Protection Module could be subordinate to a remote host (Substation or SCADA Master Station).

The IMPACS Line Protection Module may be applied in parallel with conventional protective devices on one end of a transmission line while only conventional relays are used at the opposite end of the line. Since the application of digital relaying technology is likely to parallel conventional relaying for many years, the IMPACS Protection Module is designed to be used in the relay rack of a conventional substation control building. Nothing in the architecture of IMPACS would prevent locating a Protection Module in the switchyard. To do so would, however, require a higher level of immunity to electromagnetic interference than is needed for a control house location.

The IMPACS Line Protection Module is capable of performing the following protection oriented functions.

- 3 Zone Step Distance Directional Comparison Blocking (Including Carrier Start/Stop)
- Single Phase Tripping
- Ground Time Overcurrent
- Location of Close-In Faults
- Lo-Set Trip On Breaker Reclose Into Fault
- Ultra High Speed Trip Output
- Dynamic Zones Of Protection
- Breaker Failure
- Out Of Step Blocking
- Close (Auto, Lockout, Check Sync)
- Load Compensation
- Carrier Received Delay Dropout

Using twelve samples per cycle, the IMPACS Line Protection Module can issue a trip at the time of the third data sample (1/4 cycle) for any fault within the nearest half of the transmission line. For faults in the furthest half of the line up to the Zone 1 setting, additional samples of the fault data are processed to insure an accurate decision. Most trip signals will be issued after six samples (1/2 cycle). In theory, a fault exactly on the Zone 1 boundary might not result in a trip; therefore, a time limit on the Zone 1 decision process could be set [3]. In addition to the above protection functions, the IMPACS Protection Module also provides the following:

- Current and Voltage Samples
- Watt and Var Flow Calculations
- Frequency and Rate of Change of Frequency Measurement
- Sequence of Events Recording
- Fault Data Capture
- Control Functions (i.e., Restoration Configuration, SCADA, etc)

The present line module configuration of prototype equipment at the Matt Funk Substation is illustrated in Figure 9.

IMPACS equipment can also be programmed to test protection support equipment, such as carrier, on a routine basis. Any failure of the carrier equipment would be alarmed to an operator at the SCADA Master Station.

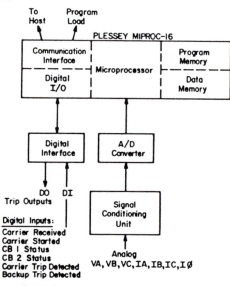

Figure 9. Present Matt Funk Configuration

Substation Test Set

This module is a recent addition to the IMPACS configuration; the first substation test set prototype was constructed in mid-1983. It is essentially a portable Substation Host Computer. Some speed of operation was sacrificed for a light weight and portable package that includes a hard copy terminal. The Substation Test Set can, as is illustrated in Figure 8, be connected to either a Protection Module or to the SCADA Master Station. The software for execution in the Substation Test Set is contained on tape and any data to be read from a Protection Module would be written on a blank tape. Upon power up, the software is automatically booted from tape and the unit is ready for operation in approximately two minutes. This seems reasonable when one considers the time necessary for a technician to follow normal maintenance setup procedures.

Advantages of the IMPACS Configuration

The IMPACS Line Protection Module, with its dedicated data acquisition and control hardware is the critical functional element. Once initialized with settings for the line to be protected, this module operates independently of the Substation Host. The advantages of an integrated architecture, however, are not realized when the Protection Module is isolated from a Host (Substation and/or SCADA Master). Therefore, to cite advantages of the IMPACS configuration, we will expect the Host levels to be functioning normally most of the time.

- IMPACS is compatible with present practice - IMPACS is installed in the existing control house like a SCADA Remote Terminal Unit. It parallels and/or works with conventional protection. IMPACS uses off-the-shelf micro-computers, SCADA communication protocol, and digital electronics typical of SCADA Remote Terminal Units. All the tools and techniques for electronic reliability engineering, pioneered by the military in the 1950's, apply to the IMPACS design.

- Functionally active at all times - By the continuous reporting of SCADA data and frequent use of diagnostics, the capability of the Protection Module to perform its primary function is known to a higher level of confidence than with conventional relays. Any failure to perform the non-protection functions, or of a diagnostic, results in maintenance service at that time, not after a failure to trip when a fault was present on the line.

- Cost Effective Maintenance Options - The diagnostic capabilities, in particular the analog calibration option, can be used to direct maintenance to the need rather than following a predefined cycle, even when no maintenance is required.

- Facilities Sharing - This feature of integration lowers both installation and operating costs. Connections to ct's and pt's provide both protection and SCADA data through the IMPACS Protection Module. The communication channel from the transmission substation to the SCADA Master Station is used for both SCADA and Disturbance Data Tables. The SCADA Master Station computer provides the data storage and computational environment for the protection related programs. As a result of the integrated sequence of events reporting and automatic disturbance data analysis, reporting, and plotting the need for an oscillograph in the substation is reduced (maybe eliminated).

- System Control Options - parameters, such as frequency and rate-of-change of frequency, provided in the IMPACS Protection Module can be used locally for protective functions or in the dispatch center to support system security applications. The capability to remotely alter protection criteria or settings can be more responsive to sudden changes in the network or generation than dispatching personnel to various substations.

Disadvantages of the IMPACS Configuration

- Unfamiliar Technology - An IMPACS Protection Module represents technology that is at least thirty years newer than that of the electromechanical relay installed on the next panel. Utility field personnel will need to be technically proficient in both technologies.

- Impact of a Failure - The failure of any IMPACS module can have a broader impact if the failure coincided with a fault on a line from that substation. Protection of the line would be available from backup schemes; however, some SCADA and disturbance data could be lost or delayed in being reported to the SCADA Master Station. The configuration is designed to store disturbance data; therefore, a temporary loss of a Host or communication channel will introduce a delay; however, data normally scanned by the SCADA Master Station could be lost.

Hardware vs. Software

The design and configuration of IMPACS uses data processing capabilities of hardware and of software. The field implementation of software is in a non-volatile media such as programmable read only memory (PROM).

IMPACS hardware "processing" is used to:

- Satisfy IEEE SWC
- provide anti-aliasing filtering
- Eliminate DC offset from analog samples

- Scale (multiply) magnitudes
- Auto range an 8 bit A/D Converter
- Perform complex arithmetic

IMPACS software processing:

- Recognizes transient initiation (e.g. fault)
- Computes Symmetrical Components using Discrete Fourier Transforms
- Calculates distance to the fault
- Implements zone and carrier logic
- Initiates trip commands
- Manages Disturbance Data
- Computes SCADA Parameters
- Performs diagnostics
- Manages communication channels

and performs everything else that needs to be done. Basically, the microcomputer is allocated those functions that are efficient in software. The dedicated functions and capabilities that can be accomplished in hardware, and would be time consuming in software are allocated to the hardware.

The Line Protection Module has been implemented in two different microcomputers. A microcomputer manufactured by Plessey of England was selected in 1977, programmed, tested in the laboratory, used for single phase tests at Kammer in 1979, and installed at Matt Funk substation in mid-1980. In 1982 the Motorola 68000 microcomputer was selected and will be used at Glen Lyn substation. The Motorola 68000 based microcomputer chips are available from multiple sources, the cost trend is downward while the performance trend of the family of chips is upward.

At the time of this writing, the final packaging of the Motorola 68000 based Line Protection Module is not complete; however, enough is known to make a reasonable cost comparison between IMPACS and present practice.

Cost Comparison IMPACS vs Present Practice

With reference to the IMPACS configuration illustrated in Figure 8, the costs associated with the integrated option will be compared to non-integrated alternatives. To make this comparison as realistic as possible, a marketplace value is assigned to the components of IMPACS. In assigning a marketplace value, the components were each compared to existing hardware and/or software modules used in similar applications. Futhermore, market value is associated with 100 copies of the Line Protection Module, 25 Substation Hosts and 5 Central Station installations. This assumption is necessary in the process of assigning a market value if a new product is to gain a share of the market. A manufacturer must base projections on some quantity, and the price of components (e.g. M68000 chips) is also influenced by quantity. No cost comparison of IMPACS, quantity one, would be realistic when the marketplace cost of non-integrated alternatives is based on hundreds of units manufactured.

Central Station Cost Comparison

In the SCADA Master Station, the IMPACS configuration requires the following resources:

- Disk space for FORTRAN programs and disturbance data tables
- Use of communication, computer and printer resources (on a priority lower than the SCADA function)

The increase in central station cost to accomadate IMPACS is estimated to be:

$5000 - License for IMPACS specific FORTRAN programs
$5000 - Additional disk space (if necessary)

Substation Host Cost Comparison

The IMPACS Substation Host, with its associated SCADA module and hardware necessary to communicate with one Master Station, is no more complex than a modern SCADA RTU and the cost is equivalent to that of the RTU that it replaces. The installation costs of the IMPACS Substation Host could be less then that of the SCADA RTU because some analog and control connections will be to the Protection Modules and not to the SCADA Module.

The marketplace cost of either the IMPACS Substation Host or the microcomputer based transmission RTU is $15,000 or more depending upon the number of data input and control output signals that are to be accommodated.

Line Protection Module Cost Comparison

The cost of the IMPACS Line Protection Module is the most difficult to estimate and an objective basis for comparison may be subject to comment.

The IMPACS Line Protection Module can be thought of as a line protection terminal. Solid state line terminals that implement the similar protection functions were quoted in 1983 at around $41,000. Alternately, a utility can assembly a panel of electromechanical relays to perform the line terminal functions of IMPACS, but not at the same speed of operation and without single phase tripping. This approach requires 11 individual relays and panel hardware with a total market price of $30,000. To this point, only protection functions have been considered; however, the IMPACS Protection Module provides disturbance data tables and sequence of event messages. Since these sequence of events and analog data are available and used for analysis of the disturbance, one might not use an oscillograph on the lines protected by IMPACS Protection Modules. The eight analog signals and the sequence of events data represent approximately 1/4 of the capacity of an installed oscillograph or $15,000.00. Therefore, the market value of state-of-the-art alternatives to an IMPACS Line Protection Module range from $45,000 to $66,000. The low end priced alternative does not have the equivalent speed of operation nor the single phase tripping capability.

The IMPACS Line Protection Module has analog input hardware for four voltages (3 line, 1 bus) and eight currents (4 each breaker) and digital input hardware for sixteen inputs and digital output hardware for eight outputs. The marketplace cost of the input/output subsystem is estimated at $18,000. The microprocessor module hardware, using a Motorola 68000 and including the Substation Host communication hardware, has an estimated marketplace cost of $11,000. The cost trends for these technological alternatives was, in 1983, downward which suggests better profit margins for the manufacturer or lower costs to the user. The firmware that implements the line protection functions discussed should also carry a marketplace value. A value of $15,000 is placed on each copy of the IMPACS Line Protection Module firmware. This value is based on the research and development man-years allocated to that portion of the work.

Based upon a 100 unit quantity, the IMPACS Line Protection Module marketplace cost is estimated to be $44000 which includes I/O Hardware, processor and power supply, and firmware

SUMMARY

AEP plans to continue its development and demonstration of IMPACS. The cost vs. performance trends for digital electronics continue to improve as does our confidence in the digital solution.

For line protection, we believe that an IMPACS Protection Module is competitive, both in performance and cost, with non-digital alternatives.

ASEA'S CO-OPERATING SYSTEM

ASEA in conjuction with the Swedish State Power Board (SSPB) has developed a microcomputer-based, bus-oriented control system which has been installed in a new substation connected to the main Swedish power grid. The station is now in commercial operation.

This microcomputer control system cooperates with the protective system, as a hybrid system. The control system can cooperate with an optional protection system.

Microcomputers have in many years been used in transmission substations as components in remote terminal units, event recorders and automatic switching equipment. They will certainly also be used in protective relays, fault locators etc. But this has not caused a basic system change. The control system structure has not changed, the microprocessor has just been used as a component inside some of the control products where it solves some limited functions to an attractive price.

However, microcomputers can be used for a complete reorganization of the control equipment. Its features can for instance be used as follows:

- Decentralized microcomputer-based units with dedicated functions and bus-communication which could save a lot of cabling.

- A microcomputer-based control system can be designed to supervise its own operation and also most of the operation of surrounding equipment.

- A microcomputer-based system can produce its own descriptive documents, thus assuring the documentation always being up-to-date.

- A microcomputer-based system can replace huge control boards.

In order to ensure that the new technology is technically & economically feasible, requirements fulfilled by traditional control equipments need to be analyzed. The life-cycle-cost including training, modification, extension, maintenance, etc. should be considered in making economic comparison.

Traditional Design

The practice for control and protection equipment varies between different utilities. Past experiences, different theoretical approaches, different educational background, etc. are factors that have influenced practices.

Figure 10 shows an example of the structure of a traditional substation control equipment. It has been used by SSPB for the last 10 years and has influenced the development of the new microcomputer-based equipment.

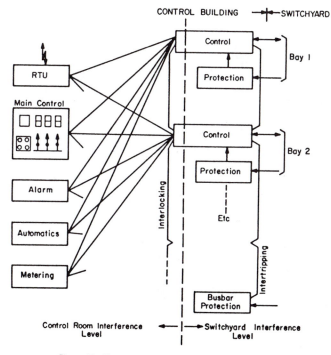

Figure 10. Example of Control Equipment Structure

The basic idea of SSPB's control system structure is to separate the switchyard interfacing equipment from other equipment in the control room. The separation is made in so called interface cubicles where all binary signals are converted from 110 V dc (switchyard) to 48 V dc (control room) and transducers are used for analog signals.

Each bay is equipped with one control interface cubicle and one or (for redundant protection) two cubicles with protective relays. The switchyard interference level is allowed only in these cubicles.

The interface cubicles contain operating and trip relays for isolators and breakers. They contain position repeat relays and supervisory equipment for all switching devices. They also contain additional relays for interlocking, voltage selection, etc. Transducers are installed for current, voltage and power measuring.

From the interface cubicles 48 V-signals are taken from and to other control equipment such as main control board, alarm system, automatic restoration equipment, event recorder and remote terminal unit.

Isolator interlocking, protection backup tripping, etc. are performed via communication buses between the interface cubicles.

The advantages of this arrangement are that standard control and protective modules can be used for most applications and that testing and modifications can be performed for a small part of the station without affecting other parts. The number of feeders and number of functions can easily be extended.

The relay protection has one primary and several secondary outputs. The primary output gives a trip impulse to one or several trip relays (one for each breaker). The secondary outputs can be start indication, trip indication, etc. The primary output must, contrary to the secondary outputs, be handled with highest speed and reliability. This way of thinking influences also the new generation of control equipment with microcomputers.

Dependability Between Functions

The introduction of microcomputers makes it possible to implement several functions into one hardware system. Therefore it is important to analyze to what degree the inter-dependability is advisable between functions that are performed in independent hardware units in traditional system:

- Protection equipment must be as independent as possible from other control equipment.

- There must be a backup for alarm transmitted to the remote control center even if the ordinary remote communication equipment is out of service.

- Each substation must be autonomous so that operation can continue without remote communication.

- The control equipment for each function such as feeders should be independent such that a single fault in the control system for a feeder will not disable other needed control functions for other feeders.

- The control equipment must be able to operate in the same environment (interference, temperature etc.) as the traditional one.

- The control equipment must be suited to the engineering organization of the user. Personnel lacking special programming knowledge must be able to write new application programs and to change existing programs.

- The control system should generate its own descriptive documents which guarantees that they always correspond to the actual design.

- The new type of equipment must be suited to the maintenance organization of the user. The system should be self-diagnostic and by help of step-by-step fault tracing manuals it must be possible to repair most of the faults at site by non-specialists.

- The hardware and software must be used in many applications (not special design for substations or still worse for one substation) which leads to better manufacturing tests, less expensive spare parts and more reliable software.

Technical Aspects on Integrating Differenct Functions

It is expensive for a bus-oriented microcomputer system to fulfill high speed requirements. It is therefore important to analyse the time requirements for all functions within the substation.

Tripping functions affecting a power system are normally very rapid. Also events and values used for later disturbance analysis must be registered with high time accuracy. Man-machine communication can on the other hand be slower.

Figure 11 shows approximate required processing and communication times for some functions in the substation. From this analysis it can be found that some functions require fast bus-communication while other functions can be slow but require a great number of data to be transmitted. A combination to the same bus can therefore hardly be economical.

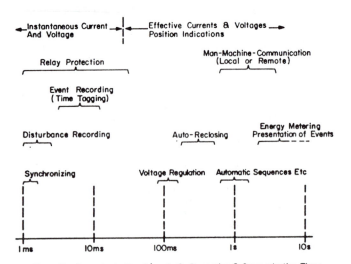

Figure 11. Approximate Requirements On Processing & Communication Times

If events are time-tagged by their input units, only protective relays, fault locator, disturbance recorder and synchronizing equipment require very fast operating times. All these functions are based on fast analysis of values from measuring transformers (currents and voltages). This is the background of a principle arrangement as illustrated in Figure 12.

There should probably be one high speed data bus for each feeder while the medium speed communication bus could cover the whole station.

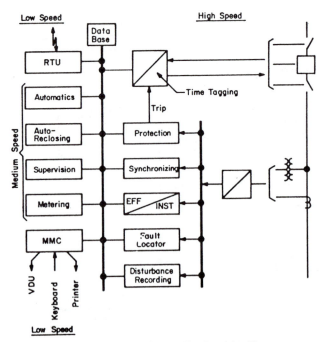

Figure 12. Design Approach Regarding Processing Times

374

Hierarchical Structure

To satisfy the above requirements, the control equipment is arranged at two levels as illustrated in Figure 13. The control and supervision of the substation is coordinated at the higher level (substation level) where also the bus protection is performed. From the lower level (unit level), individual feeders, transformers etc are controlled, protected and supervised. The functions should preferably be located at the lower level. But, if a function is common for several units, it has to be located at the higher level.

Unit control should be performed by a number of dedicated modules (including both control and protection) which means that the units are largely independent of each other. This increases the availability since any fault influences only a small part of the control equipment. Since the most important functions are located at unit level, the operation can be maintained even if a fault occurs at substation level.

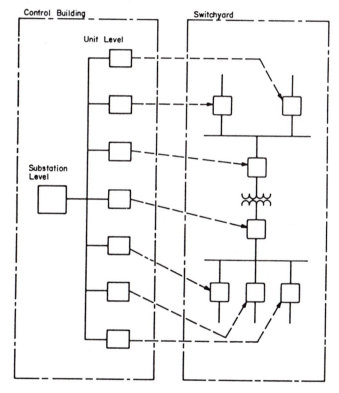

Figure 13. Two-Level Hierarchical Structure

Substation Control Level

The substation level, from the viewpoint of man-machine interface, has a lot of new facilities than traditional control boards have as illustrated in Figure 14. It is quite clear that many new functions can be implemented due to low cost. Examples of such new functions may be automatic operation sequences and double check of measured values.

But the substation level is only of small interest when discussing integration of control and protection. It is obvious that a computerized bus protection will be very complex and expensive to give the same excellent features as the traditional electronic protection. For instance the ASEA bus protective relay RADSS has an operating time of 2 ms with a modest requirements on current transformers, and is absolutely stable for external faults.

Unit Level Control

From the discussion above, it is preferred that protection and control functions on unit level should be installed in separate but cooperating hardware systems.

The medium speed equipment can easily be implemented in a microcomputer as indicated in Figure 14. However, the performance of the high speed functions cannot be economically improved by computer technique (at least with the present state of the art for microcomputer cost/performance). The features of solid state protective relays are normally good enough for today's requirements. However, the situation may be radically changed when new measuring methods with optronic have been developed, where the main savings will be for measuring transformers and transducers, switchyard layout and cabling.

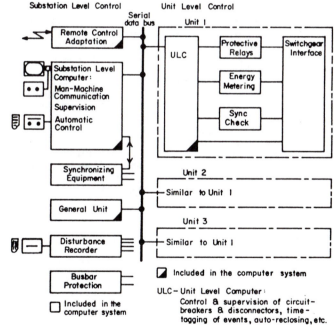

Figure 14. Allocation of Main Functions

Bus Communication

The equipment for unit level control can and should be located in separate cubicles for different feeders, transformers, etc, and connected to the substation computer by the medium speed communication bus. In this design the bus has an asynchronous serial communication, half-duplex with the polling principle and a transmission speed of 153.6 kbit/s. It has cyclic redundancy test and multi-drop connection. It is passive which means that it requires no power supply.

This design approach is particularly suitable for utilities using decentralized relay houses in the switchyards. The communication bus can then be an optical fiber.

No Need for a Double Computer Arrangement

The original design approach assumed two redundant substation computers and a double communication bus. It was compared to the single computer arrangement (see Figure 13) with a life-cycle-cost calculation for an important 400 kV substation. It was found, surprisingly, that the single computer arrangement gave such a high availability to the station that there were no need of a small increase of the availability by a more complicated and more expensive double computer arrangement.

This conclusion was possible due to the hierarchical design with protective functions decentralized to the unit level.

Experiences

The system has been carried out using components from a general purpose microcomputer system developed to be used as control equipment for power and industrial plants, but not especially designed for substations. This system has self-supervision. It has a very easy programming language and it produces its documentation automatically. It has been adapted to substation control by SSPB and installed and commissioned in a Swedish 220 kV-substation. Commissioning on site was carried out in less than two months and all tests have indicated desired performance. SSPB's design and maintenance staff have been trained to design and maintain the equipment. The hardware and basic software, programming and maintenance aids are taken from a standard ASEA-system called Asea Master.

Cost comparison indicates that the control equipment for a normal Swedish main grid substation (220 or 400 kV) would be cheaper with this microcomputer technique than with traditional equipment.

It is possible to gain advantages and avoid difficulties with the new microcomputer technique by designing a substation control system where control and protection are not integrated in the same computer. The development can be taken step by step and utilities can use control and protection systems from different suppliers.

Development of a co-operating computerized protection system should be done together with the development of new optronic measuring devices. The new type of protection must be able to communicate with the control system through the station bus.

Conclusions

This report has presented some advantages and disadvantages of integrating control and protection systems for transmission substations. First, the characteristics of the conventional system with its stand-alone subsystems and the reasons for its evolution to integrated system were reviewed, then two integrating systems and one hybrid system have been described and some specific concerns have been discussed such as security, environment, long-term availability, reliability, hardware vs. software, and cost comparison.

This report is by no means comprehensive, but it intends to encourage discussion of the implementation of this new technology. It is obvious whether integrate or not the microprocessor-based technology will emerge into control and protective equipment. While each utility user has to justify its own cost benefit for integrating, one common factor stands out; that is the cost for the input/output devices. Institution barriers have to be overcome for general acceptance of this new technology.

As many demonstration projects are gaining field experiences with positive results, it will accelerate acceptance by utilities. The instant accessibility of real time data on power systems and the capability of manipulating protective devices to suit for the changing power systems centrally will have long term impact on the availability and reliability of power supplies. While there is still a long way to go, it is recognized that any digital protective system should be developed to be compatible with future SCADA requirements and that the standardization of SCADA protocol is highly desirable now in order for the integration to be accomplished with the highest efficiency and the lowest cost.

References

1. J. S. Deliyannides and E. A. Udren, "Design Criteria for an Integrated Microprocessor-based Substation Protection and Control System," IEEE Transactions on Power Apparatus and Systems, Vol. PAS-101, pp. 1664-1673, 1982.

2. J. S. Deliyannides, "Recent Advances in Integration of Substation Digital Control and Protection - A Progress Report," Edison Electric Institute (EEI) Systems and Equipment Meeting, Atlantic City, NJ, May 10, 1983.

3. S. L. Nilsson, "EPRI Research and Development of New Substation Control and Protection Equipment," IEE (U.K.) Conference Publication 185, Developments in Power System Protection, London, June 1980, pp. 88-92.

4. E. A. Udren and M. Sackin, "Relaying Features of an Integrated Microporcessor-Based Substation Control and Protection System," Ibid, pp. 97-101.

5. J. S. Deliyannides, M. Kezunovic, and T. H. Schwalenstocker, "An Integrated Microprocessor-Based Substation Protection and Control System," IEE (U.K.) conference Publication 187, "Developments in Power System Monitoring and Control," London, June 1980, pp. 50-55.

6. "Digital Techniques for Control and Protection of Transmission Class Substations," Electric Power Research Institute (EPRI), Palo Alto, Ca, Report WS-79-184, August 1980.

7. "Substation Control and Protection Project - System Requirements Specification," EPRI, Palo Alto, CA, Report EL-1813, May 1981.

8. "Surge Withstand Capability Tests for Protective Relays," IEEE P472/ANSI C37.90a, 1983.

9. "Withstand Capability of Relay Systems to Radiated Electromagnetic Interference," IEEE p. 734/ANSI C37.90.2, 1983. (Trial Use Test Standard)

10. "Measurement and Characterization of Substation Electromagnetic Transients," EPRI, Palo Alto, CA, Report EL-2982, March 1983.

11. A. G. Phadke, M. Ibrahim, T. Hlibka, "Fundamental Basis for Distance Relaying with Symmetrical Components," IEEE Trans. on PAS, Vol. PAS-96, No. 2, March/April 1977 pp. 635-646.

12. A. G. Phadke, T. Hlibka, M. Ibrahim, M. G. Adamiak, "A Microcomputer Based Symmetrical Component Distance Relay," Proceedings of PICA, April, 1979.

13. J. S. Thorp, A. G. Phadke, S. H. Horowitz, J. E. Beehler, "Limits to Impedance Relaying," IEEE Trans. on PAS, Vol. PAS-98, No. 1, pp. 246-260 January/February, 1979.

14. A. G. Phadke, T. Hlibka, M. G. Adamiak, M. Ibrahim, J. S. Thorp, "A Microcomputer Based Ultra-High-Speed Distance Relay: Field Tests," IEEE Trans. on PAS, Vol. PAS-100, No. 4, pp. 2026-2036 April, 1981.

15. D. A. Hodges, J. C. Johnson, D. F. Koenig, R. D. Elliott "Computerized System Fault Analysis," IEEE Paper C 73 348-0.

16. S. Glazer "High-end, 16 Bit Microcomputers Gravitate Toward M C68000 chips," pp. 81-87 Mini-Micro Systems February, 1983.

17. K. P. Lau, W. R. Block, D. F. Koeing, B. D. Russell, D. M. Raikoglo, "SCADA's State-of-Art and Furture U.S. Trends in Substation Controls," CIGRE paper 23-02, 1982.

PROTECTION OF SF$_6$ GAS INSULATED SUBSTATIONS - INDUSTRY SURVEY RESULTS

A report prepared for the IEEE Power System Relaying Committee by the
Protection of Gas Insulated Substation Working Group - R. W. Dempsey, Chairman

Contributing Members: J. K. Akamine, E. A. Baumgartner, J. T. Emery, R. W. Haas, T. J. Murray

Abstract

This paper summarizes the result of an industry survey of gas insulated equipment protection practices and develops recommendations where necessary. Tables are included to show the type of gas insulated equipment located at each substation (current transformers, voltage transformers, switches, bus bars, bushings, lightning arresters, and cable end terminations), the equipment configuration (single or three conductors), the type of gas monitoring equipment used (density or pressure), the use of gas monitoring equipment (alarm and/or trip), unique relaying protection applications, and unique operating procedures. Gas insulated circuit breakers are specifically excluded from this survey.

Introduction

The use of SF$_6$ (sulfur hexafluoride) gas insulated substations by the electric utility companies has become widespread due to space limitations and economic considerations. Gas insulated substations are now used not only at EHV, but at other voltage levels, as well as at nuclear and fossil fuel plants.

Using a gas insulated substation design in no way limits the application of basic design concepts such as single bus, dual bus, breaker-and-one-half or ring bus in laying out the substations.

The use of gas insulated equipment by the electric utility companies has become more popular as the technology has evolved.

The use of gas insulated equipment has developed concerns over the application of protective devices for the gas insulated equipment. These concerns generated the development of this working group by the IEEE Power System Relay Committee.

To ascertain what the protective relaying practices are for various types of gas insulated substations, a survey was made of the electric power utilities to determine what unique protection aspects are involved in the installation of gas insulated equipment. The survey did not cover the use of gas breakers.

The purpose of this paper is to summarize the results of the survey and develop recommendations where necessary. It is not the intent of this paper to list all protective design considerations for a substation utilizing a gas bus design, but only those design considerations that are unique to the protection of gas insulated equipment.

87 WM 109-2   A paper recommended and approved by the IEEE Power System Relaying Committee of the IEEE Power Engineering Society for presentation at the IEEE/PES 1987 Winter Meeting, New Orleans, Louisiana, February 1 - 6, 1987. Manuscript submitted September 8, 1986; made available for printing November 12, 1986.

Types of Equipment

Tabulated in Table I are the type of gas insulated equipment, voltage class, BIL, current carrying capability and configurations installed in North America.

Use of Monitoring Equipment

To maintain the integrity of the gas insulation medium, all users of gas insulated equipment monitor either the pressure or density (temperature compensated pressure) within the gas enclosure (Table I).

The manufacturers recommend the pressure levels for alarming. All respondents provide for at least one low pressure/density alarm.

The most predominant setting is 90% of normal system pressure/density. If a second low pressure/density alarm is provided, it is set at approximately 80% of normal system pressure/density.

More than 50% of the respondents indicated that in addition to alarming at the second low pressure/density level, they also automatically isolate the specific equipment. While a few of the respondents reported providing a high pressure/density alarm, none indicated a practice of isolating (tripping) for high pressure/density.

In general, most have not experienced a high failure rate or unreliability associated with monitoring equipment. Comments did indicate that some (3) had mechanical relay failures and that the relay required calibration semi-annually. Another noted that temperature transducers failed due to moisture.

Based on experience, recommendation to alarm for a low gas density/pressure condition was unanimous (see Figure 1). Recommendation to trip for a low gas density condition was evenly divided among users; however, a recommendation to trip for a low gas pressure was unanimous among users. The majority indicated that high gas density/pressure monitoring is of little value.

Reclosing

A strong preference (75%), was indicated, not to reclose manually or automatically after a fault which was positively identified as being in the SF$_6$ equipment. For faults that are detected by protection that cannot differentiate between faults internal or external to the SF$_6$ equipment, about half of the respondents would reclose. Only one fourth of the group considered the magnitude of fault current in determining their reclosing policy.

Auxiliary Equipment

Most installations did not contain surge arresters inside the GIS equipment. The ones that used the arresters inside the GIS equipment used the gapped type arresters, as well as metal oxide arresters. Most companies reported experiencing no problems with SF$_6$ leaks into the gap chamber of the arrester.

Reprinted from *IEEE Transactions on Power Delivery*, vol. PWRD-2, no. 4, pp. 1053–1059, Oct. 1987.

In most applications the location of potential transformers was internal to the SF_6 equipment. All current transformers were located external to the gas environment.

Most users indicated that they require view ports to check the position of isolators, switches, and/or ground switches inside the GIS equipment. Some users required grounding switches for each SF_6 section, and others required a key interlock on all ground and line switches. Most utilities require both current and potential secondary circuits to be located outside the gas environment.

Unique Relaying/Operating Practices

Table II contains the list of unique relaying practices based on the areas of concern and Table III contains the lists of unique operating practices based on areas that are considered when employing GIS equipment.

About half of the users reported using unique relaying applications with GIS equipment. However, many things are considered when operating requirements are determined for the gas equipment. These include such things as switching visibility requirements, extensive interlocks, safety grounding switches, and special instructions/guidelines. Note that only half of the users reported using what they considered "unique" operating procedures. Half of the companies have the same procedures with GIS equipment as with standard equipment.

Conclusions

Types of Equipment

Gas insulated equipment has been applied at many different voltages, but mainly at 138kV and above. While three phase equipment has been enclosed in one pipe, the single phase conductor configuration is predominantly used.

Use of Monitoring Equipment

With the choice of monitoring pressure or density most users prefer gas density monitoring.

There is extensive application of pressure/density monitoring devices to insure that the gas environment remains within specification. However, there is no common agreement as to whether or not to automatically isolate the equipment under certain abnormal conditions. This decision is based on individual utility philosophy as well as manufacturer's recommendations.

Reclosing

Use of automatic or manual reclosing in any application is dependent upon an evaluation of (1) the possibility of whether a fault in the protected zone will be transitory or permanent; (2) the need for minimizing time of interruption to service following a fault; and (3) the potential damage to equipment and safety to personnel. In GIS installations the effect of the increased probability of "burn thru" (sic) of the outside enclosure and internal damage to equipment as a result of reclosing into a permanent fault are factors that must be taken into account when evaluating the above considerations.

It was generally found that for a fault within the gas insulated equipment, reclosing would be unsuccessful.

Auxiliary Equipment

It was generally considered inadvisable to locate gapped arresters inside the bus because of potential contamination problems.

Unique Relaying/Operations Practices

Most utilities are concerned with the protection of the equipment and if that protection is adequate.

The unique relaying practices generally deal with (1) special bus differential protection and (2) detecting loss-of-gas condition.

The use of differential protection for the equipment seems a must for all users. Solid state differential protection to provide greater speed in clearing is widely used. However, high speed clearing is not a concern among all users. The speed of the protection seems to depend more on the power system concerns, as does protection for conventional equipment.

There seems to be little difference between the way the GIS equipment is protected and the way standard equipment is protected.

The application of switching devices in the equipment raises different concerns within the industry. Some utilities require view ports to verify correct operation while others demand no view ports in order to protect against eye damage in the event the switch may operate while being observed. Again, individual utility philosophy must determine desired practice.

While the use of gas insulated equipment in substations may cause changes in operating practice, the application of standard protection practices recommended for conventional equipment seems adequate for the gas equipment.

Company #	Substation #	CT	VT	Mtr Op Sw	Disc Sw	Grd Sw	Bus Bars	Bus Bshing	L A	Cable Term	Voltage Class kV	BIL kV	Cont Curr Current kA	Avail. fault 3 Ph kA	Avail. fault 1 Ph kA	Sngl Cond	Three Cond	Normal PSIG	Low Density Alarm PSIG	Low Density Iso/Trip PSIG	High Density Alarm PSIG	High Density Iso/Trip PSIG	Low Pressure Alarm PSIG	Low Pressure Iso/Trip PSIG	High Pressure Alarm PSIG	High Pressure Iso/Trip PSIG	Note 4
4	1	X	X	X	X	X	X	X	X	X	145	650	1.25	21.9	17.1	X		47.3	45 43.5								
4	2	X	X	X	X	X	X	X	X	X	145	650	2.00	12.7	9.02	X		50									Note 1
4	3	X		X	X	X	X	X	X	X	145	450	3.00	19.9	22.6	X		47.3	45 43.5								
4	4	X	X	X	X	X	X	X	X	X	800	2000	3.00	5.86	5.33	X		47.3	45 43.5								
4	5	X		X	X	X	X	X	X	X	550	1550	3.00	12.6	10.6	X		42.2	40.7		65.5						
4	6	X	X	X	X	X	X	X	X	X	145	650	3.00	22.7	22.2	X		46.5	40.7		65.5						
13	1	X	X	X	X	X	X	X	X	X	500	1550	4.08	40.4	43.0	X		44.1	40.0 35.0								
13	2	X	X	X	X	X	X	X	X	X	500	1550	3.00	40.4			X	51.5	46.3 41.2								
13	3	X	X	X	X	X	X	X	X	X	230	900	2.00	45.2	45.0	X		44.1	39.7 35.3								
13	4	X	X	X	X	X	X	X	X	X	230	900	4.00	66.5			X	62	55.8 50.0								
13	5	X	X	X	X	X	X	X	X	X	230	900	2.00	50.2	50.0	X		51.5	46.3 41.2								
13	6	X	X	X	X	X	X	X	X	X	230	900	2.00	40.2	40.0	X		56.8	51.1 45.4								
13	7	X	X	X	X	X	X	X	X	X	230	900	2.00	15.6	15.0	X		56.8	51.1 45.4								
13	8	X	X	X	X	X	X	X	X	X	500	1550	4.00	35.8	40.0	X		56.8	51.1 45.4								
13	9	X	X	X	X	X	X	X	X	X	230	900	2.00	50.2	50.0	X		50.0	45.0 40.0								
16	1		X		X	X	X				550	1550	2.00	14.0	12.0	X		55.0					50.0 45.0				
20	1					X	X	X			345	1050	2.00	17.8	12.8	X		30.0	25.0-27.0 20.0-22.0								
32	1	X	X	X	X	X	X	X	X	X	345	1300	3.00	47.0	47.0	X		30.0	27.0 24.0								

Table I Continued

Company #	Substation #	CT	VT	Mtr Op	Disc	Grd	Buss Bars	Bush ing	LA	Cable Term	Voltage Class kV	BIL kV	Cont Curr kA	Carry ing Cap kA	Three Ph kA	Single Ph kA	Sng Conn d	Three Conn d	Normal PSIG	Low Density Alarm PSIG	Low Density Isol/Trip PSIG	High Density Alarm PSIG	High Density Isol/Trip PSIG	Low Pressure Alarm PSIG	Low Pressure Isol/Trip PSIG	High Pressure Alarm PSIG	High Pressure Isol/Trip PSIG
32	2	X	X	X	X	X	X	X	X	X	345	1300	3.00		47.0	47.0	X		30.0	27.0 24.0							
32	3	X	X	X	X	X	X	X	X	X	345	1300	3.00		47.0	47.0	X		30.0	27.0 24.0							
35	1									X	115	550	16.0		21.9	24.3	X		30					27.0 24.0			
40	1		X		X	X	X	X	X		345	1050	2.00		22.3	22.2	X		22.0	19.0							
40	2		X		X	X	X	X	X		345	1050	2.00		25.7	21.2	X		30.0	26.0							
40	3		X		X	X	X	X	X		345	1050	2.00		10.0	9.20	X		30.0	26.0							
50	1	X	X	X	X	X	X	X		X	138	650	2.00		41.7	41.6	X		50.0	47.0 45.5							
50	2	X	X	X	X	X	X	X		X	138	650	2.00		39.0	31.6	X		30.0	27.0 25.5							
55	1					X	X				242	900	3.76		53.0	35.0	X		52.0					47.0 5.0	5.0		
56	1	X	X	X	X	X	X	X	X	X	345	1050	2.00		23.5	22.0	X		30.0	26.0		22.0					
56	2	X	X	X	X	X	X	X	X	X	138	650	3.00		18.0	13.0	X		30.0	43.5		37.0					
60	1	X	X	X	X	X	X			X	138	650	1.6		40.0		X	X	30.0					15.0	10.0	45.0	
65	1					X					138	650	1.60		12.1	7.80	X		26.0	23.4		20.8					
70	1	X		X	X	X	X	X	X		500	1550	3.00		63.0			X	55	43.3 36.0							
70	2	X		X	X	X	X	X	X		230	900	3.00		63.0			X	55	43.3 36.0							
70	3	X		X	X	X	X	X	X		500	1550	3.00		63.0			X	55	43.3 36.0							
70	4	X		X	X	X	X	X	X		500	1550	3.00		63.0			X	55	43.3 36.0							
85	1				X		X				230	900	2.00 3.00		7.18	12.9	X		30	26.0 22.0							

Table I Continued

Company #	Subst. #	CT	VT	Mtr Opc Sw	Disc Sw	Grd Sw	Bus Bars	Bushings	LA	Cable Term	Voltage Class kV	BIL kV	Carry Cont Curr kA	Three Ph kA	Single Ph kA	Sng Conn	Thr Conn	Low Density Normal PSIG	Low Density Alarm PSIG	Low Density Isol/Trip PSIG	High Density Alarm PSIG	High Density Isol/Trip PSIG	Low Pressure Alarm PSIG	Low Pressure Isol/Trip PSIG	High Pressure Alarm PSIG	High Pressure Isol/Trip PSIG	
85	2	X		X		X		X			500	1550	3.00	11.7	11.3	X		45	40.0 35.0								
89	1						X	X			500	1550	3.00	11.6	9.42	X		45					36.0 32.0				
113	1	X	X	X	X	X	X	X		X	121	550	2.00	30.0	32.0	X							26.0	22.0			
113	2	X	X	X	X	X	X	X		X	362	1050	3.00	15.0	16.0	X		30.0					26.0	22.0			
120	1		X		X	X	X	X		X	138	650	2.00	48.0	53.0	X		30.0	27.0	24.0							
133	1	X	X	X	X	X	X	X		X	230	90	3.00	80.0	80.0	X		66.7	63.8	60.9							
133	2	X	X	X	X	X	X	X		X	500	1550	4.00	80.0	80.0	X		66.7	63.8	60.9							
133	3	X	X	X	X	X	X	X		X	500	1550	4.00	100.	100.	X		66.7	63.8	60.9							
137	1	X			X	X	X	X	X		500	1550	3.00	20.1	15.7	X		65.0	57.5	49.3					80.0		
137	2	X	X	X	X	X	X	X			500	1550	3.00	15.2	13.4	X		65.0	57.5	49.3					80.0		
148	1			X	X	X	X			X	138	550	1.60	9.71	9.68	X		30.0	27.0	24.0	45.0	45.0					Note 2
148	2			X	X	X	X			X	138	650	1.60	15.2	14.6	X		50.0	20.0	20.0	60.0	60.0					Note 3
149	1	X		X		X		X	X		345	1050	4.00	27.1	32.5	X		30.0	26.0	22.0							
171	1	X	X	X	X	X	X	X	X	X	500	1550	3.25	17.0	20.0	X		55.0	42.0	36.0							
181	1	X	X		X	X	X			X	145	650	2.00	29.4	29.0	X		50.0	46.5	40.0							
182	1						X	X			345	1050	1.60	13.5	15.2	X		30.0					27.0 24.0				

Note 4 (applies to Use of Gas Monitoring Equipment section)

Notes

1. Differential pressure monitoring set for alarm at 6.5 PSIG.

2. Bus has high pressure oil to SF_6 interface. Alarms are given by loss of high-pressure oil, high pressure gas in pot-head, and presence of oil in bus. Trip also on any combination of two of the above.

3. Differential density between phases and low density or high density causes trip.

4. To be consistent, all gas pressure or density measurements are given in PSIG at $20^{\circ}C$ or $68^{\circ}F$.

TABLE II
PROTECTION OF GAS INSULATED SUBSTATION, SURVEY REPORT
UNIQUE RELAYING PROTECTION APPLICATIONS
RESPONSE BY COMPANY

CO #	AREA OF CONCERN	UNIQUE RELAYING APPLICATIONS
#4	Differential	SF_6 feeder current differential and GITL (Gas Insulated Transmission Line) lead differential which detects sheath currents
#13	Differential	Extra set of high speed bus differential relays and CT's
#32	Differential	Instantaneous overcurrent relays fed by differentially connected CT's in each bushing trip and lockout portion of gas bus section within line zone protection
#56	Gas Conditions	Two-level gas density detection for alarm and trip
#60	Gas Conditions	Two-level gas density detection for alarm and trip
#70	Gas Conditions	Two-level gas pressure detection for alarm and trip
	Ferroresonance	One-ohm loading resistor on potential transformers mounted on the bus to prevent ferroresonance during switching
#113	Ampere Level	Low current level (3 amperes or less) permissive for automatic opening of disconnects
#120	Gas Conditions	Two-level gas density detection for alarm and trip
#171	Differential	Bus Differential zone within the line protection zone locks out line
	Gas Conditions	Gas Density detection for alarm and trip

TABLE III
PROTECTION OF GAS INSULATED SUBSTATIONS, SURVEY REPORT
UNIQUE OPERATING PROCEDURES
RESPONSE BY COMPANY

CO #	AREA OF CONCERN	UNIQUE OPERATING PROCEDURES
#4	Ground Switches	Ground switches will be used
	Alarms	BIO (Before improper Operation) alarm will be used
	Visibility	Disconnect and ground switch positions and all analog gauges will be visible from one location
#13	Ground Switches	Safety switches without interlocks will be used for each SF_6 section
#32	Reclosing Conditions	Reclosing of remote line breaker delayed two seconds to allow motor operated, gas-insulated disconnect switches to isolate a failed transformer on local end of line
#40	Interlocks	Key Interlock system between disconnect switches and safety grounding switches
	Visibility	Use of viewports to verify switch blade position
	Reading of Gas Conditions	Logging of gas bus temperatures and pressure periodically
#60	Switching Sequence	Electrical disconnecting of control power to motor operators
#70	Switching Sequence	Switch bus and unloaded transformer bank together to prevent ferroresonance at 230kV stations
#89	Instructions	Guidelines have been written for SF_6 equipment
#120	Interlocks	Key Interlocks on all line and ground switches
#149	Interlocks	Key Interlocks using both mechanical and electrical permissive devices
#171	Instructions	More restrictive procedures are being considered but have not been completed

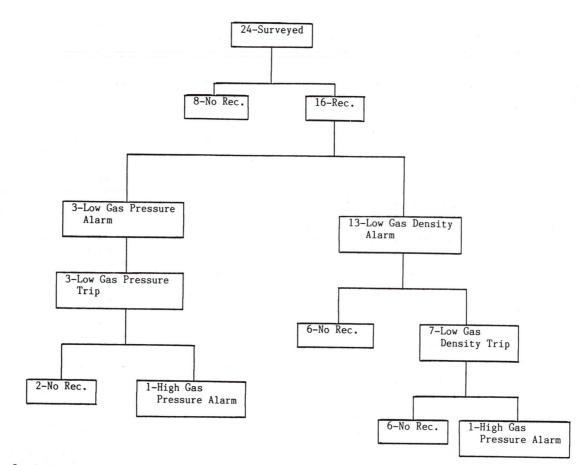

Figure 1 Graph showing the number of recommendations to alarm or trip for low gas density/pressure or high gas density/pressure conditions.

SHUNT CAPACITOR BANK PROTECTION METHODS

A working-group* report for the Power System Relaying Committee

J. A. Zulaski, Senior Member, IEEE
S&C Electric Company, Chicago, Illinois

Abstract - A "Guide for Protection of Shunt Capacitor Banks," ANSI/IEEE Standard C37.99-1980, has been prepared recently by the Power System Relaying Committee to assist in the effective application of relays for the protection of shunt capacitor banks used in substations. The various protective considerations along with recommended and alternate methods of protection for the most commonly used capacitor bank configurations are covered. The purpose of this paper is to review some of the highlights of the guide to familiarize the industry with the scope and usefulness of this new document, and to invite considered discussion for the purpose of improving this guide in the future.

INTRODUCTION

Maximum availability of a capacitor bank for service requires reliable protection which will isolate the capacitor bank from the system before it is exposed to severe damage and certainly before a fault is established on the system.

The first line of protection for a capacitor bank is the individual capacitor fuse, shown in a typical bank arrangement in Figure 1. The individual fuse senses and indicates the failure of a single capacitor unit and isolates the defective unit from the bank fast enough to prevent case rupture and subsequent damage to adjacent units, while permitting the remainder of the capacitor bank to remain in service.

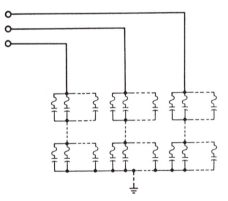

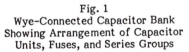

Fig. 1
Wye-Connected Capacitor Bank
Showing Arrangement of Capacitor
Units, Fuses, and Series Groups

---

*Members of the Working Group are: J. E. Stephens, Chairman, E. A. Baumgartner, R. E. Hart, W. R. Lund, M. Neumann, D. Zollman, J. A. Zulaski.

81 SM 348-2   A paper recommended and approved by the IEEE Power System Relaying Committee of the IEEE Power Engineering Society for presentation at the IEEE PES Summer Meeting, Portland, Oregon, July 26-31, 1981. Manuscript submitted March 31, 1981; made available for printing April 13, 1981.

Unfortunately, the isolation of a failed capacitor unit by its fuse results in an increase in voltage across the remaining units within that same series group. It is well known that the expected life of a power capacitor is dramatically affected by excessive overvoltage. ANSI/IEEE Std. 18-1980, IEEE Standard for Shunt Power Capacitors indicates that capacitors shall be capable of continuous operation at up to 110% of rated terminal rms voltage including harmonics, and that higher voltages should be permitted for only a short time as shown in Table 1. Extensive capacitor bank damage as a consequence of continuous excessive overvoltage should be prevented by means of protective relays which trip the bank-switching device to remove the bank from service.

Table 1
Limits of Short Time Power Frequency Overvoltage at Subzero Temperatures

Duration	Multiplying Factor Times Rated rms Voltage
0.5 c	3.0
1.0 c	2.7
6.0 c	2.2
15.0 c	2.0
1.0 s	1.7
15.0 s	1.4
1.0 min	1.3
5.0 min	1.2
30.0 min	1.15

NOTE: The short time power frequency overvoltage should be limited to the values listed here at subzero temperatures. Higher limits may be permissible with less severe conditions.

These limits are for emergency or infrequent conditions. See IEEE Std 18-1980.

Most capacitor bank installations will require an individual engineering analysis to determine the best and most economical relay protection method. The engineer will want to take a systems approach since bank design, fuse coordination, and selection of a sensing device will directly affect sensitivity and timing requirements of the protection method. Selection of the bank configuration and the bank design should include an analysis of the effect of "inherent unbalance" on the performance of the protective relay. Inherent unbalance resulting from capacitor-unit manufacturing-tolerance variations and/or system voltage unbalance may introduce errors to the voltage or current signal being monitored by the protective relay.

GENERAL UNBALANCE RELAY CONSIDERATIONS

Protective relays for capacitor-bank protection have been termed unbalance relays because they detect the isolation of an individual capacitor unit by its fuse and thus respond when an unbalance exists within the capacitor bank. In general, the capacitor-bank unbalance relay should:

(1) coordinate with fuses so that the latter provide a convenient visual means of locating defective capacitor units;

Reprinted from *IEEE Trans. Power App. Syst.*, vol. PAS-101, no. 6, pp. 1305-1312, June 1982.

385

(2) be sensitive enough to alarm on loss of one capacitor unit and trip on loss of that number of individual capacitor units that will cause a group overvoltage condition in excess of 10% over rated voltage;

(3) have time delay short enough to minimize damage due to an arcing-type fault within the bank structure and yet long enough to avoid false operations due to inrush, ground faults on the line, lightning, switching of nearby equipment, or nonsimultaneous pole operation of the energizing switch;

(4) be protected against transient voltages appearing on control wiring;

(5) include a filter to minimize the effect of harmonics;

(6) have a lockout feature to preclude automatic closing of the capacitor bank after a fault within the bank has been detected; and

(7) provide a compensating means to negate the effect of system voltage unbalance and/or capacitor-unit manufacturing-tolerance variation.

INHERENT UNBALANCE

The trend in recent years has been toward larger and larger capacitor banks at transmission voltage levels. The unbalance signal due to the loss of one or two individual capacitor units for these very large banks is such that the inherent unbalance can no longer be considered negligible. In practice, the voltage or current detected by the unbalance relay due to loss of individual capacitor units is somewhat different from the calculated value because of inherent unbalance which exists on all capacitor bank installations, and which is primarily due to system voltage unbalance and/or capacitor-unit manufacturing-tolerance variations. The inherent unbalance signal may be at such a phase angle in relation to the signal resulting from isolation of faulted capacitor units as to prevent protective-relay operation or to cause false operations. Inherent unbalance

resulting from system voltage unbalance and/or capacitor-unit manufacturing-tolerance variations for various configurations may be estimated using equations shown in Table 2. A worst-case estimate can be made by assuming the unbalance errors to be additive. When the error signal due to inherent unbalance exceeds 50% of the unbalance signal resulting from isolation of a single capacitor unit, a means of negating the influence of this error should be considered.

NEUTRAL CURRENT UNBALANCE PROTECTION METHOD: GROUNDED WYE CONNECTED CAPACITOR BANK

Figure 2 shows a neutral unbalance relay protection method for a grounded-wye capacitor bank. An unbalance in the capacitor bank will cause current to flow between the

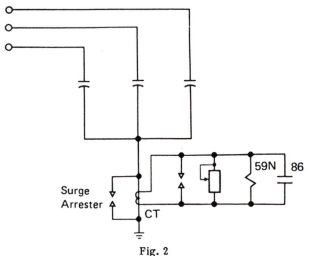

Fig. 2
Neutral Current Unbalance Protection Method —
Grounded Wye-Connected Capacitor Bank

Table 2
Effect of Inherent Unbalance on Displacement Signal

Shunt Capacitor Bank Configuration	Effect of Capacitor Manufacturing Tolerances	Effect of System Voltage Magnitude Changes	Effect of System Voltage Phase Angle Change
Grounded Y with Neutral Current Sensing	$I_N = \dfrac{\Delta C \, var_B}{3 \, V_{LG}}$	$I_N = \dfrac{(\Delta V_{LG}) P \, var_U}{S \, (V_C)^2}$	$I_N = \dfrac{2 P V_{LG} \left(\sin \frac{\phi}{2}\right) var_U}{S \, V_C^{\,2}}$
Ungrounded Y with Neutral Potential Sensing	$V_{NG} = \dfrac{\Delta C \, V_{LG}}{3}$	$V_{NG} = \dfrac{\Delta \, V_{LG}}{3}$	$V_{NG} = \dfrac{2}{3} \left(\sin \frac{\phi}{2}\right) V_{LG}$
Ungrounded Double Y with Neutral Differential Potential Sensing	$I_N = \dfrac{\Delta C \, var_B}{6 \, V_{LG}}$	$I_N = 0$	$I_N = 0$
Ungrounded Double Y with Neutral Differential Potential Sensing	$\Delta V_{NN} = \dfrac{\Delta C \, V_{LG}}{3}$	$\Delta V_{NN} = 0$	$\Delta V_{NN} = 0$
Grounded Y with Differential Potential Sensing	$\Delta V_{TG} = \Delta C \, V_{LG} \dfrac{S_T}{S^2} (S - S_T)$	$\Delta V_{TG} = 0$	$\Delta V_{TG} = 0$

See appendix for definition of symbols.

capacitor bank neutral and ground. The amount of current due to loss of individual capacitor units and the voltage across the remaining capacitor units can be determined from Figures 3 and 4 respectively.

harmonic filter. The relay may include provisions to compensate for the fixed tap-point error voltages caused by inherent capacitor-bank unbalance, fixed system voltage unbalance, and voltage-sensing device ratio errors.

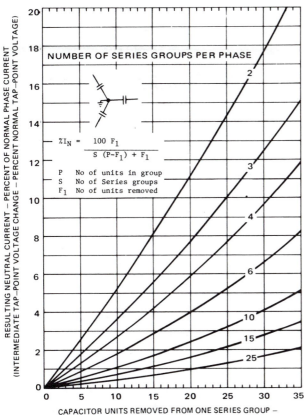

Fig. 3
Grounded-Wye-Connected Capacitor Bank:
Neutral Current (also change in intermediate tap-point voltage) Versus Percentage of Capacitor Units Removed from Series Group

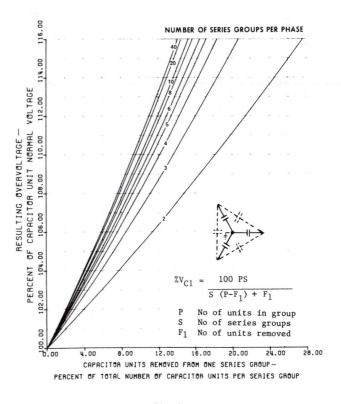

Fig. 4
Grounded Wye-Connected or Delta-Connected Capacitor Bank: Voltage on Remaining Capacitor Units in Series Group Versus Percentage of Capacitor Units Removed from Series Group

This unbalance protective method utilizes a properly burdened current transformer connected between the capacitor-bank neutral and ground, plus a time-delay voltage relay having a third-harmonic filter for reduced sensitivity to frequencies other than 60 Hertz. This voltage relay operates a latching relay to initiate the opening and to block the closing of the capacitor switch.

SUMMATION OF INTERMEDIATE TAP-POINT VOLTAGE PROTECTION METHOD: GROUNDED WYE CONNECTED CAPACITOR BANK

An unbalance voltage protection method, as illustrated in Figure 5, provides a means of detection of an unbalance in a capacitor bank by monitoring the departure from zero of the summation of voltages at the tap points of the three phase legs. The tap-voltage percent unbalance due to the loss of individual capacitor units and the voltage across the remaining capacitor units can be determined from Figures 3 & 4 respectively. This protective method utilizes three voltage-sensing devices with primaries connected between the capacitor-bank intermediate tap point of each phase and ground, and secondaries connected in broken delta, plus a time-delay voltage relay with a third-

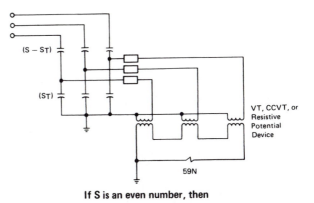

If S is an even number, then

$$S - S_T = S_T = S/2$$

If S is an odd number, then

$$S_T = \frac{S-1}{2} \text{ and } S - S_T = \frac{S+1}{2}$$

Note: Summation may also be obtained by use of a summing amplifer.

Fig. 5
Summation of Intermediate Tap-Point Voltage Protection Method — Grounded Wye-Connected Capacitor Bank

VOLTAGE DIFFERENTIAL PROTECTION METHOD: GROUNDED WYE CONNECTED CAPACITOR BANK

Another approach for grounded banks, shown in Figure 6, uses three single-phase relays, each of which senses the difference between a capacitor-bank tap voltage and the bus voltage. For each phase, the tap voltage and bus voltage signals are initially adjusted to be equal, assuming that all capacitor units are sound, and that no fuses have operated, and therefore, under these conditions, capacitor-unit manufacturing-tolerance variations and system voltage unbalance variations are compensated. If the system voltage unbalance should vary, the relay system is still compensated since a given percent change in the bus voltage of a given phase results in the same percent change on the capacitor-bank tap voltage for that same phase. Any subsequent voltage difference between capacitor-bank tap-voltage and bus-voltage signals will be due to unbalances caused by loss of capacitor units within a particular phase leg. The amount of unbalance due to loss of individual capacitor units and the voltage across the remaining capacitor units can be determined from Figures 3 and 4 respectively.

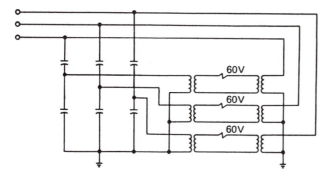

Fig. 6
Voltage Difference Protection Method —
Grounded Wye-Connected Capacitor Bank

DOUBLE WYE UNBALANCE PROTECTION

Four methods of providing unbalance protection for double wye banks are shown in Figure 7. Methods (a) and (b) are ungrounded and use either a current transformer with an overcurrent relay or a voltage transformer with an overvoltage relay connected between the two neutrals. Both of these methods are unaffected by system voltage unbalance, third harmonic voltage or current, or switching-surge currents. The current transformer or voltage transformer should be rated for system voltage. The amount of neutral current and the overvoltage on remaining capacitor units for the method illustrated in Figure 7(a) can be determined from Figures 8 and 9 respectively, while the neutral voltage and the overvoltage on the remaining capacitor units for the method illustrated in Figure 7(b) is determined from Figures 10 and 11 respectively.

In Figure 7(c) the neutrals of the two sections are grounded through separate current transformers to a common ground. The current transformer secondaries are cross-connected to an overcurrent relay so that the relay is insensitive to any outside condition which affects both sections of the capacitor bank in the same manner. The current transformers are subjected to switching-transient currents and require surge protection. They should be sized for single phase load currents if this is a possibility. The relay does not require a harmonic filter. The unbalance

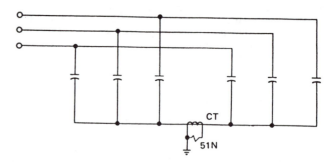

(a) Neutral Current Unbalance Detection Method — Ungrounded, Double-Wye-Connected Capacitor Bank

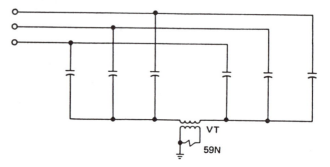

(b) Neutral Voltage Unbalance Protection Method — Ungrounded, Double-Wye-Connected Capacitor Bank (Neutrals Isolated)

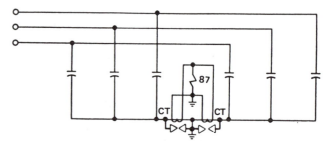

(c) Neutral Current Differential Protection Method — Grounded, Double-Wye-Connected Capacitor Bank

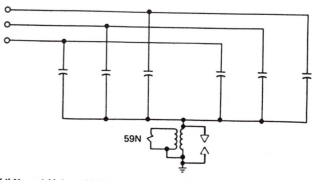

(d) Neutral Voltage Unbalance Protection Method — Ungrounded, Double-Wye-Connected Capacitor Bank (Neutrals Tied Together)

Fig. 7
Shunt Capacitor Bank Protection Methods —
Double-Wye-Connected Banks

388

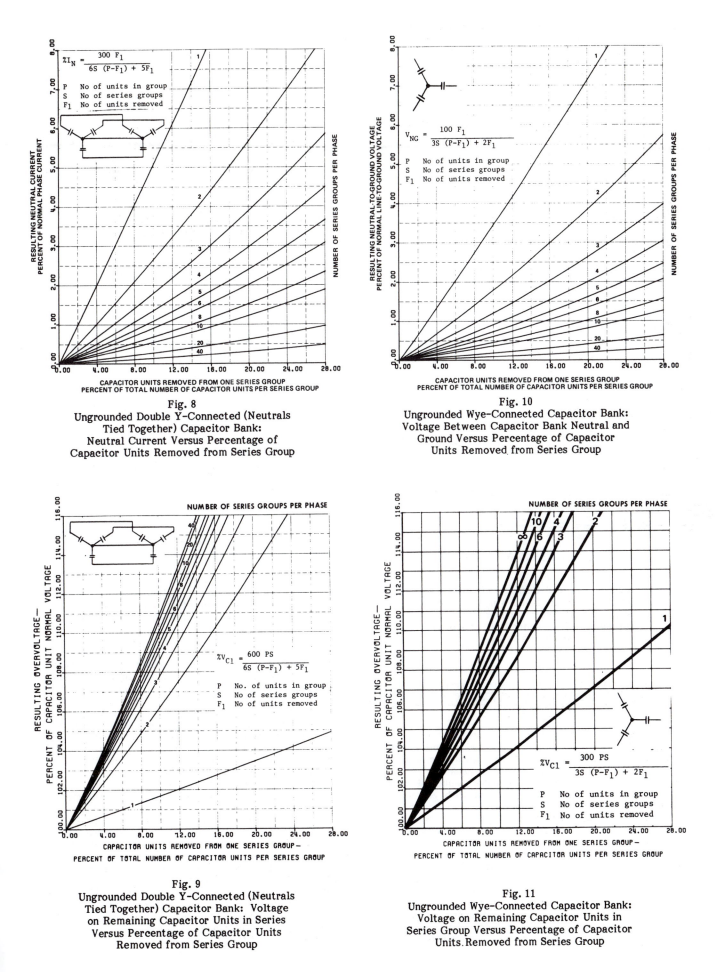

$$\%I_N = \frac{300\ F_1}{6S\ (P-F_1) + 5F_1}$$

P   No of units in group
S   No of series groups
F_1 No of units removed

RESULTING NEUTRAL CURRENT
PERCENT OF NORMAL PHASE CURRENT

NUMBER OF SERIES GROUPS PER PHASE

CAPACITOR UNITS REMOVED FROM ONE SERIES GROUP
PERCENT OF TOTAL NUMBER OF CAPACITOR UNITS PER SERIES GROUP

Fig. 8
Ungrounded Double Y-Connected (Neutrals
Tied Together) Capacitor Bank:
Neutral Current Versus Percentage of
Capacitor Units Removed from Series Group

$$V_{NG} = \frac{100\ F_1}{3S\ (P-F_1) + 2F_1}$$

P   No of units in group
S   No of series groups
F_1 No of units removed

RESULTING NEUTRAL-TO-GROUND VOLTAGE
PERCENT OF NORMAL LINE-TO-GROUND VOLTAGE

NUMBER OF SERIES GROUPS PER PHASE

CAPACITOR UNITS REMOVED FROM ONE SERIES GROUP
PERCENT OF TOTAL NUMBER OF CAPACITOR UNITS PER SERIES GROUP

Fig. 10
Ungrounded Wye-Connected Capacitor Bank:
Voltage Between Capacitor Bank Neutral and
Ground Versus Percentage of Capacitor
Units Removed from Series Group

NUMBER OF SERIES GROUPS PER PHASE

RESULTING OVERVOLTAGE—
PERCENT OF CAPACITOR UNIT NORMAL VOLTAGE

$$\%V_{C1} = \frac{600\ PS}{6S\ (P-F_1) + 5F_1}$$

P   No. of units in group
S   No of series groups
F_1 No of units removed

CAPACITOR UNITS REMOVED FROM ONE SERIES GROUP—
PERCENT OF TOTAL NUMBER OF CAPACITOR UNITS PER SERIES GROUP

Fig. 9
Ungrounded Double Y-Connected (Neutrals
Tied Together) Capacitor Bank: Voltage
on Remaining Capacitor Units in Series
Versus Percentage of Capacitor Units
Removed from Series Group

NUMBER OF SERIES GROUPS PER PHASE

RESULTING OVERVOLTAGE—
PERCENT OF CAPACITOR UNIT NORMAL VOLTAGE

$$\%V_{C1} = \frac{300\ PS}{3S\ (P-F_1) + 2F_1}$$

P   No of units in group
S   No of series groups
F_1 No of units removed

CAPACITOR UNITS REMOVED FROM ONE SERIES GROUP—
PERCENT OF TOTAL NUMBER OF CAPACITOR UNITS PER SERIES GROUP

Fig. 11
Ungrounded Wye-Connected Capacitor Bank:
Voltage on Remaining Capacitor Units in
Series Group Versus Percentage of Capacitor
Units Removed from Series Group

389

current and the overvoltage on remaining capacitor units can be determined from Figures 3 and 4 respectively.

In Figure 7(d) the neutrals of the two capacitor sections are ungrounded but tied together. A VT or potential device is used to measure the voltage between the capacitor bank neutral and ground. The relay should have a harmonic filter. The amount of neutral-to-ground voltage derived due to the loss of individual capacitor units can be determined from Figure 12 and the voltage on remaining capacitor units can be determined from Figure 9.

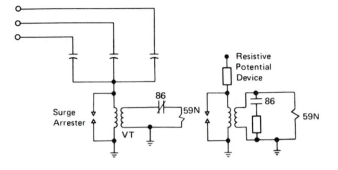

Fig. 13
Neutral Voltage Unbalance Protection Method —
Ungrounded Wye-Connected Capacitor Bank

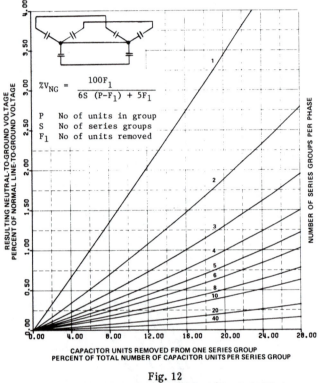

$$\%V_{NG} = \frac{100F_1}{6S\,(P-F_1) + 5F_1}$$

P No of units in group
S No of series groups
F_1 No of units removed

Fig. 12
Ungrounded Double Y-Connected (Neutrals Tied
Together) Capacitor Bank: Voltage Between
Capacitor Bank Neutral and Ground Versus
Percentage of Capacitor Units Removed
from Series Group

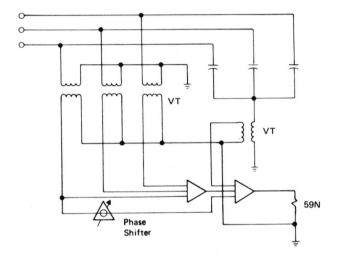

Note: Compensating voltage is system zero sequence voltage and a phase shifted voltage of one phase.

Fig. 14
Neutral Voltage Unbalance Protection Method —
Ungrounded Wye-Connected Capacitor Bank
With Compensation for Inherent Unbalance

NEUTRAL VOLTAGE UNBALANCE PROTECTION METHOD: UNGROUNDED WYE CONNECTED CAPACITOR BANK

Figure 13 shows a neutral unbalance relay protection method for an ungrounded, wye-connected capacitor bank. Unbalance sensing is accomplished by means of a voltage-sensing device connected between the bank neutral and ground. An unbalance in the capacitor bank will cause voltage to appear at the bank neutral with respect to ground. The amount of neutral voltage due to loss of individual capacitor units and the voltage on the remaining capacitor units can be determined from Figures 10 and 11 respectively.

This unbalance protective method consists of a time-delay voltage relay with third-harmonic filter connected across the potential-device secondary, and may include a means for compensation of inherent unbalance.

Figure 14 shows a method for compensation of system voltage unbalance on single ungrounded banks. Use is made of the fact that the voltage appearing at the capacitor-bank neutral due to system voltage unbalance is the zero sequence component. Another system-voltage-derived zero sequence component can be obtained utilizing three voltage-sensing devices with their high side connected from line to ground and the secondaries connected in a broken delta. The difference in voltage magnitude, due to system voltage unbalance, between the neutral signal and the broken-delta output of the system-connected voltage-sensing devices is then adjusted to zero by means of an amplitude control. Once this adjustment is made, the effect of system voltage unbalance will be compensated for all subsequent conditions of system unbalance.

To compensate for fixed inherent capacitor-tolerance unbalance it is necessary to generate an equal and opposite phasor to be summed with the fixed unbalance phasor to yield a null or zero signal output. The inherent unbalance at the neutral of the capacitor bank still exists, but the relay is no longer responsive to this fixed unbalance component.

A phase shifting network with amplitude and phase adjustment is used to generate the fixed compensating phasor. The input for the phase shifter should be bus derived to reduce the effect of nominal system voltage changes. The output of the phase shifter along with the inherent capacitor-tolerance unbalance signal can be summed by means of transformers or a summing amplifier.

OTHER CONSIDERATIONS

The Capacitor Protection Guide covers other topics related to capacitor bank protection which should at least be mentioned. An extensive list of references and a bibliography will be found useful for those engineers needing additional background material on specific subjects. Included are discussions of capacitor bank switching device requirements, inrush current during switching of parallel banks, grounding considerations, transient overvoltage phenomena, and bank design considerations that affect protection.

CONCLUSIONS

A guide for protection of shunt capacitor banks is now in print which should be of considerable value to the industry. This paper has attempted to cover the highlights of this new guide.

Discussion of shunt capacitor bank protection methods, particularly by experienced users, is encouraged to provide the Power System Relaying Committee with feedback that can be used to further improve this guide in the future.

APPENDIX
Symbol Definitions

V_{LG} = line-to-ground voltage, V (use maximum value where appropriate)

ΔV_{LG} = variation of V_{LG} between phases, per unit

S = $\dfrac{V_{LG}}{V_C}$ = number of series groups

S_T = number of series groups, sensing tap to ground

P = number of parallel capacitors per series group

F_1 = number of removed capacitors within the same series group within one phase

V_C = rated capacitor voltage, V

V_{C1} = voltage across the series group with removed capacitors, V

I_N = neutral current, A

ΔC = per unit variation of capacitance between phases

V_{TG} = voltage between intermediate tap point and ground, V

V_{NG} = neutral to ground voltage, V

V_{NN} = voltage between neutrals, V

var_U = individual capacitor reactive power rating, VA

var_B = capacitor bank reactive power rating, VA

B. Bozoki and **E. K. Christian** (Ontario Hydro, Toronto, Ontario, Canada): We commend the working group for producing a useful summary of the "Guide for Protection of Shunt Capacitor Banks" (ANSI/IEEE C37.99-1980). The Guide outlines both the complexities involved in, and the need for a systems approach to, some of the problems that can be encountered. To further emphasize the overall approach needed, we offer some supplementary information - and problems - associated with large capacitor banks.

a) *Grounding of Neutrals of Large High Voltage Banks*

The Guide states it is *desirable,* in terms of switch rating and cost, to ground the neutral of large shunt banks rated above 100kV". Two ANSI Standards for Breakers (C37.04-1979, C37.06-1979) *require* "both the shunt capacitor bank and the system to be *grounded* at voltage levels of 121 kV and above".

The matter should not be considered so clear-cut. Ontario Hydro has installed several large *undergrounded* wye banks with SF6 breakers meeting the associated, normal-system, capacitor-switching requirements. (Data for capacitor banks: 119.5kV, 96Mvar; 239kV, 192Mvar.)

The banks were purchased for installation at existing stations, some with limited space. A choice had to be made between two different sets of problems that can emanate from the choice between grounded or ungrounded banks. At the time, the consensus moved toward the use of ungrounded banks. (Some predominant ideas: possible problems from damaging transient voltages in existing switchyard wiring; location restrictions; suitable capacitor-switching breakers available; possible difficulties or extra costs with single-point or peninsula grounding.)

Our main intention is neither to contradict ANSI recommendations nor to promote the use of large ungrounded banks. It is to emphasize that high-voltage; ungrounded banks should not be left out of the considerations.

b) *Breaker-Failure Problems*

The Guide states that special capabilities are required by capacitor-switching breakers. It does not cover the case of a capacitor breaker failing to open when tripped. Assume a large capacitor bank (including breaker) is connected to a bus. If the bank protection operates but the capacitor breaker fails to open, the backup isolation breakers must be tripped. The latter are the existing line and bus-tie breakers - all of which may have inadequate capacitor-switching capabilities. ANSI Guide C37.0731-1973, section 4.10.3, discusses the subject briefly but simply indicates the last circuit breaker to clear must be adequately rated. Short of proposing the replacement of the existing backup breakers, what other alternatives can a protection engineer offer to cover breaker failure conditions? How do other utilities approach these problems?

c) *Capacitor Bank Overvoltage Protections*

As pointed out in the Guide (Table 4), capacitor banks have high overvoltage-withstand capabilities. A bank, itself, may be a cause of undesirable high voltages at some stations if specific lines are "tripped off". Overvoltage relays tripping the capacitor breaker seem to offer an obvious solution. (Section 6.4 of Guide.) The setting of these relays, however, can lead to complications:

High settings based upon capacitor *equipment* overvoltage (ov) capabilities can result in tripping breakers at voltages higher than their rated maximum values. Several ANSI breaker standards (eg C37.012-1979) state that the operating voltage for capacitor-switching breakers "should not exceed the rated maximum voltage since this is the upper limit for operation". Have other utilities established any guidelines for setting ov relays above the rated limits of capacitor-switching breakers?

Relatively *low ov settings* based on breaker or other power equipment official limitations may undesirably take system voltage control out of the hands of operators.

In establishing bank protections, one should consider a "capacitor bank" as the *complete installation* (including the breaker or switching device) - as recommended in IEEE/ANSI Standard 18-1980, section 2. The terms "capacitor equipment" and "capacitor bank" should not be synonymous terms if the different ratings and requirements of the capacitor assembly and of the associated switching devices are to be emphasized.

In summary, the ov relays should be set after considering several factors: operating practices; maximum, anticipated, emergency voltages; breaker ov tripping capabilities; effect of ov on other power system equipment; capacitor equipment ov capabilities, etc. Settings based on compromise seem to be necessary. Note: For the ov capabilities of capacitor equipment there is now a discrepancy in the data. In Table 4 of the Guide, the multiplying overvoltage factor for 30.0 min is listed as *1.15*. In the ANSI Standard 18-1980 (section 8.3.2.1) the factor is *1.25*. Which is correct?

Manuscript received August 6, 1981.

John A. Zulaski: We are in agreement with Mr. Bozoki and Mr. Christian that the guide should not lean so heavily toward grounding of the capacitor bank neutral at voltage levels of 121 kV and above. As the discussors point out, the decision whether or not to ground the bank neutral may involve considerations other than that of selection of the capacitor bank switching device.

Replacement of existing backup breakers with breakers having full capacitor switching capability is not economically feasible. Each application must be reviewed individually, but it appears that most would take the risk that the backup breaker has sufficient margin to successfully switch the bank on those infrequent occasions where backup operation is required.

Tripping of the bank switching device under overvoltage conditions usually is not a problem since the bank switching device is specifically designed for capacitor switching, and adequate margin for overvoltage usually is available.

The discussors have correctly pointed out that ANSI Standard 18-1980 shows the maximum permissible voltage multiplying factor for a duration of 30 minutes is now 1.25. Previous versions of this standard had shown the multiplying factor as 1.15.

Manuscript received September 22, 1981.

SHUNT REACTOR PROTECTION PRACTICES

A Power System Relaying Committee Report
prepared by the Shunt Reactor Protection Working Group*

Abstract

This paper has been prepared by a working group of the Power System Relaying Committee, to report on present shunt reactor protective relaying practices. Relay protection of both dry-type and oil-immersed reactors is covered.

INTRODUCTION

This paper covers protection of shunt reactors used typically to compensate for capacitive shunt reactance of transmission lines. A survey of shunt reactor protection, conducted in 1979 by the Shunt Reactor Protection Working Group of the IEEE Power System Relaying Committee, was used as a guide to determine the more common circuit arrangements and protective relaying schemes presently in use.

Other arrangements or special applications of reactors such as harmonic filter banks or current-limiting reactors are not covered, although some of the protective relaying methods may apply.

Use of Reactors

Shunt reactors are used to provide inductive reactance to compensate for the effects of high charging current of long transmission lines and pipe-type cables. For light load conditions, this charging current can produce more leading reactive kVA than the system can absorb without risk of instability or excessively high voltages at the line terminals.

Reactor Construction

The two general types of construction used for shunt reactors are dry type and oil immersed. The construction features of each type, along with variations in design, are discussed under the headings which follow.

Dry-Type Construction: Dry-type shunt reactors generally are limited to voltages through 34.5 kV and are usually applied on the tertiary of a transformer which is connected to the transmission line being compensated. The reactors are of the air-core (coreless) type, open to the atmosphere, suitable for indoor or outdoor application. Natural convection of ambient air generally is used for cooling the unit by arranging the windings so as to permit free circulation of air between layers and turns. The layers and turns are supported mechanically by bracing members or supports made from materials such as ceramics, glass-polyester, and concrete. The reactors are constructed as single-phase units, and are mounted on base insulators or insulating pedestals which provide the insulation to ground and the support for the reactor. A typical installation is pictured in Figure 1.

---

* Members of the Working Group are: J. Zulaski, Chairman; J. Bonk, D. Dawson, R. Dempsey, H. Disante, L. Dvorak, C. Gadsden, W. Lund, G. Nail, T. Niessink.

84 WM 101-2   A paper recommended and approved by the IEEE Power System Relaying Committee of the IEEE Power Engineering Society for presentation at the IEEE/PES 1984 Winter Meeting, Dallas, Texas, January 29 - February 3, 1984. Manuscript submitted July 25, 1983; made available for printing November 11, 1983.

Because the dry-type shunt reactor has no housing or shielding, a high-intensity external magnetic field exists constantly while the reactor is energized. Care is required in specifying the clearances and arrangement of the reactor units, mounting pad, station structure, and any metal enclosure around the reactor or in the proximity of the reactor. A closed metallic loop in the vicinity of the reactor produces losses, heating, and arcing at poor joints, and therefore it is important to avoid closed metallic loops and to maintain sufficient separation distances.

For the same range of application, the primary advantages of dry-type air-core reactors, compared to oil-immersed types, are lower initial and operating costs, lower weight, lower losses, and the absence of insulating oil and its maintenance. The main disadvantages of dry-type reactors are limitations on voltage and kVA ratings and the high-intensity external magnetic field mentioned above.

Oil-Immersed Construction: The two most common design configurations of oil-immersed shunt reactors are: coreless type and iron-core type. Oil-immersed reactors are designed for either self-cooling or forced cooling.

Most coreless shunt reactor designs have a magnetic circuit (magnetic shield) which surrounds the coil to contain the flux within the reactor tank. The steel core-leg that normally provides a magnetic flux path through the coil of a power transformer is replaced (when constructing coreless reactors) by insulating support structures. This type of construction results in an inductor that is linear with respect to voltage.

The magnetic circuit of an iron-core reactor is constructed in a manner very similar to that used for power

Figure 1.   Typical dry-type shunt reactor.

Reprinted from *IEEE Trans. Power App. Syst.*, vol. PAS-103, no. 8, pp. 1970–1976, Aug. 1984.

transformers with the exception that a small gap is introduced to improve the linearity of inductance of the reactor when compared to a reactor without a gap.

Oil-immersed shunt reactors can be constructed as single-phase or three-phase units, and are very similar in external appearance to that of conventional power transformers. See Figure 2.

DRY-TYPE REACTORS — APPLICATION AND PROTECTION

Reactor Connections

Dry-type reactor banks are generally connected to the tertiary of a transformer bank as shown in Figure 3. Each wye-connected, ungrounded reactor bank can be switched individually on the supply side of the reactor bank or on the neutral side, as shown in Figures 3 and 4. A grounding transformer having a grounded-wye connected primary and a broken-delta connected secondary, with a grounding resistor, as shown in Figure 3, is normally used on the tertiary circuit to provide a limited amount of ground current. The grounding transformer and the grounding resistor are sized for a continuous zero-sequence current at least equal to the zero-sequence current flowing through the tertiary-circuit capacitance to ground under ground-fault conditions. In addition, the grounding transformer must be rated for continuous application of line-to-line voltage in order to withstand a permanent ground fault on the tertiary.

Failure Modes and Types of Faults

The faults encountered in dry-type reactor installations can be categorized as follows:

1) Phase-to-phase faults on the tertiary bus, resulting in a high-magnitude phase current.

2) Phase-to-ground faults on the tertiary bus, resulting in a low-magnitude ground current, dependent upon the size of the grounding transformer and resistor.

3) Turn-to-turn faults within the reactor bank, resulting in a very small change in phase current.

Phase-to-phase faults are not likely to occur in dry-type reactors, since they are banks of single-phase units arranged with considerable separation between phases. Instances have been reported, however, where arcing from a faulted reactor contacted the tertiary bus to initiate a phase-to-phase fault.

Since dry-type reactors are mounted on insulators or supports which provide generous clearances to ground, direct winding-to-ground faults are not likely to occur without extenuating circumstances, such as when an animal bridges the insulation to ground. The damage which occurs for a winding-to-ground fault depends on how much ground current is permitted by the grounding transformer.

Winding-insulation failures in dry-type reactors may begin as tracking due to insulation deterioration or as turn-to-turn faults, but once an arc is initiated, these failures, if not detected promptly, often involve the entire winding due to the arc's strong interaction with the magnetic field of the reactor. The result is a phase-to-neutral fault which increases the current in the unfaulted phases to a maximum of $\sqrt{3}$ times normal phase current. This increase in phase current, if not detected, will cause thermal damage to the unfaulted phases of the reactor bank.

System Considerations

The transmission system is generally not disturbed by a faulted dry-type reactor, since even a shorted phase leg of an ungrounded wye-connected reactor connected to a transformer delta tertiary will have only a minor effect on the

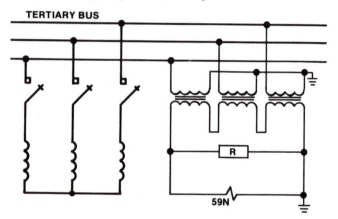

Figure 3. Typical dry-type shunt reactor connection with three-pole supply-side switching and with grounding transformer.

Figure 2. Typical oil-immersed shunt reactor.

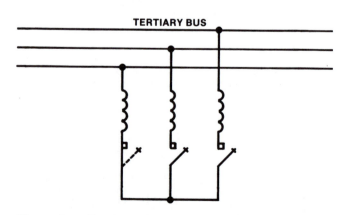

Figure 4. Dry-type shunt reactor connection with two-pole or three-pole neutral-side switching (grounding transformer not shown).

magnitude of the phase current. Unless the fault were to evolve into a phase-to-phase fault on the tertiary bus, it is necessary only to isolate the faulted reactor by tripping the reactor switching device, leaving the rest of the transmission system intact. If the reactor bank is not equipped with a switching device, the transformer bank must be tripped in the event of a reactor fault.

When a faulted dry-type reactor is isolated from the tertiary circuit, the voltage on the transmission line will increase. Studies of the system should be made to be sure that the loss of the reactor will not cause a significant over-voltage condition on the system.

Relaying Practices

Protection for Phase-to-Phase Faults: Relaying protection practice for phase-to-phase faults is generally a combination of overcurrent, differential, and/or negative-sequence current relaying. The more common schemes are illustrated in Figure 5 and are conventional schemes requiring no further explanation.

Protection for Phase-to-Ground Faults: The practice for relaying ground faults is illustrated in Figure 3. The broken-delta output of the grounding transformer is monitored by an overvoltage relay equipped with a harmonic filter. It is common practice to alarm but not trip for this fault condition.

Protection for Turn-to-Turn Faults: Turn-to-turn faults in dry-type reactors present a formidable challenge to the protection engineer. The current and voltage changes encountered during a turn-to-turn fault can be of the same order of magnitude as variations expected in normal service, and therefore, sensitive, reliable protection using the conventional relaying schemes described above is not possible. The

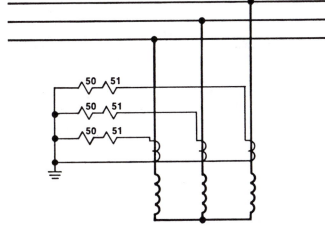

a) Overcurrent relay protection with supply-side connected current transformers

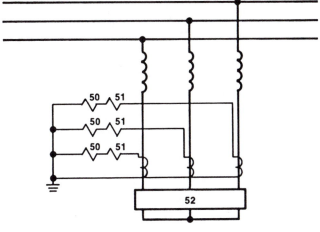

c) Overcurrent relay protection with neutral-side connected current transformers

b) Negative-sequence relay protection

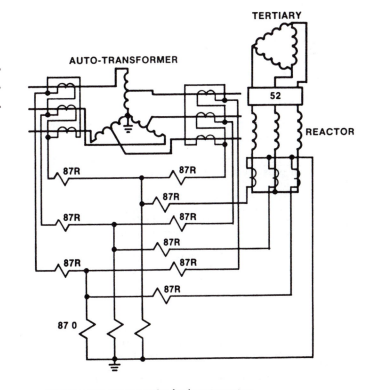

d) Differential relay protection for dry-type reactors

Figure 5. Common protective relaying schemes for dry-type reactors.

voltage-unbalance relaying scheme illustrated in Figure 6 has come into use recently. The voltage signal appearing between the neutral connection of the reactor bank and ground can be the result of 1) reactor-bank unbalance due to a faulted reactor, 2) reactor-bank unbalance due to manufacturing tolerance,* or 3) tertiary bus-voltage unbalance with respect to ground. The manufacturing tolerance tends to be a fixed error that can be negated by an equal and opposite voltage generated by means of a phase-shifting network. System-voltage unbalance may be variable; however, a given percent change in system unbalance affects both the reactor-bank neutral-to-ground voltage and the grounding-transformer broken-delta voltage to the same degree, and therefore these two signals can be used to cancel each other. The summing-amplifier signal output of Figure 6 is thus representative of the degree of unbalance due only to the faulted reactor, and hence this scheme can discriminate between a turn-to-turn fault and other sources of unbalance.

A variation of this relaying scheme involves the use of a separate phase-to-phase connected voltage transformer on the tertiary bus to derive the source for the phase-shifting network. This method avoids tripping of the reactor bank in the event of a ground fault on the tertiary bus.

OIL-IMMERSED REACTORS — APPLICATION AND PROTECTION.

Reactor Connections

Oil-immersed reactors are generally connected to one or both ends of a long transmission line (see Figure 7) and are wye-connected with a solidly grounded neutral. These reactor banks may be switched or may be permanently connected to the line.

Another reactor-bank arrangement that has recently come into use for single-pole tripping and reclosing of circuit breakers is the four-reactor scheme. In this application, a fourth reactor is connected between the reactor-bank neutral and ground to suppress the secondary arc current in a faulted and disconnected phase conductor during single-pole fault interruption.

Oil-immersed reactors may also be connected to the substation bus, and as with line-connected reactors, are generally solidly grounded and may be either switched or permanently connected to the bus.

Relaying protection methods used for bus-connected reactors and for four-reactor configured banks are basically the same as those used for line-connected, solidly grounded, oil-immersed reactors.

Failure Modes and Types of Faults

The faults encountered with oil-immersed reactor installations can be categorized as follows:

1) <u>Faults resulting in large changes in the magnitude of phase current, such as bushing failures, insulation failures, etc.</u> Because of the proximity of the winding with the core and tank, winding-to-ground failures can be expected. The magnitude of current resulting from this type of fault is dependent upon the location of the winding-to-ground fault with respect to the line terminal of the reactor. The farther the fault is away from the line terminal, the lower the fault current. Bushing failures within or external to the tank, as well as faults on the connection between the transmission line and the

reactor bank, will result in large changes in the magnitude of phase current.

2) <u>Turn-to-turn faults within the reactor winding, resulting in small changes in the magnitude of phase current.</u> Low-level faults within an oil-immersed reactor will result in a change in the reactor impedance, increased operating temperature and internal pressure, and accumulation of gas. If not detected, the turn-to-turn fault will evolve into a major fault.

3) <u>Miscellaneous failures such as low oil and loss of forced cooling.</u>

System Considerations

Clearing of Faults: Typically, the relaying practice for line-connected reactors is to trip the local line breaker and transfer-trip the remote line breaker. For a reactor fault in a direct-connected line reactor, both line breakers have to be locked out.

For a fault in a switched line reactor, where rapid reclosing is desired, both line breakers will trip, the reactor-bank switching device will open, and then the line breakers will reclose.

Another approach, when utilizing a circuit switcher as the reactor bank switching device, is a blocking or coordinated-tripping scheme. In this scheme, the circuit switcher interrupts low-level reactor faults within its rating, and the terminal breaker operates only on higher-level faults beyond the rating of the circuit switcher.

Resonance Phenomena: The distributed shunt capacitance of the transmission line forms a parallel-resonant

TERTIARY BUS

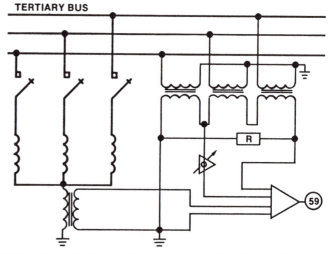

Figure 6. Voltage-unbalance relay protection for dry-type reactors.

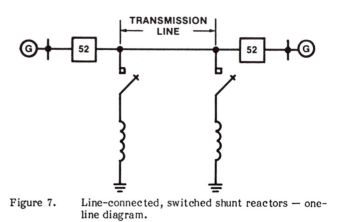

Figure 7. Line-connected, switched shunt reactors — one-line diagram.

---

* Per ANSI C57.21, in the case of a three-phase shunt reactor or a bank made up of three single-phase reactors, the maximum deviation of impedance in any one phase shall be within 2% of the average impedance ohms of the three phases. For dry-type shunt reactors without magnetic-field shielding, this tolerance applies only when units are arranged in an equilateral-triangle configuration and isolated from any external magnetic influences.

circuit with the shunt reactor(s) having a center frequency close to 60 Hz. This resonant circuit can be troublesome if not taken into account by the system planners and the protection engineer.

When a de-energized transmission line with directly connected reactor(s) is physically close enough to another energized line for the two lines to be electrically coupled, it is possible for higher than rated system voltage to be developed across the "de-energized" reactor. This problem can be prevented by isolating the reactor by means of a dedicated reactor-switching device at the same time as, or immediately following, the de-energization of the line.

Another phenomenon of concern to the relay engineer occurs when a compensated transmission line is de-energized. The parallel-resonant circuit will produce a damped sinusoidal voltage at a frequency generally less than 60 Hz which can last several seconds, with an initial voltage that can approach rated voltage. This substantial voltage with reduced frequency can cause misoperation of impedance relays used to protect shunt reactors unless the impedance relays are specifically designed for this application.

Relaying Practices

Protection for Large-Magnitude Faults: Relaying protection practice for faults producing large changes in the magnitude of phase current is generally a combination of overcurrent, differential, and/or impedance relaying. The more common schemes are illustrated in Figure 8 and are conventional schemes requiring no further explanation.

Where a reactor differential relaying scheme is used, it is essential that the current transformers on both sides of the reactor be identical (same excitation characteristic).

Protection for Turn-to-Turn Faults: It is generally agreed that overcurrent relay schemes are not sufficiently sensitive to provide much protection for turn-to-turn faults. Differential relay schemes cannot detect turn-to-turn faults. The impedance relay offers some improvement in protection, but the sudden-pressure relay and/or gas-accumulator relay are felt to provide the most sensitive means of detecting turn-to-turn faults within oil-immersed reactors.

Distance relays have been applied to detect shorted turns in iron-core shunt reactors. The use of distance relays for this type of protection is possible due to the significant reduction in the 60-Hz impedance of a shunt reactor under turn-to-turn fault conditions. The turn-to-turn fault sensitivity that can be achieved is basically a function of the apparent impedance seen by the relay during the inrush period when the reactor is energized. The relay reach must be set below the reduced impedance seen during this inrush period and should be designed so that the relay will not operate incorrectly on the natural-frequency oscillation which occurs when a compensated transmission line is de-energized.

The gas-accumulator relay is applicable on reactors which are equipped with conservator tanks and have no gas space inside the reactor tank. This relay is inserted in the pipe between the reactor and the expansion chamber (conservator). Low-energy partial discharges, creepage, and overheating caused by turn-to-turn short circuits or by high contact resistance cause the insulation at these points to slowly decompose while evolving gas. The gas rises through the oil and is accumulated in the relay. The relay will also operate for severe internal arcing or heavy-current flashovers, which force oil through the relay at a high velocity before the gases rise through the system to the device. This relay is commonly known as a Buchholz relay.

The sudden-gas-pressure relay, also known as a sudden-pressure relay, is applicable to gas-cushioned oil-immersed reactors. The relay is mounted on the reactor tank in the region of the gas space at the top of the reactor, and consists of a pressure-sensing bellows, a pressure-actuated switch, and a pressure-equalizing orifice. The relay operates on the difference between the pressure in the gas space of the reactor and the pressure inside the relay. During slow pressure variations associated with reactor temperature changes, the pressure-equalizing orifice will equalize the pressure

between the relay and the reactor, and thus prevent operation. For internal arcing that produces large amounts of gas and a sudden rise in gas pressure, the bellows will expand, causing the relay to operate.

The sudden-oil-pressure relay, also known as a fault-pressure relay, is applicable to all oil-immersed reactors. The relay is mounted on the reactor tank below the minimum de-energized liquid level. Oil fills the lower chamber of the relay housing, within which a spring-backed bellows is located. The bellows is completely filled with silicone oil. There is also silicone oil in the upper chamber, which is connected to the bellows via an equalizer hole. Should an internal fault develop, the resulting rapid rise in oil pressure, or pressure pulse, is transmitted to the bellows, causing the relay to operate. In the event of gradual increases in oil pressure due to temperature variations in the reactors, the equalizing hole stabilizes the pressure in the bellows and keeps the relay from operating.

Loss of Cooling: Oil-immersed reactors are sometimes built with forced cooling to reduce size and cost. For such reactors, the cooling is critical and must be operational any time the reactor is energized.

The loss of cooling can be detected by monitoring the oil flow with flow indicators, monitoring the ac supply voltage to the cooling fan and oil pumps, and by monitoring the temperature with temperature relays.

The oil-flow and ac supply-voltage indicators typically are connected for alarm only. The temperature relays are generally connected to trip and remove the reactor from service. To adequately protect the reactor, a combination of all the above indicators is usually necessary.

Overvoltage: Normally the reactor is installed on the line to provide inductive reactance to compensate for the effects of high charging current of long transmission lines or pipe-type cables, and thereby reduce the risk of excessively high voltages at the line terminals. Possible system overvoltage conditions should be considered when planning the installation of reactors.

Voltage relays are sometimes used to detect line overvoltage conditions which may cause damage to the reactor from operation at higher-than-rated voltage. Steps of overvoltage with associated timers can be used to provide this protection. Consultation with the reactor manufacturer is required to arrive at settings appropriate for a specific reactor.

It is more common, however, to use voltage relays to monitor the voltage on the system and to energize the reactor or, if the reactor is already energized, to protect the line against overvoltage. Because isolation of a reactor from the system in case of overvoltage would only further aggravate the overvoltage condition on the system, the voltage relays should remove both the transmission line and the reactor from the system.

SUMMARY

A working group of the Power System Relaying Committee has conducted a survey of present protective relaying practices for the protection of shunt reactors. Two basic shunt-reactor configurations are covered:

a) Dry-type, wye-connected with ungrounded neutral, and connected to the impedance-grounded tertiary of a power transformer.

b) Oil-immersed, wye-connected, with a solidly grounded or impedance-grounded neutral, and connected to the transmission system.

For dry-type reactors, protection for major faults is achieved through a combination of overcurrent, differential, and/or negative-sequence relaying schemes, while protection for low-level turn-to-turn faults is achieved through a voltage-unbalance relay scheme with compensation for inherent unbalance.

For oil-immersed reactors, protection for major faults is achieved through a combination of overcurrent and/or differential relaying, while protection for low-level turn-to-turn faults is achieved through a combination of impedance, overtemperature, gas-accumulator, and/or sudden-pressure relays.

REFERENCES

1) Requirements, Terminology, and Test Code for Shunt Reactors, ANSI Standard C57.21-1971.
2) Reactors, International Electrotechnical Commission Publication 289, 1968.
3) J. J. LaForest, et al, "Resonant Voltages on Reactor Compensated Extra-High-Voltage Lines," IEEE Transactions on Power Apparatus and Systems, vol PAS-91, pp 2528-2536, November/December 1972.
4) J. W. Copper and L. W. Eilts, "Relay for Ungrounded Shunt Reactors," IEEE Transactions on Power Apparatus and Systems, vol PAS-92, pp 116-121, January/February 1973.
5) L. Carlson, et al, "Single-Pole Reclosing on EHV Lines," International Conference on Large High-Voltage Electrical Systems, C.I.G.R.E., Paris, France, paper no. 31-03, 1974.
6) M. J. Pickett, et al, "Near Resonant Coupling on EHV Circuits," IEEE Transactions on Power Apparatus and Systems, vol PAS-87, pp 322-325, August 1967.

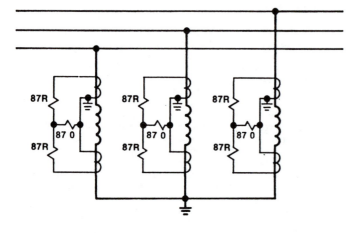

a) Differential relay protection with restraint

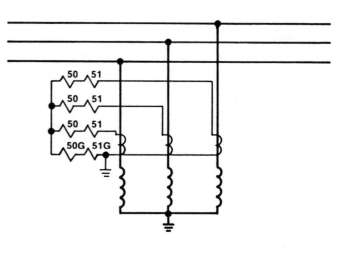

c) Overcurrent relay protection

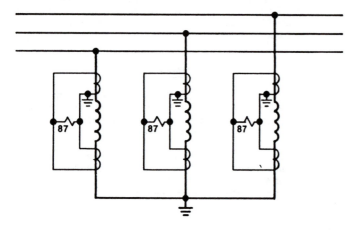

b) Differential relay protection

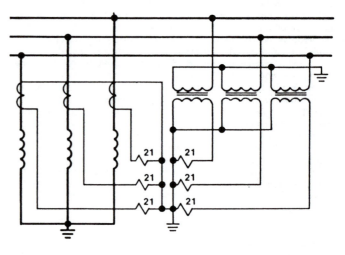

d) Impedance relay protection

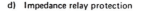

Figure 8.   Common protective relaying schemes for oil-immersed reactors.

Discussion

J. M. Crockett (Westinghouse Canada Inc., Hamilton, Ontario, Canada): Can consideration now be given to maintaining the established momentum as represented by this commendable paper and undertake preparation of a summary of experience with shunt reactor protection and ultimately, a guide to shunt reactor protection? If the experiences reported by Engelhardt[1] prove to be typical, gas detector relays are secure and reliable and significant time delays are required in sensitive electrical protections such as turn to turn fault detection.

Can the WG provide further data on the type of impedance relay used to distinguish between inrush and the impedance reduction resulting from a shorted turn? In my opinion, the use of a replica type impedance relay would be suspect because the inrush current applied to a replica would not reproduce the source voltage and I would anticipate a tendency to overreach during reactor energization. Are distance relays being used when a neutral reactor is involved?

There is no threat to system stability during low current shorted turn faults, and there is a fixed cost associated with examining and repairing a reactor no matter how minor the fault damage. Can the working group comment on experience relating to the economic benefits of high speed detection of low level faults?

REFERENCES

[1] K. H. Engelhardt, "EHV Shunt Reactor Protection—Application and Experience" 1983 Western Relay Conference, Spokane, Washington

Manuscript received February 13, 1984.

J. Zulaski, Chairman: The working group wishes to thank Mr. Crockett for his comments on our paper. Indeed, the working group is planning to recommend preparation of a guide on Shunt Reactor Protection Practices.

The specific type of impedance relays discussed in the paper are electromechanical type impedance relays, which are specifically designed for the purpose of shunt reactor protection. The replica type relay was not considered and therefore the working group has no comment on the use of this type of relay for shunt reactor protection.

To our knowledge, distance relays are not being used when a neutral reactor is involved. This may be because of the added sensitivity requirements that are imposed by this application.

We agree that there is a fixed cost associated with examining and repairing a shunt reactor, even for minor fault damage. However, we feel the answer to your question must be split into two parts by considering dry-type reactor banks and liquid immersed reactor banks separately. For dry-type reactors, even with sensitive, fast detection, it seems improbable that the damage caused by a turn-to-turn fault can be repaired. At best, it appears that all one can do is keep the fault from escalating to the point where it would involve the tertiary bus or cause damage to adjacent phases of the reactor bank. Therefore, it would appear that prompt, but not necessarily high speed, detection is important to minimize the escalation of the fault. With liquid immersed reactors it would seem that a faster relay response is important because of the cost of these larger sized reactors, as well as the consequences of a minor fault escalating into a catastrophic failure.

Manuscript received April 17, 1984.

A THREE-PHASE DIFFERENTIAL RELAY FOR TRANSFORMER PROTECTION
By

C.-H. Einvall
ASEA
Västerås, Sweden

J. R. Linders
Consultant
Cleveland, Ohio

ABSTRACT

Established harmonic restraint differential relay concepts have been enlarged upon to provide a new static multiple-winding transformer differential relay. The resulting design includes simplification of the frequency selective circuits and new techniques for rapidly determining the harmonic content of the applied currents. The relay is shown to have improved sensitivity, to be responsive to internal transformer winding faults in less than two cycles, and to the more severe type of fault in less than one cycle. Full security against inrush currents and overexcitation as well as to external faults is demonstrated. Poor performance of current transformers is shown not to adversely affect either the security or dependability of the relay. Heavy current test results are presented which confirm the relay principles.

INTRODUCTION

The concept of harmonic restraint has provided the basis for many successful high speed, sensitive transformer differential relays since its introduction some 20 years ago. The principle is sound and there is seldom any report of improper performance of relays based on this principle. However, there are limitations in the applications of these relays which justify a reappraisal of their characteristics. These limitations are becoming of concern because they involve restrictions on through-fault current magnitude, sensitivity to minor faults, speed of operation, and current transformer performance. In addition, problems of transformer overexcitation are becoming of more concern in view of the trend to flatter magnetization characteristics above the knee of the saturation curve.

Electromechanical relays inherently respond primarily to the rms value of an applied signal. While this signal can be modified by means of frequency selective filters and rectifiers, the basic rms characteristic of the measuring device remains. Such measurement limitations do not exist in static type measuring circuits. Depending on application, any of a variety of specific characteristics of a signal can be chosen as the most significant for measurement. The relay to be described utilizes these new freedoms in static circuitry.

OPERATING CONDITIONS

There are four different non-linear effects which must be considered in the design of a transformer differential relay.

These are:

a. Non-linear inrush conditions when energizing a transformer. These are of course the most obvious. "Outrush," when an adjacent transformer is energized, and "inrush" resulting from voltage recovery upon clearing an external fault are of a similar natur. These result in 2nd and higher even order harmonic distortions.

Paper T 75 187-0, recommended and approved by the IEEE Power System Relaying Committee of the IEEE Power Engineering Society for presentation at the IEEE PES Winter Meeting, New York, N.Y., January 26-31, 1975. Manuscript submitted September 4, 1974; made available for printing December 16, 1974.

b. Transformer overexcitation. This results in a different non-linear effect. This effect is becoming more severe due to the sharper curvature in the magnetization characteristic of modern transformer steel. This is largely 3rd and 5th harmonic distortion.

c. Non-linearity of current transformers. CT's can be driven into saturation by ac overcurrents, or the dc component in fault or inrush currents. CT saturation in the transformer differential system is more likely with breaker and one-half or other type configurations which can result in very large external fault currents not limited by transformer reactance. 2nd, 3rd and 5th harmonic distortions are possible.

d. Combinations of (c) with (a) or (b). These can result in additional wave shape distortions. For example, dc components which saturate CT's can result in the time to saturation varying between CT's. This can result in brief periods of further complexities in the wave shapes of the relay currents.

A harmonic restraint relay necessarily involves frequency selective circuit functions. But a frequency measurement inherently requires a minimum time to make the determination which is based on several periods of the wave in question. Thus for a high speed, harmonic restraint relay, additional techniques besides frequency segregation are likely to be needed if operating times of less than two cycles are to be attained. Furthermore, since the total operating time of such a relay may be less than the time for complete steady state conditions to be reached, the characteristics of the relay should be explainable on a transient as well as on a steady state basis.

Very large transformers can lead to a very large system X/R ratio with a large time constant. In critical cases this can sustain a dc component for a sufficient length of time to cause CT saturations. Thus, formerly acceptable relay characteristics may be inadequate at locations near such transformers. Further, relays on existing transformers may become inadequate when new, large transformers are added nearby.

When a faulty, large transformer is energized, the harmonic restraint may delay the differential relay tripping for a significant period until the harmonic current has subsided sufficiently to permit tripping. Thus, it is undesirable to use an excessive amount of harmonic restraint "to be safe." Just the opposite may occur.

Relays associated with generator unit transformers have a further operating condition when the unit is started up. Frequencies down to as low as 1/2 to 1/3 normal may exist. The differential relay should maintain reasonable sensitivity during these conditions. On the other hand, it should not indicate a transformer fault if a reasonable excess volts/hertz should occur. The low frequency characteristic of a given relay will depend on not only the arrangement of the relay filter circuits, but also on all of the CT characteristics and how they are matched to the burdens.

Reprinted from *IEEE Trans. Power App. Syst.*, vol. PAS-94, no. 6, pp. 1971–1980, Nov./Dec. 1975.

The above are, of course, in addition to the routine requirement that the relay work correctly for all types of overloads and internal and external fault conditions.

RELAY DESIGN

A static transformer differential relay based on a unique three-phase concept has been designed which effectively treats all of the operating conditions. Type designated RADSE, this relay processes these complex wave shapes in 1/2 to 2 cycles total time to breaker trip impulse. The relay is constructed on a modular basis and assembled in a 19" equipment frame. This design allows full flexibility for specifying any number of relay input circuits either initially or additionally at a later date.

The relay is shown in Fig. 1. On the left is an 18 position test switch. The next three larger modules are the individual phase units. These contain the input and variable restraint circuitry, frequency selective filters and the first stage of a two-stage level detector. These phase units are interconnected so that the harmonics in any phase provide restraint to all phases.

The fourth large module is the three-phase measuring and output unit. This includes the 2nd stage of the level detector and the "instantaneous" level detector. These both function on a three-phase basis. Threshold setting resistors, auxiliary voltage regulating circuits, operating indicator and a dry reed output are also in this module. The two smaller modules on the right are the 3 ms output tripping relay for up to four trip circuits and the optional phase indicator unit.

Fig. 1. Three phase transformer differential relay RADSE.

Fig. 2 shows the general construction of a plug-in module. Supplemental modules are used when additional inputs are desired. These modules can be used to increase the number of input circuits without practical limit. The basic relay of Fig. 1 is for up to three inputs (transformer windings or two breakers on one winding). The characteristics of the relay are not affected by the number of input units.

1. The input circuitry shown in Fig. 3 is for three inputs of one phase. It provides the following components and features:

a. Air gapped isolating CT's to reduce the dc component and to convert the current to a voltage signal. This results in only the largest of the applied currents determining output voltage. Normally, paralleled CT secondaries yield the sum of the applied currents. But when air-gapped CT's with relatively low magnetizing impedance are paralleled across a relatively higher resistance beyond a rectifier, the resulting voltage is proportional to the largest of the applied currents, not to the sum.

b. Full wave rectification in a two-diode, two-resistor bridge. This results in only the larger of the outgoing currents affecting the output. A four-diode, full-wave bridge rectifier does not have this sense of ac polarity. The two-diode, two-resistor bridge does. Thus, for an internal fault only the largest of the incoming currents will flow in R_1 of Fig. 3 on one half cycle, and in R_2 on the other half cycle. But for an external fault, with the largest of the incoming currents flowing in R_1, the largest of the outgoing currents will be flowing in R_2. On the alternate half cycle, the conditions reverse. Hence, for the external fault, voltages are developed across both R_1 and R_2 during both half cycles, but for the internal fault voltage is developed across each resistor only one-half of the time.

c. A network of zener diodes and resistors which provides the variable percentage restraint.

Mathematically, if I_x is the maximum (CT secondary) current into, and I_y the maximum current out of the protected transformer the restraining voltage developed across $R_1 + R_2$ of Fig. 3 is

$$V = k (I_x + I_y)$$

where k is the proportionality between voltage and current of the entire circuit. For internal faults, I_y will usually be zero (i.e. no current flowing out of the transformer). Thus, for a given amount of fault current into a transformer, there will not be more than one-half the restraint as for an external fault. And for a multiple-winding transformer, this ratio may be still more favorable.

Fig. 2. Three phase measuring and output unit.

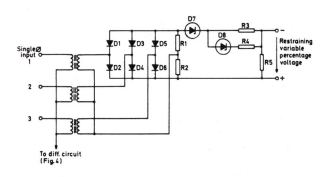

Fig. 3. Input-restraint circuitry for one phase.

2. The differential circuitry, shown in Fig. 4 for one phase, is connected in the conventional manner to receive the total of all of the currents into and out of the transformer for each phase. This circuitry provides:

(a) Current to voltage conversion again with air gapped CT's for both the operating voltage and the harmonic restraint voltages.

(b) Low pass filter to suppress components in the operating voltage above 500 Hz.

(c) 2nd harmonic and 5th harmonic filters to provide the respective inrush and overexcitation restraint voltages. The filter inductances are provided by the secondary windings of the air-gapped CT's.

(d) Two full wave, 4 diode bridge rectifiers for the operating and harmonic restraint signals.

(e) The rectified harmonic restraining voltages for each phase are paralleled and the result used for harmonic restraint for each phase. This resultant will be proportional to the sum of the applied signals. This summing action occurs because of the relatively high equivalent impedance of the filter tuned circuits. This is in contrast to the action of the Fig. 3 circuitry where the signal source is a low impedance.

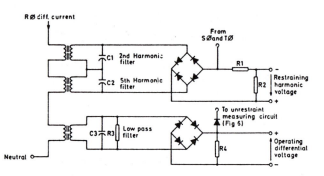

Fig. 4. Differential circuitry for one phase.

3. Phase threshold and integration circuitry shown in Fig. 5 provide three functions:

(a) The differential operating voltage and the two restraining voltages are summed in a conventional resistance summing circuit. When the resistors, $R_{1, 2, 3}$, are large compared to R_4, the voltages across R_4 will be proportional to the algebraic sum of the three applied voltages.

(b) This resultant voltage is then compared to a settable reference voltage which is applied to the 2nd input of an operational amplifier. The reference voltage is adjustable and provides the means for setting the threshold level for the minimum operating current of the relay.

(c) The integration output circuitry develops a voltage proportional to the time the signal is above the threshold level. The magnitude of the signal has no effect on this output voltage. This is accomplished by driving the amplifier to cut-off when the input signal is above the reference. This results in the R_8-C_1 integrating circuit in the amplifier output charging at a fixed rate independent of signal level. When the signal drops below the reference level, the R_8-C_1 integrating circuit discharges also at a fixed rate.

Note that there are no smoothing capacitors in any of the rectifier circuits up to the input of this phase threshold circuitry. Thus, while most of the signal is dc, it has a complex ac wave superimposed on it. The integrator constants are so proportioned that a sufficient output signal exceeds the threshold 41% of the time with a minimum duration of 3.4 ms (for a 60 Hz relay). Thus, momentary spikes, regardless of magnitude, will not cause a false output.

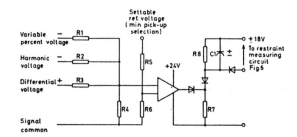

Fig. 5. Phase threshold and integration circuitry.

4. Trip measurement circuitry shown in Fig. 6 is a single unit for all three phases. This unit performs two separate measurements:

(a) The magnitude of the restrained signal as developed by the above described phase integrators is compared to a fixed reference voltage.

(b) The magnitude of the unmodified phase differential current signal, as developed in Fig. 4, is compared to a fixed reference voltage. By means of a voltage divider, the signal can be set for 8, 13 or 20 times nominal rated relay current.

In both these measurements, the signal is the composite of the three individual phase signals. This common measuring technique for all three phases is a design convenience and has no significant bearing on the relay operating characteristics.

When either of these two signals exceeds the respective reference voltage, the amplifier output is energized. This is a dry reed relay which operates in less than a millisecond. This relay then energizes a standard 3 ms, 6 contact output tripping relay. Five of these contacts are available and suitable for signalling and direct tripping of circuit breakers.

This 3 ms relay can be energized continuously, thus it can also be directly incorporated into lockout control functions. The usual ancillary functions and optional phase indicator have not been diagrammed.

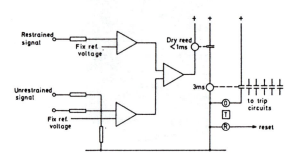

Fig. 6. Trip measurement circuitry.

5. The two stage tandem level detection in the phase units and the measuring and output unit are functionally diagrammed in Fig. 7. The unrestrained, instantaneous function has a single level measurement occurring in block E in the three-phase measuring unit, one measuring circuit for all three phases. This is not further diagrammed in Fig. 7(b).

The variable percentage, harmonic restrained and differential signals are shown in Fig. 7(a) as V_6. Until this combined signal exceeds the phase threshold setting, there is no output signal, V_a, and hence no signal at second level detector input.

When V_6 exceeds the threshold setting, an output signal V_a is developed across the integrator circuit. But unless this signal exceeds the threshold for more than 41% of the time, insufficient voltage, V_b, is developed to intiate a trip function. Thus noise spikes, regardless of magnitude, cannot create a trip signal in this high speed, sensitive differential measuring circuitry.

This double threshold technique is not necessary for the less sensitive unrestrained, instantaneous unit. However, its circuitry also has noise immunity in that no signal of less than 3 ms duration can cause a trip output.

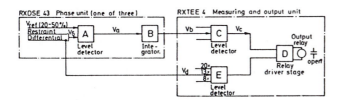

Fig. 7a. Block diagram of two-stage level detection.

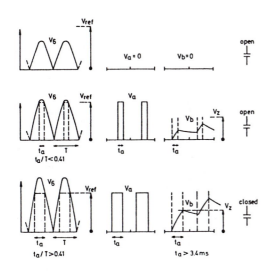

Fig. 7b. Principle functioning of two-stage level detection showing immunity to noise spikes.

OPERATING CHARACTERISTICS

Fundamental Restraint

The slope of a differential relay operating characteristic is not sufficiently defined to be meaningful without detailing the method of determination. Slope is the ratio of the required differential current to cause operation to the restraining currents. The operating current is readily agreed to, but the restraining currents are subject to a variety of interpretations. Consider a relay with an operating threshold of 10A in the differential circuit when one restraint is at 20A and the other restraint at 30A. When expressed as a percent of the lesser restraint, the slope is 10/20 or 50%. As a percent of the larger restraint it is 10/30 or 33%. And if expressed as the average of the two restraining currents it is 10 / 1/2 (20+30) or 40%. When multi-winding restraints are considered, a wider variation in apparent sensitivity may result, depending on how the slope is expressed.

As noted above, the RADSE restraint is developed from the sum of the largest incoming current and the largest outgoing current. Thus it is logical to plot its characteristics to this parameter. In practice, this number is divided by two to get an average value which is akin to a "through current." To illustrate this apparent variation in different relays, Fig. 8 is a plot of the slope of the described relay together with a second relay with a "25%" slope based on the lesser restraint current[1], and another "20-60%" variable restraint relay based on the larger restraint current[2].

Two of the three relays have a minimum sensitivity of 30% of tap, or rated current value. The third is settable between 20% and 50%. The differences in these characteristics are not obvious when stated in conventional terms of minimum sensitivity and "slope."

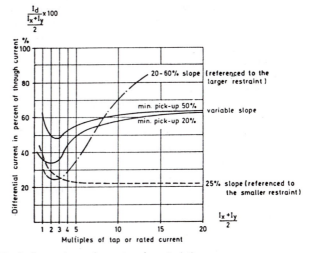

Fig. 8. Comparison of restraint characteristics.

Fig. 9 is a typical plot of operating current versus through current for the RADSE. The variable restraint is not adjustable in the field. The minimum pickup is adjustable with a range of from 20% to 50%. It has a minor influence on the slope as noted in Figs. 8 and 9. The variable percentage restraint or slope is conventional. It enhances the good security for high current external faults and at the same time assures good sensitivity to minor internal faults.

These characteristics are for single-phase, pure sine wave conditions with the restraint currents in phase. The three-phase nature of the relay with the mixing of the currents from all three phases in the harmonic restraint function results in desirable additional characteristics.

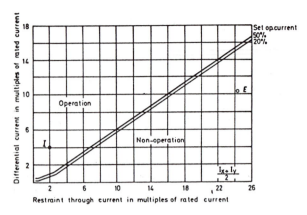

Fig. 9. *Fundamental frequency operating characteristics.*

First, there is approximately a 50% desensitizing action for balanced three-phase conditions. Thus, the relay is more secure against misoperation during severe overloads such as would be encountered during a major system disturbance. But a balanced internal three-phase fault can only occur across the conductors at full voltage; hence, there need be no concern as to the adequacy of this reduced sensitivity.

This characteristic results from designing the harmonic filters so that some fundamental frequency passes through the harmonic filters and is utilized in the three-phase restraint function. Thus, the resulting pulsating dc voltage from three-phase rectification averages larges than that from only a single-phase condition; hence, more restraint is developed.

On the other hand, heavy load currents do not significantly mask minor fault conditions. A minor fault will be (a) single-phase; (b) reactive. The fault current will thus be at a large phase angle with respect to at least one of the phase load currents. The relay has inherently a phase angle sensitivity characteristic because of the high speed response of the measuring circuits. This is shown in Fig. 10. With the fault current in phase with the load current, the standard operating curve ($I_d0°$) prevails. This is replotted from Fig. 9, but to a different scale. With a single-phase fault and a three-phase load, such in-phase condition cannot exist on both faulted phases. One of the faulted phases will have the fault current approximately at 90° to the load current. During such conditions, the increased sensitivity of the relay as shown by the $I_d90°$ curve in Fig. 10 will result. For even full load on the transformer, the minimum operating current due to a minor fault is thus increased by only a few percent.

Harmonic Restraint

The harmonic restraint characteristics have been selected so that either 20% 2nd harmonic, or 35% 5th harmonic of the fundamental will restrain the relay. This is on a single-phase basis. These are the values one observes on single-phase testing. However, the inrush to a three-phase transformer is more complex than just the sum of three individual transformer inrushes. In particular, a delta connection either of the main transformers or of the CT's will result in a less favorable ratio of 2nd harmonic current to fundamental in at least one of the phases. This is because wye-delta transformers cause different relative phase shifts to three-phase, single-phase and the harmonic currents. Since these are all complexities stemming from the three-phase environment, it is logical to resolve them with a three-phase analysis.

Thus, the described relay has interconnections between the phase units which modify the characteristics of one phase depending on conditions in the other phases. The result is an increase in restraint when all three phases contribute. The 2nd harmonics from each phase, as during inrush, will help each phase unit restrain, even though they are seldom symmetrical about the three phases.

This three-phase harmonic restraint reinforcement is important for transformers built with magnetic material which has a sharp kneepoint and a very flat magnetizing curve above the kneepoint. Fig. 11 illustrates this trend in transformer design. The energizing inrush depends significantly upon the residual magnetization in the transformer core. The maximum possible residual will be very nearly the y-axis intercept of the magnetizing curve projection as shown dotted in Fig. 11, i.e. 1.8 Wb/m^2 vs. 1.4 Wb/m^2 for older transformers. Thus, modern transformers tend to have more inrush. But the 2nd harmonic content does not necessarily increase in proportion. A wave analysis will show that if a transformer could be left with a 100% residual flux (90% is about the practical limit) the first cycle of the inrush could look very nearly like a 100% offset fault current, with no significant 2nd harmonic content.[2] During such conditions, the other two phase currents would contain 2nd harmonics. While this is a theoretical limit not reached in practice, it illustrates that the value of summing the three-phase harmonic restraints is to get the restraint when it is needed, where it is needed.

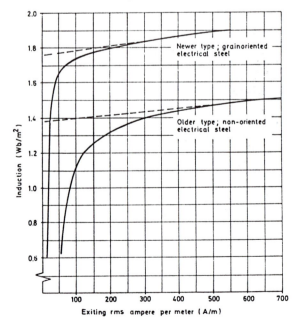

Fig. 11. *Trend in transformer steel magnetizing characteristics. Exiting rms ampere curves tested by Epstein test apparatus.*

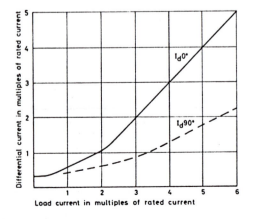

Fig. 10. *Phase angle sensitivity.*

The whole purpose of providing the least 2nd harmonic restraint possible is in the case of energizing a faulty transformer. It is of course desirable that the harmonic restraint have the least effect possible in delaying the tripping during such conditions. This is the objective in the described system without jeopardizing the other characteristics.

5th Harmonic Restraint

A 35% 5th harmonic restraint to prevent tripping on overexcitation has been determined as the most desirable through analysis of overexcitation wave shapes. While the 3rd harmonic could be used for this function, at a simplification in the filter by combining 2nd and 3rd harmonic restraints into one, it has not been used for two reasons:

a. The magnitude of the 3rd harmonic at the relay is dependent upon many factors beyond the control of the relay designer.[3]

b. The 3rd harmonic is the predominant one in the output current of a saturated CT. Thus, to use it for restraint is to hazard a blocked relay during a severe fault.

It is undesirable to trip a transformer by instantaneous differential relay action when a fault does not exist in the transformer. But, philosophically, a transformer should be tripped off before an overexcited condition can damage it, if the basic cause cannot be corrected. Thus, a separate overexcitation protection scheme is preferable to protect the transformer rather than the differential relay. However, on extreme overexcitation, the exciting current can approach full load current and a transformer can be destroyed in seconds. Should this condition ever arise, the differential relay is about the only device which could respond in time. Thus, the 35% 5th harmonic restraint provides restraint during all normal overexcited conditions, but if an overvoltage catastrophe imminent, the relay will trip the Transformer off the line.

Fig. 12 shows the composition of the exciting currents of a modern three-phase transformer. The fundamental and 5the harmonic currents will be as shown during overvoltage conditions, but the 3rd harmonic currents to the relay will in general be substantially less than shown, depending on transformer and CT connections and system characteristics.

The described relay will be blocked from tripping by this 5th harmonic exciting current for overvoltages up to at least 140%. Generator excitation systems in general cannot exceed this value even in a maloperating mode, and the exciting current of 45% of full load current at this point is about the maximum the transformer can sustain for even a brief period.

Frequency Performance

The single-phase, frequency response of the relay is shown in the curves of Fig. 13. This shows that the relay will provide good fundamental frequency protection at low frequency as during generation startup.

The 2nd harmonic restraint for inrush suppression is not usually of interest during low frequency. At 2/3rds of rated frequency any 3rd harmonic will provide restraint because this frequency coincides with the normal 2nd harmonic frequency. The 5th harmonic restraint will never be more effective that at rated frequency. At som specific low frequencies, the 5th harmonic may actually develop an operating signal due to the various filter characteristics. This will never be sufficient to cause a relay misoperation if the volts/hertz does not exceed 130% of normal.

Fig. 13 also shows that the relay develops a net operating voltage at the 3rd harmonic of the system frequency. This is desirable when considering saturated CT's.

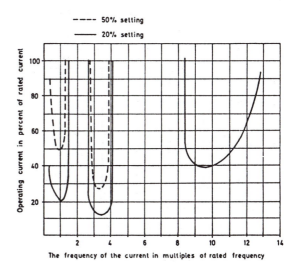

Fig. 13. Frequency response.

Fig. 14 shows the wave shapes and harmonic content in the output current of a CT during various degrees of saturation. This may be considered typical of modern low leakage reactance CT's built with modern core material. The CT burden was resistive. A reactive burden would cause less saturation effect for a given overload.

Unrestrained, Instantaneous Unit

The purpose of this unit is to provide slightly faster relay operation for severe faults. As demonstrated above, it is not needed to assure tripping on severe internal faults. Fig. 15 gives the operating time for both units.

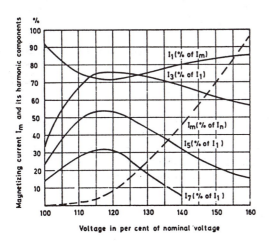

Fig. 12. Magnetizing current and 3rd and 5th harmonics in the windings of an overexcited transformer (60 MVA, 78 kV, grainoriented electrical steel).

405

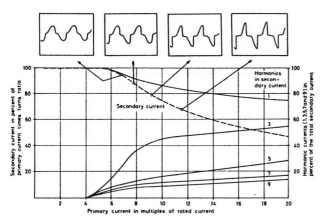

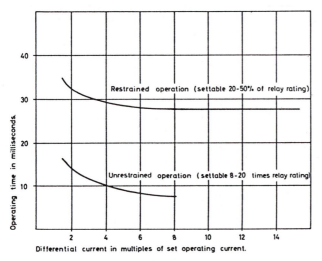

Fig. 14. Wave shapes and composition of secondary CT current during ac saturation.

Impulse Limits

A. The unrestrained unit requires an input signal for at least 3ms to develop a trip signal. Hence, typical noise spikes will not cause operation of this unit.

B. The restrained unit requires a signal above its setting for at least 15 ms (60 Hz relay) for it to complete a measurement and initiate a trip signal. This unit is thus also immune to mis-operation due to noise spikes.

Fig. 15. Operating time-current characteristics.

APPLICATION CRITERIA

Current Transformers

There are no practical restrictions on the selection of current transformer ratios, nor on the treatment of maximum fault conditions. Relay burdens are very low, thus assuring that a minimum CT will perform with standard 10% error to 20 times rating. This is more of a good design criteria than a relay requirement. Excess-ive lead burden is readily treated by use of the 1A nominal rated relay and/or utilizing the separate ratio matching CT's to minimize total burdens.

Instantaneous Setting

The unrestrained, instantaneous unit is set without regard to the setting of the restrained unit. The restrained unit will not block tripping on an internal fault due to CT saturation. The unrestrained unit is set above anticipated inrush. Its purpose is to accelerate relay operating time by about 1 cycle for very severe faults.

Minimum Pickup Setting

This setting is determined by possible tap-changer regulating range and CT ratio errors at light loading conditions. It is selected to provide about 15 percentage points margin over expected maximum ratio error. It has very little effect on the slope characteristic and is seldom chosen for security for through faults.

This setting does not affect the harmonic restraint action of the relay, hence its determination need not consider inrush nor over-excitation conditions.

HEAVY CURRENT TESTS

This relay principle has been subjected to independent laboratory high capacity tests as well as to the vendor's tests.

The security of the system to inrush when energizing a power transformer is shown in Fig. 16. The current wave shapes show the typical variations as the inrush subsides. Variations in CT performances are also evident with SØ showing earlier saturation than TØ, which has the higher peak current. The lower traces showing signals delivered to the threshold detectors have two items of interest:

1. The three signals are very similar in wave shape in spite of major differences in the three input currents. This illustrates the mixing of the harmonics between phases and confirms the three-phase nature of the relay.

2. The polarity sense is restrain on positive and trip on negative. S and T phases initially develop a signal in the tripping direction which vanishes even before the first peak of inrush is reached. It is too short to activate the 2nd level detector beyond the integrator.

Fig. 17 shows an external fault with a severe dc component which saturates the CT's in less than one cycle. The relay properly restrained during the entire period including that during the gradual CT recovery as the dc component decayed.

Fig. 18 shows a severe internal fault with a large dc component plus CT ac-saturation. The unrestrained unit has been decommissioned to demonstrate that the harmonic restrained unit performed correctly in 2.2 cycles in spite of the distorted currents from saturated CT's and a minimum pickup setting of 50%.

406

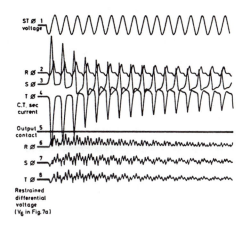

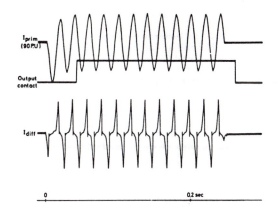

Fig. 16. Typical inrush test showing three-phase restraint action. (Inrush 10.8 P.U. on TØ).

Fig. 18. Severe internal fault with CT saturation. The unrestrained (instantaneous) measuring unit decommissioned. Operation in 2.2 cycles.

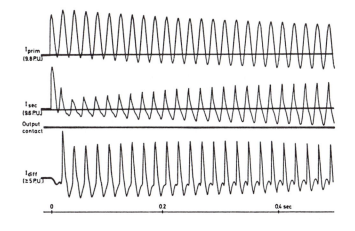

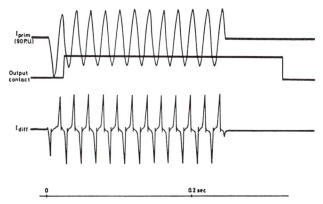

Fig. 17. Through-fault current with a dc component of 440 ms time-constant that saturates the CT's in less than one cycle. No operation.

Fig. 19. Severe internal fault with CT saturation. Instantaneous unit operation within 1.2 cycles.

Fig. 19 shows that the instantaneous unit will also properly respond to the peaked current wave shapes of saturated CT's even when set to pickup at 20 times relay rating.

CONCLUSIONS

The design of a high speed transformer differential relay requires detailed attention to all aspects of the non-linear nature of both power transformers and current transformers. Such a relay with 1/2 to 2 cyle operating time and other improved characteristics has been described. These improvements result from a three-phase evaluation of the current data, as contrasted to previous methods which evaluate each set of phase parameters separately. A further improvement has resulted from a two-stage signal level detector-integrator technique which makes the relay immune to noise spikes and which enhances the discriminating ability of the relay to recognize faults from inrush currents. Static measuring is used with an electro-mechanical output, the combination resulting in a relay which requires no new or special skills to apply.

The basic principle of the relay fulfills all demands of a modern transformer relay. The soundness of the theory has been proven by over 3,000 successful installtions throughout the world.

REFERENCES

(1) C.A. Mathews, "An Improved Transformer Differential Relay". AIEE Transactions, Vol. 73, pt III, pp. 645-50, 1954.

(2) W. K. Sonnermann, G. L. Wagner, G. D. Rockerfeller, "Magnetizing Inrush Phenomena in Transformer Banks," AIEE Transactions, Vol. 77, pt. III, pp. 884-892, 1958, and its Ref. 1.

(3) ANSI Standard C37.91-1972,"Guide for Protective Relay Applications to Power Transformers". American National Standards Institution.

(4) W. J. Smolinkski, "Design Considerations in the Application of Current Transformers for Protective Relaying Purposes". IEEE Transactions, July/Aug. 1973, Vol. PAS-92 No. 4, pp. 1329-1336.

R. E. Dietrich and **F. E. Newman** (Bonneville Power Administration, Portland, Oregon): The authors are to be commended for their presentation of a solution to a problem which has only very recently manifest itself—at least here in the west.

The paper presents a straightforward approach to the problems associated with transformer differential protection; particularly those problems associated with saturation—both in the transformer and current transformers associated with the differential relay.

The authors emphasize the three phase restraint characteristic of the relay, which is particularly effective during charging transformers built with magnetic material with a sharp kneepoint. We would appreciate their comments on the performance of this relay on clearing a close-in one line-to-ground fault external to an autotransformer, where the source voltage on one phase of the transformer goes to zero then recovers to normal, re-energizing that phase of the transformer. In this case will the 2nd harmonic restraint coming only from the phase that has been subjected to the fault be sufficient to prevent a misoperation?

We are also impressed with the modular, solid state construction and the flexibility which it provides. We believe the relay could possibly further improve if taps could be provided on the input transformers to facilitate matching dissimilar current transformer ratios. With the present design CT ratio matching must be done exclusively with auxiliary CTS mounted externally.

Manuscript received February 18, 1975.

P. A. Solanics (Cleveland Electric Illuminating Co., Cleveland, Ohio): The authors are to be congratulated for preparing a very informative and concise paper. Of particular interest is the use of 5th harmonic restraint to preclude transformer differential relay operation for normal overexcited conditions.

On generator unit transformer applications, our current practice is to provide separate overexcitation protection and provide additional restraint to the transformer differential relay for overexcited conditions. This additional restraint is provided via CT's located in the delta windings of the unit transformer. During overexcited conditions, third harmonic and their multiples will flow in the delta windings and the associated CT's. The output of the CT's supply a separate differential restraint winding and thus provide additional relay restraint. This special connection has proven satisfactory but is complicated and requires low voltage fault testing to verify the relative polarities of the inside-the-delta CT's.

The use of the 5th harmonic to provide the additional restraint is attractive in view of the complexities associated with our existing scheme. However, does the 5th harmonic content have to be determined for each application, i.e. does the harmonic content vary for 3-phase core, 3-phase shell or single phase transformer design? If so, is the relay level of 5th harmonic restraint adjustable?

REFERENCE

ANSI Stnadard C37.91 1972, "Guide for Protective Relay Applications to Power Transformers," *American National Standards Institution,* pp 21-22.

Manuscript received January 30, 1975.

M. S. Sachdev (University of Saskatchewan, Saskatoon, Sask., Canada): Some salient features of a three phase differential relay, which primarily uses solid state circuitry, are described in this paper. The authors are to be complimented for clearly presenting the information for the benefit of engineers in the power field. Will the authors care to discuss the following comments and questions?

1. It is true that electromechanical relays primarily respond to rms values of the input signals. Electronic circuits, suitable for identifying other characteristics of the inputs, can be designed and built. Analog type circuits are also subject to some limitations and in many cases these can be quite complicated. Another alternative is the use of digital technology which has its own limitations. I believe that a judicious blend of electromechanical, and analog and digital type electronic circuits can provide best results.

2. The author's statement "The basic relay of Fig. 1 is for up to three inputs (transformer windings or two breakers on one winding)," is misleading. Is the basic version of the relay designed to accept inputs from six breakers controlling a three winding transformer (two breakers for each winding)?

Manuscript received March, 6, 1975.

3. For internal faults a restraining signal proportional to the largest incoming current is generated. Consider a single phase internal fault in a two winding transformer such that currents in the faulted phase are $4.0/\underline{-60°}$ and $5.0/\underline{115°}$ p.u. The restraining signal generated in the relay will be proportional to 5.0 p.u. In a conventional type differential relay, the restraining torque will be proportional to about 0.54 p.u. current. Is the proportionality constant k adjusted to reduce the restraining voltage signal to a level comparable to that in other relays? It seems that this approach cannot be used because it will also reduce the restraining signal for through faults.

4. Second and fifth harmonic filters have been used to obtain harmonic voltages and a low pass filter to suppress components of over 500 Hz from the operating voltage. These filters are of conventional electromagnetic type. Has the use of active filters, instead of the conventional type, been investigated and/or tried for this relay?

5. "The restrained unit requires a signal above its setting for at least 15 ms for it to complete a measurement and initiate a trip signal." Does most of this delay occur in the phase threshold and integration circuitry given in Figure 5?

6. Figure 4 indicates that the operating differential voltage signal of each phase is directly applied to the trip measurement circuitry. Only two such inputs are indicated in Figure 6. Will the authors reconcile this discrepancy?

At the end, I once again congratulate the authors for a well prepared paper which brings various aspects of the problem of transformer differential protection and some feature of the RADSE relay to the notice of engineers in the power field.

C. H. Einvall and **J. R. Linders**: We thank the discussers for their comments. The several points they mention are important considerations in the application of any transformer differential relay, and warrant additional discussion. In the case of the external line to ground fault near to a set of single phase autotransformers without tertiaries, as posed by Messrs. Dietrich and Newman, the described relay will properly restrain on the voltage recovery inrush. While this is a single phase phenomenon with respect to the autotransformers as the discussers note, it is a polyphase condition to the DSE relay. By virtue of the delta connected CT's (or otherwise suppression of the zero sequence component) these recovery currents will flow in two of the relay differential circuits. Thus, the harmonic content in the single phase inrush current will also appear in two relays, and through the interconnection between the relay restraint circuits will provide additional restraint to each. This restraint will be substantially the same as for a comparable three phase inrush, because of these restraint interconnections plus the delta, or equivalent, CT connections.

The addition of ratio taps on the relay input transformers would appear to increase the convenience in the use of the relay. The present design is a trade-off judgement based largely on the additional values derived from the air gapped input transformers. These do not presently lend to multiple-tap windings with the desired accuracy. The use of separate auxiliary CT's provides a more accurate CT turns ratio adjustment and this is particularly useful when the full sensitivity of the relay is desired. Also these separate auxiliary CT's have the advantage of allowing any required delta CT connection to be made up independent of the main CT's, thus allowing them to be connected in the preferred wye configuration.

In response to Mr. Solanic's question, there is no need to make any 5th harmonic calculations to apply the relay. The relay has no adjustments for setting either the 5th or 2nd harmonic restraint. The effect of core geometry, i.e., 3 φ core form, 3 φ shell or single phase transformer, has an influence on the 3rd order harmonics, but very little effect on the 5th's. The 5th harmonic exciting current will depend on the quality of the steel used in the core and not on the core geometry. From a harmonic exciting current standpoint, there is little difference in transformer steels among various manufacturers. The small differences which do exist mainly influence the sharpness of the knee of the saturation curve. This effect is small and becomes unimportant at exciting currents above the minimum pickup current of the relay. In this higher exciting current region, the current will be largely dependent on the air core inductance of the excited winding and very little on the actual characteristics of the steel or the core geometry. There may be as much variation between the air core inductance of the high and low voltage winding of a given transformer as there is between transformer designs and manufacturers. These air core inductance differences also have little effect on the 5th harmonic currents. In other words, the 5th harmonic content in overexcited currents is due primarily to the general shape of the magnetization curve and is little effected by moderate variations in it. However, none of these overexcited transformer harmonic current characteristics can be accurately obtained from the usually available

Manuscript received April 21, 1975.

transformer magnetizing curves. These are plotted on an rms basis and must first be converted to instantaneous values before a fourier analysis can be made on the distorted exciting currents. Computer programs which are used to generate these rms curves can frequently also provide the harmonic content directly.

The statement in the paper that the relay is secure against over-excitation up to 140% rated voltage is conservative. In actual practice the relay will properly restrain significantly above this value. On a single phase basis the relay restrains on 35%–45% 5th harmonic. On a three phase basis, the restraint interconnection between relay phases increases the effectiveness of the restraint significantly. Also, on a three phase basis the pickup current of the relay increases about 50%, as noted in the paper. The net result is a three phase, 5th harmonic effective restraint of 20-25%. Thus, the relay characteristic is adequate for any foreseeable overvoltage abnormality or improvement in transformer steel. No calculations or setting adjustments are required to assure this performance. Initial installation checkout requires no more than the usual load checks to confirm the correctness of the installation wiring.

We thank Prof. Sachdev for noting the inadequacy in Figure 6. The restrained and unrestrained input signals are respectively the maximum signal from any of the three phases, selected through a simple diode auctioning circuit. The second input position shown for the unrestrained signal is a tap on a voltage divider network for selecting the sensitivity setting for this signal. It should have been so labeled in the Figure.

The described relay has the same characteristics regardless of the number of relay inputs required to satisfy the system configuration. As Prof. Sachdev notes, some two input differential relays utilize restraining cancellation for internal faults. (This is not feasible for multiple input relays.) Thus, where two breakers are used on one transformer winding these relays required a decision as to whether to parallel CT's so as to utilize this two winding characteristic. Such decisions are not required with the DSE relay. One parallels CT's or not as needed to provide the desired security to external faults, but not to get a different relay characteristic.

This restraint cancellation characteristic is not meaningful in situations where fault current does not flow into both windings for an internal fault. Also in actual practice it may not be a significant characteristic since load currents may nullify any benefit of improved sensitivity to minor internal faults. This restraining cancellation is not required in the described relay to develop the 20% minimum fault sensitivity. The threshold circuits used in the relay are quite insensitive to signal levels above the threshold values. Thus, the relay sensitivity, dependability and speed of operation would not be improved by providing these other characteristics in a two winding version.

The mentioned 15 ms is a design constant of the integrating circuit. This time interval is required for properly evaluating the complex wave form of the input signal to the first level detector. The value of this intergration is illustrated in the oscillogram of Figure 16. This circuitry also provides the needed security to noise, thus simplifying the overall design.

A direct substitution of active for the passive filters would gain little in the relay's performance. Active filters do have a place in future relay designs. When applied to any specific relay, the relay operating characteristics will be developed in a manner to take advantage of the unique characteristics of these devices. A redesign of the present relay is not contemplated at this time.

A MICROPROCESSOR BASED THREE-PHASE TRANSFORMER DIFFERENTIAL RELAY

J.S. Thorp
Senior Member, IEEE
Cornell University, Ithaca NY

A.G. Phadke
Fellow, IEEE
American Electric Power
Service Corporation, New York, NY

ABSTRACT

This paper presents an algorithm for digital protection of a three-phase, three-winding power transformer. Simple recursive expressions are given for the harmonics using a sampling rate of 12 times a cycle. These expressions are used as the basis for a one cycle algorithm which is compatible with proposed digital line protection schemes. The algorithm was tested on data obtained from a model transformer. Test results are given for 19 cases of energizations and faults.

INTRODUCTION

Digital computer based protection has been the subject of several studies during recent years. [1] It has been recognized that such protection systems are likely to be parts of a hierarchical computer system within the substation; dedicated to protection, control, alarm and data logging functions of a substation. Computer based protective devices are independent of all other subsystems, yet they are able to share certain data paths with other subsystems in order to improve the redundancy in the total system.

Work in progress at the American Electric Power Service Corporation has been reported in several earlier publications [2-5]. Effort at AEP up to this time has concentrated upon transmission line protection and the general framework for the hierarchical computer system. [6] This paper considers the problem of transformer protection within the context of the heirarchical system. The transformers considered are three phase, multi-winding transformers.

Protection of Power Transformers

Principles of transformer protection have been discussed in relaying literature for many years. [7] A brief discussion of the issues involved would be in order. Consider the power transformer shown in Fig. 1(a). The protection under consideration is the percent differential protection of the transformer. The slope of percent differential characteristic shown in Fig. 1(b) is adjusted to make the differential relay insensitive to CT and relay inaccuracies, as well as the off-nominal tap positions of the transformer. It has been known for many years that magnetizing inrush current may cause the percent differential relay to trip during transformer energization. An early solution to the inrush problem was the use of a "tripping suppressor" [7], which used a voltage relay to suppress the tripping function of the differential relay when the voltages were high. Interestingly

enough, there are possible advantages to be gained by adapting this principle in digital protection schemes for power transformers. This will be considered in a later section. By far the most common technique used for preventing false trips during energization is the use of a 'harmonic restraint' relay. [8,9] The fact that an inrush current is richer in harmonics than a fault current is key to the design of a harmonic restraint function.

The harmonic content of the inrush current depends upon many factors: residual magnetization of the core and the instant of switching are two of the more important factors. There are other phenomena which also contribute to the harmonics, and their effect on the behavior of the harmonic restraint function must also be considered. Three major sources of harmonics are as follows:

(1) Magnetizing inrush due to nonlinearities of transformer core

(2) Saturation of current transformers

(3) Overexcitation of the transformers due to a dynamic overvoltage condition.

The harmonic restraint function should be so designed that it restrains during the magnetizing inrush and overexcitation conditions, while during an internal fault the harmonics generated by a saturated CT should not restrain the differential relay.

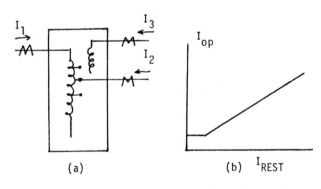

Figure 1: Transformer Current Differential

Hierarchical Substation Computer System

The computers in a substation are connected in a heirarchical network as shown in Fig. 2. The substation host computer maintains disc files of recorded events and oscillographic and other power flow data gathered within the substation. The relaying computers obtain this data from transducers and pass it on to the substation host over a serial communication link. The relaying computers are stand-alone devices and do not depend upon any other processors for their operation. The host computer also handles communication to a remote computer over the switched telephone network.

Reprinted from *IEEE Trans. Power App. Syst.*, vol. PAS-101, no. 2, pp. 426–432, Feb. 1982.

Figure 2(b) shows one bay of a breaker-and-half arrangement, where the transmission line L_1 and transformer T_1 share circuit breaker B. The current in circuit breaker B is thus used by both relaying computers R_{L1} and R_{T1}. A redundancy in data paths is provided by making the Input/Output (I/O) subsystems of R_{L1} and R_{T1} accessible to both R_{L1} and R_{T1}. Since this is the maximum extent of data sharing by any two relaying processors in a breaker-and-half schemes, this method of providing redundancy is considered to be preferable to having a common data I/O bus.

The relaying processor considered for this application is the Plessey Miproc-16 microcomputer having a 16 bit word length and available with a 250 ns or a 385 ns cycle time. Most instructions in this machine are single cycle instructions. The analog input system designed at AEP includes a Random Acess Memory to buffer the data samples. This RAM buffer also facilitates the data sharing mentioned earlier.

The design of digital protection schemes frequently involves a reexamination of the basic protection problem in order to obtain an appropriate digital solution. Digital relays, due to their inherent nature, need not be an imitation of an analog device but should, instead, offer a digital solution to the same problem. Particularly in the area of power transformer protection, it is difficult to imagine an exact digital match to existing analog devices. Existing electromechanical and static relays use analog filters to obtain trip and restraint signals with different harmonic content. It should be recognized that even the best design for an anlog filter that selects the second harmonic, for example, is imperfect. That is, the filter output contains other harmonics in addition to the second. Further the analog filter has a transient response which affects relay performance. It would seem unnecessary and undesirable to attempt to include these effects in a digital relay.

The first suggested use of digital techniques for transformer protection [10] used digital simulation of analog filters to provide the harmonic restraint function. Other authors have suggested the use of Fourier techniques [11], and Finite Duration Impulse Response Filters [12] to accomplish the same objective All of these publications have used a single phase two winding transformer for their investigations; concentrating mainly on the harmonic restraint function in the presence of magnetizing inrush phenomenon. Another paper [13] uses line voltages to enhance the speed of response of the relay in the presence of inrush currents.

This paper describes a transformer protection algorithm for a three-phase three-winding transformer which is compatible with the digital line protection schemes described in [5]. The major issue in compatibility is the ability to share samples. A fundamental restriction imposed by this requirement is that the sampling rate of 720 Hz used in [5] must be used for the transformer algorithm. It will be shown in the next section that the advantages of a 720 Hz sampling rate for line protection are even greater advantages in the transformer protection problem. A second consideration in shared samples is that the current samples used by the line relay have been filtered by an analog mimic (to match the X/R of the line). The analog mimic will be neglected in the next three sections but its impact on the algorithm will be considered in the final section. Based on experience gained in implementing a line relay on a microprocessor [6] it is important that simple recursive expressions be obtained for all quantities that are computed at each sample time. The next section describes such recursive expressions for the harmonics.

CALCULATION OF HARMONICS

The 720 Hz sampling rate makes the calculation of harmonics as straight forward as the fundamental frequency calculations used in line relaying [5]. If the signal $f(t)$ is assumed, for simplicity in presentation, to be periodic with a fundamental frequency of 60 Hz and limited to the fifth harmonic by the anti-aliasing filter then the relation between the samples

$$f_k = f(k \Delta t) \quad , \Delta t = \frac{1}{720} \text{ s}$$

and the harmonics

$$F_n \quad , n=0, \pm 1, \pm 2, \pm 3, \pm 4, \pm 5$$

is given by

$$F_n = \frac{1}{12} \sum_{k=0}^{11} f_k e^{\frac{-j2\pi kn}{12}} \qquad (1)$$

Using the conventions of [5] the phasor $\overline{F_n}$ associated with the nth harmonic is given by

$$\overline{F_n} = j \frac{(F_n + F_{-n}^*)}{2}$$

The expression in equation (1) only uses the first 12 sample values beginning at t=0. Let $F_n(r)$ represent the nth harmonic term beginning at the rth sample, i.e,

$$F_n(r) = \frac{1}{12} \sum_{k=r}^{r+11} f_k e^{-j\frac{2\pi kn}{12}} \qquad (2)$$

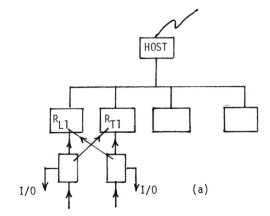

Figure 2: Hierachical Substation Computer Network

It should be observed that equation (2) is consistent in the sense that if $f(t) = F_1 e^{j\omega_0 t}$ where $\omega_0 = 2\Pi \times 60$ then $F_1^{(r)} = F_1$. After the next sample becomes available the next set of harmonics can be computed by updating $F_n^{(r)}$, viz.

$$F_n^{(r+1)} = F_n^{(r)} + \frac{1}{12}[f_{r+12} - f_r]e^{-j\frac{2\pi nr}{12}} \quad (3)$$

The correction term in equation (3) is the difference between the most recent sample, f_{r+12}, and the sample obtained one cycle in the past, f_r, multiplied by an angle which rotates more rapidly for the higher harmonics. The rotation is shown symbolically in Figure 3. Real and imaginary parts of the phasor can be computed using an expression similar to equation (3) with sin $(nr\pi/6)$ and cos $(nr\pi/6)$ instead of the exponential.

The fact that sin $(m\pi/6)$ and cos $(m\pi/6)$ involve only 0, $\pm\frac{1}{2}$, $\pm\sqrt{3}/2$, and ±1 has been exploited in the line relaying algorithm [5] by using two A/D converter channels one with unit gain and a second with an attenuation of $\sqrt{3}/2$. To obtain a sample, f_r, multiplied by $\sqrt{3}/2$, the second A/D converter channel is used. The update in equation (3) is obtained by addressing the proper A/D converter channel and multiplying by 0, $\pm\frac{1}{2}$, or ±1. One cycle of coefficients for the real and imaginary parts of the fundamental, second, and fifth harmonic is shown in Table 1. It should be noted that the n=5 entries are identical with the fundamental entries except the $\sqrt{3}/2$ channel is opposite in sign. Similar figures and tables can, of course, be produced for the third and fourth harmonics.

Each of the nine phase currents for a three-phase, three-winding transformer must be sampled simultaneoulsy (directly and through the .866 attenuator) and the desired harmonics computed. The one-cycle window implied by equation (3) is a reasonable minimum amount of data needed for a secure discrimination between a fully offset fault current and severe magnetizing inrush current.

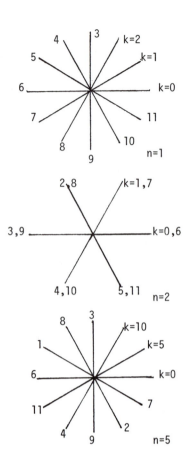

Figure 3: Rotation of first, second, and fifth Harmonic coefficients.

A TYPICAL ALGORITHM

In order to examine the computation beyond the harmonic calculation, a typical algorithm was programmed. Fundamental frequency currents must be used for trip and restraint functions to guard against tripping due to a through fault. It should also be realized that in the case of a multiwinding transformer, more than one fundamental frequency restraining signal must be examined to allow for different types of through faults that are possible. In determining the relay response these multiple restraint signals must be checked and the appropriate signal selected based upon the prevailing currents in the transformer winding.

There is evidence[14] that a secure restraint function for a three-phase transformer can be obtained if a single harmonic restraint signal is derived by combining the harmonics of the three phases. In addition it is common to think of the second harmonic as being indicative of inrush and the third and fifth as being associated with overexcitation. Arguments that the third is hazardous because of its predominace in CT saturation [14] would suggest the use of the fifth. The algorithm was programmed with two harmonic restraints per phase: one a second harmonic obtained by summing the magnitude of the second harmonic for all phases and the other a fifth harmonic restraint developed in a similar manner.

A flow chart of the algorithm is shown in Figure 4. At each sample instant one trip and two restraint samples are obtained for each phase. There are thus 18 samples in total counting the $\sqrt{3}/2$ A/D channel. Using the polarity of the one-line diagram, in Figure 1a the trip signal is

$$I_{TRP} = I_1 + I_2 + I_3$$

Harmonic	1	1	1	1	2	2	2	2	5	5	5	5
Real	x	x			x	x			x	x		
Imaginary			x	x			x	x			x	x
A/D Channel												
1.0	x		x		x		x		x		x	
.866		x		x		x		x		x		x
n												
1	1	0	0	0	1	0	0	0	1	0	0	0
2	0	1	½	0	½	0	0	1	0	-1	½	0
3	½	0	0	1	-½	0	0	1	½	0	0	-1
4	0	0	1	0	-1	0	0	0	0	0	1	0
5	-½	0	0	1	-½	0	0	-1	-½	0	0	-1
6	0	-1	½	0	½	0	0	-1	0	1	½	0
7	-1	0	0	0	1	0	0	0	-1	0	0	0
8	0	-1	-½	0	½	0	0	1	0	1	-½	0
9	-½	0	0	-1	-½	0	0	1	-½	0	0	1
10	0	0	-1	0	-1	0	0	0	0	0	-1	0
11	½	0	0	-1	-½	0	0	-1	½	0	0	1
12	0	1	-½	0	½	0	0	-1	0	-1	-½	0

Table 1: One Cycle of Coefficients

while the restraint signals are

$$I_{RST1} = I_1 - I_2 - I_3$$

and

$$I_{RST2} = -I_1 + I_2 - I_3$$

The second restraint is necessary to protect the transformer operating with the primary breaker open.

The fundamental frequency phasors for the trip and restraint signals for each phase are computed using the recursive DFT formulation (Equation (3) with n=1). In addition the second and fifth harmonic phasors for the first restraint current, I_{RST1}, are computed for each phase (Equation (3) with n=2 and n=5).

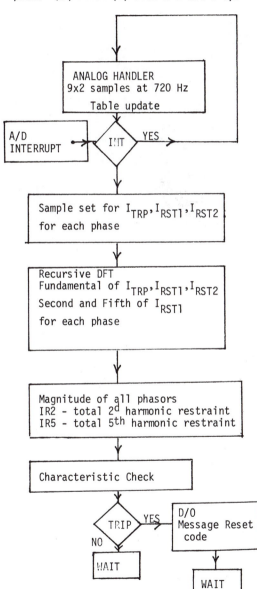

Figure 4: Flow Chart for Three Winding Transformer Protection

In total, then, ten updates of the form of equation (3) are computed for each phase (real and imaginary parts). It is clear that the simplicity in computation produced by the 720 Hz sampling rate is critical if 30 such calculations are to be made at each time step. It is also apparent that more harmonics, i.e. the 3d and 4th would add computational burden so that it is well worth determining whether the second and fifth are adequate.

The magnitude of all of the phasors are computed and the second harmonic restraint, IR2, is formed by summing the second harmonic magnitudes over the phases and the fifth harmonic restraint, IR5, is formed by summing the fifth harmonic magnitudes over the phases.

Denoting the magnitude of the fundamental frequency phasors for the trip and restraint signals by $\bar{I}_{TRP}$, $\bar{I}_{RST1}$, and $\bar{I}_{RST2}$ the relay is restrained for a given phase if

$$|\bar{I}_{TRP}| < 120 \tag{4}$$

or
$$|\bar{I}_{TRP}| < .133 \,|\bar{I}_{RST1}| \tag{5}$$

or
$$|\bar{I}_{TRP}| < .133 \,|\bar{I}_{RST2}| \tag{6}$$

or
$$IR2 > .30 \,|\bar{I}_{TRP}| \tag{7}$$

or
$$IR5 > .20 \,|\bar{I}_{TRP}| \tag{8}$$

The inequalities (4) and (5) produced the characteristic shown in Figure 1b while inequalities (4) and (6) produce a similar characteristic for the second restraint current. The inequalities (7) and (8) are the second and fifth harmonic restraints respectively. The transformer is tripped if any phase trips i.e., if none of the inequalities (4)-(8) are satisfied for some phase.

The particular numbers in the inequalities are those used for the model transformer to be described in the next section. The fifth harmonic restraint coefficient is affected by the particular anti-aliasing filters used. The threshold value of 120 is a function of the A/D converter range used in the experiment. It should be emphasized that the algorithm is only one of many possible using the DFT techniques. It is typical in the sense that it incorporates three phase percent differential operation with two restraint currents and two harmonic restraints.

ALGORITHM PERFORMANCE

The algoirithm was programmed in Fortran for off-line processing of sampled data. The nine currents were sampled in real time and the data stored in a file which was subsequently processed by the Fortran program. The transformer was simulated on the AEP Simulator with a 765 kV Y connected primary, 345 kV Y connected secondary and 138 kV Δ connected tertiary. The model transformers were single phase units connected to make a three phase unit. Each single phase transformer has five windings wound on 2 mil Deltamax tape wound core (Arnold Engineering Company) with saturating flux density of 13,500 Gauss normally operated at about 10,000 Gauss.

The model transformer exhibited inrush phenomena (see Figure 5 for an example of an energizing sequence which produced inrush currents in all three phases). The inrush currents decayed somewhat rapidly in the simulator representation. This is attributed to high losses which are usually associated with small scale models. However, the inrush currents during the period of interest for the relaying application are realistic.

A total of 19 faults and energizations were simulated to test the various features of the algorithm. The cases are listed in Table 2. Examining Case 12 to demonstrate the output, let a one in the trip column of Table 3 indicate a trip from the percent differential (inequalities (4)-(6)) while a one in the R2 column indicates that trip is restrained by the second harmonic restraint and a one in the R5 column indicates restraint from the fifth harmonic. An unrestrained trip then corresponds to entries of one under trip and zeros under R2 and R5.

Table 3 indicates that prior to sample 17 no phase had a trip from the percent differential and that the second and fifth harmonic restraints were effective. At sample 17 phase A differential indicates trip but the second and fifth harmonic restraints are still effective. At sample 22 the fifth harmonic restraint drops out and at sample 27 the second harmonic restraint drops out. The trip command would be given at sample 27, one cycle after fault initiation. Since the development of the fundamental frequency signal always lags the development of the harmonic signals during a transition period, the restraint function prevails immediately following the occurrence of a fault, until fault data of one cycle duration is available, at which time a secure relaying decision can be made. In this sense the transient response of the DFT calculations are ideal for the transformer protection problem. Some fully off-set internal faults take somewhat longer than one cycle to produce a trip decision. All other phenomena lead to a secure decision in about one cycle. The trip, restraint, second and fifth harmonics are shown in Figure 6.

CASE

1	Inrush
2	Inrush
3	Overexcitation
4	Overexcitation and Inrush
5	Energize with external fault on secondary
6	Energize with external fault on secondary
7	Phase a to ground external fault on secondary
8	3 phase external fault on secondary
9	3 phase external fault on tertiary
10	phase to phase external fault on tertiary: primary open
11	3 phase external fault on tertiary: primary open
12	phase a to ground fault internal primary
13	3 phase internal fault on secondary
14	phase a to ground on secondary internal
15	phase a to ground on primary internal
16	phase a to ground on primary internal, strong source
17	One winding of tertiary shorted
18	one half winding primary internal fault
19	one half winding secondary internal fault

Table 2: Fault Cases

Sample	Phase A			Phase B			Phase C		
	Trip	R2	R5	Trip	R2	R5	Trip	R2	R5
14	0	1	1	0	1	1	0	1	1
15	0	1	1	0	1	1	0	1	1
16	0	1	1	0	1	1	0	1	1
17	1	1	1	0	1	1	0	1	1
18	1	1	1	0	1	1	0	1	1
19	1	1	1	0	1	1	0	1	1
20	1	1	1	0	1	1	0	1	1
21	1	1	1	0	1	1	0	1	1
22	1	1	0	0	1	1	0	1	1
23	1	1	0	0	1	1	0	1	1
24	1	1	0	0	1	1	0	1	1
25	1	1	0	0	1	1	0	1	1
26	1	1	0	0	1	1	0	1	1
27	1	0	0	0	1	1	0	1	1
28	1	0	0	0	1	1	0	1	1
29	1	0	0	0	1	1	0	1	1
30	1	0	0	0	1	1	0	1	1
31	1	0	0	0	1	1	0	1	1
32	1	0	0	0	1	1	0	1	1
33	1	0	0	0	1	1	0	1	1
34	1	0	0	0	1	1	0	1	1

Table 3: Outputs for Case 12

In an abbreviated form of Table 3 the output for the 19 cases of Table 2 are summarized in Table 4. The inrush in Case 1 produces differential trip in phase B and then phase A. The fifth harmonic restraint drops out at sample 27 but the second harmonic restraint persists. In Case 3 the fifth harmonic restraint is successful for overexcitation. In cases 10 and 11 the second restraint current is used. The algorithm never misoperated: for all non-internal faults and energizations the restraint signal dominated the trip signal and for all internal faults the proper trip signal was developed (in cases 12-19 a trip was established at sample 27 in 6 cases and at sample 28 in two cases).

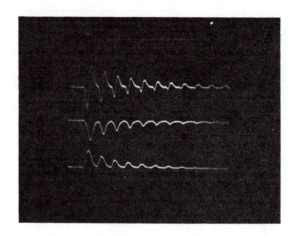

Figure 5: Model Inrush Currents

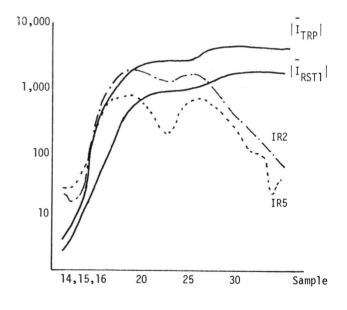

Figure 6

MODIFICATIONS AND EXTENSIONS

The current signals in the experimental model were not processed by the analog mimic that is used in line protection. The mimic has a transfer function given by $k(1+\tau s)$ where τ is matched to the line time constant. Figure 7 shows an inrush current (below) and the corresponding mimic output. The transformer algorithm must be modified if the currents are filtered by the mimic. The differential mode is unaffected but the harmonic restraints must be changed since the mimic increases the harmonic content of the currents. The fifth harmonic content of the output of the mimic is approximately 5 times the input fifth harmonic while the second harmonic is approximately twice as large. The inequalities (6) and (7) would then be modified to form

$$IR2 > .60 \ |\bar{I}_{TRP}| \qquad (9)$$

$$IR5 > 1.0 \ |\bar{I}_{TRP}| \qquad (10)$$

The factors of two and five for the second and fifth harmonics were verified experimentally. Currents from a single inrush were processed with and without the mimic filters to obtain the second harmonic verification and currents from an overexcitation case used to verify the fifth harmonic factor.

An ultra high speed version of the "tripping suppressor" can also be implemented digitally. It has been shown [5] that fundamental frequency phasors for both voltage and current can be computed in as little as one quarter of a cycle. Using the quarter cycle calculations the differential check in inequalities (4)-(6) can be computed more rapidly than the harmonic restraint is available. If the voltage phasor is computed with the same quarter cycle algorithm then the voltage restraint can be used instead of the harmonic restraint. The inequalities (7)and (8) would be replaced by a test on the voltage magnitude of the form

$$|\bar{V}| > V_0 \qquad (11)$$

where V_0 must be selected so that no internal fault is restrained by the voltage restraint. At the expense of obtaining voltage samples there could be a savings in computation (the voltage DFT is equivalent to one of the two harmonic calculations) and an increase in speed.

In addition to incorporating the mimic and investigating the performance of the digital "tripping suppressor" a number of other tasks remain. Foremost is the implementation of the program on a suitable microcomputer. Experience gained from the line relaying program [5] indicate the proposed algorithm is not too large for such implementation. In addition to the implementation a complete relaying package needs to be developed and tested. Such a package will include appropriate output signals, reset functions, backup

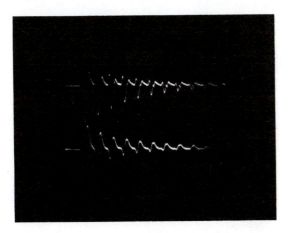

Figure 7: Mimic response to Inrush Current

CASE	SAMPLES	A			B			C		
1	1-21	0	1	1	0	1	1	0	1	1
	22-23	0	1	1	1	1	1	0	1	1
	24-27	1	1	1	1	1	1	0	1	1
	28-36	1	1	0	1	1	0	0	1	1
2	1-20	0	1	1	0	1	1	0	1	1
	21-24	0	1	1	1	1	1	0	1	1
	24-26	1	1	1	1	1	1	0	1	1
	27-36	1	1	0	1	1	0	0	1	1
3	1-11	1	1	1	1	1	1	1	1	1
	12-36	1	0	1	1	0	1	1	0	1
4	1-11	1	1	1	1	1	1	1	1	1
	12-36	1	0	1	1	0	1	1	0	1
5	1-36	0	1	1	0	1	1	0	1	1
6	1-36	0	1	1	0	1	1	0	1	1
7	1-36	0	1	1	0	1	1	0	1	1
8	1-36	0	1	1	0	1	1	0	1	1
9	1-36	0	1	1	0	1	1	0	1	1
10	1-10	0	1	1	0	1	1	0	1	1
	11-18	0	0	1	0	0	1	0	0	1
	19	0	0	1	0	0	1	0	0	1
	20	0	1	1	0	1	1	0	1	1
	21-22	1	1	1	0	1	1	0	1	1
	23-36	1	1	0	0	1	1	0	1	1
11	1-13	0	1	1	0	1	1	0	1	1
	14-15	0	0	0	0	0	0	0	0	0
	16-17	0	1	1	0	0	1	0	0	1
	18	0	1	1	0	0	1	0	0	1
	19-29	0	1	1	0	1	1	0	1	1
	30-36	0	1	1	0	1	1	0	1	0
12	1-16	0	1	1	0	1	1	0	1	1
	17-21	1	1	1	0	1	1	0	1	1
	22-26	1	1	0	0	1	1	0	1	1
	27-36	1	0	0	0	1	1	0	1	1
13	1-16	0	1	1	0	1	1	0	1	1
	17-18	1	1	1	0	1	1	1	1	1
	19-26	1	1	1	1	1	1	1	1	1
	27-29	1	1	0	1	0	0	1	1	0
	30	1	1	0	1	0	0	1	0	0
	31-35	1	1	0	1	0	0	1	1	0
	36	1	0	0	1	0	0	1	0	0
14	1-7	0	1	1	0	1	1	0	1	1
	8-19	1	1	1	0	1	1	0	1	1
	20	1	1	0	0	1	1	0	1	1
	21	1	1	1	0	1	1	0	1	1
	22-26	1	1	1	0	1	1	1	1	1
	27	1	1	0	0	1	1	1	1	1
	28-32	1	0	0	0	1	1	1	1	1
	33-36	1	0	0	0	1	1	0	1	1
15	1-17	0	1	1	0	1	1	0	1	1
	18-20	1	1	1	0	1	1	0	1	1
	21-26	1	1	1	0	1	1	1	1	1
	27	1	1	0	0	1	1	1	1	1
	28	1	0	0	0	1	1	1	1	1
	29	1	0	0	0	1	1	1	1	0
	30-33	1	0	0	0	1	1	1	1	1
	34-36	1	0	0	0	1	1	0	1	1
16	1-17	0	1	1	0	1	1	0	1	1
	18-20	1	1	1	0	1	1	0	1	1
	21-25	1	1	1	0	1	1	1	1	1
	26	1	1	0	0	1	1	0	1	1
	27-36	1	0	0	0	1	1	0	1	1
17	1-17	0	1	1	0	1	1	0	1	1
	18-19	1	1	1	0	1	1	0	1	1
	20-26	1	1	1	0	1	1	1	1	1
	27-36	1	0	0	0	1	1	1	1	1
18	1-17	0	1	1	0	1	1	0	1	1
	18-25	1	1	1	0	1	1	0	1	1
	26	1	1	0	0	1	1	0	1	1
	27-36	1	0	0	0	1	1	0	1	1
19	1-16	0	1	1	0	1	1	0	1	1
	17-20	0	1	1	0	1	1	1	1	1
	21-26	0	1	1	0	1	1	1	1	1
	27-36	0	1	1	0	1	1	1	0	0

Table 4: Summary of results

modes of operation and possible interfaces with other protection and host processors within the substation.

A final issue is the testing of the algorithm under saturation of current transformers. It was not possible to simulate saturation of current transformers in the laboratory so that this effect on both the current differential and harmonic restraint could not be adequately judged.

CONCLUSIONS:

(1) Simple recursive expressions can be obtained for harmonic phasors up to the fifth using a sampling rate of 12 times a cycle. These DFT calculations can be used as the basis for a one cycle digital relay for power transformer protection which is compatible with proposed digital line protection schemes.

(2) The transient response of the DFT calculations (the development of the fundamental lags the development of the harmonics) restrains the relay for approximately one cycle until a secure decision can be made. The algorithm did not mis-operate in 19 energizations and fault situations.

(3) An ultra-high speed digital version of the "trip-ping suppressor" has been suggested which is worthy of further investigation.

ACKNOWLEDGEMENT

The authors would like to acknowledge the assistance of M. Adamiak in carrying out the experiments on the model transformer, T. Hlibka for programming the data acquisition system and J. Jauch for the design of the A/D converter used in the experiment.

REFERENCES

(1) "Computer Relaying", IEEE Tutorial Course, 79 EH0148-7-PWR.

(2) A.G. Phadke, M. Ibrahim, T. Hlibka, "Fundamental basis for distance relaying with Symmetrical Components", IEEE Trans. on PAS, Vol. PAS-96, No. 2, March/April 1977, pp. 635-646.

(3) A.G. Phadke, T. Hlibka, M. Ibrahim, M.G. Adamiak, "A Microcomputer based Symmetrical Component Distance Relay", Proceedings of PICA, April 1979.

(4) J.S. Thorp, A.G. Phadke, S.H. Horowitz, J.E. Beehler, "Limits to Impedance Relaying", IEEE Trans. on PAS, Vol. PAS-98, No. 1, pp. 246-260 April, 1979.

(5) A.G. Phadke, T. Hlibka, M.G. Adamiak, M. Ibrahim, J.S. Thorp, "A microcomputer based ultra-high-speed Distance Relay: Field Tests" 80SM649-4 IEEE PES Summer Meeting, July 1980.

(6) A.G. Phadke, "Recent Development in Digital Computer Based Protection and Control in Electric Power Substations", presented at Conference on Power System Protection, The Institution of Engineers, Madras India, April 1980.

(7) E.L. Harder and W.E. Marter, "Principles and Practices of Relaying in the United States", AIEE Transactions, Vol. 67, Part II, pp. 1005-1022. 1948.

(8) L.F. Kennedy and C.D. Hayward, "Harmonic-Current-Restrained Relays for Differential Protection", AIEE Transactions, Vol. 57, pp. 262-266, 1938.

(9) C.A. Mathews, "An Improved Transformer Differential Relay" AIEE Transactions, Vol. 73, Part III, pp. 645-50, 1954.

(10) J.A. Sykes and I.F. Morrison, "A Proposed method for Harmonic-Restraint Differential Protection for Power Transformers", IEEE Trans. on PAS, Vol. PAS-91, No. 3, pp. 1260-1272, 1972.

(11) O.P. Malik, P.K. Dash and G.S. Hope, "Digital Protection of Power Transformer," paper No. A76 191-7 IEEE PES 1976 Winter Power Meeting, New York.

(12) E.O. Schweitzer, R.R. Larson and A.J. Flechsig Jr, "An Efficient Inrush Current Algorithm for Digital Computer Relaying, Protection of Trans-formers", IEEE paper No. A77 510-1, 1977 PES Summer Meeting Mexico, July 17-22, 1977.

(13) J.A. Sykes, "A New Technique for High-Speed Transformer Fault Protection Suitable for Digital Computer Implementation", IEEE paper No. C72 429-9, Summer Power Meeting of PES, 1972.

(14) C.H. Einvall and J.R. Linders, "A Three-Phase Differential Relay for Transformer Protection", IEEE Trans. on PAS, Vol. PAS-94, No. 6, Nov/Dec 1975.

Advances In The Design of Differential Protection for Power Transformers

by

Tony Giuliante, *GEC Measurements Division, The English Electric Corporation, Hawthorne, NY.*

and

Graham Clough, *GEC Measurements Division, The English Electric Corporation, St. Louis, Mo.*

Transformer differential relays make the comparison, in phase and magnitude, of the current entering one winding of the power transformer, with the current leaving the transformer via the other winding(s). Any difference in current resulting from this comparison passes through the operating winding of the differential relay. When the current difference reaches a predetermined level, relay operation occurs, initiating the tripping of the associated circuit-breakers. The basic principle is shown in figure 1.

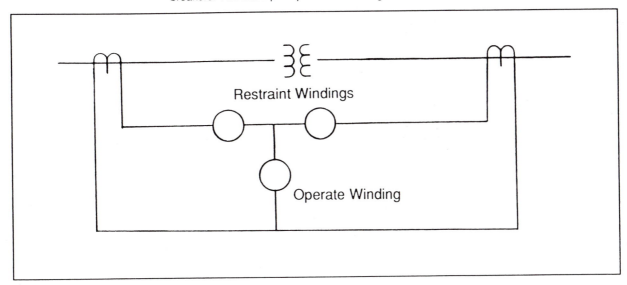

Restraint Windings

Operate Winding

Figure 1 – Percentage restrained differential transformer protection

This comparison is complicated by six factors:

* the ratios of the current transformers which connect the protection to the primary system may not match each other

* there may be a phase shift between the power transformer primary and secondary windings

* the power transformer may have a tap changer on one of its windings

* the current transformers can saturate under through fault conditions, giving an effective ratio error

* when a power transformer is energized, inrush current flows for a short time into the energized winding with no corresponding flow in the other windings

* overfluxing of the transformer can give rise to exciting current flowing in only one winding

These problems are addressed by:

* providing the protection with percentage restraint

* correct interconnection of the ct's on the power transformer inputs and outputs

* incorporating auxiliary current transformers into schemes

* providing some means of detecting the transformer inrush and overfluxed conditions

It is the means of detecting the inrush condition that is the weakest aspect of the design of most transformer differential relays. Traditionally relays have been stabilized against inrush by using the second (and other) harmonic content of the inrush waveform to apply extra restraint during transformer inrush. This has the disadvantage, however, of causing the relay to become very slow on some internal faults.

When a wye connected power transformer with a low level pre-existing fault on one phase is energized, inrush currents from the unfaulted phases will flow into the relay on the faulted phase because of the delta connected ct's. The second harmonic content of this inrush can restrain the relay. Very long operating times also result when a dc offset in the fault waveform causes the line ct's to saturate. The saturation induces high levels of second harmonics into the ct secondary current.

Reprinted with permission from *1991 Georgia Tech Protective Relaying Conference*, Atlanta, GA, pp. 1–12, May 1–3, 1991.

TRANSFORMER INRUSH CURRENTS

Under steady state, unloaded, conditions the power transformer draws exciting current sufficient to produce the flux level in the core that will produce a back-emf which exactly opposes the applied voltage. The magnitude of the exciting current depends on the permeability of the iron core. The working flux density is about 1.6 teslas (1 tesla = 10,000 gauss) and is produced by an exciting current of less than 1% of the full load rated current. If the flux level is increased, however, the iron will saturate and this demands a much greater exciting current to produce the flux. Saturation occurs at about 2 teslas in modern transformers. An example of a core magnetization characteristic is shown in figure 2.

Inrush current flows into a transformer winding as a result of transient saturation of the core when the winding is energized. This occurs because the iron core can be left in a magnetized condition when the transformer is de-energized.

The exciting current lags the applied voltage by almost 90° but when load current is also present, the total current flowing is the vector sum of the exciting current and the load. The load current can have a phase relationship to the output voltage over a very wide range, so that when the transformer is de-energised, although the load current will be broken at or near a current zero, the exciting current component of this could be at a current maximum. This would leave a remnant flux in the power transformer core near the peak working flux, perhaps as high as 1.3 teslas.

Re-energization of the transformer at a voltage zero will cause substantial saturation if the remnant flux is in the right direction. The inductance of the transformer core during saturation will be the air-cored value, about 0.15 henry, with a time constant of about 100ms.

Figure 3a shows the steady state relationship between the applied voltage and the exciting current and figure 3b shows the associated flux in the core. In figure 3c the remanent flux is at a maximum and the flux level present if the transformer is energized at a positive-going voltage zero is shown. The flux will rise so as to generate the required back-emf, but when the core enters the saturated region, the current needed to generate this flux rapidly increases as the permeability of the magnetic circuit approaches that of air. This continues until the flux level falls, later on in the cycle, to a value below the saturation level and the exciting current then falls to the lower value. By using the magnetizing characteristic shown in figure 3d, the amount of exciting current required to produce the flux can be found.

This figure is, of course, a simplified case. Saturation does not occur suddenly but takes place gradually. The simplification used can be considered to be a worst-case.

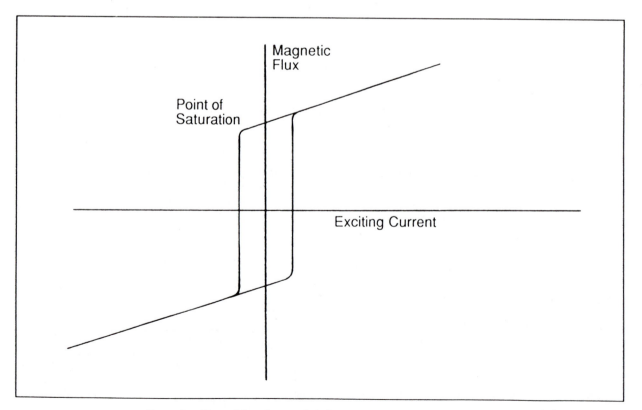

Figure 2 – Magnetizing characteristic for a typical power transformer core

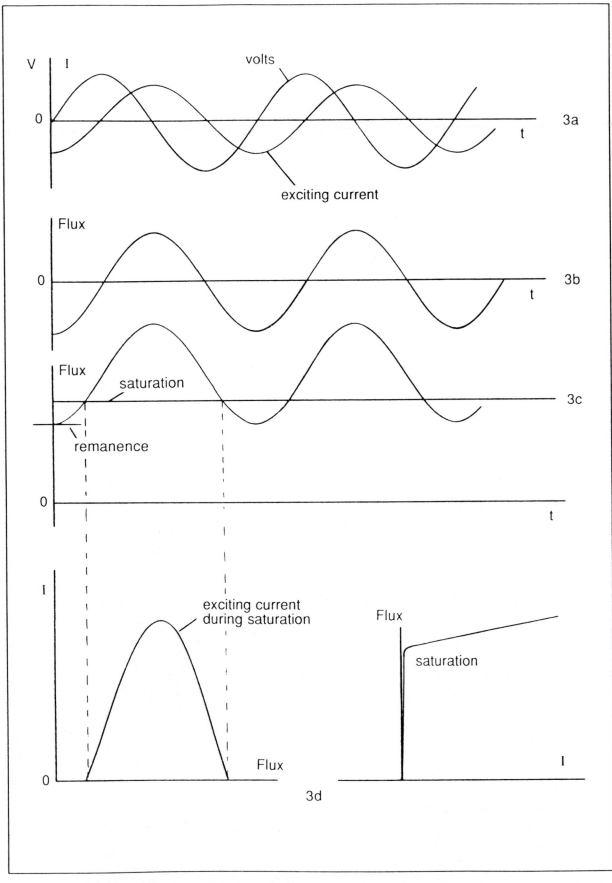

Figure 3 – Relationship between the applied voltage, exciting current and magnetic core flux

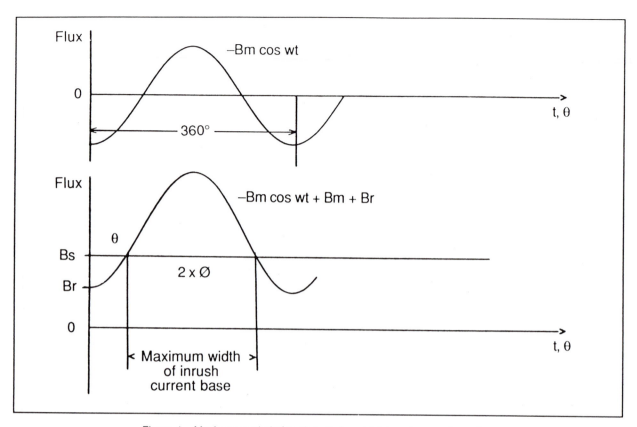

Figure 4 – Maximum period of each cycle for which inrush current can flow.

Figure 4 shows how the maximum proportion of each cycle for which inrush current flows, can be determined. If the remnant flux is Br, the peak working flux is Bm and the saturation flux for a particular steel is Bs, then when the circuit is closed at a voltage zero, the equation of the flux will initially be:

$$flux \ = \ -Bm \ coswt + Bm + Br$$

The core will go into saturation when the flux = Bs and this point will be determined by:

$$Bs \ = \ -Bm \ coswt + Bm + Br$$

$$thus \ coswt \ = \ \frac{Bm + Br - Bs}{Bm} = cos \ \theta$$

Angle θ is the initial saturation angle

From this, the maximum proportion of each cycle for which inrush current flows, can be found as follows.

In half a cycle, the core is unsaturated for θ degrees therefore in half a cycle, the core is saturated for 180 − θ = Ø degrees.

$$Ø \ = \ 180 - \theta$$

$$cos \ Ø \ = \ cos(180 - \theta)$$

$$cos \ (180 - \theta) \ = \ cos180 \ x \ cos\theta + sin180 \ x \ sin\theta$$

$$= \ -cos\theta + 0$$

Therefore, $cos \ Ø \ = \ -cos\theta$

and $Ø \ = \ arc \ cos \ (-cos \ \theta)$

$$Ø \ = \ arc \ cos \ \frac{(Bs - Bm - Br)}{Bm}$$

The total angle of inrush per cycle = 2 x Ø

This assumes that the inrush starts suddenly at saturation and thus is pessimistic.

Substituting the figures for the fluxes given on page two,

$$2 \ x \ Ø = 2 \ x \ arc \ cos \ \frac{(2 - 1.6 - 1.4)}{1.6} = 257°, \ or \ 0.72 \ cycle.$$

The decrement in the waveform will reduce this, a decrement having an initial time constant of 100ms, decreasing the period to 255 degrees or 0.71 cycle.

Thus the inrush current will flow for a maximum of 0.71 of a cycle: during 0.29 of a cycle, the current flow will be the normal value of exciting current.

The foregoing has assumed that the power transformer core has maximum remanence. How can this occur?

It must be remembered that if the power transformer is de-energized when the magnetic flux and hence the exciting current are at a maximum value, the remanence will be a maximum. The load is usually disconnected before the transformer primary winding is de-energized. If however, a transformer is feeding a load with a power factor near unity and the feeder supplying the transformer is faulted, the fault current will be interrupted near a zero crossing. As the exciting current is almost in quadrature with the load current, it will be interrupted at a peak value, giving maximum remnant flux in that limb of the transformer core.

Figure 5 shows a computer simulation of the inrush waveform with a superimposed decrement.

In practice the waveforms are more complex than this, because of the mutual interaction effect between phases. The flux in the core of a three phase transformer is a product of the effects of all three phases and in a wye delta transformer or a transformer with a delta tertiary winding,

420

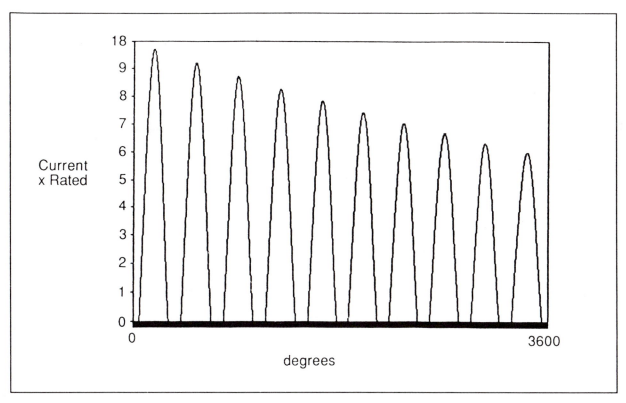

Figure 5 – Computer simulation of the inrush current with a decrement

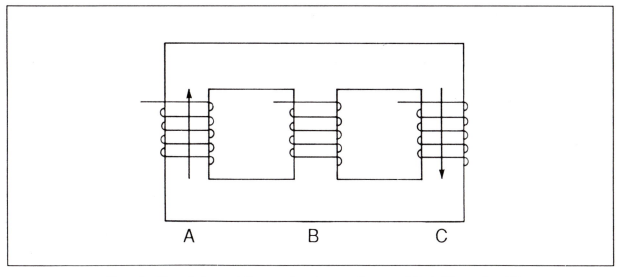

Figure 6 – Three limb wye-delta power transformer with remnant flux in Phases A and C

the current circulating in the delta contributes to the current in all phases. The waveform is not necessarily the unidirectional type previously discussed, but frequently has both positive and negative loops. A detailed discussion on this is outside the scope of this paper, so only an elementary treatment is given.

Figure 6 shows a three limb wye-delta transformer with remnant flux in opposite directions in phases A and C.

In order to understand the inrush waveform in phase B, we must use basic electromagnetic theory. The excitation current in phase B will depend upon the magnetic reluctance of the magnetic circuit of which it is a part. When the outer limbs of the core are not saturated, the exciting

current in phase B will be small. because magnetic flux will be easily produced in the highly permeable material. When either of the outer limbs approaches saturation because of the combination of the remnant flux and the direction of the exciting current in that phase, the effective reluctance of the magnetic circuit seen by phase B will increase as the permeability starts to fall.

When both outer limbs are in saturation, the reluctance of the circuit seen by the center limb will rise to a maximum and thus the exciting current in phase B will also rise. In figure 7, the three flux traces are shown 120' apart and the periods of inrush for phases A and C are shown. It is during these periods that core saturation is occurring and hence inrush current will flow in phase B as shown in the diagram.

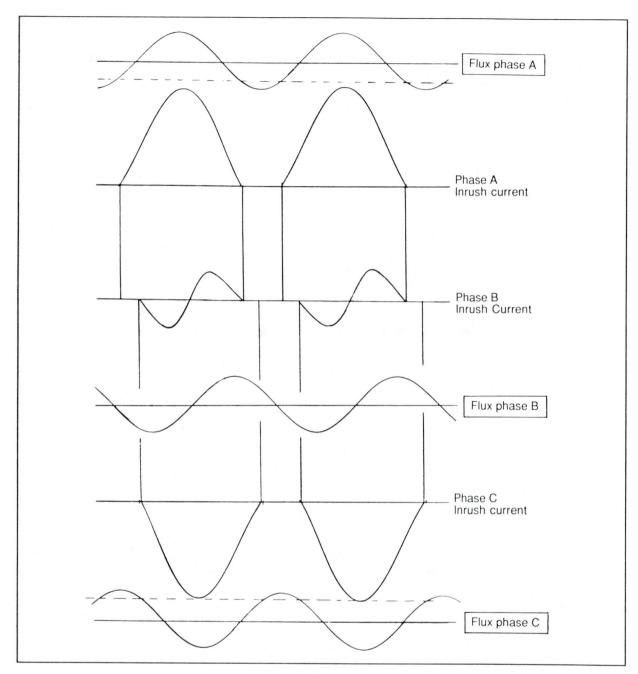

Figure 7 – Mutual interaction in a three limb transformer with remanence.

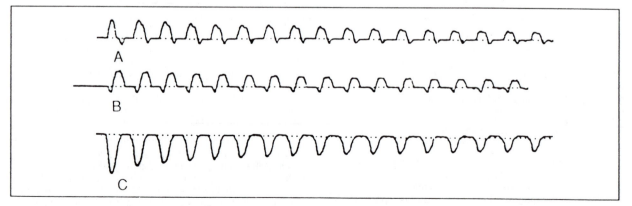

Figure 8 – Inrush waveforms of a typical power transformer.

422

The presence of the circulating current in the delta winding complicates the phenomenon but the basic principle remains. The relay current is further modified by the delta connection of the ct's on the wye side of the power transformer.

Figure 8 shows the inrush waveforms taken from a typical 345/23.7 kV wye-delta power transformer with a 645 MVA rating. The currents are flowing in the relay circuit on the 345 kV side and come from delta connected ct's. The time constant of this inrush is about 10 seconds. This shows how bi-directional waveforms can be produced.

HARMONIC ANALYSIS OF INRUSH WAVEFORMS

The inrush current in phase C shown in figure 8 is the unidirectional waveform that was shown in computer simulation in figure 5 and the width of the base varies with the degree of saturation as was shown previously. A Fourier analysis on these waveforms reveals the following composition:

Harmonic content %
Waveform base (degrees)

	260°	250°	240°	230°	220°	210°
Fundamental	48	48	47	47	46	45
Second harmonic	5	6.5	8	10	11	13
Third harmonic	3.5	4	4	4	4	3.4
Fourth harmonic	1.5	1	1	0.2	0.6	1.6
Fifth harmonic	0.1	0.4	1	1.3	1.5	1.7
DC	38	37	35	34	33	31
Ratio: $\frac{\text{second harmonic}}{\text{fundamental}}$	10%	13.5%	17%	21%	24%	29%

From the table it can be seen that sufficient second harmonic is probably present to restrain a harmonic restrained relay for most conditions although with more severe inrush the ratio of second harmonic to fundamental does fall below the 15% setting used on some relays. This indicates that relays that rely on harmonic restraint will mis-operate sometimes, on inrush.

The presence of the bi-directional waveform does substantially increase the proportion of second harmonic, this proportion in phase B in figure 7, being over 60%. A technique is sometimes used in which the total second harmonic content in all phases is used as an overall restraint. This has the severe limitation however, of preventing clearance of incipient faults for several hundred milliseconds.

HARMONIC ANALYSIS OF FAULT WAVEFORMS

During internal faults, the saturation of the line ct's will also produce comparable values of second harmonics.

The harmonic analysis of a fully offset wave shows the presence of almost no second harmonic. Thus if there is no line ct saturation, the second harmonic content is negligible. However, if a typical transformer installation is considered, the effect of the dc offset on the line ct for an internal fault is substantial.

The example here is the computer simulation of an installation with a power transformer of 100MVA, 345/13.8kV and with a 400/5A ct having a 240V knee point. The ct secondary resistance is 0.28 ohm and the lead and relay burden is 0.35 ohm. The primary X/R is 27.

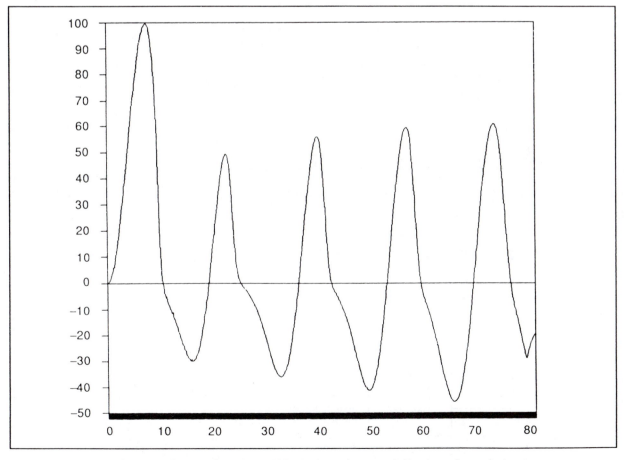

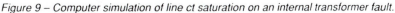

Figure 9 – Computer simulation of line ct saturation on an internal transformer fault.

423

Figure 9 shows a computer simulation of the secondary current waveform for a fault at maximum offset and value of 8.25 times rated current (3300 A) with ct remanence of 50%.

Harmonic analysis of this on a cycle by cycle basis shows:

	Cycle 1	Cycle 2	Cycle 3	Cycle 4
Fundamental	60.8	32.5	40	46
Second harmonic	12.7	12.8	12	11
Third harmonic	8.8	5.25	6	5.6
Fourth harmonic	5.6	2.9	2.3	2.4
Fifth harmonic	1.75	1.5	2	1.5
DC	22.1	1.2	1	1.23
Ratio: second harmonic / fundamental	21%	39%	30%	24%

A typical harmonically restrained relay has a harmonic setting of about 15%, that is, the relay is restrained as long as the proportion of second harmonic to fundamental is higher than 0.15. After 4 cycles, the proportion of second harmonic, relative to the fundamental present, still exceeds the proportion which can be present in inrush; the saturation of the line ct has produced a signal which will restrain the relay.

The conditions chosen in this example are not extreme. The ct application is taken from a practical case and although a condition of maximum offset has been assumed, a remnant flux level of only 50% has been considered. A fault could occur under better or worse circumstances than have been described.

The use of an instantaneous element would not necessarily be of value either, because its setting must be above the level of peak inrush and the magnitude of the fault current seen by the relay, especially when the ct saturates, can be less than this. The relay will have an operating time of well over 300ms under some conditions.

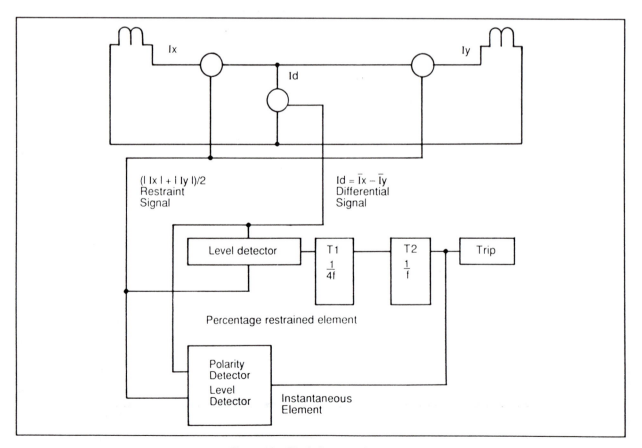

Figure 10 – Block diagram of the relay.

424

DESIGN OF A NEW RELAY

Modern electronics provides a practical solution to these problems.

The inrush wave is always distinguished from a fault wave by the period in each cycle during which normal exciting current flows, when the core is not in saturation (figures 5 and 7).

It is this characteristic that can be used to prevent a percentage restrained relay from operating during inrush without affecting its performance during internal faults.

Figure 10 is a block diagram of a relay that will be insensitive to inrush and yet have an operating time of between 12 and 25 ms for most internal faults. Figures 11 and 12 illustrate the principle of operation.

The relay has three current inputs. Two are in series with the instrument ct's and develop the restraint signal which is the mean scalar value of the currents from each instrument ct. The third is in the differential circuit and detects the vector sum of the instrument ct secondary currents. The differential signal is fed to a level detector which also has an input from the restraint signal. Here the relay setting is

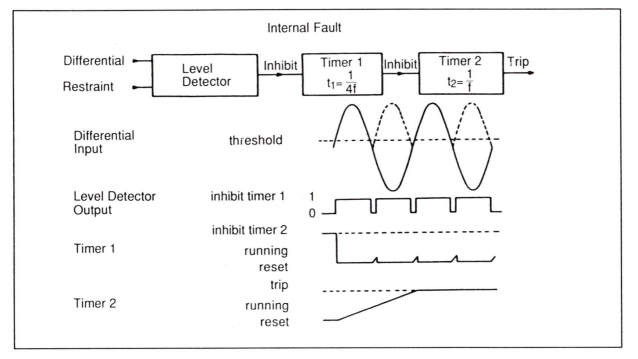

Figure 11 – Operation for an internal fault

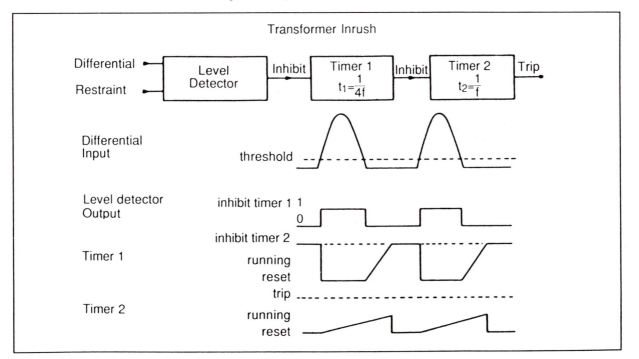

Figure 12 – Restraint under inrush conditions.

425

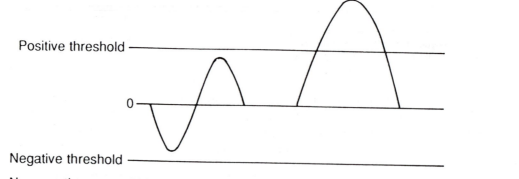

Positive threshold
0
Negative threshold

No operation on – – bidirectional or unidirectional inrush

Figure 13a

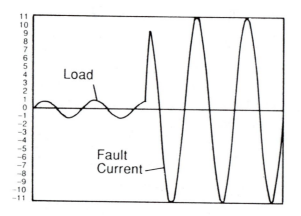

11
10
9
8
7
6
5
4
3
2
1
0
−1
−2
−3
−4
−5
−6
−7
−8
−9
−10
−11
Load
Fault
Current

Internal fault waveform (computer simulated)

Figure 13b

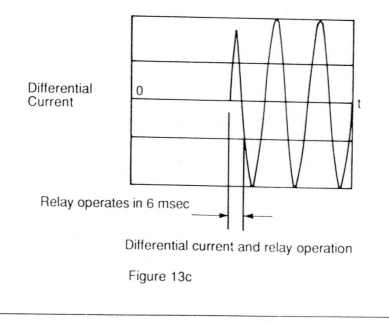

Differential
Current
0
t
Relay operates in 6 msec

Differential current and relay operation

Figure 13c

Figure 13 – Operation of the instantaneous element.

determined and the percentage restraint characteristic is generated. The output of this level detector is connected to two timers, T1 and T2. It is these timers that identify the difference between fault current and inrush.

Timer 1 has a delay of a quarter of a cycle and timer 2 has a delay of one cycle. When there is no differential current, the output of the differential level detector is low and this allows timer 1 to time out and its output prevents timer 2 from operating. Timer 2 is thus held reset.

When the differential current is above the threshold level of the differential level detector, its output resets timer 1 instantaneously and this allows timer 2 to start to time out. If the comparator has operated on fault current, timer 1 will not time out because there will be no gaps greater than a quarter of a cycle in the current waveform.

Timer 2 will therefore time out and the relay output contacts will close. This is shown in figure 11. However, during inrush, the characteristic gap of more than a quarter of a cycle will be present in each cycle of differential current and although timer 2 will start to time out, every time a gap appears in the current waveform, timer 1 will time out preventing timer 2 from operating. This is shown in figure 12.

However, the performance of the relay can be improved beyond this. The problem with a traditional instantaneous element is that it must be set higher than the value of the unidirectional inrush current.

If a solid state instantaneous element is used however, as shown in figure 10, it can be given two operating criteria. One criterion is that the differential current must exceed a threshold value and the second is that the current must contain both positive and negative loops which are above the threshold as shown in figure 13. The setting of this element can then be made just higher than the maximum value of the bi-directional inrush current, typically 3.5 times rated current. This element will not operate on uni-directional inrush, which can be as much as 10 times rated current and will not operate on bi-directional inrush because it is not big enough. Under through fault conditions the restraint signal increases its setting to cope with ct saturation and for internal faults, operation can take place in less than three quarters of a cycle, including the operating time of the output element. A composite time-current curve is shown in figure 14.

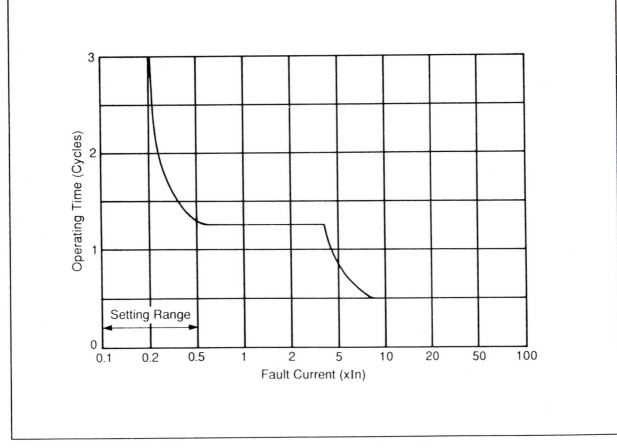

Figure 14 – Composite time – current characteristic.

427

CONCLUSION

Harmonically restrained transformer protection has been used effectively for many years. However, the use of harmonic restraint will not always guarantee immunity to mal-operation on inrush. Even techniques, such as using the third harmonic as an operate quantity to overcome excessively long operating times, when ct saturation occurs on internal faults, are only a compromise.

Modern electronics, however, allows techniques to be used to overcome all of these disadvantages, providing a relay which is stable on the severest inrush and yet will give very fast operation for internal faults. Since its introduction about 6,000 of these units have been successfully put into service world-wide.

DEVELOPMENT AND APPLICATION OF FAULT-WITHSTAND STANDARDS FOR POWER TRANSFORMERS

C. H. Griffin
Senior Member, IEEE
Georgia Power Company
Atlanta, Georgia

ABSTRACT

After five years of effort, a working group of the IEEE Transformers Committee completed work on one of the most important new electrical standards to be developed in recent years, the Transformer Through Fault Current Duration Guide. This document defines the exposure limits of <u>liquid-immersed</u> transformers to short circuit currents, and establishes fault withstand guidelines for all conditions of service. This paper provides a description of these new standards and a brief history of how and why they were developed.

Simultaneously with the development of the current duration standards, the IEEE Power System Relaying Committee was engaged in a complete revision of the IEEE Guide for Protective Relay Applications to Power Transformers, ANSI C37.91. A valuable feature of this new revision is Appendix A, which gives detailed instructions on how to apply relays and power fuses so as to properly protect transformers by coordinating overcurrent devices with the transformer fault-withstand curves. A summary of these application guidelines is included here, in order to assist protection engineers in employing the new standards.

INTRODUCTION AND HISTORICAL DATA

During the latter half of 1977, the members of the Transformer and Bus Protection Subcommittee of the IEEE Power System Relaying Committee became aware of the fact that the IEEE Transformers Committee had adopted a revision of ANSI C57.12.00, Section 10, "Short Circuit Characteristics" that would severely limit power transformer short circuit requirements for through faults. Under these new standards, transformers were divided into the following categories:

	Minimum Nameplate KVA	
Category	Single Phase	Three Phase
I	5 to 500	15 to 500
II	501 to 1,667	501 to 5,000
III	1,668 to 10,000	5,001 to 30,000
IV	Above 10,000	Above 30,000

For Category I (primarily distribution transformers), the duration of the short-circuit current was determined by the formula:

$$t = \frac{1250(f)}{I^2(60)}, \text{ or } I^2t = 1,250 \text{ (for 60 Hz)}$$

where t = duration in cycles
f = frequency in hertz
I = symmetrical short-circuit current in times-normal base current as determined under various methods, depending on the category

For all other categories, the new standard stated that "the duration of the short-circuit current is limited to two seconds, unless otherwise specified by the user."

It is interesting to compare these requirements with those of the old C57.12.00. Under Section 10.1.1 "Permissible Short Circuit," we find the

85 WM 128-4   A paper recommended and approved by the IEEE Power System Relaying Committee of the IEEE Power Engineering Society for presentation at the IEEE/PES 1985 Winter Meeting, New York, New York, February 3 - 8, 1985. Manuscript submitted July 16, 1984; made available for printing November 19, 1984.

following: "The duration of the short circuit is limited to the time periods shown below. Intermediate values may be determined by interpolation."

rms Symmetrical Current in Any Winding	Time Period in Seconds
25 times base current	2
20 times base current	3
16.6 times base current	4
14.3 or less times base current	5

Directly conflicting with this standard was ANSI C37.91-1972, which is the IEEE "Guide for the Protective Relay Applications to Power Transformers," which states on page 12, in referring to transformer protection by phase overcurrent relays, that "the time setting is usually chosen for coordination with relays on adjacent equipment. However, an upper limit to the overcurrent relay time setting exists in the time-overcurrent characteristic of the transformer windings. For proper backup protection, the relays may (must?) operate for an external fault before the windings are damaged by the heavy currents. Many such transformer "heat curves" are available, but as all seem based on USA Standard C57, one may use the curve in Appendix Figure 1."

This curve is reproduced as Figure 1 of this paper. Note that it allows fault clearing times of literally hundreds of seconds for low level faults, as opposed to the 5 second limit under the old standard, or the 2 second limit of the new revision. The reason for this conflict is because the curve of Figure 1 is a thermal damage curve, whereas the C57.12 limits are essentially concerned with the mechanical strength of transformers, and their ability to withstand the disturbing forces produced by through-fault currents.

Since the Figure 1 curve appeared in an IEEE publication published by and for protective relay engineers, it appears that most transformer over-current protection in the United States has been based on this curve. Although this curve provides an accurate description of the ability of a liquid immersed power transformer to withstand low level through faults, it is not a truly valid curve, since it does not correctly portray the mechanical stresses imposed on a transformer by high fault currents.

Due to a breakdown in the internal communications of the IEEE, the Power System Relaying Committee did not receive its ballot for the "proposed revision of ANSI C57.12.00—1973" until October 1977, and the PSRC negative vote was too late to stop publication. However, members of the Transformers Committee immediately began a cooperative effort with the PSRC engineers to resolve what now appears to have been a breakdown in communications. This is well set forth in a letter dated November 4, 1977 from Mr. W. J. McNutt of the Transformers Committee, who was Chairman of the Subcommittee which developed the revised standard, to Mr. R. W. Haas, then Chairman of the Transformer and Bus Protection Subcommittee:

"I think that transformer people did not recognize that the material on current duration in the old standard was used by relaying engineers for protection coordination. There was considerable misunderstanding about the times normal-current duration table and at times it was misused. The philosophy adopted for power transformers in the revised standard is "self-protecting in a prescribed environment," rather than a times normal limit. The two-second fault duration limit was adopted in consideration of mechanical wear effects, rather than thermal heating, and was influenced by results of an EEI survey which indicated no fault durations exceeding one second. I expect that this survey really dealt with duration of worst magnitude faults. Certainly we in the IEEE Transformers Committee were not thinking of reduced level faults when we set the two-second limit."

FORMATION OF THE WORKING GROUP

In order to resolve the situation, and develop a standard that would be accurate and acceptable to both transformer and protection engineers,

Reprinted from *IEEE Trans. Power App. Syst.*, vol. PAS-104, no. 8, pp. 2177–2188, Aug. 1985.

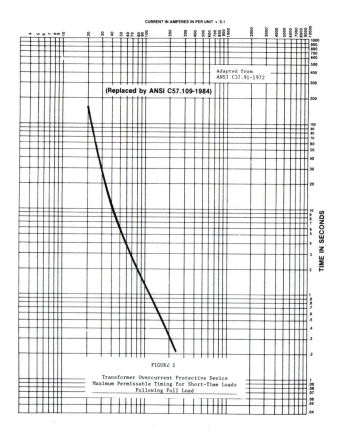

CURRENT IN AMPERES IN PER UNIT × 0.1

Adapted from
ANSI C37.91-1972

(Replaced by ANSI C57.109-1984)

TIME IN SECONDS

FIGURE 1

Transformer Overcurrent Protective Device
Maximum Permissable Timing for Short-Time Loads
Following Full Load

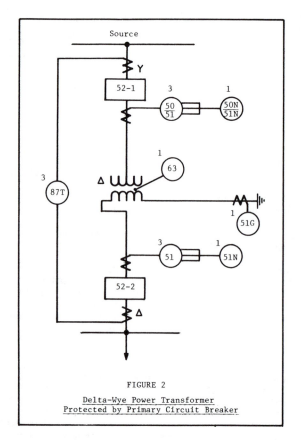

FIGURE 2

Delta-Wye Power Transformer
Protected by Primary Circuit Breaker

a working group was formed by Mr. Leonard Long, Chairman of the Performance Characteristics Subcommittee of the IEEE Transformers Committee. This working group was titled the "Working Group on Short Circuit Duration," and was under the chairmanship of Mr. W. F. Griffard of Commonwealth Associates. Members included knowledgeable transformer engineers, representing utilities, consultants, and manufacturers; Mr. Jack Allachi, of Westinghouse Electric Corporation, representing the IEEE Industry Application Society; and the writer, representing the IEEE Power System Relaying Committee.

The first meeting of the Working Group was held on October 30, 1978 at Chattanooga, Tennessee. At this meeting, Mr. McNutt proposed that the two-second rule for Category II, III, and IV transformers be modified as follows:

For through faults ranging from maximum (1.0 p.u.) to 0.5 p.u., fault withstand time shall be calculated in the following manner:

1. Determine maximum through fault. Express in times-normal full load.

2. Calculate constant K using formula $I^2t = K$, where t is 2 seconds.

   Example: For transformer whose maximum through fault is 20 times full load, I = 20, and K = (20)(20)(2) = 800.

   Then, for 0.5 p.u. fault, t = 800/(10)(10) = 8 seconds

Note that full load is always calculated on the minimum nameplate rating of the transformer. Also note that although the standards provide that the source impedance may be included in calculating the faults on Category III and IV transformers, an infinite bus is assumed for the calculations in this paper.

For faults below 0.5 per unit, Mr. McNutt proposed that the old thermal limit curve (Figure 1) be used. When the two portions are combined, a discontinuous curve is created. However, the result is a considerable improvement over all previous C57.12 requirements as far as allowing for proper coordination of transformer overcurrent protection with downstream protective devices.

The principal concern of protection engineers with the "2 second" and "5 second" rules involved the difficulty of providing proper transformer protection at low fault current levels, particularly for single-phase-to-ground faults on the low-voltage "wye" side of delta-wye transformers. This is particularly important for Category II transformers, which are usual-

ly protected by fuses on the high-voltage "delta" side. Since the fuses will only receive 57.7% of the low-side ground fault current, it becomes apparent that for stuck-breaker faults, very long clearing times will be required.

INITIAL INVESTIGATION OF PROPOSED STANDARD MODIFICATIONS

In order to determine whether the proposed "McNutt Curve" could be applied under real-world conditions, a number of studies were undertaken by the System Protection Section of Georgia Power Company. These studies were based on actual transformers installed on the Georgia Power Company system, using the actual relay settings or high-side fuse curve, as appropriate. Figure 2 generally shows the protection provided for large units (above 10 MVA), while small units are usually fused. The principal consideration was the clearing time for through faults, including those outside the transformer zone of protection. No attempt was made to modify the existing protection to make it "fit" the McNutt Curve.

Curves Nos. 1 and 2

These curves show typical backup protection for three-phase and phase-to-ground faults for power transformers protected by relays. They indicate that it should be possible to adequately protect any transformer constructed under Mr. McNutt's proposal if proper protective relaying is applied, and if the relays are properly set in accordance with generally accepted industry standards. The curves shown on these figures provide room for coordination with overcurrent protection that may be installed on the substation feeders.

Curve No. 3

This curve shows overcurrent relay protection for three-phase and phase-to-phase faults on a 115/46-kV forced-cooled transformer. On this sheet is plotted for comparison the McNutt Curve, the old C37.91 Curve (Figure 1), the old C57.12 Curve, and the new C57.12 "2-second" Curve, along with the highside bank backup overcurrent relay curve. The following points are of interest:

*Transformers constructed according to the proposed McNutt curve can be adequately protected.

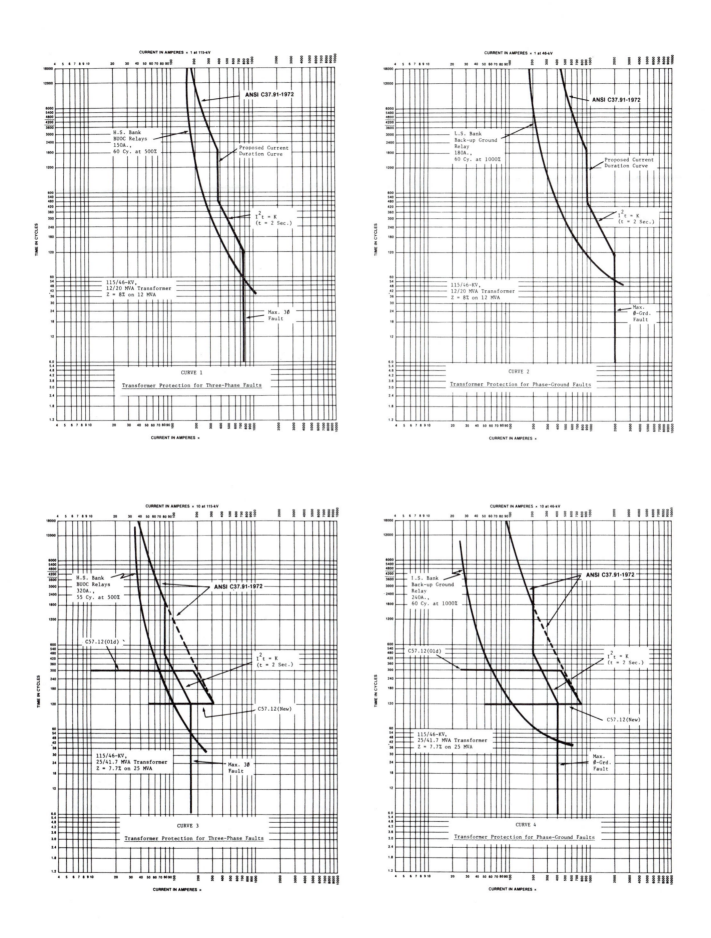

431

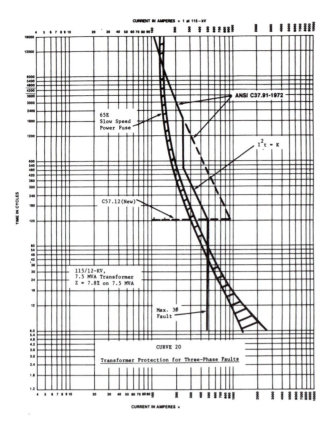

CURVE 20

Transformer Protection for Three-Phase Faults

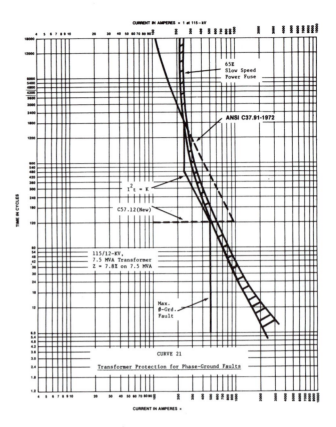

CURVE 21

Transformer Protection for Phase-Ground Faults

*Transformers constructed according to the old C57.12 curve can be protected for faults above 700 amperes (115-kV base). This is about 40% of the maximum through fault.

*Transformers constructed according to the "2-second" rule can be protected for faults above about 1000 amperes.

*The relay curve is well below the old short-time overload curve for all values of fault current.

Curve No. 4

This curve shows the overcurrent protection for ground faults on the wye side of the 115/46-kV transformer of Curve 3. Again it should be noted that transformers constructed in accordance with the McNutt Curve can be adequately protected, since the ground relay curve is faster than the proposed withstand curve for all values of fault current. Banks constructed according to the 2-second rule cannot be protected for faults below about 1100 amperes.

Curve No. 20

This curve indicates that fused Category III self-cooled transformers, with 8% impedance, can be properly protected for three-phase faults if constructed according to the McNutt curve. They cannot be protected under the 2-second criteria.

Curve No. 21

This curve indicates that a fused delta-wye 8% self-cooled transformer cannot be adequately protected for phase-to-ground faults under the McNutt rule. This result is obtained because the fuse curve must be shifted to the right by the factor of 1/0.577 to allow for the fact a fuse on the delta side will receive only 57.7% of the ground-fault current flowing on the wye side. Note, however, that fuse protection is provided for heavy faults for transformers with through-fault capability equivalent to that shown for the old overload curve.

Curve No. 22

This is an interesting set of curves. When forced cooling is added to the 7.5 MVA transformer of Curves 20 and 21, the increased KVA capacity requires an 80 ampere fuse instead of the 65 ampere fuse previously ap-

plied. Protection is poor for all criteria, except for comparison to the overload curve. It appears from an examination of Curve 22 that no 8% forced-cooled transformer should be fused if constructed in accordance with the McNutt curve. This is certainly true for the "2-second" curve (not plotted). Note that the old C57.12 curve is superior to the McNutt curve for faults above about 320 amperes (115-kV base). Note also that these curves are plotted for three-phase faults. Protection for ground faults would be substantially worse.

Curve No. 23

These curves are for a larger forced-cooled transformer, and were plotted at the specific request of a member of the WG. The result is essentially the same as for the previous curves, and shows that protection would be inadequate for nearly all faults. The fuse protection of high-impedance forced-cooled units is inherently poor, since the damage curve for a particular transformer is the same, whether it has forced cooling or not; whereas the higher MVA forced-cooled rating usually requires that the fuse size be increased, thus increasing the clearing time for all types of faults.

Curve No. 24

These curves are for a 46/12-kV transformer, and show that when the impedance is reduced to the 6% range, even FAC transformers can be properly fused for three-phase and phase-to-phase faults, if constructed in accordance to the McNutt proposal. They can only be poorly protected under the 2-second rule. This plot is an excellent example of an apparent conflict in IEEE/ANSI standards. Note the typical cross-over of the ANSI Standard "E" fuse curve, and the transformer thermal damage curve. This is because of the extreme inverseness of the former compared to the latter. The reason and/or necessity for this conflict should probably be investigated and resolved.

Curve No. 25

The IEEE Industry Application Society has been concerned for several years over the fact that Category II transformers, which are nearly always fused, have been lumped with Category III and IV transformers, which are generally provided with protective relays and high-side switch. The reason is apparent upon an examination of Curve No. 25. This diagram shows that fuses on small 6% transformers will not coordinate with the McNutt curve for ground faults on the low voltage (wye) side. That is, the same

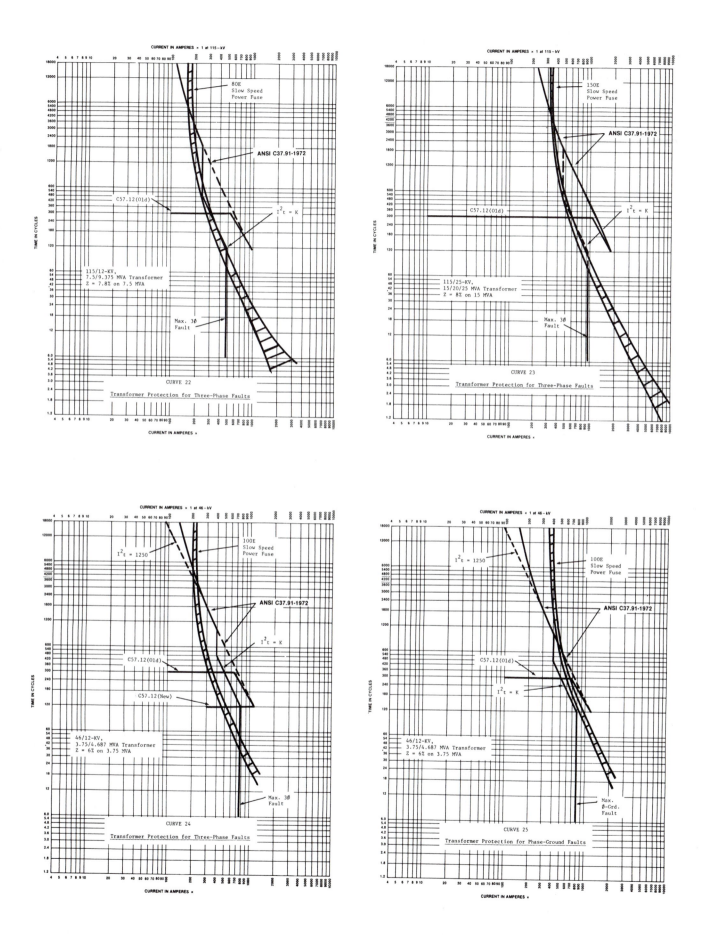

433

transformer that can be satisfactorily fused for three-phase and phase-to-phase faults is exposed to damage for ground faults, again because the 1/0.577 factor must be applied.

STUDIES AND DISCUSSIONS BY THE WORKING GROUP

The information derived from the curve plots was presented to the Working Group on Short Circuit Duration at a meeting in San Diego on March 26, 1979. From the standpoint of protection engineers, these curves raised several questions:

1. Was it realistic (or technically correct) to have one through-fault withstand standard for three-phase, phase-to-phase, and phase-to-ground faults on delta-wye power transformers?

2. What are the relative mechanical stresses imposed by a single-phase-to-ground fault on the wye side of a three-phase power transformer, versus the stresses imposed on one single-phase transformer in a bank of three?

3. How should a protection engineer interpret the old "Short-Time Overload Curve"? This curve is very generous as far as permissible fault currents are concerned, especially compared to the old and new C57 standards. It appeared that as far as fault currents standards are concerned, there should be a single curve, as proposed by Mr. McNutt, that tells the protection engineer exactly what protection he must provide for all excess currents, from moderate overloads to maximum bolted faults.

With the above questions in mind, the following findings appeared to be valid:

4. The modification to the existing short circuit duration standard proposed by Mr. W. J. McNutt was a substantial and significant improvement over the "2-second maximum" introduced in C57.12.00—1980.

5. The McNutt proposal allowed all power transformers to be properly protected by protective relays.

6. The McNutt proposal allowed fused transformers with impedances of approximately 6% and below to be adequately protected against three-phase through faults.

7. The McNutt proposal allowed fused self-cooled transformers above 6% impedance to be adequately protected against three-phase and phase-to-phase faults.

8. Forced-cooled transformers with impedances higher than 6% may not be adequately protected with fuses. This is because the higher impedances force the damage curve to the left, and the higher MVA ratings force the fuse curve to the right, beyond the damage curve.

9. Fuses on the delta side of the Category II, III, and IV transformers that were studied may not provide adequate protection against line-to-ground faults that may occur on the wye side. This not only applies to the McNutt proposal, but to transformers built under previous standards.

After the San Diego meeting, Mr. Griffard, the WG Chairman, worked very hard to reconcile the sometimes conflicting demands of the various members. Several letter ballots were taken, and additional meetings were held in Houston and Williamsburg. A December 1979 letter ballot generally proposed that the McNutt curve be adopted for Category II, III, and IV transformers, "provided that the rate of occurrence of faults is no greater on the average than one fault per day." This proposal received several negative votes. One reason was the fact that some members felt strongly that it was a mistake to lump Category II banks with the large transformers. Another objection was the "average of one fault per day" limitation. Average of what? Average number in a year? Could you go 364 days with no faults, then have 365 faults in one day and still meet the requirement?

DEVELOPMENT OF PROPOSALS IN LETTER BALLOT OF FEBRUARY 1980

In October 1979, this writer made the following proposals for consideration by the working group:

1. Adopt the "McNutt Curve" for Category III and IV transformers. These banks are generally protected by relays that trip a high-side

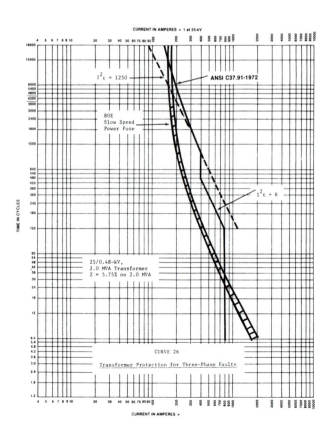

CURVE 26

Transformer Protection for Three-Phase Faults

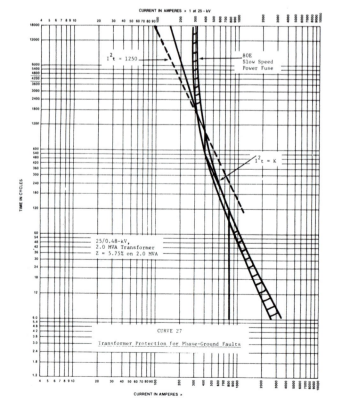

CURVE 27

Transformer Protection for Phase-Ground Faults

switch. Curves 1 and 2 show that these categories can be adequately protected for through faults if equipped with a standard protection scheme, such as that shown in Figure 2.

2. Place Category II transformers in the same fault-withstand class as Category I, as requested by Jack Alacchi, R. W. Haas, and others, since both categories are usually protected by high-side power fuses. Curves 26 and 27 show a Category II transformer built according to two different standards. The solid curve is the McNutt curve, while the dashed curve is the Category I curve. Here, Curve No. 27, for a phase-to-ground fault, contains the significant data, which is this: The protection for a "McNutt" curve transformer is questionable (or completely lacking) for all faults, whereas we do find adequate protection for a "Category I" curve bank for high current faults. Protection for low current faults is still poor, however.

3. In order to improve protection for low current faults, consider a new "composite" withstand curve for both Category I and II transformers. This curve can be constructed using the present Category I curve from 1.0 to 0.5 p.u. of maximum fault, and the transformer thermal damage (overload) curve below 0.5 p.u. Curves 28 and 29 indicate that the "composite" curve allows for considerably better fuse protection for all faults than any curve proposed so far. The apparent deficiency at the lower current values is probably due to the shape of the "E" fuse curve, and does not represent any weakness in transformer design.

In February 1980 Mr. Griffard issued another letter ballot that substantially followed these recommendations. Significant changes included the following:

* Category I and II transformers were grouped together with fault withstand defined by the "composite" curve.

* The McNutt Curve was proposed for Category III and IV transformers.

* The transformer fault-withstand curves were presented as "low risk" curves, not as definitive "go-no go" damage curves.

This ballot failed of adoption, principally because some working group members felt that Category I (distribution) transformers should not be grouped with Category II, and others considered that neither the McNutt curve nor the "composite" curve could be properly applied to medium power transformers under certain conditions of service.

DEVELOPMENT OF THE FINAL THROUGH-FAULT STANDARDS

It soon became apparent to the working group that in order to develop acceptable curves for some categories that it would be necessary to define the "fault frequency" to which a transformer might be subjected; and this would permit a more lenient (slower) curve to be applied for faults that would occur only rarely during the service life of the transformer. Faults between the transformer low-voltage terminals and its associated low-side bank breaker would be typical of this classification.

After considerable discussion, and a good deal of hard work, Mr. Griffard presented a compromise proposal to the working group. This proposal, which was designated P784/D7, was approved by the working group on October 25, 1982. After approval by the Transformers Committee, it was transmitted to the IEEE Standards Board, where it was adopted as an IEEE Standard on June 23, 1983. The title of this new standard is the Transformer Through Fault Current Duration Guide, ANSI C57.109. This Guide defines the damage curves for each classification as follows:

Category I Transformers

The "composite" curve was adopted for this category. The guide states that this curve "reflects both thermal and mechanical damage considerations and should be applied as a protection curve for faults which will occur frequently or infrequently."

Category II Transformers

Two curves were adopted for Category II transformers (Figure 3), depending on "fault frequency." This is defined as "the number of faults with magnitudes greater than 70 percent of maximum."

The left-hand curve of Figure 3 is a modified McNutt curve. According to the Guide, this curve "reflects both thermal and mechanical damage considerations, (and) should be applied as a protection curve for faults which will occur frequently (typically more than 10 in a transformer lifetime). It is dependent upon impedance of the transformer for faults above 70% of maximum possible and is keyed to the I^2t of the worst-case mechanical duty (maximum fault current for 2 seconds)."

The right-hand curve is the old thermal damage curve of Figure 1. The Guide states that this curve "may be applied as a protection curve for faults which will occur only infrequently (typically not more than 10 in transformer lifetime). This curve may also be used for backup protection where the transformer is exposed to frequent faults normally cleared by high speed relaying."

Category III Transformers

Two curves were also adopted for Category III transformers (Figure 4). In this category, fault frequency "refers to the number of faults with magnitude greater than 50% of maximum."

The left-hand curve of Figure 4 is the McNutt curve, and the Guide states that this curve "reflects both thermal and mechanical damage considerations, (and) should be applied as a protection curve for faults which will occur frequently (typically more than 5 in transformer lifetime). It is dependent upon impedance of the transformer for fault current above 50% of maximum possible and is keyed to the I^2t of the worst-case mechanical duty (maximum fault current for 2 seconds)."

As in Category II, the right-hand curve is the old thermal damage curve, and "may be applied as a protection curve for faults which will occur only infrequently (typically not more than 5 in transformer lifetime)." This curve may also be used when applying backup protection.

Category IV Transformers

The McNutt curve was adopted for this category (Figure 5). Since these large transformers are generally protected by high-speed relays and circuit breakers, and are frequently subjected to very high fault currents, it is appropriate that more stringent fault-duty standards be applied.

APPLICATION OF TRANSFORMER OVERCURRENT PROTECTION

The publication C57.109 provides information that permits coordination of transformer overcurrent protection with the new through-fault withstand curves. This is not difficult to do, but does require some care. Also, it will soon become apparent that instances will occur where a transformer overcurrent device that properly coordinates with the transformer fault-withstand curve, will not coordinate with upstream and/or downstream protection, and vice-versa. In such cases a compromise will be required that will call for careful and informed judgement by the protection engineer.

Fortunately for the industry, the IEEE Power System Relaying Committee has completed an extensive revision of the Guide for Protective Relay Applications to Power Transformers, ANSI C37.91. This Guide, which has also been approved by the Transformers Committee and the Industrial Applications Society, was approved by the IEEE Standards Board in March 1984. It should be published in early 1985 and is required reading for all protection engineers in North America.

"Appendix A," which is titled "Application of the Transformer Through-Fault Current Duration Guide to the Protection of Power Transformers" is particularly valuable. This appendix was developed in close cooperation with the Transformers Committee and includes detailed instructions on how to apply the new fault duration curves. Protection engineers will find Figure (A)5 of this appendix to be very useful. This diagram, reproduced here as Figure 6, was developed by the PSRC working group to assist engineers in the application of the "frequent" and "infrequent" curves of Categories II and III.

TYPICAL COORDINATION EXAMPLES

Detailed instructions on how to apply overcurrent relay protection and power fuses to each transformer category are given in C37.91. The following examples will provide additional information and guidance.

Relay Protection—Phase Faults

Figure 7 shows the coordination curves for phase faults for a 230/25-kV, 30/50 MVA transformer with an impedance of 10% on 30 MVA base. This is a Category III transformer.

To construct the "frequent" fault-withstand curve, proceed as follows:

Normal base current at 230-kV is 75 amperes.
Maximum through-fault with infinite source is 10 x 75 = 750 amperes. This establishes the 2-second point (1).

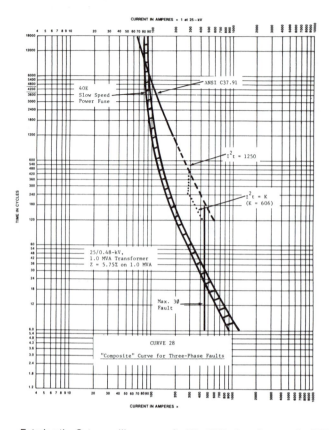

CURVE 28

"Composite" Curve for Three-Phase Faults

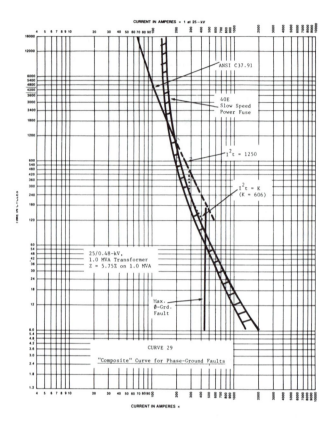

CURVE 29

"Composite" Curve for Phase-Ground Faults

Entering the Category III curve, we find the 50% of maximum point (375 amperes) to be at 8 seconds. This establishes point (2). We then connect (1) and (2).

Next, trace the top half of the Category III curve from five-times normal (375 amperes), point (3), to the 1000 second crossing, point (4). Connect (2), (3), and (4) and the "frequent" fault-withstand curve (A) is completed.

Primary protection for this transformer is provided by high-speed differential relays and a fault pressure relay. Backup protection for phase faults is provided by three high-side overcurrent relays. These relays should be set with the following requirements in mind:

* Setting should be high enough to permit the transformer to carry an emergency overload when required. A setting of 150% of the 55 degree forced-cooled rating of 50 MVA should be sufficient.

* Setting should coordinate with the transformer fault-withstand curve.

* Setting should clear a maximum three-phase fault in approximately 60 cycles (1 second).

* Setting should coordinate with the maximum 25-kV feeder relay setting, for both three-phase and phase-to-phase faults.

Curve (B), for the 230-kV bank overcurrent relays, is plotted with a setting of 200 amperes (80 MVA), 42 cycles at 500% of pickup. This curve coordinates with the transformer "frequent" fault-withstand curve and crosses the maximum three-phase fault line at one second.

Curve (C) is plotted for a phase-to-phase fault on the 25-kV feeder with the highest relay setting, in this case 720 amperes, 36 cycles at 500%. This setting is converted to a 230-kV base as follows:

Relay pickup = (25/230)(720) = 78.26 amperes.
Equivalent pickup for phase-to-phase fault = 1.15 x 78.26 = 90 amperes.
Calibration point at 500% of pickup = 5 x 90 = 450 amperes at 0.6 second.

The minimum coordination time interval (CTI) between Curve (B) and Curve (C) is 0.63 second, which is satisfactory. It is not necessary to plot the 25-kV feeder curve for a three-phase fault, since it would be to the left of Curve (C), and the CTI would therefore be larger.

Note: Since protection has been established for all faults using the "frequent" fault curve, the "infrequent" curve is not required. It is shown here as a dotted line for reference only.

Fuse Protection—Ground Faults

Figure 8 shows the coordination curves for phase-to-ground faults for a 115/12.5-kV 7.5/9.375 MVA transformer with an impedance of 8% on 7.5 MVA base. This is a Category III transformer.

To construct the "frequent" fault-withstand curve, proceed as follows:

Normal base current at 12.5-kV is 347 amperes.
Maximum through-fault with infinite source is 12.5 x 347 = 4338 amperes.
This establishes the 2-second point (1).

Entering the Category III curve, we find the 50% of maximum point (2169 amperes) to be at 8 seconds. This establishes point (2). We then connect (1) and (2).

Next, trace the top half of the Category III curve from 6.25 times normal (2169 amperes), point (3), to the 1000 second crossing, point (4). Connect (2), (3), and (4) and the "frequent" fault withstand curve (A) is completed.

The "infrequent" fault curve should always be plotted for fused transformers. This curve is indicated by the dotted line.

Protection for this transformer is provided by a set of 80E slow speed power fuses. Since this is a delta-wye transformer, and since this plot is for phase-to-ground faults on the 12.5-kV side, the current values for the 115-kV fuses will be 1/0.577 times the values for a three-phase fault. Thus, the 300-second point is not 160 amperes times the ratio of transformation (9.2), but 160 x 9.2 x 1.73 = 2547 amperes, and the fuse curves (B) are moved substantially to the right.

Curve (C) is plot of the high-set ground relay on the 12.5-kV feeders, which is 240 amperes, 72 cycles at 1000 % of pickup. This setting is designed to coordinate with a 140/280 ampere oil circuit recloser (not shown).

An examination of Curves (A), (B), and (C) reveals the following:

* The transformer is well protected for "frequent" phase-to-ground faults that may occur on the 12.5-kV feeders.

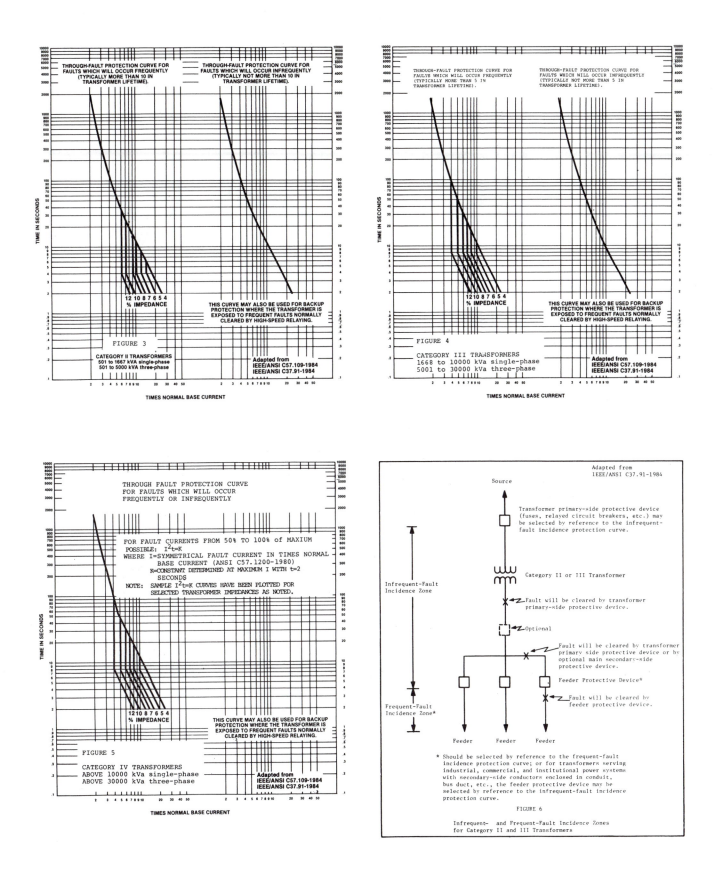

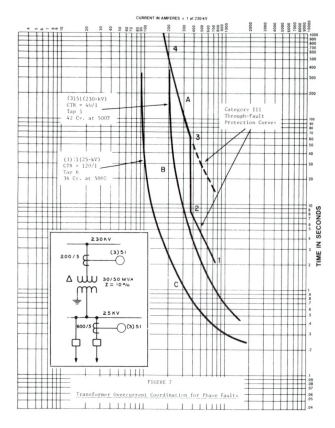

FIGURE 7

Transformer Overcurrent Coordination for Phase Faults

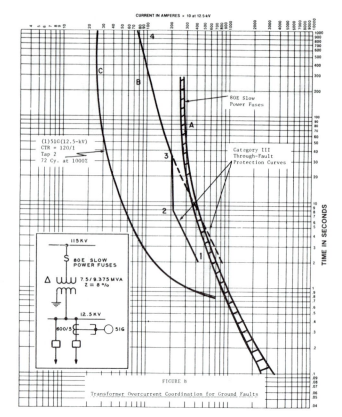

FIGURE 8

Transformer Overcurrent Coordination for Ground Faults

*The coordination between the feeder ground relays and the 115-kV fuses is excellent.

*Coordination of the fuse curves with the "infrequent" fault-withstand curve is marginal. This occurs because the fault-withstand curve must be plotted on the transformer self-cooled rating of 7.5 MVA, while the 80 ampere fuse is selected on the basis of the forced-cooled rating of 9.375 MVA. Better protection can be provided by selecting a 65 ampere fuse, at the risk of poor coordination for some phase-to-phase faults, and a reduction in the emergency overload capability of the transformer.

CONCLUSION

Development of the IEEE Transformer Through Fault Current Duration Guide, and its companion, the revised IEEE Guide for Protective Relay Applications to Power Transformers are significant achievements in the continuing efforts by electrical engineers to improve the reliability and protection of power transformers. It is very important that both manufacturers and users understand how and why these new standards were developed, and how to apply them to the various transformer categories, as defined by the IEEE Transformers Committee. Proper application of these guides can be expected to greatly improve the performance of the utility and industrial electrical systems in North America.

REFERENCES

1. Applied Protective Relaying, Chapter 8. Westinghouse Electric Corporation, Pittsburgh, PA. 1976.

2. General Requirements for Liquid-Immersed Distribution, Power, and Regulating Transformers. ANSI Standard C57.12—1980. Available from the Institute of Electrical and Electronic Engineers, Inc., New York, NY.

3. Transformer Through Fault Current Duration Guide. ANSI Standard C57.109—1984. Available from the Institute of Electrical and Electronic Engineers, Inc., New York, NY.

4. Guide for Protective Relay Applications to Power Transformers. ANSI Standard C37.91—1984. Available from the Institute of Electrical and Electronic Engineers, Inc., New York, NY.

Discussion

L. S. McCormick (Westinghouse Elec. Corp., Pittsburgh, PA): Users of the new Transformer Through Fault Current Duration Guide, C57.109 when applying the rules to transformers rated above 100 MVA should do so with caution. The reason for this statement is apparent from a close inspection of the curves for category IV transformers given in Fig. 5. At fault currents of 50 percent or less of maximum possible, the curves leave the I^2t range and immediately enter what should be termed the overload range. Operation in this range does not conform to the new IEEE Trial-Use Guide for Loading Mineral-Oil-Immersed Power Transformers Rated in Excess of 100 MVA, IEEE Std. 756 which was issued in May of 1984. Table 3 of this guide limits all overloads of system tie and substation transformers to 150 percent and of generator step-up transformers to 110 percent of maximum ratng. C57.109 attempts to recognize this inconsistency through the use of a caution note added to the table listing the transformer categories and corresponding protection curves.

How critical is this caution note? Perhaps an example will be the best explanation. Assume any transformer with an FOA rating above 100 MVA, 65 °C average winding rise and operating in an ambient of 30 °C. Conservatively, this unit would typically have a top liquid temperature rise of 45 °C and the leads would be operating at 25 °C above this which is approximately 10 °C below the allowable hottest spot rise of 80 °C. Assume also that the fault protection equipment allows a 3 P.U. fault to last for 5 minutes — as allowed by Fig. 5. A typical time constant for the leads is 8 minutes. Thus, in 5 minutes the 25 °C temperature difference has risen to approximately 115 °C which when added to the 45 °C oil rise and 30 °C ambient yields a copper temperature of 190 °C. This would surely result in degradation of the paper insulation and gas generation within the transformer.

Perhaps we are all fortunate that most, if not all, power transformers in this rating range are protected by breakers which react much faster than fuses. But the inconsistency in the documents does exist.

It should be noted that there is a typographical error in Fig. 1. The time seal should be multiplied by 10 to be correct.

Manuscript received March 25, 1985

W. J. McNutt (General Electric Co., Pittsfield, MA): Mr. Griffin has provided a valuable service in documenting the course of events which occurred during the resolution of a very difficult technical issue within the power industry. His paper also illustrates the progress which can be made through cooperation between IEEE technical committees. For me the one lesson which this experience has taught is that industry standards should not remain too long unattended, principally because the state-of-the-art knowledge advances with time, but also because the basis for some provisions in the standard can be forgotten and the standard may be applied in fashions not envisioned by those who prepared it.

The earliest transformer document of which I am aware that provides guidance for application of overcurrent protective deivces is the 1956 ASA Appendix C57.92, Guide for Loading Oil-Immersed Distribution and Power Transformers. Tucked away on the last page of this document, and perhaps earlier issues of it as well, is a section on Overcurrent Protective Devices. A curve was provided which was qualified by a statement "For use . . . when specific information applicable to individual transformers is not available." The curve starts at the short-time end with the traditional 25 tmes base current for 2 seconds and progresses through 11.3 times base current for 10 seconds and 4.75 times base current for 60 seconds to 2 times base current for 30 minutes. The two end points of this curve seem to be the same as those in Fig. 1 of Mr. Griffin's paper, but the intermediate values differ. Evidently the engineers who prepared the two guides had different shaped French curves. At any rate, each of the end points appears to have been based on thermal criteria. The two-second point was based on a limiting conductor temperature of 250 °C with all heat stored, and the 30-minute point was set based on limiting the thermal degradation to the conductor insulation during a severe overload.

Transformer engineers within IEEE then seem to have lost sight of the need for or the existence of such guidelines for application of overcurrent protective devices. Certainly most never thought of the table in the Short Circuit Characteristics section of ANSI C57.12.00 (times normal base current vs current duration) as being associated with that purpose. Thus when the mechanical performance of transformers became unsatisfactory in the 1960's, work was started to completely rewrite short circuit requirements. Among other changes which appeared in the final draft in 1977 was a uniform fault current duration limit of 2 seconds for power transformers, introduced solely to control the amount of mechanical wear and tear which could result from each *worst case* fault event. No consideration was given to acceptable durations for fault current magnitudes less that *worst case*.

Mr. Griffin accurately relates the course of events after the oversight was uncovered. I believe we have arrived at a reasonably sound consensus positon on transformer fault current duration capability in the new ANSI/IEEE C57.109-1984. This position was based on the best informed engineering judgment available in the industry today, which is more enlightened in many respects than the judgment of the engineers in the 1940's and 1950's by virtue of the advances in knowledge which have taken place over the intervening time interval. However, the new guide should not be viewed as a sacred document which proclaims the final word. It should be periodically reviewed and tested for soundness, and be updated when new facts demonstrate the need.

Mr. Griffin certainly deserves a great deal of credit for his persistence in pushing for a realistic position which satisfies the needs of both transformer engineers and protection engineers. He was instrumental in the development of the new guide.

Manuscript received February 14, 1985

W. A. Elmore (Westinghouse Elec. Corp., Coral Springs, FL): Mr. Griffin has once again demonstrated his ability to reduce a very complex problem to simple understandable terms, and for this he deserves our appreciation.

One point in the paper may be somewhat misleading, and I would like Mr. Griffin's comments on this. In going from curve 20 to curve 21 the transformer protection curve stays fixed, intimating that it has the same withstand capability whether it be for a three-phase fault or a phase-to-ground fault. While this is true in terms of wye current and high side winding current, it appears in curve 21 to be true of 115 kv line current. While Mr. Griffin's treatment of coordination is accurate, it should not be misunderstood that the transformer can tolerate 500 amperes for 2 seconds at 115 kv with a phase-to-ground fault on the 12 kv winding.

It would be more accurate to hold the fuse characteristic to be the same in curves 20 and 21 and to shift the transformer protection curve to the left on curve 21.

Manuscript received February 19, 1985

John C. Appleyard (Wisconsin Power & Light Co., Madison, WI): I appreciate Mr. Griffin's efforts both to provide historical data and insight into the interpretation and application of transformer fault-withstand standards. We have recently reviewed our application practices and find this paper most helpful.

One area I found difficult to follow concerned the application of the damage curve for three-phase and phase-to-ground faults. Specifically I refer to either curves 20 and 21 or curves 24 and 25. The author has chosen to shift the fuse curve and maximum single phase to ground fault to the right by 1/0.577 to compensate for the delta-wye transformer connection. Initially I found this confusing. I was expecting the maximum single phase high side fault to be 57.7 percent of the three-phase value.

It is my understanding that three approaches may be used to evaluate fuse/damage curve applications for single phase to ground faults on delta-grounded wye transformers.

1. Current reference-transformer high side

   Shift both the fuse curve and maximum single phase to ground fault to the right by 1/0.577.

   Damage curve is referenced to base rated high side load current.
2. Current reference—transformer high side.

   Shift the damage curve to the left by 0.577.

   Use calculated high side single phase to ground current. Fuse curve is not shifted.
3. Current reference—transformer low side

   Damage curve is referenced to base rated low side load current.

   Use calculated low side single phase to ground current.

   Shift the high side fuse curve to the right by the transformer winding ratio multiplied by 1.73.

439

I also have difficulty understanding the additional fuse curve shift of 1.73 indicated in approach three. With the damage curve referenced to the low side of the transformer winding and line current are identical. It would seem to me that the fuse curve should be shifted to the right by the winding ratio only. The author uses approach three in discussing ground fault protection in Fig. 8.

Also in this example (Fig. 8) the statement is made: "Since this is a delta-wye transformer, and since this plot is for phase to ground faults on the 12.5kV side, the current values for the 115kV fuses will be 1/0.577 times the values for a three phase fault." Shouldn't this value be 0.577 times the three-phase fault?

Manuscript received March 1, 1985

C. H. Griffin: The discussers have brought forth several important subjects which point up the need to correct and clarify certain sections of the paper. First, Mr. McCormick is correct that the time scale in Fig. 1 should be multiplied by 10. I can't comment on his discussion of the apparent conflict between C57.109 and the new IEEE trial-use loading guide for large power transformers. These are both Transformers Committee documents and this problem will have to be solved by that group. Appendix A of C37.91 is based solely on C57.109. I doubt if many members of the Power System Relaying Committee know of the existence of the other guide.

I wish to thank Mr. McNutt for his kind remarks, and for giving us additional background as to how the new transformer standards were developed. I certainly agree with his statement that standards "should be periodically reviewed and tested for soundness and be updated when new facts demonstrate the need."

Both Mr. Elmore and Mr. Appleyard are correct in stating that Curve 21 (also 25, 27, and 29) should not be used to calculate transformer fault withstand capability for ground faults. These curves were plotted when we were investigating the relative performance of the early "McNutt" curves for three-phase and phase-to-ground faults and are valid for that comparison only. The curves of Fig. 8 correctly portray transformer fault withstand capability and coordination for ground faults, and it will be noted that they are referenced to the low-side voltage. This follows the basis rule for plotting ground fault coordinaiton curves for delta/wye power transformers, and corresponds to Mr. Appleyard's "approach number three."

To Mr. Appleyard's question, "shouldn't this value be 0.577 (instead of 1/0.577) times the three-phase fault?," the answer is *No*. Remember, we are plotting coordination curves here, and we must constantly be aware of the amount of current in the transformer and in each series protective device for any particular fault. This means, for example, that a 500 ampere low-side ground fault will be viewed as a 500 ampere fault by the 51G relay of Fig. 8, but only a 500/1.73 ampere fault by the high-side fuses, if we consider a one-to-one transformer. So, in order to plot the curves on the same sheet, we must move the fuse curve to the right by multiplying all values by 1.73. Think of it this way: For a one-to-one transformer, we must compare the relay speed for a 500 ampere fault to the fuse speed for a fault of 500/1.73 amperes. In other words, it takes 1.73 times as much low-side ground fault current as it does three-phase fault current to give the same fuse melting time.

Manuscript received March 25, 1985

Part 5
Input Devices

I. General Discussion

MAGNETIC instrument transformers have been very reliable sources of inputs to protective relays. They are capable of monitoring primary power system conditions and reproducing secondary magnitudes that faithfully reflect the primary conditions. As primary power system voltages and protective relay speeds have increased, the transient behavior of the instrument transformers has become of greater and greater concern. At higher primary power system voltages, the prohibitive cost of magnetic voltage transformers dictates that the relay voltage source will become a cascade-type magnetic voltage transformer or a coupling capacitor voltage transformer.

The coupling capacitor voltage transformer is a power-line carrier coupling capacitor with a magnetic voltage transformer connected to an appropriate tap brought out from the capacitor divider that makes up the carrier coupling capacitor. The burden (magnetic voltage transformer, connected load, etc.) on the tap of the carrier coupling capacitor must operate at 60 Hz series tuned circuit in order for the secondary voltage to remain stable. The transient response of the coupling capacitor voltage transformer is not very good, particularly during large, fast, voltage excursions, (e.g., a phase-to-ground fault or energizing the line). These steep wave fronts produce oscillations (ringing) that must be damped. A number of components and protective gaps in the device already exists. Now, additional components must be added to improve the transient response and prevent ferroresonant-suppression. The coupling capacitor voltage transformer is not as large as the cascade voltage transformer, is less expensive, but is not as reliable.

High voltage circuit breakers equipped with grading capacitors across multiple breaks in each phase, can be involved in resonance problems when deenergizing magnetic voltage transformers or cascade voltage transformers. The power system serves as a source of supply to the capacitance of the grading capacitors across the open circuit breaker in series with the inductance of the voltage transformer. If the values of the capacitance and inductance are close enough to produce resonance, the voltage at the circuit node between the capacitor and the inductor will be quite high and will be sustained. The result will be damage to the voltage transformer.

At primary power system voltages, the prohibitive cost of "dead" tank high voltage circuit breakers dictates that the high voltage circuit breaker be a "hot," or "live," tank design. The use of bushing-type (window or bar) current transformers is no longer feasible and therefore the current transformer will be a freestanding, oil, or SF$_6$ insulated type. If primary power system voltages continue to increase, cascade-type magnetic current transformers may be used. Seis-mic tests may also pose a problem to the very large freestanding current transformers. The freestanding, oil insulated current transformers have experienced a very high failure rate. Disruptive failures have resulted in major damage to adjacent high voltage equipment because the explosive failure will hurl large chunks of porcelain long distances. The SF$_6$ insulated current transformers can contain the damage to a much smaller area, but these current transformers are more expensive and have a different set of problems unique to the SF$_6$ insulating medium. One additional solution to the freestanding current transformer problem is to buy only "dead" tank high voltage circuit breakers. Of course, the circuit breakers are considerably more expensive to purchase and install. They also require more switchyard space (they have a larger "footprint") as well as a considerably larger volume of space to ensure proper electrical clearances.

Gas insulated substations (GISs), with their own set of unique problems, might be a solution to the current transformer problem because the use of bushing transformers is permissible. Switching disconnects in GIS (equipped with voltage transformers) can result in internal flashovers to ground caused by the very steep wave front from restrikes impinging on the circuit discontinuity of the voltage transformer.

Because of the existing problems with input sources presently in use today, it is easy to justify research and development projects to produce better input sources. Considerable success has been achieved utilizing a Faraday sensor for a current transformer and a Pockels sensor for a voltage transformer. Development of these devices to replace the Ct and voltage transformer and their acceptance for use on a power system could result in a number of benefits:

(1) Eliminates oil/paper insulation systems and their catastrophic failures.
(2) Does not contain iron; therefore eliminates saturation, residual flux, and ferroresonance.
(3) Greatly reduces size and weight (could even be combined with a bus insulator).
(4) Passive devices with fiber-optic cable to the switchhouse provide EMI immunity.
(5) Increases accuracy (only limited by the electronics).
(6) Greatly increases the dynamic range of the Ct.

II. Discussion of Reprints

The first paper "Transient Response of Coupling Capacitor Voltage Transformers," represents an important contribution to this difficult analytical problem. The PSRC has provided examples and appropriate tables to allow the relay engineer to determine how relays will respond in the presence of transients. The paper, "Transmission Line Protection with Magneto-Optic Current Transducers and Microprocessor-Based Relays," by Udren and Cease, describes a current transducer with an improved optic path using the Faraday principle of the rotation of light polarization in the presence of a magnetic field. In addition, a voltage sensor consisting of a capacitor divider and an adaptation of the current transducer provides the input to the protection package of a 161 kV transmission line. The last paper, "An Analytical Approach to the Application of Current Transformers for Protective Relaying," by Zocholl, Kothheimer, and Tajaddadi, presents a PC program that allows relay engineers to analyze the Ct performance and optimize the relay application for critical situations.

III. Associated Standards

"*Guide for Field Testing of Relaying Current Transformers*," ANSI/IEEE C57.13.1—1981.

"*Standard Conformance Test Procedures for Instrument Transformers*," ANSI/IEEE C57.13.2—1986.

"*Guide for the Grounding of Instrument Transformer Secondary Circuits and Cases*," ANSI/IEEE # C57.13.3—1983.

"*Requirements for Instrument Transformers*," ANSI/IEEE C57.13—1978.

"*Requirements for Power-Line Coupling Capacitor Voltage Transformers*," ANSI/IEEE C93.2.

"*Transient Response of Current Transformers*," IEEE Special Publication 76 CH 1130-4 PWR, 1976.

IV. Supplementary Bibliography

M. Kanoi, G. Takahashi, T. Sato, M. Higaki, E. Mori, and K. Okumura, "Optical voltage and current measuring system for electric power systems," *IEEE Trans. on Power Delivery*, vol. PWRD-1, no. 1, Jan. 1986, pp. 91–97.

T. W. Cease and P. Johnston, "A magneto-optic current transducer," *IEEE Trans. on Power Delivery*, vol. 5, no. 2, April 1990, pp. 548–555.

TRANSIENT RESPONSE OF COUPLING CAPACITOR
VOLTAGE TRANSFORMERS IEEE COMMITTEE REPORT

Transient Response of CCVTs
Working Group of the Relay Input Sources Subcommittee
of the Power System Relay Committee

Abstract - The results are presented of the effort to standardize the limits of the transient errors in the output of a CCVT, or CVT, permitted during a subsidence of the input voltage. Various proposed methods of describing and measuring these deviations are discussed relative to their use with protective relays. The conclusion is that no standard can be written at this time.

INTRODUCTION

Capacitor Voltage Transformers (CVTs) not equipped for carrier coupling and Coupling Capacitor Voltage Transformers (CCVTs) have been used successfully for many years as an economical way to transform transmission voltages (138,345kV, etc.) to voltages at which relays are designed to operate (66.4, 115V). Ideally, the low voltage applied to the protective relays should be an exact replica of the transmission voltage. CCVTs are good, but not ideal, and lose some fidelity in reproducing transient voltage variations because of inductive, capacitive, and non-linear elements in the device. Electromagnetic voltage transformers (VTs) perform better with respect to transient response than CCVTs, but they are more costly at high voltages. A subsidence transient occurs following a fault and as a result of a fault. It should be noted that the "Transient response" of CTs refers to the ability of the CT to handle the DC component in an asymmetrical current wave, but "transient response" of a CCVT refers to its ability to control the tendency to create extraneous frequencies in the output.

Industry experience has indicated few, if any, problems with electromechanical relays due to CCVT subsidence transients; however there is a concern with the effect of the subsidence transient on fast, solid state impedance relays. The solutions to the anticipated false operations have been for the relay manufacturers to add input filters with a resultant delay or for the user to reduce the relay reach below that used with electromechanical relays. Reducing the reach does not solve the problem when there is an error in the detection of direction.

The relay engineer's problem is: "How does one choose a CCVT for a particular relay application?" At this time, the engineer can be sure of no problems with respect to relay response only if he uses CCVTs which have been factory tested with his particular relays. Although several standards exist concerning transient response of CCVTs they were written before solid state relays were developed, and none appears to insure satisfactory relay operation. The International Electrotechnical Commission (IEC) Standard 186 Section 20,

paragraph 43, "Transient Response" says: "Following a short circuit of the supply at the primary and earth terminals, the secondary output voltage of a capacitor voltage transformer shall decay within one cycle of rated frequency to a value of less than 10% of the peak value before short circuit". The Australian standard on CVTs requires any residual voltage oscillations to be outside of a band, for instance, they must be below 12Hz or above 300Hz.

In the interest of having a CCVT standard, ANSI C93.2 was approved in 1976 without specific transient response requirements.

ANSI Standard C93.2-1976 specified test circuits and burdens for transient response measurements but states that the required response for the user's particular application is a matter to be resolved between the user and manufacturer.

In preparation for the reissuance of the standard, the WG assignment adopted September 18, 1973 was to "Establish transient response classifications for coupling capacitor voltage transformers for use with protective relays".

A previous WG of the Relay Input Sources Subcommittee had proposed to the Standards Committee that three classes of transient response be established based on residual voltage and decay time. These were: Class I-following a short circuit of the primary voltage, the secondary output voltage of the CCVT shall decay, within 8 milliseconds and thereafter, to a value of 6% or less of the peak value before short circuit. Class II-Following a short circuit of the primary voltage, the secondary output voltage of the CCVT shall decay within 8 milliseconds and thereafter to a value of 12% or less of the peak value before short circuit. Class III-Following a short circuit of the primary voltage, the secondary output voltage of the CCVT shall decay within one cycle of rated frequency to a value of less than 10% of the peak value before short circuit.

These classes were rejected for the C93.2-1976 standard on the basis that they are not meaningful in relation to the ability for a CCVT - protective relay system to provide the desired function. However, this WG has failed to agree on a proposal for the next revision of the standard. The purpose of this report is to summarize what was learned during the past seven years.

Subsidence Transient

A subsidence transient is an error voltage appearing at the output terminals of a CCVT resulting from a sudden significant drop in the primary voltage, typically produced by a nearby phase-to-ground fault. This produces a transient voltage in the secondary that may be damped oscillatory or decaying unidirectional depending upon the design of the CCVT, the connected burden, and the incidence point on the voltage wave. The apparent impedance to a relay may include errors in both magnitude and phase angle.

Figure 1 shows the manner in which the secondary voltage may collapse in response to a sudden reduction to zero in the primary voltage.

81 SM 457-1   A paper recommended and approved by the IEEE Power System Relaying Committee of the IEEE Power Engineering Society for presentation at the IEEE PES Summer Meeting, Portland, Oregon, July 26-31, 1981. Manuscript submitted March 27, 1981; made available for printing May 11, 1981

Reprinted from *IEEE Trans. Power App. Syst.*, vol. PAS-100, no. 12, pp. 4811-4814, Dec. 1981.

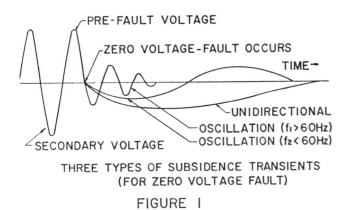

PRE-FAULT VOLTAGE

ZERO VOLTAGE-FAULT OCCURS

TIME→

UNIDIRECTIONAL

OSCILLATION ($f_1 > 60Hz$)

OSCILLATION ($f_2 < 60Hz$)

SECONDARY VOLTAGE

THREE TYPES OF SUBSIDENCE TRANSIENTS
(FOR ZERO VOLTAGE FAULT)

FIGURE I

Factors Affecting CCVT/Relay Performance

In this discussion, comments on the effects of various factors are based on references (1) and (2). Comments regarding the effects of various parameters on the performance of a protective relay/CCVT system are necessarily of a rather general nature. Reference (1) gives an excellent discussion of the effects of various CCVT parameters on the transient waveform output of the CCVT. Relating this to relay performance is another problem entirely. This is due to the complexity of the CCVT and relay response and the large number of relay types (2).

The discussion which follows on the factors as they affect CCVT/relay performance is intended as an introductory guide only. For specific individual cases the user should consult the equipment manufacturer for advice in the evaluation of possible transient response problems.

CCVT Burdens

Burden Magnitude - Most CCVT designs give better transient performance for burdens lower than the rated burden.

Burden Power Factor - In general, the transient response becomes worse as the power factor decreases, either leading or lagging, and generates a low frequency oscillation.

Composition and Connection of the Burden - Reactive and resistive elements of the burden are usually complex combinations of series and parallel elements connected in one phase, between phases and across a broken corner delta connection as the source for triple zero sequence voltage. Some relay types introduce coupling between the current and voltage inputs in the relay replica impedances. Another factor can be the presence of surge filters in the voltage input to solid state relays, which introduce additional capacitance. Any relay installation can be quite complex and it is only possible to make the following general remarks: 1. High Q inductive elements in the burden tend to make transient response worse. 2. Surge capacitors have only a minor effect on the transient response except for very low burdens where they may accentuate high frequency oscillations. 3. Series RL burdens for the same volt-ampere and power factor give better transient response than parallel RL circuits.

Incidence Angle of Fault (Point on Voltage Wave)

In general the worst CCVT response is obtained when the fault occurs at the zero point on the voltage waveform because the capacitance has maximum stored energy at this time and a longer time constant than the inductance. The inductance stores an equal amount of energy at voltage crest, but the discharge time constant is much smaller (1).

Because faults occurring at zero voltage will also have maximum offset fault current, it seems likely that large CCVT transient errors and CT transient errors may occur at the same time, at least on one of the phases.

CCVT Design

Capacitance of the Voltage Divider - Higher capacitance CCVTs produce subsidence transients of smaller magnitude but of longer duration.

Ferroresonance Suppression - This protection is necessary on all CCVTs and does have a bearing on the transient response. The method used may or may not degrade relay performance but it is not beneficial.

Design of Voltage Transformer - Higher turn ratios produce subsidence transients of smaller magnitude but of longer duration. Higher excitation currents are a source of low frequency oscillations and so the transformer is designed for low excitation current.

Tuning Reactor Design - A low Q is desired for this component to improve the transient response by dissipating energy; however, this conflicts with the requirement for a low series resistance to meet steady state accuracy requirements.

Relay Design

Protection Function - There are many different types of protective relays that utilize voltage input. Their purposes differ widely and therefore it is to be expected that for some the transient response of the CCVT will be of no consequence, whereas for others it will have a major impact on performance. The following table illustrates a few examples:

Protective Relay	Effect of CCVT Transient
(a) Synchronizing	Not important—uses steady state waveform.
(b) Frequency	Depends upon speed of operation. Normally a time delay is acceptable.
(c) Undervoltage/ Overvoltage	Response to voltage changes not appreciably affected. Time delay, when used, overrides any effect.
(d) Volts/Hertz	Normally supplied by VTs. CCVTs are not recommended because of their tuned circuit.
(e) Solid State First Zone Distance, Mho, Impedance, etc. using basic phase to phase or phase to ground quantities and with no intentional filtering or time delay.	CCVT transients can cause errors in measurement, both overreach and underreach. Slower responding relays are less affected. Momentary directional error can occur but memory action tends to override this error. Error can be

444

<table>
<tr><td>(f)</td><td>Solid State Distance as in (e) with filtering or time delay.</td><td>The filtering or other added delay is the method normally used to prevent errors. In some designs the filter is not used unless the fault is severe.</td></tr>
<tr><td>(g)</td><td>First Zone Distance using sequence quantities from sequence filters.</td><td>Less affected than (e) because symmetrical components are derived from three phase voltages, which results in a smoothing effect and fewer errors in the signals.</td></tr>
<tr><td>(h)</td><td>Second Zone Distance</td><td>Time delay overrides transients.</td></tr>
<tr><td>(i)</td><td>Directional Comparison relay system</td><td>Momentary directional error possible for relays without time delay. Transient errors can occur beyond first zone reach when source impedance is high.</td></tr>
<tr><td>(j)</td><td>Traveling Wave</td><td>User must specify required transient response of CCVT to meet the requirements of the traveling wave relay system.</td></tr>
</table>

<u>Effect of Transient Error on Speed of a Relay</u> – The effect of the CCVT transient error on the operating time of a relay depends upon the type of measurement that is being made and on the operating principle of the relay. The general effect is to delay operation by a few milliseconds because the CCVT output changes voltage levels more slowly than the primary.

In the test experience of one manufacturer who used a model power system and a model CCVT, there is only a very slight increase in average operating times of distance relays applied in an overreaching mode in a directional comparison scheme. However, there is a more significant increase for faults between 80 and 100% of the reach setting of a first zone relay.

<u>Solutions Considered and Evaluation</u>

Certain CCVT designs are modeled and tested along with a relaying system on analogue power system models to assess the reliability of the composite system. In considering alternative CCVT designs which are not tested with specific relays, the user needs to determine if the CCVT candidate is approximately as good as or superior to the CCVT tested as to transient response. This equivalency criterion initially appeared to be less difficult to specify than the objective of the WG which was to establish transient response classifications. The following approaches were proposed and considered.

Proposal #1 CCVT Equivalency by Circuit Parameters

Simplify the CCVT to a series RLC Thevenin equivalent per Figure 4. Use C_E as the figure of merit on the basis that as C_E increases, the transient response improves. For a given burden, energy storage decreases in a linear manner with an increase in C_E. Since L_E and C_E are approximately in series resonance at rated

frequency, C_E essentially defines L_E. The ANSI Standard C93.2 describes the burden for transient response rating as, "one impedance shall be a pure resistance (Rp) and the other (R_s plus X_s) shall have a lagging power factor of 0.5. The inductive reactor shall be an air core type," (3). These symbols are shown in Figures 3 and 4.

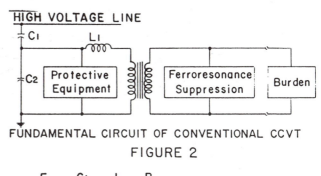

FUNDAMENTAL CIRCUIT OF CONVENTIONAL CCVT
FIGURE 2

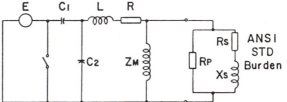

APPROXIMATE EQUIVALENT DIAGRAM
FIGURE 3

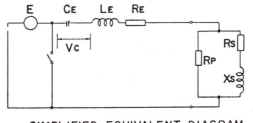

SIMPLIFIED EQUIVALENT DIAGRAM
FIGURE 4

Ferroresonance suppression circuits are either "active" which store energy or "passive" which do not. The active circuits may materially modify the CCVT response. Where these circuits are separate elements, they could be excluded from the equivalent circuit representation and evaluated separately by specifying a maximum deviation in overall time response from that of the same design without the ferroresonance suppression circuit.

Doubts were raised about the practicality of reducing all CCVT designs to a common form of equivalent circuit. Probably two or three circuit configurations would be needed to properly represent all available designs. Also it was felt that the equivalent circuit is not necessarily indicative of relaying performance. For example, an increase in C_E tends to prolong the transient, on the other hand this "slow" transient can be readily rejected by the relays.

445

Proposal #2 Fourier Transform Classifications.

The Fourier Transform of the CCVT time response for a zero voltage fault over the 0-200 Hz band looks meaningful to the relays, where the relays have the most difficulty rejecting "noise"!

A computer analysis of the time responses could yield a voltage vs. frequency curve as in Fig. 5. The analysis takes the non-periodic wave and assumes it to be periodic with a period T (e.g. 100 ms). It yields a pair of coefficients, a & b, for each frequency $\frac{n}{T}$; where a_n is the sine coefficient; b_n the cosine coefficient. If the time response is measured each 0.5 ms, for example, 200 values would be given for a 100 ms period. These would yield voltage magnitudes at $n=1$ to $n=99$ (plus the average value for $n=0$), which correspond to frequencies of 10, 20...990Hz. Alternatively, the frequency response could be characterized by separate sine and cosine plots, which would define the phase angle relations amoung the frequency components.

Each transient classification would limit the magnitude at specified frequencies, and would be so chosen as to properly encompass existing CCVT designs.

Each transient classification could consist of several characteristics. For example, it could specify both a time domain and a frequency domain: The magnitude of voltage must get below and stay below some value and not exceed a certain set of values at specified frequencies.

The Working Group could not, from experience, evaluate the adequacy of this proposal, nor develop specific magnitude limits.

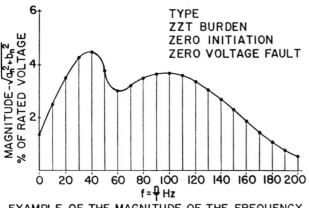

TYPE
ZZT BURDEN
ZERO INITIATION
ZERO VOLTAGE FAULT

$MAGNITUDE = \sqrt{a_n^2 + b_n^2}$
% OF RATED VOLTAGE

$f = \frac{n}{T}$ Hz

EXAMPLE OF THE MAGNITUDE OF THE FREQUENCY COMPONENTS NEEDED TO REPRESENT A CCVT SUBSIDENCE TRANSIENT WITH A PERIOD OF T=100ms WITH TIME FUNCTION VALUES TAKEN EACH 0.5ms
FIGURE 5

Proposal #3 Envelope Restriction

Restrict the magnitude of the peak of the envelope of the remnant voltage existing at, for example, 8 ms, rather than restricting the magnitude of the response at a particular time as does the IEC Standard. Use of an envelope would tend to overcome the objection that the response can be oscillatory and may reach a higher magnitude after the specified time. The IEC Standard does not say that the voltage must remain below 10%, although some believe that is what is intended. However, an envelope restriction does not define the wave form (e.g. oscillatory vs. exponentially decaying), which appears crucial to relay response.

Proposal #4 Tracking Requirement

When an abrupt change in primary voltage magnitude or phase angle occurs as the result of a fault, the output must track the primary voltage after 8 ms within the limits: (a) Zero crossing deviation $\pm$ 0.5 ms and no additional crossings. (b) The magnitude of the secondary voltage at each instant when the primary voltage has a zero slope shall be within +20% and -5% or +10V and -2.5V, whichever is greater and no polarity reversal.

This is a relay oriented requirement which is probably conservative. It may not be practical to design a CCVT that meets these requirements.

Proposal #5. Time - Voltage Bands.

Develop a band on each side of the time response of existing CCVT designs which would be grouped into three of four categories. Each of these bands would define a classification based on ANSI C93.2-1976 tests.

As with Proposal #3, the band limit fails to define the wave form; the response could be oscillatory near 60Hz, for example.

Summary, Conclusions, Recommendations

This Working Group has found no serious industry problem due to a lack of standard classifications for the transient response of CCVTs being used with protective relays. This conclusion and the collected information tends to justify their work. Even though the investigation was limited to the subsidence transient response, no standard was written.

The readers are encouraged to express their opinion by a Discussion or to the Relay Input Sources Subcommittee as to the existence of problems, the need for a standard and what form a standard should take. The responses will be helpful in evaluating the need to continue this Working Group.

Working Group Members and Past Members

J. Berdy, R. Brookes, J.W. Chadwick, Jr., D.H. Colwell, M.B. DeJarnette, J.B. Dingler, W.A. Elmore, C.W. Fromen, L.L. Green, F.B. Hunt, Chairman, J.L. Koepfinger, W.C. Kotheimer, R.W. Pashley, G.D. Rockefeller (Past Chairman) A. Sweetana, K. Winick (Deceased).

References

1. A. Sweetana, "Transient Response Characteristics of Capacitive Potential Devices", IEEE Transactions on Power Apparatus and Systems. Vol PAS-90, pp. 1989-2001, September/October 1971 (71TP 197-PWR).
2. G.A. Gertsch, F. Antolic, F. Gygax, "Capacitor Voltage Transformers and Protective Relays," CIGRE Report 31-14 1968.
3. AIEE Committee Report. The "Effects of Coupling Capacitor Potential Device Transients on Protective Relay Operation". Vol 70 Part II pp. 2089-96, disc 2096 1951.

Transmission Line Protection with Magneto-Optic Current Transducers and Microprocessor-Based Relays

E.A. Udren
ABB Power T&D Co.
Coral Springs, Florida

T.W. Cease
Tennessee Valley Authority
Chattanooga, Tennessee

ABSTRACT

TVA has installed three-phase magneto-optic current transducer (MOCT) sets at the two ends of a 161 kV transmission line, along with microprocessor-based numerical transmission-line relays which can accept the inputs from the MOCTs or conventional cts. The equipment includes multiple MOCT sets and multiple relays at each end; current-signal switching and test facilities; and data communications networks for monitoring performance. The paper discusses reasons for pursuing MOCT technology; MOCT operating principles; and the microprocessor-based line relays, with focus on the interfacing to the MOCT. Current-signal switching and input testing facilities are explained. The paper also describes the data communications configuration and experience.

INTRODUCTION

The Magneto-Optic Current Transducer (MOCT) promises to become one of the rare technological advances which brings benefits in performance and application, without the drawbacks often associated with change to a radically different technology. Compared to a conventional iron-cored magnetic ct, the MOCT offers a host of benefits listed below. The drawbacks are those associated with the birth and introduction of any new technology, and are fading with successful demonstrations of the system in projects like the one described in the present paper.

The Tennessee Valley Authority (TVA) has promoted and sponsored pioneering development of practical MOCT systems, working with the Electric Metering Systems Division of ABB Power T&D Company. Initial work focused on the use of MOCTs for high-accuracy metering. As successful single-phase and then three-phase metering demonstrations have been carried out at TVA, a number of other utilities have installed similar MOCT-based metering systems. Previous papers [1,2,3] have described the MOCT metering work. All of these activities have advanced the MOCT from a laboratory curiosity to a practical transducer.

More recently, the researchers have tackled the problem of performing protective relaying with current signals from the same MOCTs. The present paper describes a joint effort of TVA, the MOCT development group of ABB EMS Division, and ABB Relay Division to field and demonstrate practical protective relaying using MOCTs.

Reprinted with permission from *1991 Georgia Tech Protective Relaying Conference*, pp. 1–21, May 1–3, 1991.

The following sections present a summary of MOCT operating theory and construction practice, highlighting the benefits and limitations of MOCTs versus those for familiar iron-core cts. The program for demonstration of relaying is described, and the equipment design is explained. The testing program and field trial experience to date are summarized. We conclude with a discussion of future prospects and issues for broadened MOCT use in the utility industry.

BACKGROUND - CT FEATURES AND PROBLEMS

Ferromagnetic current transformers have served the utility industry faithfully since its inception. Since these cts provided the only practical method of current measurement, engineers simply accepted and worked with their performance limitations - notably, difficulty in maintaining accuracy over the full range of operating conditions, and most particularly their tendency to suffer saturation of the iron core during severe faults, with accompanying severe ratio error or loss of output. The difficulties with saturation were circumvented by using a larger core, and by inventing and applying relays whose characteristics provided robustness and security in the face of heavy faults with saturation. For example, differential relays were developed with restraint based on fault current magnitude, and percentage-differential operating characteristics which desensitize the relay during heavy faults.

As the size and importance of power networks grew, ct designs evolved to meet the ever-higher demands placed upon them. The gradual nature of the change may have failed to highlight some of the ct design features which have come into existence:

1. As the complexity of relaying has increased, the number of relays connected to each ct winding has grown. The size of the ct assemblies has grown to carry the burden. More secondary windings have been added to independently drive primary and backup relays for two or more adjacent zones which overlap at the ct. Some functions, such as metering, have different performance requirements and may require yet more cts which are independent of those used for relaying. The amount of iron and wire in a typical EHV ct installation is staggering.

2. As fault current levels have increased, there has been yet more need for larger cores and heavier wire in the secondary circuit to limit ratio error problems.

3. Ct wiring is prominent among the mass of switchyard-to-control house wiring which conducts intense and destructive surges to connected, vulnerable electronic relays and control equipment.

4. Increasing system voltages have imposed the fundamental requirement for more elaborate insulating systems. Insulation for HV or EHV cts requires a massive porcelain column which can support all of the iron and wire described above, plus a primary-to-secondary insulating system which uses oil or SF6 gas.

5. Older dead-tank circuit breakers provided convenient mounting of bushing cts on both ends of the breaker. The higher voltages have brought about the use of live-tank circuit breakers, which do not provide a convenient mechanical means for mounting the cts. Accordingly, the cts are installed in a separate, free-standing insulating assembly which is

costly. Because of the cost, all of the cts formerly installed in bushings on both sides of the older breakers are lumped together in just one free-standing ct assembly. This has eliminated the very desirable overlapping of relaying zones, leaving some breaker internal faults unprotected, and has added to the complexity of relaying schemes.

Most of these problems have been solved by spending more money - supplying the bigger cores with more windings, large free-standing insulating structures, shielding and surge suppression for the wiring, and extra breaker-fault cts and relays. The internal insulation problem, however, is less tractable. At The Tennessee Valley Authority (TVA) and at other utilities, failures of oil insulation systems have caused dramatic explosions of 161 kV cts, with serious physical damage, fire, hazard of injury to personnel in the vicinity, and interruption of electric service. Gas insulation systems bring new maintenance problems of their own, and in TVA's view do not provide an adequate long-term solution.

The MOCT Solution

The MOCT concept, based on the Faraday-effect rotation of light polarization in glass in the presence of a magnetic field, has been advanced to the state of a practical, high-performance current transducer with none of the above-listed inherent drawbacks of cts.

We must observe here that iron-core cts are able to deliver the high levels of operating energy required by electromechanical relays. Modern solid-state and microprocessor-based relays don't require much energy and allow consideration of new approaches such as the MOCT. However, many existing reliable electromechanical relays will remain in service for many more years yet, and a large number are still manufactured and applied today. For some applications, users do not have a superior electronic replacement available for trusted electromechanical designs. An MOCT can drive an electromechanical relay only if equipped with an expensive electronic high-current output amplifier, and today it appears unlikely that such amplifiers will be economical or reliable enough for widespread use. Thus we can see that MOCTs and conventional cts must coexist for the foreseeable future.

Also, no one will forget that the iron-core ct, with all of its drawbacks, has proven itself to be among the most reliable of components in the relaying chain. Most of the performance problems have been overcome through relay design and application engineering, and by paying the price of high-capability cts.

MOCT SYSTEM CONFIGURATION

Figure 1 shows how the MOCT-based current sensing system is installed in the substation. A machined-glass Faraday sensor, described just below, is installed around the power conductor. Three are required for three-phase measurement, and additional sets of three may be installed for redundancy as with conventional ct secondaries. The rotators built to date are about 4 inches by 4 inches, and about 1 inch thick, with a large window through which the power conductor passes.

Connected to each sensor is a pair of optical fibers. The fibers conduct light between the sensor and an electronic module in the control house, perhaps several hundred meters away, as shown in the Figure. The module launches light to the sensor, and interprets modulation of the returned light to develop an output voltage which is proportional to current in the power conductor.

Figure 1 shows the sensors with dedicated insulating columns to support the sensors and to shield the optical fibers from the environment. Note, however, that the sensing system comprises all dielectric materials, weighs under a kilogram, and does not require its own insulating system. The Faraday sensors and fibers can be embedded in existing structures which are always installed by the utility in any case—live-tank circuit-breaker bushings, transformer bushings, or even bus support insulators.

Also, note that the switchyard components are simple, entirely passive, and unlikely to drift over time. The active electronic portion of the system - the light source and the signal-processing electronics - are all installed in a rack in the easily-accessed, environmentally-benign control house.

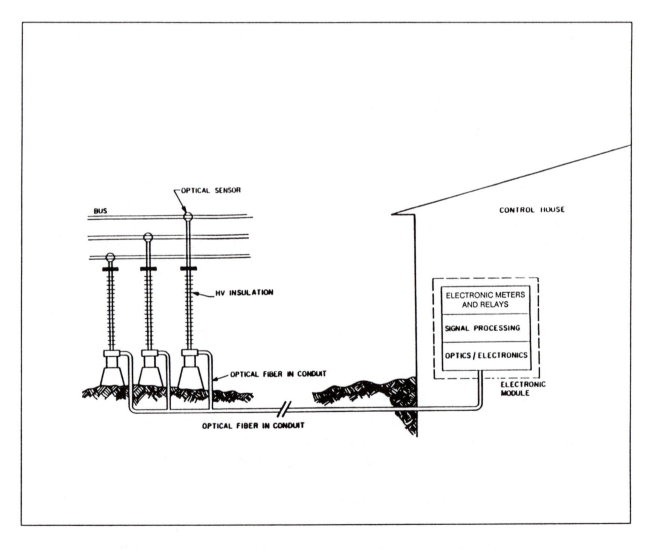

Figure 1. Configuration of MOCT Installation

MOCT THEORY OF OPERATION

An MOCT system develops an analog output voltage signal whose instantaneous value is proportional to the instantaneous current flowing through the primary power conductor. The waveshape of the current in the primary circuit is reproduced with large bandwidth, and with very high accuracy and dynamic range. The output voltage is low—in the range of ±10V for the largest excursions of fault current in the primary circuit. Available energy is minimal without an amplifier, and thus the MOCT interfaces most readily to electronic measurement and relaying systems. Modern microprocessor-based systems normally sample and digitize the analog wave according to the needs of the measuring device.

The key element in the system is the Faraday effect sensor, also referred to as a rotator in the following. The sensor is a block of precisely-machined optical glass, with a hole through which the primary power conductor passes. Input and output optical fibers are joined to the block through lensing assemblies which collimate and polarize the light beam. As emphasized above, this passive sensor is the only component in the transducer system which is installed at high voltage, making this design much simpler than other electronic techniques for current sensing.

The Faraday effect can be observed when a light beam is passed through a substance which becomes optically active in the presence of a magnetic field. The angle of polarization of the light beam is shifted in proportion to the direction and intensity of the field. If a scheme is devised to measure the polarization shift, the field strength can be measured. Stated simply, the MOCT uses the measurement of this shift to determine the instantaneous magnetic field strength surrounding the power conductor, which in turn is proportional to the current.

Figure 2 shows the implementation of this measurement principle. A light-emitting diode (LED) in the control house electronic interface system launches light into a multimode optical fiber, which passes into the switchyard, up through an available insulating structure, to the rotator which surrounds the power conductor. As mentioned above, the fiber-to-rotator attachment includes a polarizing filter and collimating lens arrangement, so that a narrow light beam of specific polarization passes into the rotator.

The machined faces of the rotator reflect the light at each corner, so that the light travels in a loop around the conductor. Finally, the light beam strikes a second collimator and polarizer assembly as it exits the glass rotator. Light which passes through the second polarizer is carried by the second fiber back to the electronic interface in the control house.

The two polarizing filters are positioned with intentional misalignment of polarization so that, with no magnetic field, the light exiting the Faraday sensor is partially attenuated by the second polarizing filter. If current flows in the power conductor, the resulting field shifts the polarization of the light beam in the rotator. For current in one direction, the polarization axis of the beam shifts to become even less well aligned with the exit polarizing filter, so that the attenuation at the exit point increases. If the current reverses, the beam polarization becomes more closely aligned with the exit polarizing filter, so that the attenuation decreases. Thus, the light beam exiting the Faraday sensor is amplitude-modulated by the current and its resultant field. A photodetector at the control-house end of the return fiber responds to the amplitude-modulated light,

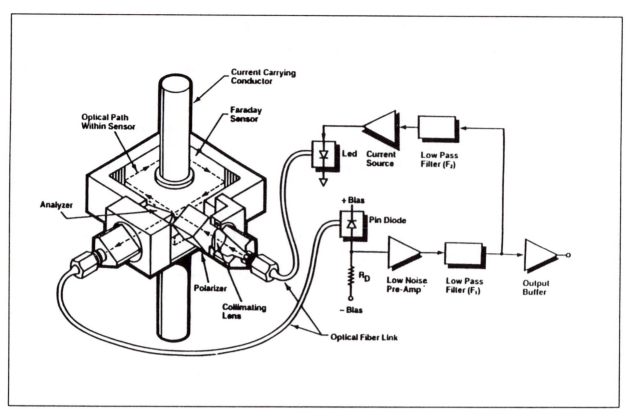

Figure 2. Schematic Diagram of MOCT System.

and drives an electronic amplifier circuit. The amplifier produces a low-level voltage output which is proportional to the instantaneous current flow in the power conductor.

It is important to note that the Faraday rotator, and the light beam it guides, loops completely around the high voltage conductor. Fields originating outside the loop produce no net effect on the beam as it traverses the entire path, so that the Faraday sensor does not respond to external or environmental fields, even if fairly intense or close by. In fact, it can be shown mathematically that for an arbitrary closed path around the current-carrying conductor, the modulation of the angle of polarization will depend only upon currents within the closed path. Furthermore, the modulation will not depend upon the location or position of the conductor within the looped light path. This means that the mounting or centering of the sensor on the power conductor does not affect accuracy or calibration.

Sensitivity

The relationship between the power-conductor current and the angular shift of polarization is expressed as:

$$\Theta = 2VI$$

where I is the current flowing in the conductor; Θ is the angle of rotation; and V is a sensitivity constant for the particular rotator material, called the Verdet constant. Material selection for the rotator element becomes a trade-off involving optical characteristics, operating range, and temperature stability. The selection of the glass used in this system was guided by metering system requirements of stability and minimized optical attenuation. Within these constraints, the chosen material had to provide the dynamic range needed for protective relaying. This has led to the selection of a material with a small Verdet constant - one which shows relatively low sensitivity to magnetic fields.

Use of a material with a small Verdet constant makes the task of recovering high current signals relatively easy. The challenge is then to detect small signals, where the designer faces the noise figure limitations of the photodetector and amplifier circuit. To provide a useful revenue-metering function, the combination of the sensor and the external electronic meter should maintain accuracies within 0.2% to 0.5% of reading over a dynamic range of 0.01 per unit (pu) to 2.0 pu current levels. Maintaining this accuracy on an instantaneous basis with the small rotation angles is difficult due to the noise generated by the photodetector. However, the integrating function of the watt-hour meter reduces the effect of this shot noise dramatically. This smoothing action allows the accuracy of the MOCT-based sensor/meter system to easily exceed performance specifications for existing 0.1% metering systems connected to conventional metering cts, with the additional error budget these cts provide.

Fault Currents

The MOCT operates on the basis of intensity modulation of the injected light source. As a result, the transfer function is a cosine squared curve. Depending upon the linearity restrictions placed upon the system, this transfer function approximates a linear function over a range of rotation angles as wide as ±25 degrees.

For large load or fault currents in the primary circuit, noise in the signal processing circuits does not contribute significant errors. As the magnitude of the current rises, the larger rotation angle means the approximation of the cosine squared transfer function by a linear transfer function is increasingly less valid. Even with the chosen low-sensitivity material (small Verdet constant), very large fault currents over 50,000 A may lead to instantaneous ratio errors of several percent. Such errors are far less than those typically seen even with high quality conventional cts, and exceed the needs of the relays for such intense fault conditions. It is important to note, furthermore, that this error does not appear as saturation or unpredictable loss of output. Instead, it appears only as a benign, slight, predictable attenuation of the highest waveform peaks. Correction is possible, although probably not justified.

Electronics

The light source for the optical portion of the system is a light-emitting diode (LED), and not a laser diode. The LED is far more reliable, has a far longer life such that replacement isn't normally needed, and is much lower in cost. The LED has much less light output, but still more than adequate considering the total attenuation of the MOCT optical system.

The LED emitter and photodetector interface to the rotator via multimode fibers of 400 μm core diameter. These are relatively large, easy to terminate, have small aperture losses, and are relatively low in cost. The light transmission losses are acceptably low for runs of several hundred meters in the switchyard.

Figure 2 shows a filtered feedback connection from the photodetector amplifier circuit to the LED emitter driver. This closed-loop arrangement controls the quiescent light intensity level in the optical portion of the sensing system. Low-frequency filtering in the feedback loop allows only fixed attenuation and long-term variations to influence LED output - the normal power-system current waveforms are not fed back to the LED emitter. The closed loop light-output control automatically adjusts to the total attenuation in the optical path, as well as to variations of LED output and photodetector sensitivity with time and temperature.

One emitter-detector sensing module is provided for each phase; a microprocessor monitors operation of the entire closed-loop system.

MOCT ADVANTAGES AND CHARACTERISTICS

We now summarize from the preceding discussion the benefits of the MOCT design presented here:

1. All-dielectric construction requires no elaborate insulation system, and cannot explode.

2. The small, light MOCT Faraday sensor is a candidate for embedding in other power apparatus, eliminating the heavy and expensive free-standing ct assembly or bushing cts. Even live-tank breakers and bus-suspension insulators can carry MOCT sensors. Seismic withstand capability of substation apparatus may be substantially improved.

3. MOCTs can be installed as needed to restore desirable overlapping of protection zones, previously lost when live-tank breakers with separate ct assemblies came into use.

4. Optical-fiber connections do not conduct surges from the switchyard to the control house.

5. Switchyard elements of the MOCT system are entirely passive. Electronics are in the control house for environmental protection and easy maintenance.

6. With its very wide dynamic range, a single MOCT can serve both high-accuracy revenue-metering and high-current protective relaying requirements, thus eliminating two expensive ct sets at some switchyard locations.

7. The MOCT has no memory or remanence effects, and cannot saturate or lose output, even for the most intense fully-offset fault-current waves. The worst distortion is a slight peak reduction - a few percent - for the most severe fault currents available on utility systems.

8. The MOCT has wide, flat frequency response from 1 Hz to an upper limit imposed only by the electronics, and not by the Faraday effect. This response can be extended into the Megahertz region, orders of magnitude beyond even the most broadband transient monitoring systems previously installed on power networks. Dc current measurement is possible, although with a different system design than the one described here.

Table 1 indicates the target specifications for the sensor.

TABLE 1	
MOCT System Specifications	
Rated Current	2000A = 1 pu
Output at Rated Current	200 mV
Thermal Rating Factor	2 pu
Fault Current Rating, Symmetrical	20 pu
Fault Current Rating, Fully-Offset	40 pu
Short-Term Mechanical Rating	80 kA
Short-Term Thermal Rating	80 kA
Operating Temperature Range	-40 to +85°C
Ratio Error at Rated Current	0.2%
Ratio Error, Max. Symmetrical Fault	5%
Ratio Error, Max. Fully-Offset Fault	10%
Maximum Phase Angle Error	1 degree

PROGRAM DESCRIPTION

Previous MOCT Development and Demonstration - Metering Systems

The project described in this paper is one in a series of efforts to develop and demonstrate the capabilities of the MOCT. The first was a single-phase metering system [1,2], followed by a three-phase metering system with optical current and voltage sensing [3], and leading to the present relaying demonstration [4]. All of these have been installed at 161 kV. The voltage levels at which other utilities are testing MOCTs range from 23 kV to 345 kV [5]. The next system to be implemented by TVA will be a 500 kV relaying installation. The design of the MOCT optics has remained relatively constant from the outset, with changes mainly in the collimating-lens arrangements; however, the electronics package has evolved considerably.

First Metering Installation

The first application for the MOCT was as a metering class transducer. The objective of this field test was to establish a figure of merit for the stability of the sensor when operating in a substation environment. As part of a metering system, the optical sensor system was instrumented and monitored on a continuous basis for comparison with equipment which is accepted by the utility industry as accurate and stable.

The test method employed was to connect two metering systems to the same phase of a 161 kV transmission line and to record the measurements from both. System parameters such as attenuation of the optical system and load level, as well as environmental parameters such as temperature, were recorded. From this data base, performance levels could be determined and problem areas corrected. The data acquisition system, simultaneously recording other parameters along with the metered quantities, provided a way to view performance as a function of several variables and to analyze deviations, in order to refine the MOCT design.

The MOCT was placed in series with a conventional free-standing oil-filled revenue-metering accuracy ct on one phase of a transmission line at TVA's Chickamauga Dam switchyard. Both sensors fed solid-state revenue meters. One conventional potential transformer was connected to both meters. This arrangement was monitored for a little over two years. The system was modified several times at the site. At the end of the trial, the MOCT had operated continuously for one year. Over this period, the accumulated deviation between the MOCT system and the conventional system was 0.08 percent. The installation was decommissioned after the conventional oil-filled ct developed an oil leak for the third time.

Three-Phase Metering Installation

The first MOCT project was so successful that the next step was a big one. TVA installed a three-phase metering system that used not only optical current sensing, but also optical voltage sensing. To sense voltage, an additional MOCT was connected to sense the current flowing through a capacitor stack connected from line to ground. This configuration has been referred to as the Magneto-Optic Voltage Transducer (MOVT). The MOCT/MOVT metering system was installed on the high side of a transformer bank delivering power to the Electric Power Board of Chattanooga. The existing revenue-metering system is also on the high side, allowing direct comparison with the MOCT/MOVT system. A multi-input data acquisition system like that described just above monitored the installation.

For this project, a microprocessor was added to the electronics; this introduced a number of problems which were solved. Just as serious assessment of the installation was getting under way, a bird flew into the capacitor bank on the low side of the power transformer; the resulting fault damaged the transformer, which is now being replaced. The optical sensors in this system have operated reliably. However, due to problems with other components and with the failed power apparatus, the data collected so far falls short of expectations.

Project for Line Relaying with MOCTs

The project reported in the present paper is designed to integrate the MOCT into a conventional modern relaying scheme, and to allow TVA to evaluate the performance of the MOCT in a relaying application. Also, the test has allowed evaluation of some of the latest digital relaying systems, with their ability to communicate fault and performance data over a local area network.

MOCTs and new microprocessor-based relays with dedicated carrier sets are installed on the 30 mile, 161 kV line between Widows Creek Generating Station and Oglethorpe Substation, as shown in Figure 3. The transmission line runs over a mountain, and experienced 14 faults in 1987.

A total of five new relays are installed. At each terminal, one digital pilot relaying system is connected to either of two installed three-phase MOCT sets through a selector switch arrangement described further below. A second pilot terminal at each end operates from existing conventional cts. Thus, pilot relaying performance with the two types of current transducers can be directly compared. The fifth relay, at Oglethorpe, operates from an MOCT in the stepped-distance mode.

Distance-relaying functions are supplied from existing station voltage transformers. All relays are connected to trip the line circuit breakers.

This project will last for at least two lightning seasons. To determine the acceptability of performance, digital transient recorders on both ends capture fault and relay data. The digital transient recorders monitor not only the system parameters, but also the times of trip or block signals. This will allow direct comparison of tripping times and action taken for a particular disturbance. A comparison of the fault current waveforms from conventional cts and MOCTs can be made.

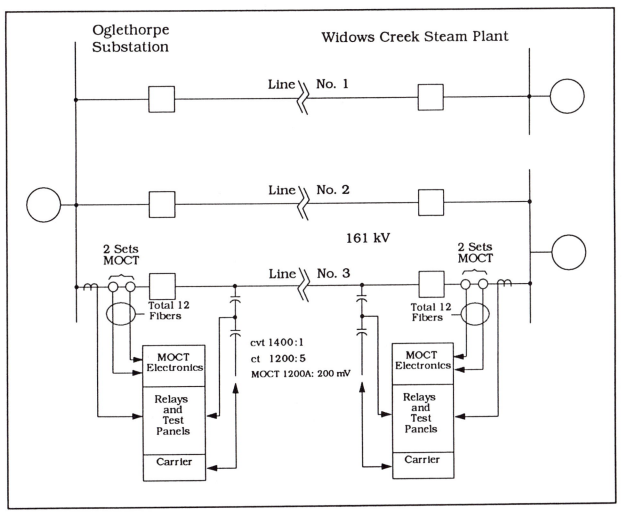

Figure 3. TVA Relaying Installation—One Line Diagram.

THE DIGITAL LINE RELAYING SYSTEM

The MOCT is demonstrated in conjunction with a stand-alone, compact microprocessor-based numerical line protection system whose design stands in contrast to the larger line relaying configurations used in the past [7]. The MDAR relaying system provides all protection functions for one transmission line terminal in a compact four rack-unit package. It was originally designed to work with conventional wound ct, but in the present project, relay samples have been modified for connection to the MOCT electronic interface unit. Figure 4 shows the overall functional arrangement of hardware within the relay. Line voltages and currents from potential devices and ct are connected to isolating transformers, surge suppression, and antialiasing filters. The ac inputs are connected through a multiplexer to the analog conversion subsystem. An A/D converter places instantaneous samples of these ac signals in the microprocessor memory eight times per power cycle. Status or contact inputs are also scanned.

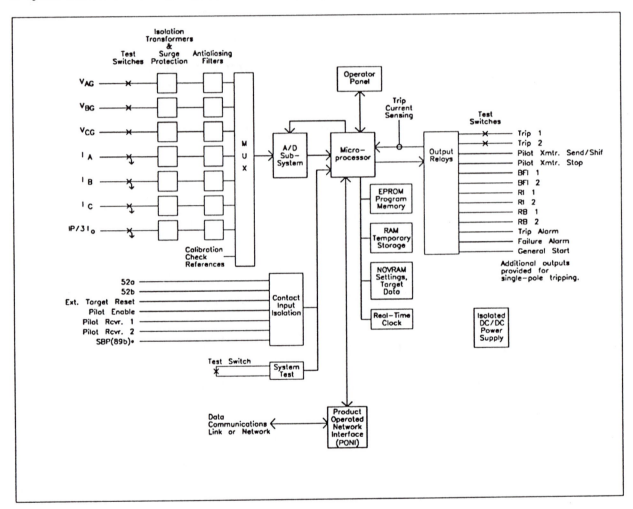

Figure 4. Relay System Block Diagram

For use with MOCTs, the conventional current-isolating transformers are replaced with special units which match the MOCT's low analog voltage output to the relay's analog input subsystem. Provision for connecting a conventional 5A ct or a high-current relay test set is provided in the test panel described below. Although the Figure shows a 3Io input, the modified relays compute residual current in software from the three phase-current input signals. All relaying measurements and logic are performed by software executed in the microprocessor. The programs also handle operator interface and self-checking functions. The hardware performs tripping or other control outputs through a series of contacts as shown in Figure 4. A data port for remote communications is included as explained below.

The MDAR relaying measurements and logic support all of the popular distance-based or directional-comparison pilot and non-pilot line protection schemes, plus ancillary measurements, and some novel relaying options not available before. All of the protection options are included in each relay. TVA selects the carrier-blocking (phase-distance, ground directional overcurrent) logic and stepped-distance backup protection it prefers by way of settings.

The operator's panel consists of two vacuum-fluorescent alphanumeric displays, a row of indicator lights, and a set of pushbuttons. The operator can check or change settings, or check on the status of the relay hardware. More frequently, the display is used to review data for recent or previous faults including trip targets, fault type, time of fault, relaying units which operated, physical fault location, and fault voltage and current values.

Fault location estimates in miles or kilometers are also provided, as part of a comprehensive log of response data for each fault, on the front-panel display or via the data-communications network described next.

INTERFACE DESIGN BETWEEN MOCT AND RELAY

The operating principle of the MOCT is such that the output is necessarily an analog (amplitude modulated) optical signal on a fiber as explained above. That signal must first be converted to a low-level analog electrical signal; the electrical signal is then digitized for processing by front-end hardware in the digital relaying system. Light signals on the six inbound fibers from the two three-phase MOCT sets at each line terminal are converted by the electronic interface package to six low-level analog signals, representing the two redundant three-phase current sets. The two signal sets are compared, for security checking. The test panel is installed in the same cabinet with the digital relaying systems to which it connects. Wired connections also run from the MOCT system analog outputs to a digital fault recorder, in the same room or the control house.

The test panel is installed adjacent to the relays and the MOCT electronics packages as shown schematically in Figure 5. Switching is provided so that the low-level analog signals from the MOCT interface are normally fed directly into the voltage isolating transformers of the relay analog interface; these are the low-voltage input transformers which replace the normal 5A input transformers, as described above. However, the original 5A isolating transformers are retained as part of the separate test panel. The primaries are connected to the conventional ct, or can be fed from a relay test set through test switches. The secondaries are provided with burden resistors such that the output voltage corresponds to that which the MOCT electronics would produce for a given primary current value.

One test panel is provided for each of the three modified line relays. TVA engineers can use a three-position toggle switch on each test panel to select the source of current information for that relay. The actual switching is performed by a group of telephone-type relays, and the switched signals are low voltages only. The switch choices are MOCT No. 1 (the primary MOCT set); MOCT No. 2 (the redundant set); or conventional ct (the voltage which is proportional to current from the ct, or a relay test set connected through test switches).

459

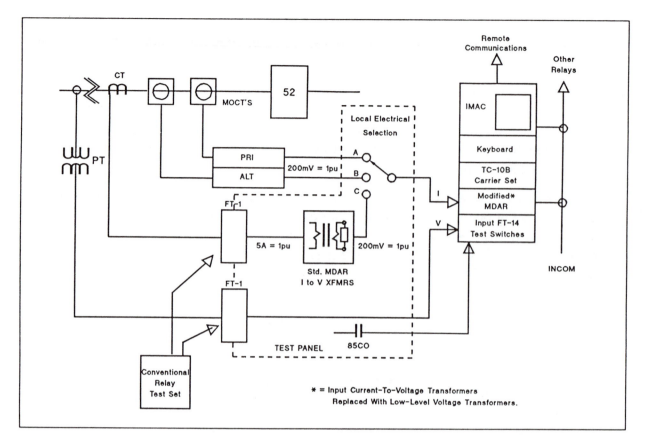

Figure 5. Interfacing of MOCT to Line Relay.

SUBSTATION LOCAL-AREA NETWORK

The TVA trial installation also includes demonstration of the WRELCOM Local-Area Network (LAN) [8] for interconnection of relays in substations. The LAN provides consolidated substation handling of relay data. The data of interest includes event or target records (fault type, units which operated, breakers which were tripped) with time tags; power-system ac value readings obtained during any nonfault time, ac values saved during faults (voltage, current, angle); fault locations, hardware diagnostic information, and tables of relay settings.

WRELCOM is a low-cost, moderate-speed local-area network for substations. The backbone of the network is a shielded small-gauge twisted-pair cable which conveys a carrier signal. The carrier is modulated with frames of serially-transmitted data at 1200 bits per second. All the devices in the network are attached in parallel via spliced-on drops. The network addressing scheme can accommodate thousands of drops.

The [1]INCOM-based Multi-Access Controller (IMAC) is a 4 rack-unit microcomputer system which serves as the host or master of the network. It communicates with the other drops by using a unique address assigned to each; a particular drop responds only when addressed by the host.

1. INCOM is a registered trademark of Westinghouse Electric Corporation.

The IMAC provides a single point of connection for remote and/or optional local operator interfaces. Locally, the IMAC can include printing and display functions, plus mass archival data storage. For communications with the control center or the relay engineer's office, a modem output is provided. Thus, the single off-site communications channel ties the engineer to the network, which can connect to every relay in the substation. In this project several relays are attached to WRELCOM network drops. Other new relays are available with the same interface.

A special network-interface device also exists for connecting conventional relays into the remote automatic data-gathering process. The Existing-Relay Network Interface (ERNI) senses trip outputs of conventional relays, and other contact operations of substation apparatus, with isolated wide-range inputs. Each operation is time-tagged and stored in a memory for up to 50 such operations. The operating-time data is periodically read by the IMAC and saved, for later transmission to the protection engineer. Each ERNI is compact, mounts on the rear of the control panel, and monitors 4 contact inputs. With ERNIs on the existing electromechanical line relays, the performance those relays can be compared to that of the ct- and MOCT-connected digital line relays.

For the demonstration at TVA, the WRELCOM Network comprises one to three standard or MOCT-modified MDAR relays, up to 4 ERNI contact-sensing interfaces, plus an IMAC host at each substation, which also provides remote data communications. The IMAC constantly scans the network, collecting fault and operating data as the relays and ERNIs generate it. The relay engineer communicates from his office to IMAC using a personal computer, modem, and WRELCOM communications software designed for handling data from the installed relays. The WRELCOM Relay Communications Program is provided by the relay manufacturer.

TESTING PROGRAM

High-Current Tests

Characterization of the sensor at high currents is in its early stages - only a few measurements have been made to date. One test was directed at verifying that the Faraday rotator was immune to external fields; a second test was run to verify the mechanical withstand capabilities of the sensor mounting assembly. In both tests the current waveforms were symmetrical sinusoids. These tests were performed on the latest version of the sensor with the most recent collimating-lens arrangement.

During the two tests described above, the MOCT system was energized and output waveforms were monitored. These tests were intended to locate problems associated with mechanical stress on the rotator, and to indicate any gross deviations from linearity. The verification of short-term mechanical ratings was a significant result. It was also shown that the MOCT output does properly reproduce the waveform of very large primary currents. The mechanical withstand test was performed at 80 kA. The test was to have been for a symmetrical waveform. However, due to an error in the current generator setup, the initial test was performed with the high-current wave fully offset. The test was repeated after the current generator was set up properly.

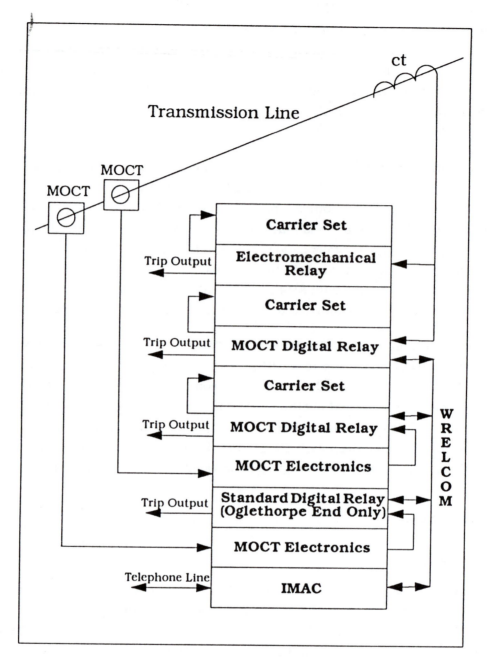

Figure 6. Equipment Configuration.

Model Power System Testing of Relays

Cabinets containing the digital line relays, carrier sets, MOCT electronics, test panels, and data-communications equipment were thoroughly exercised on an analog high-energy Model Power System at ABB Relay Division. To fully test the relay-MOCT interface, one three-phase set of MOCTs was actually installed on the MPS at one line terminal. A cable carrying multiple optical fibers ran from the MPS power equipment area to the relay and mimic-board location.

The MPS operates at a nominal current level of 5 A, whereas the TVA target line operates at 1200 A. To obtain adequate MOCT output for MPS faults, multiple-turn primaries were wound onto the MOCT rotators installed at the MPS.

The MPS was configured to simulate the ultimate field installation site. The tests were carried out with relays at one line terminal operating from the MOCTs, and those at the other end from the conventional MPS ct, as selected through the test panel. At one point in the testing, the inputs to the two panels were switched, so that both terminals could be tested with MOCTs and with the conventional ct set.

Initially, interfacing difficulties between the subsystems appeared and were corrected. Power-supply noise affected the MOCT signal quality through coupling in the cabinet wiring, which was rerouted. Balance of the MOCT channels had to be adjusted with consideration of response of some algorithms in the relays. The relay software itself was also modified to deal with the different characteristics of the MOCTs. Notably, the new software computes residual current samples from the three phase inputs, since the MOCT has no residual current output or return circuit from which a direct measurement can be made as with a conventional ct set. The changes are all compatible with conventional ct use.

Close attention was paid to characterizing the differences in transient response and phase shift of the MOCT versus conventional ct. The wideband precision of the MOCT forced the designers to recheck results, and illuminated the errors and shortcomings of the conventional ct.

The MOCT-energized relays demonstrated their ability to properly relay the MPS line for a wide variety of fault conditions. The IMAC and other data-communications equipment helped greatly with the documentation and analysis of the tests.

INSTALLATION AND FIELD EXPERIENCE

The field installation was straightforward; the work was handled by the TVA area electricians, with testing performed by the area test personnel. The only special provision was that the manufacturer supplied an experienced person to attach connectors to the ends of the fiber optic cables.

These cabinets were installed in an existing, older substation. There was insufficient space in the control and protection area of the control house; the new equipment was installed in the communications room. Figure 7 shows the installed electronic equipment. Figure 8 is a view of the MOCT switchyard installation.

As with any new type of system, there have been a few difficulties, both with the MOCT and with the relaying portion of the system.

The MOCT portion of the system experienced noise problems. Rerouting of some cabinet wiring cured the problem as mentioned above.

On one occasion, a digital relay tripped incorrectly on a current reversal caused by a parallel line fault. This problem has been resolved by changing the pilot logic; it is unrelated to the MOCT aspect of the installation.

Figure 7a. MOCT Electronics and Relaying
Equipment in Control House.

Figure 7b. Fiber Optic Connections to
Electronics.

The high-set overcurrent tripping threshold of one digital line relay was set by an operator from a remote location, inadvertently, to 10% of its correct value; this caused an undesired trip, which was correct considering the setting. Subsequently, the callback scheme of remote access control was instituted in the IMACs for the dial-in telephone line; increased setting-change security features are being added to the remote communications software as well.

TVA electricians, while changing out the station battery, caused the 125 Vdc supply to rise far higher than the 140 Vdc maximum rated value (actual level is not known, but may have approached or exceeded 200 Vdc). This caused the power supplies in the carrier sets to fail. A protective varistor at the input of the IMAC power supply was also damaged. All power supplies have been repaired and are back in service.   Relays and MOCT electronics were not damaged.

The contacts of one telephone relay in a test panel became dirty, causing the current input signal to the relay to be incorrectly low. Cleaning the relay contacts solved the problem, and it has not reoccurred.

This transmission line was selected because it historically has experienced a high number of faults as mentioned above. During the last year, since installation of this system, the number of faults has been less than half of normal. For those faults which have occurred, all systems have responded as expected, except for the above-noted misoperation for a fault on a parallel line.

RECENT PROGRESS IN OPTICAL SENSOR TECHNOLOGY

Calibration and Maintenance

There is still much to be done in the field of optical sensors. A fundamental need is for practical field calibration. The National Institute of Standards and Technology (NIST) (formerly the National Bureau of Standards) has been working to develop the methodology for traceable calibrations. This work has been going on for some time and will continue as new types of optical sensors become available. Field calibration units which insure metering levels of accuracy are being designed.

Figure 8. MOCT Installation in Switchyard.

To maintain relaying accuracy if MOCT electronics should need to be replaced, calibration standards for the optical elements are being established which can be coordinated with settings made in the replacement electronics. Thus, no replacement or testing at high-voltage levels is needed to restore relaying or other current-based functions. Repairs of MOCT electronics can be fast, as with the modern digital relays themselves.

System Voltage Sensing

One of the new optical sensors that is in the early stages of development is the Electro-Optic Voltage Transducer (EOVT). The EOVT sensor is a small crystal (about 6 inches long) that has full line-to-neutral voltage impressed along its length. For a 161 kV system, 93 kV is applied directly to the crystal. The EOVT employs the Pockels effect, rather than the Faraday effect of the MOCT.

The physical principals involved in modulating the light are very different. However, the result is similar - an amplitude–modulated light beam.

A proof-of-concept model, rated for 20 kV line-to-ground, has been built and is undergoing test. More work is needed in critical areas at this point. However, it is expected that a unit suitable for optical measurement of full transmission-system voltage will be available within two years. No capacitor stacks are employed as with the previous MOVT mentioned above.

TVA is planning to install a free-standing 500 kV MOCT system this fall. Meanwhile, another utility will be installing a 23 kV MOCT system on a cogeneration tie. Thus, the range of application is extending in both directions. At the same time, work is underway to extend the metering accuracy to 0.3 percent of reading at 0.01 pu.

Other utilities have ordered or installed MOCT systems like the ones described above (e.g., see [5]). Most of them operate at HV or EHV levels.

Interfacing Standards

Industry committees are looking at the possibilities for standard interfacing of MOCTs and other new transducers to the electronic systems - relays, meters, and others - which use the measurements.

There is discussion of the possibility of digital interface standards - the analog-to-digital conversion is performed once by the sensor, and sent as a data stream to using equipment. The data stream is well suited to optical transmission, and to sharing by a number of systems which have each been designed to conform precisely to the standard. Technical requirements for digitizing - speed and dynamic range - must serve the most demanding uses, and lesser applications will need to accept the full volume of resulting data. The economics of design for integrated relaying and control systems in substations will lead to such standards and practices, but at some future time which cannot be accurately predicted now.

Meanwhile, the low-level analog interfacing scheme of the present project has demonstrated its ability to drive a variety of existing proven relays, meters, and fault recorders; while meeting or exceeding technical performance requirements. The MOCT, and devices using its output, have been operating reliably in the substation environment. All-new digital designs, incompatible with conventional ct use, have not been required.

CONCLUSION

The magneto-optic current transducer has already been demonstrating its capabilities for accurate revenue metering in HV systems for several years. The present paper has described a project in which the range of application has been extended to protective relaying of transmission lines.

The benefits and limitations of the MOCT have been compared to those of the familiar iron-core ct. Experiences with the metering and relaying installations at TVA have been summarized. The design of the relaying system, with its arrangements for redundancy, testing, and data communications, has been presented.

Technical glitches, expected with any new approach so different from the standard practice of decades, have been surprisingly minimal, and have been remedied or have known solutions. The benefits are substantial—the MOCT solves serious ct problems—notably insulation failures, explosions, and saturation during faults. Performance specifications are impressive.

In spite of this, the widespread application of MOCTs is likely to be pushed by the cost benefits. While existing demonstration systems are hand-built, more efficient production is possible. The method is inherently simple. However, the ability to embed the sensors in other existing insulating structures offers a compelling promise of savings.

ACKNOWLEDGEMENT

The authors, and others associated with the work, are grateful to The Tennessee Valley Authority for its continuing financial and human support of the development. The series of projects is demonstrating benefits for the entire electric utility industry, and for the advance of basic technology. The authors also acknowledge the continuing accomplishments of scientists presently and formerly associated with the Westinghouse Science and Technology Center in the successful conception and development of optical transducer technology.

REFERENCES

1. J. C. Stites, C. L. Carter, "The Magneto-Optic Current Transducer", Georgia Tech Relay Conference, May 1986.

2. T. W. Cease, P. M. Johnston, "A Magneto-Optic Current Transducer", IEEE Trans. 89 SM 732-9 PWRD.

3. T. W. Cease, J. G. Driggans, S. J. Weikel, "Optical Voltage and Current Sensors Used In a Revenue Metering System", IEEE Trans. 91 WM 245-1 PWRD.

4. T. W. Cease, P. M. Johnston, E. A. Udren, "A Digital Distance Relay Using a Magneto-Optic Current Sensor as the Input", CIGRE Symposium on Digital Technology in Power Systems (1989).

5. T. D. Maffetone, T. M. McClelland, "345 kV Substation Optical Current Measurement System for Revenue Metering and Protective Relaying", IEEE Trans. 91 WM 171-9 PWRD.

6. J. R. Boyle, "TVA's Experience and Action Plans With Freestanding Oil-Filled Current Transformers," National Electric Reliability Council, Southeast Region, Operating Committee Meeting, Chattanooga, TN (1985).

7. E.A. Udren and H.J. Li, "Transmission Line Relaying Using Microprocessors", Georgia Tech. Relaying Conference, May 1989.

8. J.P. Garitty, "Data Collection and Control Techniques for Protective Relays", Georgia Tech. Relaying Conference, May 1990.

An Analytic Approach To The Application Of Current Transformers For Protective Relaying

S. E. Zocholl   W. C. Kotheimer

ABB Power Transmission Inc
Protective Relay Division
Allentown, Pa.

F. Y. Tajadoddi

Questar Engineering Inc.
Norwalk, California

Abstract

When applying current transformers for protective relaying the goal is to insure undistorted sine wave operation for the most severe faults encountered. Yet in nearly all cases involving cts in metal clad switchgear the maximum fault current guarantees ct saturation and a distorted secondary current. The dilemma for the relay engineer is in predicting the operation of the relays when subjected to a saturated waveform. This paper analyzes the effects of ct saturation on the performance of transformer differential and overcurrent relays in actual case histories. The applications are reviewed and performance of the relays are analyzed aided by computer simulation of the current transformer.

Introduction

Generally protective relays are designed for sine wave operation and their performance is not specified for other wave forms. Therefore, in a protective relay application, the rating and burden of the ct should be specified to insure undistorted secondary current for the maximum fault condition.

The steps in the application procedure which insure this condition are:

1. Current transformers in switchgear are mounted either on the primary disconnects in the breaker compartment, or on the bus work behind the breaker compartment. Determine the mounting location and choose the ct with the highest C-rating that will can be mounted in that location.

Reprinted with permission from *43rd Annual Georgia Tech Protective Relaying Conference-1989*, pp. 1–21, May 3–5, 1989.

2. Determine the X/R ratio of the faulted primary circuit which produces maximum fault current. Calculate the symmetrical voltage rating of the ct by dividing the C-rating by (1+X/R).

3. Calculate the allowable burden impedance by dividing the voltage calculated in step 2 by the secondary current produced by the maximum fault.

4. Calculate the maximum lead resistance by subtracting the relay impedance from the burden impedance calculated in step 3.

5. Determine the length of the secondary leads and choose a wire size to guarantee a resistance not exceeding the maximum lead resistance calculated in step 4.

In switchgear, limited space is allocated for mounting current transformers. Because ANSI standards [1] list low ratio, low accuracy cts to accommodate the space, maximum C-ratings are not supplied in standard assemblies. Specifying the larger C-rating required in step 1 incurs added cost. With undersized cts, the burden impedance determined by steps 2 and 3 are not achievable. In addition in differential relaying one set of cts are remote from the relay cabinet adding to the difficulty of satisfying step 5. Consequently, the conditions of the above procedure are rarely achieved in practice and ct saturation is inevitable.

The problem, then, is to minimize saturation and determine its effect. This paper discusses the use of computer simulation as a means to analyze performance in such applications.

Application of a Transformer Differential Relay

Figure 1 shows a 4160 volt power plant auxiliary bus supplied by a 5000 kva delta-wye resistance grounded transformer. The transformer has a 4.95% impedance on a 5 mva base and is fed by a standby and an emergency bus.

The bus is protected by phase and ground directional overcurrent relays and the transformer by percentage differential relays with harmonic restraint and a ground overcurrent relay fed by a ct in the transformer neutral.

A rating C-200, 1200/5 was selected for the low side cts and a C-50, 600/5 rating for the high side cts. The taps for the 87T relay were set at 2.9 amps for the high side winding and at 8.7 amps for the low side winding leaving a ratio mismatch of 4.4%. The percentage differential for tripping was set at 25%.

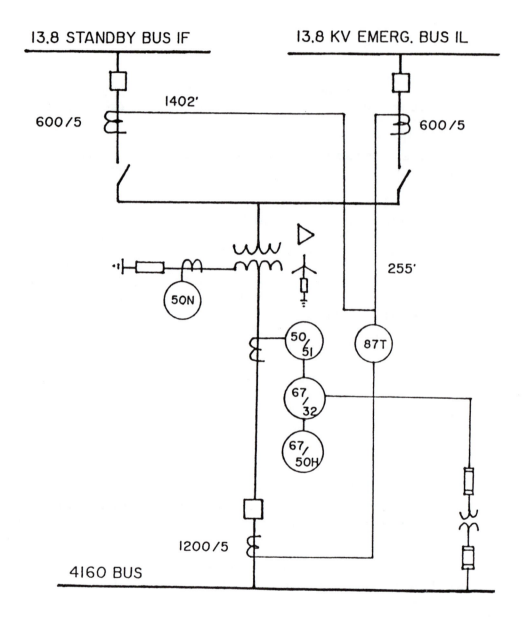

Figure 1. Power Plant Auxiliary

The complication in this application is in the fact that the 600/5 cts are located 1400 feet from the switchgear requiring at least 2800 feet of leads. Only 25 feet of leads were required for the 1200/5 cts mounted in the switchgear with the relay. The long run was installed with #10 gauge wire which has a calculated resistance of 6.72 ohms including the totally resistive relay burden. The 1200/5 ct was installed with a calculated total burden of 0.236 ohms. The maximum through fault current was calculated as 12,312 amps on the 4160 bus (3711 amps at 13.8 kv) with an X/R ratio of 11.

RMS Calculation

All the data is available to check this application using the above procedure with the following results:

1200/5 ct

```
Step 1. Rating = C-200
Step 2  Vrms    = 200/(1+X/R)= 200/12     = 16.67  volts

Step 3  Maximum secondary I = 12,312/240 = 51.3   amps.
        Allowable burden  Z = 16.67/51.3 =   .325 ohms

Step 4  Allowable Lead R   = .325 - .04 =   .285 ohms
Step 5     Actual lead R               =   .236 ohms  OK
```

600/5 ct

```
Step 1 Rating = C-50
Step 2 Vrms    = 50/(1+X/R)= 50/12       =  4.17 volts

Step 3 Maximum secondary I = 3711/120  =30.9   amps
       Allowable burden  Z = 4.17/30.9 =  .135 ohms

Step 4 Allowable lead R    = .135 - .04 =  .095 ohms
Step 5     Actual lead R               =  6.7  ohms  ??
```

The calculations show the 1200/5 to be sized to produce undistorted secondary current for assymetrical faults. However, the procedure shows the 600/5 ct at C-50 comes nowhere near to meeting the criteria. With the a burden of 6.72 ohms, 50 volts, the threshold of saturation occurs at 7.44 amps or 893 primary amps.

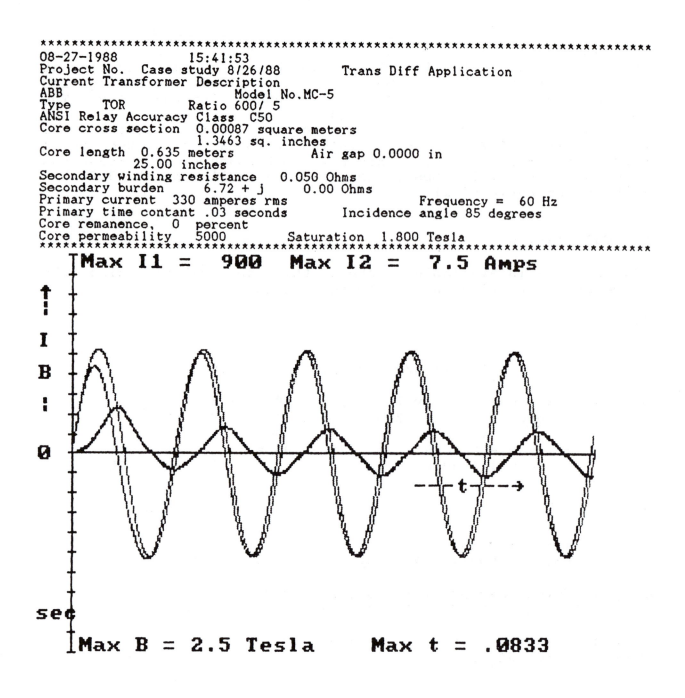

```
********************************************************************
08-27-1988        15:41:53
Project No.  Case study 8/26/88        Trans Diff Application
Current Transformer Description
ABB                            Model No.MC-5
Type    TOR          Ratio 600/ 5
ANSI Relay Accuracy Class  C50
Core cross section  0.00087 square meters
                    1.3463 sq. inches
Core length  0.635 meters          Air gap 0.0000 in
             25.00 inches
Secondary winding resistance   0.050 Ohms
Secondary burden      6.72 + j      0.00 Ohms
Primary current  330 amperes rms               Frequency =  60 Hz
Primary time contant .03 seconds     Incidence angle 85 degrees
Core remanence,  0  percent
Core permeability  5000          Saturation  1.800 Tesla
********************************************************************
```

Max I1 =   900   Max I2 =   7.5 Amps

Max B = 2.5 Tesla   Max t = .0833

Figure 2. Computer Simulation showing primary, secondary
           and magnetizing current. C-50 600/5 ct, 330 amps.

472

Postmortem by Computer Simulation

The installation of a C-50 rating, 600/5 ct resulted in nuisance tripping of the 87T relay during the starting of a 1500 hp motor. It is clear from the rms calculations that the cts are under sized for the 3711 amp. through fault condition. However, since the ct can support 893 amps. what caused the trip for currents of the order of 400 amperes?

The answer is seen in Figure 2 which is a computer simulation of the ct performance for symmetrical current of 330 amps. with the 6.72 ohm burden. The simulation is the time plot of the primary, and secondary current of the ct scaled to give the same plot magnitude. The scaling allows the plot of the difference current also shown in Figure 2.

The simulation program used in this and subsequent example is given as reference [2] and is explained in detail in reference [3]. The program allows the user to input the follow data which is then printed above the plot of each case as shown Figure 2:

        Manufacturer and Model
        Type of Current Transformer (Toroid or wound)
        Initial Core permeability
        Saturation Flux Density
        Mean Length of Core
        C-rating, Core Area, or Knee Point Voltage
        Burden Impedance
        Primary Current
        Primary Time Constant
        Current Incident Angle

In Figure 2 the primary current plot can be taken as the secondary current in the low side winding of the transformer differential relay. This is valid since the 1200/5 ct was shown to be ideally sized and introduces no distortion. The current in the high side winding of the relay is the 600/5 ct secondary current shown. The difference or operating current is then seen to be the magnetizing current of the undersized ct which exceeds the differential setting and causes the trip condition. The example shows that the simulation serves the application engineer as would an oscillogram of an actual full scale test.

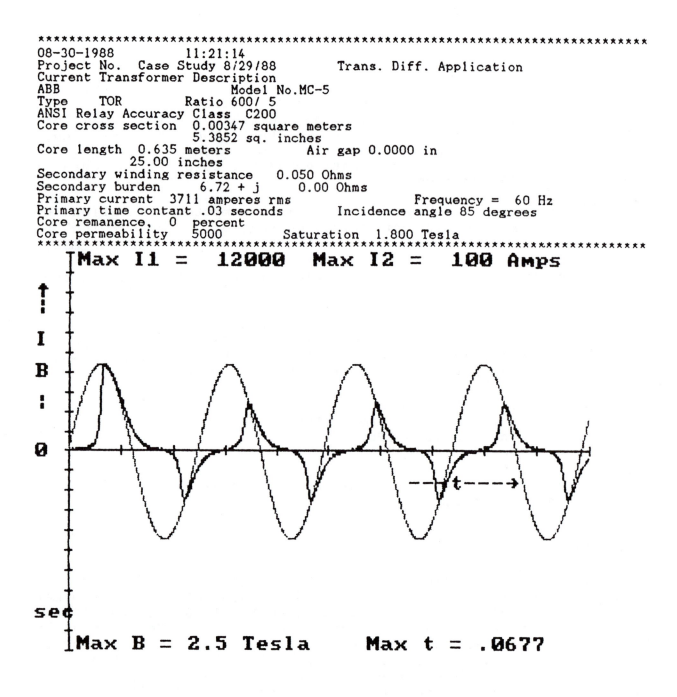

```
*****************************************************************************
08-30-1988        11:21:14
Project No.  Case Study 8/29/88        Trans. Diff. Application
Current Transformer Description
ABB                      Model No.MC-5
Type     TOR          Ratio 600/ 5
ANSI Relay Accuracy Class  C200
Core cross section  0.00347 square meters
                    5.3852 sq. inches
Core length  0.635 meters          Air gap 0.0000 in
             25.00 inches
Secondary winding resistance    0.050 Ohms
Secondary burden        6.72 + j      0.00 Ohms
Primary current  3711 amperes rms            Frequency =  60 Hz
Primary time contant .03 seconds     Incidence angle 85 degrees
Core remanence,  0  percent
Core permeability  5000          Saturation  1.800 Tesla
*****************************************************************************
```

Max I1 =   12000   Max I2 =   100 Amps

I B : 0

sec

Max B = 2.5 Tesla   Max t = .0677

Figure 3. Computer Simulation showing the primary and
         magnetizing current. C-200 600/5 ct, 3711 amps.

474

Analyzing Field Modifications

The field engineers were faced with replacing the high side ct and possible modification of 1400 ft. of leads to lower the burden. It was determined that a C-200 rating at 600/5 was the largest ct that could be mounted in place of the original C-50 rating. Calculations show that the 30.9 secondary amps produced by the maximum fault causes 207 secondary volts across the 6.72 ohm burden, barely matching the C-rating.

Is the change sufficient for secure operation? Under the circumstances is the cost of additional wiring justified? These difficult questions can be answered with the aid of the ct simulation.

Figure 4 is the computer simulation of the C-200, 600/5 ct with the maximum assymetrical fault. The plot shows only the primary current and the magnetizing current plotted to the same scale. As before, the primary current can be taken as the undistorted low side secondary current and the magnetizing current is the difference or operating current for the differential relay.

In this case, the magnetizing current during saturation caused by the asymmetrical current is of the same wave shape and is due to the same phenomenon as the inrush current of the power transformer itself. Consequently, the second harmonic restraint will prevent tripping during the offset.

Figure 3, however, shows a large steady state magnetizing current for the case for a maximum symmetrical fault. In this case it may be possible to prevent tripping by a maximum (40%) differential setting but the call is close.

Figure 5 shows the simulation when a second set of leads are paralleled to half the burden impedance. This case shows a dramatically reduced magnetizing current and increased security for the maximum through fault condition. In this case saturation, as indicated by the C-rating, occurs at twice the available fault current.

CT Application For Overcurrent Relaying

ANSI standards allow the use of low ratio current transfomers in switchgear. Normally a range of fault currents of up to 20 times ct rating would be expected. However, on power plant auxiliaries, overcurrent relays have been applied for motor protection using

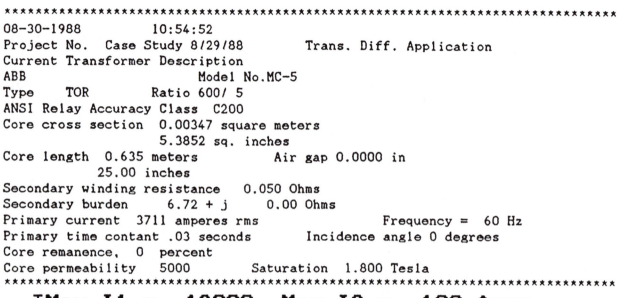

```
********************************************************************
08-30-1988          10:54:52
Project No.  Case Study 8/29/88      Trans. Diff. Application
Current Transformer Description
ABB                    Model No.MC-5
Type    TOR        Ratio 600/ 5
ANSI Relay Accuracy Class  C200
Core cross section  0.00347 square meters
                    5.3852 sq. inches
Core length  0.635 meters        Air gap 0.0000 in
          25.00 inches
Secondary winding resistance   0.050 Ohms
Secondary burden      6.72 + j      0.00 Ohms
Primary current  3711 amperes rms           Frequency =  60 Hz
Primary time contant .03 seconds      Incidence angle 0 degrees
Core remanence,  0  percent
Core permeability  5000        Saturation  1.800 Tesla
********************************************************************
```

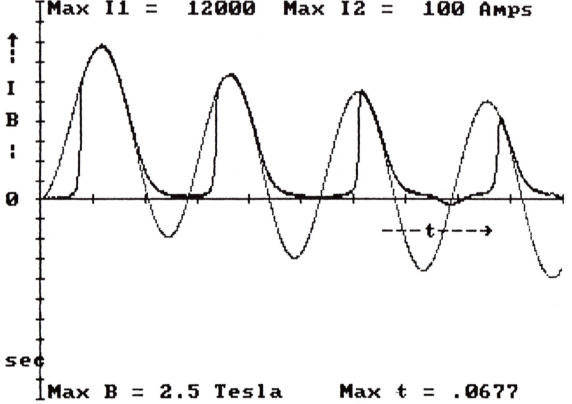

Max I1 = 12000   Max I2 = 100 Amps

Max B = 2.5 Tesla   Max t = .0677

Figure 4. Computer Simulation showing the primary and
          magnetizing current. C-200 600/5 ct, 3711 amps.

low ratio cts where the fault current can be 200 to 500 times the ct rating. The extremely high primary current causes severe saturation in the current transformer and produces a secondary current as shown in Figure 7. Figure 7 shows a 37000 amp symmetrical primary current and the resulting secondary current produced by a C70, 75/5 ct subjected to this extremely high value of fault current. As the figure shows, the secondary current can be described as a large spike of current lastings less than 4 ms each half cycle.

Full scale high current tests have identified performance anomalies in various types of overcurrent relays subjected to spiked waveforms of extremely short duration. Eddy currents, armature inertia and saturation affect the response of plunger type instantaneous elements where the wave form of less than 6 milliseconds duration. Consequently, a working knowledge of the effects of extreme saturation is needed in order to avoid inadvertent applications of this type.

Air Core Coupling at Extremely High Current

Normally, a solid state overcurrent relay would be expected to peak detect the short duration high current spike and produce an instantaneous trip. However, the small sine wave secondary current caused by the air core coupling at extremely high current can cause a relay using a saturating input transformer to produce a signal corresponding to the small current rather than to the high magnitude portion of the waveform. When this occurs, the relay produces a signal which may be less than the instantaneous pickup setting. In this case a time delayed trip rather than the expected instantaneous trip occurs and coordination is lost. The conditions producing this phenomenon are analyzed below.

In self powered relays, saturating input transformers are used as a means to present a low burden, to limit signal to prevent circuit stress and also to reduce the size of the relay input transformers.

Saturation of the input transformer occurs just above one perunit of tap current and the last linear peak is produced for a current approximately two perunit of tap setting. The signal across the relay burden resistor on the output of the saturating input transformer is shown in Figure 6. As shown in Figure 6, the relay operates on the peak signal determined by the point on the current waveform at which saturation occurs. Since the relay uses a resistive burden, the voltage and the secondary current of the input transformer have the same waveform.

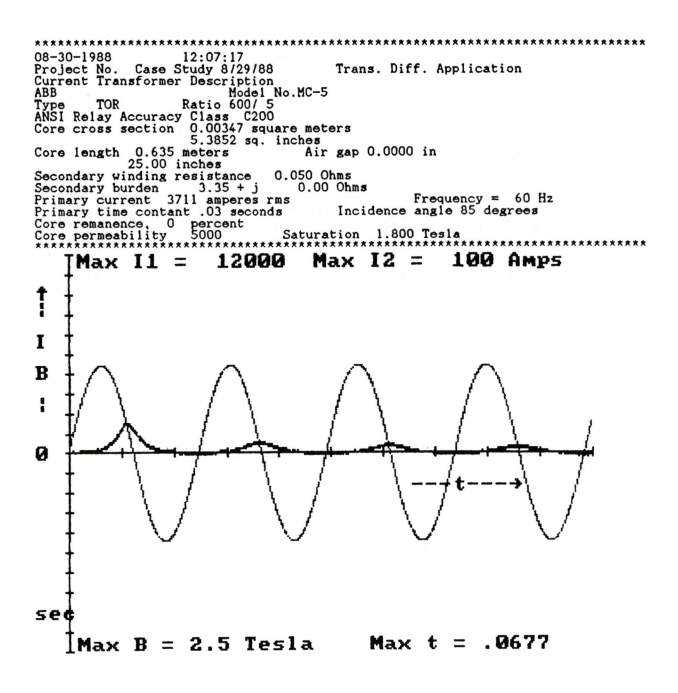

```
*****************************************************************************
08-30-1988          12:07:17
Project No.  Case Study 8/29/88      Trans. Diff. Application
Current Transformer Description
ABB                      Model No.MC-5
Type    TOR        Ratio 600/ 5
ANSI Relay Accuracy Class  C200
Core cross section  0.00347 square meters
                    5.3852 sq. inches
Core length 0.635 meters          Air gap 0.0000 in
            25.00 inches
Secondary winding resistance   0.050 Ohms
Secondary burden      3.35 + j      0.00 Ohms
Primary current  3711 amperes rms           Frequency =  60 Hz
Primary time contant .03 seconds     Incidence angle 85 degrees
Core remanence,  0  percent
Core permeability   5000        Saturation  1.800 Tesla
*****************************************************************************
```

Figure 5. Computer Simulation showing the primary and
          magnetizing current. C-200 600/5 ct, 3711 amps.
          with reduced burden.

478

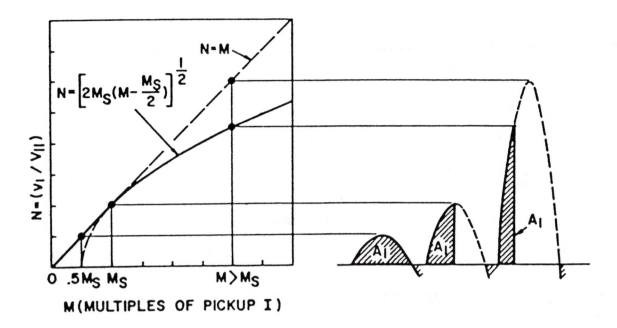

Figure 6   Resistive Saturation Peak Volts and Wave Forms

The point of saturation and the peak signal for any perunit current is can be calculated using the equation shown in Figure 6 and given below:

$$N=[2M_S(M-M_S/2)]^{0.5} \qquad (1)$$

where   N is the peak output signal
        M is the input in perunit of relay tap setting
        M_S is the last linear peak value and is 2.299

Low Ratio CT Application

Figure 9 shows the same waveforms as in Figure 7 with the scale amplified 100 times. The secondary current due to the air core coupling of the 15 turns of the ct are seen as a sine wave of 18 amps peak superimposed on the waveform due to the saturation of the iron core.

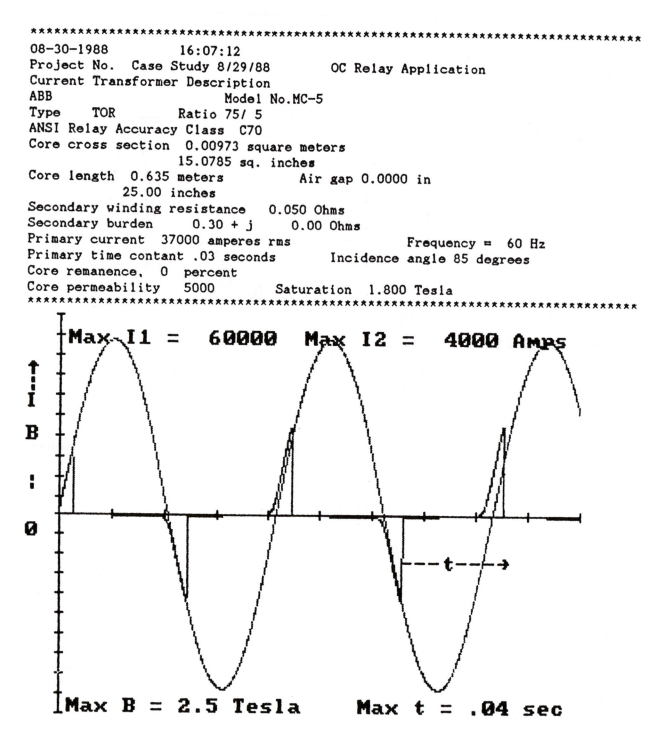

```
************************************************************************
08-30-1988           16:07:12
Project No.  Case Study 8/29/88       OC Relay Application
Current Transformer Description
ABB                      Model No.MC-5
Type    TOR         Ratio 75/ 5
ANSI Relay Accuracy Class  C70
Core cross section  0.00973 square meters
                    15.0785 sq. inches
Core length  0.635 meters          Air gap 0.0000 in
              25.00 inches
Secondary winding resistance   0.050 Ohms
Secondary burden      0.30 + j      0.00 Ohms
Primary current  37000 amperes rms          Frequency =  60 Hz
Primary time contant .03 seconds     Incidence angle 85 degrees
Core remanence,  0  percent
Core permeability  5000         Saturation  1.800 Tesla
************************************************************************
```

Max I1 =   60000   Max I2 =   4000 Amps

I

B

0

t

Max B = 2.5 Tesla   Max t = .04 sec

Figure 7. Computer Simulation showing primary and
          secondary current. C-70 75/5 ct, 37000 amps.

If the overcurrent relay subjected to this waveform has a five amp tap setting its input transformer will saturate and produce a peak calculated using equation (1) as follows:

$$I = 18 \text{ amps (peak of the aircore waveform)}/1.414 = 12.72 \text{ amps}$$

$$M = 12.72 \text{ amps}/ 5 \text{ amp tap} = 2.54 \text{ perunit}$$

$$N = [2(2.299)(2.54 - 2.299/2)]^{0.5} = 2.53 \text{ perunit}$$

Saturation of the relay input transformer terminates its secondary current at:

$$\emptyset = \arcsin(N/M) = \arcsin(2.53/2.54) = 84.4 \text{ degrees}$$

Because the instantaneous characteristic has a small RC delay, at least two pulses of the wave form in Figure 9 must be measured to produce the trip signal. As can be seen in Figure 9, the relay measures the initial peak properly. However, after the initial peak the relay measures a point 84.4 degrees from a zero crossing which occurs on the small aircore sine wave and the high current peaks are missed.

Where Figure 7 shows a symmetrical fault current, Figure 8 shows the 37,000 amp fault current with dc offset and Figure 10 shows the offset case with the scale amplified 100 times. It is important to note in Figure 10 that the dc offset causes the high magnitude portion of the waveform to occur closer to the zero crossing on the positive going wave. In this case, the saturation of the relay input transformer would occur at a point of high magnitude and total saturation on the small sine wave is less likely.

Solutions

Although there is only a small probability that this extremely high current phenomenon could occur, the message is that as a matter of practice one set of cts or relays cannot be expected to cover such an extreme range of fault current. The following remedies are suggested:

1.   Add a relay operated by the same cts that has at least a 12 amp tap setting. This relay will not saturate in the small sine wave portion of the waveform caused by air core coupling and consequently will measure a point of high magnitude on the major portion of the secondary current for the case above.

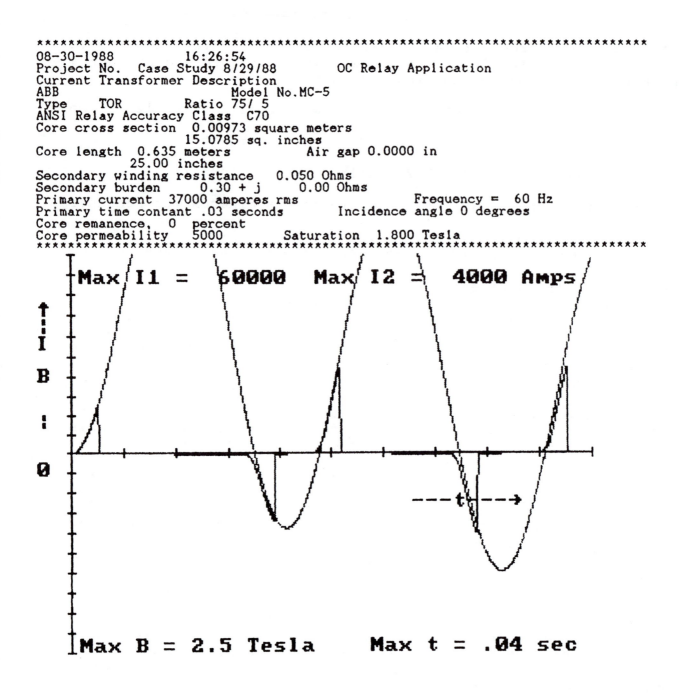

```
*****************************************************************
08-30-1988          16:26:54
Project No.  Case Study 8/29/88       OC Relay Application
Current Transformer Description
ABB                      Model No.MC-5
Type    TOR        Ratio 75/ 5
ANSI Relay Accuracy Class  C70
Core cross section  0.00973 square meters
                15.0785 sq. inches
Core length  0.635 meters          Air gap 0.0000 in
                25.00 inches
Secondary winding resistance   0.050 Ohms
Secondary burden       0.30 + j     0.00 Ohms
Primary current  37000 amperes rms            Frequency =  60 Hz
Primary time contant .03 seconds      Incidence angle 0 degrees
Core remanence,  0   percent
Core permeability  5000          Saturation  1.800 Tesla
*****************************************************************
```

Max I1 =   60000   Max I2 =   4000 Amps

I

B

0

--- t --->

Max B = 2.5 Tesla   Max t = .04 sec

Figure 8. Computer Simulation showing primary and secondary
          current. C-70 75/5 ct, 37000 amps. asymmetrical

2.   Add 0.1 ohm reactance in series with the relays to store the energy of the pulse and deliver it over a longer time interval as shown in Figure 11. Figure 12 shows the same waveforms magnified 100 times as in Figure 9. It can be seen that the relay would measure a high magnitude at its point of saturation on this waveform.

3. Solutions 1 and 2 may be considered as expedients for modifying existing installations. Adding a set of high ratio cts and relays to cover the extreme faults should be considered in this type of application.

4. The operation principle of specific overcurrent relays [3] insure reliable tripping over this range of current. Comprehensive full scale high current test data for such relays should be available to certify operation.

Conclusions:

Applications common to power plant auxiliary systems and similar industrial and cogeneration projects have been reviewed. In these cases involving 4.16-13.8 kv switchgear, maximum fault conditions produced ct saturation which affected protective relay performance. The cases described in this paper indicate that the method used in selecting cts for transformer differential protection as given in most guides and relay literature is not adequate to produce proper relay behavior without determining the transient characteristic of the ct in question. Cts should be chosen to accomodate the transient flux swings without saturation to assure proper relay behavior under the worst fault condition.

It should be emphasized that the space provided for cts in medium voltage switchgear, especially at 13.8 kv, is inadequate for the purpose. These situations have been encountered on many occasions with undesirable results. It should also be emphasized that overcurrent relays have a wide but limited range of operation. Consequently, overcurrent relays should not be applied to clear fault currents hundreds of times the ct rating in magnitude without adequate analysis and supporting test data to prove operation.

In the analysis of such cases, computer simulation provides an effective aid in determining performance of a relay under any fault condition allowed by switchgear standards. It is especially useful in relay performance analysis with severe saturation caused by extremely high fault current.

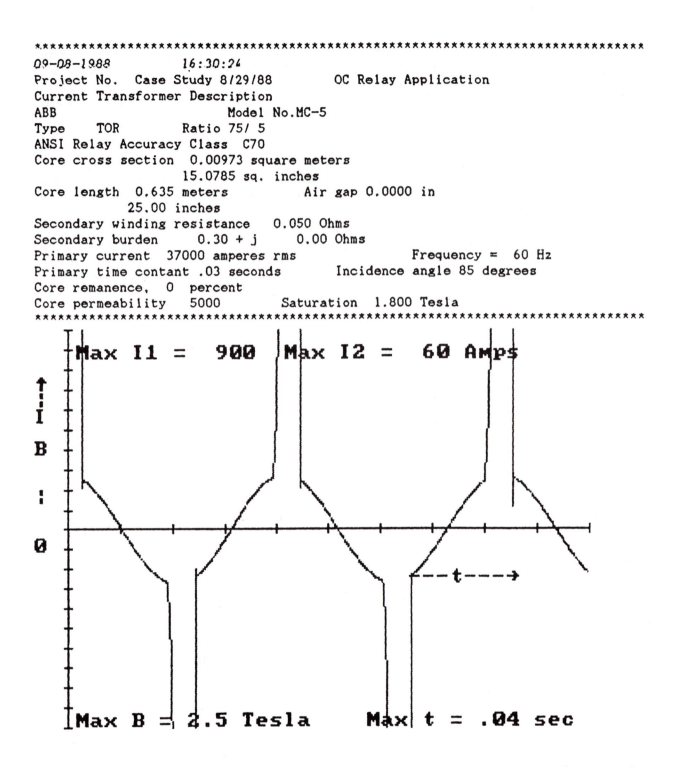

```
*****************************************************************
09-08-1988          16:30:24
Project No.  Case Study 8/29/88        OC Relay Application
Current Transformer Description
ABB                     Model No.MC-5
Type    TOR        Ratio 75/ 5
ANSI Relay Accuracy Class  C70
Core cross section  0.00973 square meters
                15.0785 sq. inches
Core length  0.635 meters        Air gap 0.0000 in
            25.00 inches
Secondary winding resistance   0.050 Ohms
Secondary burden      0.30 + j      0.00 Ohms
Primary current  37000 amperes rms        Frequency =  60 Hz
Primary time contant .03 seconds      Incidence angle 85 degrees
Core remanence,  0  percent
Core permeability   5000        Saturation  1.800 Tesla
*****************************************************************
```

Max I1 =   900   Max I2 =   60 Amps

↑
┆
I

B

┆

0

--- t ---->

Max B = 2.5 Tesla   Max t = .04 sec

Figure 9. Computer Simulation showing primary and secondary
         current. C-70 75/5 ct, 37000 amps. showing aircore
         coupling.

484

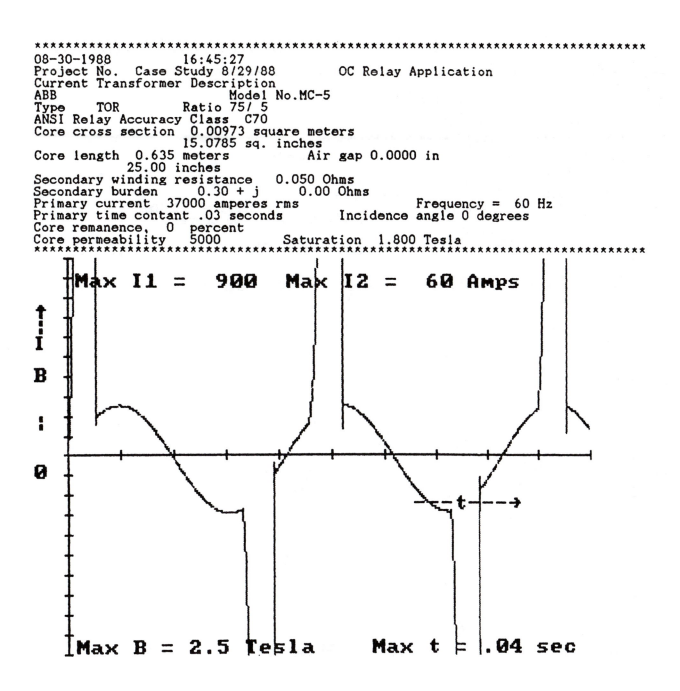

```
*********************************************************************
08-30-1988          16:45:27
Project No.  Case Study 8/29/88        OC Relay Application
Current Transformer Description
ABB                        Model No.MC-5
Type    TOR          Ratio 75/ 5
ANSI Relay Accuracy Class  C70
Core cross section  0.00973 square meters
                    15.0785 sq. inches
Core length  0.635 meters          Air gap 0.0000 in
             25.00 inches
Secondary winding resistance   0.050 Ohms
Secondary burden      0.30 + j     0.00 Ohms
Primary current  37000 amperes rms           Frequency =  60 Hz
Primary time contant .03 seconds      Incidence angle 0 degrees
Core remanence,  0  percent
Core permeability   5000          Saturation  1.800 Tesla
*********************************************************************
```

Max I1 =   900   Max I2 =   60 Amps

I

B

0

← - t - - →

Max B = 2.5 Tesla   Max t = .04 sec

Figure 10. Computer Simulation showing primary and secondary
           current. C-70 75/5 ct, 37000 amps. showing aircore
           coupling, asymmetrical case.

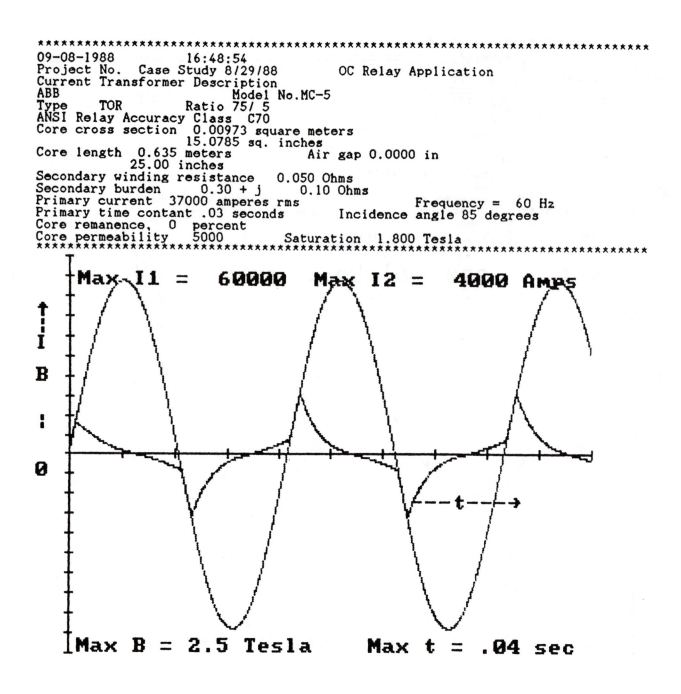

```
*******************************************************************
09-08-1988          16:48:54
Project No.  Case Study 8/29/88          OC Relay Application
Current Transformer Description
ABB                         Model No.MC-5
Type    TOR          Ratio 75/ 5
ANSI Relay Accuracy Class  C70
Core cross section  0.00973 square meters
                    15.0785 sq. inches
Core length  0.635 meters          Air gap 0.0000 in
             25.00 inches
Secondary winding resistance   0.050 Ohms
Secondary burden      0.30 + j      0.10 Ohms
Primary current  37000 amperes rms          Frequency =  60 Hz
Primary time contant .03 seconds     Incidence angle 85 degrees
Core remanence,  0  percent
Core permeability   5000          Saturation  1.800 Tesla
*******************************************************************
```

Figure 11. Computer Simulation showing primary and secondary
current. C-70 75/5 ct, 37000 amps. Inductance added
to burden.

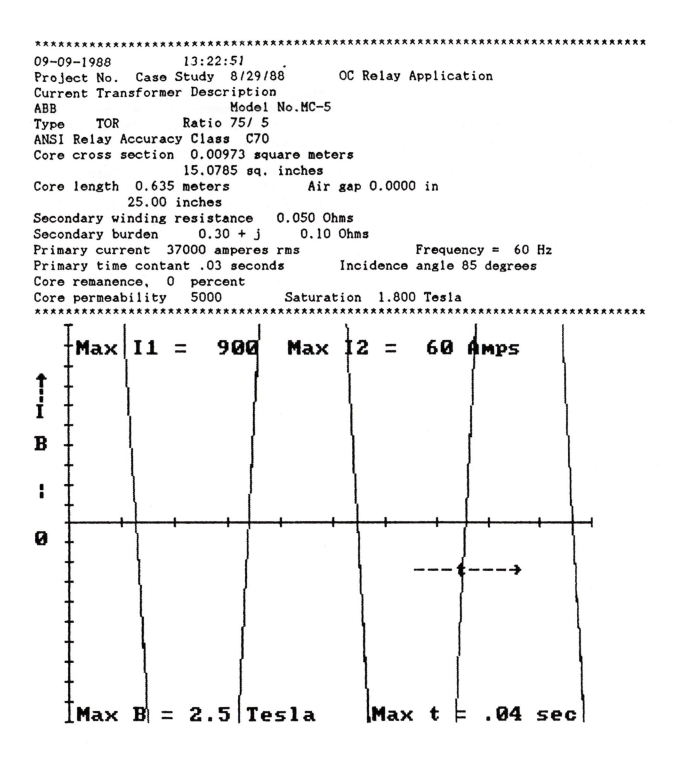

```
********************************************************************************
09-09-1988          13:22:51      .
Project No.  Case Study  8/29/88        OC Relay Application
Current Transformer Description
ABB                      Model No.MC-5
Type    TOR        Ratio 75/ 5
ANSI Relay Accuracy Class  C70
Core cross section  0.00973 square meters
                    15.0785 sq. inches
Core length  0.635 meters          Air gap 0.0000 in
            25.00 inches
Secondary winding resistance    0.050 Ohms
Secondary burden      0.30 + j      0.10 Ohms
Primary current  37000 amperes rms              Frequency =  60 Hz
Primary time contant .03 seconds        Incidence angle 85 degrees
Core remanence,  0  percent
Core permeability   5000         Saturation  1.800 Tesla
********************************************************************************
```

Max I1 = 900   Max I2 = 60 Amps

I B

0

Max B = 2.5 Tesla   Max t = .04 sec

Figure 12. Computer Simulation showing secondary current C-70
            75/5 ct, 37000 amps. showing detail at zero crossings
            with inductance added to burden.

487

References

1.ANSI C37.20 IEEE Standard for Switchgear Assemblies.

2. W. C. Kotheimer, "SEETEE-B.BAS", A PC Compatible Current Transformer Simulation Program, ABB Allentown Pa.

3. R. M. Garrett, W. C. Kotheimer, S .E. Zocholl, "Computer Simulation of Current Transformers and Relays for Performance Analysis",14th annual Western Relay Conference, Spokane, WA, OCtober 1987

Part 6
Digital Relays

I. General Discussion

DIGITAL relays have "come of age." In the past two decades, most manufacturers of relays have started producing extensive lines of digital relays. These vary from simple stand-alone relays meant to replace aging analog or electromechanical relays, all the way to advanced integrated substation computer systems, which include relaying functions. As with most engineering endeavors, the driving force behind these developments has been the economic advantage of designing, building, and maintaining digital relays. At the same time, the success of early digital relays in field installations has earned them an acceptance and a loyalty on the part of power system engineers. After the long debates and discussions about the practicality of digital computer-based relays, it is indeed gratifying that the acceptance of these devices is no longer an issue, and the research and development activity can concentrate on the implementation of traditional and novel relaying ideas with digital relays.

In this part, we are concerned with the fundamental concepts of digital relays. The papers included here discuss theory and technology as they apply to the broad spectrum of computer-based protection. The implementation and applications of microprocessor-based devices to specific power apparatus is covered in related parts of this volume.

The most widespread use of digital relays has been as stand-alone devices, either as replacements for conventional relays or as special protection systems for unusual relaying problems. In the early days of this technology, high expectations were placed on the integrated substation computer system. It was thought that the full flowering of this technology would be realized when relays became part of a substation-wide integrated computer system in the following manner: sharing data and functions; providing redundancy through the computer network, and, in general, bringing order into the myriad metering, recording, relaying, and controlling functions in a substation. Although some notable systems of this type have been developed, it must be noted that integrated (or coordinated) substation computer systems have been slow to emerge. This may be partly due to the reduced construction activity in the power industry in general. Or, it may be due to the cautious attitude of relay engineers toward this new technology—it may be more appealing to wait until there is more evidence of the success of digital relays on an individual basis before the more complex integrated substation computer is accepted. In either case, a wider acceptance of integrated substation computer systems is certainly anticipated.

A highly beneficial outcome of computer-based relays is the ability to measure power system quantities with precision, and, further, to make these measurements available to several other substation functions. This aspect of computer relaying developments is discussed in Part 7: Monitoring.

In the field of digital relaying, new ways of looking at the relaying process have been proposed in the recent technical literature. For example, considerable literature has evolved from the subject of Kalman filtering for implementing relaying algorithms. Another concept has been to implement relays as statistical decision-theoretic devices. Knowledge-based systems (e.g., expert systems, artificial intelligence systems) have also been proposed as appropriate vehicles to frame the relaying algorithms. However, these ideas have not been fully accepted as yet, and far more convincing demonstrations must be furnished before they will be accepted as having a practical value. For now, the algorithms developed in the early years of computer relaying remain dominant in the field.

There is no question that a large proportion of the newer developments in the field of relaying utilize the computer as a relaying medium of choice. Engineers are beginning to realize the potential of this technology, and the possibilities it offers for innovative relaying practices. The coming years will show us what these devices can do.

II. Discussion of Reprints

The six papers on digital relaying presented here span a period of approximately 20 years. They represent a body of work that has influenced the digital relay developments thus far, as well as some novel theoretical ideas which are stimulating discussions among relay engineers, and which could influence relay designs of the future.

The first paper, "Computer Relaying: Its Impact on Improved Control and Operation of Power Systems," by Phadke, provides an overview of the field of computer relaying, and introduces two new concepts; synchronized measurements in power systems, and adaptive relaying. As computer relays become more common in field installations, their ability to make direct measurements of power system states becomes increasingly more important. This subject is dealt with in greater detail in Part 7. Adaptive relaying introduces the idea that as power system conditions change, relays may also change, and adapt to the changing power system conditions. These subjects are addressed more fully in this volume and future papers on relaying are anticipated.

The second paper, "The Use of Digital Computers for Network Protection," by Poncelet, published in 1972, provided, for the first time, an excellent account of the differential equation algorithm for transmission line protection. This paper has influenced a number of researchers, and its proposed techniques have become part of many practical line relay implementations. This paper is an important technical and historical reference.

The third paper, "A Digital Computer System for EHV Substations: Analysis and Field Tests," by Phadke, Hlibka, and Ibrahim, describes one of the early field installations of computer-based line protection systems by an electric utility. A complete protection system, consisting of step distance primary and backup (overcurrent) protection, reclosing functions, carrier interfaces, and breaker failure protection were all included. The paper was the first report of a completed system, with operational experience from a field installation. It should be noted that at present most commercially available line relays offer all these features.

III. Supplementary Bibliography

A. G. Phadke and J. S. Thorp, *Computer Relaying for Power Systems*, London: Research Studies Press, New York: Wiley, 1988.

T. Sakaguchi, "A Statistical Decision Theoretic Approach to Digital Relaying," *IEEE Trans. on Power Apparatus and Systems*, vol. PAS-99, no. 5, Sept./Oct. 1980, pp. 1918–1926.

A. A. Girgis and D. G. Hart, "Implementation of Kalman and Adaptive Kalman Filtering Algorithms for Digital Distance Protection on a Vector Signal Processor," *IEEE Trans. on Power Delivery*, vol. 4, no. 1, Jan. 1989, pp. 141–156.

Computer Relaying: Its Impact on Improved Control and Operation of Power Systems

Arun G. Phadke

Introduction

This article describes a development in the field of substation monitoring, protection, and control, which can have significant impact on all aspects of power system engineering. A hierarchical system of computers is being installed, which can provide direct access to all relays, control devices, and measuring systems throughout the power system. Although the field installations of these systems have been slow, the pace is now accelerating and we are poised at the threshold of a new era of sophisticated control of power systems. It is our intention to make this development known to people working in all branches of power system engineering.

The name "relaying" given to the entire field of protection is an unfortunate choice, but it is too late to change. It seems unfortunate because it conjures up an image of a solenoid pulling in a plunger when energized. A modern relay is nothing like a solenoid, although some relays may have plungers and solenoids and electromagnets in them. A modern relay is a very sophisticated control element: it uses the most modern circuitry to achieve extremely high speed control. In fact, "feedback control system" is a far more descriptive phrase for relays. More recently, the relays have become computer-based, and there is no longer any resemblance to the old electromagnetic devices. Of course, these devices do exist on present-day power systems—and in large numbers—but that is because they have been working so well; it is difficult to discard something that is working well. So we still have electromagnetic relays, but the times are changing. It is with computer relays that we are bringing in a revolution. My comment about the name "relaying" being unfortunate is of interest for another reason: as computer-based relays become active in the traditional area of power system control, we begin to see some institutional difficulties within the utilities as well as with manufacturers of power equipment. Who will be responsible for this new "relaying-control"? Organizational problems are appearing, and, often, these are far more difficult to resolve than the technical problems. It would have been simpler if relaying was always called

by its rightful name—"control"; none of the organizational problems would have been created. Of course, I am only half serious. This field is relaying, and relaying it will remain.

Relays are supposed to protect the power system from those conditions that we consider to be undesirable. These conditions may be harmful to the power apparatus—such as a line, a transformer, or a bus—but, more often, the potential is for harm to the power system. It is the system that may go unstable, split up, and cause a blackout. It is to prevent these happenings that the relays must act quickly and accurately. *And this, in fact, is the central dilemma of relaying: how to be quick and accurate at the same time.* If you can be slow, you can think things over and be more accurate. Quickness of action is an invitation to make mistakes. So, relays will make mistakes. They can err on the side of being too cautious: the relays will fail to trip when they should have—in the relaying jargon, this is failure of dependability. Or they can err on the other side: they will trip when they should not have—again, in the jargon, this is loss of security. Both errors are bad, but we have to accept both. The relay designer strikes a balance between the two errors, and depending upon what the protection engineer did with his balancing act, the system will be prone to overtripping, thus bringing the power system down—or it will fail to trip soon enough and may bring the system down anyway. It is a tough balancing act.

Considering the choices a relay engineer makes and how many relays exist on a power system, it is indeed reassuring that the relay systems work exceedingly well. We should also recognize that often, most of the major system disruptions have protection system malfunction as a contributing cause. If we examine the annual NERC (North American Electric Reliability Council) reports on system disturbance analysis, we will find that, on an average, in 80 percent of the cases analyzed, protection systems must shoulder the blame. Of course, the relay engineering community does not always agree with these conclusions, but one must admit that a very significant number of power system disturbances are caused by—or made worse by—an unwanted relay op-

Reprinted from *IEEE Computer Applications in Power*, pp. 5–10, Oct. 1988.

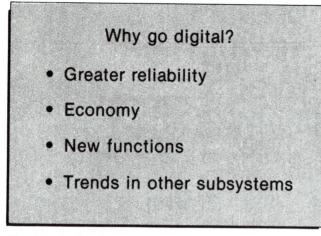

Fig. 1 Motivation for developing and using computer relays.

eration. In my view, this brings protection system design philosophy to center stage. Every power engineer should appreciate the significance of relaying decisions made on his or her system: it is the relays that determine what contingencies are feasible. It is the relays that determine whether the power system can recover from an outage. It is the protection system that is in the front line when we are fighting to contain the damage resulting from a major outage. And computer relaying has the potential to alter some of the key elements of the protection philosophy (see Fig. 1).

Relaying with Computers

The field of computer relaying is about 25 years old. The practical implementation of computer-relaying techniques has been slow for many reasons; the energy shock of the mid-'70s and the resulting slowdown in utility construction are probably the most important ones. Just now, we are beginning to hear of computer-relay installations taking place in the field. Although a vast number of traditional relays will continue to exist for the foreseeable future, the computer relays are here. We are going to see them in service in increasing numbers.

At the heart of a computer relay is one or several high-performance microcomputers. Sixteen-bit computers with 100- to 200-nsec instruction execution time, 1 Mbyte of random-access memory, an analog data-acquisition system, and a digital input-output system make up the essential elements of the relaying computer. The retail cost of this hardware today is about $5,000. To this must be added filters of various kinds for EMI (electromagnetic interference) shielding, anti-aliasing, and galvanic separation. The complete package can be put together for under $8,000 retail. This is not cheap, but it is only a fraction of the cost of a comparable analog relay on the market. To this basic cost must be added the software development costs, retraining of personnel, etc. The fact remains that a computer relay is cost-competitive with most of the existing analog relays, and this cost advantage is continually becoming more favorable to the computer relay as digital hardware prices continue their downward trend.

Reduced cost is an important (but not the only) reason for the move toward computer relays. Computer relays can monitor themselves and, when defective, can log a service alarm at a remote location. This self-diagnostic ability is what makes them so attractive to the relay engineer. Most relay misoperations occur not because the protection engineer designed a faulty protection system but because the relay was in a failed state, and no one knew about it. More frequent maintenance (even if it was economically possible) is not the solution to this problem. Indeed, many of the relay failures can be traced directly to some maintenance activity. Therefore, although we expect the computer relay to fail in service from time to time just as a conventional relay does, we expect to be informed about its failure immediately and repair it.

There are several other perceived advantages. The computer relays make possible electronic CTs (current transformers) and CVTs (capacitine voltage transformers). Several functions within the substation could be combined in a relay. Redundancy in protection functions can be built up in many novel ways. Relays can be made to respond to control decisions made on a systemwide basis.

This last is an issue of major philosophical change in the protection field: adaptive relaying. Should relays adapt to changing system conditions? Should they be under control of a remote center? Where is the responsibility? If relays are going to change as the system changes, what becomes of relay coordination? The relaying community is actively debating these and other issues dealing with computer relays. All power engineers should be aware of this debate—and take part in it.

Measurement with Relays

Let us consider the computer relay as an intelligent device sitting idle in the substation most of the time. It gets extremely busy in short bursts of time when a fault occurs nearby, but, otherwise, the relay is coasting, and this means for more than 99 percent of the time. What else can it do in its spare time?

One of the important tasks that a relay does all the time is the measurement of system currents and voltages. It does this very quickly (remember that a relay must act in just a few milliseconds—and accurately). It can, therefore, serve as a high-accuracy short time-constant measurement device for power system voltages and currents. Most relays extract the fundamental frequency quantities, and some relays measure symmetrical components. We, thus, have devices in the substation that can measure positive sequence current and voltage phasors and that can track changes taking place in the power system by tracking these measure-

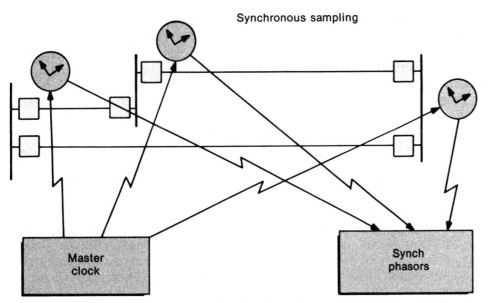

Synchronous sampling

Master clock

Synch phasors

Fig. 2 Synchronous sampling of currents and voltages in a substation with a master satellite clock.

ments. (The relays can also—and often do—measure negative and zero sequence quantities, as well as the harmonic content of waveforms.) In addition, a relay can measure the local power system frequency and the rate of change of frequency with high precision and with a minimum of time delay. From these basic measurements, a number of derived measurements could be obtained: real and reactive power, power factor, etc. Models of the power system can also be constructed in real time.

How accurate are these measurements? In addition to the errors introduced by the transducers (CTs and CVTs), we must now take into account the errors introduced by the electronics in the data-acquisition system of the relaying computer. It is well known that the relaying transducers are not as accurate as the measurement system transducers. But is this really so? If we examine this statement closely, we find that the relaying CTs have greater errors at fault current levels, and the error may be affected by the burden placed in the secondary circuit. If the burden is fixed, and we are interested in the accuracy at the load current levels, the answer may be quite different. In fact, we could calibrate the relaying transducer with its connected burden and correct any errors inside the relay computer through software by using the calibration curves. Potentially, the transducers could be made as accurate as needed. The thermal variations need to be checked in the field, and some experiments in this direction are in the planning stage at this time. The A/D converters are typically 12-bit devices. This is sufficiently accurate if the inputs are scaled for full-scale deflection at full load current. If the 12-bit converter must accommodate fault currents as well, the accuracy at load current magnitudes becomes quite poor. The solution is to use a dynamic-ranging A/D converter or to use two channels (of different gains) for each current input. The

load current channels will saturate when a fault occurs, but no measurements (except those obtained for relaying functions through the low-gain input channels) should be needed during a fault transient, anyway. It seems certain that with some care, the input system of a relay can be made to provide the same accuracy as that given by a measurement system.

If we consider that a substation may have several computer relays, the measurements made by all the computers could be used to obtain a more robust measurement set from the substation. Bad data can be detected within the substation and rejected. If several relays measure the same quantity—for example, all the relays in a substation measure the same voltage—accuracies greater than those of any single measurement could be achieved through simple averaging or through the more sophisticated weighted-least-squares technique.

When we talk about phasor measurements, we imply a common reference for the phasors. This is easy to achieve within a substation. All that is needed is the synchronization of all the sampling clocks used by the computer relays, as shown in Fig. 2. If we wish to put systemwide phasor measurements on a common reference, clock synchronization throughout the system is a necessity. Several experiments to achieve this are in progress at this time.

Uses of Measurements

What uses can one make of these measurements? Assuming that sampling clock synchronization has been achieved throughout the system, what we have in the positive sequence voltage measurements is the (static) state vector of the power system. If all the bus voltages were measured in this fashion, and the accuracy of the measurements was good, there would be no need for the state estimation programs in the control centers. We would, of course, make use of the positive sequence current measurements also, and, in fact, some state-estimation-like computation is necessary if the current measurements are used. If the positive sequence voltage and current phasors are denoted by the vector Z, then the state estimate is given by

$$G\hat{E} = BZ$$

where $\hat{E}$ is the state estimate, and G and B are constant matrices. The constant gain matrix G is specially welcome, as no repeat factorization of the gain matrix is

493

needed to solve this equation. Furthermore, the state estimation process is linear—i.e., no iterative solutions are needed. One pass of computation on one set of measurements and we are finished. Estimation becomes very simple. And to remind ourselves of what was said above, *no estimation is necessary if voltages are measured with sufficient accuracy and are known to be error-free.*

This last aspect is important not just because it eliminates the need to perform a state estimation calculation at the center; the advantage goes much deeper. The whole concept of data flow used in traditional state estimation systems (transmission of all the system data to the center, processing it there, and then getting the bus voltage vector at the center) is no longer necessary. The state vector is available where the measurement is made. It is not necessary to invest time and money in this substantial data collection process. If the state vector can be used at places other than at the system center, then the state vector can be sent there directly. If only a few elements of the state vector are needed to arrive at a new control strategy, you can get them directly from the measurement points. These may be only one bus away from the decision-making center. You can get the measurements quickly, and you can get them selectively. With this direct measurement technique, we position ourselves in an ideal position for designing control systems that need state vector feedback.

Even if the synchronization experiments currently under way prove successful, we are not proposing that we abandon the existing state estimation systems. Fortunately, the new measurements can be grafted onto an existing state estimator. Thus, we have an ideal method of expanding the coverage of existing state estimators into portions of the system where no measurement systems exist, and it is desirable to go in with incremental investments. We can add these measurements where we want them and when we want them. The complete measurement system need not be in place before *any* states can be obtained.

With these measurements, we can talk of dynamic state estimation. To estimate the dynamic state vector of the power system, we need measurements that exhibit the dynamics. A quick responding measurement of the bus voltage vector is just what makes this possible. A few papers dealing with this aspect of voltage vector measurement have been published in recent years.

We would like to discuss one aspect of these measurements that was mentioned before but was not emphasized sufficiently. The measurements made within the relays are (or should be) those of the positive sequence quantities. SCDFT (symmetrical component discrete Fourier transform) is used over a data window lasting one or more cycles. This measurement is exactly what is meant by the state vector of a power system. The system model used in state estimation is the one-line diagram, which is the positive sequence representation of the system. The currents and voltages that exist on this model are the positive sequence quantities, and those are what we are measuring with computer relays. In addition, we compute the fundamental frequency components of the input signals with the DFT (discrete Fourier transform). Estimates of phasors based upon this technique are far better than those that derive phase angle based upon zero-crossing instants of the waveforms. These instants are affected by the harmonic content, and if two waveforms have different harmonic contents, phase angle differences calculated with zero crossing of waveforms are in error. Analog filtering of the input data would be one way of getting around this difficulty, but analog filtering is not perfect, and, furthermore, it slows down the response time of the measuring system. A variable window DFT is the best available method of estimating the phasors.

Relaying and Control

Opening, closing, and reclosing of network switches and circuit breakers is one of the most used controls in a power system. As the relays control the circuit breakers, it is natural that such control takes place through the computer-relaying hierarchy. Reclosing is traditionally considered to be a relaying task, and we will defer our discussion of newer reclosing techniques to the following section "Adaptive Relaying." Other control systems for the power network consist of transformer taps, series and shunt capacitor switch-

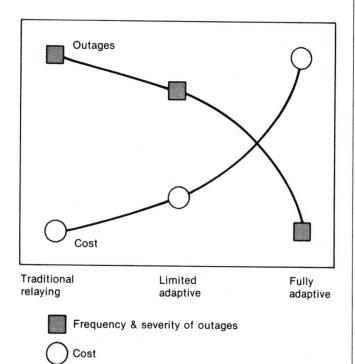

Fig. 3 Qualitative representation of costs and benefits of adaptive relaying. The phrase "limited adaptive" implies minimal local communication links to the relay. The term "fully adaptive" implies complete systemwide high-speed communication capability.

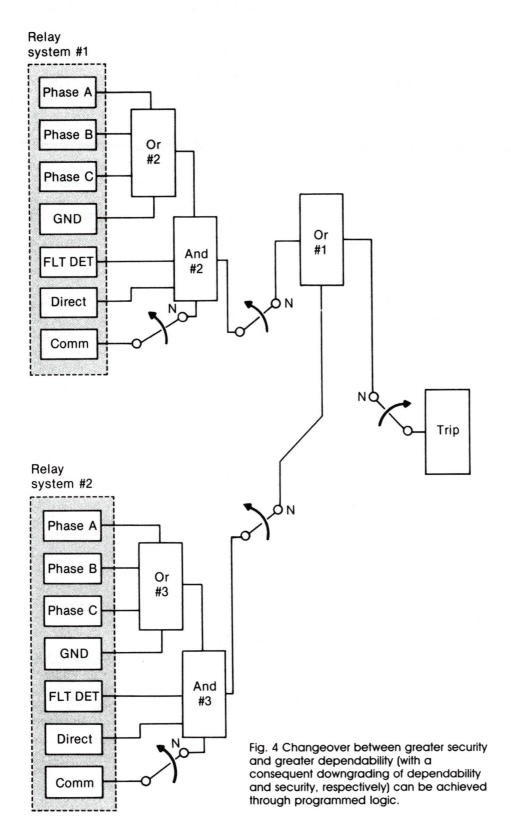

Relay
system #1

Phase A
Phase B
Phase C
GND
FLT DET
Direct
Comm

Or
#2

And
#2

Or
#1

Trip

Relay
system #2

Phase A
Phase B
Phase C
GND
FLT DET
Direct
Comm

Or
#3

And
#3

Fig. 4 Changeover between greater security
and greater dependability (with a
consequent downgrading of dependability
and security, respectively) can be achieved
through programmed logic.

ments for feedback. The only exception may be the voltage control at a remote bus through tap changing, through excitation control, or through static var control. The term *remote bus* usually implies a bus within the substation—perhaps on the high side of a transformer. Physically, the measurement can still be considered local. It is recognized that a control action at one place on the power system will change conditions throughout the system; certainly, the influence of the control will be felt in the neighborhood of the control element. It would be far more appropriate to use as the feedback signal all the quantities that are influenced by the control. Ideally, this demands a state vector feedback. Complete state vector feedback should not be necessary in most cases, but measurement of several voltage phasors at key locations would go a long way in improving the efficacy of the control system. Analytical studies of such control systems have begun, and some interesting results pertaining to state vector feedback optimal controllers, which handle system nonlinearities in a novel fashion, have been reported in the literature. Computer relays as measuring elements are expected to provide the feedback signal required in such control systems.

Adaptive Relaying

This is a new development in the relaying field, and, quite simply, it is relaying with state vector feedback. Several papers have been published on this subject, and this topic is being considered by working groups in IEEE and CIGRÉ. It is recognized that many—*not all*—features (settings) of relays are based upon

ing, static var control, and phase shifting transformer. The generating station controls consist of excitation control and turbine control, although these latter controls are somewhat too slow-acting in the present context.

Most of these control systems use local measure-

certain assumptions about the state of the power system. If the state of the system changes significantly, the setting is no longer the best that it could be, and, occasionally, it could be wrong. Figure 3 shows the relation of cost to outage minimization.

For example, the zone settings of multiterminal line relays are based upon the assumption that all terminal breakers are closed. If some of them should be open, the zone settings are usually too short to provide high-speed protection for the entire line. A transformer percentage differential relay is set based upon the assumption that its tap setting may change, and, consequently, the slope of the differential characteristic is set high enough to cover all possible tap positions. This makes the relay insensitive to low-grade fault currents. Reclosing relays are provided with time delays, interlocks, and several check synchronizing features, which are appropriate only if the system conditions happen to be what was assumed in devising the strategy. If the system reaches a state not anticipated during the reclosing logic design, undesirable reclosing performance would result.

There is an even more basic feature of a relay system design that could be responsive to changes in system state. Recall the earlier discussion about the dependability and security of the protection system. Dependability of a relay is the property by which a relay will perform as expected under all circumstances. Security is the property that it will not take any action for events for which it was not supposed to. A relay setting is a compromise between the requirements of dependability and security. Most relay systems are biased toward dependability; the perceived need is to get a fault off the system under all circumstances. Consequently, the bias is toward a loss of security; i.e., from time to time, there may be a number of false (unwanted) trips. This is an appropriate position to take when all the transmission and generation facilities are in service. Unnecessary loss of a line or two is not too serious. A fault remaining too long on the system because of a loss of dependability is not acceptable. However, when the power system is in a state where generation and transmission are loaded to the limit, the possibility of a false trip is far more threatening. One should now accept a certain loss of dependability in order to reduce the possibility of a false trip. This type of adaptive relaying offers many exciting opportunities, but it requires careful and long study before it can be implemented.

Indeed, all adaptive relaying must be studied carefully. What is the fallback position if the feedback signals are lost? In a coordinated protection system, what happens if one element changes in adapting to the prevailing power system state and the other does not? These and other issues are being discussed by relay engineers now and will continue to be discussed in the coming years. All power system engineers should be aware of this development and should take part in this discussion. Figure 4 shows trade-offs between dependability and security.

Future Prospects

Computer relays are a fact of life. All manufacturers of power system relays have production lines with computers playing a vital role in the relaying tasks. Computers have made possible the entry of small (and new) manufacturers in this field. Computer relays are being designed with extensive communication capabilities that are a prerequisite for the kinds of things we have been talking about. Synchronization of clocks in all the substations already has been done in most systems for sequence-of-event analysis. What we have proposed here is that this synchronization be made much tighter—within 1 μs of each other throughout the power system. This is technically feasible today, and both the utilities and the manufacturers should make room for such a capability in the stations and in the relaying hardware. Apart from this, we must await the accumulation of substantial field experience with computer relays. This cannot be—and should not be—bypassed, but a resumption of load growth and the attendant construction projects by the utilities will accelerate this process.

For Further Reading

Publications on this subject have appeared in the *Transactions of the Power Engineering Society*, Proceedings of the CIGRÉ general assemblies, CIGRÉ colloquia, and journals of many other engineering societies. IEEE has two tutorial publications, one from 1979 and the other from 1988, that contain very extensive lists of references.

THE USE OF DIGITAL COMPUTERS FOR NETWORK PROTECTION

by

R. PONCELET

Director of Studies, Université Libre de Bruxelles

(Belgium)

SUMMARY

The characteristics and features of process calculators are briefly reviewed. The sequence pattern of the programmes necessitates a reappraisal of the conception of network protection and in particular protection of a substation. This report sets out the formulation of protection programmes for a complete substation and the associated lines with a single computer. Distance protection is dealt with in particular. The method proposed employs techniques for the identification of parameters evaluating resistance and inductance from the outgoing point of a line with the occurrence of a fault. Modelling on a scientific computer employing simulated data or data obtained from a network model shows that it is possible to obtain correct R and L values in less than 10 msec for a three-phase line. Present day process computer performances are inadequate for handling all the programmes but, on the one hand the performances are continually improving and, on the other hand the principles can be used for setting up independent relays.

Digital computer, Distance, High speed protection, Line, Mathematical modelling, Substation, Transient.

REPORT

1. INTRODUCTION

Digital process computers are nowadays finding application in substations, substituting a single piece of equipment for independently wired analog and logical circuits for functions such as telemeasurement, maintaining a log, circuit-breaker reclosure and checking any failure etc. This process is helped by the lower cost of the equipment (hardware) due to its non-specialized nature, to some extent balanced by the cost of drawing up programmes (software).

Up to the present use has been made of independent equipment in order to ensure for each unit of a substation the protective functions : detection and location of faults and circuit-breaker control. Continuous network development has necessitated a higher operating speed and increased selectivity of protection with a consequent increase in complexity and cost of the relays. This investment, although essential, is only little used, since fortunately the protection functions only on rare occasions. This again always leaves a doubt as regards future sound functioning of a system which is continuously idle. In contrast to this, the majority of the circuits of a digital computer are continuously in service and the periodic test routines bring to light any faulty circuit which is quickly replaced.

The construction of current transformers of the non-magnetic type necessary at extra high voltages is fraught with difficulties due to the high-power heavy-current supplies in the input circuits of conventional relays, even when transistorised. A more elegant solution is to provide the transmission of information in digital form.

It is logical therefore to examine the possibilities of computers for protection since this function occupies very little time and, as indicated above, requires only a marginal increase in outlay.

The sequential pattern of programmes for a computer necessitates reconsideration of the conception of network protection and, in particular, substation protection. This report sets

Reprinted with permission from *CIGRE, Paper No. 32-08*, pp. 1–15, Aug. 28–Sept. 6, 1972.

out the formulation of protection programmes for a substation and associated lines with a single computer. The method is not fundamentally different from that proposed by Rockefeller in his basic article [1]. A practical and simple solution has been aimed at.

Line protection, provided by distance measurement, has been particularly developed. Rockefeller proposed simple transposition of analog techniques. Mann and Morrison have proposed a number of methods. The first, "table look-up" is not very accurate [2]. The second, "peak determination" assumes that the aperiodic component is eliminated and requires evaluation of the current derivative [3]. These authors have in mind immediate application with an existing computer. Slemon et al. apply Fourrier analysis techniques to extract the fundamental current and voltage waves from which inductance and resistance have to be deduced [5]. The calculations appear fairly tedious. The method proposed employs parameter identification techniques (least squares methods, degenerate where appropriate) in order to evaluate the instantaneous resistance and inductance from the point of departure of a transmission line. Knowing R and L, it is possible to construct all imaginable types of characteristics. It is also possible to accelerate the remote back-up protection but this feature will not be considered here.

Considering the rated frequency (50 c/s), inductance L and reactance X are expressed by the same number in the form of reduced variable. Impedance can be defined as follows :

$$\underline{Z} = R + jX = Z \underline{/\xi}$$

2. STATE OF RESEARCH AND PROSPECTS

In order to maintain a high level of reliability of supply, extreme prudence is necessary when introducing new techniques. Computer study of protection must be carried out in a number of stages, at first by simulation on the scientific computer employing initially completely simulated data, then obtained from a network model and from the actual network. The programmes are then written out for a process computer and the necessary interface is prepared to allow real time operation on a network model. In this way it is possible to apply the programmes in turn. Application to the actual network, at first evidently with open loop, is the final stage, achieved progressively, one line and finally the entire substation.

In this way a substantial data file was drawn up giving the voltages and currents with the occurrence of a fault on a simple single phase network, varying the different parameters. Employing this data file, programmes written out for the scientific computer enabled a comparison to be made of the various algorithms as regards the effect on precision on the R and L calculations of the sampling frequency, measurement accuracy, aperiodic current component, any nonlinear characteristics (arc, current transformer saturation), etc. This comparison is facilitated by calculation of the mean value, of the maximum positive and negative deviation from this mean, and the standard error in R, X, Z and ξ for a certain number of measurements carried out during the fault. It emerges from the study that it is possible to obtain correct R and L values in 10 msec for a three-phase line. These programmes are also fed from data obtained from a small number of short-circuit tests carried out on a network model. The results are very encouraging but pursuit of the study requires the construction of a high speed analog-digital converter.

In order to evaluate the time necessary for calculation in actual operation, the appropriate programmes were transposed onto a small process computer. In the present state of the technique this type of computer can in practice protect only one line. It is evident however that technological advances and advances in computer organisation will rapidly overcome this impediment. On the other hand it is possible that techniques developed here may be no longer used by a single computer but, in the manner of existing relays, by simplified computers comprising perhaps parallel digital circuits in dialog with the central computer.

3. OVERALL SUBSTATION PROTECTION

3.1 <u>Brief summary of the operation of a process computer</u> - The essential characteristic of digital computers is the sequential mode of performing programmes. It can be said in practice that a computer carries out only one thing at a time while the conventional protection units for example function simultaneously and independently.

There is little essential difference in the structure of process computers from scientific computers. There is a central arithmetic and control unit, high speed and low speed memories etc. The following features must however be emphasised :

1. The process computer functions in real time, that is to say it takes the necessary information directly from the installation in operation to which it is coupled by an interface. It is also in service continuously including during maintenance.

2. The process computer possesses a hierarchical interruption system. It would be inconceivable that the high computing speeds achieved should be restricted by the slow performance of the peripheral equipment (10 characters/sec typewriter). The main programme is restricted to issuing an instruction to one input/output unit and proceeds accordingly. When the operation is completed (letter struck, measurement carried out) a signal is transmitted which interrupts the progress of the main programme, secures the working resisters and passes control to a programme processing the entering or leaving information with final return to the main programme. The time loss is thus minimum. This procedure is used for exchanges with all the peripheral equipment (analog/digital converters, multiplexers etc.). Since the computer performs only one operation at a time a hierarchical interrupting system controls the treatment sequence of the peripheral equipment simultaneously occupying the computer.

Control of these inputs/outputs imposes too big a load on the central unit if the peripheral equipment is too numerous or too rapid. In this case the computer is fitted with a direct access unit to the storage (DMA) functioning independently of the central unit and sharing with it access to the high-speed storage. In the extreme case the DMA undertakes complete control of the peripheral equipment and transmits to the central unit only one signal "available data" or receives an instruction "data to be transmitted".

3. The process computer has a "real time monitor" (MNT), that is to say a programme for controlling all the separate programmes but communicating through a common part of the storage, set up for a particular application. The order in which the programmes are performed is not fixed but depends on the external signals or results of previous calculations. Functioning of the monitor is shortly described below. All the programmes have an index indicating their level of priority. The real time monitor issues an "authorisation" activating all or certain programmes. The routine performed is the one which, when activated, possesses the lowest levels of priority. Programmes situated on the same level are performed in a pre-determined or arbitrary order. Following completion, the programme is inactivated and returns control to the real time monitor. The operation proceeds with the active programme on the level immediately above if the real time monitor has not cancelled the general authorisation. An example is given in the next chapter.

3.2 <u>Organisation of substation protection programmes</u> - For simplification the account will be restricted to substation protection at one voltage level but with several sets of busbars and the associated lines (fig. 1). The method also applies to a substation with several voltage levels but, apart from the fact that this eventuality is not very realistic, the presence of transformers introduces a number of complications which would be too tedious at this point.

At precise instants, separated by a constant period of time called DT and equal for example to 0.8 msec, all the substation current and voltage measurements are sampled and classified by the DMA channel in a central storage. For the substation shown in Figure 1 this means three times 3 = 9 voltage values (which can be reduced to 6 and even 3 if the busbars are coupled together) and to 11 times 3 = 33 current values (3 per outgoing line and 1 or 2 times 3 per circuit-breaker on the busbars). These measurements are obtained either from digital pickups or more probably at the start from analog/digital converters. The latter are supplied from conventional step-down units and the problems which these present cannot be disregarded (for example saturation of current transformers). The computer must hold

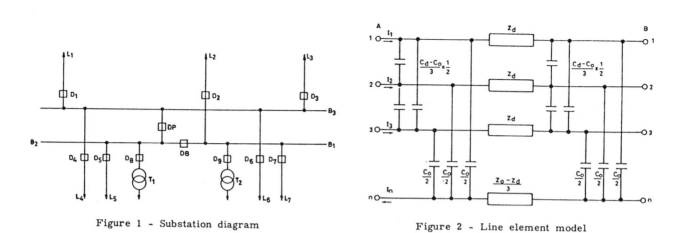

Figure 1 - Substation diagram   Figure 2 - Line element model

within the storage the values obtained during the previous 20 msec. This necessitates $42 (20 (1/0.8) + 1) \simeq 1100$ words of storage. "Filling" of the storages is organised in a cyclic manner. When the signal "available data" is transmitted to the central unit the real time monitor issues the instruction for performance of the fault detection routine.

The three main principles employed in the formulation of protection programmes are :

1. The programmes, a list of which is given in table 1, are designed to provide location of the defective element with increasing precision. For example, while the fault detection routine simply signals the existence of a fault, SELECT determines the defective phase or phases, DIFPOST localises the fault inside or outside the substation and DISTANi determines the defective line and controls the delay.

2. The performance of the programmes must not be stopped when one fault has been localised. The programmes are classified in accordance with a system of priority levels and continues successively so long as the time is available, attempting to localise any other faults.

3. When a programme is thought to have localised a fault it varies the level of priority of certain programmes so as to be certain that they proceed to the next step. Lack of time is actually found to be the main difficulty in designing a computerised protective system. All the programmes in table 1 cannot be performed in the period (0.8 msec) separating each sampling. Hence, if the DISTAN1 programme is thought to detect a fault in the first step on line i, it will bring its priority level to 4 so as to be certain to proceed to the subsequent steps and to bring about disconnection after a certain number of positive steps. In actual fact there is only one DISTANi programme performed with the measurements and characteristics of the line i. It is preferable here to assume as many programmes as lines and that the level of priority of each of these is modified. The same comment applies to DIFPOST and DIFBAR, DIRPOST and DIRBAR.

3.3 <u>Fault detection and selection of defective phases</u> - The fault detection routine (DDT) is initiated indefinitely every 0.8 msec by the real time monitor. It should therefore be as simple as possible, while taking care not to diminish its sensitivity and selectivity. It is difficult to conceive a simple programme giving a positive indication only during the period of a fault. It is simpler to detect its appearance and to adopt a different principle to establish its elimination.

A detection criterion based on voltage measurements is preferable since these are fewer and can be subject to discontinuities. Thus, for the substation or Figure 1, it is sufficient to examine a set of three simple voltages and the homopolar voltage when the busbars are coupled together. The method chosen is similar to that proposed by Mann and Morrison [4], and consists in comparing the actual voltage on each phase with the value measured 20 msec earlier. For each phase, if the difference exceeds a fixed tolerance, a counter steps up by one unit, otherwise it diminishes, without falling below zero. This type of counter functions for the homopolar voltage but comparison is made with the instantaneous value and not the difference. When at least one counter reaches the maximum value MAXNO and another reaches or passes MAXNO - 1, a fault is detected. It is clear that this detection occurs, at the earliest (MAXNO - 1) DT msec after the occurrence of the fault.

Table 1

Protection Routines

Routines	Function	Initial Level	Effect if Positive	Negative
DDT	Fault detection	(2)	call in SELECT	return to real time monitor
SELECT	Selection of faulty phases	(2)		
DIFPOST	differential, substation	(6)	DIRPOST (11 $\longrightarrow$ 7)	
DIRPOST	directional, substation	(11)	DIFBARi Group (12 $\longrightarrow$ 8)	DIRPOST (7 $\longrightarrow$ 11)
DIFBARi	differential, busbars i	(12)	DIFBARi ($\longrightarrow$ 4) DIRBARi ($\longrightarrow$ 5) DIFPOSi ($\longrightarrow$ 11) DIRPOST ($\longrightarrow$ 11)	DIFBARi (4 $\longrightarrow$ 8) (8 $\longrightarrow$ 12)
DIRBARi	directional, busbars i	(14)	after app. 5 msec, DECLi ($\longrightarrow$ 1)	DIRBARi ($\longrightarrow$ 14)
DISTANi	line i, distance	(9) Z3 Z1	DISTANi ($\longrightarrow$ 10) DISTANi ($\longrightarrow$ 4) after app. 4 msec, DECLi ($\longrightarrow$ 1)	DISTANi ($\longrightarrow$ 13) DISTANi ($\longrightarrow$ 9)
DECLi	disconnection	without	RATDISJ ($\longrightarrow$ 3) DECLi ($\longrightarrow$ without)	
RATDISJi	circuit-breaker failure	without		
VIDDI	changeover DISTANi routines from (10) to (9) when the (9) level is vacant.			

Two versions of the programme have been tested. In the first, comparison was carried out with the absolute value of the difference and in the second with the difference of the absolute values. In the latter case a positive indication is provided only if the voltage is falling. Both methods have disadvantages, the extent of which depends on local conditions and which are impossible to assess without many test runs on actual and simulated networks.

All the routines of table 1 are single phase and therefore require previous determination of the defective phase or phases. The following chapter indicates how the single phase quantities are set up on the basis of three-phase quantities when using the DISTANi routines. The SELECT routine employs the fault detection routine counter reading for carrying out selection. The "defective phases" are defined as the two phases for which the counter reaches the highest value, the "homopolar" counter being considered in the same way as the others but, for equal readings, one of the three phases is chosen rather than "earth".

If the fault continues after 20 msec the fault detection routine no longer provides an indication. Selection of the phases is then eliminated and the programmes are applied successively to each of the phases unless one of the programmes can determine which phases are defective. In this case the choice is valid for the next time interval.

3.4 Differential and directional protection - This type of protection is exactly similar in principle for the whole of the substation and for each of the sets of busbars but a knowledge of internal couplings is essential. A positive indication is provided when, for the defective phase or phases the instantaneous differential current exceeds a fixed limit. This criterion, which is very rapid, is only rough and not very selective. A differential pseudocurrent appears for example in the event of an external fault with saturation of a current transformer. Use of a more reliable routine based on a directional criterion is essential in order to justify disconnection. The algorithm for the DIRPOST and DIRBARi routines employs only the direction of the currents and does not introduce voltages which may disappear completely in the event of a dead short at the busbars. The principle is similar to that described in reference [6].

Table 1 allows to foresee the priority movements of the routines from the initial level according to whether the response is positive or negative, that is to say whether a defect is being detected within or outside a protected zone.

A positive indication of the DIRBARi routine is necessary during 5 msec approximately in order to bring about disconnection of the set of busbars i. This is obtained by bringing priority of a DECLi programme to the level 1, bringing about disconnection of the circuit-breakers for the busbars and checking any failure of one of these by the RARDISJ programme.

3.5 Distance protection - All the DISTANi routines are initially at level 9. They are carried out in an arbitrary sequence. If the impedance, reckoned from the outgoing point of the line k, is within the first step, the priority for the DISTANk routine moves to 4 so as to occur with certainty during the following passages. Keeping the impedance in the first step for approximately 4 msec brings about disconnection by the DECLk. The lines for which the measured impedance is not in the third step have their programme priority transferred to 13, following that of the DIFBARi routines. On the other hand, the lines for which the impedance is in the third step without being in the first have their programme priority transferred to 10. Provision must be made so that all the DISTANi routines cannot be performed in one DT interval, allowing the successive examination of each.

After a certain time the range of the first step can be modified to obtain the stepped delay characteristic of distance relays.

3.6 Return to normal - Return to normal is detected by negative response of all the localising routines, that is to say by their presence at a certain priority level. This routine will restore the initial state of priorities.

4. DISTANCE PROTECTION

4.1 Principles of protection

4.1.1 Calculation procedures - It is known that a symmetrical line can be represented by a succession of three-phase cells as shown in Figure 2 [7]. For a no-resistance fault between phases i and j, disregarding capacitances, we can write :

$$u_{ij} = (R_d + L_d d/dt) i_i - (R_d + L_d d/dt) i_j = R_d[i_i - i_j] + L_d d/dt[i_i - i_j]$$

for a single-phase fault :

$$u_i = (R_d + L_d d/dt) i_i + (R_t + L_t d/dt) i_n = R_d[i + (R_t/R_d) i_n] + L_d d/dt [i_i + (L_t L_d)i_n]$$

In both cases these are equations of the form :

$$u = Ri_A + L \, di_B/dt \tag{1}$$

writing $u = u_{ij}$ and $i_A = i_B = i_i - i_j$ in the polyphase case and $u = u_i$, $i_A = i_i + (R_t/R_d)i_n$ and $i_B = i_i + (L_t/L_d)i_n$ in the single-phase case. Determination carried out by SELECT provides calculation of the single-phase quantities.

1. The differential equation method (abbreviated to DIFF).

In equation (1) the quantities u, i_A and i_B are obtained by measurement and are therefore known. The unknowns are R and L. On account of measurement errors, phenomena disregarded etc., equation (1) is not strictly proved and should be replaced by the relationship :

$$R i_A + L \, di_B/dt - u = \varepsilon(t) \tag{2}$$

The principle of the method consists in calculating the R and L values minimising the error function $\varepsilon(t)$ over a period of time T (from t_n to t_o). The instantaneous variable time value is represented by t_o with $t_n = t_o - T$. The commonest criterion consists in rendering the square deviation integral $\varepsilon(t)$ minimum over the period T. We obtain the system :

$$A R + B L = C$$
$$D R + E L = F \tag{3}$$

where

$$A = \int_{t_n}^{t_o} i_A^2 dt, \quad B = \int_{t_n}^{t_o} i_A di_B/dt, \quad C = \int_{t_n}^{t_o} ui_A dt, \quad D = B, \quad E = \int_{t_n}^{t_o} (di_B/dt)^2 dt,$$

$$F = \int_{t_n}^{t_o} u \, di_B/dt \, dt$$

This system can be readily resolved knowing the coefficients A, B, ..., requiring calculation at each instant of the product or squares and integration over the period T. Since the u and i values are known only at the moments of sampling, integration is carried out by the conventional approximation method (trapeze or Simpson). The number of successive samples entering the calculations is equal to NECH = (T/DT) + 1. Various expressions exist for evaluating the derivative of i_B [8] such as central derivative, first, second order rear derivative etc. The errors involved can be calculated at least in theory, but the classification drawn up is scarcely valid in practice on account of the noise always affecting the actual measurements.

2. Integral equation procedure (abbreviation INTEG).

By dividing the internal (t_n, t_o) into N parts and integrating equation (1) over each of these the differential equation is substituted by a system of N integral equations of the type :

$$R \int_{t_i}^{t_j} i_A dt + L \int_{t_i}^{t_j} di_B/dt \, dt = \int_{t_i}^{t_j} u \, dt \tag{4}$$

This system is incompatible unless N deviation variables ε_j are introduced. We now obtain a system of N equations of the type :

$$R U_j + L V_j - W_j = \varepsilon_j$$

where

$$U_j = \int_{t_i}^{t_j} i_A dt, \quad V_j = (i_{Bj} - i_{Bi}) \quad \text{and} \quad W_j = \int_{t_i}^{t_j} u \, dt.$$

Minimalising $\sum_{j=1}^{N} \varepsilon_j^2$ in relation to R and L we obtain the same system of two equations as above (3) where :

$$A = \sum_{j=1}^{N} U_j^2, \quad B = \sum_{j=1}^{N} U_j W_j, \quad D = B, \quad E = \sum_{j=1}^{N} V_j^2, \quad F = \sum_{j=1}^{N} V_j W_j$$

Evaluation of the coefficients is simpler than for the previous method. Only the case where N = NECH - 1 has been examined.

3. Mean procedure (abbreviation MEAN).

In the previous method, considering only two intervals designated $T_2(t_2, t_1)$ and $T_1(t_1, t_o)$ the system (4) is compatible and calculation of the coefficients is now very simple :

$$A = \int_{t_2}^{t_1} i_A dt, \quad B = i_B(t_2) - i_B(t_1), \quad C = \int_{t_2}^{t_1} i_A dt, \quad E = i_B(t_1) - i_B(t_o), \quad F = \int_{t_1}^{t_o} u \, dt$$

Since $t_1 = t_o - T/2$ and $t_2 = t_o - T$, the integration intervals are displaced at each sampling while t_o increases. Various methods have been tested in order to reduce further the calculations by fixing the limits of one of the two intervals, thus maintaining three of the six coefficients constants, but experience indicates excessive errors.

4.1.2 <u>Characteristics</u> - Knowing the instantaneous R and L values provides all imaginable characteristics. The simplest are doubtless those comprising straight segments. Thus, by combining the characteristics shown in Figures 3a, b, c, the characteristics of Figure 3d showing zones one and three can be readily constructed.

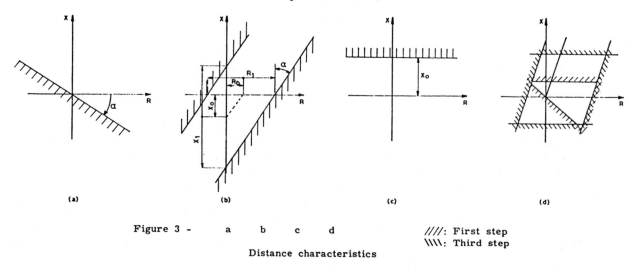

Figure 3 - a b c d

Distance characteristics

////: First step
\\\\: Third step

4.1.3 <u>Procedure</u> - For the various methods proposed, to yield coherent results, all the measurements employed must be taken after the inception of the fault. The results obtained by integrating strictly from the instant of the fault are of no significance. As the fault is detected by the fault detection routine only MAXNO elementary instants after its inception, it is sufficient for MAXNO and NECH to be equal. It is found however, as will be seen later, that methods employing the mean squares criterion yield results tending more regularly towards the end value than the mean method.

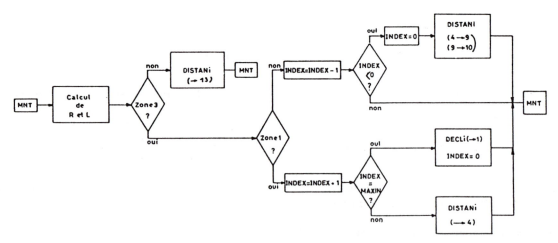

Figure 4 - DISTANi organigram

Figure 4 clarifies the functioning of the DISTANi programme. The instantaneous R and L values are first calculated. Disconnection is not decided with the first appearance of the impedance representative point in the first step but after a certain number fixed by the index counter, the functioning of which is identical with those introduced in the fault detection routine. Increasing the MAXIN limit of the counter reduces the possibility of incorrect disconnection but delays disconnection particularly in the event of saturation of the current transformer. 4 msec in the first step to bring about disconnection appears to be a good compromise.

4.2 <u>Comparison by simulation</u> - A number of programme variants employing the principles set out above have been written for a scientific computer. A big series of tests were carried out using data obtained by calculation on the basis of a single-phase network, sketched in Figure 5. The results are summarised below.

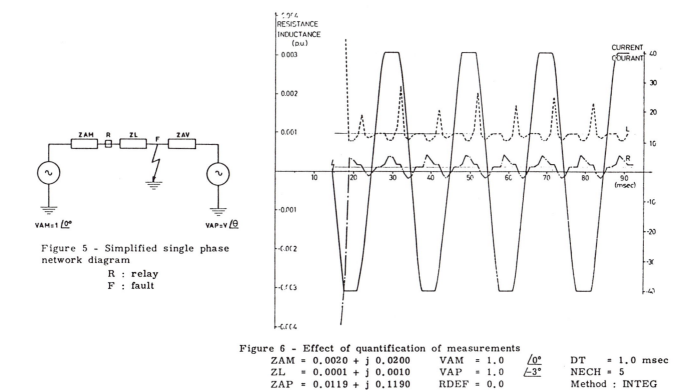

Figure 5 - Simplified single phase network diagram

R : relay
F : fault

Figure 6 - Effect of quantification of measurements

ZAM = 0.0020 + j 0.0200	VAM = 1.0 $\underline{/0°}$	DT = 1.0 msec
ZL = 0.0001 + j 0.0010	VAP = 1.0 $\underline{/-3°}$	NECH = 5
ZAP = 0.0119 + j 0.1190	RDEF = 0.0	Method : INTEG

4.2.1 <u>Exact data. Linear system</u> - These tests aimed at applying the programmes and eliminating errors. For sinusoidal conditions the coefficients A, B, C ... can be readily calculated.

4.2.2 <u>Errors in the u and i measurement</u> - If a relative error (of linearity) of $\varepsilon_i(\%)$ is committed in the current measurement and $\varepsilon_u(\%)$ in the voltage measurement, the error in the R and L values is approximately $\varepsilon_u - \varepsilon_i(\%)$. It is more difficult to provide a simple expression setting out the effect of an absolute error (or zero shift).

4.2.3 <u>Deviations in the sampling instants</u> - In sinusoidal conditions a constant deviation between the sampling instants of the current and voltage waves can be assimilated to a parasitic phase shift between the two waves, the impedance argument being converted to an angle value corresponding to the time shift. Thus 100 µsec corresponds to 0,9°. In the presence of an aperiodic current component the theoretical values in a particular case oscillate around their mean from 1 to 2 % for R, L and Z and from 1 to 2° for ξ .

Deviation in the frequency of sampling does not affect the value of R but results in an equal deviation in the L value.

4.2.4 <u>Quantification of measurements and saturation of the converters</u> - All the present analog/digital converters (ADC) provide information in the form of integer variable. The measurement is thus quantified and the range is limited. It appears difficult to exceed 4,000 measurement points, i.e. a possible variation of ± 2,000 points (11 bits + sign).

For voltage the quantum represents 0.1 % of the nominal peak value, providing correct measurement at twice the rated voltage.

The range of current variation is very wide. It is impossible to measure to the same precision a normal current representing a fraction of the rated current and a completely asymmetric short-circuit current equal to 100 times this value. It is necessary therefore to locate the ADC measurement range at the best position within this range and to determine the SATAM current value corresponding to saturation of the ADC, which must be avoided for faults at the end of the line where precision is essential. As a precaution the completely asymmetric short-circuit peak current is taken for the SATAM, calculated for a fault towards the middle of the line. Under these conditions it is rarely necessary to exceed 30 to 40 times the rated peak current, corresponding to a quantum of 2 %. These values were taken for the following tests. For faults nearer the substation the ADC may evidently be saturated but a signal indicating this state is interpreted as an indication of fault in the first step by the programme, which suspends calculation of R and L. It is easy to ensure that ADC does not become saturated for near faults situated before the ADC. When the ADC is desaturated, the calculations can be resumed from the immediately preceding measurements or with those preceding saturation. In order to assume the worst conditions and to simplify programming, the first method was chosen for the test shown in Figure 6. For a near fault for which the residual voltage represents 5 % of the rated voltage, the representative point of the Z measurement in the R X diagram is maintained in the semi-plane confined by a straight −45°, resulting in disconnection. It is necessary to fall to 0.5 % to obtain a delay in disconnection.

4.2.5 Integration period and frequency of sampling - A study of table 2 shows that increasing the integration period scarcely affects the deviation of the mean of the measurements in relation to the exact value but substantially improves the variance. Increasing the frequency of sampling yields inverse results.

Table 2
Results of the DISTANi routines

T msec	DT msec	METHOD	MEAN ERROR				STANDARD ERROR			
			R %	L %	Z %	ξ_o	R %	L %	Z %	ξ_o
4	1.0	MEAN	0.7	0.6	0.6	0	26.1	1.0	1.0	0.7
		INTEG	0.7	0.6	0.6	0	21.2	0.9	0.9	0.6
		DIFF	23.1	−9.4	−10.3	−0.6	421.8	8.1	9.9	7.6
8	1.0	MEAN	0.7	0.6	0.6	0	7.6	0.4	0.4	0.2
		INTEG	0.5	0.6	0.6	0	4.2	0.3	0.3	0.1
		DIFF	−0.6	−2.3	−2.3	0	34.7	0.8	0.8	1.0
4	0.5	MEAN	0.4	−0.1	−0.1	0	25.2	0.9	0.9	0.7
		INTEG	0.7	0.0	0.0	0	19.7	1.0	1.0	0.6
		DIFF	1.3	−1.9	−1.9	−0.1	67.0	1.4	1.5	1.8
8	0.5	MEAN	0.2	−0.1	−0.1	0	7.0	0.4	0.4	0.2
		INTEG	0.2	−0.0	−0.0	0	4.2	0.3	0.3	0.1
		DIFF	−0.2	−0.7	−0.7	0	9.4	0.4	0.4	0.3
4	1.0	MEAN	0.4	0.5	0.5	0	28.9	1.2	1.2	0.8
		INTEG	−1.2	0.5	0.5	0	25.4	1.1	1.1	0.7
		DIFF	5.5	−12.2	−13.6	−0.2	470.6	14.6	15.3	10.3
8	1.0	MEAN	−2.5	0.6	0.5	0.1	20.4	0.5	0.5	0.6
		INTEG	−0.7	0.6	0.5	0	11.0	0.5	0.5	0.3
		DIFF	3.9	−3.3	−3.3	−0.2	59.9	1.8	1.8	1.6
4	0.5	MEAN	−1.4	0.1	−0.1	0	34.1	1.3	1.2	1.0
		INTEG	−2.1	0.1	0.1	0.1	27.1	1.1	1.1	0.8
		DIFF	−2.2	−2.2	−2.3	0	77.0	2.2	2.2	2.2
8	0.5	MEAN	−0.7	−0.0	−0.0	0	22.0	0.5	0.5	0.6
		INTEG	−0.5	0.0	0.0	0	9.9	0.4	0.4	0.3
		DIFF	0.0	−0.9	−0.9	0	18.7	0.5	0.5	0.5

a : without aperiodic current component
b : with aperiodic current component

ZAM = 0.010 + j 0.200   VAM = 1.0   /0°
ZL = 0.005 + j 0.100   VAP = 1.0   /−3°
ZAP = 0.010 + j 0.200   RDEF = 0.0

All the tests carried out confirm these results.

4.2.6 _Aperiodic component. Angle and frequency deviation_ - Comparison of the results shown in table 2 indicates that a maximum aperiodic component in the fault current does not have an appreciable effect on the accuracy of the calculations.

A network frequency deviation has no effect, like an angle deviation between the voltage at the two ends other than, in the latter case, the familiar effect of the current due to the source below on the common fault resistance.

4.2.7 _Arc_ - An arc was simulated by an equation of the form [9] :

$$|u_{arc}| = K_1 + K_2/(1 + K_3 |i|)$$

where K_1, K_2 and K_3 are constants. The effect of this is slight for faults at the end of the line but is evidently greater for near faults, without however bringing about incorrect disconnection or delay even when the arc voltage represents half the peak voltage at the relay.

4.2.8 _Presence of a saturable current transformer_ - Several studies have shown that correct transmission of the aperiodic component of the fault current requires considerable over-dimensioning of the conventional current transformer. Otherwise the transformer is saturated, but it has been demonstrated that it is exceptional to reach saturation less than 8 to 9 msec after inception of the fault and that the current transformer provides a desaturation period and therefore correct signal transmission at each wave period.

For the purpose of our study the current transformer is assumed to feed an ohmic load. It does not display either magnetic losses nor hysteresis.

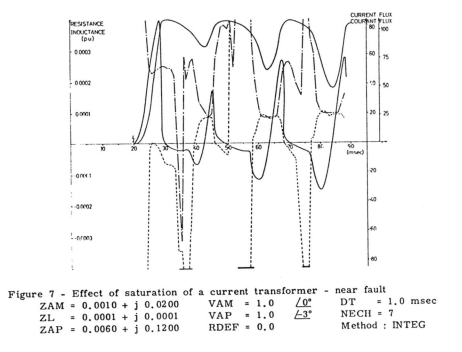

Figure 7 - Effect of saturation of a current transformer - near fault

ZAM = 0.0010 + j 0.0200	VAM	= 1.0	$\angle 0°$	DT	= 1.0 msec
ZL = 0.0001 + j 0.0001	VAP	= 1.0	$\angle -3°$	NECH	= 7
ZAP = 0.0060 + j 0.1200	RDEF	= 0.0		Method : INTEG	

By way of example, for a near fault with maximum aperiodic component, saturation is reached after 8 msec (Fig. 7). However, the representative impedance point leaves the characteristic of the first step only 14 msec after the fault. The periods where this occurs are indicated by a thick line at the bottom edge of the drawing. The saturation flux of the current transformer is chosen so that it is almost reached in the event of a fault without aperiodic component at the same position, for which the maximum angle deviation is 13° by the INTEG method.

For the same current transformer and a fault at the limit of the first step, saturation is reached only at the 5th period. In order to make the test more severe the saturation flux was reduced to half. Figure 8 demonstrates how the programme operates. Saturation is reached

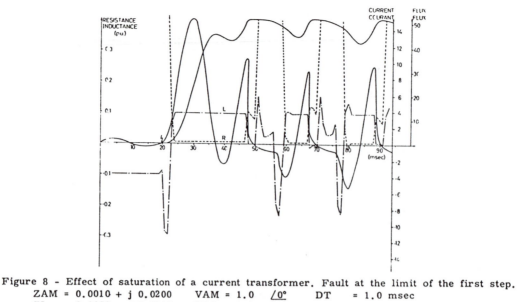

Figure 8 - Effect of saturation of a current transformer. Fault at the limit of the first step.

ZAM = 0.0010 + j 0.0200	VAM = 1.0 $\underline{/0°}$	DT = 1.0 msec
ZL = 0.0050 + j 0.1000	VAP = 1.0 $\underline{/-3°}$	NECH = 5
ZAP = 0.0010 + j 0.0200	RDEF= 0.0	Method : INTEG

27 msec after inception of the fault but desaturation periods and therefore correct measurements are obtained if necessary between 41 and 47. 61 and 68 msec after the fault.

4.2.9 <u>Comparison of methods and conclusions</u> - Comparing all the tests carried out, it appears that the DIFF methods are less precise. As they are often more complicated they can be rejected. The INTEG method provides a gain in precision of the mean but in particular an increase in coherence of results (reduction in variance) as compared with the MEAN method, the improvement being particularly marked with non-linearities (arc, quantification, ...). It is more complicated and therefore more tedious. Experience will have to distinguish the two methods.

The gain in precision as the result of using an integration method different from the trapeze is not generally significant.

Figures 9 and 10 show that the INTEG method provides a smoother transition between the impedance values before and after the fault.

As regards simulated tests it is feasible to hope for the development of a computerised protective system. The interval of 0.8 msec between two samplings is a good compromise. With values MAXNO = 5, NECH = 5, or 7, MAXIN = 5, a disconnection time would be obtained of the order of 7 msec. A delay of two to three msec in selection of the correct DISTANi routine is tolerable.

4.3 <u>Network model test</u> - The Université Libre de Bruxelles possesses a network model specially designed for testing conventional relays. Unfortunately it does not have an analog/digital converter, essential for computer protection studies. A very small number of tests have been carried out. The tedious labour of conversion is carried out on the basis of an oscillograph record employing human labour.

It will be remembered that the theory of sampling functions assumes that the spectrum of the signal does not extend beyond half the sampling frequency (Shannon). The necessary filtering is obtained to an adequate extent by inertia of the galvanometric loops (1200 Hz). Figure 11 shows the voltage record from a three-phase short circuit (DT = 0.5 msec). Saturation of a voltage transformer during voltage build-up explains the presence of a homopolar component after elimination of the fault.

Performance of the fault detection and SELECT routines is satisfactory with the reservations indicated above. In the event of a short-circuit affecting a number of phases, selection carried out by SELECT may vary in the course of time. This does not delay functioning of the protection and has the advantage of more precise measurement if the fault develops for example from single-phase to two-phase-earth.

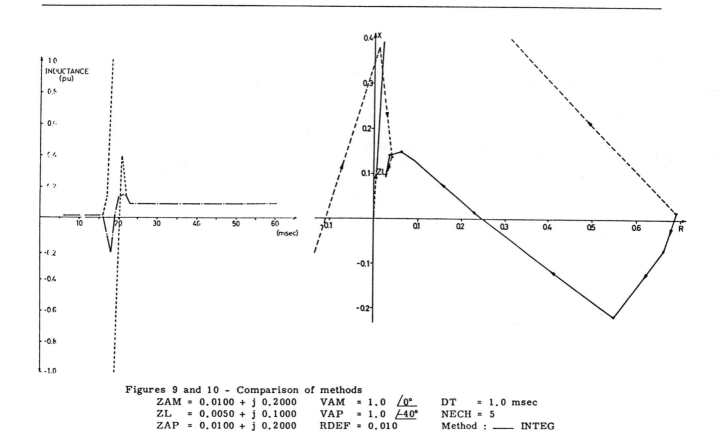

Figures 9 and 10 - Comparison of methods

ZAM = 0.0100 + j 0.2000	VAM = 1.0 ∠0°	DT = 1.0 msec	
ZL = 0.0050 + j 0.1000	VAP = 1.0 ∠-40°	NECH = 5	
ZAP = 0.0100 + j 0.2000	RDEF = 0.010	Method : —— INTEG	
		---- MEAN	

Table 3

Results of the DISTANi routine

T msec	DT msec	METHOD	MEAN ERROR				STANDARD ERROR			
			R %	L %	Z %	ξ_o	R %	L %	Z %	ξ_o
4	0.5	MEAN	6.1	−0.9	0.4	1.2	37.9	8.3	7.7	6.9
		INTEG	5.4	−1.4	−1.2	1.3	27.4	7.4	6.7	5.4
	1.0	MEAN	9.6	−2.0	−0.0	1.8	42.7	8.3	7.7	7.9
		INTEG	7.3	−2.0	−0.3	1.6	31.8	7.6	6.8	6.1

Network model test
Direct-on   3-phase fault
Phases 1,3.

Figures 12 and table 3 show the results of impedance measurements carried out between phases one and three. The lower accuracy observed is evidently attribuable to the factors disregarded in the calculations (capacitances, etc.), and even more so to the measurement errors. More than half of the error can be explained by inaccuracy of measurement recording and chart speed. It is evidently essential to design an automatic ADC.

4.4 Process computer : duration of a routine - DISTANi routines have been written out for a small process computer (words of 16 bits, cycle time = 2 μsec. Excluding the use of the real time monitor, input/output operation, determination of the step, the time taken for the R and L calculation is ∼ 1.1. msec by the MEAN method and ∼ 2.9 msec for the INTEG method, for an interphase fault. These periods are evidently too long at present. However, the continuously increasing computer speed, the development of Read Only Memories providing specialised instruction programmes and the development of an arithmetic unit with speed as high for floating numbers as integer numbers, promise well for the future.

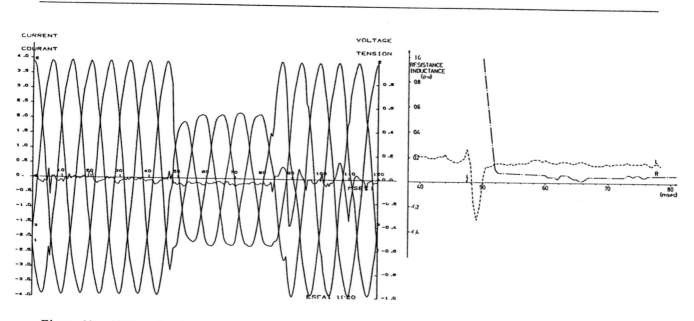

Figure 11 - Measured voltages on the network model with 3-phase fault.

Figure 12 - Impedance with 3-phase fault on the model.
Phases : 1.3   DT = 0.5 msec
NECH = 9   Method : INTEG

5. CONCLUSIONS

A digital computer substation protection procedure is presented with emphasis on distance protection. Advantages to be anticipated are : high speed and selectivity of protection by reason of the principles employed, reliability due to continuous operation with regular automatic testing of the computer circuits, and savings from the use of marginal elements mainly performing other tasks.

Despite difficulties encountered in programming, the accumulation of the substantial volume of data necessary and treatment of the mass of calculations, the prospects appear interesting.

REFERENCES

[1] G.D. ROCKEFELLER - Fault Protection with a Digital Computer. (IEEE Transactions, Power, Apparatus and Systems, Vol. PAS-88, pp 438-464, April 1969).

[2] B.J. MANN - Real-Time Computer Calculation of the Impedance of a Faulted Single-Phase Line. (IFAC Symposium, Sydney, August 1968, Paper No. 2502).

[3] B.J. MANN and I.F. MORRISON - Digital Computer Calculation of Impedance for Transmission Line Protection. (IEEE Power Meeting, Paper 70TP165-PWR, Jan. 1970).

[4] B.J. MANN and I.F. MORRISON - Relaying a Three-Phase Transmission Line with a Digital Computer. (IEEE Power Meeting, Paper 70TP547-PWR, July 1970).

[5] G.R. SLEMON, S.D.T. ROBERTSON and M. RAMAMOORTHY - High-speed protection of Power Systems based on improved Power System models (CIGRE Paper No. 31-09, 1968).

[6] H. HAUG and M. FORSTER - Electronic Bus Zone Protection. (CIGRE Paper, No. 31-11, 1968).

[7] J.P. BARRET - Etude à l'aide d'un modèle analogique des phénomènes transitoires liés à l'appareillage d'interruption. (R.G.E., Vol. 74, No. 5, pp. 441-456, May 1965).

[8] E. ISAACSON and H.B. KELLER - Analysis of numerical Methods. (John Wiley & Sons, 1966, pp 245-299).

[9] E. WALCZUK - Über die Lichtbogenenergie im Kontaktspalt beim Ausschalten von Wechselstrom. (Archiv für Elektrotechnik, Vol. 54, pp 43-55, 1970).

Extrait de la *Conférence Internationale des Grands Réseaux Electriques.*
Session de 1972.
Imprimerie LOUIS-JEAN — 05 - GAP

A DIGITAL COMPUTER SYSTEM FOR EHV SUBSTATIONS: ANALYSIS AND FIELD TESTS

A. G. Phadke   T. Hlibka   M. Ibrahim

American Electric Power Service Corp.
New York, N. Y.

Abstract

This paper describes the research and development work being done at the AEP Service Corp. on digital computer applications in EHV substations. Relaying of transmission lines and other power system equipment, station alarm monitoring and data logging, oscillography, and supervisory control functions are considered for possible computer implementation. A new algorithm for computing the impedance as seen from the terminals of a faulted transmission line is presented. Results obtained with the new algorithm for faults simulated with a model transmission line are included in the paper. Field experience obtained from operating a minicomputer in an EHV substation is also described.

INTRODUCTION

The development of process control computers and continuous advances made in the field of microprocessors have made it possible to consider digital computer applications in areas which traditionally are the domains of dedicated (most commonly of the analog type) devices. Relaying of power system equipment and in particular of EHV transmission lines is an area in which computer applications were first considered in the late 1960's. (1) This reference and later work (2-5) broke much new ground in this field, and since then a number of analytical papers dealing with computer relaying of transmission lines have been published (6-8).

Computer hardware technology has considerably advanced since early 1960's. Newer generations of mini and micro computers tend to make digital computer relaying a viable alternative to the traditional relaying systems. Indeed it appears that a simultaneous change is taking place in traditional relaying systems, which are beginning to use solid state analog and digital subsystems as their building blocks. It is quite possible that the next major step in the evolution of protection systems will include many of the concepts being discussed here and in the references cited earlier.

Although most of the work reported in the literature concentrated on development of digital computer algorithms for relaying of EHV transmission lines, it should be noted that a computer in the substation presents opportunities to take on many of the other substation tasks. There is an obvious economic motivation for assigning multiple tasks to the computer. On the other hand reliability and integrity needs of the protection functions demand that back-up computers be available for certain prescribed modes of equipment failure. As a result of these two considerations, it becomes necessary to consider a system of substation computers or processors, rather than a single substation computer. These ideas, which are the basis of our future efforts in this field are not discussed further in this paper.

Paper F 75 543-9, recommended and approved by the IEEE Power System Relaying Committee of the IEEE Power Engineering Society for presentation at the IEEE PES Summer Meeting, San Francisco, Calif., July 20-75, 1975. Manuscript submitted February 6, 1975; made available for printing May 12, 1975.

The first phase of our work at AEP, which is the subject of this paper, is a feasibility study of a substation computer system using a single computer. A major portion of this effort concentrated on developing programs for various substation tasks, investigating the available algorithms for distance relaying of transmission lines and developing new ones where necessary, and conducting a field test of computer hardware and software in an EHV substation. The three sections that follow summarize the results of this phase of our work.

IMPEDANCE COMPUTATION FROM SAMPLED DATA

List of Symbols:

i, e	: Instantaneous values of current and voltage signals.
$y(t)$, y_k	: A time dependent variable and its k'th sample.
y_s, y_c $= \sqrt{2}\,(Y_s \cdot Y_c)$	: Real and imaginary components of phasor representation of the fundamental frequency component of $y(t)$.
$N = 2M$ $= 4P$	: Number of samples taken in one fundamental period.
$\Sigma_j\, y_k$	: A j − order partial sum of y_k defined in equations (10)-(11).
T	: Sampling interval.
ω	: $120\,\pi$ radians/sec.
$\propto$	: Decrement factor for the decay of dc offset terms in fault current waveforms.
γ	: Decrement ratio for the dc offset between consecutive samples.
R_{ml}	: Residual harmonic components of order ml, assumed negligible.
$Z_f = (R_f + jX_f)$	: Positive sequence impedance from a terminal of the transmission line to the fault location.

Phasor representation from sampled data of waveforms:

High speed relaying, and in particular distance relaying of EHV transmission lines is the most demanding computational task faced by the substation computer. Its main effort is to determine the phasor representations for a current and a voltage signal from sampled data of their instantaneous values. Given the sample sets $\{e_k\}$ and $\{i_k\}$, their phasor representations are determined:

$$\{e_k\} <=> E_s + jE_c$$

$$\{i_k\} <=> I_s + jI_c$$

When the signal pair is selected properly for a given fault (10), the ratio of the two phasors is a measure of the distance of the fault from the transmission line terminal.

$$Z_f \equiv R_f + jX_f$$

$$= \frac{E_s I_s + E_c I_c}{I_s^2 + I_c^2} + j\,\frac{E_s I_c - E_c I_s}{I_s^2 + I_c^2} \tag{1}$$

Location of the impedance to the fault - point Z_f on the complex Z plane (on which a relay characteristic is defined) is the key finding upon which the distance relaying program bases its decision.

Reprinted from *IEEE Trans. Power App. Syst.*, vol. PAS-95, no. 1, pp. 291–301, Jan./Feb. 1976.

Sampling rate and data window:

Since the digital computer must spend some time in acquiring the samples of a signal, it is clear that the signal sampling rate should be kept as low as possible. The phasor representation desired is a 60 Hz result and consequently the lowest permissible sampling rate must be greater than 120 Hz. (11) One of the two methods of obtaining the phasor representation from data samples described below uses Fourier Transform (or digital harmonic filter) equations, and in that case it is advantageous to have the sampling rate a multiple of the fundamental frequency. Thus nx60 Hz (where n = 3, 4...) are the acceptable sampling rates. The other technique described below is based upon least-square polynomial smoothing of the data, and does not require sampling at synchronous rates.

Data window is the time span covered by the sample set needed to execute the computation procedure. The smaller the data window, the faster is the response of the computation procedure. In most Fourier Transformer computations a data window of one period of the fundamental frequency is considered to be desirable, although half-cycle data windows are also acceptable. As will be explained in the next section, in the presence of a decaying dc offset component in the signal, a data window of one sample period plus one-half cycle of the fundamental frequency enables phasor computations to be performed with acceptable accuracy. (It should be remembered that an additional delay of one sample period may take place if a sampling instant falls immediately before the occurrence of a fault.) Table I lists required data windows computed on this basis (which is equivalent to the response time of the relay algorithm following a fault) for several synchronous sampling rates.

As mentioned previously, the least square polynomial filter equations do not require a synchronous sampling rate. These methods will be discussed in more detail later. It turns out that these techniques are efficient on relatively narrow data windows (of the order of one-sixth of a fundamental period or less), and require about five or more points within the data window. The sampling rates which produce satisfactory computations with these techniques should therefore be faster than about 30 times per period of the fundamental frequency. These algorithms have a potential of having a faster response time. They suffer from serious errors of computation in the presence of decaying dc offsets and harmonics in the sample set.

TABLE I

Minimum Data Windows for Reliable Phasor Computation with Synchronous Sampling

Sampling Rate (Hz)	Min. Data Window (Response Time) (MS)
240	12.5
360	11.1
480	10.4
720	9.7
960	9.4

Considering the response time shown in Table I, any sampling rate above 480 Hz should be equally satisfactory from the point of view of the speed of response. Among these, the rate of 720 Hz is particularly attractive as the corresponding Fourier Coefficients are $0, \pm 1, \pm 1/2$ and $\pm \sqrt{3}/2$. All the coefficients except for the last two are easily obtained in integer arithmetic. Even the last coefficient has an adequate binary representation of five terms, and the digital harmonic filter equations corresponding to this sampling rate are executed quite rapidly in integer arithmetic.

Data smoothing with least square polynomial fits:

A number of papers dealing with computer relaying of transmission lines (1-5) use smoothing procedures for the data samples

and subsequent estimation of time derivatives of the signal. A formal procedure for smoothing the data is to fit a polynomial of degree m to 2n+1 samples of the data. If m < 2n+1, a least square fit can be obtained (12). The necessary time derivatives are then estimated by differentiating the polynomial. Clearly, many such fits can be obtained depending upon the value of m and n chosen subject to the condition m < 2n+1. An example of this procedure (taken from reference 12) is a third degree polynomial fit to a seven point data set.

Given the data set $\{y_k\}$ where $K = -3, -2, ..., +2, +3$, the smoothed values of y_0 y'_0 and y''_0 are given by

$$\bar{y}_0 = \frac{-2(y_{-3}+y_3) + 3(y_{-2}+y_{+2}) + 6(y_{-1}+y_1) + 7y_0}{21}$$

$$\bar{y}'_0 = \frac{22(y_{-3}-y_3) - 67(y_{-2}-y_2) - 58(y_{-1}-y_1)}{252}$$

$$\bar{y}''_0 = \frac{5(y_{-3}+y_3) - 3(y_{-2}+y_2) - 4y_0}{42} \qquad (2)$$

These estimates are than used to determine the phasor representation for $\{y_k\}$:

$$y^2 = y_s^2 + y_c^2 = 2(Y_s^2 + Y_c^2)$$

$$\cong (\bar{y})^2 + \frac{1}{\omega^2}(\bar{y}')^2$$

$$\cong \frac{1}{\omega^2}(\bar{y}')^2 + \frac{1}{\omega^4}(\bar{y}'')^2 \qquad (3)$$

or,

or,

$$\cong \frac{(\bar{y}'_1)^2 + (\bar{y}'_2)^2 - 2(\bar{y}'_1)(\bar{y}'_2)\cos(k\omega t)}{\omega^2 \sin^2(k\omega t)}$$

and

$$\frac{y_c}{y_s} \cong \omega(\bar{y})/(\bar{y}')$$

or,

$$\cong -(\bar{y}'')/\omega(\bar{y}') \qquad (4)$$

or,

$$\cong \frac{(\bar{y}'_1)\cos k\omega t - \bar{y}'_2}{(\bar{y}'_2)\sin k\omega t}$$

where the subscripts 1 and 2 refer to estimates at sample times t_1 and t_2 such that $t_1 - t_2 = k\omega t$, where k is an integer. It shoulb be repeated that equations (2) above are representative of a fairly large group of polynomial fits, and consequently many different expressions for Y_c and Y_s in terms of $\{y_k\}$ are possible.

We recall that a meaningful least-squares procedure requires data redundancy, which implies that the inequality m < 2n + 1 should have a wide margin. A cubic fit over seven points or a quadratic fit over five points are perhaps necessary to satisfy the redundancy requirement. A minimum data window for methods of this type should therefore span four to six sample periods. Furthermore, a low order polynomial fit over the data window would be meaningful only if the fundamental frequency component of the signal does not change radically within the data window. Perhaps one sixth of the fundamental period is the upper limit on the data window from this point of view. These conclusions have been borne out by a number of off-line studies performed at AEP.

The exponentially decaying dc offset present in some of the signals gives rise to fairly large errors (in excess of ten percent) in the phasor estimates unless the offset terms are removed prior to the execution of the smoothing equations.

Digital Harmonic Filter equations

(a) Signals with constant dc offsets:

Considering the case of signals containing a constant dc offset first, the Fourier Transform of N data samples taken over one period of the fundamental frequency produces the phasor estimate for the fundamental frequency

$$y_c = \sqrt{2}\,Y_c = \frac{2}{N} \sum_{k=1}^{N} y_k \cos \frac{2k\pi}{N}$$

$$y_s = \sqrt{2}\,Y_s = \frac{2}{N} \sum_{k=1}^{N} y_k \sin \frac{2k\pi}{N} \tag{5}$$

The case considered in this paper is N=12, and the corresponding formulas for y_c and y_s are

$$6\,y_c = -(y_6 - y_{12})$$
$$+ \frac{1}{2}(y_2 - y_8 - y_4 - y_{10})$$
$$+ \frac{\sqrt{3}}{2}(y_1 - y_7 - y_5 - y_{11}) \tag{6}$$

$$6\,y_s = (y_3 - y_9)$$
$$+ \frac{1}{2}(y_1 - y_7 + y_5 - y_{11})$$
$$+ \frac{\sqrt{3}}{2}(y_2 - y_8 + y_4 - y_{10})$$

A somewhat poorer estimate may be obtained with six data samples (a data window of half cycle)

$$6\tilde{y}_c = -2y_6 + (y_2 - y_4) + \sqrt{3}\,(y_1 - y_5)$$
$$6\tilde{y}_s = 2y_3 + (y_1 + y_5) + \sqrt{3}\,(y_2 + y_4) \tag{7}$$

The transform $(\tilde{y}_c, \tilde{y}_s)$ is the same as (y_c, y_s) if no even harmonics are present in the signals. If even harmonics are present, an error proportional to their magnitude is introduced in $(\tilde{y}_c, \tilde{y}_s)$. This is an example of leakage phenomena in Finite Fourier Transform computations [11].

(b) Signals with decaying offsets:

The fault currents of a transmission line may often contain a dc offset term which decays exponentially with time. A general waveform for such signals is

$$y(t) = y_0\, e^{-\alpha t} + \sum_{l=1}^{\infty} y_l \sin(l\omega t + \phi_l) \tag{8}$$

where (y_0/y_1) depends upon the instant of fault occurrence, and α depends upon the X/R ratio of the faulted circuit. The sample set $\{y_k\}$ is given by

$$y_k = y_0\,(\gamma)^k + \sum_{l=1}^{\infty} y_l \sin(l\omega t + \phi_l) \tag{9}$$

If the number of samples in one period of the fundamental frequency is $N = 2M = 4P$, a partial sum $\Sigma_1 y_k$ may be defined by

$$\Sigma_1 y_k = y_k + y_{k+M}$$
$$= y_0\,(\gamma)^k \{1 + (\gamma)^M\} + R_{21} \tag{10}$$

The residual R_{21} is dependent upon harmonics of order 21, 1 being

any positive integer. Similarly a second partial sum $\Sigma_2 y_k$ may be defined

$$\Sigma_2 y_k = \Sigma_1 y_k + \Sigma_1 y_{k+p}$$
$$= y_0\,(\gamma)^k \{1 + (\gamma)^M\} \{1 + (\gamma)^P\} + R_{41} \tag{11}$$

where the residual R_{41} depends upon harmonics of order 41. Clearly additional partial sums of higher order can be defined in this manner if N, the sampling rate, has a higher power of 2 as a factor. For the case of N = 12, M = 6 and P = 3, the following three second order partial sums can be constructed from N data samples:

$$\Sigma_2 y_1 = y_1 + y_4 + y_7 + y_{10}$$
$$\Sigma_2 y_2 = y_2 + y_5 + y_8 + y_{11} \tag{12}$$
$$\Sigma_2 y_3 = y_3 + y_6 + y_9 + y_{12}$$

Making the assumption that R_{41} in equation (11) is a negligible quantity

$$\Sigma_2 y_1 \cong y_0\,\gamma\,\{1 + (\gamma)^6\} \{1 + (\gamma)^3\}$$
$$\Sigma_2 y_2 \cong (\Sigma_2 y_1) \cdot (\gamma) \tag{13}$$
$$\Sigma_2 y_3 \cong (\Sigma_2 y_1) \cdot (\gamma)^2$$

Parameters y_0 and γ of the decaying dc offset term can now be estimated from equations (13). A new sample set

$$\tilde{y}_k = y_k - y_0\,(\gamma)^k \tag{14}$$

is then computed and used in the filter equations (6).

Equations (13) refer to a data window of one cycle duration. A faster response time is possible if certain assumptions are made. A data window of seven samples (one half cycle plus one sample) is sufficient for the computation of $\Sigma_1 y_1$. Assuming R_{21} in equation (10) to be small,

$$\Sigma_1 y_1 \cong y_0\,(\gamma)\,\{1 + (\gamma)^6\} \tag{15}$$

If γ is further assumed to be determined solely by the X/R ratio of the transmission line it is a known parameter and y_0 can be estimated from equation (15). Equations (14) and (7) can then be solved to estimate the phasor representation of $\{y_k\}$.

(c) Numerical considerations:

Although equations (12) through (15) and equations (6) and (7) seem to require a great deal of computation, the actual computational burden is kept at a minimum by using recursive relationships, integer arithmetic, and truncated binary expansion of constants in multiplications and divisions. In addition, a considerable portion of the arithmetic work is scheduled to be performed while waiting for additional samples of data. On the computer used in this project (IBM S/7), the additional delay introduced by the computations after the last sample is obtained has been of the order of 1 millisecond. Results obtained by using these algorithms to compute transmission line impedance are, included in the next section.

LABORATORY DEVELOPMENT OF A SUBSTATION COMPUTER SYSTEM

A simulator laboratory was designed and built to develop and test various substation computer programs. This simulation laboratory, which has recently been expanded, has the capability of modeling several three phase EHV transmission lines, multi-winding three phase transformers and source impedances. An electronic synchronous switch (9) and other available instrumentation make this laboratory a useful tool for real-time simulation of power system faults and related transients.

The programs developed in the laboratory performed these substation functions:

(a) Alarm monitoring and data logging
(b) Control
(c) Oscillography
(d) Relaying

(a) Alarm monitoring and data logging:

This function requires scanning of contact status and certain analog signals on a periodic basis, as well as upon a change of status. Certain alarms require annunciation at a remote location. These functions are relatively slow speed, and pose no special problem to the computer. It is generally necessary to provide adequate isolation and protection of the computer contact sense circuits. It is also possible to furnish the data required by the central computer state-estimator programs where such data are available within the substation. This last function was not considered in our work.

(b) Substation control:

The extent of control actions (initiated from a remote location, through stored programs, or on a local decision basis) performed at a substation are dependent upon the operating philosophy of a utility. Generally, substation control actions as they are understood today, require a relatively moderate speed of response. The functions programmed during this project included initiation of high speed reclosing of circuit breakers, automatic reclosing with resynchronization test, and breaker failure protection schemes.

(c) Oscillography:

The traditional oscillography functions have been performed locally with automatic optical recorders which are triggered by the system disturbance. These generally lose about one-quarter cycle of the fault data, and have an initial time scale distortion. Recently certain magnetic devices with memory capability for pre-fault data have been used in substations to eliminate many drawbacks of the commonly used optical recorders.

The substation computer, with disc and teleprocessing capabilities, was programmed to store pre- and post-fault data and transmit it to a central computer for processing and plotting.

(d) Relaying:

Among the various relaying functions, the most challenging computational function is the multi-zone distance relaying of an EHV transmission line. Consequently, a program to perform this function was developed, and includes these features:

1) Three zone distance relaying and high set current relaying for phase faults.

2) Directional overcurrent, instantaneous, and inverse-time protection for ground faults.

3) Breaker failure protection.

4) High speed and automatic reclosing logic with appropriate synchronization checks.

5) Carrier logic.

Input signals to the program are the three line currents, three phase-neutral voltages on the line, and phase 1 voltage on the bus side of the circuit breaker. The program flow chart is shown in Figure 1. An internal timer presents an interrupt to the computer twelve times per cycle of the 60Hz signals. Following the interrupt, a data vector consisting of the three phase currents and the three phase to neutral voltages is sampled and stored in a storage ring of about 10 cycle duration. Upon detection of a disturbance, the active ring is reserved and transferred to the disc, while continuous data storage is shifted to another ring. The two rings are used alternately, assuring that no data is lost while oscillography data is transferred to the disc.

The sampled voltages and currents are combined to form zero sequence quantities which are also stored in the data rings. The currents are tested against high set limits to provide High Set trip functions. These limits must be exceeded for three successive samples to produce a trip signal. The voltages are compared with their one cycle old values, and upon detecting a significant variation which lasts for three samples, a data ring is reserved for oscillography. Should the variation persist for 6 samples, a fault is suspected and an attempt is made to classify the fault. A scheme designed to discount switching and other high frequency transients from triggering the fault detection mode was proposed by Mann and Morrison (2, 3), and is used in the fault detection algorithm. A counter K is assigned to each voltage and is incremented whenever a significant variation is found. (Refer to Figure 2). It is also decremented if it is greater than zero whenever the variation is found insignificant. By changing the threshold value of K at which a disturbance is detected, the scheme can be adjusted to be more or less sensitive to transients.

Fault classification is based upon the relative magnitude of the counters K, and is supplemented by current magnitude information. Following the classification, a calculation of the impedance is made using the appropriate relaying voltage-current pair. Carrier, zone detection, and other relaying programs are executed in the conventional manner based upon the value of the fault impedance. The clear-fault program determines whether the fault persists or has been

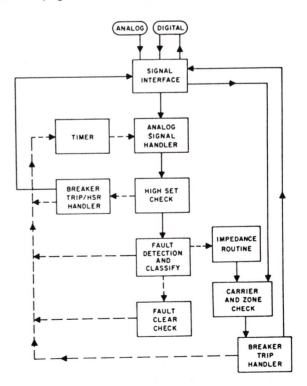

Fig. 1. Transmission Line Relaying Program Flow Chart.

515

cleared by breaker action, and provides appropriate blocks to various tripping programs.

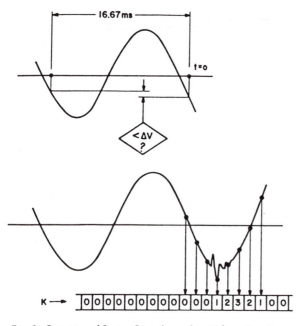

Fig. 2. Detection of System Disturbance from Voltage Waveforms.

The heart of the transmission line relaying program is the impedance computation procedure discussed in the previous section. A number of laboratory tests of this program were made to determine the effect of dc offset terms, line loading, line charging current, switching surges, etc. upon the computational error. A few of these results are shown in Figures 3 through 5. Figure 3 is obtained by putting line-line faults at various points on a transmission line having negligible line charging current and the fault currents were fully offset. The error in computation of the impedance at any point is expressed as a percentage of the impedance for the entire line. The maximum error increases approximately linearly with fault location-or the error expressed as a percentage of the measured quantity is approximately constant. A typical voltage and current signal pair for these cases is shown in Figure 6.

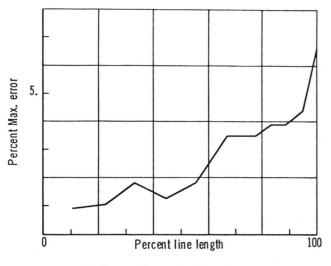

Fig. 3. Maximum Error in Impedance Computation. No Line Charging Capacitance.

Figure 4 shows the effect of dc offset on errors of computation for line to line faults at the end of a line. The second curve on Figure 4 corresponds to a 225 mile 765 kv line model having very signifi-

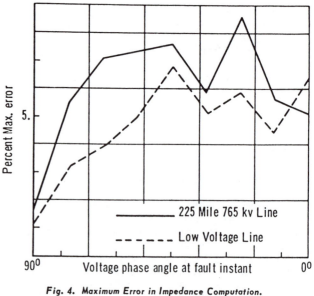

Fig. 4. Maximum Error in Impedance Computation. Variation Due to Varying dc Offset.

cant line charging current. It can be seen from this Figure that the switching surges associated with long EHV lines increase the computational errors. Figure 5 shows the error for faults at different points on the EHV line, one curve showing no offset in the fault current, and the other showing a fully offset fault.

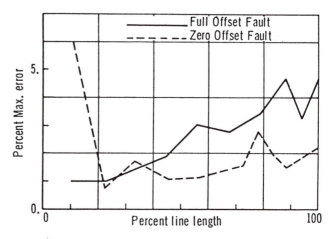

Fig. 5. Maximum Error in Impedance Computation. 225 Mile 765 kv Line.

These results are obtained from a series of runs made to evaluate the performance of the algorithm. The first computation is performed as soon as the seventh sample after the fault becomes available. The computations are repeated with every new sample as it becomes available, and the largest of the errors in each of these series of computation (corresponding to a line section) is plotted in Figures 3 through 5. In general, the largest error is made at the seventh sample computation, and the average error for a series is much smaller than the maximum.

Although the accuracy of the algorithm seems adequate, there is room for improvement in its performance. A detailed study of the errors and their causes was not made at this time since many of these errors are computer hardware dependent, and it is not certain that the computer used for these tests would be a final choice for a substation computer system. It may be mentioned that it is entirely possible to use scaling of signals and integer arithmetic routines in such a manner that the error is a minimum at the balance point of the first zone of the relay.

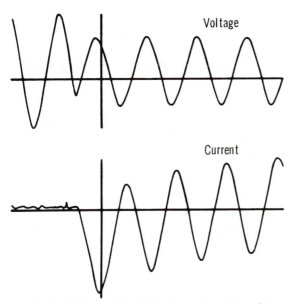

Voltage

Current

Fig. 6. *Model Generated Fault Waveforms. Low Voltage Line.*

FIELD TESTS

To test many of the programs developed in the laboratory, a minicomputer (IBM S/7) was installed in the Matt-Funk substation near Roanoke, Virginia. A station one line diagram for Matt-Funk is shown in Figure 7.

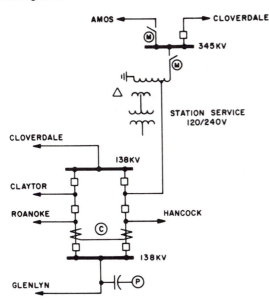

Fig. 7. *Matt-Funk 138 kv Station One Line Diagram.*

The computer and its peripheral hardware configuration is shown in Figure 8. The computer was connected to monitor the Matt-Funk Glen Lyn line. Various programs tested during the field test included alarm and monitoring programs, teleprocessing programs, oscillography programs, and certain parts of the relaying programs.

It was decided that this first field test would be used primarily to gain experience with the operation of a computer system in an unattended EHV substation. The computer input power was supplied form the AC station service during this initial research phase. Clearly this cannot be a practical mode of operation for a relaying computer. However, in view of the limited objectives for this phase of field testing, it was decided that an uninterruptible power supply would not be used at this time, and consequently the computer was

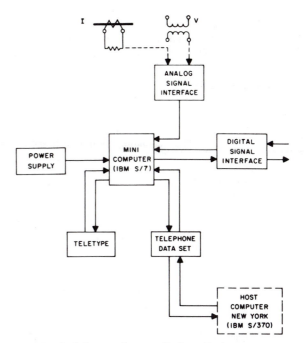

Fig. 8. *Substation Computer Hardware Configuration.*

expected to be inoperative for several types of faults. This, coupled with some early software errors, caused the need for operator intervention before the computer could resume operation following a power supply interruption. Similar loss of the computer resulted from some teleprocessing system failures, again requiring operator intervention for restart. Since the substation was an unattended substation, manual restarts involved fairly long delays. Needless to say, after the software errors were corrected, no manual intervention was needed and the programs restarted automatically following a failure. There was only one confirmed computer hardware problem requiring correction in the field. It is felt that this preliminary test of the computer hardware and software systems was quite encouraging.

During this period, the computer recorded one hundred system transients, while the station oscillograph recorded forty-two. The larger number of computer recordings is due to the higher sensitivity of its threshold as designed into the program. Twelve events were correlated between the station oscillograph output and the computer

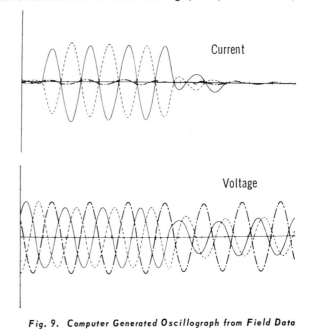

Fig. 9. *Computer Generated Oscillograph from Field Data*

generated oscillograms. As mentioned earlier, several events, including the 345 kv faults were lost to the computer because of the interruptions to its power supply. Samples of the computer generated oscillograms are shown in Figure 9.

Several relaying program components including the fault detection and classification modules were field tested; and in all events recorded by the computer their operation was found to be correct. The impedance calculation algorithm discussed above and tested in the laboratory was not operational during the field test and consequently no estimates of its performance in the field are available at this time. However, from the laboratory tests, and also from some off-line checks made on the data collected from the field it appears that the algorithms will prove to be satisfactory.

A brief summary of the failure history of programs and computer hardware during the field test is shown in Table II.

TABLE II

Substation Computer Field Test. Failure Analysis.
Total Days Under Test: 316

Cause of Failure	Events	Days Out of Service
Substation Computer Hardware	1	1
Host Computer Hardware & TP	4	32
Substation Computer Software	8	34
Power Failure	5	8
Unknown	2	16

CONCLUSIONS

(1)   Our analytical and experimental work up to this time supports the view that distance relaying of EHV transmission lines with a digital computer is technically a viable concept.

(2)   Response time of the computational algorithms is somewhat greater than one-half cycle after the occurrence of a fault; with the accuracy of distance estimates improving with one-cycle or greater elapsed time after the fault.

(3)   Decaying dc offsets, when present in a waveform, create substantial errors in the estimates and must be dealt with specifically by the computational algorithms.

(4)   In general, techniques using polynomial smoothing of the signals and their derivatives operate well when high sampling rates and narrow data windows are used. They are susceptible to errors caused by the decaying dc offset and other frequency components present in the signal waveform. Fourier Transform techniques (digital harmonic filters, digital notch filters) are very satisfactory at relatively low sampling rates provided that the decaying dc offsets are pre-processed and removed.

(5)   A multi-task substation computer system consisting of mini and micro processors connected in a network at the substation is an attractive arrangement from the economic, securtiy and reliability view point. We intend to study such systems in the future.

REFERENCES

(1)   G. D. Rockefeller, "Fault protection with a digital computer," IEEE Trans. on PAS, vol. PAS-88, pp. 438-464, April 1969.

(2)   B. J. Mann and I. F. Morrison, "Relaying a three phase line with a digital computer," IEEE Trans. on PAS, vol. PAS-90, No. 2, pp. 742-750, March 1971.

(3)   B. J. Mann and I. F. Morrison, "Digital calculation of impedance for transmission line protection," IEEE Trans. on PAS, vol. PAS-90, No. 1, pp. 270-279, January 1971.

(4)   G. B. Gilcrest, G. D. Rockefeller and E. A. Udren, "High speed distance relaying using a digital computer, I - system description," IEEE Trans. on PAS, vol. PAS-91, No. 3, pp. 1235-1244, May 1972.

(5)   G. D. Rockefeller, E. A. Udren, "High speed distance relaying using a digital computer, II - test results," IEEE Trans. on PAS, vol. PAS-91, No. 3, pp. 1244-1258, May 1972.

(6)   A. M. Ranjibar and B. J. Cory, "An improved method for the digital protection of high voltage transmission lines," IEEE PES Summer Power Meeting, IEEE paper no. T74-380-2, July 14-19, 1974.

(7)   M. Soulliard, P. Sarquiz and L. Mouton, "Development of measurement principles and of the technology of protection systems and fault location systems for three phase transmission lines," CIGRE paper no. 34-02, August 21-29, 1974.

(8)   J. Carr and R. V. Jackson, "Frequency domain analysis applied to digital transmission line protection," paper no. T75-055-9, presented at the 1975 Winter Power Meeting.

(9)   J. R. Ziegler, L. N. Walker and A. G. Phadke, "Design of twelve phase synchronous switch," 1973 Midwest Power Symposium, Cincinnati, Ohio, October 22-23, 1973.

(10)   W. A. Lewis and L. S. Tippett, "Fundamental basis for distance relaying on 3-phase systems," AIEE Proceedings, vol. 66, 1947.

(11)   E. Oran Brigham, "The fast fourier transform," Prentice-Hall, Inc. 1974.

(12)   A. Savitzky and M. J. E. Golay, "Smoothing and differentiation of data by simplified least squares procedures, "Analytical Chemistry, vol. 36, No. 8, July 1964.

Discussion

J. L. Koepfinger (Duquesne Light, Pittsburgh, Pa.): The multi task capability of a digital computer is a stimulus for developing its application to power systems substation. The authors have made a contribution to this ultimate goal. As the authors recognize, the prime deterrent to a more expanded use of computer based system in substation has been reliability. An availability of 99.997% is generally considered good for many single CPU systems. Our experience with mini CPU used for Sequence of Events shows that the performance is near this level of availability. Such experience indicates to us that application of digital computers to protective relaying function must be approached with care.

An equally important consideration is the maintainability of the system. Today's single function relay designs generally affect only small part of the substation equipment. This allows maintenance to be performed with relatively minor disturbance to the substation operation.

Recent experience with digital oscillographs confirms the authors finding that a data sample rate of four times the fundamental frequency of the analog information being sampled is desirable.

This paper has investigated the affect of an assymmetrical signal upon the impedance measuring algorithm. Both current transformers and voltage coupling devices can exhibit distorted wave form in their output signal. Would the authors care to discuss how such a phenomenon will affect the algorithm?

Manuscript received July 15, 1975.

R. J. Marttila (Ontario Hydro, Toronto, Ontario, Canada): The authors are to be complimented for an informative paper on the application of the Fourier Transform to the calculation of phasor quantities from waveform samples.

The authors' preprocessing algorithm to reduce the effect of dc offset in the current signal, assumes that the offset has a particular time constant. In your test results of the algorithm, how does the time constant in the current signal compare with that assumed in the algortihm? What is the sensitivity of the algorithm to varying dc constants? Also, it would be of interest to see how the calculated current magnitude and phase of a current signal, with 100% dc offset, behaves between pre-fault and post-fault steady states.

A deteriorating A/D conversion resolution is indicated by Figures 3 and 5 of the paper which show errors increasing with faults farther down the line. What is the resolution of the A/D conversion system used in the study?

As suggested by the authors, from the standpoint of reliability, a substation requires a host of processors and it would seem that each EHV transmission line should have a dedicated processor which is capable of the multiple tasks discussed by the authors. To execute these functions, what are the authors' estimate of the minimum computer requirements with respect to speed and memory size?

Manuscript received July 21, 1975.

G. S. Hope and **O. P. Malik** (The University of Calgary, Calgary, Alberta, Canada): We would like to congratulate the authors for an interesting paper on digital protection. It is interesting to see that the authors have been successful in limiting the window of the Fourier-type analysis to half to a 60-Hz period. Since three years ago we tried without success to use dc filtering techniques, we are in a position to appreciate a successful attempt.

However we would like the authors to comment on general-analysis procedure; e.g., if one were to do spectral analysis using the FFT, one would expect to calculate the magnitude of a 2-Hz waveform by analysis over its entire period or say perhaps half its period. This means that the minimum window to detect a low-frequency component say at 2 Hz would be some 30 times the 60-Hz window. Would the authors comment on a general technique which allows one to overcome this fundamental restraint in rejecting low-frequency components.

Manuscript received July 25, 1975.

The results of our studies previously reported before the PES meetings show that 8 samples per cycle, i.e., a sampling rate of 480 Hz is satisfactory for digital impedance relaying purposes. The authors of this paper have come to the same conclusion but they suggest the rate of 720 Hz as being particularly attractive. We consider 960 Hz to be particularly attractive.

W. A. Lewis (Systems Control, Inc., Sunnyvale, CA.): My interest in the subject of accurate distance relaying of transmission systems has continued over a long period. This is attested by the fact that I was a co-author of Reference 10 of the paper, presented as an AIEE conference paper in 1931 and upgraded to transactions status in 1947. At that time digital computers did not exist and the concept that computers fast enough to be used for relaying purposes could not even be dreamed of.

I am greatly impressed by the care and thoroughness with which the authors have approached their task. I am particularly gratified by two of their conclusions:

1. That satisfactory accuracy requires a data window of a little more than a half-cycle, in order to eliminate the effects of dc offset.

2. That the computer system should consist of several processors, so that an individual processor can be assigned to the impedance calculation of each terminal.

With regard to the first, my physical intuition has been horrified by the claims which have appeared several times in the literature that the effect of offset can be eliminated by using as few as three data samples taken in less than one-quarter cycle. The work of the authors should lay this claim to rest.

With regard to the second, the proposal made several times that a single computer should replace all the relays in a station has seemed to be utter folly from the standpoint of reliability and the sudden burden applied to the computer by the needs for relay determinations for all circuits that have to be made at the same time, when a disturbance occurs. A single computer would require at least one full duplicate, to provide for instant backup in case of computer failure. The complication of keeping both fully operable simultaneously for all functions appears likely to offset any saving made by using an overall computer. The development of economical minicomputers offers, to me, the much more attractive alternative of parallel processing of the relaying functions, with the station computer being used to analyze the results of relay action (by separate minicomputers) and develop restoration strategy, rather than to perform all relay functions by one processor. It seems probable that one or two spare minicomputers could serve as backup for all the minicomputers performing relaying functions, thus substantially reducing the total complement of computer power in a large station.

I have two questions. The complete description of how the distance measuring function is to be performed is not described, but it is suggested that an end result of the sampling computations is to develop R_f and X_f in equation 1. At present I do not see what useful purpose the determination of R_f serves. With the possibility of unknown fault resistance being present, it does not seem reasonable to use R_f for any distance measuring function, and I do not see any other useful purpose. Is there a use for it?

It seems to be indicated that the input to the computer is derived directly from sampling of voltage and current from instrument transformer secondaries. However, analog processing of the inputs is certainly permissible. I believe that if the current were passed through an inductive shunt having approximately the same X/R ratio as the line, the offset in the voltage drop across the shunt, now the "current" input, would largely disappear. The time constant of the local transient for the loop comprising the inductive shunt and the input sensor could be much shorter than the dc time constant of the system, and thus would have little effect after the first sample or so. The phase shift resulting from the shunt will be known and can be allowed for in the computer logic. I would like to ask the authors if they have examined this possibility, and if so, what results were obtained.

Manuscript received July 30, 1975.

J. Carr (Saskatchewan Power Corporation, Regina, Saskatchewan, Canada): This description of field tests is very worthwhile and much appreciated. AEP's commitment to the tests should encourage efforts by others.

Manuscript received August 1, 1975.

It is interesting to note that oscillography has been included as a tested function. I believe that such functions will add greatly to the value of switching station computers while adding little to their cost. It would be interesting to have details of the smoothing or interpolating technique used in the oscillograph generation. Details of the instrument transformers and any filtering used would also be interesting. Such details might reveal why the fault waveforms appear so smooth.

The waveform handling algorithm is based on there being exactly 12 samples per cycle. If the clock used to determine sampling instants is synchronized to the power system, such sampling will be possible. If the clock is unsynchronized, frequency and time tolerances will make exactly 12 samples per cycle impossible. Which timing technique did the authors use? If the unsynchronized system was used, what tolerances were involved and what, if any, adverse effects due to non-integral sampling were noticed?

John T. Tengdin (General Electric Company, Philadelphia, Pa.): The authors have presented a very interesting paper which goes beyond the narrow scope of digital relaying. If digital techniques are to be applied in substation functions, many believe that the resultant performance must be at least equal to existing analog approaches. In the case of oscillography, Figure 9 shows oscillograms which appear to be pure 60 Hertz with no offset. What sampling rate do the authors recommend to obtain frequency response results comparable to oscillograms from analog equipment? Others active in digital relaying have proposed front end filtering to remove both harmonics and dc offset, and, thus, avoid the algorithm problems associated with unfiltered input data. Would the authors comment on the future acceptance of oscillograms which do not reproduce such data, and contain phase shifts as result of the filtering?

Manuscript received August 8, 1975.

A. M. Ranjbar and **B. J. Cory** (Imperial College, London, England): We wish to compliment the authors for the useful work which they have done on digital protection and we would like to give a few comments on their paper.

1. From figures (3) to (5) it seems that the authors have used the modulus of impedance and (probably) its angle as a criterion for measurement thus providing the modulus of impedance with direction. It is well known that such a characteristic does not provide immunity against balanced system conditions such as heavy loads and power swings. By calculating the real and imaginary parts of the voltage and current it is possible to build an ideal quadrilateral characteristic[A] which has the necessary immunity against extreme loads and power swings. The authors have mentioned this (equation 1) but they have not used it throughout the paper.

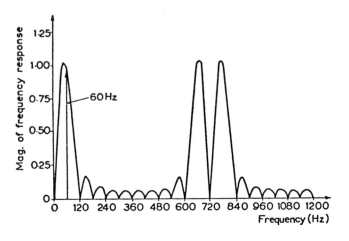

Fig. 1 Spectrum of equation (6) (one cycle data window)

Manuscript received August 25, 1975.

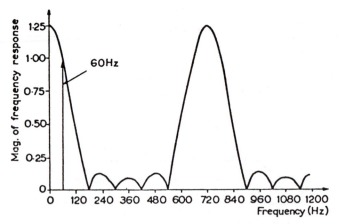

Fig. 2 Spectrum of equation (7) (half a cycle data window)

2. In Table I of the paper, the authors have given the minimum data windows for different sampling rates. In determining these windows, one important factor which has been neglected is the spectrum of equations (5). By reducing the data window, the spectrum becomes worse and less unwanted components are filtered out. For example, the spectrum of equations (6) for one cycle data window and equation (7) for half a cycle data window have been plotted in figures 1 and 2 and it can be seen that with half a cycle data window, even a constant dc offset cannot be filtered out.

3. In equations (10) and (11), R_{21} and R_{41} are not dependent only upon harmonics of order 21 and 41. They are also functions of non-harmonic components and so we cannot agree with the authors that R_{41} is a negligible quantity. We can neglect R_{41} if the current and voltage waveforms consist only of odd harmonics, but, in general, this is not true. Our computations on transmission lines have shown that there is no discernable pattern for voltage and current spectrums and they might contain almost any harmonic or non-harmonic components.

4. The computation of $\hat{Y}_k$ from equations (12) to (15) involves several multiplications. $\hat{Y}_k$ is then used in equation (5) which, in turn, needs further computation, producing the real and imaginary parts of only one current. The whole process must now be repeated for at least two other currents (assuming that the dc offset of voltage is negligible) and then the computed real and imaginary parts are used in equation (1) for the resistance and reactance calculation. This is a tedious procedure which takes some time on a mini-computer. Can the authors say if they have given any consideration to reducing the computations necessary for fault determination?

REFERENCE

A: A. M. Ranjbar and B. J. Cory: "Algorithms for distance protection" IEEE Conference on modern developments in protection, March 1975, Publication No. 125.

R. W. Pashley (Long Island Lighting Company, Hicksville, N. Y.): My congratulations to the authors for presenting a clear and well organized paper on this complex subject. The mathematical techniques employed for the impedance computation, the laboratory test and the subsequent field test are detailed thoroughly and demonstrate a solid approach to the problem. The authors conclusions are thought provoking to all of us, in this area of utility expansion.

My comments are as follows:

The concept of the digital computer in substations has been studied for several years with minimal progress. The approach of using a computer strictly for relaying is limited because the computer cannot provide high speed clearing for 100% of the transmission line without a response from the remote terminal as our present individual solid state systems require and its installation costs are high because of the hostile substation environment. The authors have recognized the economics of such a limited approach and have suggested several periphery duties which could be performed by a substation computer and make them economically justifiable.

Manuscript received September 29, 1975.

I would like to add:

Supervisory (Remote) Control
Relay Target Analysis and Report
Environmental Monitoring
Automatic Meter Reading
Open Conductor Detection
Pattern Recognition
Load Shedding

Dual matched computers would be required and would become part of a network of satellite installations for the main central operations computer complex.

I visualize the relaying application described in this paper being used as common back-up for phase faults and a pattern recognition system as common back-up for ground faults. The latter settings would be programmable, based on the latest system fault studies performed by the main central operations computers. The individual high speed solid state systems would remain as the primary relaying.

Finally, I congratulate the American Electric Power Service Corporation for this pioneering effort. Field experience with substation computers is sorely needed to solve installation, maintenance, programming and the general interface problems between utility users and the manufacturers.

A. G. Phadke, T. Hlibka, and M. Ibrahim: We appreciate these contributions by the discussers which help clarify a number of important questions. Certain points raised by Messers. Hope, Malik, Ranjbar, Cory and Lewis are somewhat related and will be considered first. We are in total agreement with Prof. Lewis's remark that a secure relaying decision must await a significant evolution of the fault waveforms. Thus it is possible to calculate estimates of fault current and voltage phasors with very narrow data windows but the confidence limits accompanying

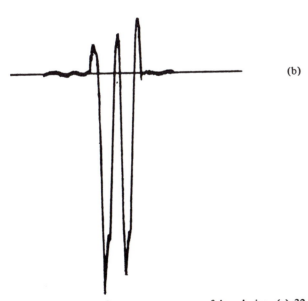

(a)

(b)

Fig. 1   Fault current waveforms upon unsuccessful reclosing. (a) 32 mile, 138 KV line, (b) 203 mile, 765 KV line.

Manuscript received September 29, 1975.

those estimates are quite poor. We have found that a half-cycle window with half-cycle of in-fault data performs very well indeed. There is of course steady improvement in the quality of the estimate as the window is enlarged, which bears out the intuitive feeling that a slower responding relay is potentially a more accurate relay.

The ability of the relaying algorithm operating within a fixed data window to reject the non-harmonic frequencies is affected by the sampling rate. Messers. Cory and Ranjbar point out that with a half-cycle window the dc component cannot be rejected. This is of course true. However, as pointed out in the paper, the dc offset is removed before the Fourier Transform equations are executed. They also raise the question of non-harmonic components (excluding the exponentially decaying dc offset). Should there be significant amounts of non-harmonic components (which lie within the pass-band of the analog filter), these will contribute to the error in the estimation of the impedance. There is no easy solution to this problem (short of increasing the sampling rate), and indeed this must be done if fault waveforms exhibit this characteristic. We have examined numerous fault waveform records obtained from our 138 KV, 345 KV, and 765 KV transmission lines, and find no cases where this would be a serious problem. Figures 1 (a), and (b) for example show a typical fault current (following unsuccessful reclosing) on a 32 mile 138 KV line and on a 203 mile 765 KV line respectively. Note the smoothness of the waveforms in both cases. Also note the significant dc offsets which are often present in the current waveforms of a reclosing operation on a faulted line. Messers. Hope and Malik also bring up a similar question of very low frequency components. Again in the absence of any such components in the observed fault currents, we do not feel that we need to be concerned with them. In our opinion the approach to the relaying problem should be pragmatic, one must balance between the conflicting requirements of a secure and a fast estimation procedure.

Messers. Cory and Ranjbar have commented on the advantages of using rectangular vs. polar formula for the impedance for zone determination. We agree with them and have ourselves used the rectangular form in zone checks. The reason for using the impedance magnitude in Figures 3 to 5 is that the error of the estimate of a complex quantity is most naturally expressed as a magnitude of the error, which defines a disc in the complex plan centered on the true value and encompasses all the erroneous estimates. Prof. Lewis has questioned the need for estimating R_f. It is computed primarily because one may enter the fault algorithms without there actually being a fault on the system (possibly as a consequence of some switching operations), and the complex number $R_f + jX_f$ must then be used for zone-checking purposes. Furthermore, R_f is obtainable with very few extra computations.

The point raised by Messers. Cory and Ranjbar about the excessive number of calculations involved in obtaining $\tilde{Y}_k$ from Y_k is not valid. The computations are performed for only one current as indicated by the classification algorithm. Using these techniques on the IBM System/7, one entire impedance computation is completed in slightly under one millisecond.

Messers. Tengdin, Carr and Koepfinger have commented on various aspects of digital oscillography. We agree with these discussers that including functions such as these at little or no overhead in computational requirements is an asset of the substation computer system. In response to Mr. Koepfinger's question, the transient components of voltage waveforms are given as much attention as those of the current waveform. The reason for discussing the currents only in such detail in the paper is the exponentially decaying dc offset component, which requires special handling. On the question of suppressing the dc offset in the current waveform with the use of mimic circuits (or transient shunts) as pointed out by Prof. Lewis, we have been considering such a scheme for our future work. One of the reasons that such circuits pose a problem is that for modern EHV lines with high X/R ratios they act like a differentiator, thereby amplifying the higher frequencies. We will report our findings on this topic in a future paper. Mr. Tengdin raises a valid question about the quality of the oscillography generated by digital samplers operating at 720 Hz; and the related question of effects of various signal conditioning circuits used on the waveforms. Certainly, if mimic circuits and low pass filters are used the character of the waveforms being recorded is changed. For many users such a modified waveform would be of no use and in that case one must use the traditional oscillographs. However, in a number of instances where oscillography is to be used for sequence of event analysis, the waveforms produced by the low speed sampler may provide adequate resolution. Perhaps the oscillograph generated by the digital relay should be viewed as an important supplement to the conventional oscillograph — where often the number of recording channels is severely limited. It is of course possible to dedicate a processer to the oscillography function and obtain greater resolution through higher sampling rates.

Mr. Carr has asked about the smoothing algorithm used to generate the off-line plot of Figure 9. A modified quadratic interpolation is used to generate three additional points between every pair of observed points. This supplemented data array is then drawn with connecting straight-line segments. As to the signal conditioning used, usual CT's and a two stage RC filter with a cut-off frequency of 360 Hz were used for all signals. We would like to point out again that most of the field oscillograms we have observed are indeed quite smooth. (See Figure 1). Waveforms generated by computer simulation programs using distortion-less modal propagation equations (or other similar approximations) tend to be very rich in high frequencies, and should be used with caution as factual waveforms for fault phenomena lasting several cycles.

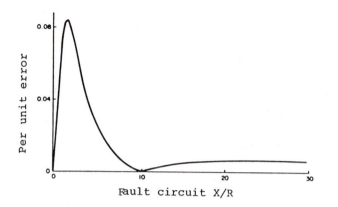

Fig. 2 Error in Phasor estimate due to errors in the X/R assumed by the algorithm.

We have considered the effect of non-synchronous sampling as a result of a drift in computer clocks or in power system frequency. Within the context of observable frequency drifts on a large interconnected power system, the errors are likely to be quite small. On systems with wide frequency excursions, this effect may be significant.

In response to Mr. Martilla's question, Figure 2 for example shows the per unit error of a phasor estimate made with an algorithm using an X/R of 10 while the fault path X/R ratio is varied between 0-30. Interestingly enough, the performance of this algorithm is somewhat similar to that of a mimic circuit using a fixed X/R ratio. Figures 3 and 5 of the paper show the error expressed as a fraction of total line impedance. The error as a fraction of the nominal result of the computation is substantially independent of fault position. We therefore do not agree that the A/D degradation is indicated by these Figures. The A/D converter used had a 14 bit plus sign configuration.

The comment by Mr. Pashley that a computer relay cannot provide high-speed relaying for 100% of the line applies equally well to all relays which have no communication with the remote terminal. The economics of the substitution of a computer relay for a multi-zone distance relay is not yet clearly established — either pro or con — and must await the evaluation of other technical considerations. We therefore do not agree that the computer relay may only be applicable as a back-up relay. It certainly has the speed and accuracy needed for primary and back-up relaying functions.

Messers. Koepfinger, Pashley and Martilla have also touched upon the question of reliability and maintainability of a computer system. We have considered these aspects of a substation computer system and it seems to us that a micro-computer network with dedicated primary responsibility for a relaying task offers many practical benefits. We intend to pursue this subject in greater detail in the future.

Part 7
Monitoring

I. General Discussion

A DIGITAL relay, by its very nature, first makes a measurement of power system quantities of interest (e.g., voltages and currents) and then makes a relaying decision based on these measurements. This is in contrast to conventional relays, where the operating characteristic is inherent in the relay design. The relay acts when a combination of its input signals produces conditions that are satisfactory for its operation. Since the field of digital relaying was first studied, it has been recognized that the measurement made by a digital relay, by itself, can be of great value in many applications.

By monitoring the voltages and currents during nonfault conditions, for example, the relay could act as an ammeter, a voltmeter, or a power meter. Although the question of accuracy must be noted, the potential is still there. Similarly, knowing the currents and voltages during a fault, the distance to a fault could be estimated. Thus, the relay measurements could lead to a fault locator. A number of other measurement possibilities can also be enumerated, e.g., harmonics, unbalances, power factor, etc.

All the measurement tasks a relay can perform can be viewed as a monitoring function. Measuring a system, i.e. monitoring it, is the first step to better utilizing and controlling it. This concept is most fully developed in the area of synchronized phasor measurements made by digital relays and measuring instruments.

The idea of measuring phase angles between voltages at different buses of a power system has an instinctive appeal to the power engineer. To measure phase angles and phase angle differences, is to monitor the parameter that controls the flow of synchronizing power between the buses. Stability is defined in terms of the phase angles, and many control systems depend on the detection of instability. Also, many relaying functions depend on phase angle differences. Undoubtedly, the measurement of phasors and phase angles of AC quantities is a very important function.

To measure phase angle differences between remote buses, synchronization of the measuring instant must be achieved. Many techniques have been undertaken to achieve this synchronization. Recently, the advent of the global positioning system (GPS) of satellites has made the synchronization problem amenable to an economical solution. Some of the papers collected in this part describe this technology.

If synchronized measurements can be achieved, it becomes possible to form a consistent picture of the power system at that instant. This has always been the goal of the state estimation functions in the power system control centers. However, in the absence of the truly synchronized measurements, at best a quasi steady-state estimate was possible. With the high degree of precision in synchronization possible

with GPS transmissions, and with the high speed of response of the measurement process, it is now possible to construct a dynamic picture of the power system. Thus, many dynamic monitoring and control functions can also be undertaken.

These, and other future developments, are certain to revolutionize the monitoring and control functions in power system substations and control centers.

II. Discussion of Reprints

The first paper, ''An Accurate Fault Locator with Compensation for Apparent Reactance in the Fault Resistance Resulting from Remote-End Infeed,'' by Eriksson, Saha, and Rockefeller, describes an excellent technique for using prefault load current data to correct for ground fault resistance effects on apparent impedance seen by a ground distance relay. This technique is an example, wherein the development of a monitoring technique (fault locator) has led to an improved protective relay function. It is likely that this type of mutual interaction between monitoring and protection functions will increase as experience with computer relays is gained.

The next two papers, ''Application of Fault and Disturbance Recording Devices for Protective System Analysis,'' a summary paper by the PSRC, and ''Field Experience with Absolute Time Synchronism between Remotely Located Fault Recorders and Sequence-of-Events Recorders,'' by Burnett, point to the importance of fault recorders in analyzing the performance of modern power systems and their associated protective systems. Fault recorders from various substations can now be synchronized, and a coherent picture of what took place on a power system, and in what sequence, is gained. As the power systems and their protection and control equipment become more sophisticated, the station recording equipment will be even more crucial to our understanding of the power system behavior.

The next papers, ''Phase Angle Measurements with Synchronized Clocks—Principle and Applications,'' by Bonanomi, and ''A New Measurement Technique for Tracking Voltage Phasors, Local System Frequency, and Rate-of-Change of Frequency,'' by Phadke, Thorp, and Adamiak, have ushered in a new era in precise monitoring of power systems through synchronized sampling of voltages and currents. Most of the digital computer-based devices in substations use sampling clocks, and the measurements made by these devices are traceable to the sampling instants. If the sampling clocks in the various substations are synchronized to a certain degree of accuracy, then the measurements made in those substations, particularly phasor measurements, are also accurate to the same level of accuracy. With modern GPS systems, it is possible to achieve sampling clock synchronization to within 1 microsecond. For the first time we now have the ability to measure phase angle differences across the system to a fraction of an electrical degree. This technology is opening the door to improved monitoring, control, and protection of power systems.

III. ASSOCIATED STANDARDS

"*Application of Fault and Disturbance Recording Devices for Protective System Analysis*," IEEE Special Publication 87 TH 0195-8-PWR, 1987.

IV. SUPPLEMENTARY BIBLIOGRAPHY

T. Takagi, Y. Yamakoshi, M. Yamamura, R. Kondow, and T. Matsushima, "Development of a New Type Fault Locator Using the One-Terminal Voltage aand Current Data," *IEEE Trans. on Power Apparatus and Systems*, vol. PAS-101, no. 8, August 1982, pp. 2892–2898.

G. Missout, J. Béland, G. Bédard, and Y. Lafleur, "Dynamic Measurement of the Absolute Voltage Angle on Long Transmission Lines," *IEEE Trans. on Power Apparatus and Systems*, vol. PAS-100, no. 11, November 1981, pp. 4428–4434.

AN ACCURATE FAULT LOCATOR WITH COMPENSATION FOR APPARENT REACTANCE
IN THE FAULT RESISTANCE RESULTING FROM REMOTE-END INFEED

Leif Eriksson,
ASEA AB, Vasteras, Sweden

Murari Mohan Saha, Member, IEEE
ASEA AB, Vasteras, Sweden

G. D. Rockefeller, Fellow, IEEE
Rockefeller Associates, Inc.
Morris Plains, New Jersey

Abstract - A microprocessor based fault locator is described, which uses novel compensation techniques to improve accuracy. It displays the distance to the fault in percent of transmission line length, for facilitating repair and restoration following a permanent fault. Also, it pinpoints weak spots following transient faults.

This new method for fault location on electric power transmission lines uses recorded phase currents and voltages at the near end. The main feature of the method is that it considers the influence of the remote-end infeed of the transmission line by using a complete network model.

A microprocessor filters the ac currents and voltages from the protective relaying instrument transformers to extract the fundamental components of the signals. It then computes the distance to the fault point, compensating for the apparent reactance in the fault resistance resulting from load current and the variations in impedance angles in the power-system network.

The design has undergone field tests and evaluation. The outcome of the field tests is presented and has confirmed the validity of the concept of the fault locator, showing an accurate display of the distance to fault.

INTRODUCTION

Distance relays for transmission-line protection provide some indication of the general area where a fault occurred, but they are not designed to pinpoint the location. Since power circuit breakers are only installed at the terminals, it is immaterial where the line faulted for isolating the flashover. However, with immediate knowledge of the location, the nature (type and measuring data) of the fault can be determined quickly, facilitating repair and restoration. A locator is also useful for transient faults, pointing to a weak spot that is threatening further trouble.

Various methods were developed during the recent years to detect the location of fault on a transmission line. They are mostly based on analog techniques[2]. Some fault locator designs are in service which are reliable in detecting permanent faults. New methods and systems of fault location based on both analog and digital techniques seem to offer good prospects to obtain a precise location of transient faults also. The method reported in [1] is an approximate method. The method of fault location technique described here seems more advantageous, since it takes into consideration the effects of both ends of the line. This novel approach is accomplished by using a complete network model, where the infeed from the network beyond the remote end point is rigorously taken into consideration.

84 SM 624-3   A paper recommended and approved by the IEEE Power System Relaying Committee of the IEEE Power Engineering Society for presentation at the IEEE/PES 1984 Summer Meeting, Seattle, Washington, July 15 - 20, 1984. Manuscript submitted August 30, 1984; made available for printing May 4, 1984.

Pre-fault load-current samples are stored and used for compensation to eliminate a substantial effect on accuracy. Also, representative values for the source impedances are stored to compensate for variations in impedance angles. This novel approach is described as well as the system design and performance.

PURPOSE FOR FAULT LOCATORS

Even where helicopters are immediately available for patrol following unsuccessful reclosing, fault locators perform a valuable service. Trouble cannot always be found with a routine patrol with no indication of where the fault occurred. For example, tree growth could reduce clearances, resulting in a flashover during severe conductor sagging. By the time the patrol arrives, the conductors have cooled, increasing the clearance to the tree. The weak spot is not obvious.

The importance of fault locators is more obvious where foot patrols are relied upon, particularly on long lines, in rough terrain. Also, locators can help where maintenance jurisdiction is divided between different companies or divisions within a company.

Fault locators are valuable even where the line has been restored either automatically or non-automatically. In this category are faults caused by cranes swinging into the line, brushfires, damaged insulators and vandalism. The locator allows rapid arrival at the site before the evidence is removed or the "trail becomes cold". Also, the knowledge that repeat faults are occurring in the same area can be valuable in detecting the cause. Weak spots that are not obvious may be found because a more thorough inspection can be focussed in the limited area defined by the fault locator.

FAULT LOCATION FUNDAMENTALS

The fault location computations determine the apparent fault impedance with novel compensation for the fault resistance drop, eliminating the errors inherent in conventional reactance-type measurements. For a fault-protection relay these errors are tolerable because a safety margin is inserted in its setting. However, for the fault locator a more precise measurement is quite desirable. The Appendix derives the equation for the apparent impedance seen by a reactance-type measurement. Figs. A1 and A2 illustrate the apparent reactance effect in the fault-resistance term.

Fig. 1 shows the connection of the fault locator at station A. Using the ac quantities available at station A, it is not possible to determine the total fault current I_F unless p and Z_{SB} are known; conventional devices tolerate the errors resulting from the infeed current I_B flowing through the fault resistance R_F, out of phase with respect to I_A.

In the method reported in Reference 1, the current distribution factors for the parts of the network, located on either side of the fault point, are assumed to have the same arguments, or else it is assumed that the difference between the arguments is known and constant along the line segment

Reprinted from *IEEE Trans. Power App. Syst.*, vol. PAS-104, no. 2, pp. 424-436, Feb. 1985.

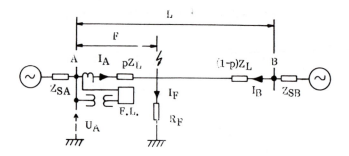

Fig. 1 Power system one-line diagram with fault.

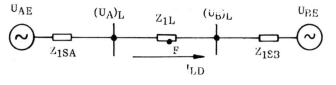

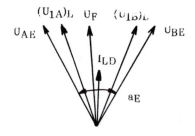

Fig. 2 Prefault conditions.

in question. This simplified assumption can give rise to substantial errors in the calculated fault distance. The difference between the two arguments mentioned generally varies with the distance to the fault.

The fault locator program described here utilizes representative values of the source impedances to determine a correct description of the network. The value of R_F is unknown; however, it is not needed — only the angle of $I_F R_F$, the fault point voltage, is used.

The other key aspect of this locator's algorithm is its use of pre-fault current memory to determine the change in line current caused by the fault: the actual fault current minus the prefault value. Eq. (A9) from the appendix, repeated here as Eq. (1) states that the station A voltage is the sum of the drops in the line to the fault point plus the fault point voltage:

$$U_{1A} = I_{1A} \, p \, Z_{1L} + I_{1F} R_F \qquad (1)$$

Eq. (1) is written for a 3-phase fault, using the positive sequence notations. Eq. (2) expresses the same concept in general terms applicable to any fault:

$$U_A = I_A \, p \, Z_L + I_F R_F \qquad (2)$$

To consider load current effects, the actual voltages and currents can be considered to be composed of the prefault values plus the changes caused by the fault. The appendix discusses this concept. The load currents are generated in Fig. 2 by the angular difference between the source voltages. Voltage U_F exists at the fault point prior to the fault and is the driving voltage in Fig. 3 producing the changes caused by the fault.

From Fig. 3, the change in positive-sequence current in the line at A, ΔI_{1A}, is related to the total positive-sequence current by the current distribution factor D_{1A}:

$$\Delta I_{1A} = D_{1A} \, I_{1F} \qquad (3)$$

Writing a similar, but general expression:

$$I_{FA} = D_A I_F \qquad (4)$$

where I_{FA} is the current change produced by the fault, equal to the actual fault current less the prefault current. The expression for I_{FA} varies with the fault type; Table I defines these. Changes in phase currents are used except for single-line-ground faults. The latter use the change in faulted phase current less the zero sequence current I_{0A}. The zero-sequence current is extracted because the zero-sequence distribution factor D_{0A} is not known as reliably as the positive-sequence factor D_{1A}. The 3/2 factor in Table I provides heavier weighting to compensate for the removal of the zero-sequence current.

The I_{FA} expression for a single-line-ground (SLG) fault will now be derived with the objective of eliminating the

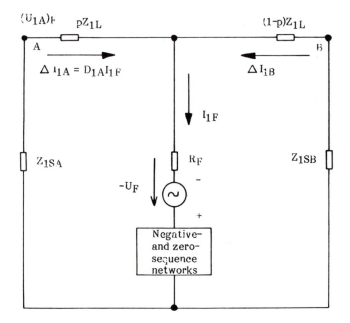

Fig. 3 Faulted network (Thevenin equivalent).

processing of the zero-sequence components to obtain I_F. Fig. 4 shows the current relations in the fault for an SLG fault:

$$\frac{I_F}{3} = I_{1F} = I_{2F} = I_{0F} \qquad (5)$$

Eliminating the zero-sequence current I_{0F}:

$$I_F = 3/2 \ (I_{1F} + I_{2F})$$
$$= 3/2 \ (\frac{\Delta I_{1A} + \Delta I_{2A}}{D_{1A}}) \qquad (6)$$

The distribution factors for the positive- and negative-sequence currents may be assumed, with great accuracy, to be equal (i.e., $D_{1A} = D_{2A}$).

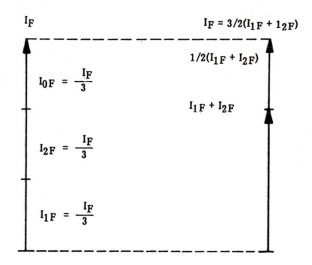

Fig. 4   Current relations for a single-line-to-ground fault.

At station A in the line:

$$\Delta I_A = \Delta I_{1A} + \Delta I_{2A} + \Delta I_{0A} \tag{7}$$

or

$$\Delta I_{1A} + \Delta I_{2A} = \Delta I_A - \Delta I_{0A} \tag{8}$$

Neglecting pre-fault zero-sequence current:

$$\Delta I_{1A} + \Delta I_{2A} = \Delta I_A - I_{0A} \tag{9}$$

Eq. (9) in (6):

$$I_F = 3/2 \; \frac{\Delta I_A - I_{0A}}{D_{1A}} \tag{10}$$

Let $I_{FA} = 3/2 \; (\Delta I_A - I_{0A})$ (11)

The general expression of Eq. (2) can be rewritten as:

$$U_A = I_A \, p \, Z_L + (\frac{I_{FA}}{D_A}) \, R_F \tag{12}$$

From Fig. 1:

$$D_A = \frac{(1-p) \, Z_L + Z_{SB}}{Z_{SA} + Z_L + Z_{SB}} \tag{13}$$

Substituting Eq. (13) in (12) and rearranging, yields:

$$p^2 - p \, K_1 + K_2 - K_3 \, R_F = 0 \tag{14}$$

where:

$$K_1 = \frac{U_A}{I_A Z_L} + 1 + \frac{Z_{SB}}{Z_L} \tag{15}$$

$$K_2 = \frac{U_A}{I_A Z_L} \, (\frac{Z_{SB}}{Z_L} + 1) \tag{16}$$

$$K_3 = \frac{I_{FA}}{I_A Z_L} \, (\frac{Z_{SA} + Z_{SB}}{Z_L} + 1) \tag{17}$$

The complex expression of Eq. (14) contains the unknowns p and R_F. However, Eq. (14) can be separated into two simultaneous equations, one real and one imaginary. By eliminating R_F, a single expression results with the single unknown p. This is solved by the program, using the peak values and their phase position, taken from the Fourier analysis routine which yields the fundamental components of the signals.

Fig. 5 provides an alternative view of how the fault locator determines p without requiring the scalar value of the fault-point voltage. Fig. 5 is a triangulation problem to determine the intersection point F, knowing U_A and I_A by measurement and computing a_D and a_L. While the computer is not specifically using this algorithm, nevertheless the conclusions reached using the concept of Fig. 5 are valid: viz. that the magnitude of $I_F R_F$ is not required, but only its angle a_D.

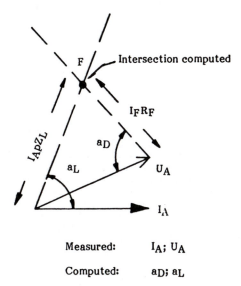

Measured: I_A; U_A

Computed: a_D; a_L

Fig. 5   Fault-resistance-drop compensation.

Parallel Line Cases

The locator can be finished with a modified algotithm for a parallel-line application. The positive-sequence network is completely described by re-defining equation 13 as:

$$D_A = \frac{(1-p) \, (Z_{SA} + Z_{SB} + Z_L) + Z_{SB}}{2 \, Z_{SA} + 2 \, Z_{SB} + Z_L} \tag{13A}$$

Eq. (13A) assumes an identical parallel line. Use of the equation will improve accuracy with the parallel line in service.

Zero-sequence mutual coupling can be compensated for by interconnecting two fault locators. With a fault locator on the parallel line its residual current reading can be input to its companion locator via the local printer loop circuit; the mutual resistance and reactance are input as additional setting parameters.

ALGORITHMS

The fault location algorithm uses the pre-fault and fault currents and voltages at the near end of the transmission line to determine the distance to fault. Current and voltage samples are continuously measured. Following a signal from the line protection at the instant it initiates breaker tripping, current and voltage samples for six cycles are frozen until the completion of the distance-to-fault computation.

527

TABLE I QUANTITIES PROCESSED FOR VARIOUS FAULT TYPES

Type of Fault	U_A	I_A	I_{FA}
RN	U_{RA}	$I_{RA} + K_N \times I_{NA}$	$3/2\,(\Delta I_{RA} - I_{0A})$
SN	U_{SA}	$I_{SA} + K_N \times I_{NA}$	$3/2\,(\Delta I_{SA} - I_{0A})$
TN	U_{TA}	$I_{TA} + K_N \times I_{NA}$	$3/2\,(\Delta I_{TA} - I_{0A})$
RST RS RSN	$U_{RA} - U_{SA}$	$I_{RA} - I_{SA}$	ΔI_{RSA}
ST STN	$U_{SA} - U_{TA}$	$I_{SA} - I_{TA}$	ΔI_{STA}
TR TRN	$U_{TA} - U_{RA}$	$I_{TA} - I_{RA}$	ΔI_{TRA}

The program then determines the loop of analog data (currents and voltages) on which to base the computation. For example, the RS loop is selected if the R and S phase selectors operate. Table I shows the normal loop selection. For double-line-ground faults only, a phase-to-ground loop may be selected instead of the normal phase-phase loop, if desired. Table II shows the loop options. The program next determines the fault inception point within the 6 cycle data, looking at the selected loop ac quantities, choosing the current(s) first. If necessary, the voltage(s) is also processed to find the fault inception point.

Reading 24 samples apart are compared for a significant change, starting with oldest samples. The required threshold for the current change is adaptive, depending on the prefault current level. The voltage threshold is fixed. If no change is found, a one sample advance occurs and the procedure is repeated.

TABLE II LOOP SELECTION FOR DOUBLE-LINE-TO-GROUND FAULTS

Phase Input	Loop Selected Normal	Cyclic	Acyclic
RSN	RS	SN	RN
STN	ST	TN	TN
TRN	TR	RN	RN

If unable to find the fault-inception point, the program reverts to a "slow start" algorithm, which eliminates the load-current compensation. This would occur for a switch-into-the-fault case or a time delay trip for an end-zone fault, where in such cases no prefault samples are available. In both cases, there would be no load flow to compensate, because at the time of tripping the far end of the line would be open.

When a two-phase loop is selected (e.g. R and S for an RSN selector input), two different fault inception points will generally occur; the program selects the latter of the two fault points.

The program selects 24 samples from the two periods spanned by SL1 to SF1 in Fig. 6, where SF1 is the third sample following the fault point. For the fault data, 24 samples immediately following SF1 in Fig. 6 are selected.

The pre-fault and fault currents and voltages on all phases are filtered by Fourier analysis to yield the fundamental component. The selected loop quantities are then processed for the distance-to-fault determination, while the complete set is available for printing for user analysis.

The fundamental components are determined by multiplying each sample by the appropriate instantaneous sine value and integrating over a full period, then repeating the process by multiplying by cosine values. The result provides the scalar value and argument for each ac quantity. The Fourier filtering effectively attenuates the dc offset component and power system harmonics plus CCVT transients and CT saturation distortion.

The program uses the peak and angle outputs from the filtering routine to compute the distance to the fault.

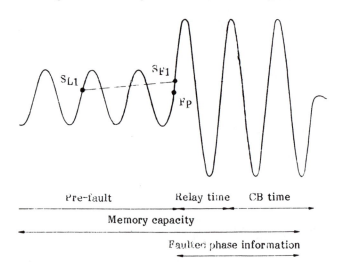

Fig. 6 Samples are selected from two periods before and after fault inception point F_p.

HARDWARE

Fig. 7 shows the main elements of the hardware. The digital inputs consist of the start signal from the line protection relay breaker-trip output and the phases R, S, T (i.e., A, B, C) and ground phase selection outputs from the line protection or from the integral phase underimpedance and ground overcurrent phase-selector units. The start input initiates a distance-to-fault computation.

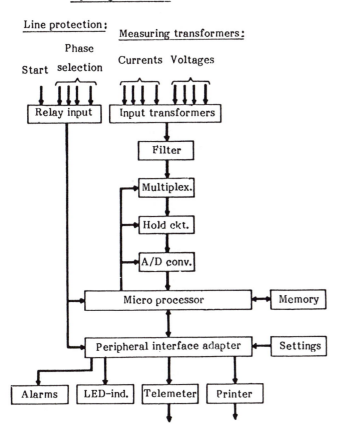

Fig. 7   Hardware configuration.

The 3-phase currents and voltages enter through input transformers which provide galvanic isolation from the instrument transformers, as well as transforming the signals to a level suitable to the electronics. A screen between the windings minimizes common-mode surge coupling. The analog signals feed through low-pass filter for signal conditioning, using a 500 Hz cut-off frequency. The filter outputs are switched in sequence by the multiplexer and fed into the hold circuit in preparation for conversion to a digital value proportional to the instantaneous value of the ac wave. Both the digitized signals and the relay input status are stored in a 6-cycle circular file in the memory with the aid of the microprocessor (MC 6803 microprocessor, 8 bit, 814 ns memory cycle).

The microprocessor processes the measuring values according to the fault location algorithm and presents the distance to fault in percentage of the line length on the LED indicator or remote connected equipment (printer, telemeter, etc.). The microprocessor continuously executes a monitoring routine. If the computer fails to periodically output an "all's well" signal, a peripheral circuit operates an alarm relay. During normal service, a pushbutton is used for

initiating test programs to the memory and for resetting the LED indicator.

Table III shows the parameters to be set by the user. See GLOSSARY OF SYMBOLS. Parameters 1 to 4 are the line constants; 5 to 8 are the source constants. The source impedances will change with system conditions, so a representative value is selected. The inclusion of the source impedances in the fault location algorithm is a novel approach, which allows compensation for different impedance angles in the power system network.

The parameters are set with the help of thumbwheels. A pushbutton operation initiates an EPROM (electrically programmable read-only memory, 22 Kbytes used) burn-in. The digits of parameter No. 9 determine four different items listed in Table III.

TABLE III: SETTING & PROGRAMMING PARAMETERS

1. R_{1L}	7. R_{1SB}
2. X_{1L}	8. X_{1SB}
3. R_{0L}	9. – Type of phase selection (Normal, cyclic, acyclic)
4. X_{0L}	– CT polarity
	– Printout (A,B,C)
5. R_{1SA}	– Line number
6. X_{1SA}	

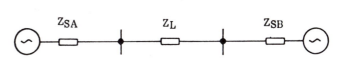

DESIGN

Fig. 8 shows the fault locator assembly with phase selector and printer, which is suitable for 19" rack mounting. The assembly consists of test unit, power supply unit, transformer unit and measuring shunt unit, which are directly mounted to and held together by two apparatus bars. The remaining plug-in units are connected to a mother-board, fed from the shunt unit. The bottom half of the assembly consists of the fault-type selection underimpedance and ground overcurrent units and the printer.

The result of the distance-to-fault calculation is shown on the LED indicator. Fig. 9 shows an example of output results from the printer. The relative distance to fault is expressed in percentage of the line length. "Phase = RN" indicates A and ground inputs from the phase selectors (i.e., a phase R to N fault). "LOOP = RN" indicates that the computer processed phase A to ground voltage and phase A and residual currents to determine the fault location. The computed r.m.s. values of the fundamental component of ac signals in polar form prior to and during the fault are also shown in Fig. 9.

DESIGN TESTS

Model-line tests were performed on six fault locator units at the factory under dynamic conditions. A 100 km, 170 kV line was modelled using CT's and PT's or a CCVT model. The line impedances referred to 380 V were: $Z_{1L} = 2.2 + j\,17.0$ and $Z_{0L} = 16.4 + j\,68.0$ ohms. The ohmic values

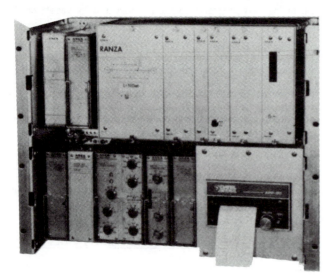

Fig. 8 Fault locator assembly with
phase selector and printer.

Line number 1

Relative distance to
fault p = 75%

Phase = RN, Loop = RN

IR	=	AMPL. =	002,213 A
		ARG. =	028,4 DEG.
IS	=	AMPL. =	000,292 A
		ARG. =	308,7 DEG
IT	=	AMPL. =	000,831 A
		ARG. =	212,0 DEG
IRØ	=	AMPL. =	000,519 A
		ARG. =	103,1 DEG
ISØ	=	AMPL. =	000,514 A
		ARG. =	343,3 DEG
ITØ	=	AMPL. =	000,524 A
		ARG. =	223,1 DEG.
IN	=	AMPL. =	001,475 A
		ARG. =	015,1 DEG
UR	=	AMPL. =	052,157 V
		ARG. =	093,0 DEG.
US	=	AMPL. =	062,611 V
		ARG. =	332,9 DEG.
UT	=	AMPL. =	063,192 V
		ARG. =	210,5 DEG.
URØ	=	AMPL. =	063,510 V
		ARG. =	092,0 DEG.
USØ	=	AMPL. =	063,672 V
		ARG. =	331,7 DEG.
UTØ	=	AMPL. =	063,774 V
		ARG. =	211,9 DEG.

IRØ, ISØ, ITØ — Pre-fault

URØ, USØ, UTØ — Pre-fault

Fig. 9   Printer output.

stated below are also referred to 380 V, the primary of the system simulator. Faults were applied with varying incidence angles. Three series of tests were conducted.

Series No. 1 tests were run with zero fault resistance and rated load current, exported power from station A; the errors were less than 2 % of the line setting. Series No. 2 were run with 50 % load, with fault resistance variations from 5 to 40 ohms; the errors were less than 4 % of the line setting. Series No. 3 tests checked for errors resulting from variations in source impedance from set values; errors were less than 2 % of the line setting. Source A was varied from 2 to 32 ohms, B from 6 to 23 ohms. All ohmic values are on a 1 A rated base.

The locator was also run through the normal test series applied to protective relays, including disturbance (surge) tests: a 4-8 kV fast transient test per SEN361503 (a fast rise test simulating inductive interruptions in the control circuits) and the IEC 255-4 1 MHz, 2.5 kV test.

FIELD RESULTS

1)   Two prototype units were installed for evaluation in Swedish utilities. No misoperations or component failures have been experienced. One of the prototypes was installed on the 30th of June 1982, on a 76 km, 130 kV transmission line between Nybro and Vaxjo, on the Sydkraft transmission system in southern Sweden.

On July 2nd 1982 , a disturbance was detected on the line between Nybro and Vaxjo. The fault locator display showed 89 % (corresponding to 67.6 km). The type of fault that was printed out was a single-phase ground fault on phase T. An aerial inspection of the transmission line was made on July 6th 1982 and the point of disturbance was found in phase T at a distance of 67.0 km from Nybro. The flashover occurred from conductor to arcing horns, across a tower insulator string. The distance to fault, calculated by the fault locator deviated from the actual distance to fault by less than 1 %.   Load current was 238A and the calculated fault resistance was 7.2 ohms. The parameter settings matched those existing in the network at the time of the fault.

The other prototype was installed on a 400 kV Swedish State Power transmission line. However, no fault has occurred on that line.

2)   A staged fault test was conducted jointly by Texas Power & Light Co. and Texas -New Mexico Power Co. on the described fault locator. A fault locator was installed at Northwest Carrollton sub as shown in Fig. 10.

NW Carrollton   TI (Lewisville)   Highlands
138kV

3.1 ∠80   2.6 ∠80
10.6 ∠76   8.7 ∠76

FLT. LOC.   2.7 ∠74

Source: 3.4 ∠80   Source:
0.8 ∠83   11.3 ∠75   6.9 ∠79
1.4 ∠82   19.8 ∠75

Imp. in % on 100 MVA base

Fig. 10 Staged faults were applied at TI (Lewisville).

There is mutual coupling over 59 % of the Northwest Carrollton - TI (Lewisville) line with the Northwest Carrollton - Highlands line. It was the intention to estimate the performance of the fault locator on that line <u>without any compensation for the mutual impedance.</u>

Five faults, two phase-to-phase and three phase-to-ground faults were placed at the Lewisville substation during the early morning of October 25th 1983. The faults were applied at the TI (Lewisville) substation where both lines loop into the substation. The actual distance to fault was 55 % of the line. The load current was about 100A. Estimated fault resistance was 1.6 ohms for the ground faults.

The locator functioned properly for all five faults. The highest deviation of the results between the distance to fault shown on the fault locator display and the actual distance to fault was 3 %.

ACCURACY ANALYSIS

Results of overall accuracy during dynamic model-line tests are described under DESIGN TESTS. This section will discuss the accuracy of this locator in comparison with those using other approaches.

The fault locator described above computes the distance to the fault point by compensating for the apparent reactance in the fault resistance resulting from load current and variations in the impedance angles in the power system network. However, minor errors (see Fig. 11) can occur if the settings are not adequate. These errors should be moderate for practical situations. For one, the value of the fault locators increases with line length; for these applications the line impedance tends to mask the effect of source impedance variations. Again for longer lines, the fault resistance, particularly for single-phase-ground faults, becomes a smaller percent of the line impedance.

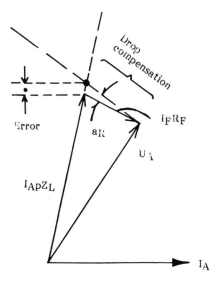

Fig. 11: Location error resulting from error in computing argument of $I_F R_F$.

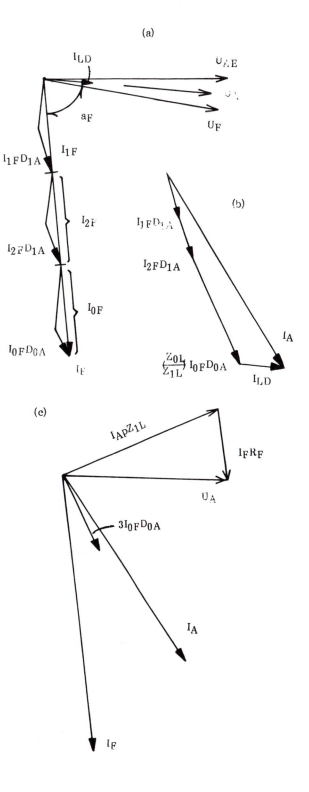

Fig. 12: System quantities for a single-line-ground fault with load <u>export</u> from station A.

The next series of illustrations shows qualitatively the fault-resistance infeed effects on two alternative methods of determining fault location: the reactance and the zero-sequence current-reference techniques. Two examples are shown: one for load export; the second, import.

Fig. 12(a) shows the station A line currents out of phase with the total fault current I_F. Stating it another way, source B produces out-of-phase infeeds in the fault

resistance. The total phase current at A is developed in Fig. 12(b) and is the sum of the currents resulting from the fault and driven by U_F, and the pre-fault current I_{LD}. Fig. 12(c) extracts from Figs. 12(a) and (b) the essentials for developing the error diagrams of Figs. 13 and 14.

The reactance type of Fig. 13 uses the same basis of measurement as does the conventional reactance relay. The angular shift of $I_F R_F$ produces an error which can become

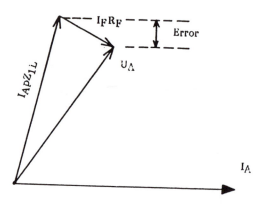

Fig. 13: Reactance-type error resulting from conditions in Fig. 12.

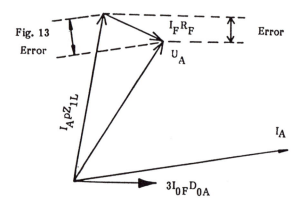

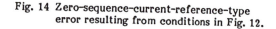

Fig. 14 Zero-sequence-current-reference-type error resulting from conditions in Fig. 12.

significant for the fault location function where greater precision is needed than for a protective relay.

The type shown in Fig. 14 differs from the reactance type in its use of residual current, $3I_0FD_0A = 3I_0A$ as the reference rather than using I_A. For the example of Fig. 12, the two currents $3I_0FD_0A$ and I_A are nearly in the same phase position, so the error in Fig. 14 is almost as much as the Fig. 13 error.

In Fig. 15(a), power is being imported to Station A so I_{LD} is about 180° from its position in Fig. 12. The fault currents here are identical to those of Fig. 12; however, the superposition of load current results in a different magnitude and angle for I_A. Compare 16 and 17: the latter now has a larger error because its reference current, $3I_0FD_0A$, leads the I_A position.

The load compensation method [1] will experience errors generally less than those shown in Figs. 14 and 17, depending upon the amount of out-of-phase infeed into the fault from the far end. This can be seen in the following numerical examples.

Single-Line Examples

Table IV shows the location errors for three cases, comparing four computational methods, using an off-line computer simulatation. Table V defines the power system conditions for two different transmission lines. These were

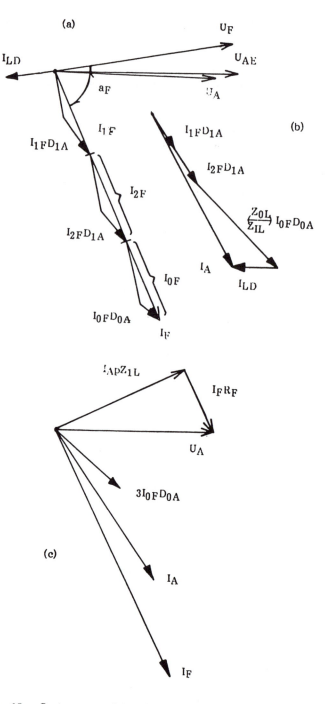

Fig. 15: System quantities for a single-phase-ground fault with load import into station A.

then faulted as specified in Table IV. Method 1 has zero error because the source impedance values were input, describing the complete system. Method 4 shows substantial errors, although smaller than both methods 2 and 3.

Case 3 in Table IV yields greater errors than cases 1 and 2 because the zero-sequence line impedance argument of 76.7 degrees is lower with respect to the zero-sequence source impedance argument. The corresponding angle for cases 1 and 2 is 81.7 degrees. Thus, the zero-sequence out-of-phase infeed is much greater in case 3, causing greater errors for methods 2, 3 and 4.

With smaller fault resistances the errors in Table IV will be correspondingly smaller. On the other hand, lower voltage lines with lower arguments will produce larger

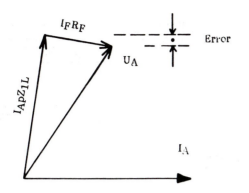

Fig. 16: Reactance-type error resulting from conditions in Fig. 15.

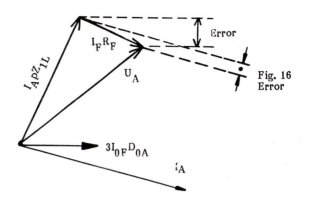

Fig. 17 Zero-sequence-current-reference-type error resulting from conditions in Fig. 15.

TABLE IV COMPARISON OF FAULT LOCATOR ERRORS FOR THE APPLICATIONS IN TABLE V

					CALCULATED RESULTS							
	Power	Fault	Fault	R_F	Method 1		Method 2		Method 3		Method 4	
Case	System	Locator At:	Point	Ohms	p	Error	p	Error	p	Error	p	Error
1	Ashe-Marion	Ashe	0.9	50	0.9	0	0.87	−3.1	0.62	−28.4	0.88	−1.9
2	Ashe-Marion	Ashe	0.9	100	0.9	0	0.86	−4.5	0.54	−36.7	0.87	−2.7
3	Stenkullen-Borgvik	Stenkullen	0.8	125	0.8	0	0.72	−7.9	0.39	−41.0	0.76	−4.2

Method 1   As described in this paper.

Method 2   Zero-sequence-current-reference type (see Fig. 14)
K_N as defined in the Glossary

Method 3   Reactance type (see Fig. 13)
K_N as defined in the Glossary.

Method 4   Load compensated [1] .

Errors are in percentage of total length of line.

errors where the remote source argument is higher, such as would be the case near a generating station.

Parallel-Line Examples

Table VI shows computer-simulation results for two parallel-line cases. In both instances the parallel line is identical to its companion listed in Table V. The errors in Table VI for method 1 would occur if parallel-line compensation were not included. With both the positive - and zero-sequence effects included in the algorithms, the computed errors were zero. The uncompensated errors for the Lieto-Forsa application are greater because of the lower line arguments compared to the Ashe-Marion lines. The parallel-line effect upon accuracy will be the greatest for far-end faults, because these produce the maximum parallel-line current.

APPLICATION CONSIDERATIONS

The unit is suitable for use with lines or sources with secondary impedances in the range of 0-1000 ohms based on a 1 A CT (0-200 ohms for 5 A CT's). Signal current range is 0.1-20 times rated; voltage, 0.01-1.5 rated. It is available with 1, 2 and 5 A ratings, withstanding 3 times rated current continuously.

It is designed for use with the relaying instrument transformers and introduces a burden of 1 VA per phase at rated voltage or current.

The unit is designed and tested just as if it were a protective relay.

Operating temperature range is 0 to 55° C.

It requires 1.5 cycles of fault duration for an accurate measurement.

It is not necessary to apply one locator at each line terminal. Four units can share one optional printer.

The distance to the fault can be remotely logged using the built-in telemetering outputs from the locator.

TABLE V SYSTEM DESCRIPTIONS FOR EXAMPLES IN TABLES IV AND VI

		Ashe-Marion (USA)	Stenkullen-Borgvik (SWEDEN)	Lieto-Forsa (FINLAND)
Voltage – kV		525	400	110
Z_{1SA}	mag.	17.6	21.3	5.8
	arg.	90	90	85
Z_{0SA}	mag.	6.2	24.4	70
	arg.	90	90	85
Z_{1SB}	mag.	42.2	25.6	15
	arg.	90	90	85
Z_{0SB}	mag.	29.5	27.8	130
	arg.	90	90	80
Z_{1L}	mag.	118.5	53.5	27.6
	arg.	87.0	86.4	73.7
Z_{0L}	mag.	463.0	247.6	108.7
	arg.	81.7	76.7	74.3
U_{AE}	mag.	320.4	225.2	63.5
	arg.	-17.0	0	0
U_{BE}	mag.	334.9	240.2	63.5
	arg.	-67.0	-29.4	-15
I_{LD}	mag.	1555	1184	
	arg.	-37.0	-5.7	

Impedances in ohms
Voltages in kV
Currents in A
Arguments in degrees

SUMMARY

1   The fault locator displays the percent distance to the fault, using novel compensation techniques for the errors resulting from remote-end source infeed into the fault resistance. This is accomplished by using a complete network model.

2   Compensation for fault-resistance voltage drop utilizes prefault current and representative values for source resistance and reactance. The inclusion of source impedances is a novel approach which provides improved accuracy.

3   The microprocessor-based system is designed and built to protective-relaying standards, including all surge withstand requirements.

4   It is suitable for use with the relaying instrument transformers, introducing negligible burden.

5   Extensive software filtering minimizes the effects of power-system and instrument-transformer transients.

6   The microprocessor monitors itself and built-in functional-test facilities can pinpoint a particular chip or analog-system failure.

7   The locator can function either with the line-relay phase selectors or with the optional built-in relay units.

8   Only one end of the line needs to be equipped.

TABLE VI FAULT LOCATOR ERRORS (METHOD 1) FOR PARALLEL LINE APPLICATIONS WITHOUT COMPENSATION

Fault locator installed at Ashe and Lieto.

		Fault Location	R_F Ohms	Ashe Marion	Lieto Forsa
Z_{0M}	mag	-	-	332.1	75.8
	arg.	-	-	73.6	73.5
I_{LD}	mag	-	-	1165	240
	arg.	-	-	-37.8	4.3

Percent Error:

		Fault Location	R_F Ohms	Ashe Marion	Lieto Forsa
		0.3	50	0.1	
		0.3	100	0.1	
		0.4	4.3		0.4
		0.4	20		0.4
		0.8	50	3.2	
		0.8	100	2.9	
		0.8	4.3		15.8

REFERENCES

1)   T. Takagi, Y. Yamakoshi. M. Yamaura, R. Kondow, T. Matsushima: "Development of a new type fault locator using the one-terminal voltage and current data", IEEE Trans., VOL. PAS-101, No. 8, August 1982, pp 2892-2898.

2)   M. Souillard, Ph. Sarquiz, L. Mouton: "Development of measurement principles and of the technology of protection systems and fault location systems for three-phase transmission lines", CIGRE No. 34-02, August 1974.

GLOSSARY OF SYMBOLS

a_D angle of fault resistance drop, $I_F R_F$, with reference to U_A

a_E angle between generated voltages U_{AE} and U_{BE}

a_F angle of total fault current, with reference to U_F

a_L argument of line impedance

a_R angular error in computing $I_F R_F$

D_A general expression for current distribution factor at station A.

D_{1A}, D_{2A}, D_{0A}
positive-, negative- and zero-sequence current distribution factor in line at station A

I_A, I_B line current on faulted phase(s) at station A and B, respectively. General, i.e. not referring to a specific symmetrical component sequence

I_{1A} total positive-sequence current in line at station A.

$\Delta I_{1A}, \Delta I_{1B}$
positive-sequence current change as a result of fault, in line at station A and B, respectively

I_F total fault current

I_{FA} change in line current at station A, as a result of fault. General – see Table I for specific definition depending upon fault type

I_{1F}, I_{2F}, I_{0F} positive-, negative- and zero-sequence currents in fault

I_{LD} load current in line

I_{NA} residual current ($3 I_{0A}$) in the line at station A

I_{0A} zero-sequence line current at station A

I_{RA}, I_{SA}, I_{TA} current in line at station A, phases R, S and T, respectively

ΔI_{RA}, etc.—total current minus pre-fault current, in the line at station A, phase R

K_N $\dfrac{Z_{0L} - Z_{1L}}{3Z_{1L}}$

N ground, when referring to fault type

p proportion of the total line from station A (fault locator location) to the fault

R, S, T power system phases (alternative: A, B, C)

R_{1L}, R_{0L} positive-and zero-sequence line resistances

R_{1SA}, R_{1SB} positive-sequence source resistance at station A and B, respectively

R_F fault resistance

U_A, U_B voltage on faulted phase(s) at station A and B, respectively. General, i.e. not referring to a specific symmetrical component

U_{AE}, U_{BE} source voltage behind equivalent impedance at station A and B, respectively

U_{1A} actual positive-sequence voltage at station A

$(U_{1A})_F$ positive-sequence change in voltage at station A resulting from fault

$(U_{1A})_L, (U_{1B})_L$ positive-sequence voltage prior to fault at station A and B, respectively

U_F voltage at fault point prior to fault

U_{RA}, U_{SA}, U_{TA} voltage to ground at station A, phases R, S and T, respectively

X_{1L}, X_{0L} positive-and zero-sequence line reactance

X_{1SA}, X_{1SB} positive-sequence source reactance at station A and B, respectively

Z_A apparent impedance on the line at station A

Z_{1SA}, Z_{1SB} positive-sequence source impedance at station A and B, respectively

Z_{0SA}, Z_{0SB} zero-sequence source impedance at station A and B, respectively

Z_{1L}, Z_{0L} positive- and zero-sequence line impedance

Z_{SA}, Z_{SB} source impedance behind station A and B, respectively. General, i.e. not referring to a specific symmetrical component sequence

Z_L total line impedance. General, i.e. not referring to a specific symmetrical component sequence

Z_{0M} mutual impedance of line for the zero-sequence network

APPENDIX

Apparent Impedance for Three Phase Fault

The apparent impedance seen by a conventional reactance relay or fault locator will be derived for a 3-phase fault, including load current effects.

Load current flow in the line will be developed by phase angle a_E between two sources, per Fig. 2. Shunt capacitances and reactances will be neglected.

The derivation will use the superposition principle with Thevenin's theorem, where the total voltage and current at any point in the network is the sum of the prefault values plus those produced by short circuiting the voltage sources and inserting U_F, the open-circuit voltage at the fault point, as shown in Fig. 3. The polarity of U_F in the equivalent circuit of Fig. 3 is such as to cancel the pre-fault voltage except for the fault resistance drop. For the 3-phase fault case, the negative- and zero-sequence voltage drops are zero.

The positive direction of the load current I_{LD} is arbitrarily chosen as shown in Fig. 2.

The positive-sequence prefault voltage at station A is:

$$(U_{1A})_L = U_{AE} - I_{LD} Z_{1SA}$$
$$= U_F + I_{LD} p Z_{1L} \tag{A1}$$

The positive-sequence change in voltage at A produced by the fault is:

$$(U_{1A})_F = - D_{1A} I_{1F} Z_{1SA}$$
$$= D_{1A} I_{1F} p Z_{1L} + I_{1F} R_F - U_F \tag{A2}$$

The actual voltage at A is:

$$U_{1A} = (U_{1A})_L + (U_{1A})_F$$
$$= U_F + I_{LD} p Z_{1L} + D_{1A} I_{1F} p Z_{1L}$$
$$+ I_{1F} R_F - U_F$$
$$= I_{LD} p Z_{1L} + (D_{1A} p Z_{1L} + R_F) I_{1F} \tag{A3}$$

The actual current in the line at A is:

$$I_{1A} = I_{LD} + \Delta I_{1A}$$
$$= I_{LD} + D_{1A} I_{1F} \tag{A4}$$

From (A3) and (A4):

$$Z_A = U_{1A}/I_{1A}$$

$$= \frac{p\,Z_{1L}\,(I_{LD} + D_{1A}\,I_{1F}) + R_F\,I_{1F}}{I_{LD} + D_{1A}\,I_{1F}}$$

$$= p\,Z_{1L} + R_F\left(\frac{I_{1F}}{I_{LD} + D_{1A}\,I_{1F}}\right)$$

$$= p\,Z_{1L} + \frac{R_F}{D_{1A} + \dfrac{I_{LD}}{I_{1F}}} \qquad\text{(A5)}$$

From Fig. 3:

$$I_{1F} = \frac{U_F}{R_F + D_{1A}\,(p\,Z_{1L} + Z_{1SA})} \qquad\text{(A6)}$$

where

$$D_{1A} = \frac{(1-p)\,Z_{1L} + Z_{1SB}}{Z_{1SA} + Z_{1L} + Z_{1SB}} \qquad\text{(A7)}$$

Eq. (A6) in (A5):

$$Z_A = p\,Z_{1L} + \frac{R_F}{D_{1A} + I_{LD}\,\dfrac{R_F + D_{1A}\,(p\,Z_{1L} + Z_{1SA})}{U_F}}$$

$$\text{(A8)}$$

Eq. (A8) is a quadratic in p, the unknown fault location. R_F is also unknown.

Note that the argument for the R_F term is influenced by both the load current angle and by the impedances.

For a long line and a fault near station B, D_{1A} in Eq. (A7) approaches:

$$D_{1A} = \frac{Z_{1SB}}{Z_{1L}}$$

Assuming that Z_{1SB} has a larger argument than Z_{1L}, D_{1A} will have positive argument.

Neglecting load current, Eq. (A8) reduces to:

$$Z_A = p\,Z_{1L} + \frac{R_F}{D_{1A}}$$

This is plotted in Fig. A1 with the locus for varying R_F; the curve with the load effect included is also plotted.

The apparent-reactance effect can also be seen on a phasor diagram such as Fig. A2, with the current plotted horizontally as the reference.

Fig. A2 can be derived by rearranging Eq.(A3):

$$U_{1A} = p\,Z_{1L}\,(I_{LD} + D_A\,I_{1F}) + I_{1F}\,R_F$$

$$= p\,Z_{1L}\,(I_{1A}) + I_{1F}\,R_F \qquad\text{(A9)}$$

where I_{1A} = actual line current at station A.

Thus, a conventional reactance distance relay or fault locator using the same principle is subject to an error because of fault resistance, load current and differences in system-impedance arguments. The error shown in Figs. A1 and A2 is negative; however, for an

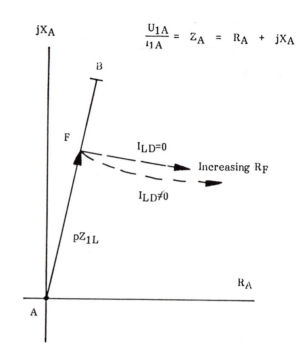

$$\frac{U_{1A}}{I_{1A}} = Z_A = R_A + jX_A$$

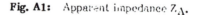

Fig. A1: Apparent impedance Z_A.

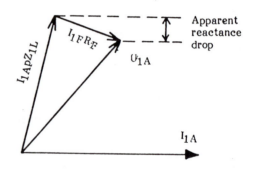

Fig. A2: Phasers showing apparent reactance effect in fault point potential.

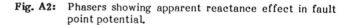

import condition at station A (U_{AE} lags U_{BE}), the error will be positive.

536

Discussion

J. M. Crockett (Westinghouse Canada Inc., Hamilton, Ontario, Canada): The authors have outlined an interesting concept in fault locator design but some additional information would be useful.

The derivation of the equivalent source impedances Z1SA and Z1SN could be described in greater detail. It would appear that they are not actual source impedances but based on a system reduction to a two line machine equivalent and then the delta star conversion of the equivalent parallel line and source impedances. If the resulting mutual branch is eliminated, positive sequence source impedances result which yield correct distribution factors for all fault locations. Is this the approach used? It is not exact in the zero sequence network if mutual coupling is involved.

The positive sequence equivalent is not precisely rigorous if shunt loads at the line terminals are ignored or variable, and the degree of precision as regards fault locator accuracy may be very important even though the resulting source impedance argument variation is less than 3 degrees.

As an example of errors, consider source to line impedance ratio in the order of 0.5, shunt loads in the order of 12 or more times the source impedance, a substantial phase shift between sources, a high resistance fault to ground such as a tree, and a fault locator at the power receiving line terminal. I believe substantial changes in measured fault location would be observed as the shunt load, fault location, and fault impedance vary. Would the authors agree?

The ability to locate a high impedance fault to foliage would indeed be useful but has this capability been demonstated? Slow relaying would be expected for such a fault and hence apparent current and voltage deviations could be zero when the line was tripped.

Finally, I believe that the power-sending terminal is the preferred fault locator position. Would the authors agree?

Manuscript received August 2, 1984.

L. Eriksson, M. M. Saha, and **G. D. Rockefeller:** The authors thank Mr. Crockett for his interest in this paper and for his relevant comments and questions. His discussion can be segregated into four areas:

1. derivation of the current distribution factor for the parallel line case
2. effect of shunt loads
3. availability of prefault current information for a high-impedance fault
4. preferred spot for the locator.

These item numbers will be used below.

1. The current distribution factor as stated in Eq. 13A is rigorous assuming an identical parallel line. It is independent of any zero-sequence effect, including mutual impedance, because only the positive-sequence factor is needed to derive the total fault current I_f.

Measured voltage will contain any mutual induction from a parallel line—this can be compensated for by inputting the zero-sequence mutual impedance to the locator and providing the program with the measured parallel-line zero-sequence current via the printer loop, from the companion locator.

The Eq. 13A expression is derived using a delta-star conversion, where the source impedances are merged into the equivalent impedances. However, the *actual* source impedances are input just as for the single line algorithm which uses Eq. 13 for the distribution factor.

2. Shunt impedance effects at the two ends of the line can be rigously canceled by including these impedances in the source impedances. This is the case for either the single line or the parallel line algorithm. The load and source impedances are merely paralleled. Mr. Crockett is correct that changes in these, coupled with fault resistance, will introduce errors. The magnitude of the load impedance will have little effect, but its angular difference compared to the other impedances will have an effect as the impedance varies from the nominal value used to determine the source impedance input. The following example illustrates this effect.

A computer simulation on a 400 kV, 150 km line with a shunt load applied at Station A of the proportions stated by Mr. Crockett produced these results:

Case	Fault Point	R_f Ohms	Source Impedance Scalar	Angle	Error %
1	0.5	20	23	90	1
2	0.5	20	22.9	85.2	0
3	0.5	100	23	90	3
4	0.5	100	22.9	85.2	0
5	0.8	100	23	90	4
6	0.8	100	22.9	85.2	0

All six faults were phase to ground. Cases 2, 4 and 6 have zero error, because the load impedance has been rigorously accommodated. Cases 3 and 5 represent extreme conditions, with 100 ohms of fault resistance, particularly case 5 where the fault is near the far end. By choosing a representative value for the load impedances variations in this value will have a limited error effect.

The parameters for this example are:

	magnitude	angle-deg.
Z_{1SA}	23	90
Z_{0SA}	23	90
Z_{SH}	276	0
Z_{1SB}	20	90
Z_{0SB}	40	90
Z_{1L}	46.5	84.4
Z_{0L}	156.7	77.5
U_{AE}	230.9	0
Z_{BE}	230.9	-38.5
I_{LD}	2312	-14.3
Z_{SP}	22.9	85.2

Impedances are in ohms, voltages in kV, currents in amperes and angles in degrees. Z_{SF} is the positive-sequence equivalent of the paralleled source Z_{SA} and load impedance Z_{SH}. Note that the effect of the load impedance at Station A is to shift the actual source angle from 90 deg. to 85.2 deg. for the paralleled equivalent of the source and load impedances. The load has a negligible effect on the magnitude of this equivalent.

3. If the normal high-speed proteciton fails to detect a high-resistance fault, such as a mid-span flashover to foliage, two possibilities exist for the locator:

    a) The fault-locator terminal opens following the remote terminal. In this case, there is no load current flow during the fault period used by the locator to compute. Thus, no error occurs because of load current.

    b) The fault-locator terminal opens first. The prefault data are not available in memory if the line protection trip is delayed. This situation will be recognized by the program and an alternative "slow trip" routine is processed. Errors will occur because of the load current component.

4. The direction of power flow is not of concern to the fault locator, since its algorithm eliminates load flow effects. That is, there is no bias towards installing the locator at the "power sending terminal" unless this describes the *stronger* source of fault current. The stronger source installation reduces errors because the weaker end produces less of an infeed effect.

5. The following additional field results are noteworthy. Six phase-ground faults occurred during very high winds on a 400 kV, 135 km line. The precise location of these temporary faults is not known. The locator operated in each case, indicating in the range of 93 to 99%.

Manuscript received September 4, 1984.

SUMMARY OF THE SPECIAL PUBLICATION
"APPLICATION OF FAULT AND DISTURBANCE RECORDING
DEVICES FOR PROTECTIVE SYSTEM ANALYSIS"

Paper 7.2

A Working Group of the Relaying Practices Subcommittee of the IEEE Power System Relaying Committee: C. W. Barnett, Chairman; J. R. Boyle, J. D. Brandt, J. A. Bright, R. O. Burnett, A. D. Cathcart, D. M. Clark, D. H. Colwell, C. W. Fromen, E. A. Guro, R. W. Haas, E. A. Hauptman, R. W. Hirtler, D. Hollands, S. R. Jack, T. L. Kaschalk, D. K. Kaushal, L. E. Landoll, M. J. Lang, E. R. Mendrysa, J. Miller, J. J. Murphy, T. Niessink, R. W. Ohnesorge, A. Politis, H. S. Smith, J. E. Stephens, J. Tudor, C. L. Wagner, M. A. Xavier

Abstract

The special publication "Application of Fault and Disturbance Recording Devices for Protective System Analysis" was printed and made available in the Summer of 1987. Identified as IEEE Publication No. 87TH0195-8-PWR, this report is intended to supplement previous papers to provide additional guidance in justifying, utilizing and specifying fault recording equipment for the electric power system. Although most of the information was drawn from experiences with light-beam galvanometer paper-record applications, much of the material will be generally applicable to digital fault recorder systems now rapidly growing in popularity.

Following are brief summaries of the major sections of the 35-page report.

Introduction

Analysis of problems that occur during the protective system (instrument transformers, relays, circuit breakers, etc.) operation requires the application of high speed fault and disturbance recording devices. Detailed information on pertinent quantities and events will permit definitions of such problems as:

1. failure of relay systems to operate as intended.
2. fault location and possible cause of fault.
3. incorrect tripping of terminals for external zone faults.
4. determination of the optimum line reclose delay.
5. determination of the magnitude of station ground mat potential rise (gpr) that influences design of communication circuit protection and station ground grids or assists in the quantification of gpr and its dc offset.
6. determination of optimum preventive maintenance schedules for fault interrupting devices.
7. deviation of actual system fault currents significantly from calculated values.
8. impending failure of fault interrupting devices and insulation systems.

Practical, specific solutions to these problems can then be devised to prevent unnecessary tripping of load, loss of revenue, and incorrect relay system operation.

89 WM 216-3 PWRD   A paper recommended and approved by the IEEE Power System Relaying Committee of the IEEE Power Engineering Society for presentation at the IEEE/PES 1989 Winter Meeting, New York, New York, January 29 - February - 3, 1989. Manuscript submitted August 16, 1988; made available for printing November 17, 1988.

The sequence of event recorder (SER), although not thought of as a fault recorder, is now frequently being used in conjunction with fault recording systems to provide a multitude of event (device change-of-state) data. Because of this closely related application, SERs will be referenced in this report when appropriate. Appendix D of the paper will provides an outline for an SER specification.

Equipment cost generally dictates that fault recording and SER devices are normally installed to monitor generators and step-up transformers, transmission lines (69 kV and above), bulk system switching stations, and system tie transformers. Distribution or sub-transmission systems are not normally monitored except for study of special problems or research projects.

1 Power System and Protective Equipment Performance Evaluation

To the untrained eye, a fault-recorder record (oscillogram) is just a conglomeration of meaningless "wiggles". Notwithstanding that most technically oriented people recognize the familiar 60 Hz voltage and current traces, the bits of data superimposed on the 60 Hz waveform in the form of spikes, "pips", "dips", higher or lower frequency waveforms, etc. can be difficult to evaluate or to relate to known high voltage equipment operations. Section 1 of the report illustrates many of these phenomena, or "signatures", and relates them to protective equipment performance. Examples are shown in Figure 1.

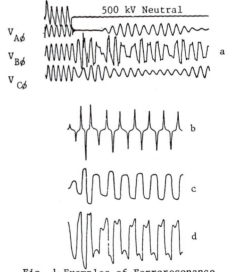

Fig. 1-Examples of Ferroresonance

Figure 1a shows the actual measured unfaulted B-phase 500 kV line voltage during ferroresonance experienced by a line-connected shunt reactor. Note in Figures 1b, 1c, and 1d the "typical" different waveforms of currents and voltages during ferroresonance.

Reprinted from *IEEE Transactions on Power Delivery*, vol. 4, no. 3, pp. 1625–1630, July 1989.

Figure 2 illustrates how the pattern or "signature" on a given voltage or current trace during system faults will be affected by the type of fault and the system configuration at the time of fault. Consequently, analysis of fault recorder data will be made easier if one first obtains an accurate system diagram. For example, Figure 2a shows selected traces from a record of a fault, located as shown in Figure 2b, that was phase-to-phase for 2 cycles and then became 3-phase. With no additional information, the approximate 0.6 cycle of residual current ($3I_0$) in CB 2550 would probably be interpreted as unequal pole opening of CB 2550 or CB 2530. However, the Main Transformer 6 neutral current indicates something else is happening. Further investigation revealed that Figure 2b should have been detailed as shown in Figure 2c. An air switch that opened under load had flashed A-phase to B-phase. About 2 cycles later, C-phase was contacted by the arc but only on the circuit 1 side of C-phase. In reality, the fault was a simultaneous combination of a 3-phase fault and an open conductor, the 0.6 cycle of residual current being the time span between the flash to C-phase and the opening of CB 2550. Figure 2c made this much easier to visualize than did Figure 2b.

Since multiple grounds on instrument transformer secondary circuits can result in recorded data that misrepresents the true primary events, Figures 3 and 4 are included as important examples of the need for caution in applications.

Figure 3a shows selected phase-neutral voltage traces for an A-phase to ground fault (verified from other fault recorder records). Note that C-phase voltage trace also indicates a fault. This is caused by multiple grounds on the vt neutral as shown in Figure 3c. The A-phase primary fault current flowing from "m" to "n" will produce a voltage drop in the vt neutral that will lag Va-n by some 60° to 80°. This neutral conductor voltage when added vectorially to the unfaulted phase voltages will decrease Vc-n and increase Vb-n as measured by the fault recorder (see Fig. 3b.) The errors in the measured voltages could cause incorrect relay operation since these voltages also drive the relay circuits.

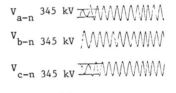

(a)

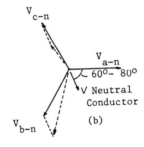

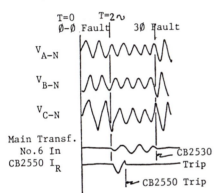

(a) Osc. from Permian Basin

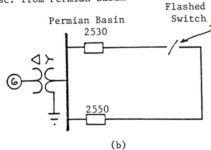

(b)

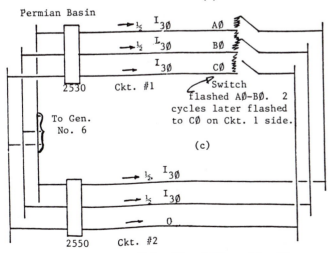

Fig. 2 - Comparison of One Line & Three Line Diagrams for Use with Fault Recorder Records

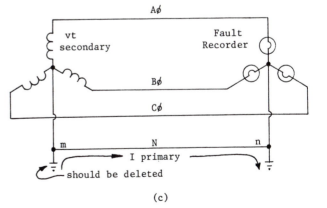

(c)

Fig. 3 - Phase Voltage Error Because of VT Neutral Multiple Grounds

Figure 4 shows the results of breaker 620 of a 69 kV ring bus reclosing unintentionally into a C-phase 69 kV fault. All current transformers are identical with circuit burdens well below the ct capability. Multigrounding of the current circuit neutrals and differences in voltage across the ground mat due to fault current flow caused over-excitation of one of the paired cts of each circuit and the resulting unbalanced current in the line relaying and metering circuit. The instantaneous ground overcurrent relay of line 6613 falsely operated because of the unbalance current $(3I_o)$.

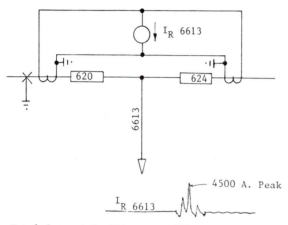

Total Current in CB's 620 & 624 - 10,400 A.

Fig. 4 - Current Circuit Error Because of CT Neutral Multiple Grounds

Fault recorders should be used to monitor generator quantities to determine the generator performance for both external and internal faults. Figure 5 shows generator pre-, during-, and post-fault information. A phase-to-ground fault on the generator (top trace) quickly evolved into an inter-phase fault.

Sufficient quantities are monitored (generator neutral current - or optionally, voltage - generator phase 3 current, phase 1-2 generator voltage, and exciter field voltage) to determine the performance of the generator breaker and field breaker. It is also apparent that even after the field breaker opened, the residual field voltage was sufficient to re-fault the generator phase 3 approximately 8 cycles after both the generator and field breaker had opened. Note the presence of a third harmonic current in the generator neutral during the prefault. The presence of the third harmonic neutral current (or optionally, voltage) can be extremely useful in assessing the continuity of the neutral circuit during normal operating conditions.

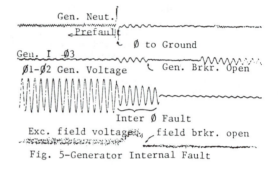

Fig. 5-Generator Internal Fault

Figure 6 illustrates how a fault recorder record can clearly indicate the sequence of events for a multi-event fault. Initially, a 345 kV B-phase to ground failure occurred. Even after the 345 kV system was disconnected from the fault at about 2 cycles, the decaying generator field continued to sustain voltage sufficient to maintain a significant current contribution from the generator. AT 70 cycles, the arc also involved the 345 kV A-phase, and then about 3x cycles later involved one phase of the generator as indicated by the No. 2 Gen. Vn voltage trace. This last bit of data prompted a careful inspection which revealed damage to the isophase bus. This fault record also permitted determination of the $(I_2)^2 t$ duty imposed on the generator rotor to be about 13% of its capability. Consequently, no inspection of the rotor was required.

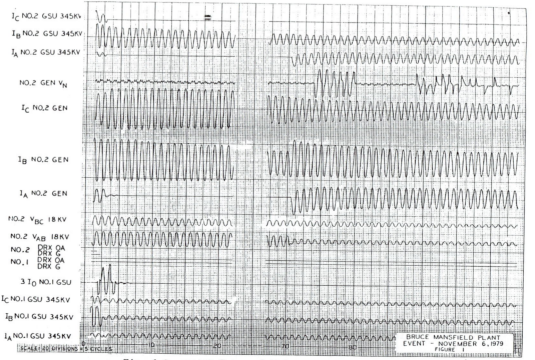

Fig. 6-Generator External Multi-event Fault

2 Basis for Selection of Power System and Protective Device Parameters for Monitoring Items in Section 1

Section 2 discusses the normally preferred system quantities or parameters that should be monitored by the fault or disturbance recorder to provide the analysis benefits or results illustrated in Section 1.

In addition, because of the high frequency operation of a converter and sometimes complex failure modes, high frequency fault recorders have found widespread application in converter stations for fault analysis. The fault recorders have proven to be a valuable supplement to regular relay targets and sequence of event recorders. Typically the recorded quantities are:

1. ac converter bus voltages.
2. converter currents on the dc side of the converter transformer.
3. dc pole to ground voltage.
4. converter dc current.
5. selected key control system analog quantities or event indicators.

In a bipolar scheme, the neutral current and neutral bus voltage may be valuable to record. Some of these measuring points are indicated in Figure 7.

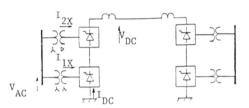

Fig. 7 - Single Pole DC Tie

3 Parameter Monitoring & Initiating Methods

Section 3 provides a summary review of monitoring and recording device initiation options for analog and event data. A contrast between the requirements for generating stations and transmission stations is emphasized.

4 Comparison of State-of-the-Art Fault Recording Equipment

There are three basic types of systems (quick start, prefault, continuous) which are in widespread use for fault and disturbance recording. These are indicated in Figure 8. Although both magnetic tape and digital fault recorder systems provide "continuous" monitoring of the system, it should be noted that the recording intervals of the two systems are greatly different. Consequently, the presently available digital fault recorders are considered in this report to be microprocessor-based prefault fault recording systems. The summary found in Table I (Appendix B) shows some of the characteristics of each type of system.

Discussion of fault recorder accessory features and related considerations covers sensors, signal conditioning equipment, power supplies and inverters, clocks, event markers, transducers, filters, and recording control devices.

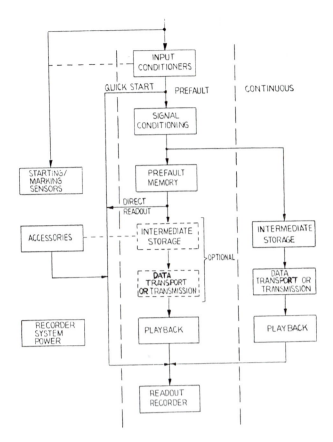

Fig. 8-Basic Flow Diagram Fault Recording Systems

5 Recommended Monitoring Installations

Physical, electrical, and recording system requirements that must be considered in locating the fault recording system are discussed. Concerns for monitoring and time coordination between fault recorders and sequence-of-event recorders (SER) are pointed out. Suggested parameter monitoring is included for a typical station configuration.

6 Typical Parameter Labels

To encourage consistency in identifying monitored system parameters, several are listed with their generally used labels. Included are phase, neutral, and residual currents and voltages and communication channel quantities.

7 Definitions

To further improve communications among suppliers and users of fault recording equipment, seventeen pertinent technical terms are defined and illustrated; such as amplitude resolution, dc offset, dc trip offset, circuit breaker clearing time, frequency/transient response, line dead time, re-ignition, re-strike, etc.

8 Calibration

The extremes between normal and abnormal values of voltages and currents must be considered to ensure that the recorder channel dynamic range (including input conditioning) is sufficient to prevent saturation. For currents, the preferred full scale deflection should be twice the maximum symmetrical current. The factor of two is used to provide for the maximum possible current offset of 100%. This ideal calibration may result in unsatisfactory noisy traces on normal faults and prohibit recognition of distant faults unless the dynamic range of the equipment is adequate.

The signal to noise (S/N) ratio specification of the particular fault recording apparatus will establish the dynamic range for the recording channels. Figure 9 illustrates the comparison of signal-to-noise ratio characteristics of the several identified systems.

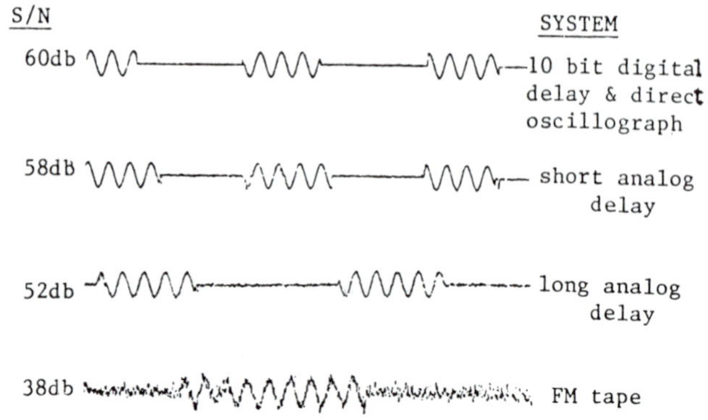

S/N		SYSTEM
60db	WV————\/\/\/——\/\/\/—	10 bit digital delay & direct oscillograph
58db	WV—————/\/\/\——\/\/\—	short analog delay
52db	-\/\/\/————\/\/\/\—	long analog delay
38db	—————\/\/\/\/\/\/\/————	FM tape

Fig. 9-Comparison of S/N Ratios
Deflection of 4A/cm
Full Scale of 75 Amperes

A convenient method to determine current and voltage magnitudes from fault recorders is through the use of scale factors. Sometimes, there may be confusion in utilizing scale factors to convert recorded waveforms to primary quantities. It must be made clear whether the scale factor is given as an rms equivalent or an instantaneous value. Because of the difficulty in identifying the zero line position on the recorded waveform, the scale measurement of an ac quantity is most easily obtained by scaling from the extremities of the negative peak to the positive peak. Therefore, the scale multiplying factor would be expressed as "primary units per unit peak-peak of deflection;" such as, 100 amperes/inch peak-peak. Figure 10 illustrates this procedure. For an asymmetrical current waveform as shown in the solid line of Figure 11, the rms value of the ac component cannot be determined because the zero reference line of the ac component cannot be located. However, note that the <u>instantaneous</u> value of the total current (I_{ac} + I_{dc}) can be evaluated at any instant using the peak-peak rms calibration factor multiplied by 2.828.

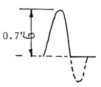

Scale Multiplier=100 Amps
RMS/1" Pk-Pk
Pri. Amps =0.7" x
2.83 x 100=198 Amps Inst.

Fig. 11-Measurement of an
Asymmetrical Waveform for
Conversion to Primary Val.

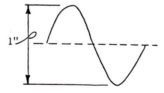

Scale Multiplier=100 Amps
RMS/1"P-P. Pri. Amps=1"
x 100=100 Amps RMS

Fig. 10-Measurement of a
Symmetrical Waveform for
Conversion to Primary Values

Methods for determining scale factors for magnetic tape playback machines and direct write systems are outlined.

Appendix
The appendices includes:
1. An outline for a fault recorder specification,
2. a. A table of comparison of fault recording techniques (included in this summary),
   b. A table on input conditioners
   c. A table on signal conditioning
   d. A table on prefault memory methods
   e. A table comparing features of on-site recorders
   f. A table of fault recorder initiating device characteristics
3. A typical calibration sheet for an FM tape system, and
4. An outline for a sequential events recorder specification.

TABLE I. COMPARISON OF FAULT RECORDING TECHNIQUES

Type of Recording	Quick Start	Prefault		Continuous Magnetic Tape
		Prefault System	Digital Fault Recorder	
Recording Start	Generally auto o'graph 4 msec to paper movement 1 msec to lamp start (fault current peak readable)	O'graph for immed. readout. Magnetic tape for storage/ playback. 3 to 10 cycles prefault	Automatic for fault Manual/Auto/Timed from Master or Remote Station	Data recorded continuously
Frequency Response Lower Upper	$\pm$ 10% DC Std - 800 to 1600Hz Upper 10KHz	(10%) DC or 0.25Hz(-6db) 400, 800 or 1600Hz 1500-1600Hz (-3db)	$\pm$ 10 % DC or 0.25 Hz (-6db) Up to 100 kHz available. Actual depends on scan rate memory, input conditioning. 10 - 20 kHz normal maximum	DC or 0.25Hz(-6db) 1250Hz(-6db) at 3 3/4 i.p.s. for disturbance recording 15/16 i.p.s. used @ ¼ above response
No. of Channels (Maximum) Analog Events Time Code	 32 16 1	 40 48 1	Theoretically unlimited. Function of Master Sta.Capacity Groups of 8 to 32 Groups of 8 or 16 1 channel dedicated or interleaved	Total - 32 in IRIG format analog & digital -may substitute 8 events per analog -uses 1 of 32 above for clock
Estimated Relative Cost Factor - Recorder Only	1	1.2 to 1.4 x	1.3 * assumes Master Station available	1.3 to 1.6 x *assumes playback unit available
Disadvantages	Possible record clutter No prefault data Loss of data possible (UV)	Possible record clutter Increased cost & complexity Signals conditioned	Initial Cost - Master Station Required Local Readout may be necessary.	Cost Higher Maintenance Playback mandatory Optional local playback may be req.
Maintainability Relative Rating	1 - Simplest system Calibration, test Possible 1 year calibration cycle	2 - 6 to 12 mos. cycle + 1 to 3 hours	1 - Minimum mechanical maintenance - Self/auto test can replace scheduled maintenance. Some 3rd party maintenance.	3 - Most difficult mechanical Most frequent periodic mtce. e.g.4-6 wk. head cleaning recom.
Operating Supply	Recording Paper	Recording Paper where applicable	Printer paper - ribbons floppy or hard disks	Reusable tape
Major Advantages	System Simplicity Lowest Cost Fault current amplitude Max. freq. respon. (limited by paper speed) No conversion errors Immediate/on-site readout Isolation by high voltage insulation Req. o'graph channel isolation 1500Vac	Shows prefault conditions Reduces recorder quick start requirements On-site read-out Possible data storage Playback for auto analysis, remote read-out	Maximum Data Gathering and Manipulating Capability. Superior Field expandability, flexibility, auto data analysis. Re-run and presentation variations	Max. data gather Delayed data save possible Max.(6 hrs. poss.) prefault Ability to run & re-run record with record expansion capability Maximum post fault recovery data Future remote (playback read-out) Flexible trace positioning

543

FIELD EXPERIENCE WITH ABSOLUTE TIME SYNCHRONISM
BETWEEN REMOTELY LOCATED FAULT RECORDERS
AND SEQUENCE OF EVENTS RECORDERS

R. O. Burnett, Jr.
Associate Member, IEEE
Georgia Power Company
Atlanta, GA

ABSTRACT

This paper addresses the problem of maintaining absolute real time synchronism between many remote substation locations containing fault and sequence of events recording equipment. Several methods are described which have been under field evaluation for over four years and are in current use at Georgia Power Company.

These methods include: (1) utilizing a master time source for transmission via microwave of a time code data stream (IRIG-B) to selected remote locations. (2) utilizing the National Bureau of Standards 60 KHz transmission (WWVB) from Ft. Collins, Colorado to provide time data via commercial WWVB receivers. (3) utilizing the GOES satellite transmission to synchronize time data at selected locations via commercial GOES receivers.

Advantages and disadvantages of each method are presented along with field data to support these methods.

Synchronism to absolute real time provides a common time base for analyzing system faults and disturbances recorded at many remote locations over a large geographical area with typically one millisecond resolution. This approach also allows the user to compare events and fault data between various utilities where reference is made to absolute real time.

INTRODUCTION

Since the first fault and event recording systems were placed in service years ago, some method of "time tagging" the recorded data was provided. These methods have included a physical stamping of the time of day on oscillographic records, flash tube photographic techniques, L.E.D. strobe techniques, and recording time data on magnetic tape.

With the advent of higher transmission line voltages, larger capacity generating plants, and more sophisticated high speed relaying systems, an increasing burden was placed on the fault data equipment to be able to properly resolve events and disturbances over a large area.

The increasing requirement to be able to analyze data on a real time basis from several locations for any one disturbance has led to the investigation of improved methods of time tagging the recorded data.

This paper will address the methods investigated and used by Georgia Power Company to significantly improve the time synchronism between various locations. The systems in use have improved the synchronism accuracy by three orders of magnitude over previous methods. These improvements have greatly enhanced the ability to properly analyze, document, and take corrective action on problems occurring with power system transmission equipment and protective relaying systems.

BACKGROUND

In any fault recording system, some method must be used to identify each record so it can be related to system conditions and aid in documenting and analyzing various events and disturbances. Usually, time of year information (DAYS, HRS, MINS, SECS) is recorded along with power system data at the time of fault occurrence. This time of year data can have a resolution of one second without too much difficulty. Usually these clocks providing the time data are synchronized to the 60 Hz line frequency and contain some type of back up oscillator to insure continued operation during AC power outages. While the 60 Hz line method can provide very good long term stability (months), its short term stability (minutes—hours) is grossly inadequate to achieve one millisecond accuracy between several recorders. Even if the above problem is not considered, all too often these recorder clocks (fault recorder and sequence of events recorder) are set by personnel using local station clocks, wristwatches, or pure guesswork as a reference. While the individual recorder clocks can then maintain a fairly accurate elapsed time over the long term, the accuracy between different clocks and any relationship to absolute real time can be horrendous at best.

Continuous monitoring thirty-two channel magnetic tape fault recorders provide the majority of all fault data for Georgia Power. With forty-two of these systems in service, tremendous amounts of pre-fault, fault, and post-fault data is available for analysis on any one given disturbance. Additionally, seventy-five sequence of events recorders are located throughout the power system and also provide data during disturbances.

Originally the above recorders were synchronized to the 60 Hz line frequency. The time resolution was very poor as was the short term accuracy. Eventually electronic time code generators with digital read-outs and improved code outputs were installed and resolution improved by a factor of six. However, we were still at the mercy of local clocks, wristwatches, and guesswork.

As the power system grew, with more large critical EHV substations and generating plants coming on line, more stringent requirements were necessary to properly document faults, outages, and disturbances.

Companies within the Southern Company (Alabama Power, Georgia Power, Gulf Power and Mississippi Power) were in need of a system for synchronizing clocks on fault recording equipment to some master source.

IRIG-B SYSTEMS

Several possibilities were investigated with the final determination made to use equipment located at Southern Company Services in Birmingham, Alabama to generate the master time signals. Since the Southern Company Communications and Data Center is at this location and very extensive company microwave and communications equipment is in place to all sister operating companies, this made an ideal location from which to transmit time code data to the various company headquarters. Once this time code was transmitted to the Georgia Power General Office in Atlanta, it could be further routed over company owned microwave facilities to individual substations and fault recording equipment.

The time code format to be used was the Inter-Range Instrumentation Group Type B (IRIG-B) and consisted of a 1000 Hz carrier that was amplitude modulated with the time code data in BCD format. This format has adequate resolution (1 millisecond), transmission capabilities that fit available data grade channels, and can easily be recorded on magnetic tape. Refer to Figure 1 for an example of the IRIG-B format. Time information is BCD coded by use of the modulated carrier with two cycles of modulated data (large cycles) representing a "0" bit and five cycles representing a "1" bit. Frame position identifiers are shown as eight cycles of modulation. This format is one second long and is a continuous time code updated every second. Complete time of year data is contained in this time code from day of year through seconds with "time of occurrence" being resolved to within one millisecond. Additional data bits for any special needs or control functions are also provided in the time code and may be used as necessary.

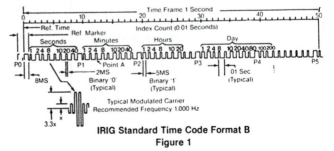

IRIG Standard Time Code Format B
Figure 1

Reprinted from *IEEE Trans. Power App. Syst.*, vol. PAS-103, no. 7, pp. 1739–1742, July 1984.

The basic idea behind the master time synchronizing project was to have all fault data clocks on the same time base and have that time base referenced to absolute real time. If only relative time is important, savings can be experienced and a much simpler system installed. However, be aware that the ability to use other time sources such as WWV, WWVB and the GOES satellite in conjunction with the relative time system would not be feasible.

The system currently being used at the master station consists of the following equipment: one primary Cesium atomic time standard, one secondary Rubidium atomic time standard, one WWV receiver, one Loran-c receiver and three independent digital clocks feeding a majority gate arrangement to insure agreement between standards. Sufficient power supply redundancy is provided to insure continuous operation.

Since all time transmitted from the master is tracable, and synchronized to the National Bureau of Standards, any WWV, WWVB, or GOES equipment may also be used along with the IRIG-B data with accurate synchronism possible. However, the time code format and sequence of events recorder scan time limits the practical resolution to one millisecond.

Certain locations within our system are not served by company owned microwave facilities and leased circuits are not a viable option. In these cases, WWVB or GOES Satellite receivers are being used to achieve synchronism to the master clock. Additional data will be provided later for information on these auxiliary devices.

Data from the master clock is transmitted via numerous microwave sites to the Georgia Power General Office communications center. Here the signal is rerouted and split as necessary for retransmittal to the many remote locations throughout the state. Various synchronized time code generators are used to decode the incoming IRIG-B, adjust for propagation delay, display, and buffer the data for recording on the magnetic tape during fault conditions. This equipment has proven to be immune to noise, switching transients, and signal polarity reversals while providing reliable operation.

No matter what type of equipment is used, if a real time system is desired, this equipment must update the display and all outputs at the beginning of the second rather than at the end. Since it would take a second to read the incoming data, the outputs would be in error by this time. With accuracy and resolution of 1ms desired, attention must also be given to the propagation delay times from the master to each remote.

A portable Rubidium atomic time standard is used to measure all propagation delay times. This standard is periodically carried to the Birmingham master clock for resynchronism. Typical drifts have been on the order of 2 milliseconds over a twelve month period. Once the portable clock has been synchronized to the master, the portable is transported to the remote location and the signal delay is observed on a scope and measured on a counter with time interval measurement capability. Within Georgia, propagation delays range from 3.75 milliseconds to 7.70 milliseconds from the Birmingham master. Attention must also be given to observing proper scope triggering techniques whether on the leading or trailing edge of the 1PPS signal in the various equipment to insure accurate measurements. The propagation delay offset switches in the time code generator (TCG) can then be adjusted so the output of the TCG exactly matches the time of the master. If a portable standard is not available, various loop-back/computational methods can be used to derive propagation delay times.

This master time is also used to synchronize the substation sequence of events recorder (SER) clock. The best time resolution of the SER is 1 millisecond, therefore, it is not very critical to maintain real time accuracy better than that. Experience indicates that time synchronism of both the SER and the fault recorder to be of great benefit during fault conditions. Times can be compared from both ends of the faulted line. Sequential tripping and associated reclosing can now be observed without the previous guesswork of what happened first. In the past, many events happened but it was always practically impossible to correlate relay trip outputs and breaker operations from several remote SER's. Many times, operations would occur when faults were not involved and the SER was the sole source of data from each end of the line. With time synchronism between these SER's, data can be compared on a real time basis. Experience has shown over the years that it is becoming more and more critical to install not only fault recording systems but also sequence of events recording systems to gather all possible data for analysis of power system operating conditions.

The necessary synchronizing circuits to interface the fault recorder clock to the SER clock are usually added as a field modification and may consist of sync pulses of 1PPS, 1PPM, and/or 1PPH depending upon the equipment involved. Certain SER's are also capable of reading the parallel BCD time code and applying that time directly to the internal clock for automatic reading and updating. This feature greatly enhances and simplifies the automatic synchronism of SER systems to absolute real time.

Most SER systems employ some type of internal digital filtering to eliminate contact bounce. This procedure can introduce errors where absolute real time recordings are required. In some cases, these offsets are corrected in software. Many other types require the addition of at least four milliseconds to the printed time to allow real time comparisons to the fault recorder times.

Another method that may be used to record time code does not involve the use of a TCG at all. The IRIG-B signal as received at each remote location may be applied directly to the fault recorder channel input. While an obvious advantage is saving the cost of a TCG, several disadvantages are also realized. One, propagation delay times cannot be easily and automatically accounted for, if that is desired. Two, no on-site display is available to indicate the presence of time code data and show the time of year information. Three, if the master fails or any communication channel fails, there is no time code data available for recording and usually no way of knowing the channel is inoperative.

Since the IRIG-B signal drives the TCG input, and the TCG output drives the recorder, all of the above problems are solved. Even if the input signal from the master clock fails, the TCG continues to function as a stand-alone time code generator. As soon as the master clock signal is received again, the TCG will automatically correct any drift and continually update time once more. An internal oven controlled oscillator in the TCG then serves as the time base. Its stability is approximately 1 part x 10 -9/day. During propagation delay measurements, using the atomic standard, this internal oscillator is fine-tuned for minimum drift.

While requiring extensive communications circuits, the IRIG-B system has proven to be extremely reliable.

The desired end results and user needs will determine the most feasible approach for recording the time information.

The user must also be aware of an operating problem when using a master clock system with magnetic tape or any other device having synchronous motors as a driving medium.

If these synchronous motors are driven from an inverter source and the inverters throughout the system are not synchronized to each other, the recorded oscillographic data can not easily be physically aligned between machines. The recorded times will be correct but will not necessarily match if overlayed. The use of stabilized and synchronous inverters would eliminate this problem. Using the IRIG-B carrier to sync the inverters would be one method. Consideration should also be given to the playback machine itself if extreme accuracy and trace alignment is a rigid requirement.

While this shortcoming exists, it usually does not present a problem unless several oscillograms are overlayed on each other for comparisons. In these cases, times must be indicated individually on each oscillogram for comparisons.

Tape speed differences can be observed that are translated into distances on the oscillogram on the order of .25 to .50 MM (.01 to .02 in) for a record 30 cm (12 in) in length.

WWVB SYSTEMS

Several locations in this system are not served by company owned microwave sites, yet it is desired to have the fault and event recording equipment at these locations synchronized to the master clock. Cost estimates and reliability evaluations have resulted in considering methods other than leased circuits.

Initially, WWVB receivers were installed at certain 500KV substations to synchronize the local TCG's. While the equipment purchased has proven to be reliable, several operational problems have been experienced. One of the major problems is the location of the WWVB receivers approximately 2250 km (1400 mi) from the NBS transmitter in Ft. Collins, Colorado. Sites within a few hundred kilometers of the transmitter may experience signal strengths of 2500 microvolts/meter while those in Georgia usually receive 100 microvolts/meter or less. If the user is only interested in measuring frequency, using the WWVB signal as the reference, no major problems are encountered even at distant locations. However, complete phase lock and accurate time reception is desired 100% of the time. Data has been gathered over a period of years on the feasibility of extensive use of WWVB equipment. This data has shown predictable and expected signal drop-outs usually twice a day at times of local sunrise and sunset. This phenomena is caused by the "dinural shift" and is caused by the day/night line progressing between the transmitter and receiver. Usually this loss of signal lasts for less than one hour and its occurrance varies as the time of local sunrise and sunset varies every day. Realize that even though the receiver loses phase lock it continues to function on its back-up internal oscillator but a drift from real time is experienced. This drift rate has been measured to be approximately an average of 10 milliseconds per hour after final oscillator adjustments have been made.

A second problem in using WWVB timing has been noticed during times of severe weather disturbances. Thunderstorm activity has a detrimental effect in not allowing noise cancellation on the received signal with the approaching storms. Since numerous thunderstorms and related bad weather during the spring and summer months are a common occurrence, many signal dropouts have been recorded. Unfortunately, most faults also occur during these times. Then, a true synchronized time system for these WWVB installations as related to our IRIG-B sites does not exist. The IRIG-B sites are in sync over 99.9% of the time day in and day out. The WWVB sites approach 85% signal lock at their best. Again, this is mainly because of our distant location from the transmitter and the location of the receiving equipment. An electric transmission substation has not proven to be the best environment for receiving the 60 KHz WWVB signal for time code reception. Excessive corona from switches and various buswork has made reception extremely difficult at several locations. The physical arrangement of the substation and practical limits of antenna height and distance from the receiver was also a limiting factor.

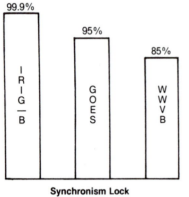

Synchronism Lock
Figure 2

During laboratory tests away from the substations this equipment performed as expected. In all cases the antenna consisted of an active system using a 30 cm (12 in) ferrite rod.

Two WWVB installations remain in service for continuing evaluations but no others have been scheduled for future additions due to the above mentioned problems.

Propagation delay measurements are made at these WWVB sites just as they are at the IRIG-B sites. Using the portable atomic clock the 1PPS received via the WWVB receiver can be exactly aligned with the 1PPS from the master clock in Birmingham thereby maintaining true real time sync between the WWVB receiver and the master clock.

WWV equipment has not been considered for use since its code format does not provide a continuous time code broadcast but only 1PPS ticks. This prevents the automatic clock setting capability from being used.

GOES SATELLITE SYSTEMS

Several years ago commercial equipment became available for receiving and decoding the time code data from the GOES satellite system. These are the Geostationary Operational Environmental Satellites placed in synchronous orbit approximately 22,300 miles above the equator. As such they continuously remain in the same place above earth. There are three working satellites in orbit. The East Satellite at 75 degrees West Longitude, the West Satellite at 135 degrees West Longitude and a spare at 105 degrees West. Excellent coverage is obtained over all of North and South America with these satellites. Experiments were conducted with both the East and West satellites with reliable synchronism obtained from both. However, the East satellite was used for normal operations mainly because of its higher angle above the horizon (52 degrees vs 24 degrees) and its operating frequency.

Commercial satellite receivers have been under evaluation for approximately two years and five different sites are in service. This equipment is being monitored by sequence of events recorders to determine the time and duration of signal outages. Basically, field data indicates GOES provides much more reliable operation (95%) than with the WWVB system. The satellite signal reception is not susceptible to the "dinural effect" and thus far thunderstorm activity has not had a detrimental effect on the received signal. One location has the antenna "looking" through a 230KV bus structure with significant corona and no detrimental effects have been

seen on the received signal. A WWVB receiver would not work at all at this site. Even though we have experienced no satellite data problems with radio interference, the user should be aware the operating frequency is approximately 468 MHz. This is the spectrum used by certain "land-mobile" services.

A few equipment problems have been experienced but the biggest reason for signal disturbances on the East satelite has been due to the spring and autumn solar eclipse period. These eclipses can last up to approximately one hour per day at approximate local midnight from March 1 to April 15 and again from September 1 to October 15. Usually the spare satellite is used during this time to insure continuous operation but the predetermined propagation delay times on each receiver are then in error. This could cause a difference of up to 20 milliseconds between the GOES receiver and our master clock. Readjustment of the propagation delay or switching to the West satellite during these times would eliminate these problems. Since the time transmitted via the satellites is maintained to NBS time by time standards at Wallops Island, Virginia, direct comparison to our IRIG-B master source can be maintained.

To insure all time is synchronized, the portable atomic clock is transported to each GOES receiver and the time delay measured and adjusted on each receiver to bring all 1PPS signals together. After analyzing oscillograms from IRIG and GOES receivers for the same fault, very good results within the desired 1 millisecond requirement are obtained. While evaluations are still underway, it appears that the satellite receivers will be a cost effective alternative to leased circuits. This system could also provide excellent synchronizing capabilities for certain individual remotely located sequence of events recorders. By merely pointing the rather small antenna at the satellite and plugging the receiver output into the event recorder, accurately synchronized time can be automatically maintained. The antenna for this system is approximately 30 cm (12 in) square containing an active microstrip antenna. Field tests have indicated the requirement to mount this antenna outside the control house is one shortcoming. However due to other advantages previously mentioned, additional GOES timing systems will be installed in the future.

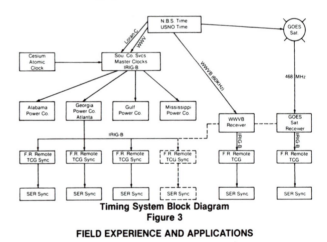

Timing System Block Diagram
Figure 3

FIELD EXPERIENCE AND APPLICATIONS

Several techniques may be employed to record the time code data on the magnetic tape. Usually this data is recorded only for approximately ten seconds when the fault or disturbance occurs. This data can be in the form of a DC level shift TTL output voltage that would appear as a square wave pulse train coded in BCD format. Resolution can be interpolated to within one millisecond.

We have chosen to use the true 1 KHz amplitude modulated IRIG-B signal from the TCG rather than the DC level shift code. One advantage is the one millisecond time base is much easier to resolve since each cycle of the carrier represents one millisecond. By using this format, a time code reader can be used on the tape playback system to automatically display the recorded time data on a digital clock. Instead of having to manually decode the time data during playback, the code reader can be used to do this automatically. This saves considerable time and oscillograph paper since the desired fault can be located before any paper has to be processed. With the costs of direct print paper increasing, it doesn't take long in saved time and paper to pay for the reader. The particular commercial model has the capability of not only reading the recorded time, but also allows one to dial in a selected time and have the search unit control the tape deck until that time is located.

As previously mentioned, the time code data is usually recorded for only approximately ten seconds. We have chosen to lengthen this to sixty seconds. Two advantages are realized. One, the longer time frame makes locating the fault information much easier. Two, the longer time frame is necessary for proper time references during system disturbances and faults involving generators.

At certain installations even sixty seconds is not adequate for the type data we desire to analyze. For these applications, we dedicate an entirely separate channel for recording continuous time code. This method better utilizes the continuous data recording available on the fault recorder. Every cycle of recorded data can be identified with a time resolution of one millisecond and also be identified relative to real time. If we desire to locate data that occurred, for example, 17 minutes 28 seconds prior to some event or fault, that can be easily and rapidly done with the combination of continuous time code and time code reader. It may be then determined that the preceding event really occurred at 17 minutes 28 seconds and 373 milliseconds in absolute real time. Data from several other recorders may also be compared to each other since all clocks are synchronized to a master source. This allows viewing vast amounts of prefault as well as post fault data from several recorders and identify each cycle on each recorder with a real time tag having one millisecond resolution. Initiating events, and "cause and effect" type phenomena can be easily resolved over a large geographical area.

In one case we were able to determine the initiating event for a serious generator problem was recorded some two minutes prior to the actual trip. Two isolated recorders over 200 km (125 mi) apart had prefault data that could be viewed on a real time basis and allowed the true initiating events to be identified.

Many times distant problems have occurred before, during, and after system faults. Previously, trying to relate these problems to the system faults was difficult to do. Now everything is time synchronized and all recorded data is easier to relate to each other in real time.

Often problems at first were thought to occur simultaneously. After reviewing real times, it was determined this was not the case. Two different problems occurring 12 milliseconds apart had occurred. A more direct troubleshooting approach was then initiated and the problem solved much quicker. One must also realize we are considering the sequence of events recorder as well as the fault recorder at locations remote from the fault.

There have also been cases where the data revealed the problems actualy did occur simultaneously during a fault, reclose, etc. Previously this may have gone undetected but now corrective action can be taken before additional problems occur.

Presently automatic reader high speed search capabilities on the fault recorder are not possible since the time code data is recorded using the FM saturation recording technique. Once the recorder is modified to allow time code recording using the direct analog method, this problem will be eliminated. Then the user can fully utilize the reader/search unit to automatically search the tape looking for some specific predetermined time and locate that exact spot in time. The second channel previously used for continuous time code would not be necessary.

When time code data is decoded to retrieve the time of year information, the question that always arises is, "Was the time code generator synchronized to the master clock at the time of the disturbance or fault?". This information is vital if many recorders are to be compared with any degree of accuracy. If the recorder clock was not in synchronism, the internal drift could cause a minimum of several milliseconds error.

Modifications have been completed to allow a previously unused control bit in the time code format to indicate if the remote clock was synchronized to the master. This "sync bit" is easily readable on the oscillogram and is also displayed on the time code reader. One can tell at a glance if the recorded data was syncrhonized to the master. Knowing this, all data can be compared with confidence. The presence or absence of the "sync bit" is also monitored on the substation sequence of events recorder to provide additional data on channel reliability.

CONCLUSIONS

In summary, this synchronized timing system consists of a master station transmitting continuous time code (traceable to NBS) to various remote locations. The received time data is displayed and buffered for recording on the magnetic tape and events recorders with propagation delay accounted for. If no company owned communications facilities are available, equipment is installed for receiving the time data as transmitted from the GOES satellites or NBS station WWVB. This essentially takes the place of the microwave channel equipment. A portable atomic clock is carried to each location to measure time delays and insure the 1PPS signal is syncrhonized to all other 1PPS signals. The goal is to have all recorded times within 1 ms of each other at all locations.

Field experience has indicated the above goals are obtainable on a consistent basis. This timing system not only allows accurate data comparisons between Georgia Power Company recorders but also provides common time base data for three other operating companies within the Southern Company.

While considerable progress has been made since this project conception, newer methods and improved equipment should allow future improvements as requirements change. Possible one area of even more accurate time and frequency dissemenation to be used in the future will utilize the Gobal Positioning Satellites (GPS) as a master time base.

Presently the combination of continuous monitoring magnetic tape fault recorders, sequence of events recorders, and a real time clock synchronizing system provides extremely valuable fault data while allowing determination on a real time basis of events happening over a very large transmission system.

REFERENCES

1. Publication TFS-602, January 1978, **NBS Time Via Satellites,** National Bureau of Standards, Boulder, Colorado.

2. Application Note 52-1, **Fundamentals of Time and Frequency Standards,** Hewlett-Packard, Palo Alto, California.

3. Application Note 52-2, **Timekeeping and Frequency Calibration,** Hewlett-Packard, Palo Alto, California.

4. Application Note 191, **Precision Time Interval Measurement Using an Electronic Counter,** Hewlett-Packard, Palo Alto, California.

System Comparisons
Figure 4

	IRIG-B	WWVB	GOES
Installed Cost	High	Moderate	Moderate
Sync Lock	99.9%	85%	95%
Reliability	Very High	Dinural Effect	Eclipse Periods
Installation Time	High	Very Low	Very Low
Turn-On Times	Immediate	Long 5-60 Mins.	Moderate 5 Min.

547

PHASE ANGLE MEASUREMENTS WITH SYNCHRONIZED CLOCKS - PRINCIPLE AND APPLICATIONS

P. Bonanomi
Brown, Boveri & Co., Ltd.
CH-5405 Baden, Switzerland

ABSTRACT - A simple and economical technique for measuring voltage phase angles and generator shaft angles is reviewed. The principle is briefly exposed and its practical operation is illustrated by an experiment in a real power system. Due to the importance of accurate time for this application a short survey on time dissemination systems is included. The main part of the paper is devoted to the possible applications of phase angle measurements for the control and monitoring of power systems. The discussion deals with various topics and makes ample reference to available literature. The paper is intended as an information survey for assessing the practical significance of phase angle measurements.

INTRODUCTION

The voltage phase angles and rotor angles in a power system are important variables for describing the state of the system. When the voltage magnitudes, voltage angles, impedances, and boundary injections are known, for example, real and reactive power and current can readily be calculated. The rotor angles between generators are equally important because they represent state variables which are significant for the control of power systems.

Phase angles and rotor angles are often computed from other measured variables in the power system. Using state estimation, for example, the voltage phase angles can be computed from voltage magnitude, power and current measurements. In spite of these possibilities it may be advantageous for certain applications to measure these angles directly.

Phase angle measurements have already been performed in practice. The system described in [1] for example uses microwave channels to transmit AC voltage waveforms to the control center. The angle between the voltages is determined by digital transducers and displayed for the dispatcher. In [2], a measuring system is presented which alleviates the burden on the data transmission system. The method requires accurate clocks at the measuring points.

Experimentation with phase angle measurements is relatively new, and there is only limited information available to make a judgment on this matter. The present paper makes two contributions in this respect: first, the results of a detailed field test are presented, using a measuring technique similar to [2,43]; second and perhaps more important, a survey of examples is given, which illustrates the possible applications of phase angle measurements.

MEASURING PRINCIPLE

The proposed technique which is similar to the one described in [2] is characterized by the synchronous sampling of local voltage angles. Assuming that the instantaneous angle of the voltage sine wave or generator shaft is continuously available at every bus, the angle deviations between the different locations are simply determined by sampling every local angle simultaneously. The sampled values are then transmitted to some central processing station and the angle differences between the buses can be computed. The principle is illustrated in Fig. 1 for the simple case with two buses.

The simultaneous sampling is achieved with accurate clocks for which a very high accuracy is obviously required. This point will be discussed in later sections.

When computing the phase angle differences between the buses, one must be sure that the samples belong to the same sampling time. This may be achieved by attaching the exact time information to every sample. Data acquisition and data processing can then operate according to their own speed without risk of confusion.

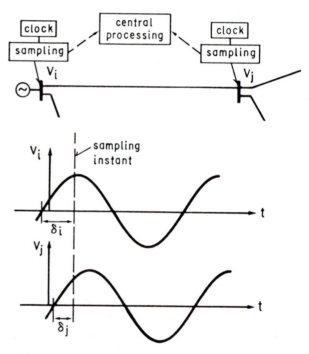

Fig. 1   Measuring the phase angle $\delta_{ij} = \delta_i - \delta_j$ between bus voltages V_i and V_j

Several circuit principles are possible for sampling the phase angle of the voltage. A simple solution using a counter and zero-crossing detection was found to be adequate during the field test. The counter is started at the zero-crossings and the value is latched at the sampling times (see Fig.1). This value is then divided by the total count obtained for the preceding period of the AC voltage. The same principle may also be used for rotor angles, in which case the counter is started by position detectors on the generator shaft.

The sampling frequency for the angle measurements is chosen according to the needs of the application. For power system monitoring, a sampling interval in the range of seconds is appropriate. For control applications, faster rates may be required. The device used in the field test is designed for sampling rates up to 50Hz (line frequency).

81 SM 405-0   A paper recommended and approved by the IEEE Power System Engineering Committee of the IEEE Power Engineering Society for presentation at the IEEE PES Summer Meeting, Portland, Oregon, July 26-31, 1981. Manuscript submitted January 28, 1981; made available for printing April 28, 1981.

Reprinted from *IEEE Trans. Power App. Syst.*, vol. PAS-100, no. 12, pp. 5036-5043, Dec. 1981.

SOURCES OF ERRORS

There are several sources of errors affecting the angle measurements. The most significant is probably the time inaccuracy of the clocks. If one clock is off by one millisecond, for example, an error of 360/20 =18°/ms (50Hz) is produced. Considering that the largest phase angle difference in a typical power system is in the order of 10 to 20 degrees, an error of 1 ms would be unacceptable. If the error between the clocks is constant, however, a constant bias appears in the angle measurements, which can be compensated as explained later.

A certain error is also due to the voltage transducers. This error is negligible for the classical transformer-type transducers, but not quite insignificant for the capacitive devices. Due to the resonant circuit characteristic of these transducers, the error depends on several factors, such as burden, line frequency and voltage magnitude. The deviations as a function of frequency and burden characteristic is typically given in the form of a chart, which is provided by the manufacturer. As an example for the order of magnitude of these deviations, it may be found that a line frequency which is 0.75Hz off nominal can cause an angle deviation between 0 and 20min depending on the burden.

The effect of nonideal conditions in the 3-phase voltage is significant when making phase angle measurements. For the measurements to be useful, the waveforms must be very near the ideal sinusoidal and symmetrical condition. In case this is not true, due to noise, harmonics or unbalance, the effect would have to be investigated. The result depends not only on the type of distortion, but also on the way the phase angle of the AC waveform is extracted. The device used in the field test, for example, is based on the zero-crossings of the voltage and uses only one phase. The measuring device could be more sophisticated, for instance including filters and measuring all 3 phases of the AC system.

SYNCHRONIZED CLOCKS

Very accurate clocks are required for the phase angle measurements. A precision in the order of a few microseconds must be achieved (1μs corresponds to an error of 360/20'000 = 0.018° at 50Hz). Accurate clocks are also required for other power system applications, but the requirements are less stringent; precision in the order of milliseconds is usually sufficient. Due to the general interest in time dissemination systems for power system applications, this topic is discussed in the appendix. A short survey is given which goes beyond the immediate needs of the present application for phase angle measurements.

In summary, it may be said that the desired accuracy for phase angle measurements can be obtained with radio time signals for synchronizing the clocks.

FIELD TEST

Experiment setup

The principle of the phase angle measurements has been verified in a field test in the Swiss network. A 220kV transmission line was used for this purpose. The length of the line is 35km and the parameters are X=15.0Ω and R=2.05Ω. The voltage phase angle δ_{ij} across the line is measured using two identical units as in Fig.1. The zero-crossings of a line-to-ground voltage are used to evaluate the phase angle at each local bus. The measurement and sampling is performed by a digital circuit and has been implemented in a standard system for data acquisition and remote control. A digital processor is included in each unit for the purpose of monitoring the experiment. Instead of transmitting the sampled data to a central processing station, the samples, together with their time identification, are

stored on a local disc. The phase angle differences are obtained off-line by computing the difference between the corresponding samples from the two discs. The experiment can thus be carried out without real-time data transmission.

The clocks are of the type described in [3], using the HBG radio time signal. Synchronization with the radio signal is performed in the following way: first, coherent detection of the second- and minute-impulses is used to obtain an approximate synchronization. The oscillators are then locked on the carrier of the HBG signal so that the required time stability in the order of a few microseconds is achieved.

The phase angles at the two stations were sampled according to the following pattern: every 10 minutes, starting on the hour, a set of 15 samples was recorded with 20ms sampling intervals.

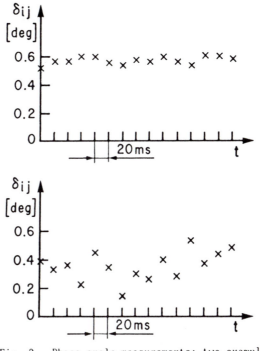

Fig. 2   Phase angle measurements: two examples. (Bias in the measurements = -0.95°)

Results

The two recordings in Fig.2 illustrate the rapid time fluctuations in the measurements of δ_{ij}. The amount of the fluctuations is observed to be different in the two cases. No correlation between this amount and the time of the day has been established, however.

In Fig.3, the phase angle variations are compared with the measured power flow on the transmission line. For every angle measurement δ_{ij}, the corresponding line power is calculated using the formula for a lossless transmission line:

$$P_{ij} = \frac{V_i V_j}{X_{ij}} \sin\delta_{ij} \qquad (1)$$

where P_{ij}, X_{ij}, V_i and V_j are the line power, the line reactance and the voltage magnitudes at the two buses respectively. For each set of 15 samples, the smallest and the largest value are considered. The calculated power values are then superposed on the chart recording for the line power (Fig.3). An agreement is thus obtained up to a constant bias between the curves. This bias is due to a constant but very small time shift between the two clocks, as explained in the appendix of the paper. By shifting the curves until the agreement in Fig.3 is obtained, the bias is found to be -65MW in our case, which corresponds approximately to -0.95° for the

549

angle and -53μs for the time shift between the clocks.

The power variations have been calculated from the angle measurements using equation 1. It may be noted that a certain approximation has been made, due to the bias in the measurements and the nonlinear nature of this equation. This approximation is quite acceptable because of the small value of the angles.

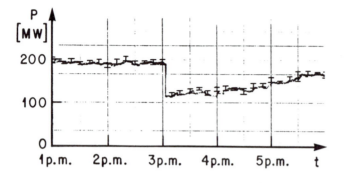

Fig. 3   Power values calculated from phase angle measurements, superimposed on the power recording for the transmission line.

Interpretation of the results

The angle deviations measured in this field test are in the order of one degree (1° corresponds to about 68MW). Despite these very small values, the relationship between line power and phase angle measurements is clearly established (Fig.3). The validity of such measurements has also been demonstrated in [2] for angles over 30° on a long transmission line.

The small errors observed in Fig.3 may be produced by the following effects:
- The response time of the phase angle measurements is very fast, whereas the chart recording of the line power is affected by sluggish time constants and friction effects.
- Nonideal conditions in the 3-phase system may have been responsible for some error. In this respect, it may be mentioned that the quality of the results remained unchanged whether using line-to-line or line-to-ground voltages.
- The phase locked loop in one of the clocks may have been defective during the experiment, causing unwant-ed hunting about the phase of the carrier signal.

APPLICATIONS IN POWER SYSTEM MONITORING

The voltage magnitudes and angles in the power system are important data for monitoring the operation of the system. In modern control centers, these voltages are always known, being continuously computed and updated by the state estimator. In this section, the physical interpretation of the phase angles and the corresponding applications in power system monitoring will be discussed. The different topics are introduced without considering how the phase angles are obtained. The discussion is then devoted to the measurement and data acquisition system.

Monitoring of power flows

The basic relationship between phase angle and power flow over a transmission line has been given in equation 1. Since the bus voltage magnitudes are usually held constant, the power flow is basically determined by the angle difference δ_{ij} between the buses. This property leads to a certain intuition about the load flows in a power system, whereby the phase angles at the buses are regarded as levels of potential and the power flows on the lines as a result of the potential differences. This interpretation is common among power system operators.

Phase angles are closely related to the transient stability margin of a power system. A simple example is given by the 1-machine infinite bus system. In this case, first swing stability is determined by the equal-area criterion, and the stability margin is clearly visible in terms of the angle distance to the stability limit. In multimachine systems, a similar interpretation may be applied. Small angles between the buses are regarded as "relatively secure", for example, as opposed to large angles approaching 90°. Such observations lead to an intuitive security assessment and help finding preventive or corrective actions in case of inadequate security.

Two examples may be mentioned in this respect. In reference [4], for instance, the average difference in bus voltage phase angles is minimized in order to obtain the generation dispatch during alert status operation. In reference [5], the phase angles are used in a pattern recognition algorithm for assessing tran-sient stability margins.

Due to the intuitive relationship between phase angles and power flows, such information would be valuable to the operators in a control center. If the phase angles at the buses were displayed on the power system diagrams, for example, certain features of the load flow situation could be determined at a short glance. Areas with predominant generation and predomi-nant load could be identified, and the main directions of power transfer and interchanges would be visible. Some feeling for the stability margin would be provided as well. Such a monitoring facility has been considered in some cases [1,6]. In reference [7] an interesting example for an unusual phase angle distribution in a real power system may be found.

Reactive power requirements

The reactive power injections which are required for controlling the voltage magnitudes in the power system depend basically on two factors: the load characteris-tics and the reactive losses in the network. Phase angles are important quantities in this respect because they influence the reactive losses in transmission lines, especially for large angle values δ_{ij}. The reac-tive power absorbed by a transmission line is given by $Q_{ij}+Q_{ji}$, which is the sum of the powers flowing into the line from the two ends. the reactive power flowing into the line from node i is given by:

$$Q_{ij} = -\frac{V_i V_j}{X_{ij}}\cos(\delta_{ij}-\theta_{ij}) + \frac{V_i^2}{X_{ij}}\cos\theta_{ij} - V_i^2 B_{ij}/2 \qquad (2)$$

where V_i and V_j are the voltage magnitudes at the end of the line and δ_{ij} is the angle between V_i and V_j. θ_{ij} is defined by $\tan\theta_{ij}=R_{ij}/X_{ij}$, which is usually a small value (0.1 for example). R_{ij}, X_{ij} and B_{ij} are the line resistance, reactance and susceptance respectively.

For small angles δ_{ij}, the reactive loss $Q_{ij}+Q_{ji}$ is seen to depend primarily on the voltage magnitudes V_i and V_j. For larger phase angles, the influence of the phase angle itself becomes also significant. In fact, the sensitivity $\partial Q_{ij}/\partial \delta_{ij}$ increases rapidly as a func-tion of δ_{ij}, according to $\sin(\delta_{ij}-\theta_{ij})$.

Due to this dependence of the reactive losses on the phase angles, the reactive power requirements for voltage control are also a function of these angles. This relationship is a major factor in the process of voltage collapse as it happened in France in 1978 [8]. Abnormally large angles are initially produced by high power transfers, which demand increased reactive compen-sation [9]. The process then becomes unstable if the compensation is insufficient, resulting in voltage collapse.

Due to the relationship between phase angles, reactive power requirements and the associated sensitivity, the phase angles in the power system are useful indicators for the security with respect to reactive control and voltage stability.

Parameters of the Dynamical System

For many applications, the dynamical equations (differential equations) of the power system are linearized about the steady state operating point. This simplifies analysis and design procedures. The parameters of the resulting linear model are of course dependent on the operating point. Since a major nonlinearity in the power system is produced by the trigonometric functions in the power equations ((1) and (2) for example), the parameters of the linear model are strongly related to the phase angles. There is therefore a certain inclination to specify the operating point by the phase angles in the system.

A typical example for the relationship between dynamical parameters and phase angles is found in the dynamic stability problem: it is known from theory and practice that the dynamic stability (small oscillations) of a generator with voltage regulator is significantly affected by the steady state load angle [10]. Since unstable operation must be avoided as much as possible, real-time monitoring of dynamic stability is often required in practice [11]. The analysis is usually performed by determining the eigenvalues of the linearized system. Knowledge of certain phase angles across the power system would provide an intuitive feeling for this dynamic stability margin.

Synchronism, Stability

During emergency operating conditions, phase angles represent a key information about the state of the system. As such, they may be useful for various monitoring functions to limit the spread of the disturbance and facilitate system restoration. System sectioning, load shedding and system resynchronization are a few examples for such functions.

Comparison with State Estimation

It has been mentioned earlier, that the voltage magnitudes and angles are important data in a modern control center. The information is provided by the state estimator, which computes the complex voltages from other quantities like voltage magnitudes and power flows. In this situation, there is no apparent motivation for making phase angle measurements, since the information is already provided. State estimation suffers from various limitations, however, suggesting that some improvements are possible. Phase angle measurements may thus be proposed as a backup for state estimation, or as an additional monitoring information in case no state estimation is provided. The following points may be mentioned in this respect:

- State estimation requires a computer system for its implementation. If such a facility is not available, or if the computer is out of function, phase angles may still be available as measurements. A few selected measurements in the power system would be a helpful monitoring facility in such cases.
- The response time of state estimation is relatively slow (in the order of seconds up to a minute). On the other hand, the speed of direct phase angle measurements is fast, being only limited by data transmission delays and by the time resolution of the local angle measurements (in the order of milliseconds).
- Due to the previous two properties, phase angle measurements may be used in event recorders. The response of certain phase angles and rotor angles during transients is informative for a-posteriori analysis of power system disturbances and for validation of power system models [43].

- Neighbouring power systems which are outside the range of the control center are usually not processed by the state estimator. It is important, however, to gather certain on-line information about their configuration and operating state. This information could be favourably complemented by a few selected phase angle measurements. These measurements would show the overall load flow pattern and the synchronism between the power systems.
- State estimation may fail to be successful in certain situations. This happens particularly when the monitoring and control functions are needed most: during overloaded operation of computer and power system, when oscillations and other perturbations are present, and when the integrity of the system is in danger. State estimation may fail due to the following reasons:
  - The state estimator requires consistent sets of measurements. Due to the lack of synchronism in the sampling process, this is usually not possible during perturbed and oscillatory conditions in the power system.
  - When certain measurements are missing, the state estimator may not be able to reconstruct the entire state of the system. In such cases, observable islands are identified and processed separately. The phase angles and power flows between the islands remain unknown.
  - The results of state estimation may be spoiled by the presence of bad data in the measurements.
  - Convergence problems may occur when the phase angles across the system are large, and when bad (inconsistent) data is present. The difficulties tend to be more frequent with large networks and during perturbed operating conditions.

Finally, a basic restriction on the application of angle measurements should be mentioned. The problem arises when phase angle measurements are performed between tightly coupled buses. In such cases when the line impedances are small, the relationship between phase angles and power flows is very sensitive (equation 1), which means that small errors in the angle lead to large deviations in the power flow. In state estimation, a similar sensitivity problem exists, but the problem can be eliminated by selecting a suitable placement of the measurements in the network.

Combination with State Estimation

The purpose of state estimation is primarily to detect, identify and correct gross measurement errors and to compute a good estimate of the bus voltage magnitudes and angles. Phase angle measurements may enhance the performance of the state estimator in the following respects:

- Additional data validation may be achieved by comparing the phase angle measurements with the results from state estimation. The test is most beneficial in networks with low measurement redundancy, such as radial portions of a power system. Conversely, state estimation is also useful for verifying the phase angle measurements.
- Phase angle measurements may be used to initialize the state estimator. Iterations can be saved as compared to the "flat start" where all voltage angles are initially zero. The advantage would be more apparent in power systems with large angles.
- The phase angle measurements may be considered as part of the measurement set for the state estimator. In this case, the equations of the estimator need to be expanded, so as to include the new measurements. Assuming the usual model [13] $z=h(x)+v$ for the traditional measurements, the following expanded model is obtained:

$$\begin{bmatrix} z \\ z_\delta \end{bmatrix} = \begin{bmatrix} h(x) \\ H_\delta x \end{bmatrix} + \begin{bmatrix} v \\ v_\delta \end{bmatrix} \qquad (3)$$

where z and z_δ denote the vectors containing the usual measurements and the angle measurements respectively, v and v_δ the corresponding measurement errors and x the unknown state vector in polar coordinates. h(.) is the function relating the state x to the measurements z, and H_δ is the matrix relating the state to the angle measurements. Since the state vector x consists of the bus voltage magnitudes and angles, the matrix H_δ turns out to be simple and very sparse, consisting of a single "1" in every row. A linear relationship as in the lower part of equation 3 may be obtained for the entire measurement model by performing a certain transformation of the measurements z. This is suggested in reference [14], as a means to increase the speed of the solution.

CONTROL APPLICATIONS

In the previous section, applications in a relatively slow process were considered, namely power system monitoring. The following examples deal with control systems in a faster time frame, such as excitation control, HVDC control, etc.

An important requirement for the control systems in this area is reliability. The requirement is much more stringent for this type of apparatus than it is for power system monitoring equipment. In this respect, two points appear to be critical in the proposed technique for phase angle measurements:
- The reliability of the clock system (time signal availability, quality of reception and hardware reliability).
- The reliability of the data links.

These two aspects must be kept in mind when evaluating the applicability of the phase angle measurements in this section.

Original Properties

Phase angle measurements provide original features in the following aspects:
- Due to the absence of long time delays, phase angle measurements are quite suitable for control applications. Calculations using state estimation would be too slow in most cases. Ultra fast state estimation techniques [14] may also be considered, however. Such techniques seem to be attractive for small portions of a network, but their performance declines as the number of elements increases.
- Phase angles can be measured across asynchronous links. The computation of such phase angles from other measured variables would be difficult in this case.
- Angle measurements can also be performed on the rotor of generators instead of AC voltages. This allows to measure relative shaft angles between generators.

The phase angle over a transmission line may also be determined by measuring the power flow and using equation 1 [24,36]. Another solution is to use the principles of protective relaying (out-of-step relaying) [15] for evaluating the angle. Strictly speaking, these techniques are only valid for individual transmission lines, however.

Excitation Control

Modern excitation systems have a good potential for improving the transient stability of power systems (preventing AC system breakup) [10]. The regulators can be designed accordingly, making optimal use of the available feedback variables. Angles between generator shafts play an important role in this respect, because they are "state variables" of the power system, which means that they are necessary for achieving optimal control. Optimal excitation control has been a popular research topic, for which an interesting discussion and conclusion is given in [16].

In some cases, the shaft angles may be obtained approximately by integrating the frequency or the speed [17]. This idea has been applied in a practical system [18] for improving the transient stability. After the disturbance, the excitation voltage is pushed to the ceiling as a function of the angle signal, which prevents the system from breaking apart (first swing stability).

The influence of the rotor angle on the parameters of the linearized generator equations has been mentioned earlier. This influence is nicely illustrated by the principle of the adaptive regulator proposed in [12]. In this paper, the settings of the regulator are adapted as a function of the rotor angle, and it is shown that the region of stability of the generator can thus be extended.

HVDC Power Modulation

Due to the fast controllability of the DC power, HVDC systems are powerful tools for improving the transient stability of AC systems. The control may be continuous in nature, using feedback variables or it may be discrete, consisting of predetermined changes in the DC power. Rotor and phase angles in the AC system appear to be useful variables for this type of control. Typical examples based on a DC link in parallel with an AC system may be found [19-23]. For certain configurations, the required phase angle may be determined from AC power measurements (equation 1). In [24] for example, the first swing (transient) stability is improved by using a power measurement on an AC line parallel to the DC link. The damping of oscillations is obtained by forming the rate of change of this power, which corresponds to the relative speed deviation [25].

In contrast to an AC line, the power transfer over a DC link does not depend on the voltage phase angle between the ends of the line. This property may be regarded as an advantage, because it provides an additional degree of freedom for control. The intrinsic synchronizing effect of the transmission system is lost, however. In some applications, a synchronizing characteristic would be desirable [26,25]. A P-δ relationship between transmitted power and phase angles helps controlling the load flow during steady state and after disturbances. The phase angles in the AC system play an important role in this respect.

Supplementary control for providing synchronizing power has already been applied in practice: In the pacific HVDC intertie, a fast increase of DC power is initiated as a control action in case of a sudden loss of transmission on the parallel AC system. The loss is compensated within the limit of the DC transmission capacity.

Dynamic Braking

Dynamic braking is used to prevent AC system separation (transient instability) by briefly removing kinetic energy at some generators. The control schemes are usually discrete in nature [18], consisting of on-off insertions of ohmic resistors. The essence of the control strategy lies in the criteria for the switching times.

Due to their importance for transient stability, phase angles (rotor angles) are useful variables for this control. The beginning of braking is usually triggered by detecting a disturbance in the power system, and the duration is determined by some frequency or angle response. In [27] for example, the resistors are proposed to be controlled using angle information extracted from local measurements. Control units are assumed to be available at every generator. In [28], the required angle information is calculated from local measurements, assuming a simplified model for the power system. The control suggested in [29] relies on direct phase angle measurements.

Capacitor Switching

Series capacitors are used to increase the power transfer capacity of long transmission lines. The capacitors may be in service continuously or switched according to a control rule [18]. Phase angles in the power system provide useful information for this control. A scheme proposed in [30] for example uses a phase angle signal to determine the switching instants. A control scheme which has been implemented in practice uses phase angle information in the form of an out-of-step relay signal.

Phase Shifting Transformers

Thyristor controlled phase shifting transformers have been proposed for the fast control of phase angles in the AC system [31-33]. Several control applications emerge from this new component. A certain similarity exists with the control of HVDC links in AC systems.

Angle measurements may be used for the phase shifter in a similar way as for a DC link. The two basic aspects are steady state control (control of load flow [7], dynamic stability) and control during transients (preventing overloads and transient instability) [32,33]. Intelligent reclosure of transmission interties is another application.

Load Shedding, Generator Tripping, System Separation

Load shedding, generator tripping and system separation are aids to maintain stable operation or prevent further collapse of the power system after major disturbances [18]. The aim is to avoid critical situations such as out-of-step conditions, underfrequency (causing unwanted trips of thermal generation) and overfrequency. The control may be based on two principles:
a) The action is taken immediately after the disturbance, using a detection mechanism.
b) The control operates as a function of the dynamical response of the power system, as in underfrequency and out-of-step relays, for example.

In case a) knowledge of certain phase angles may be useful for the following reason: The best protective action depends on the nature of the disturbance as well as on the state of the power system before the disturbance. In automatic relays, some information about the state before the disturbance may indeed be provided [11]. The purpose is to adapt the triggering level, for example. In this case, knowledge of certain phase angles can be useful. Other relaying functions such as line protection may also benefit from phase angle measurements in a similar way.

In case b), an important task is to detect out-of-step conditions or to anticipate such situations. The devices which are typically used for this purpose are out-of-step and swing relays [15,34,36]. In complex power systems the detection of stability disturbances using such relays may be complicated, however [35]. Phase angle or rotor angle measurements could be used as an alternative [36].

CALIBRATION OF THE MEASUREMENTS

It has been mentioned earlier that the phase angle measurements may be affected by constant offsets due to time shifts in the clocks. These offsets may be compensated in two ways: either by calibrating the clocks using a highly accurate portable clock or by comparing the measurements with state estimation results. The first solution is only adequate when a very high reliability can be achieved for the clocks (good signal reception and availability, emergency power supplies, etc.). The second calibration method can be performed whenever required by making a full state estimation or localized state calculations [14] using the available power system measurements.

CONCLUSION

A technique for measuring voltage phase angles by the synchronous sampling of bus voltage angles has been illustrated. The approach is basically independent of the data transmission delays, but it requires highly accurate clocks to control the sampling times. The significance of phase angles and rotor angles in the power system has been illustrated by several examples and applications in power system monitoring and control. The reliability of the clocks and data links was found to be a significant factor in evaluating the applications of the measurements.

ACKNOWLEDGEMENTS

The author would like to thank Mr.P.Storrer of the management of the Bernische Kraftwerke who made the field test possible, and Mr.R.Hofmann for his valuable assistance during the experiments. He also acknowledges the help of Dr.Th.Lalive d'Epinay and Mr.R.Kesenheimer for design and implementation of the measuring devices.

APPENDIX

Synchronized Clocks

Accurate clocks are needed for many applications in power systems. For precisions in the order of 10 ms, suitable solutions may be found easily. If the requirements are more stringent, some difficulties may arise. The highest precision in present day applications is required by event recorders. In these devices, a precise identification of the particular cycles of the AC waveforms is desired. A precision in the order of a few milliseconds is therefore needed [37]. For evaluating HVDC disturbances and control phenomena, the precision must be even better [38]. The use of accurate clocks has also been proposed for achieving synchronous sampling of certain power system variables, as a means to improve load-frequency control and state estimation, for example. In the following discussion, time dissemination systems are considered which allow a precision of 1 millisecond and better.

From the point of view of planning requirements, two cases may be distinguished:
1) Radio signals like HBG, WWVB or Loran-C are available for synchronizing the clocks to a central time reference.
2) The clocks are synchronized without the help of a central radio signal.

Case 1) will be discussed first. Accurate time may be obtained by synchronizing the local clocks to a central time reference using radio signals. Each clock possesses its own time base, for example a quartz oscillator, and the radio signal is used to make appropriate corrections. The constant term in the propagation delays is easily compensated in the receiving system. This concept is attractive with respect to both reliability and accuracy. The local time base ensures continued operation in the case of poor reception or temporary absence of the radio signal, and it permits a high immunity against noise perturbations in the radio signal.

Several types of transmitters may be used for time reference. The emitted signals usually consist of a carrier modulated by square impulses. In the RF range long wave signals exhibit the most reliable propagation characteristics. Short waves are inadequate due to fading and because the frequency band tends to be overcrowded. FM signals have the disadvantage of a limited range. Furthermore, the VLF range and the L-band are also appropriate for time dissemination. A few major transmitters are given below:
- The time signal transmitters like HGB (75 kHz) in Switzerland, DCF (77.5 kHz) in Germany, MSF (60 kHz) in England and WWVB (60 kHz) in the United States are well suited for low cost receiver-clock systems. Accuracies better than 1 ms can be attained with a good design of receiver, signal detector and phase locked loop [3]. Among the factors which restrict the attainable accuracy, the following may be the most significant: the stability of the local time base (quartz oscillator), the noise in the received signal

(lightning and man made noise), the fluctuations due to sky-wave propagation and the rise time of the modulated impulses resulting from the limited bandwidth of the transmitters.

A much higher accuracy in the sense of time stability may be obtained by locking the local clock on the carrier of the radio signal using a phase locked loop. If the carrier is controlled by a highly accurate time standard, the same long term stability is available at the local clock. The error due to variations in the propagation delay is in the order of one microsecond for distances up to 1000 km.

- The global navigation System Omega [39], which consists of 8 VLF transmitters distributed around the globe can also be used for time reference. The attainable accuracy is lower than with the previous type of transmitters, however. The rise time of the signal envelope is slow and is subject to significant disturbances. The good navigational accuracy is obtained by locking on the carrier signal.

- A much higher precision may be obtained with the navigation system Loran-C (100 kHz) [2]. Beside its navigational purpose, the system is also used for time coordination between the western countries [40]. The accuracy which can be better than 1 μs is made possible by the wide bandwidth and high power of the transmitters. This allows precise and secure detection of the signal edges from the first incoming wave (ground wave propagation).

Due to the increased bandwidth and complexity of the Loran-C signals, the corresponding receiver-clock systems are more expensive than the HBG- or WWVB-type devices.

- The Navstar Global Positioning System [41] is a space-based radio navigation network in development. The L-band signals will allow a high precision in the order of 1 μs, at the cost of a relatively expensive antenna.

In case a certain area has no suitable time signal available, it is possible to set up a time dissemination system at one's own initiative. Cooperation with government, utilities or other companies may be advantageous for such a project. A multipurpose long wave transmitter can be attractive in this respect. As an example, the Swiss time signal transmitter HBG may be mentioned, which will be used concurrently for a long-range paging system starting 1981. In this arrangement, the entire transmitter is expected to be financially self-supporting from the paging system alone. The time synchronization is also a benefit for the paging system, because it allows improved performance of the data transmission and can be used to synchronize any associated clocks.

In case 2), in which an autonomous system without central radio signal is required, one of the following two solutions is available.

- Highly accurate and stable cesium beam or rubidium frequency standards can be used. This solution has the disadvantage of relatively high cost, low reliability and limited life time of the components. Its decentralized nature may be viewed as an advantage for the reliability, on the other hand. Synchronism between the different clocks can be achieved by periodic corrections using a portable time standard.

- Master-slave time sychronization [38,42] is another way of synchronizing geographically disseminated clocks. Such a scheme requires a data transmission network with stringent constraints on the transmission delays. In evaluating the cost of such a system, significant effort and expenses may be required for data processing, interfacing equipment and for the design of the entire system.

Phase angle measurements demand an accuracy in the order of a few microseconds. Only two of the above schemes satisfy this requirement - the navigation systems Loran-C and Navstar. For some applications, however, time signals like HBG or WWVB may also be used. As explained

earlier, a good time stability in the order of a few microseconds can be obtained when the local clocks are locked on the carrier of the radio signal. An initial synchronization is necessary at the start, since the particular cycle on which the clock is locked cannot be identified. Corresponding calibration procedures have been outlined at the end of the main text. In order not to lose the cycle during poor reception or temporary shutdowns of the transmitter, the signal of a second transmitter can be used, on which to lock the oscillator during absence of the first signal and thus improve the reliability of the clocks.

An important remark should be made at this point: High noise levels in the low-frequency and VLF range exist in the vicinity of high voltage power lines, due to coronal discharges and carrier frequencies for telemetry and remote control. For good signal quality it is therefore advisable to place the antenna at least 100m away from high potentials, using shielded cable.

REFERENCES

1. W.A. Flowers, "Phase angles displayed for dispatcher", Electrical World, pp. 48-49, May 1973.

2. G. Missout, P. Girard, "Measurement of bus voltage angle between Montreal and Sept-Iles", IEEE Trans. on Power Apparatus and Systems, Vol. PAS-99, No.2, pp. 536-539, March/April 1980.

3. P. Schumacher, "Horloges de précision synchronisées par l'émetteur HBG", Procès verbal du Congrès International de Chronométrie, Genève, pp. 91-95, 1979.

4. J.G. Blaschak, G.T. Heydt, J.M. Bright, "A generation dispatch strategy for power systems operating under alert status", IEEE Paper No. A 79 474-8, presented at the IEEE PES Summer Meeting, Vancouver BC, Canada, July 15-20, 1979.

5. O. Saito, et al., "Security monitoring systems including fast transient stability studies", IEEE Trans. on PAS, Vol. PAS-94, No.5, pp. 1789-1805, Sept./Oct. 1975.

6. M.J. Britt, G.H. Couch, "Power system control center display of voltage phase angle", IFAC Symposium on Automatic Control and Protection of Electric Power Systems, Melbourne, Australia, Preprint of Papers, pp. 410-413, February 1977.

7. L.J. Rubino, W.J. Tishinski, "System interconnection using special phase shifting transformers", IEE Conference on High Voltage DC and/or AC power transmission, London, Conf. Publication No.107, pp. 183-188, 1973.

8. "L'incident du 19 Décembre 1978", Revue Générale de l'Electricité, Tome 89, No.4, Avril 1980.

9. Committee Report, "EHV operating problems associated with reactive control", IEEE Paper No. 80 SM 513-2, presented at the IEEE PES Summer Meeting, Minneapolis, Minnesota, July 13-18, 1980.

10. "IEEE Guide for Identification, Testing and Evaluation of the Dynamic Performance of Excitation Control Systems", IEEE Std 421A-1978.

11. A. Shalaby, V.F. Carvalho, I.A. Findlay, "On-line computer monitoring of complex power system stability limits", Proceedings of the Power Industry Computer Applications Conference (PICA), Cleveland, Ohio, pp. 64-72, May 1979.

12. M. Venkata Rao, M.P. Dave, "Extension of stable operating regions of synchronous machines by non-

linear excitation control", IEEE Paper No. A 78 099-4, presented at the IEEE PES Winter Meeting, New York, N.Y., January 29 - February 3, 1978.

13. S.A. Arafeh, R. Schinzinger, "Estimation algorithms for large-scale power systems", IEEE Trans. on PAS, Vol. PAS-98, No.6, pp. 1968-1977, Nov./Dec. 1979.

14. J. Zaborszky, K.W. Whang, K.V. Prasad, "Ultra fast state estimation for the large electric power system", IEEE Trans. on Automatic Control, Vol. AC-25, No.4, pp. 839-841, August 1980.

15. W.A. Elmore, "System Stability and Out-of-Step Relaying" in Applied Protective Relaying, Westinghouse Electric Corporation, Newark, N.J., 1976.

16. J.H. Anderson, et al. "Microalternator experiments to verify the physical realisibility of simulated optimal controllers and associated sensitivity studies", IEEE Trans. on PAS, Vol. PAS-97, No.3, pp. 649-658, May/June 1978.

17. P. Bonanomi, R. Bertschi, "On-line identification of an equivalent reactance for stability applications", IEEE Paper No. 80 SM 502-5, presented at the IEEE PES Summer Meeting, Minneapolis, Minnesota, July 13-18, 1980.

18. IEEE Committee Report, "A description of discrete supplementary controls for stability", IEEE Trans. on PAS, Vol. PAS-97, No.1, pp. 149-165, Jan./Feb. 1978.

19. D.B. Goudie, "Steady state stability of parallel h.v.a.c. - d.c. power-transmission systems", Proc. IEE, Vol.119, No.2, pp. 216-224, February 1972.

20. P.K. Dash, et al. "Transient stability and optimal control of parallel A.C.-D.C. power systems", IEEE Trans. on PAS, Vol. PAS-95, No.3, pp. 811-820, May/June 1976.

21. A.M. El-Serafi, H.S. Khalil, "Effect of HVDC control signals on stability of AC/DC parallel system", Canadian Communications & Power Conference, Montreal, Canada, pp. 445-448, October 20-22, 1976.

22. J. Käuferle, E. Rumpf, "The influence of HVDC transmission systems and their control on the stability of associated A.C. networks", CIGRE Report No. 32-15, Paris, 1970.

23. H. Kobayashi, K. Ichiyanagi, "Improvement of the transient stability by optimal switching control of parallel AC-DC power systems", IEEE Trans. on PAS, Vol. PAS-97, No.4, pp. 1140-1148, July/Aug. 1978.

24. N.G. Hingorani, N. Sato, J.J. Vithayathil, "Use of fast power control on Pacific Northwest-Southwest D-C intertie operating in parallel with two A-C interties", IEE Conference on High Voltage DC and/or AC power transmission, London, Conf. Publication No.107, pp. 212-220, 1973.

25. R.L. Cresap, et al., "Operating experience with modulation of the Pacific HVDC intertie", IEEE Trans. on PAS, Vol. PAS-97, No.4, pp. 1053-1059, July/Aug. 1978.

26. R. Jötten, et al., "Control in HVDC systems. The state of the art. Part 1: Two-terminal systems", CIGRE Report No. 14-10, Paris, 1978.

27. J. Zaborszky, K. Prasad, K.-W. Whang, "Operation of the large interconnected power system by decision and control", IEEE Trans. on PAS, Vol. PAS-99,

No.1, pp. 37-45, Jan./Feb. 1980.

28. J. Meisel, "Transient stability augmentation using a hierarchical control structure", IEEE Trans. on PAS, Vol. PAS-99, No.1, pp. 256-267, Jan./Feb. 1980.

29. V.M. Gornshtein, Ya. N. Luginskii, "The use of repeated electrical braking and unloading to improve the stability of power systems", Electric Technology USSR (England), Vol.2, pp. 292-302, 1963.

30. O.J.M. Smith, "Power system transient control by capacitor switching", IEEE Trans. on PAS, Vol. PAS-88, No.1, pp. 28-35, January 1969.

31. R.M. Mathur, R.S. Basati, "A thyristor controlled static phase-drifter for AC power transmission", IEEE Paper No. 80 SM 660-1, presented at the IEEE PES Summer Meeting, Minneapolis, Minnesota, July 13-18, 1980.

32. C.P. Arnold, R.M. Duke, J. Arrillaga, "Transient stability improvement using thyristor controlled quadrature voltage injection", IEEE Paper No. 80 SM 631-2, presented at the IEEE PES Summer Meeting, Minneapolis, Minnesota, July 13-18, 1980.

33. H. Glavitsch, R. Baker, G. Güth, "Stability improvement by static phase shifting or tap changing transformers", Power Systems Computation Conference (PSCC), Lausanne, Switzerland, July 12-17, 1981.

34. F.R. Schleif, et al. "A swing relay for the East-West Intertie", IEEE Trans. on PAS, Vol. PAS-88, No.6, pp. 821-825, June 1969.

35. V.L. Mansvetov, "Discovery of disturbance of stability in complex power sytems", Electric Technology USSR (England), No.3, pp. 13-26, 1977.

36. L.A. Boguslavsky, et al. "Assembly of general application devices for control of power systems under fault condition", CIGRE Report No. 34-04, Paris 1980.

37. V. Arcidiacono, et al. "On-line detection and recording of major disturbances in the ENEL power system", Proceedings of the PICA Conference, Cleveland, Ohio, pp. 273-283, May 1979.

38. D.D. Hofferber, S.L. Nilsson, "Integrated data acquisition system for HVDC converter stations and large AC substations", IEEE Trans. on PAS, Vol. PAS-99, No.2, pp. 540-548, March/April 1980

39. J.F. Kasper Jr., Ch.E. Hutchinson, "Omega: global navigating by VLF fix", IEEE Spectrum, pp. 59-63, May 1979.

40. P. Schumacher, "Beim letzten Ton ist es genau..." Schweizer Elektronik-Magazin rte, No.12, pp. 54-63, 1979.

41. IEEE Spectrum, Januar 1980, p.80

42. G. Missout, W. Le Francois, L. La Roche, "Time Dissemination in the Hydro Quebec Network", 11th Precise Time and Time Interval (PTTI) applications and planning meeting, Nasa Goddard Space Flight Center, Greenbelt, Maryland, 1979.

43. G. Missout, et al., "Dynamic measurement of absolute voltage angle on long transmission line", submitted for presentation at the IEEE PES Summer Meeting, Portland, Oregon, July 26-31, 1981.

A NEW MEASUREMENT TECHNIQUE FOR TRACKING VOLTAGE PHASORS, LOCAL SYSTEM

FREQUENCY, AND RATE OF CHANGE OF FREQUENCY

A. G. Phadke
Fellow, IEEE
VPI & SU
Blacksburg, VA.

J. S. Thorp
Senior Member, IEEE
Cornell University
Ithaca, New York

M. G. Adamiak
Member, IEEE
AEP Serv. Corp.
New York, N.Y.

Abstract: With the advent of Substation Computer Systems dedicated to protection, control and data logging functions in a Substation, it becomes possible to develop new applications which can utilize the processing power available within the substation. The microcomputer based Symmetrical Component Distance Relay (SCDR) described in the references cited at the end of this paper possesses certain characteristics which facilitate real-time monitoring of positive sequence voltage phasor at the local power system bus. With a regression analysis the frequency and rate-of-change of frequency at the bus can also be determined from the positive sequence voltage phase angle. This paper describes the theoretical basis of these computations and describes results of experiments performed in the AEP power system simulation laboratory. Plans for future field tests on the AEP system are also outlined.

INTRODUCTION

Positive sequence voltage phasor at a power system bus is a parameter of vital significance. The collection of all positive sequence voltage phasors constitutes the state-vector of a power system. It is also one of the output variables of such steady-state analysis programs as the load flow and state-estimation. The idea of measuring these voltage phasors directly has appeared in the technical literature from time to time. However, no entirely satisfactory practical method for measuring voltage phasors in real-time has yet been proposed.

It is reasonable to assume that a measurement system capable of measuring positive sequence voltage phasors directly and with adequate accuracy has the potential to simplify and improve state-estimation algorithms as well as other real-time analysis programs. In addition, accurate measurement of phasor voltages in real-time may lead to development and implementation of adaptive control procedures which can be based on direct measurement of key system variables. Use of local frequency and rate of change of frequency for load-shedding and load restoration functions is yet another well-known application of such measurements. In fact, the measurement of phase angle,

82 SM 444-8   A paper recommended and approved by the IEEE Power System Relaying Committee of the IEEE Power Engineering Society for presentation at the IEEE PES 1982 Summer Meeting, San Francisco, California, July 18-23, 1982. Manuscript submitted February 2, 1982; made available for printing May 3, 1982.

frequency and rate of change of frequency (which are quantities of "system" significance) with substation computers can be viewed as a building block for the computer based distributed-processor, hierarchical control systems which have been proposed for real-time monitoring and control of power systems.

The steady-state analysis programs such as load-flow, state-estimation and optimal power flow use the positive sequence bus voltage as the state variable, and the network model uses the positive sequence representation of the power system. It should therefore be a matter of some concern that many of the phase angle measurement schemes proposed in the literature [1] measure the magnitude and phase angle of voltage of one phase. In the presence of system unbalance (which exists on almost all high voltage systems) such a measurement is an approximation at best and does not represent the positive sequence voltage at the bus accurately. Furthermore, almost all of these schemes measure the phase angle of the voltage by timing the zero-crossing instant of the voltage waveform with respect to a standard reference pulse. Since the zero-crossing instant of a voltage wave is affected by the magnitude of the non - 60 Hz components present, this is yet another source of error as far as the measurement of positive sequence fundamental frequency voltage measurement is concerned. In contrast, the technique described in this paper measures a digitally filtered positive sequence voltage phasor at the bus directly.

Recently, papers [2,9] have described techniques for measuring the frequency and rate-of-change of frequency with a microcomputer. This technique uses the aggregate magnitude of all non - 60 Hz components in a waveform as a measure of the input waveform frequency. It has been observed that this measurement principle (based on "leakage effects" in the Fourier Transform calculation) is not a particularly sensitive measure of frequency deviations, and may in fact go completely astray if a non - 60 Hz component is present in the input waveform. In contrast, this paper describes a robust computation technique for phase angle, frequency, and rate-of-change of frequency.

Following a theoretical derivation of the necessary equations, a laboratory experiment using a small scale power system model will be described and the results of the experiment discussed. Plans for field tests of these ideas on the AEP system will also be presented.

Reprinted from *IEEE Trans. Power App. Syst.*, vol. PAS-102, no. 5, pp. 1025–1038, May 1983.

FILTERED PHASORS FROM SAMPLED DATA

Consider a sinusoidal input signal of frequency ω given by

$$x(t) = \sqrt{2} X \sin(\omega t + \phi) \tag{1}$$

This signal is conventionally represented by a phasor (a complex number) $\bar{X}$

$$\bar{X} = X e^{j\phi} = X \cos\phi + jX \sin\phi \tag{2}$$

Assuming that x(t) is sampled N times per cycle of the 60 Hz waveform to produce the sample set $\{x_k\}$

$$x_k = \sqrt{2}\, X \sin\left(\frac{2\pi}{N} k + \phi\right) \tag{3}$$

The Discrete Fourier Transform of $\{x_k\}$ contains a fundamental frequency component given by

$$\bar{X}_1 = \frac{2}{N} \sum_{k=0}^{N-1} x_k\, e^{-j\frac{2\pi}{N}k}$$

$$\tag{4}$$

$$= \frac{2}{N} \sum_{k=0}^{N-1} x_k \cos\frac{2\pi}{N}k - j\frac{2}{N} \sum_{k=0}^{N-1} x_k \sin\frac{2\pi}{N}k$$

$$\equiv X_c - jX_s \tag{5}$$

where X_c and X_s are the cosine and sine multiplied sums in the expression for $\bar{X}_1$ Substituting for x_k from equation (3) in equations (4) and (5) it can be shown that for a sinusoidal input signal given by equation (1),

$$X_c = \sqrt{2}\, X \sin\phi$$

$$X_s = \sqrt{2}\, X \cos\phi \tag{6}$$

From equations (2), (5) and (6), it follows that the conventional phasor representation of a sinusoidal signal is related to the fundamental frequency component of its DFT by

$$\bar{X} = \frac{1}{\sqrt{2}} j\, \bar{X}_1 = \frac{1}{\sqrt{2}} (X_s + j X_c) \tag{7}$$

In the preceding development it was assumed that the input signal is a pure sine wave of fundamental frequency. When the input contains other frequency components as well, the phasor calculated by equation (7) is a filtered fundamental frequency phasor. The input signals must be band-limited to satisfy the Nyquist criterion[3] to avoid errors due to aliasing effects. It is therefore assumed that the input signals are filtered with low-pass analog filters having a cut-off frequency of ωN/4π Hz. The effect of anti-aliasing analog filters on the fundamental frequency signals has been discussed in reference [4].

Another point to note is that equation (4) assumes data collected over one complete cycle of the fundamental frequency. Although the filter equations (4) are particularly simple for the case of a one cycle data window, similar filter equations can be formulated for any other window length. [5] The consequence of using other window lengths is to affect the accuracy of the phasor computation. A more detailed discussion of this aspect will be found in reference [5]. The remainder of this paper will assume the use of one cycle data window, it being a simple matter to modify one cycle data-window results to reflect the effects of other data window lengths.

RECURSIVE PHASOR COMPUTATION

The input signal of equation (1) is shown in Figure (1). Data window 1 produces the sample set $\{x_k, k=0,\ldots N-1\}$ and the phasor representation obtained from this sample set is given by equation (7). A new sample is obtained after an elapsed time corresponding to the sampling angle $2\pi/N$ radians. At this time the data window 2 becomes operative with sample set $\{x_k, k=1,\ldots N\}$. The phasor computations using data window 2 are performed with equations (1), (2) and (7) as follows:

$$x(t) = \sqrt{2} X \sin\left(\omega t + \phi + \frac{2\pi}{N}\right) \tag{8}$$

$$\bar{X}^{(new)} = X\, e^{j(\phi + \frac{2\pi}{N})}$$
$$= \bar{X}^{(old)}\, e^{j\frac{2\pi}{N}} \tag{9}$$

$$\bar{X}^{(new)} = \frac{1}{\sqrt{2}} j\, \bar{X}_1^{(new)}$$
$$= \frac{1}{\sqrt{2}} \left(X_s^{(new)} + j X_c^{(new)} \right) \tag{10}$$

where the superscripts 'new' and 'old' signify computations from data windows 2 and 1 respectively. Equation (9) shows that the use of equation (7) for calculating the filtered phasor of an input signal produces a phasor which rotates in a counterclockwise direction in the complex

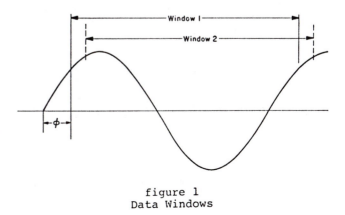

figure 1
Data Windows

557

plane by the sampling angle $2\pi/N$. This phenomenon is illustrated in Figure 2. The angular velocity of the phasor computed from a 60 Hz input signal is thus $120\pi \approx 377$ radians per second, although the phasor is available only at discrete angles.

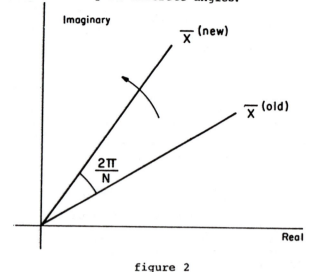

figure 2
Phasors from Different Data Windows

In general, when sample data from the rth window is used,

$$X_c^{(r)} = \frac{2}{N} \sum_{k=0}^{N-1} x_{k+r-1} \cos \frac{2\pi}{N} k \qquad (11)$$

$$X_s^{(r)} = \frac{2}{N} \sum_{k=0}^{N-1} x_{k+r-1} \sin \frac{2\pi}{N} k \qquad (12)$$

$$\bar{X}^{(r)} = \frac{1}{\sqrt{2}} (X_s^{(r)} + j X_c^{(r)})$$

$$= \bar{X}^{(r-1)} \cdot e^{j \frac{2\pi}{N}} \qquad (13)$$

Clearly the procedure described by (11)-(13) is non-recursive, and requires 2N multiplications and 2(N-1) additions to produce the phasor $\bar{X}^{(r)}$. (The factor 2 is of no consequence, and is usually suppressed). It should however be noted that in progressing from one data window to the next, only one sample (x_o) is discarded and only one sample (x_N) is added to the data set. It is therefore advantageous to develop a technique which retains 2(N-1) multiplications and 2(N-1) sums corresponding to that portion of the data which is common to the old and new data windows.

A recursive computation of the type described above is made possible by the fact that the DFT computation is arbitrary to the extent of its phase angle. Consider the calculation of $X_c(\theta)$, $X_s(\theta)$ and $\bar{X}(\theta)$ with Fourier coefficients having an arbitrary phase angle θ:

$$X_c^{(\theta)} = \frac{2}{N} \sum_{k=0}^{N-1} x_k \cos \left(\frac{2\pi}{N}k + \theta\right) \qquad (14)$$

$$X_s^{(\theta)} = \frac{2}{N} \sum_{k=0}^{N-1} x_k \sin \left(\frac{2\pi}{N} k + \theta\right) \qquad (15)$$

$$\bar{X}^{(\theta)} = \bar{X} e^{-j\theta} \qquad (16)$$

The phasor representation of the input signal in equation (16) contains as much information as does the one described by equations (2) and (7), and can therefore be used without any loss of generality. It is advantageous to calculate the phasor for data window 1 with equations (2) and (7), and that for data window 2 with equations (14) - (16):

$$X_c^{(new),\theta} = \frac{2}{N} \sum_{k=0}^{N-1} x_{k+1} \cos \left(\frac{2\pi}{N} k + \theta\right) \qquad (17)$$

$$X_s^{(new),\theta} = \frac{2}{N} \sum_{k=0}^{N-1} x_{k+1} \sin \left(\frac{2\pi}{N} k + \theta\right) \qquad (18)$$

$$\bar{X}^{(new),\theta} = \bar{X}^{(old)} \cdot e^{-j\theta} \qquad (19)$$

If θ is now made equal to $\frac{2\pi}{N}$, equations (17) and (18) become

$$X_c^{(new),2\pi/N} = \frac{2}{N} \sum_{k=o}^{N-1} x_{k+1} \cos \frac{2\pi}{N}(k+1) \qquad (20)$$

$$X_s^{(new),2\pi/N} = \frac{2}{N} \sum_{k=o}^{N-1} x_{k+1} \sin \frac{2\pi}{N}(k+1) \qquad (21)$$

Equations (20) and (21) are recursion relations, since they can be re-written as

$$X_c^{(new),\frac{2\pi}{N}} = \frac{2}{N} \sum_{k=0}^{N-1} x_k \cos \frac{2\pi}{N}k$$

$$+ \frac{2}{N} \cos 2\pi (x_N - x_o) \qquad (22)$$

$$X_s^{(new),\frac{2\pi}{N}} = \frac{2}{N} \sum_{k=0}^{N-1} x_k \sin \frac{2\pi}{N} k$$

$$+ \frac{2}{N} \sin 2\pi (x_N - x_o) \qquad (23)$$

or

$$X_c^{(\text{new}),\frac{2\pi}{N}} = X_c^{(\text{old})} + \frac{2}{N}\cos 2\pi(x_N - x_o) \quad (24)$$

$$X_s^{(\text{new}),\frac{2\pi}{N}} = X_s^{(\text{old})} + \frac{2}{N}\sin 2\pi(x_N - x_o) \quad (25)$$

If it is understood that the angle θ is always set equal to 2π/N, this factor can be dropped from the superscript on the left hand side of equations (24) and (25), and the new phasor is given by

$$\bar{X}^{(\text{new})} = \bar{X}^{(\text{old})} + j\frac{1}{\sqrt{2}}\cdot\frac{2}{N}(x_N - x_o)e^{-j\frac{2\pi}{N}} \quad (26)$$

In general, the rth phasor is computed from the (r-1)th phasor by

$$\bar{X}^{(r)} = \bar{X}^{(r-1)} + j\frac{1}{\sqrt{2}}\frac{2}{N}(x_{N+r} - x_r)\,e^{-j\frac{2\pi}{N}(r-1)} \quad (27)$$

Recursive equations (24) and (25) are comparable to the non-recursive equations (11) and (12). With the recursive procedure only two multiplications need be performed at each new sample time; making this a very efficient computational algorithm.

It is interesting to note that when the input signal is a pure sine wave of fundamental frequency $x_{N+r} = x_r$ for all r; and consequently for this case equation (27) becomes

$$\bar{X}^{(r)} = \bar{X}^{(r-1)} \quad \text{for all r.} \quad (28)$$

Equation (28) shows that when a recursive computation is used to calculate phasors, it leads to stationary phasors in the complex plane when the input signal is a pure sine wave of fundamental frequency. Recall that non-recursive phasor computation leads to phasors which rotate in the complex plane with an angular velocity ω.

RECURSIVE SYMMETRICAL COMPONENT COMPUTATION

Recursive computation of symmetrical components of three phase signals has been described in literature [4,6,7]. A few salient points of that procedure are summarized here for ready reference. A sampling rate of 720 Hz (N=12) has been found to be most beneficial for real-time computation of symmetrical components [4]. The sines and cosines of multiples of 30° (2πk/N) are shown arranged around the circumference of a circle in Figure 3. Computation of X_c and X_s of equation (5) can be visualized as an inner product of the data vector $\{x_k, k=o.... N-1\}$ and the constants of Figure 3 starting at constant no. 1 and constant no. 4 respectively.

It will be recalled that the positive sequence voltage can be calculated from three phase voltages by the relation

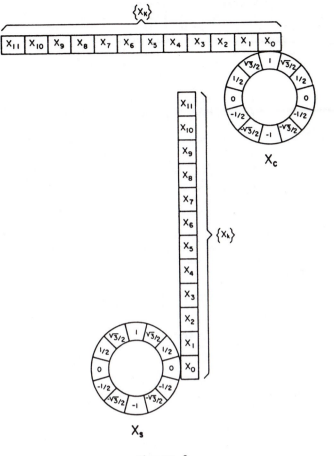

figure 3
Phasor from Samples

$$\bar{V}_1 = \frac{1}{3}(\bar{V}_a + \alpha\bar{V}_b + \alpha^2\bar{V}_c) \quad (29)$$

where α and α^2 are phase shifts of 120° and 240° respectively. Equations (14) and (15) provide the necessary phase shifts for phase b and phase c voltages. Since the sampling rate selected (720 Hz) corresponds to a phase angle of 30° between samples, the necessary phase shifts are obtained simply by introducing an offset of 4 and 8 respectively in using the constant table of Figure 3. Thus $\alpha\bar{V}_b$ is produced by calculating X_c and X_s starting with constants no. 5 and 9 on data samples of $v_b(t)$, while $\alpha^2\bar{V}_c$ is obtained when X_c and X_s are calculated starting with constants no. 9 and 12 on data samples of $v_c(t)$. The recursive relation for the positive sequence voltage is similar to equation (27)

$$\bar{V}_1^{(\text{new})} = \bar{V}_1^{(\text{old})}$$
$$+ j\frac{2}{N}\cdot\frac{1}{\sqrt{2}}\cdot\frac{1}{3}[(v_{a,r+N} - v_{a,r})e^{-j\frac{2\pi}{N}(r-1)}$$
$$+ (v_{b,r+N} - v_{b,r})e^{-j\frac{2\pi}{N}(r+5-1)} \quad (30)$$
$$+ (v_{c,r+N} - v_{c,r})e^{-j\frac{2\pi}{N}(r+9-1)}]$$

Other symmetrical components of voltages and currents are calculated in a similar manner. Several practical considerations for computing symmetrical components will be found in references [4,6,7]. It should also be reiterated that although equation (30) assumes data window of one cycle, other data windows can be readily accommodated in this procedure.

It will be assumed that a positive sequence voltage phasor is computed recursively at each sampling instant (12 times a cycle). This computation is done by the Symmetrical Component Distance Relay as part of its relaying algorithm, and if such a relay exists at the substation this phasor is available for use at no additional computational burden. In the absence of a Symmetrical Component Distance Relay at the substation, equation (30) must be programmed on a microprocessor and installed at the substation as a stand-alone measurement system.

CALCULATION OF LOCAL FREQUENCY

Assume that the sampling clock used for obtaining the sampled data from the three phase voltage inputs operates precisely at 720 Hz while the power system frequency is precisely 60 Hz. When a recursive relation such as equation (27) is used to calculate the phasors, the resultant phasors will remain stationary in the complex plane. The equations derived in this section will consider phasor computations from a single input signal, although it may be verified readily that the results of this section apply directly to the positive sequence voltage calculated from three input voltage signals according to equation (30).

If the input signal frequency is now assumed to change slightly from 60 Hz by an amount Δf, while the sampling clock frequency remains at 720 Hz, it can be shown that the recursive relation of equation (27) changes into

$$\bar{X}_{60+\Delta f}^{(r)} = \bar{X}_{60}^{(o)} \cdot \frac{\sin \frac{\Delta f}{60} \pi}{N \sin \frac{\Delta f}{60} \frac{\pi}{N}} \cdot e^{j \frac{\Delta f}{60} \frac{2\pi}{N} r} \quad (31)$$

where $\bar{X}_{60}^{(o)}$ is the initial computation of the phasor from a 60 Hz input signal having the same magnitude as the $(60+\Delta f)$ Hz signal, r is the recursion number, and N is the number of samples in a period of the 60 Hz wave. Equation (31) shows that when the input signal frequency changes from 60 Hz to $(60 + \Delta f)$ Hz, the phasor obtained recursively undergoes two modifications:

a magnitude factor of $(\sin \frac{\Delta f}{60}\pi / N \sin \frac{\Delta f}{60} \frac{\pi}{N})$;

and a phase factor of $\exp(j\frac{\Delta f}{60} \frac{2\pi}{N}r)$. The magnitude factor is independent of r, and is relatively small for small changes in frequency. The magnitude factor is a manifestation of the "leakage effect", and has been proposed as a measure of the frequency deviation Δf [2].

However, the phase angle effect is far more sensitive to the frequency Δf, and provides a most direct measure of Δf.

Denoting the phase factor by $\exp(j\psi_r)$

$$e^{j(\frac{\Delta f}{60} \frac{2\pi}{N} r)} \equiv e^{j\psi_r} \quad (32)$$

$$\psi_r = \frac{\Delta f}{60} \frac{2\pi}{N} r \quad (33)$$

and thus the phase angle at rth recursive computation directly depends upon the frequency deviation and the recursion order r. Since r increases by 1 in each iteration, the recursive relation for ψ_r becomes

$$\psi_r = \psi_{r-1} + \frac{\Delta f}{60} \cdot \frac{2\pi}{N} \quad (34)$$

Further, the time interval between two iterations is 1/60N seconds: and therefore the angular velocity of ψ is given by

$$\frac{d\psi}{dt} = \frac{\psi_r - \psi_{r-1}}{(1/60N)}$$

$$= 2\pi \Delta f \text{ radians/second} \quad (35)$$

The rate of change of the complex phasor angle is thus directly related to the input signal frequency. For example, an input signal with frequency (60 ± 1) Hz would produce a phasor that turns one complete circle per second in the complex plane. When the input signal frequency is 61 Hz, the phasor rotates in the counterclockwise direction, whereas for an input signal frequency of 59 Hz the phasor rotates in a clockwise direction. There is a striking resemblance between this phenomenon of a rotating phasor and the principle of a power system synchroscope so familiar to most power system engineers.

Just as the frequency is calculated by calculating $d\psi/dt$, the rate of change of frequency can be calculated by computing $d^2\psi/dt^2$. From equation (35)

$$f = 60 + \Delta f$$

$$= 60 + \frac{1}{2\pi} \frac{d\psi}{dt} \quad \text{Hz} \quad (36)$$

$$\frac{df}{dt} = \frac{1}{2\pi} \frac{d^2\psi}{dt^2} \quad \text{Hz/second} \quad (37)$$

The actual computations for f and df/dt are performed with the help of regression formulas as explained below.

IMPLEMENTATION ON A MICROCOMPUTER

It is well known that a frequency meter (or a frequency measurement technique) has a longer time constant (i.e. takes longer to make the measurement) for input signals with low frequencies. It can therefore be expected that when input signals have frequencies which differ from 60 Hz by small amounts, the technique presented in this paper would take longer to make that measurement. This effect is linked to the accuracy requirement of the frequency measurement, and is also present in the algorithm developed to implement equations (36) and (37).

The real-time computation of phase angle, frequency and rate-of-change of frequency on a microcomputer requires careful consideration of the numerical technique to be selected. The first step in the algorithm is to calculate the phase angle between two phasors. Consider the two phasors OA and OB in Figure 4. The phase angle ψ between OA and OB is given by

$$\tan \frac{\psi}{2} = \frac{\left| \frac{1}{2}(OB-OA) \right|}{\left| \frac{1}{2}(OB+OA) \right|} \tag{38}$$

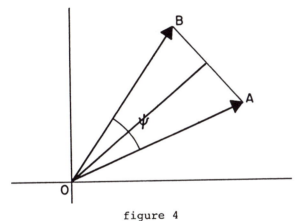

figure 4
Angle between two Phasors

The ratio on the right hand side of equation (38) is calculated first, and then ψ obtained from a stored table in the microcomputer. The ratio calculation can be simplified greatly by noting that (OB-OA) is perpendicular to (OB+OA); and for two phasors $\bar{X}$ and $\bar{Y}$ which are perpendicular to each other,

$$\frac{|\bar{X}|}{|\bar{Y}|} \equiv \frac{|X_r + jX_i|}{|Y_r + jY_i|}$$

$$= \frac{|X_r| + |X_i|}{|Y_r| + |Y_i|} \tag{39}$$

Equation (39) is far simpler than equation (38), and is used in the on-line program to calculate ψ.

The denominator of equation (38) is the magnitude of positive sequence voltage at the substation. During non-fault periods, this magnitude is relatively unchanging. The Symmetrical Component Distance Relay developed at AEP uses internal scaling factors so as to produce a normal positive sequence

voltage magnitude of about 25,000. It is well to design an algorithm such that the numerator in equation (38) turns out to be about 6000 in all cases so that the effect of truncation error in the numerator is negligible (less than 0.02%). A numerator of 6000 (with a denominator of 25,000) corresponds to $\psi/2$ of about 0.25 radian; or ψ, the angle between the two phasors of about 0.5 radian. Recall that the phasor computed from an input signal of frequency $(f+\Delta f)$ Hz rotates at an angular velocity of $2\pi\Delta f$ in the complex plane. Thus to generate a phase angle of 0.5 radians between two phasors, an input signal of $(f+\Delta f)$ Hz frequency must rotate for a time T given by

$$T = \frac{0.5}{2\pi\Delta f} \simeq \frac{0.08}{\Delta f} \quad \text{seconds} \tag{40}$$

Assuming that at least 4 raw phase angle measurements would be used to produce one smoothed frequency result (for example, by a simple linear regression formula), the time required to obtain the frequency reading of $(60 + \Delta f)$ Hz would be

$$T_{(60+\Delta f)} = 4T = \frac{0.32}{\Delta f} \quad \text{seconds} \tag{41}$$

A frequency of 60.1 Hz would thus be measured in 3.2 seconds, while a frequency of 65 Hz would be measured in 0.064 second. Assuming that $\pm$ 5 Hz is the limit of frequency deviation that may have to be measured, four phasor computations must be made available in about 0.064 seconds. A phasor computation frequency of about one computation per cycle is thus sufficient to measure frequencies from 55 Hz to 65 Hz with inaccuracies of numerical computation less than 0.02%.

Equation (41) shows the relationship between the frequency and its measurement time. It is clear that smaller Δf would require longer times to make the measurement. Notice however that this particular relationship is based upon specifying that the numerator of equation (38) be of the order of 6000 in order to limit the errors of computation to less than 0.02%. If greater errors are permissible (and in many application this may well be the case) the measurement time can be shortened in an inverse proportion. Thus if it takes 3.2 seconds to measure 60.1 Hz for a given error limit, it would take 1.6 seconds to measure the same frequency if the allowed error is doubled. Based upon arguments of this nature, an adaptive algorithm could be designed which measures frequencies to differing accuracy specification in different times.

It should also be noted that errors discussed here are those of computation alone; other sources of error - such as the effects of temperature variation on electronic component values etc. - have not been considered. In a complete system, all these errors must be accounted for in specifying an overall accuracy capability of the proposed measurement scheme.

Figure 5 is a flow chart of an algorithm which would measure frequencies in the range of 55 Hz to 66 Hz in measurement times as per equation (41).

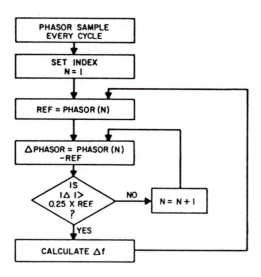

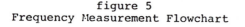

figure 5
Frequency Measurement Flowchart

EXPERIMENTAL RESULTS

The experiments described here were performed in the AEP power system simulation laboratory [8]. In the first series of experiments, the positive sequence voltage at a system bus having a nominal supply frequency of 60 Hz (local utility supply) was monitored by the Symmetrical Component Distance Relay. (See Figure 6)

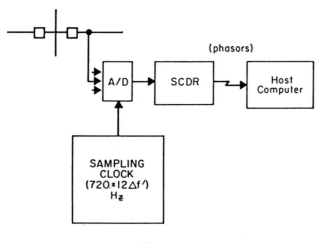

figure 6
Positive Sequence Voltage Measurement

The three phase voltage input was sampled at a variable clock frequency of $(720 + 12\Delta f')$ Hz, and the calculated positive sequence voltage phasors were stored in real time once every 12 samples in a table within the SCDR. These were then moved to the host computer where the algorithm of Figure 5 was executed in an off-line mode. Integer arithmetic was used in all key operations. The power supply frequency remained at a nominal frequency of 60 Hz, and differed from the fundamental frequency of the sampling clock by $\Delta f'$. Hz. As this experiment was meant to

simulate a variable power system frequency and a fixed clock frequency of 720 Hz, the frequency difference $\Delta f'$ is related to the frequency difference Δf being simulated by the relation

$$\Delta f = \Delta f' \frac{60}{60 + \Delta f'} \qquad (42)$$

Table I shows the sampling clock setting converted to Δf according to equation (42), the measured Δf according to the algorithm of Figure 5, the average time to make that measurement, and the error of the measurement.

Note that the measurement time for a given Δf is approximately inversely proportional to Δf in keeping with the result of equation (41). The Δf in Table I could not be adjusted to a simple desired setting precisely because the sampling clock could only be adjusted to the nearest tenth of a microsecond.

TABLE I

Case No.	Δf Setting	Δf Measured	Error	Average Measuremt Time(Sec)
1	-5.002	-5.014	.012	0.097
2	-4.501	-4.495	-.006	0.089
3	-3.999	-3.994	-.005	0.088
4	-3.499	-3.490	-.009	0.133
5	-3.002	-2.995	-.007	0.133
6	-2.501	-2.491	-.01	0.136
7	-1.999	-1.990	-.009	0.167
8	-1.002	-0.992	-.01	0.308
9	-0.751	-0.736	-.015	0.421
10	-0.501	-0.489	-.012	0.587
11	-0.250	-0.233	-.017	1.22
12	0.251	0.257	-.006	1.46
13	0.502	0.509	-.007	0.787
14	0.748	0.754	-.008	0.417
15	0.998	1.010	-.012	0.395
16	1.249	1.269	-.02	0.253
17	1.499	1.509	-.01	0.218
18	2.000	2.015	-.015	0.16
19	2.502	2.512	-.01	0.133
20	2.999	3.010	-.011	0.16
21	3.499	3.526	-.027	0.133
22	4.001	3.998	.003	0.09
23	4.502	4.501	.001	0.083
24	4.999	4.998	.001	0.066

It is the authors belief' that the errors in Table I, although quite small, may be due to the normal frequency excursions of the power system; and in fact the measurement accuracy of the algorithm presented here is much greater than that indicated by Table I.

The second experiment consisted of measuring the positive sequence voltage phasors at two ends of a transmission line, while the generator at one end was going through power swings. The experimental set-up is shown in Figure 7.

The analog input data was sampled at sampling pulses generated by a common sampling clock. The positive sequence voltage phasors from the two ends were computed by a single SCDR computer, although in a field installation two separate SCDR's (and also two separate

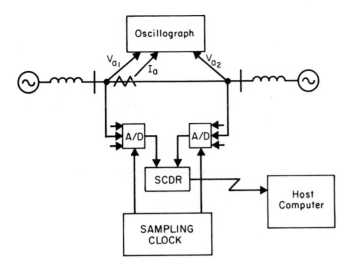

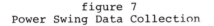

figure 7
Power Swing Data Collection

sampling clocks) would be needed. The phasor samples were transmitted to the host computer for off-line processing. The observed phase a voltages from the two ends of the line and phase a current in the line are shown in Figure 8(a). Figure 8(b) shows the calculated phase angle difference between the positive sequence voltages of two ends of the line and the computed frequency at each end.

In general, regression relations for computing $\dot\psi$ and $\ddot\psi$ can be developed from observed values of ψ. However, in practice it has been found adequate to use simple averages over a data window to estimate these quantities:

$$\dot\psi_0 = f_0 \approx \frac{1}{2N} \sum_{k=-N}^{N} \frac{\psi_{k+1} - \psi_k}{T_0} \tag{43}$$

and

$$\ddot\psi_0 = \dot f_0 = \frac{1}{2M} \sum_{k=-M}^{M} \frac{f_{k+1} - f_k}{T_0} \tag{44}$$

where T_0 is the sampling interval, and the window for computing f_0 and $\dot f_0$ is $(2N+1)$ and $(2M+1)$ samples respectively.

CONSIDERATIONS FOR FIELD INSTALLATION

Experiments similar to those described above will be performed in the field on AEP system in the coming months. Two Symmetrical Component Distance Relays will be installed at two substations at the two ends of a transmission line. Computation of local phase angles, frequency and rate of change of frequency would be carried out as described above. However, an attempt will also be made to synchronize the sampling clocks at the two substations so that the phase angles measured at the two substations will be with respect to a common reference. One scheme capable of achieving this coor-

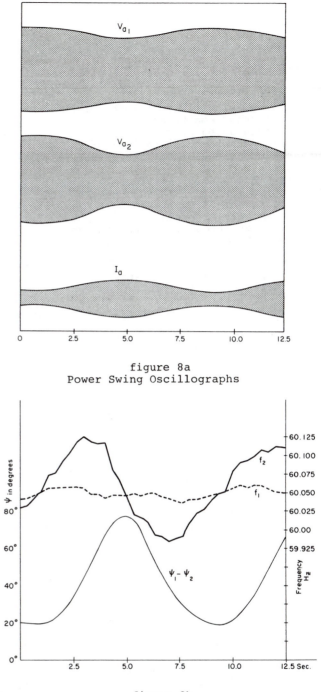

figure 8a
Power Swing Oscillographs

figure 8b
Calculated Frequency and Phase Angle

dination has been described in Reference [1]. The authors have started a preliminary investigation of the possibility of using a commercially available WWVB receiver as the source for the sampling clock synchronization. Certain on-line applications of phase angle measurement may require high accuracy in the synchronization of sampling clocks. The forthcoming field trials at AEP will be directed towards determining the limits of accuracy that can be achieved with a commercially available WWVB receiver. Other schemes for synchronization of sampling clocks will also be investigated. [9]

CONCLUSIONS

1) An algorithm for measuring phase angle and frequency of ac power system signals has been described. The ac signals are also filtered by the algorithm.

2) The measurement of positive sequence voltage motivated by developments in the digital relaying field is much more useful than the measurements made on one phase of a three phase ac power system.

3) This technique can be implemented as a stand-alone computer based measurement system, or can be made part of a Symmetrical Component Distance Relay used for transmission line protection.

4) The technique of phase angle measurement presented here is not dependent upon zero-crossing detection of a waveform, and hence is not subject to errors created by harmonic components.

5) The laboratory experiments performed on the AEP power system simulator have confirmed the findings of the paper.

6) Making local computations (in a sub-station) which have significance for system-wide estimation and control procedures is a concept which has attracted the interest of power system engineers for a number of years. Techniques presented here may be viewed as a step in a direction of developing one element of a distributed-processing hierarchical computer system for real-time monitoring and control of power systems.

REFERENCES

[1] G. Missout, P. Girard, "Measurement of Bus Voltage angle between Montreal and Sept-Iles", IEEE Trans. on PAS, March/April 1980, pp. 536-539.

[2] A. A. Girgis, F. M. Ham, "A new FFT-based digital frequency Relay for load Shedding", Proceedings of PICA, 1981, Philadelphia.

[3] E. Oran Brigham, "The Fast Fourier Tranform" (book), Prentice Hall, 1974.

[4] A. G. Phadke, M. Ibrahim, T. Hlibka, "Fundamental basis for Distance Relaying with Symmetrical Components", IEEE Trans. on PAS, Vol. PAS-96, No.2, pp 635-646, March/April 1977.

[5] J. S. Thorp, A. G. Phadke, S. H. Horowitz, J. E. Beehler, "Limits to Impedance Relaying", IEEE Trans. on PAS, Vol. PAS-98, No.1, Jan/Feb 1979, pp 246-260.

[6] A. G. Phadke, T. Hlibka, M. Ibrahim, M. G. Adamiak, "A Microcomputer Based Symmetrical Component Distance Relay", Proceedings of PICA, 1979, Cleveland. pp 47-55.

[7] A. G. Phadke, T. Hlibka, M. G. Adamiak, M. Ibrahim, J. S. Thorp, "A Microcomputer Based Ultra High Speed Distance Relay: Field Tests", IEEE Trans. On PAS, Vol. PAS-100, No. 4, April 1981, pp 2026-2036.

[8] A. G. Phadke, M. Ibrahim, T. Hlibka, "Computer in an EHV Substation: Programming Considerations and Operating Experience", CIGRE SC. 34 Colloquium, Philadelphia, October 1975.

[9] P. Bonanomi, "Phase Angle Measurements with Synchronized Clocks-Principle and Application", IEEE Trans. On PAS, Vol. PAS-100, No. 12, December 1981, pp 5036-5043.

Discussion

R. J. Marttila (Ontario Hydro, Toronto, Canada): The authors have presented a very detailed description of an interesting approach to tracking power system quantities.

The authors make the point that derivation of the positive sequence voltage from the power system phase voltages will lead to more accurate correspondence with positive sequence voltages calculated from state-estimation, load flow, and optimal power flow programs. If the load flow and the other programs use the positive sequence equivalent of the power system, the positive sequence voltage obtained from these programs will not, in the presence of unbalances, correspond to the actual positive sequence voltage in the power system. In order to take full advantage of the improved capability of measuring the actual positive sequence system voltages, as proposed in the paper, the state-estimation, load flow and optimal power flow programs should have the capability of correctly representing unbalance effects in the power system. Would the authors comment?

The authors make a comparison of the relative quality of the bus voltage phase angle measurement as achieved by the digital approach discussed in the paper and by the detection of zero-crossings of the voltage wave. If any non-60 Hz components which could influence the accuracy of the measurement were filtered out from the signal, fundamentally it seems that accuracies comparable to the digital approach should be achievable with the detection of the zero crossings. The digital approach does provide more flexibility and convenience in the implementation, but ultimately, after digital filtering of the signal, this approach also depends on correctness of the zero crossings of the fundamental 60 Hz signal for correct results. Would the authors comment?

In the context of Table I, the authors state that the errors in excess of the expected between the actual and calculated frequency deviation, is probably due to normal frequency excursions of the 60 Hz supply used in the tests. If this source is considered to be predominant, then the results suggest that during the test period, encompassing Case No's 1-24 in Table I, the supply frequency (taking 60 Hz as a reference) varied from 60.012 Hz (Case #1) to 59.973 Hz (Case #21). Is variation in this range expected over the period of time during which the tests were performed? Could another source of error be the stability of the clock frequency (720 Hz ± variation) for the A/D converter? What is the specified stability of the clock to achieve the desired accuracy?

Manuscript received December 22, 1982.

William Premerlani (GE CR&D, Schenectady, NY): The method for measuring frequency and rate-of-change of frequency with a microcomputer described by the authors is one of the most accurate methods described to date. It becomes most sensitive as the frequency approaches the frequency assumed for establishing the sampling rate. As the authors point out, the method bears a "striking resemblance" to the "principle of a power system synchroscope." The advantages of the scheme over other methods include:

1. Using all three phase voltages, it is less sensitive to individual phase voltage variations than methods based on one phase.
2. Using more signal information than methods based on zero-crossing times, it is more immune to noise and harmonics.
3. Because it is based on sampled data, the method can be added for low incremental cost to microcomputer-based devices that are performing other functions, without additional hardware.
4. The method described by the authors is more sensitive to frequency deviation and less sensitive to noise and harmonics than methods based on "leakage effects" in the Fourier Transform calculation.

I believe there is a term missing from the equation given for the rotation of the phasor. It can be found as follows.
The phasor estimate of the signal at time sample r is given by:

$$\bar{X}(r) = \frac{2}{N} \sum_{k=r-N}^{r-1} x(k\Delta t) e^{-j\frac{2\pi k}{N}} \tag{1}$$

$$\Delta T = \text{time between samples}$$

A real valued sinusoidal signal can be expressed as:

$$x = \text{Real}(\bar{X}e^{j2\pi ft}) = \frac{\bar{X}e^{j2\pi ft} + \bar{X}*e^{-j2\pi ft}}{2} \tag{2}$$

564

X = phasor value of x
f = signal frequency
* denotes complex conjugate

Substituting Eq. 2 into Eq. 1, the following is obtained:

$$\bar{X}_{60+\Delta f}^{(r)} = \frac{\bar{X}}{N} e^{+j2\pi(\frac{r}{N}-1)\frac{f}{60}} \sum_{k=0}^{N-1} e^{j\frac{2\pi k}{N}(\frac{\Delta f}{60})}$$

$$+ \frac{\bar{X}}{N} e^{-j2\pi(\frac{r}{N}-1)\frac{f}{60}} \sum_{k=0}^{N-1} e^{-j\frac{2\pi k}{N}(\frac{\Delta f}{60}+2)}$$

(3)

Eq. 3 can be simplified using the following identity:

$$\sum_{i=0}^{N-1} (e^{j\theta})^i = \frac{\sin\frac{N\theta}{2}}{\sin\frac{\theta}{2}} e^{j(N-1)\frac{\theta}{2}}$$

(4)

Using Eq. 4 in Eq. 3, the following is obtained:

$$\bar{X}_{60+\Delta f}^{(r)} = \frac{\bar{X}}{N} \frac{\sin\pi\frac{\Delta f}{60}}{\sin\frac{\pi\Delta f}{N60}} e^{-j\pi\frac{\Delta f}{60}(1+\frac{1}{N})} e^{j\frac{\Delta f}{60}\frac{2\pi}{N}r}$$

$$+ \frac{\bar{X}^*}{N} \frac{\sin\pi\frac{\Delta f}{60}}{\sin(\frac{2\pi}{N}+\frac{\pi}{N}\frac{\Delta f}{60})} e^{j\pi(1+\frac{1}{N})\frac{\Delta f}{60}} e^{j\frac{2\pi}{N}} e^{-j\frac{\Delta f}{60}\frac{2\pi}{N}r}$$

(5)

Let $X_{60}^{(0)}$ be denoted:

$$\bar{X}_{60}^{(0)} = \bar{X} e^{-j\pi\frac{\Delta f}{60}(1+\frac{1}{N})}$$

(6)

Then Eq. 5 becomes:

$$\bar{X}_{60+\Delta f}^{(r)} = \bar{X}_{60}^{(0)} \frac{\sin\frac{\Delta f}{60}\pi}{N\sin\frac{\Delta f}{60}\pi} e^{j\frac{\Delta f}{60}\frac{2\pi}{N}r}$$

$$+ \bar{X}_{60}^{*(0)} \frac{\sin\frac{\Delta f}{60}\pi}{N\sin(\frac{\Delta f}{60}\frac{\pi}{N}+\frac{2\pi}{N})} e^{j\frac{2\pi}{N}} e^{-j\frac{\Delta f}{60}\frac{2\pi}{N}r}$$

(7)

It can be readily verified that this is the equation of an ellipse in the complex plane. The orientation of the ellipse depends on the phase angle of $X_{60}^{(0)}$. For small frequency deviations, the second term of the equation vanishes, and the equation reduces to that of a circle. As the frequency deviation increases, the eccentricity of the ellipse increases. Fortunately, the second term in the equation is not very large for frequencies near 60 Hz. Would the authors care to comment on the effect of this term on the overall accuracy of the computed frequency using the method they described?

Manuscript received August 2, 1982.

S. Zocholl (Brown Boveri Electric, Inc.): The authors have presented a useful and novel digital technique producing vector rotation which identifies frequency change.

The authors restrict the sampling rate to 720 hertz to conveniently obtain symmetrical component values. It should be noted at a slower sampling rate, say for example, 240 hertz; the symmetrical components may be obtained for a small software burden, using a rotation matrix.

In addition, the slower rate requires less bandwidth.

Regarding the introduction, the authors imply that harmonics effect the zero crossings when measuring frequency. Harmonics have no effect when measured over a full period.

Manuscript received August 4, 1982.

Adly A. Girgis (North Carolina State University, Raleigh, NC): The authors are to be complimented for their application of the discrete Fourier transform in measuring the frequency and its rate of change at a local bus. We have the following comments and questions for the authors' consideration.

Most of the time, engineers are interested in the discrete Fourier transform (DFT) only because it approximates the continuous Fourier transform (CFT). Most of the problems in using the discrete Fourier transform (DFT) are caused by neglecting of what this approximation involves [A]. When the continuous Fourier transform (CFT) is used to analyze a constant-amplitide waveform of a single frequency f, the result is a phasor of a constant amplitude rotating at f cycles/sec. Therefore, when the frequency of the input signal is $60 \pm \Delta f$ Hz, the resultant phasor of the CFT turns an angle of $\pm 2\pi\Delta f$ radians per second relative to the phasor of a 60 Hz waveform.

When the discrete Fourier Transform is used with N points in a window of 1/60 sec, it is assumed that the signal to be analyzed has a constant amplitude and is periodic in a period of 1/60 sec. Any deviation from this assumption will cause errors in calculating the magnitude and the phase angle of the fundamental frequency of the of the DFT (60 Hz component).

Therefore, the error in the phase angle of the fundamental frequency component is function of many parameters, some of these parameters are:

(i) magnitude of frequency deviation $|\Delta f|$
(ii) sign of frequency deviation (positive or negative)
(iii) starting point of the window
(iv) the presence of nonperiodic functions
(v) the larger frequency deviations of the different harmonics ($n\Delta f$ for the nth harmonic), etc.

Aside from the presence of a nonperiodic function or the effect of harmonics, the change in the phase angle of the fundamental frequency component of the DFT is a nonlinear function of the frequency deviation and its sign and the starting point on the window. To show this relation, consider the fundamental frequency component of the DFT (X_1) of a signal x(t) where

$$x(t) = A \sin(2\pi(60 + \Delta f)t + \phi)$$

(D-1)

then

$$\bar{X} = \frac{1}{\sqrt{2}} j\bar{X}_1$$

$$= \frac{A}{\sqrt{2}} \frac{2}{N} \left[e^{j\phi} \sum_{k=0}^{N-1} e^{j\frac{\Delta f}{60}\frac{2\pi}{N}k} - e^{-j\phi} \sum_{k=0}^{N-1} e^{-j(\frac{4\pi}{N}+\frac{\Delta f}{60}\frac{2\pi}{N})k} \right]$$

(D-2)

thus

$$\bar{X} = \bar{X}_{60} \frac{\sin(\frac{\pi\Delta f}{60})}{N\sin(\frac{\Delta f}{60}\frac{\pi}{N})} e^{j\Theta_0(\Delta f)}$$

$$- \bar{X}_{60}^* e^{j\frac{2\pi}{N}} \frac{\sin(\frac{\pi\Delta f}{60})}{N\sin(\frac{2\pi}{N}+\frac{\pi}{N}\frac{\Delta f}{60})} e^{-j\Theta_0}$$

(D-3)

where

$$\Theta_0(\Delta f) = -\frac{\pi\Delta f}{60}(1-\frac{1}{N})$$

Eq. (D-3) can be rewritten in the form

$$\bar{X} = \bar{X}_{60} F_1(\Delta f) e^{j\Theta_0(\Delta f)} - \bar{X}_{60}^* e^{j\frac{2\pi}{N}} G_1(\Delta f) e^{-j\Theta_0(\Delta f)}$$

(D-4)

565

where (for Δf within 5 Hz):

$$F_1(\Delta f) \simeq 1.0$$

$$G_1(\Delta f) = \sin\left(\frac{\pi\Delta f}{60}\right)/N\left(\sin\left(\frac{2\pi}{N} + \frac{\Delta f}{60}\frac{\pi}{N}\right)\right)$$

or

$$G_1(\Delta f) \simeq \left(\pi\frac{\Delta f}{60}\right)/\left(N\sin\frac{2\pi}{N}\right) \qquad \text{(D-5)}$$

As the change in the phase angle of the fundamental frequency component is considered to be $\frac{2\pi\Delta f}{60}$ radians per cycle of the 60 Hz, the percentage error due to $G_1(\Delta f)$ varies between $-\frac{100}{2N\sin(2\pi/N)}$ and $+\frac{100}{2N\sin(2\pi/N)}$. For N = 12, this error will be between -8.3% to $+8.3\%$ depending on the sign of Δf and the starting point of the DFT window. That is in addition to the error due to the initial angle $\theta_0(\Delta f)$. The error due to $\theta_0(\Delta f)$ will not be eliminated by choosing an arbitrary angle θ in the DFT (the angle θ described in Eqs. 14 - 18). That is simply because the arbitrary angle is constant, but the angle $\theta_0(\Delta f)$ is function of the frequency deviation.

Also the effect of the second term of equation (D-4) will not be eliminated by considering the positive sequence voltage. To simplify the mathematical proof of this statement, consider the fundamental frequency component of a sinewave, x(t), and then for a cosinewave, y(t).

$$x(t) = A\sin(2\pi(f + \Delta f)t) \qquad \text{(D-6)}$$

$$y(t) = A\cos(2\pi(f + \Delta f)t) \qquad \text{(D-7)}$$

It can be shown from Eq. (D-2 - D-4) that

$$\overline{Y}_1 = F_1(\Delta f)\,e^{j\theta_0(\Delta f)} + e^{j\frac{2\pi}{N}}\,G_1(\Delta f)\,e^{-j\theta_0(\Delta f)} = \overline{F}_1 + \overline{G}_1 \qquad \text{(D-8)}$$

and

$$j\overline{X}_1 = F_1(\Delta f)\,e^{j\theta_0(\Delta f)}\,e^{j\frac{2\pi}{N}} - e^{j\frac{2\pi}{N}}\,G_1(\Delta f)\,e^{-j\theta_0(\Delta f)} = \overline{F}_1 - \overline{G}_1 \qquad \text{(D-9)}$$

The phasors Y_1 and jX_1 are shown graphically in Fig. (D-1) which clearly indicates that Y_1 will never be equal to jX_1 since Δf is not equal to zero.

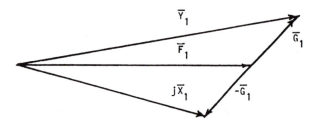

Fig. D-1: Phasor representation of Y_1 and jX_1

Therefore, the voltage of the three phases will not be displaced by 120° in the frequency domain, when there is a frequency deviation Δf.

The measurement time tabulated in Table 1 is calculated according to Eq. 41 ($T = 0.32/\Delta f$), that implies that the measurement time is constant for a constant frequency deviation. However, there is a large difference due to the change of the sign of the frequency deviation. Consider, for example, the frequency deviation of -0.501 and $+.502$, the measurement times are 0.587 and 0.787 sec respectively, which differ by -8.3% and 23% from the values given by Eq. 41. Can this error be considered as a measure of the degree of accuracy of the technique?

The experimental test on a power system model does not show how the power swing condition was experimentally obtained. However, Fig. 8.a implies that the results were obtained by having the two machines of Fig. 7 running at two slightly-different-constant speeds. If that is the case, f_1 and f_2 are supposed to be constants. Examining Fig. 8.b indicates that the frequencies f_1 and f_2 fluctuate according to the fluctuation of the magnitudes of Va_1 and Va_2. Also, comparing Fig. 8.a with Fig. 8.b shows that the fluctuation in f_2 is higher than the fluctuation in

f_1, which appears to be due to the higher fluctuation in Va_2; therefore, can these fluctuations in f_1 and f_2 be considered as a measure of the error due to the change of the signal amplitude?

It is claimed that the method is not affected by harmonics, that is based on the assumption that the frequency of the harmonics is not affected by the frequency deviation of the fundamental. But when the frequency of the fundamental component changes by Δf, the frequency of the nth harmonic changes by $n\Delta f$.

The fundamental frequency (the 60 Hz component) of a signal of frequency $n(60 \pm \Delta f)$ will be a phasor quantity whose magnitude and phase angle are functions of the frequency deviation (Δf) of the fundamental. It should be also noted that the effect of this harmonic will not be diminated by considering the positive sequence voltage. The reason is that the fundamental frequency components of the nth harmonic in the three phases will not be displaced by an angle of $2\pi\frac{n}{3}$.

The technique will be affected by any nonperiodic signal that may be present in the waveform such as dc offset or voltage spikes. Do the authors intend to use an analog filter to filter such nonperiodic functions?

The leakage coefficient method for detecting the frequency deviation is based on the fact that due to a frequency deviation of the 60 Hz waveform by $\pm\Delta f$, the magnitude of any of the harmonics in the DFT is not zero. We found that the sum of the magnitudes of the harmonics in the DFT is linearly related to the magnitude of the frequency deviation, provided that the waveform to be analyzed is a sinewave [2, A], therefore, a zero crossing detector is used to have a fixed starting point on the waveform to be analyzed. As the sum of DFT components will be also affected by the presence of non 60 Hz component, the first step in the leakage coefficient scheme is to band-pass filter the waveform. The band-pass filter used in a 4-pole Butterworth filter with a center frequency of 60 Hz and a bandwidth of 30 Hz [B, C]. The leakage coefficient technique is not affected by the change of the amplitude of the waveform to be analyzed. Furthermore, any frequency deviation is calculated in two cycles (33 msec). The change of the angle of DFT fundamental component is also used in the leakage coefficient method to detect only the sign of the frequency deviation (position or negative).

Finally, I congratulate the authors for an interesting paper.

REFERENCES

[A] Adly A. Girgis and F. Ham, "A Quantitative Study of Pitfalls in the FFT," *IEEE Transactions on Aerospace and Electronic Systems*, Vol. AES-16, No. 4, July 1980, pp. 434-438.
[B] Adly A. Girgis and F. Ham, "Frequency Measuring and Monitoring Apparatus, Methods and Systems" U.S. Patent No. 4,319,329.
[C] F. Ham and Adly A. Girgis, "Measurement of Power Frequency Fluctuations Using FFT", to be published.

Manuscript received August 13, 1982.

J. G. Gilbert and **N. P. Marino** (Pennsylvania Power & Light Company, Allentown, PA): The authors have made a useful contribution to the art of computer protection and control of power systems. We would appreciate having the author's thoughts on the following to clarify concepts they have proposed:

1. A fixed frequency sampling clock is used with the SCDR computer described by the authors. Had a frequency tracking sampler been employed, power system frequency could not be measured in the manner described in this paper. What issues lead to the author's choice of a fixed frequency sampling clock? Do the authors believe it worthwhile and practical to compensate results of frequency sensitive calculations based on computed power system frequency?

2. The authors assume four or more raw phase angle measurements would be used to produce one smoothed frequency result. Did the authors consider using phase angle measurements made from a longer data window? Would the authors discuss benefits of smoothing several raw phase angle measurements instead of using a correspondingly longer data window?

3. Fig. 8b graphically shows that precautions are required to prevent misoperation of load shedding schemes during power swings. Time delay has been used to prevent misoperation of conventional relay schemes. Have the authors investigated any alternatives to intentional time delay for this application?

We believe some of the issues raised by this paper point to the need for early establishment of guidelines and standards. A nonexhaustive list of subjects which comes to mind is: establishment of standard sampling rate(s); selection of fixed frequency versus frequency tracking sampling clocks; standard electrical interfaces to digital protection subsystems; man-machine interface standards; identification, modularity, and residency of standard protection and control functions; standards for test and calibration of digital protection subsystems; etc. This list can be made much longer. Early development of guidelines and standards for issues such as these could prevent many future concerns and difficulties.

Manuscript received August 2, 1982.

A. G. Phadke, J. S. Thorp, and **M. G. Adamiak:** We thank the discussors for the interest they have shown in our paper.

Mr. Martilla observes correctly that to form an entirely consistent model of the power system, its measurements, the model and the computation procedure used should all use a positive-sequence frame of reference. In most cases, the positive sequence network model used in load flow or state estimation programs simply neglects the negative and zero sequence networks and their coupling to the positive sequence network. A model which included the coupled network would be the most accurate model, and when combined with sequence voltage measurements would produce an accurate representation of the system. In our view, measuring the positive sequence voltage on the positive sequence network model (neglecting the couplings) is a step in the direction of improving model consistency. In many cases, this improvement may be all that is called for. We believe this procedure would be better than using phase voltage measurements with the positive sequence network model. Perhaps additional investigation in this subject is justified where system unbalances are a significant factor.

Another point raised by Mr. Martilla discusses the efficacy of using zero crossing information for frequency measurement. Analog filters, at least in their practical realizations are not as sharp as digital (Fourier type) filters in rejecting harmonics. Further, as pointed out by Mr. Premerlani in his discussion, since the Fourier type technique uses more data (including the non-zero sample values), it forms a more reliable estimate of the frequency. Ultimately, if the analog filters used were able to produce a pure fundamental frequency component, then its zero-crossing frequency measurement would be as accurate as that produced by the digital technique. However, such an analog filter would not function when the power system frequency changes from the nominal power frequency.

The last point made by Mr. Martilla concerns the frequency excursions observed in our area. Although it was not mentioned in the paper, the data collection was carried out over several days, and consequently the observed frequency excursions are well within normal variations of the power system frequency. The crystal oscillator clock used had a stability of 10 parts per million. It seems to us that such accuracies being readily available, the stability of the clock should be of no concern in the present context.

Mr. Zocholl is quite right in his statement that a sampling rate of 240 Hz can also be used to produce symmetrical components. In fact, almost any sampling rate would be acceptable, although the computational burden associated with the transformation from phasors to symmetrical components depends upon the sampling rate chosen. A sampling rate which is a multiple of 180 Hz has particularly advantageous properties, as pointed out in our Ref. (4). Also, as the digital relays we have developed use a sampling rate of 720 Hz, the technique presented allows a computation of frequency with little overhead. We are not sure about the import of Mr. Socholl's comment about the bandwidth. It seems to us that in terms of presently available analog components and digital computers, there is not much difference between bandwidths associated with sampling rates of 180 Hz, 240 Hz or 720 Hz.

Mr. Zocholl is quite right in pointing out that a stationary harmonic content does not affect the zero crossing of a waveform. However, in our observations, we have noted a constant movement of the harmonic components, and consequently the zero-crossing does change due to changing harmonic components in system voltage waveforms. It should be noted that several successive zero-crossing points must be used to create a single frequency measurement.

Mr. Gilbert and Mr. Marino have correctly pointed out the reasons for using sampling clocks that are independent of power system frequency. Without such clocks, frequency measurements would not be possible. As a practical matter, such clocks are also much simpler to construct than clocks which are synchronized to the power system waveform.

In some cases, relaying parameters are dependent upon frequency. Although the present authors' technique using symmetrical components is immune to power system frequency deviations, other impedance relays may depend upon the frequency. Since the frequency is known (obtained by using the technique described in this paper), the frequency correction could be applied to the relay characteristic where appropriate. Individual cases must be considered separately to determine specific frequency compensation to be applied.

In general, either raw phase angle measurements or averaged phase angle measurements could be used to calculate the frequency, as long as at least two phase angle measurements are used at sufficiently separated instants of time. It should be pointed out that averaging of data from a rotating phasor could lead to serious errors if the averaging is done over a large angle of rotation. To take an extreme example, the average of two phasors which are obtained from a rotating phasor when the amount of rotation is 180° is zero, a result which is of no use in any frequency calculation.

We have not studied the problem of load shedding relays. This clearly is an application problem. If engineering analysis shows that higher order derivatives of frequency should be used as inputs to the load shedding relay, then it would be a simple matter to compute the necessary frequency derivatives by using the regression technique outlined in the paper.

We agree with Messrs. Gilbert and Marino that there is a great need for standardization in the areas they mention. Perhaps, in time, such standards will be developed by technical subcommittees and working groups of IEEE.

Mr. Premerlani has given an excellent summary of the advantages of a frequency measurement technique based on recursive Fourier Transforms. We are also grateful to him and to Mr. Girgis for pointing out that Eq. (31) of our paper is an approximation. The analysis given by Messrs. Premerlani and Girgis gives the correct form, and it is instructive to study the effect of this approximation. Taking Eq. (7) of Mr. Premerlani (which is essentially the same as Eq. (D-4) of Mr. Girgis) it should be noted that the phasor at $(60 + \Delta f)$ inscribes an ellipse in the complex plane. The equation for the ellipse is of the type

$$\bar{X} = A e^{j\Delta\omega t} + xA^* e^{j\frac{2\pi}{N}} e^{-j\Delta\omega t}$$

where x is a scale factor given by

$$x = \frac{\sin \frac{\pi\Delta f}{60N}}{\sin \left(\frac{\pi\Delta f}{60N} + \frac{2\pi}{N} \right)}$$

The two phasors A and xA^* rotate in opposite directions in the complex plane, and at $t = 0$ the resultant phasor is

$$\bar{X} (t=0) = A + xA^* e^{j\frac{2\pi}{N}}$$

This relationship is illustrated in Fig. 9 for an arbitrary phasor A using a somewhat magnified value for x.

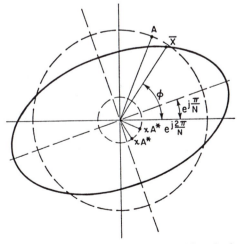

Fig. 9. Phasor of frequency $(60 + \Delta f)$ at $5 = 0$

567

The major axis of the ellipse is at an angle π/N with the real axis, and the minor axis is perpendicular to it. This angle results from the placement of the sampling instants with respect to the time zero point, and does not depend upon $X(0)_{60}$ or any other paramter. Disregarding this shift in angle (or assuming that the sampling instants are evenly placed around t = 0, leads to an ellipse placed with its major axis along the real axis.

One could calculate the angular velocity of the resultant phasor which describes the ellipse. With some manipulations, it can be shown that the angular velocity $\frac{d\phi}{dt}$ of the phasor $X_{(60+\Delta f)}$ is given by

$$\frac{d\phi}{dt} = \Delta\omega \frac{1 + \tan^2 \Delta\omega t}{\frac{1+x}{1-x} + \frac{1-x}{1+x} \tan^2 \Delta\omega t}$$

Clearly the angular velocity differs from $\Delta\omega$ when x is not equal to zero as pointed out by Messrs. Premerlani and Girgis. The variation of $\frac{d\phi}{dt}$ is cyclic around $\Delta\omega$, varying between $\frac{1+x}{1-x}\Delta\omega$ and $\frac{1-x}{1+x}\Delta\omega$ as shown in Fig. 10.

For a frequency difference of 1 Hz, x is 0.00866, and consequently the measured frequency would vary between 60 ± (0.9828) Hz to 60 ± (1.017) Hz depending upon the instant of measurement.

This phenomenon suggests a remedy to the problem: If $\frac{d\phi}{dt}$ is averaged over a half period, the effect of this variation will be removed for small values of Δf. Indeed, if the frequency is to be measured from a single phasor, this averaging procedure should be used.

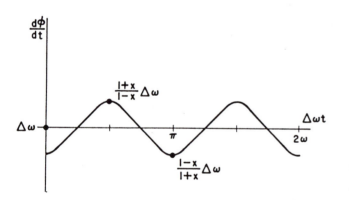

Fig. 10. Variation of computed frequency

However, when positive sequence quantities are used, there is an automatic filtering effect of this variation. We will now consider the positive sequence computation from three phase voltage measurements where the power system frequency is $(60 + \Delta f)$ Hz. (In a brief conversation at the Summer Power Meeting, we had mentioned this phenomenon to Mr. Girgis. We are not certain whether we explained our idea in sufficient detail at that time. Perhaps this explanation will help answer Mr. Girgis' question on symmetrical components.)

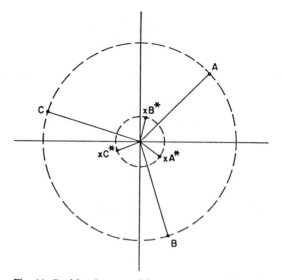

Fig. 11. Positive Sequence Measurement at $(60 + \Delta f)$ Hz

Each of the three phase voltage phasors at frequency $(60 + \Delta f)$ Hz has two components, one rotating in clockwise direction, and the other in a counterclockwise direction. As the axes of the ellipse are fixed by the sampling instants, they are identical for the three phasors. See Fig. 11: (The axes of the ellipse are drawn coincident with real and imaginary axes for convenience. As pointed out earlier, this choice is immaterial to the present discussion.)

The phasors A, B, C rotate in the clockwise direction, whereas phasors xA^*, xB^* and xC^* rotate in the counterclockwise direction. Note that xA^* is obtained as a mirror image of A in the real axis with a scale factor x, and similarly for xB^* and xC^*. It is clear from Fig. 11 that while A, B, C are positive sequence quantities, xA^*, xB^* and xC^* are negative sequence quantities. Consequently, when the symmetrical component transformation is applied to $(A + xA^*)$, $(B + xB^*)$, $(C + xC^*)$, the positive sequence component of the result will be the same as though it was computed from (A, B, C). The frequency computed from the positive sequence voltage phasor is thus free of the cyclic variation shown in Fig. 10; and reads $\Delta\omega$ correctly. No averaging is necessary, and the correction term omitted from Eq. (31) of the paper is indeed unnecessary.

To answer the questions of Mr. Girgis regarding the experiment, the machines were *not* turned at constant speeds, indeed one machine was oscillated around the synchronous speed. The measured frequency change is the actual frequency change, and is immune to the changes of voltage amplitude.

The measurement times do not follow Eq. (41) of the paper exactly, it being a continuous variable result, whereas the actual measurement time jumps in units of the data window (one cycle). Eq. (41) should be viewed as a general guideline for the kind of meaurement times to be expected.

The harmonic effect mentioned by Mr. Girgis is very interesting, but we have not studied it in any detail. Perhaps such a study should be made. However, we would make the observation that the leakage effect method will fail completely in the presence of harmonics, in fact we find this to be one of the main drawbacks of the leakage effect methods.

We would like to conclude by saying that the method presented in the paper is extremely sensitive and accurate for small frequency deviations. Since this is the nature of frequency excursions in a power system, the method is particularly attractive. Also, by using phasors which are computed for digital relaying applications, the method adds frequency measurement function to a relay at no additional burden on the relaying processor.

Manuscript received October 28, 1982.

Author Index

A

Adamiak, M. G., 556
Allguren, B. J., 360
Aucoin, M., 164

B

Beehler, J. E., 108
Begovic, M. M., 77
Blackburn, J. L., 87
Bonanomi, P., 548
Bowler, C. E. J., 244
Brown, P. G., 244
Buckley, G. W., 202
Burnett, Jr., R. O., 544

C

Cease, T. W., 447
Chadwick, Jr., J. W., 338
Clough, G., 417

D

Damborg, M. J., 47
Deliyannides, J. S., 350

E

Einvall, C.-H., 400
Elmore, W. A., 87, 270
Eriksson, L., 525

F

Feero, W. E., 20
Fuchs, E. F., 29
Fuller, J. F., 29

G

Gish, W. B., 20
Giuliante, A. T., 143
Giuliante, T., 417
Griffin, C. H., 210, 130, 429

H

Haddad, M. V., 260
Hicks, Sr., K. L., 54
Hlibka, T., 512
Horowitz, S. H., 67, 77, 108

I

Ibrahim, M., 512
IEEE PSRC Report, 3, 9, 38, 122, 173, 181, 187, 222, 235, 293, 302, 314, 378, 393, 443, 538

J

Jones, R. H., 20

K

Khunkhun, K. J. S., 260
Kimbark, E. W., 96
Koenig, D. F., 360
Koepfinger, J. L., 260
Kotheimer, W. C., 343, 468
Kramer, C. A., 270

L

Linders, J. R., 54, 400

M

Mankoff, L. L., 343
McPherson, G., 202

N

Nilsson, S. L., 360

P

Phadke, A. G., 67, 77, 108, 410, 491, 512, 556
Poncelet, R., 497
Pope, J. W., 193, 210

R

Ramaswami, R., 47
Ray, R. E., 154
Rizy, D. T., 54
Rockefeller, G. D., 54, 525
Roesler, D. J., 29
Russell, B. D., 164

S

Saha, M. M., 525
Schlake, R. L., 202
Stranne, G., 143
Sun, S. C., 154
Sutton, H. J., 325

T

Tajaddodi, F. Y., 468
Thorp, J. S., 67, 77, 108, 410, 556

U

Udren, E. A., 350, 360, 447

569

Subject Index

Editor's Biography

Stanley H. Horowitz (M'50–SM'63–F'79) is a graduate of the City College of New York where he received his B.S.E.E.

Mr. Horowitz retired from American Electric Power Service Corporation in 1989 after thirty-seven years of service. During that time he was the Relay Section Manager, responsible for relay protection and control of transmission lines and generating units; Assistant Chief Electrical Engineer, responsible for protection, communication, metering, and electrical research; Assistant Division Manager, responsible for electrical aspects of nuclear and fossil power plants; and Consulting Electrical Engineer, reporting to the Senior Vice-President of Electrical Engineering.

Mr. Horowitz is a Fellow of the IEEE, a past member of the IEEE Executive Board, past Chairman of the IEEE Power System Relaying Committee and past International Chairman of CIGRE's Study Committee 34-Protection and Control. He has co-authored the textbook *Power System Relaying* (London: Research Studies Press; New York: Wiley, 1992), has authored more than a dozen papers covering transmission system and generator protection and control, and has edited the first volume of the IEEE Press Book *Protective Relaying for Power Systems*. Mr. Horowitz was given a Power System Relaying Committee prize paper award on ultra high speed fault detection as well as the Power System Relaying Committee Distinguished Service Award. He has lectured throughout the world on the fundamentals of power system protection, adaptive relaying, and the application of microprocessors to protective relaying.